Developments in Environmental Modelling, 1

Energy and Ecological Modelling

Developments in Environmental Modelling

Series Editor: S.E. Jørgensen
Langkaer Vaenge 9,
3500 Vaerløse,
Copenhagen,
Denmark

Developments in Environmental Modelling, 1

Energy and Ecological Modelling,,

Edited by

Dr. W.J. MITSCH and Dr. R.W. BOSSERMAN
Systems Science Institute, University of Louisville, Louisville, Kentucky 40292, U.S.A.

and

Dr. J.M. KLOPATEK
Oak Ridge National Laboratory, Oak Ridge, Tennessee, U.S.A.

Proceedings of a symposium held from 20 to 23 April 1981 at Louisville, Kentucky.

Sponsored by the International Society for Ecological Modelling (ISEM)

in cooperation with:
    Ecological Society of America,
    Oak Ridge National Laboratory,
    University of Louisville,
    Environmental Protection Agency.

ELSEVIER SCIENTIFIC PUBLISHING COMPANY
Amsterdam — Oxford — New York   1981

Published jointly by

INTERNATIONAL SOCIETY FOR ECOLOGICAL MODELLING (ISEM)
Langkaer Vaenge 9, 3500 Vaerløse, Copenhagen, Denmark

and

ELSEVIER SCIENTIFIC PUBLISHING COMPANY
1, Molenwerf 1014 AG Amsterdam
P.O. Box 330 1000 AH Amsterdam, The Netherlands

Distributors for the United States and Canada:

ELSEVIER/NORTH-HOLLAND INC.
52, Vanderbilt Avenue
New York, N.Y. 10017

ISBN 0-444-99731-8 (Vol. 1)
ISBN 0-444-41948-9 (Series)

Printed in Denmark by Fair-Print AS, Roskilde

PREFACE

The papers in this volume were presented at the international symposium "Energy and Ecological Modelling," held in Louisville, Kentucky (U.S.A.) on April 20-23, 1981. The conference was co-sponsored by the International Society for Ecological Modelling (ISEM), the University of Louisville, and the Oak Ridge National Laboratory. The Ecological Society of America and the U. S. Environmental Protection Agency also co-operated in early preparations for the conference.

We have attempted to use the common themes of energy and modelling to join two seemingly disparate fields of study into one volume. The first section of the volume describes the use of ecological modelling as a tool for dealing with energy-environment issues such as CO_2 increases in the atmosphere, oil spills, coal mining pollution, acid rain, thermal power plant impacts, and hydroelectric dams. In the second section, papers discuss the use of energetics and other systems analysis techniques to explain the economics of ecosystems, regions, energy alternatives, and national and international systems. Many investigators today deal in research presented in both sections under such disciplinary headings as ecological modelling, systems ecology, ecoenergetics, systems analysis, regional ecology, and energy economics. We hope that this volume will contribute to continued development of these fields.

This volume and the symposium, which can trace their roots to a discussion in Liege, Belgium in April, 1980, are the result of tireless contributions from many people. Thelma Goldstein, of the Systems Science Institute at the University of Louisville, served enthusiastically and loyally as technical coordinator of the symposium and manuscript editing. Karen Cozine, Marty Ising, Beth Dillon, and Cheryl Hartley all provided major technical assistance to the conference planning and manuscript preparation. Graduate students at the University of Louisville, particularly Debbie Cool, Lynne Dauffenbach, Paul Hill, Christopher Lind, Kathy Lowry, Francie Smith, and Jan Taylor, assisted in many technical, yet indispensable, details. Ronald M. Atlas and Robert Costanza assisted in planning for the conference. John A. Dillon, Jr., Director of the Systems Science Institute, and X.J. Musacchia, Dean of the Graduate School, both of the University of Louisville, were most generous in their support for this endeavor. Sven E. Jorgensen of ISEM gave the support of the society to the symposium and this publication. The Oak Ridge National Laboratory provided printing and other professional services to the conference. Finally, and most importantly, we extend our thanks to the authors for their timely and provocative contributions for this rewarding, academic experience.

William J. Mitsch
Robert W. Bosserman
Louisville, Kentucky

Jeffrey M. Klopatek
Oak Ridge, Tennessee

July 1981

LIST OF CONTRIBUTORS

Miguel F. Acevedo L. Universidad de Los Andes, Escuela de Ingenieria de Sistemas, Merida, Venezuela.

Austin K. Andrews. Western Energy & Land Use Team, U.S. Fish & Wildlife Service, Fort Collins, Colorado 80526.

Ronald M. Atlas. Biology Department, University of Louisville, Louisville, Kentucky 40292.

Gregor T. Auble. Western Energy & Land Use Team, U.S. Fish & Wildlife Service, Fort Collins, Colorado 80526.

M. C. Axelrod. Environmental Sciences Division, Lawrence Livermore National Laboratory, Livermore, California 94550.

Steven W. Ballou. Energy and Environmental Systems Division, Argonne National Laboratory, Argonne, Illinois 60439.

Steven M. Bartell. Oak Ridge National Laboratory, Union Carbide Corporation—Nuclear Division, Oak Ridge, Tennessee 37830.

Mitchell Berger. Cornell University, Ithaca, New York 14850.

G.E. Bingham. Environmental Sciences Division, Lawrence Livermore National Laboratory, Livermore, California 94550.

R.J. Blackwell. Jet Propulsion Laboratory, California Institute of Technology, Pasadena, California 91103.

M. Blake-Jacobson. Systems Ecology Research, San Diego State University, San Diego, California 92115.

P. Bogdonoff. Cornell University, Research Park, Ithaca, New York 14850.

S. Bollinger. Horn Point Environmental Laboratory, University of Maryland, Cambridge, Maryland 21613.

Robert W. Bosserman. Systems Science Institute, University of Louisville, Louisville, Kentucky 40292.

Walter R. Boynton. Chesapeake Biological Lab, University of Maryland, Solomons, Maryland 20688.

Robert W. Brocksen. Ecological Studies Program, Electric Power Research Institute, Palo Alto, California, 94303.

H. Brohl. Institut fur Physik n. Theor. Chemie der Universitaet Frankfurt, Robert-Mayer-Schr. 11, 6 Frankfurt 1, West Germany.

Joan Browder. National Marine Fisheries Service, Miami, Florida 33176.

J. M. Brown. Eastern Energy & Land Use Team, U.S. Fish & Wildlife Service, Kearneysville, West Virginia 25430.

Mark T. Brown. Department of Urban & Regional Planning, University of Florida, Gainesville, Florida 32603.

Sandra Brown. Department of Forestry, University of Illinois, Urbana, Illinois 61801.

Mike Burnett. Western Solar Utilization Network, Portland, Oregon 97205.

Thomas J. Butler. Center for Environmental Research, Cornell University, Ithaca, New York 14853.

John Cairns, Jr. Center for Environmental Studies, Virginia Polytechnic Institute and State University, Blacksburg, Virginia 24061.

Donald S. Cherry. Department of Biology, Virginia Polytechnic Institute and State University, Blacksburg, Virginia 24061.

Edward F. Cheslak. Wildlife Science Department, U.M.C. 52, Utah State University, Logan, Utah 84322.

Wilson Clark. California Governor's Office, Washington, D.C.

Cutler J. Cleveland, Jr. Department of Marine Sciences, Louisiana State University, Baton Rouge, Louisiana 70803.

Robert Costanza. Coastal Ecology Lab and Department of Marine Sciences, Center for Wetland Resources, Louisiana State University, Baton Rouge, Louisiana 70803.

Bernard Coupal. Dept. de Genie Chimique, University of Sherbrooke, Sherbrooke, Quebec, Canada.

C.T. Cushwa. Eastern Energy & Land Use Team, U.S. Fish & Wildlife Service, Kearneysville, West Virginia 25430.

Lynne Dauffenbach. Systems Science Institute, University of Louisville, Louisville, Kentucky 40292.

P.A. Dauzvardis. Argonne National Laboratory, Argonne, Illinois 60439.

John W. Day, Jr. Coastal Ecology Laboratory, Center for Wetland Resources, Louisiana State University, Baton Rouge, Louisiana 70803.

Edward H. Dettmann. Environmental Impact Division, Argonne National Laboratory, Argonne, Illinois 60439.

R. P. Detwiler. Cornell University, Research Park, Ithaca, New York 14850.

J.L. Dodd. Natural Resources Ecology Laboratory, Colorado State University, Fort Collins, Colorado 80523.

Thomas W. Doyle. Environmental Sciences Division, Oak Ridge National Laboratory, Oak Ridge, Tennessee 37830.

Calvin Dubrock. Eastern Energy & Land Use Team, U.S. Fish & Wildlife Service, Kearneysville, West Virginia 25430.

Richard A. Ellison. Western & Energy Land Use Team, U.S. Fish & Wildlife Service, Fort Collins, CO 80526.

W.R. Emanual. Environmental Sciences Division, Oak Ridge National Laboratory, Oak Ridge, Tennessee 37830.

A.H. Eraslan. Department of Engineering Science and Mechanics, University of Tennessee, Knoxville, Tennessee 37916.

Katherine C. Ewel. School of Forest Resources and Conservation, University of Florida, Gainesville, Florida 32611.

Linda B. Fenner. Energy Policy Studies, Incorporated, Omaha, Nebraska 68152.

Véronique Garçon. Electricite de France, Chatou, France.

George Gertner. Department of Forestry, University of Illinois, Urbana, Illinois 61801.

James D. Giattina. Corvallis Environmental Research Laboratory, U.S. Environmental Protection Agency, Corvallis, Oregon 97331.

John P. Giesy. Aquatic Toxicology, Michigan State University, East Lansing, Michigan 48823.

Martha W. Gilliland. Energy Policy Studies, Inc., 11109 N. 61 Street, Omaha, Nebraska 68152

D.N. Gladwin. Eastern Energy & Land Use Team, U.S. Fish & Wildlife Service, Kearneysville, West Virginia 25430.

Robert Goldstein. Electric Power Research Institute, Palo Alto, California 94303.

Phillipe Gosse. Direction des Etudes et Recherches, Electricite de France, Chatou, France.

Robert T. Haar. Center for Quantitative Science, University of Washington, Seattle, Washington 98195.

Charles A. S. Hall. Cornell University, Research Park, Ithaca, New York 14850.

David B. Hamilton. Western Energy & Land Use Team, U.S. Fish & Wildlife Service, Fort Collins, Colorado 80526.

Bruce Hannon. Energy Research Group, Office of Vice Chancellor for Research, University of Illinois, Urbana, Illinois 61801.

Archie Arkadius Harms. McMaster University, Hamilton, Ontario, Canada.

John E. Heasley. NREL, Colorado State University, Fort Collins, Colorado 80523.

George W. Heitzman. Center for Ecological Modeling, Rensselaer Polytechnic Institute, Troy, New York 12180.

Brian Henderson-Sellers. Department of Civil Engineering, University of Salford, Salford, England.

Robert A. Herendeen. Energy Research Group, Office of Vice Chancellor for Research, University of Illinois, Urbana, Illinois 61801.

Albert J. Hermann. Horn Point Environmental Lab, University of Maryland, Cambridge, Maryland 21613.

Paul Hill, Jr. Systems Science Institute, University of Louisville, Louisville, Kentucky 40292.

Mark Homer. Department of Environmental Health Sciences, University of South Carolina, Columbia, South Carolina 29208.

Charles Hopkinson. University of Georgia Marine Institute, Sapelo Island, Georgia 31327.

Richard J. Horwitz. Academy of Natural Sciences of Philadelphia, Philadelphia, Pennsylvania 19103.

Kevin J. Hussey. Jet Propulsion Laboratory, Pasadena, California 91103.

Barry Indyke. Center for Ecological Modeling, Rensselaer Polytechnic Institute, Troy, New York 12180.

Bengt-Owe Jansson. Asko Laboratory, University of Stockholm, 11386, Stockholm, Sweden.

Ann-Mari Jansson. Asko Laboratory, University of Stockholm, 11386, Stockholm, Sweden.

Sven Erik Jorgensen. The Pharmaceutical University of Copenhagen, Langkaer Vaenge 9, DK-3500 Vaerloese, Copenhagen, Denmark.

J.R. Kahn. Department of Economics, State University of New York, Binghampton, New York 13901.

Ekrem V. Kalmaz. Engineering Science and Mechanics, University of Tennessee, Knoxville, Tennessee 37916.

Prahlad Kasturi. Department of Agriculture & Resource Economics, University of Hawaii, Honolulu, Hawaii 96822.

Robert K. Kaufmann. Complex Systems Group, University of New Hampshire, Durham, New Hampshire 03824.

W. Michael Kemp. Horn Point Environmental Laboratory, University of Maryland, Cambridge, Maryland 21613.

R. Kendall. Systems Ecology Research, San Diego State University, San Diego, California 92115.

James R. Kercher. Environmental Sciences Division, Lawrence Livermore National Laboratory, Livermore, California 94550.

George G. Killough. Health and Safety Research Division, Oak Ridge National Laboratory, Oak Ridge, Tennessee 37830.

Thomas B. Kirchner. NREL, Colorado State University, Fort Collins, Colorado 80523.

W. Kitchens. National Coastal Ecosystem Team, U.S. Fish & Wildlife Service, Slidell, Louisiana 70458.

Jeffrey M. Klopatek. Oak Ridge National Laboratory, Oak Ridge, Tennessee 37830.

Robert L. Knight. Department of Environmental Engineering Sciences, University of Florida, Gainesville, Florida 32611.

G. H. Kohlmaier. Institut fur Physik n. Theor. Chemie der Universitaet Frankfurt, Robert-Mayer-Shr. 11, 6 Frankfurt 1, West Germany.

G. Kratz. Institut fur Physik n. Theor. Chemie der Universitaet Frankfurt, Robert-Mayer-Shr. 11, 6 Frankfurt 1, West Germany.

E.M. Krenciglowa. McMaster University, Hamilton, Ontario, Canada L8S 4M1.

J.R. Krummel. Environmental Sciences Division, Oak Ridge National Laboratory, Oak Ridge, Tennessee 37830.

Peter F. Landrum. National Oceanic & Atmospheric Administration, Ann Arbor, Michigan 48109.

Vincent A. Lamarra. Wildlife Science Department, Utah State University, Logan, Utah 84322.

William K. Lauenroth. NREL, Colorado State University, Fort Collins, Colorado 80523.

Mitchell J. Lavine. Center for Environmental Research, Cornell University, Ithaca, New York 14853.

Clara H. Leuthart. University College, University of Louisville, Louisville, Kentucky 40292.

Gordon J. Leversee. Savannah River Ecology Laboratory, Aiken, South Carolina 29801.

Karin Limburg. Department of Environmental Engineering Sciences, University of Florida, Gainesville, Florida 32611.

Christopher G. Lind. Systems Science Institute, University of Louisville, Louisville, Kentucky 40292.

Stephen C. Lonergan. Department of Geography, McMaster University, Hamilton, Ontario, Canada.

Orie L. Loucks. The Institute of Ecology, 4600 Sunset Boulevard, Indianapolis, Indiana 46208.

Ariel E. Lugo. Southern Forest Experiment Station, Institute of Tropical Forestry, Rio Piedras, Puerto Rico.

James W. Male. Department of Civil Engineering, University of Massachusetts, Amherst, Massachusetts 01003.

James B. Mankin, Jr. Union Carbide Corporation, Nuclear Division, Oak Ridge, Tennessee 37830.

David R. Marmorek. Environmental & Social Systems Analysts, Ltd., Vancouver, British Columbia, Canada.

W.T. Mason. Eastern Energy & Land Use Team, U.S. Fish & Wildlife Service, Kearneysville, West Virginia 25430.

Hank McKellar. Department of Environmental Health Sciences, University of South Carolina, Columbia, South Carolina 29208

Daniel H. McKenzie. Ecological Sciences Department, Battle Pacific Northwest Laboratories, Richland, Washington 99352.

Christopher McVoy. Ecology & Systematics, Cornell University, Ithaca, New York 14850.

Henning F. Mejer. Copenhagen Engineering School, Copenhagen, Denmark.

Ane D. Merriam. Institute of Science and Public Affairs, Florida State University, Tallahassee, Florida 23206.

Michael A. Miller. Center for Wetlands, University of Florida, Gainesville, Florida 32611.

Philip C. Miller. Systems Ecology Research, San Diego State University, San Diego, California.

William J. Mitsch. Systems Science Institute, University of Louisville, Louisville, Kentucky 40292.

Kenneth A. Morrison. Dept. de Genie Chimique, University of Sherbrooke, Sherbrooke, Quebec, Canada.

John K. Munro, Jr. Computer Science Department, Oak Ridge National Laboratory, Oak Ridge, Tennessee 37830.

Ishwar P. Murarka. Environmental Physics & Chemistry Program, Electric Power Research Institute, Palo Alto, California 94303.

Christopher Neill. Coastal Ecology Laboratory, Center for Wetland Resources, Louisiana State University, Baton Rouge, Louisiana 70803.

James Novak. Department of Forestry, University of Illinois, Urbana, Illinois 61801.

Howard T. Odum. Environmental Engineering Sciences, University of Florida, Gainesville, Florida 32611.

Hisashi Ogawa. Department of Civil Engineering, University of Massachusetts, Amherst, Massachusetts 01003.

Jerry S. Olson. Environmental Sciences Division, Oak Ridge National Laboratory, Oak Ridge, Tennessee 37830.

R.V. O'Neill. Environmental Sciences Division, Oak Ridge National Laboratory, Oak Ridge, Tennessee 37830.

T.T. Palgar. Martin Marietta Environmental Center, Baltimore, Maryland 21227.

Richard A. Park. Center for Ecological Modeling, Rensselaer Polytechnic Institute, Troy, New York 12181.

Bernard Patten. Institute of Ecology, University of Georgia, Athens, Georgia 30602.

Leonard Pearlstine. Department of Environmental Health Sciences, University of South Carolina, Columbia, South Carolina 29208.

Donald B. Porcella. Tetra Tech, Inc., Lafayette, California 94549.

Alician V. Quinlan. Department of Mechanical Engineering, Massachusetts Institute of Technology, Cambridge, Massachusetts 02139.

Philip J. Radford. Institute for Marine Environmental Research, Prospect Place, The Hoe, Plymouth PL1 3DH, United Kingdom.

P.F. Ricci. Ecological Studies Program, Electric Power Research Institute, Palo Alto, California 94303.

John Richardson. Environmental Engineering Sciences, University of Florida, Gainesville, Florida 32603.

James E. Roelle. Western Energy & Land Use Team, U.S. Fish & Wildlife Service, Fort Collins, Colorado 80526.

R. Nicholas Rouse. Eastern Energy & Land Use Team, U.S. Fish & Wildlife Service, Kearneysville, West Virginia 25430.

H.M. Runke. Environmental Research Group, St. Paul, Minnesota 55112.

T.R. Schueler. Horn Point Environmental Laboratory, University of Maryland, Cambridge, Maryland 21613.

Robert Schware. National Center for Atmospheric Research, Boulder, Colorado 80307.

Kristin Shrader-Frechette. Philosophy Department, University of Louisville, Louisville, Kentucky 40292.

Herman Shugart. Environmental Sciences Division, Oak Ridge National Laboratory, Oak Ridge, Tennessee 37830.

J.E. Shultz. Consolidation Coal Company, Stanton, North Dakota 58571.

Lalit Sinha. Illinois Environmental Protection Agency, Springfield, Illinois 62704.

E.O. Sire. Institut fur Physik n. Theor. Chemie der Universitaet Frankfurt, Robert-Mayer-Shr. 11, 6 Frankfurt 1, West Germany.

Francie Smith. Systems Science Institute, University of Louisville, Louisville, Kentucky 40292.

Hugh T. Spencer. Chemical and Environmental Engineering, University of Louisville, Louisville, Kentucky 40292.

Gary Spiller. University of Sherbrooke, Sherbrooke, Quebec, Canada.

Kenneth Staver. Horn Point Environmental Laboratory, University of Maryland, Cambridge, Maryland 21613.

J.C. Stevenson. Horn Point Environmental Laboratory, University of Maryland, Cambridge, Maryland 21613.

J. Kevin Summers. Martin Marietta Environmental Center, Baltimore, Maryland 21227.

Edward T. Sullivan. Forest Resources & Conservation, University of Florida, Gainesville, Florida 32611.

Gordon Swartzman. Center for Quantitative Science, University of Washington, Seattle, Washington 98195.

S. Tartowski. Cornell University, Research Park, Ithaca, New York 14850.

Jan R. Taylor. Systems Science Institute, University of Louisville, Louisville, Kentucky 40292.

Normand Thérien. Department of Chemical Engineering, University of Sherbrooke, Sherbrooke, Quebec, Canada.

Robert E. Ulanowicz. Chesapeake Biological Laboratory, University of Maryland, Solomons, Maryland 20688.

Thomas Veselka. Energy and Environmental Systems Division, Argonne National Laboratory, Argonne, Illinois 60439.

Robert Walker. Regional Science Department, University of Pennsylvania, Philadelphia, Pennsylvania 19104.

Flora C. Wang. Center for Wetland Resources, Louisiana State University, Baton Rouge, Louisiana 70803.

Darrell C. West. Environmental Sciences Division, Oak Ridge National Laboratory, Oak Ridge, Tennessee 37830.

R.T. Willey. Horn Point Environmental Laboratory, University of Maryland, Cambridge, Maryland 21613.

Thomas Zaret. Department of Zoology, University of Washington, Seattle, Washington 98195.

G.W. Zimmerman. Department of Engineering Sciences & Mechanics, University of Tennessee, Knoxville, Tennessee 37830.

Jim Zucchetto. Regional Science Department, University of Pennsylvania, Philadelphia, Pennsylvania 19104.

CONTENTS

5. Modelling of Air Pollution Impact

SECTION II: ENERGY AND SYSTEMS

6. Regional Systems

MODELLING OF ENERGY/ENVIRONMENT ISSUES

1. Global Carbon Models

I. Global Carbon Models

CARBON DIOXIDE DYNAMICS OF THE BIOSPHERE[1]

Sandra Brown[2], George Gertner[2], Ariel E. Lugo[3], and James Novak[2]

Abstract.--The seasonal oscillation in atmospheric CO_2 records measured at four worldwide stations were converted to monthly and annual rates of uptake (U) and release (R) of CO_2 to see if new information about the dynamics of the biosphere could be obtained. Our preliminary results show that annual rates of R and UR ratio (from the observed Mauna Loa data) appear to be responding to fossil fuel produced CO_2 inputs; annual U is not, suggesting carbon is accumulating in the biosphere. The amount of fossil fuel produced CO_2 remaining in the atmosphere is about 51-55% at Mauna Loa and the Southern Hemisphere stations, but about 25% at Point Barrow. However, the degree of uncertainty about these estimates is large. The consistency in the timing and magnitude of the monthly U and R rates in the Northern Hemisphere stations suggest no major human impact on the biota. Until more is known about the pattern of CO_2 exchange with the oceans, other sinks or sources of CO_2, and what the airsheds for the various monitoring stations are, we cannot use atmospheric CO_2 data to determine with certainty whether the biota is a source or sink for CO_2.

INTRODUCTION

The concentration of atmospheric CO_2 measured at several worldwide stations (Keeling et al. 1976a and 1976b, Lowe et al. 1979) has attracted the attention of many disciplines, each trying to relate the observations and their particular paradigms to the carbon balance of the world (Broecker et al. 1979, Hall et al. 1973, Junge and Czeplak 1968, Machta et al. 1977, Pearman and Hyson 1980, and Stuiver 1978). The long-term atmospheric CO_2 concentration data (fig. 1a-d) has two trends: a secular rise and a seasonal oscillation. The secular rise corresponds to about one-half of the CO_2 released from the burning of fossil fuels (major source) and the manufacture of cement remaining in the atmosphere, assuming that any releases from the world's biota are insignificant.

The seasonal oscillation is attributed to the metabolism of the biosphere (Bolin and Bischof 1970, Hall et al. 1975, Junge and Czeplak 1968, Keeling et al. 1976a and 1976b, Lowe et al. 1979, Machta et al. 1977, Woodwell et al. 1973), and its amplitude varies with latitude from about 15 ppm at Point Barrow (71°N latitude) to about 1 ppm at the South Pole (c.f. Machta et al. 1977 for summary of long-term and short-term measurements). The decrease in amplitude of the seasonal oscillation generally parallels the decreasing ratio of land area to total area in a given latitudinal belt from north to south.

Our interest in the atmospheric CO_2 data was to analyze the seasonal oscillations at the four different stations (fig. 1) by converting CO_2 concentrations into monthly rates of net uptake (U) or net release (R) using established procedures in limnology and terrestrial ecology for analyzing diurnal O_2 or CO_2 measurements. We then applied modeling tools to these rates, with particular emphasis on the extensive Mauna Loa data base, to take out the fossil fuel effects (secular rise) and to see if new information can be obtained about the CO_2 dynamics of the biosphere. Specifically, we raised the following questions:

1. Are the uptake and release rates of CO_2 changing in response to fossil fuel produced CO_2 inputs?

[1] Paper presented at the international symposium Energy and Ecological Modelling, sponsored by the International Society for Ecological Modelling. (Louisville, Kentucky, April 20-23, 1981)

[2] Sandra Brown and George Gertner are Assistant Professors in Forestry and James Novak is a graduate student in Forestry, all at the University of Illinois, Urbana, Illinois, U.S.A.
[3] Ariel Lugo is a project leader of the Southern Forest Experiment Station's Institute of Tropical Forestry, Rio Piedras, Puerto Rico.

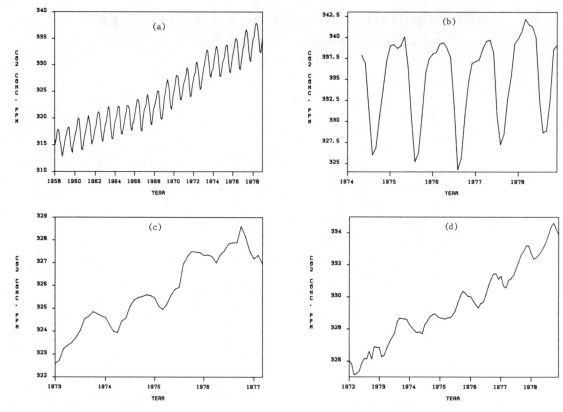

Figure 1.--Atmospheric CO_2 concentration at (a) Mauna Loa[4]
(20ºN latitude), (b) Point Barrow[5] (71ºN latitude),
(c) New Zealand (42ºS latitude, Lowe _et al._ 1979),
and (d) South Pole (90ºS latitude, Keeling _et al._
1976a).

2. Is the amount of fossil fuel-produced CO_2 remaining in the atmosphere the same at all monitoring stations?

3. Do the timing and magnitude of the rates of uptake and release of CO_2 relate to what we know about the functioning of terrestrial ecosystems?

4. Can the atmospheric CO_2 measurements help us determine whether the terrestrial biosphere is a sink for part of the fossil fuel produced CO_2 or is it itself a source?

Estimates of whole ecosystem metabolism have been made from atmospheric inversions (Woodwell

and Dykeman 1966); however, only total respiration (release rate) could be estimated by this method. Others have used the Mauna Loa data to establish "net hemispheric metabolism" (Hall _et al._ 1975). They, however, looked only at annual rates of uptake and release calculated as the difference between peak to trough ("semiannual net photosynthesis") and trough to peak ("semiannual net respiration") after correcting for a presumed 42% uptake by the oceans. Other statements about biospheric activity have been made using atmospheric concentration data (Lowe _et al._ 1979, Woodwell _et al._ 1973), but the authors have confused concentration with rates. For example, Woodwell _et al._ (1973) showed that the lowest CO_2 concentration occurred in September and the highest in December-January. The authors commented that the largest decrease in CO_2 concentration occurred "in the fall as photosynthesis in the northern hemisphere stores carbon". By calculating the rate of change from their figure 4, however, we find that the peak _rate_ of CO_2 uptake occurred earlier in June-July.

---

[4]Bacastow, R. 1979. Personal correspondence. Scripps Institute of Oceanography, La Jolla, California, U.S.A.

[5]Komhyr, W. D. 1979. Personal correspondence. NOAA-US Dept. Commerce, Boulder, Colorado, U.S.A.

Rust and Kirk (1978) also confused rates with concentrations in their analysis of the Mauna Loa data. They found that the waveform resulting from the detrended and normalized atmospheric CO_2 <u>concentration</u> measurements was strikingly similar to a waveform resulting from the normalization of the monthly <u>rates</u> of heat storage in the atmosphere of the Northern Hemisphere. Because they confused rates with concentrations, one must question their statement that "most of the seasonal variation of atmospheric CO_2 at Mauna Loa might be caused by the transport of CO_2 molecules across that (ocean-atmosphere) interface."

EFFECTS OF FOSSIL FUEL INPUTS ON ANNUAL RATES OF CO_2 UPTAKE AND RELEASE

The analysis in this section will deal only with the Mauna Loa data because it has the longest period of continuous record (1958-1978) and it is roughly representative of the Northern Hemisphere. Updated atmospheric CO_2 concentration data for Mauna Loa[4] was used for the analysis. These data are tentative because the measurement system is still being calibrated. However, we assumed that the errors are systematic and would not affect our analysis because we deal with differences in concentrations (rates). Data for the CO_2 production from burning of fossil fuel and manufacture of cement were obtained from Rotty (1979), and we assumed that most of this CO_2 production occurred in the industrial Northern Hemisphere.

The first step in our analysis was to calculate the monthly rate of change of CO_2 concentration. We did this by subtracting the mean monthly concentration at month m from the mean of month m+1 and, assuming that the mean concentration occurred in the middle of the month, we assigned the monthly rate value to the middle of month m+1. Positive rates of change represent net release of CO_2 (R) and negative rates of change represent net uptake of CO_2 (U). We next integrated the monthly rates of change to calculate the annual uptake and release of CO_2 for the period of record.

Annual R is always greater than annual U, and their shapes are very similar (fig. 2). Annual R varied between 5.1 and 8.0 ppm/yr and annual U between 4.6 and 6.1 ppm/yr. We applied ordinary least squares regression (OLSR) to estimate simple linear models of annual U vs. annual CO_2 input from fossil fuel burning (FFI), annual R vs. FFI, and annual UR ratio vs. FFI. We found: 1) no significant relationship between U and R and FFI at the 0.05 level, however, if we accept a significance level of 0.15, we found that R was increasing with increasing FFI, and 2) a significant (P = 0.05) negative relationship between UR ratio and FFI (fig. 3). Because the secular rise of atmospheric CO_2 represents a net source to the atmosphere, we should expect the annual U rates to be masked somewhat, resulting in an observed decrease in U, an observed increase in R, and an observed decrease in the UR ratio. This can be best explained by

the following analogy: if CO_2 uptake of two identical plants is measured under identical conditions, except that the soil is isolated in one and not in the other, it would appear that the plant with the exposed soil was fixing CO_2 at a lower rate than the one with the isolated soil. In reality, both plants fixed the same amount of CO_2. At the global scale, production of CO_2 from fossil fuel is analogous to the production of CO_2 by the soil and biospheric CO_2 uptake (U) is analogous to plant CO_2 uptake. We did find that the observed annual UR ratio and annual R followed the expected trend, but not annual U. The fact that annual U is not changing with FFI suggests that 1) the storage of carbon in the Northern Hemisphere is increasing by either the rate of storage increasing (CO_2 enrichment) or the area of vegetated land increasing (more forest land) or 2) the oceans are taking up more in the summer than in the winter or 3) the rate of CO_2 accumulation is increasing in other unknown, but significant, processes.

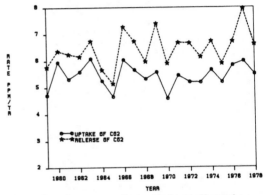

Figure 2.--The course of observed annual uptake and release of CO_2 (fossil fuel effects not removed) at Mauna Loa for the period of record.

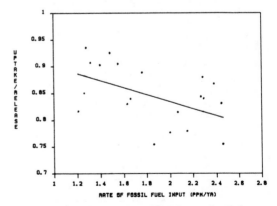

Figure 3.--Simple linear model of annual uptake to release ratio (U/R) vs. annual rate of fossil fuel produced CO_2. The equation is Y = 0.97-0.66 X (significant at P = 0.05).

AMOUNT OF FOSSIL FUEL CO_2 REMAINING IN THE ATMOSPHERE

Estimates of the amount of fossil fuel produced CO_2 remaining in the atmosphere at Mauna Loa range from 49 ± 12% (Bacastow and Keeling 1973) to 52 ± 4% (Broecker et al. 1979) and 54% (Keeling et al. 1976a); at the South Pole the airborne fraction is estimated to be 52% (Keeling et al. 1976a). These values were obtained either from the results of models or by determining the secular rise over a period of years by fitting cubic trend to the data (by OLSR) using time as the independent variable and expressing the rise as a fraction of the total input during the same period. These results assume that the fraction remaining airborne is constant over time. An analysis by Machta et al. (1977), however, indicated that the fraction remaining airborne at Mauna Loa was not constant from year to year but varied between 20-100% during the period 1961-74.

Our approach to determine the fraction remaining in the atmosphere was to fit a linear model to the CO_2 concentration data as a function of cumulative FFI using OLSR for the four monitoring stations shown in figure 1. The results of this analysis (table 1) show that the fraction remaining airborne at Mauna Loa and South Pole is within the range of values found by other investigators (see above). Our analysis also shows that the amount remaining airborne at New Zealand is similar to that at Mauna Loa and South Pole. Point Barrow, however, appears to be responding differently. We did not get a significant relationship at this station, suggesting that considerably less remains airborne at Point Barrow, or the amplitude of the seasonal oscillation is large (about 15 ppm) compared to the magnitude of the FFI, and/or the period of record was too short, which prevent us from obtaining a significant slope parameter (b_1). The standard errors of b_1 (table 1) should be viewed with caution. To use OLSR requires the assumption to be made that the observations of CO_2 concentration be independent; this requirement is violated when cumulative data are used. When the assumption of independence is violated, the estimate of the slope parameter is unbiased, while the standard error estimate of the slope parameter is biased downward, i. e., the standard error estimate is smaller than it should be (Mandel 1957). Therefore, the estimates of the fraction remaining airborne (table 1) are unbiased, but the estimated degree of precision of the parameter estimates are greater than they should be.

Converting cumulative data into rates helps partially overcome some of the problems of using OLSR, and as stated earlier, we are interested in rates rather than concentrations. We used, therefore, OLSR to fit a linear model to the monthly rates of change of CO_2 concentrations (ΔCO_2) and the monthly FFI rates (ΔFFI), assuming that the ΔFFI was constant throughout the year. The results of this analysis produced no meaningful estimates for the slope parameter for Point Barrow, New Zealand, or South Pole stations. This was probably

Table 1.--Ordinary least squares regression parameter estimates of the linear model $[CO_2] = b_0 + b_1$FFI.

Data base	b_0	b_1[1] ± 1SE	r^2
Mauna Loa (1958-1978)	315.21	0.55 ± 0.011	0.90
Point Barrow (1974-1978)	327.10	0.25 ± 0.210	0.03
New Zealand (1973-1977)	310.64	0.53 ± 0.027	0.89
South Pole (1972-1978)	314.45	0.51 ± 0.013	0.95

[1]Fraction remaining airborne.

caused by a combination of the small number of points (short record) and the large variation in the ΔCO_2 relative to the ΔFFI (which ranged from about 0.1-0.2 ppm/mo). The more extensive data base at Mauna Loa produced more meaningful results with a linear model (table 2), with an estimate of the slope comparable to that in table 1, but standard error indicated that this estimate was not significant. Because of the large amplitude in oscillation of the ΔCO_2 relative to the magnitude of ΔFFI, we tried fitting more complex nonlinear models to remove the oscillation effects (table 2). These models gave lower estimates of the fraction remaining airborne than did the linear model, and both had large standard errors showing that these estimates were insignificant, although the 12 and 6 mo model gave a smaller standard error than the 12 mo model.

Table 2.--Ordinary least squares regression parameter estimates of ΔCO_2 as a function of ΔFFI only and ΔFFI and time combined.

Model	b_0	b_1	±	1SE
Linear[1]	0.02	0.62	±	1.94
Non-linear				
12 mo sine[2]	0.02	0.42	±	1.03
12 mo and 6 mo sine[3]	0.01	0.48	±	0.58

[1]$\Delta CO_2 = b_0 + b_1 \Delta$FFI

[2]$\Delta CO_2 = b_0 + b_1 \Delta$FFI $+ b_3 \sin(\frac{2\pi}{12}(t + b_4))$

[3]$\Delta CO_2 = b_0 + b_1 \Delta$FFI $+ b_3 \sin(\frac{2\pi}{12}(t + b_4)) + b_5 \sin(\frac{2\pi}{6}(t + b_6))$

In conclusion, our analysis suggests that the fraction of FFI remaining airborne at Mauna Loa ranges from 42-62%, with a large degree of uncertainty. Our analysis produced values for Mauna Loa, New Zealand, and South Pole that were comparable to those published elsewhere. The fraction remaining airborne at Point Barrow appears to be considerably less than at the other three monitoring stations.

MAGNITUDES AND PHASING OF CO_2 UPTAKE AND RELEASE RATES

Monthly Uptake and Release Rates

The first step in this analysis was to remove the effects of fossil fuel produced CO_2 from the rate data. Our approach was first to fit a linear model using OLSR to the monthly ΔCO_2 and ΔFFI for all four stations shown in figure 1. We then plotted the residuals of this model against time as shown in figure 4a-d, with U rates plotted above the zero line and R rates plotted below the zero line. These results show the following:

. the pattern of monthly U rates in the two Northern Hemisphere stations is smoother than for the two Southern Hemisphere stations. This smoother pattern in the Northern Hemisphere possibly reflects the influence of the larger land masses.

. the magnitudes of the monthly U and R rates decrease latitudinally from Point Barrow to South Pole.

. maximum U rates are generally higher than maximum R rates at Point Barrow and Mauna Loa (monthly U=6-8 ppm/mo and 2 ppm/mo, respectively; monthly R=4-5 ppm/mo and 1.5 ppm/mo, respectively), whereas at New Zealand and South Pole maximum U rates are similar to maximum R rates (approximately 0.5 ppm/mo for both stations).

. all four stations show a minimum monthly R rate in their winter months that coincides with the period of peak monthly U in the opposite hemisphere, suggesting that not only do the air masses of the two hemispheres mix, but that they do so relatively quickly (on the order of a month). At the two Southern Hemisphere stations there is also a decrease in the monthly U rate in most years coinciding with peak R in the Northern Hemisphere; we cannot detect a similar trend in the Northern Hemisphere stations. We suspect, however, that Northern Hemisphere rates are similarly affected and this is one line of analysis we are presently pursuing. If they are affected by activity in the Southern Hemisphere, the timing of maximum monthly U may occur at some other time of the year than is indicated in figure 4a and 4b.

. the length of the growing season (the length of time U $\geq$ 0, i.e., above the zero line) varies from 131 ± 15 (SE) days at Point Barrow to 142 ± 2 (SE) days at Mauna Loa and 195 ± 5 (SE) days at New Zealand (we did not calculate the length for South Pole because the pattern was too erratic). The small standard errors for the length of the growing season, particularly for Mauna Loa measurements, may be surprising when one considers that the world is composed of a variety of ecosystems, all adapted to different sets of environmental factors.

. peak monthly U occurred at approximately mid-August at Mauna Loa, late June at Point Barrow, and mid-January at New Zealand (table 3). This timing, of course, depends upon the validity of our assumption that the monthly rate of change between any two adjacent months occurs at the middle of month m + 1 (see first section). Regardless of the assumption, our analysis does show that peak U at Point Barrow occurs approximately 1.5 mo earlier than at Mauna Loa.

Table 3.--Timing of maximum monthly rate of uptake of CO_2.[1]

	b_4 ± 1SE	Month of year[2]
Mauna Loa	7.16 ± 0.08	7.84
Point Barrow	8.75 ± 0.27	6.26
New Zealand	1.89 ± 0.37	1.11

[1]Determine from fitting the model by OLSR: $\Delta CO_2 = b_1 + b_2 \Delta FFI + b_3 \sin(\frac{2\pi}{12}(t + b_4))$.

[2]Monthly rate was assigned to the middle of the months (see first section), e.g., month 7 = mid-July.

In summary, monthly U and R rates in the Northern Hemisphere are more consistent and of greater magnitude than those in the Southern Hemisphere. The length of the growing season is shorter, the magnitude of U is higher, and the timing of peak U is earlier at Point Barrow than at Mauna Loa. The consistency in timing of peak U, the consistency of the annual pattern of monthly U and R, and the consistency in the magnitude of U and R for the periods of record at Point Barrow and Mauna Loa suggest that CO_2 exchange in the Northern Hemisphere does not reflect any impact by humans on the biota during the last two decades. The absence of a uniform pattern of monthly U and R at South Pole and New Zealand possibly reflects a lesser influence from large land masses (approximately 5% of the total surface area south of latitude 30°S and not including Antarctica is land; Pearman and Hyson 1980), interference from northern air masses; and a greater oceanic influence.

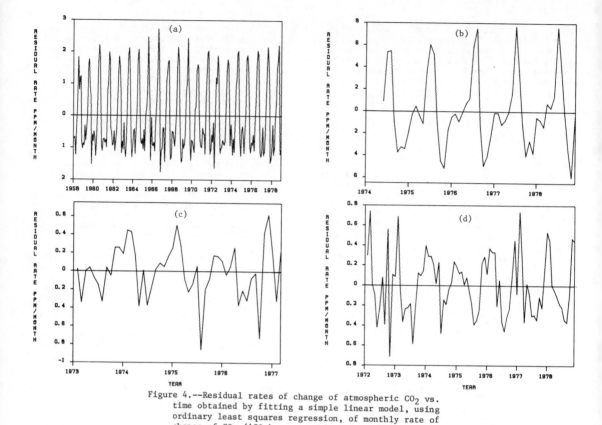

Figure 4.--Residual rates of change of atmospheric CO_2 vs. time obtained by fitting a simple linear model, using ordinary least squares regression, of monthly rate of change of CO_2 (ΔCO_2) vs. monthly fossil fuel produced CO_2 input rate (ΔFFI) for the following stations: a) Mauna Loa, b) Point Barrow, c) New Zealand, and d) South Pole. Monthly uptake rates are plotted above the zero line and release rates below the zero line.

Annual Uptake and Release Rates

In addition to monthly rates we were also interested in the annual rates of U and R. We tried fitting a Fourier series to the monthly rates because fitting a sine wave assumes a symmetry of U and R which would prevent us from seeing if there were differences between them in any given year. The model produced by fitting the Fourier series did not produce satisfactory results, so instead we used the empirical data. The annual U and R rates were calculated using the same method as described in the first section; we did not use South Pole data because of the erratic nature of the monthly rates (fig. 4d).

Annual U and R are, respectively, approximately 14 ppm and 14.5 ppm at Point Barrow, 5.5 ppm and 6 ppm at Mauna Loa, and 1.5 ppm for both U and R at New Zealand (fig. 5). Using the value of 2.12×10^{15} g C/ppm CO_2 (Verniani 1966) these rates translate to annual U and R of 29 and 31 x 10^{15} g C/yr for Point Barrow, 11.5 and 13 x 10^{15} g C/yr at Mauna Loa, and 3 x 10^{15} g C/yr for both U and R at New Zealand. If we assume that Mauna Loa represents the Northern Hemisphere, net ecosystem production there (NEP = net annual U) is approximately 11.5×10^{15} g C/yr. This is double the value obtained by Pearman and Hyson (1980) from their model of CO_2 exchange in the atmosphere.

During three out of the four years of data, annual R was greater than annual U at both Point Barrow and New Zealand; R was greater than U for about half the period of record at Mauna Loa (fig. 5). When R is greater than U, is the net release from the biosphere really greater than net uptake, or does it mean that more fossil fuel produced CO_2 is remaining in the atmosphere (i.e., more than the average amount estimated by the model, e.g., = 62% at Mauna Loa, c.f. table 2) as was suggested by Machta et al. (1977)? Similarly, when U is greater than R, is the net uptake by the biosphere greater than net release, or is less remaining in the atmosphere and maybe more being removed by the oceans and other sinks? Until more is known about the pattern of CO_2 uptake by ocean processes and what the airsheds for the various stations are, we cannot answer this question at this time.

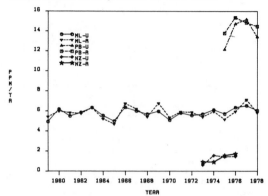

Figure 5.--The course of annual rates of CO_2 uptake (U) and release (R) with fossil fuel effects removed. ML = Mauna Loa, PB = Point Barrow, and NZ = New Zealand.

Conformity of Atmospheric CO_2 Rates With Ecosystem Function

Are the observed trends in the rates of U and R consistent with what we know about the functioning of terrestrial ecosystems? This question implies that all the uptake and release of CO_2 are due to the activity of the terrestrial biota. This may be a valid assumption for the Northern Hemisphere because total U during the growing season varies between 5-6 ppm at Mauna Loa to 13-15 ppm at Point Barrow (fig. 5) and results of ocean models suggest that only 1 ppm (40% of FFI) is being sequestered by the oceans during the whole year (Broecker et al. 1979). In addition, the ratio of land area to total area is approximately 40% (Pearman and Hyson 1980). In the Southern Hemisphere the situation is somewhat different because of the reasons given earlier. Therefore, the question posed at the beginning of this section will be answered in reference to the Northern Hemisphere stations only.

There are few studies that have measured CO_2 exchange from whole ecosystems for comparison with atmospheric exchange rates. Two studies in northern latitutde ecosystems (Coyne and Kelley 1975, Woodwell et al. 1973) did show peak net CO_2 uptake in June-July and net uptake approaching zero by the end of August, two trends exhibited by the Point Barrow data. There are no similar types of studies for lower latitude ecosystems with which to use as a reference for Mauna Loa, but peak net CO_2 uptake in early August is consistent with our expectations. We would also expect the length of the growing season to be longer for the Northern Hemisphere as a whole than

for higher latitudes alone; this is supported by our analysis (fig. 4a and 4b). However, our analysis shows that net CO_2 uptake starts earlier at Point Barrow (generally April-May, when temperatures are close to zero) than at Mauna Loa (generally May-June). Our experience tells us that for the Northern Hemisphere, in general, plants break dormancy earlier than May-June. Because no net CO_2 uptake is observed until later suggests that processes in ecosystems that release CO_2 are more active than those that fix it. In high latitude ecosystems the reverse trend appears to be true. There is evidence that tundra plants can uptake CO_2 at temperatures as low as $0^{\circ}C$ (Coyne and Kelley 1975). However, all the processes of CO_2 uptake and release at high latitudes may not be biological (see below).

The high annual U rate at Point Barrow appears to suggest that net ecosystem production is greater in high latitude ecosystems than in those typical for the Northern Hemisphere (represented by Mauna Loa). It has been suggested that these higher annual rates result from the large land areas in higher latitudes (Keeling et al. 1976b). However, if Mauna Loa is assumed to represent the whole Northern Hemisphere, then the total land area of its airshed is greater than for Point Barrow. We suggest that the difference in land area alone cannot account for the more than two-fold higher annual U and R at Point Barrow than at Mauna Loa. Factors other than the metabolism of high latitude ecosystems must be responsible for the high annual U and R rates. For example, Broecker et al. (1979) showed the North Atlantic Ocean (at 70° N latitude) to be a strong sink for CO_2 based on differences in the partial pressure of CO_2 (pCO_2) exerted by the ocean and atmosphere along the GEOSECS track, and they stated that large seasonal changes in ocean pCO_2 are to be expected at high latitude locations. However, the seasonal pattern of pCO_2 in high latitude oceanic waters is unknown (Broecker et al. 1979). This factor and possibly others may explain why annual U and R are high and why there is a lower fraction of the fossil fuel produced CO_2 remaining in the atmosphere at Point Barrow.

Assuming Mauna Loa does represent the Northern Hemisphere, we divided the average annual U rate (11.5×10^{15} g C/yr) by the area of land surface[6] and produced an estimate of net ecosystem production (NEP) during the growing season (annual U rate) of 115 g C/m². This is about half the average value of net primary production (NPP) estimated by Pearman and Hyson (1980, who used data from Whittaker and Likens [1973]) for four latitudinal belts which represent major ecosystems types found in the Northern Hemisphere. Because our estimate from atmospheric CO_2 data is in the right order of magnitude (NEP should be lower than NPP, and it should be greater than zero during the growing season so that

---

[6]Surface area of Northern Hemisphere = 2.56×10^{14}m², ratio of land area to total surface area = 0.39 (estimated from data in Pearman and Hyson 1980), land area = 9.88×10^{13}m².

some organic production is available for the remainder of the year) we suggest that the annual rates of CO_2 U and R measured at Mauna Loa do represent production and respiration of the biosphere in the Northern Hemisphere.

ANNUAL RATES OF UPTAKE AND RELEASE WITH VARYING AMOUNTS OF FOSSIL FUEL PRODUCED CO_2 REMOVED

We tested the sensitivity of annual U, R, and UR ratio of the Mauna Loa data to varying amounts of FFI removed because the degree of uncertainty in the amount of fossil fuel produced CO_2 remaining in the atmosphere is large.

As the percent removed decreases (or percent remaining increases) corrected annual U increases, corrected annual R decreases and UR ratio changes from less than to greater than one (table 4). However, the annual rates of U and R and UR ratio appear not to be very sensitive to changes in the percent removed because none of them change by more than 2% around the average value and the annual course of U and R with, e.g., 40% removed (fig. 6) is the same as that shown in figure 5.

We can use the results in table 4 as another means of estimating the average amount of FFI remaining in the atmosphere. Because the secular rise (percent remaining) is the difference between all known and unknown sources and sinks of CO_2, removal of the exact amount of percent remaining must result in an average UR ratio of one; this occurs when 50-55% of FFI remains (or 45-50% is removed). This estimate is similar to the others discussed earlier.

Table 4.--Sensitivity of annual uptake (U), release (R), and uptake to release ratio (U/R) to varying amounts of fossil fuel produced CO_2 removed from the atmosphere.[1]

% Removed[2]	Rate of CO_2		U/R
	U	R	($\pm$1SE)
	(ppm/yr $\pm$ 1SE)		
60	5.70 $\pm$ 0.10	5.93 $\pm$ 0.15	0.97 $\pm$ 0.02
55	5.74 $\pm$ 0.10	5.88 $\pm$ 0.15	0.98 $\pm$ 0.02
50	5.77 $\pm$ 0.10	5.83 $\pm$ 0.15	0.99 $\pm$ 0.02
45	5.81 $\pm$ 0.10	5.78 $\pm$ 0.15	1.01 $\pm$ 0.02
40	5.85 $\pm$ 0.10	5.73 $\pm$ 0.15	1.03 $\pm$ 0.02
35	5.88 $\pm$ 0.11	5.67 $\pm$ 0.15	1.04 $\pm$ 0.02

[1]Using the simple linear model:
(Residual ΔCO_2) = ΔCO_2 - $b_1 \Delta FFI$, where

ΔCO_2 = rate of change of CO_2

ΔFFI = monthly rate of fossil fuel produced CO_2

b_1 = fraction remaining.
[2]Percent removed = 100 (1 - b_1).

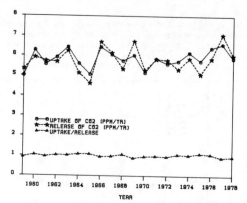

Figure 6.--The course of annual CO_2 uptake, release, and uptake to release ratio with 40% of the fossil fuel produced CO_2 removed (approximately the amount removed by oceans, Broecker et al. 1979) for the period of record at Mauna Loa.

To determine whether the biosphere is a sink or source of CO_2, we need to know the magnitude of the other global sinks. If we assume that the oceans remove between 35-40% of the FFI as suggested by Broecker et al. (1979), the biosphere is a small sink of CO_2 of the order 0.12 - 0.21 ppm/yr or 0.3 - 0.4 x 10^{15} g C/yr (table 4). This is contrary to other estimates based on models of the world's terrestrial ecosystems, which suggest that the biosphere is a source of 4 - 8 x 10^{15} g C/yr (Woodwell et al. 1978) or a source or sink of magnitude $\pm$ 2 x 10^{15} g C/yr (Seiler and Crutzen 1980).

The course of annual U, R, and UR ratio for 40% of FFI removed is shown in figure 6. There was no significant change (P = 0.05) in U, R, or UR ratio over time. Annual R was considerably larger than annual U during the three years 1966, 1969, and 1977; it was slightly larger than U in 1961. For the remainder of the period of record annual U was greater than R or approximately equal to it. Bacastow (1976) suggested that during those years when the Southern Oscillation Index (SOI = difference in the average monthly barometric pressure between Easter Island, near the center of a high pressure cell, and Darwin, representative of the equatorial low pressure zone) was high there was an increased loss of CO_2 from the atmosphere attributed to an increased uptake by either the oceans or biota, but most likely the oceans. When the SOI was low the converse was true. The SOI exhibited low values during 1961, 1966, 1969, and 1973 (Bacastow 1976). We found annual R greater than annual U during three of these years (fig. 6); no data were available for 1977 when R was also greater than U. The apparent relationship between those years when R is greater than U and the presence of a low SOI appears to indicate that less fossil fuel produced CO_2 was removed by the oceans rather than net R from the biosphere being greater than net U.

SUMMARY AND CONCLUSIONS

By converting atmospheric CO_2 concentration records into rates of uptake and release we have been able to determine the following about the CO_2 dynamics of the biosphere:

- observed annual R rate and UR ratio determined from the Mauna Loa data are changing significantly with increasing inputs of fossil fuel produced CO_2. The observed annual U rate has remained constant suggesting that carbon is accumulating in the biosphere.

- 42-62% of fossil fuel produced CO_2 appears to be remaining in the atmosphere, but the degree of uncertainty is large. Point Barrow appears to be much less affected by fossil fuel inputs than the other monitoring stations.

- the constancy in timing and magnitude of the monthly rates of U and R for the two Northern Hemisphere stations suggest little human impact on the biosphere during the last two decades.

- the timing of peak monthly U and the length of the growing season at Point Barrow and Mauna Loa are generally consistent with what we know about ecosystem processes.

- the magnitudes of monthly and annual U and R rates at Point Barrow suggest processes other than biological are responsible for the high rates obtained there.

- annual U and R rates for Mauna Loa are comparable to those reported in the literature for terrestrial ecosystems of the Northern Hemisphere.

- annual U, R, and UR ratio are fairly insensitive to varying fractions of fossil fuel produced CO_2 remaining in the atmosphere.

- if the fraction of fossil fuel produced CO_2 removed by the oceans is approximately 40%, the biosphere is a sink of about 0.3-0.4 x 10^{15} g C/yr.

Our analysis confirms the earlier ones that the seasonal oscillation in atmospheric CO_2 measurements is due mainly to biological processes in the biosphere. We also suggest that oceanic and other large scale atmospheric processes affect the annual rates of CO_2 exchange. Until more is known about the annual variation and seasonality in oceanic rate of uptake and release of CO_2 and the factors that affect these rates, we cannot exploit fully the atmospheric CO_2 measurements to address the question of whether the biosphere is a sink or source of carbon dioxide.

LITERATURE CITED

Bacastow, R. 1976. Modulation of atmospheric carbon dioxide by the Southern Oscillation. Nature 261:116-118.

Bacastow, R. and C.D. Keeling. 1973. Atmospheric carbon dioxide and radiocarbon in the natural carbon cycle: 2. Changes from A.D. 1700 to 2070 as deduced from a geochemical model. p. 86-135. In G.M. Woodwell and E.V. Pecan (eds.), Carbon and the biosphere. U.S. Atomic Energy Commission. CONF-720510, NTIS, Springfield, Virginia, USA.

Bolin, B. and W. Bischof. 1970. Variations of the carbon dioxide content of the atmosphere in the northern hemisphere. Tellus 22:431-442.

Broecker, W.S., T. Takahashi, H.J. Simpson, and T.H. Peng. 1979. Fate of fossil fuel carbon dioxide and the global carbon budget. Science 206:409-418.

Coyne, P.I. and J.J. Kelley. 1975. CO_2 exchange over the Alaskan arctic tundra: meterological assessment by an aerodynamic method. J. Appl. Ecol. 12:587-611.

Hall, C.A.S., C.A. Ekdahl, and D.E. Wartenberg. 1975. A fifteen year record of biotic metabolism in the Northern Hemisphere. Nature 255:136-138.

Junge, C.E. and G. Czeplak. 1968. Some aspects of the seasonal variation of carbon dioxide and ozone. Tellus 20:422-434.

Keeling, C.D., J. Adams, Jr., C.A. Ekdahl, Jr., and P.R. Guenther. 1976a. Atmospheric carbon dioxide variations at the South Pole. Tellus 28:552-564.

Keeling, C.D., R.B. Bacastow, A.E. Bainbridge, C.A. Ekdahl, P.R. Guenther, and L.S. Waterman. 1976b. Atmospheric carbon dioxide variations at Mauna Loa Observatory, Hawaii. Tellus 28:538-551.

Lowe, D.C., P.R. Guenther, and C.D. Keeling. 1979. The concentration of atmospheric carbon dioxide at Baring Head, New Zealand. Tellus 31:58-67.

Machta, L., K. Hanson, and C.D. Keeling. 1977. Atmospheric carbon dioxide and some interpretations. p. 131-144. In fate of fossil fuel CO_2 in the oceans. Plenum Press, New York, USA.

Mandel, J. 1957. Fitting a straight line to certain types of cummulative data. J. Amer. Stat. Assoc. 52:552-566.

Pearman, G.I. and P. Hyson. 1980. Activities of the global biosphere as reflected in atmospheric CO_2 records. J. Geophys. Res. 85(C8):4457-4467.

Rotty, R. 1979. Data for global CO_2 production from fossil fuels and cement. Unpublished manuscript. 3 pp., Institute of Energy Analysis, Oak Ridge Associated Universities, Oak Ridge, Tennessee, USA.

Rust, B.W. and B.L. Kirk. 1978. Inductive modelling of population time series. p. 154-192. In H.H. Shugart, Jr. (ed.), Time series and ecological processes. Proceedings of SIAM, Inst. for Mathematics and Society, Philadelphia, Pennsylvania, USA.

Seiler, W. and P.J. Crutzen. 1980. Estimates of gross and net fluxes of carbon between the biosphere and the atmosphere from burning biomass. Climate Change 2:204-247.

Stuiver, M. 1978. Atmospheric carbon dioxide and carbon reservoir changes. Science 199:253-258.

Verniani, S. 1966. The total mass of the earth's atmosphere. J. Geophys. Res. 71:385-391.

Whittaker, R.H. and G.E. Likens. 1973. Carbon in the biota. p. 281-302. In G.M. Woodwell and E.V. Pecan (eds.), Carbon and the biosphere. U.S. Atomic Energy Commission. CONF-720510, NTIS, Springfield, Virginia, USA.

Woodwell, G.M. and W.R. Dykeman. 1966. Respiration of a forest measured by CO_2 accumulation during temperature inversions. Science 154: 1031-1034.

Woodwell, G.M., R.A. Houghton, and N.R. Tempel. 1973. Atmospheric CO_2 at Brookhaven, Long Island, New York: patterns of variation up to 125 meters. J. Geophys. Res. 78:932-940.

Woodwell, G.M., R.H. Whittaker, W.A. Reiners, G.W. Likens, C.S. Delwiche, and D.B. Botkin. 1978. The biota and the world carbon budget. Science 199:141-145.

SIMULATION OF SOME ECOLOGICAL CONSEQUENCES
OF ENERGY DEVELOPMENT IN NORTHERN
ECOSYSTEMS[1]
P. C. Miller, M. Blake-Jacobson, and R. Kendall[2]

Abstract.--Two simulation models, one simulating processes
within a 1- to 2-year period and the other simulating processes
in a 50-year period, were used to evaluate potential impacts
associated with the use of fossil fuel energy resources in the
Arctic. Increased carbon doxide had less effect than increased
temperature on the accumulation of organic matter because of the
different responses of underlying processes.

INTRODUCTION

Arctic regions of Alaska, Canada, northern Europe, and the U.S.S.R. have large reserves of minerals and fossil fuels and have large accumulations of organic carbon. In Alaska, known coal reserves underlie about 40% of the 0.22×10^6 km^2 of tundra north of the Arctic Circle, all of which also may be underlain by oil reserves (Selkregg 1975). Most of these energy reserves are covered by Eriophorum vaginatum tussock tundra which is one of the most widespread tundra types in the circumpolar arctic, covering about 16% of the 5.7×10^6 km^2 ice-free land area. The increasing exploitation of these mineral and energy reserves is impacting the Arctic in diverse ways. Most obvious are the immediate changes in the vegetative cover or organic mat caused by off road vehicles, dust deposition, or oil spills. More subtle and long lasting are changes in the depth of the thawed layer and in nutrient availability. Even more subtle, long lasting, and far reaching are changes induced by global increases in atmospheric carbon dioxide and temperature. Past experiences indicate that the Arctic is extremely sensitive to impact and slow to recover because of the slow plant growth and the subsidence and erosion following melting of permafrost (Webber and Ives 1978). The increasing human activities in the Arctic and the desire to preserve its wilderness character, as mandated in the Alaska National Interest Lands Conservation Act PL 96-487, dictate that

management alternatives be evaluated from an understanding of how arctic ecosystems function. This understanding of ecosystem function should be formalized in a framework which is quantitative, predictive, and integrated. The short- and long-term effects of perturbations in the Arctic require a framework, or frameworks, operative over both time periods.

In 1974, at the conclusion of the International Biological Program (IBP), unconnected models of photosynthesis and growth of vascular plants and mosses and preliminary models of decomposition and mineral cycling existed (Miller and Tieszen 1972, Miller et al. 1976, 1978, Lawrence et al. 1978, Stoner et al. 1978a, b, Bunnell and Tait 1974). The models developed during the IBP were a major step toward a quantitative and integrative framework but lacked sufficient detail and breadth to be important in general predictions. Also, more research was needed in the areas of plant growth, plant demography, and plant-soil nutrient relations. Following the IBP, research funded by the National Science Foundation continued on plant-herbivore interactions at Meade River, Alaska, while the research reported here, funded by the Department of Energy, Office of Environmental Programs, concentrated on plant-soil interactions at Eagle Creek, Alaska and along the Trans Alaskan Oil Pipeline. Early in the DOE sponsored research a simulation model of the decomposition of dead organic matter and mineralization of nitrogen, phosphorus, and calcium was completed based on the research conducted during the IBP in coastal wet meadow tundra (Barkley et al. 1978). This model was later expanded to include a moss layer, and existing models of plant and soil processes were added to it. The resulting detailed mechanistic computer program was then simplified at several workshops involving investigators and a simpler model, the Arctic Tundra Simulator (ARTUS), was defined based on experimental data and field observations.

[1]Paper presented at the international symposium Energy and Ecological Modelling, sponsored by the International Society for Ecological Modelling. (Louisville, Kentucky, April 20-23, 1981)

[2]Philip C. Miller is the Director and Martha Blake Jacobson and Richard Kendall are Systems Modellers for Systems Ecology Research Group, San Diego State University, San Diego, California U.S.A.

A second model, the Northern Ecosystems Carbon Model (NECS), was based on the understanding and quantification of arctic ecosystems developed from the earlier, more process based simulation models and on the results of a workshop held during March 1980 at which the potential long-term effects of increased atmospheric carbon dioxide and temperature on arctic and taiga ecosystems were clarified (Miller 1981).

ARTUS is a detailed process oriented model which operates in 1-day time steps. It is used to understand the short-term effects of perturbations in the Arctic. NECS is less detailed and considers the effect of perturbations over a time scale of 1-50 years in 1-year time steps.

BACKGROUND ON MODELS

The DOE research program on primary production, decomposition, and mineralization processes in arctic tundra focused on E. vaginatum tussock tundra. The operational objective of this research is to develop the ARTUS simulation model of plant and soil processes in arctic tundra on the basis of an understanding of fundamental underlying physical, biochemical, and physiological processes (1). These processes are

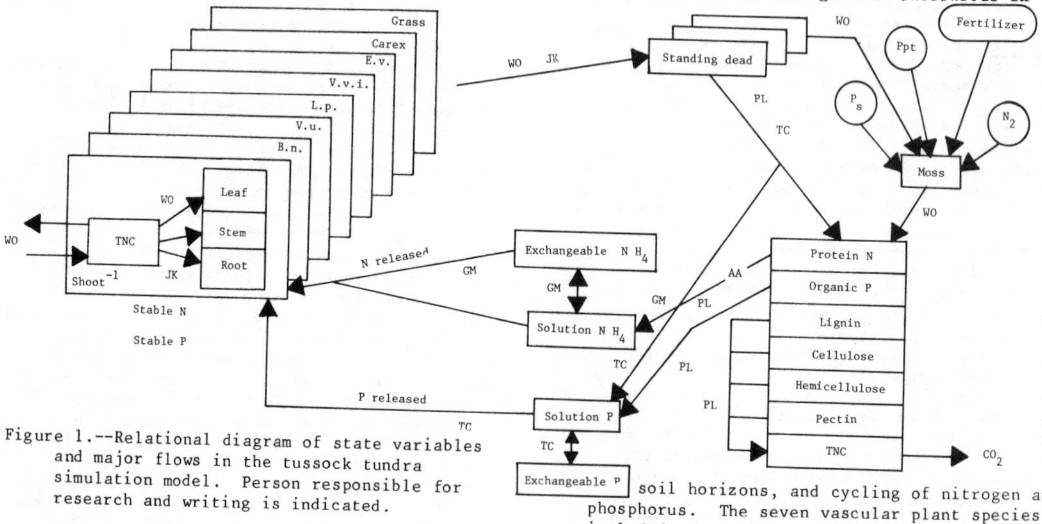

temperature and water status. The processes affecting the growth and development of the vascular plants and mosses are critical in maintaining or restoring the natural species composition, which preserves the wilderness character of the Arctic. The species composition in turn affects the site thermal a water balance and nutrient cycling. The understandings gained from the simulation mode are generally applicable and can be applied to diverse problems that arise due to energy exploration and exploitation. The principal investigators are each responsible for developing the understanding encoded in different parts of the model and for suggestin alternatives to the code or model. ARTUS is being validated as funds permit at several sit along the Trans Alaskan Oil Pipeline in north-central Alaska and at Cape Thompson in northwest Alaska.

ARTUS is structured to calculate processe during a 2- to 5-year period with the ambient conditions and following various perturbations at several geographic locations in Alaskan tussock tundra that are susceptible or are now receiving impacts from energy development. The model calculates thermal and water balances of the site, carbon incorporation, growth and dea of seven vascular plant species and four moss types, decomposition of dead organic matter, mineralization of nitrogen and phosphrous in

Figure 1.--Relational diagram of state variables and major flows in the tussock tundra simulation model. Person responsible for research and writing is indicated.

important in stabilizing and restoring disturbed areas. The physical processes of energy and water exchange affect the temperature of all organisms, the rate of all metabolic processes, and the stability of the permafrost. The processes affecting the cycling of nitrogen and phosphorus, including their input and losses, are critical to the long-term maintenance of species composition and are affected by site

soil horizons, and cycling of nitrogen a phosphorus. The seven vascular plant species included are: the tussock forming sedge Eriophorum vaginatum; the deciduous shrubs Betula nana and Vaccinium uliginosum; the evergreen shrubs Ledum palustre and Vaccinium vitis-idaea; the rhizomatous sedge Carex _____ and a generalized grass species tha in the field, is often a species of Arctagrost or Calamagrostis. The four moss types include are: Polytrichum, Dicranum, Sphagnúm, and other.

Of the possible perturbations related to energy development, five were selected for study and simulation. These impacts, ranked from most to least tractable for experimental research and listed in the order in which they have been studied, are: fertilizer addition, altered soil water drainage, off-road vehicle tracks, oil spill, and fire. The potential effects of increasing carbon dioxide and temperature have been considered. The perturbations were included by changing state variables or processes according to the current understanding of their impact.

The Northern Ecosystems Carbon Simulator calculates plant and soil processes in 1-year time steps through a 50-year period. The model includes nine ecosystem types: polar desert, semidesert, wet meadow tundra, tussock tundra, low shrub tundra, tall shrub tundra, northern taiga, middle taiga, and southern taiga. It calculates rates of photosynthesis, respiration, growth, death, decomposition, mineralization, and freezing of organic carbon and nutrients into the permafrost. The state variables are plant carbohydrates, biomass, thawed dead organic matter, and frozen dead organic matter. Provision for the changing areal extent of each ecosystem type is made.

In NECS four basic compartments, storage carbohydrates, total plant biomass, dead organic matter in the active layer, and dead organic matter frozen in the permafrost, are considered for each of the nine northern ecosystems types (fig. 2). The carbon dioxide content of the air

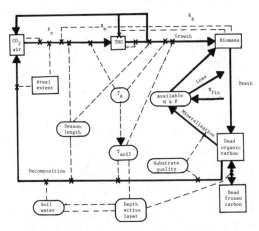

Figure 2.--Relational diagram of variables and processes affecting the carbon balance of northern ecosystems.

is updated according to the predicted global change (Keeling et al. 1976) plus or minus the amount of carbon released or fixed by northern ecosystems. The change in the areal extent of

each ecosystem type is calculated according to area lost and gained as boundaries shift. Season length, soil water content, depth of the active layer, available nitrogen and phosphorus, and air, soil, and plant temperature are calculated as intermediate variables that depend on various environmental conditions and affect the fluxes between compartments.

The fluxes directly affecting the ecosystem carbon balance include photosynthesis, maintenance and growth respiration, growth, death, decomposition, mineralization, and changes between the unfrozen and frozen dead organic matter compartments. The processes are calculated separately for nine arctic and taiga ecosystem types. The state variables are updated annually, although photosynthesis is calculated for shorter time intervals.

Photosynthesis depends on a maximum rate, light intensity, leaf senescence, leaf biomass, length of the active season, temperature, and storage carbohydrate. A daily rate of photosynthesis is calculated for each month of the active season by integrating the product of a Michaelis-Menten curve for the solar irradiance response and a temperature function through an average day. This rate is adjusted by multiplying by leaf biomass and functions of leaf senescence and carbon dioxide concentration. The yearly rate is obtained by summing the products of the daily rates per month and the numbers of days in that month in which the air temperature exceeds $-4°C$.

Maintenance respiration is calculated separately for woody stems and roots and for leaves. The maintenance respiration depends on a base rate, temperature, biomass of the appropriate plant part, and season length. If storage carbohydrate increases above a maximum level, the excess is assumed to be lost in maintenance respiration.

Growth is defined as the synthesis of new tissue from storage carbohydrates. The biochemical composition of this tissue determines the growth respiration percentage. The annual growth rate is calculated to use all of the storage carbohydrate, available nitrogen, or available phosphorus, whichever yields the smallest growth rate and therefore would be the limiting factor. This rate is then adjusted according to a temperature response. Growth is limited by nitrogen and phosphorus such that the nitrogen and phosphorus content of new tissue is effectively about 0.01 g N/g C and 0.001 g P/g C. Death is set equal to growth, based on the assumption that the biomass is in a steady state.

Decomposition depends on a basic rate, season length, and temperature. Currently, several temperature relations for decomposition can be used interchangeably. Using the temperature relation of Heal and French (1974) gives extremely high decomposition values. The temperature relations of Linkins[3] and Marion and Miller[4] give more reasonable values. Decomposition is insensitive to temperatures below 10°C and increases above 10°C.

Mineralization is related to the decomposition rate, the substrate quality, and a constant representing the net release or the inorganic nutrient per unit of carbon mineralized. The transfer of organic matter between frozen and dead is controlled by the change in the depth of the active layer, which is related to the annual damping depth, the mean temperature, and the temperature amplitude. The damping depth is calculated from the thermal conductance and the volumetric heat capacity of the soil. The two thermal properties of the soil depend on the organic matter and water contents. Material is transferred between thawed and frozen organic matter compartments depending on the bulk density of the soil material and the organic matter content at the bottom of the thawed layer.

The seasonal course of air temperature is calculated for each of the nine ecosystem types with a sine curve which is defined by the mean temperature and the temperature amplitude characteristic of each ecosystem type. The soil temperature is set above the air temperature by several degrees; air temperature varies by ecosystem type. A plant temperature is calculated as a weighted mean of the air and soil temperatures. The weighting factors vary with the growth form of the plants in the ecosystem. Photosynthesis depends on this calculated plant temperature. Respiration and growth depend on soil temperature modified slightly by the air temperature because of the large fraction of biomass found belowground in most northern ecosystems.

The length of the active season is calculated as the period during which air temperatures are above -4°C. Photosynthesis, growth, and decomposition occur when temperatures are above this minimum temperature.

VALIDATION OF MODELS

ARTUS was validated by comparing calculated values with measured annual totals of processes; peak season biomasses and nutrient contents; and seasonal progressions of biomass, carbohydrate, and nutrient contents. Specifically, the

[3]Linkins, Arthur E. 1981. Unpublished data. Biology Department, Virginia Polytechnic Institute, Blacksburg, VA.

[4]Marion, Giles M., and Philip C. Miller. 1981. Unpublished data. Systems Ecology Research Group, San Diego State University, San Diego, Calif.

validation variables are early and peak season new leaf and stem biomass by species, early and peak season nitrogen and phosphorus contents by species, nitrogen and phosphorus uptake by species, ecosystem respiration, and peak season depth of thaw. Several sets of observations have been completed to validate ARTUS. No one data set provides a rigorous independent validation. The best set of data includes shoot populations, shoot growth, and nutrient contents measured several times in the season along the Trans-Alaskan Oil Pipeline. The measured trends indicated that production per shoot on the site north of the Brooks Range was about half that on the south slope of the Brooks Range. Nitrogen and phosphorus contents of new biomass were usually slightly higher at the north slope site. Thaw depths did not decrease in the northward direction probably because of the variable organic matter and ice contents. Ecosystem respiration did not vary consistently in the northward direction. In the simulations using ARTUS, production by B. nana was consistently low, perhaps because short shoot production was not included in the peak leaf mass. Production per shoot by all species was about half the measured values at all sites. Simulated nitrogen contents of B. nana, L. palustre, and E. vaginatum were low and decreased in the northward direction. Contents of V. vitis-idaea and Carex were similar to measured values but were variable. Simulated values for ecosystem respiration were higher than the measured values at Timberline and Sagwon, perhaps due to error in the number of active shoots per square meter or the moss cover. Simulated thaw depths were deeper than those measured at sites along the haul road.

The relations used in NECS were parameterized based on information summarized in Miller (1981). NECS has not been validated because the data needed for validation are not available. Unmeasured parameters were adjusted so that the net carbon flux in 1973-1975 was 10-20 g C $\cdot$ m^{-2} $\cdot$ yr^{-1}, which is in agreement with available data.

SIMULATION RESULTS AND DISCUSSION

Some of the more interesting results of the simulation, with regard to new field experiments, follow. In the simulation with ARTUS a pulse of inorganic nitrogen is released into the soil solution when the soil solution is thawed (fig. 3). Mineralization proceeds relatively rapidly and uses all the potentially mineralizable nitrogen within 3 weeks after thaw in each horizon. These pulses are delayed by about 2 weeks as each horizon thaws. The amount of nitrogen in the frozen soil solution is relatively large, 0.7 g N/m^2, and can supply a large fraction of the annual nitrogen uptake when mineralization is stopped. Mineralization in the mineral soil is exceedingly low. The mineralization rates for the C horizon at Eagle Creek were used to calculate mineralization at

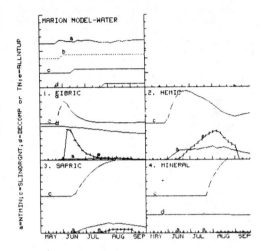

Figure 3.--Simulated seasonal progression of
nitrogen mineralization in organic soil
horizons and the thawed mineral soil with
ambient conditions.

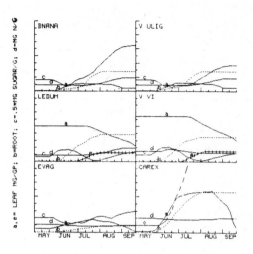

Figure 4.--Simulated seasonal progression of the
leaf and root biomasses and of the sugar
and nitrogen fractions of six vascular
plant species with ambient conditions.

the timberline site. The calculated
mineralization rates appear low relative to
required plant uptake rates at the timberline
site. Probably, the mineralization rate for the
C horizon cannot be extrapolated to mineral
soils supporting vegetation. Of the two
mineralization models included in ARTUS,
Linkin's decomposition-mineralization model
based on soil enzyme activities gives lower and
more realistic rates than Marion's model based
on incubation stands. Nitrogen is taken up by
V. vitis-idaea only early in the season because
its shallow root system exploits only early
thawing soil. Deeper rooted shrubs have access
to a more continuous nitrogen supply.

The carbohydrate balance of mosses and
vascular plants is out of balance. Stem and
root respiration measurements allow too much
latitude. Plant sugar and storage polymer
levels generally rise to unrealistic levels,
indicating that respiration is underestimated or
that photosynthesis is overestimated (fig. 4).
Unfortunately, the balance between carbon gain
and loss is very delicate and depends on very
difficult measurements. The onset and
progression of growth is reasonable.
Mycorrhizal relations of the species should be
clarified relative to the relative magnitudes of
the mycorrhizal factors. The rooting depth
relations of the species should be clarified
because it appears that rooting depth is
important in the nitrogen uptake pattern of the
species. Because V. vitis-idaea has such a
short time period for nitrogen uptake, its

mycorrhizal factor is large to give it a large
uptake rate during a short period relative to
the other species. The increased uptake rate in
late summer recorded by Stuart and Miller
(unpubl. data) are not simulated. With carbon
dioxide induced climate change sugar-storage
polymer levels rose faster and the growth of
evergreen shrubs increased more than the other
species (fig. 5). Temperature had a greater
influence than the increased carbon dioxide
content of the air.

Simulations with NECS indicated that
throughout the simulated period from 1970 to
2020, the net annual accumulation of carbon in
northern ecosystems ranged from 5 to 10% of the
carbon added to the atmosphere each year
(fig. 6). With no change in climate and no
atmospheric input of nutrients, less than 8 GT C
accumulated by year 2020. Atmospheric
deposition of nutrients by itself raised the
accumulation to 9 GT. With climatic change and
nutrient deposition carbon accumulation
continued at slightly increasing rates, so that
by year 2020 the carbon in northern ecosystems
increased by about 35 GT (table 1). Doubling
the carbon dioxide content but maintaining
temperatures constant, such as might occur with
a combination of carbon dioxide and dust
pollution, had little effect on the accumulation
of carbon. Increasing temperatures but
maintaining carbon dioxide contents constant
increased the accumulated carbon by 25 GT.
Thus, most of the effect of increasing carbon
dioxide and temperature was due to temperature.

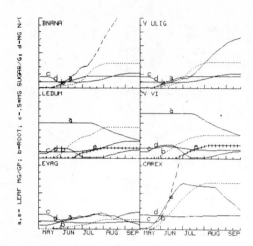

Figure 5. Simulated seasonal progression of the leaf and root biomasses and of the sugar and nitrogen fractions of six vascular plant species with temperature increased by 4°C and photosynthesis doubled.

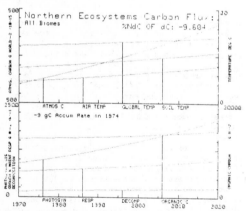

Figure 6. A preliminary projection of the carbon flux of northern ecosystems to the year 2020, including atmospheric carbon, arctic summer and global mean air temperature, percentage change in area, photosynthesis, respiration, decomposition, and organic carbon. The percentage of the global increase in atmospheric carbon dioxide which is due to carbon flux in northern ecosystems is calculated every 5 years and given in the upper right (-9.604).

Table 1.--Total organic carbon and rate of accumulation of organic carbon simulated for all northern ecosystems with different assumptions. Values are GT and GT/year. The change per year is calculated from the value in 2020 minus the value in 1980 divided by 40 years.

Assumption	Total organic carbon			Change per year
	1970 (GT)	1980 (GT)	2020 (GT)	1980-2020 (GT/year)
No climate change, no atmospheric input	230.2	231.7	237.9	0.155
No climate change, atmospheric input	230.2	231.9	239.2	0.183
CO_2 doubled by 2020, atmospheric input	230.2	233.4	265.2	0.795
Tmp rise but no CO_2 rise	230.2	233.3	255.0	0.543
CO_2 doubled but no tmp rise	230.2	232.0	239.5	0.188
Death maintained at 1970 rates	230.2	233.7	263.1	0.735

Assumption	Rate of accumulation			
	(GT/yr)	(GT/yr) [a]	(GT/yr)	(GT/yr²) [b]
No climate change, no atmospheric input	0.158	0.152	0.160	0.0002
No climate change, atmospheric input	0.158	0.178	0.186	0.0002
CO_2 doubled and Tmp up 4°C by 2020, atmospheric input	0.158	0.323	1.406	0.3271
Tmp up 4°C but no CO_2 rise, atmospheric input	0.158	0.297	1.051	0.0189
CO_2 doubled but no Tmp rise, atmospheric input	0.158	0.182	0.195	0.0003
Death maintained at 1970 rates, dead	0.158	-0.379	-0.300	-0.0170
live	0.000	0.760	1.379	0.0155

[a] Current rates of accumulation 0.15 to 0.19 based on gas analysis and dated peat or 0.01 to 0.28 based on Glacier Bay.

[b] Expected change in the rate of accumulation is about 0.004 from dated peat and 0.001 to 0.004 from current distribution of carbon.

These simulation results are a consequence of assuming that growth is more temperature and nutrient sensitive than is photosynthesis and that excess photosynthate is used in respiration or decomposition. In another set of simulations, annual death of biomass was assumed to be equal to growth and to be constant at the 1970 rate. These assumptions had little effect on the total organic carbon by the year 2020, although when death was maintained at a constant rate, biomass increased while dead organic carbon decreased.

Simulations were run varying the assumptions about the stability of the permafrost. In these simulations, half the area of northern taiga and of tussock tundra was assumed to develop a decomposition rate 8 times the current rate as a result of lower permafrost and better soil drainage. Northern ecosystems became a carbon source with the assumption of better soil drainage. Most of the effect was because of the northern tagia. The amount of carbon frozen in the permafrost was underestimated in these simulations (Everett[5]); thus northern ecosystems could become an even greater carbon source.

[5]Everett, Kaye. 1981. Agronomy Department, Institute of Polar Studies, Ohio State University, Columbus, Ohio.

The relative amounts of carbon in the different ecosystem types remained almost constant. The low shrub type gained organic carbon relatively more and the southern taiga relatively less. Most of the carbon continued to be in the taiga. The polar desert gained more carbon on a percentage basis because it had a low initial carbon content although the lack of data on decomposition in the polar desert makes the polar desert carbon balance problematic. Total organic carbon in the separate ecosystems increased by 0.1-5.0%/yr.

The processes differed slightly in their response to carbon dioxide induced climatic change (table 2). The annual rates of

Table 2.-- Rates of carbon fluxes in different years simulated for the different ecosystems with CO_2 content doubled and 4°C temperature rise. PD = polar desert; SD = semidesert; WM = wet meadow tundra; TT = tussock tundra; LS = low tundra; TS = tall shrub tundra; NT = northern taiga; MT = middle taiga; and ST = southern taiga. Rates are in g C m^{-2} yr^{-1}.

Year and treatment	PD	SD	WM	TT	LS	TS	NT	MT	ST
Photosynthesis									
1970	2	109	591	1786	354	793	671	1378	2038
1980	2	139	741	2163	431	965	794	1603	2391
2020	6	363	1833	4539	875	1967	1622	2951	4559
g C m^{-2} yr^{-1}	0.08	5.1	24.8	55.1	10.4	23.5	19.0	31.5	50.4
% 1970/yr	4.0	4.7	4.2	3.1	2.9	3.0	2.8	2.3	2.5
Respiration									
1970	1	89	430	1649	169	360	545	918	1691
1980	1	113	532	1994	191	479	645	998	1957
2020	2	310	1395	4230	246	1162	1383	1921	3985
g C m^{-2} yr^{-1}	0.02	4.4	19.3	51.6	1.5	16.0	16.8	20.1	45.9
% 1970/yr	2.0	5.0	4.5	3.1	0.9	4.5	3.1	2.2	2.7
Growth									
1970	1	14	108	90	135	315	90	340	250
1980	1	19	144	114	176	355	107	450	317
2020	3	39	311	219	468	595	176	775	424
g C m^{-2} yr^{-1}	0.04	0.5	4.1	2.6	6.7	5.6	1.7	8.7	3.5
% 1970/yr	4.0	3.6	3.8	2.9	4.9	1.8	1.9	2.6	1.4
Decomposition									
1970	0	14	93	75	130	315	75	320	250
1980	0	15	117	91	166	326	85	410	307
2020	1	20	173	127	313	393	116	613	380
g C m^{-2} yr^{-1}		0.1	1.6	1.0	3.7	1.6	3.8	5.9	2.6
% 1970/yr		0.9	1.7	1.4	2.8	0.5	1.1	1.8	1.0
Accumulation									
1970	0	0	15	15	5	0	15	20	0
1980	0	4	27	23	10	29	22	40	10
2020	2	18	138	91	156	201	60	160	43
g C m^{-2} yr^{-1}			2.5	1.5	3.0		0.9	2.8	
% 1970/yr			16.4	10.1	60.4		6.0	14.0	

photosynthesis, respiration, and growth increased between 2 and 5% annually because of increased carbon dioxide and temperature. The rate of decomposition increased 0.9-2.8%. The accumulation of organic matter increased 6-60%. Thus, changes in the rates of photosynthesis, respiration, growth, and accumulation due to increased carbon dioxide and temperature should be easier to measure than changes in rates of decomposition. The rates of photosynthesis and respiration changed more than rates of the other processes when carbon dioxide was doubled without temperature increases. Growth and accumulation rates increased more when temperatures were increased without carbon dioxide increase than when both were increased together.

ACKNOWLEDGEMENTS

The research and development associated with ARTUS was supported by the Ecological Research Division, and the syntheses and development associated with NECS was supported by the Carbon Dioxide and Climate Division of the U.S. Department of Energy, Office of Health and Environmental Research. We thank Beth Sigren and Patsy Miller for their help on this paper.

LITERATURE CITED

Barkley, S. A., D. Barel, W. A. Stoner, and P. C. Miller. 1978. Controls on decomposition and mineral release in wet meadow tundra--a simulation approach. p. 754-778. In D. C. Adriano and I. L. Brisbin, Jr. (eds.) Environmental chemistry and cycling processes. U.S. Department of Energy Symposium Series CONF-760429, Technical Information Center, Washington, D.C.

Bunnell, F. L., and D. E. N. Tait. 1974. Mathematical simulation models of decomposition processing. p. 207-225. In A. J. Holding, O. W. Heal, S. F. MacLean, Jr., and P. W. Flanagan (eds.) Soil organisms and decompositon in tundra. Tundra Biome Steering Committee, Stockholm.

Heal, O. W., and D. D. French. 1974. Decomposition of organic matter in tundra. p. 279-308. In A. J. Holding, O. W. Heal, S. F. MacLean, Jr., and P. W. Flanagan (eds.) Soil organisms and decompositon in tundra. Tundra Biome Steering Committee, Stockholm.

Keeling, C. D., R. B. Bacastow, A. E. Bainbridge, C. A. Ekdahl, P. R. Guenther, L. S. Waterman, and J. F. S. Chin. 1976. Atmospheric carbon dioxide variations at Mauna Loa Observatory, Hawaii. Tellus 28:538-551.

Lawrence, B. A., M. C. Lewis, and P. C. Miller. 1978. A simulation model of population processes of arctic tundra graminoids. p. 599-619. In L. L. Tieszen (ed.) Vegetation and production ecology of an Alaskan arctic tundra. Springer-Verlag, New York.

Miller, P. C. 1981. Carbon dioxide effects research and assessment program: Carbon balance in northern ecosystems and the potential effect of carbon dioxide induced climatic change. Report of a workshop. [San Diego, Calif., March 7-9, 1980] U.S. Department of Energy, CONF-8003118, 109 p. Washington, D.C.

Miller, P. C., and L. L. Tieszen. 1972. A preliminary model of processes affecting primary production in the arctic tundra. Arct. Alp. Res. 4:1-18.

Miller, P. C., W. C. Oechel, W. A. Stoner, and B. Sveinbjornsson. 1978. Simulation of CO_2 uptake and water relations of four arctic bryophytes at Point Barrow, Alaska. Photosynthetica 12:7-20.

Miller, P. C., W. A. Stoner, and L. L. Tieszen. 1976. A model of stand photosynthesis for the wet meadow tundra at Barrow, Alaska. Ecology 57:411-430.

Selkregg, L. L. 1975. Alaska regional profiles, arctic region. Divison of planning and research, State of Alaska. Juneau, Alaska.

Stoner, W. A., P. C. Miller, and W. C. Oechel. 1978a. Simulation of the effect of the tundra vascular plant canopy on the productivity of four moss species. p. 371-387. In L. L. Tieszen (ed.) Vegetation and production ecology of an Alaskan arctic tundra. Springer-Verlag, New York.

Stoner, W. A., P. C. Miller, and L. L. Tieszen. 1978b. A model of plant growth and phosphorus allocation for Dupontia fischeri in coastal wet meadow tundra. p. 559-576. In L. L. Tieszen (ed.) Vegetation and production ecology of an Alaskan arctic tundra. Springer-Verlag, New York.

Webber, P. J., and J. D. Ives. 1978. Damage and recovery of tundra vegetation. Environ. Conserv. 5:171-182.

A BOX-DIFFUSION MODEL OF THE GLOBAL
CARBON CYCLE INCLUDING
MARINE ORGANIC COMPARTMENTS AND A POLAR OCEAN[1,2]
John K. Munro, Jr.[3] and Jerry S. Olson[4]

A box-diffusion model of the global carbon cycle has been extended to include a polar ocean and marine organic compartments corresponding to the mixed layer, intermediate, and deep ocean. Both stable carbon and radiocarbon are coupled throughout. Measured ocean CO_2 profiles are used to infer carbon flows into the individual ocean layers due to respiration of organic carbon and advection.

INTRODUCTION

Our primary objective is to extend recent modelling of organic matter of land ecosystems to the large pools of (mostly dead) organic matter in the oceans, in order that both parts of the biosphere may find a place in global models of the carbon cycle. A second objective is to explore how the recycling rates of this organic material may compare to other fluxes (e.g., from diffusion, major eddies, and advection processes) in the larger marine cycle of inorganic carbon.

Some additional aspects of several ocean models are treated in a companion paper by Killough and Emanuel (1981) and therefore need not be duplicated here. Earlier simplified models (cf. Keeling 1973a, Bacastow and Keeling 1973, Oeschger et al. 1975, Killough 1977, Broecker et al. 1971, 1979, Bjorkstrom 1979,

Emanuel et al. 1980a, Bolin et al. 1981) and ours all seem compatible with projections of substantial further increases of atmospheric CO_2, presumably well beyond these already well documented.

REVIEW OF GLOBAL AND LAND BALANCES

A brief review of work published recently on the terrestrial cycle will suggest that land parts of the biosphere could have been important as temporary sources and perhaps (very recently?) as sinks for atmospheric carbon dioxide. A classic paradigm asserts that the world's atmospheric CO_2 is balanced by a rate of photosynthetic exchange to plants that is nearly equal to releases by fires plus respiration by plants, animals, and decomposers. This has been the starting point for numerous models showing linear and nonlinear departures from such a balance since Eriksson and Welander (1956). The relatively large gross and net primary production of ecosystems on land and large respiration and nonfossil burning rates (Olson 1970, Olson et al. 1978, Bolin et al. 1979) lead us to a continuing re-evaluation.

Relatively small differences between these large opposing rates would be sufficient to contribute 1 to 3 Pg yr^{-1} (footnote 5) of C to the atmosphere (Baes et al. 1976, 1977). Some authors (Woodwell et al. 1978) even contemplated the possibility that such nonfossil sources of CO_2 would exceed the 5 Pg yr^{-1} currently released from burning of oil, natural gas, and coal. A critical review of that problem is being presented elsewhere (Olson, to be published) and shows that the latter figure, or a higher one, is conceivable

[1] This work was supported jointly by the CO_2 and Climate Program, U.S. Department of Energy, under contract W-7405-eng-26 with the Union Carbide Corporation and by the U.S. National Science Foundation's Ecosystems Studies Program under interagency agreement No. DEB77-26722.

[2] Paper presented at the International Symposium on Energy and Ecological Modelling, sponsored by the International Society for Ecological Modelling. (Louisville, Kentucky, April 20-23, 1981).

[3] John K. Munro, Jr. is Computational Physicist, Computer Sciences Division at Oak Ridge National Laboratory, Union Carbide Corporation, Nuclear Division, Oak Ridge, Tennessee, U.S.A.

[4] Jerry S. Olson is Senior Ecologist, Environmental Sciences Division, Oak Ridge National Laboratory, Oak Ridge, Tennessee, U.S.A.

[5] 1 Pg = 10^{15}g = 1 Gton.

as a biospheric source of CO_2 only if unreasonably high estimates of global closed-forest area, forest <u>biomass density</u>, and extraordinary disturbance <u>rates</u> for forest and soil are combined.

Partly because of regrowth of forests already depleted by earlier clearing, and possibly with some modest future stimulus of photosynthesis by enhanced atmospheric CO_2 concentrations, land ecosystems could temporarily be storing organic carbon. Either as a fluctuating source or as a temporary sink, these interactions of the land biosphere with the atmosphere will continue to attract modelling efforts but are beyond the scope of this report. Subdividing functionally distinct components of the whole land biosphere or its regional parts into categories that separate both living and dead and pools of rapid and slow transit times is an obvious step toward refinement (Olson and Killough 1977, Chan and Olson 1980, Emanuel <u>et al.</u> 1980b, Chan <u>et al.</u> 1981, Emanuel <u>et al.</u> 1981, Moore <u>et al.</u> 1981).

This discussion, however, shifts attention to the larger pools of marine organic matter, and simplifies the land pools deliberately (as did SCEP 1970 and Keeling 1973a) into one rapidly exchanging and one slowly exchanging pool. Unlike Machta (1972) and Oeschger <u>et al.</u> (1975), we do not envision a discrete transit time for each pool, but essentially exponential distribution of residence times. Thus leaves, plant reproductive parts, almost all animals, their dead remnants, and the main decomposers of such fresh detritus are pooled as rapidly exchanging components for the global system (here; cf. Olson <u>et al.</u> 1978, Chan <u>et al.</u> 1980 for a broad regional treatment). The slowly exchanging pool includes woody materials of forests and other landscapes and those humus pools that are derived from the foregoing materials that undergo decay over periods of decades and centuries. Additional deep peat and humus that are buried or otherwise out of circulation with current atmospheric exchange were not included in the reviews by Baes <u>et al.</u> (1977) which we use here to define the initial conditions of 1860 in our simulations.

MARINE ORGANIC AND INORGANIC POOLS

A simplified diagram of model structure given in figure 1 associates one organic pool each with the mixed layer of the ocean (taken as averaging 75 m in depth), thermocline or other intermediate waters [averaging 75 to 1000 m, following Oeschger <u>et al.</u> (1975), for convenient comparison], and deeper waters (below 1000 m). Naturally this means that organic materials near the source, in sunlit surface waters, include most of the living organisms (only 1 to 4 Pg C) and recently formed debris and dissolved residues (~60 Pg C); this is relatively susceptible to

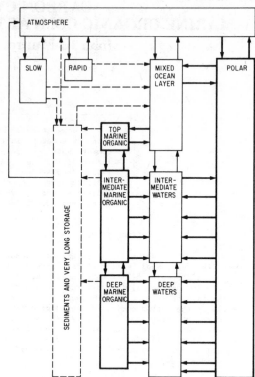

ORNL-DWG 81-2841   FED

Figure 1.--Simplified diagram of model structure. New compartments and flows added to the box-diffusion model of Killough (1980) are shown by heavy lines. Flows not included yet in the program described here are shown with broken lines. The flow shown from the sediment pool is the fossil fuel input to the atmosphere, currently treated as an external source.

early recycling to CO_2. In the deep ocean, a surprisingly uniform residue of relatively resistant material is found. Menzel and Ryther (1970) estimated 3 to 10 µg liter^{-1} of particulate carbon and 350 to 700 µg liter^{-1} of dissolved carbon. We followed Mopper and Degens (1979) for a recent judgment of representative radiocarbon ages. The model's self-consistency favored accepting the larger values within the range of pool sizes which they considered consistent with data from recent cruises and improvements in analytical techniques. Intermediate layers were assumed to have a wide mixture of fresh and resistant organic residues, with residence times averaging between those of surface and deep waters (figure 1).

MODEL IMPLEMENTATION

We extended the box-diffusion model of the global carbon cycle, as originally described by Oeschger et al. (1975) and implemented by Killough (1977, 1980) by adding marine organic and polar ocean compartments. We used the profile for the Pacific Ocean (Craig and Weiss 1970) for testing, but provide for the alternative of using regionally or globally averaged profiles of CO_2 concentration as a function of depth. Thus, this model will permit us to close the cycle for a self-consistent treatment of organic forms of carbon. The next phase of development will follow with the addition of a sediment pool and the flow of carbon down rivers to the sea. In its current form the model can be used to study the processes which transport organic carbon from shallow to deep waters and convert organic forms to inorganic ($CO_2 + HCO_3^- + CO_3^=$) at each depth of the ocean. Addition of a polar ocean allows explicit treatment of net circulation of CO_2 from warm surface waters to the atmosphere and from the atmosphere into cold surface waters.

In this model the polar ocean size is determined by the fraction of the surface area of the world's oceans which make up the Arctic and Antarctic waters, about 13%. This is a very simplified description since the polar waters have a very complicated structure at each depth with tongues of cold water moving toward low latitudes beneath overlying layers of warmer water, as for example in the Atlantic where the Antarctic Intermediate tongue of water is defined down to about 1100 m and extends far northward. The intent in this model is to treat what features we can, if only in a much simplified fashion.

Figure 1 shows these additions and some further developments planned for this model. The large sediment pool was omitted for the present time because it is not expected to be significant in the ocean dynamics as long as the periods of interest in the simulations are less than a few centuries (Broecker et al. 1980).

Since this model is an extension of earlier work, it retains most of the features introduced previously (Oeschger et al. 1975, Killough 1980). The latter reference explains the use of the globally averaged ^{14}C specific activity profile for the pre-industrial ocean, the historical record of CO_2 injection into the atmosphere due to fossil fuel burning through 1974 and a suitable logistic function to project anticipated future use, and the historical record of atmospheric release of ^{14}C from atmospheric tests of nuclear weapons. The treatment of stable carbon and radiocarbon is coupled throughout for each compartment. One can use a constant CO_2 concentration profile for the oceans as a function of depth to make comparisons of results with previous work since

this shuts off all advective flows through the polar ocean and transfer of carbon from marine organic compartments to the intermediate and deep waters.

For the results of calculations reported here, we used a CO_2 concentration profile of Craig and Weiss (1970) for the Pacific. We used ^{14}C ages given by Mopper and Degens (1979), which they base on an evaluation of many measurements, for the intermediate and deep marine organic compartments. Mopper and Degens (1979) also gave a range of values for the concentration of organic forms of carbon for different depths. We found it necessary to use the maximum values they gave to obtain consistent results for the pre-industrial balance.

Measured CO_2 concentration profiles typically will vary as a function of depth. The resulting vertical gradients, if they are to be maintained, require the presence of sources or sinks of CO_2. Under conditions of equilibrium, flows between adjacent vertical layers must be determined by the diffusion equation (in the vertical direction)

$$J = -D \frac{\partial C}{\partial z} , \qquad (1)$$

where

$J = J(z,t)$ = flux in the vertical direction,
D = diffusion coefficient, and
$C = C(z,t)$ = carbon concentration.

Source or sink information is incorporated through the continuity equation

$$\frac{\partial J_z}{\partial z} + \frac{\partial C}{\partial t} = S , \qquad (2)$$

where

$S = S(z,t)$ is the source ($S > 0$)
or sink ($S < 0$) function.

The value of the eddy diffusion constant of Oeschger et al. (1975) was used for the coefficient D. The Craig and Weiss (1970) data used in the work reported here are shown in figure 2 together with the vertical gradient to show more clearly the regions of CO_2 sources and sinks. We assume the pre-industrial ocean CO_2 concentration profile is in approximate steady state so that there must be sources and sinks of CO_2 to support the observed gradient. The sinks appear to be quite weak compared to the sources, so we consider only the sources in our calculations. Injection of CO_2 at a depth z can occur through biological and chemical processes that convert organic forms of carbon to CO_2 at that depth and through physical processes in which CO_2-rich waters flow from one depth to another where the CO_2 concentration is less.

In the absence of globally averaged values of oxidation rates of organic material in the ocean as a function of depth, we used an exponentially decreasing dependence for the rate in the top few hundred meters of the ocean. We also linearly extrapolated the CO_2 production rates from marine organic carbon material in the upper layers of the deep ocean up through the intermediate waters. Flows required to support the CO_2 concentration in the deep layers are assumed to come from the deep marine organic compartment only. Some justification for the exponentially decreasing

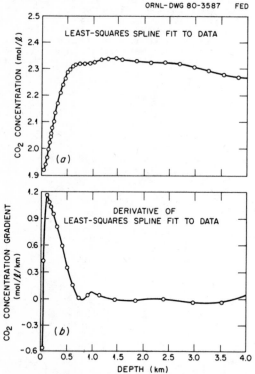

ORNL-DWG 80-3587   FED

Figure 2.--Least-squares spline fit to Craig and Weiss CO_2 concentration profile for the Pacific Ocean, used as the basis for calculations reported here, and the CO_2 concentration vertical gradient obtained from this profile. The graph of the spline fit to the data does not have enough resolution to show an upturn for very shallow depths which gives rise to the values of the derivative which are negative for these depths.

rates near the surface can be given in terms of exponentially decreasing light levels and the typical correlation between light intensity and respiration rate (Clarke and Denton 1962). The extrapolation upward through intermediate waters of transfer rates in the upper deep waters is somewhat arbitrary and is an attempt to explore consequences of likely upper limits on these rates.

While it is not clear whether the pre-industrial distribution of carbon was close to an equilibrium, we used an equilibrium condition as a means of determining various unknowns in order to have a well-defined set of initial conditions. We used the year 1860 as the initial time because the fossil fuel burning estimates before that time give very low CO_2 injection rates into the atmosphere (Keeling 1973b).

MODEL RESULTS

Every effort was made to use results from experiments to specify model parameters and to minimize the necessity for making arbitrary choices. Such choices had to be made for the net flow of carbon from the atmosphere to the surface waters of the polar ocean (the maximum the model would allow was used) and for the partitioning of flows from the marine organic compartments and polar waters into the diffusive layers of the nonpolar waters as discussed in more detail in the previous section. Freedom to adjust values to improve agreement with observation existed for only a few of the model parameters. All such possibilities for making adjustments are discussed explicitly in this paper.

The model structure and choice of initial conditions impose bounds on the range of values which may be used for some parameters, while other parameters may be adjusted without constraints. One may use whatever logistic for future fossil fuel CO_2 injection into the atmosphere seems appropriate. The return flow of CO_2 from the polar surface waters to the atmosphere may also be adjusted as required by data. However, for the model structure shown in figure 1, upper bounds exist for the net flow of CO_2 from the atmosphere to polar ocean (1.02 Pg yr^{-1}) and for the direct injection of carbon from the polar ocean into the deepest layer of the ocean (0.23 Pg yr^{-1}). The oxidation rate of marine organic carbon apparently has some upper bound which is less than the ocean CO_2 gradient would imply to be possible. We have not yet found a way to determine directly from existing data how much of the gradient is due to oxidation of marine organic carbon and how much should be attributed to advective processes. Lower bounds exist for the carbon contents in intermediate and deep marine organic compartments (about 700 and 2000 Pg, respectively) in order to obtain results consistent with the ages of marine organic carbon in them. The model also implies a lower bound for the size of the polar ocean pool (3940 Pg). The upper bound could be adjusted according to preference for narrower or broader geographic coverage.

Figure 3 shows results for the atmospheric concentration of CO_2 with and without a modest CO_2 "fertilization" response by the terrestrial biosphere. The accompanying table summarizes the compartment contents and net carbon changes for major compartment groups for the same period of time shown in the figure and for a biota growth factor value $\beta = 0.1$. The net change in the marine organic pools is a factor of 2 less than the change in the terrestrial biosphere pools.

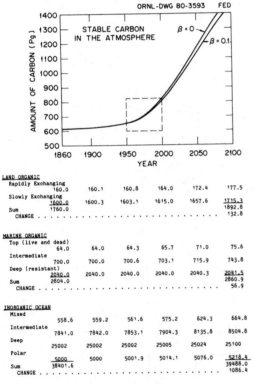

ORNL-DWG 80-3593   FED

LAND ORGANIC					
Rapidly Exchanging					
160.0	160.1	160.8	164.0	172.4	177.5
Slowly Exchanging					
1600.0	1600.3	1603.1	1615.0	1657.6	1715.3
Sum					
1760.0					1892.8
CHANGE					132.8
MARINE ORGANIC					
Top (live and dead)					
64.0	64.0	64.3	65.7	71.0	75.6
Intermediate					
700.0	700.0	700.6	703.1	715.9	743.8
Deep (resistant)					
2040.0	2040.0	2040.0	2040.0	2040.3	2041.5
Sum					
2804.0					2860.9
CHANGE					56.9
INORGANIC OCEAN					
Mixed					
558.6	559.2	561.6	575.2	624.3	664.8
Intermediate					
7841.0	7842.0	7853.1	7904.3	8135.8	8504.8
Deep					
25002	25002	25002	25005	25024	25100
Polar					
5000	5000	5001.9	5014.1	5076.0	5218.4
Sum					
38401.6					39488.0
CHANGE					1086.4

Figure 3.--Simulated atmospheric concentration of CO_2 with ($\beta = 0.1$) and without ($\beta = 0$) a modest CO_2 fertilization response by the terrestrial biosphere. The Mauna Loa record covers less than half the period shown by the dashed rectangle. The accompanying table summarizes simulated compartment contents and net carbon changes for major compartment groups for $\beta = 0.1$.

Some interesting responses emerged from this work. When the value taken from Mopper and Degens (1979) for carbon flow from the top marine organic compartment to the mixed layer (2.25 Pg yr^{-1}) was used with the average mixed layer depth (of 75 m) to evaluate the constants for the decreasing exponential dependence for oxidation of organic carbon, the calculated extinction coefficient was in the range of values for light extinction in clear temperate waters (Clarke and Denton 1962). Significant polar advection is suggested for intermediate waters, especially for depths of 450 to 800 m, as illustrated in figure 4 which shows the initial contents and flows. The Suess effect, reflecting the dilution of ^{14}C in the atmosphere resulting from injection of fossil fuel CO_2, is simulated quite well (-1.8% for 1950 compared to the measured -2.0 ± 0.3%) and the simulated $\Delta^{14}C$ values for the mixed layer compare well with observed values as shown in figure 5.

CONCLUSION

We extended the box-diffusion model implemented by Killough (1977, 1980) to include treatments of marine organic and polar ocean compartments. The model is designed to use a globally averaged CO_2 concentration profile for the ocean as the basis for calculating flows of carbon by advection from polar to nonpolar waters and the transfer of carbon from the marine organic compartments to inorganic compounds. We used evaluated values based on many measurements for the the contents of stable carbon in the marine organic compartments and for their ages in the intermediate and deep waters. Values for the contents of stable carbon in the intermediate and deep marine organic compartments had to be taken from the upper limits of the range of values given by the experimental errors in order to get results consistent with the inferred ages.

The extinction coefficient for the exponentially decreasing dependence of oxidation of organic carbon in the top few hundred meters of the ocean was calculated to be 0.036 m^{-1}, a value within the range of values of the light extinction coefficient for warm temperate waters. Most of the advective flows from the polar ocean were calculated to occur in the middle third of the intermediate waters.

Some of the model parameters were found to have constraints, in some cases lower bounds (sizes of the polar ocean, intermediate, and deep marine organic compartments) and in others upper bounds (net flow of carbon from atmosphere to polar ocean and flow of carbon from polar ocean to bottom of deep ocean by direct injection).

42

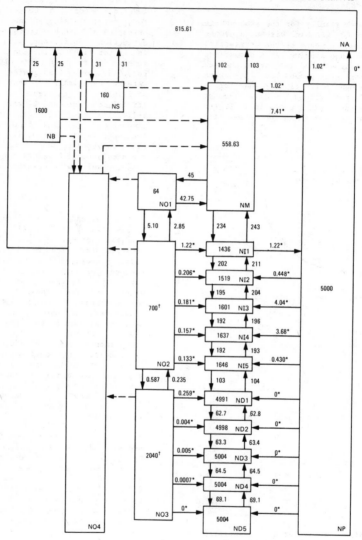

Figure 4.--Diagram of model structure showing initial
compartment contents and flows. Flow values
identified by asterisks may have to be adjusted
when more data are available. Contents of
intermediate and deep marine organic compartments
(identified by daggers) may have to be changed if
errors in measurements are reduced.

This model will permit a self-consistent treatment of the entire cycle of organic forms of carbon when treatment of the ocean sediment pool is added and the flow of carbon down rivers to the sea is included. One should now be able to include possible nutrient limits on the levels of marine organic carbon due to the levels of phosphorus and nitrogen in the oceans (Baes, to be published). A model such as this, which focuses on organic carbon, needs a more detailed treatment of the terrestrial biosphere to distinguish between carbon in soil and biota and to show effects of agricultural practices and changes in forest size. These topics should be considered in future development.

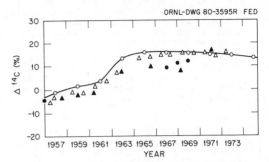

Figure 5.--Simulated $\Delta^{14}C$ values for the mixed ocean layer (open circles) compared to observed values. of Nozaki et al. (1978) (solid triangles), Druffel and Linick (1978) (open triangles), and Nydal et al. (1976) (solid circles).

LITERATURE CITED

Bacastow, R. B. and Keeling, C. D. 1973. Atmospheric Carbon Dioxide and Radiocarbon in the Natural Carbon Cycle: II Changes from A.D. 1700 to 2070 as Deduced from a Geochemical Model. p. 86-135. AEC Symposium Series 30: In Carbon and the Biosphere, Proceedings of the 24th Brookhaven Symposium in Biology. G. M. Woodwell and E. V. Pecan (Eds.), [Upton, N.Y., May 16-18, 1972] CONF-720510, P. 392.

Baes, C. F., Jr., Goeller, H. E., Olson, J. S., and Rotty, R. M. 1976. The Global Carbon Dioxide Problem. ORNL-5194. Oak Ridge National Laboratory, Oak Ridge, Tennessee.

Baes, C. F., Jr., Goeller, H. E., Olson, J. S., and Rotty, R. M. 1977. Carbon Dioxide and Climate: The Uncontrolled Experiment. American Scientist 65: 310-320.

Baes, C. F., Jr. (To be published). The Response of the Oceans to Increasing Atmospheric Carbon Dioxide. Institute of Energy Analysis, Oak Ridge Associated Universities, Oak Ridge, Tennessee.

Bjorkström, A. 1979. A Model of the Carbon Dioxide Interaction between Atmosphere, Oceans, and Land Biota. Chapter 15, p. 403-457. In The Global Carbon Cycle. SCOPE 13. p. 491. B. Bolin, E. T. Degens, S. Kempe, and P. Ketner (Eds.). Proceedings of a Workshop, Ratzeburg, German Federal Republic. [March 21-26, 1977] John Wiley and Sons, New York, N.Y.

Bolin, B., Degens, E. T., Kempe, S., and Ketner, P. (Eds.). 1979. The Global Carbon Cycle. SCOPE 13. p. 491. Proceedings of a SCOPE Workshop, Ratzeburg, German Federal Republic. [March 21-26, 1977] John Wiley and Sons, New York, N.Y.

Bolin, B. et al. (Eds.) 1981. Global Carbon Modeling. SCOPE 16. John Wiley and Sons, Inc. New York City, N.Y. (in press).

Broecker, W. S., Li, Y.-H., and Peng, T.-H. 1971. Carbon Dioxide - Man's Unseen Artifact. p. 297-324. In Impingement of Man on the Oceans, D. W. Hood (Ed.) John Wiley and Sons, Inc., New York, N.Y.

Broecker, W. S., Takahashi, T., Simpson, H. J., and Peng, T.-H. 1979. Fate of Fossil Fuel Carbon Dioxide and the Global Carbon Budget. Science 206: 409-418.

Broecker, W. S., Peng, T.-H., and Engh, R. 1980. Modeling the Carbon System. Radiocarbon 22 (3): 565-598.

Chan, Y.-H. and Olson, J. S. 1980. Limits on the Organic Storage of Carbon from Burning Fossil Fuels. Journal of Environmental Management 11: 147-163.

Chan, Y.-H., Olson, J. S., and Emanuel, W. R. 1980. Simulation of Land-Use Patterns Affecting the Global Carbon Cycle. ORNL/TM-6651, Oak Ridge National Laboratory, Oak Ridge, Tennessee.

Clarke, G. L. and Denton, E. J. 1962. Light and Animal Life. Chapter 10. p. 456-468. In The Sea, Volume 1: Physical Oceanography. M. N. Hill (Ed.). Interscience Publishers, New York, N.Y.

Craig, H. and Weiss, R. F. 1970. Geosecs 1969 Intercalibration Station: Introduction, Hydrographic Features, and Total Carbon Dioxide - Oxygen Relationships. Journal of Geophysical Research 75 (36): 7641-7647.

Druffel, E. M. and Linick, T. W. 1978. Radiocarbon in Annual Coral Rings of Florida. Geophysical Research Letters 5 (11): 913-916.

Emanuel, W. R., Olson, J. S., and Killough, G. G. 1980a. The Expanded Use of Fossil Fuels by the U.S and the Global Carbon Dioxide Problem. Journal Environmental Management 10:37-49.

Emanuel, W. R., Post, W. M., and Shugart, H. H., Jr. 1980b. In Proceedings of the 1980 Pittsburgh Conference Modeling and Simulation. Instrument Society of America, Pittsburgh, Pennsylvania.

Emanuel, W. R., Killough, G. G., and Olson, J. S. 1981. In Global Carbon Modeling. SCOPE 16. John Wiley and Sons, Inc., New York, N.Y. (in press).

Eriksson, E. and Welander, P. 1956. On a Mathematical Model of the Carbon Cycle in Nature. Tellus 8 (2): 155-175.

Keeling, C. D. 1973a. The Carbon Dioxide Cycle: Reservoir Models to Depict the Exchange of Atmospheric Carbon Dioxide with the Oceans and Land Plants. In Chemistry of the Lower Atmosphere. S. I. Rasool (Ed.). Plenum Press, New York, N.Y.

Keeling, C. D. 1973b. Industrial Production of Carbon Dioxide from Fossil Fuels and Limestone. Tellus 28 (2): 174-198.

Killough, G. G. 1977. Diffusion-Type Model of the Global Cycle for the Estimation of Dose to the World Population from Releases of Carbon 14 to the Atmosphere. ORNL-5269. Oak Ridge National Laboratory, Oak Ridge, Tennessee.

Killough, G. G. 1980. A Dynamic Model for Estimating Radiation Dose to the World Population from Releases of ^{14}C to the Atmosphere. Health Physics 38 (March): 269-300.

Killough, G. G. and Emanuel, W. R. 1981. Distributions of Transit Time and Age for Several Models of Carbon Turnover in the Ocean. (This symposium).

Machta, L. 1972. Mauna Loa and Global Trends in Air Quality. Bulletin of the American Meteorological Society 53 (5): 402-420.

Menzel, D. W. and Ryther, J. H. 1970. Distribution and Cycling of Organic Matter in the Oceans. p. 31-54. In Organic Matter in Natural Waters. D. W. Hood (Ed.). Proceedings of a Symposium, University of Alaska, College, Alaska, [September 2-4, 1968] University of Alaska, Institute of Marine Science, College, Alaska.

Moore, B., Boone, R. D., Hobbie, J. E., Houghton, R. A., Melillo, J. M., Peterson, B. J., Shaver, G. R., Vorosmarty, C. J., and Woodwell, G. M. 1981. A Simple Model for Analysis of the Role of Terrestrial

Ecosystems in the Global Carbon Budget. In Global Carbon Modeling. SCOPE 16. Bolin et al. (Eds.). John Wiley and Sons, Inc., New York, N.Y. (in press).

Mopper, K. and Degens, E. T. 1979. Organic Carbon in the Ocean: Nature and Cycling. In The Global Carbon Cycle. SCOPE 13. p. 491. B. Bolin, E. T. Degens, S. Kempe, and P. Ketner (Eds.). Proceedings of a Workshop, Ratzeburg, German Federal Republic [March 21-26, 1977] John Wiley and Sons, New York, N.Y. Chapter 11.

Nozaki, Y., Rye, D. M., Turekian, K. K., and Dodge, R. E. 1978. A 200 Year Record of Carbon-13 and Carbon-14 Variations in a Bermuda Coral. Geophysical Research Letters 5 (10): 825-828.

Nydal, R., Lovseth, K., and Gulliksen, S. 1976. A Survey of Radiocarbon Variation in Nature Since the Test Ban Treaty. Paper presented at 9th Int. Radiocarbon Conf., Univ. of Calif., Los Angeles and San Diego [June 20-26, 1976].

Oeschger, H., Siegenthaler, U., Schotterer, U., and Gugelmann, A. 1975. A Box-Diffusion Model to Study the Carbon Dioxide Exchange in Nature. Tellus 27 (2): 168-192.

Olson, J. S. 1970. Carbon Cycles and Temperate Woodlands. p. 226-239. In Analysis of Temperate Forest Ecosystems. Ecological Studies 1. D. E. Reichle (Ed.). Springer-Verlag, New York, N.Y.

Olson, J. S. and Killough, G. G. 1977. Atmospheric Carbon Exchange Model With Subdivided Biospheric Pools and a Diffusive Ocean. Eos 58:808.

Olson, J. S., Pfuderer, H. A., and Chan, Y.-H. 1978. Changes in the Global Carbon Cycle and the Biosphere. ORNL/EIS-109. Oak Ridge National Laboratory, Oak Ridge, Tennessee. 169 pp.

Olson, J. S. .To be published. Earth's Vegetation as a Carbon Pool, Source, and Potential Sink for Carbon Dioxide. (Submitted to Nature); and The Role of the Biosphere in the Carbon Cycle. In Beyond the Energy Crises: Opportunity and Challenge. Pergamon Press, New York, NY.

SCEP (Study of Critical Environmental Problems). 1970. Man's Impact on the Global Environment. MIT Press, Cambridge, MA.

Woodwell, G. M., Whittaker, R. H., Reiners, W. A., Likens, G. E., Delwiche, C. C., and Botkin, D. B. 1978. Biota and the World Carbon Budget. Science 199;141-146.

DISTRIBUTIONS OF TRANSIT TIME AND AGE FOR SEVERAL MODELS OF CARBON TURNOVER IN THE OCEAN[1]

G. G. Killough and W. R. Emanual[2]

Abstract.--Five mathematical models of distribution by depth of carbon in the world ocean are compared with respect to their steady-state distributions of carbon transit time and age. In two of the models, the ocean is represented as two well mixed compartments, namely, a surface layer and a deep-water compartment. The remaining three models divide the ocean depth into 19 well mixed layers and differ among themselves in the patterns of transfer of carbon among the layers. All five models are calibrated to the same preindustrial depth profile of natural carbon-14, which in each case uniquely determines the transfer coefficients. The models show quite different distributions of transit time, and their carbon age distributions deviate significantly from the "apparent age" deduced from the calibrating profile by the radioactive carbon-14 decay rate.

INTRODUCTION

Within the past decade the global carbon cycle has come under intense scientific scrutiny as a result of concern about rising levels of CO_2 in the atmosphere, due at least in part to exponentially increasing utilization of fossil fuels. Carbon dioxide gas is transparent to incoming ultraviolet radiation from the sun and relatively opaque to scattered and reflected radiation in the infrared band. The result is a net trapping of heat (the so-called "greenhouse effect"). The rising atmospheric level of CO_2 is therefore believed to be a potential stress for planetary warming, with resultant disturbances of climate patterns that could have profound consequences for all human activities. Baes et al. (1977) give an overview of this problem.

Quantification of the effects of future scenarios of world-wide fossil fuel utilization is dependent upon our ability to simulate the dynamic response level of CO_2 in the atmosphere. Such simulations require dynamic models of the global carbon cycle that estimate exchanges of carbon among the major reservoirs, namely the atmosphere, the ocean, and land biota (fig. 1).

In the following somewhat simplified view of the carbon cycle, we assume that before the industrial revolution and up to mid-19th century the carbon cycle was essentially in steady state with respect to each of its isotopes, the most abundant of which is carbon-12 or ^{12}C (98.89%; nonradioactive). Most of the remainder is nonradioactive ^{13}C. Radioactive ^{14}C, which is produced in the upper atmosphere at an assumed constant rate, is present from this natural source in an abundance of about one part in 10^{12}. The radioactive half-life of ^{14}C is 5730 yr, and we assume that the natural rate of formation exactly balances the radioactive decay rate from the world system.

Man's activities in the past 150 years have impinged on the carbon cycle in important ways. Two are of primary interest here: (1) the rising atmospheric level of CO_2 due partly to combustion of fossil fuels and partly to forest clearing and land-use conversion is a significant perturbation of the cycle from its steady state; (2) nuclear and thermonuclear weapons testing since the 1940s has introduced about 27 million grams of ^{14}C into the system (Killough and Till 1978), so that by 1963

[1]Paper presented at the international symposium Energy and Ecological Modelling, sponsored by the International Society for Ecological Modelling (Louisville, Kentucky, April 20-23, 1981). Research sponsored by the National Science Foundation through Interagency Agreement AG-199, BMS 76-00761.

[2]G.G. K. is in Health and Safety Research Division, and W.R. E. is in Environmental Sciences Division, Oak Ridge National Laboratory, Oak Ridge, Tennessee, 37830 U.S.A. Oak Ridge National Laboratory is operated by Union Carbide Corporation for the U.S. Department of Energy under Contract No. W-7405-eng-26.

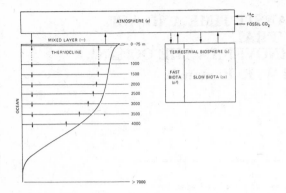

Figure 1.--Schematic presentation of the global carbon cycle.

in the high northern latitudes, the specific activity of atmospheric ^{14}C had reached a level about 100% in excess of its natural value.

The ocean contains about 60 times the carbon content of the atmosphere and perhaps 19 times that of the land biota. It is not well mixed, and a carbon atom that enters the surface waters from the atmosphere faces a wide distribution of possible transit times before it returns to the atmosphere. Each carbon atom in the ocean may be assigned as its "age" the time since it last entered the ocean from the atmosphere. The aggregate of carbon atoms in the ocean possesses a distribution function for age. Explicit determination of transit time and age distributions for several models of carbon in the ocean forms the principal part of this presentation.

Five one-dimensional structures for carbon turnover in the ocean will be considered (fig. 2). All have been calibrated to a mass distribution of natural radiocarbon specific activity estimated from measurements taken in the Atlantic, Pacific, and Indian Oceans in the late 1950s and early 1960s, so that in steady state their natural radiocarbon distributions (insofar as their structural resolution permits) are the same.

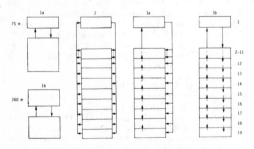

Figure 2.--Structures of the five models of carbon turnover in the ocean.

This calibration determines the transfer coefficients. For each model so calibrated we have deduced the steady-state distributions of transit time and age. Details and results are given in the sections that follow.

STRUCTURAL DEFINITION OF THE MODELS

The five models considered here all divide the ocean into a surface layer and a deeper portion, with all carbon that is exchanged with the atmosphere passing through the former compartment. Two of the models treat the deeper ocean as a single well mixed compartment which returns carbon to the surface layer by first-order kinetics. The remaining three models subdivide this deeper part into 18 layers; these three models differ from each other only in patterns of exchange of carbon among the layers. All five models are subject to the requirement that at steady state they predict the same mass distribution of natural ^{14}C specific activity; details of this distribution and the resulting calibration are given in the next section.

For convenient reference, the models are designated 1 (a and b), 2, and 3 (a and b) (fig. 2). The two-box models 1a and 1b are distinguished from each other only by the fact that the surface layer of 1a is considered to be 75 m deep, in conformity with the layered models, whereas 1b has a 260-m surface layer. The deeper surface layer corresponds to a calibration made by some investigators to bring the two-box model into better agreement with observed responses (Bacastow and Keeling 1979).

Of the layered group, Model 2 injects carbon from the surface layer directly into each of the deep-water layers, and each deep layer returns it directly to the surface layer. Thus the model might be interpreted to represent an extreme view that the total downward transport of carbon is by the sinking of cold surface water and its return to the surface is exclusively due to unwelling.

Layered Models 3a and 3b move carbon upward through the deep water reservoir by linear donor-controlled transfers from layer to layer. Model 3b represents downward transport in the same way, with passage through successive layers, whereas Model 3a uses direct injection from the surface layer to the deep water layers as does Model 2. Model 3a might be designated as "advective" and Model 3b as "diffusive."

Four of the five models are simplified prototypes of others proposed and studied by various investigators of carbon-cycle dynamics (Craig 1957; Bolin and Eriksson 1959; Keeling 1973; Bacastow and Keeling 1979; Oeschger et al. 1975; and Björkström 1979).

In each compartment or layer (indexed by j) we consider the dynamic quantities Y_j and X_j, which stand for total grams of nonradioactive carbon ($^{12}C + ^{13}C$) and ^{14}C, respectively, with initial (steady-state) values Y_j^0 and X_j^0. In the preindustrial steady state, we write $A_j = X_j^0/(Y_j^0 + X_j^0) \cong X_j^0/Y_j^0$ for the natural $^{14}C/(total carbon)$ mass ratio or _specific_ activity. We assume that the Y_j^0 and the A_j (and hence the X_j^0) are known.

Mass balance is expressed by differential equations that relate rates of change of compartment levels to fluxes in and out of the compartments; two equations are written for each compartment: one for Y_j, the other for X_j. These equations are given below, followed by definitions of symbols used in them.

Surface layer, j=1:

$$\frac{dY_1}{dt} = k_{a,m}Y_a - \left(\mu + \sum_{j=2}^{N} k_{i,j}\right)\cdot Y_1 + \sum_{j=2}^{N} k_{j,1}Y_j \qquad (1)$$

$$\frac{dX_1}{dt} = k_{a,m}\alpha_{a,m}X_a - \left(\alpha_{m,a}\mu + \sum_{j=2}^{N} k_{i,j} + \lambda\right)\cdot X_1 + \sum_{j=2}^{N} k_{j,1}X_j \qquad (2)$$

Deep waters, layer j = 2, ..., N:

(Note: [] means the enclosed term is omitted if j = 2; {} is omitted if j = N.)

$$\frac{dY_j}{dt} = [k_{1,j}Y_1] + k_{j-1,j}Y_{j-1} - (k_{j,1} + \{k_{j,j+1}\} + [k_{j,1}])Y_j + \{k_{j+1,j}Y_{j+1}\} \qquad (3)$$

$$\frac{dX_j}{dt} = [k_{1,j}X_1] + k_{j-1,j}X_{j-1} - (k_{j,1} + \{k_{j,j+1}\} + [k_{j,1}] + \lambda)X_j + \{k_{j+1,j}X_{j+1}\} \qquad (4)$$

(For N = 2 eqs. (1)-(4) define Models 1a and 1b; for Models 2, 3a, and 3b, N = 19. The numbering j = 1, ..., N of the layers increases with depth.) In eqs. (1)-(4) $k_{a,m}$ = total carbon transfer-rate coefficient from atmosphere to ocean surface layer (yr^{-1}); Y_a = total grams of atmospheric carbon; μ = transfer-rate coefficient for evasion of carbon from the ocean surface water to the atmosphere (yr^{-1}; calculated from steady-state conditions); $k_{i,j}$ = transfer-rate coefficient from compartment i to compartment j (yr^{-1} calculated from steady-state conditions); X_a = total grams of radioactive ^{14}C in the atmosphere; $\alpha_{a,m}$, $\alpha_{m,a}$ = isotope fractionation coefficients for ^{14}C relative to total carbon in transfers between atmosphere and ocean surface waters ($\cong$0.972, 0.955, respectively); and λ = radioactive decay-rate coefficient for ^{14}C (= 1.21 x 10^{-4} yr^{-1} based on a half-life of 5730 yr).

For the steady state, the derivative terms in eqs. (1)-(4) are zero, $Y_j = Y_j^0$, and $X_j = X_j^0 = Y_j^0 A_j$. The preindustrial atmospheric levels Y_a^0 and $X_a^0 = Y_a^0 A_a$ are assumed known. Table 1 gives numeric values of these initial values and other parameters. When these substitutions are made in eqs. (1)-(4), we may solve uniquely for the transfer coefficients $k_{a,m}$, $k_{i,j}$, and μ for each of

the model structures under study, although in the general case with N = 19 there would be too many unknowns (4N - 4) for the 2N linear equations to determine. For N = 2 the 4 equations uniquely determine $k_{a,m}$, μ, $k_{1,2}$, and $k_{2,1}$; for larger N the special structure of Model 2 and Models 3a and 3b imposes the condition $k_{i,j} = 0$ on the excess transfer coefficients. Explicit solutions for the nonzero $k_{i,j}$, $k_{a,m}$, and μ are possible and are given elsewhere (Killough and Emanuel, in press), or alternatively the values may be obtained directly with a linear equation solver by setting the derivative terms in eqs. (1)-(4) equal to zero; see Table 1 footnote.

MASS DISTRIBUTION OF NATURAL RADIOCARBON SPECIFIC ACTIVITY IN THE OCEAN

Our goal is to estimate the distribution function F(A'), where

$$M \cdot F(A') = \text{carbon mass in the ocean of natural } ^{14}C \text{ activity} < A' \qquad (5)$$

By A' we mean the normalized quantity A/A_a, where A and A_a denote $^{14}C/C$ ratios in the ocean and atmosphere, respectively. We note that A_j' must

Table 1.--Distribution of nonradioactive carbon and natural ^{14}C for the five ocean models

Layer j	Depth (m)	Nonradioactive carbon Y_j^0 (10^{15} g)		^{14}C activity*	
		1a	1b	1a	1b
1	0	659		.960	
2	75	432		.959	
3	125	643	2,251	.957	.960
4	200	862		.953	
5	300 [260]	880		.939	
6	400	895		.925	
7	500	895		.914	
8	600	895		.909	
9	700	895		.904	
10	800	895		.900	
11	900	37,542 893	35,950	.839 .896	.839
12	1,000	4,437		.885	
13	1,500	4,328		.864	
14	2,000	4,187		.837	
15	2,500	4,015		.814	
16	3,000	3,678		.795	
17	3,500	3,177		.781	
18	4,000	2,506		.770	
19	4,500	3,029		.723	
	ΣY_j^0	38,201			

*The numbers shown are fractionation-corrected $^{14}C/^{12}C$ ratios normalized to the preindustrial atmospheric value. Because the effect of isotopic fractionation has been removed, the fractionation coefficients $\alpha_{m,a}$ and $\alpha_{a,m}$ in eq. (2) should be set equal to one when these numbers are substituted into steady-state versions of eqs. (1) - (4) to calculate the transfer coefficients μ and $k_{i,j}$.

decrease with increasing depth: $A'_j < A'_\ell$ for $j > \ell$. This fact is a consequence of loss by radioactive decay and the absence of sources of radiocarbon within the ocean. M is the total mass of carbon in the ocean (Table 1).

Samples of inorganic carbon taken from surface waters of the ocean since the late 1950s show increasing contamination by ^{14}C produced by nuclear weapons tests. Comparison of depth profiles of ^{14}C activity taken in the Pacific over time suggests that as of the middle 1960s the excess radiocarbon had not migrated below 200 m in significant amounts (Bien and Suess 1967). Systematic sampling of radiocarbon in the oceans does not predate the weapons testing, but Broecker et al. (1978) have given an estimate for the surface waters that we approximate in the A' scale as A' = 0.96.

Therefore we proceed as follows:

(A) Use data from the ocean to estimate the distribution $F_{200}(A')$ of A' below 200 m

(B) With the result of step (A) and the assumption of decreasing A' with increasing depth, assign values for the A'_j for those layers at depths $\geq$ 200 m

(C) Assign A' = 0.96 to the surface layer (j = 1) and the two intervening layers down to 200 m by linear interpolation.

Figure 3 shows an empirical distribution of A' for depths below 200 m overlaid by a coarser step function that adapts the distribution to the resolution permitted by the 16 corresponding layers of the model. The indicated cumulative fractions are arrived at by summing the Y^0_j of Table 1 from the deepest layer upward and dividing by the total carbon mass for 200 m and below. The empirical distribution of A' is based on 191 carbon samples from various depths and locations in the Atlantic (Broecker et al. 1960) and Pacific and Indian Oceans (Bien et al. 1965); we eliminated all samples taken above 200 m and did not include more recent sampling (post 1965) by other investigators.

Table 1 exhibits the A'_j and Y^0_j, with the former determined by steps (A)-(C) above. The latter are based on water volumes at various depths (Sverdrup et al. 1942) and depth profiles of dissolved carbonate (Broecker 1974).

Our estimate of the distribution function F(A') [eq. (5)] may now be stated for our one-dimensional models as

$$F(A') = \frac{\sum_{\{j:\, A'_j < A'\}} Y^0_j / M}{}$$ (6)

where M is the total mass of carbon in the ocean.

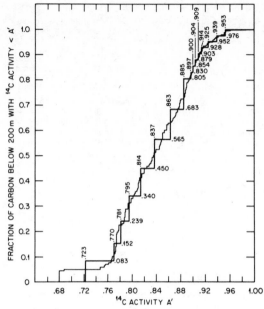

Figure 3.--Cumulative carbon mass distribution of natural ^{14}C activity A' in the ocean below 200 m. The finer step function is the empirical distribution function and is based on measurements in the Atlantic, Pacific, and Indian Oceans in the late 1950s and early 1960s. The coarser step function is a calibration fit based on the resolution of the three layered models (2, 3a, and 3b).

DISTRIBUTION OF CARBON TRANSIT TIME AND AGE IN THE MODELS

Let us consider a unit pulse of carbon introduced at time t = 0 into the surface layer (j = 1) of the ocean. If we denote the response in layer j by n_j, then we have the perturbation equations

$$\frac{dn_1}{dt} = -\left(\mu + \sum_{i=2}^{N} k_{1,i}\right) \cdot n_1 + \sum_{i=2}^{N} k_{i,1}\, n_i$$ (7)

$$\frac{dn_j}{dt} = -\left(\sum_{\substack{i=1 \\ i\neq j}}^{N} k_{j,i}\right) \cdot n_j + \sum_{\substack{i=1 \\ i\neq j}}^{N} k_{i,j}\, n_i ,$$

$$j = 2, \ldots, N$$ (8)

with initial conditions

$$n_1(0) = 1, \quad n_j(0) = 0, \quad j = 2, \ldots, N$$ (9)

We assume that the $k_{i,j}$ and μ have been determined for the model under study as indicated previously. We define the unit response function

$$R(t) = \sum_{j=1}^{N} \eta_j(t), \qquad t \geq 0 \qquad (10)$$

which may be interpreted as the fraction of carbon atoms entering the ocean at time zero that remain at time t. Alternatively, $R(t)$ is the probability that a carbon atom will remain in the ocean for a time $\geq t$, so that $1 - R(t)$ is the cumulative distribution function for the transit time of carbon atoms entering the ocean:

$$\Phi(\tau) = 1 - R(\tau) \qquad (11)$$

where the transit (or residence) time τ is defined as the time elapsed between the entry of a carbon atom into the ocean from the atmosphere and its return to the atmosphere.

When the system is in its preindustrial steady state, the distribution of the age of carbon in the ocean may be derived. Let $\psi(T)$ denote the normalized age density. Then $\int_T^{T+\Delta T} \psi(t)$ dt is the fraction of carbon mass that has been in the ocean for a time between T and T + ΔT. If I denotes the total invasion rate, the mass of this carbon is $\int_T^{T+\Delta T} I \cdot R(t)$ dt, so that ψ is proportional to R. Therefore, with normalization

$$\Psi(T) = \int_0^T R(t)\, dt \bigg/ \int_0^{+\infty} R(t)\, dt \qquad (12)$$

is the cumulative distribution function of the age T of carbon in the ocean at steady state.

Figure 4 shows the cumulative distribution curves for transit times of carbon entering the ocean — one curve for each of the five ocean models. All of these distributions have the same mean: $\bar{\tau}$ = 393 yr. Yet the distribution curves discriminate rather sharply among the models, with the exception that the curves for Models 3a and 3b remain fairly close together. The medians range from 5.6 yr (Model 1a) to 39 yr (Model 3a), and it is important to realize that the curves deviate most from each other in the region of transit times $<\sim 100$ yr — a region that is of greatest importance for discerning the recent history of the system and making projections into the next century.

The cumulative distributions of carbon age in the steady-state system are shown in Fig. 5 for the five models. While the model curves do not offer much contrast among themselves, they differ strongly from the "apparent age" histogram that is calculated directly from radiocarbon depletion of each layer relative to the atmosphere:

$$\text{apparent age of layer } j = -\lambda^{-1} \ln A_j' \qquad (13)$$

Each model curve shows a theoretical "true" distribution of age, whereas the step function is the result of information loss due to mixing processes. The means of the age distribution are as follows:

Model	1a	1b	2	3a	3b
Mean age $\bar{T}$ (yr)	1564	1514	1611	1663	1607

The median age varies from 854 yr (3a) to 1104 yr (1a).

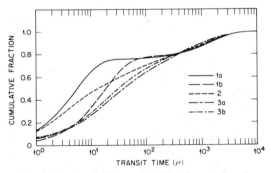

Figure 4.--Cumulative distribution of the transit time of carbon entering the ocean at steady state, for each of the five models.

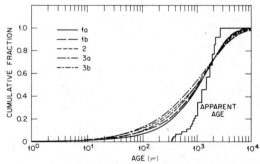

Figure 5.--Cumulative distribution of the age of carbon in the ocean at steady state, for each of the five models. The step function represents the "apparent age" distribution for the resolution of the layered models (2, 3a, and 3b).

CONCLUSIONS

Distributions of age and transit time for material in a dynamic reservoir have long been recognized as useful concepts for the interpretation of tracer measurements. Our results show quite clearly that for a fixed mass distribution of $^{14}C/C$ in the ocean (and therefore a fixed mean transit time), structural changes that result in relatively small variations in the distribution of age induce important perturbations in the transit-time distribution for τ less than about 100 years. Simulations reported elsewhere (Killough and Emanuel, in press) demonstrate a considerable impact of these perturbations on the interpretation of dynamics of the global carbon cycle, and particularly the role of terrestrial ecosystems in the cycle. Therefore, conclusions drawn about dynamics of the global carbon cycle on the basis of estimates of the age of carbon in ocean water are subject to question. Information in other tracer distributions (e.g. tritium, ^{226}Ra, ^{39}Ar) must be used to narrow the uncertainty.

Finally, it is proper to mention the essential role played by modern computer software in investigations of the kind reported here. We have had to perform three types of calculations: (1) solution of linear algebraic equations to estimate the mean age for each model; (2) evaluation of $\exp(At)$, A a constant matrix, in order to plot the transit time and age distributions; and (3) discrete-variable solution of systems of nonlinear ordinary differential equations in order to perform the simulations referred to in this section. For (3) the IBM CSMP3 system was used, and (2) was based on subroutines from EISPACK (Smith et al. 1974) for calculating the characteristic roots and vectors of A. Linear algebraic systems were solved with local library subroutines adapted from Forsythe et al. (1977). All calculations were quite routine with no evidence of numeric difficulties.

LITERATURE CITED

Bacastow, R. and Keeling, C. D. 1979. Models to predict future atmospheric CO_2 concentrations. p. 72-90. In Global effects of carbon dioxide from fossil fuels (ed. W. P. Elliott and L. Machta), CONF-770385. National Technical Information Service, Springfield, VA.

Baes, C. F., Jr., Goeller, H. E., Olson, J. S., and Rotty, R. M. 1977. Carbon dioxide and climate: The uncontrolled experiment. Am. Sci. 65:310-320.

Bien, G. and Suess, H. 1967. Transfer and exchange of ^{14}C between the atmosphere and the surface water of the Pacific Ocean. p. 105-115. In Radioactive dating and methods of low-level counting. International Atomic Energy Agency, Vienna

Bien, G. S., Rakestraw, N. W., and Suess, H. 1965. Radiocarbon in the Pacific and Indian Oceans and its relation to deep water movements. Limnol. Oceanogr. 10 (Suppl.):R25-R37.

Björkström, A. 1979. A model of CO_2 interaction between atmosphere, oceans, and land biota. p. 403-451. In The global carbon cycle — SCOPE 13 (ed. B. Bolin, E. T. Degens, S. Kempe, and P. Ketner). J. Wiley, New York.

Bolin, B. and Eriksson, E. 1959. Changes in the carbon dioxide content of the atmosphere and sea due to fossil fuel combustion. p. 130-142. In The atmosphere and sea in motion (ed. B. Bolin). The Rockefeller Institute Press, New York, N. Y.

Broecker, W. S. 1974. Chemical oceanography. Harcourt Brace Jovanovich, New York, N. Y.

Broecker, W. S., Peng, T. H., and Stuiver, M. 1978. An estimate of the upwelling rate in the equatorial Atlantic based on the distribution of bomb radiocarbon. J. Geophys. Res. 83:6179-6186.

Broecker, W. S., Gerard, R., Ewing, M., and Heezen, B. C. 1960. Natural radiocarbon in the Atlantic Ocean. J. Geophys. Res. 65: 2903-2931.

Craig, H. 1957. The natural distribution of radiocarbon and the exchange time of carbon dioxide between atmosphere and sea. Tellus 9:1-17.

Forsythe, G. E., Malcolm, M. A., and Moler, C. B. 1977. Computer methods for mathematical computations. Prentice-Hall, Englewood Cliffs, N. J.

Keeling, C. D. 1973. The carbon dioxide cycle: reservoir models to depict the exchange of atmospheric carbon dioxide with the oceans and land plants. p. 251-328. In Chemistry of the lower atmosphere (ed. S. I. Rasool). Plenum Press, New York, N. Y.

Killough, G. G. and Till, J. E. 1978. Scenarios of ^{14}C releases from the world nuclear power industry from 1975 to 2020 and the estimated radiological impact. Nucl. Saf. 19:602-617.

Killough, G. G. and Emanuel, W. R. In press. A comparison of several models of carbon turnover in the ocean with respect to their distributions of transit time and age, and responses to atmospheric CO_2 and ^{14}C. Tellus.

Oeschger, H., Siegenthaler, U., Schotterer, U., and Gugelmann, A. 1975. A box diffusion model to study the carbon dioxide exchange in nature. Tellus 27:168-192.

Smith, B. T., Boyle, J. M., Garbow, B. W., Ikehe, Y., Klema, V. C., and Moler, C. B. 1974. Matrix eigensystem routines — EISPACK guide. Springer-Verlag. Berlin, Heidelberg, and New York.

Sverdrup, H. V., Johnson, M. W., and Fleming, R. H. 1942. The oceans — their physics, chemistry, and general biology. Prentice-Hall, New York, N. Y.

ENERGY DEVELOPMENT AND MAJOR CARBON DIOXIDE
PRODUCERS: MODELLING THE CUMULATIVE
RESPONSIBILITY FOR FUTURE CLIMATE CHANGE[1]

Robert Schware[2]

Abstract -- Inevitably, questions may arise concerning responsibility for the costs of the adverse climatic impacts that might be suffered by regions or nations as a consequence of increased atmospheric carbon dioxide. In practice this means that nations would need to agree on measures to determine the proportional indemnity to be charged to each region or country. Such measures should be calculated in terms of cumulative emissions. Beyond the turn of the next century, developed countries might have to bear much of the responsibility for new climatic regimes, as they will become the principal suppliers of fuel to the world, the major consumers, and the major cumulative contributors of carbon dioxide to the atmosphere.

INTRODUCTION

One global effect of our increasing burning of fossil fuels is the accumulation of atmospheric carbon dioxide. As a result, a significant average warming of the earth's surface may occur within the first half of the next century. It is possible that this so-called "greenhouse effect" will cause a temperature change and shifts of precipitation patterns without precedent in human history.

But knowledge about the timing, intensity, and effects of possible carbon dioxide-induced impacts is still poorly understood. Assessments to date have been aggravated by great uncertainties in: the behavior of the natural carbon cycle; the rate of uptake by the oceans and biomass of anthropogenic carbon dioxide; the accuracy of general circulation models that simulate the atmospheric response to an increased loading of carbon dioxide; and the expansion, conservation, and fuel switching in the energy supply that may occur over the next few decades (Machta 1973; Chervin 1978; Schneider and Chen 1980).

[1]Paper presented at the international symposium Energy and Ecological Modelling, sponsored by the International Society for Ecological Modelling. (Louisville, Kentucky, April 20-23, 1981).
[2]Robert Schware is with the Advanced Study Program, National Center for Atmospheric Research, Boulder, Colorado, USA. The National Center for Atmospheric Research is sponsored by the National Science Foundation.

In light of the wide range of uncertainties about estimates and impacts of possible climatic futures, an important, and less uncertain, policy aspect of the global carbon dioxide problem can be addressed now: Who are the major carbon dioxide-emitting nations? To what extent do they contribute to the carbon dioxide rise?

In this paper, the countries producing most of the carbon dioxide (presently or in the recent past) are identified and their importance to the future of the carbon dioxide problem is discussed. There are, as we will see, important differences for carbon dioxide emissions between countries and groups of countries. These might be critical in the long term if responsibility for the costs of adverse carbon dioxide-induced impacts is to be assessed (Kellogg and Schware 1981).

HISTORICAL BUILDUP OF ATMOSPHERIC CARBON DIOXIDE

Just before the Industrial Revolution the natural sources and sinks of carbon dioxide must have been in near-equilibrium. Undoubtedly there have been large changes in carbon dioxide concentration during the earth's geological history. In fact there have been changes of a factor of two in just the past 20,000 years (Berner et al. 1980; Delmas et al. 1980). One hundred years ago the concentration of carbon dioxide was probably between 270 and 300 parts per million by volume (ppmv). But at that time a new source was introduced. Mankind began taking carbon locked in the earth's crust in the form of coal, petroleum, and natural gas and burning it in large quantities, thereby producing carbon dioxide. Since the turn of this century the rate of increase in fossil fuel burning has averaged about 4.3 percent per year, a steady rise that slackened only during the two world wars and the 1930s depression (Rotty 1979). Carbon dioxide concentration is now almost 336 ppmv, a 17 percent increase over the assumed pre-1900 value.

The United Nations has collected data on how much fossil fuel has been used worldwide (United Nations 1976; United Nations 1978). These figures make it possible to compare the amount of carbon dioxide produced with the amount that has remained in the atmosphere. If no other source were present, it appears that about 55 percent of the fossil fuel produced carbon dioxide is still airborne. Generally, it has been assumed that the rest was taken up by the oceans, or possibly by the living biota -- that is, primarily the forests of the world. This issue is controversial, however (Schneider and Chen 1980).

Before we consider the rate at which carbon dioxide emissions from fossil fuel combustion might be increased over the next several decades, it is instructive to look at what has happened in the past thirty years. The worldwide pattern of carbon dioxide production is more certain from 1950, since United Nations data are not available before then.

A first step in disaggregating carbon dioxide producers is to look at countries separately. The relative carbon emissions (in the form of carbon dioxide) from the consumption of fossil fuels for the six largest producing countries appear in figure 1 and cover the period 1950-1976.[1] The largest contributors have been the U. S. A., USSR, & China. Figure 2 illustrates the recent history of rate of growth of carbon dioxide production (percent per year) for the countries shown in the previous figure.

The overall picture of the major geopolitical sectors (as defined by the United Nations) producing carbon dioxide between 1950 and 1976 is shown in figure 3 (including World totals). All of these sectors have gradually been

---

[1]The following carbon release coefficients per terrawatt-year (in units of 10^9 tons of carbon in the form of carbon dioxide) have been used in this analysis:
  i) Solid fuels (coal and lignite)
      0.76×10^9
 ii) Liquid fuels (crude petroleum and natural gas liquids)
      0.6×10^9
iii) Natural gas
      0.43×10^9
See Keeling (1973), Perry and Landsberg (1977), Rotty (1978), Woodwell et al. (1979), Zimen et al. (1977).

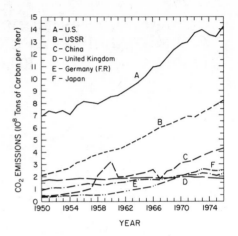

Figure 1.--Historical trend of carbon dioxide emissions by the six largest carbon dioxide producing countries.

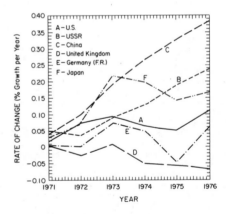

Figure 2.--Rate of changes in carbon dioxide emissions for the six largest carbon dioxide producing countries.

increasing their carbon dioxide emissions during the period for which data are available. The annual growth for the remaining seven geopolitical sectors of the world is shown in figure 4. (Note that the vertical scale has been expanded by a factor of ten.) Their relative contribution to world emissions is considerably outweighed by the major geopolitical sectors comprised of developed countries.

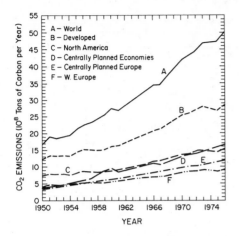

Figure 3.--Historical trend of carbon dioxide emissions by the largest carbon dioxide geopolitical sectors (see United Nations 1978).

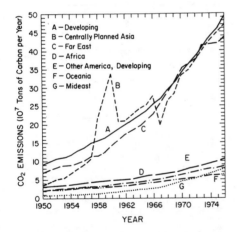

Figure 4.--Historical trend of carbon dioxide emissions by the smallest carbon dioxide producing geopolitical sectors (see United Nations 1978).

FUTURE PRODUCERS

There are major uncertainties as to which countries or groups of countries will be the dominant contributors of atmospheric carbon dioxide emissions to the year 2000. Attempting to determine the rate of additions to the atmospheric concentration of carbon dioxide by

countries after 2000 is highly speculative. In both cases, estimates must assume (somewhat arbitrarily) growth rates that will depend on numerous political, economic, and demographic variables. These growth rates must also take into account past carbon dioxide emissions as well as the transient response of the earth-atmosphere-ocean-biota system to annual carbon dioxide inputs.

An illustration of the uncertainty that surrounds most estimates of past and future fossil fuel consumption is presented in table 1. It is apparent that major errors in past projections of the future have included both large overestimates and underestimates.

Figure 5 provides for a wide range of future possibilities regarding the carbon dioxide doubling date and the relative fraction of the added carbon dioxide contributed by developed and developing countries. For example, if we assume fossil fuel growth rates of 0.02 and 0.03 (approximately 2 and 3 percent per year) in developed and developing countries respectively, the carbon dioxide doubling of the assumed preindustrial value of 290 ppm will occur around 2050, and 75 percent of it will have been contributed by developed nations. If we make a conservative assumption that fossil

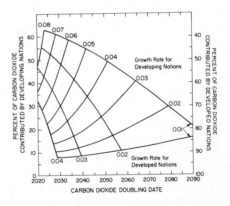

Figure 5.--Percent of the doubled airborne carbon dioxide contributed by developed and developing countries. Note that growth rates are used in an exponential carbon dioxide release model for dates after 1976 (Schware and Friedman 1981).

Table 1. Review of past energy consumption projections (Adapted from Darmstadter and Schurr, 1974)

Region	Actual Data		Projections		
	Period	Average Annual Growth Rate (%)	Reference*	Period	Average Annual Growth Rate (%)
		Energy Consumption			
World	1960-70	5.6	(A)	1960-70	4.6
	1970-78	3.2	(G)	1970-80	4.8
			(B)	1960-80	5.0
Western Europe	1960-70	6.3	(C)	1955-75	2.8
			(A)	1960-70	4.4
			(A)	1970-80	4.0
			(B)	1964-70	4.2
			(B)	1970-80	4.1
United States	1960-70	4.2	(D)	1960-70	2.9
	1973-79	1.0	(G)	1970-80	2.8
			(A)	1960-70	3.6
			(A)	1970-80	3.3
Japan	1960-70	11.9	(A)	1960-70	9.1
			(A)	1970-80	7.0
			(B)	1964-70	10.0
			(B)	1970-80	6.9
		Oil Consumption			
Western Europe	1962-72	10.5	(B)	1964-80	4.1
United States	1962-72	4.6	(A)	1960-70	3.5
			(A)	1970-80	2.7
			(E)	1965-75	3.4
			(E)	1975-80	2.9
			(F)	1965-80	3.1
Japan	1962-72	17.5	(B)	1964-70	14.3
				1970-80	8.3

*Symbols refer to the following references:
(A) European Coal and Steel Community. 1966. Review of the Long Term Energy Outlook of the European Community, Luxembourg;
(B) O.E.C.D. 1966. Energy Policy, Paris;
(C) O.E.C.D. 1960. Towards a New Energy Pattern in Europe, Brussels;
(D) Landsberg, H., L.L. Rischman and J.L. Fisher. 1963. Resources in America's Future. Resources for the Future, Johns Hopkins, Baltimore;
(E) Schurr, S.H., P.T. Homan, et al. 1971. Middle Eastern Oil and the Western World: Prospects and Problems. Elsevier, New York;
(F) U.S. Department of the Interior. 1968. United States Petroleum through 1980. U.S. Department of the Interior, Washington, D.C.;
(G) U.S. Department of Energy. 1980. Monthly Energy Review (May), U.S. Department of Energy, Washington, D.C.

fuel growth rates of 0.01 and 0.02 (approximately 1 and 2 percent per year) will occur for developed and developing countries respectively, then the doubling would be reached in 2080, and 70 percent of the carbon dioxide airborne at that date will have been contributed by developed countries.

It seems likely that this latter low-growth scenario would delay the impacts of a carbon dioxide-induced warming trend. The time available and needed for policy makers to consider implications of the climate change in a deliberate, purposeful policy design would then be extended considerably. It may even be sufficient for the transition in energy production to non-fossil fuels without massive adjustments toward non-fossil supply expansion, lower economic growth, or higher fossil fuel prices. Nonetheless, developed countries might have to bear much of the responsibility for the new climatic regimes, should they occur: they are likely to be the major cumulative contributors of carbon dioxide to the atmosphere.

LITERATURE CITED

Berner, W., H. Oescherger, and B. Stauffer. 1980. Information on the CO_2 cycle from ice core studies. Proceedings of the 10th International Radiocarbon Conference, Bern and Heidelberg, August 1979 (in press).

Chervin, R.M. 1978. The limitations of modelling: The question of statistical significance. p. 191-201. In Climate Change, ed., J. Gribbin, Cambridge University Press, Cambridge, England

Darmstadter, J., and S.S. Shurr. 1974. World energy resources and demand. Phil.Trans.-Roy.Soc.London A276, No. 1261: 413.

Delmas, R.J., J.-M. Ascencio, and M. Legrand. 1980. Polar ice evidence that atmospheric CO_2 20,000 years BP was 50% of present. Nature 284: 155-157.

Keeling, C.D. 1973. Industrial production of carbon dioxide from fossil fuels and limestone. Tellus 25: 174-198.

Kellogg, W.W., and R. Schware. 1981. Climate change and society: consequences of increasing atmospheric carbon dioxide. Westview Press, Boulder, Colorado.

Machta, L. 1973. Prediction of CO_2 in the atmosphere. In Carbon and the Biosphere, ed., G.M. Woodwell and E.V. Pecan, U.S. Atomic Energy Commission CONF - 720510: 21-31.

Perry, H., and H. Landsberg. 1977. Projected world energy consumption. p. 35-50. In Energy and Climate, National Academy of Sciences, Washington, D. C.

Rotty, R.M. 1979. Energy demand and global climate change. p. 269-283. In Man's Impact on Climate, ed., W. Bach, J. Pankrath, W. W. Keellog, Development in Atmospheric Science 10, Elsevier, Amsterdam.

Rotty, R.M. 1978. The atmospheric CO_2 consequences of heavy dependence on coal. In Carbon Dioxide. Climate and Society, ed., J. Williams, Pergamon Press, New York.

Schneider, S.H. and R.S. Chen. 1980. Carbon dioxide warming and coastline flooding: Physical factors and climatic impact. Ann.Rev. Energy 5: 107-140.

Schware, R. and E.J. Friedman. 1981. Anthropogenic climate change: assessing the responsibility of developed and developing countries. Bull.Atomic Scientists (in press).

United Nations. 1978. World Energy Supplies 1972-1976. Statistical papers, Series J, No. 21.

United Nations. 1976. World Energy Supplies 1950-1974. Statistical papers, Series J, No. 18.

Woodwell, G.A., G.J. MacDonald, and C.D. Keeling. 1979. The carbon dixoide report. Bull.Atomic Scientists, October 56-57.

Zimen, K.E., P. Offerman and G. Hartmann. 1977. Source functions of CO_2 and the future CO_2 burden in the atmosphere. Zeits Naturforschung 32a: 1544-1554.

THE SOURCE-SINK FUNCTION OF THE TERRESTRIAL BIOTA WITHIN THE GLOBAL CARBON CYCLE[1]

G. H. Kohlmaier, G. Kratz, H. Bröhl and E. O. Siré[2]

Abstract.--A five biome model describing the carbon exchange within tropical and non-tropical forests and grassland as well as deserts and CO_2 exchange with the atmosphere has been constructed under consideration of vegetation changes. For the time range from 1860 to 1980 the calculations yield a cumulated CO_2 net input into the atmosphere from biogenic sources of about 100 - 110 Gt C, the differential net input being about 2 Gt C/yr at present. Good agreement with the observed atmospheric C-13 data for the time range between 1956 and 1978 is obtained.

INTRODUCTION

Whether today's biosphere is to be regarded as a source or a sink for additional atmospheric carbon dioxide is still not known. On the one hand estimates have been made concerning the areas of changes in the vegetation cover (e.g. forest clearings). Starting from these figures together with data for living and dead biomass per unit area in the affected biomes and under consideration of processes such as regrowth (Broecker et al. 1979; Hampicke 1980), incomplete combustion (Seiler and Crutzen 1980), accelerated humus oxidation (Buringh 1979) and erosion one can arrive at estimates for the CO_2 actually released as a consequence of man's impact on the biota. Some more recently published values concentrate within the range of 1 - 4 Gt C/yr for the present time (Hampicke and Bach 1979; Kohlmaier et al. 1980) Calculations based on the measurements of the C-13/C ratio in tree-rings (Stuiver 1978) suggest a cumulated input up to 120 Gt C for the period from 1860 to 1920 and a neutral function of the biota in recent times.

On the other hand, numerical ocean models calibrated with the stationary and bomb C-14 distribution with depth have been constructed (Oeschger et al. 1975) with such an uptake capacity for additional atmospheric CO_2 that, taking into account the carbon dioxide release from fossil fuel combustion and the increase of the atmosphere's CO_2 content, the biosphere has to be regarded as a CO_2 sink. The significant model parameters can at most be varied in such a way as to account for a weak biospheric source (Oeschger et al. 1980). Recently an analytical ocean model which had been calibrated with the stationary C-14 distribution in the ocean has been presented (Kohlmaier et al. 1980; Siré et al. 1981). This model's version with depth dependent eddy diffusivity suggests that the biosphere is neither a source nor a sink. If one, however, additionally takes into consideration downwelling of polar waters (advection) a biospheric CO_2 source can probably be accounted for.

In this paper we shall present a numerical model of the biota which are subdivided into five major biome types: tropical forests and grasslands, non-tropical forests and grasslands as well as deserts. After removal of the vegetation cover of a certain area the biome type on this area may change or regrowth of the old vegetation type may start. The humus content is correspondingly changed as well. Loss of humus by immediate oxidation and erosion as well as incomplete combustion is accounted for. Thus the calculations should be able to give a gross survey on the present state and changes of the biota

---

[1]Paper presented at the international symposium Energy and Ecological Modelling, sponsored by the International Society for Ecological Modelling. (Louisville, Kentucky, USA, April 20-23, 1981)

[2]Institut für Physikalische und Theoretische Chemie, Universität Frankfurt, Robert-Mayer-Str. 11, D-6000 Frankfurt/Main, F.R.G.

and the CO_2 released from biogenic sources. The model is also run for C-13 in order to verify the measured atmospheric data, because the correct reproduction of these constitutes one condition for the model reliability.

DESCRIPTION OF THE MODEL

General system behaviour

Each of our model biomes consists of living phytomass and dead organic matter in the soil (humus).(For the more detailed description of the world's forests see the subsequent paragraph.) The uptake of CO_2 from the atmosphere is modeled on the net primary production level; it is assumed to be proportional to the δ-th power ($0 \leq \delta < 1$) of the biomass (assuming a Bertalanffy (1960) type kinetics)

$$f_{ab} = k_{ab} \cdot b^{\delta}. \qquad (1.a)$$

f_{ab}: net primary production per unit area.
b: biomass per unit area or biomass density (The total fluxes and the total C-masses in a certain compartment can be obtained by simply multiplying the biomass and flux densities by the corresponding areas.) All other fluxes involved are assumed to be linearly donor controlled:

$$f_{ba} = k_{ba} \cdot b , \qquad (1.b)$$

$$f_{bh} = k_{bh} \cdot b , \qquad (1.c)$$

$$f_{ha} = k_{ha} \cdot h . \qquad (1.d)$$

Thus this model (fig. 1) can be described by the following system of differential equations:

$$db/dt = k_{ab} \cdot b^{\delta} - (k_{ba} + k_{bh}) \cdot b , \qquad (2)$$

$$dh/dt = k_{bh} \cdot b - k_{ha} \cdot h , \qquad (3)$$

h: soil organic carbon mass per unit area. Stability analysis of such systems have been made by Eriksson and Welander (1956) and by Kohlmaier et al. (1979). Eq. (2) can be solved analytically to give:

$$b(t) = b_{max} \cdot \left\{ 1 - [1 - (b_o/b_{max})^{1/d}] \cdot \right.$$
$$\left. \exp[(\delta - 1) \cdot (k_{ba} + k_{bh}) \cdot (t - t_o)] \right\}^d \qquad (4)$$

with $d = 1/(1-\delta)$,

which is a function for regrowth on an area where the vegetation cover has been removed at time t_o. b_o may be identified with the mass of newly imported seeds or the first layer of herbaceous plants (per unit area). For our calculations its magnitude does not have a significant influence on regrowth dynamics as long as it is greater than zero and less than 10 % of the climax biomass density, b_{max}. δ is the climax parameter

taken to be in the range between 0.2 and 0.5.

In contrast to previously assumed uptake kinetics (for a summary and an analysis see e.g. Kohlmaier et al. 1979) stimulation of growth by increased atmospheric CO_2 content is not considered in the numerical calculation Although the effect is well established on the photosynthesis level it is, however, difficult to estimate on the biome level. (Lemon 1977, Hampicke and Bach 1979).

Within a first order approximation the additional flux of CO_2 from the atmosphere to the biota may be formulated as:

$$F_{\gamma} = \gamma \cdot F_{ab}^o \cdot (\Delta N_a/N_{ao}) , \qquad (5)$$

where $0 \leq \gamma \leq 1$ is the fertilization parameter or biota growth factor, F_{ab}^o is the total net primary production and $\Delta N_a/N_{ao}$ is the relative change of the atmospheric CO_2 content. From all experimental evidence (Hampicke and Bach 1979) we believe that it may be realistic to assign to γ a value between 0 and 0.15, which when considering

$F_{ab}^o \approx 60$ Gt C/yr and $\Delta N_a/N_{ao} = 0.15$ in 1980 yields an annual flux of F_{γ} (1980) = 0 – 1.35 Gt C/yr. The cumulated sink strength due to CO_2 fertilization may also be calculated with respect to the reference year 1860. Assuming an exponential growth rate ($r = 0.033$/yr) for ΔN_a, one can calculate:

$$\int_{1860}^{1935} F_{\gamma} \, dt = 0 - 9.5 \text{ Gt C} ,$$

$$\int_{1860}^{1980} F_{\gamma} \, dt = 0 - 24.2 \text{ Gt C} ,$$

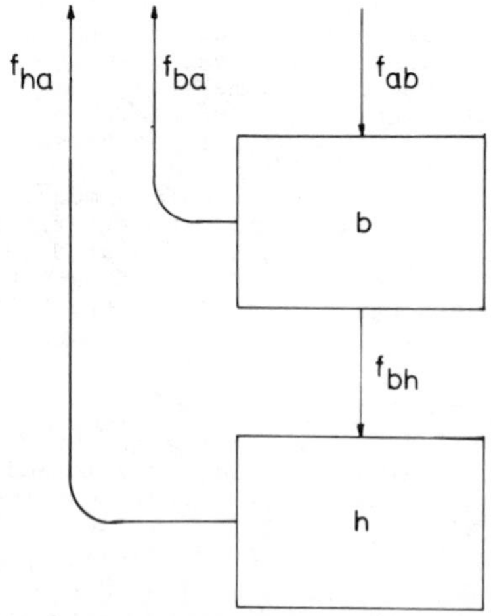

Figure 1.--Compartment structure of the model biomes.

$$\int_{1860}^{2000} F_\gamma \, dt = 0 - 87 \text{ Gt C} \, .$$

It may be seen that the reduction of the net carbon input caused by CO_2 fertilization up to 1935 falls within the uncertainty range of the total cumulated input (50 Gt C) whereas for the year 2000 its influence may be highly significant.

Analytical results: theoretical basis for monobiome and multibiome approach

Starting with the Bertalanffy expression (2) the stationary state condition of the climax state yields:

$$\frac{db}{dt} = f_{ab}^{max} \cdot [\, (\frac{b}{b_{max}})^\delta - \frac{b}{b_{max}}] \, , \tag{6}$$

where f_{ab}^{max} is the net primary productivity and b_{max} is the biomass density at climax state. Under the assumption of a homogeneous biosphere, the net ecosystem production may be obtained by multiplying eq. (6) with A, the total land area, if f_{ab}^{max} is identified with the average net primary productivity and b_{max} with the average biomass density at climax state:

$$A \cdot \frac{db}{dt} = A \cdot f_{ab}^{max} \cdot [\, (\frac{b}{b_{max}})^\delta - \frac{b}{b_{max}}] \tag{7.a}$$

or

$$A \cdot \frac{db}{dt} = A \cdot f_{ab}^{max} \cdot [\, (1 + \frac{\Delta b}{b_{max}})^\delta$$

$$- (1 + \frac{\Delta b}{b_{max}})] \, , \tag{7.b}$$

if $\Delta b \equiv b - b_{max}$. δ is the climax parameter which has been estimated to lie between 0.2 and 0.5 (Kohlmaier 1981) with the stability bounds $0 \leq \delta < 1$. Assuming in addition a homogeneous reduction of all biota by e.g. 10 %, 20 % or 30 %, the sink strength of the biota with respect to atmospheric CO_2 may be calculated to be 3.9 Gt C/yr, 7.7 Gt C/yr and 11.3 Gt C/yr. These results deviate very little from the linearized eq. (7.b):

$$A \cdot \frac{db}{dt} = - A \cdot f_{ab}^{max} \cdot [\, (1 - \delta) \cdot \frac{\Delta b}{b_{max}}] \tag{8}$$

where the corresponding fluxes are given by 4 Gt C/yr, 8 Gt C/yr and 12 Gt C/yr. If the reduction is not homogeneous but different in different regions A_i (with $\sum_i A_i = A$), then the following set of equations in accordance with eq. (7) may be studied:

$$A_i \frac{db_i}{dt} = A_i \cdot f_{ab,i}^{max} \cdot [\, (1 + \frac{\Delta b_i}{b_{i,max}})^\delta$$

$$- (1 + \frac{\Delta b_i}{b_{i,max}})] \, . \tag{9}$$

If the biota themselves are still considered homogeneous, then i may be just a placeholding index: $f_{ab,i}^{max} = f_{ab}^{max}$ and $b_{i,max} = b_{max}$. $\Delta b_i / b_{max}$ may now be different for each region A_i. The Δb_i obey the weighted average normalization condition:

$$\sum_i A_i \frac{\Delta b_i}{b_{max}} = A \frac{\Delta b}{b_{max}} \, . \tag{10}$$

An example may help to illustrate this effect. If, e.g. $A_i = A/10$ for all i and if $\Delta b / b_{max} = -0.2$ (20 % averaged total reduction) then

$$\frac{1}{10} \sum_i \frac{\Delta b_i}{b_{max}} = -0.2 \, .$$

Depending on the distribution of $\Delta b_i / b_{max}$ any sink strength value from 0 ($\Delta b_1 / b_{max} = \Delta b_2 / b_{max} = -1$, $\Delta b_i / b_{max} = 0$ for i = 3,...,10) up to the maximum for homogeneous reduction may be obtained. In the linearized version corresponding to eq. (7), the distribution of reduction is of no importance which stresses the fact that the linearization may introduce serious errors.

Model dynamics of the numerical simulation

In order to reproduce the single successional states more realistically than can be done by a simple one-box phytomass model the forest biomes have been subdivided into 20 compartments representing different age classes. The distribution in the stationary initial state (constant area cleared per year) is obtained as follows: From eq. (4 the time t_{95} which the biome needs to regrow until it has reached 95 % of its climax biomass density, b_{max}, is calculated. t_{95} multiplied with the annually cleared area gives the area then in use which is divided into 19 regions the succession states of which are calculated from eq. (4) as well. The 20th compartment exhibits climax vegetati

Within each simulation run the clearing rate may vary with time. Furthermore not the complete area cleared is regrown by forests again, but part of it is transformed to permanently used agricultural land. For each compartment the differential eq. (2) is numerically solved to simulate regrowth.

The humus of each biome type is modeled by a single compartment. Following total clearings 40 % of the humus content of the affected area is assumed to be liberated, one part of which is immediately oxidized and released to the atmosphere, another part of which is eroded. Now depending on whether the area is regrown by the original vegetation type again or whether it is transformed into another biome type the remaining humus is left in its original compartment or transferred to the succeeding biome's humus box. Then its time variation is simulated according to eq. (3).

DATA BASIS

Data for the model pools and fluxes

To estimate the contents of the chosen model compartments and the fluxes between them we use data for biome areas, net primary productivity, biomass, litter and humus density, litterfall and humification given by various authors.

Table 1 shows the estimated areas of different biomes at a preagricultural state, at the beginning of industrialisation (1860) and at present (1970) as given by Olson et al. (1978). Though it can be seen that great changes in biome areas occur up to 1860, we start from a hypothetical steady state with the 1860 conditions, assuming that the transformation rates had been small before this time. Taking the given sums for the wood complexes for our tropical and non-tropical forest biomes, dividing the grasslands in tropical and non-tropical as done by Rodin et al. (1975) and the crops and fringe areas as done by Hampicke and Bach (1979) and adding the area of buildings etc. to desert and semi-desert, we arrive at the data given in table 4, column 1.

For net primary production and biomass density we adopt the values given by Whittaker and Likens (1975) (table 2). Compared with more recent studies (Ajtay et al. 1979; Olson 1980; Brown and Lugo 1980) their npp-values might be somewhat low, while their biomass density values are biased on the high side, thus perhaps leading to an overestimation of carbon released by human impact and an underestimation of the accumulation capability of the biota. From these data we compute the figures in table 4, column 2-3, taking the area weighted means for our model biomes.

For the humus compartment we use the data given by Schlesinger (1979) (table 3). Similar values are reported by Ajtay et al. (1979), while Brown and Lugo (1980) give a significantly lower value for tropical forests (calculated as a mean of 6 kg C/m^2).

We did not find sufficient data for the flux into the humus compartment for all biomes. Ajtay et al. (1979) give estimates of litterfall and litter (table 3) but do not state which part of this is converted to humus and

Table 1.--Changes in biome areas from a pre-agricultural state to 1970 (after Olson et al. 1978).

Reservoir	Estimated Area (10^6 km^2)		
	preagricultural	1860	1970
WOODS COMPLEXES			
Boreal + Temperate			
Boreal (taiga)	10.10	9.5	9
Semiboreal	6.91	5.6	5
Cordilleran	3.77	3.5	3
Other cool temperate	3.76	3	2
Warm temperate	5.76	4.2	3.8
Semiarid	3.83	2.5	2
Arid moistland	1.07	0.4	0.2
TOTAL	35.2	28.7	25.0
Tropical + Subtropical			
Wet site, rainforest	4.56	4.3	3.3
Other tropical moist	8.83	7	5.3
Montane, seasonal	1.18	1	0.5
Montane, humid	2.42	2.2	2
Arid moistland	0.32	0.2	0.1
Woody savanna, scrub	14.05	13	11
TOTAL	31.36	27.7	22.2
WOODS TOTAL			47.2
NONWOODS COMPLEXES			
Agro-urban			
Crops	0	5	12
Fringe areas	1	3	7
Buildings, etc.	0	1	3
Other land			
Tundra-like, bogs	13.53	13	12
Grasslands	22.96	21	20
Desert, semidesert	29.35	30	29
NONWOODS TOTAL	67	73	83
TOTAL			130.2

which part is rapidly oxidized. From case studies (Kitazawa 1977; Kawahara 1977; Bormann and Likens 1979; Numata 1975) and carbon cycle models on the biome or ecosystem level (Olson 1973, 1975; Harris et al. 1975; Young 1976; Kira 1978) we calculate the humus production, expressed as a percentage of net primary productivity in table 4, column 4, in such a way that total humus production is in accordance with recent estimates (as quoted above).

Table 2.--Net primary production (NPP) and biomass for various
ecosystem types (after Whittaker and Likens 1975).

Ecosystem type	NPP (g DM m^{-2}yr^{-1})		Biomass (g DM m^{-2})	
	Normal range	Mean	Normal range	Mean
Tropical rain forest	1000 – 3500	2200	6000 – 80000	45000
Tropical seasonal forest	1000 – 2500	1600	6000 – 80000	35000
Temperate forest: evergreen	600 – 2500	1300	6000 –200000	35000
deviduous	600 – 2500	1200	6000 – 60000	30000
Boreal forest	400 – 2000	800	6000 – 40000	20000
Woodland and shrubland	250 – 1200	700	2000 – 20000	6000
Savanna	200 – 2000	900	200 – 15000	4000
Temperate grassland	200 – 1500	600	200 – 5000	1600
Tundra and alpine	10 – 400	140	100 – 3000	600
Desert and semidesert scrub	10 – 250	90	100 – 4000	700
Extreme desert: rock, sand, ice	0 – 10	3	0 – 200	20
Cultivated land	100 – 4000	650	400 – 12000	1000
Swamp and marsh	800 – 6000	3000	3000 – 50000	15000
Lake and stream	100 – 1500	400	0 – 100	20
TOTAL CONTINENTAL		782		12200

Table 3.--Estimated litter and litterfall (after Ajtay et al.
1979) and soil carbon content (after Schlesinger 1979).

Ecosystem type	Litterfall (g DM m^{-2}yr^{-1})	Litter (g DM m^{-2})	Mean Soil Profile Carbon (g C m^{-2})
Forests: Tropical rain	1850	650	9800 (TF lowland)
Tropical seasonal	1300	850	28700 (TF montane)
Mangrove	600	10000	
Temperate	850	3000	13400
Boreal closed	600	3500	20600
Boreal open	550		
Forest plantations	875	500	
Temperate woodlands	1220	2500	6900 (Woodland)
Chaparral, maquis, brush land	1000	500	
Savanna: Grass dominated savanna	1500	350	4200 (Tropical grassland)
Savanna forest	800		
Temperate grassland: wet	900	500	18900
dry	550	325	
Tundra arctic/alpine: polar desert	20	30	
herb lichen tundra	145	500	20400
scrub tundra	200	5000	
Desert and semidesert	125	100	5800 (Desert scrub)
Extreme desert	15	15	170 (ED, rock, ice)
Perpetual ice	0	0	
Lake and stream	?	?	
Bog, swamp and marshes	600	2500	72300 (Swamp, marsh)
Cultivated lands: annual	450	50	7900
perennial	150		
Human area	300	300	
TOTAL	634.3	795.7	

Table 4.--Area, net primary productivity (npp), biomass
density (b), humus productivity (hp) and humus density (h) for the model biomes, as derived from
tables 1 - 3.

	Area (10^6 km^2)			npp $(g \text{ C m}^{-2}\text{yr}^{-1})$	b $(kg \text{ C m}^{-2})$	hp (% npp)	h $(kg \text{ C m}^{-2})$
	preagri-cultural	1860	1970				
Tropical forest	31.4	27.7	22.2	834	16.8	22	11.1
Non-tropical forest	35.2	28.7	25.0	441	10.4	45	15.4
Tropical grassland	11.5	14.0	20.5	410	1.8	30	5.5
Non-tropical grassland	25.0	28.0	30.5	300	1.18	45	16.5
Desert	29.4	31.0	32.0	18	0.014	20	2.6

Data for the transformation of the biomes

The carbon released from biogenic
sources due to human impact is assumed to
be composed (1) of the gaseous products of
combustion of living phytomass (CO_2, CO etc.)
and (2) humus oxidized immediately after
the removal of the original vegetation cover.
Part (1) is estimated as the product of the
affected area, the biomass density, the percentage of the aboveground biomass, and the
burning efficiency of the respective ecosystem. On the average about 40% of the
original humus content is eliminated after
clearing of forests from which one has to
subtract the eroded fraction in order to
obtain part (2).

Wong (1978) has given figures of the
forest areas burned by wildfires in the USA
and Canada from 1920 to 1965. We took 1.5
times these values in order to account for
wildfires in other parts of the non-tropical forests as well. From 1960 on the annual
burning rate is left constant ($0.065 \cdot 10^6$
km^2/yr, Seiler and Crutzen 1980). All these
areas are to be covered by forests again,
whereas another fraction is intentionally
cleared for agriculture and thus transformed
to extratropical grassland. Taking the data
for the increase in agricultural land in
the developped world (including China) as

given by Revelle and Munk (1977) and assuming that half of this area had originally
been forest the corresponding data in fig.
2a can be gained. The aboveground biomass
is assumed to be 73% (Seiler and Crutzen 1980)
and the burning efficiency shall be 35%.
Furthermore we estimate that about 12.5%
of the humus of the affected area is eroded.

We now turn to vegetation changes in
the tropics considering three different
types of impact.
(1) Shifting cultivation means periodical
clearing of a certain (tropical) forest
area which is abandoned after some years
of agricultural use and which then may be
regrown by forests again. According to
Seiler and Crutzen (1980) the area cleared
each year for shifting cultivation amounts
to $a_{sc}(1970) = 0.248 \cdot 10^6$ km^2/yr. Since this
area is proportional to the rural population
in the developping countries which up to
the year 1970 increased with a growth coefficient of $k_{sc} = 1.18\%/yr$ an exponential
increase of the annually cleared area is
assumed:

$$a_{sc}(t) = a_{sc}(1970) \cdot \exp[k_{sc} \cdot (t - 1970)] \quad (11)$$

It is suggested here that after 1970 the
exponential growth expressed in eq.(11) will
be modified by a term which takes into
account the reduced available tropical
forest area (A_{tot}) as well as its diminishing

quality. Both effects together are summarized as $(1 - A_{loss}/A_{tot})$, where A_{loss} is the area permanently transformed to grassland (by increase of permanent agriculture and commercial clearings).

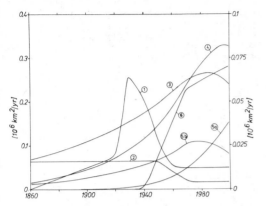

Figure 2a.--Areas affected by different types of vegetation changes (in 10^6 km^2/yr):
curve 1: wildfires in nontropical forests (left scale)
curve 2: clearings in nontropical forests (right scale)
curve 3: shifting cultivation in tropical forests (left scale)
curve 4: increase of permanent agriculture-- tropical forest (right scale)
curve 5a: commercial clearings in tropical forests - low scen. (left scale)
curve 5b: commercial clearings in tropical forests - high scen. (left scale)
curve 6: desertification (right scale)

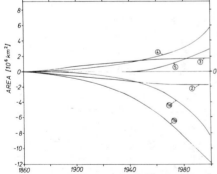

Figure 2b.--Changes of biome areas according to fig.2a (in 10^6 km^2):
curve 1a: tropical forests (low scenario)
curve 1b: tropical forests (high scenario)
curve 2: nontropical forests
curve 3: nontropical grassland
curve 4: tropical grassland (low scenario)
curve 5: deserts

(2) The contribution of increase of permanent agriculture to tropical forest clearings amounts to about a_{pa} (1970) = $0.05 \cdot 10^6$ km^2/yr. The population growth to be considered here must relate to the total population of the developing countries (and not only to its rural part). We assume a growth coefficient of k_{pa}= 2.4%/yr (Revelle and Munk 1977).

(3) The areas affected by commercial clearings in the tropics are mainly used for purposes like cattle raising and timber production. Within the model these areas are assumed to be changed into tropical grassland as are the additional areas for permanent agriculture. The range of respective data here is even wider then in the other cases such that it seems justified to set up two scenarios: (1) high scenario: the area of tropical forests cleared in 1970 (a_{tc} (1970)) may be as high as $0.1 \cdot 10^6$ km^2. Again assuming an exponential increase until 1970 and taking a value of $5 \cdot 10^6$ km^2 of tropical forests cleared until that year for commercial purposes this yields an increase of 2%/yr. (2) low scenario: Taking the data given by Seiler and Crutzen (1980) and by Buringh (1979), i.e. a_{tc}(1970)=0.047· 10^6 km^2/yr and a cleared area of $0.9 \cdot 10^6$ km^2 a growth rate of 5.22 %/yr is obtained such that in this case the impact up to 1970 is only one fifth of that in case (1) but there results a higher growth rate. According to Seiler and Crutzen (1980) the aboveground biomass is taken as 78% and the burning efficiency is assumed to be 30%.

From the data given by Mensching(1979) we conclude that since 1930 the deserts of the world have been expanding at an increasing rate which in 1970 might have been about $0.057 \cdot 10^6$ km^2/yr. The desertification map presented at the UN conference on desertification at Nairobi in 1977 (Mensching 1979) exhibits regions threatened by desertification to a different extend. We assume that until the year 2000 all regions with a very high degree of desertification hazards will be desert ($3.4 \cdot 10^6$ km^2).

MODEL TREATMENT OF C-13 DATA

The release of fossil and biogenic CO_2 into the atmosphere leads to a decline of the atmospheric C-13/C-12 or C-13/C ratios which should also be reflected in tree-rings (Stuiver 1978; Freyer 1978, 1979; Wagener 1978). Such a decrease, however, has more recently been questioned by Francey (1981) for southern hemisphere trees. Because of the uncertainties involved in tree-ring data interpretation it seems more appropriate to try to reproduce the atmospheric C-13 data as given by Keeling et al. (1979).

In most cases the C-13/C-12 ratio is given in terms of the relative deviation from a standard (subscript "std"):

$$\delta^{13}C\ [\text{‰}] = (\frac{R^{13}}{R^{13}_{std}} - 1)\cdot 1000 \qquad (12)$$

$$R^{13} = n^{13}/n^{12} \qquad (13.a)$$

n^{12}, n^{13}: mole numbers of C-12 and C-13, resp. We now introduce the C-13/C ratio

$$Q^{13} = n^{13}/n \qquad (n = n^{12} + n^{13}) \quad (13.b)$$

which is related to R^{13} as follows:

$$Q^{13} = \frac{1}{\frac{1}{R^{13}} + 1} \qquad (13.c)$$

(Although the absolute values of R^{13} and Q^{13} do not differ much it is these small diffe- rences which can be significant.) All C-13 fluxes from a compartment i to a compartment j (fig. 3) are formulated as:

$$F^{13}_{ij} = \alpha^{13}_{ij}\ Q^{13}_i\ F_{ij} \qquad (14)$$

α^{13}_{ij}: fractionation factor

F_{ij}: fluxes of total C in moles per year

The α^{13}_{ij} in general can be set equal to 1, except α^{13}_{ab}, which is calculated assu- ming stationary conditions at simulation starting time (1860, subscript "o"):

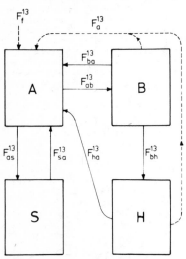

Figure 3.--C-13 fluxes as described in the text.

$$F^{13}_{abo} = F^{13}_{bao} + F^{13}_{hao}, \ F_{abo} = F_{bao} + F_{hao}$$

$$\alpha^{13}_{ab} = Q^{13}_{bo}/Q^{13}_{ao}, \ Q^{13}_{bo} = Q^{13}_{ho} \qquad (15.a-d)$$

Thus the variation of C-13 in each biotic compartment can be calculated by adding up the C-13 influxes and outfluxes. After numerical integration the Q^{13} values can be obtained from eq. (13.b).

The first time derivative of the atmos- pheric C-13 content is given in the following way:

$$\frac{d}{dt}n^{13}_a = F^{13}_f + F^{13}_a + F^{13}_{ba} + F^{13}_{ha}$$
$$- F^{13}_{ab} - F^{13}_{as} + F^{13}_{sa} \qquad (16)$$

F^{13}_f: C-13 flux to atmosphere from fossil fuel combustion

F^{13}_a: C-13 flux to atmosphere from biogenic sources due to man's impact

The exchange of atmospheric total carbon and C-13 with the sea (index "s") is described by a two-reservoir-ocean model (see Bacastow and Keeling 1973, and Keeling 1979) which allows to reproduce the atmospheric carbon dioxide partial pressure as given by the Mauna Loa measurements (Keeling et al. 1976). The mole number of C-13 as well as the total carbon mass in the atmosphere are then obtained by numerical integration of eq. (16) and of

$$\frac{d}{dt}n_a = F_f + F_a + F_{ba} + F_{ha}$$
$$- F_{ab} - F_{as} + F_{sa} \qquad (17)$$

leading to the δ^{13}_aC-values (eqs. 12 - 13).

Now we have to determine F^{13}_{ba}, F^{13}_{ha} and F^{13}_a. We assume that only newly formed parts of the plants are heterotrophically respired. These parts can be assumed to completely exhibit the actual Q^{13}_b at the time of their formation. F^{13}_{ha}, however, is regarded to consist of such material only to an extend of 10 %, the rest dating back from preindustrial times (Q^{13}_{bo}). The respective ratio between recently formed material and fractions being older than 1860 which is released to the atmosphere by vegetation changes is assumed to be 1:1.

The C-13 values to be compared with the tree-ring data for a certain year are determined as

$$Q_b^{13}(t) = \alpha_{ab}^{13} Q_a^{13}(t) \qquad (18)$$

Little is known about α_{ab}^{13} which also expresses fractionations during growth, which themselves depend on temperature and other factors, such that in a more refined study α_{ab}^{13} may be considered time dependent.

RESULTS AND DISCUSSION

As can be seen from fig. 4 the tropical forest biome has been a source for CO_2 over the whole time range considered. The loss of area and of living as well as dead organic matter is greater in the scenario for enhanced commercial clearings. The computation results show that the total organic mass of the tropical forests in 1980 decreases by about 4.3 Gt C/yr or by about 5.5 Gt C/yr for the high clearing rate scenario, resp. In both scenarios approximately the same amounts of total organic mass are released (2.4 and 2.7 Gt C/yr, resp.) The differences between released biomass and total organic mass decrease is due to erosion, charcoal formation and the changing of the vegetation type. Other data published are 1.3 - 4 Gt C/yr (Hampicke and Bach 1979) and 3.5 Gt C/yr (Woodwell et al. 1978). From living phytomass only 1.2 Gt C/yr are released in 1980 according to our calculations; Seiler and Crutzen (1980) assume 1.09 Gt C/yr. The humus oxidation immediately after clearing in tropical forests is given as 1.35 and 1.41 Gt C/yr, resp.

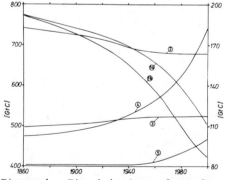

Figure 4.--Time behaviour of total organic
    carbon masses in the model biomes:
curve 1a: tropical forests, low scen. (left)
curve 1b: tropical forests, high scen. (left)
curve 2 : non-tropical forests (left scale)
curve 3 : non-tropical grassland (left scale)
curve 4 : tropical grassland, low scen.(right)
curve 5 : deserts (right scale)

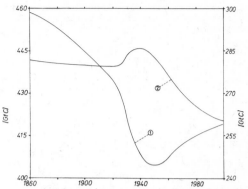

Figure 5.--Calculated living biomass(1 -
    right scale) and humus (2 -left scale)
    for the non-tropical forest biome.

As has been said in a previous section, tropical forests can be converted into agricultural land, and, on the other hand, area of tropical grassland is transformed to desert. The latter process seems to be of minor importance as far as biomass loss is concerned. Thus the total organic mass of tropical grasslands is increasing over the whole time range considered (see fig. 4).

The main impact on the extratropical forests has been wildfires which reached a maximum in about 1935. Although, due to regrowth, the living phytomass increases after 1950, this is not the case for soil organic matter which responds only with large delay (fig. 5). Because humus decline predominates regrowth according to our model this biome as a whole has a source for atmospheric CO_2 in the time range from 1860 to 1980. This is in clear contrast to Hampicke and Bach (1979) who assume a net uptake of CO_2 by the temperate forests of 0.5 to 1 Gt C/yr as well as to Woodwell et al. (1978) who regard the boreal and temperate forests as major source for atmospheric CO_2 (2.2 Gt C/yr). According to our calculations the non-tropical forests' total organic mass decreases at a rate of 0.11 Gt C/yr (1980) which mostly will remain in the atmosphere. The cumulated carbon loss from this biome amounts to about 90 Gt C from 1860 to 1980.

Extratropical grasslands expand by clearings in non-tropical forests; its total organic mass increases by about 27 Gt C in the above time range.

Fig. 6a,b show the differential and cumulated global transfers to and from the biosphere as a whole. The net ecosystem production of the global system first increases due to regrowth in non-tropical forests and later (after about 1960) decreases because at that time the main impact is brought about in the tropics and large parts of areas affected here are not allowed to be regrown by forests but are converted to grassland.

A comparison of the amount of CO_2 released from living biomass and from humus by vegetation changes reveals that nowadays these are of the same size summing up to about 3 Gt C/yr (1980). (Buringh (1979) found 4,6 Gt C/yr humus loss only). The difference between the CO_2 released from living biomass and humus and the NEP is equal to the net biogenic input into the atmosphere which according to our calculations amounts to about 2 Gt C/yr for 1980. As fig. 6b shows the cumulated biogenically released CO_2 which is not taken up again by the biota amounts to about 108 Gt C (1980) which is about

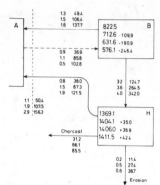

Figure 7.--Global biota carbon pools and fluxes for the years 1935, 1980 and 2000. Pool contents are also given for the initial steady state (1860). Small numbers indicate cumulated changes with respect to 1860. Solid arrows indicate fluxes as induced by human impact (per year and cumulated), while the dashed arrow indicates the net ecosystem production. Values at the dashed line give the net input to the atmosphere (all values in Gt C and Gt C/yr, resp.).

2/3 of the fossil input until that year. (see also fig. 7).

For the years 1956 and 1978 our calculations yield $\delta^{13}C$-values for the atmosphere of -6.66 % and -7.25 % , resp., which are in good agreement with the experimental results of -6.69 % and -7.24 % , resp. (Keeling et al. 1979). We think that this agreement lends support to the assumed model biome structure and the anthropogenic impact for the period of the last 25 years. Our model calculations show a smaller decrease of $\delta^{13}C$ in tree-rings than the measured data given e.g. by Stuiver (1978), Wagener (1978) and Freyer (1978, 1979) for the period from 1860 to 1935.

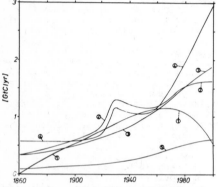

Figure 6a.--Global transfers of carbon between atmosphere and biota (low scen.):
curve 1: net ecosystem production
curve 2: CO_2 release from living biomass due to vegetation changes
curve 3: CO_2 release from humus due to vegetation changes
curve 4: net input into the atmosphere
curve 5: erosion

CONCLUSIONS

The model calculations presented show that the biota since 1860 until nowadays have always been a source for atmospheric CO_2 resulting in a net flux of biogenic carbon into the atmosphere of about 2 Gt C/yr at present, or a cumulated net input of about 100 - 110 Gt C until 1980, resp. The corresponding figures for the year 2000 are calculated to be about 3 Gt C/yr for the net input and 150 - 160 Gt C for the cumulated net biogenic input. Agreement with atmospheric C-13 data for the time range 1956 to 1978 can be obtained. It was interesting to see that the more detailed treatment of the multi-biome approach with transitions from forested areas to grassland and from grass-

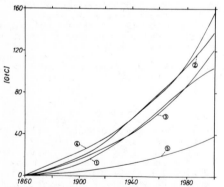

Figure 6b.--Cumulated fluxes within the atmosphere-biota-humus system (explanation of curves corresponding to those given in fig. 6a.).

land to desert as well leads to a lower NEP and thus a higher net transfer of biogenic carbon to the atmosphere than is obtained by means of the mono-biome approach.

ACKNOWLEDGEMENT

We thank C.D. Keeling (Scripps Institution of Oceanography, La Jolla) for a valuable discussion on C-13 modelling.
Financial support of this project by the Stiftung Volkswagenwerk and the Deutsche Forschungsgemeinschaft is gratefully acknowledged.
The calculations have been carried out on the university computers at Frankfurt/Main and Darmstadt.

LITERATURE CITED

Ajtay, G.L., P. Ketner, and P. Duvigneaud. 1979. Terrestrial primary production and phytomass. In The global carbon cycle. Eds. B. Bolin, E.T. Degens, S. Kempe and P. Ketner. Wiley, New York

Bacastow, R., and C.D. Keeling. 1973. Atmospheric carbon dioxide and radiocarbon in the natural carbon cycle: II. Changes from A.D. 1700 to 2070 as deduced from a geochemical model. In Carbon and the biosphere. Eds. G.M. Woodwell and E.V. Pecan. USAEC CONF 720510, NTIS, Springfield, Va.

Bertalanffy, L.v. 1960. Principles and theory of growth. In Fundamental aspects of normal and malignant growth, Ed. W.W. Nowinski, Elsevier, Amsterdam.

Bormann, F.H., and G.E. Likens. 1979. Pattern and process in a forested ecosystem. Springer, New York, N.Y.

Broecker, W.S., T. Takahashi, H.J. Simpson, and T.-H. Peng. 1979. Fate of fossil fuel carbon dioxide and the global carbon budget. Science 206: 409-418.

Brown, S. and A. Lugo. 1980. Preliminary estimate of the storage of organic carbon in tropical forest ecosystems. In The role of the tropical forest on the world carbon cycle. Eds. S. Brown, A. Lugo and B. Biegel. US Dept. of Energy, CONF-800350 UC/11.

Buringh, P. 1979. Decline of organic carbon in soils of the world. presented at the SCOPE conference: Role of the terrestrial vegetation in the global carbon cycle, Woods Hole, May 6-11.

Eriksson, E., and P. Welander. 1956. On a mathematical model of the carbon cycle in nature. Tellus 8: 155-175.

Francey, R.J. 1981. Tasmanian tree rings belie suggested anthropogenic $^{13}C/^{12}C$ trends. Nature 290: 232-235.

Freyer, H.D. 1978. Preliminary evaluation of past CO_2 increase as derived from ^{13}C measurements in tree rings, In

Carbon dioxide, climate, and society. Ed. J. Williams, Pergamon Press, Oxford.

Freyer, H.D. 1979. On the ^{13}C record in tree rings. Part I.: ^{13}C variations in northern hemisphere trees during the last 150 years. Tellus 31: 124-137.

Hampicke, U. 1980. The role of the biosphere. In Interactions of energy and climate. Eds. W. Bach, J. Pankrath and J. Williams, Reidel, Dordrecht.

Hampicke, U., and W. Bach. 1979. Die Rolle terrestrischer Ökosysteme im globalen Kohlenstoffkreislauf. Bericht im Auftrag des Umweltbundesamts, FE-Vorhaben Nr. 10402513.

Harris, W.F., P. Sollins, N.T. Edwards, B. E. Dinger, and H.H. Shugart. 1975. Analysis of carbon flow and productivity in a temperate deciduous forest ecosystem. In Productivity of world ecosystems. Proc. Symp. Seattle, August 31 - September 1, 1972, National Academy. of Sciences, Washington, D.C.

Kawahara, T. 1977. Dynamics of soil organic matter and carbon cycling. In Primary productivity of japanese forests, Eds. T. Shidei and T. Kira, JIBP Synthesis 16, University of Tokyo Press, Tokyo.

Keeling, C.D. 1979. The Suess effect: $^{13}Carbon$ -$^{14}Carbon$ interrelations. Env. Int. 2: 229-300

Keeling, C.D., R.B. Bacastow, A.E. Bainbridge, C.A. Ekdahl, P.R. Guenther, L.S. Waterman, and J.S. Chin. 1976. Carbon dioxide variations at Mauna Loa observatory, Hawaii. Tellus 28: 538-551.

Keeling, C.D., W.G. Mook, and P.P. Tans. 1979. Recent trends in the $^{12}C/^{13}C$ ratio of atmospheric carbon dioxide. Nature 277: 121-123.

Kira, T. 1978. Carbon cycling. In Biological production in a warm temperate evergreen oak forest of japan. Eds. T. Kira, Y. Ono and T. Hosokawa. JIBP Synthesis 18, University of Tokyo Press, Tokyo

Kitazawa, Y. 1977. Ecosystem metabolism of a subalpine coniferous forest of the Shigayama IBP area, central Japan. JIBP synthesis 15, University of Tokyo Press, Tokyo.

Kohlmaier, G.H. 1981. Possible self-consistent paths of the terrestrial biota-humus-atmosphere system in response to man's impact. Radiat. Environ. Biophys. 19: 67-78.

Kohlmaier, G.H., U. Fischbach, G. Kratz, E.O. Siré, J. Hirschberger, and W. Schunck. 1979. Modelling man's impact on the subsystem atmosphere-biosphere of the global carbon cycle. In Man's impact on climate. Eds. W. Bach, J. Pankrath and W.W. Kellogg. Elsevier, Amsterdam

Kohlmaier, G.H., U. Fischbach, G. Kratz, H. Bröhl, and W. Schunck. 1980. The carbon cycle: sources and sinks for atmospheric CO_2. Experientia 36: 769-776.

Lemon, E. 1977. The land's response to more

68

carbon dioxide. In The fate of fossil fuel CO_2 in the oceans. Eds. N.R. Andersen and A. Malahoff, Plenum, New York, N.Y.

Mensching, H. 1979. Desertification. Geogr. Rundsch. 31: 350-355.

Numata, N. (Ed.), 1975. Ecological studies in Japanese grasslands, JIBP synthesis 13, University of Tokyo Press, Tokyo.

Oeschger, H., U. Siegenthaler, U. Schotterer, and A. Gugelmann. 1975. A box diffusion model to study the carbon dioxide exchange in nature. Tellus 27: 168-192.

Oeschger, H., U. Siegenthaler, and M. Heimann. 1980. The carbon cycle and its perturbation by man. In Interactions of energy and climate. Eds. W. Bach, J. Pankrath and J. Williams, Reidel, Dordrecht.

Olson, J.S. 1973. Carbon cycles and temperate woodlands. In Analysis of temperate forests ecosystems. Proc. Symp. Seattle, August 31 - September 1, 1972, Nat. Acad. Sci., Washington, D.C.

Olson, J.S. 1980. Earth's vegetation as a carbon pool, source, and potential sink for carbon dioxide. Preprint to appear in Nature.

Olson, J.S., H.A. Pfuderer and Y.-H. Chan. 1978. Changes in the global carbon cycle and the biosphere. Environmental Sciences Division Publication No. 1050 ORNL/EIS-109, Oak Ridge National Laboratory, Oak Ridge, Tennessee.

Revelle, R., and W. Munk. 1977. The carbon dioxide cycle and the biosphere, In Energy and climate. Nat. Acad. Sci., Washington, D.C.

Rodin, L.E., N.I. Bazilevich and N.N. Rozov. 1975. Productivity of the world's main ecosystems, In Productivity of world ecosystems. Proc. Symp. Seattle, August 31 - September 1, 1972, Nat. Acad. Sci., Washington, D.C.

Rotty, R.M. 1980. Past and future emission of CO_2. Experientia 36: 781-783.

Schlesinger, W.H. 1979. The world carbon pool in soil organic matter: a source of atmospheric CO_2? In The role of terrestrial vegetation in the global carbon cycle: method for appraising changes. Ed. G.M. Woodwell, Wiley, New York, N.Y.

Siré, E.O., G.H. Kohlmaier, G. Kratz, U. Fischbach, and H. Bröhl. 1981. Comparative dynamics of atmosphere-ocean models within the description of the perturbed global carbon cycle. Z. Naturforsch. 36a: 233-250.

Stuiver, M. 1978. Atmospheric carbon dioxide and carbon reservoir changes. Science 199: 409-418.

Wagener, K. 1978. Total anthropogenic CO_2 production during the period 1800 - 1935 from carbon-13 measurements in tree rings. Radiat. Environ. Biophys. 15: 101-111.

Whittaker, R.H., and G.E. Likens. 1975. The biosphere and man, In Primary productivity of the biosphere, Eds. H. Lieth and R.H. Whittaker, Springer, Berlin.

Wong, C.S. 1978. Atmospheric input of carbon dioxide from burning wood. Science 200: 197-200.

Woodwell, G.M., R.H. Whittaker, W.A. Reiners, G.E. Likens, C.C. Delwiche, and D.B. Botkin. 1987. The biota and the world carbon budget. Science 199: 141-146.

Young, A. 1976. Tropical soils and soil survey. Cambridge University Press, Cambridge.

THE ROLE OF TROPICAL LAND USE CHANGE IN THE GLOBAL CARBON CYCLE: DETAILED ANALYSIS FOR COSTA RICA AND PANAMA AND PRELIMINARY ANALYSIS FOR PERU AND BOLIVIA[1]

R. P. Detwiler[2] Charles A. S. Hall[3] Philip Bogdonoff[4] Christopher McVoy[4] Sandy Tartowski[2]

Abstract.--Our group, composed of modelers working in conjunction with tropical ecologists, has produced a simulation model that quantifies the net carbon exchanges between vegetation and the atmosphere resulting from land use changes in the tropics. The model uses Holdridge's concept of Life Zones. Our results indicate that the Life Zone approach produces lower rates of carbon additions to the atmosphere than some previous approaches because most of the tropics have lower carbon storages than once thought, and because human impact may be concentrated in Life Zones of moderate carbon storage.

INTRODUCTION

Uncertainties associated with assessing the global carbon cycle remain. Summaries of the cycle and problems with its interpretation can be found in Woodwell et al. 1978; Baes et al. 1977; Broecker et al. 1979; and Detwiler and Hall 1980. Land use changes in the tropics have the potential to change significantly the carbon content of the atmosphere due to the large area of tropical ecosystems, the large difference between the carbon stored in natural systems and human disturbed systems, and the purportedly large rates of development in tropical countries (e.g. Myers 1980). Estimates of the exchange between the earth's biota and the atmosphere range from an annual addition to the atmosphere of 4 to 8 billion metric tons (BMT) (Woodwell et al. 1978)

to an annual removal from the atmosphere of 2 BMT (Bacastow and Keeling 1973; Seiler and Crutzen 1980). The range is in large part a function of the lacunae in our knowledge of tropical systems and tropical land use change.

Carbon is added to the atmosphere when vegetation with large carbon storage is replaced by vegetation with little carbon storage, and is removed from the atmosphere by the growth of vegetation. In the past, it was assumed that the release of carbon due to the destruction of vegetation occurred completely and simultaneously with the change in land use. Recent work by others as well as our own studies of tropical agricultural systems indicate that this is probably not true (Seiler and Crutzen 1980; Appendix A). In shifting cultivation--which is an important land use in many countries--the land often is not completely cleared. Large trees may be left to provide shade for crops or because felling them would be too difficult (Manshard 1974). Wong (1978) estimates that 25% of the above-ground vegetation is not burned and instead decays slowly. Seiler and Crutzen (1980) estimate that up to 75% of the above-ground biomass is left to decay after burning. In addition, they believe that a significant portion of the material is converted to charcoal, which may persist for many years. As it now appears that both carbon

[1]Paper presented at the International symposium Energy and Ecological Modelling, sponsored by the International Society for Ecological Modelling. (Louisville, Kentucky, April 20-23, 1981).

[2]R. P. Detwiler and Sandy Tartowski are graduate students;
[3]Charles A. S. Hall is Assistant Professor, and;
[4]Philip Bogdonoff and Christopher McVoy are research assistants; Ecology and Systematics, Cornell University, Ithaca, New York U.S.A.

uptake by the recovery of vegetation on disturbed sites and carbon release by land use change occur in complex patterns over periods of years, we believe that the calculation of the annual exchange of carbon between the atmosphere and the tropical biota requires the use of computer simulation combined with as complete a data set as is available.

METHODS

We developed an empirically-driven computer model that combines estimates of land use change and carbon storage in vegetation to determine the exchange of carbon between tropical vegetation and the atmosphere. A detailed discussion of our model structure and rationale is given in Detwiler and Hall (1980). We have organized the model structure, information on the biomass of tropical ecosystems, and rates of land use change in tropical countries according to the Holdridge Life Zone system (Holdridge 1947; Holdridge 1967; Holdridge et al. 1971). Drs. S. Brown and A. Lugo, who are members of our research group, have compiled biomass estimates from the literature by Life Zone. The estimates give an average biomass for each Life Zone with a relatively small standard error (Brown and Lugo 1980; table 1). It should be noted that these averages are significantly lower than the averages most often used, those of Whittaker and Likens (1975). The numbers we use include estimates of carbon in below-ground vegetation but do not include carbon in soil and litter.

Members of our research group determined the area of each Life Zone in Bolivia, Costa Rica, Panama, and Peru by planimetering Life Zone maps (Tosi 1980; Tosi and Brown 1980). We excluded from this study the portions of Bolivia which have vegetation characteristic of the Warm Temperate Life Zone (sensu Holdridge). The excluded area is about one-third of the total land area of Bolivia.

Table 1.--Estimates of organic carbon storage in tropical
Life Zones. One standard error is given in parenthesis
(from Brown and Lugo 1980)

Life Zone[a]	n	Above Ground	Below Ground	Total Plant	n	Soil	n	Litter	Total Forest
						(mtC/ha)[b]			
T-Rain and Wet Forest	8	145 (18.6)	20 (3.1)	164 (21.0)	36	106.5 (20.3)	5	5.0 (1.5)	276
T-Moist Forest	18	152 (12.7)	31 (4.6)	182 (11.1)	19	73.5 (7.6)	17	4.8 (0.8)	260
T-Dry Forest	-	-	-	91[c]	7	55.3 (14.5)	-	5.5[c]	152
ST-Rain & Wet Forest	4	92 (19.6)	40 (9.1)	132 (27.5)	5	104.6 (37.8)	2	2.4	239
ST-Moist Forest	4	98 (13.5)	11 (2.1)	110 (14.8)	8	56.1 (8.8)	4	2.6 (0.7)	169
ST-Dry Forest	3	49 (22.4)	8 (4.0)	57 (26.0)	-	24.2	5	3.0 (0.5)	84

[a] T = Tropical
ST = Subtropical

[b] Assumed carbon content = 0.5 x organic matter

[c] We found no plant biomass data for Tropical Dry Forest. We assumed the ratio of the organic carbon storages of the T-Dry Forest to the T-Moist Forest was the same as the ratio of organic carbon storages for the ST-Dry to ST-Moist Forests of 0.5. Organic carbon storage T-Dry Forest = 0.5 x 182 = 91. Carbon storage in litter was calculated by a similar method.

The rates of land use change also were determined by Life Zone for each country based on remote sensing and ground observations supplemented by demographic statistics (Tosi 1980; Tosi and Brown 1980). Tosi found that the intensity of land use change varies as a function of the Life Zone. Life Zones of intermediate wetness (where the ratio of precipitation to potential evapotranspiration is near unity) are in general the ones first developed for agriculture. If more land is needed, development expands into wetter and drier regions. It is important to know specific land use, particularly whether it is a permanent or temporary use (e.g. permanent agriculture or shifting cultivation). In the past, others often have assumed that all estimates of forest conversion were complete clearing for permanent use (Bolin 1977; Brunig 1977; Woodwell et al. 1978). We believe that this assumption leads to overestimates of carbon release from tropical ecosystems.

The land use information used in our analyses came from three sources: Tosi (1980), Tosi and Brown (1980), and agricultural statistics for each country (e.g. Censos Nacionales de 1960 for Panama; see also literature cited). We give Panama as an example of the methods used to adapt the statistical data on agriculture to the model because the data base for Panama, like that for Costa Rica, is relatively comprehensive.

Example of Agriculture Calculations:
Panama, 1950-1980

We used information from the National Agricultural Censuses of Panama, years 1950, 1960, and 1970, supplemented by several secondary sources including the Area Handbook of Panama (1972) and personal communications to estimate net land use changes in permanent (i.e. non-shifting) agriculture.

The above reports enabled us to compile figures on land area in Panama for three land use categories: temporary crops, permanent crops[5], and pastures. The data are only for land holdings greater than one hectare and hence probably exclude most shifting cultivation. The data are given by province, of which there are nine, ranging in total area from 240,000 to 1,700,000 hectares. Our model requires the assignment of agricultural data to Life Zones. For a first approximation we assumed that agriculture in each province is distributed according to the area of each Life Zone in that province. This assumption and planimetry of combined Life Zone and political boundary maps provided us with a first estimate of agricultural land use as a function of Life Zone (Appendix B). Our future

---

[5]Not to be confused with "Permanent Agriculture" as defined in the following section.

calculations will use any available information on agricultural patterns to more accurately distribute land use among Life Zones. A detailed analysis of agricultural patterns in Panama is presented as Appendix B.

Trends Over Time: 1950-1980

Permanent Agriculture. The following definitions, based upon the National Census data, were used to estimate areas in Permanent Agriculture, comprised of: (1) "Temporary Crops", those crops which are sown every year, e.g. rice, corn, vegetables and tobacco--including land in fallow, as well as land in sugarcane, even though the latter may be harvested for several years (this category does not include land in shifting cultivation); (2) "Permanent Crops", those which are not sown every year, but are harvested from long-lived shrubs or trees--coffee, bananas, cacao, citrus fruits; and (3) "Pastures" those areas, natural or improved, used for grazing dairy and beef cattle.

Land area for Permanent Agriculture has increased by approximately 70% between 1950 and 1970 based on the national statistics, going from 1,002,000 to 1,683,000 hectares, an average increase of 34,000 hectares per year (37,000 hectares per year for 1950-1960; 31,000 hectares per year for 1960-1970). About 87% (589,000 hectares) of this 20 year increase in Permanent Agriculture was due to an increase in the area of pastures. Temporary and Permanent Crops together declined slightly (10,000 hectares) between 1960 and 1970. Pastures increased from 552,000 hectares in 1950 to 818,000 hectares in 1960 and to approximately 1,141,000 hectares in 1970. The annual rate of increase was 27,000 hectares per year in the decade 1950-1960, and 32,000 hectares per year in the subsequent decade.

Trends for 1950 to 1970 were based on three national agricultural censuses. The fourth such census is scheduled for May 1981, so nationwide data will not be available for at least a year. Therefore, we extrapolated the 1950, 1960, and 1970 data to 1980 using a second-order polynomial.

Similar analyses were done for Costa Rica using data from Censo Agropecuario for the years 1950, 1963, and 1973 and similar province by province planimetry of Life Zones. We also give preliminary results for Peru for the nation as a whole. We do not have agricultural data for Bolivia but used for our preliminary analysis Tosi's (1980) estimate of total land area with herbaceous vegetation.

Shifting Cultivation. Quantitative information on shifting cultivation in Latin America is both rare and of unknown accuracy. Our assessments of the area of land in shifting cultivation in 1950 and 1980 were based upon three major types

of information: government estimates of the land
area in shifting cultivation made in association
with national censuses; forestry estimates of the
amount of deforestation due to shifting cultivation; and government estimates of the population
size, proportion of the population engaged in
swidden agriculture and the amount of land cleared
per capita per year by each family of swidden
agriculturalists. The distribution of shifting
cultivation by Life Zone was inferred from information on the geographic distribution of: agriculture, population, forest, deforestation and
qualitative descriptions of shifting cultivation.
Generally, shifting cultivators prefer Moist Lower
Montane sites in old secondary forest (Manshard
1974). A more detailed analysis of shifting
cultivation and the three procedures by which we
derived our estimates of the area in shifting
cultivation in Panama are presented in Appendix
A. Similar methods were employed to estimate
shifting cultivation parameters for Costa Rica,
Peru, and Bolivia.

The Model

The model is essentially an exercise in
bookkeeping based on the conservation of mass and
area (fig. 1). The area in each of the land
classifications--natural system vegetation (NS),
mature secondary vegetation (2F), swidden area in
cultivation (SB(1) and SB(2)), swidden area in
fallow (RG(1) through RG(FLW), where FLW is the

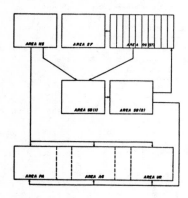

Figure 1.--Land use flow chart for the model
GLOBC6. Boxes represent land areas in
various categories of use: NS is land with
"natural system" vegetation; 2F is land with
mature secondary vegetation; SB(1) and SB(2)
areas in the two years of cultivation
in swidden agriculture; RG(ST) are areas in
the fallow portion of the swidden agriculture
cycle; PA is pasture land; AG is land with
permanent agriculture; and UR is area with
settlements. PA, AG, and UR were combined
into one area as their vegetation has
similar carbon content.

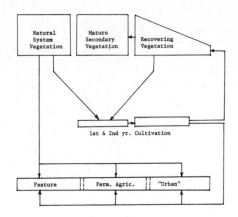

Figure 2.--Quantity of carbon stored in the
characteristic vegetation of each land use.
The height of each box is roughly proportional to the carbon content of the vegetation per unit area.

number of years of the fallow cycle)--is assigned
based upon the user's information and assumptions
about land use in the tropics. For the simulations
discussed in this paper, we assigned these areas
based on data from Tosi (1980), Tosi and Brown
(1980), and national agricultural censuses for
the 1950's and our crude estimates of land area in
shifting cultivation (see Appendix A and B). Next,
the carbon content of the vegetation associated
with each of these classifications is specified.
To determine the carbon content of the fallow
vegetation, the model requires the entry of the
number of years necessary for secondary vegetation
to mature, and the shape of the curve of carbon
plotted against time--linear, logistic, or
exponential. Given these values, the model interpolates the carbon content of fallow vegetation
for each age class (fig. 2). Our estimates of the
carbon content of vegetation came from Brown and
Lugo (1980). The user must then enter rates of
area transfer among land use classifications for
the first and last years of the simulation. The
model determines the rates for the intervening
years by interpolation, again using linear,
logistic, or exponential functions as specified by
the user. Finally, the user enters in known values
or assumptions for the length of the fallow cycle
in swidden, the percentage of cleared vegetation
that is left after burning to decay slowly, the
decomposition rate of that vegetation, and the
percentage of cleared vegetation that is permanently
stored as wood products, charcoal, and/or additions
to the soil organic matter. We apportioned cleared
vegetation among these categories based on information in Seiler and Crutzen (1980).

Given the above inputs, the model runs as
follows. At the beginning of a year of the simulation, the total carbon in living vegetation, dead
and decaying vegetation, and permanent storage

(wood products, charcoal, and additions to soil organic matter) is determined. For each region modeled (e.g. the Tropical Moist Life Zone in Panama), the carbon stored in living vegetation is the summation of the products of the area in each land use and the carbon storage per hectare of its characteristic vegetation. The area in each land use then is changed according to the data supplied by the user. On areas converted to human use, the next step is determining the fate of the vegetation. Some portion is left alive; some is cleared. Of the cleared portion, some is burned, some is left to decay, and some is permanently stored. On areas released from human use, the vegetation may grow. In areas under little pressure from increasing populations, swidden cycles may have fallow periods of 15-20 years (Watters 1971; Gross et al. 1979). Some areas may be permanently abandoned. Areas released from use may act as carbon sinks for tens of years until the vegetation reaches some steady-state. Even if the vegetation does not return to its former carbon content, the area can act as a sink for a portion of the carbon released by the original clearing.

After some areas are cleared and the vegetation assigned to decaying or permanent storage, and after other areas are abandoned and the vegetation allowed to grow, the amount of carbon stored in the living vegetation is determined. To this is added material permanently stored or yet to decay. The sum is all the carbon stored as vegetation, decaying material, and lumber, charcoal, and new soil organic matter within the Life Zone. From this sum is subtracted the same sum calculated at the beginning of the year. A negative result is a net transfer from the biota to the atmosphere. A positive result is the reverse.

Because the determination of net transfers is based on a number of assumptions (see Detwiler and Hall 1980, p. 144), we felt it necessary to incorporate as much flexibility as possible into the model structure. Most values of variables-- carbon content of vegetation, fallow periods, conversion rates--can be entered by the user. The model will interpolate between values of a variable known at two or more different times using linear, logistic, or exponential functions. This flexibility allows the model to adapt to changes in the quantity and quality of data.

RESULTS

A principal result of our research to date (table 2) is that it is very difficult to make our model produce the very high estimates of CO_2 release from forests that have sometimes appeared in the literature (e.g. Brunig (1977) or Woodwell et al.'s (1978) higher estimates) unless the tropics as a whole are being cut at rates at least as great as our present estimates of land use changes in Costa Rica and Panama--countries that we believe have considerably greater rates of

forest clearing than the average for the tropics as a whole (table 2).

A second important result is that for Costa Rica, Panama, and presumably other regions where shifting cultivation is may be important, the net exchange of carbon is dependent on the proportion of forested land that was in fallow in past decades. In other words, the model results are sensitive to whether increases in the area of permanent agriculture have come from areas once in shifting cultivation rather than areas with undisturbed vegetation. If shifting cultivation was less important in the past than we have assumed, our present estimates (see, e.g. fig. 3a and 4a) may be substantial underestimates of the carbon released from the land.

Our results for Peru and Bolivia, if extrapolated to the entire tropics, give considerably lower estimates of carbon release from the vegetation than do the extrapolated results for Panama and Costa Rica because the rates of land clearing in Peru and Bolivia appear to be much smaller than for the Central American countries we studied (table 2). This conclusion is based on a less adequate data base and less comprehensive data analysis and so is preliminary.

Panama.--Since our most comprehensive data base is for Panama we will discuss the results for that country first and in most detail. All of Panama is in the Tropical Life Zone region so that there are only three Life Zone categories to consider. Figures 3a, 4a, and 5a show our best estimates of land use change in the three Life Zone categories for Panama based on the official agricultural statistics adjusted to Life Zones, Tosi's (1980) estimate of forest and bush cover in 1980, and a linear extrapolation back in time of Tosi's contemporary forest cutting rates. Figure 3b, the result of a computer run forced to fit the land areas and rates of change of figure 3a, shows that if the land use changes occurred as in figure 3a, 0.0034 billion metric tons of carbon would be released per year in 1980. An alternative extrapolation of this data set is given as figure 3c in order to examine the potential importance of having had different quantities of land in shifting cultivation fallow. The results are given for that Life Zone within that country (0.0008 BMT per year) and also as the net exchange with the atmosphere if the tropics as a whole were behaving in a fashion similar to the land area in question, i.e. by multiplying the results by the ratio of tropical non-savanna areas to the land area in question (0.536 BMT per year; table 2 and figure 3c).

Figures 4a and 5a are data, and figures 4b-d and 5b simulation results, for the Tropical Moist and Dry Life Zones in Panama. In brief, carbon exchange ranged from a low of 0.0001 BMT per year in 1980 for the Tropical Dry Life Zone to a high of 0.007 BMT for the Tropical Moist Life Zone with the assumption that there was essentially no land in

Table 2.--Summary of simulation results for Panama, Costa
Rica, Bolivia, and Peru.

Country/Life Zone	Area[1]	CUMCEX[2]	EXCH80[3]	Tropics[4]
I. Panama				
Tropical Rain & Wet	3 871 900	−0.1168	−0.0034	−2.269
Tropical Moist	3 316 900	−0.1169	−0.0035	−2.719
Tropical Dry	354 800	−0.0041	−0.0001	−0.794
Total	7 543 600	−0.2378	−0.0070	−2.413
Panama, alt. runs				
Tropical Rain & Wet[5]	3 873 000	−0.0689	−0.0008	−0.536
Tropical Moist[5]	3 316 900	−0.0531	−0.0017	−1.334
Tropical Moist[6]	3 316 900	−0.2371	−0.0070	−5.451
II. Costa Rica				
Tropical Rain & Wet	3 337 600	−0.1474	−0.0045	−3.490
Tropical Moist	1 373 420	−0.1061	−0.0032	−6.087
Tropical Dry	375 500	−0.0078	−0.0003	−1.745
Total	5 086 520	−0.2613	−0.0080	−4.089
III. Peru				
Total	109 065 900	−1.0207	−0.0323	−0.770
IV. Bolivia				
Tropical Rain & Wet	1 709 890	−0.0041	−0.0001	−0.202
Tropical Moist	4 777 000	−0.0149	−0.0006	−0.334
Tropical Dry	299 100	−0.0005	−0.00002	−0.149
Subtrop. Rain & Wet	7 128 100	−0.0594	−0.0023	−0.828
Subtrop. Moist	41 361 740	−0.1381	−0.0045	−0.280
Subtrop. Dry	1 205 000	−0.0214	−0.0007	−1.487
Total	56 480 830	−0.2383	−0.0082	−0.378

[1]Area of the Life Zone or country in hectares.
[2]Cumulative carbon exchange from 1950 to 1980. Numbers
are in billion metric tons (BMT) of carbon.
[3]Annual exchange for 1980, in BMT C. Negative numbers
indicate a net release of carbon to the atmosphere.
[4]If the whole of the tropics (26 x 10[8] ha.) looked like
the Life Zone or country, both in biomass on the land and in
cutting rates, this would be the net exchange of carbon.
[5]These simulations were set up so that all the land cut
for shifting cultivation or permanent agriculture comes from
the recovering fallow areas; no "new" natural system
vegetation is cut.
[6]This simulation was set up so that all the land cut for
shifting cultivation or permanent agriculture comes from the
natural system vegetation; there is very little fallow land
in 1950.

shifting cultivation in 1950. Shifting cultivation
appears to be unimportant in the Tropical Dry Life
Zone both now and in the recent past.

Our estimate of carbon exchange for Panama
as a whole is the weighted (by land area) average
of the land use changes depicted as figures 3a, 4a,
and 5a, and is about 0.01 BMT carbon per year, or
3.65 BMT per year if extrapolated to the tropics
as a whole.

Costa Rica.--We also have a reasonably good
data base for determining land use changes in
Costa Rica although our present analysis is
limited by lack of data on the area of early
successional forest. Figures 6 through 8,
analogous to figures 3 through 5 for Panama,
summarize our information base and simulation
results for land use changes in Costa Rica, based
on official agricultural statistics and Tosi's
(1980) forest clearing estimates. Table 2

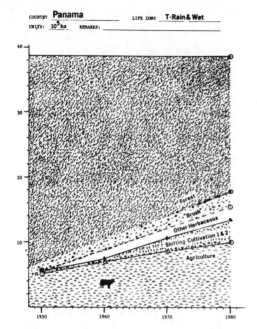

Figure 3a.--Estimates of land use in the Tropical Rain and Wet Life Zone in Panama during the period of 1950 to 1980.

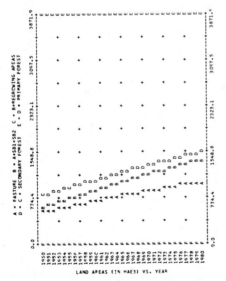

Figure 3b.-- Results of the model using data shown in figure 3a. This run, for an area of 3872 x 10³ ha, had a carbon exchange of -0.0034 x 10⁹ metric tons of carbon per year (mTC/yr) in 1980. If extrapolated to the entire tropics (2600 x 10⁶ ha), the annual exchange would have been -2.27 x 10⁹ mTC.

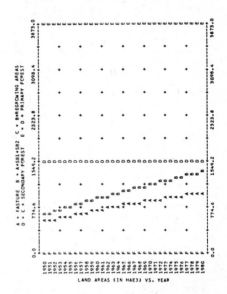

Figure 3c.--Results of the model using data shown in figure 3a except that the initial conditions were changed so that natural system area remained constant and all additions to pasture and permanent agriculture areas came from areas in the fallow portion of the shifting cultivation cycle. (Note: the computer graphing routine plots the A-line first, then B, etc., so, for example, if there is no land in secondary forest, the D-line will plot on top of the C-line. The F-line is not used and can be ignored.) This run, for an area of 3872 x 10³ ha, had a carbon exchange of -0.0008 x 10⁹ mTC/yr in 1980. If extrapolated to the entire tropics, the annual exchange would have been -0.54 x 10⁹ mTC.

also includes an extrapolation of these results to the tropics giving our estimate of the maximum possible carbon exchange from tropical land clearing, since Costa Rica has recently undergone especially intensive development (Tosi, personal communication).

    Bolivia.--Our analysis for Bolivia is more uncertain as we have not been able to find extensive agricultural information on Bolivia. We have guessed at the quantities of land in permanent agriculture and shifting cultivation based on Tosi's (1980) estimates of land with herbaceous vegetation. The assumptions and the simulation results are summarized in figure 9a through f and table 2. New forest clearing appears very low (Tosi 1980), and as a result our preliminary model estimate of the next exchange of carbon with the atmosphere is low.

Peru.--Peru is probably the most complex nation in the world from the perspective of geographic and ecological variability (Tosi 1980). Since it has not yet been possible for us to convert the agricultural information (given by province) to Life Zones, and because human impact in Peru appears relatively stable (Tosi 1980), we have lumped all Life Zones of Peru into one and have made a preliminary model run based on the nation as a whole. The results of the analysis for Peru (table 2, and figs. 10a and b) indicate that there may be very little, if any, net exchange of carbon between the vegetation and the atmosphere as a result of forest clearing.

DISCUSSION

Our estimates for Bolivia and Costa Rica, if extrapolated to the entire tropics, gives a range of 0.4 to 4.1 BMT of carbon released annualy due to land use changes. The range is consistent with our earlier assessment (Detwiler and Hall 1980) that tropical vegetation is a source of atmospheric carbon dioxide of the same order of magnitude as, but significantly smaller than, fossil fuel combustion, which is about 5.1 BMT per year (Rotty 1977). The range is also consistent with some other estimates found in the literature. Seiler and Crutzen (1980) estimated a net release of 0.8 to 1.5 BMT of carbon per year due to "deforestation" in tropical ecosystems. Hampicke (1979)

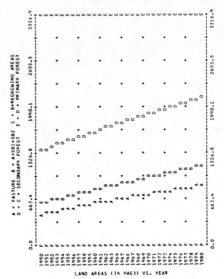

Figure 4b.--Results of the model using data shown in figure 4a. This run, for an area of 3317 x 10^3 ha, had a carbon exchange of -0.0035 x 10^9 mTC/yr in 1980. If extrapolated to the entire tropics, the annual exchange would have been -2.72 x 10^9 mTC.

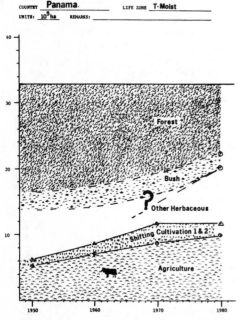

Figure 4a.--Estimates of land use in the Tropical Moist Life Zone in Panama, 1950-1980.

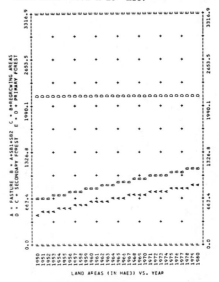

Figure 4c.--Results of the model using data shown in figure 4a, except that initial conditions were changed as in figure 3c. This run, for an area of 3317 x 10^3 ha, had a carbon exchange of -0.0017 x 10^9 mTC/yr in 1980. If extrapolated to the entire tropics, the annual exchange would have been -1.33 x 10^9 mTC.

gives an estimated net release of 2.6 BMT per
year from "forest clearing in the Third World."
Wong's (1978) estimated net annual release is
1.5 BMT as a result of "new tropical forest
clearings." Bolin's (1977) range of net annual
release is 1.1 ± 0.6 BMT due to "deforestation and
use of fuel wood in developing countries." But
our range is lower than Brunig's (1977) estimate
of 6 BMT per year from "tropical forest clearing"
and Woodwell et al.'s (1978) range of 1 to 7 BMT
released annually because of "clearing and harvest
in tropical forests." Our estimated range is lower
for several reasons: (1) the estimates of the
carbon content of vegetation we used are lower;
(2) rates of land use change are smaller; (3) we
do not assume that all of the original vegetation
is cleared and immediately converted to carbon
dioxide; and (4), we include a significant uptake
of carbon by growing vegetation on abandoned areas,
a process which may be important where shifting
cultivation is a primary land use in terms of area.

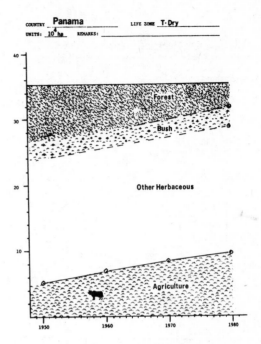

Figure 5a.--Estimates of land use in the Tropical
Dry Life Zone in Panama, 1950-1980.

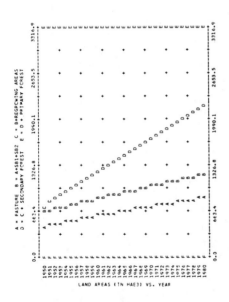

Figure 4d.--Results of the model using data shown
in figure 4a, except that initial conditions
were changed such that there was little area
in the fallow portion of swidden agriculture
cycle and all increases in pasture and permanent agriculture areas came from areas with
natural system vegetation. This run, for an
area of 3317 x 10³ ha, had a carbon exchange
of −0.0070 x 10⁹ mTC/yr in 1980. If extrapolated to the entire tropics, the annual
exchange would have been −5.45 x 10⁹ mTC.

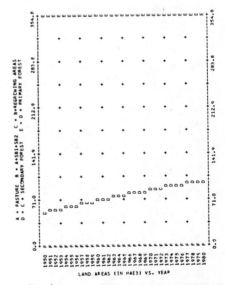

Figure 5b.--Results of the model using data shown
in figure 5a. This run, for an area of 355
x 10³ ha, had a carbon exchange of −0.0001
x 10⁹ mTC/yr in 1980. If extrapolated to
the entire tropics, the annual exchange would
have been −0.79 x 10⁹ mTC.

78

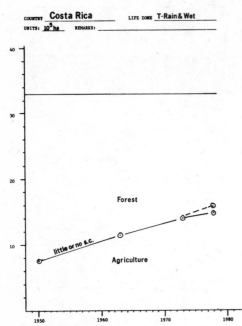

Figure 6a.--Estimates of land use in the Tropical
Rain and Wet Life Zone in Costa Rica during
the period of 1950 to 1980.

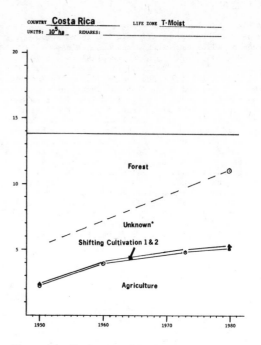

Figure 7a.--Estimates of land use in the Tropical
Moist Life Zone in Costa Rica, 1950-1980.

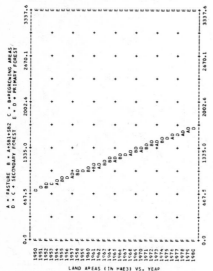

LAND AREAS (IN HAE3) VS. YEAR

Figure 6b.--Results of the model using data shown
in figure 6a. This run, for an area of 3338
x 10^3 ha, had a carbon exchange of -0.0045
x 10^9 mTC/yr in 1980. If extrapolated to
the entire tropics, the annual exchange would
have been -3.49 x 10^9 mTC.

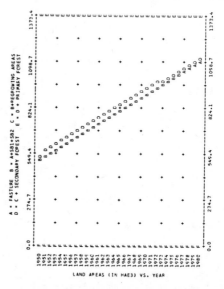

LAND AREAS (IN HAE3) VS. YEAR

Figure 7b.--Results of the model using data shown
in figure 7a. This run, for an area of 1373
x 10^3 ha, had a carbon exchange of -0.0032
x 10^9 mTC/yr in 1980. If extrapolated to
entire tropics, the annual exchange would have
been -6.09 x 10^9 mTC.

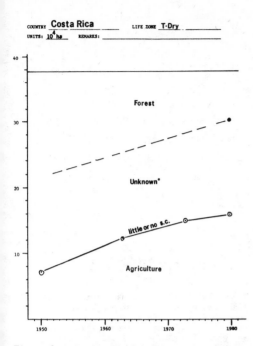

Figure 8a.--Estimates of land use in the Tropical Dry Life Zone in Costa Rica, 1950-1980.

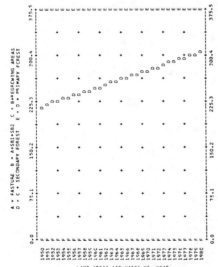

Figure 8b.--Results of the model using data shown in figure 8a. This run, for an area of 376 x 10³ ha, had a carbon exchange of -0.0003 x 10⁹ mTC/yr in 1980. If extrapolated to the entire tropics, the annual exchange would have been -1.75 x 10⁹ mTC.

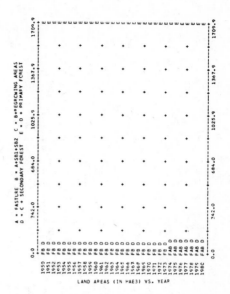

Figure 9a.--Results of the model using Tosi's (1980) data for the Tropical Rain and Wet Life Zone in Bolivia. This run, for an area of 1710 x 10³ ha, had a carbon exchange of -0.0001 x 10⁹ mTC/yr in 1980. If extrapolated to the entire tropics, the annual exchange would have been -0.20 x 10⁹ mTC.

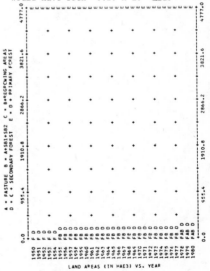

Figure 9b.--Results of the model using Tosi's (1980) data for the Tropical Moist Life Zone in Bolivia. This run, for an area of 4777 x 10³ ha, had a carbon exchange of -0.0006 x 10⁹ mTC/yr in 1980. If extrapolated to the entire tropics, the annual exchange would have been -0.33 x 10⁹ mTC.

80

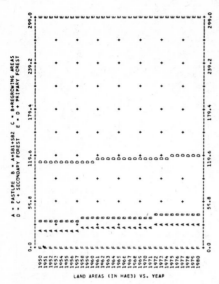

Figure 9c.—Results of the model using Tosi's
(1980) data for the Tropical Dry Life Zone in
Bolivia. This run, for an area of 299 x 10^3
ha, had a carbon exchange of -0.00002 x 10^9
mTC/yr in 1980. If extrapolated to the
entire tropics, the annual exchange would
have been -0.15 x 10^9 mTC.

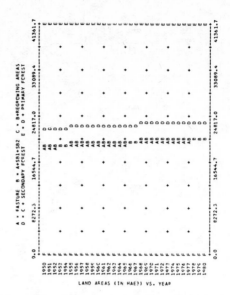

Figure 9e.—Results of the model using Tosi's
(1980) data for the Subtropical Moist Life
Zone in Bolivia. This run, for an area of
41362 x 10^3 ha, had a carbon exchange of
-0.0045 x 10^9 mTC/yr in 1980. If extrapolated
to the entire tropics, the annual exchange
would have been -0.28 x 10^9 mTC.

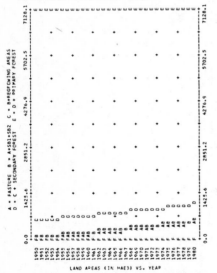

Figure 9d.—Results of the model using Tosi's
(1980) data for the Subtropical Rain and Wet
Life Zone in Bolivia. This run, for an area
of 7128 x 10^3 ha, had a carbon exchange of
-0.0023 x 10^9 mTC/yr in 1980. If extrapolated
to the entire tropics, the annual exchange
would have been -0.83 x 10^9 mTC.

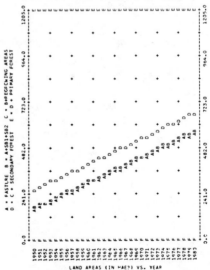

Figure 9f.—Results of the model using Tosi's
(1980) data for the Subtropical Dry Life
Zone in Bolivia. This run, for an area of
1205 x 10^3 ha, had a carbon exchange of
-0.0007 x 10^9 mTC/yr in 1980. If extrapolated
to the entire tropics, the annual exchange
would have been -1.49 x 10^9 mTC.

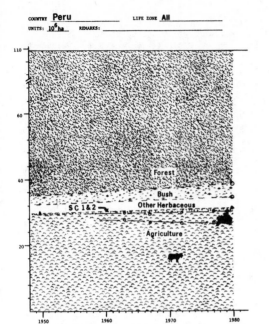

Figure 10a.--Estimates of land use in all of Peru during the period of 1950 to 1980.

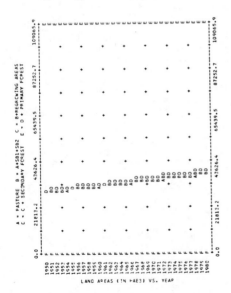

Figure 10b.--Results of the model using data shown in figure 10a. This run, for an area of 109065 x 10³ ha, had a carbon exchange of -0.0323 x 10⁹ mTC/yr in 1980. If extrapolated to the entire tropics, the annual exchange would have been -0.77 x 10⁹ mTC.

We emphasize the tentative nature of our estimates. Obviously, no one country, or even four countries, can be expected to accurately represent the tropics as a whole. We chose these countries to illustrate the large differences that exist in rates of land use change. Until we can analyze larger areas with at least the precision of our analyses for Panama and Costa Rica, and until we can better document past land use patterns, our estimates will remain reasoned guesses. Especially important is the past and present extent of shifting cultivation, because land cleared for this purpose may sequester in recovering vegetation at least a portion of the carbon dioxide released when it was cleared. In addition, if area in the fallow portion of the shifting cultivation cycle is a large portion of the "forested" area of a country, there is significantly less carbon released from cutting this smaller-statured vegetation than if these areas had "mature" or "climax" vegetation. For these reasons our model results were quite sensitive to the quantity of land that was assumed to be in swidden agriculture in the past (figs. 3b and c, 4b, c, and d). Unfortunately, even the present area in shifting cultivation can be estimated only from very incomplete data (see Appendix A) as shifting cultivators are not normally included in agricultural and economic statistics of their country due to the small size of their holdings, the frequent absence of legal title to the holdings, and the individually small contributions to markets.

We are reluctant to equate our estimates of net release from the cutting of tropical vegetation to net additions to the atmosphere from that region. For soils, our model allows only additions of carbon from cleared vegetation. Carbon can accumulate in soils as vegetation recovers following abandonment. Human activity may also reduce the carbon content of tropical soils, but we do not have sufficient information on the carbon content of various soil types, the areal extent of these types, and the changes in carbon content as a function of land use to improve upon present estimates of the release from tropical soils (see Schlesinger 1977). We believe that soil processes should be modeled explicitly; to do this, information on soils will have to be collected and organized by Life Zones. In addition, we do not have sufficient information to assess the importance of suggested biotic sinks of atmospheric carbon dioxide, such as enhanced photosynthesis due to increased atmospheric carbon dioxide (see Goudriaan and Atjay 1979), or sequestering of carbon by mature systems in soils or ground water (Holdridge 1980). If these processes are occurring to a significant degree in tropical or other ecosystems, they may offset to some greater or lesser extent the release of carbon dioxide from land use change.

LITERATURE CITED

Bacastow, R., and C. D. Keeling. 1973. Atmospheric carbon dioxide and radiocarbon in the natural carbon cycle. II: Changes from A.D. 1700 to 2070 as deduced from a geochemical model, p. 86-135. In G. M. Woodwell and E. V. Pecan (eds.). Carbon and the Biosphere, U.S.A.E.C., Washington, D.C.

Baes, C. F., Jr., H. E. Goeller, J. S. Olson, and R. M. Rotty. 1977. Carbon dioxide and climate: The uncontrolled experiment. American Scientist 65:310-320.

Bolin, B. 1977. Changes of land biota and their importance for the carbon cycle. Science 196:613-615.

Broecker, W. S., T. Takahashi, H. J. Simpson, and T.-H, Peng. 1979. Fate of fossil fuel carbon dioxide and the global carbon budget. Science 206:409-418.

Brown, S., and A. E. Lugo. 1980. Preliminary estimate of the storage of organic carbon in tropical forest ecosystems, p. 65-117. In S. Brown, A. E. Lugo, and B. Liegel (eds.). The Role of Tropical Forests on the World Carbon Cycle. CONF-800350, U.S. Department of Energy, Washington, D.C.

Brunig, E. F. 1977. The tropical rain forest-- A wasted asset or an essential biospheric resource? Ambio 6:187-191.

Comision para la Reforma Agraria y la Vivienda. 1961. La reforma agraria en al Peru; Documenta Lima. 62 p. Comision para la Reforma Agraria y la Vivienda, Lima, Peru.

Consejo Nacional de Reforma Agraria. 1975. El proceso de reforma agraria en cifras 1975. 163 p. Departmento de Estadistics, La Paz, Bolivia.

Council on Environmental Quality and the Department of State. 1980. The Global 2000 Report to the President. Volume Two: The Technical Report. 766 p. U.S. Government Printing Office, Washington, D.C.

Departmento de Censos. Censo Nacional de 1950. Departmento de Censos, Lima, Peru.

Departmento de Censos. Censo Nacional de 1970. Departmento de Censos, Lima, Peru.

Departmento de Censos. Primer Censo Nacional Agropecuario 2 de Julio 1960. Departmento de Censos, Lima, Peru.

Departmento de Censos. Segundo Censo Nacional Agropecuario de 1972. Departmento de Censos, Lima, Peru.

Detwiler, R. P., and C. A. S. Hall. 1980. The development of an empirically-driven simulation model of carbon exchange between human-impacted tropical ecosystems and the atmosphere, p. 140-156. In S. Brown, A. E. Lugo, and B. Liegel (eds.). The Role of Tropical Forests on the World Carbon Cycle. CONF-800350, U.S. Department of Energy, Washington, D.C.

Direccion de Estadistica y Censo. Censo Nacionales de 1950: Primer Censo Agropecuario. Direccion de Estadistica y Censo, Panama City, Panama.

Direccion de Estadistica y Censo. Censo Nacionales de 1960: II Censo Agropecuario. Direccion de Estadistica y Censo, Panama City, Panama.

Direccion de Estadistica y Censo. Censo Nacionales de 1970: III Censo Agropecuario. Direccion de Estadistica y Censo, Panama City, Panama.

Direccion General de Estadistica y Censos. 1953. Atlas estadistico de Costa Rica. 114 p. Seccion de Cartografica y Divulgacion, San Jose, Costa Rica.

Direccion General de Estadistica y Censos. 1974. Censo Agropecuario 1973. 286 p. Seccion de Publicaciones, San Jose, Costa Rica.

Food and Agriculture Organization. 1971. Mapa Ecologico de Panama, 1:500,000. FAO, Rome. (Map drawn by J. A. Tosi, Jr., 1967).

Food and Agriculture Organization. 1976. Proceedings of the 12th Session of the Latin American Forestry Commission. 427 p. FAO, Rome.

Food and Agriculture Organization. 1978. Proceedings of the 8th World Forestry Congress. 374 p. Secretariat of the 8th World Forestry Congress, Jakarta, Indonesia.

Goudriaan, J., and G. L. Atjay. 1979. The possible effects of increased CO_2 on photosynthesis, p. 237-250. In B. Bolin, E. T. Degens, S. Kempe, and P. Ketner (eds). The Global Carbon Cycle. John Wiley and Sons, Chichester, England.

Gross, D. R., G. Eiten, N. M. Flowers, F. M. Leoi, M. L. Ritter, and D. W. Werner. 1979. Ecology and acculturation among native peoples of Brazil. Science 206:1043-1050.

Hampicke, U. 1979. Net transfer of carbon between the land biota and the atmosphere, induced by man, p. 219-236. In B. Bolin, E. T. Degens, S. Kempe, and P. Ketner (eds.). The Global Carbon Cycle. John Wiley and Sons, Chichester, England.

Holdridge, L. R. 1947. Determination of world plant formation from simple climatic data. Science 105:367-368.

Holdridge, L. R. 1967. Life Zone Ecology Revised edition. 206 p. Tropical Science Center, San Jose, Costa Rica.

Holdridge, L. R. 1980. A new look at atmospheric carbon dioxide. p. 19-29. In S. Brown, A. E. Lugo, and B. Liegel (eds). The Role of Tropical Forests on the World Carbon Cycle. CONF-800350, U.S. Department of Energy, Washington, D.C.

Holdridge, L. R., W. C. Grenke, W. H. Hatheway, T. Liang, and J. A. Tosi. 1971. Forest environments in tropical Life Zones: A Pilot Study. 747 p. Pergamon Press, New York.

Krieslory, M. 1970. Costa Rican Agriculture. 55 p. Foreign Economic Development Service, USDA, Washington, D.C.

Manshard, W. 1974. Tropical Agriculture. 226 p. Longham, London.

Ministerio de Agricultura. 1961. Agricultura boliviana epoca 1961. 85 p. Ministerio de Agricultura, La Paz, Bolivia.

Myers, N. 1980. Conversion of Tropical Moist Forests. 205 p. National Academy of Sciences, Washington, D.C.

Nye, P. H., and D. J. Greenland. 1960. The Soil Under Shifting Cultivation. Technical Communication No. 51. 156 p. Commonwealth Bureau of Soils, Harpenden, London.

ONERN. 1976. Mapa Ecologico del Peru, 1:1,000,000. Republica del Peru, Oficina Nacional de Evaluacion de Recursos Naturales del Peru, Lima.

Pacific Science Association. 1958. Proceedings of the 9th Pacific Congress. Vol. 20. Special Symposium on Climate Vegetation and Rational Land Utilization in the Humid Tropics. 169 p. Secretariat of the 9th Pacific Congress, Bangkok, Thailand.

Persson, R. 1974. World Forest Resources: Review of the World's Forest Resources in the Early 1970's. Department of Forest Survey Research Notes No. 17. 261 p. Royal College of Forestry, Stockholm, Sweden.

Revelle, R., and W. Munk. 1977. The carbon dioxide cycle and the biosphere, p. 140-158. In Energy and Climate, National Academy of Sciences, Washington, D.C.

Rotty, R. M. 1977. Global carbon dioxide production from fossil fuels and cement, A.D. 1950 - A.D. 2000, p. 167-181. In The Fate of Fossil Fuel CO_2 in the Oceans. Plenum Press, New York and London.

Ruthenberg, H. 1976. Farming Systems in the Tropics. Second edition. 313 p. Oxford University Press, Oxford.

Schlesinger, W. H. 1977. Carbon balance in terrestrial detritus. Annual Review of Ecology and Systematics 8:51-81.

Seiler, W., and P. J. Crutzen. 1980. Estimates of gross and net fluxes of carbon between the biosphere and the atmosphere from biomass burning. Climatic Change 2:207-247.

Sommer, A. 1976. Attempt at an assessment of the world's tropical moist forests. Unasylva 28(112&113):5-24.

Tosi, J. 1969. Mapa Ecologico de Costa Rica (1:750,000). Centro Cientifico Tropical, San Jose, Costa Rica.

Tosi, J. 1975. Mapa Ecologica de Bolivia (1:1,000,000). Republica de Bolivia, Ministerio de Asunto Campesinos y Agropecuarios, La Paz, Bolivia.

Tosi, J. 1980. Vegetational cover and rates of conversion of forest and bush cover in Panama, Peru, and Bolivia with commentary on the methodology, p. 75-106. In Models of Carbon Flow in Tropical Ecosystems with Emphasis on Their Role in the Global Carbon Cycle. [U.S. Department of Energy Report DOE/EV/06047-1] U.S. Department of Energy, Washington, D.C. (see also J. Tosi. 1980. Life Zones, land use, and forest vegetation in the tropical and subtropical regions, p. 44-64. In S. Brown, A. E. Lugo, and B. Liegel (eds.). The Role of Tropical Forests on the World Carbon Cycle. CONF-800350, U.S. Department of Energy, Washington, D.C.

Tosi, J., and S. Brown. 1980. Areas of Life Zones and land use, and organic carbon storage in ten tropical countries, p. 107-150. In Models of Carbon Flow in Tropical Ecosystems with Emphasis on Their Role in the Global Carbon Cycle. [U.S. Department of Energy Report DOE/EV/06047-1] U.S. Department of Energy, Washington, D.C.

Wadsworth, F. 1978. Deforestation--Death to the Panama Canal, p. 22-24. In U.S. Department of State and U.S. Agency for International Development. Proceedings of the U.S. Strategy Conference on Tropical Deforestation, U.S. Printing Office, Washington, D.C.

Watters, R. F. 1971. Shifting Cultivation in Latin America. FAO Forestry Development Paper No. 17. 303 p. FAO, Rome.

Watters, R. F. 1975. Shifting Cultivation: Its Past, Present, and Future. 239 p. IUCN Secretariat, Morges Switzerland.

Weil, T. E. 1972. Area Handbook for Panama. 415 p. U.S. Government Printing Office, Washington, D.C.

Whittaker, R. H., and G. E. Likens. 1973. Carbon in the biota, p. 281-302. In G. M. Woodwell and E. V. Pecan (eds.). Carbon and the Biosphere. U.S.A.E.C., Washington, D.C.

Whittaker, R. H., and G. E. Likens. 1975. The biosphere and man, p. 305-328. In H. Leith and R. H. Whittaker (eds). Primary Productivity of the Biosphere. Springer-Verlag, New York.

Wong, C. S. 1978. Atmospheric input of carbon dioxide from burning wood. Science 200: 197-200.

Woodwell, G. M., R. H. Whittaker, W. A. Reiners, G. E. Likens, C. C. Delwiche, and D. B. Botkin. 1978. The biota and the world carbon budget. Science 199:141-146.

Appendix A: Shifting Cultivation Estimates for Panama

Three different patterns of shifting cultivation occur in the neotropics. The traditional form of swidden agriculture involves cutting and burning a small (less than 1 hectare), carefully selected area of forest, growing mixed food crops for one to three years, and then leaving the land fallow for 20 or more years while cutting and cultivating other plots (Ruthenberg 1976). With increasing population pressure the fallow period may decrease, eventually leading to permanent or semi-permanent agriculture, or land abandonment due to declining yields (Watters 1971). Yields may decline for a variety of reasons including: weed competition, insect pest attack, and decreasing soil nutrient availability (Nye and Greenland 1960).

Another type of shifting cultivation occurs as an advancing front of deforestation. As new roads make forests more accessible, shifting cultivators clear land, grow subsistance and/or cash crops for 1-2 years, and then move on to another, newly accessible site. After cultivation the land is converted to other uses such as pasture, plantation agriculture, or human settlement instead of returning to forest fallow.

A third type of marginal shifting cultivation occurs in regions which have been settled for some time. Landless peasants cultivate roadsides, field edges, and other temporary sites for 1-2 years then move to other similar sites when evicted or yields decline. The land often remains in a scrub or bush fallow for only 2-4 years before it is cultivated once again.

There are virtually no reliable estimates of land areas in shifting cultivation available for most countries at this time. Until a comprehensive remote sensing analysis can be undertaken our crude estimates will continue to be limited by this lack of information.

Our estimates of land area devoted to shifting cultivation include the traditional and advancing front forms of shifting cultivation but do not include marginal shifting cultivation in settled areas because of little available information and because the land use changes are unlikely to significantly alter carbon storages.

The first method for estimating the area of land devoted to shifting cultivation in a region is based on the amount of land necessary to support the estimated number of shifting cultivators. The following equation was used:

$$A_{(y)} = [P_{(y)} * (R_{(y)} * S_{(y)}) * C_{(y)} * W_{(y)}] \div F_{(y)}$$

Where,

A = area (in hectares) covered by shifting cultivation crops
P = population of region (country or smaller)
R = proportion of population which is rural
S = proportion of rural population engaged in shifting cultivation
C = amount of land cleared per year per family
W = number of years a field is cultivated
F = size of family
y = year of calculation.

As an example of the calculations involved, the population of Panama in 1950 was 863,000 (Censo Nationales de Panama 1950). Sixty-five percent of the population was rural and of those approximately 20% were engaged in shifting cultivation (Censo Nationales de Panama 1950). Estimates of the proportion of the rural population engaged in shifting cultivation are no more than best guesses of observers familiar with the area. In Panama, rural families consist of approximately seven persons and each family of shifting cultivators clears approximately one hectare of forest each year (Manshard 1974; Myers 1980). Since fields are usually cultivated for 2 years before fallow or abandonment, each year a family has 1-2 hectares in crops. In 1950, this would have been 32,054 hectares in shifting cultivation crops and a forest clearing rate of 16,027 hectares a year.

$$863{,}000 * (0.65 * 0.20) * 1 * 2/7 = 32{,}054 \text{ hectares in 1953.}$$

Similarly, in 1979 the population of Panama was 1,855,000 (Myers 1980, p. 29). Since the rate of population increase was 0.023 in 1979 (Myers 1980) we estimate the 1980 population of Panama as 1,900,000. In 1970, fifty percent of the 1980 population was considered rural and of these approximately 20% were engaged in shifting cultivation (Watters 1971). Assuming that these proportions did not change significantly, the 1980 population would have 54,285 hectares in crops and clear 27,143 hectares a year of primary and secondary forest.

The second method of estimating land area devoted to shifting cultivation is based upon forest clearing rates multiplied by the area of forest. It has been proposed that approximately 2% of the closed forest in Latin America is cleared each year (Myers 1980; FAO 1978; Persson 1977). Extrapolating Persson's 1974 estimate of forest area of Panama to 1980 we obtain 3,616,000 hectares of forest being cleared at a rate of 72, 320 hectares a year. Similarly, using Tosi's 1980 estimate of 3,190,100 hectares of forest the clearing rate is 63,802 hectares a year. Since nearly all forest cleared in Latin America is first cleared by shifting cultivators (Watters 1971) these rates of deforestation suggest that approximately 127,604 to 144,640 hectares were covered with shifting cultivator's crops in 1980.

The third method of estimation depended upon government reports of the amount of forests cleared by shifting cultivators. The government of Panama estimates that 75,000 hectares of forest were cleared by shifting cultivators in 1972 (Area Handbook of Panama 1972). If this clearing rate remained unchanged it would indicate approximately 150,000 hectares in shifting cultivator's crops in 1980.

Overall, estimates of land area devoted to shifting cultivation crops in 1980 range from 54,285 hectares to 150,000 hectares. Although we used an estimate of 100,000 hectares in our simulation we believe this value to be quite conservative. Government descriptions of the importance of shifting cultivation are prone to underestimation because of the difficulty of censusing often inaccessable traditional shifting cultivators and locating rapidly moving advancing front shifting cultivators not attached to previously recognized settlements. In addition, the general lack of consideration shifting cultivators receive due to their small social, political, and economic impact virtually insures the unavailability of sufficient

resources to conduct thorough, accurate investigations. Furthermore, many shifting cultivators probably identify themselves to interviewers as legal title holders practicing permanent agriculture. If their land holdings were large than one hectare these shifting cultivators would be counted as part of the national agricultural production census. The government census however does not include any holdings smaller than one hectare.

In addition, although we might expect foresters to present inflated estimates of deforestation in an effort to generate action in what they see as a crisis situation, the usual mode of expressing deforestation as a percent loss per year of the total closed forest in some reference year most likely results in an underestimate when extrapolated into the future. As the forest area in Latin America decreases through time the rate of deforestation as a proportion of remaining forest probably will continue to increase. The only apparent limit to deforestation in Central America has been the exhaustion of the forest resource as in El Salvador (Myers 1980). In comparison to these sources of underestimation of the area in shifting cultivation potential sources of overestimation seem insignificant.

Table 1-A presents the estimates of land area in shifting cultivation used in our simulation. Estimates for Costa Rica were calculated in the same way as those for Panama. However, less information is availble for Peru and Bolivia, thus, these estimates are based upon only the first two methods of estimation described for Panama.

Table 1-A--Land Area in Shifting Cultivation

Country	1950	1980	References*
Panama			Manshard 1974; Wadsworth 1978;
T-Wet	10,500	50,000	Weil 1972; Direccion de Estadistica y
T-Moist	21,500	50,000	Censo 1950, 1960, 1970
T-Dry	0	0	
Total	32,000	100,000	
Costa Rica			Direccion General de Estadistica y
T-Wet	0	30,000	Censos 1953, 1974; Manshard 1974;
T-Moist	6,000	20,000	Kreislory 1970
T-Dry	0	0	
Total	6,000	50,000	
Bolivia			Ministerio de Agricultura 1961;
Total	1,000	7,000	Consejo Nacional de Reforma Agraria y la Vivienda 1961
Peru			Departmento de Censos 1950, 1960, 1970;
Total	300,000	500,000	Comision para la Reforma Agraria y la Vivienda 1961

*
The following references were used for all countries: FAO 1976, FAO 1978, Myers 1980, Persson 1974; Sommer 1976.

Areas in shifting cultivation were assigned to Life Zones based upon the location of: remaining forest, previous deforestation, cities and settlements, intensive permanent agriculture, new access roads, indigenous populations, national parks and preserves, and the general preference of shifting cultivators for the Tropical Moist Life Zone (Manshard 1974; Watters 1971). For example, the Gatun Lake watershed west of the Canal Zone in Panama was 85% forested in 1952 and less than 10% forested in 1978 (Wadsworth 1978). The initial clearing of this area was accomplished almost entirely by shifting cultivators (Wadsworth 1978). According to Tosi's (1971) Life Zone map of Panama this area lies in the Tropical Moist Life Zone. In contrast, the Darien province in eastern Panama, which lies in the Tropical Wet Life Zone, is almost completely forested. Darien has no large settlements, no premanent agriculture, and only a few primitive access roads. The indigenous people living in the Darien forest probably practice shifting cultivation; however, their population is so small that less than 10% of all estimated shifting cultivation in Panama takes place in Darien (Myers 1980; Area Handbook of Panama 1975). In addition, the dry southwestern penninsula and seacoast support many settlements, extensive permanent agriculture, and contains only a small portion of coastal forest (Censo Nacionales 1970). Therefore, we assume that almost no shifting cultivation occurs in the Tropical Dry Life Zone of Panama.

The integration of the above information and other such specific indicators resulted in the assignment of areas of shifting cultivation presented in table 1. Obviously, such a subjective estimate is no more than an educated guess.

Appendix B: General Description of Agriculture and Derivation of Data Used in Model for Panama

Panama has approximately an area of 7.6 million hectares. According to the Area Handbook of Panama (Weil 1972), about one-fifth of the land is above 750m and is largely unsuitable for agriculture as it is steep and heavily forested. These high regions, essentially unaffected by agriculture, include: the Cordillera Central, running from the Costa Rican border through Bocas del Toro and Chiriqui into Cocle; a group of low mountains found in the southwest corner of Veraguas; another range on the Azuero peninsula; the Serrania de San Blas; the Serrania Darien; the Serrania de Maje; and the mountains in southwest Darien province. The Area Handbook indicates that most crops are grown below 300m of elevation, except for coffee which is more common in the 300-750m regions. Livestock grazing is also concentrated in the 300-750m regions. With this information, the lowlands (and thus potential agricultural regions) can be divided into four groups: (I) the Atlantic and Pacific coasts of western Panama (Atlantic-northern Bocas del Toro; Pacific-southern Chiriqui, southern Veraguas); (II), the Golfo de Parita/Golfo de Panama Coastal Region (Los Santos, Herrera, and Cocle); (III),

the Central Region (Colon, western Panama); and (IV), the Inland Plain Region (eastern Panama and Darien). Of these regions, the Atlantic coastal part of Group I (Bocas del Toro) has been isolated by lack of roads and is largely unused for agriculture except for limited banana and cacao production. Group IV (the inland plains of Darien and eastern Panama provinces) also lacks roads, contributes a minute portion to the censused agriculture, and contains only small indigenous Indian populations. This leaves as the main agricultural regions: (A) Gulfo de Panama coastal area (Los Santos, Herrera, and Cocle provinces), (B) the western and Pacific area (Chiriqui and Veraguas provinces), and (C) the central area around the Canal Zone (Colon and Panama provinces).

The Gulfo de Panama area includes nearly all of the Tropical Dry Life Zone area in Panama. The area is said to be most favorable for agriculture; this is reflected in the high percentage of these provinces in agriculture. Chiriqui, and to a lesser extent Veraguas, are also important agriculturally, particularly for pastures and permanent crops.

Agriculture appears to extend to higher altitudes in Chiriqui and Veraguas than in the Golfo de Panama provinces. We found little information on the agriculture of the lowlands around the Canal Zone, other than census statistics which indicate the area is not extensively farmed. This may be a result of the efforts of the United States to maintain forest cover so that silt inputs into the canal do not increase.

The main staple crops in Panama are rice, corn, and beans, with bananas grown as the major export crop. Exports of sugar and beef have been increasing. Bananas and coffee are the main permanent crops. Area in permanent crops (i.e. shrubs and trees) has been about 1/4 to 1/5 of the area in temporary (i.e. herbaceous) crops. Pasture land has been 1 to 3 times the area of permanent and temporary crops combined. Both the 1960 and 1970 census data divide area into croplands (temporary croplands) and into areas actually planted in crops and areas in fallow, i.e. that which has been used for temporary crops and presumably will be used again, but which was not being cultivated at the time of census. It is of interest that the area in fallow has been approximately as much as the area actually growing crops; about half the cropland at any time is uncultivated. This may relate to the rapid exhaustion of some tropical soils or difficulties with insects and weeds in fields cultivated for many years. These problems also encourage the use of shifting cultivation, but it is not possible to determine directly from the statistics how much of the temporary cropland actually is used for shifting cultivation.

The agricultural censuses give information on ownership of the land worked: these indicate that approximately half the farm land is held without legal title ("sin titulo"). The census statistics also give information on the use of fertilizer, animal power, mechanization, and farm size. All of these seem to indicate that much of Panamanian agriculture during the 1950-1970 period occurred on a small scale using simple and traditional methods. The exceptions are certain large-scale enterprises growing bananas and coffee, and to a lesser extent, sugarcane and rice.

Calculation of Area in Permanent Agriculture

The area in Permanent Agriculture used in the simulations was determined by the following formula:

AREAPA (R, LZ, YR) = p(R, LZ) * PA(R, YR)

where,

AREAPA = the area, in hectares, of permanent agriculture in a region R, the Life Zone LZ, for the year YR

p(R,LZ) = is our ratio of the planimetered area of the Life Zone LZ in the region R to the total planimeterd area of that region

PA(R,YR) = is the area of land in Permanent Agriculture for that year and region, according to the official statistics or our extrapolation for 1980 (see tables 1-B and 2-B).

Sample Calculation:

AREAPA (Bocas del Toro Province, Tropical Moist Life Zone, 1960)

= 0.19 * 29,304

=5,568 hectares.

The planimetered areas of each Life Zone of each province in Panama and our calculated estimates of permanent agriculture by province and Life Zone are given as tables 1-B and 2-B.

Appendix C: Deriving Coefficients and Initial Conditions for the Model

Three types of data are required by our model to calculate the carbon transfer between the atmosphere and tropical ecosystems due to land use changes:

A. The amount of land in the following classifications (state variables):

1. Natural System (NS): the undisturbed vegetation of the Life Zone,

2. Secondary Vegetation (2F): the climax vegetation of the Life Zone after disturbance. Also needed is the number of years (MAT) required for succession to grow back to steady-state biomass.

3. Permanent Agriculture (PA): temporary crops, permanent crops, and pasture. We have combined the land area in these three categories with the assumption that each has about the same quantity of carbon per hectare.

4. Shifting Cultivation:

a. Cultivated Land (SB(1) and SB(2)): the first two years of the cycle when crops are grown, and

b. Fallow Land (RG(1) to RG (FLW)): land that has been abandoned to lie fallow for FLW years. Also needed is the shape of the curve of recovered biomass over time (RSS2F).

B. The rates of land use change in hectares per year (fluxes; see fig. 1):

1. The rate at which land is taken from the Natural System (NS) for swidden (shifting) cultivation (SW): NS->SW.

2. The rate at which NS land is converted directly to pasture and agriculture (PA): NS->PA.

3. The rate at which fallow land (RG) is cut for swidden: RG->SW.

4. The rate at which land cleared by shifting cultivators (SB(1) and SB(2)) is converted to pasture and permanent agriculture, thereby removing it from the swidden cycle: SW->PA.

C. The carbon content per hectare of vegetation characteristic of the following classifications:

1. Natural System (TCHANS).

2. Secondary Vegetation (TCHA2F).

3. Vegetation at the start of the swidden cycle: after clearing and before cultivation (TCHASS).

4. Permanent Agriculture: temporary crops, permanent crops, and pasture (TCHAPA).

Data Preparation:

A. We establish areas in each of the land uses by combining several sources of information:

1. We use the numbers derived from agricultural statistics and planimetry of Life Zone maps to obtain the total area in agriculture in each Life Zone.

Table 1-B—Area in Permanent Agriculture in Panama

Region / Life Zone	(areas in hectares) Total Area	%	Perm. Agriculture 1950	1960	1970
All Panama	7565000.0	1.000	1002262.0	1371726.0	1682602.0
T-Dry	378250.0	0.050	50113.1	68586.3	84130.1
T-Moist	3252950.0	0.430	430972.6	589842.2	723518.8
T-Wet	3933799.0	0.520	521176.2	713297.4	874953.0
Bocas	891700.0	1.000	26663.0	29304.0	35097.0
T-Dry	0.0	0.000	0.0	0.0	0.0
T-Moist	169422.9	0.190	5066.0	5567.8	6668.4
T-Wet	722277.0	0.810	21597.0	23736.2	28428.6
Cocle	503500.0	1.000	103489.0	145460.0	189957.0
T-Dry	138462.4	0.275	28459.5	40001.5	52238.2
T-Moist	150042.9	0.298	30839.7	43347.1	56607.2
T-Wet	214994.4	0.427	44189.8	62111.4	81111.6
Colon	746500.0	1.000	28463.0	43547.0	54589.0
T-Dry	0.0	0.000	0.0	0.0	0.0
T-Moist	156764.9	0.210	5977.2	9144.9	11463.7
T-Wet	589735.0	0.790	22485.8	34402.1	43125.3
Chiriqui	875800.0	1.000	303094.0	364870.0	421585.0
T-Dry	0.0	0.000	0.0	0.0	0.0
T-Moist	289013.9	0.330	100021.0	120407.1	139123.0
T-Wet	586786.0	0.670	203072.9	244462.9	282461.9
Darien	1680300.0	1.000	14429.0	23524.0	22035.0
T-Dry	0.0	0.000	0.0	0.0	0.0
T-Moist	890558.9	0.530	7647.4	12467.7	11678.5
T-Wet	789741.0	0.470	6781.6	11056.3	10356.4
Herrera	242700.0	1.000	109265.0	132527.0	145867.0
T-Dry	89799.0	0.370	40428.0	49035.0	53970.8
T-Moist	104361.0	0.430	46983.9	56986.6	62722.8
T-Wet	48540.0	0.200	21853.0	26505.4	29173.4
Los Santos	386700.0	1.000	145680.0	193480.0	242202.0
T-Dry	100541.9	0.260	37876.8	50304.8	62972.5
T-Moist	235887.0	0.610	88864.8	118022.8	147743.2
T-Wet	50271.0	0.130	18938.4	25152.4	31486.3
Panama	1129200.0	1.000	109053.0	173040.0	219683.0
T-Dry	27100.8	0.024	2617.3	4153.0	5272.4
T-Moist	658323.6	0.583	63577.9	100882.3	128075.2
T-Wet	443775.6	0.393	42857.8	68004.7	86335.4
Veraguas	1108600.0	1.000	162126.0	265974.0	351587.0
T-Dry	0.0	0.000	0.0	0.0	0.0
T-Moist	576471.9	0.520	84305.5	138306.4	182825.2
T-Wet	532128.0	0.480	77820.4	127667.5	168761.8

Table 2-B—Area in Permanent Agriculture in Costa Rica

Region / Life Zone	(areas in hectares) Total Area	%	Perm. Agriculture 1950	1963	1973
Costa Rica	5090000.0	1.000	980336.0	1686572.0	2048512.0
T-Dry	371570.1	0.073	71564.5	123119.8	149541.4
T-Moist	1185969.0	0.233	228418.3	392971.2	477303.3
T-Wet	3532460.0	0.694	680353.2	1170480.0	1421667.0
San Jose	520000.0	1.000	134088.0	204443.0	225357.0
T-Dry	0.0	0.000	0.0	0.0	0.0
T-Moist	62400.0	0.120	16090.6	24533.2	27042.8
T-Wet	457599.9	0.880	117997.4	179909.8	198314.1
Alajuela	950000.0	1.000	188985.0	341577.0	414704.0
T-Dry	0.0	0.000	0.0	0.0	0.0
T-Moist	218500.0	0.230	43466.6	78562.7	95381.9
T-Wet	731499.9	0.770	145518.4	263014.3	319322.1
Cartago	260000.0	1.000	74889.0	101873.0	94669.0
T-Dry	0.0	0.000	0.0	0.0	0.0
T-Moist	20800.0	0.080	5991.1	8149.8	7573.5
T-Wet	239200.0	0.920	68897.9	93723.1	87095.4
Heredia	290000.0	1.000	27901.0	56558.0	63078.0
T-Dry	0.0	0.000	0.0	0.0	0.0
T-Moist	8700.0	0.030	837.0	1696.7	1892.3
T-Wet	281300.0	0.970	27064.0	54861.3	61185.7
Guanacaste	1040000.0	1.000	334595.0	517722.0	676443.0
T-Dry	353599.9	0.340	113762.3	176025.4	229990.6
T-Moist	572000.0	0.550	184027.3	284747.1	372043.6
T-Wet	114400.0	0.110	36805.5	56949.4	74408.7
Puntarenas	1100000.0	1.000	156929.0	336953.0	449572.0
T-Dry	22000.0	0.020	3138.6	6739.1	8991.4
T-Moist	242000.0	0.220	34524.4	74129.6	98905.8
T-Wet	835999.9	0.760	119266.0	256084.3	341674.7
Limon	930000.0	1.000	62949.0	127447.0	124690.0
T-Dry	0.0	0.000	0.0	0.0	0.0
T-Moist	83699.9	0.090	5665.4	11470.2	11222.1
T-Wet	846300.0	0.910	57283.6	115976.8	113467.9

2. We use the information in Appendix A to obtain the area in the first and second year of shifting cultivation. We assume a 15 year fallow period for 1950 in Panama and Costa Rica, and 12 years in Bolivia and Peru. We estimate that it takes 20 years for cleared areas to grow into mature Secondary Vegetation in Panama and Costa Rica and 30 years in Bolivia and Peru. We assume that during vegetative recovery carbon content plotted against time yields a linear function.

3. Tosi's 1980 paper on Panama, Bolivia, and Peru enables us to establish a rough estimate of the land in fallow areas in 1980 by: RG_{1980} = "Bush" + ("Total Herbaceous" - Permanent Agriculture). The fallow area for 1950 is then calculated as:

$$RG_{1950} = FLW_{1950} / FLW_{1980} * (SB(1) + SB(2))_{1950} / (SB(1) + SB(2))_{1980} * RG_{1980}$$

4. We assume that there is no Secondary Vegetation (2F) in 1950 through 1980 or that this area is indistinguishable, and hence contained within, Natural System vegetation. The area in Natural System (NS) vegetation is then calculated as:

$$NS = Total\ Area - (PA + SB(1) + SB(2) + RG + 2F)$$

5. The model also requires values for:

   a. the percentage of cleared vegetation which does not burn and decays over time (PRVNEC).

   b. the percentage of cleared vegetation which does not burn and is stored as wood products, charcoal, and/or additions to soil organic matter (PRVCHR).

   c. the rate at which cleared vegetation left after burning decays (KDECSW).

We estimated these coefficients using information found in Seiler and Crutzen (1980), Wong (1978), and from discussions with C. Jordan (personal communication). In the simulations for this paper, 70% of the cleared vegetation was left after burning to decay at a rate of 20% per year, and 5% of the cleared vegetation was stored as wood products, charcoal, and/or additions to soil organic matter.

B. Changes in land use for the period 1950-1980 were calculated from Tosi (1980), the official agricultural statistics and the information reported in Appendix A.

   1. $RG->SW_{1950}$ and $RG->SW_{1980}$: annual conversion of fallow land to first year of swidden cultivation. Rates determined from data in Appendix A. The model interpolates rates for the years between 1950 and 1980.

   2. $SW->PA_{1950,1980}$: annual transfer of land in the second year of cultivation in swidden agriculture to permanent agriculture and pasture. Rate set to annual increase in total agricultural land unless information about swidden agriculture was known. See Appendix A.

   3. $NS->PA_{1950,1980}$: annual conversion of land with natural system vegetation to pasture and permanent agriculture. Rate was zero unless increase in PA was greater that could be accounted for by the transfer of swidden land to pasture.

   4. $NS->SW_{1950,1980}$: annual conversion of land with natural system vegetation to the first year of cultivation in the swidden cycle. Rate set by increases in PA and swidden cycle fallow areas, RG.

C. The carbon content of the natural system vegetation (NS) and the mature secondary vegetation (2F) were assigned according to the Life Zone averages found in Brown and Lugo (1980). If the area of the simulation contained more than one Life Zone, a weighted average was used. The carbon content of the vegetation at the start of swidden (SS) was estimated at 10 metric tons of carbon per hectare for all simulations. Similarly, the carbon content of permanent agriculture and pasture was set at 5 metric tons of carbon per hectare until we can get better estimates for these parameters from the literature from research we have proposed.

Extrapolation to the Whole Tropics:

Net carbon exchanges between the vegetation and the atmosphere for a given region were extrapolated to the tropics as a whole by assuming that the land area being modeled was characteristic of the tropics as a whole. Our estimate for the appropriate land area of the whole tropics, 26 x 10^6 km^2, was derived as 44 x 10^6 km^2 (mean of four estimates of 'total land area of tropics', Brown and Lugo 1980) minus 15 x 10^6 km^2 of savannas, minus an arbitrary estimate of 3 x 10^6 km^2 of barren land. The amount of carbon which would be released annually by the whole tropics was calculated by multiplying the amount of carbon released per hectare by the country or Life Zone being simulated by the figure for the area of the whole tropics.

2. Ecological Models of Fossil Fuel Development

ENVIRONMENTAL IMPACTS OF COAL CONVERSION PLANTS AND MINE ACTIVITIES[1]

Ekrem V. Kalmaz[2]

Abstract.--The advent of the energy crisis during the last decade has spurred the consideration of alternative energy sources to oil and natural gas as staple fuels. Coal conversion is considered one of the best solutions for the energy shortage problems, currently encountered by the Western industrialized countries. Coal conversion plants produce emissions not commonly found in conventional energy producing facilities. In order to evaluate the environmental impacts of coal conversion plants, preconstruction environmental impact studies are designed to encompass the probable emissions from these plants. In this paper, potential environmental impacts associated with operation of these plants and mine activities are discussed.

INTRODUCTION

The increased energy demand in the United States has brought about a major effort committed to examining coal conversion processes (CCP) under current economic conditions. Coal represents a large percentage of this nation's energy resources and in all likelihood, will be a primary source for meeting future energy needs. The economic utilization of the vast coal reserves for the production of clean-burning gaseous and liquid fuels would be one of the more attractive and challenging contributions toward solving the national energy problem, particularly since the major transportation and commercial sectors depend primarily on such fuels.

In the last several years, considerable efforts has been committed toward developing the necessary technology for the conversion of coal to liquid and gaseous fuels. Through chemical reactions involving coal, conversion processes produce a broad range of materials that

contain toxic, carcinogenic, and mutagenic organic and inorganic compounds, including aromatic amines, aromatic and heterocyclic compounds, highly substituted phenols, and other compounds. As an integral part of the synthetic fuels technology development program, unacceptable degradation of the quality of the environment must be prevented by appropriate control technology implemented concurrently with the coal conversion technology. Indeed, from an engineering viewpoint, any potential environmental impact problems should be determined before conversion technology is commercialized. The technical framework of overall synthetic fuel and environmental impacts characterization programs need to be factored into the design of various stages of process development from bench-scale to demonstration and ultimately, to commercial-scale operations. Although information and materials are frequently available, interpretative engineering analysis must be performed to project commercial-scale characteristics from data obtained on subcommercial-scale systems. In some cases, technical limitations may affect the design of environmental impact preventive measures which can be developed in advance of commercial-scale operations. Most of these limitations can be minimized with careful equipment design and judicious choice of subcommercial-scale studies.

The primary goal for environmental operational strategy is the construction of a comprehensive environmental impacts data base which will provide (a) a quantitative basis for analysing identified environmental

[1]Paper presented at the international symposium of Energy and Ecological Modelling, sponsored by the International Society for Ecological Modelling. (Louisville, Kentucky, April 20-23, 1981).

[2]Ekrem V. Kalmaz is Research Assistant Professor of Engineering Science and Mechanics, The University of Tennessee, Knoxville, TN 37916 U.S.A.

risks; (b) a quantitative basis for determining environmental management needs options and performance requirements; (c) a quantitative basis for the selection of control systems and for preparing required environmental impacts statement; (d) indicators of the scope of significant environmental issues; and (e) broad substantive evidence of the environmental acceptability of coal conversion technology.

COAL CONVERSION PROCESS

The coal conversion processes planned for commercial development in the United States are: (a) coal liquefication, (b) coal gasification, and (c) solvent-refined coal (SRC). The conversion of coal to high-BTU gaseous and liquid fuels require major chemical and physical changes in coal. Both gaseous and liquid fuels are produced as a result of decreasing the carbon-to-hydrogen weight ratios of solid coal. The carbon-to-hydrogen weight ratio of coal ranges from 12 for lignite to 20 for bituminous grade coal. By the rejection of carbon and/or the addition of molecular hydrogen, the weight ratios of coal can be lowered to 10 to produce synthetic liquids. When hydrogen content is increased, methane gas can form at a carbon-to-hydrogen ratio of three. Gasification processes can convert solid coal to low- and medium-BTU gases for use as utility or industrial fuels or chemical feedstocks, or to high-BTU gases for use as natural gas substitutes. Solid coal gasifies in reactions at temperatures approximately from 500-1000°C (1000-2000°F) and pressure from 500-1000 pounds per square inch gauge pressure. The amount of oxygen and/or air added controls reaction rate and temperature. In the process, oxygen concentration is controlled to facilitate the proper partial combustion of the coal to produce the desired final product. The general chemical reaction for coal gasification is rather simple:

$$\text{Solid coal} + H_2O \longrightarrow CH_4 + CO_2 \qquad (1)$$

This reaction process takes place in several different steps. Most gasification technologies under development incorporate the concept of hydrogasification in which the incoming coal is initially reacted with a hydrogen-rich gas to form methane. The conditions inside the gasifier produce only partial combustion of the coal. The result is a strong reducing state in which the carbon bonds in the coal are severed. In the system, the rate of reactions and degrees of conversion are functions of temperature, pressure, oxygen concentration, and the nature of the particular coal employed. Gasification processes consume hydrogen in quantities too great

to be economically produced on site. Hydrogen may be produced by a partial oxidation gasifier which reacts with a carbon source and a limited quantity of oxygen in the presence of steam to produce carbon monoxide and hydrogen. The carbon monoxide is then reacted with steam in a shift conversion reaction to produce hydrogen and carbon dioxide in accordance with the following reaction:

$$CO + H_2O \longrightarrow H_2 + CO_2 \qquad (2)$$

Coal and oil can be utilized as gasifier feed in this reaction system. Most coal gasification processes can be categorized according to the methods of contacting gas and liquid streams including suspension (entrained-bed reactor), fixed-bed reactor, moving-bed reactor, fluidized-bed reactor, and molten-bath reactor.

Liquid fuels produced from conversion processes have potential use in the firing of industrial and electric boilers and gas turbines. The major advantages of liquefication are that a broad range of possible products (such as fuel oil, gasoline, jet fuel, and diesel oil) can be produced from coal by varying the type of catalyst used as well as other physical and chemical processing conditions. In contrast to gasification, liquefication processes involve fewer chemical changes have a significantly higher energy conversion efficiency. Major components of coal conversion processes are carbon monoxide (CO), hydrogen (H_2), methane (CH_4) and carbon dioxide (CO_2). A large number of other inorganic and organic derivatives are also produced, although in smaller quantities.

ENVIRONMENTAL IMPACTS

The possible environmental impacts of the coal conversion process are a principle concern in the development of synthetic fuels technology. Unfortunately, our knowledge of the environmental health effects of human exposure to the primary effluent streams of air, water, and solid waste from coal conversion processes is currently fragmentary and unclear. The potential effects of CCP effluents must be methodically examined and evaluated before the various technologies are employed on a commercial scale in order to prevent major environmental and health problems possibly associated with such processes. Because of the chemically complex nature of coal and because of the reaction conditions under which it is converted to gas and liquid hydrocarbons, a broad range of organic and inorganic compounds are produced that may be hazardous to human health. Certain CCP intermediates, products, by-products and

wastes may contain trace elements and or-
ganometallic compounds. When released to
the environment, trace elements are normally associated with the particulate e-
missions. Selected streams may also contain a wide range of complex organic compounds, including polynuclear aromatic
hydrocarbons (PAHs), aromatic amines that
are known and/or suspected to be mutagenic
or teratogenic in microbial systems and
carcinogenic or highly toxic in experimental animals (Cowser 1980). In coal
conversion most of these complex organic
compounds are released primarily during
the hydrogenation reaction. Although
little is known about commercial-scale
conversion facilities, investigations on
related processes and from measurements on
process streams of bench scale or pilot
scale facilities have aroused legitimate
health concerns. Experience from petroleum refineries and related petrochemical
industries, however, has furnished considerable aid in the planning and implementing of environmental health protection
programs (fig.1).

to minimize the release of hazardous organic and inorganic chemicals and to furnish
maximum protection from exposure to both
the human and environment. Pilot plant
studies have demonstrated that the qualitative and quantitative compositions of
coal conversion products and wastes vary,
and are sensitive to changes in the nature
of coal and such process conditions as
hydrogen pressure, temperature, and resistance time (U.S. Department of Energy, Division of Coal Conversion 1979; Battelle
Pacific Northwest Laboratory 1979). This
variability, combined with the fact that
only a conceptual coal conversion plant
design exists at the present time, reduces
the ability to predict the specific chemical species and exact concentrations that
may be present in products, intermediate
streams, wastes, and the working environment.

In general, exposure in modern coal
conversion facilities are expected to be,
at worst, much lower than occurred with the
coal utilization technologies of past years.
Since most coal conversion processing is
carried out in closed systems, emissions
would result primarily from leaks, accidents, or the emergency venting of tar,
chars, soots, or other toxic chemicals.
Nevertheless, potential exposures vary
widely from process to process and continual
monitoring is essential in the working
environment.

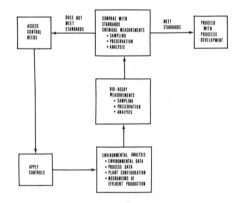

Figure 1.--Environmental impact research
related to coal conversion processes
is concerned with evaluating the potential ecological effects arising
from effluent and product releases
and with determining the ultimate
fate of released materials in the
environment.

Although present and projected environmental quality standards and emission
limitations require application of state-
of-the-art pollution control techniques,
standards and limitations do not yet exist
for the most classes of compounds that are
of a major concern. Compliance with existing environmental requirements alone may
not ensure complete protection of the
workers, the general public, and the environment. Thus, the design of pilot and/or
demonstration plants must include the means

Environmental impact assessments of
coal conversion plants, can be conducted
in an iterative and simultaneously with
technology development, may help to identify potential problems and provide a
framework for accumulating necessary data
for the decision making process in the face
of uncertainties (fig. 2).

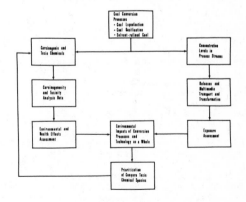

Figure 2.--Decisions made on risks to human
health must include consideration of
the data on biological effects at every
level of biological organisms.

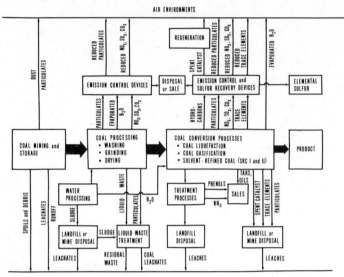

Figure 3.--Overview of the waste streams and primary emissions from coal conversion processes.

Coal conversion technology usually consists of a schematic diagram showing process, effluent, and emissions streams (fig.3) for each component of the fuel cycle. Identification of the list of toxic materials of potential human health concern is the first step before the development and/or soon after the operational facilities design is completed.

ATMOSPHERIC EMISSIONS

Atmospheric emissions of criteria pollutants such as sulfur dioxide, nitrogen oxides, carbon monoxide, total suspended particulate (TSP), and hydrocarbons must be controlled in compliance with the Clean Air Act regulations. The aim of air pollution control legislation is to protect and enhance public health and welfare, regardless of location. The Clean Air Act of 1970 and its 1977 amendments authorize programs to meet this goal. In polluted areas, regulations have to be set to improve air quality through the National Ambient Air Quality Standards (NAAQS).

Atmospheric emissions of unidentified organic derivations are not criteria pollutants but may be suspected of having geotoxic or other biological activity which can be minimized by the use of several engineering controls. The most important of these is a controlled combustor

to convert hydrocarbon emissions efficiently to carbon dioxide and water vapor. Except during the most severe emergencies, hydrocarbon emissions from process equipment can normally be completely contained and any emissions can be directed to a controlled combustor. Emissions from the venting of safety valves, evaporative emissions from tanks and filling operations at product shipping points, fumes from product solidification, and all other relatively continuous or small point sources can be treated in this fashing.

Fugitive emissions to the atmosphere should not be a major source of occupational exposure for CCP-plant workers involved in maintenance operations. Dermal exposure would likely be the most significant factor. There is also a need to have detailed information available on the fugitive materials which must be monitored continuously to ensure adequate health protection.

The basic areas of concern with regard to engineering controls for CCP include those materials produced from furnaces and coal dryers, sulfur plant tail, gas, cooling towers, and carbon dioxide from acid gas liquor (phenols, sulfur and nitrogen compounds, oil, cresols). Although various control techniques have been developed and have application, numerous control technology gaps on coal gasification and liquefication do exist. A principle objective in control technology for CCP is the removal of sulfur and other organic and inorganic substances from gas streams at high temperatures and pressures.

Sulfur Oxides

Most sulfur removal units efficiently scrub the gas with water and/or organic absorbents to remove this pollutant. Approximately 85 to 98% of the sulfur compounds in the CCP feed are reduced during reaction to hydrogen sulfide and organic sulfur species. In CCP, fired heaters are potentially the major source of sulfur oxides but the fuel normally fired is a process-generated, low-sulfur fuel gas. The alternate fuel is oil which is also process-generated, but contains greater amounts of sulfur. The fuel gas from calciner can be incinerated and scrubbed to minimize SO_2 and particulates. Other alternative scrubbing technologies include a dry system utilizing a sodium carbonate spray drier, fabric filters, and a wet Venturi scrubber system utilizing sodium carbonate absorption and cyclone-type separation of liquids and particulate from fuel gas.

Other sulfur species are first removed from process streams in acid -gas removal areas, then are converted to elemental sulfur. Some residual reduced sulfur compounds remain in the tail gas and in the CO_2 streams. These may be vented from a common stack resulting in reduced sulfur in the vent stream.

Particulates

Particulate emissions from coal conveying can be significantly reduced by wetting the coal with chemical dust-suppressing agents and by partially enclosing conveyors and transfer points with hoods and chutes. After coal pulverizing, all coal conveyors and processing equipment can be totally enclosed and vented through single dust collectors or in an integrated bag house system. Fugitive emissions from coal piles can be minimized by spraying the piles with a chemical wetting agent or a polymer crusting agent. Vent gas from a coal pulverizing/drying operation can also be filtered in a bag house. Particulates can be removed from the calciner flue gas by the SO_2 scrubbing system. The molten element sulfur from the sulfur-recovery area can be stored in a closed insulated tank to prevent emission of fugitive dust from open piles and solid sulfur handling. The anode coke from calciner can be stored in silos handled by a closed conveyer equipped with dust control system.

Solid coal contains numerous trace elements of different concentrations. In the emission process, most of these trace elements are normally associated with the particulate emission and their quantities are quite low except for such highly volatile elements as selenium, mercury, and flourine.

Hydrocarbons

Another major environmental concern is the possible emission of organic compounds such as polycyclic aromatic hydrocarbons (PAH's). Because some of these compounds are highly toxic and/or carcinogenic, an accurate baseline concentration for these compounds at the proposed plant site must be obtained.

Hydrocarbon emission control is achieved by utilizing storage tanks with fixed roofs, where most of the vapors can be condensed and returned to storage. Uncondensed vapors can be combusted in the flare system. Some liquid storage tanks are equipped with a floating roof, and vapors are purged from the space between double seals and combusted in the flame system. During process, upset and emergencies pressure relief streams containing hydrocarbons be flared. In CCP, most hydrocarbons can be condensed, and the uncondensible portion can be flared. All hydrocarbons should be combusted before release to the atmosphere. Hydrocarbons volitalized from the coal-drying operation can be vented to the atmosphere. The extent of volitalization may be minimized by controlling the drying temperature at less than 200° C and designing the pulverizer/drier to minimize residence time.

Although CCP units do not emit hydrocarbons, (other than fugitive releases), process upsets and emergencies may require immediate disposal of combustible gases and vapors to relieve internal pressure. Since some of these combustible organic chemicals may contain toxic and carcinogenic substances, flaring from an elevated flare stack may not adequately destroy these organic compounds. The separation of light from heavy materials by cooling, condensing, and flaring the easily combustible light vapors while incinerating the heavy liquids may provide complete combustion of all hydrocarbons. A controlled combustion system can be utilized to incinerate heavy hydrocarbons.

Nitrogen Oxides

Nitrogen oxides are principally a product of fuel in process-fired heaters, boilers, the calciner, and control combustors. Their quantities can be reduced significantly by applying such combustion techniques as staged combustion, flue gas, recirculation, and the minimizing of excess air. Emissions of nitrogen oxides can be totally eliminated by utilizing process-derived fuel gas which contains no fuel-bound nitrogen. Thermal NO_x formation can be controlled by utilizing low-NO_x burner design and combustion techniques, which do not compromise fuel efficiency to the point where the amount of plant derived gas would be insufficient to meet heat requirements. When plant gas is unavailable, plant-derived fuel oil can be used in the heaters.

Although coal conversion plants produce many potentially toxic gases, the plants do not regularly emit these compounds. When operating properly, carbon dioxide and nitrogen comprise the principal fraction of atmospheric emissions. Since a large quantity of carbon monoxide is produced in the first stage gasifier, the escape of carbon monoxide is a possibility. For this reason, an accurate baseline of ambient carbon monoxide concentration is necessary. Special emphasis should be placed on the monitoring of trace elements, SO_2, NO_2, O_3, and the particulate size distribution of total suspended particulates (TSP).

The cooling tower is another potential source of atmospheric emissions which could result in public exposure to the biologically active chemical species and substances. These towers also are a potential source of atmospheric emissions. If organic and inorganic compounds invade the cooling water, they could be released to the air through codistillation, aerosol formation, or entrainment in drift. From a health stand point, the major concern involves the processing equipment which contains the heavy hydrocarbon (boiling point greater than 260°C). Usually this type of equipment includes a coal dissolver, hydrogen recovery unit, expanded-bed hydro-cracker, and a coker-calciner. Most units can be equipped with aerial coolers where hot process fluids are cooled by flowing through finned tubes with atmospheric air circulated around them by motor driven fans. driven fans.

WASTE WATER

The waste water disposal system of CCP-plants can be mitigated by a plant design which incorporates a sophisticated tertiary treatment process capable of removing most toxic metals and organic chemicals. Waste water treatment systems can collect and treat all process waste waters, process area runoff, runoff from product storage and shipping areas, leachates from coal and product storage areas, as well as the slag disposal and hazardous-waste landfill areas, and blowdown from the cooling towers and boilers. The system can be

designed for total water recycling and zero discharge, but should be flexible enough to treat and discharge all waste water when operating conditions do not allow recycling. Major spills occurring at the main process, storage, and shipping areas can be contained to prevent surface runoff. Residues from clean-up operations from such spills should be recycled whenever possible and contained at the plant site.

If waste water is discharged into the aquatic environment, a monitoring program must be instituted to assess the impact. This has to be emphasized both in the case of emissions monitoring in the waste water stream itself and monitoring within the river for bioaccumulation of specific chemical species with potential for adverse health effects. Although coal conversion plants will be design to meet existing effluent limitation standards, some water quality deterioration will be unavoidable. Effluents from CCP may ultimately reach major water course, with large capacities for pollutant dispersion and dilution. At present time, the exact constituent make-up of commerical scale plant effluent streams is not known. Table 1 depicts the worst-case, treated wastewater discharges from SRC-1 coal liquefaction demonstration plant.

SOLID WASTE DISPOSALS

Solid waste disposal from coal conversion processes is expected to have only a minor impact on the environment. In most locations, solid wastes from pilot plants can be disposed of on site, in compliance with the provisions of the Resource Conservation and Recovery Act. Organic and inorganic substances of concern can be contained on the site in a landfill designed to minimize transport into surface and subsurface soils. The landfill must be lined with low permeability clay underlaid with a plastic liner, with a leachate collection system above and below the double liner to collect any leachate that might escape. Most leachate chemicals can be recycled to the waste water treatment plant.

In coal conversion processes, the gasifier will be the major solid waste producing system. However, solid waste produced by the gasifier may not be toxic. Gasifier slag may contain a small quantity of benzene-soluble organic compounds which may be biologically active. However, leaching of these compounds from gasifier wastes may have a very minor environmental impact. Nevertheless, these compounds must be treated during processing as hazardous materials and should be disposed of accordingly.

Table 1.--Worst-case, treated wastewater discharges from SRC-1 coal
liquefaction plants

(100,000 bbl/d oil equivalent)

Parameters	Average concentrations (mg/liter)	Maximum concentrations (mg/liter)	Average discharge rate [(tons x 10^3)/year]	Maximum discharge [(tons x 10^3)/year]
BOD	20.0	40.0	457.0	914.0
COD	50.0	100.0	1142.0	2284.0
TSS	20.0	35.0	457.0	800.0
Oil and grease	10.0	20.0	228.0	456.0
NH_3 (as N)	5.0	20.0	114.0	456.0
Phenols	0.1	0.6	2.3	14.0
H_2S	0.04	0.09	0.9	2.0

Source: U.S. Department of Energy, Division of Coal Conversion (1979)

The method of disposal for such residuals depend on their physical and chemical characteristics as well as for their potential biological activities and the environmental mobility of leachate components within the waste material. Physical, chemical and biological data from unquenched gasifier bottom ash derivative has shown a low potential for a short-term but acute hazard to aquatic or terrestrial environments when proper landfill disposal procedures are not used.

The most of the produced solid wastes are in the form of bottom ash, fly ash, chars, and sulfur residues. Residual material (sludges) are also generated during the treatment of aqueous wastes. Most waste consists of organic compounds, which have questionable value and would ultimately be buried in landfills. Since, the inorganic content (ash) of most types of coal consists of about 10% of total weight, a conversion process producing 6000 tons of coal per day could be expected to have 600 tons/day of disposable material.

The fate of various chemicals present in coal during the conversion process is of concern since a large amount of some toxic chemicals would be generated in a large-scale coal conversion industry.

The adverse effects of leaching landfill on regional ground water systems occurs when the leachates reach the water table located below the landfill. Deeper aquifer water quality may also be affected by the leaching of chemical contaminants in water. Leaching processes depend mainly on the chemical and physical properties of soil and the chemical constituents of the receiving water. In certain localities, the low water storage capacity of soil coupled with high hydrolic conductivity allows rapid percolation of excess water. In contaminated water, some of the chemicals

are more soluble than are others. The potential exists for leaching large amounts of soluble chemical species toward the water table. Distribution of wells and springs with sources in the shallow ground water changes the flow path of contaminated surface and ground waters. Such alterations in the regional flow pattern could change the hydrologic system and high concentrations of toxic chemicals could accumulate in the water of deep aquifers.

The monitoring of surface and ground water quality, carried out in the vicinity of the disposal site would determine the impact of pollutant concentration levels. The programs can be designed to monitor the impact of leachates both at the point of maximum concentration and at significant receptors even though pollutants are dispersed before reaching the latter. The use of this type of monitoring helps to ensure compliance with water quality standards for criteria pollutants, particularly those potentially toxic chemical species, which must be closely watched in the vicinity of the disposal site. Dispersion modeling is a means by which the greatest exposure for surrounding unsaturated and saturated media can be predicted.

The chemical impact of the disposed material on ground water can be realistically determined by taking into consideration all soil parameters which could affect water infiltration and runoff. These include soil, structure, surface tension, forces which influence moisture movement, soil moisture content, soil permeability, soil minerology, vegetative cover, tillage effects, temperature of water and soil, entrapped air, sorption characteristics (physical, chemical, and biological), adsorption, absorption, ion exchange capacity, complex formation, dissolution, and topography. By combining leachability, transport, and dispersion data with information from site monitoring, the potential environmental impact of solid waste disposal areas can be evaluated.

POTENTIAL IMPACT FROM COAL MINING

Surface mining frequently has a significant impact on land and water resources. In the surface mining activities, the most immediate effect is a reduction of vegetation and animal population in the area. The restoration of surface mining areas depends on the physical and meteorological conditions as well as on the efforts invested by man. Reclamation of strip-mined land is certainly possible. Although underground mining has less impact on land and vegetation, solid wastes produced from underground mines can have impacts on land. However, by backfilling underground mines, the amount of solid wastes and the associated acidic drainage can be decreased if proper sealants are used.

The quality of surface and ground water can be degraded during mining activities by acid mine drainage (AMD) and siltation. AMD results when sulfuric acid forms from the oxidation of pyrite and marcasite coal inclusions in the presence of bacteria, oxygen, and water. In some cases, AMD waters contain high levels of iron, manganese, sulfate ions, and many trace elements.

Potential Impacts To Surface And Ground Water

Surface mining can change the land's drainage pattern to the point that the flow of receiving streams is significantly reduced. In turn, flow reduction can affect aquatic life. The erosion of exposed spoils by wind and rain contributes greatly to stream siltation. Streams in the vicinity of mining activities usually receive high concentrations of trace elements leached from freshly exposed spoils. As the mined land is reclaimed and vegetation cover is established, the damage resulting from erosion and spoil leaching diminishes and the stress on the aquatic life in the receiving stream ceases.

The adverse effect of site specific leaching and strip mines on regional ground water systems occur during the mining operation when the shallow ground water is drawn down by dewatering operations at the strip mine site. Generally, dewatering is achieved through some combination of well points trenches, and sumps that locally lower the water table below the coal bed. This distruption of wells and springs with sources in the shallow ground changes the path of the contaminated surface and groundwaters. In the process, deeper aquifer water quality is also effected by leaching of the inorganic and organic chemical contaminants in water. Hence, alteration of the regional flow pattern may change the hydrologic system and accumulates high concentrations of toxic chemical compounds in water of deep aquifers.

CONCLUSION

Many of the potential environmental impacts associated with CCP can be mitigated with the careful use of available technology. Monitoring and control technology can fulfill the needs to prevent occupational and environmental health impacts that may be attendant to major CCP plants. Pollutants product end-use may involve environmental and occupational exposure. Generally, occupational exposure may result primarily through stain contact while environmental exposure may be due to combustion emissions (assuming that combustion is the only end-use for the plant product). The chemical composition of effluents from coal conversion processes are not yet fully classified. However, pilot plant monitoring programs presently being implemented for these facilities are an attempt to establish a reliable base line for future emissions.

The development of any technology involved in chemical processing or organic and/or inorganic emission in conversion processes requires serious investigation of the toxicological and environmental impacts of that particular technology. This concept is also true for coal-conversion processes because of the diverse nature of the raw material used. A major effort should be made, before technological developments proceed beyond the demonstration-plant stage, to identify and eliminate potential hazardous materials that may pose a potential health hazard. The expansion of world coal production and coal conversion processes projected by the World Coal Study (Vilson 1980), predicts that each country will need to consider the resulting environmental impact issues. In recent years, extensive experience has been gained with the mining, transportation, and use of coal and the application of environmental controls in the industrialized countries. Also, knowledge about environmental control strategies and technology improved significantly in the last several years. Many industrialized countries have adopted detailed legislative and regulatory systems or other less formal systems, for controlling the environmental effects of coal production and use. However, the magnitude of long-term environmental effects of coal combustion emissions on global climate and the environmental impacts posed by coal conversion plants, have not yet been determined.

Most environmental risk factors emanating from coal conversion processes are amenable to technological control. Emission release and other occupational and environmental effects can usually be reduced to acceptable levels by applying current available control techniques. In the environmental control processes, each increment of reduction increases cost and as total control is approached, these costs become very high. Within what can be expected of ambient air standards, the coal can be transported, produced, and converted to liquid and gaseous states at costs competitive with those of other fuels. Environmental concerns or control costs will probably preclude the development of certain coal resources and/or sites. However, possible sites and resources remaining in this country and worldwide are so many that such exclusions should not be a limiting factor for expanding the utilization and conversion processes of coal.

National perception of values differ on such things as exposure of the general population to health risks or visibility reduction in air environments. For example, the possible magnitude of environmental impact and health effects of emissions from coal conversion processes has aroused considerable controversy. In general, environmental impacts significantly differ between possible plant sites because regional characteristics such as topography, meteorology, population density, and resource distribution. Therefore, different nations take different positions on the kind and extent of environmental control measures to be required as the use of coal in conjunction with conversion processes increase. Recent studies (Vilson 1980) indicate that most nations expect to apply measures that will ensure compliance with their own particular national environmental standards. On the other hand, some environmental issues will doubtlessly require joint action among the nations involved in the expansion of coal use and conversion processes.

Agreement on the application of existing control technology in the interest of neigboring nations despite individual national interests may be difficult but necessary. For example, long-range transport, distribution, and disposition of acid rain in several countries is receiving increased attention and may require joint action. Also, improved understanding of the effects of pollutants on air and aquatic environments may lead to continuing international cooperation. The need for understanding, integration and coordination of several ecological actions at global, regional, national and local levels is becoming increasingly apparent.

Because technical solutions for controlling carbon dioxide (CO_2) emissions are costly and because large increases in the amount of atmospheric CO_2 may alter global climate, CO_2 emissions are one of the most important environmental problems resulting from the combustion of all carbon fuels, including oil, gas, coal, and wood. In spite of its low concentration in the atmosphere, CO_2 does influence atmospheric temperature. Carbon dioxide, while transparent to sunlight, absorbs infrared radiation emitted from the earth's surface and reradiates part of it, thereby reducing the rate of surface cooling. As a result, any increase in atmospheric CO_2 contributes to a rise in the earth's temperature referred to as the greenhouse effect. Currently, little is known about CO_2 inputs from various sources, the absorption of CO_2 by various sinks, and the consequences of rising CO_2 concentration in the atmosphere. If the effects prove as serious as some investigators predict, the resulting situation would call for extraordinary international cooperation to control world fuel combustion and/or the use of alternative energy sources.

A report issued recently by the Department of Energy outlines a broad range of needed research on the impact of increasing concentrations of atmospheric CO_2 and calls for a multidisciplinary, international effort on the problem. The significance of the problem lies in the fact that the impacts would probably be incremental over a long period of time.

LITERATURE CITED

Battelle Pacific Northwest Laboratory, 1979. Biomedical studies on solvent refined coal (SRC-II) liquefaction materials: A status report, PNL-3183, October. Richland, WA.

Cowser, K.E. 1980. Synthetic fossil fuel technology-potential health and environmental effects. p.288. Ann Arbor Science Publishers, Inc., Ann Arbor, Michigan.

U.S. Department of Energy, Division of Coal Conversion, 1979. Demonstration plants quarterly report, April-June 1978, DOE/ET 0068/2.

Vilson, C.L. 1980. Coal-bridge to the future. Ballinger Publishing Company Cambridge Massachusetts.

MODELS OF WETLANDS AMID SURFACE COAL MINING REGIONS OF WESTERN KENTUCKY[1]

by

William J. Mitsch, Robert W. Bosserman, Paul L. Hill, Jan R. Taylor and Francie Smith[2]

Abstract.--Conceptual models are developed in a hierarchical fashion for management of wetlands in the Western Kentucky coal field. The models are developed for regional, watershed and ecosystem levels. Three specific wetland sites are chosen for ecosystem models. Each model emphasizes pathways that are affected by major environmental impacts such as coal mining, logging, agricultural development and drainage.

INTRODUCTION

Conceptual and diagrammatic models can be useful tools for the identification and summary of environmental impact on ecosystems (Odum 1972; Hall and Day 1977; Farnworth et al. 1979). Even if they are not developed further into mathematical simulation models, these models 1) serve as guides for identifying critical ecological processes, 2) focus subsequent field and laboratory research on pertinent measurements, and 3) summarize environmental impacts on one or a few pages. This paper deals with conceptual models that are being developed as part of a study entitled "Wetland Identification and Management Criteria for the Kentucky Western Coal Field." Our emphasis is on defining the function of wetland ecosystems in the western coal field of Kentucky in a hierarchical fashion with conceptual models and on identifying the impacts of surface coal mining and other management options on the structure and function of these wetlands.

[1]Paper presented at the international symposium Energy and Ecological Modelling, sponsored by the International Society for Ecological Modelling. (Louisville, Kentucky, April 20-23, 1981)

[2]William J. Mitsch is Associate Professor of Systems Science and Environmental Engineering, Robert W. Bosserman is Assistant Professor of Systems Science and Biology, Paul L. Hill, Jan R. Taylor and Francie Smith are Graduate Research Assistants. All at University of Louisville, Louisville, Kentucky, U.S.A.

Western Kentucky Coal Field

Surface mining of coal is expected to increase in western Kentucky and elsewhere as synthetic fuel plants and electric power plants are built and conversion to coal burning by industry is encouraged. The western coal field of Kentucky (fig. 1), one of two major coal fields in the state, is part of the Eastern Interior Coal Basin that extends through southwestern Indiana and much of Illinois.

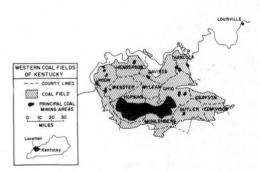

Figure 1.--Map of Western Kentucky Coal Field with Intensively Mined Areas.

The field comprises about 12,000 square kilometers and has experienced heavy strip mining since the technique was first introduced in Muhlenberg County in 1829. In 1979, the total amount of coal mined in western Kentucky was 44 million metric tons, with about 23 million metric tons obtained through surface mining (Kentucky Department of Mines and Minerals 1979). Total coal mined in 1979 was down about 20 percent from 1976. However, several synthetic fuel plants have recently been proposed for four locations on the Ohio River adjacent to the Kentucky western coal fields (Louisville Courier-Journal, February 1, 1981). These plants, if built, could increase the annual coal use in western Kentucky by 35 million metric tons per year with much of this coal coming from the Western Kentucky Coal Field.

Wetlands in Western Kentucky Coal Field

The Western Kentucky Coal Field is located in the Shawnee Hills Section of the Interior Low Plateaus Physiographic Province. Most of the region is undulating to hilly with wide, flat, silt-filled, alluvial valleys along the Green, Rough, Pond, Tradewater, Barren, and Ohio Rivers (Harker et al. 1980). The relatively flat topography supports a wide variety of wetlands. We have tentatively identified about 40,000 hectares of wetlands in the Western Kentucky Coal Field. Many of these wetlands are over 2,000 hectares in size.

Coal mining, particularly coal surface mining, have affected many of the wetlands in the region. Acid drainage and sedimentation from mining has adversely impacted the water quality, hydroperiod, and vegetation of some wetlands. Other wetlands have been formed as a direct result of mining activity and thus display characteristics of developing ecosystems. Several wetlands in the region are seemingly not influenced by mining at all. Logging, channelization, conversion to agriculture, and highway construction are other major influences on wetlands in the region.

Little published information has existed until recently on wetlands in the Western Kentucky Coal Field. Harker et al. (1980) compiled a notable floristic and faunistic survey of many of the wetland areas in the coal field. Aerial photo and LANDSAT imagery were used by Whinnery (1977) to map Clear Creek Swamp and nearby strip mining areas. Water quality and biotic parameters were examined in a marsh adjacent to Clear Creek Swamp (Neichter 1972; Leuthart 1975). Grubb and Ryder (1972) described the effects of acid drainage on the water resources of the Tradewater River Basin.

Intensive Study Areas

Three major wetland areas in the Western Coal Field were chosen for detailed study because of significant or unique features. Much of the following information comes from Harker et al. (1980).

Clear Creek Swamp

This wetland, over 3400 hectares in size, is the largest extant wetland system in the Western Kentucky Coal Field. It is located in Hopkins County west of Madisonville. The wetland has been severely affected by a combination of logging, altered hydroperiod and, notably, surface mining of coal. Major wetland communities include young forest thickets of Betula nigra (river birch) and Acer rubrum (red maple), open water communities interspersed with Cephalanthus occidentalis (buttonbush), and stands of Liquidambar styraciflua (sweetgum).

Cypress Creek Wetlands

These riparian wetlands are found along 15 kilometers of Little Cypress and Cypress Creeks in Muhlenberg County. A major feature is the presence of Taxodium distichum (bald cypress) communities. The cypress stands are the most extensive and best developed in the interior of western Kentucky and represent some of the eastern-most cypress in the Mississippi Basin. Swamp forest communities of Betula nigra and Acer rubrum and marshes of Typha sp. (cattail) are also found in this study area. There are signs of mine drainage although channelization and levee construction on all of the major streams appear to have caused the most impact. Cypress Creek wetlands are an important waterfowl area, especially for mallards and wood ducks.

Henderson Sloughs

These wetlands are located along the Ohio River in Henderson and Union Counties and are annually flooded by the river. These wetlands are elongated sloughs divided by low ridges and roughly parallel to the direction of the river. Much of the region has been drained for agriculture and significant environmental impact may result from oil drilling in the region. No coal mining impact is apparent. The major community found in the seasonally-inundated sloughs are dominated by Acer saccharinum (silver maple) and Betula nigra. This wetland is one of only two Kentucky wetlands cited by Goodwin and Niering (1975) in their survey of wetlands in the United States.

MODELLING

The Hierarchical Framework

Modern techniques for the management and evaluation of wetlands require that decisions be made using disparate variables at different levels of organization. Such variables can best be dealt with if they are arranged so they interact within the same spatial and temporal context. In order to accomplish this organizational task effectively, hierarchical models are often useful. The upper level of such a hierarchical model involves variables that change over a broad spatial extent and a long time period while the lower levels involve more resolved variables that change over a smaller spatial and a quicker time scale. For example, while the variables that describe an ecosystem can vary over several hectares on a week by week basis, the variables that describe an organism can vary over several square meters on a minute by minute basis. Generally there is more than one way to model a hierarchical system; however, during research activities the questions asked by observers provide constraints for model development. Hypotheses about system behavior can be developed in conjunction with the hierarchical model and then can be tested with an appropriate experimental design (O'Neill 1975). In this research project a hierarchical model provides a framework by which management strategies for particular wetlands and their most important impacts can be developed and assessed. Techniques will then be developed which provide an integrated management strategy that incorporates each level of organization.

Within our hierarchical framework, several levels of organization have been identified as being important to wetland assessment. We have decided to use three levels of organization as shown in figure 2: regional, watershed and ecosystem. The upper, most-inclusive level that we identified is the regional level which encompasses the entire Western coal field. Among variables which are relevant to wetland management at this level are the location, number and area of wetlands and other land categories. Man-induced impacts such as agriculture, mining, and lumbering tend to cause changes in these variables. In our conceptual model of the regional level, land categories in the Western Kentucky coal region are the compartments and transfers of area between compartments are the interactions.

At the second level of the hierarchy, the entire region is regarded as being partitioned into the watersheds which make up the region and contain the wetlands. Watersheds have been identified as being an important unit in ecological studies because they are well defined and greatly affect inputs to aquatic ecosystems. A watershed is a hydrological unit which is identified by the topography of the region; therefore, water movement, is closely related to the relevant variables of interest. Relevant variables include surface runoff, stream flow, water level, and flow volume. Because many inorganic and organic materials are carried in the water, they are closely tied to watershed characteristics and can be examined at this level. Chemical and

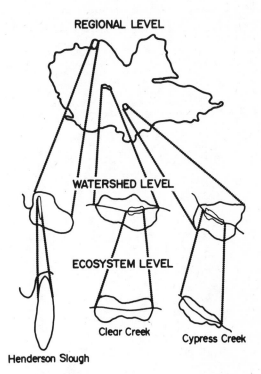

REGIONAL LEVEL

WATERSHED LEVEL

ECOSYSTEM LEVEL

Clear Creek

Cypress Creek

Henderson Slough

Figure 2.--Diagram of Hierarchical Framework for This Study.

hydrologic budgets can be constructed for each study site in order to properly characterize wetland behavior at this level of organization. Impacts at this level may include levee construction, channelization, and land use in the uplands. Many of these impacts affect the inputs and outputs of wetlands and thereby help to determine wetland characteristics.

Specific wetland ecosystems occur at the lowest level of our hierarchical framework. The variables of interest that we have chosen to examine are community structure and the energy and nutrient flows among various ecosystem components. Impacts include the effects of acid mine drainage on primary production and other ecosystem processes. Each of the intensive study sites has a significantly different community structure and is affected by different impacts; therefore they are modelled separately.

Conceptual Modelling Languages

Conceptual models can be depicted by several modelling languages. Each of these languages is appropriate for describing different aspects of system organization and behavior. Energy and matter storages and flows are often represented with Energese, the energy circuit language of H. T. Odum (1971)and Odum and Odum (1976). On the other hand, material storages and flows and changes in land use status can be represented with the DYNAMO language of J. W. Forrester (1961). The symbols used in this paper are shown in figure 3 where the analogous Energese and DYNAMO symbols are presented side by side. In Energese, each

are compatible with the hierarchical framework that we have developed. In DYNAMO, state variables which represent storages are shown as boxes.

CONCEPTUAL MODELS OF WESTERN KENTUCKY WETLANDS

Regional

On the regional level a conceptual model was developed to describe transfer of land area between categories of land use (fig. 4).

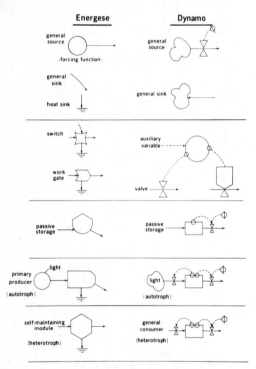

Figure 3.--Major Symbols Used in This Study Including Energy Symbols of Odum (1971) and DYNAMO Symbols of Forrester (1961).

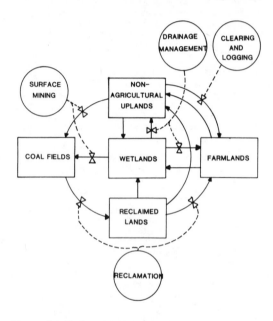

Figure 4.--Regional Model of Western Kentucky Coal Field.

energy module represents a dominant mode of processing energy such as autotrophy, heterotrophy, or passive storage. Solid lines demonstrate energy flows between model components, and information interactions. Circles depict factors such as energy inputs, perturbations, or management efforts which originate outside the boundary of the system. Effects of external factors such as mining, logging, and agriculture are represented from the circles to work gates on the various energy flows. Components of the energy circuit language are hierarchical because each module can be broken into smaller modules and can be contained within a larger module; therefore, they

The main land use divisions found in the Western Kentucky Coal Fields are 1) non-agricultural upland ecosystems, 2) wetland ecosystems, 3) farmland, 4) stripped coal fields and 5) reclaimed land. Non-agricultural uplands are defined as land with predominantly mesophytic or xerophytic vegetation, and with mainly nonhydric soil that is not flooded or saturated at some time each year. Stripped coal fields include active mines and abandoned mines that have not yet been reclaimed.

The major management impacts that affect the region are mining operations, logging and clearing activity, reclamation operations and drainage manipulation. These impacts work in a unidirectional manner causing wetlands to become non-agricultural uplands, farmlands, or coal fields; coal fields to become reclaimed land;

reclaimed land to become farmland; and non-agricultural uplands to become coal fields or farmlands. Any flows opposing these are due to natural ecosystem succession (i.e., farmlands to non-agricultural uplands) and are not directly controlled by management activities.

Watershed

Water flow through a watershed of the Western Kentucky Coal Field is described by a conceptual hydrologic model (fig. 5). The main

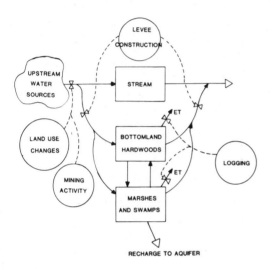

Figure 5.—Watershed Model of Western Kentucky Showing Major Hydrologic Compartments.

compartments through which water moves are the stream, the bottomland hardwood forests, and the marshes and swamps.

Streams include flowing, open water within a channel. The stream systems excludes trees, shrubs and emergent vegetation. Bottomland hardwood forests consist of areas that are flooded or saturated with water at some time of year. The vegetation can be any number of flood-tolerant hardwoods. Marshes and swamps are characterized by emergent hydrophytes and standing water for most of the growing season (Cowardin et al. 1979).

The major impacts on water flow are channelization and land use, mining activity, logging activity and levee building. Mining activity can

affect water flow by actual disruption of the terrain or by increasing sediment loads.

Ecosystems

Clear Creek

Energy flow through a coal mine impacted wetland is shown in a conceptual model of Clear Creek Swamp in Western Kentucky in figure 6. The

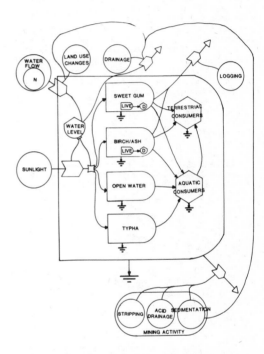

Figure 6.—Ecosystem Model of Clear Creek Swamp, Hopkins County, Kentucky.

main autotrophic communities found in the Clear Creek system are a sweet gum community, birch/ash thicket, open water and Typha marsh.

The sweet gum communities are dominated by mature monospecific stands of Liquidambar styraciflua (sweet gum). There are three associations within the birch/ash thicket community. One association is predominantly Betula nigra (river birch) and Acer rubrum (red maple). The second association is characterized by Fraxinus pennsylvanica (green ash) and Acer rubrum. The third distinct association is composed of Liquidambar styraciflua and Acer rubrum. Open water areas are edged by the sedges Eleocharis quadrangulata (spike rush) and Rhynchospora corniculata. The dominant species of cattail found in the marsh areas is <u>Typha latifolia</u>.

The main forcing functions acting on the system are water flow (including nutrients) and sunlight. Mining, drainage management and logging activities are the major impacts on the system. Mining can be divided into three types of impact: sedimentation, acid mine drainage and removal of ecosystem by strip mining.

Upstream land use changes also affect the flow of water and nutrients through the Clear Creek system. Inflow of water, which controls the water level of the wetland, affects the proportion of the system that makes up the various communities. This is shown in the diagram through a switching function. A successional change due to water level change is important in the flow of energy through the system. A storage of dead trees is shown in both the sweet gum and birch/ash communities due to the longevity of dead standing timber when water levels rise and kill the trees.

Cypress Creek

A conceptual model of the Cypress Creek ecosystem (fig. 7) demonstrates the effects of upstream mining activity on energy flow through the system. The communities found in this area are <u>Typha</u> marsh, scrub cypress, bottomland hardwood and two creeks: Cypress and Little Cypress. This wetland area is considered a high quality area which support populations of restricted or declining wildlife species (Harker <u>et al.</u> 1980).

The main canopy trees of the scrub cypress community are <u>Taxodium distichum</u> (bald cypress) and <u>Fraxinus profunda</u> (red ash). The bottomland hardwood association, where flooding is only for a short period, is predominantly <u>Acer rubrum</u>, <u>Betula nigra</u> and <u>Cephalanthus occidentalis</u> (buttonbush).

Major impacts on this wetland are channelization, agricultural activities and mining operations. Both Cypress Creek and Little Cypress Creek have been channelized. Mining operations occur upstream from this system, causing some acid and sediment additions to the systems. Cypress Creek may act as a biological filter that damps the effect of the upstream mining activity. Direct effects on the wetland has been through clearing for farmland and channelization of the streams.

Henderson Sloughs

The conceptual model of Henderson Sloughs (fig. 8) shows the influence of the annual flooding of the Ohio River on riparian wetlands.

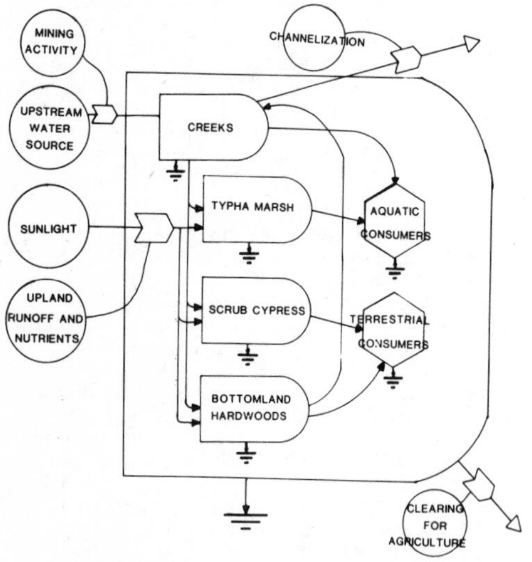

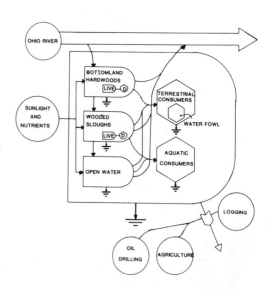

Figure 7.--Ecosystem Model of Cypress Creek Wetlands, Muhlenberg County, Kentucky.

Figure 8.--Ecosystem Model of Henderson Sloughs, Henderson and Union Counties, Kentucky.

The flooding river contributes nutrients and organics to the elongated sloughs. The undulating terrain is flooded annually with the Ohio River spring overflow. This typical river-bottom hardwood habitat consists of bottomland hardwood forests, wooded sloughs and open water areas. The bottomland hardwood forest of slightly raised elevation is primarily composed of oak-hickory (Quercus sp., Carya ovata) and red maple-ash (Acer rubrum, Fraxinus pennsylvanica, F. americana) associations (Goodwin and Niering 1975; Quarterman and Powell 1978). In the wooded sloughs, Acer saccharinum (silver maple), Betula nigra (river birch), Carya illinoensis (pecan) and Liquidambar styraciflua (sweet gum) are the dominant species. Both of these areas contain both a living population and a standing dead crop of the above species. The open water areas may potentially support rare Kentucky plant species. The rare species are Decodon verticillatus (swamp loosestrife), Echinodorus rostratus (burhead), Pontederia cordata (pickerel weed) and Utricularia gibba (humped bladderwort) (Harker et al. 1980).

Henderson Sloughs are valuable wildlife areas, supporting a variety of terrestrial and aquatic animals. Goodwin and Niering (1975) noted the importance of the sloughs as a late winter, prebreeding season conditioning area for ducks and geese. Foxes, gray squirrels, and raccoons were reported as common. Rare animal species in the sloughs are the great blue heron and the masked shrew.

Encroachments on the area consist primarily of oil drilling operations, agricultural activities and timber harvesting. Oil operations cause chemical and mechanical damages to the area and wildlife through spills and overflows of crude and brine (Goodwin and Niering 1975). Public-funded agricultural practices and subsidies encourage drainage and clearing. Timber harvesting also has an impact on the area.

Ecosystem Impact Summary

At the ecosystem level, three distinct wetland areas were designated for models and intensive study. Clear Creek, Cypress Creek and Henderson Sloughs all have unique ecological communities and different hydrologic conditions. The energy flows in each of these systems are affected by different impacts. The impacts on each wetland ecosytem are summarized in table 1.

DISCUSSION

The Hierarchical Approach and Wetland Management

Management techniques and evaluation procedures are being developed that will correspond to the levels of organization identified in our hierarchical model (fig. 2). Careful coordination of these techniques is required in order to achieve a harmonious balance between natural and human needs.

Regional

At the regional level, which encompasses the entire western coal field, impacts involve the destruction or creation of wetlands through human activities. Managing the coal region should involve control of such variables as number, location, areal extent, and types of wetlands in the region. Certain wetland types are valuable

Table 1.--Summary of major impacts on wetland study sites in the Western Kentucky Coal Field

Wetland	Stress						
	Coal Mining			Agriculture	Timber Operations	Oil Operations	Drainage
	Acid Drainage	Sedimentation	Stripping				
Clear Creek	X	X	X		X		X
Cypress Creek	X	X		X			X
Henderson Sloughs				X	X	X	

components of the regional landscape because of recreational use, wildlife production, flood protection, water quality control, and lumbering. Other wetland types may contribute little to environmental quality of the region. The values of the various wetland types must be assessed before their management can be accomplished. A number of techniques have been developed for evaluating wetlands (Wharton 1970; Gosselink et al. 1974; Mitsch et al. 1979; Schamberger et al. 1980). These methods will be adopted or new techniques will be developed in order to properly assess the western Kentucky wetlands. The values of different wetland types can be incorporated into a scheme which examines the entire mosaic of ecosystem types in the region. Even if a particular wetland type is valuable in terms of wildlife production, an excess of such acreage can lower its value by adversely affecting the development of other land uses. Maintenance of an optimal acreage and distribution of each wetland type is a useful goal to be pursued at this level of organization. Maps of the coal field and conceptual models such as shown in figure 4 will be useful for generating such management strategies.

As with other biological phenomena, value with respect to the amount of total acreage of a particular wetland type in a region can best be described with a hump-backed curve (fig. 9). The value of a wetland type tends to increase with total acreage until an optimum is reached and then the value tends to decrease. An area which is too small provides suboptimal conditions for maintenance of wildlife diversity and production, recreation, and timber production (Odum 1977; Smith 1980). On the other hand, if the total area is too large, that wetland category competes with other useful land categories. Each land use type can be represented by a similar Value/Total Acreage curve. A strategy of management at the regional level would be to obtain an optimal mosaic of all land types, including various wetland categories. A curve which describes the value of a wetland category with a broad peak would be desirable in order to maximize flexibility in decision-making and to accommodate the diverse opinions of interest groups. An assessment of the wetland value in terms of desired acreage would be useful and should involve various government agencies, coal companies, researchers, and community leaders.

Another management strategy which can be pursued at the regional level is to reclaim wetlands from mined areas. Wetlands tend to develop in any location where a depression has been created in the ground and where water can accumulate. Indeed, many wetlands in the Western Kentucky coal fields were originally established by man's activities in mining, highway construction, and water control. By applying simple reclamation techniques to mined lands, wetlands can be generated in an effective and economical fashion. Such techniques would include the establishment of depressions and holding ponds,

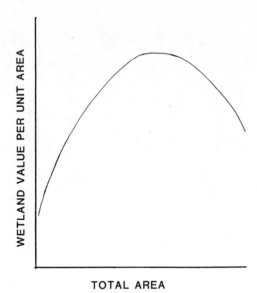

Figure 9.--Possible Relationship Between Wetland Value and Total Acreage of Wetland.

introduction of wetland species, modification of drainage patterns through channelization and levee construction, and maintenance of wetland/ upland ecotones. These techniques would probably be more economical than traditional techniques of restoring the land to original contours and revegetating with a few species. A diverse mosaic of reclaimed ecosystem types would be more stable and more valuable than the undiverse ecosystems created by present reclamation techniques.

Watershed

Management techniques at the watershed level involve the modification of runoff from surrounding lands and mined areas, of rivers associated with wetlands, and of water levels and water flow rates in the wetlands. Runoff from the surrounding watershed is a major source of water, nutrients, sediments, and organic materials for many wetlands. The magnitude of these sources affects productivity and diversity in aquatic situations. Sediments, for example, increase turbidity and abrasion and can thereby decrease primary and secondary production. Nutrients, on the other hand, tend to increase production and successional rates but decrease the diversity of species. Impacts of coal mining, lumbering, and highway construction are varied and include changes in sediment load and type, elemental concentrations, water level and water flow characteristics. Acid mine drainage, a common problem in mining areas has deleterious effects on many organisms and processes in wetlands. Many of these effects can be reduced by diverting water to holding ponds where it can be held while sedimentation and biotic processes reduce the sediment load, and harmful elemental concentrations. Such techniques are important if the potentially affected wetland is a valuable wildlife management area which is sensitive to this impact. However, rather than diverting water from wetlands, some wetlands can be used and maintained as interface ecosystems between man and other aquatic ecosystems in order to improve water quality before it goes further downstream (Mitsch 1977). Wetlands tend to slow water movement, reduce sediment load, take up nutrients and in other ways reduce the magnitude of effects that are transmitted through hydrological processes. The location of wetlands in relation to disturbed areas is important in considering their use in this regard. Again, evaluation of wetlands must be made in order to assess which wetland should be protected from particular impacts, and which ones should be used to buffer impact. The conceptual model used in this paper (fig. 5) will be useful in making such an evaluation.

Ecosystem

At the ecosystem level, the various impacts affect flows between ecosystem components (figs. 6-8). Certain components of a wetland ecosystem are more tolerant to such impacts than others. Organisms which are sensitive may be eliminated or substantially reduced, while other may become more abundant because of increased nutrients, reduced competition, or reduced predation. The sensitivities of various wetland components and processes must be assessed in order to make decisions about the management of wetlands at this level. Those organisms which are most sensitive and valuable should be protected, while those which are insensitive and invaluable may not need protection. Wetlands which are insensitive to impacts can be established and located in order to buffer the impact of activities. To an extent, many of these wetlands are self-designing and require little management; however, plants and animals can be introduced which tolerate the modified conditions in impacted marshes. Typha, the cattail, and other aquatic plants can be introduced successfully into a wetland. Other reclamation techniques which should encourage colonization of organisms are the addition of treated sewage wastewater and fertilizers or lime to wetlands and other reclaimed areas or the creation of corridors through which immigrating organisms can pass. Addition of sewage materials would stimulate productivity, alter successional rates, and indirectly reduce effects of acid mine drainage; it would also provide a use for some of the municipal wastes that are accumulating in Kentucky's cities. Sewage recycling with wetlands has already been successfully demonstrated in Florida (Odum et al. 1975) and Michigan (Kadlec and Tilton 1979).

Management strategies should also include pathways and rates of succession in wetlands. Ecosystems are not static objects but will eventually change into another ecosystem type through successional processes. By identifying the successional sequences and the factors affecting them, management strategies can be adopted which accelerate, or set back successional states in order to maintain the most valuable types of habitat and species composition.

Models as Aids for Data Collection

Models such as those developed here can be used as guides to data gathering and hypotheses generation for the three levels chosen. Models can identify key variables and pathways for measurement and can help prevent a common malaise in ecological modelling--trying to measure everything.

Regional Data Collection

The regional model (fig. 4) serves as our guide for the gathering of data on land-use categories in the coal field. We have obtained recent (1979-80) color aerial photography at 1:24,000 scale of the Western Kentucky Coal Field from the Kentucky Nature Preserves Commission and similar scale black and white photography from the early 1950's from the Kentucky Department of Transportation. These maps will enable us to determine the cumulative effects of coal mining and other activities on wetlands over more than 25 years. These changes are represented by the pathways in the model shown in figure 4.

Watershed Data Collection

The watershed model (fig. 5) serves as a guide for data collection on hydrologic conditions of selected wetlands. Streamflow, where applicable, and soil and standing water conditions in selected wetlands will be collected routinely. Climatalogical data from nearby weather stations will be used to determine several other components of the hydrologic cycle. Flooding frequency and duration important variables for specific wetlands, can be determined where hydrologic data have been collected for several years. There are a number of U.S.G.S. streamflow gauges in various watersheds in the study region. As with the other level, our data watershed studies will emphasize easily obtainable data.

Ecosystem Data Collection

Our ecosystem models have identified those pathways that are most susceptible to impacts by coal mining and other management strategies. In figure 6 of Clear Creek Swamp, water level controls the selection of the ecosystems that develop in the wetland. The water level, in turn, is controlled by mining activities and other upstream land use changes. Our literature surveys and possible field and laboratory measurements will attempt to elucidate the threshholds of water level that cause a succession from one plant community to another.

The three models of ecosystems, drawn in similar fashion, will also facilitate comparative ecosystem studies on functions such as primary productivity and nutrient cycling. In figures 6 through 8, the models have been drawn with similar functions located in similar parts of the diagram. This further facilitates the comparative ecosystem approach to data collection.

In our study, our ultimate goal is to ascertain the implications of energy development and other management practices on the health of selected wetland ecosystems. We will also attempt to determine watershed and regional scale measurements in support of the "next-larger-system" approach of Odum (1977). Some of the critical measurements anticipated in our studies at the ecosystem and watershed levels are summarized in table 2.

Table 2.--Ecosystem, Watershed and Regional Measurements Anticipated for Wetlands in Western Kentucky.

Scale	Measurement	
	Function	Structure
Ecosystem	Primary Productivity Hydrology Decomposition Nutrient Cycling	Biomass Water Chemistry Sediments
Watershed	Steamflow Flooding Frequency Precipitation/ Evapotranspiration	Groundwater Water Levels Flooding Duration
Regional	Change in Land Use	Land Use

SUMMARY

In order to successfully manage the wetlands of the Western Kentucky Coal Fields, strategies that are developed at each level of organization must be integrated into a comprehensive scheme. Establishment or use of a particular wetland should take into account the number of wetlands in the regional landscape and also the watershed in which it occurs. Such integration must involve several government agencies, coal mining companies, and community groups. Values of various land categories must be assessed in order to provide the necessary information required for decisions about wetlands. Our study will not only continue to provide techniques for evaluating wetlands but will also suggest management strategies which will increase the value of wetlands and the surrounding environment.

LITERATURE CITED

Cowardain, L.M., V.G. Carter, F.C. Golek, and E.T. Laroe. 1979. Classification of wetlands and deepwater habitats of the United States. U.S. Fish and Wildlife Service 79/31, 103 pp. Washington, D.C.

Farnworth, E.G., M.C. Nichols, C.N. Vann, L.G. Wolfson, R.W. Bosserman, P.R. Hendrix, F.B. Golley, and J.L. Cooley. 1979. Impacts of sediments and nutrients on biota in surface waters of the United States. USEPA, Environmental Research Laboratory, EPA 600/3-79-105, 315 p. Athens, Ga.

Forrester, J.W. 1961. Industrial dynamics. MIT Press, Cambridge, Mass.

Goodwin, R.H., and W.A. Niering. 1975. Inland wetlands of the United Stated. National Park Service National History Theme Studies, No. 2. 550 p. National Park Service Washington, D.C.

Gosselink, J.G., E.P. Odum, and R.M. Pope. 1974. The value of tidal marsh. 20 p. Center for Wetlands, Louisiana State University, Baton Rouge.

Grubb, H.F., and P.D. Ryder. 1972. Effects of coal mining on the water resources of the Tradewater River Basin, Kentucky. Geological Survey Water-Supply Paper 1940, Kentucky Geological Survey, Lexington, Ky.

Hall, C.A.S., and J.W. Day, Jr. 1977. Ecosystem modeling in theory and practice. 673 p. John Wiley and Sons, New York, N.Y.

Harker, D.F., R.R. Hannan, M.L. Warren, L.R. Phillippe, K.E. Camburn, R.S. Caldwell, S.M. Call, G.J. Fallo, and D. Van Norman. 1980. Western Kentucky coal field: preliminary investigations of natural features and cultural resources. Kentucky Nature Preserves Commission. Frankfort, Ky.

Kadlec, R.H., and D.L. Tilton. 1979. The utilization of a freshwater wetland for nutrient removal from secondarily treated wastewater effluent. J. Environ. Qual. 8:328-334.

Kentucky Department of Mines and Minerals. 1979. Annual Report. Lexington, Ky.

Leuthart, C.A. 1975. Reclamation of orphan strip mined land in southern Illinois and western Kentucky. Doctoral Dissertation, University of Louisville, Louisville, Ky.

Mitsch, W.J. 1977. Energy conservation through interface ecosystems p. 875-881. In Proceedings of International Conference on Energy Use Management. [Tucson, Ariz., October 18-24, 1977] Pergamon Press.

Mitsch, W.J., M.D. Hutchinson, and G.A. Paulson. 1979. The Momence wetlands of the Kankakee River in Illinois-an assessment of their value. Illinois Institute of Natural Resources Document No. 79/17. 55 p. I.I.N.R. Chicago, Ill.

Neichter, P.L. 1972. Water quality survey of the Weirs Creek levee system, Kentucky. Master's Thesis. 66 p. University of Louisville, Louisville, Ky.

Odum, E.P. 1977 The emergence of ecology as a new integrative discipline. Science 195:1289-1293.

Odum, H.T. 1971. Environment, power and society. Wiley-Interscience, Inc., New York, N.Y.

Odum, H.T. 1972. Use of energy diagrams for environmental impact statements. Tools for Coastal Management, Proceedings of the Conference, Feb. 14-19, 1972, Marine Technology Society, Washington, D.C.

Odum, H.T., K.C. Ewel, W.J. Mitsch, and J.W. Ordway. 1975. Recycling treated sewage through cypress wetlands in Florida. p.35-

67. In Wastewate renovation and reuse. F.M. D'Itri (ed.). Marcel Dekker, Inc., New York, N.Y.

Odum, H.T., and E.C. Odum. 1976. Energy basis for man and nature. 296 p. McGraw Hill, New York, N.Y.

O'Neill, R.V. 1975. Modeling in the eastern deciduous forest biome p. 49-72. In Systems analysis and simulation in ecology (B.C. Patten, ed.). Academic Press, New York, N.Y.

Quarterman, E., and R.L. Powell. 1978. Potential ecological/geological natural landmarks on the Interior Low Plateaus. U.S. Department of the Interior, Washington, D.C.

Schamberger, M.L., C. Short, A. Farmer. 1980. Evaluation of wetlands as wildlife habitats. p. 74-83. In wetland functions and values: the state of our understanding (P.E. Greeson, J.R. Clark, and J.E. Clark, eds.). American Water Resources Association. Minneapolis, Minn.

Smith, R.L. 1980. Ecology and field biology, 3rd edition. 835 p. Harper and Row, Publishers, New York, N.Y.

Wharton, C.H. 1970. The southern river swamp-a multiple use environment. 48.p. Bureau of Business and Economic Research, School of Business Administration, Georgia State University, Atlanta, Ga.

Whinnery, W.N. 1977. Remote sensing of a western Kentucky acid mine drainage swamp. Master's Thesis. 64 p. University of Louisville, Louisville, Ky.

PREDICTING THE FATE OF COAL-DERIVED POLLUTANTS IN AQUATIC ENVIRONMENTS[1]

Richard A. Park, Barry H. Indyke, and George W. Heitzman[2]

Abstract.--The PEST model predicts the fate of toxic organic materials (TOM) in aquatic ecosystems. The model is capable of simulating the time-varying concentration of TOM in as many as sixteen biotic and abiotic "carrier" compartments. Process equations represent all important modes of degradation and transport.

INTRODUCTION

The impact of coal-derived contaminants on aquatic ecosystems is of current concern. Mathematical models can provide an evaluative and, eventually, a predictive tool for site-specific risk assessment; these models can also aid in the formulation of optimal mitigation strategies. One such model is PEST (Park et al. 1980), which represents chemical kinetics and bioenergetics in a user-oriented, interactive program that is well suited to the needs of the power industry.

PEST is a dynamic simulation model for evaluating the fate of potentially toxic organic materials (TOM) in aquatic environments. It represents the time-varying concentration of a given TOM in each of as many as sixteen carrier compartments typical of aquatic ecosystems. Possible carriers include phytoplankton, macrophytes, zooplankton, waterbugs, zoobenthos, fish, particulate organic matter, floating organic matter, clay, and water (with TOM in the dissolved phase) as shown in figure 1.

It simulates degradation of TOM by acid- and base-catalyzed hydrolysis, oxidation, sensitized and unsensitized photolysis, microbial metabolism, and biotransformation by higher organisms. It simulates TOM transfer by

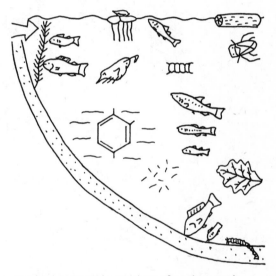

Figure 1.--Possible carriers of toxic organic material represented by the PEST model.

solution, volatilization, sorption, absorption onto gills, consumption, excretion, defecation, biodeposition, mortality, and throughflow (fig. 2). These are subject to time-varying environmental factors such as pH, temperature, dissolved oxygen, wind, solar radiation, and biomass and condition of organisms.

The model has been partially verified for several compounds and sites. Although more than fifteen man-years of effort have gone into its development, it is still being tested and improved. PEST requires more data than most models, but it has proven to be quite useful in linking process-level laboratory studies and

[1] Paper presented at the international symposium Energy and Ecological Modelling, sponsored by the International Society for Ecological Modelling. (Louisville, Kentucky, April 20-23, 1981).

[2] Richard A. Park is Professor of Geology and Director of the Center for Ecological Modeling, Rensselaer Polytechnic Institute, Troy, New York U.S.A. Barry H. Indyke is a student in Computer Science and George W. Heitzman is a student in Chemical and Environmental Engineering at Rensselaer.

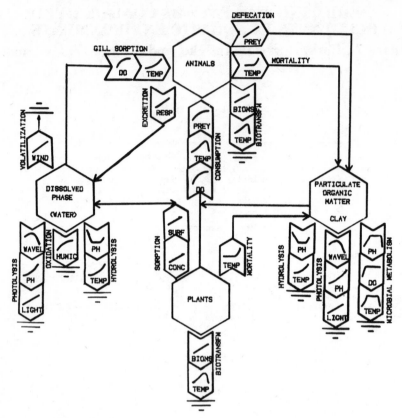

Figure 2.--Processes of transfer and degradation represented
by the PEST model.

larger microcosm studies and in providing an
extrapolation to specific site conditions. This
paper describes one such application involving
the fate of p-cresol and other coal-derived
contaminants.

PROCESS REPRESENTATION

Smith et al. (1977) present an extensive
summary of laboratory studies on p-cresol. Both
the SRI model that they developed and parameterization of our PEST model show photolysis to
be an important degradation process for normal
light intensities (fig. 3). However, this
process is sensitive to depth and turbidity of
water. Volatilization is primarily a function of
wind speed and is relatively unimportant (fig.
4). Oxidation is negligible and hydrolysis does
not occur.

Microbial degradation is quite important
(Smith et al. 1977). As represented in PEST, it
is a function of bacterial biomass, temperature
(fig. 5), pH, and dissolved oxygen (fig. 6).

MICROCOSM STUDIES

Boling and Cooper (1981) have studied the
fate of p-cresol in an experimental stream at
Monticello, Minnesota. The experimental channel
received two dosages during the summer of 1980.
Total loadings of p-cresol were not measured,
but the level in the dissolved phase in the
channel was held at 8 ppm for 24 hours and 48
hours respectively. PEST was run using observed
site conditions and parameter values taken from
the literature. In the absence of independent
data, the depuration (clearance) parameters were
calibrated so that the resulting process response
curve matched the residue data obtained during
the experiment; this was the only calibration
that was performed.

A plot of the state-variable values shows
that the model comes reasonably close in
simulating the time-varying levels of p-cresol in
the dissolved phase (fig. 7), but is off by a
factor of 5 in representing the concentration in
largemouth bass on the first day (fig. 8).
Predicted residue levels in the primary producers

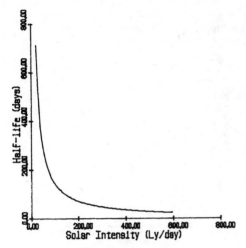

Figure 3.--Photolytic degradation of p-cresol as
a function of solar intensity in
Langleys/day.

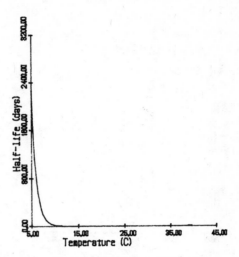

Figure 5.--Biodegradation of p-cresol as a
function of temperature.

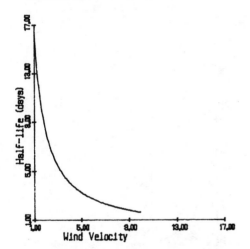

Figure 4.--Volatilization of p-cresol as a
function of wind velocity (m/sec).

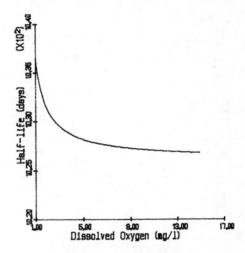

Figure 6.--Biodegradation of p-cresol as a
function of dissolved oxygen.

(fig. 9) and in other consumers (fig. 10) seem
reasonable but cannot be confirmed until residue
analyses are completed. The differences in
predicted levels among the carriers are due to
differences in surface area, lipid content, and
bioenergetics.

HYPOTHETICAL APPLICATION

Assume that one is faced with the problem of
disposal of effluent containing p-cresol and
other coal derivatives typical of the coal
liquefaction process. PEST can be used to gain
insights into the fate of those materials under
natural and semi-natural conditions. For
example, based on environmental sensitivities
demonstrated by the model and considering the
responses of aquatic ecosystems, a holding pond
could be designed to take advantage of the
natural pathways of degradation.

The task is made easier when PEST is used in
conjunction with an ecosystem model such as
MINI.CLEANER (Park et al. 1980; MacLeod et al.
in press). In that way the interactions between

118

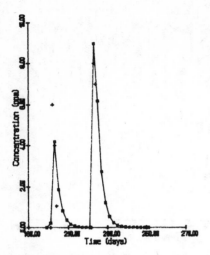

Figure 7.--Simulated (solid line) and observed
(+) concentrations of p-cresol (in ppm) in
the dissolved phase, Monticello (MN) stream
facility.

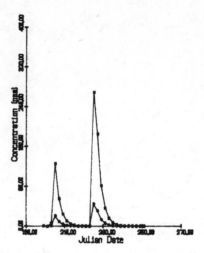

Figure 9.--Simulated concentrations of p-cresol
in algae (top line) and macrophytes.

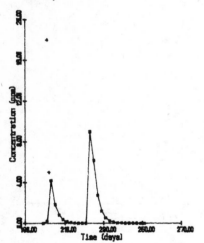

Figure 8.--Simulated (solid line) and observed
(+) concentrations of p-cresol in largemouth
bass.

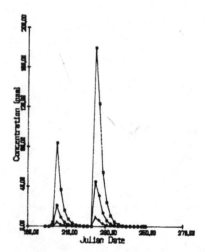

Figure 10.--Simulated concentrations of p-cresol
in zooplankton (top line), zoobenthos
(middle line), and smallmouth bass (bottom
line).

the biota and the physical and chemical environments can be determined, and the effects on the
xenobiotic compounds can then be assessed. At
present there is no model capable of representing
the complex feedbacks from the toxic material to
the ecosystem; when such a toxic-effects model is
available - and we are developing one now - it
will be even easier to design an optimal system
in harmony with nature.

In our hypothetical example we are assuming
that the holding pond is being constructed near

St Louis, Missouri, and that the effluent will
contain p-cresol, benz-thiophene, and quinoline.
Not only do we wish to promote rapid degradation
of the compounds, we also wish to minimize
volatilization.

Based on the process-response curves for the
various compounds (for example, fig. 11), we
conclude that photolysis and microbial
degradation are the most important degradative
pathways and that these are enhanced by clear
water and abundant organic substrate. However,

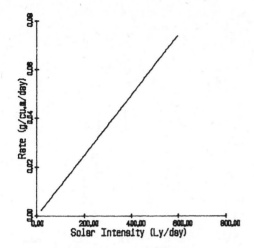

Figure 11.--Photolysis of quinoline as a function of solar intensity (Langleys/day), expressed as a rate (g/m /day).

microbial activity *is* greatly decreased by low dissolved oxygen levels and low temperatures. Volatilization is a problem with benz-thiophene at higher wind velocities, as shown by both half-life and rate plots (figs. 12 and 13); volatilization of p-cresol is a lesser problem.

The best results were obtained by simulating a series of shallow ponds with abundant submerged macrophyte growth and low phytoplankton biomass; wind velocity was decreased in accordance with

the effect of surrounding trees serving as windbreakers. Ice cover decreased both volatilization (fig. 14) and photolysis (fig. 15), and low winter temperatures decreased microbial degradation rates (fig. 16), causing a build-up of the xenobiotics, such as p-cresol, in winter (fig. 17). Occasional anaerobic conditions during the growing season also sharply reduced biodegradation (fig. 16). Aeration and heated effluent gave satisfactory results (Fig. 18).

In summary, the PEST model was shown to be useful in linking laboratory and microcosm studies and in allowing the results to be extrapolated to varying site conditions. By means of such detailed fate models, environmental engineers should be able to assess the persistence of energy-related xenobiotics and design cost-effective treatment facilities that utilize the functioning of natural and semi-natural ecosystems.

ACKNOWLEDGMENTS

The p-cresol data from the Monticello Ecological Research Station were kindly furnished on short notice by William E. Cooper and Robert H. Boling. We are very grateful for their generosity. Their research was funded with federal funds from the Environmental Protection Agency under Grant No. R807555010.

Development of the PEST model was financed with federal funds from the Environmental Protection Agency under Grant No. R80482003. The contents of this paper do not necessarily reflect the views and policies of that agency.

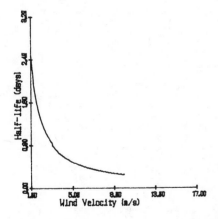

Figure 12.--Volatilization of benz-thiophene as a function of wind velocity (m/sec), expressed as half-life.

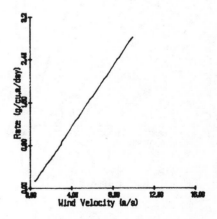

Figure 13.--Same as figure 12 but expressed as rate; note discontinuity at 3 m/sec as a result of wave set up.

120

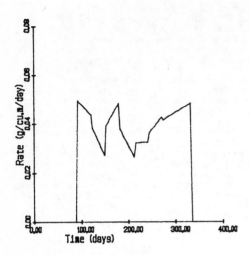

Figure 14.--Volatilization of p-cresol in
hypothetical holding pond; note influence of
ice cover and wind.

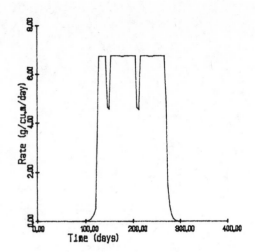

Figure 16.--Biodegradation of p-cresol in
hypothetical holding pond; note influence of
temperature in winter and anaerobic
conditions in summer.

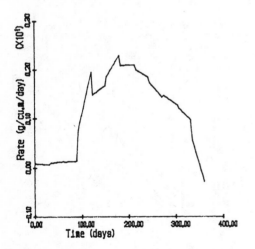

Figure 15.--Photolysis of p-cresol in
hypothetical holding pond.

BIBLIOGRAPHY

Boling, Robert H., Jr. and William E. Cooper. 1981.
Assessment of energy-related toxicant stress
in an engineered ecosystem. 26 p. Interim
Report for EPA Grant No. R807642010. Michigan
State University, West Lansing, Mich.

MacLeod, Bruce B., Richard A. Park, Carol D.
Collins, James R. Albanese, and Diana
Merchant. In press. Documentation of the
aquatic ecosystem model MINI.CLEANER. 71 p.
Final Report for EPA Grant No. R806299020.
Rensselaer Polytechnic Institute, Troy, NY.

Park, Richard A., Christopher I. Connolly, James
R. Albanese, Lenore S. Clesceri, George W.
Heitzman, Harry H. Herbrandson, Barry H.
Indyke, Joseph R. Loehe, Sydney Ross,
Dineschand D. Sharma, and William W.
Shuster. 1980. Modeling transport and
behavior of pesticides and other toxic
organic materials in aquatic environments.
Center for Ecological Modeling Report 7.
163 p. Rensselaer Polytechnic Institute,
Troy, NY.

Smith, J.H., W.R. Mabey, N. Bohonos, B.R.
Holt, S.S. Lee, T.-W. Chou, D.C.
Bomberger, and T. Mill. 1977.
Environmental pathways of selected chemicals
in freshwater systems. EPA Report
600/7-77-113. 81 p. EPA Environmental
Research Laboratory, Athens, Ga.

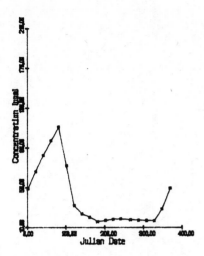

Figure 17.--Concentration of p-cresol in
dissolved phase in hypothetical holding
pond.

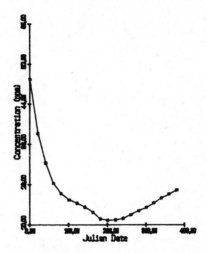

Figure 18.--Concentration of p-cresol in
dissolved phase in hypothetical holding pond
with ΔT of 15° and aeration; initial
condition is same as in figure 17.

EFFECTS OF A MAJOR OIL SPILL ON THE LIFE CYCLES OF THREE AQUATIC INSECTS: A MODEL OF WILSON CREEK, NELSON COUNTY KENTUCKY[1]

Francie L. Smith[2]

Abstract.--Models of Ephemeroptera, Plecoptera and Trichoptera populations were developed to assess the effects of an oil spill on these insect orders. Results of simulations corresponded with field observations of communities in oil affected areas of Wilson Creek, Kentucky. The primary stabilizer of simulated and natural communities was nymphal drift.

INTRODUCTION

Kentucky is becoming increasingly important to the production, storage and transportation of fossil based fuels due to its abundant resources of coal and oil shale. Yet, these resources pose a threat to Kentucky's streams, another major resource. Kentucky, with more stream miles than any state except Alaska, is threatened with environmental destruction by the growing fossil fuel industry. Studies of potential impacts to the environment are necessary in order to preserve energy and nutrient flows in natural ecosystems while developing an efficient fossil fuel industry.

On October 24, 1979 a major crude oil spill affected Wilson Creek, a previously uncontaminated stream of northcentral Kentucky. A crack in a pressurized pipeline used for transporting crude oil across the state resulted in the spillage of 252,000 gallons of South Louisiana crude oil. Approximately four miles of Wilson Creek were affected. A study of the benthic insect community, a vital link in energy flow through the stream ecosystem, showed that a drastic effect had occurred.

A modelling approach was used to study the effects of such impacts on aquatic insect life

cycles. Models have been used to describe the structure of ecosystems since Lindeman discussed the flow of energy through a food web (Lindeman 1942). In recent years, as the techniques of modelling and computer simulation have been refined, numerous models of stream ecosystems have been developed (Bender 1973, McIntire & Colby 1978, Park et al. 1975). All ecosystem models to date treat insects, in fact all benthic invertebrates, as a single compartment. This compartment actually contributes a great deal to the stream system. As Cummins (1973) pointed out in regards to nutrients and energy, "a significant portion of such cycling and flow involves the processing of various forms of organic matter by freshwater invertebrate animals, especially insects." It is important to have a better knowledge of how this benthic compartment is functioning.

One approach to studying aquatic insects is to examine their life cycles. Lehmkuhl (1979) observed that environmental disturbances which have no immediate affect may, in the long run, have significant effects on life histories of insects. These long term effects may even lead to the eventual extinction of an insect in an affected area.

An attempt to model an insect life cycle was made by Rotenberg (1977). His model dealt with a general holometabolous insect. The main assumptions were that all eggs hatched, all pupae emerged and that change from one life stage to another was dependent on age and weight.

The models proposed here deal with three orders of aquatic insects. The Ephemeroptera (mayflies) and Plecoptera (stoneflies) are hemimetabolous, while the Trichoptera (caddisflies) are holometabolous insects. These three orders

[1]Paper presented at the International Symposium on Energy and Ecological Modelling, sponsored by the International Society for Ecological Modelling. (Louisville, Kentucky, April 20-23, 1981).

[2]Francie L. Smith is a Graduate Research Assistant in the Systems Science Institute, University of Louisville, Louisville, Kentucky. 40292. U.S.A.

represent approximately 80% of the insect community of undisturbed ares of Wilson Creek. The
Plecoptera are of particular interest because
they are considered indicator organisms that
exhibit extreme sensitivity to polluted conditions.

The main assumptions under which these
models function are that insects overwinter in
the nymphal stage, that they exhibit univoltine
life cycles and that temperature is the signal
for the change from one life stage to another.
The two assumptions are generalizations about
insect life histories. Many aquatic insects do
have univoltine life cycles, though there are
some exceptions. Overwintering in the nymphal
stage is also very common, yet, again there are
exceptions to this rule. The third assumption is
based on numerous ecological studies that indicate temperature and daylength may function together as a key for life form changes (Lehmkuhl
1974, Ulfstrand 1968).

An important facet of the models is the
inclusion of nymphal drift. Drift has been shown
to be an important aspect of benthic community
ecology (Elliott 1967, Elliott 1968, Ulfstrand
1968). Drift also functions in the recolonization of denuded areas of streams (Waters 1964).

Of major interest is the reaction of the
system to environmental disturbances such as oil
spills. In the model proposed here the oil spill
acts as a modifier of the pathways between
compartments. The characteristics of system behavior that are important are initial reaction to
impact and rate of recovery of the stream. These
characteristics are analogous to two ecological
aspects of system stability. Initial reaction is
related to system resistance and rate of recovery
is related to system resilience. These concepts
are discussed more fully by Waide and Webster
(1976), Holling (1973) and Long (1974).

The hypothesized effect of the oil on the
benthic community is a scouring effect. Initially there is complete eradication with subsequent
recovery. Yet there may also be lingering effects as traces of oil remain in the sediments.
The time of impact in relation to the insect's
life cycle is also an important aspect of the
system.

Study of the dynamics of the models, using
techniques of sensitivity analysis, aids in understanding the functioning of the ecosystem.
Sensitivities of pathways in stable systems compared to those in impacted systems may lead to a
better idea of ecosystem response mechanisms.

MODELLING METHODOLOGY

Community structure is the parameter of
primary interest in a study of environmental
impact on benthic organisms (Cairns 1974). A
main characteristic of insect community structure
is the distribution of individuals among various
life stages of the insect in question. The
models developed during this study are based on
generalized life histories of three aquatic insect orders: Ephemeroptera, Plecoptera and
Trichoptera. Storages and flows between compartments of the models deal with numbers of individuals.

Energese

Three models were utilized to establish
stable populations of the insects. Conceptual
models of insect life cycles were designed to
delineate possible paths by which an oil spill
affects populations of aquatic insects. These
models were developed using H. T. Odum's Energese
language for ecological models (Odum 1972). The
symbols used are listed and defined in table 1.

The various life stages of the insects were
considered to be storages where insects remain
until they die or move to the next stage. These
storages and their connecting pathways were acted
upon by forcing functions that were external to
the system. Forcing functions included in the
model acted as system input/outputs and pathway

Table 1.-- Energese symbols that are used in the
conceptual models of insect life cycles.

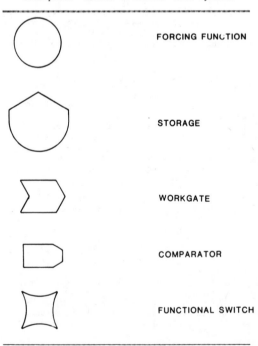

FORCING FUNCTION

STORAGE

WORKGATE

COMPARATOR

FUNCTIONAL SWITCH

regulators. Temperature, nymphal drift and oil spills were considered as forcing functions. The mechanisms through which these forcing functions act were designated by workgates, comparators and functional switches. The specific mathematical characteristics of these mechanisms were unique for each model.

Conceptual Models

Figure 1 shows the model of the Ephemeroptera segment of the community. Three storages represent the main life stages (egg, nymph and adult) of the mayflies. The subimago stage, a sexually immature winged stage unique to the Ephemeroptera, was not considered separately because it is a very short stage. The transformations from egg to nymph and nymph to adult were switched on and off by temperature cues. This temperature dependence was supported by studies of environmental effects on aquatic insect life histories (Lehmkuhl 1974). Nymphal mortality was also temperature dependent due to lowered predation rates in colder conditions. Adults begain to lay eggs as soon as they emerged.

Nymphal drift, a forcing function, was dependent on the number of nymphs present in the system. Thus, drift acted as an input during times of low population and as an output during times of overpopulation. Numerous studies have related the phenomenon of drift to community density (Hildebrand 1974, Muller 1974, Waters 1964).

The conceptual model for the Plecoptera is shown in figure 2. Conceptually the Plecoptera community functioned like the Ephemeroptera community because both orders are considered hemimetabolous. Major differences between the two orders were inherent in temperature cues for life form changes and in sensitivity to environmental disturbance.

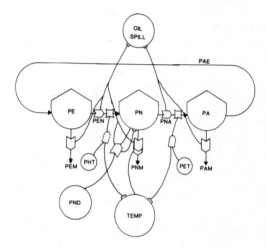

Figure 2.-- Conceptual model of Plecoptera population. PE = eggs, PN = nymphs, PA = adults, PEN = hatch, PNA = emergence, PAE = fecundity, PEM = egg mortality, PNM = nymph mortality, PAM = adult mortality, PND = nymphal drift, HT = hatch temperature, ET = emergence temperature.

Figure 3 shows the model developed for the Trichoptera. Trichoptera is an order of holometabolous insects so a fourth storage was added to accomodate the pupal stage. The onset of pupation, nymphal drift and adult emergence were considered temperature dependent.

The three models worked under similar assumptions. A complete cycle of the models was accomplished in one year, thus describing univoltine life cycles. The main factors affecting pathways were temperature and the impact of the oil spill. The only unaffected parameter was the number of eggs laid per adult in each model.

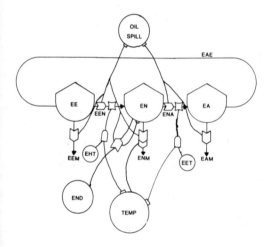

Figure 1.--Conceptual model of Ephemeroptera population. EE = eggs, EN = nymphs, EA = adults, EEN = hatch, ENA = emergence, EAE = fecundity, EEM = egg mortality, ENM = nymph mortality, EAM = adult mortality, END = nymphal drift, HT = hatch temperature, ET = emergence temperature.

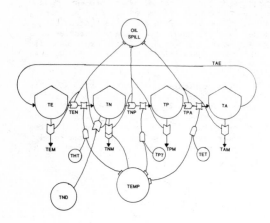

Figure 3.-- Conceptual model of Trichoptera
population.   TE = eggs, TN = nymphs, TP =
pupae, TA = adults, TEN = hatch, TNP =
pupation, TPA = emergence, TAE = fecundity,
TEM = egg mortality, TNM = nymph mortality,
TPM = pupa mortality, TAM = adult mortality,
TND = nymphal drift, HT = hatch temperature,
PT = pupation temperature, ET = emergence
temperature.

Mathematical Structure

The following basic differential equations
were developed from the conceptualizations for
simulation purposes.

$$dEE/dT = EAE*EA - EEM*EE - EEN*E \qquad (1)$$

$$dEN/dT = EEN*EE + END - ENM*EN*T - ENA*EN \qquad (2)$$

$$dEA/dT = ENA*EN - EAM*EA \qquad (3)$$

$$dPE/dT = PAE*PA - PEM*PE - PEN*PE \qquad (4)$$

$$dPN/dT = PEN*PE + PND - PNM*PN*T - PNA*PN \qquad (5)$$

$$dPA/dT = PNA*PN - PAM*PA \qquad (6)$$

$$dTE/dT = TAE*TA - TEM*TE - TEN*TE \qquad (7)$$

$$dTN/dT = TEN*TE + TND - TNM*TN*T - TNP*TN \qquad (8)$$

$$dTP/dT = TNP*TN - TPM*TP - TPA*TP \qquad (9)$$

$$dTA/dT = TPA*TP - TAM*TA \qquad (10)$$

The equations were based on donor controlled
pathways. This made for a simple linear model.

Pathway rates are given in table 2. Values
for parameters were based on a generalized insect
life table (Price 1975).   Parameters were adjusted and the model calibrated to allow for
given survivorship ratios in unaffected populations.

Table 2.-- Values used for rates in simulations.

PATHWAY	RATE
EEM	.4
EEN	.002
END	2000
ENM	.00025
ENA	.025
EAM	.05
EAE	60
PEM	.3
PEN	.00025
PND	1000
PNM	.0005
PNA	.045
PAM	.3
PAE	200
TEM	.4
TEN	.005
TND	1000
TNM	.00005
TNP	.8
TPM	.4
TPA	.8
TAM	.05
TAE	60

Temperature was described as a sine wave
with values from 0 to 90 degrees Fahrenheit.
Temperature cues were thus signals of seasonal
change.   For the Ephemeroptera hatching occured
between temperatures of 70 and 80 degrees while
Plecoptera hatched at temperatures above 70 degrees and Trichoptera hatched at temperatures
between 71 and 75 degrees.   Emergence of adults
occured at temperatures above 80 degrees in the
Ephemeroptera, above 55 degrees in the Plecoptera
and between 75 and 85 degrees in the Trichoptera.
Trichoptera pupation occured above a temperature
of 85 degrees.

The effect of the oil spill was described by
a function that varies between 0 (no effect) and
1 (maximum impact).   When a spill occured it
remained at a maximum impact level for 7 days.
The oil effect then dissipated according to a
time dependent exponential decay. This function
was based on field observations of the spill at
Wilson Creek, Kentucky.   Small amounts of oil in
the sediments continued to enter the creek ecosystem over a year after the spill had been
cleaned up.   The rate of effect decay was less
for the Plecoptera than for the Trichoptera and
less for the Trichoptera than for the Ephemeroptera.

Simulation

The models were simulated using CSMP (Continuous Systems Modeling Program) (IBM 1972). This program was compatible with the use of switches and comparators. It also allowed for user defined MACRO's (Fortran subroutines) which gave the simulator a great degree of flexibility in the simulation process.

Simulations of stable insect life cycles were contrasted with simulations of life cycles affected by an oil spill. The effects of the oil were programmed as a MACRO function. The MACRO was defined in such a way that severity of impact and day of occurrence could be manipulated. This allowed for testing hypotheses concerning effects of oil spills that occur at various times of year. Day of occurrence was set to approximate the date of the occurrence of the Wilson Creek oil spill. The MACRO program describing the effect of the oil spill is given in figure 4.

```
MEMORY SPILL
MACRO OUT=SPILL(A,B,C)
PROCEDURAL
IF(TIME.LT.A) OUT=B
IF((TIME.GE.A).AND.(TIME.LT.G)) OUT=D
G=A+7
C=(G-TIME)/X¹
IF(TIME.GE.G) OUT=2**C
ENDMAC
```

[1]X varies with sensitivity of Order.

Figure 4.--MACRO function for effect of oil.

Sensitivity Analysis

Model dynamics were examined using the techniques of sensitivity analysis. Absolute sensitivities in the sense of Brylinsky (1972) were calculated. Due to the cyclical nature of the models a method of analysis similar to that of Glass (1971) was employed. Parameters were varied individually by 10%. Corresponding variations were calculated for each state variable. The sensitivities of the compartments to parameter changes were of particular interest. Sensitivities were calculated for models with and without the oil effects. Comparison of these sensitivities provided a method for examining the effects of an oil spill on the aquatic insect community.

RESULTS AND DISCUSSION

Simulations were run to compare stable life cycles with cycles affected by an oil spill. Because of the date of occurrence of the Wilson Creek oil spill the life stage of major interest was the nymphal stage. In both the Ephemeroptera and Plecoptera models all adults had emerged and all eggs that were going to hatch had already

hatched. In the Trichoptera model pupation, emergence and egg hatch had been completed. Collections of nymphs and adults at Wilson Creek verified the assumptions about the relation of the oil spill to emergence, pupation and hatch.

Plots of nymphs from the Ephemeroptera simulations are given in figure 5. Simulations were based on a population of 2000 nymphs. The stable mayfly life cycle showed a fairly constant number of nymphs with a minimum of 1240 during the period of emergence and a peak of 2750 at the time of hatch. The number of nymphs was kept stable by the input through nymphal drift.

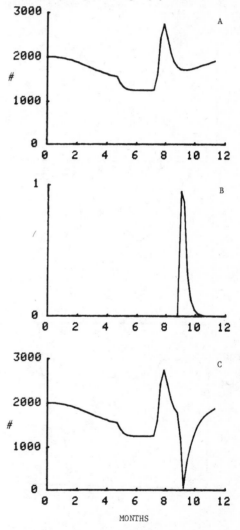

Figure 5.--A. Seasonal variation of nymphs from simulation of stable Ephemeroptera population B. Effects of oil spill on Ephemeroptera. C. Variation of nymphs in affected system.

When the oil effect was added to the system, an immediate response was seen with subsequent recovery within two months. The initial effect of the oil reduced the number of nymphs to 56. Recolonization by nymphs that drifted into the system began after two weeks. Field observations at Wilson Creek verified the rapid recovery of the Ephemeroptera.

Simulations of Plecoptera nymphs are shown in figure 6. These simulations were based on populations of 1000 nymphs. In simulations without oil the number of nymphs showed double emergence minimums, each of 660 nymphs. Hatch rate caused a gradual increase back to 1000 nymphs.

The oil impact caused an initial reduction to 69 nymphs. Recovery rate of the Plecoptera was much slower than recovery rate of the Ephemeroptera. Drift was the primary mechanism of recolonization, yet the Plecoptera were not capable of colonizing affected areas as readily as the Ephemeroptera. The study of Wilson Creek showed that Plecopterans had begun to return two weeks after the spill but had not yet successfully recolonized after seven months.

The Trichoptera fell between the Ephemeroptera and Plecoptera in regard to sensitivity to impact. Figure 7 shows the results of simulations of the Trichoptera. The baseline populations of the Trichoptera was 1000 nymphs.

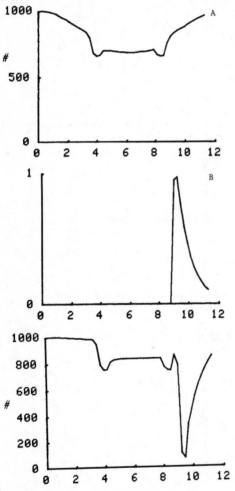

Figure 6.-- A. Seasonal variation of nymphs from simulation of stable Plecoptera population. B. Effect of oil spill on Plecoptera. C. Variation of nymphs in affected system.

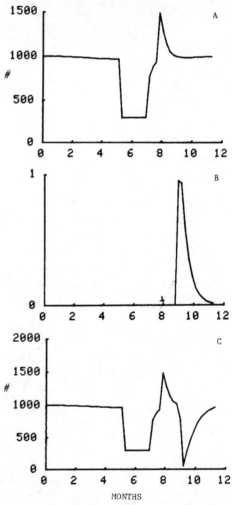

Figure 7 A.-- Seasonal variation of nymphs from simulation of stable Trichoptera population. B. Effect of oil spill on Trichoptera. C. Variation of nymphs in affected system.

A minimum of 290 occurred during emergence and a maximum of 1480 during hatch period of the stable Trichoptera population.

The initial response to the oil impact produced a drop to 50 nymphs. A fairly complete recovery to a stable population of 1000 occurred within three weeks. At Wilson Creek the Trichoptera had begun to return within ten days of the occurrence of the oil spill.

All three models demonstrated relatively rapid recovery from a major impact. This result was generally verified by observations of the communities at Wilson Creek. In the models nymphal drift was responsible for the recovery of the systems. This effect of drift has been observed to function in natural systems.

The importance of nymphal drift was evident in the results of sensitivity analysis of the models. Sensitivities of Ephemeroptera life stages to pathway changes are presented in table 3. Sensitivities are given in percent change of variable. A positive value indicates an increase in numbers, whereas a negative value indicates a decrease in numbers. Sensitivities for oiled communities are given under those for unoiled communities for comparison.

Table 3.-- Sensitivities of Ephemeroptera eggs, nymphs and adults to changes in population parameters in oiled and unoiled systems. Values represent percent change in variable due to a 10% increase in parameter.

		ENA	EAE	EEN	ENM	END	EAM	EEM
EE	Unoiled	+8.5	+10	0	-1.3	+7.3	-8.5	0
	Oiled	+8.5	+10	0	-1.3	+7.3	-8.5	0
EN	Unoiled	+.9	+1.2	+1.0	-2.2	+9.2	-1.1	-1.0
	Oiled	+.9	+1.4	+1.3	-2.3	+10.1	-1.1	-1.1
EA	Unoiled	-1.3	-1.4	0	-1.3	+7.4	-9.0	0
	Oiled	+1.3	+4.1	+4.1	-1.3	+7.2	-9.0	0

Changes in the nymphal drift parameter caused the largest overall changes in state variable values. In unaffected populations a 10% increase in drift caused a 9.2% increase in nymphs, a 7.4% increase in adults and a 7.3% increase in number of eggs. In effect, changes in drift rate dictated changes in allowable density of nymphs which changed overall population size.

The least effect was seen in changes of egg mortality. Increased mortality of eggs produced a slight decrease in the number of nymphs (-1.0) and no change in the number of adults.

The only variable that exhibited a difference in sensitivities between unoiled and oiled was the nymph variable. This is reasonable since the nymphal stage was the only one affected. Variations were slight yet the oil effect appeared to increase sensitivities. The greatest increase was again seen in the effect of the drift parameter. Sensitivity to changes in drift increased 0.9%, while there was no change in sensitivity to the nymph to adult pathway.

Plecoptera sensitivities are given in table 4. Again the parameter of drift had the greatest overall effect on the system. The effect of drift was more evident in analysis of the Plecoptera model. The inclusion of drift created a damping effect on other parameter changes as the population went through the nymphal stage. Drift also caused the nymph variable to be the least sensitive to parameter changes. Drift enabled the population to compensate for variations in pathways.

Table 4.--Sensitivities of Plecoptera eggs, nymphs and adults to changes in population parameters in oiled and unoiled systems. Values represent percent change in variable due to 10% increase in parameter.

		PNA	PAE	PEN	PNM	PND	PAM	PEM
PE	Unoiled	+8.5	+11.2	+1.1	-.2	+8.4	-8.5	-.9
	Oiled	+8.5	+11.2	+1.1	0	+8.4	-9.9	-.9
PN	Unoiled	-.2	0	+<.1	-.2	+9.0	0	-<.1
	Oiled	-1.9	0	0	-1.0	+9.1	0	0
PA	Unoiled	+8.8	+1.6	+1.6	-.4	+8.8	-10.4	-1.6
	Oiled	+8.8	+1.1	+1.1	-.4	+8.8	-10.4	-1.6

The Plecoptera sensitivities reflected the greatest effect of the oil spill on the population. The greatest increase in sensitivity was seen in the nymph variable as affected by the nymph to adult pathway. An increase in sensitivity of 1.7% was seen. Again, though, only the nymph variable was affected by the oil.

Table 5 shows the sensitivities of the Trichoptera variables to changes in pathway parameters. The same general pattern was seen in the Trichoptera sensitivities as in the other two models. Changes in nymphal drift had the greatest effect while changes in the hatch rate had the least effect. The addition of the pupal stage added an interesting aspect to the Trichoptera model. A 10% increase in the rate of pupation caused a 5.5% increase in adults, 5.8% increase in eggs, a 5.5% increase in nymphs and a 5.7% increase in pupae. These changes were close to those caused by increasing nymphal drift.

Table 5.--Sensitivities of Trichoptera eggs, nymphs, pupae and adults to changes in population parameters in oiled and unoiled systems. Values represent percent change in variable due to 10 % change in parameter.

		TNP	TPM	TAE	TPA	TEN	TNM	TND	TAM	TET
TE	Unoiled	+5.8	-12.0	+10.0	+3.1	0	0	+5.7	-9.1	-<.1
	Oiled	+5.8	-12.0	+10.0	+3.1	0	-<.1	+5.7	-7.9	0
TN	Unoiled	+5.5	-<.1	+<.1	0	+<.1	0	+6.8	-.1	-<.1
	Oiled	+5.2	-<.1	+<.1	+<.1	+<.1	-.4	+7.0	-<.1	-<.1
TP	Unoiled	+5.7	0	0	0	0	0	+5.9	0	0
	Oiled	+5.7	0	0	0	0	-<.1	+5.9	0	0
TA	Unoiled	+5.5	-12.	0	+3.0	0	-.3	+5.6	0	-<.1
	Oiled	+5.6	-12.2	0	+3.0	0	-.3	+5.6	0	0

Yet, the pupal variable showed no sensitivity to any parameters but pupation rate and drift rate. Pupal mortality had dramatic effects. An increase in pupal mortality of 10% caused a 12.2% decrease in adults and a 12% decrease in eggs.

The addition of the oil spill to the Trichoptera model had very little effect on sensitivities. Slight increases in sensitivity of the nymphal compartment occurred. All changes in sensitivity were less than 1%.

Basic patterns of sensitivity were similar in all three models though magnitudes of changes varied. The factor of nymphal drift played an important role in stabilization of the systems. Nymphal drift effectively turned the nymphal compartment into a buffering agent. Any changes in population parameters were compensated for as the populations cycled through the nymphal compartment. This result would seem valid because drift is an important ecological phenomenon of benthic communities.

CONCLUSION

The models developed here simulated stable insect population cycles for the orders Ephemeroptera, Plecoptera and Trichoptera. When an oil spill was added to the model in order to simulate the spill at Wilson Creek, Kentucky only the nymphs were affected. Recovery of the populations was rapid. Recovery was consistent with field observations of the Wilson Creek spill.

The primary factor affecting the stability of the populations and their rapid recovery was nymphal drift. The input of nymphs from a source external to the system created nonautonomous systems which were inherently more stable than autonomous systems.

The modelling effort and analysis of model dynamics aided in understanding the workings of the natural systems. Though generalizations about the insects were necessary in order to develop the models, the simulations of the models were consistent with field observations. Because of the generality of the models, they could be used to simulate the behavior of insects in similar streams throughout Kentucky and in other regions. Other impacts besides those resulting from oil spills could be added easily.

Predicting the effects of environmental impacts such as oil spills is important in this time of fossil fuel resource developement. Study of spills such as the one at Wilson Creek provide much information about impact effects. But such incidents are isolated in time and location. Quantitative predictions of the magnitudes of environmental impact can be made successfully through the use of mathematical models and computer simulations.

Acknowledgements.--I would like to thank Dr. Robert Bosserman for reviewing this manuscript. This study was supported in part by a grant from the Committee for Academic Excellence, University of Louisville, Louisville, Kentucky.

LITERATURE CITED

Bender, M. E. 1973. Water quality models and aquatic ecosystem status, problems, and prospectives. In Models for Environmental Pollution Control. R. A. Deininger (ed). Ann Arbor Science Pub. Inc. 448p. Ann Arbor, Mich.

Brylinsky, M. 1972. Steady-state sensitivity analysis of energy flow in a marine ecosystem. p. 81-101. In Systems Analysis and Simulation in Ecology. B. C. Patten (ed). Vol. II. Academic Press. 592 p. New York.

Cairns, J. Jr. 1974. Indicator species vs the concept of community structure as an index of pollution. Water Resources Bulletin. 10:338-347.

Cummins, K. W. 1973. Trophic relations of aquatic insects. Annual Review of Entomology. 18:183-206

Elliott, J. M. 1967. The life histories and drifting of the Plecoptera and Ephemeroptera in a Dartmoor stream. J. Animal Ecology. 36:343-362

------------ 1968. The life histories and drifting of Trichoptera in a Dartmoor stream. J. Animal Ecology. 37:615-625.

Glass, N. R. 1971. Computer analysis of predation energetics in the largemouth bass. p. 325-363. In Systems Analysis and Simulation in Ecology. B. C. Patten (ed). Vol I. Academic Press. 607 p. New York.

Hildebrand, S. G. 1974. The relation of drift to benthic density and food level in an artificial stream. Limnology & Oceanography. 19:951-957.

Holling, C. S. 1973. Resilience and stability of ecological systems. p. 1-23. In Annual Review of Ecology and Systematics, Vol. 4. R.F. Johnston (ed). Palo Alto, Calif.: Annual Reviews.

IBM. 1972. Systems/360 Continuous Systems Modeling Program -User's Manual. Fifth ed. International Business Machine Corp. White Plains, New York. 76 p.

Lehmkuhl, D. M. 1974. Thermal regime alteration and vital environmental physiological signals in aquatic organisms. p. 216-222. In Thermal Ecology. J. W. Gibbons & R. R. Scharits (eds). AEC Symposium Series.

-------------- 1979. Environmental disturbance and life histories: principles and examples. J. Fisheries Research Board of Canada. 36:329-334.

Lindeman, R. L. 1942. The trophic-dynamic aspect of ecology. Ecology 23:399-418.

Long, G. E. 1974. Model stability, resilience, and management of an aquatic community. Oecologia. 17:65-85.

McIntire, C. D. & J. A. Colby. 1978. A hierarchical model of lotic ecosystems. Ecological Monographs. 48:167-190.

Muller, K. 1974. Stream drift as a chronobiological phenomenon in running water ecosystems. Annual Review of Ecology and Systematics. 5:309-323.

Odum, H. T. 1972. An energy circuit language for ecological and social systems: Its physical basis. p.139-211. In Systems Analysis and Simulation in Ecology. Vol.II. B. C. Patten (ed).Academic Press. 592 p. New York.

Park, R. A., D. Scavia & N. L. Clesceri. 1975. CLEANER: The Lake George Model.p.49-82.In Ecological Modeling in a Resource Management Framework. C. S. Russell (ed).Washington D.C.

Price, P. W. 1975. Insect Ecology. John Wiley & Sons. New York. 514 p.

Rotenberg, M. 1977. Mathematical description of holometabolous life cycles. J. Theoretical Biology. 64:333-353.

Ulfstrand, S. 1968. Benthic animal communities in Lapland streams. Oikos Supplement 10.120 p.

Waide, J. B. & J. R. Webster. 1976. Engineering systems analysis: applicability to ecosystems. p. 329-371. In Systems Analysis and Simulation in Ecology Vol. IV. B. C. Patten (ed). Academic Press. 593 p. New York.

Waters, T. F. 1964. Recolonization of denuded stream bottm areas by drift. Transactions American Fisheries Society. 93:311-315.

SIMULATED TRANSPORT OF POLYCYCLIC AROMATIC HYDROCARBONS IN ARTIFICIAL STREAMS[1]

Steven M. Bartell, Peter F. Landrum, John P. Giesy, and Gordon J. Leversee[2]

Abstract.--A model was constructed to predict transport of polycyclic aromatic hydrocarbons (PAH) in artificial streams. Model processes included volatilization, photolysis, sorption to sediments and particulates, and net accumulation by biota. Simulations of anthracene transport were compared to results of an experiment conducted in the streams. The model realistically predicted the concentration of dissolved anthracene through time and space. Photolytic degradation appeared to be a major pathway of anthracene flux from the streams.

INTRODUCTION

Polycyclic aromatic hydrocarbons (PAH) are potentially hazardous by-products of the synthetic fuels industry and the generation of electricity by combustion of fossil fuels. Toxic, mutagenic (LaVoie et al. 1979), and carcinogenic (Norden et al. 1979) characteristics of PAH require estimation of human health risks posed by introduction of these chemicals into the environment.

Assessment of health risks associated with introduction of PAH into the environment depends in part upon quantification of environmental transport and subsequent dose (Crawford and Leggett 1980). Thousands of different species of PAH are chemically possible. Therefore, application of elaborate screening protocols (e.g., Duthie 1977) to identify major processes of transport, accumulation, and degradation of specific PAH is impractical for purposes of risk assessment.

This paper reports an attempt to predict the pattern of flow and accumulation of three PAH (anthracene, naphthalene, and benzo(a)pyrene) in artificial streams located on the Savannah River Plant near Aiken, South Carolina. Predictions were based upon the premise that the fundamental chemistry of individual PAH contains useful information for predictive purposes.

Predictions of transport of PAH were made for streams because energy industries that produce PAH require large volumes of water for production or cooling purposes. Thus, they are often located near streams and rivers where the probability of accidental or chronic addition of PAH to adjacent waters might be relatively high.

MODEL DESCRIPTION

Simulation of PAH transport in lotic systems requires an understanding of the basic structure and function of streams and rivers, as well as of the physical, chemical, and biological processes that determine transport of PAH compounds. Basic ecological information concerning species composition, standing crop, nutrients, and energy flow has accumulated for a variety of lotic systems (Coffman et al. 1971, Fisher and Likens 1972, McIntire and Phinney 1965, Minshall 1978, Odum 1957, Teal 1957, Tilly 1968). Yet streams and rivers have not received the attention from ecological modelers that other aquatic systems have (e.g., Patten 1968). Models of lotic systems have classically been the purview of sanitation engineers. State variables in these models included dissolved oxygen, biological oxygen demand, and total solids. Model structure represented a stream or river as a series of segments (reaches), which can be individually very different in dimensions and hydraulic characteristics, but which are assumed internally homogeneous.

[1]Paper presented at the International Symposium on Energy and Ecological Modelling, sponsored by the International Society for Ecological Modelling (Louisville, Kentucky, April 20-23, 1981).

[2]Steven M. Bartell, Research Associate, Oak Ridge National Laboratory, Oak Ridge, Tennessee, U.S.A.; Peter F. Landrum, Research Associate, National Oceanic and Atmospheric Administration, Ann Arbor, Michigan U.S.A.; John P. Giesy, Coordinator of Aquatic Toxicology, Michigan State University, East Lansing, Michigan U.S.A.; Gordon J. Leversee, Associate Director, Savannah River Ecology Laboratory, Aiken, South Carolina U.S.A.

134

Recently, some ecologists have begun to simulate energy and material flow through lotic systems, borrowing the reach-model approach (Chen and Wells 1976, Knowles and Wakeford 1978, McIntire and Colby 1978, Sandoval et al. 1976, Zalucki 1978). Our strategy was to adopt this approach.

Specific processes believed to influence PAH flux included photolytic degradation, volatilization, sorption to suspended particulates and sediments, and uptake and depuration by stream biota. These processes were incorporated into a Fates of Aromatics Model (FOAM). Figure 1 illustrates model pathways of PAH flux through a single reach.

An overall mass balance approach was taken to model the change in PAH concentration through time and space:

$$\partial P_{(t,j)} = \frac{\partial P_{(t,j)}}{\partial t} + U_{(t,j)} \frac{\partial P_{(t,j)}}{\partial j} \quad , \qquad (1)$$

where

P = PAH concentration in reach j at time t,

j = distance downstream (m), and

U = current velocity (m/t).

Equation (1) was solved by numerical approximation after conversion to difference equations (Sandoval et al. 1976). $P_{(t,j)}$ is the sum of time dependent change and advective change. Dispersion was not considered in the model. The time step (Δt) was determined by a necessary relationship between reach length and current velocity: $U \cdot \Delta t = \Delta j$ (Bella and Dobbins 1968).

Hydrological Submodel

Application of the model was constrained to conditions of constant reach geometry and constant discharge. For modeling purposes, the 91.4-m long artificial stream channel was conceptually divided into five equal reaches. Reach length was 18.28 m, width was 0.63 m, and depth was 0.20 m. The continuity equation calculated current velocity in relation to constant cross sectional area and discharge:

$$V = Q/A \quad , \qquad (2)$$

where

V = current velocity (18.28 m/h),

Q = discharge (4.53 m³/h), and

A = reach cross sectional area (0.12 m²).

ORNL-DWG 81-10842 ESD

SIMULATION OF PAH FLUX THROUGH STREAMS

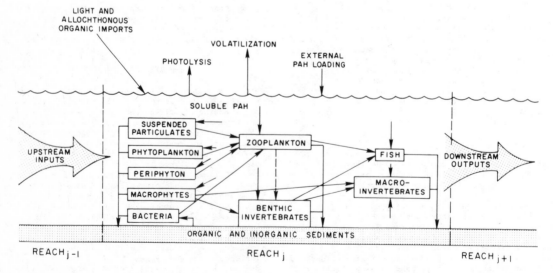

Figure 1.--Flow chart indicating components, pathways, and processes for PAH transport within a stream segment.

This submodel, while simplistic, was representative of the artificial streams.

Temperature Dependent Processes

In FOAM, water temperature can influence transport of PAH through the foodweb by regulating rates of photosynthesis and respiration according to an empirically derived function (Shugart et al. 1974). Temperature gradients can be simulated using hourly input temperatures for each reach. However, for simulations reported in this study, water temperature was constant, 22°C, for all reaches.

Solar Radiation Submodel

Incident solar radiation drives photolytic degradation of PAH and photosynthesis in FOAM. A modification of Satterlund and Means' (1978) model was incorporated in FOAM to simulate daily noon light intensities used in a 12-h light:dark cycle:

$$I_t = (I_{ds} + I_sF) \cdot (1 - C) + (I_{do} + I_s) \, Ce^{-d} F, \quad (3)$$

where

I_t = light intensity at reach surface (ly/h),

I_{ds} = slope dependent, direct beam radiation (ly/h),

I_s = scattered clear sky radiation (ly/h),

I_{do} = direct radiation on horizontal surface (ly/h),

C = fractional cloud cover,

F = slope correction for scattered radiation, and

d = regression parameter for seasonal cloud effects.

Light intensity at the reach surface was attenuated with water depth. Hyperbolic equations generated partial extinction coefficients for suspended particulate matter and phytoplankton.

Photolysis Submodel

Light-dependent degradation of aromatic hydrocarbons can be important in the physical-chemical transformation of PAH in aquatic systems (Southworth 1979a). Direct photochemical breakdown of dissolved PAH and degradation of photosensitized PAH sorbed to suspended particulates both occur; however, Miller and Zepp (1979) concluded that indirect photolytic degradation of sorbed PAH was minor in comparison to direct photolysis. In FOAM only direct photolysis of dissolved PAH was modeled.

In the photolysis submodel, PAH was degraded according to an equation adapted from Zepp and Cline (1977):

$$\frac{dPAH_d}{dt} = \phi \sum_\lambda k_\lambda \, PAH_d \quad , \quad (4)$$

where

ϕ = a molar yield coefficient (unitless),

k_λ = light absorbed at wavelength λ, and

PAH_d = concentration of dissolved PAH (μmoles/L). ·

Reported yield coefficients ranged between 0.001 and 0.01 for polycyclic aromatic hydrocarbons (Zepp and Schlotzhauer 1979). A linear regression based upon data in Zepp and Schlotzhauer (1979) related molecular weight to yield coefficient:

$$\phi = 0.0235 - 8.38 \times 10^{-5} \cdot MW \quad . \quad (5)$$

Light absorbance for specific PAH compounds was summed over 10-nm wavelength increments between 300 and 500 nm, a region of maximum absorbance for PAH. Each k_λ was calculated as the product of light intensity at lambda and the molar extinction coefficient for lambda, e_λ. Compound-specific values of e_λ were supplied as model input.

Light intensity, Ro, incident on each reach surface was simulated by equation (3). Light intensity at each wavelength was attenuated as a function of depth according to a regression derived for 12 southeastern United States rivers (Zepp and Cline 1977):

$$\alpha_\lambda = 0.42e^{(-0.0044 \cdot \lambda)} \quad , \quad (6)$$

where α_λ has units of 1/cm.

A depth-specific rate of photolysis was calculated for 1.0-cm intervals until: (1) the depth of the reach had been equaled or (2) the depth specific rate was less than 10% of the rate calculated for the 0 to 1.0 cm depth interval. Depth-specific rates were integrated over the water column and converted to units of $gPAH/m^2$. The mass of degraded PAH was added to the dissolved metabolite pool.

Modeled differences in rate of photolysis of specific PAH resulted from differences in molecular weight that translated to different molecular yield coefficients (eq. 5), and from differences in absorbance spectra.

Sorption Submodel

Karickhoff et al. (1979) modeled sorption as a first-order relationship between the concentration of dissolved PAH and an equilibrium-partition coefficient, K_p. The partition coefficient increased linearly as a function of the organic content of the sorbent material. The partition coefficient also correlated with the octanol/water partition coefficient (K_{ow}) of the PAH.

Depending upon available data, rates of sorption were calculated in one of three ways. First, if K_{ow} and the organic fraction of the sorbent were known, K_p was calculated from:

$$LOG(K_{oc}) = LOG(K_{ow}) - 0.21 \quad , \qquad (7)$$

and

$$K_p = K_{oc} \cdot oc \quad , \qquad (8)$$

where

oc = fractional organic content of the sorbent, and

K_{oc} = coefficient for partitioning into organic fraction of sorbent.

Second, if K_{ow} was not known, K_{oc} was estimated from the solubility of the particular PAH:

$$LOG(K_{oc}) = -0.54 * LOG(S) - 0.44 \quad , \qquad (9)$$

where

S = solubility of the PAH, as a mole fraction, and K_p was estimated from (8).

Third, if neither K_{ow} nor the organic fraction of the sorbent was known, a partition coefficient was estimated directly from:

$$LOG(K_p) = 4.16 - 0.51 \cdot LOG(S) \quad , \qquad (10)$$

where

S = solubility of the PAH, (ppm).

Equation (10) resulted from analysis of data in Karickhoff et al. (1979).

As a first approximation, K_p values were divided by the duration of the Karickhoff et al. (1979) experiments to convert to units of inverse time. We assumed that these values provided first estimates of a maximum rate of sorption per unit mass of sorbent, S_{max}. At each iteration of the model, specific rates of sorption were calculated for suspended particulate matter and sediments according to:

$$S = S_{max} \cdot SBATE \cdot SBENT \cdot oc/[SBATE + (SBENT \cdot oc)] \quad , \qquad (11)$$

where

S = rate of sorption (gPAH m^{-2} h^{-1}),

S_{max} = maximum sorption rate (gPAH g^{-1} h^{-1}),

SBATE = soluble PAH (g/m^2), and

SBENT = sorbent (g/m^2).

With (11), the rate of sorption was proportional to the concentration of dissolved PAH when the concentration of the sorbent was large by comparison. Conversely, when sorbent concentration was low, sorption primarily reflected the concentration of the sorbent.

Desorption of PAH from sorbent material was not included in the sorption submodel. Net suspension of suspended particulate matter and sediments provided pathways for advective loss of sorbed PAH from each reach in the modeled stream.

Volatilization Submodel

The submodel that simulated losses of PAH from the artificial streams as a result of volatilization was taken directly from Southworth (1979b). The rate of volatilization was modeled as a function of current velocity, wind velocity, reach depth, and molecular weight of the PAH according to a hyperbolic equation involving Henry's Law:

$$K_L = H \cdot K_g \cdot K_1/[(H + K_g) \cdot K_1] \quad , \qquad (12)$$

where

H = molar concentration of PAH in air divided by molar concentration of PAH in water,

K_g = gas phase exchange coefficient (cm/h),

K_1 = liquid exchange coefficient (cm/h), and

K_L = overall transfer coefficient (cm/h).

The logarithms of H, K_g, and K_1 were estimated from regressions defined in Southworth (1979b). Current velocities required by the regression equations were calculated by the hydrological submodel. Wind velocity was specified as an input parameter and was a constant 1.0 m/s in all simulations.

Biological Production Submodel

Rates of biomass change were determined by first-order, mass-balance, differential equations. The equation for producer i (phytoplankton, periphyton, macrophytes) in reach j at time t was:

$$\frac{dP_i}{dt} = (FI_i - FO_i) + P_i (PS_i - R_i - M_i - U_i - S_i) - G_i \quad , \qquad (13)$$

where

P_i = biomass of producer i (g dry wt/m^2),

FI_i = inflow of P_i from reach j-1 (g m^{-2} h^{-1}),

FO_i = outflow of P_i to reach j+1 (g m^{-2} h^{-1}),

PS_i = rate of gross photosynthesis of P_i (g g^{-1} h^{-1}),

R_i = respiration rate (g g^{-1} h^{-1}),

M_i = non-grazing mortality rate (1/h),

U_i = secretion-excretion rate (1/h),

S_i = sinking rate (1/h), and

G_i = loss of P_i to grazing (g m^2 h^{-1}).

The mass balance equation for changes in consumer biomass (zooplankton, bacteria, benthic invertebrates, fish) was:

$$\frac{dN_i}{dt} = (FI_i - FO_i) + {}_j C_{ij} - N_i (R_i + F_i + U_i + M_i) - G_i \ , \qquad (14)$$

where

N_i = biomass of consumer i in reach j at time t (g dry wt/m^2),

C_{ij} = consumption of prey j by consumer i (g m^{-2} h^{-1}),

FI_i, FO_i = same as in equation (13), but for consumer i,

R_i = respiration rate of consumer i, (g g^{-1} h^{-1}),

F_i = fraction of consumption that was egested (1/h),

U_i = excretion rate for consumer i (1/h),

M_i = non-predatory mortality rate for consumer i (1/h),

G_i = predatory losses of consumer i (g m^{-2} h^{-1}).

Photosynthesis.--Gross photosynthesis was modeled after Smith (1936):

$$PS_{(ijt)} = f(T) \cdot P_{max} \cdot I \cdot (I_k + I^2)^{1/2} \ , \quad (15)$$

where

PS_{ijt} = grams photosynthesized per gram producer at time t in reach j by producer i,

P_{max} = light saturated photosynthetic rate for i (g ly^{-1} h^{-1}),

I = light intensity from solar submodel (ly/h),

I_k = light saturation constant for producer i (ly/h), and

$f(T)$ = temperature dependence of photosynthesis.

Respiration.--For producers, a fraction of gross photosynthesis, or for consumers, a fraction of consumption, was lost through respiration.

Consumption.--PAH moved through the foodweb by the feeding activities of the consumer components. Consumption was modeled as a function of predator and prey biomass (DeAngelis et al. 1975):

$$C_{i,j} = f(T) \cdot \frac{C_{max\ i} \cdot N_i \cdot w_{ij} \cdot N_j \cdot {}_j}{N_i + {}_j (w_{ij} \cdot N_j \cdot {}_j)} \ , \quad (16)$$

where

$C_{i,j}$ = biomass of prey j consumed by consumer i (g m^2 h^{-1}),

$C_{max\ i}$ = maximum feeding rate of consumer i, (g g^{-1} h^{-1}),

N_i = biomass of consumer i (g/m^2),

N_j = biomass of prey j (g/m^2),

w_{ij} = Bayesian preference of consumer i for prey j (unitless),

$f(T)$ = temperature dependence of predation (eq. 4), and

$_j$ = PAH concentration of prey j (gPAH/g dry wt).

This equation has been usefully applied to models of feeding of zooplankton, benthic invertebrates and fish (Kitchell et al. 1974, O'Neill 1976, and Smith et al. 1975).

Excretion.--At each iteration a fraction of biomass and its PAH content were excreted from each primary producer and consumer component. It was assumed that 25% of the excretory products was a metabolic product that was shunted to the metabolite pool. The remainder was excreted into the dissolved PAH pool.

Mortality.--Nonpredatory mortality was modeled in a linear, donor-dependent manner. Biomass of dead planktonic components was added to the suspended detritus pool. Nonplanktonic mortality was added to the sediments. Loss of PAH from biota was the product of biomass lost to mortality and the PAH concentration of the biomass.

Sinking Losses.--Losses of PAH from the water column due to sinking phytoplankton were modeled in a linear fashion. In FOAM the sinking rate was not coupled to the current velocity.

Egestion Losses.--A fraction of ingested food was egested by the consumer components of the model. Losses due to egestion were added to the suspended particulate pool. Egested PAH was simulated as the product of egestion rate and the PAH concentration of the food item egested.

Direct PAH Uptake.--Uptake of dissolved PAH by the primary producers in FOAM was modeled as a second-order process:

$$PU_i = P_i \cdot UP_i \cdot PAH_d/(KP_i + PAH_d) \quad , \quad (17)$$

where

PU_i = uptake by producer i (gPAH m^{-2} h^{-1}),

P_i = biomass of producer i (g dry wt/m^2),

UP_i = maximum uptake rate (gPAH g dry wt^{-1} h^{-1}),

PAH_d = dissolved PAH concentrations (gPAH/m^2), and

KP_i is analogous to half the saturation constant (gPAH/m^2).

Direct uptake was assumed independent of rates of photosynthesis or respiration.

Direct uptake of PAH by consumers was coupled to the rate of respiration, assuming that uptake occurred across respiratory membranes and was an active process:

$$CU_j = N_j \cdot UC_j \cdot R_j \cdot PAH_d/(KC_j + PAH_d) \quad , \quad (18)$$

where

CU_j = uptake of dissolved PAH (gPAH m^{-2} h^{-1}),

NC_j = biomass of consumer j (g dry wt/m^2),

UC_j = maximum uptake per unit respiration,

R_j = g dry wt respired per g dry wt,

PAH_d = dissolved PAH (gPAH m^2), and

KC_j is analogous to half saturation constant (gPAH m^2).

As an initial assumption, values of KP_i and KC_j were all defined as half of the maximum solubility of the particular PAH.

Degradation of PAH.--Metabolic degradation of assimilated PAH was modeled as a function of respiration rate and was therefore indirectly affected by water temperature.

$$D_i = R_i \cdot P_i \cdot h_i \cdot D_{max\ i} \quad , \quad (19)$$

where

$D_{max\ i}$ = gPAH degraded h^{-1} g dry wt^{-1} respired by component i

P_i = g dry wt/m^2 of component i, and

h_i = fractional PAH content of P_i.

Metabolically degraded PAH was added to the metabolite pool.

External Loading of PAH and Solubility

Polycyclic aromatic hydrocarbons were added to the headwaters of the artificial streams. FOAM structure reflects this practice; dissolved PAH entered the first reach in the modeled stream as the product of a predefined discharge regime (m^3/h) and the PAH concentration of the influent (gPAH/m^3).

A relationship between PAH molecular weight and solubility of specific compounds was derived from data in Braunstein et al. (1977):

$$LOG(PAH_d) = 6.50 - 0.04 \cdot MW \quad , \quad (20)$$

where

PAH_d = dissolved PAH (mg/ml), and

MW = molecular weight.

This regression was used to estimate solubility for individual PAH compounds and to ensure that solubility was not exceeded during the course of a simulation. PAH in excess of solubility was added to the sediments.

Model Parameters

Parameters for physical/chemical processes of PAH flux were estimated from data presented in papers referenced in sections that describe physical/chemical submodels.

Parameters for biological pathways of uptake and depuration result from experiments in laboratory microcosms with ^{14}C-PAH tracers and various components of the stream biota (Leversee et al. 1981).

Initial Conditions

Initial biomass conditions for the simulations were estimated from samples collected from the artificial stream or measured directly before organisms were added to the stream. Periphyton biomass was estimated as 0.18 g dry wt/m^2 in all reaches. Biomass of papershell clams (Anodonta imbecilis) was 9.31 g dry wt/m^2 in reach 1 and 8.80 in reach 5. Clams were added only to

reaches 1 and 5. Bluegill sunfish were
constrained by cages to reaches 1 and 5; biomass
was 0.64 and 0.66 g dry wt/m². Bacteria biomass
was estimated as 0.1 g dry wt/m² for all reaches.
Biomass of settled detritus was arbitrarily defined as 0.5 g dry wt/m². The mass of active
sediments was estimated as 1.0 g dry wt/m².

RESULTS

Simulated and Observed Transport of Anthracene

During simulated daylight hours of the
12L:12D cycle, the model predicted a downstream
gradient of dissolved anthracene. Between
reaches 1 and 5, a distance of 81 m, the modeled
anthracene concentration decreased from 0.056 to
0.036 µmoles/liter. In the simulation, this
gradient was absent at night. FOAM predictions
agreed with measured concentrations in water
samples collected at midday and at dawn from the
artificial streams (table 1). After termination
of the addition of anthracene, predicted concentrations of dissolved anthracene decreased to
zero within eight hours. Measured concentrations
of anthracene decreased to below detection limits
(250 nmol/L) in the channel within 24 h.

Predicted uptake of anthracene by periphyton
approached a steady-state concentration of approximately 0.17 µmol/g dry weight by day 3 of the
simulation (figure 2). Concentration of anthracene in periphyton exhibited a downstream gradient in the simulation. Simulated concentrations were consistently higher at night. The
predicted concentrations were nearly four times
higher than concentrations (0.04 µmol/g dry
weight) measured in periphyton samples collected
from the stream. Half-life of anthracene calculated from the simulation after the end of the
infusion period was approximately 17 h, which was
within 2 to 3 h of the value calculated from measurements made on periphyton in the stream.

Table 1.--Observed and predicted concentrations
(µmol/L) of dissolved anthracene in
artificial stream channels.

Time	Stream reach		
	1	3	5
Dawn			
Observed	0.066	0.065	0.063
Predicted	0.056	0.056	0.055
Midday			
Observed	0.067	0.045	0.028
Predicted	0.056	0.045	0.036

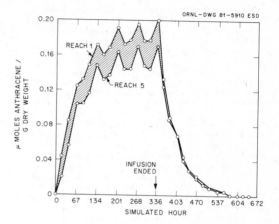

Figure 2.--Simulated concentration of anthracene
in periphyton through time in reaches 1 and
5. Stippled area contains transient concentrations for reaches 2 to 4.

Figure 3 illustrates comparison of predicted and measured concentration of anthracene
in the papershell clam (A. imbecilis). FOAM predicted the general pattern of the time-dependent
uptake and depuration of anthracene. However,
the model underestimated the actual rates of
uptake and depuration by a factor of approximately 1.5. The downstream gradient predicted by
the model was not supported by the data.

The model predicted a linear rate of
increase in anthracene concentration in stream
sediments during the course of the simulation.
Estimated steady-state values approximated 0.090,

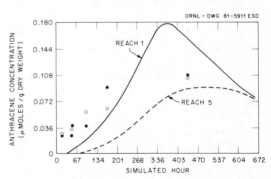

Figure 3.--Simulated concentration of anthracene
in the papershell clam (Anodonta imbecilis)
in reaches 1 (solid line) and 5 (dashed
line). Solid dots are measured concentrations for clams collected from reach 1.
Open circles are measured concentrations
for clams collected from reach 5.

0.004, and 0.003 μmol/g dry weight in reaches
1, 3, and 5, respectively (table 2). FOAM failed
to predict the pattern of uptake and release of
anthracene from the sediments as measured concentrations decreased after termination of the
anthracene infusion to the stream. FOAM also
underestimated the rate of uptake. The model
did, however, predict a downstream gradient in
sedimented anthracene similar to the measured
gradient.

Figure 4a illustrates the relative importance of non-advective physical/chemical processes versus biological processes in determining
the flux of anthracene through the simulated
lotic system. The periodic behavior in figure 4a
corresponded to the light:dark cycle and emphasizes the importance of photolyic degradation of
anthracene in overall transport. As more anthracene entered the food web during the simulation,
the biological processes became increasingly
important. Of the modeled processes, egestion
and defecation of anthracene by consumers dominated the biological flux of compound in the
stream. After the end of the infusion period,
anthracene flux was completely dominated by
biological processes.

Naphthalene and Benzo(a)pyrene Simulations

The initial premise stated that the transport and fate of specific PAH compounds could be
predicted from basic chemistry of the compounds.
Given some agreement between simulated and measured fate of anthracene in the artificial stream
channel, the experiment was simulated again, this
time with equal infusion rates for naphthalene and
benzo(a)pyrene. Initial conditions and parameter
estimates were identical to those for the anthracene simulation. Only values of molecular weight
and absorbance spectra were changed to correspond
to either naphthalene or benzo(a)pyrene.

Table 2.--Comparison of observed and predicted concentrations
(μmol/g dry weight) of anthracene in stream sediments.

Time (h)	Stream reach					
	1		3		5	
	Obs	Pre	Obs	Pre	Obs	Pre
24	0.0024	0.00026	0.0023	0.00020	0.0021	0.00017
48	0.0044	0.00059	0.0040	0.00045	0.0034	0.00037
96	0.0070	0.0014	0.0060	0.00096	0.0053	0.00080
168	0.0096	0.0027	0.0086	0.0018	0.0073'	0.0015
336	0.0149	0.0073	0.0134	0.0040	0.0123	0.0033
360	0.0137	0.0079	0.0118	0.0041	0.0106	0.0034
384	0.0108	0.0084	0.0104	0.0043	0.0104	0.0035
432	0.0113	0.0092	0.0116	0.0044	0.0107	0.0035
528	0.0092	0.0096	0.0076	0.0044	0.0065	0.0035
672	0.0072	0.0092	0.0065	0.0042	0.0053	0.0034

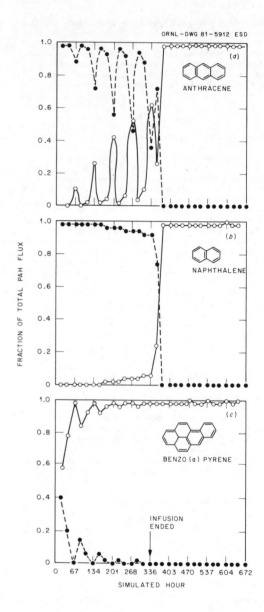

Figure 4.--Relative importance of non-advective
physical/chemical processes (solid dots,
dashed line) and biological processes (open
circles, solid line) in total transport of
(a) anthracene, (b) naphthalene, and
(c) benzo(a)pyrene through artificial
streams.

Without data for comparison with FOAM simulations, discussion of the dynamics of individual state variables becomes rather meaningless. However, it is reasonable to examine the relative importance of various pathways of simulated flux in light of available information concerning observed behavior of these compounds in natural or experimental systems. The simulated fate of naphthalene was similar to that of anthracene with respect to the relative importance of physical/chemical versus biological processes. Figure 4b shows that physical/chemical processes accounted for more than 90% of naphthalene flux through the model system. However, the absence of periodic behavior corresponding to the light-dark cycle indicated that, unlike anthracene, photolysis was not the dominant process. Closer inspection of the data summarized by figure 4b revealed that volatilization was the major physical/chemical pathway of naphthalene loss from the simulated stream. Lee et al. (1978) concluded that evaporation was an important process in the overall loss of naphthalene from large pelagic enclosures. Similar to the anthracene simulation, the processes of consumer egestion and defecation assumed importance in the overall flux of naphthalene following the termination of naphthalene infusion to the simulated stream.

The transport of benzo(a)pyrene (BaP) was dominated by biological processes, in contrast to the lighter weight anthracene and naphthalene (figure 4c). Following an initial 8-h period of infusion where photolytic degradation was important, uptake by primary producers, movement through the foodweb, and consumer processes of egestion and defecation played the determining role in the non-advective transport and fate of BaP in the modeled stream. Lu et al. (1977) observed biomagnification of BaP by biota in a simple aquatic microcosm that contained algae, mosquito larvae, Daphnia, snails, and fish. Lee et al. (1978) found 40% of BaP added to marine pelagic enclosures in samples of sediments. Of the physical/chemical processes simulated, sorption to suspended particulates and settled sediments was the most important vector simulated for BaP by FOAM.

DISCUSSION

The modeling effort attempted to evaluate the usefulness of molecular weight for predicting transport of individual PAH compounds. Molecular weight was correlated with water solubility, eq. (20), photolytic molar yield coefficient, eq. (5), rate of sorption (indirectly), eq. (10), and gas-phase exchange coefficient in volatilization, eq. (12) (Southworth 1979b).

The absence of rate coefficients for biological transport processes in relation to molecular weight or other fundamental chemical properties of PAH reflected the limited amount of data available for estimation of such correlations. Therefore, the model depended on extrapolation of rate constants measured in the

laboratory or published in the literature. As the number of necessary rate parameters that must be directly measured in the system of interest increases, specific application of transport models becomes more difficult (e.g., Baughman and Lassiter 1978).

Results of initial simulations indicated that FOAM predicted the observed spatial-temporal dynamics of dissolved anthracene with reasonable accuracy (table 1). This agreement might simply reflect the comparative sophistication of the photolytic submodel and the susceptibility of anthracene to photolytic degradation (Zepp and Schlotzhauer 1979, Southworth 1979a).

Lack of close agreement between simulated and measured concentrations of anthracene in periphyton, clams, and sediments indicates the difficulties in extrapolating laboratory determined rate constants to the streams. For example, rate of uptake by periphyton was overestimated. The cultures of algae used in the laboratory experiments to determine uptake did not resemble the species composition of the flora that was growing in the experimental stream.

While data from the stream were not available for comparison, simulations of naphthalene and BaP transport demonstrated that the model was sensitive to changes in parameters that determined PAH transport in relation to molecular weight and light absorbance. Furthermore, the general patterns of flux predicted by the model for these two compounds agreed qualitatively with observations made on other aquatic systems (Lee et al. 1978, Lu et al. 1977).

Results of this initial modeling effort suggest that:

(1) rate of physical/chemical processes of PAH transport in lotic systems can be predicted from molecular weight of individual compounds,

(2) light absorbance spectra between 300 and 500 nm contain information for predicting rates of photolytic degradation of PAH and

(3) parameters for biological transport and accumulation of PAH can be extrapolated to artificial streams with limited success.

The modeling effort suggested that future work is required in the following areas:

(1) collection of data to examine possible relationships between PAH compound structure and rates of biological uptake, degradation, and depuration, and

(2) collection of data in additional monitoring experiments for comparison with model predictions.

Results of this suggested research will increase our ability to predict exposure and dose in an

overall assessment of risk associated with introduction of PAH into aquatic systems.

ACKNOWLEDGEMENTS

This work continues under Interagency Agree-
ment EPA-79-d-X0290-1 with the Environmental Protection Agency and the Department of Energy, under
contract EY-A-09-0943 between the University of
Georgia and the Department of Energy, and by Oak
Ridge National Laboratory operated by Union
Carbide Corporation under contract W-7405-eng-26
with the Department of Energy. Publication
No. 1769, Environmental Sciences Division. Without the help of T. Fannin, S. Gerould, J.
Haddock, M. Bruno, J. Bowling, J. Cheetam, S.
Giddings, and R. Schloesser in the experimental
phase of this project, the modeling effort would
have been impossible. Comments made by G. R.
Southworth, J. M. Giddings, and W. Van Winkle on
earlier drafts were greatly appreciated.

LITERATURE CITED

Baughman, G. L., and R. R. Lassiter. 1978. Prediction of environmental pollution concentration p. 35-54. In Cairns, J., K. L.
Dickson, and A. W. Maki, (eds.), Estimating
the hazard of chemical substances to aquatic
life. ASTM STP, American Society for Testing and Materials. 657 p.

Bella, D. A., and W. E. Dobbins. 1968. Difference modeling of stream pollution. J.
Sanitary Engineering ASCE 94:995-1016.

Braunstein, H. M., E. D. Copenhaver, and H. A.
Pfuderer, editors. 1977. Environmental,
health, and control aspects of coal conver-
sion: an information overview. ORNL/EIS-94,
Oak Ridge National Laboratory, Oak Ridge,
Tennessee.

Chen, C. W., and J. T. Wells, Jr. 1976. Boise
River ecological modeling. p. 171-203. In
R. P. Canale (ed.), Modeling biochemical
processes in aquatic ecosystems. 389 p.
Ann Arbor Science Inc., Ann Arbor, Michigan.

Coffman, W. P., K. W. Cummings, and J. C.
Wuycheck. 1971. Energy flow in a woodland
stream ecosystem: tissue support structure
of the autumnal community. Arch. Hydrobiol.
68:232-276.

Crawford, D. J., and R. W. Leggett. 1980.
Assessing the risk of exposure to
radioactivity. Amer. Sci. 68:524-536.

DeAngelis, D. L., R. A. Goldstein, and R. V.
O'Neill. 1975. A model for trophic interaction. Ecology 56:881-892.

Duthie, J. R. 1977. The importance of sequential
assessment in test programs for estimating
hazard to aquatic life. pp. 17-35. In
Mayer, F. L. and J. L. Hamelink (eds.),
Aquatic toxicology and hazard evaluation.
307 p. ASTM STP 634, American Society for
Testing and Materials.

Fisher, S. G., and G. E. Likens. 1972. Stream
ecosystem: organic energy budget. BioScience 22:33-35.

Karickhoff, S. W., D. S. Brown, and T. A. Scott.
1979. Sorption of hydrophobic pollutants to
natural sediments. Wat. Res. 13:241-248.

Kitchell, J. F., J. F. Koonce, R. V. O'Neill,
H. H. Shugart, Jr., J. J. Magnuson, and
R. S. Booth. 1974. Model of fish biomass
dynamics. Trans. Amer. Fish. Soc.
103:786-798.

Knowles, G., and A. C. Wakeford. 1978. A mathe-
matical deterministic river-quality model.
Part I: Formulation and description. Wat.
Res. 12:1149-1153.

LaVoie, E., L. Tulley, V. Bedenko, and D. Hofman.
1979. Mutagenicity, tumor initiating
activity and metabolism of tricyclic polynuclear aromatic hydrocarbons (abstract).
In Jones, P. W. and P. Leber (eds.), Polynuclear Aromatic Hydrocarbons. 892 p. Ann
Arbor Sciences, Ann Arbor, Michigan.

Lee, R. F., W. S. Gardner, J. W. Anderson,
J. W. Blaylock, and J. Barxwell-Clarke.
1978. Fate of polycyclic aromatic hydrocarbons in controlled ecosystem enclosures.
Environ. Sci. Technol. 12:832-838.

Leversee, G. J., J. P. Giesy, P. F. Landrum,
S. Bartell, S. Gerould, M. Bruno, A. Spacie,
J. Bowling, J. Haddock, and T. Fannin. 1981.
Disposition of benzo(a)pyrene in aquatic
systems components: periphyton, Chironomids,
Daphnia, fish. In Cooke, M. and A. J. Dennis
(eds.), Chemical analysis and biological
fate: polynuclear aromatic hydrocarbons.
Proceedings of symposium, Battelle Labora-
tories, Columbus, Ohio, October 19-22, 1980.

Lu, P., R. L. Metcalf, N. Plummer, and
D. Mandel. 1977. The environmental fate of
three carcinogens: benzo(a)pyrene, ben-
zidine, and vinyl chloride evaluated in
laboratory model ecosystems. Arch. Environ.
Contam. Toxicol. 6:129-142.

McIntire, C. W., and J. A. Colby. 1978. A
hierarchical model of lotic ecosystems.
Ecol. Monogr. 48:167-190.

McIntire, C. D., and H. K. Phinney. 1965.
Laboratory studies of periphyton production
and community metabolism in lotic environments. Ecol. Monogr. 35:237-258.

Miller, G. C., and R. G. Zepp. 1979. Effects of
suspended sediments on photolysis rates of
dissolved pollutants. Wat. Res. 13:453-459.

Minshall, G. W. 1978. Autotrophy in stream
ecosystems. BioScience 28:767-771.

Norden, B., U. Edlund, and S. Wold. 1979.
Carcinogenicity of polycyclic aromatic
hydrocarbons studied by SIMCA pattern recognition (abstract). In Jones, P. W. and P.
Leber, (eds.), Polynuclear Aromatic Hydrocarbons. 892 p. Ann Arbor Sciences, Ann
Arbor, Michigan.

Odum, H. T. 1957. Trophic structure and
productivity of silver springs. Ecol.
Monogr. 27:55-112.

O'Neill, R. V. 1976. Ecosystem persistence and
heterotrophic regulation. Ecology
57:1244-1253.

Patten, B. C. 1968. Mathematical models of
plankton production. Int. Rev. Ges. Hydrobiol. 53:357-408.

Sandoval, M., F. H. Verhoff, and T. H. Cahill. 1976. Mathematical modeling of nutrient cycling in rivers, p. 25-232. In Canale, R. P. (ed.), Modeling biochemical processes in aquatic ecosystems. 389 p. Ann Arbor Science, Inc., Ann Arbor, Michigan. p. 389.

Satterlund, D. R., and J. E. Means. 1978. Estimating solar radiation under variable cloud conditions. For. Sci. 24:363-373.

Shugart, H. H., R. A. Goldstein, and R. V. O'Neill. 1974. TEEM: A terrestrial ecosystem energy model for forests. Oecol. Plant. 25:251-284.

Smith, E. L. 1936. Photosynthesis in relation to light and carbon dioxide. Proc. Nat. Acad. Sci. 22:504-511.

Smith, O. L., H. H. Shugart, R. V. O'Neill, R. S. Booth, and D. C. McNaught. 1975. Resource competition and an analytical model of zooplankton feeding on phytoplankton. Nat. 109:571-591.

Southworth, G. R. 1979a. Transport and transformations of anthracene in natural waters. p. 359-380. In L. L. Marking and R. A. Kimerle (eds.), Aquatic Toxicology. 480 p. ASTM STP 667, American Society for Testing and Materials.

Southworth, G. R. 1979b. The role of volatilization in removing polycyclic aromatic hydrocarbons from aquatic environments. Bull. Environ. Contam. Toxicol. 21:507-514.

Teal, J. M. 1957. Community metabolism in a temperate cold spring. Ecol. Monogr. 27:283-302.

Tilly, L. J. 1968. The structure and dynamics of Cone Spring. Ecol. Monogr. 38:169-197.

Zalucki, M. P. 1978. Modeling and simulation of the energy-flow through Root Spring, Massachusetts. Ecology 59:654-659.

Zepp, R. G., and D. M. Cline. 1977. Rates of direct photolysis in aquatic environments. Environ. Sci. Technol. 11:359-366.

Zepp, R. G., and P. F. Schlotzhauer. 1979. Photoreactivity of selected aromatic hydrocarbons in water p. 141-148. In Jones, P. W. and P. Leber (eds.), Polynuclear Aromatic Hydrocarbons, 892 p. Ann Arbor Sciences, Ann Arbor, Michigan.

A COMPUTER SIMULATION MODEL OF THE FATE OF CRUDE PETROLEUM SPILLS IN ARTIC TUNDRA ECOSYSTEMS[1]

Lynne Dauffenbach[2], Ronald M. Atlas[3], and William J. Mitsch[4]

ABSTRACT.--A computer simulation model is being developed to investigate the fate of petroleum hydrocarbon pollutants in Arctic tundra soils. The model consists of five interrelated submodels: hydrology, nitrogen, phosphorus, non-petroleum carbon and petroleum hydrocarbons. The model emphasizes tundra ecosystem processes affecting the biodegradation of petroleum hydrocarbons. Development of the model has been aimed at more clearly elucidating factors which determine the fate of petroleum hydrocarbons in tundra soils, and at providing predictive capabilities for developing spill management strategies which will maximize removal of oil and minimize ecological damage.

INTRODUCTION

Despite recent efforts toward the development of alternative energy sources, the industrial world still relies heavily upon fossil fuels. As the demand for energy increases, so do problems associated with the extraction, transportation and use of fossil fuels, particularly petroleum products. Accidental spillages are inevitable wherever petroleum products are handled. The development of Alaskan North Slope oil has already been accompanied by a number of major spills.

Oil spills on tundra soil present unique difficulties for cleanup and ecological recovery. Cleanup operations are costly and the disturbances resulting from cleanup activities may be more harmful ecologically than the spill itself. Field activities are restricted by the harsh climate and high transportation costs. In developing a model of the fate of crude oil

spills on tundra, we hope to identify areas in which additional field research is most needed or seems most promising. We have tried to keep the model general enough to apply to tundra areas other than the field study site at Barrow, Alaska, so that it could eventually be used to determine the best strategies for handling any particular spill.

One of the goals of the International Biological Programme's Tundra Biome studies was to generate the necessary data base for ecosystem modelling (Miller et al. 1973). Modelling was an important part of the Tundra Biome studies, and a number of models resulted. Two of these, ABISKO II, a model of carbon flux (Bunnell & Scoullar 1975), and DECOMP, a model of soil decomposition processes (Bunnell & Tait 1974), helped to provide some of the groundwork for this modelling effort.

Models ot oil spills on tundra to date have largely dealt with physical and chemical processes (Deita 1972; Davis et al. 1972; Mackay et al. 1975). The models primarily consider phenomena such as evaporation, leaching, flow into and out of the soil, leaching, flow into and out of the soil, dispersion-diffusion effects and transport by ground water. In some models, biodegradation is included, but simply as a proportional loss rate; these models are a good complement to a model of biological processes affecting the fate of oil.

One exception to this general trend was a model by McGill (1975) of microbial decomposition of oil. This model focussed on the effects of adding fertilizer nitrogen to oiled soils to enhance biodegradation, and was aimed at predicting the amount of fertilizer nitrogen required to achieve a desired rate of decomposition.

[1] Paper presented at the international symposium Energy and Ecological Modelling, sponsored by the International Society for Ecological Modelling. (Louisville, Kentucky, April 20-23, 1981).

[2] Lynne Dauffenbach is a graduate student in Interdisciplinary Studies, University of Louisville, Louisville, Kentucky, U.S.A.
[3] Ronald M. Atlas is Professor of Biology, University of Louisville, Louisville, Kentucky, U.S.A.
[4] William J. Mitsch is Associate Professor of System Science and Environmental Engineering, University of Louisville, Louisville, Kentucky, U.S.A.

Although a number of studies have been conducted on the fate and effects of oil spills on tundra, very little is actually known about the ways in which a spill affects ecosystem processes in the tundra, or the mechanisms by which the rate of oil biodegradation is controlled. Most studies have been directed at determining the effects of spills on vegetation and investigating means for enhancing vegetation recovery. Another area of emphasis has been the determination of decomposition rates of crude oil as a whole and of various factions, in situ and in the laboratory. Everett (1978) studied the effect of oil on the physical and chemical characteristics of wet tundra soils, a crucial step in understanding the interaction of the spill with ecosystem processes. A review of the literature and the preliminary formulation of the conceptual model, including the initial form of the mathematics for the petroleum submodel, were presented by Dauffenbach et al. (1981). This paper describes the development of the equations for the carbon and petroleum submodels of the simulation model, and the forcing functions used to drive the model.

MODEL DEVELOPMENT

Since our primary goal is to examine the fate of crude oil spills rather than simply the ecological effects, we have made an effort to model processes which might be expected to play a role in controlling the fate of oil in tundra soils. The major emphasis of this stage of the model development has been to identify possible points of interaction between ecosystem processes and spilled oil, and to develop hypotheses about the nature of these proposed interactions. The exercise of formulating mathematical expressions to represent these hypotheses has forced us to express our assumptions about the system processes in a more rigorous and more testable form than in earlier versions. Throughout the modelling effort we have tried to design the model so that existing data or easily generated new data could be used for parameter estimation. This is crucial in view of the high cost of research in the tundra and the scarcity of data on some processes.

The conceptual model describes the interaction of spilled oil with carbon, nitrogen, phosphorous and hydrologic submodels. The simulation model currently involves the carbon and petroleum submodels only. The time frame for this model is five years. A future long-term version of the model will have a time frame of 50-100 years. Figure 1 defines the symbols used in the conceptual models presented here.

The Carbon Submodel

Figure 2 shows the carbon and petroleum submodels and their interactions. The carbon

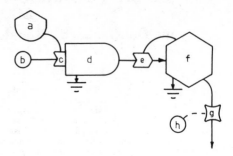

Figure 1.--Symbols used in model diagrams. a-passive storage; b and h-forcing functions or independent variables; c and e-work gates or functional relationships; d-autotroph; f-heterotroph; g-logic switch.

submodel is essentially a vegetation and detritus model. Structurally, it is similar to, but much simpler than, the Tundra Biome study group's ABISKO II model (Bunnell and Scoullar 1975). Since we are not attempting to model carbon cycling in tundra per se, we have chosen to sacrifice some sophistication in the interest of simplicity, while still attempting to retain the salient features of the overall behavior of carbon cycling in the tundra. We have assumed animal biomass and the effects of grazing by herbivores to have a negligible effect on processes effecting the rate of oil decomposition. Bunnell and his coworkers (1975) report that in non-peak years in the lemming population cycle, lemmings consume less than 0.1% of the above ground net primary production. We are not aware of any studies of the effects of oil spills on lemming grazing patterns, but it seems unlikely, even in peak population years, that lemmings would consume oiled vegetation if uncontaminated vegetation were available in the vicinity. Invertebrate herbivores are low in numbers and diversity and we assume them to have a negligible effect on the rate of oil decomposition as well.

The carbon submodel consists of six compartments:

C_1 -- above ground live vegetation: all above-ground vascular plant biomass.

C_2 -- belowground live vegetation: roots, rhizomes and stem bases.

C_3 -- standing dead: plant tissues which have died but have not yet fallen into the litter layer.

C_4 -- belowground dead, litter, and soil organic matter: organic residues on

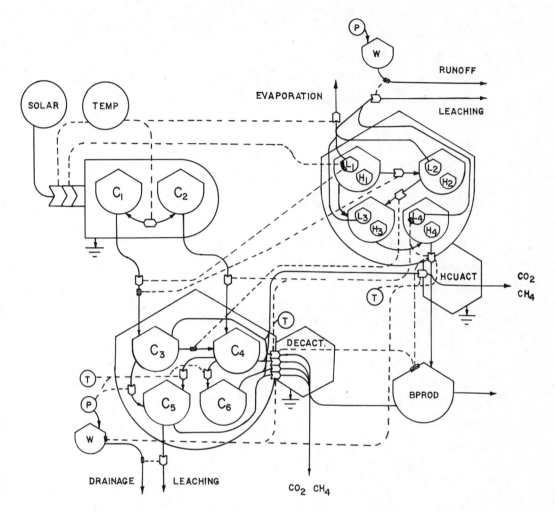

Figure 2.--Carbon and petroleum submodels.   C1 - aboveground live vegetation; C2 - belowground live vegetation; C3 - standing dead vegetation; C4 - belowground dead vegetation,litter, and soil organic matter;
C5 - soluble soil carbon; C6 - soil humus; L1 and H1 - LMW and HMW fractions,respectively, of oil on
aboveground live vegetation; L2 and H2 - LMW and HMW fractions of oil on standing dead; L3 and H3 - LMW
and HMW fractions of oil in surface pool; L4 and H4 - LMW fractions and HMW fractions of oil in soil; SOLAR
- solar radiation; TEMP or T - temperature; HCUACT -activity of hydrocarbon-utilizing microorganisms;
DECACT- activity of other decomposers; BPROD - toxic or inhibitory byproducts of decomposition; W - hydrology submodel; P - precipitation.

or in the soil which still have a
recognizable origin.

C5 -- soluble soil carbon: dissolved carbon
in the soil.

C6 -- soil humus: organic matter whose origin
is no longer recognizable; the more
recalcitrant detrital component.

The mass balance equations for the carbon
sector are given in Table 1.

Growth of the aboveground live vegetation
is assumed to be function of light, temperature,
nitrogen and phosphorus availability, spilled oil,
and the plant biomass itself.   We assume a maximum photosynthetic rate when light, temperature,
nitrogen   and   phosphorus   are   not

Table 1.--Mass balance equations for carbon submodel.

C1--aboveground live vegetation

$\overset{.}{C1}$ = PHSYN + TRANSU - TRANSR - RESPA - DEATHA

    PHSYN = (SOLAR/SOLMAX)*(TEMP/TEMOPT)*(C1/C1+C3)*KC1*C1*(1-((L1+H1)/(ALPHS1+L1+H1)))
    TRANSU = TDOT*(KC2*C2*COMPAR(TDOT,0)+(KC4*C1*(1-COMPAR(TDOT,0)))
    TRANSR = KC3*TEMP*C2*COMPAR(PHSYN+TRANSU-RESPA,KC3*TEMP*C2)
    RESPA = KC5*TEMP*C1
    DEATHA = C1*KC14+(L1/L1+ALPHS2)) C1

C2--belowground live vegetation

$\overset{.}{C2}$ = TRANSD + TRANSR - RESPB - DEATHB

    TRANSD = -TRANSU
    RESPB = KC3*TEMP*C2
    DEATHB = KC6*C2+(L4/(ALPHS3+L4))*C2

C3--standing dead vegetation

$\overset{.}{C3}$ = DEATHA - KC7*C3 - KC8*C3*RAIN - KD1*C3*TEMP

C4--belowground dead, litter, soil organic matter

$\overset{.}{C4}$ = KC7*C3 + DEATHB - DECC4

    DECC4 = KC9*C4*DECACT*(Q1**((TEMP-10)/10))
    DECACT = K14*(1-K15*L4-K22*BPROD)*COMPAR(1,K15*L4+K22*BPROD)

C5--soluble soil carbon

$\overset{.}{C5}$ = KC8*C3*RAIN + LEACHB - LEACHO - DECC5

    LEACHB = KL*(DECC4+DECC6)
    LEACHO = KC10*C5*DRAIN
    DECC5 = KC13*C5*DECACT*(Q2**((TEMP-10)/10))

C6--soil humus

$\overset{.}{C6}$ = KC11*DECC4 - DECC6

    DECC6 = KC12*C6*DECACT*(Q3**((TEMP-10)/10))

limiting. This maximum rate is then multiplied by a set of terms, each ranging between 0 and 1, which represent the availability of light, temperature, nitrogen and phosphorus. Light and temperature are assumed to have a linear effect on photosynthetic rate while nitrogen and phosphorus availability are assumed to follow saturation curves. A shading term is included to account for the shading effect of standing dead. We assume that the presence of oil affects photosynthetic rate through the blockage of light resulting from a coating of oil on the leaves. This effect is modelled by a term which decreases to zero along an inverse saturation curve as the amount of oil on above ground live increases.

Because snowmelt occurs just prior to the summer solstice, it is imperative that tundra plants be able to respond with rapid growth immediately after meltoff. This is accomplished by the use of stored reserves in the belowground portions of the plant which are translocated upward and used for rapid growth early in the season. When temperatures are rising, the model allows upward translocation to take place at a rate proportional to the amount of stored reserves and to the rate of temperature increase. When temperatures are falling, we assume aboveground live vegetation storages to be translocated downward in the same manner.

Respiratory losses are assumed to be a function of temperature, and it is assumed that if respiratory activities exceed photosynthetic input tissue death results. Winter kill is modelled as occurring at the same rate throughout the winter, rather than primarily during freezeup. This should not affect the behavior of simulated oil decomposition, since we assume for simplicity that biological processes such as biodegradation only take place during periods when the temperature is above freezing. We have also included a term in the equation for death of aboveground live vegetation to allow simulation of the hypothesis that oil on aboveground live increases winter kill. The LMW fraction appears in this term since we assume that most of the toxic effects of the spill are due to the LMW (low molecular weight) fraction.

Growth of belowground live vegetation is accomplished through downward translocation of aboveground live storages. Respiratory losses are assumed to be a function of temperature. Death of belowground live vegetation is assumed to occur at a rate proportional to belowground live biomass. The model also allows for death rate of belowground live to be increased by the presence of LMW hydrocarbons in the soil.

Inputs to the standing dead vegetation compartment consist of winter kill from aboveground live, plus tissue death of aboveground live which occurs when respiratory needs and downward translocation exceed photosynthetic imputs. It is assumed that standing dead vegetation falls into the litter layer at a rate proportional to the amount of standing dead present. This is a simplifying assumption, since it does not take into account the "age structure" of standing dead. However, we feel that it is a reasonable first approximation. Likewise, leaching from standing dead vegetation is assumed to be simply proportional to the amount of standing dead and the rate of incoming rainfall. Decomposition of standing dead vegetation is assumed to be a simple linear function of temperature. Decomposition of standing dead is modelled more simply than decomposition of the other storages because standing dead does not undergo much decomposition until it falls into the litter layer. Thus, in this instance, the relationship used allows for a simpler model without the danger of sacrificing realism.

Inputs to the belowground dead vegetation, litter and soil organic matter storage are from the attrition of standing dead and death of belowground live. Decomposition of belowground dead vegetation, litter and soil organic matter is modelled by a modified version of a relationship developed by Bunnell & Tait (1974) for the model DECOMP. Decomposition is assumed to be a function of soil moisture and temperature. Q_{10} values were initially drawn from the ABISKO II model, as were maximum decomposition rates. However, as this model involves fewer compartments, the rate constants and Q_{10} values used here generally represent weighted averages of those given for the corresponding ABISKO II compartments. In addition, we have multiplied the decomposition rate by a term relating decomposer activity to the level of LMW hydrocarbons in the soil. This term allows for decomposer activity to decline linearly as a function of LMW hydro-

carbons in the soil, reaching zero at a specified level of contamination.

Inputs to the soluble soil carbon compartment are leaching from standing dead and from belowground dead, litter and soil organic matter. Outputs are leaching of soluble soil carbon and decomposition. The leaching terms are assumed to be proportional to the amount of leachable material and the volume of water, in the form of rainfall or groundwater drainage acting on the substrate. Decomposition is modelled in the same manner as for belowground dead, litter and soil organic matter by substituting the appropriate rate constants.

Input to the soil humus compartment is the recalcitrant portion of belowground dead, litter and soil organic matter which is left after the more easily decomposed material has been utilized. Output is through decomposition, which is modelled as for belowground dead, litter and soil organic matter and soluble soil carbon.

Two hypotheses have been developed relating the biodegradation of oil to its effects on the decomposition of detritus. Plant residues may compete with petroleum hydrocarbons as a substrate for the growth of those miocroorganisms which can utilize hydrocarbons. If this occurs, an increased input to the detrital compartments, resulting from plants killed by the spill, would slow oil decomposition. Also, it is possible that the decomposition of plant residues results in by-products which inhibit the activity of hydrocarbon-utilizing microorganisms. Once again, this would result in a decrease in rate of oil decomposition. See Dauffenbach et al. (1981) for a further discussion of these hypotheses.

The Nitrogen and Phosphorus Submodels

The nitrogen and phosphorus submodels are shown in figures 3 and 4, respectively. We have postulated that interactions between nutrient cycling and the spilled oil are through alterations in carbon cycling, and through the limiting effect of low nutrient availability on the rate of decomposition of any substance in tundra soils. As McGill (1975) pointed out in his model, the high C:N ratio of crude oil makes the limiting effect of low nutrients even more pronounced.

Low nutrient availability also means that any perturbation which kills aboveground live vegetation before nutrients have been translocated downward to the roots and rhizomes at the end of the growing season would have a drastic effect on the growth of vegetation in subsequent seasons. Smaller standing crops in recovery years would mean lower inputs to detritus compartments. Also, because decompositional processes control the release of

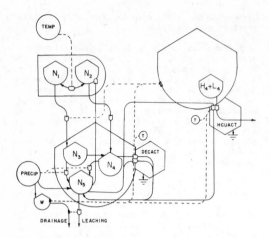

Figure 3.--Nitrogen submodel. N1 - nitrogen in aboveground live vegetation; N2 - nitrogen in belowground live vegetation; N3 - nitrogen in standing dead vegetation; N4 - nitrogen in belowground dead vegetaion, litter, soil organic matter and soil humus; N5 - soluble and exchangeable soil nitrogen; PRECIP - precipitation.

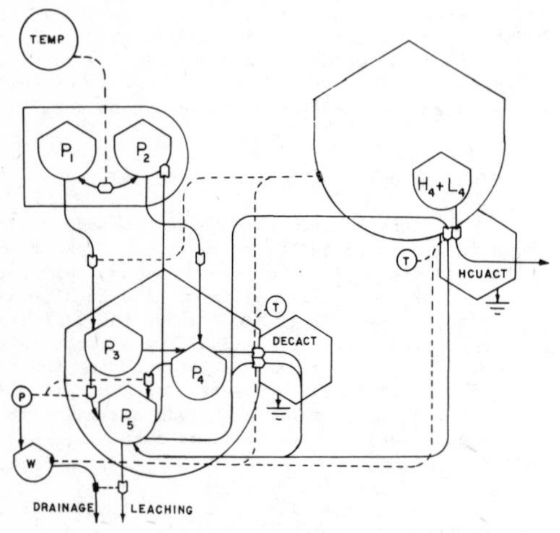

nutrients, any disturbance which alters decomposition rates alters nutrient release rates. This, in turn, alters vegetation growth rates and availability of nutrients in the soil. If decomposition, and thus nutrient release rates, were decreased by a toxic or inhibitory effect of oil, this would exacerbate the problem of nutrient limitation on the rate of oil decomposition.

Also, Everett (1978) noted that spilled oil reduced the available soil phosphorus. We have not yet postulated a mechanism by which this effect takes place; however, we expect to incorporate it into the model later.

The Hydrology Submodel

The hydrology submodel is included primarily as a means of driving the rest of the model. Interactions between the hydrology and petroleum carbon submodels may be seen in Figure 5. While the soils at our study site at Barrow were found to have relatively constant high moisture levels, there are some significant hydrologic events affecting the fate of spilled oil on tundra.

Figure 4.--Phosphorus submodel. P1 - phosphorus in aboveground live vegetation; P2 - phosphorus in belowground live vegetation; P3 - phosphorus in standing dead vegetation; P4 - phosphorus in belowground dead vegetation, litter, soil organic matter and soil humus; P5 - soluble and exchangeable soil phosphorus

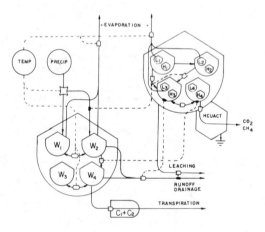

Figure 5.--Hydrology submodel. W1 - snowcover and frozen surface water; W2 - standing water; W3 - soil moisture, solid phase (frozen); W4 - soil moisture, liquid phase.

The time of a spill relative to spring melt-off will determine whether the oil is carried in surface runoff, resulting in a larger area being affected, but by a lower concentration of oil. This could have a significant effect on subsequent biodegradation if the concentration is reduced enough to ameliorate the toxic effects of the spill on the system. Standing water or water-saturated soil may prevent the spill from penetrating the soil until conditions are drier, but the presence of the spilled oil may also retard evaporation or reduce the wettability of the soil. Obviously, transport by ground water and surface runoff, and leaching of soluble components play a major role in determining the fate of spilled oil, but for the most part we leave these aspects to the extant spatial models mentioned earlier. The model treats these losses as simply proportional to the volume of water flow involved.

Another point of interaction between spilled oil and the hydrologic cycle is the increase in depth of thaw resulting from a spill. This is accompanied by an increase in soil temperature, which would tend to increase decomposition rates

of both the spilled oil and of the detritus.

The Petroleum Submodel

The petroleum submodel and its interactions with the rest of the model have already been seen in figures 1-5. This submodel consists of four main compartments, each divided into a high molecular weight (HMW : MW > 200) and a low molecular weight (LMW : MW ≤ 200) fraction. We have assumed for simplicity that the low molecular weight fraction is volatile, leachable and toxic while the high molecular weight fraction is relatively inert. We do not, at present, postulate any difference in rate of biodegradation for the two fractions. There is still debate over whether microorganisms degrade different fractions at significantly different rates in situ.

The mass balance equations for the petroleum submodel are given in Table 2. Spilled oil is

Table 2.--Mass balance equations for petroleum submodel.

H1--oil on aboveground live, high molecular weight fraction

$\dot{H}1$ = INT1*(1-KC)*SPILL - LOSSH1 - DRIP1*H1

    INT1 = KS1*(C1/KS2)*((KS3*C1)/(ALPHS4+C1))
    LOSSH1 = (H1/C1)*DEATHA
    DRIP1 = KS4*((H1+L1)-ASAT*C1)*COMPAR((H1+L1)-ASAT*C1,0)

L1--oil on aboveground live, low molecular weight fraction

$\dot{L}1$ = INT1*KC*SPILL - LOSSL1 - OEVAP*L1 - RLEACH*L1 - DRIP1*L1

    LOSSL1 = (L1/C1)*DEATHA
    OEVAP = KS6*(TEMP+273)
    RLEACH = KSL*RAIN

H2--oil on standing dead, high molecular weight fraction

$\dot{H}2$ = LOSSH1 + INT2*(1-KC)*SPILL - (H2/C3)*KC7*C3 - DRIP2*H2

    INT2 = KS1*(C3/KS2)*((KS7*C3)/(ALPHS5+C3))
    DRIP2 = KS4*((H2+L2)-DSAT*C3)*COMPAR((H2+L2)-DSAT*C3,0)

L2--oil on standing dead, low molecular weight fraction

$\dot{L}2$ = LOSSL1 + INT2*KC*SPILL + (L2/C3)*KC7*C3 - DRIP2*L2 - OEVAP*L2 - RLEACH*L2

H3--oil in surface pool, high molecular weight fraction

$\dot{H}3$ = (1-INT1-INT2)*(1-KC)*SPILL + DRIP1*H1 + DRIP2*H2 + (H2/C3)*KC7*C3 - SOAK*H3 - FLOATH

    SOAK = KS8*(SAT-W1)
    FLOATH = KS9*H3*RUNOFF

L3--oil in surface pool, low molecular with fraction

$\dot{L}3$ = (1-INT1-INT2)*KC*SPILL + DRIP1*L1 + DRIP2*L2 + (L2/C3)*KC7*C3 - SOAK*H3 - FLOATL
    - OEVAP*L3 + RLEACH*(L1+L2) - KS11*L3*INFLTR

    FLOATL = KS10*L3*RUNOFF

H4--oil in soil, high molecular weight fraction

$\dot{H}4$ = SOAK*H3 - DEGRAD*H4

    DEGRAD = HCUACT*(QOIL**((TEMP-10)/10)
    HCUACT = K16*(1-K17*L4+K18*(L4+H4)-K23*BPROD)*COMPAR(1+K18*(L4+H4),K17*L4+K23*BPROD)

L4--oil in soil, low molecular weight fraction

$\dot{L}4$ = SOAK*L3 + KS11*L3*INFLTR - KS12*L4*DRAIN - DEGRAD*L4

---

intercepted and absorbed by aboveground live and standing dead vegetation at a rate which follows a saturation curve to some maximum value as the amounts of aboveground live and standing dead vegetation increase. Oil on aboveground live vegetation is transferred to the oil on standing dead compartment by the death of aboveground live. Likewise, oil on standing dead is transferred to the surface pool by attrition of standing dead.

We assume that oil on aboveground live and standing dead vegetation drips into the surface pool at a rate proportional to the excess oil over that which can be absorbed by the plant materials. Evaporation and leaching of low molecular weight fractions are assumed to be linear functions of temperature and rainfall, respectively.

Once in the surface pool, spilled oil enters the soil at a rate inversely proportional to the degree of saturation of the soil. We assume that oil in the soil reduces the saturation point by physical displacement of soil water, but we have not as yet modelled any changes in the we wettability of the soil.

Once in the soil, we assume that the high molecular weight fraction is removed only through biodegradation, and the low molecular weight fraction only through leaching or biodegradation.

Auxiliary Variables and Forcing Functions

The auxiliary variables and forcing functions are given in Table 3. The temperature

Table 3.--Forcing functions and auxiliary equations.

PRECIP--total precipitation

PRECIP = table function of 29-yr average precipitation with linear interpolation

SOLAR--incoming solar radiation

SOLAR = table function of 14-yr average solar radiation with quadratic interpolation

TEMP--temperature

TEMP = BASE + AMP*SIN(FREQ+SHIFT)

TDOT = derivative of temperature function

DEPTH--depth of thaw

DEPTH = (K12*TEMP + K13*(H4+L4))*COMPAR(TEMP,0)

RUNOFF--surface runoff

RUNOFF = (IMPULS(23,52)+IMPULS(24,52))*KR + KRR*RAIN

RAIN--rainfall

RAIN = PRECIP*COMPAR(TEMP,0)

SPILL--oil spill

SPILL = PULSE(SWEEK)*KSPILL

BPROD--inhibitory byproducts

BPROD = KB1*(DECC4+DECC6) + KB2*(DEGRAD*(H4+L4)) - KB3*BPROD

COMPAR--comparator (CSMP built-fn macro function)

COMPAR(A,B) = 1 if A $\geq$ B, 0 if A $<$ B

IMPULS--impulse function (CSMP built-in macro function)

IMPULS(C,D) = 0 if TIME $\neq$ kD, 1 if TIME = kD, where k is a positive integer

PULSE--pulse function

PULSE(E) = 1 if TIME = E, 0 otherwise

and precipitation functions were generated from 29-year averages from the National Weather Service as reported by Bunnell and others (1975). The solar radiation function was generated from 14-year averages from the same source. Rainfall is generated from precipitation by the use of a switch which converts precipitation to rainfall when the temperature is above   Runoff is assumed to be proportional to rainfall and includes a pulse input of 2 weeks duration representing snowmelt in the early summer.

Depth of thaw is assumed to be function of temperature and the presence of oil. Perhaps a more realistic arrangement would be to allow soil temperature to increase in the presence of spilled oil. An increase in soil temperature following a spill is presumably due to a decrease in the shading effect of vegetation and to an increase in albedo. We hope to incorporate temperature into the model as an auxiliary variable as soon as vegetation dynamics seem reliable.

Decomposer activity and activity of hydrocarbon utilizers are at a maximum in the absence of spilled oil, and then modified in a linear fashion to represent the toxic inhibitory effects of low molecular weight hydrocarbons in the soil, the stimulatory effect of the presence of oil on on hydrocarbon utilizers and the postulated inhibitory effects of by-products of decomposition of detritus or spilled oil. These by-products are assumed to accumulate as a linear function of the respective decomposition processes and to undergo a loss proportional to the amount present.

Simulations of the forcing functions and auxiliary environmental variables are shown in figure 6. Solar radiation and precipitation functions necessarily fit the data, as they were generated directly from the data as table functions. Temperature was fit to a sine curve by nonlinear regression. The use of a single temperature function is somewhat simplistic, as some processes in the model depend upon air temperatures while others depend upon soil temperatures. Depth of thaw could probably be modelled more realistically as a function of soil temperature than air temperature.

Gaming with the carbon and petroleum submodels has indicated that while parameter estimates drawn from the literature for the carbon sector produce simulated results of a reasonable order of magnitude, additional calibration is required for some model parameters. In particular, the model is sensitive to the values chosen for maximum photosynthetic rate and for translocation rates between belowground live and above ground live. It is also possible that simulating translocation by assuming linear dependence upon the rate of change of temperature is inappropriate. This mechanism was chosen for convenience rather than from any theoretical rationale, and

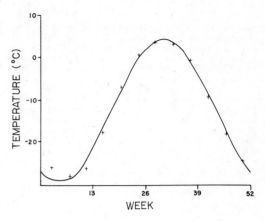

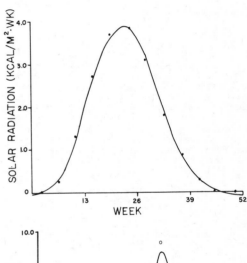

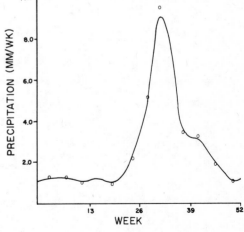

Figure 6.--Forcing functions. Solid lines are simulations.

rationale, and may be inadequate to describe the behavior of translocation. This rate will be a critical feature of the nitrogen and phosphorus submodels, and thus merits close scrutiny before those submodels are coupled to the rest of the model.

Decomposition rates postulated for detrital compartments are crucial to the interactions between the carbon and petroleum submodels. Fortunately, the model appears much less sensitive to these values. The actual values produced are altered only slightly by changes in the parameters in the case of the belowground dead, litter and soil organic matter compartment and the soil humus compartment. Soluble soil carbon is more strongly affected, but as this compartment is not directly coupled to the petroleum submodel, degradation rate of soluble soil carbon is less crucial to the simulation of oil decomposition.

Calibration and Stability

The petroleum submodel requires a good deal of additional calibration. Due to the lack of data on some processes, this submodel includes a number of parameters for which no estimates yet exist. Some parameters present difficulties in estimation due to the influence of confounding factors in the field data. Often it is impossible to separate the process of interest from the other factors influencing data values from the field. For example, it is difficult to infer biodegradation rates from field data on the amount of oil present in the soil over a sampling period, as biodegradation is not the only process affecting the amount of oil in the soil. A waterlogged or flooded soil may show very low amounts of oil until the soil has dried sufficiently to allow the spill to infiltrate the soil. Lack of data on soil temperatures for the experimental sites presents some difficulties, although soil temperatures might reasonably be inferred from depth-of-thaw measurements and from air temperature data, which does exist for part of the study period.

It has not yet been determined whether instabilities exhibited by the petroleum submodel are the result of too simplistic or erroneous mathematical assumptions or from parameter values which are in error by several orders of magnitude. More calibration and gaming is required before this question can be answered.

CONCLUSION

Much more work is required before the model will be suitable for drawing inferences about the behavior of the fate of oil spilled on arctic tundra soils. However, the exercise of developing the model has resulted in the generation of several hypotheses (see Table 4) about interactions between the spill and the tundra ecosystem which might influence biodegradation of spilled oil. The first three of these were developed early in the modelling effort in an attempt to

Table 4.--Hypotheses about controls on biodegradation rate to be investigated with the model.

1. Additional input of plant material to the soil as a result of oil toxicity provides a competitive substrate for hydrocarbon-utilizing microorganisms.

2. Increased activity of other decomposers results in competition for nitrogen and phosphorus in the soil.

3. Byproducts of oil decomposition are produced which inhibit further decomposition.

4. Byproducts of detritus decomposition inhibit oil decomposition.

5. Oil in soil inhibits decomposition of detritus and thus inhibits nutrient release.

6. Spilled oil in soil reduces availability of nutrients in soil (see Everett, 1978).

7. Initial elevation in soil temperatures following a spill increases biodegradation rate, but biodegradation rate then drops off as soil tempereatures return to normal, causing the appearance of a slowdown in biodegradation a few years after the spill.

8. Oil in soil reduces soil moisture availability for microorganisms by reducing the wettability of the soil.

9. Saturation of upper soil layers with oil interferes with water and gas movement in the soil and creates an anaerobic environment not conducive to rapid biodegradation.

account for an observed cessation or extreme slowdown in biodegradation in the second summer following the experimental spills. The rest were developed in the course of expressing the conceptual model in mathematical form. We intend to use these latter hypotheses to modify the simulation model to produce the most parsimonious (and tractable) simulation model which still includes all of the interactions we consider important for investigating the fate of oil spills in arctic tundra ecosystems.

ACKNOWLEDGMENTS

Portions of this work were funded by the U.S. Department of Energy. We thank Arthur Linkens, Kaye Everett, Alan Sexstone and Ellie Pesek for their contributions in generating data for this modelling effort.

LITERATURE CITED

Bunnell, F.L., and K.A. Scoullar. 1975. ABISKO II: A computer simulation model of carbon flux in tundra ecosystems. p. 425-448. In Rosswall, T., and O.W. Heal (eds.) Structure and function in tundra ecosystems. Ecol. Bull., Stockholm, Sweden.

Bunnell, F.L., and D.E.N. Tait. 1974. Mathematical simulation models of decompositional processes. p. 207-225. In Holding, et al. (eds.) Soil organisms and decomposition in tundra. Tundra Biome Steering Committee, Stockholm, Sweden.

Bunnell, F.L., S.F. MacLean, Jr., and J. Brown. 1975. Barrow, Alaska, USA. p. 73-124. In Rosswall, T., and O.W. Heal (eds.) Structure function in tundra ecosystems. Ecol. Bull., Stockholm, Sweden.

Dauffenbach, Lynne, William J. Mitsch, and Ronald M. Atlas. 1981. Modeling the fate of crude petroleum spills in Arctic tundra ecosystems. In Dubois, Daniel M. (ed.) Progress in ecological engineering and management by mathematical modelling. Proceedings of the conference. [Liege, Belgium, April 18-24, 1980] Centre Belge d'etudes et de Documetation, Liege, Belgium.

Davis, J.B., V.E. Farmer, R.F. Dreidner, A.E. Staub, and K.M. Reese. 1972. The migration of petroleum products in soil and groundwater: principles and counter measures. 36 p. American Petroleum Institute, Washington, DC.

Deitz, D.N. 1972. Behavior of components from spilled oil on their way through the soil. A.I.M.E.--Soc. Pet. Eng., Paper No. SPE 4077, Annual Fall Meeting, SPE, Oct. 1972.

Everett, K.R. 1978. Some effect of oil on the physical and chemical characteristics of wet tundra soils. Arctic 32(3):260-276.

McGill, W.B. 1975. A preliminary model of oil decomposition in soils. p. 291-312. In Proceedings of the conference on oil and salt water spills on land. 237 p. University of Calgary and Alberta Research Secretariat, Edmonton, Alberta, Canada.

Mackay, D. 1975. The physical behavior of oil spills on northern terrian. p. 81-110. In Proceedings of the conference on oil an salt water spills on land. 327 p. University of Calgary and Alberta Environment Research Secretariat, Edmonton, Alberta, Canada.

Mackay, D., P.J. Leinonen, J.C.K. Overall, and B.R. Wood. 1975. The behavior of crude oil spilled on snow. Arctic 23(1):9-20.

Miller, P.C., Boyd D. Collier, and Fred Bunnell. 1973. Development of ecosystem modelling in the U.S. IBP Tundra Biome Report, 73-2. 16 p.

AQUATIC TRANSPORT OF SINKING PARTICULATES: MODEL RESULTS AND IMPLICATIONS FOR DESIGN OF PLUME SAMPLING PROGRAMS AT OFFSHORE OIL AND GAS WELLS AND OTHER DISCHARGES[1]

Edward H. Dettmann[2]

Abstract.--Transport of particulate wastes by vertically-shearing currents in lake and marine environments is modelled. Veering of current vectors with depth causes particulate plumes to veer and foreshortens horizontal particle transport distance. These effects have implications for spatial design of plume sampling programs.

INTRODUCTION

Particulate wastes are often discharged into large lakes and marine environments, for example, in disposal of drill cuttings and muds from offshore gas and oil drilling rigs, as well as dumping of dredge spoils and sewage sludge. Patterns of movement and ultimate fate of discharged particulates are often of interest because of direct effects on water turbidity, physical covering of benthic communities by deposition on the bottom, and the dispersal of potentially toxic elements and compounds.

In this paper, a mathematical model is used to investigate the behavior of particulate plumes and spatial patterns of deposition on the bottom in the presence of vertical current shear (changing current speed and/or direction with depth). Implications of the results for optimal design of spatial sampling patterns in the study of particulate plumes are considered.

Most previous modelling studies of particulate waste transport, for instance those of Edge and Dysart (1972), Koh and Chang (1973) and Krishnappan (1975), have focused upon dump or jet discharges of barged wastes. These models have been reviewed by Johnson (1974) and Bowen (1976). The near-field behavior of the discharge jet or descending cloud of negatively buoyant material

was considered in detail by these models, although dispersion theory was used by some (Edge and Dysart 1972; Koh and Chang 1973) to describe far-field transport. Use of the dispersion equation to simulate particle transport requires numerical solution of model equations on a case-by-case basis, and may also require simplification of data on particle-size distributions and the current field in order to reduce a problem to manageable proportions.

THE MODEL

The model used in this paper focuses specifically on the effects of vertical current shear on particulate transport. In the analysis presented here, certain simplifications or idealizations are made in order to reduce the problem to its essentials. The model describes the transport of particulates introduced into the water column at low concentrations with zero momentum, or alternatively, the transport of particles once the momentum of a jet discharge has dissipated. Trajectories for particles with settling times small compared with the time required for significant diffusion to occur are examined in the assumed absence of turbulent or molecular diffusion.

This approach yields closed-form solutions of the model equations, and permits ready analysis of the functional relationships of both plume positional parameters and spatial deposition patterns to current field characteristics and the particle-size distribution. Closed-form solutions also simplify calculation of results for specific cases. By neglecting diffusional effects, however, this methodological approach eliminates the possibility of calculating particulate concentrations in the water column and

[1]Paper presented at the International Symposium Energy and Ecological Modelling, sponsored by the International Society for Ecological Modelling. (Louisville, Kentucky, April 20-23, 1981.)

[2]Edward H. Dettmann is Environmental Scientist, Division of Environmental Impact Studies, Argonne National Laboratory, Argonne, Illinois 60439 U.S.A.

predicting the lateral spread of material deposited on the bottom. In effect, the plume is assumed to be collapsed laterally onto the plume centerline. The model is being expanded to take diffusion into account, but this extension is unimportant to the analysis presented here.

With these assumptions, the equations of motion for an individual particle may be written

$$\frac{dx}{dt} = u(z) \tag{1a}$$

$$\frac{dy}{dt} = v(z) \tag{1b}$$

$$\frac{dz}{dt} = w(z) , \tag{1c}$$

where x, y, z are the particle coordinates (z positive upward), t is time, u(z) and v(z) are the x and y components of current velocity at depth z, and w(z) is the particle sinking velocity at depth z. These equations may be combined to give the following equations for the trajectory:

$$\frac{dx}{dz} = \frac{u(z)}{w(z)} \tag{2a}$$

$$\frac{dy}{dz} = \frac{v(z)}{w(z)} . \tag{2b}$$

For any given current field, defined by u(z), and v(z), these equations may be integrated to find particle trajectories. Solutions of these equations for a range of current fields have been described by Dettmann (1980, 1981).

While the detailed configuration of the current field depends upon governing mechanisms and driving forces, current speed in general falls off with depth. The direction of the current vector may also be depth-dependent, as described by Ekman (1905) for drift (wind-driven) and gravity currents. For drift currents, the current vector rotates uniformly, and decreases exponentially in magnitude with depth as in equations 3a and 3b, where z_o is the depth of release, ϕ_o is the angle between the current vector and the x-z plane, U_o is the current speed, and D_f is the depth of frictional influence.

$$u(z) = U_o \, e^{\frac{\pi}{D_f}(z-z_o)} \cos\left(\phi_o + \frac{\pi}{D_f}(z - z_o)\right) \tag{3a}$$

$$v(z) = U_o \, e^{\frac{\pi}{D_f}(z-z_o)} \sin\left(\phi_o + \frac{\pi}{D_f}(z - z_o)\right) \tag{3b}$$

MODEL RESULTS

If the functional dependence of u(z) and v(z) on z is known, equations 2a and 2b may be solved for particle trajectories (Dettmann 1980, 1981). More generally, however, the equations may be solved to give a general expression (equation 4) for the net horizontal transport distance of particles as a function of sinking speed and mean current components over the depth interval of interest (Dettmann 1981).

$$D = \left[\frac{z - z_o}{w}\right] \sqrt{(\bar{u})^2 + (\bar{v})^2} \tag{4}$$

One also finds that, if the sinking speed of a given particle is considered constant over the depth interval of interest, the displacement vectors of all particles along the plume centerline at any given depth make the same angle θ with the vertical plane containing the current vector at depth z_o where θ is given by equation 5 (Dettmann 1981):

$$\theta = \arctan\left(\frac{\bar{v}}{\bar{u}}\right) \tag{5}$$

This behavior is illustrated in figure 1. And, finally, if p(w) is the frequency distribution giving the fraction of particles with settling velocities between w and dw, the total mass settling rate M(D) in the horizontal distance interval D to D + dD is

$$M(D)dD = C \, p(w(D)) \left(\frac{dw(D)}{dD}\right) dD , \tag{6}$$

where C is the total loading rate, and where w(D) and dw(D)/dD are determined from equation 4 (Dettmann 1980, 1981).

Various cases of model behavior are illustrated in figures 2-4 for particles sinking through a depth interval of 10 meters. In each case, the current speed at the depth of release is chosen to be 0.1 m/s.

The dependence of plume veer angle on depth is illustrated in figure 2 for two current configurations. In one case a current vector of constant magnitude is assumed to rotate 0.2 radians/meter (11.5 degrees/meter) in a clockwise sense. In this case, the plume rotates 0.1 rad/m uniformly with depth. Total plume veer angle at 10-m depth is 1 radian (57.3 degrees). In the second case, the current vector again rotates 0.2 rad/m, and the current speed falls off exponentially with a decay constant of 0.2. This describes Ekman shear (eq. 3a, 3b) with $D_f = \pi/0.2 \sim 15.7$ m. In this case, the plume angle initially changes at the same rate as in the previous case, but changes more slowly at greater depth. At 10-m depth the total plume veer angle is 0.670 radians (38.4 degrees).

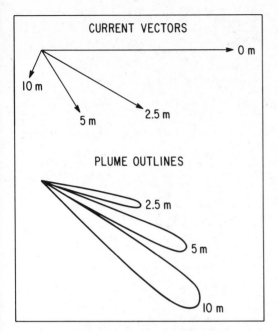

Figure 1.--Behavior of a particulate plume in the presence of a veering current. The plume veers in the same sense as the current, but more slowly.

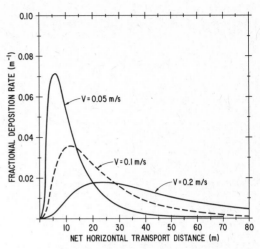

Figure 3.--Effect of the parameter V on the spatial deposition pattern for a medium to coarse sand ($\bar{\phi}$ = 1.0 on Wentworth scale).

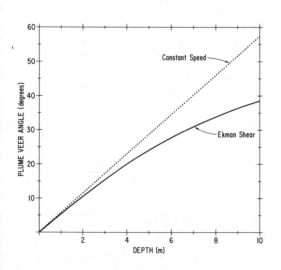

Figure 2.--Variation of plume veer angle for two sample current configurations.

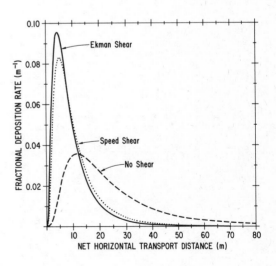

Figure 4.--Effect of current shear on the spatial deposition pattern for a medium to coarse sand ($\bar{\phi}$ = 1.0 on Wentworth scale).

It is apparent from equation 4 that the net horizontal transport distance is inversely proportional to particle settling speed, and directly proportional to the parameter V, where

$$V \equiv \sqrt{(\bar{u})^2 + (\bar{v})^2} \text{, and} \qquad (7)$$

V is equal to the mean current speed over the depth increment of interest only for unidirectional currents. For veering currents V is proportional to, but smaller than the mean current speed. Therefore, veering of the current vector restricts net horizontal transport distance.

The effects of mean current velocity and vertical shear on spatial deposition patterns are illustrated in figures 3 and 4 for a particle-size distribution that is log-normally distributed in size. The effects of a change in the magnitude of parameter V on the deposition pattern are shown in figure 3. Since V is proportional to the mean current speed in the depth interval of interest, the three curves in this figure may be interpreted as demonstrating the effects of factor-of-two changes in current speed. The ordinate in figure 3 is the fractional deposition rate, that is, the fraction of the total settling rate of material deposited per unit distance interval at the indicated distance. As V increases, the mean transport distance and the width of the settling distribution also increase.

The specific effects of vertical shear on spatial deposition patterns are shown in figure 4. The dashed line gives the deposition pattern for a current which is unidirectional and constant in speed (0.1 m/s) for the 10-meter depth increment of interest. The dotted line is for a unidirectional current having a speed of 0.1 m/s at the depth of particle release, but a mean speed of 0.0432 m/s over the 10-meter depth interval. In this case, the deposition pattern is more strongly skewed toward lower horizontal transport distances than for the case without shear, and a higher peak deposition rate is observed. The solid curve represents the deposition pattern for a current field with the same mean current speed over the depth interval of interest as the last example, but which veers with depth. The current pattern assumed is an Ekman Spiral, with $D_f = \pi/0.2 = 15.7$ m. The deposition pattern is slightly more strongly skewed toward smaller horizontal transport distances than in the previous example, and higher peak deposition rates are calculated. In addition, the path along which the particles settle is rotated by 38.4 degrees from the direction of the current vector at the point of release.

IMPLICATIONS FOR THE DESIGN OF SAMPLING PROGRAMS

A procedure sometimes used in sampling plumes is the following: the position of the plume at or near the surface is located, the horizontal concentration distribution is determined by sampling along horizontal transects perpendicular to the plume centerline, and the vertical concentration distribution is determined by sampling along vertical transects at selected points along the plume centerline.

The above analysis shows, however, that some current configurations will lead to complex behavior for particulate plumes. Samples taken along vertical transects from the centerline of the plume at 2.5-m depth in figure 1, for instance, may be far removed from the plume centerline at greater depths, and may miss the plume entirely.

One approach to sampling in a manner including the plume within the collection grid would be to sample at points along a series of arcs of circles concentric with the discharge. The lengths of these arcs would be chosen to maximize the likelihood of including the plume at all depths. This method requires collection of a large number of samples, and the length of time required to collect the samples increases the likelihood that changes in the velocity field or discharge conditions will result in changes of plume configuration during the sampling period.

Sampling requirements may be considerably reduced by use of a plume-location technique in order to determine optimal sampling points. For example, acoustic sounding, transmissometry, and nephelometry may be used for plume location. Each of these techniques requires towing a sensor through the water along a transect of interest. Of these three techniques, acoustic sounding (Proni et al. 1976) is possibly the most efficient since relative particulate abundance may be determined for all depths along a given horizontal transect in a single tow, while transmissometry and nephelometry require a tow at each sampling depth.

An alternative method for reducing sampling requirements would be to monitor the current-velocity field before and during the sampling program, and to calculate the expected plume position at any given sampling depth. This method requires use of a current-meter array spanning the depth interval of interest, and capable of being interrogated rapidly before and during the sampling program. If this method is to be practical, currents would need to remain reasonably constant during the sampling program.

LITERATURE CITED

Bowen, Stuart P. 1976. Modeling of coastal dredged material disposal. P. 202-225. In Proceedings of the Speciality Conference on Dredging and Its Environmental Effects. [Mobile, Alabama, January 26-28, 1976.] American Society of Civil Engineers, New York, N.Y.

Dettmann, Edward H. 1980. Transport of Particulate Matter by Shearing Currents in Lake Erie. Argonne National Laboratory Report ANL/ES-86, 21 p. Argonne, Ill.

Dettmann, Edward H. 1981. Aquatic transport of sinking materials by vertically shearing currents. Manuscript in preparation, to be published.

Edge, Billy L. and Dysart, Benjamin C., III. 1972. Transport Mechanisms Governing Sludges and Other Materials Barged to Sea. Civil Engineering and Environmental Systems Engineering, Clemson University.

Ekman, V. Walfrid. 1905. On the influence of the earth's rotation on ocean-currents. Arkiv för Matematik, Astronomi och Fysik 2, paper no. 11, 52 pp.

Johnson, Billy H. 1974. Investigation of Mathematical Models for the Physical Fate Prediction of Dredged Material. Technical Report D-74-1, 54 p. & app. U.S. Army Engineer Waterways Experiment Station, Vicksburg, Miss.

Koh, Robert C.Y., and Y.C. Chang. 1973. Mathematical Model for Barged Ocean Disposal of Wastes. U.S. Environmental Protection Agency Report EPA-660/2-73-029, 178 p. & app. Washington, D.C.

Krishnappan, B.G. 1975. Dispersion of Granular Material Dumped in Deep Water. Scientific Series No. 55, Inland Waters Directorate, Canada Centre for Inland Waters, 114 p. Burlington, Ontario.

Proni, John R., Fred C. Newman, Ronald L. Sellers, and Charles Parker 1976. Acoustic Tracking of Ocean-Deumper Sewage Sludge. Science 193:1005-1007.

3. Aquatic Ecosystem Models
and Power Generation

ECOSYSTEM MODELS TO ASSESS IMPACTS OF POWER PLANT COOLING SYSTEMS[1]

Ishwar P. Murarka, Donald B. Porcella, and Robert W. Brocksen[2]

Abstract.--The Electric Power Research Institute sponsored research to develop and demonstrate the use of ecosystem models to assess effects of cooling system operations at one or multiple power plants located on a single waterbody. The research has developed methods for ecosystem and population analysis.

The methods have been tested by carrying out a demonstration application to a lake ecosystem. In simulating these results, researchers were constrained by an inadequate understanding of certain ecosystem functions. Future contributions in this methodology should, therefore, be in the form of improved quantitative definition of ecosystem dynamics to increase precision and sensitivity in assessing ecosystem change.

INTRODUCTION

Since 1975, the Electric Power Research Institute (EPRI) has funded several research projects to advance methods for assessing aquatic ecological effects of cooling water uses (Goldstein 1979). The projects have ended in advancing the state-of-the-science in ecosystem modelling.

The overall assessment methodology consists of models and computational procedures designed to simulate the major features of ecosystem behavior and their interaction with power plant operations. The methodology allows utilities, regulatory agencies, and researchers to analyze power plant impacts on aquatic communities and to design facilities and operational procedures that mitigate those impacts.

[1]Paper presented at the international symposium Energy and Ecological Modelling, sponsored by the International Society for Ecological Modelling. (Louisville, Kentucky, April 20-23, 1981).

[2]Ishwar P. Murarka is Project Manager, Environmental Physics & Chemistry Program, Electric Power Research Institute, Palo Alto, California U.S.A.; Donald B. Porcella is Principal Scientist, Environmental Systems Engineering, Tetra Tech Inc. Lafayette, California U.S.A.; and Robert W. Brocksen is Program Manager, Ecological Studies Program, Electric Power Research Institute, Palo Alto, California U.S.A.

The issue is to determine the potential for aquatic ecosystem damage and how to at least minimize damage. Frequently, an issue is related to recognized problems in such a way that the approach and understanding cannot be tested; in other words, no hypotheses can be tested and processes cannot be measured. Because of the cost of environmental regulation and the social and ecological costs of environmental damage, the most important part of environmental research is this need to be able to test approaches for their scientific validity. In addition to their analytical role, models are useful for framing hypotheses for later testing in the field. The EPRI research essentially took the task of developing and testing approaches so that potential for ecosystem damage can be properly estimated. In this paper we will describe four projects and report the results from two completed projects which have contributed to the science of ecosystem modelling.

Under EPRI Research Project entitled: "Development of a Methodology for Assessing Integrated Effects of a Number of Thermal Power Stations on a Single Body of Water" (Research Project 878) the Contractor (Tetra Tech, Inc.) developed an ecosystem simulation model that was test demonstrated to assess multiple power plant effects on the Lake Cayuga, New York, ecosystem. In a companion project entitled: "Development of Methodology for Assessing Ecosystem Effects Related to Intake of Cooling Waters" (Research Project 876) the Contractor (Lawler, Matusky and Skelly, Engineers) catalogued the methods for assessing population and ecosystem effects of impingement and entrainment on major fish and invertebrate species and carried out a test demonstration of the methodology restricted to population-level assessment of effects due to multiple plant development on Lake Cayuga. While projects 876 and 878 resulted in ecosystem or population models for assessing ecological effects, a somewhat different approach was taken in studying Lake Sangchris and Lake Shelbyville by the Illinois Natural History Survey (INHS) under EPRI Research Project 573 entitled: "Evaluation of Cooling Lake Fishery." Instead of first developing a model and simulating the effects on an ecosystem, the Illinois Project made use of extensive field investigations to formulate a Cooling Lake Ecosystem Model as a tool to manage viable sport fisheries. The output of this research effort was a Lake Sangchris fisheries evaluation model.

Cooling water use effects on lake's fisheries were examined by a case history approach under EPRI Research Project 880 entitled: "Synthesis and Analysis of Cooling Pond and Lake Information" by Battelle Pacific Northwest Laboratory. Fisheries data for 14 cooling lakes were evaluated to determine established levels of populations over 10 or more years in use as a cooling water body.

In the following sections we will elaborate on the ecosystem models that have been developed by Tetra Tech, Inc. and the Illinois Natural History Survey.

ECOSYSTEM ASSESSMENT MODEL - TETRA TECH, INC.

EPRI Research Project 878 was initiated to partially fulfill a need to develop methods for qualitative and quantitative assessment of ecological effects of multiple power plants on a single water body. The Contractor (Tetra Tech, Inc.) approached this objective by advancing a set of model-oriented computational procedures which incorporate the direct effects of power plant cooling system operation with ecosystem behavior to allow an assessment of ecosystem impacts.

There are two components to the Ecosystem Assessment Model (EAM): (a) Ecosystem Simulation Model, which comprises ecosystem functions and hydrodynamic modeling and (b) Power Plant Simulation Model. There is a five volume EPRI report (Tetra Tech, Inc. 1979a, 1979b, 1979c, 1980a, 1980b) which has documented the details on model formulation and use.

Ecosystem Simulation Model

The Ecosystem Simulation Model represents the prototype by a system of elements to provide various levels of spatial detail. The waterbody being modelled is divided into a one-, two-, or three-dimensional system of elements, depending on the complexity. Each element is assumed to be well mixed to give a uniform constituent concentration within each element. Appropriate

hydrodynamic models can be used to simulate advective flows and heat balances in one, two, or three dimensions. Flow data and temperatures are used by an interface program which can integrate over time and space for use in the ecosystem model.

The ecosystem model simulates the transport and reactions of 46 state variables as listed in table 1. The ecosystem model computes the changing concentration of each variable on a daily time step.

Table 1.--List of state variables in the ecosystem model.

- fish (four groups, five life stages)
- benthic animals
- zooplankton (three groups)
- phytoplankton (four groups)
- attached algae
- suspended organic detritus
- organic sediment
- ammonia nitrogen
- nitrite nitrogen
- nitrate nitrogen
- phosphate phosphorus
- dissolved silica
- suspended particulate silica
- sediment silica
- total inorganic carbon
- alkalinity
- pH
- dissolved oxygen
- biochemical oxygen demand
- total dissolved solids
- toxic substances
- coliform bacteria

The concentrations of all biotic variables are expressed in terms of dry weight biomass. Differential equations for each variable express the rate of change in concentration as a function of advection, dispersion, inflows, outflows, and sources and sinks. The terms for advection, dispersion, inflows, and outflows account for the hydraulic transport mechanisms which move the constituents throughout the waterbody, while the source and sink terms account for all biological, chemical and physical reactions which change the constituent concentrations within each element.

The general trophic relationships included in the model are illustrated in figure 1. The mass balance equations for each state variable are described in detail in Section 3 of Volume II of the EPRI report (Tetra Tech. Inc. 1979b). The model output includes the value of each state variable at each time step in each element.

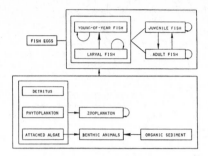

Figure 1.--Potential trophic relationships in ecosystem model.

Power Plant Simulation Model

The power plant simulation model calculates numbers of organisms entrained and impinged based on specified formulations, cooling water flow rates, and constituent concentrations computed by the ecosystem model for the element in which the intake is located. The power plant model allows for withdrawal from one element of the ecosystem, passage through a cooling system and discharge plume, and then release into another element of the ecosystem. The various portions of the model are shown in figure 2. Computation of the direct effects of plant operation has been divided into three categories: uptake, plant passage, and plume. The intake model provides for an approximation of impingement rate and modification of ambient concentrations of other biotic constituents to account for organism behavior, intake location, and physical characteristics. Biological functions included are: zooplankton and meroplankton vertical migration, juvenile and adult fish reaction to bypass louvers, juvenile and adult fish survival of screens and screen wash return systems. The plant passage model provides for an approximation of mortality rates for all organisms entrained in passing through each cooling system. Total mortality for each group of organisms is computed based on individual estimates of mortality due to biocide use, mechanical effects and thermal stress. The organisms surviving plant passage are discharged back to the appropriate element of the ecosystem. Further details of this model can be found in Section 4 of Volume II (Tetra Tech, Inc. 1979b).

Test Application to Lake Cayuga

In this example we illustrate the use of the Ecosystem Assessment Model by applying it to Cayuga Lake in New York.

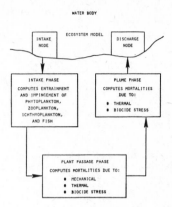

Figure 2.--Relationship of power plant simulation model to ecosystem model.

The overall approach is based on a description of alternative scenarios, prescreening using simple hand calculation procedures, followed by application of the ecosystem model. The alternative scenarios depend on vertical and horizontal location of intakes and discharges, cooling water flows and condenser residence times, biocide usage, and other important factors. Some scenarios can be eliminated during prescreening or consideration of site constraints but most will require evaluation by the model.

Scenarios

The scenarios (table 2) for Cayuga Lake were based on the present situation having an existing power plant (No. 2), a baseline situation before any power plant was located on the lake (No. 1), and three hypothesized situations (No. 3-5) where additional power plants would be added cumulatively to Cayuga Lake. These power plants were selected to increase significantly the thermal loading (from 0.015 up to 0.24 MWe/acre) and the hydraulic cycling of cooling water (cooling water residence time relative to lake volume increases up to 0.1 percent/day). Generally, the hypothesized power plant characteristics and operating conditions were defined to be the same as that of the existing plant--Milliken station, Units 1 and 2, New York State Electric and Gas Company.

Critical variables were intake locations (depth and lake area) condenser flow rates, ΔT, and biocide use. The intakes were all located below the summer thermocline at 12 meters depth. Discharges were at shoreline locations. Condenser flow rates and ΔT were based on typical values for fossil fuel plants. Biocide use followed the schedule of chlorination for the Milliken Station.

Scenario Number	Plants	Location	m³/sec 24-hour Average Cooling Water Withdrawn	At	Intake Depth, m (Node #)	Discharge Depth, m (Node #)	Total MWe
1	None	None	None	None	None	None	0
2	Existing	Mideast shore	10.65	8.3	12 (9)	1 (10)	270
3	E + FFI	Mideast shore	39.0	9.0	12 (9)	1 (10)	1270
4	E + FFI	Mideast shore	39.0	9.0	12 (2)	1 (3)	1270
	FFII	Southwest	28.36	10.0	12 (2)	1 (3)	1000
5	E + FFI	Mideast shore	39.0	9.0	12 (15)	1 (16)	1270
	FFII	Southwest	28.4	10.0	12 (15)	1 (16)	1000
	FFIII	Northeast	56.6	10.0	12 (15)	1 (16)	2000

Table 2.--Scenarios for Cayuga Lake multiple power plant cooling system effects.

Normally, the plant location would be selected to minimize visual and ecosystem impacts. Two of the hypothetical plants (scenario 4 and 5) were selected to be placed in critical habitats; a spawning area for lake trout was used for the first plant (scenario 4) and a summer feeding area was used for the second (scenario 5).

Computations and Results

After a review of data and publications on morphology, climate, hydrology, circulation, water chemistry, and biology of Cayuga Lake, the initial conditions were determined. Then simulations were carried out to compute the behavior of selected state variables (temperature, D.O., nutrients, phyto- and zooplankton and fish groups) to determine the differences in ecosystem effects due to the various scenarios of multiple power plant development on Lake Cayuga. No calibration or validation were performed and only tentative conclusions were drawn. Phosphorus concentrations for each state variable are similar to field measurements. Phosphorus flux values are reasonable and are similar to results reported by Scavia (1981). These results support the conclusion that the ecosystem model represents ecosystem processes and state variables accurately.

Near field and far field effects of power plant operation can be compared. Annual average concentrations of green algae, cladocerans, adult lake trout, and adult clupeids (mostly, alewives) for two water columns impacted by hypothetical power plants showed progressive changes with cumulative addition of power plants (table 3).

Scenario	Green Algae Intake	Green Algae Discharge	Cladocerans Intake	Cladocerans Discharge	Lake Trout Intake	Lake Trout Discharge	Clupeids Intake	Clupeids Discharge
1	0.104	0.108	0.125	0.111	0.00194	0.00190	0.0722	0.0301
2	0.118	0.154	0.131	0.127	0.00199	0.00232	0.0347	0.0435
3	0.236	0.230	0.124	0.120	0.00214	0.00289	0.0449	0.0568
4	0.352	0.306	0.118	0.116	0.00208	0.00276	0.0468	0.0568
5	0.227	0.221	0.119	0.119	0.00194	0.00257	0.0402	0.0508

Table 3.--Annual average of selected biological variables in two water columns impacted by hypothetical power plants.

Compared with the no power plant scenario (Number 1) the scenarios with power plants had increased green algae and clupeid biomass and no change in cladocerans or lake trout for the intake water column. The discharge water column had a somewhat different pattern (increased biomass of green algae, clupeids and lake trout) and differed quantitatively from the intake.

Near field biomass is similar to the lakewide results (far field effects). Results in table 4 show that nutrients are increased by power plant operation resulting in increased productivity and biomass of the different trophic groups. The increased productivity results from increased vertical transport of nutrients from the deeper waters of Cayuga Lake where regeneration of nutrients occurs. The additional nutrients increased growth, compensating in part for increased mortality due to entrainment and impingement. As expected, higher phosphorus flux was associated with phytoplankton, zooplankton and fish for those scenarios having more power plants. Although greater phytoplankton activity occurred, lakewide dissolved oxygen (DO) concentrations were less because of lower DO saturation at the higher lakewide temperatures occurring with added power plants.

Scenarios	Water Quality Variables Temperature °C	DO, mg/l	pH	Biomass of Biota (Total of Groups) mg/l Phytoplankton	Zooplankton	kg/ha Fish
1	7.85	10.2	7.96	0.084	0.097	18.3
2	8.04	10.3	8.00	0.101	0.120	18.5
3	8.18	10.3	7.99	0.116	0.129	19.6
4	8.33	10.2	7.99	0.130	0.141	20.5
5	8.51	10.1	7.98	0.126	0.140	20.3

Scenarios	Nutrients, mg/l PO_4-P	$NH_3 + NO_3$-N	CO_2	Phosphorus Flux µg P/l day Phytoplankton uptake	Zooplankton Respiration	Fish Respiration
1	0.020	0.67	0.76	0.076	0.013	0.013
2	0.025	0.82	0.86	0.080	0.013	0.013
3	0.025	0.82	0.86	0.094	0.015	0.014
4	0.024	0.81	0.87	0.113	0.016	0.014
5	0.024	0.81	0.90	0.103	0.016	0.014

Table 4.--Comparison of lakewide annual average concentrations of selected ecosystem variables.

Findings

The application of the multiple power plant methodology to a prototype system, Cayuga Lake, tested its role as a tool for

- evaluating the regulatory implications of single and multiple power plant operation on lake ecosystems (entrainment, impingement and near-field effects).

- evaluating ecosystem impacts (both positive and negative) of multiple power plants on a lake ecosystem (far-field effects) in an objective manner.

- comparing alternative designs, sites, and mitigating technology.

- assessing the potential of using a power plant to manage aquatic resources through deep mixing (or not mixing) operations.

The overall assessment of this illustration suggests that up to a certain level, an increase in power plants on Cayuga Lake would have little or no detectable effect on the lake ecosystem community composition or biomass.

Added linear effects of single power plants would result in higher predicted impacts than were predicted by this ecosystem simulation. Some processes are density dependent (e.g., probability of being impinged) and simple linear addition of effects would result in an overestimate of impacts.

Most of the increased biomass resulted from transport of recycled nutrients from the deep layer to the surface layer. The location of the intakes in deep layers had the beneficial effect of more efficient power plant operation and less entrainment and impingement than would be expected for a surface intake.

In this respect compensation by fish populations is an inherent result of the modeling approach and a mechanism for compensation need not be hypothesized. Thus positive and negative effects on fish biomass are inherent in the modeling approach because an attempt has been made to include all fundamental processes of fish growth and decay. Therefore, designs to mitigate impacts can be tested at the ecosystem level by use of the model, such as effects of placement of intakes on fish biomass.

LAKE ECOSYSTEM MODEL - INHS

EPRI Research Project 573 was initiated to evaluate the Lake Sangchris fishery and to develop a model (or conceptual understanding) of those aspects that appear important for managing a cooling lake fishery. The Contractor (Illinois National History Survey) approached the research by examining a cooling lake (Lake Sangchris) and a nearby unheated lake (Lake Shelbyville) to focus on aspects of the waterbodies which support viable fish populations. Fish growth, reproduction, food habits, movements, primary production and water quality variables were extensively examined. The Lake Sangchris Cooling Lake Ecosystem Model (CLEM) represents an attempt to merge the general nature of a purely theoretical model with the specific applicability of a more empirical simulation model. The model structure has coupled the outside lake ecosystem forces (management practices, climatological effects, and power plant influences) to the dynamics of the fishery through all relevant pathways, direct or indirect. There is a four-volume EPRI report (Illinois Natural history Survey 1979a, 1979b, 1979c, 1980) which

has documented the details on model formulations and uses.

Cooling Lake Ecosystem Model (CLEM)

The Cooling Lake Ecosystem Model consists of a number of submodels as shown in table 5. Each submodel is basically a system of differential equations that describe mass transfer and, in the fish model, changes in numbers of individuals. Physiological processes provide the primary level of description in the biological models. The fish submodel has the most complex structure because a high degree of specificity has been included. Readers are referred to Volume 2 of the EPRI report (Illinois Natural History Survey 1979a) for the details of the Model.

Table 5.--Lake Sangchris cooling lake ecosystem model contents.

State Variables	Components (Numbers, description)
Mass	
Phosphorus	1, orthophosphate
Detritus	1, organic detritus
Phytoplankton	2, bluegreen, non-bluegreen
Periphyton	1, diatom-like
Macrophyton	1, submerged and floating leaf
Zooplankton	1, herbivorous cladocera and copepoda
Benthos	1, based on chironomids
Numbers	
Fish	5 species
	8 life stages (eggs, larvae, postlarvae, 3 subadult sizes, 2 adult sizes)

Results and Findings

Because of the objective to develop an ecosystem model specific to Lake Sangchris, the researchers used extensive field investigations as the primary path to arriving at CLEM. Field investigations concerned with age, sex products, and enzyme composition were used to estimate the effects of the thermal effluent on the growth, life span, age at maturity, spawning periods, and genetic selection of various fishes.

The distribution and movements of various fishes were examined by long-term radiotelemetry and tagging studies as well as analyses of time-effort samples collected for Commonwealth Edison Company. The composition and population density of the reservoir were evaluated using standing crop estimates and the time-effort samples. A comprehensive extension of the creel survey program provided reliable estimates of the economic value and public utilization of the fishery. Dynamic interactions between the fish and fish food resources were determined in a food habits investigation of all important fishes of Lake Sangchris.

The important fish food resources, particularly benthos, zooplankton, and algae, were also examined. The composition and abundance of all groups in Lake Sangchris were compared with data from Lake Shelbyville and the literature. In addition, seasonal fluctuations and life histories of principal benthic macroinvertebrates from Lake Sangchris were examined. Profundal and littoral populations of both benthos and zooplankton in Lake Sangchris were compared. Seasonal variability of zooplankton from various areas of the cooling lake were also investigated. Primary and secondary (zooplankton) production were measured and evaluated in heated and ambient areas of the lake and in Lake Shelbyville for comparison. Phytoplankton and periphyton algal communities were examined with particular emphasis on the persistence of blue-green algae in those communities. The cell sizes of various algal species were measured to determine if they were typical of midwestern flora. An ecosystem model was constructed using the data from field investigations as a final part of the project.

The greatest value of the model was its applicability as a tool for examining interactions within the ecosystem. It was constructed to permit evaluation of the effects of manipulation of physical or biological parameters on the fishery of Lake Sangchris.

As an example of the usefulness of the CLEM to analyze management alternatives on fisheries, simulations were carried out to see how spawning success affects the biomass of gizzard shad (fig. 3) and the zooplankton (fig. 4), which is the principal source of food for the shad. Resultant biomass of gizzard shad shows significant variations with greatest yearend biomass occurring at lowest spawning success. This may have resulted because of overpredation of the food resource (zooplankton biomass) by the greater populations of gizzard shad that occur at higher spawning success.

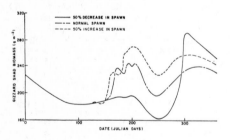

Figure 3.--Simulations demonstrating the effects on gizzard shad biomass of a 50% increase or decrease in spawning success.

The effects of variable fishing effort on fish standing crop is important because fishing effort is controlled by regulating fishing habits with various management tools (fishing season and hours, bag limits, fishing methods). As shown in figure 5, the expected standing crop of game species decreases sharply as fishing effort is increased for largemouth bass and channel catfish while the forage fishes, bluegill and gizzard shad (prey for game species) increase. White bass could sustain a greater fishing effort and the manager might encourage fishermen to fish differently to capture a greater percentage of this species.

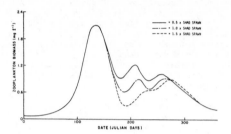

Figure 4.--Simulation results indicating the effect of changes in gizzard shad spawning success on zooplankton biomass.

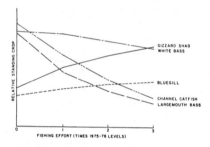

Figure 5.--The expected standing crops of gizzard shad, white bass, bluegill, channel catfish, and largemouth bass as predicted from one-year simulation runs at fishing efforts of 0-, 1-, 2-, and 3-times the estimated 1975 through 1976 levels.

STATUS OF THE ECOSYSTEM MODELS

In the preceding sections, we have summarized the two major ecosystem models that were developed and tested. The Tetra Tech, Inc. work illustrated that complex ecosystem models are important to make a priori assessment of effects as a result of multiple power plants on a single waterbody.

The model as defined provides a way of estimating the effects of power plants on aquatic ecosystems. Because the entire lake is simulated, all significant biological groups are evaluated, and all seasons are considered, the integrated effects of power plants can be estimated on the aquatic community. The model conceptual and computational constraints may reduce the accuracy of prediction as do field sampling procedures. The results must therefore be applied with judgment and experience.

For the model contained in this methodology, some qualitative estimates can be made as to the accuracy of the various predictions. Predictions are probably most accurate for chemical processes, i.e., DO, pH, nutrients, conservative constituents and lower food chain organisms (phytoplankton, zooplankton); fish are less accurately described than the plankton. Attached organisms (benthic invertebrates, Cladophora) are not well modeled nor are sediments.

In general, temperature distributions, mixing, and circulation patterns can be adequately simulated with existing hydrodynamic models. However, more spatial resolution (more nodes) and a much shorter computational time step is typically required than for an analagous ecosystem simulation. As a result, hydrodynamic models tend to be relatively expensive to apply and, therefore, may not be cost-effective. A more simplistic approach was chosen to estimate the long-term average circulation in Cayuga Lake for this illustration of the methodology. This provides an example of the type of analysis one may perform with a minimum of data and effort. In this case, the configuration of the lake and the locations of the major inflows and outflows simplified the analysis. For a more complicated situation, more specific information on the existing circulation patterns would probably be required to estimate the average net flows in the absence of a suitable hydrodynamic model. Alternatively, the user may select an appropriate hydrodynamic model and obtain the necessary data to apply it.

The Illinois Natural History Survey work provided an illustration of using a reverse approach in modeling an ecosystem. The researchers developed the insights of ecosystem processes by collecting and analyzing data. This naturally followed by a mathematical abstraction of an essential structure to simulate production of a sport fishery in a cooling lake.

Some improvements in the model could be made if larval fish dynamics were better understood. Fish migration and transport of plankton could be incorporated if greater resolution were required. Although improved modeling of physical chemical processes could be incorporated to increase the applicability of the model, it is a useful tool to investigate the dynamics of cooling lakes.

The results of simulations, although only indicative of trends, could be applied to impact analysis and appraisal of management alternatives for Lake Sangchris as well as other systems. The greatest value of the model may be in investigating the compatibility of power generating facilities in general and aquatic ecosystems in cooling lakes. There is great potential for utilizing the energy of heated water to increase the total production and, consequently, the recreational value of reservoirs used for cooling water.

LITERATURE CITED

Goldstein, Robert A. 1979. Development and implementation of a research program on ecological assessment of the impact of thermal power plant cooling systems on aquatic environments. p. 117-130. In Environmental Biomonitoring, Assessment, Prediction, and Management - Certain Case Studies and Related Quantitative Issues, ed., J. Cairns, Jr., G. P. Patil, and W. E. Waters, International Cooperative Publishing House, Fairland, Maryland.

Tetra Tech, Inc. 1979a. Methodology for evaluation of multiple power plant cooling system effects. Volume I: General description and screening. Electric Power Research Institute Report No. EA-1111, 98 p. Palo Alto, CA.

Tetra Tech, Inc. 1979b. Methodology for evaluation of multiple power plant cooling system effects. Volume II: Technical basis for computations. Electric Power Research Institute Report No. EA-1111, 130 p. Palo Alto, CA.

Tetra Tech, Inc. 1979c. Methodology for evaluation of multiple power plant cooling system effects. Volume III: Data Requirements. Electric Power Research Institute Report No. EA-1111, 179 p. Palo Alto, CA.

Tetra Tech, Inc. 1980a. Methodology for evaluation of multiple power plant cooling system effects. Volume IV: User's guide to model operation. Electric Power Research Institute Report No. EA-1111, 264 p. Palo Alto, CA.

Tetra Tech, Inc. 1980b. Methodology for evaluation of multiple powerplant cooling system effects. Volume V: Methodology application to prototype-Cayuga Lake. Electric Power Research Institute Report No. EA-1111, 101 p. Palo Alto, CA.

Lawler, Matusky & Skelly Engineers. 1979. Methodology for assessing population and ecosystem level effects related to intake of cooling waters. Electric Power Research Institute Report No. EA-1238, 164 p. Palo Alto, CA.

Lawler, Matusky & Skelly Engineers. 1980. Methodology for assessing population and ecosystem level effects related to intake of cooling waters. Volume 1: Handbook of methods population level techniques. Electric Power Research Institute report No. EA-1402, 391 p. Palo Alto, CA.

Lawler, Matusky & Skelly Engineers. 1980. Methodology for assessing population and ecosystem level effects related to intake of cooling waters. Volume 2: Handbook of Methods Community Analysis Techniques. Electric Power Research Institute Report No. EA-1402, 306 p. Palo Alto, CA.

Battelle, Pacific Northwest Laboratories. 1979.
Synthesis and analysis of ecological infor-
mation from cooling impoundments. Volume 1.
Electric Power Research Institute Report No.
EA-1054, 103 p. Palo Alto, CA.

Battelle, Pacific Northwest Laboratories. 1979.
Synthesis and analysis of ecological infor-
mation from cooling impoundments. Volume 2:
Appendix A - study site histories and data
synopsis. Electric Power Research Institute
Report No. EA-1054, 283 p. Palo Alto, CA.

Illinois Natural History Survey. 1980. Evalu-
ation of a Cooling Lake fishery Volume 1:
Introduction, Water Quality and Summary.
Electric Power Research Institute Report No.
EA-1148, 70 p. Palo Alto, CA.

Illinois Natural History Survey. 1979a.
Evaluation of a Cooling-Lake Fishery Volume
2: Lake Sangchris Ecosystem Modelling.
Electric Power Research Institute Report No.
EA-1148, 205 p. Palo Alto, CA.

Illinois Natural History Survey. 1979b.
Evaluation of a Cooling-Lake fishery. Volume
3: Fish population studies. Electric Power
Research Institute Report No. EA-1148, 304
p. Palo Alto, CA.

Illinois Natural History Survey 1979c.
Evaluation of a Cooling-lake fishery.
Volume 4: Fish food resources studies.
Electric Power Research Institute Report No.
EA-1148, 749 p. Palo Alto, CA.

Scavia, D. 1981. Use and interpretation of
detailed, mechanistic models of phytoplankton
dynamics. In Phytoplankton - Environmental
Interactions in Reservoirs. U.S. Army Corps
of Engineers. Waterways Experiment Station,
Vicksburg, MS. In press.

EVALUATION OF THE USEFULNESS OF ECOLOGICAL SIMULATION MODELS IN POWER PLANT IMPACT ASSESSMENT[1]

Gordon L. Swartzman, Robert T. Haar, Daniel H. McKenzie and Thomas Zaret[2]

Abstract.--Comparisons were made of the equations, ration-ale, data sources and parameter values of 26 simulation models of fish and zooplankton population dynamics and energetics and results were compared in standard notation and units, process by process. The major process categories considered were con-sumption, predation, metabolic processes, assimilation, growth, fecundity, recruitment and mortality. A model simulation lan-guage, AEGIS (Aquatic Ecosystem General Impact Simulator) was built to compare model equations process by process allowing convenient interchange of model equations for any process "module". This simulator was parameterized to a test site, Lake Keowee, South Carolina, on which resides the Oconee Nuclear Power Station. Model parameter estimation and comparison of these models with biological monitoring data allows evaluation of ecosystem models from the standpoint of 1) prediction of behavior under normal and perturbed conditions, 2) organization of data into an ecosystem framework, and 3) evaluation of data to address impact questions.

INTRODUCTION

Within the field of impact assessment simulation models play an ambivalent role. On the one hand, it is recognized that simulation models, with their focus on changes in biota through time, and the mechanisms which link these changes to environmental conditions, and to direct perturbations on the biota, present an ideal medium for exploring the possible consequences of various types of impacts. On the other hand, no simulation models at present have proven effective at giving accurate quantitative predictions of the effects of impacts (Swartzman et al. 1978).

Power plant impacts on aquatic ecosystems offer a good case in which to study the usefulness of simulation models in impact assessment. The

direct impacts from power plants are quantifiable in terms of number of fish and other biota killed through power plant entrainment and impingement and the increase in temperature of the water body (ΔT) at different distances from the plant over time (seasonal changes in ΔT).

Nuclear power plants represent an additional opportunity for ecological modeling since a large amount of monitoring and baseline data are collect-ed on biota both before and after plant operation is initiated. These data represent a biological sketch of the ecosystem under consideration and may be useful to estimate model parameter values and to compare with model behavior.

Simulation models have been applied to power plant impact assessment at the single species level (Lawler 1975, Saila 1976, Van Winkle et al. 1974, Warsh 1975, Eraslan et al. 1976), and at the multi-species level (McKellar 1975). While these models purport to predict the effect of power plant impact on relevant biota, in fact their predictions are very sensitive to parameter values which are not well known and to process equations which are empirical at best (Swartzman et al. 1978).

This paper reports on a preliminary review, comparison and evaluation of a large number (26) of simulation models of aquatic ecosystems built for a variety of purposes. This review was

[1]Paper presented at the international symposium Energy and Ecological Modeling, sponsored by the International Society for Ecological Modeling (ISEM), Louisville, Kentucky, April 20-23, 1981

[2]Gordon L. Swartzman is Research Associate Professor, Center for Quantitative Science, Univer-sity of Washington, Seattle, Washington 98195 U.S.A.; Robert T. Haar is Statistician, Center for Quantita-tive Science, University of Washington, Seattle, Washington 98195, U.S.A.; Daniel H. McKenzie is Senior Research Scientist, Ecological Sciences Department, Battelle Pacific Northwest Laboratories, Richland, Washington, 99352, U.S.A. Thomas Zaret is with the Department of Zoology, University of Washington, Seattle, WA. 98195

conducted with the explicit objective of evaluating these models' usefulness in impact assessment, not necessarily the purposes for which they were built.

The three major areas of potential usefulness for simulation models considered are:

1) Impact prediction - the ability of models to give an accurate assessment of changes in "important" biota over time due to changes in environmental conditions and to power plant operations. This also includes the ability of a model to generate testable hypotheses concerning changes to an ecosystem under stress.

2) Data organization - the ability of the model to organize the data available on a site and on similar sites into a whole ecosystem framework using such abstracting methods as annual budgets for biomass flow, time traces of seasonal change in biota and diet data.

3) Data evaluation - the ability of the model to assess the usefulness of monitoring and baseline data for model development. What additional data are needed and what do they represent in terms of increased model reliability and additional addressable questions?

While the primary use of models of interest is impact prediction, the other two take on increased importance when we consider the long term perspective under the hypothesis that with a proper data base, models may eventually prove more accurate than they are at present. Also, at present, models are our only tool offering the potential for impact prediction over a wide range of scenarios which allow peer review of how these predictions were made. With the above focus in mind, this paper reviews our progress to date on evaluating the usefulness of simulation models reviewed for power plant impact assessment relative to the three criteria cited above. The methods we adopted, in brief, were:

1) Review published models for fish and plankton population dynamics, energetics, feeding and growth.

2) Compare the model equations, rationale, parameter values, and data sources on a process by process basis - these being included in standard notation and units in a process notebook (Swartzman et al. 1980).

3) Choose a site(s) for simulation comparison of the process equations under analogous environmental conditions. This choice was made based on criteria of a) frequency and replication of samples, b) completeness of data representing dominant species, c) existence of life history studies on important fish species, and d) existence of a power plant. The first ecosystem chosen was Lake Keowee in South Carolina, site of the Oconee Nuclear Power Station managed by the Duke Power Company (DPC).

4) Develop a preliminary test model framework (FISH1) for the model comparison. This model is a composite of most of the models reviewed, with compartments (major species, functional groups) based on data from the comparison site.

5) Assemble and organize the data to be used for model parameterization (e.g., fecundity), validation (e.g., plankton biomass change through time, annual biomass budget), and simulation (e.g., temperature, initial conditions.)

6) Conduct simulations using the test model framework including comparison with validation data. Critique the current model structure (choice of compartments) based on test model evaluation.

7) Evaluate underlying model assumptions using site specific data. These include spatial homogeneity and randomness (versus patchiness) in the distribution of biota.

8) Develop a model compatibility chart in preparation for a process by process comparison of the models on the test site.

9) Simulation comparison of models. Summarize the various equations' usefulness relative to data needs for parameter estimation and impact questions addressable.

Work to date on this project has progressed through the middle of step 8) above - although we are returning to earlier steps with a second site comparison - Lake Ontario. We have chosen to skim over the surface of findings from steps 1) to 7) giving examples where appropriate and then giving our major conclusions relative to model usefulness. We want to state at the outset that we found the test model wanting in the predictive arena and that, in fact, the data available, in its present form, would be insufficient to develop a model to realistically mimic dynamic patterns of the major biota on Lake Keowee.

MODEL REVIEW AND COMPARISON

The Process Notebook for Aquatic Ecosystem Simulation (Swartzman et al. 1980) represents a distillation of the equations, rationale and parameters from 26 simulation models for (primarily) fish population dynamics, feeding and energetics. The models are documented on a process by process basis, the major processes of concern being ingestion, respiration, growth, natural mortality, fecundity, feeding selectivity, predation mortality, and gut contents dynamics. The models reviewed here range from multi-trophic level models (Andersen and Ursin 1977) to models of fish feeding dynamics (Kerr 1971), from natural ecosystem models to fish-tank experiment models. Our aim was to bring together in a common framework the major modeling approaches representing the mechanisms of fish growth, population change, and feeding.

The process notebook contains a listing, in standard notation, of all of the different approaches known to the authors to representing a given process (e.g., all equations for representing the effect of temperature on maximum ingestion rate). These are accompanied by tables listing parameter values and data sources from each model as well as notes on units conversion, etc.

The format used here is similar to that followed in Swartzman and Bentley (1979). Also included for most processes is a chart showing the historical precedents for each of the equation forms. Various groups of processes fit together (e.g., energetics processes, feeding processes, mortality processes) and for those groups compatibility charts were drawn (as suggested in step 8 above), to show which equations from one model are compatible with those from other models. Thus, when a model is constructed for comparison purposes having, for example, equations from one model for temperature effect on feeding and from another for weight effect on feeding, the compatibility chart for feeding processes tells which combinations are compatible and which are not. The process notebook also contains a general comparison of the models with respect to trophic structure, spatial and temporal assumptions, processes considered and driving variables. There is also a translation section which gives standard notation and its corresponding notation from the models reviewed, enhancing access to the original papers.

Besides serving as a resource document for our model comparison the process notebook has also proven useful in a number of other ways: 1) as a review of available models for future model development, 2) as a repository for parameter values, to check the realism of parameter values commonly used for fish (e.g., assimilation efficiency, maximum ingestion rate), 3) for the review of process information in one area and to see its relationship to other processes, 4) as an educational tool to see differences and similarities in modeling approach, 5) to set standards for future model development and to make explicit the relationship between models and their predecessors, 6) to find errors or discrepancies in model equations and parameter values, and 7) to connect the various process equations with the data needed to estimate their parameters.

CHOICE OF TEST SITE FOR MODEL COMPARISON

While model equations can to some extent be compared relative to their equations, rationale, parameter values, and data needs, an evaluation of these models usefulness for impact assessment is incomplete without a substantive comparison of the models on a test site (preferably several test sites). With this in mind, and recognizing our need to examine power plant impacts, we reviewed monitoring data abstracts from about 40 sites and settled on Lake Keowee as the best. A lake site was chosen because we wanted to work initially with a system that was hydrodynamically simple and where fish populations (our primary focus) are most easily estimable. Lake Keowee was chosen partly for its

'adequate' monitoring data set, partly for the existence of an independent group, the Southeast Reservoir Investigation (SERI) group, studying the reservoir and partly for the availability of a large amount of comparative data from nearby reservoirs (Leidy and Jenkins 1977). An abstract of the available monitoring data is given in table 1. Summarial details are given in Haar and Swartzman (in press).

LAKE KEOWEE TEST MODEL - FISH1

FISH1, a test model, was a composite of the models reviewed in the process notebook and in Swartzman and Bentley (1979). The model assumes spatial homogeneity throughout the reservoir except that it can be run with or without the power plant operating and the power plant operating case can be looked at as near field. The model time step is daily and it is represented as a series of differential equations solved numerically using the Euler method. The physical environment is represented by driving variables input to the model on a daily basis including water temperature (deg C), surface solar radiation (ly/day), day length as a fraction of 24 hours, nitrate concentration (gm N/m^3), and ammonia concentration (gm N/m^3). Phosphate concentration, which is important on Lake Keowee since it is phosphate limited (NES 1975) was omitted because reported concentrations in the DPC monitoring data (DPC 1975) were less than the tolerance limits of the method used. Thermal loading is simulated by increasing temperature by the desired ΔT. The model state variables include 5 functional groups of phytoplankton (greens, blue-greens, diatoms, dinoflagellates and chrysophytes), 2 functional groups of zooplankton including cladocera (represented by *Bosima* sp.) and copepods (represented by *Diaptomus mississipiensis*), and 4 species of recreationally important fish including largemouth bass, bluefill sunfish, black crappie and yellow perch.

The phytoplankton are represented by functional groups having representative (constant) average size with numbers varying through time. Zooplankton and fish are represented by cohorts having numbers, average weight, average age, and developmental class, state variable attributes. Other attributes of these cohorts are intermediate variables such as growth rate or ingestion rate and a species, or functional group identifier. This identifier labels the cohort as associated with a particular species or functional group object which contains as its attributes species-specific parameters. This linked list structure is organized as part of our model simulator called the Aquatic Ecosystem General Impact Simulator (AEGIS). AEGIS allows us to obtain any attribute or collection of attributes from any object or collection of objects. Thus we can print out the biomass of a single cohort of fish or all fish cohorts of a specific species (a collection of cohorts sharing a specific attribute), or of all fish (a collection of cohorts of similar type) at any time step during a model run.

AEGIS also has an event oriented simulation structure patterned after SIMULA, a simulation

Table 1.—Baseline and monitoring data available from Lake Keowee.

Variables	Gear or Method	Taxonomic Detail	Sampling Season	Frequency	Number of Stations	Replicates	Remarks
Abiotic							
Dissolved Oxygen	Hydrolab Surveyor 6-D				13	1	In situ measurements were made at a number of depths from the surface to 10 m above the bottom.
Temperature	Thermistor				13	1	
Solar Radiation	Photometer	not applicable	all	monthly	12	1	
Euphotic Depth	Photometer				12	1	
NO₂-NO₃ Nitrogen					13	1	
NH₃ Nitrogen	All whole water samples were analyzed with a Technichron Autoanalyzer II				13	1	Whole water samples were taken at 1/2 meter intervals from 0.3 m below the surface to 1.0 m above the bottom.
Soluble Ortho-phosphate					13	1	
Total Phosphorus					13	1	
Phytoplankton							
Biovolume	Van Dorn water bottle	species	all	every other month	6(+6)	1	Two samples were taken at each station; a euphotic zone composite and a sample from near the bottom. 6 stations were added halfway through the sampling program.
Concentration							
Zooplankton							
Concentration	0.5 net with 76 micron mesh	species	all year	monthly	10	2	
Benthos							
Concentration	Modified Peterson grab	Genus or family	all year	1/qtr	6	3	
Fish							
CPUE	Shoreline seining	species with size and age	June-Sept.	every other week	5	4	4 stations in each of 5 areas. 10 stations in each of 4 areas. For both these, all stations in an area are treated as replicates.
CPUE	Gill-netting		all year	2/qtr		10	
Concentration	Cove Rotenoning		August	1/year	3	1	
CPUE	Electro-shocking		all year	1/qtr	3	1	Each of these stations is a 2 mile stretch of shore. These were paired samples; one on the surface and one at 20 ft.
CPUE	Icthyoplankton trawling		March-Sept.	weekly	5	2	

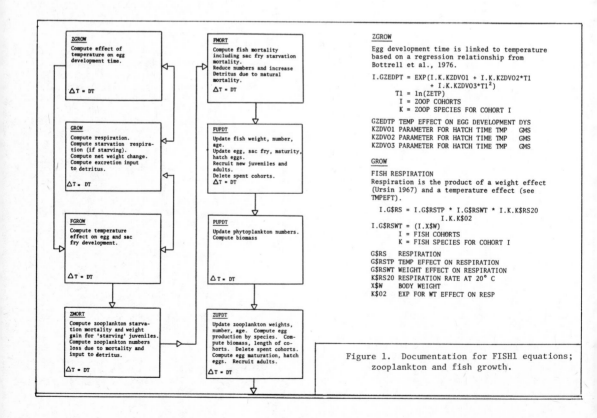

ZGROW

Compute effect of temperature on egg development time.

ΔT = DT

GROW

Compute respiration. Compute starvation respiration (if starving). Compute net weight change. Compute excretion input to detritus.

ΔT = DT

FGROW

Compute temperature effect on egg and sac fry development.

ΔT = DT

ZMORT

Compute zooplankton starvation mortality and weight gain for 'starving' juveniles. Compute zooplankton numbers loss due to mortality and input to detritus.

ΔT = DT

FMORT

Compute fish mortality including sac fry starvation mortality. Reduce numbers and increase Detritus due to natural mortality.

ΔT = DT

FUPDT

Update fish weight, number, age. Update egg, sac fry, maturity, hatch eggs. Recruit new juveniles and adults. Delete spent cohorts.

ΔT = DT

PUPDT

Update phytoplankton numbers. Compute biomass.

ΔT = DT

ZUPDT

Update zooplankton weights, number, age. Compute egg production by species. Compute biomass, length of cohorts. Delete spent cohorts. Compute egg maturation, hatch eggs. Recruit adults.

ΔT = DT

ZGROW

Egg development time is linked to temperature based on a regression relationship from Bottrell et al., 1976.

$$I.GZEDPT = EXP(I.K.KZDVO1 + I.K.KZDVO2*T1 + I.K.KZDVO3*T1^2)$$

$$T1 = ln(ZETP)$$

I = ZOOP COHORTS

K = ZOOP SPECIES FOR COHORT I

GZEDTP TEMP EFFECT ON EGG DEVELOPMENT DYS
KZDVO1 PARAMETER FOR HATCH TIME TMP   GMS
KZDVO2 PARAMETER FOR HATCH TIME TMP   GMS
KZDVO3 PARAMETER FOR HATCH TIME TMP   GMS

GROW

FISH RESPIRATION

Respiration is the product of a weight effect (Ursin 1967) and a temperature effect (see TMPEFT).

$$I.G\$RS = I.G\$RSTP * I.G\$RSWT * I.K.K\$RS20$$
$$I.K.K\$02$$

$$I.G\$RSWT = (I.X\$W)$$

I = FISH COHORTS

K = FISH SPECIES FOR COHORT I

G\$RS   RESPIRATION
G\$RSTP   TEMP EFFECT ON RESPIRATION
G\$RSWT   WEIGHT EFFECT ON RESPIRATION
K\$RS20   RESPIRATION RATE AT 20° C
X\$W   BODY WEIGHT
K\$02   EXP FOR WT EFFECT ON RESP

Figure 1. Documentation for FISH1 equations; zooplankton and fish growth.

language which allows scheduling of discrete model events (e.g., spawning), continuous model events (e.g., computation of process rates), input-output events (e.g., graphs of diet for selected species), and user controlled events (e.g., power plant operating schedule), all within one framework. AEGIS is an interactive programming language implemented on a PRIME 300 minicomputer. Complete documentation of FISH1 is given in Haar and Swartzman (in press). An example page of this documentation is shown in figure 1. Our documentation scheme combines a model flow chart with equations, and their sources and rationale, plus relevant notation. Each section of the flow chart is accompanied by documentation for the equations contained in that part of the model. Model parameter values and data sources for FISH1 are given separately in Haar and Swartzman (in press).

DATA BASE FOR MODEL COMPARISON AND PARAMETER ESTIMATION

Parameter values for FISH1 are given in Haar and Swartzman (in press). These values were either a) obtained from a regression relationship (e.g., effect of fish weight on fecundity), b) a commonly accepted physical quantity (e.g., carbon to dry weight ratio), c) a literature value based on direct measurement in the laboratory or field,

(e.g., O_2 consumption due to respiration), d) deduced from data (e.g., egg weight), or e) calibrated by running the model.

A general description of lake trophic dynamics can be obtained by constructing an annual budget for biomass flow. A biomass budget for Lake Keowee is given in table 2. Assumptions made in constructing this budget are discussed in detail in Haar and Swartzman (in press). The major difficulties we encountered in the data were 1) lack of reliable value for average phytoplankton production and production to biomass (P/B) ratio, 2) lack of any zooplankton or fish production estimates which necessitated an approach based on trophic efficiency (zooplankton production/phytoplankton production) for zooplankton and on an assumed P/B value for fish.

Comparisons of production (P) and standing crop (B) for phytoplankton and zooplankton in our Lake Keowee budget with those from other lakes (Winberg 1972, Schindler 1972) indicate that our calculated zooplankton to phytoplankton standing crop ratio is unrealistically high.

SIMULATION AND VALIDATION

Using driving variable and initial conditions data from Lake Keowee from 1974, calibration of

Table 2.—Annual budget for Lake Keowee, 1974.

Group	B	P	P/B	R	TC	Consumption by Food Type									
						1 %	Amt	2 %	Amt	3 %	Amt	4 %	Amt	5 %	Amt
Phytoplankton	14.20	2100	147.9	-----	------										
Zooplankton	27.80	210.0	7.55	315.00	875.00										
Benthos	4.00	10.00	2.50	23.30	55.50										
Fish:															
Shad	4.25	2.16	0.51	5.03	8.99	90	8.09	10	0.90	--	-----	--	-----	--	-----
Carp	55.32	28.20	0.51	65.71	117.38	60	70.43	30	35.21	10	11.74	--	-----	--	-----
Shiners	0.48	0.24	0.51	0.56	1.00	--	-----	15	0.15	85	0.85	--	-----	--	-----
Catfish & Bullheads	4.85	2.47	0.51	5.76	10.28	25	2.57	55	5.65	--	-----	20	2.06	--	-----
Sunfish	15.84	8.08	0.51	18.83	33.63	5	1.68	65	21.86	--	-----	10	3.36	20	6.73
Basses	9.09	4.64	0.51	10.81	19.31	--	-----	10	1.93	--	-----	85	16.41	5	0.97
Crappie	9.82	5.01	0.51	11.67	20.85	5	1.04	20	4.17	15	3.13	55	11.47	5	1.04
Perch	3.90	1.99	0.51	4.64	8.28	--	-----	20	1.66	20	1.66	60	4.97	--	-----
Total Fish	118.64	60.49	0.51	140.95	251.77		109.45		73.13		22.19		38.27		8.74

B – average standing crop (kg ww/ha)
P – production (kg ww/ha/yr)
P/B – production to biomass ratio (1/yr)
R – metabolic costs (kg ww/ha/yr)
TC – total consumption (kg ww/ha/yr)
Consumption by Food Type – per cent of total consumption and amount in kg ww/ha/yr

Food types:
1 – phytoplankton and detritus
2 – benthos
3 – zooplankton
4 – fish
5 – terrestial organisms

the model was done by adjustment of various parameters until stable and realistic behavior resulted.
No field data or annual budget data were used in
this preliminary calibration phase. Changes were
also made during calibration to several process
equations. For example we had to include a refuge
for plankton, i.e., a density below which no grazing
or predation occurs. Otherwise, relatively rare
groups are eliminated due to high grazing rates
resulting from relatively high abundances of other
groups of plankton. An 'other' food category was
added to the model because insufficient food was
available for fish to meet their energetic needs
unless their ingestion rates were increased to such
a level that they overgrazed their food base.
Benthos were obviously important as a food source.
However, since not enough was known about their
population dynamics to explicitly include them, they
were implicitly included as an 'other' food source
following the population dynamics of those prey
modeled explicitly in the model. Another essential
modification was the addition of a starvation
mortality term for larval fish. In the model larval
fish were unable to meet their high expected growth
rate with average food abundance. In fact, many
larvae in natural ecosystems do not get enough food
and starve to death while others see more than
average (Hunter, personal communication). To simulate this phenomenon we added a starvation mortality
term that adjusted mortality so that those larvae
that survive would get enough food to grow at known
growth rates.

The model, after the initial parameterization
phase, was compared with field data from Lake Keowee
for 1974 using the following components: annual
biomass budgets, time traces (mean density plus or
minus one standard deviation) for phytoplankton and
zooplankton functional groups, and fish diets. This
comparison, and the simulation work and data analysis
leading to it, resulted in a critique and suggestions
for improvement in both the model and the data. As
examples of this comparison, an annual biomass budget
for 1974 was constructed from the model (table 3)
and was compared with table 2, our data based biomass
budget. The following observations were made:

1) Zooplankton P/B in the model is much
   higher than for the data.

2) Benthos is important in the data and
   is not included explicitly in the model.

3) Fish production and P/B is much higher
   in the model than in the data.

Table 3.   Budget calculated from model run.

Group	Biomass kg ww/ha	Production kg ww/ha yr	P/B 1/yr
Phytoplankton	11.44	1056.00	95.30
Zooplankton	6.02	1038.00	172.40
FISH			
Largemouth bass	21.17	95.46	4.51
Black crappie	4.07	6.25	1.54
Bluegill sunfish	10.60	11.80	1.11
Yellow perch	1.54	3.08	1.99
Total Fish	37.38	116.60	3.12

4) Dominant fish species in Lake Keowee
   (e.g., carp) are missing in the model.

From these observations we concluded that all
dominant fish species need to be included in FISH1.
Benthos are not included explicitly in FISH1,
although an 'other'category comprising up to 50%
of all food eaten by fish and implicitly representing fish was included. In a reservoir like Lake
Keowee with a high shoreline development index
and many shallow coves, benthos probably are
important, especially as food for littoral bottom
feeding fish (such as carp and bluegill sunfish)
and must be explicitly included in a model. As
many of the models reviewed were developed for
larger water bodies, apparently in most of those
cases benthos could be omitted, without seriously
affecting model performance, which explains their
absence in most of these models. Zooplankton
production in FISH1 seems unrealistically high.
The cause probably lies in overly high fecundity
- an area about which little is known on Lake
Keowee. Information on size (weight) distribution
of zooplankton and egg production for the dominant
species on Lake Keowee is needed to better model
zooplankton dynamics. The relatively high fish
production in the model may be due either to high
zooplankton production and/or to the inclusion of
larvae and juveniles to compute fish production.
The data base P/B ratio of .51 is based on annual
production of adult fish. Much of the fish production in FISH1 is by ichthyoplankton and juveniles,
many of whom do not reach adulthood because of high
mortality rates.

Time traces of organism density were also
compared. Figure 2 shows, for example, model output for cladoceran density for 1974 while figure 3
shows monitoring data from three stations on Lake
Keowee from 1974-1976. The three stations were
considered as replicates and their means ± 1
standard deviation bars are plotted for each sampling data.

Similar comparisons were made for photoplankton
and copepods. Observations from these comparisons
are:

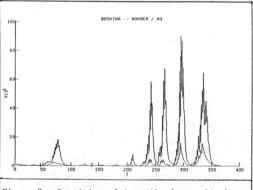

Figure 2.   Densities of juvenile (upper line)
cladocerans from FISH1 - Lake Keowee 1974.

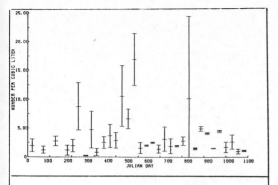

Figure 3.--Cladoceran density on Keowee River
arm, 1974-1976.

1) Plankton seasonal dynamics agree fairly
   well between model and data although
   timing of blooms for some groups are
   different or average values differ by
   as much as a factor of 2.

2) At times of the year when the plankton
   data values have high variance the model
   output often shows oscillation. Details
   are given in Haar and Swartzman (in press).

The unexpectedly close relationship for plankton between model and data time traces suggests
that the plankton dynamics of the model are fairly
realistic. However, there still is the important
question of whether the oscillations shown in the
model are really occurring in Lake Keowee or
whether they are artifacts of the model. A data
collection frequency of bimonthly is insufficient
to evaluate such oscillations since the period of
oscillation in FISH1 (see figure 2) is often less
than a month. Edmondson (personal communication)
in biweekly phytoplankton samples on Lake Washington
found indications of oscillation in total chlorophyll a during the summer months. These oscillations, if they exist on Lake Keowee, may be important to understanding the ecosystem because they
indicate a biotic coupling between plankton and
predator dynamics. They may also result in phase
lags between zooplankton and fish hatchings. Such
lags could lend credence to Cushing's critical
period hypothesis and could possibly give a mechanistic basis to year class strength in fish, though
we believe that other mechanisms such as wind speed
and water level fluctuations are also important.

Few substantive conclusions can be derived
from fish diet data. There were several studies
of larval and adult yellow perch diets on Lake
Keowee (Oliver, unpublished manuscript). These
data do not provide a quantitative basis for model
validation but do indicate that model fish diets
are reasonable. Fish feeding in FISH1 is based on
the size selective feeding hypothesis of Andersen
and Ursin (1977). Model behavior was very sensitive
to the parameters in this equation and dominance
changes between copepods and cladocera resulted

from relatively small changes in these parameters,
the values of which are not well known. The model
sensitivity to these parameters, the lack of much
information on fish feeding, and the lack of diet
studies taken in conjunction with water column
sampling lead us to question the predictive value
of multi-species, multi-trophic level models.
Another feature of this, the possible implications
of patchily distributed prey and spatial variability of prey and predator density over different
habitats will be discussed in the nest section.

UNDERLYING MODEL ASSUMPTION EVALUATION

A number of structural and spatial assumptions
inherent in the model may be evaluated in light of
available data from Lake Keowee and from simulation
work.

The major spatial assumption is that of
spatial homogeneity. All biotic and abiotic components of the model are presented in FISH1 by
spatially averaged quantities. FISH1 also assumes
that the process equations are representative of
average lake dynamics. There are several factors
observed in many aquatic ecosystems which argue
against the homogeneity assumption. These are:
1) fish habitat selection and the existence of
separable habitats on a lake like Lake Keowee,
2) prey densities tend to be different in these
different habitats (i.e., they are spatially
heterogeneous), 3) plankton may be patchily distributed, 4) power plant effects are point sources,
yet are treated as spatial averages, and 5) diel
migration of prey makes average densities erroneous.
We first use available data, where possible, to
ascertain whether spatial heterogeneity exists,
then discuss the type and magnitude of the error
resulting from the spatial homogeneity assumption,
given the lake is actually heterogeneous. Data
were available on a number of stations in the
Keowee River arm and discharge area of Lake Keowee.
Unfortunately, the DPC data did not have multiple
observations from each sampling station, making
it impossible to test for differences between
stations in the same area.

All the stations in the same area of the lake
sampled at the same time were treated as replicates.
Unfortunately, the only areas comparable are the
Keowee River arm and the discharge area. There
were no replicates on the one station in the other
arm of the lake. It appears from the data that
there is a difference in average abundance and to
a lesser extent in seasonal pattern between the
Keowee River arm and the discharge area, although
the large variances make the difference of
questionable statistical significance. Most phytoplankton and zooplankton densities tend to be lower
around the discharge area than in the Keowee River
arm although relative species composition is different
in the two areas. Some replicated data are available
on zooplankton (Hudson, unpubl. data) and have a
coefficient variation of 25%. Thus on Lake Keowee,
no evidence for zooplankton patchiness at the spatial
scale of the samples was found.

Diel migration of zooplankton is striking on Lake Keowee (Hudson and Nichols 1978) with significant differences in depth distribution between day and night.

Conclusions from the above comparisons are:

1) Though differences in plankton abundance and species compositions between the areas near and close to the plant are apparent, these differences are not statistically separable from the high station to station variability within each area.

2) Patchiness either is not important or exists at a spatial scale larger than the sampling region over which replicates were taken.

3) Diel migration of zooplankton is seasonal and marked.

4) The assumption of spatial homogeneity of the fish distribution could not be investigated from the available data but life history evidence indicates spatial heterogeneity to be an important feature of fish feeding patterns.

Errors in model output resulting from incorrect spatial assumptions are difficult to delineate. Some factors, however, are evident. Computing fish feeding rate as a function of an average density probably leads to erroneous estimates, since fish probably choose habitat areas with high concentrations of their preferred prey. Other factors are probably also important including temperature preferences, schooling characteristics, and the fact that most fish feed only during part of the day presumably due to diel migration of their prey or predator avoidance. These combinations of spatial and temporal factors make the fish feeding and growth model based on spatially averaged (and temporally averaged over the day) prey abundances highly suspect. Furthermore, Beyer (1976) and Vlymen (1977) have shown in theoretical and simulation studies that larval growth depends strongly on spatial aggregation of prey. Vlymen shows that a certain degree of prey patchiness is necessary for larvae to grow at observed growth rates.

Two structural features of FISH1 that bear further discussion are zooplankton cohort structure and identification of species-specific parameters for fish. Organizing zooplankton into cohorts and allowing spawning as frequently as every other day created a plethora of cohorts. This large number of cohorts produced a size distribution of zooplankton within the model. While this cohort division realistically represented zooplankton population dynamics and allowed differential predation on different aged (and sized) cohorts, it had the unpleasant effect of greatly increasing simulation time. Also, since zooplankton fecundity and size selection of fish preying on zooplankton are not well understood on Lake Keowee, having such size detail is inappropriate. We plan

to replace the zooplankton cohorts by size classes and have dynamic pool aging and recruitment between size classes. Having a single species-specific parameter in order to achieve parameter parsimony was unacceptable for certain fish parameters, especially ingestion related parameters such as the effect of weight on ingestion rate and the size selective feeding parameters. We had difficulty fitting reasonable growth rates to both adult and larval fish with mechanisms based on a single parameter value.

FURTHER CRITIQUE OF MODEL AND DATA

One omission of the model was not including phosphorous as a driving variable despite our knowledge that Lake Keowee is phosphate limited (NES 1975). This occurred because the phosphate values given for the summer (DPC 1975) was at the detection limit for the method used. In order to compensate for phosphate limitation in the model we had to assign a maximum biomass for each species to prevent explosive blooms which are prevented in the real system by limited summer phosphate reserves. Thus, while the general plankton dynamics were reasonable, an important process could not be included in the model.

Also, life history studies were only done on recreationally important fish. This made inclusion of carp and forage fish such as shad and shiners difficult even though these species are the dominant species based on biomass. These fish need to be included in a model purporting to be realistic.

There may be important environmental factors affecting fish year class strength which are not included in FISH1. For example, Miller and Kramer (1971) found that spawning success of centrachids (all fish species in the model except yellow perch) could be seriously reduced by high winds and/or fluctuating water levels since these would cause the adults to abandon their nests. Analysis of available data on these abiotic variables and larval abundance from tow netting samples could be used to investigate the importance of this mechanism as a determinant of year class strength in these fish.

A major potential impact of power plant operation on fish could be related to the effect of thermal loading on the timing of cycles in larval food base. This hypothesis is demonstrated in FISH1 by oscillations in plankton biomass and sensitivity of larval growth to the density of prey following spawning. In order to check for this phenomenon in Lake Keowee, samples are needed on a biweekly basis with at least three replicates at each station - these to be taken over the period of larval growth (April to August on Lake Keowee).

SUCCESS OF MODELS FOR IMPACT ASSESSMENT

We mentioned earlier three criteria for model evaluation. The first, impact prediction, or at least accurate quantitative prediction of system change under impact, is beyond the scope of the models reviewed. We did mention that part of this could include the ability of a model to formulate testable hypotheses. Here FISH1 has done better - indicating that there are oscillations in plankton biomass and hypothesizing that changes in the timing of these cycles or in the timing of spawning (presumably due to thermal loading) could create a critical stress on a fish population.

FISH1 did serve as a very useful tool for organization of available data in a variety of formats, including biomass budgets, diets, population size structure, plots of biomass change with time, and parameter values related to specific process mechanisms. Some aspects of this data organization are given here; the body of it is in Haar and Swartzman (in press).

The third criterion, a model's ability to evaluate the available data and suggest further data needs, is by no means straightforward to apply to a model. We see model functioning as a data screener and evaluator on three distinct levels. The first level evaluates the data base as a resource file for delineating the structural components and process mechanisms necessary to connect a particular set of impacts of concern (e.g., thermal loading and impingement resulting from power plant operating) with a particular output criteria of interest (e.g., year class strength of sport fish). The second level concerns the use of available data to product a quantitative sketch of the important features of the particular ecosystem (e.g., biomass budgets and system behavior) under study and to check underlying model assumptions. The third level involves simulation study and assesses the sufficiency of the available data to test hypotheses suggested by the simulation and to validate or invalidate model behavior.

DATA USE FOR DELINEATING MODEL STRUCTURE,
MECHANISMS, AND PARAMETER VALUES

The larger question of both the data and the model is, "To what use are they to be put? Or, what questions do we want to ask, which kinds of impacts do we expect to be able to detect or at least consider? With this perspective, models and data can be seen to go hand in hand; the data collection is designed to detect certain types of possible impacts, and the model deals with the mechanisms expected to result in those kinds of impacts. Modeling is then seen as a translator between data and impact questions relative to types of expected impacts. Data evaluation becomes a measure of how well the data speaks through the model - whether data is available to substantiate and test a model that relates to the impact that we wish to detect with a monitoring program.

The types of impact questions that arise are usually specific to a water body type in a region. For example, Lake Keowee, being a Southeastern U.S. reservoir, would be expected to have impacts 'typical' of water bodies of this type in this area. Other factors which might vary between lakes in the same region may also be important (nutrient status,

volume, turnover rate, fish introductions, shore-line development index, water clarity, etc.). Let us consider some example impact questions in Lake Keowee as a prelude to connecting them with models and data.

In a discussion of fish cove rotenoning data for seven years from Lake Keowee, Jenkins (1977) hypothesized that there had been a drop in fish abundance from 1972-1976 primarily due to the aging process of the reservoir. After inundation, reservoirs tend to be high in nutrients and these drop off as the reservoir ages, reaching equilibrium after about six years. However, he also observed that the ratio of predator to prey (forage fish) also dropped in the cove rotenoning samples from 1973 to 1976. This he attributed not to a real drop in this ratio but rather to the reduced preference of predators for the cove habitat due to warmer waters (reduction in the hypolimnion habitat in the cove caused by the drop in thermocline) from mixing of hypolimnion and epilimnion waters by the power plant. The focal question arising from this discussion is how to separate the impact of the power plant from the effect of reservoir aging.

Other possible questions are: What are the effects on total fish production? On total fish production? On fish habitats? On long term catch of sports fish? Primary focus might be on short term or long term effects; noxious blue-green algal blooms might also be of concern.

Figure 4 outlines the process of structuring the model by connecting expected direct impacts to expected direct impacts to desired output criteria.

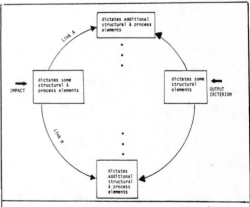

Figure 4. Model structuring connecting impacts with output criterion.

Both the impact of concern and the output criterion of interest necessitate certain structural components of a model. Each link in figure 4 is one pathway considered as a possible connection between the impact and the output criterion. Data assessment during this process aids in the decision about which links are to be included in the model – a decision based on data availability and the a priori judgement of the modeler about the importance of a particular link.

Suppose for example, we are concerned with the effects of power plant operation on the year class strength of sports fish in Lake Keowee. A useful model in this case would be similar to FISH1 in its structural aspects. Direct effects of the power plant on the lake are increased temperatures, increased epilimnion-hypolimnion mixing, lowered thermocline, and entrainment/impingement.

To model the potential direct impingement and entrainment effects, which are size and species selective, size classes of fish are needed for young-of-the-year fish. Separate compartments for those kinds of fish most easily impinged (usually small schooling fish) must also be included. To implement possible effects of thermal loading, mechanisms must be considered for how temperature affects the various fish processes – ingestion, matabolism, mortality, fecundity, timing of spawning. To consider thermocline and mixing effects, a model must have two layers and a hydrodynamic submodel.

Next, we move to the outcome of concern – year class strength of sports fish. Obviously all sports fish must be included either as species or, depending on their functional similarity, as functional groups. Direct impingement/entrainment effects on sports fish are already considered by having size classes of all impinged fish as a structural detail in the model. Focus on young-of-the-year fish necessitates having age classes or at least separating adults from young-of-the-year for the sports fish of concern.

Turning to other possible links between direct impacts and output criterion, one connection is through direct impact of the power plant on prey species (zooplankton, forage fish, benthos) of the sports fish. Another linkage is through sports fish predators which could affect predation mortality. Further links are through the direct effects of entrainment and thermal loading on the food of the prey of the sports fish and so on down the trophic food chain, or through the direct effect of nutrient mixing on nutrient status and phytoplankton growth.

For each of the possible links we assess the expected importance of this link and the availability of data to support a model having this link. A useful rule is that simpler, more direct links are more important to include than indirect links. When an important link does not have sufficient data to support its inclusion, then the data is deemed insufficient to build a realistic model to study that impact on the particular output criterion. For example, on Lake Keowee, we decided that a major weakness of the data base was that no information was available on life history of the major forage fish on Lake Keowee.

Some links are eliminated or simplified due to data availability or the lack of it. For example, to consider the effect of lake mixing due to power plant operation on nutrient status in a model, we must quantify this mixing on a lake-wide basis, using a hydrodynamic model, and separate the effect of mixing from those of nutrient uptake, phytoplankton sinking, and nutrient regeneration through excretion. Alternatively, we can measure the changed nutrient status in the field and use this as a driving variable in a model. Since phytoplankton-nutrient interactions are far removed from fish year class strengths, the latter approach seems preferable. However, if algal blooms were the output criterion of interest, this decision would be more difficult.

The decision about the amount of detail to include for each process along a particular link is often difficult. An example is the effect of temperature on sports fish fecundity. We must weigh the cost of obtaining information about temperature effects along with the plausibility that fecundity will change sufficiently under thermal loading to have an important or even measurable influence on year class strength.

DATA USE IN AN ECOSYSTEM SYNTHESIS

The second level of data evaluation involves using data to give a sketch of the ecosystem that parallels the model and to examine the realism of several model assumptions, for example, spatial homogeneity. The data are evaluated here for their sufficiency to test model assumptions and to construct a system sketch, as well as by the believability of the numbers. An example on Lake Keowee of data evaluation on this level was the insufficiency of the monitoring data resulting from lack of replicates at each station with which to evaluate the existance of patchiness and spatial heterogeneity. Another example was our finding that the zooplankton standing crop was unrealistically higher than the phytoplankton standing crop on Lake Keowee when compared with data from other lakes. Information on this level may bear on decisions about model structure - for example, our discovery of the importance of benthos as a fish food source from the budget in table 2.

DATA USE IN SIMULATION COMPARISON

The third level of data evaluation involves simulation. Data may be required to test hypotheses resulting from observed model behavior. An example is the hypothesis based on approximate 30 day oscillations of zooplankton abundance. Here, the possibility of fish larvae hatching into a low zooplankton period and starving is high – the timing and period of oscillations possibly being affected by temperature or mixing. First, we must establish whether oscillations in fact exist - a sampling program requiring at least biweekly instead of the present bimonthly sampling. Thus, present sampling frequency was not sufficient to evaluate this hypothesis.

Regarding model validation or invalidation, also involved in third level data evaluation, the output criterion of primary interest sets a priority for the type and frequency of validation data needed. If that output criterion is year class strength of sports fish, a good set of validation data would be several years of townetting data at multiple stations compared with a multiple year run of the model. Priorities on other validation data should be based on which link (figure 4) the data is in and how important that link is deemed to impact on sports fish. There should be more replicates and a higher sampling frequency of validation data along the important links.

Another means to aid in this level of data evaluation is sensitivity analysis using newly developed methods which involve an experimental design, or Monte Carlo procedure, for generating simulation experiments. This is done in such a way as to ascertain the relative importance of various parameters (and hence the data supporting them) to the output criteria of interest. These new methods consider interactions between parameters and are therefore to be preferred to the classical Tomovic, or single-parameter-at-a-time perturbation experiments which do not (Rose and Swartzman 1981).

THE VALUE OF ON SITE-EXPERIENCE

In some ways our method of approaching a site with the idea of parameterizing an existing model (or a synthesis from existing models) to it is like putting the cart before the horse. We were hampered by lack of familiarity with the system under study and lack of accessibility to it. True, a great deal was learned about how the model and data could be improved through the process of comparing model assumptions, parameters and output with available data. However, we are always anxious that our lack of on-site familiarity with the system under study will lead us to overlook some major (and to an on-site person, obvious) process which makes the rest of our model peregrinations fantasies. We have observed this failing in other modeling studies and are aware of how valuable field experience is to the success of a modeling process. We think there must be a balance. Quantitative skill and a theoretical model framework are valuable tools for gaining perspective but so are on-site observation and rapid access to a system to test hypotheses soon after they are formulated. With more on-site experience we never would have left out benthos in FISH1. However, with stronger data synthesis, quantitative skills and a supporting model, the expanded field data effort by DPC would have addressed some issues that might possibly have led to insight about the possible mechanisms for the power plant's impact.

LITERATURE CITED

Andersen, K.P., and E. Ursin. 1977. A multi-species extension of the Beverton and Holt theory of fishing, with accounts of phosphorous circulation and primary production. Medd. fra Danmarks Fisk.-og Havund. 7:319-435.

Beyer, J.E. 1976. Ecosystems---An Operational Research Approach. The Institute of Mathematical Statistics and Operations Research, The Technical University of Denmark, Lyngby.

Duke Power Company. 1974a,b. Oconee Nuclear Station. Semi-Annual Reports for 1974.

Eraslan, A.H., W. VanWinkle, R.D. Sharp, S.W. Christensen, C.P. Goodyear, R.M. Rush, and W. Fulkerson. 1976. A computer simulation model for the striped bass young-of-the-year population in the Hudson River. ORNL/NUREG-8, ESD-766, Oak Ridge National Laboratories, Oak Ridge, TENN.

Haar, R.T.,and G.L. Swartzman. (in press). Evaluation of simulation models in power plant impact assessment: a case study using Lake Keowee. U.S. Nuclear Regulatory Commission.

Hudson, P.L., and S.J. Nichols. 1978. Relations between zooplankton migration and entrainment in a South Carolina cooling reservoir. Waste Heat Management Utilization Conf., Miami Beach, FLA.

Jenkins, R.M. 1977. Prediction of fish biomass, harvest, and prey-predator relations in reservoirs. p. 282-293. In Proceedings of the Conference on Assessing the Effects of Power-Plant_induced Mortality on Fish Population, Pergamon Press, New York, N.Y.

Kerr, S.R. 1971. Prediction of fish growth efficiency in nature. J. Fish. Res. Board, Canada. 28(6):809-814.

Lawler, Matusky and Skelly Engineers. 1975. Report on development of a real-time, two dimensional model of the Hudson River striped bass population. LMS Project No. 115- 49.

Leidy, G.R., and R.M. Jenkins. 1977. The development of fishery compartments and population rate coefficients. Army Waterways Experiment Station, Vicksburg, MISS.

McKellar, H. 1975. Metabolism of esturine bay ecosystems affected by a coastal power plant. University of Florida, Dissertation.

Miller, K.D., and R.H. Kramer. 1971. Spawning and early life history of largemouth bass, *Micropterus salmoides*, in Lake Powell. p. 73-83. In G. E. Hall (ed.) Reservoir Fisheries and Limnology, Spec. Publ. No. 8, American Fisheries Society.

National Eutrophication Survey. 1975. Lake Keowee, Oconee and Pickens Counties, South Carolina. National Technical Information Service, PB-260 500.

Oliver, J.L. (unpublished manuscript) Food of larval yellow perch in a reservoir receiving heated effluent. 16 pp.

Rose, K., and G. Swartzman. 1981. A review of parameter sensitivity methods applicable to ecosystem models. Technical Report NUREG/CR-2016, Center for Quantitative Science, University of Washington, Seattle.

Saila, S.B. 1976. The effects of power plant entrainment on winter flounder populations near Millstone Point. Final Report, NE Utilities Serv. Co., URI-NUSCO Rep. 5.

Schindler, D.W. 1972. Production of phytoplankton and zooplankton in Canadian Shield lakes. p. 311-331. In Proceedings of the IBP-UNESCO Symp. on Productivity Problems of Freshwaters, Z. Kajak and A. Hillbricht-Ilkowska (eds.), Poland.

Swartzman, G., and R. Bentley. 1979. A review and comparison of plankton simulation models. ISEM Journal, 1(1-2):30-81.

Swartzman, G., E. Smith, D. McKenzie, B. Haar and D. Fickeisen. 1980. Process notebook for aquatic ecosystem simulation. Technical Report NUREG/CR-1182, College of Fisheries and Center for Quantitative Science, University of Washington, Seattle.

Swartzman, G., R.B. Derise, and C. Cowan. 1978. Comparison of simulation models used in assessing the effects of power-plant-induced mortality on fish populations. Technical Report NUREG/CR-0474, Center for Quantitative Science, University of Washington, Seattle.

VanWinkle, W., B.W. Rust, C.P. Goodyear, S.R. Blum, and P. Thall. 1974. A striped bass population model and computer programs. ORNL/TM-4578, ESD-643, Oak Ridge National Laboratories, Oak Ridge, TENN.

Vlymen, W.J. 1977. A mathematical model of the relationship between anchovy, Engraulis mordax, growth prey microdistribution, and larval behavior. Env. Biol. Fish. 2(3): 211-233.

Warsh, K.L. 1975. Hydrological-biological models of the impact of entrainment of spawn of the striped bass, Morone saxatilis, in proposed power plants at two areas in the upper Chesapeake Bay, Johns Hopkins University, Applied Physics Laboratory.

Winberg, G.G., V.A. Babitsky, S.I. Gavrilov, G.V. Gladky, I.S. Zakharenkov, R.Z. Kovalevskaya, T.M. Mikheeva, P.S. Nevyadomskaya, A.P. Ostapenya, P.G. Petrovich, J.S. Potaenko, and O.F. Yakushko. 1972. Biological productivity of different of lakes. p. 383-404. In Proceedings of the IBP-UNESCO Symp. on Productivity Problems of Freshwaters, Z. Kajak and A. Hillbricht-Ilkowska (eds.) Poland.

THE DIRECT AND INDIRECT IMPACTS OF ENTRAINMENT ON ESTUARINE COMMUNITIES-TRANSFER OF IMPACTS BETWEEN TROPHIC LEVELS[2]

Richard J. Horwitz[2]

Abstract.--A model of the impact of the entrainment of planktonic groups on an estuarine ecosystem is discussed. The transfer of impacts from entrained groups onto predators or prey is emphasized. The model suggests that the greatest impact may occur on the predators of an entrained group, though the direct negative effect on the entrained group is more consistently present. Impacts do not accumulate on top-level carnivores, primarily because of a slight shift toward a food chain based on detritus and benthos.

INTRODUCTION

In 1928 Vito Volterra (Volterra 1928) published an extension of the Lotka-Volterra predator-prey equations which considered the effects of additional mortality on a species. He showed that identical additional mortality on both populations leads to an increase in equilibrial prey densities and a decrease in equilibrial predator populations (Scudo 1971). The concordance of fisheries data with Volterra's model has been called the first instance on which an ecological model matched empirical data (Scudo 1971). The issue of additional mortality is more general, being applicable to fishery management, pest control, nutrient removal and estimation of impacts of entrainment and impingement. Barclay and Van der Driessche (1977) generalized these results to include quadratic self-regulatory terms for both populations and arbitrary functional response formulations. Under generally realistic constraints on the parameters the equilibrium abundance of the predator always decreases, even with no added mortality to itself. The prey may or may not increase; an increase is likely when predation is strong and predator self-regulation is weak.

These results imply that the effects of additional mortality on the predators or prey of a population may be extremely important to the dynamics of the population. This may be emphasized by comparing two pairs of very simple single species and multispecies models of the effects of additional mortality. Both show that the prediction of an impact of additional mortality on a prey population depends on whether or not predator dynamics are modeled.

The first pair models systems controlled by the rate of input of resources F. Consider a species P, feeding on a resource F.

$$\frac{dF}{dt} = i - uPF \qquad (1)$$

$$\frac{dP}{dt} = (-a + vF)P - mP \qquad (2)$$

Where i is the rate of input of resource F
u and v reflect rates of uptake and assimilation of resource F
a is the death rate of P

At equilibrium

$$F = \frac{a + m}{v} \qquad P = \frac{vi}{u(a + m)} \qquad (3)$$

and the proportional reduction of P by the additional mortality is $\frac{a}{a + m}$.

Suppose P is preyed upon by Z:

$$\frac{dF}{dt} = i - uPF \qquad (4)$$

$$\frac{dP}{dt} = (-a_1 + vF - xZ) P - mP \qquad (5)$$

$$\frac{dZ}{dt} = (-a_2 + yP) Z \qquad (6)$$

Then,

$$\hat{F} = \frac{iy}{ua_2} \qquad \hat{P} = \frac{a_2}{y} \qquad \hat{Z} = \frac{viy}{ua_2 x} - \frac{a_1 + m}{x}$$

[1]Paper presented at the International Symposium, Energy and Ecological Modelling, sponsored by the International Society for Eclogical Modelling. (Louisville, Kentucky, April 20-23, 1981).

[2]Richard J. Horwitz is Assistant Curator of the Division of Limnology and Ecology, Academy of Natural Sciences of Philadelphia, Philadelphia, Pennsylvania.

Here, the prey suffers no reduction from the additional mortality; it is the predator density that decreases.

The same contrast between single species models and multispecies models can be shown in systems controlled by self-regulation. Consider a species which follows the logistic growth equation with added entrainment mortality, m:

$$\frac{dP}{dt} = rP \left(1 - \frac{P}{k}\right) - mp \qquad (7)$$

r is the intrinsic growth rate and k is the asymptotic equilibrium in the absence of entrainment. The equilibrium is $k(1 - \frac{m}{r})$; it is smaller than the corresponding equilibrium without entrainment by a factor of $(1 - \frac{m}{r})$. Now suppose that P is preyed upon by predator Z; let the growth equations be:

$$\frac{dP}{dt} = rP \left(1 - \frac{P}{k}\right) - mP - \frac{xZP}{P+S} \qquad (8)$$

$$\frac{dZ}{dt} = dZ + \frac{yZP}{P+S} \; . \qquad (9)$$

Here d is the death rate of the predator and x, y and s are parameters determining the rate and efficiency of the predator.

$$\hat{P} = \frac{ds}{y-d} \qquad \hat{Z} = \frac{r(1 - \frac{\hat{P}}{k})P}{dx}$$

$\hat{P}$ is unaffected by entrainment mortality, but $\hat{Z}$ decreases by a factor $m/[r(1-(\hat{P}/k))]$. Note that this is greater than $\frac{m}{r}$, the reduction of the entrained species in the one-species model. Thus there is both a transfer and amplification of impact from additional mortality on prey to impact on the predator. Like a number of single species models with "density-dependent mortality" (cf. Ricker 1954; Hess 1975; DeAngelis *et al*. 1977), this model examines the effects of density-dependent intraspecific factors and interspecific factors on reduction of an entrained species. It shows the same decrease in impact from these compensatory factors, but it adds something new: the compensation at one level involves a new loss on another level. This loss may be relatively greater than the non-compensated loss in the absence of the interspecific effects.

I have extended these models to chains of 3 and 4 species, with self-regulatory quadratic terms in either the bottom trophic level, the top level, or all levels. The tendency for impact to be transferred away from self-regulated levels is general. However, when more than one level is self-regulated, the transfer is not complete as it is in the 2-species models described above. The general hypotheses about system behavior predicted by the model with several levels self-regulated are:

1) All levels are adversely affected by entrainment on their own level and all lower levels.
2) Levels are enhanced by entrainment of predators,

3) All species are adversely affected by the entrainment of predators of predators, etc.

The general result of these models is that additional mortality on one level may be translated in whole or part into impact at other levels. The implication for modeling is that single species models, even with self-regulatory mechanisms, are likely to be inadequate. The transfers of impact are not necessarily second-order effects dwarfed by the primary impact: they can represent significant impacts. Single-species models may fail to include the main impact of a given source of additional mortality. Although this point is well-known and obvious in the simple context so far presented (cf. May *et al*. 1979), trophic complexity is often sacrificed in applications of biological models. For example, many models of water quality and entrainment include spatial and hydrologic complexity at the expense of trophic complexity; fishery models have often been more concerned with age-class structure than community structure. While these decisions are understandable considering the sketchy information about many interspecific interactions, ignoring these effects completely may jeopardize even the qualitative results of the models.

While the type of simple models discussed above can demonstrate the need for multispecific models, they cannot fulfill that need. Their simplicity limits their validity to equilibrial conditions, which may never be reached. They include no seasonal variation, nutrient cycling and have a very simple trophic structure. All these factors are likely to affect system dynamics and the response to additional mortality. I have developed a model which includes these factors. Its main application is the estimation of impacts of entrainment and impingement from power plant once-through cooling on a Chesapeake Bay estuarine ecosystem.

The model is based on material flow and follows the fluxes of two substances through 11 compartments. Compartments and fluxes are quantified as the density of ashfree dry weight; units are mg dry weight/1000 m^3. The two substances, N and C, represent limiting nutrient and non-nutrient material. Thus, the model can generally be applied to any situation with a single limiting nutrient. For application to Chesapeake Bay estuaries, nitrogen was considered limiting, so that N is identified with nitrogen and C with the remaining material. N is conserved. C is not conserved; this reflects the fixation of CO_2 by phytoplankton and the respiration of organics into CO_2.

The 11 compartments are highly aggregated, each compartment containing a number of species and size classes. In some cases this aggregation simplifies the modeling. Eppley's (1972) demonstration of temperature response by phytoplankton is the classic example of this--the whole community maintains an exponential response to increasing temperature as different groups, each with a complicated peaked response to temperature, replace each other. In other cases the heterogeneity makes parameter estimation difficult. The following aggregations were chosen to provide a detailed representation of trophic function without requiring a huge number of parameters:

PHYT3   phytoplankton

ZOOP4   zooplankton; all species and age-classes are grouped together.

PFISH8   adult and juvenile planktivorous fish, smaller benthic-feeding fishes and some macroplankton. Representative species in this group are the bay anchovy, *Anchoa mitchilli*, the Atlantic menhaden, *Brevoortia tyrannus*; the anadromous clupeids; the silversides, *Menidia* spp.; the spot, *Leiostomus xanthurus*, the hogchoker, *Trinectes maculatus*; and the naked goby, *Gobiosoma bosci*; the shrimp *Neomysis* and the blue crab *Callinectes sapidus*.

PLARV5   larvae of PFISH8.

FFISH9   adults and juveniles of the larger, piscivorous and benthic feeding fishes. The major species and white perch, *Morone americana*; striped bass, *Morone saxatilis*; the eel, *Anguilla rostrata*; and the white catfish, *Ictalurus catus*.

FLARV6   larvae of FFISH9.

CTEN7   ctenophores and jellyfish. *Mnemiopsis leidyi* and *Chrysaora quinquecirrha* are the major species in the Chesapeake Bay estuaries.

BENTB   benthos; including polychaetes, molluscs, amphipods and isopods.

REFA   refractory organic materials, consisting largely of slowly decomposing plant material.

LAB1   labile organic material and the associated bacterial component.

DISS2   dissolved nutrient, including nitrate, nitrite, ammonia and urea.

The proportion of N in all compartments except PHYT3 and DISS2 is variable, so that the main variables of the model are the 11 standing crops and 9 proportions of nutrient.

The system is open, with several types of sources and sinks. Active migrations of fish and ctenophores are modelled by seasonal inputs and outputs. Phytoplankton, dissolved nutrient and the two types of detritus enter the system with spring and fall peaks; this represents inputs from marshes and runoff (Heinle *et al*. 1973).

There is an additional first-order loss of zooplankton, phytoplankton and detritus mimicking advective river transport and sedimentation. Fixation and respiration of non-nutrients also represent net inputs and losses to the system. Since the main interest is the effect of trophic interactions on impact, the system is modeled to be spatially homogeneous. Tidal dilution at the boundaries is modeled by daily exchange of a fraction of the study volume with an adjacent system, which can be considered a farfield (no impact) zone. This farfield is assumed identical with the nearfield in all respects except that its compartments suffer no additional mortality. Thus, in the absence of additional mortality, the daily exchange does not affect system dynamics; when additional mortality is present, the exchange reduces the impact.

The interactions between compartments are shown in figure 1. The system is analagous to the set of models in equations (3-5) in that abundances are controlled by the rates of inputs and losses to the system rather than by explicit density-dependent processes.

The basic form of the system is two interwoven feeding-nutrient cycles. One is based on dissolved nutrient and consists of nutrient-phytoplankton-zooplankton-fish larvae, planktivorous fish and ctenophores-piscivorous fish. The second is detritally based, consisting of detritus-benthos-fish. There are two main paths of nutrient regeneration: a fast cycle through excretion and a slow cycle through bacterial processing and mineralization.

The main physical forcing functions are temperature, day-length and insolation. All are modeled as annual sinusoidal cycles, with a mimimum temperature on 1 February (day 1 in the model). Temperature affects respiration rates, detrital processing rates and feeding rates. Day-length and insolation affect rates of carbon fixation and nitrogen uptake by phytoplankton.

All calculations are performed in double precision. Simulation follows day-by-day iteration from initial conditions. The simulation converged quickly to repeated annual cycles—the cycles typically repeated to three significant figures within 3-5 years of simulation. The cycles are independent of initial conditions within reasonable ranges of initial conditions. Impact assessments were based on the 10th year of simulation. The ratios of mean, maximum and minimum annual values for impact conditions relative to non-impact conditions were the primary metrics. The ratios of total annual recruitment of impact to non-impact conditions of the two fish compartments were also considered.

The day-to-day iteration procedure is three step, as follows. In the first step, most of the internal fluxes—fixation, uptake, feeding and defecation, respiration, density-independent mortal-

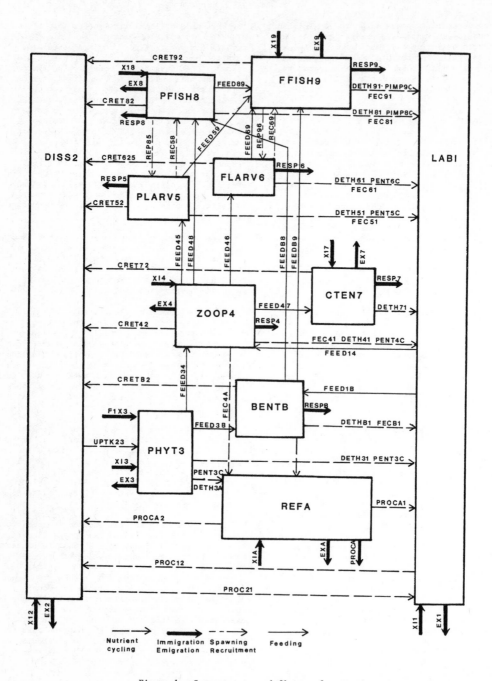

Figure 1.--Compartments and fluxes of estuarine ecosystem model.

ity, detrital processing, entrainment and impinge-
ment--are computed. In the second step, excretion
is calculated to balance nutrient to non-nutrient
ratio. Recruitment, reproduction and immigration
and emigration are also calculated in this step.
In the third step, exchange with the far field is
computed. The individual fluxes for the first
step are calculated as proportional parts of a total loss generated by the integration over the day
of the sum of the instantaneous rates of mortality
from each cause. Losses of nutrient and non-nutrient are calculated separately. This formulation
accounts for loss of each donor compartment over
the course of the day. It does not include changes
in instantaneous predation rate from changes in predator or prey density. (cf. Hassell 1978; Kremer
and Nixon 1978). If daily fluxes are small, the
error introduced by this approximation is small.

The individual fluxes are described below.

Feeding

Feeding of zooplankton and benthos are modeled by a multispecific analogue of the Holling type
II functional response, a hyperbolic response with
an asymptotically approached maximal feeding rate
per predator (Holling 1959, Hassell 1978).
Feeding by larval and adult fishes and ctenophores
follows a Holling type III response, a sigmoid response. The type III response is used to account
for several aspects of predation which reduce predation at low prey density (cf. Hassell et al.
1977), e.g., spatial refuges, cessation or curtailment of feeding at low prey density, and habitat
selection. All of these have the effect of producing a sigmoid functional response.

In all groups, feeding rate is exponentially
temperature-dependent. This reflects the low feeding activity in winter and increased feeding rates
as metabolic demands rise with increasing temperature.

Defecation

The benthos and the four fish compartments assimilate a constant fraction of the food killed.
Zooplankton and ctenophores have variable assimilation fractions reflecting lower assimilation at
high food intake. Assimilation and defecation
occur in the same time step as the feeding which
generated them.

Respiration; Excretion

Each animal compartment has an exponentially
temperature-dependent respiration rate.

Excretion of nitrogenous material is used to
balance the C:N ratio, keeping it above a prescribed
minimum for each compartment. Since the primary excretory products of aquatic organisms are ammonia
and urea (Conover 1978), which are used directly by
phytoplankton (McCarthy et al. 1977, McCarthy 1972
a,b, Webb and Haas 1976), it is justified to consider all excretory products as going directly into
the dissolved nutrient pool (DISS2).

Uptake and Photosynthesis

Limitations on phytoplankton growth are represented by multiplicative limitation of maximum nutrient uptake by light and a single nutrient N.
The maximum uptake ratio is exponentially temperature
dependent. Nutrient limitation is represented by
the Monod model of asymptotic saturation. Light
limitation is based on a deterministic analogue of
the algorithm of Kremer and Nixon (1978), based on
work by Steele (1962) and Ditoro et al. (1971).
This model assumes that over a year acclimatization maintains maximum uptake rates for all light
levels above some mimimum; approximate intergrals
over dept. and time of day yield a proportion of
maximum uptake for each day; this proportion depends on the depth of the estuary, the extinction
coefficient of light, and the daily insolation
and day length.

Given a daily level of nutrient uptake, carbon
fixation is assumed to balance respiration so as to
maintain a constant C:N ratio.

Detrital Processing

The pathways of nutrient cycling are shown in
figure 1. These fluxes are all donor-controlled,
first-order processes. Each rate is exponentially
temperature dependent.

Reproduction and Recruitment

Explicit reproduction is necessary only for
the fish compartments, since life-stages are not
separated for the other groups. A proportion of
the adult biomass goes into reproduction; reproduction is spread over a number of days, with
peak reproduction in the middle of the reproductive period. The piscivorous fishes reproduce
for a short period in early spring; the other
fishes spawn over much of the spring and summer.

Recruitment of fish larvae into the adult
classes occurs after a fixed number of days after
reproduction. This scheme does not incorporate
one of the commonly cited sources of compensation--
increased growth rate at lower densities leading
to shorter larval period and shorter period of
vulnerability to predators.

Density-Independent Mortality

The animal compartments each sustain a fixed
proportional loss from density-independent losses.
The principal source is fishing mortality on the
fishes. This category also includes losses from
a number of sources, such as disease, injury, sinking, stress and senescence. While some of these
may be density-dependent in reality, not enough is
known about these to allow more complex formulations.

190

Immigration and Emigration

Active immigration is modeled as a constant annual input spread over a span of days. Emigration is modeled as a proportion of the compartment which leaves the estuary each year; this emigration is also spread over a number of days. For the standard run ctenophores immigrated in mass in midsummer and emigrated in entirety in late summer, and "forage" fishes immigrated in early spring and most of the population (typically 80%) emigrated in late fall. Tests was also done which included movements of larval fish.

Additional Mortality

Additional mortality, in this application resulting from entrainment and impingement, was modeled as a seasonally constant, density-independent mortality rate for each biological compartment except benthos. These rates are the product of an entrainment rate (constant for all entrained groups) or impingement rate multiplied by a probability of mortality given entrainment or impingement. Based on real studies (ANSP), the latter probabilities were taken to be .10 for phytoplankton, .40 for zooplankton, 1.0 for ctenophores, .75 for larval fish, and 1.0 for impinged adult fish.

Estimation of Parameter Values

The primary application of the model is to a 10 nautical mile section of the middle Patuxent estuary, on the Maryland Western shore. To model this system, parameters were estimated from the literature and from studies by the Academy of Natural Sciences of Philadelphia (ANSP 1981). Whenever possible, data from the Patuxent River were used; when unavailable, preference was given to studies on dominant Patuxent River species in the laboratory or in other parts of the Chesapeake Bay system, and to studies on the same species outside the Chesapeake and, finally, to literature studies of related species. A number of parameters are poorly known; this, and the simplifications of the model necessitated adjustments of some of the parameter estimates. In particular, the parameters of feeding rates required considerable adjustment. A detailed discussion of model formulations and sources of parameter values may be obtained from the author (Horwitz and Odhner 1981).

RESULTS

Standard Run

The model converged quickly into repeated annual cycles (fig. 2 and 3); the annual means, maxima and minima are shown in table 1. The cycles illustrate many of the features typical of estuarine systems. Phytoplankton standing crops increased from a later winter minimum, following increasing temperature, day-length and nutrients. Around day 120, phytoplankton stocks declined about an order of magnitude, despite increasing nutrient uptake resulting from an increase in nutrient concentrations. The drop was due to rapidly increasing predation. Phytoplankton stocks remained rather constant through the summer, in balance between high levels of production (uptake is about 75% of standing crop of nutrient in this period) and predation. Zooplankton density rose sharply with increasing phytoplankton and detritus until early spring, when cropping by returning forage fish, larvae of piscivorous fishes, and, later, larvae of forage fish, sharply reduced zooplankton densities. The densities of the two groups of fish larvae and the adult forage fish peaked soon after; the ensuing decrease is attributable to increased predation by piscivorous fish, decreased food following the zooplankton crash, and larval recruitment. Zooplankton started to increase again in midsummer, 16 days after the peak in adult forage fish, but densities remained low from predation by the remaining fish planktivores, and, in late summer, by ctenophores. Zooplankton increased steadily over fall and winter, after recruitment of larval fishes and emigration reduced feeding of ctenophores and forage fish.

Forage fish increased sharply in spring from increased feeding and immigration. There was a short decline after the zooplankton crash, but the fish population subsequently increased through feeding on expanding benthic populations. The fish population peaked around day 125, dropped to a plateau in late summer and declined and emigrated in fall. This decline was due to predation by piscivorous fish and declining populations of benthos. The piscivorous fish showed a slight increase in spring from recruitment and the immigration of for-

Table 1.--Annual mean, maximum and minimum densities-- standard (non-impact) run.

	LAB1	DISS2	PHYT3	ZOOP4	PLARV5	FLARV6	CTEN7	PFISH8	FFISH9	REFA	BENTB
Mean=	.103E+7	.893E+4	.299E+7	.488E+4	.301E+3	.210E+4	.188E+4	.919E+6	.108E+7	.103E+7	.971E+5
Max.=	.214E+7	.502E+5	.508E+7	.498E+5	.578E+4	.540E+5	.620E+5	.342E+7	.388E+7	.149E+7	.141E+7
Min.=	.969E+5	.750E+3	.399E+6	.389E-3	.000	.000	.000	.213E-1	.292E+5	.536E+6	.618E-1

TREC8= .47104E+5; TREC9= .22429E+4

Values are reported in exponential, base 10, notation. For example, .103E+7=1,030,000.

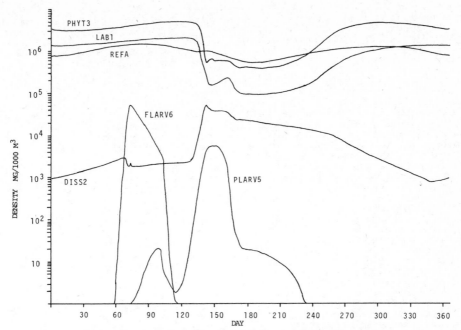

Figure 2.--Annual cycle of standard run of estuarine ecosystem model: zooplankton, benthos, ctenophores and
adult fish. Day 1 = February 1.

Figure 3.--Annual cycle of standard run of estuarine ecosystem model: phytoplankton, detritus, dissolved
nutrient and larval fish. Day 1 = February 1.

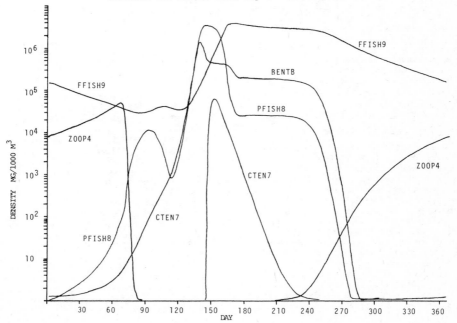

age fishes, and a sharp increase in summer from predation on the expanded forage fish and benthic
populations. The piscivorous fish decreased about
20 days after the peak of these two prey populations, and declined slowly until the following
spring.

The larvae of the piscivorous fish increased
as the zooplankton crashed, and then decreased from
recruitment and starvation. The abundance of PLARV5
followed the abundance of the parent population.
Since the zooplankton crashed before the main period
of spawning, these larvae fed relatively little and
declined from starvation; there were moderate levels
of recruitment. Similarly, ctenophores enter the
study area only in the summer, when zooplankton populations are low.

Refractory nutrient was strongly controlled by
input from outside the system, reaching a peak during the April influx and a minimum in midsummer,
when processing rates were high and import low.
Refractory material, as expected, showed a relatively narrow range of variation. The labile detritus
increased in spring from increasing breakdown of
refractory material and increasing production of
fecal material, triggered by increasing temperature and immigration of the larger organisms.
This increase in flux of labile material and the
increasing production of phytoplankton triggered
the spring burst in benthos: labile detritus and
phytoplankton increased little in the spring while
the benthos increased manyfold. Both the labile
detritus and phytoplankton decreased sharply around
day 130, as a result of increasing predation by
benthos. Dissolved nutrient increased slowly
through the spring, primarily from the increase in
breakdown of labile material. This breakdown peaked
around day 120 (after the decline in labile standing crop) but dissolved nutrient showed a sharp increase from excretion by larger organisms. During
much of the year, much of the available pool of dissolved nutrient was taken up daily by phytoplankton
but nutrient was recycled rapidly through excretion
by the dominant consumer (alternately zooplankton,
forage fishes, piscivorous fishes and benthos at
different times). Input and mineralization were
also important sources in the warmer months.

While the general pattern of annual abundances
was realistic, there were several atypical aspects
of the model. Phytoplankton showed a relatively
narrow range of variation--about one order of magnitude and did not show the observed winter crash.
The zooplankton also did not show a late winter
crash; furthermore, the decline from the spring
bloom occurred too early and was too severe. The
piscivorous fish were generally more abundant than
usually observed, though measurements of fish
standing crop are notably inprecise. The summer
increase in benthos and smaller fishes was probably more extreme than usually observed.

These discrepancies are interdependent; for
example, high piscivorous fish densities maintain
high phytoplankton productivity through excretion

and contribute to the high amplitude cycles of
other fishes and benthos. The zooplankton crash
also contributes to high summer phytoplankton levels. These discrepancies may be due to several
structural oversimplification of the model.

The high density of piscivorous fish probably
results from the lack of any movement into and out
of the area. It is probable that these fish (primarily white perch) move out of the area at high
densities. However, in the absence of any population data on movements of these fish, this movement would be difficult to model.

The multistage life cycle of the common copepod species may strongly affect zooplankton dynamics. By considering all stages as part of a
single compartment, these dynamics were lost.
Kremer and Nixon (1978) found that separating zooplankton into stages had a major effect; in particular, the zooplankton cycles were more stable than
under a lumped configuration. Thus, the high
amplitude of zooplankton cycles may arise from
this modeling artifact.

Test runs were made changing various parameters of the model - e.g., adding immigration of
fish larvae, changing feeding preferences, and
adding herbivory by planktivorous fishes. While
these changes affected seasonal dynamics, they
did not cause significant changes in the effects
of additional mortality. Thus, the model is
appropriate for its main purpose, the study of
effects of additional mortality.

Effects of Additional Mortality·

Runs with real estimate of entrainment, impingement and dilution rates generally showed only
trivial changes from the standard run. Since the
major interest here is the relative effects of
additional mortality on different compartments,
runs with no dilution (an unrealistic situation)
will be emphasized - these display the impacts
more clearly.
In order to sort out the effects of mortality
at different levels, a series of runs with entrainment at only one level was made. These used rates
of .025, .05, .071, .082, .10 and .15. These values span real estimates for entrainment (levels 3-
7); while they are much greater than the observed
rates of impingement (levels 8, 9), these values
were used for purposes of comparison. Runs were
made at three exchange rates: 0 (i.e., no exchange), .30, and .68. Where effects were observed,
their direction was almost always the same in comparable runs at the high and no exchange rates.
The few differences between the two sets of runs
(6 of 147 comparisons) always involved impacts of
mortality of PHYT3, PFISH8 or FFISH9 on PHYT3 or
BENTB.

The model is robust to changes in entrainment
rates. That is, the qualitative behavior did not
change over the range of entrainments modeled (.025-
.15)--direction of impacts were similar and the

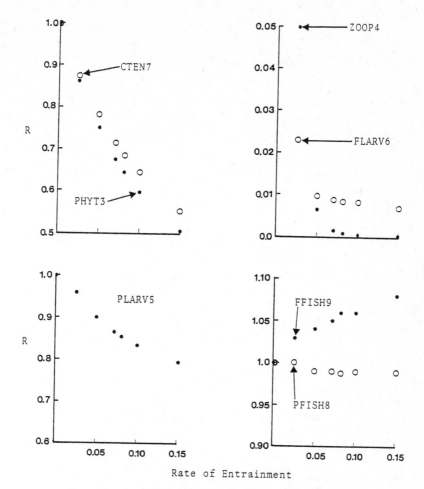

Figure 4.--Sensitivity of model results to rate of entrainment. Ratios (R) of mean biomass with entrainment to
mean without entrainment, as a function of rate of
entrainment. Levels PHYT3, ZOOP4, PLARV5, FLARV6
and CTEN7 entrained. Rate of impingement of PFISH8=
0.0002, and of FFISH9=0.00009. Rate of exchange =0.0.
Note differences in scale of y-axes.

magnitude of impact did not change greatly with
small changes in entrainment rate. As expected,
in virtually all cases, impact is approximately
linear with increasing mortality for small reductions (the realistic cases) and exponential for
drastic reductions (fig. 4 and 5). (The insensitive
response of PFISH8 and FFISH9 reflects the fact
that they are not entrained and are relatively insensitive to change at lower levels.) Thus, the
model is well-behaved with respect to entrainment.
For this reason, only the cases for m (mortality
from entrainment) = .071 are discussed in detail.
The qualitative behavior--direction of impact--
will be discussed first, followed by analysis of
quantitative aspects.

Many of the predicted effects of single-level
entrainment are seen in this model as well (table
2). In all cases, entrainment (or impingement) on
a level reduced the mean, maximum and minimum of
that level. The amplitude decreased for entrainment of phytoplankton and zooplankton, but increased with impingement of adult fish.

As hypothesized, transfers of impact between
levels were apparent, especially for mortality of
phytoplankton, zooplankton and adult fish. These
were often in the form discussed in earlier sections, with predators decreasing with entrainment of lower levels, especially their prey, and
prey increasing with predator mortality. (tables

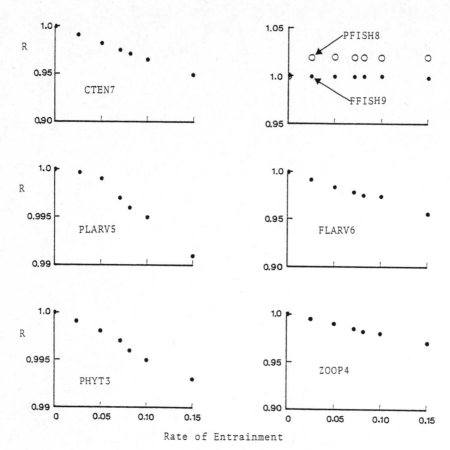

Rate of Entrainment

Figure 5.--Sensitivity of model results to rate of entrainment. Ratios (R) of mean biomass with entrainment to
mean without entrainment, as a function of rate of
entrainment. Levels PHYT3, ZOOP4, PLARV5, FLARV6
and CTEN7 entrained. Rate of impingement of PFISH*=
0.0002, and of FFISH9=0.00009. Rate of exchange =0.68.
Note differences in scale of y-axes.

2 and 3) Thus, benthos increased when the two
fish groups, its predators, were entrained; forage fish increased when piscivorous fish were
impinged and piscivorous fish decreased when
forage fish and zooplankton were entrained; phytoplankton increased under zooplankton entrainment and zooplankton decreased under phytoplankton entrainment; and zooplankton increased under
entrainment of forage fish.

However, this is not a general phenomenon.
For example, forage fish increased in response
to entrainment of zooplankton, its prey. The
annual amplitude of forage fish increased under
entrainment of zooplankton and other trophic in-

teractions may affect forage fish in these cases.
Benthos, an alternate prey of forage fish, increased under entrainment of phytoplankton (benthos preferentially eats detritus over phytoplankton) and piscivorous fish, the main predator of the forage fish, decreased under entrainment of zooplankton--the increased amplitude of
forage fish here may cause a decline in piscivorous fish which allows a greater subsequent
increase in forage fish. Thus, while the general predictions of the equilibrium food chain
models are consistent with the predictions of
this model, the consideration of alternate prey
(i.e., of a food web rather than a chain) and
seasonal dynamics changes the predictions to
some degree.

Table 2.--Top: Effect of entrainment of single
levels on mean annual densities of other levels. + indicates an increase under entrainment, - a decrease. R is the ratio of density in impact run to density in standard run.
Middle: Histogram of signs in above matrix.
Bottom: Effect of entrainment of single levels on amplitudes of annual cycles of other
levels. / indicates that the level is not
present during some part of the year (from
emigration or recruitment).

Effect on Level	Entrained Level						
	3	4	5	6	7	8	9
3	--	+	0	-	-	+	--
4	--	--	+	+	++	--	--
5	--	+	-	+	-	--	++
6	--	--	+	--	++	--	--
7	0	0	0	0	--	0	0
8	-	+	-	+	-	--	++
9	+	-	0	0	0	--	--
B	+	0	0	0	0	++	++

++   R≥1.05
+   1.05>R>1.00
0   R=1.00
-   .95<R<1.00
--   R≤.95

FREQUENCY OF SIGNS

Minimum Difference Between Trophic Levels*	++	+	0	-	--	Predicted by Equilibrium Models
-2	1		1	2	2	(-)
-1	5	3		1		(+)
0			1	6		(-)
1		3	3		3	(-)
2	1		1	2	2	(-)

* -2 - impact of entrainment of i on prey of prey of i
  -1 - impact of entrainment of i on prey of i
   0 - impact of entrainment of i on i
   1 - impact of entrainment of i on predator of i
   2 - impact of entrainment of i on predator of
      predator of i

Effect on Level	Entrained Level						
	3	4	5	6	7	8	9
3	--	++	0	--	-	++	--
4	--	--	0	++	++	-	--
5	/	/	/	/	/	/	/
6	/	/	/	/	/	/	/
7	/	/	/	/	/	/	/
8	++	++	0	0	--	++	-
9	0	++	0	++	--	++	++
B	--	++	0	++	++	++	--

The transfer of impacts can be seen in the
magnitude of the impact ratios. The impact on an
unentrained predator is often greater than that
on an entrained prey (table 3); the impact of entrainment of phytoplankton on zooplankton is
greater than that of phytoplankton. The transfer
may also be downward. One of the most dramatic
impacts modeled is that of entraining piscivorous
fish on zooplankton; the zooplankton are drastically reduced by the resultant rise in forage
fish.

For application to the effects of the Chalk
Point S.E.S. on the Middle Patuxent Estuary the
most realistic conditions were estimated as entrainment of levels 3 through 7 at a rate of .071,
impingement of levels 8 and 9 at a rate of .0002 or
.00009, and exchange at a rate of 0.68 (table 3,
fig. 5). The estimated levels of impingement have
little or no effect on any component. The general
trends are very small decreases in phytoplankton,
small decreases (less than 3%) in zooplankton, fish
larvae and ctenophores, and an increase in amplitude of variation of both fish compartments, with
small increases in mean density of piscivorous fish
and benthos. The total recruitment of the two groups
of fish larvae are moderately reduced (7-9%), but
this does not affect adult fish densities strongly. This
implies that there is no severe accumulation
of impacts on any one group, although the depletion
of larvae of piscivorous fish and ctenophores is
most severe. One effect noted is a slight shift
from a zooplankton-based food chain toward a benthic
chain. Phytoplankton and zooplankton decrease, but
the unentrained benthos increase under entrainment
(the increase in detrital food outweighing losses of
phytoplankton) and this allows the fish groups to at
least maintain former abundance.

DISCUSSION AND CONCLUSIONS

Thus, the general conclusions of this ecosystem
model are that entrainment impacts are apt to be
transferred between predators and prey, but that the
occurrence of this effect will depend on the dynamics
and other prey of the groups. These effects may
create rather complicated webs of impacts (cf. the
simulation of zooplankton entrainment on forage fish,
benthos and adult fish). In the model with added
mortality at all levels, small decreases were noted
in the phytoplankton-based food chain, while small
increases were noted in benthos and some fish.
This indicates a possible slight shift toward a detritus-based system. In comparing the results of
this model with the general food chain models discussed earlier, the main difference is the occurrence of occasional enhancement in some levels
(particularly PFISH8) and the absence of accumulation of impacts on higher levels. These differences
are likely the result of factors in the ecosystem
model not included in the equilibrium models--a
more complicated trophic structure, including alternate food pathways; nutrient cycling, including return of material from entrained organisms to the
system; and the effect of seasonal variation on the
abundance of organisms.

In comparing the models, the ecosystem model shares some similarities in behavior with the equibrium models with self-regulation in the top level. In particular, impacts appear to be transferred down the food web; zooplankton, not piscivorous fish, showed the greatest sensitivity to entrainment. As discussed earlier, patterns of abundances suggest that the model system shows greater predator dominance than may really occur. Furthermore, in modeling adult and larval fish separately,

the model includes a self-regulatory feedback for fish populations through cannibalism. While a similar effect may occur with zooplankton, it was not explicitly modeled. Thus, the appearance of predator domination may in part be the result of model structure and parameterization. Because the direction of impacts depended on the predator dominance, some of the numerical results of the simulation might change with a more realistic simulation.

Table 3.--Ratios of mean, maximum or minimum density of compartment in impact run to corresponding density in standard run. RATR8 and RATR9 are the ratios of total annual recruitment to PFISH8 and FFISH9 in impact runs to recruitment in standard runs.

Entrained Level	Rate		LAB1	DISS2	PHYT3	ZOOP4	PLARV5	FLARV6	CTEN7	PFISH8	FFISH9	REFA	DENTB
		RATE OF EXCHANGE = 0.00											
PHYT3	.071	Mean=	.148E+1	.998	.677	.349E-1	.890	.168E-1	NI	.984	.105E+1	.102E+1	.103E+1
		Max.=	.151E+1	.997	.733	.126E-1	.827	.794E-2	NI	.955	.981	.102E+1	.102E+1
		Min.=	.949	.110E+1	.914	.144E+1	.000	.000	.000	.814	.971	.956	.132E+1
		RATR8= .8977; RATR9= .4021E-1											
ZOOP4	.071	Mean=	.977	.990	.103E+1	.675E-2	.103E+1	.138E-1	NI	.103E+1	.995	.997	NI
		Max.=	.961	.101E+1	.104E+1	.210E-2	.103E+1	.706E-2	NI	.101E+1	NI	.991	.101E+1
		Min.=	.101E+1	.883	.101E+1	.853	.000	.000	.000	.290	.956	.101E+1	.992
		RATR8= 0.1033E+1; RATR9= 0.3792E-1											
PLARV5	.071	Mean=	NI	NI	NI	.101E+1	.972	.101E+1	NI	NI	NI	NI	NI
		Max.=	NI	NI	NI	.101E+1	.971	.101E+1	NI	NI	NI	NI	NI
		Min.=	NI	NI	NI	.101E+1	.000	.000	.000	NI	NI	NI	NI
		RATR8= .9218; RATR9=.1011E+1											
FLARV6	.071	Mean=	.999	NI	.998	.102E+1	.101E+1	.690	NI	.101E+1	NI	NI	NI
		Max.=	NI	.977	.999	.102E+1	NI	.931	NI	NI	NI	NI	.101E+1
		Min.=	.998	NI	NI	.101E+1	.000	.000	.000	NI	.991	NI	NI
		RATR8= 0.1010E+1; RATR9= .1807											
CTEN7	.071	Mean=	NI	NI	.995	.119E+1	.995	.117E+1	.716	.995	NI	NI	NI
		Max.=	.101E+1	.990	.993	.118E+1	.995	.117E+1	.869	.998	NI	NI	.101E+1
		Min.=	.998	.102E+1	.998	.111E+1	.000	.000	.000	.104E+1	.101E+1	.998	NI
		RATR8= 0.9945; RATR9= .1190E+1											
PFISH8*	.071	Mean=	NI	.108E+1	.101E+1	.509	.905	.535	NI	.914	.971	NI	.123E+1
		Max.=	.983	.116E+1	.102E+1	.522	.785	.525	NI	.806	.966	.996	.105E+1
		Min.=	.102E+1	.942	NI	.582	.000	.000	.000	.636	.905	.104E+1	.994
		RATR8= .9069; RATR9= .5125											
FFISH9*	.071	Mean=	.117E+1	.249E+1	.756	.210E-5	.499E+1	.686E-6	NI	.127E+2	.236	.115E+1	.325E+1
		Max.=	.121E+1	.912	.952	.626E-6	.142E+1	.362E-6	NI	.151E+1	.323	.110E+1	.823
		Min.=	.145E+1	.108E+1	.156E+1	.650	.000	.000	.000	.110E+2	.339E-4	.124E+1	.260E+2
		RATR8= .5230E+1; RATR9= .1658E-5											
PFISH8	.0002	Mean=	NI	NI	NI	NI	NI	NI	NI	NI	NI	NI	NI
		Max.=	NI	NI	NI	NI	.999	NI	NI	.999	NI	NI	NI
		Min.=	NI	NI	NI	.998	.000	.000	.000	.999	NI	NI	NI
		RATR8= .9996; RATR9= NI											
FFISH9	.00009	Mean=	NI	NI	.999	.974	NI	.974	NI	NI	NI	NI	NI
		Max.=	.999	.994	NI	.973	.999	.974	NI	NI	NI	NI	NI
		Min.=	.101E+1	.995	NI	NI	.000	.000	.000	.999	.984	NI	NI
		RATR8= NI; RATR9= .9748											
3-7	.071	Mean=	.148E+1	.996	.678	.168E-2	.867	.898E-2	.716	.986	.105E+1	.102E+1	.103E+1
		Max.=	.151E+1	.995	.733	.857E-3	.804	.508E-2	.869	.955	.983	.102E+1	.102E+1
		Min.=	.949	.108E+1	.914	.127E+1	.000	.000	.000	.235	.970	.955	.132E+1
		RATR8= .8297E; RATR9= .1431E-1											
3-7	.071	Mean=	.148E+1	.998	.677	.164E-2	.867	.885E-2	.716	.988	.105E+1	.102E+1	.104E+1
8	.0002	Max.=	.150E+1	.987	.733	.836E-3	.802	.500E-2	.869	.955	.983	.102E+1	.102E+1
9	.00009	Min.=	.954	.108E+1	.914	.128E+1	.000	.000	.000	.230	.956	.955	.132E+1
		RATR8= .8299; RATR9= .1413E-1											
		RATE OF EXCHANGE = 0.68											
3-7	.071	Mean=	.101E+1	.994	.997	.985	.992	.978	.975	.994	.998	NI	.995
		Max.=	.101E+1	.988	.997	.986	.992	.982	.978	.996	.998	NI	.992
		Min.=	.101E+1	.986	.999	.993	.000	.000	.000	.968	.995	NI	.998
		RATR8= .9402; RATR9= .9256											
3-7	.071	Mean=	.101E+1	.996	.997	.985	.997	.978	.975	.102E+1	NI	NI	.101E+1
8	.0002	Max.=	.101E+1	.998	.997	.986	.997	.982	.978	.102E+1	.101E+1	NI	.102E+1
9	.00009	Min.=	.101E+1	.984	.999	.994	.000	.000	.000	.974	.977	NI	NI
		RATR8= .9443; RATR9= .9257											

NI - impact value equals non-impact value, i.e., impact value equals 1.00.
Values are reported in exponential, base 10, notation. For example, .103E+7=1,030,000.

ACKNOWLEDGEMENTS

I would like to thank Michael X. Odhner for much effort in reviewing the estuarine literature in search of reliable parameter estimates and Sandy De Van for preparation of the manuscript. This work was funded by the Potomac Electric Power Company and by the Baltimore Gas and Electric Company.

LITERATURE CITED

ANSP. 1981. Chalk Point 316 Demonstration of thermal, entrainment and impingment impacts on the Patuxent River in accordance with COMAR 08.05.04.13, 6 vol. Academy of Natural Sciences of Philadelphia.

Barclay, H., and P. van den Driessche. 1977. Predator-prey models with added mortality. Can. Entomol. 109:763-768.

Clark, C. W. 1976. Mathematical bioeconomics: the optimal management of renewable resources. Wiley, New York.

Conover, R. J. 1978. Transformation of organic matter. p. 221-498. In O. Kinne, ed. Marine ecology IV. John Wiley and Sons, New York.

DeAngelis, D. L., S. W. Christensen, and A. G. Clark. 1977. Responses of a fish population model to young-of-the-year mortality. J. Fish. Res. Bd. Canada, 34:2124-2132.

DiToro, D. M., D. J. O'Connor, and R. V. Thomann. 1971. A dynamic model of phytoplankton populations in the Sacramento-San Joaquin Delta. Adv. Chem. Ser. 106:131-180.

Eppley, R. W. 1972. Temperature and phytoplankton growth in the sea. Fish. Bull. 70:1063-1085.

Hassell, M. P. 1978. Dynamics of arthropod predator-prey systems. Princeton Univ. Press, Princeton.

Hassell, M. P., J. H. Lawton, and J. R. Beddington. 1977. Sigmoid functional responses by invertebrate predators and parasitoids. J. Animal Ecol. 46:249-262.

Heinle, D. R., D. A. Flemer, J. F. Ustach, R. A. Murtagh, and R. P. Harris. 1973. The role of organic debris and associated microorganisms in pelagic estuarine food-chains. Technical Report 22, Chesapeake Biol. Lab., Solomons, Md.

Hess, K. W., M. P. Sissenwine, and S. B. Saila. 1975. Simulating the impact of the entrainment of winter flounder larvae. p. 1-29. In S. B. Saila, ed. Fisheries and energy production: a symposium. Lexington Books, D. C. Heath and Co., Lexington, MA.

Holling, C. S. 1959. The Components of predation as revealed by a study of small mammal predation of the European pine sawfly. Can. Ent. 91:293-320.

Horwitz, R. J., and M. X. Odhner. 1981. Ecosystem model and definitions of model parameters. p. 329-464. In vol. 6 of Chalk Point 316 Demonstration of thermal, entrainment and impingement impacts on the Patuxent River in accordance with COMAR 08.05.04.13. The Academy of Natural Sciences of Philadelphia.

Kremer, J. N. and S. W. Nixon. 1978. A coastal marine ecosystem, simulation and analysis. Ecological Studies, Vol. 24. Springer-Verlag, New York.

May, R. M., J. R. Beddington, C. W. Clark, S. J. Holt, and R. M. Laws. 1979. Management of multispecies fisheries. Science 205:267-277.

McCarthy, J. J. 1972a. The uptake of urea by marine phytoplankton. J. Phycol. 8:216-222.

_____. 1972b. The uptake of urea by natural populations of marine phytoplankton. Limnol. Oceanogr. 17:738-748.

McCarthy, J. J., W. R. Taylor, and J. L. Taft. 1977. Nitrogenous nutrition of the plankton in the Chesapeake Bay. I. Nutrient availability and phytoplankton preferences. Limnol. Oceanogr. 22:996-1011.

Ricker, W. E. 1954. Stock and Recruitment. J. Fish. Res. Board. Can. 11:559-623.

Scudo, F. M. 1971. Vito Volterra and Theoretical Ecology. Theor. Pop. Biol. 2:1-23.

Steele, J. H. 1962. Environmental control of photosynthesis in the sea. Limnol. Oceanogr. 7:137-150.

Volterra, V. 1928. Variations and fluctuations of the number of individuals in animal species living together. J. Conseil Int. Explor. Mer 3:1-51.

Webb, K. L., and L. W. Haas. 1976. The significance of urea for phytoplankton nutrition in the York River, Virginia. p. 90-102. In M. Wiley, ed. Estuarine Processes, Vol. 1. Academic Press, New York. 541 pp.

A PROCEDURE FOR DETERMINING AND EVALUATING POTENTIAL POWER PLANT ENTRAINMENT IMPACT OR A MODELER REVISITS HIS ASSUMPTIONS[1]

J. Kevin Summers and T. T. Polgar[2]

Abstract.--A procedure is presented that assesses the effects of power-plant entrainment on aquatic populations and the ecosystem. This methodology compromises between strict modeling and empirical approaches by using field data in the framework of several conceptual submodels. While the model's estimates of potential impacts are relatively gross, the procedure provides regulatory agencies with holistic impact estimates. The major assumptions of the submodels are reviewed and discussed in relation to basic ecological principles and current studies concerning population and community dynamics.

INTRODUCTION

In the last decade, substantial effort has been directed towards assessing and evaluating the impact of power plant entrainment losses on aquatic ecosystems (Hoss et al. 1974, Horst 1975, 1977, Boreman et al. 1978, Marcy 1977, Schubel et al. 1977, Van Winkle 1978, Polgar et al. 1979, Martin Marietta Corporation 1980, Summers and Jacobs 1981, and many others). Even after these years of study, it is not completely clear which type of analytical approach (generic and site-specific data or forecasting models) is best for estimating entrainment impact (Lauer et al., in press, Christensen, in press). In its broadest sense, entrainment refers not only to the passage of organisms through power plant condenser systems but also to the exposure of aquatic organisms to the plant's thermal discharge.

Power plant entrainment acts directly on a group of organisms within a system and subsequently has indirect consequences on other organisms through system linkages.

Generally, entrainment impact has been evaluated on four levels (organismal, population, community, and ecosystem). To be interpreted

effectively, impact due to entrainment must be considered as an integrated hierarchical system, with effects seen at all levels -- from physiological processes to energy relationships in the ecosystem. While this hierarchy can be interpreted as resulting in either additive or ameliorating levels of effect (Lauer et al., in press), the resulting impacts of simple population losses can only be put into perspective at higher organizational levels (i.e., community or ecosystem).

Investigations into the impact of entrainment losses at any of these hierarchical levels have taken many forms, all of which can be essentially categorized into three primary types of analyses: 1) measurement and statistical analysis of generic or site-specific data that describe the existing effects of entrainment 2) simulation modelling based on generic and limited site-specific data that forecasts the impact of entrainment at high organizational levels or at some future date; and 3) statistical or empirical modelling based on generic and extensive site-specific data to quantitatively describe the immediate effects of entrainment and to project short-term effects on higher organizational levels.

These three approaches span a methodological spectrum from best data fit to mathematical abstraction. The basic tool of any entrainment analysis is data. Often, entrainment data are densities of organisms entrained into cooling systems and abundances of local species along a gradient of thermal exposure. While these data provide important information relating to the

[1]Paper presented at the international symposium Energy and Ecological Modelling (Louisville, Kentucky, April 20-23, 1981).

[2]J.K. Summers and T.T. Polgar are with Martin Marietta Environmental Center, 1450 S. Rolling Road, Baltimore, Maryland U.S.A.

types and magnitudes of entrainment effects, they are of little use for evaluating the impact of those effects, except through personal judgment, regardless of the hierarchical effects level.

Data evaluation is a dynamic and continuous process that can use these preliminary measurements in many ways to estimate impact. Green (1979) discussed five statistical scenarios (sequences) for impact assessment based on three basic questions (fig. 1). In essence, these scenarios can be used to describe most of the site-specific analytical studies designed for addressing the extent and magnitude of entrainment effects.

Studies at several power plants (e.g., Martin Marietta Corporation 1980) have approached the evaluation of entrainment impact at a population level by comparing preoperational and operational data (fig. 1, sequence 1). If the investigators knew that 1) an entrainment effect would be seen and where, 2) an adequate spatial control was available, and 3) the populations under investigation were at a stable equilibrium prior to plant operation, then an analytical design of optimal impact could be implemented to directly evaluate the extent and magnitude of entrainment effects at the population level and, possibly, at the community level. Unfortunately, biologists are generally first consulted to ascertain the importance of an impact only after the impact has occurred. In such cases, the investigator is left to use statistical techniques to infer impact from differences in spatial distributions without the benefits of temporal information. Other than examining and testing specific phenomenological hypotheses, these studies and the subsequent statistical analyses cannot assist much in the evaluation of entrainment impact beyond the population level,

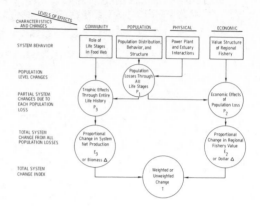

Figure 2.--General scheme for estimating entrainment and evaluating its impact.

if even to that level. Some community response parameters (e.g., clustering, diversity, discriminant analyses) can provide additional information, which is strongly based on empirical data, to estimate higher level effects. However, the utility of these techniques to evaluate impact directly is often limited.

Modelling, of necessity, often makes evaluations beyond the straight-forward estimation of the extent and magnitude of the effect. Entrainment modelling, in this sense, has taken several forms. Some modelling exercises have extended classical analytical model formulations (e.g., stock recruitment or Leslie matrix relationships) to evaluate the effect of entrainment losses on fish population levels (e.g., Lawler and Englert 1978, Horst 1975). Simulation models (e.g., Swartzman et al. 1978) have been used with limited site-specific data but extensive generic data to integrate the effects of entrainment losses and generate an evaluation of the spatial and/or temporal effects of these losses through an ecosystem. Obviously, models of this magnitude require numerous assumptions, most of which are not generally met by conditions at the site under investigation. But, at least if the model is presented correctly, the assumptions are explicitly stated and open to public and scientific scrutiny. Finally, empirical modelling allows the user to investigate somewhat beyond the available site-specific data, but generally without the myriad of assumptions required for simulation techniques. Models of this type are generally closely tied to data and statistical analyses. They are often "accounting" models that are primarily based on distributions and/or abundances derived from sampling data (Boreman et al. 1978, Polgar et al. 1979).

For regulatory purposes, real data must be addressed, corresponding to the specific concerns

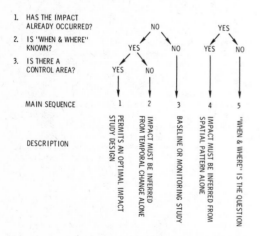

Figure 1.--Decision key for design of impact analysis studies (from Green 1979).

of environmental managers and agencies. While these data often are not available for rigorous quantitative treatment, definitive answers are needed. This paper presents a linearized impact model for estimating relative population losses in a community ensemble due to power plant entrainment in the absence of compensation, which helps provide these answers. The approach uses empirical field data in the framework of several mathematically simple, conceptual models. It estimates potential loss of equivalent adults due to entrainment of early life stages and evaluates the impact of these losses in terms of a biologically meaningful measure of consequent loss in ecosystem productivity and in terms of potential value changes in the regional fishery (fig. 2).

This approach and its application have been extensively described elsewhere (Polgar et al., in press), and the methodology will only be generally presented here. The primary purpose of this paper is to examine some of the model assumptions, particularly those of the ecological submodel, in light of ecological principles concerning population and community dynamics.

ENTRAINMENT LOSS ESTIMATION

At a gross level, the entrainment loss submodel estimates relative population losses due to entrainment as the integrated products over time of entrainment mortality probabilities and life-stage densities normalized to the size of a regional life-stage population. This procedure accounts for developmental times of specific life stages over the entire period in question and accounts for diel, vertical, migratory behavior of organisms in a given life stage. The proportion (P_{ij}) of a regional life-stage-specific population j of estuarine species i potentially lost due to entrainment is:

$$P_{ij} = \left(\{1-\exp[(RQ_p-Q_p)/Q_T]\} \cdot Q_K \cdot L_{ij} \cdot T_{ijK} \cdot \overline{D}_{ij}* \cdot \right.$$
$$\left. \cdot M_{ijE}\right)/(\overline{D}_{ij} \cdot V_{ij} \cdot N_{ij}) \qquad (1)$$

where

R = probability of cooling water recirculation

Q_p = cooling system withdrawal rate

Q_T = total water transport available for entrainment

Q_K = discharge rate in water layer K where cooling water is withdrawn

L_{ij} = length of time that life stage j of species i occurs in exposed water column

T_{ijK} = proportion of diurnal time-span that life stage j of species i spends in water layer K

$\overline{D}_{ij}*$ = mean local density of life stage j of species i over L_{ij}

M_{ijE} = mortality rate due to cooling system entrainment of life stage j of species i (assumed to be 100% for all stages)

$\overline{D}_{ij}$ = mean regional density of life stage j of species i over L_{ij}

V_{ij} = volume of region inhabited by life stage j of species i that is potentially vulnerable to entrainment

N_{ij} = number of generations of life stage j of species i during period of susceptibility (L_{ij}).

The proportional loss of each life stage j of species i to the regional population due to plume entrainment (W_{ij}) can be expressed as:

$$W_{ij} = \{[(A_C/A_T) \cdot Q_D \cdot L_{ij} \cdot T_{ijD} \cdot \overline{D}_{ij}* \cdot$$
$$\cdot M_{ijP}]/(\overline{D}_{ij} \cdot V_{ij} \cdot N_{ij})\} \qquad (2)$$

where

A_C = mean maximal 24-hour cross-sectional area of the 2°C excess temperature isotherm

A_T = total estuarine cross-sectional area where A_C occurs

Q_D = transport rate in water layer D in which the plant discharge plume is located

T_{ijD} = proportion of diurnal time-span that life stage j of species i spends in water layer D

M_{ijP} = mortality rate due to plume entrainment of life stage j of species i (assumed to be 10% for egg and larval stages and 1% for juveniles).

Entrainment losses, regardless of source, can be combined for all life stages (z) to estimate an equivalent adult population loss (P_{1i}) as:

$$P_{1i} = 1 - \prod_{j=1}^{z} [1 - (P_{ij} + W_{ij})]. \qquad (3)$$

Several key assumptions are made in this estimation of population losses due to entrainment. All planktonic life stages are assumed to be uniformly distributed within specific water layers, and only their vertical migratory behavior patterns are assumed to alter their susceptibility to entrainment. Foremost among the assumptions of the population loss estimator is that no density-dependent compensation is invoked at the loss levels anticipated and that annual losses are not cumulative. The implications of these last two assumptions are described in the DISCUSSION section below.

IMPACTS OF ENTRAINMENT LOSSES

Economic Impact

The potential economic impact associated with the estimated population losses due to entrainment is calculated from historical and present economic values of the individual commercial and recreational fisheries and the estimated population losses. The proportional economic impact (P_{2i}) to commercial and recreational fisheries due to population losses of species i through entrainment is:

$$P_{2i} = [(P_{1i} \cdot V_i)/\Sigma V_i] \qquad (4)$$

where

P_{1i} = computed equivalent adult population losses of species i

V_i = mean annual dollar value of species i in the regional fishery in question

ΣV_i = total mean annual dollar value of the regional fishery in question

The economic impact of all species losses due to entrainment (E_2) may then be expressed proportionately as:

$$E_2 = \sum_i P_{2i} \qquad (5)$$

or as a dollar figure as:

$$E_2^* = \Sigma (P_{2i} \cdot V_i).$$

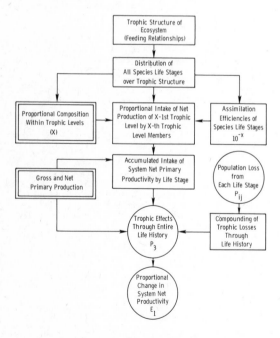

Figure 3.--Computational logic for evaluation of ecological impact due to population entrainment losses.

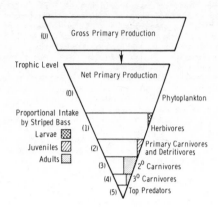

Figure 4.--Potomac River energy-flow, trophic dynamics for striped bass life stages.

The primary assumptions inherent in this estimation of economic impact, besides those related to the calculation of the estimated population losses (P_{1i}), are that the patterns of exploitation of species i have not changed significantly over the historical period used and that the regional fishery represents basic economic and stock units for the species considered.

Ecological Impact

Removal of portions of life-stage populations from any aquatic system will alter the energy flow pattern and trophic dynamics of that system. Impact at the ecosystem level due to entrainment losses can be simplistically estimated as the proportion of the system's net primary production that would go "unutilized" if portions of life-stage populations were eliminated.

Figure 3 delineates the logic used to estimate this quantity, and thus, the ecological impact of the population losses due to entrainment. In this analysis, individual life stages are used like pseudo-species, with each life-stage population composing a part of the system's trophic structure. The key variables necessary for calculating the proportional net primary productivity of each life-stage population are the trophic structure and feeding dynamics in the aquatic system, the distributions (densities) of the life-stage populations over the trophic structure, and the estimated assimilation efficiencies of each trophic level. For example, the inverted energy-flow pyramid in figure 4 illustrates the proportional intake of net primary production from the Potomac River system by the life stages of striped bass. The trophic structure of the Potomac estuarine ecosystem is shown as six trophic levels (0-5), representing

phytoplankton and other primary producers, herbivores, detritivores, and various carnivores.
The life stages of striped bass are partitioned
as larvae, juveniles, and adults. Eggs can be
generally ignored in this scheme because they
account for only a negligible portion of the
system's biomass and energy flow. In this
example, striped bass larvae feed primarily on
the herbivorous zooplankton and comprise about
2% of the larval fish in the primary carnivore
trophic level. Based on this percentage and
their ingestion rates, striped bass larvae would
ingest approximately 1.3% of the total net
production of the preceding trophic level (i.e.,
herbivores). Similarly, striped bass juveniles
ingest an estimated 9.9% of the net production
of primary carnivores and detritivores (i.e.,
0.09% of system's net production), while adults
remove about 50% and 35% of the net production of
secondary and tertiary carnivores, respectively.
Similar representations for all major components
of the system's trophic structure are necessary
to estimate potential "losses" of system productivity.

The proportion of "unutilized" net production (P_{3i}) in the system resulting from entrainment losses may be determined as:

$$P_{3i} = \sum_j \left\{ [1 - \prod_{j=k}^{z} (1 - P_{ij} - W_{ij})] \cdot \right. \tag{6}$$

$$\left. \cdot \sum_x \sum_{j=k}^{z} (I_{ijx} \cdot A^{-x} \cdot NPP) \right\} / \sum_{x=1}^{t} (A^{-x} \cdot NPP)$$

where

j = life stage (larvae, juvenile, adult)
k = life stage under investigation
(larvae, juvenile, or adult)
x = index defining the position of the
life stage in the aquatic trophic
structure
I_{ijx} = proportional intake of system net
productivity of trophic level ($x-1$)
by life stage j of species i at
trophic level x
A^{-x} = assimilation efficiency of trophic
level x (assumed to be 10^{-x})
NPP = system net primary production
t = number of trophic levels in aquatic
system.

The ecological impact of all species losses due
to entrainment (E_3) may then be expressed as:

$$E_3 = \sum_i P_{3i}. \tag{7}$$

Two major assumptions impinge on the validity of this estimate of ecological impact. All
phytoplankton and zooplankton losses due to entrainment are assumed to be rapidly and completely
compensated by mixing and high reproductive rates.
In addition, all "unused" net production in the
system will not be rapidly assimilated by other
components of the system's trophic structure
(i.e., predator switching or opportunism).

DISCUSSION

As with all model conceptions, the underlying assumptions governing the model structure
and applicability control the validity of the
modeler's abstraction of reality. Model assumptions are often shrouded in mathematical jargon
or are not specifically mentioned, making the
user feel that the model is safe to use. This
type of obscurity often lowers the esteem of the
modeler in the view of ecological researchers.
Model assumptions are sometimes presented in a
straight-forward manner, and the constraints they
place on the model's application are openly
admitted and constructively attacked. In such a
spirit of exchange and constructive criticism,
we discuss below several of the major assumptions
of this procedure for assessing entrainment
impact and their general applicability to general
estuarine situations (e.g., the Chesapeake Bay).
The assumptions discussed are 1) the applicability
of a linear approximation to entrainment losses,
2) the noncumulative (i.e., nonadditive year-to-
year) nature of entrainment losses, 3) the lack
of compensatory mechanisms at low levels of population loss, and 4) the lack of predator
switching to prey items made more available by
the entrainment destruction of the prey's primary
predator.

All computations estimating the population
losses due to entrainment were linearized. In
reality, most (if not all) of the relationships
involved in entrainment losses to life-stage
populations and their impacts are complex and
generally nonlinear. Linear approximations of
these relationships are acceptable if the loss
rates are generally small in relation to the size
of the base population. This assumption is
reasonable for many power plant facilities
located on large water bodies with widely distributed populations. Location on water bodies
characterized by complex flow regimes (e.g.,
tidal) enhances the acceptability of a linearized approximation. Even within the valid range
of linearization, population losses can result
in nontrivial economic impacts (e.g., small
losses to an important fishery) and ecological
impacts (e.g., small losses at important lower
trophic levels). The primary purpose for model
linearization is to provide managers with rapid,
inexpensive, yet quantitative estimates of potential population losses (i.e., without complex
programming and extensive computer use).

Linearized approximations are particularly
applicable to entrainment estimates for plants on
large estuaries such as the Chesapeake Bay and
its tributaries because the estimated losses are
not cumulative from year to year. This is not a
compensatory mechanism such as losses being
ameliorated within a year by changes in mortality

rates of a given life stage. Finfish generally recover slowly from population perturbation (e.g., from entrainment losses) since they normally spawn once per year. Thus, plant-induced mortalities can only be recouped in the following year. Entrainment losses are not cumulative because natural variation in environmental parameters is the primary controller of the introduction of new organisms to the various early life-stage populations. In other words, such factors as year-to-year variability in water temperature or flow play a significant role in the survival rate of eggs and larvae. The small proportional losses due to entrainment of early life stages of anadromous and resident fish are insignificant compared to the high levels of natural variability seen in abundances of early life-stage populations.

Generally, linearized estimates of regional population losses do not exceed 5-8% of a life-stage population. Losses of this magnitude are generally significantly less than the annual variability, even when associated with juvenile finfish populations. For example, the juvenile indices of populations of dominant ageclass fish such as alewives and striped bass in the Upper Chesapeake Bay have varied from -94% to 600% and -94% to 215%, respectively, from their long-term (1955-1980) mean abundance levels (e.g., Boone 1976). Reductions of less than 10% to an existing population will hardly affect that population beyond a single year when the subsequent year's abundance can range from 10% to 600% of the present standing stock.

Use of a highly variable fish stock such as striped bass may not provide a convincing argument for the noncumulative nature of entrainment losses in the Chesapeake Bay, but similar arguments can be advanced for resident fish stocks. While the annual range of variability of Maryland resident fish stocks such as spottail shiner (-66% to 120%), bay anchovy (-83% to 180%), and Atlantic silversides (-73% to 190%) is much less than the annual variation associated with anadromous stocks (Boone 1976), the annual reductions in population size attributed to entrainment (<10%) is still easily assimilated by the natural variability of the resident stocks. Thus, all population entrainment losses appear to be ameliorated or "compensated for" annually as a result of the variation of physical environmental parameters, and consequently, the impacts associated with these losses also should be short-term.

While these losses appear to be "compensated" for in the long term by natural variation in stock abundance, classical compensation (i.e., the tendency of populations to experience increases or decreases in natural mortality rates in response to environmental perturbation) over the short term probably does not occur. The contention here is not to show that compensatory mechanisms do not operate in fish populations; the continued existence of fish stocks despite commercial exploitation is ample evidence of such a mechanism. However, little is known about the degree to which compensatory mechanisms can offset increased mortalities and the critical level that additional mortalities must reach before compensation is initiated. We contend that the increased mortality levels, within the valid range of linearization, from power plant entrainment alone are not large enough to invoke compensatory mechanisms. The level of entrainment losses for early life stages are relatively trivial when compared to natural mortality rates for eggs or larvae (Hoss et al. 1974, Hettler and Clements 1977, Schubel et al. 1977, Van Winkle et al. 1979). While density-dependent mortality relationships most likely exist in nature, their exclusion from the estimation of the magnitudes and effects of losses for early-life-stage populations does not severely restrict the estimates of impact.

A major assumption for the estimation of the ecological impact of entrainment losses is the absence of predator switching. In other words, opportunistic species do not switch from prey A to prey B when 10% entrainment losses to a predator of prey B increases the availability of prey B (this increase is also assumed to be small). Predator switching, as used here, implies that the proportion of a prey taken will change from less than to greater than expected as the abundance of the available prey increases (Murdoch 1969, 1977, Lawton et al. 1974, Murdoch and Oaten 1975, Murdoch et al. 1975) rather than switching due to prey sizes (O'Brien et al. 1976) or habitat switching due to interspecific competition of predators (Werner and Hall 1977, 1979). Experimentation has revealed that predator switching occurs primarily in cases where the predator shows a highly variable and weak preference between two preys at equal abundances (Murton 1971, Manly et al. 1972, Lawton et al. 1974, Murdoch et al. 1975) and that no switching occurs when the predator shows a consistent strong or weak preference between two prey at equal abundances (Murdoch 1969, Murdoch and Marks 1973, Reed 1969). Only two of these studies have investigated potential switching in finfish predators. Reed (1969) found no evidence of switching in bluegills (Lepomis macrochirus) feeding on midge and mosquito larvae, while guppies (Poecilia reticulatus) readily switched preys between Drosophila larvae and tubificid larvae, depending on their relative abundances (Murdoch et al. 1975). Strong evidence of predator switching in guppies (Murdoch et al. 1975) and Notonectids (Lawton et al. 1974) was not evident until relative abundances of the prey items differed by greater than 40% and 30%, respectively.

In the entrainment model presented here, population losses are consistently less than 10% In an extreme case, where one prey is increased 10% and the other prey is unaffected, the change in relative abundance, assuming initial relative equality, would be altered between two prey from

50:50 to about 47:53. To reach the apparent, experimental, relative abundance thresholds that would result in switching, entrainment losses would have to be approximately 50% of a population. While there may be evidence of large local entrainment losses in some cases, consistent losses of the magnitude of 50% are highly unlikely. It is also unlikely that the abundance of only one prey item would be affected without other prey items being affected in a like manner. In short, at the small shifts in relative abundances that would be associated with entrainment losses, rapid predator switching would be unlikely. Thus, entrainment losses at the level described here could result in at least short-term alterations in the energy flow dynamics of an aquatic ecosystem, with little possibility that the changes (i.e., unused net productivity) would be ameliorated rapidly. These changes would be stabilized eventually, although the time needed for recovery is unclear. Vincent and Anderson (1979) have suggested from model evidence that restabilization of perturbed food chains is inversely related to trophic complexity.

The ultimate decision on the acceptability of the model assumptions and the projected levels of economic and ecological impact due to entrainment lies with the environmental manager. Regulatory action on the manager's part, while often based on scientific advice, is often a value judgement. It is essential, therefore, to provide managers with sound biological, statistical, and modelling information so they can make the most responsible value judgements possible. This can only be accomplished if biologists and modelers work hand-in-hand with managers and candidly reveal the limitations (i.e., limiting assumptions) of their observations.

LITERATURE CITED

Boone, J.G. 1976. Estuarine fish recruitment survey – 1958 to 1976. Maryland Department of Natural Resources Internal Report F-27-R-2.

Boreman, J., C.P. Goodyear, and S.W. Christensen 1978. An empirical transport model for evaluating entrainment of aquatic organisms by power plants. FWS/OBS-78/90. Prepared for U.S. Fish and Wildlife Service by Oak Ridge National Laboratory.

Christensen, S.W. In press. The best approach to impact assessment is to use empirically based or simulation models to forecast impacts. In Fifth National Workshop on Entrainment and Impingement, L.D. Jensen (ed.). Ecological Analysts, Inc., Baltimore, MD.

Green, R.H. 1979. Sampling Design and Statistical Methods for Environmental Biologists. John Wiley and Sons, New York. 257 pp.

Hettler, W.F., and L.C. Clements. 1977. Effects of acute thermal stress on marine fish embryos and larvae. p. 171-190 In Fourth National Workshop on Entrainment and Impingement, L.D. Jensen (ed.). Ecological Analysts, Inc., Baltimore, MD.

Horst, T.J. 1975. The assessment of impact due to entrainment of ichthyoplankton. p. 107-118 In Fisheries and Energy Production: A Symposium, S.B. Saila (ed). D.C. Heath and Company, Lexington, MA.

Horst, T.J. 1977. Use of the Leslie matrix for assessing environmental impact with an example for a fish population. Trans. Am. Fish. Soc. 106:253-257.

Hoss, D.E., W.F. Hettler, and L.C. Coston. 1974. Effects of thermal shock on larval estuarine fish -- Ecological implications with respect to entrainment in power plant cooling systems. p. 357-371 In Early Life History of Fish, J.H.S. Blaxter (ed.). Springer-Verlag, New York.

Lauer, G.J., J.R. Young, and J.S. Suffern. In press. The best way to assess environmental impacts is through use of generic and site-specific data. In Fifth National Workshop on Entrainment and Impingement, L.D. Jensen (ed). Ecological Analysts, Inc., Baltimore, MD.

Lawler, J.P., and T.L. Englert. 1978. Models useful for the estimation of equilibrium population reduction due to power plant cropping. p. 103-113 In Fourth National Workshop on Entrainment and Impingement. L.D. Jensen (ed.). Ecological Analysts, Inc., Baltimore, MD.

Lawton, J.H., J. Beddington, and R. Bonser. 1974. Switching in invertebrate predators. p. 141-158 In Ecological Stability, M.B. Usher and M.H. Williamson (eds.). Chapman and Hall, London.

Marcy, B.C. 1977. Entrainment of organisms at power plants, with emphasis on fishes -- An overview. p. 89-106 In Fisheries and Energy Production: A Symposium, S.B. Saila (ed.). D.C. Heath and Company, Lexington, MA.

Martin Marietta Corporation. 1980. Summary of findings: Calvert Cliffs Nuclear Power Plant aquatic monitoring program. Volumes I-III. Prepared for Maryland Department of Natural Resources by Martin Marietta Corporation. Ref. No. PPSP-CC-80-2.

Manly, B.J.F., P. Miller, and L.M. Cook. 1972. Analysis of a selective predation experiment. Am. Nat. 106:719-736.

Murdoch, W.W. 1969. Switching in general predators: Experiments on predator specificity and stability of prey populations. Ecol. Monogr. 39:335-354.

Murdoch, W.W. 1977. Stabilizing effects of spatial heterogeneity in predator-prey systems. Theor. Pop. Biol. 11:252-273.

Murdoch, W.W., and R.J. Marks. 1973. Predation by Coccinellid beetles: Experiments on switching. Ecology 54:160-167.

Murdoch, W.W., and A. Oaten. 1975. Predation and population stability. Adv. Ecol. Res. 9:1-131.

Murdoch, W.W., S. Avery, and M.E.B. Smyth. 1975. Switching in predatory fish. Ecology 56: 1094-1105.

Murton, R.K. 1971. The significance of a specific search image in the feeding behavior of the wood pigeon. Behavior 40: 10–42.

O'Brien, W.J., N.A. Slade, and G.L. Vinyard. 1976. Apparent size as the determinant of prey selection by bluegill sunfish (Lepomis macrochirus). Ecology 57:1304–1310.

Polgar, T.T., J.K. Summers, and M.S. Haire. 1979. Evaluation of the effects of the Morgantown SES cooling system on spawning and nursery areas of representative important species. Prepared by Martin Marietta Corporation, Environmental Center, Baltimore, MD, for Maryland Department of Natural Resources. Ref. No. PPSP-MP-27.

Polgar, T.T., J.K. Summers, and M.S. Haire. In press. Assessment of potential power plant entrainment impact at Morgantown Steam Electric Station. In Fifth National Workshop on Entrainment and Impingement, L.D. Jensen (ed.). Ecological Analysts, Inc., Baltimore, MD.

Reed, R.C. 1969. An experimental study of prey selection and regulatory capacity of bluegill sunfish (Lepomis macrochirus). M.A. Thesis, University of California, Santa Barbara.

Schubel, J.R., C.F. Smith, and T.S.Y. Koo. 1977. Thermal effects of power plant entrainment on survival of larval fishes: A laboratory assessment. Chesapeake Sci. 18:290–298.

Summers, J.K., and F. Jacobs. 1981. Estimation of the potential entrainment impact on spawning and nursery areas near the Dickerson Steam Electric Station. Prepared by Martin Marietta Corporation, Environmental Center, Baltimore, MD, for Maryland Department of Natural Resources. Ref. No. PPSP-D-81-1.

Swartzman, G.L., R.B. Deriso, and C. Cowan. 1978. Comparison of simulation models used in assessing the effects of power-plant-induced mortality on fish populations. Prepared by University of Washington for the Nuclear Regulatory Commission. NUREG/CR-0474.

Van Winkle, W. (ed.). 1978. Assessing the effects of power-plant-induced mortality on fish populations. National Technical Information Service, Springfield, VA.

Van Winkle, W., S.W. Christensen, and J.S. Suffern. 1979. Incorporation of sublethal effects and indirect mortality in modeling population level impacts of a stress, with an example involving power plant entrainment and striped bass. Prepared by Oak Ridge National Laboratory, Oak Ridge, TN. ORNL/NRED/TM-288; ORNL-1295.

Vincent, T.L., and L.R. Anderson. 1979. Return time and vulnerability for a food chain model. Theor. Pop. Biol. 15:217–231.

Werner, E.E., and D.J. Hall. 1977. Competition and habitat shift in two sunfishes (Centrarchidae). Ecology 58:869–876.

Werner, E.E., and D.J. Hall. 1979. Foraging efficiency and habitat switching in competing sunfishes. Ecology 60:256–264.

CORRESPONDENCE BETWEEN BEHAVIORAL RESPONSES OF FISH IN LABORATORY AND FIELD HEATED, CHLORINATED EFFLUENTS[1]

John Cairns, Jr., Donald S. Cherry, and James D. Giattina[2]

Abstract.--The purpose of this study was to determine the correspondence of laboratory and field data using fish prefer- ence-avoidance of heated, chlorinated waste discharges from a steam electric power plant. Such evidence is essential if the abundant information generated with single species laboratory tests is to be used in ecological modelling. The correspon- dence of laboratory and field data for most of the fish spe- cies used in this study was quite good.

INTRODUCTION

Biological assessment of water pollution has depended historically on observing effects directly in natural systems. This is unsatisfactory because: (a) best management practices require prevention rather than confirmation of damage, and (b) direct observation of an event in one system does not necessarily permit one to make claims about events in other systems. Therefore, an urgent need for enlightened water quality management is the devel- opment of a predictive capability coupled with a means of validating the accuracy of the predictions which will in turn enable corrections to be made when the predictions are wrong. Steam electric power plants are capable of producing major impacts in the aquatic environment through: (a) discharge of heated wastewater, (b) addition of potentially toxic or hazardous chemicals such as chlorine, and (c) introduction of fly ash or water exposed to fly ash into aquatic systems. This paper focuses on the first two activities which might displace aquatic ecosystems in both structure and function.

[1]Paper presented at the international sympo- sium Energy and Ecological Modelling, sponsored by the International Society for Ecological Modelling. (Louisville, Kentucky, April 20-23, 1981).

[2]John Cairns, Jr. is University Distinguished Professor, Department of Biology, and Director, University Center for Environmental Studies, Virginia Polytechnic Institute and State University, Blacksburg, Va.; Donald S. Cherry is Associate Professor of Zoology, Department of Biology, Virginia Polytechnic Institute and State University, Blacksburg, Va.; James D. Giattina is Research Aquatic Biologist, Corvallis Environmental Research Laboratory, Western Fish Toxicology Station, U. S. Environmental Protection Agency, Corvallis, Ore.

Most predictions of displacements or effects in aquatic ecosystems are based on single species laboratory toxicity tests. Rarely are these pre- dictions validated in natural systems in a scien- tifically justifiable way. This is rather aston- ishing in view of the fact that most single species laboratory tests are carried out in rather simple steady state environments, sometimes even without the benefit of renewal of test solutions. Even the simplest natural system is far more complex than the environment in most standard laboratory toxicity tests which have few of the ecological "compartments" present in natural systems. Not only is this lack of correspondence in complexity of the test system likely to affect the biological response, but its absence will also have major effects upon the analysis of the chemical and physical components eliciting the biological response (Cairns in press). Finally, both the discharges of the power plant and the environmental conditions in natural systems fluctuate markedly and sometimes violently, which is a sharp contrast to the steady state conditions maintained in most laboratory toxicity tests. However, simple toxicity tests are unlikely to be abandoned because they are: (a) inexpensive, (b) easily replicated, and (c) widely used and accepted by both industry and regulatory agencies. Since the present widespread utilization of such tests is unlikely to change in the foreseeable future, determination of the correspondence between the predictions made on the basis of data generated with these tests and the actual response of organisms to the same stresses in field situations seems prudent.

One might think that direct observation of an environmental impact on a particular system could be the basis for the development of a predictive capability. For this to be possible, transfer of direct observations in other similar systems and development of a theoretical understanding of the interactions which occur under stress situations that permit predictions to be made is essential.

The study of the factors altering the effect or the biological response is essential to the development of the predictive capabilities. Direct observation of pollutional effects has been popular with biologists because it involves minimal dependence on theoretical constructs which, in a newly developed field such as pollution ecology, may often be wrong. The major weakness is that this type of evidence has not proven adequate for extrapolation to ecosystems other than the one in which the event occurred, except in the most general way. Transferring direct observations from one field situation to another perhaps will markedly improve in the future. Until this happens, most decisions will be made on the basis of single species laboratory tests extrapolated to field situations. Therefore, this paper evaluates the correspondence of single species laboratory tests with an array of fish species to the response of these same species to similar stresses in field situations from power plant effluents. Few papers exist that address this topic of correspondence of field and laboratory responses of species. This approach should help facilitate the use of the abundant laboratory data of this type in the field of ecological modelling.

MATERIALS AND METHODS

Glen Lyn Power Plant and New River Sites

The Glen Lyn Plant, a fossil-fueled facility of Appalachian Power Company (APCO), is located at New River kilometer 153 in southwestern Virginia adjacent to West Virginia (Dickson et al. 1976). The Glen Lyn Plant has an operating capacity of 350 Mw and discharges heated, intermittently chlorinated water into the New River and a smaller tributary (East River) that shortly confluences with the New River (fig. 1). The maximum theoretical rise in temperature in the New River discharge is $7.8^\circ C$ with a weekly maximum of 0.5 mg/ℓ free residual chlorine (FRC) and an average of 0.2 mg/ℓ FRC (Giattina et al. in press). The New and East River drainages consist of a shallow series of pool and riffle areas. Of the two units currently operating at the Glen Lyn Plant, 14.9 m^3/sec of cooling water is discharged by the plant; 5.35 m^3/sec of this is directly released into the New River from one unit (105 Mw). The second unit (245 Mw) discharges 7.6 m^3/sec via a 275-m effluent channel into the East River upstream from its confluence with the New River, and 2.0 m^3/sec directly into the New River. Maximum and minimum flow rates of the New River from 1915 to 1965 ranged from 4361 m^3/sec and ~ 32 m^3/sec, respectively (Virginia Division of Water Resources 1967).

Site-Specific Glen Lyn Field Laboratory Responses

Temperature Preference Procedures

Tests were conducted in an epoxy-coated, stainless steel trough with a battery of twelve 250 watt infrared lamps located underneath (Cherry et al. 1975a). These lamps heated the test water to various temperatures and produced a gradient

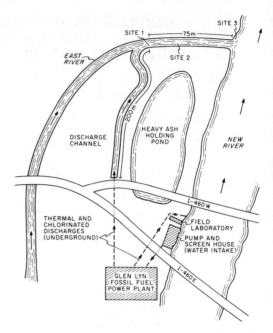

Figure 1.--Schematic representation of the Glen Lyn area showing the relationship of the field sampling sites in the East River relative to the power plant and field laboratory.

ranging from cold to hot extremes. New River water, which was cooled by a water bath prior to entry into the preference trough, was pumped into the upper end to expand the temperature gradient (e.g., $6-24^\circ C$). This combination of lamps and cooling water bath produced a gradient from cold to hot extremes and made it possible for fish to avoid the extremes and select a temperature between the undesirable ends. Fish were tested individually, in pairs, or in groups of four for each trial with at least one replication to as many as eight (depending upon the responsiveness of each species), at acclimation temperature of 6 to $36^\circ C$ at $6^\circ C$ increments. After an initial adjustment period (~60 min minimum), observations were made on the position of the fish within the gradient at 2-min intervals for 40 min. The mean temperature was then calculated from these observations and was considered to be the selected or preferred temperature. The final temperature preferendum was calculated as the point in which the acclimation and preferred temperatures coexisted using either a linear or curvilinear regression analysis (Cherry et al. 1977a).

Temperature Avoidance Procedures

The procedure used in determining the first temperature avoided by fish above the acclimation or test temperature was similar to that reported earlier (Cherry et al. 1975a,b). The trough system consisted of two distinct temperature inputs from opposite ends, one on each side of the central drains (fig. 2). A control trial run consisted of both sides of each trough being equal to the acclimation temperature in each of two troughs located in parallel but opposite in treatment to minimize position effects. For the control run, both halves of each trough received water at the acclimation temperature. This was followed by a 10-min continuous observation period in which residence time of fish was measured in the pretreated side. Then the upper half of each trough was heated to a higher temperature in 3°C temperature increments followed by 10-min observation periods at each increment. the 10-min observation period was monitored using stop watches plus a closed television monitor. New River water was consistently elevated by 3°C temperature increments from two water baths (one contained water held at the acclimation temperature and the other bath was used to raise the water temperature) until fish significantly avoided the heated half of each trough. Fish were tested either individually or in pairs depending upon the behavioral characteristics of each species. A complete description of this format is presented in Cherry et al. (1977a). Statistical avoidance was determined from a two-way factoral analysis of variance (Sokal and Rohlf 1969) when fish first avoided a 3°C temperature increment above the acclimation temperature. A Duncan's new multiple range test was conducted to determine when a significant ($P \leq 0.05$) decline in the residence time occurred between the elevated temperature and the original control run (e.g., when temperature on both sides of the trough equalled the acclimation temperature).

Chlorine Avoidance Procedures

Avoidance to chlorine was carried out in the same system described in this test for temperature

(fig. 2). Each trough received chlorinated and nonchlorinated water at opposite ends (600 mℓ/run) and drained at the center through four bottom and side drains. New River water from two water baths, each with a 67.8-liter capacity, supplied half of each trough. One bath was chlorinated by addition of calcium hypochlorite solution, while an additional solution from a 18.9-liter chlorinated carboy was continuously pumped (290 mℓ/min by peristaltic pump) into that bath to maintain a relatively constant concentration. By periodically increasing the concentration in the bath and carboy, successively higher chlorinated levels were obtained in the trough. Chlorine concentrations were specifically monitored in four regions of the troughs by drawing water by gravity flow through tygon tubing. Chlorine concentrations were made before and after each target dose by amperometric titration according to the American Public Health Association (1976) utilizing a Wallace and Tiernan titrator. The trough was completely enclosed once a trial began, and fish movement was observed through the closed-circuit television as discussed earlier. Fish were exposed to several successively doubling chlorinated doses (0.02, 0.05, 0.10 . . . 0.80 mg/ℓ total residual chlorine (TRC) in half of each trough. Following a 20 to 120-min adjustment period, depending upon the time necessary for each species to establish random movement in the troughs, fish residency or time spent in each half of the trough was continuously monitored for 10 min. The first significant ($P \leq 0.05$) avoidance to TRC, as measured by a reduction in residence for the control run, was determined to be the avoidance concentration to TRC. Statistical procedures used were the same as those for the temperature avoidance described earlier in the text.

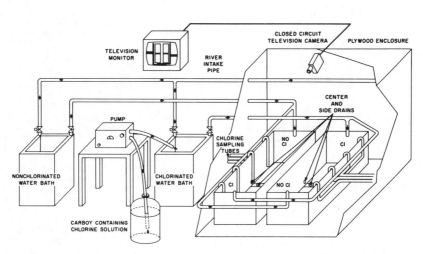

Figure 2.--Schematic diagram of the temperature and chlorine avoidance apparatus.

New River Field Responses

Temperature Preference and Avoidance Procedures

Fish were collected from six sampling stations of the New and East River drainage systems in areas above, within, and below the thermal effluent of the Glen Lyn Plant. A more detailed description of the research area can be found in Stauffer et al. (1976). Fish were collected with seine and rotenone using a block net to ensure a more adequate collection of the areas sampled. Collections were made during the spring to fall (19.4 to 35.5°C) to ensure adequate use of rotenone as a sampling technique. All fish were sorted, identified to species, and stored at the Virginia Polytechnic Institute and State University Fish Museum. A stepwise regression technique was used to evaluate the percentage abundance versus temperature (corrected for sampling effort) for each fish species relative to the determination of preference temperature from the field data (Stauffer et al. 1976). The final field temperature preference was obtained by using a stepwise regression technique that determined if a large percentage (>50%) of a particular species occurred within a selected, narrow temperature range (1.1 to 1.7°C) when other variables tested (e.g., photoperiod, flow rate, stream gradient, time prior to chlorination) did not statistically influence the field response of fish. Highest field temperature occurrences of selected fish species in the New River drainage system were used as the temperature in which fish approached "threshold" high temperature conditions.

Chlorine Avoidance Studies

The area most commonly sampled was located in the thermal discharge in the East River (Site 1 in fig. 2). Chlorine (TRC and FRC by a portable field Fisher-Porter amperometric titrator), temperature, pH (portable Markson meter), dissolved oxygen (ozide modification of the Winkler technique), and flow rate (General Oceanics Inc., Model 2030 flow meter) were measured in the field at each sampling effort. Through cooperation with power plant officials, a range of chlorine concentrations from 0.00 to 0.50 mg/ℓ FRC was applied to the in-plant cooling system. The dosing schedule was randomized to eliminate any bias due to environmental parameters which were not measured. A large (0.48 to 0.64 cm) mesh bag seine was used to sample areas ~12 by 12 m. Each sample was taken after a peak TRC concentration was measured in the field with fish preserved to establish verification of species composition. The Wilcoxon rank sum test was used to compare differences in the number of fish sampled between control and dose days in the East River. This was done by jointly ranking (from lowest to highest) the measurements for the fish samples and examining the sum of the ranks for measurements from the group with the smallest sample size (T). The data were divided into three groups: (1) the number of fish collected on control days (TRC - 0.00 mg/ℓ), (2) the number of fish collected when the TRC was measured in a range <0.15 mg/ℓ (low dose days), and (3) the number of fish collec-

ted when the TRC was measured greater than or equal to 0.15 mg/ℓ (high dose days).

Fish Species Studied

The fish species tested in the Glen Lyn Laboratory and resulting New River drainage for temperature responses included the Cyprinidae: Campostoma anomalum (stoneroller), Notropis rubellus (rosyface shiner), N. spilopterus (spotfin shiner), N. galacterus (whitetail shiner), Pimephales notatus (bluntnose minnow). Comparison of laboratory to field avoidance concentration of TRC included Cyprinus carpio (carp), Ictalurus punctatus (channel catfish), Notemigonus crysoleucas (golden shiner), Micropterus punctulatus (spotted bass), M. salmoides (largemouth bass), M. dolomieui (smallmouth bass), Notropis atherinoides (emerald shiner), N. rubellus (rosyface shiner), N. cornutus (common shiner), N. spilopterus, N. galacturus and Catostomus commersoni (white sucker). Some of these species have been identified as being either important or representative (most abundant, sport, or forage fishes) of the ichthyofauna in the New River receiving system.

RESULTS AND DISCUSSION

Laboratory and Field Temperature Responses

A high degree of similarity was evident between laboratory and field preference and avoidance responses for the five fish species studied (table 1). Channel catfish (Ictalurus punctatus) had the greatest degree of final temperature preferenda, 33.8 to 34.4°C between laboratory and field, respectively. The greatest difference between final temperature preferenda was 5°C or less, with preferenda being more highly estimated in the laboratory for the stoneroller (Campostoma anomalum) and ~2°C higher in the field for rosyface shiner. No field preferendum was available for the spotfin shiner since the euryacious nature of the species precluded this calculation (Stauffer et al. 1976).

The highest temperatures avoided between the laboratory responses to those in the heated plume of the field during summer conditions showed a greater degree of similarity than did the preference data. For the stoneroller, rosyface shiner, and bluntnose minnow, the highest temperature avoided in the laboratory was 1.4, 2.0, and 2.0°C lower, respectively, than the highest temperature in which these species were sampled in the thermal plume. The percent composition for these three species (respectively, as listed above) represented <1, 4, and 8% of the total sample; whereas at lower field collection temperatures (~21, 30, and 23°C, respectively), the same species comprised as much as 61, 68, and 44% of the fish collected (Stauffer et al. 1976). For the spotfin shiner and channel catfish, substantially higher numbers of fish were collected at 35°C in the thermal plume (25 and 45%, respectively), which was the highest temperature for sampling in the field. Laboratory avoidance

Table 1.—Range of preferred temperature with (in parentheses) 95% confidence limits of five fish species from field laboratory data at acclimation temperatures from 12-36°C compared to the final laboratory temperature preferendum and upper avoidance temperatures with the field temperature preferendum and highest temperature at which fish were collected in the field

Fish species	Acclimation temperature (°C)								
	12	15	18	21	24	27	30	33	36
	Range of laboratory temperature preferences								
Campostoma anomalum[1] (stoneroller)	16.5 (14.2-18.2)	17.0 (16.7-19.8)	21.0 (19.2-21.6)	22.4 (21.4-23.6)	25.1 (23.4-25.9)	28.2 (25.2-28.3)	27.4 (26.9-30.8)	--	--
Notropis rubellus[1] (rosyface shiner)	20.8 (18.7-22.2)	21.7 (20.2-23.0)	22.2 (21.7-24.0)	22.5 (23.0-25.0)	25.8 (24.2-26.2)	28.1 (25.2-29.5)	28.0 (26.2-29.0)	27.7 (27.0-30.5)	--
Notropis spiloterus[1] (spotfin shiner)	21.4 (19.3-24.4)	21.8 (21.0-25.2)	24.1 (22.7-26.2)	26.4 (24.3-27.2)	27.3 (25.7-28.4)	30.6 (26.9-29.8)	31.8 (28.0-31.4)	31.0 (28.9-33.1)	29.2 (29.8-34.8)
Pimephales notatus[1] (bluntnose minnow)	19.3 (18.0-20.0)	20.9 (19.9-21.5)	21.9 (21.7-23.0)	23.2 (23.5-24.6)	26.4 (25.2-26.4)	27.9 (26.7-28.8)	29.0 (28.2-30.2)	--	--
Ictaturus punctatus[2] (channel catfish)	19.9 (19.9-22.3)	21.7 (21.8-23.8)	22.9 (23.4-25.3)	26.1 (24.9-26.9)	29.4 (26.4-28.8)	29.5 (27.8-30.6)	30.5 (29.1-32.6)	--	--

Summary of laboratory and field temperature responses

	Lab final temp. pref. (°C)[1]	Highest lab avoid. temp. (30°C Acclim. Temp.)[1]	Field temp. pref. (°C)[3]	Highest field temp. that fish were collected
stoneroller	26.2 - 28.8	33	22.8 - 23.8	34.4
rosyface shiner	26.0 - 28.4	33	28.3 - 30.0	35.0
spotfin shiner	29.4 - 31.9	36	(-)[4]	35.0
bluntnose minnow	28.1 - 29.3	33	26.7	35.0
channel catfish	33.8[2]	35[2]	34.4	35.0

[1] from Cherry et al. 1977a

[2] from Cherry et al. 1975b

[3] from Stauffer et al. 1976

[4] could not be determined from field data

responses for these two species still compared well to the field data of $35^{o}C$ with significant avoidance in the laboratory being either equal to or $1^{o}C$ higher than the temperature of highest field occurrence. It should also be noted that additional laboratory trials conducted at a higher acclimation temperature ($33^{o}C$) for the channel catfish resulted in avoidance occurring at $37^{o}C$.

The comparative laboratory preference responses of the five fish species tested were species specific with the highest preferred temperature being observed for the channel catfish and spotfin shiner. The stoneroller preferred the lowest temperature ($16.5^{o}C$) when acclimated and tested at $12^{o}C$ (table 1). The highest temperature at which the stoneroller could be held for a 7-day period without suffering mortality was $31^{o}C$, which was also lower than the other species tested ($32-36^{o}C$). The general trend for all fish species tested in the laboratory showed an increase in preferred temperature with an increase in the acclimation temperature from winter ($12^{o}C$) to early summer ($27^{o}C$). However, at higher summer acclimation temperatures ($30-36^{o}C$, depending upon the species tested), a decline in temperature preference was observed. For example, the stoneroller preferred a lower temperature at $30^{o}C$ ($27.4^{o}C$) than at $27^{o}C$ ($28.2^{o}C$). For the spotfin shiner, the decline in preference at the highest acclimation temperature tested ($36^{o}C$) was substantially lower than that measured at $30^{o}C$ (29.2 and $31.8^{o}C$, respectively). All species preferred a temperature greater than the acclimation temperature tested from $12-24^{o}C$, but at higher acclimation temperatures, preferred temperatures in the laboratory temperature gradient (e.g., from $18-36^{o}C$ when tested at $24^{o}C$) fell below the acclimation temperature. The widest differential between the acclimation and preference temperatures occurred at winter test conditions ($12-15^{o}C$) which suggested that a greater attraction for warmer water was evident during winter instead of spring and early summer conditions. Similar trends were noted for at least 10 other fish species conducted by these investigators (Cherry et al. 1975b, 1977a) and others (Doudoroff 1938, Fry 1947, Pitt et al. 1956, Brett 1952, Ferguson 1958, Meldrim and Gift 1971, Stauffer et al. 1976, and others). The reduction in preferred temperatures below acclimation temperature limits in the laboratory and the close similarity between laboratory and field responses supports the premise that these fish species possess the ability to select optimal temperatures and to avoid potentially hostile temperature limits.

Laboratory and Field Chlorine Avoidance Studies

The first significant avoidance concentration to TRC for 10 fish species tested at 12, 18, 24 and $30^{o}C$ in the laboratory varied from 0.05 to 0.41 mg/ℓ (table 2). Species which avoided the lowest TRC concentration (0.05-0.20 mg/ℓ) included spotted bass, followed by the rosyface shiner, whitetail shiner, stoneroller, bluntnose minnow, and smallmouth bass (0.10-0.22 mg/ℓ). Channel catfish avoided the highest concentrations of TRC (0.20-0.41 mg/ℓ) regardless of the acclimation temperature. No general trends between fish avoidance behavior and fish

families, trophic level interaction, food habits, ecological requirements, etc., were discernible from the 10 species studied.

Field avoidance concentrations were only available for two species (spotfin and whitetail shiners) which occurred most frequently in the sampling site used (fig. 1). These two species were readily available in the thermal plume during $24-34^{o}C$ and generally absent when temperatures declined below $24^{o}C$. Comparisons between field avoidances (≥ 0.18 mg/ℓ) at discharge temperatures between $24-29^{o}C$ agreed favorably with the laboratory avoidance concentrations (0.11-0.22 mg/ℓ) for the spotfin and white shiners (table 2). Avoidances to field TRC concentrations 0.04-0.12 mg/ℓ for the same species of shiners were lower than those obtained in the laboratory at $30^{o}C$ (0.21-0.22 mg/ℓ). The overall avoidance threshold or lower detection limit of occurrence for the major fish species (fish community response which included seasonally abundant number of the stoneroller, rosyface shiner, mimic shiner [N. voluccellus], bluntnose minnow, telescope shiner [N. telescopus], sand shiner [N. stramineus], striped shiner [N. chrysocephalus], silver shiner [N. photogenis], and smallmouth bass) was 0.15 mg/ℓ at temperatures between $7-36^{o}C$. At field temperatures $\leq 18^{o}C$, little fish movement was noted as most species were located beneath the banks of the East River upstream of thermal influence which were accompanied by a canopy of leaves and branches for cover or refuge.

A comparison of the laboratory avoidance results of the 12 fish species tested compared favorably with the field-determined community thresholds, which were within the 95% confidence limits established for the summer field avoidance thresholds to TRC, and indicated a strong correlation between these data (fig. 3). At $24^{o}C$, 8 of the 12 species tested had laboratory threshold concentrations which were within the 95% confidence limits established for the summer field thresholds; while at $30^{o}C$, 10 of the 12 fell within this range. Comparisons between field avoidance thresholds ($12-18^{o}C$) to laboratory avoidance at the same acclimation temperatures were difficult to discern due to the absence of measurable TRC concentrations <0.20 mg/ℓ TRC and to lack of adequate fish numbers in the field for evaluation.

Based on previous work, Sprague and Drury (1969) concluded that avoidance reactions may be expected in polluted receiving systems at concentrations which cause fish in laboratory tests to spend 90% of their time in untreated water. The residence of both the spotfin and whitetail shiners in untreated water at the first avoidance concentration to TRC ranged from 72 to 93%. Considering the fairly close correlation for the field and laboratory avoidance thresholds determined in our study, the results appear to support the conclusion of Sprague and Drury (1969).

Determining field avoidance behavior of fish to TRC has been a difficult task since several environmental variables are present and many aquatic receiving systems are not readily managed or able

Table 2.--Laboratory and field avoidance responses of fish species acclimated and tested at 12-30°C and exposed to total residual chlorine (TRC) at the Glen Lyn field laboratory and East River thermal-intermittently chlorinated discharge sampling sites

Fish species	Acclimation temperature tested (°C)			
	12	18	24	30
	First	Significant condition to TRC	Laboratory (mg/ℓ)	Avoidance
Notropis spilopterus[1] (spotfin shiner)	0.21	0.22	0.21	0.21
Notropis galacturus[1] (whitetail shiner)	0.11	0.11	0.11	0.22
Notropis rubellus[2] (rosyface shiner)	0.10	0.10	0.20	0.20
Notropis cornutus[3] (common shiner)	0.21	0.12	0.20	0.20
Campostoma anomalum[4] (stoneroller)	0.11	0.11	0.19	0.22
Pimephales notatus[4] (bluntnose minnow)	0.12	-	0.21	-
Micropterus punctulatus[2] (spotted bass)	0.05	0.20	0.20	0.20
Micropterus dolomieui[5] (smallmouth bass)	0.10	0.20	0.20	0.20
Ictalurus punctatus[4] (channel catfish)	0.40	0.41	0.40	0.20

	Field Discharge Temperature (°C) and Avoidance Concentration to TRC (mg/ℓ)		
	7 – 36	24 – 30	>30 – 34
spotfin shiner	-	0.18	0.12
whitetail shiner	-	0.18	0.04
overall fish community	≥0.15 (lower detection limit for fish occurrence)		

[1] from Giattina et al. in press
[2] from Cherry et al. 1977c
[3] from Cherry et al. 1978
[4] data generated in this study
[5] from Cherry et al. 1977c

to be sampled. Still avoidance behavior by two fish species (spotfin and whitetail shiners) was demonstrated at the Glen Lyn Plant receiving system of the East River. The favorable comparisons of the field avoidances at specific field temperatures (when adequate species numbers were available) to the laboratory determined avoidance thresholds at these same temperature ranges provided some evidence that the current chlorination practices are not having a demonstrable, adverse effect on the resident populations. These data, along with the chlorinated discharge site, confirms this viewpoint. At a time when considerable controversy exists over either the recommended wastewater chlorine standards by Region V of the U.S. EPA or various state water quality control regulations, the use of intermittent chlorination practices at the Glen Lyn Plant do not have any demonstrable effect upon the ichthyofauna present in the New River drainage system at the Glen Lyn Plant.

CONCLUSIONS

The correspondence between the laboratory tests and the evidence obtained from the field was surprisingly good. This does not mean that on other sites or for other types of stresses the correspondence between field and laboratory studies will be this good. This data base shows that, at

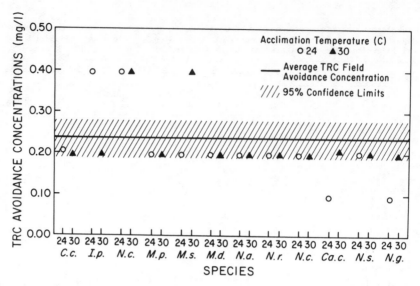

Figure 3.--Comparisons of laboratory avoidance thresholds for various species at 24 and 30°C acclimation temperatures with summer avoidance thresholds for the field data (C.c. = Cyprinus carpio; I.p. = Ictalurus punctatus; N.c. = Notemigonus crysoleucas; M.p. = Micropterus punctulatus; M.d. = M. salmoides; M.d. = M. dolomieui; N.a. = Notropis atherinoides; N.r. = N. rubellus; N.c. = N. cornutus; Ca.c. = Catostomus commersoni; N.s. = N. spilopterus; N.g. = N. galacturus).

least under certain circumstances, the correspondence may be rather good. Additional studies at other sites and with other stresses are essential so that one can identify how the laboratory tests were lacking in environmental realism when the correspondence was not particularly good. The most important modelling information needed, however, is practically nonexistent, namely the determination of correspondence between the very common single species tests and the response in natural systems at higher levels of organization, such as multispecies, community, or system level tests. Additional evidence is also needed to determine the correspondence between laboratory tests carried out at higher levels of organization than single species (e.g., multispecies tests) and responses of the same levels of organization in natural systems. Understanding the correspondence between the artificial systems with good control of the variables and the natural systems with little or no control will certainly improve modelling capabilities because studies can then be carried out in the laboratory that could not possibly be carried out in natural systems. Replication problems common to studies of natural systems can be diminished by use of laboratory systems that are more easily replicated and a variety of other advantages to modellers will ensue.

ACKNOWLEDGMENTS

The research was supported by the American Electric Power Service Corporation, Canton, Ohio 44701.

LITERATURE CITED

American Public Health Association. 1976. Standard methods for the examination of water and waste-water. 14th edition. 1193 p. Washington, D. C.

Brett, J. R. 1952. Temperature tolerance in young Pacific salmon, genus Oncorhynchus. Journal of the Fisheries Research Board of Canada 9:265-323.

Cairns, J., Jr. In press. Biological monitoring, Part VI:future needs. Water Research.

Cherry, D. S., K. L. Dickson, and J. Cairns, Jr. 1975a. The use of a mobile laboratory to study temperature response of fish. Purdue University Engineering Bulletin 145:129-140.

Cherry, D. S., K. L. Dickson, and J. Cairns, Jr. 1975b. Temperature preferred and avoided by fish at various levels of acclimation. Journal of the Fisheries Research Board of Canada 32:485-491.

Cherry, D. S., K. L. Dickson, J. Cairns, Jr. and J. R. Stauffer. 1977a. Preferred, avoided, and lethal temperatures of fish during rising temperature conditions. Journal of the Fisheries Research Board of Canada 34:239-246.

Cherry, D. S., R. C. Hoehn, S. S. Waldo, D. H. Willis, J. Cairns, Jr., and K. L. Dickson. 1977b. Field-laboratory determined avoidances of the spotfin shiner (Notropis spilopterus) and the bluntnose minnow (Pimephales notatus) to chlorinated discharges. Water Resources Bull. 13:1047-1055.

Cherry, D. S., S. R. Larrick, K. L. Dickson, R. C. Hoehn, and J. Cairns, Jr. 1977c. Significance of hypochlorous acid in free residual chlorine to the avoidance response of spotted bass (Micropterus punctulatus) and Rosyface shiner (Notropis rubellus). Journal of the Fisheries Research Board of Canada 34:1365-1372.

Cherry, D. S., S. R. Larrick, J. D. Giattina, J. Cairns, Jr., and K. L. Dickson. 1978. The avoidance response of the common shiner to total and combined residual chlorine in thermally influenced discharges. In Energy and environmental stress in aquatic systems. 854 p. Technical Information Center and U. S. Department of Energy, Washington, D. C.

Dickson, K. L., J. Cairns, Jr., D. S. Cherry, and J. R. Stauffer. 1976. An analysis of the EPA requirements for heated wastewater discharge - a site specific case history evaluation. p. 316-326. In Thermal Ecology II. Technical Information Center, Springfield, Virginia.

Doudoroff, P. 1938. Reactions of marine fishes to temperature gradients. Biological Bulletin 75:494-509.

Ferguson, R. G. 1958. The preferred temperature of fish and their midsummer distribution in temperature lakes and streams. Journal of the Fisheries Research Board of Canada. 15:607-624.

Fry, F. E. J. 1947. Effects of the environment on animal activity. University of Toronto Studies Biological Series 55, Ontario Fisheries Research Laboratory 68:1-62.

Giattina, J. D., D. S. Cherry, J. Cairns, Jr., and S. R. Larrick. In press. Comparison of laboratory to field avoidance behavior of fish in heated, chlorinated water. Transactions of the American Fisheries Society.

Meldrim, J. W., and J. J. Gift. 1971. Temperature preference, avoidance, and shock experiments with estuarine fishes. Ichthyology Associates Bulletin 7:1-75.

Pitt, T. K., E. T. Garside, and R. L. Hepburn. 1956. Temperature selection of the carp, Cyprenus carpio (Linn.). Canadian Journal of Zoology 34:555-557.

Sokal, R. R., and F. J. Rohlf. 1969. Biometry. 776 p. W. H. Freeman Co., San Francisco, Calif.

Sprague, J. B., and D. E. Drury. 1969. Avoidance reactions of salmonid fish to representative pollutants. p. 169-197. In Advances in Water Pollution Research. 936 p. Pergamon Press, London.

Stauffer, J. R., Jr., K. L. Dickson, J. Cairns, Jr., and D. S. Cherry. 1976. The potential and realized influences of temperature on the distribution of fishes in the New River, Glen Lyn, Virginia. Wildlife Monographs 40(50): 1-40.

Virginia Division of Water Resources. 1967. New River basin. In Hydrolic Analysis, Vol. III, Planning Bulletin 203, Department of Conservation and Economic Development, Richmond, Va.

KINETICS MODEL AND COMPUTER SIMULATION FOR PREDICTING THE COMPOSITION AND CONCENTRATION VARIATIONS OF PRIMARY AND SECONDARY HYDROLYSIS PRODUCTS FORMED IN CHLORINATED MARINE WATERS[1]

Ekrem V. Kalmaz[2], Arsev H. Eraslan[3] and Gary W. Zimmerman[4]

Abstract.--Chemical kinetics model is presented for simulating the compositions and concentration variations of the primary and secondary hydrolysis products formed in marine water under the influence of chlorinated water discharges from power plants located on estuaries and coastal regions. Kinetics model uses the basic principle of "law of mass" and considers the chemical reactions in terms of forward and reverse reaction rates and simulates the chemical species based on their initial diluted discharged concentrations. The results of this study indicate that the bromide derivates in marine water are less stable than similar chlorine derivates. It is apparent that the chloride concentrations shift some of the chemical reactions to higher concentrations of free molecular chlorine and repeat the cycle of toxic chlorinated compounds.

INTRODUCTION

Chlorine has been used for preventing slime formation in once-through cooling systems and cooling towers of electric generating power plants. Added to cooling systems intermittently or continuously, chlorine (Cl_2) prevents biofouling on the walls of intake conduits and water boxes and on condenser tube surfaces.

Although much of the biocide chlorine added to water may rapidly dissipate by reaction with organic and inorganic

substances in cooling water, detectable levels are usually present in the power plant effluents. The persistance of residual chlorine in aquatic environments is studied by Polgar and associates (1976), show significant modification in the distribution of two copepod species induced by chlorine reaction products in power plant cooling water effluents.

The release of chlorinated water is producing effects that are slowly being better documented as a result of continuing research. The formation of even relatively low concentrations of residual chlorine has been proven to be toxic to aquatic life, (Gehrs et al. 1974; Roberts et al. 1975). Chloramines and chlorinated organic compounds have formed in low yields during the chlorination of potable water, industrial, municipal waste waters and power plant cooling water effluent (Jolly et al. 1976). During the last decade, there has been increasing concern about the possible human health and ecological effects of residual chlorine and chloro-organics produced as a result of utilization of chlorine in water treatment practices. Although there is an ample amount of documantation from laboratory experiments about the toxic effects of chlorination on aquatic life (Kalmaz 1978), the possible ecological and health effects related to the residual chlorine

[1]Paper presented at the international symposium of Energy and Ecological Modelling, sponsored by the International Society for Ecological Modelling. (Louisville, Kentucky, April 20-23, 1981).

[2]Ekrem V. Kalmaz is Research Assistant Professor of Engineering Science and Mechanics, The University of Tennessee, Knoxville, Tennessee 37916 U.S.A.
[3]Arsev H. Eraslan is Professor of Engineering Science and Mechanics, The University of Tennessee, Knoxville U.S.A.
[4]Gary W. Zimmerman is graduate student at the Department of Engineering Science and Mechanics, The University of Tennessee Knoxville, Tennessee 37916 U.S.A.

and chlorinated organic compounds that are produced in the process are not fully understood (Trabalka and Burch 1979). The uncertainty about the toxicity and/or carcinogenicity in human and ecological effects of chlorine usage has led to the establishment of strict regulations to control the levels of residual chlorine in fresh and marine waters.

Chloramines at concentrations below 0.1 mg/l produced fish-kills in laboratory studies (Zillich 1972), and in stream trials with caged specimen (Basch et al., 1971). In studies of cage specimens, fish-kills were observed at distances up to 0.8 mile below three of the four municipal outfalls tested. In two instances even after thorough mixing with receiving water, the chlorinated effluent was toxic up to 0.7 mile farther downstream (Arthur and Eaton 1971). It has been observed that fish populations tend to avoid toxic materials even at concentrations much less than necessary for toxic symptoms to occur. Therefore, the greatest effect of chlorinated effluent waters probably is to render broad reaches of receiving water unavailable to fish (Tardiff 1975). Blockage of upstream migration of certain fish during the spawning season occurs in chlorinated, wastewater-receiving rivers and estuaries (Tsai 1975). The results of the recent studies show avoidance and reproductive failure in fresh water vertebrates and fish at chlorine concentrations of 0.003 mg/l to 0.005 mg/l (Brungs 1973; Gehrs et al. 1974).

Results of recent investigations show that the massive fish-kill in the cooling waters of an electric generating station on the shore of Saginow Bay has been attributed to lethal concentration of residual chlorine in the water (Ingersoll 1952). High concentrations of halogen residuals in chlorinated wastewater effluence have also been responsible for the large quantities of fish-kill in the estuary of the James River (Bellance and Bailey 1975).

In trying to access the true impact of residual chlorine on marine organisms it is important that we be able to identify the reaction products of chlorine and change in their concentrations with respect to time in estuarine and/or seawater, as the potential toxicity of these compounds is not yet known completely.

MECHANISM OF THE CHEMICAL REACTIONS

Chlorine is one of the three elements (bromine and iodine are the others) that react reversibly with water (fig. 1). In such rapid equilibria, the oxidizing

power of the parent element, chlorine is preserved mainly in hypochlorous acid; other oxidizing agents are formed in minute quantities as hypochlorite ion, chloride monoxide, trichloride ion, and hypochlorous acidium ion. Under appropriate conditions in aqueous solutions, irreversible reactions occur whereby additional oxychlorine species may be formed (Connick and Chia 1959; Carpenter and Macalady 1978).

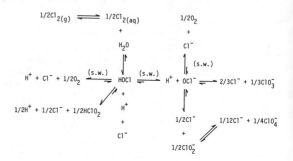

Figure 1.--Chlorine reactions in seawater.

Weil and Morris (1949) showed that the rate of formation of monochloramine is a function of pH with the maximum rate occurring at pHs where the concentration of the molecular species hypochlorous acid (HOCl) and ammonia (NH_3) is highest. At pHs greater or less than 8.4, the rate of formation is reduced considerably (Berliner 1931).

Monochloramine (NH_2Cl) either reacts directly with itself to form dichloramine ($NHCl_2$) or hydrolyzes to HOCl (Fair et al. 1947). The HOCl formed by hydrolyzing either reacts with ammonia to form monochloramine or forms dichloramine by reacting with NH_2Cl. The presence of excess ammonia favors the direct N-chlorination reaction by monochloramine by decreasing the hydrolysis of NH_2Cl to HOCl.

In chlorinated seawater, NH_2Cl hydrolyzes, yielding HOCl and NH_3 in a relatively fast reaction. This reaction controls the concentration of HOCl which is a powerful chlorinating agent. Hypochlorous acid partly dissociates into hydrogen and hypochlorite ions (OCl^-) the extent of dissociation being a function of both pH and temperature. The hypochlorous acid readily dissociates into OCl^- and H^+ in seawater with pH in the neighborhood of 8.

The three forms of free available chlorine involved in these reactions-molecular chlorine unionized hypochlorous acid and the hypochlorite ion-may exist together in

equilibrium. Their relative proportions
depend on the pH values, temperature, and
dissolved solids. Furthermore, the proportions are the same whether the chlorine
is introduced as chlorine gas or a OCl^-
into the aquatic system. In seawater, pH
is the most important factor governing the
relative proportions of these three species discussed by Duursma and Parsi (1976).

A side reaction producing hypochlorite acid ($HClO_2$) always occurs when hypochlorous acid is generated in natural pH.
At low pH $HClO_2$ is unstable and decomposes
by a complex chain mechanism discussed by
Thomann and Sobel (1964).

Bromine (Br_2) and chlorine have similar chemistries in aqueous solution
(fig. 2). The primary Br_2 species are molecular bromine, hypobromous acid (HOBr)
and hypobromite ion (OBr^-). The major difference in the two systems is the formation
of a greater predominance of Br_2 compared
to the low concentrations of Cl_2 in aquatic environment. In addition to the formation of primary species, Br_2 is also similar to Cl_2 in its reaction with ammonia
(Rook 1974; Mills 1977).

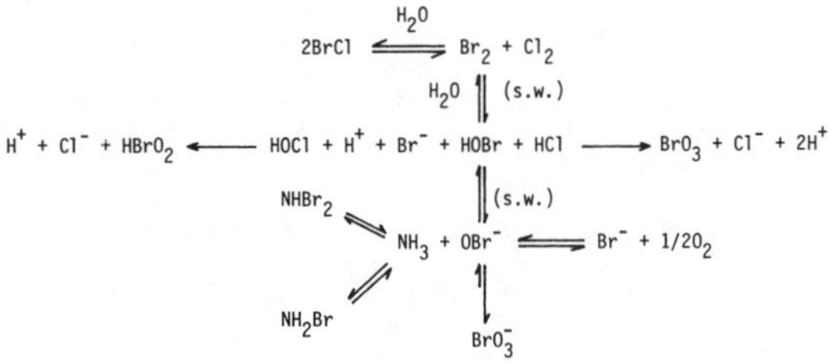

Figure 2.-- Bromine reactions in seawater.

Hydrolysis of Br_2 in seawater at pH
approximately 8 produces HOBr and OBr^- and
hydrogen ion (H^+). Both HOBr and bromide
(Br^-) are unstable with respect to decomposition and disproportionation. In seawater HOBr, Br^-, and Br_2 are photochemically active, and formation of $HOBr^-$ depends on the NH_3 concentration, pH and
temperature. Bromide is a potential reac-
tant for HOCl in seawater. It has a relatively high concentration (64 mg/l as
Br) which causes the reduction of all the
chlorine to chloride (Cl^-) with the formation of HOBr and hypobromite ($HBrO_2$).
However, if ammonia nitrogen levels are
sufficiently high, the formation of monochloramine may compete with the bromide
oxidation (Johnson et al., 1972; Floyd et
al. 1978 and Trofe et al. 1980). Free
Br_2 when present, exists almost entirely

as HOBr since the dissociation of HOBr to
hypobromite ion ($HBrO_2^-$) occurs near pH 8,
or 1.5 pH units higher than the reaction
for the formation of OCl^-.

In seawater at pH 8, the hypobromous
acid readily dissociates into hypobromate
(OBr^-) and hydrogen ion, the extent of dissociation being a function of pH, temperature, and sunlight. Both HOBr and OBr^- are
very photochemically active (Carpenter and
Macalady 1978).

In chlorinated seawater, HOCl reacts
with HOBr and forms $HBrO_2^-$, Cl^- and H^+ at pH
7-10. This reaction is also highly dependent on the ammonia nitrogen level and
photochemically active. At lower nitrogen
levels, the bromamines are formed but are
unstable in comparison to monochloramines
(Keeslar 1975). Hypobromous acid dissociates into Br^-, O_2, and H^+ under the influence of sunlight (Engel et al. 1954;
Macalady et al. 1977).

In seawater, consumption of free chlorine by oxidation of Br^- to HOBr occurs
rapidly with respect to the organic substitutions that lead to formation of volatile halocarbons. The very small yields
of bromoform ($CHBr_3$) in seawater may be
limited by the chlorine concentration, not
by the availability of organic carbon.
Haloform production requires multiple halogen substitution at one carbon atom. If
the halogen concentration is too low in
relation to the reactive carbon, multiple
substitution is unlikely and halocarbon
yields are low. Despite the low yield of
volatiles, humic-like macromolecular material in water reacts extensively with halogen (Pressley et al. 1972).

Chlorination of NH_3 to achieve the
total substitution of all the attached hydrogen atoms by chlorine has been of interest to investigators in water chlorination.
Kinetic studies on the formation and inter-
conversion of monochloramine and dichloramine have been undertaken by Weil and
Morris (1949). Other studies included formation and disproportionation reactions of
NH_2Cl and $NHCl_2$ and the hydrolysis of NH_2Cl
but very little information is available in
the literature about the hydrolysis rate of
these two species.

The literature is replete with data
on the properties and preparation of $NHCl_2$
(Everett and Jones 1941) but there is no
mention of the rate of the hydrolysis reactions. Moreover, many data are not relevant to marine water chlorination since
the concentrated reactants were used rather
than a few parts per million of aqueous
chlorine. Observations on the conditions
favorable to formation and decomposition
of $NHCl_2$ have been reported by Granstram

220

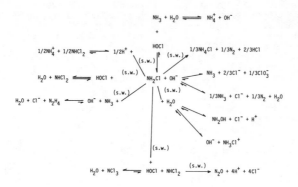

Figure 3.-- Reactions of chlorination of ammonia
in seawater (s.w.).

(1954) in connection with attempts to solve
formation   problems encountered during
water treatment, but these reactions have
not been studied quantitatively on a hydrolysis kinetic basis.

The concentration of $NHCl_2$ is rather
very low in chlorinated seawaters, and not
detectable by the existing measurement
techniques. However, hydrolyzing of $NHCl_2$
yields NH_2Cl (fig. 3), which is also an
important chlorinating agent and highly
toxic to organisms in aquatic environment
(Zillich 1972).

In aquatic environments, the species
of organic amines (RHN_2) are the major
sink for the hydrolyzing products of chlorinating agents. Most of the HOBr and
HOCl are consumed by this reaction. Many
water-soluble chlorine-containing organic
compounds of low volatility have been
found in samples of chlorinated cooling
water from electric power plants and chlorinated effluents from domestic sanitary
treatment plants. The halogenated organic
compounds occur throughout estuarine and
chlorinated seawater (Jolley et al. 1976),
but their concentrations have not been
fully determined.

MATHEMATICAL FORMULATION OF COMPLEX

CHEMICAL REACTIONS

Mathematical formulation of the complex chemical reactions of differential
equations reduced to equations for single-
stage nonequilibrium kinetics process in
chlorinated marine water (Emanuel and
Knorre 1973). Complex chemical reactions
involving the primary and secondary hydrolysis reactions and uptake of chlorine in

satisfying the chlorine demand of the water
and interaction of bromine species with
predominant residual chlorine species in
seawater that receive chlorinated discharges
from power plants located on estuaries
and coastal regions classified in three
catagories.

a)   Principal chlorine species involving
     the primary and secondary hydrolysis
     processes and reactions with ammonia
     and ammonium ion in chlorinated sea-
     water.

b)   Principal bromine species involving
     the hydrolysis processes and reactions
     with the predominant residual chlorine
     species, ammonia, and ammonium ion in
     chlorinated seawater.

c)   Known reactions involving the organic
     chemicals with predominant chlorine and
     bromine species in chlorinated seawater.

The values of the parameters of reactions and the species initial diluted concentrations were determined based on measured data and theoretical considerations
of the chemistry of the saline waters.
Most of the chemical reactions that are
known to occur when residual chlorine is
introduced into the saline water were included in the model. Chemical species included in the model is given in table 1.

Model uses the basic principal of
"law of mass" and considers the chemical
reactions in terms of forward and backward
reaction rates and predicts the variation
of net rate of formation and consumption
of the chemical species in the aquatic
system based on their initial diluted discharge concentrations (fig. 4). Code calculates the net rate of formation and

Figure 4.--Computer printout of chemical species and concentrations.

depletion of the chemical species as a function of forward and backward reaction rates, pH and temperature, and simulates the variations of the concentrations of the substances as a function of time. There is a correspondence between the kinetic and stoichiometric equations of the reactions. In complex reversible reactions systems the equilibrium constant is the ratio of the rate constant of the forward reaction to the rate constant of the backward reactions.

Chemical Reaction Model

In seawater, residual chlorine consisting of A_k species, $k=1,2,...,K$, undergoes first order chemical reactions, $r=1,2,...,R$, of the form

$$\sum_{k=1}^{K} \alpha_{k,r} A_k \underset{\overset{*}{K}_{bwd,r}}{\overset{\overset{*}{K}_{fwd,r}}{\rightleftharpoons}} \sum_{k=1}^{K} \beta_{k,r} A_k \quad (1)$$

where $\alpha_{k,r}$ and $\beta_{k,r}$ represent the stoichiometric coefficients of the reactants and the products, respectively. The general forms of forward and backward reaction potentials are given by

$$\Phi_{fwd,r} = \overset{*}{K}_{fwd,r} \prod_{k=1}^{K} [A_k]^{a_{k,r}}$$

$$\Phi_{bwd,r} = \overset{*}{K}_{bwd,r} \prod_{k=1}^{K} [A_k]^{b_{k,r}} \quad (2)$$

The stoichiometric forms of forward and backward reaction potentials are given by

$$\Phi_{fwd,r} = \overset{*}{K}_{fwd,r} \prod_{k=1}^{K} [A_k]^{\alpha_{k,r}}$$

$$\Phi_{bwd,r} = \overset{*}{K}_{bwd,r} \prod_{k=1}^{K} [A_k]^{\beta_{k,r}} \quad (3)$$

The forward molal rate of formation of chemical species k in reaction r is given by

$$[\dot{A}_k]_{fwd,r} = (\beta_{k,r} - \alpha_{k,r})$$

$$\Phi_{fwd,r} = (\beta_{k,r} - \alpha_{k,r})$$

$$\overset{*}{K}_{fwd,r} \prod_{k=1}^{K} [A_k]^{a_{k,r}} \quad (4)$$

The backward molal rate of formation of chemical species k in reaction r is given by

$$[\dot{A}_k]_{bwd,r} = (\alpha_{k,r} - \beta_{k,r})$$

$$\Phi_{bwd,r} = (\alpha_{k,r} - \beta_{k,r})$$

$$\overset{*}{K}_{bwd,r} \prod_{k=1}^{K} [A_k]^{b_{k,r}} \quad (5)$$

In seawater, pH is assumed constant (in the neigborhood of 8) and the forward and backward reaction rate constants, $\overset{*}{K}_{fwd,r}$ and $\overset{*}{K}_{bwd,r}$ respectively, for reaction r are expressed as function of temperature (T) as:

$$\overset{*}{K}_{fwd,r} = A_{fwd,r} T^{a_{fwd,r}} exp(-\epsilon_{fwd,r}/kT)$$

$$\overset{*}{K}_{bwd,r} = A_{bwd,r} T^{a_{bwd,r}} exp(-\epsilon_{bwd,r}/kT)$$

The necessary data for the constants $\epsilon_{fwd,r}/k$, $\epsilon_{bwd,r}/k$ are taken from the references (Lietzke 1977; Weast and Astle 1979).

Table 1.-- Known chemical species involved in the primary and secondary hydrolysis reactions of chlorine and bromine in chlorinated marine water.

Components of the Kinetics Model		
1. Cl_2	13. CH_3NH_2	25. $HOBr$
2. Cl^-	14. CH_3NHCl	26. $CH_3CONBrCH_3$
3. OCl^-	15. $CH_3CONHCH_3$	27. $CH_3CONClCH_2COO^-$
4. $HOCl$	16. $NH_2CONClCH_3$	28. CH_3NBr_2
5. NH_3Cl^+	17. $(CH_3)_2NH_2$	29. CH_3NHBr
6. NH_2Cl	18. CH_3NCl_2	30. $(CH_3)_3NH^+$
7. $HClO_2$	19. $CH_3CONHCH_2COO^-$	31. $(CH_3)_2N$
8. $NHCl_2$	20. Br^-	32. $CH_2CONHCl$
9. NH_3	21. Br_2	33. CH_2CONH_2
10. NH_4^+	22. OBr^-	34. RNH_2
11. CH_3NH_3	23. $HBrO_2$	35. $RNHCl$
12. $(CH_3)_2NH^+$	24. $HBrO_3$	36. H_2O

Some of the organic compounds reactions with chlorine and bromine known to occur in chlorinated marine waters included in the model (table 1). No rate constant data are available for many organochlorine species reactions and for other reactions rate constants have not been measured, but considerable variability exists in the absolute k values obtained by different investigators. In addition, reliable concentration data are not available for organochlorine compounds and all the marine transient chemical species with which organochlorine compounds might react.

The research in the identification and quantitative analysis of organochlorine compounds will increase the knowledge of the identities and distribution of these compounds in chlorinated seawater. The studies of rate of formation and kinetic data of organochlorine compounds will be useful for modelling and computer simulation and prediction of concentration variations with respect to time.

RESULTS AND DISCUSSION

The kinetics study shows that the chemical nature and lifetimes of compounds formed on the addition of residual chlorine estuarine and/or seawater depend mainly on the concentrations of ammonia and bromide as well as on pH and temperature.

Concentration-versus-time curves for the hydrolysis of the principal species of chlorine during a three-hour period (fig.1) show rapid loss of chlorine in total chlorine species (Cl_2, Cl^-, $HOCl$, OCl^-). This loss of chlorine is attributable to the

formation of hypochlorous acid and chloride (secondary hydrolysis processes).

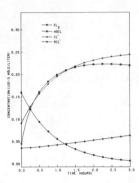

Figure 5.--Concentration variations of the principal chlorine species resulting from hydrolysis, during three-hour period.

The most striking result of this study is the high accumulation of secondary hydrolysis products such as OCl^-, $HOCl$, OBr^-, and $HOBr$ in marine water during and after one-hour period following the discharge of chlorinated effluents. Some of these secondary hydrolysis products and intermediate compounds that play an important role in the formation of some of the principal chlorine and bromine species that may be toxic to marine organisms.

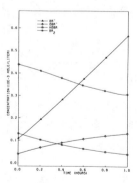

Figure 6.--Concentration variations of the principal bromine species as a result of dissociation, during one-hour period, after addition of 0.5 mg/l residual chlorine into the marine water.

Hypobromite ion concentration is increasing when the HOBr concentration is decreasing in one-hour period. There is a sharp increase in concentration of Br$^-$ in one-hour period. The secondary hydrolysis processes in the presence of residual chlorine, Br$_2$ concentration is gradually decreasing in the system.

Results of one-to-three hour simulations show that, concentration of the NH$_3$ decreasing when ammonium ion is increasing in seawater after introduction of 0.5 mg/l residual chlorine (fig. 7). There is a continuous reduction in the ammonia and methylamine (CH$_3$NH$_2$) concentrations, due to the chlorination of CH$_3$NH$_2$ and ionization of NH$_3$ in the chlorinated marine water. In chlorinated marine water secondary hydrolysis products such as OCl$^-$, NH$_2$Cl, HOCl and methylchlorite (CH$_3$Cl$^+$) are not stable HOCl and NH$_2$Cl concentrations decrease sharply during one-hour period after addition of 0.5 mg/l residual chlorine (fig. 8).

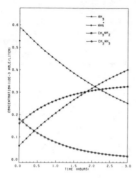

Figure 7.-- Concentration variations of the ammonia, ammonium ion and methylamine during three-hour period after addition of 0.5 mg/l residual chlorine into the marine water.

Although much of the chloramines dissipate in a short period of time low levels are present in chlorinated marine water after one-hour of hydrolysis processes. Most of the nature of the reactions of chlorine species with organic material are not fully understood and reaction rate constants are not available in the literature. However, at marine pH, rate of chlorination and disappearance of organic compounds are very slow. Chloro-organic compounds may form series of equilibria with chlorine species in chlorinated marine water. As a result of competitive reactions, intermediates form such as trimethylamine, methychloramine and methylaminedichloride. Concentrations of trimethylamine carbomide, acetoethylamine are unchanged during a three-hour period.

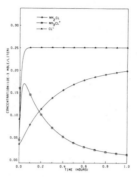

Figure 8.--Concentration variation of monochloramine, methylchlorite and chloride ion after addition of 0.5 mg/l residual chlorine into marine water.

This kinetics study reveal that small concentration of dichloramine is formed in the secondary hydrolysis process in chlorinated marine water (fig. 9) which may not be detectable by the conventional measurement techniques.

Reactions involving the irreversible uptake of residual chlorine in satisfying the chlorine demand of water show that the discharged residual chlorine has a significant effect on the concentration of the total organic matter and their chlorination in the system. Concentrations of organic amines are decreasing when chloroorganic amines are increasing during a three-hour period (fig. 10).

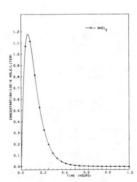

Figure 9.--Concentration variation of dichloramine after addition of 0.5 mg/l of residual chlorine into seawater in one-hour period.

224

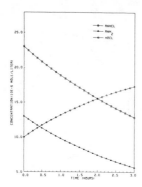

Figure 10.--Concentration variations of
RNH$_2$ and RNHCl and hypochlorous acid
during three hour period after addition of 0.5 mg/l residual chlorine
into the marine water.

Presence of high amount of amino and
organic compounds in sea and/or estuary
waters reduce the resistance of total
chlorine showing almost logarithmical loss
of total chlorine. On the other hand, the
concentration of the chlorinated organic
compounds increased significantly in chlorinated marine water. The half-time of
the consumption of total chlorine is between 1 and 16 hours in chlorinated marine
water. Some of the loss of total chlorine
in chlorinated seawater may be attributed
to the loss in dissolved organic carbon
and the deficit in the Cl$_2$ and Cl$^-$ formed
chlorophenols and chloroform.

CONCLUSIONS

The result of this study of chemical
kinetics of chlorinated marine water show
that bromide concentration have significant
influence on the concentration levels of
toxicants, such as brominated and chlorinated derivates which are formed by the
action of chlorine on bromide. The result
of this study also show that the bromine
derivates are less stable than the similar
chlorine derivates in chlorinated marine
water. The exact reason for the differences in persistance of the bromine and chlorine forms is not known, as loss of halogen
is also related to the oxidation of available reduceable material in marine water.
It is apparent from the results that the
chlorine concentration in chlorinated seawater shift some of the chemical reactions
occuring to higher concentrations of free
molecular chlorine and repeat the cycle of
production of toxic chlorinated compounds

that usually escape the conventional measurement methods of detecting for total and
combined chlorine. It appears that the
formation of chloramines in chlorinated
marine water differ significantly from that
in fresh-water. It is postulated from the
results of this study that upon chlorination of marine water containing ammonia,
bromamines regulate certain chemical reactions.

There are serious analytical difficulties in only measuring the residual chlorine
in chlorinated marine water. Thus, on-
line detection and feed-back control of low-
level residuals and dosing is not yet practically possible. The purpose of this
study is to characterize distribution and
formation of predominant species in reactions of residual chlorine with ammonia,
organic material, and other halogens in
chlorinated marine water. This together
with further characterizations of various
reactions may lead to a better understanding of the mechanism of haloform formation
to better knowledge of the identity of
other unidentified halogen-containing compounds present, and to a more general knowledge of what, in fact, remains in marine
water after certain period of time of discharge of chlorinated effluents.

This model will be further developed
into a transport system that will be able
to give chemical species concentration at
any point in water, for example, related
to ammonia concentration, temperature and
salinity (if pH is taken constant).

In the theory, a complete answer to
this question can be given only by continuously tracking the chemical species discharged from the individual sources and
computing the concentration of each species
at every point as they are transported down
the estuary and spread by diffusion and
dilution.

ACKNOWLEDGEMENTS

The authors gratefully acknowledge
T.H. Row for offering many valuable and
constructive comments throughout the coarse
of this study. We are also grateful to
Dr. A.O. Bishop, Jr., Associate Director
of the University of Tennessee Computing
Center, for his assistance and cooperation
in making all the computing facilities
available to us. We also wish to thank
Dr. M.H. Lietzke for his valuable suggestions and discussions.

This study was financed in part by a
grant obtained from the Health and Environmental Research Branch of Nuclear Regulatory Research in the U.S. Nuclear Regulatory Commission (B0169).

LITERATURE CITED

Arthur,T.W. and T.G. Eaton. 1971. Chloramine toxicity to the amphipod and fathead minnow. Journal of Fish Research Board Canada.

Basch, R.E., M.R. Newton, J.G. Truchan and C.M. Fetterolf. 1971. Water Pollution Control Res. Series EPA, WQO Grant No. 18050 G22.

Bellance, M.A. and D.S. Bailey. 1975. A case history of some effects of chlorinated effluents on the aquatic ecosystem of the lower James River in Virginia. Abstract, 48th Annual Conference, Water Pollution Control Federation, Washington, D.C.

Berliner, J.F.T. 1931. The chemistry of chloramines. Journal of American Water Works Association. 23:1320.

Brungs, W.A. 1973. Effects of residual chlorine on aquatic life. Journal of Water Pollution Control Fedaration 45:2180.

Carpenter, J.H. and D.L. Macalady. 1978. Chemistry of halogens in seawater. p. 161. In Water Chlorination: Environmental Impact and Health Effects. Ann Arbor Science Publisher Inc., Michigan.

Connick, R.E. and Y. Chia. 1959. The hydrolysis of chlorine and its variation with temperature. Journal of American Chemical Society 81:1280.

Duursma, E.V. and P. Parsi. 1976. Persistence of total and combined chlorine in seawater. Netherlands Journal of Sea Research 10(2):192.

Emanuel, N.M. and D.G. Knorre. 1973. Chemical Kinetics. 447 p. John Wiley and Sons, New York, NY.

Engel P., A. Oplakta, and B. Perlmutter-Hayman. 1954. The decomposition of hypobromite and bromite solutions. Journal of American Chemical Society 76:2010.

Everett, D.H. and W.F.K. Jones. 1941. Proceedings of Royal Society London 177A:499.

Fair, G.M., J.C. Morris and S.L. Chang. 1947. The dynamics of water chlorination. Journal of New England Water Works Association 61:285.

Floyd, R., D.G. Sharp and J.D. Johnson 1978 Inactivation of single poliovirus particles in water by hypobromite ion, American Chemical Society 12(9):1031.

Gehrs, C.W., L.D. Eyman, R.L. Jolley and J.E. Thompson. 1974. Effects of stable chlorine- containing organics on aquatic environments. Nature (London) 249:675.

Granstram, M.L. 1954. The disproportionation of monochloramine. Ph.D. Dissertation, Harvard University.

Ingersoll. B. 1952. Tie chlorine discharges fish die-off. Sun-Times, January 9, p. 5. Chicago.

Johnson, J.D., Y. Hsu, and K. Lui 1972. Bromine residual chemistry. Annual Process Report to the U.S. Army Research and Development Command.

Jolley, R.L., C.W. Gehrs and W.W. Pitt 1976. Chlorination of cooling waters: A source of chlorine containing organic compounds with possible environmental significance. p.21. In Radiology and Energy Resources. Dowden, Hutchinson and Ross, Inc. Stroudsburg, Pennsylvania.

Kalmaz, E.V. 1978. Added chlorine in environments. Journal of Environmental Sciences 21(3):30.

Keeslar, F.L. 1975. A selective amperometric membrane electrode for measurement of chlorine and bromine residuals in water. M.S. Thesis, University of North Carolina.

Lietzke, M.H. 1977. A kinetic model for predicting the composition of chlorinated water discharged from power plant cooling systems, ORNL-NUREG-13, April. Oak Ridge National Laboratory, Oak Ridge, TN.

Macalady, D.L., J.H. Carpenter and C.A. Moor 1977. Sunlight-induced bromate formation in chlorinated seawater. Science 195:1335.

Mills, J.F. 1977. Interhalogens and halogen mixtures and disinfectants. p.113. In Disinfection:Water and Wastewater. Ann Arbor Science Publishers Inc., Michigan.

Polgar, T.T., L.H. Bongers and G.M. Krainak. 1976. Assessment of power plants effects on zooplankton in the nearfield. p. 358. In Thermal Ecology II, Technical Information Center, ERDA. Washington, D.C.

Pressley, T.A., D.F. Bishop and S.G. Roan 1972. Ammonia-nitrogen removal by breakpoint chlorination. Environmental Science and Technology 6:622.

Roberts, M.H., R.J. Diaz, M.E. Bender, and R.J. Huggett 1975. Acute toxicity of chlorine to selected estuarine Species. Journal of Fish Research Board of Canada 32:2525.

Rook, J.J. 1974. Formation of haloforms during chlorination of natural waters. Water Treatment Exam 23:234.

Tardiff, R.G. 1975. Halogenated organics in tap water: A toxicological evaluation. Proceedings of the Conference on the Environmental Impact of Water Chlorination, October 22-24. Oak Ridge, TN.

Thomann, R.V. and M.J. Sobel. 1964. Journal of Sanitary Engineering American Civil Engineering Society 90(5):9.

Trabalka, J.R. and M.B. Burch. 1979. Effects of water-soluble chlorine-containing organics on aquatic environments: Another perspective. Toxicology Letters 3:201.

Trofe, T.W., G.W. Inman, Jr. and J.D. Johnson, 1980. Kinetics of monochloramine decomposition in the presence of bromide. Environmental Science and Technology 14(5):544.

Tsai, C. 1975. Effects of sewage treatment plant effluents on fish: A Review of Literature. Publication No. 36, Chesapeak Research Consortium, Inc.

Weast, R.C. and M.J. Astle, 1979. CRC-Handbook of chemistry and physics. CRC Press, Inc., Baca Raton, Florida.

Weil, I. and J.C. Morris, 1949. Kinetic studies on the chloramines. I. The rate of formation of monochloramine, N-chloromethylamine, and N-chloro-dimethylamine. Journal of American Chemical Society 71:1664.

Zillich, J.A. 1972. Toxicity of combined chlorine residuals to freshwater fish. Journal of Water Pollution Control Federation 44:212.

ECOLOGICAL RISK UNCERTAINTY ANALYSIS[1]
Robert A. Goldstein and Paolo F. Ricci[2]

Abstract.--It is important to formulate ecological
assessment in a quantitative probabilistic context that
explicitly incorporates different sources of uncertainty in
order to evaluate alternative control and mitigation strategies, and judge the desirability of proposed research. A
simple exponential decay model is used to illustrate this
concept. For certain situations, the relationship between
the width of the probability density function of the decay
constant and the probability that the variable will exceed a
critical threshold can lead, depending on the relative
magnitude of the threshold, to diametrically opposite management decisions.

INTRODUCTION

At present, most ecological assessments are
qualitative. It is important to formulate quantitative ecological assessments so that net
adverse effects can be compared to costs of control and mitigation strategies. Mathematical
modeling is a valuable tool for estimating quantitative effects. However, in instances where
modeling is applied, it is usually deterministic. Probabilistic ecological models need to be
developed that explicitly incorporate the different sources of uncertainty inherent in assessing
ecological effects. Sources of uncertainty
include variation in design amongst experiments,
uncertainty in model structure relative to reality, measurement error, intra and interspecific
variation, and spatial and temporal variation in
environmental conditions.

If an appropriate probability function for
the magnitude of an ecological effect can be
estimated, then an analysis of how the sources of
uncertainty affect the probability function will
result in a more realistic assessment and an
understanding of how additional research can

decrease the uncertainty in the assessment. By
placing assessments in the above context, an
adaptive tool is created that is not only capable
of evaluating alternative control and mitigation
strategies, but in addition the desirability of
additional research. The objective of this paper
is to explain this concept, which we call "ecological risk uncertainty analysis," and to illustrate it using a simple hypothetical mathematical
example.

BACKGROUND

Goldstein (1977) identified several major
factors that limit the ability of mathematical
ecological simulation models to predict the
effects of perturbations of specific real systems. Looked at from another perspective, these
factors can be considered sources of uncertainty
in the prediction. If output from mathematical
models are to be used by policy makers to guide
them in making management decisions, then the
models should convey to the policy maker an estimate of the uncertainty inherent in the predictions of the models. Knowledge of the uncertainty can be considered a necessary, although
not sufficient, condition for making a rational
policy decision. However, the subject of uncertainty and error associated with ecological
impact assessment is rarely addressed.

Recently, O'Neill, Gardner, Mankin, and
their colleagues at Oak Ridge National Laboratory
have given significant attention to analysis of
error in ecological models (O'Neill and Gardner
1978, O'Neill 1979, O'Neill et al. 1980, Gardner
et al. 1980a, Gardner et al. 1980b). The major
focus of their research has been the calculating

---

[1]Paper presented at the International Symposium on Energy and Ecological Modelling, sponsored by the International Society for Ecological
Modelling. (Louisville, Kentucky, April 20-23,
1981.)
[2]Robert A. Goldstein and Paolo F. Ricci are
Project Managers, Environmental Assessment
Department, Electric Power Research Institute,
P.O. Box 10412, Palo Alto, CA 94303, U.S.A.

research has been the calculation of "error terms" in model variables arising from such things as aggregation of system components, selection of alternative mathematical formulations, and neglect of distributions of parameter values. Their ultimate objective is to develop general rules of error estimation for ecological models. This is an extremely complex and difficult problem. Their strategy is to perform individual error analyses for many simple and complex models and to search for general patterns of error behavior (O'Neill and Gardner 1980).

Ecological risk uncertainty analysis, although concerned with similar subject matter as the error analysis discussed in the previous paragraph, approaches the subject from a different perspective. From the perspective of risk uncertainty analysis, the magnitude of effect resulting from an environmental perturbation can never be predicted with 100% certainty. The principal objective then is to estimate the magnitude of effect in terms of probability functions and relate the probability explicitly to the different sources of uncertainty. Hence, the emphasis is on placing the behavior of the variables in a probabilistic context; whereas in error analysis, the emphasis is on calculating differences in variable behavior between alternative models or alternative implementations of the same model. Related to this overall subject is a discussion by Shaeffer (1979) of the progation of statistical uncertainty through multiplicative chain models.

What a policy maker needs to know is the probability that an effect will be a certain magnitude, or perhaps more importantly, the probability that an effect will exceed a certain magnitude or the magnitude of effect for which there is a certain probability that the effect will be greater. Let us refer to the above as information needs of Types 1, 2, and 3, respectively. In order to give a specific example of these three types, let us assume that we are concerned about the effect of sulfur dioxide on soybean yield. Then an example of a question for Type 1 information is, "What is the probability that an exposure of Y parts per million (ppm) of SO_2 over a period of time T will produce an X% yield reduction?" An example for Type 2 is, "What is the probability that an exposure of Y ppm over a period of time T will produce a yield reduction greater than X%?", and a example for Type 3 is, "For what yield reduction will the probability be P that there will be a greater yield reduction as a result of an exposure of Y ppm over a period of time T?"

If x is the variable by which the impact is being measured, then the probabilities that the policy maker is interested in are given by the probability density function, $p(x)$, and the probability distribution function, $P(X \geqslant x)$. The first function being applicable to Type 1 questions and the second to Type 2 and 3 questions. The two functions are related by the equation:

$$P(X \geqslant x) = \int_x^\infty p(\sigma) d\sigma , \qquad (1)$$

where σ is a variable of integration.

$$P(X \geqslant x) = 1 - P(X \leqslant x) , \qquad (2)$$

since

$$\int_{-\infty}^\infty p(x)dx = 1 . \qquad (3)$$

SOURCES OF UNCERTAINTY

Factors that introduce uncertainty into ecological assessments include variation in design amongst experiments, uncertainty in model structure relative to reality, measurement error, intra and interspecific variation, and spatial and temporal variation in environmental conditions. If the probability distribution function can be explicitly related to these factors, then questions can be addressed regarding how control of these factors or additional research can influence the prediction of impact. Using the case of SO_2 fumigation of soybean crops, an example of an important management question influencing policy decisions is, "If x_t is some critical threshold of effect that is considered undesirable to exceed, how can the system be managed to decrease $P(X \geqslant x_t)$?" Management can include either altering the perturbation, for instance, lowering the exposure concentration or shortening the duration of exposure, or altering the system, for instance, by planting an injury resistant soybean cultivar.

An example of a research-related question would be, "If uncertainties regarding the nature of the mechanisms through which injury is produced can be reduced, how will $P(X \geqslant x_t)$ be effected?" Analyses of how P will vary as the magnitudes of the different sources of uncertainty change are very complicated, as can be inferred from the simple analytical example in the next section.

EXAMPLE

Model

To illustrate the concept of ecological risk uncertainty analysis, a simple model

$$x = x_o e^{-\alpha t} \qquad (4)$$

will be used. This function has also been used by O'Neill (1979) to illustrate ecological error analysis.

Equation (4) is the solution for $t \geqslant 0$ of the differential equation

$$\frac{dx}{dt} = i(t) - \alpha x , \qquad (5)$$

where $i(t) = 0, \quad t \geqslant 0$,

and $x_o = x(t=0)$.

This model is frequently used to represent a situation where an individual has been exposed to a harmful substance at or for some time $t \leqslant 0$. At time $t=0$, as a result of the exposure, the individual has an absorbed amount, x_o, of the substance in his system. For $t \geqslant 0$, the individual receives no further exposure and metabolizes the absorbed substance according to a first-order kinetic process that is characterized by the time constant α.

The model can also be assumed to be a simple representation of an exotic pest population (e.g., Mediterranean fruit fly), x, that has been recently introduced, $t \leqslant 0$, into an agricultural area. The life cycle history of the pest is such that during winter, environmental conditions do not allow reproduction and the size of the population is governed solely by a density independent mortality rate α. The size of population, including all life stages, at the start of the nonreproductive period, $t=0$, is x_o.

Let us assume that at some time t*, there exists a critical threshold x_t. In the case of the harmful substance, if $x(t^*) \geqslant x_t$, then the individual will die. In the case of the exotic pest, t* marks the end of the nonreproductive period. If $x(t^*) < x_t$, then the surviving population is not large enough to be viable or can easily be controlled by management techniques such as the dissemination of sterile adults. On the other hand, if $x(t^*) \geqslant x_t$, then the population is uncontrollable. It will increase to epidemic size and produce major crop losses.

The problem to be solved is: Given the probability density function for $\alpha, p(\alpha)$, what is the probability distribution function for $x, P(X \geqslant x)$? For the example, we assume that α has a uniform probability density (fig. 1). Hence

$$p(\alpha) = \begin{array}{lll} 0 & ; & 0 \leqslant \alpha \leqslant a-\Delta \\ 1/2\Delta & ; & a-\Delta < \alpha \leqslant a+\Delta \quad (6) \\ 0 & ; & a+\Delta < \alpha \end{array}$$

where "a" is both the mean and median of the density function and 2Δ is the width. It would probably be more realistic to assume some type of bell shaped curve within the interval $(a-\Delta, a+\Delta)$; however, to keep the example as straightforward and analytically simple as possible a uniform density was chosen. Furthermore, it will be shown that this simple example yields insight into the solution for bell shaped and other density functions.

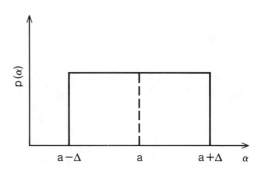

Figure 1.--Uniform probability density function of the parameter α. "a" is both the mean and median of the function. The width is 2Δ .

The existence of a distribution, instead of a single value, for α can result from multiple factors, some of which will now be identified. The distribution can result from the genetic diversity of the population or statistical uncertainty in previous measurements of α. If α is an implicit function of environmental factors, then the distribution can result from variation in the environment of an individual or components of the population with space and time. For instance, the individual who has been exposed to the harmful substance may experience different environments during the period that he is metabolizing the substance. Different clusters of the pest population may be located in different microenvironments. Even if the total pest population is in a spatially homogeneous environment, the environment and hence α will change with time. The distribution of α can also result from the fact that x does not represent a specific age class or life stage, and different age classes and life stages may have different characteristic metabolic and mortality rates. In addition, the distribution of α can result from any combination of the above factors.

Analysis

The probability distribution function, $P(A \leqslant \alpha)$, can be derived using equations (1) and (2).

$$P(A \leqslant \alpha) = \begin{array}{lll} 0 & ; & 0 \leqslant \alpha \leqslant a-\Delta \\ (1/2\Delta)(\alpha-a+\Delta) & ; & a-\Delta \leqslant \alpha \leqslant a+\Delta \quad (7) \\ 1 & ; & a+\Delta \leqslant \alpha \end{array}$$

We can now solve for the probability distribution function for $x, P(X \geqslant x)$.

$$P(X \leqslant x) = P(f(A) \leqslant x)$$

$$= P(A \geqslant f^{-1}(x)) , \qquad (8)$$

where $\quad x = f(\alpha) = x_o e^{-\alpha t}$, $\qquad$ (9)

and $\qquad \alpha = f^{-1}(x) = (1/t)\ln(x_o/x)$. $\qquad$ (10)

    The nature of the inequality in the right-hand side of equation (8) results from $f(\alpha)$ being a monotonically decreasing function of α. If $f(\alpha)$ was monotonically increasing, then the inequality would be in the opposite direction. If $f(\alpha)$ is not monotonic, then the situation becomes more complex and the function must be partitioned into separate intervals in which behavior is monotonic.

    Substituting equations (8), (9) and (10) into (2),

$$P(X \geqslant x) = \begin{cases} 1 & ; \ 0 \leqslant x \leqslant x_\ell \\ (1/2\Delta)((1/t)\ln(x_o/x)-a+\Delta); & x_\ell \leqslant x \leqslant x_u \quad (11) \\ 0 & ; \ x_u \leqslant x \end{cases}$$

where $\qquad x_\ell = x_o e^{-(a+\Delta)t}$, $\qquad$ (12)

and $\qquad x_u = x_o e^{-(a-\Delta)t}$. $\qquad$ (13)

    Note that x_ℓ is the lower bound and x_u is the upper bound for x. $P(X \geqslant x)$ (fig. 2) is the probability distribution function that X is greater or equal to x.

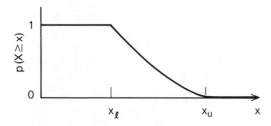

Figure 2.--Probability distribution function of the variable x(t). The upper and lower bounds of x(t) are $x_u(t)$ and $x_\ell(t)$, respectively.

    If the mean, a, of the density function of α is increased, then both x_ℓ and x_u are decreased. If we have two different situations corresponding to two values of a, a_1, and a_2, where $a_1 < a_2$, and $P_1(X \geqslant x)$ and $P_2(X \geqslant x)$ are the two respective distribution functions, then for any given threshold x_t (such that $x_{\ell 2} \geqslant x_t \geqslant x_{u1}$), $P_1 > P_2$. Hence the completely intuitive conclusion for the case of uniform distribution of α, that to decrease the probability that the threshold will be exceeded, a management strategy should be implemented that increases the mean metabolic rate in the example of the harmful substance or the mean mortality rate in the example of the pest species. The

above discussion assumes that only the mean and not the width of the density function is changed. Note that the mean value of x does not correspond to f evaluated at the mean value of α, although the median of x is so related to the median of α.

    It is obvious that the relation between $P_2(X \geqslant x)$ and $P_1(X \geqslant x)$ discussed in the previous paragraph will hold for any density function of α, defined in the interval $(a-\Delta, a+\Delta)$, when the function is simply translated to the right on the alpha axis, by a distance of $d=a_2-a_1$, without changing shape; that is, $P_2(\alpha+d)=P_1(\alpha)$. In this case, "a" is not assumed to designate either the mean or median of $p(\alpha)$ as it does for the uniform density function, but simply is the midpoint of the interval for which $p(\alpha)$ is defined.

    Let us now examine the effect on the probability distribution $P(X \geqslant x)$ of changing the width or uncertainty, as characterized by Δ, of the probability density function $p(\alpha)$. Assume that Δ is increased from Δ_1 to Δ_2.

$$\Delta_2 = \Delta_1 + \varepsilon \qquad (14)$$

Let us define $P_1(X \geqslant x) = P(X \geqslant x)$ for $\Delta = \Delta_1$,

and $\qquad P_2(X \geqslant x) = P(X \geqslant x)$ for $\Delta = \Delta_2$. (15)

    From equations (12) and (13), we can see immediately that x_ℓ will be decreased and x_u will be increased (fig. 3).

$$x_{\ell 2} = x_o e^{-(a+\Delta_2)t} < x_{\ell 1} .$$
$$\qquad (16)$$
$$x_{u2} = x_o e^{-(a-\Delta_2)t} > x_{u1} .$$

    From equation (11), we know that P_2 and P_1 are one when x is less than $x_{\ell 2}$ and $x_{\ell 1}$, respectively, and P_2 and P_1 are zero when x is greater than x_{u2} and x_{u1}, respectively, Hence P_2 and P_1 must cross at some point, x_c, where $x_{\ell 1} < x_c < x_{u1}$, and $P_2 < P_1$ when $x_{\ell 2} < x < x_c$, and $P_2 > P_1$ when $x_c < x < x_{u2}$.

    Since the median of $p(\alpha)$ remains fixed at "a" regardless of the magnitude of Δ, then based on the general property of the distribution function as given in equation (8), the point where $P(X \geqslant x)=1/2$ will also remain fixed. Hence the point at which P_2 and P_1 cross is where they equal 1/2; that is,

$$x_c = x_o e^{-at} . \qquad (17)$$

    One can also derive equation (17) algebraically by using equations (11) and (14) to equate P_1 and P_2. Doing that yields

$$\varepsilon((1/t)\ln(x_o/x_c)-a) = 0 , \qquad (18)$$

and since $\varepsilon > 0$, then the expression in the parentheses that multiplies ε must be 0. Setting that expression equal to zero yields equation (17).

The relation between P_1 and P_2 can be written

$$P_1(X \geqslant x) > P_2(X \geqslant x) > 1/2 \quad ; \quad x_{\ell 2} < x < x_c ,$$
$$P_1(X \geqslant x) < P_2(X \geqslant x) < 1/2 \quad ; \quad x_c < x < x_{u2} , \qquad (19)$$
$$P_1(X \geqslant x) = P_2(X \geqslant x) = 1/2 \quad ; \quad x = x_c .$$

In fact, it is clear from the discussion that precedes equation (17), that equation (19) is valid not only for the uniform density probability function of α, but for any density probability function of α that is defined by the independent parameters Δ and "a", where "a" is the median of the function and $(\Delta-a, \Delta+a)$ is the interval in which $p(\alpha)$ is defined. The assumption of independence means that changing Δ will not change "a".

Discussion

What is the interpretation of (19) with respect to the simple model and problem that is being analyzed? Since the location of the critical threshold, x_t, is independent of the model and assumed probability function $p(\alpha)$, x_t may be either larger or smaller than x_c. If x_t is larger than x_c, then the probability that it will be exceeded increases if the width of $p(\alpha)$ (uncertainty in α) increases. If x_t is smaller than x_c, then the probability that it will be exceeded decreases if the uncertainty in α increases. (Of course if x_t does not fall between the upper and lower bounds of x as determined by the largest Δ being considered, then the probability of it being exceeded cannot be affected by changing the width.)

If the region to the left of x_c (fig. 3) is considered to represent a situation that is "favorable" for exceeding the threshold, since $P(X \geqslant x)$ is always greater than 1/2, and the region to the right of x_c "unfavorable", then increasing the uncertainty in α increases the probability of exceeding the threshold under unfavorable conditions and decreases the probability under favorable conditions.

Let us consider the case of the insect pest. Assume that identical numbers of the pest have been introduced into two separate areas. Both areas are identical with respect to the median mortality rate but the areas have different degrees of spatial heterogeneity resulting in different diversities of microenvironments and hence different widths for $p(\alpha)$. Then in the situation where conditions favor a pest outbreak $(x_t < x_c)$, the area with greater microenvironmental diversity will have the lower probability of having a pest outbreak, while in the situation where conditions are unfavorable for pest outbreak, the area with greater microenvironmental diversity will have the higher probability of having a pest outbreak.

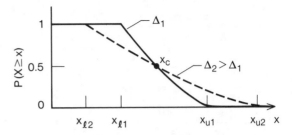

Figure 3.--Probability distribution function of the variable $x(t)$ for two uniform density functions of α. The density functions have the same median "a" but different half widths, Δ_1 and Δ_2. The two distribution functions intersect at x_c, where $P(X \geqslant x_c) = 1/2$.

A policy maker interested in preventing outbreak in both areas, could use the probability distribution functions for allocating his resources. In the first situation, more resources could be allocated to the area having lower spatial heterogeneity and in the second situation, more resources could be allocated to the area having higher spatial heterogeneity.

Ecologists tend to attach theoretical significance to the concept of "diversity" and equate increased "diversity" with increased "stability". It is interesting to note that for the simple model of the pest population, increasing the diversity of the population, as represented by the width of the density function of alpha, can either increase or decrease the probability of survival.

CONCLUSION

It is clear even from the simple hypothetical example presented in this paper that the explicit analysis of the relationships between key system variables and sources of uncertainty is complicated. For only a few simple models will analytic approaches be feasible. Complicated multivariable, multiparameter models, such as a model that predicts how a lake-watershed ecosystem is acidified by the deposition of atmospheric acids (Goldstein et al. 1980, Chen et al. 1979), will require analyses based on Monte Carlo procedures. However, such analyses are necessary if we hope to address decisions about control and mitigation of ecological impacts and needed ecological assessment research in a rational context.

ACKNOWLEDGMENTS

The authors thank Hung-Po Chao for his helpful and insightful comments and the time that he spent discussion the subject with them.

LITERATURE CITED

Chen, C. W., S. A. Gherini, and R. A. Goldstein. 1979. Modeling of lake acidification process. p. 5:1-5:43. In Proceedings of international workshop on ecological effects of acid precipitation. [Gatehouse-of-Fleet, Scotland, September 4-7, 1978] Electric Power Research Institute Report EA-79-6-LD, Palo Alto, CA.

Gardner, R. H., R. V. O'Neill, J. B. Mankin, and D. Kumar. 1980a. Comparative error analysis of six predator-prey models. Ecology 61:323-332.

Gardner, R. H., D. D. Huff, R. V. O'Neill, J. B. Mankin, J. Carney, and J. Jones. 1980b. Application of error analysis to a marsh hydrology model. Water Resources Research 16:659-664.

Goldstein, R. A. 1977. Reality and models. p. 207-215. In Mathematical models in biological discovery. ed., D. L. Solomon and C. Walter. Springer-Verlag, Berlin.

Goldstein, R. A., C. W. Chen, S. A. Gherini, and J. D. Dean. 1980. A framework for the integrated lake-watershed acidification study. p. 252-253. In Proceedings of the international conference on the ecological impact of acid precipitation. [Sandefjord, Norway, March 11-14, 1980] SNSF Project Report, Oslo-As.

O'Neill, R. V. 1979. Natural variability as a source of error in model predictions. p. 23-32. In Systems analysis of ecosystems. ed., G. S. Innis and R. V. O'Neill. International Co-operative Publishing House, Fairland, Maryland.

O'Neill, R. V., and R. H. Gardner. 1979. Sources of uncertainty in ecological models. p. 447-463. In Methodology in systems modelling and simulation. ed., B. P. Ziegler, M. S. Elzas, G. J. Klir, and T. I. Oren. North-Holland, Amsterdam.

O'Neill, R. V., R. H. Gardner, and J. B. Mankin. 1980. Analysis of parameter error in a nonlinear model. Ecological Modelling, 8:297-311.

Shaeffer, D. L. 1979. A model evaluation methodology applicable to environmental assessment models. 48 p. Oak Ridge National Laboratory Report 5507, Oak Ridge, TN.

4. Aquatic Ecosystem Models –
Streams, Reservoirs, and Hydropower

MODELLING THE IMPACT OF A TIDAL POWER SCHEME
UPON THE SEVERN ESTUARY ECOSYSTEM[1]

Philip J. Radford[2]

Abstract.--A General Ecosystem Model of the Bristol Channel and Severn Estuary (GEMBASE) has been used to predict the impact of a tidal power scheme upon estuarine biota. The proposed semi-permeable barrage, incorporating electric power generators would halve the natural tidal range and greatly reduce the dispersion and the turbidity up-estuary of the structure. Results are presented which demonstrate the sensitivity of GEMBASE to these factors and indicate the implications of such changes to the ecosystem.

INTRODUCTION

The steeply rising cost of energy from conventional sources such as from oil and coal, coupled with increasing public concern over the true economic and social costs associated with the disposal of nuclear waste, have heightened the search for alternative energy sources. In certain key locations around the world (Hoare and Haggett 1979) exceptionally high tidal amplitudes make the possibility of tidal power generation, at an economically competitive price, feasible. In 1978 the U.K. Government allocated £1.5 million for studies to assess tidal barrage schemes in the Severn Estuary (Department of Energy U.K. 1978), and over the subsequent two years increased this R & D expenditure to a total of £2.5 million. Recent estimates (Financial Times 1981) predict that by the turn of the century the cost of generating electricity using Severn tidal power will be £0.025 - £0.030/kWh compared with £0.035p/kWh from coal and £0.022/kWh from nuclear fuel. Other predictions, researched under contract to the U.K. Government (Radford 1980, Radford and Young 1980), concern environmental impact considerations, viz water quality and ecosystem effects, and these form the basis of this paper.

THE BRISTOL CHANNEL AND SEVERN ESTUARY

The Severn Estuary is one of the world's most suitable sites for tidal power generation, having a high natural tidal range of between 8-14 m, second only in magnitude to that propagated in the Minas Basin of the Bay of Fundy (16 m), Canada. One of the most favoured Severn sites for a barrage (fig. 1) is within the general area of the mouth of the estuary near Cardiff and Weston-super-Mare. Seaward of this line the estuary opens out into the Bristol Channel which extends a further 125 km to the Celtic Sea. There are numerous possible sites for tidal power generation within the Inner Channel Region, some of which are discussed in the European Energy report (Financial Times 1981) but Government sponsored ecological impact studies (Radford 1980, Radford and Young 1980) have only considered two ebb generation schemes, one across the boundary of the Outer Estuary and the Inner Channel and the other across the boundary of the Inner Channel and the Central Channel South. Results are presented here for the most landward of these two locations only.

The unmodified Severn Estuary is typified by very high levels of suspended particulate matter (table 2) kept in suspension by the strong tidal currents induced by the large tidal excursions. In the Outer Estuary total particulates as high as 1000 g/m^3 have been measured and they rarely fall below 100 g/m^3. The resultant high turbidity of the water column restricts the photic zone to less than 0.1 m on most occasions, so greatly restricting primary production. A further ecological effect of elevated particulate load is to prevent colonization by benthic suspension feeders (Warwick and Uncles 1980), whose filtering mechanisms would become overloaded and inoperative under

---

[1]Paper presented at the International Symposium-Energy and Ecological Modelling, sponsored by the International Society for Ecological Modelling (Louisville, Kentucky, April 20-23, 1981).
[2]Philip J. Radford is head of the Systems Analysis and Statistics section of the Natural Environment Research Council's Institute for Marine Environmental Research (IMER), Plymouth, Devonshire, UK.

such conditions and which require a stable substrate on which to grow. The construction of a tidal barrage would undoubtedly reduce bed stress, turbidity, and total particulate load in the water column so improving the potential of the estuary for increased primary and secondary production up-estuary of the structure.

THE GENERAL ECOSYSTEM MODEL

The General Ecosystem Model of the Bristol Channel and Severn Estuary (GEMBASE) is a dynamic simulation model which represents the ecological interactions between the component animal and plant communities typically found in estuaries (see Radford 1979). The interrelationship between the components of the estuarine ecosystem are expressed as 15 ordinary differential equations which combine 150 separate processes and are driven by six environmental forcing functions (table 1) and five exogenous variables. The biomass of each of the higher carnivores (07 - CARNIVORES, table 1) is not generated by the model but supplied as time series data obtained from empirical estimates. The seasonal development of these components is dominated by immigration and emigration rather than by interactions within the ecosystem. In GEMBASE, the higher carnivores interact fully with the rest of the model through processes such as feeding, respiration and mortality but the development of their

biomass is considered to be largely dependent upon external factors.

All of the interrelationships between the state variables in the model are assumed to occur in the context of the specific hydrodynamic regime of the Bristol Channel and Severn Estuary and the model has been simultaneously applied to six of the seven regions (2-7) shown in figure 1. The Inner Estuary Region (region 1) and the Celtic Sea are not modelled within GEMBASE but empirical time series data relating to all of their dissolved, particulate and planktonic variables are used as boundary conditions. The double-headed arrow symbol used in figure 1 represents the exchange of such material across all regional boundaries. The magnitude of these transfers are computed by a sub-model named HYDROB which is an empirical dispersive-advective model, similar in concept to HYDROBASE (Radford, Uncles and Morris 1981). The dispersion coefficients used by HYDROB were calculated using salinity measurements over the whole estuary from the tidal river to the Celtic Sea (for computational method see Uncles and Radford 1980).

The simulation results generated by GEMBASE have been verified against a three-year time series of field data (1973 - 1975), and preliminary comparisons for carnivorous and omnivorous

Table 1.--List of state variables, forcing functions and exogenous variables recognised by GEMBASE.

Variable Type	No. - Name		Code	
State variables	01 - NUTRIENTS	-	NI	- Nitrates, nitrogen only
	02 - DOM	-	OI	- Dissolved organic matter
	03 - PHYTOPLANKTON	-	P1	- Edible (by zooplankton)
	PI	-	P2	- Non-edible
	04 - PHYTOBENTHOS	-	A1	- Macroalgae
	AI	-	A2	- Benthic diatoms
	05 - ZOOPLANKTON	-	Z1	- Carnivores
	ZI	-	Z2	- Omnivores
	06 - ZOOBENTHOS	-	B2	- Suspension feeders
	BI	-	B3	- Deposit feeders
	08 - SUSPENDED POM	-	DW	- Particulate organic matter
	09 - DEPOSITED POM	-	DS	- Particulate organic matter
	10 - HETEROTROPHS	-	HW	- Planktonic bacteria
	11 - HETEROTROPHS	-	HS	- Benthic bacteria
	12 - SALINITY	-	SA	- Salinity $^o/oo$
Exogenous variables	07 - CARNIVORES	-	C1	- Birds
			C2	- Man
	CI	-	C3	- Pelagic Fish
			C4	- Demersal Fish
		-	C5	- Benthic Invertebrates
Forcing functions	13 - RADIATION	-	SR	- Hourly solar radiation
	14 - RUN-OFF	-	QI	- Daily river flows
	15 - TEMPERATURE	-	TW	- Seasonal water temperature
	"	-	TA	- Seasonal air temperature
	16 - TIDAL RANGE	-	TR	- Daily predicted range
	17 - PARTICULATES	-	TP	- Total suspended particulates

zooplankton have been published by Radford and Joint (1980). The measure of agreement between observed and simulated results is promising for all model variables for which adequate data exist but difficulty has been experienced in obtaining realistic seasonal cycles for some variables such as planktonic heterotrophs and benthic diatoms which cannot be measured so readily. The exercise of interaction between model verification and model improvement still continues as more data come to hand and as our understanding of some key processes improves. In the near future it is hoped to validate the model against a completely independent data set obtained from a series of monitoring cruises which have recently been completed (1978-1981). In the mean time it is clear that GEMBASE is capable of simulating most components of the system, reproducing realistic seasonal cycles, reasonable annual mean values, and significant differences between the six geographical regions.

THE APPLICATIONS OF GEMBASE

The urgent need of the U.K. Department of Energy (1978) to predict the effect of a tidal power scheme upon the Severn Estuary ecosystem precipitated the use of GEMBASE before full validation of the model was possible. Thus in the pre-feasibility phase of this study it has been possible to demonstrate something of the strength and limitations of the use of an ecosystem model as an aid to environmental impact assessment. It is judged that the experience gained by pursuing this policy will enable validated models to be used and accepted with greater confidence should the Severn Tidal Power Project pass into an acceptability study phase.

GEMBASE has been used to simulate both the natural estuary, and the system as modified by a notional tidal power scheme, driven by the environmental conditions which prevailed over the period November 1972 to December 1975. The initial conditions for all the state variables for all six regions (Regions 2-7 in figure 1) were taken from measurements made on a standard survey cruise which took place during November 1972. The model assumes that the direct impacts of a barrage near Cardiff / Weston-super-Mare would be:

i – The proposed volume of water above the barrage would coincide with the volume presently attained at high water level on an average neap tide.

ii – The proposed depth of water above the barrage would be consistent with the proposed new volume.

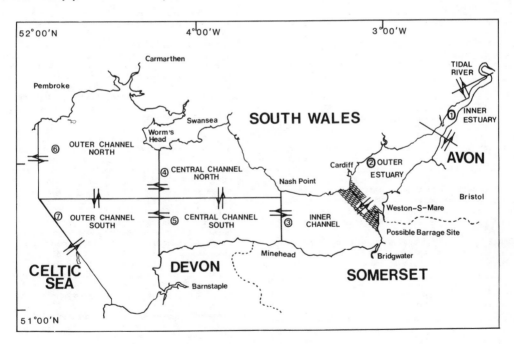

Figure 1.--Chart of Bristol Channel and Severn Estuary showing the seven regions recognised by GEMBASE and the area designated as a possible site for a tidal power scheme.

iii - The dispersion coefficients landward of the barrage would be reduced to one quarter of their present values.

iv - The turbidity of the water column above the barrage would be reduced to that presently attained in the Central Channel North (Swansea Bay). Table 2 gives the range of total particulates expected in all six regions in an unmodified estuary.

v - The mortality rate of phytoplankton above the barrage would be reduced from 10% to zero due to the greatly reduced suspended particulates and the associated increase in the photic zone.

vi - The littoral region would be halved due to the reduced tidal range above the barrage.

Each of the six assumptions are based upon realistic predictions obtained from separate studies of the hydrodynamic and engineering aspects of construction of a barrage, and where appropriate, interpreted in the light of ecological research.

The simulations predict the effect of a barrage had it been built instantaneously on 1 November 1972; the results demonstrate the subsequent development of the ecosystem over the three year period 1973-1975. No results are presented for the Inner Estuary (Region 1 in figure 1) because measured data are used in this region as boundary values for the Outer Estuary.

SIMULATION RESULTS

GEMBASE generates a large matrix of simulation results incuding time-series (1973-1975) for 15 state variables (table 1), for each of 6 regions (regions 2-7 in figure 1), for the cases both with and without a barrage (180 time-series in all). Also, the time-series for the 150 rate variables may be summarized into 40 meaningful groups of processes, again for six geographical regions, with and without a barrage, resulting in a total of 480 options. Only a few of these results can be presented in a brief paper such as this, and those selected for publication have been

Table 2.--Total suspended particulates (g/m^3) under different tidal regimes and river flow conditions.

TIDAL RANGE						
FRESHWATER PERCENTILE FLOWS	10%	50%	90%	10%	50%	90%
OUTER ESTUARY	96	154	227	399	457	531
INNER CHANNEL	12	19	27	52	59	67
CENTRAL NORTH	.1	6	14	13	19	27
CENTRAL SOUTH	.9	4	7	15	18	21
OUTER NORTH	.1	2	7	5	9	14
OUTER SOUTH	.2	2	5	4	6	9

Table 3.--Averaged simulated salinities (^{o}/oo) with seasonal standard deviations, as generated by the GEMBASE subroutine HDROB (1974-75) vs. the fully validated model HYDROBASE (1972-78).

Region MODEL	BARRAGE	NO BARRAGE
Outer Estuary		
HYDROBASE	21.8 ± 3.8	25.1 ± 3.2
GEMBASE	21.8 ± 3.1	25.2 ± 3.1
Inner Channel		
HYDROBASE	29.8 ± 1.0	29.7 ± 0.9
GEMBASE	30.7 ± 1.0	29.4 ± 1.4
GEMBASE		
Central North	32.9 ± 0.3	32.4 ± 0.4
Central South	33.2 ± 0.3	32.6 ± 0.4
Outer North	33.7 ± 0.2	33.6 ± 0.2
Outer South	34.5 ± 0.1	34.4 ± 0.1

chosen to illustrate the most important features of the system, and the most radical changes caused by the superimposition of a barrage.

Figure 2 shows simulation results for three state variables for six regions for the case of the unmodified estuary. The results for salinity reflect the effectiveness of the subroutine HYDROB in simulating the advection and dispersion terms of the system. Both the means and the seasonal standard deviations of this variable (table 3) agree well with the output of the fully validated model HYDROBASE (Radford, Uncles and Morris 1981) for the two regions (Outer Estuary and Inner Channel) simulated by both models.

The simulations show the increasing salinity of the estuarine water from east to west and from north to south and the associated decrease in seasonal variability. The inter-regional transfers of all dissolved and planktonic matter in the estuary are computed from the dispersive and advective flows predicted by subroutine HYDROB.

Phytoplankton levels are restricted in the Outer Estuary and Inner Channel regions due to the high turbidity of the water column but for the remainder of the Bristol Channel, average biomass per unit area is high and very similar for all regions except for the summer peaks when the two Northern Regions exhibit a much higher biomass which persists for about one month. The between year variability is a function of the different environmental forcing functions (solar radiation and river run-off) which are applied using the appropriate historic data. In contrast, the levels of carnivorous zooplankton are distinctive to each region, the biomass normally increasing with increasing salinity from region to region. As there is no direct link between

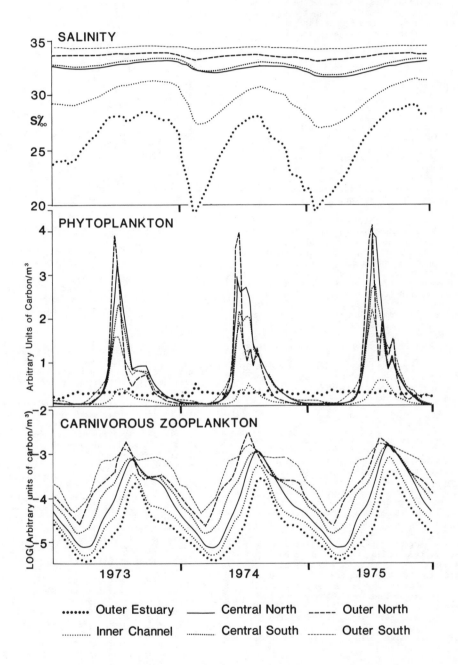

Figure 2.--Simulation results for three state variables for six regions for the case of the unmodified estuary.

salinity and the biomass of carnivorous zooplankton programmed into GEMBASE, one must assume that this correlation is caused by the interaction of a number of different biological processes which are not self evident.

Effects of a tidal power scheme on state variables

General Effects

No results are presented in this paper which apply to the Outer Channel (Regions 6 and 7 in figure 1) because the simulations do not reveal any significant effects of a barrage on the flow of carbon or nutrients in these regions at the levels of sub-system and spacial discrimination used in GEMBASE. The following discussions refer to the remaining four regions within which a barrage would be sited, Outer Estuary, Inner Channel, Central Channel North and Central Channel South.

It is recognized that the predicted initial impact of introducing a barrage instantaneously is unreal because closure would occur progressively over a period of years. Nevertheless, the immediate response of the model and the subsequent equilibration are of more than theoretical importance; they point to the numerical stability of the model, the resilience of the system and the ultimate pattern of probable changes and trends. The results generated by GEMBASE as an unvalidated model must be regarded as tentative indications; it would be unwise, at this stage, to assign absolute quantitative values to the biological components of the system. The figures, therefore, are plotted with an arbitrary y axis; they should be used to make qualitative comparisons between the regions of the present unmodified system and the effects of a barrage.

Salinity

The predicted effect of a barrage built across the boundary between the Outer Estuary and Inner Channel Regions is to reduce average salinities up-estuary of the structure by about 3 ^{o}/oo (table 3) but to slightly increase salinities in the Inner Channel, immediately below the barrage. Salinity in the remainder of the Bristol Channel including the North and South Central Regions would be very slightly but consistently increased (< 0.5 ^{o}/oo). These changes would be caused by a combination of the reduced dispersion and the increased average volumes which it is assumed would occur above the barrage. The associated retention times would be increased from 79 days to 139 days under average river run-off conditions and by a similar percentage for above or below average flows. These changes in retention times can be critical in determining the population sizes of planktonic species and in the concentrations of conservative pollutants (Radford 1980, Radford and Young 1980).

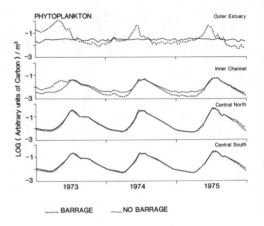

Figure 3.--Simulated levels of Phytoplankton, assuming a barrage had been built in November 1972 vs. the unmodified system, for four geographical regions.

Phytoplankton

Phytoplankton is relatively sparse in the Outer Estuary and Inner Channel (Regions 2 and 3 in figure 1) of the unmodified estuary, although measurements of chlorophyll are often enhanced by dead plant material from freshwater sources and from mudflats. Estimated annual primary production based on monthly measurements in the Inner Channel during 1974 was as low as 6.9 g C/m^2 (Joint and Pomroy (in press)) compared with 48.5 g C/m^2 in the Central Channels. These production figures agree reasonably well with GEMBASE simulated rates of 4.1 g C/m^2 and 42.3 g C/m^2 per annum for the Inner and Central Channels respectively.

The assumed reduction in suspended particulate matter in the Outer Estuary which would follow the construction of a barrage (see assumption iv and table 2) would increase the average euphotic zone in that region from about 0.5 m to 5.0 m which coupled with the presence of an ample supply of nutrients would be expected to considerably enhance the growth of phytoplankton. Simulation by GEMBASE, of the introduction of a barrage in November 1972 produces a very large peak of phytoplankton in the Outer Estuary during the summer of 1973 (fig. 3). In subsequent years the system equilibrates to a moderate seasonal cycle of phytoplankton, the same order of magnitude as that presently experienced in the Central Channel Region, although the season is considerably shorter probably because of the different hydrodynamic regimes of the two regions. Levels of phytoplankton immediately below the barrage, in the Inner Channel, remain virtually unchanged.

Plant nutrients

In GEMBASE, plant nutrients represent the
sum of nitrates plus nitrites derived from terrestrial sources, and ammonia and organic nitrogen recycled through in situ biological process.
There is a strong seasonal cycle of nutrients in
the Outer Estuary (fig. 4) due to dilution by
river run-off. The highest levels occur in
autumn and the lowest in early spring. In all
other regions the seasonal cycle is less marked.
A barrage would have two conflicting effects on
nutrient levels at positions up-estuary of the
structure.

i - Nutrient levels would be increased
      due to reduced dispersion and inc-
      reased retention times for the
      estuary.

ii - Nutrient levels would be reduced
      by increased primary production.

The resultant effect may be seen in figure 4
which demonstrates that a barrage would cause
an increase in nutrients in the Outer Estuary during the spring months but a decrease in the summer
when phytoplankton production is highest. In all
other regions of the Bristol Channel nutrient
levels would be reduced because of the combined
effects of increased nutrient uptake by phytoplankton and increased retention time in the
system above the barrage.

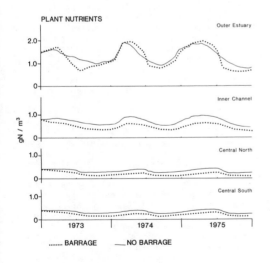

PLANT NUTRIENTS

.... BARRAGE   —— NO BARRAGE

Figure 4.--Simulated concentrations of Plant
Nutrients, assuming a barrage had been built
in November 1972 vs. the unmodified system.

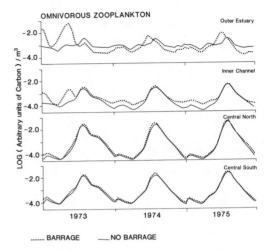

OMNIVOROUS ZOOPLANKTON

.... BARRAGE   —— NO BARRAGE

Figure 5.--Simulated levels of Omnivorous Zooplankton, assuming a barrage had been built
in November 1972 vs. the unmodified system.

Zooplankton

Omnivorous zooplankton is sparse in the
Outer Estuary and the Inner Channel; seaward of
these regions, it is an order of magnitude more
abundant (fig. 5). The increased primary production caused by a barrage has the immediate effect
of increasing zooplankton levels up-estuary of the
barrage (Outer Estuary) to levels previously only
attained in the Central and Outer Channel regions.
This effect is transient, persisting for only one
year, after which a seasonal cycle of omnivores
is established which represents only about twice
or three times the level of the unmodified system.
The impact on zooplankton outside the barrage is
negligible, with the possibility of a slight reduction in zooplankton in consequence of the
predicted reduction in phytoplankton which itself
is caused by a reduced nutrient regime. The
dependence of carnivorous zooplankton upon omnivorous zooplankton as its major food source results in a distribution of carnivores which is
correlated in abundance with omnivores but at a
much lower biomass (an order of magnitude lower).
In the presence of a barrage the populations of
carnivores fluctuate in direct response to those
of omnivorous zooplankton. The transient effect
immediately following the inclusion of a barrage
(1973) produces a very high peak of carnivores
above the structure but by 1974 the system settles
down to a seasonal pattern similar to that of the
unmodified system. This suggests that the increased primary production which follows decreased
turbidity in the Outer Estuary is not permanently
utilized by an increased carnivorous zooplankton
biomass.

Benthic suspension feeders

The high levels of suspended particulate matter in the Outer Estuary and the Inner Channel
Regions prevent the growth and development of
benthic suspension feeders because their filtering
mechanisms are unable to cope with the large volume of non-biological particles present in the
water column. Under these ambient high particulate concentrations, suspension feeders stop filtering altogether so they can survive only in
backwaters where much of the dense abiotic material has fallen out of suspension. In figure 6
it is evident that in the unmodified Outer Estuary,
levels of suspension feeders are too low to appear
on the chosen scale of the plot, although much
greater biomasses exist in the less turbid regions
of Swansea Bay (Central Channel North), where a
seasonal cycle of production is possible. When a
barrage is introduced suspension feeders experience
two environmental changes which enable them to
exploit the Outer Estuary habitat.

i   -   Levels of suspended particulates are
        greatly reduced (table 2) enabling
        them to filter at all states of tide,
        provided they remain submerged.

ii -   The nutritional value of the remain-
        ing suspended particulate matter is
        much higher due to the increased
        primary production.

GEMBASE predicts (fig. 6) that the small native
population of suspension feeders in the Outer
Estuary would readily exploit these changed
conditions and expand to a concentration several
times that which presently occurs in Swansea Bay.
In the first year after barrage construction they

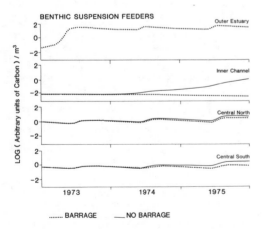

Figure 6.--Simulated levels of Benthic Suspension
Feeders, assuming a barrage had been built
in November 1972 vs. the unmodified system.

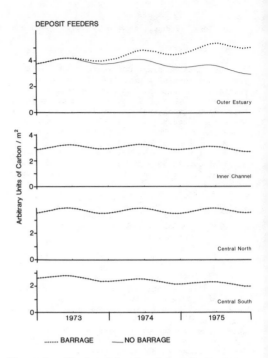

Figure 7.--Simulated levels of Deposit Feeders,
assuming a barrage had been built in November
1972 vs. the unmodified system.

would utilize the large biomass of phytoplankton
to increase their numbers and in subsequent years
this enhanced community would control the growth
of phytoplankton to a level comparable with that
experienced in Swansea Bay (see figure 3, Central
Channel North). In the Central Channel North the
presence of a barrage would slightly reduce levels
of suspension feeders because of the related reductions in nutrients and, therefore, phytoplankton

Benthic deposit feeders

It is assumed in GEMBASE that benthic deposit
feeders are able to grow in only a small fraction of the total area, because of the absence
of a suitable substrate in the tidally scoured
sub-littoral area. The biomass of these organisms
is fairly constant, in the areas where they can
establish themselves and their seasonal fluctuations are quite small (fig. 7). After construction of a barrage, the biomass of deposit
feeders per unit area is predicted to increase
steadily over a three-year period in the region
above the structure (Outer Estuary). This is
due to the increased particulate carbon which
would accumulate in the substrate following death
and decay of the increased biomass of primary
and secondary producers in the water column. The
biomass of deposit feeders in the remainder of
the Bristol Channel is predicted to remain virtu-

ally unchanged.

Other components of the ecosystem

The concentration of dissolved organic matter in the unmodified Outer Estuary is relatively high, due to terrestrial and riverine inputs, with small seasonal fluctuations caused by changes in the magnitude and quality of riverflows. The introduction of a barrage causes a general increase in levels of dissolved organics in the Outer Estuary (above the barrage) and a strong seasonal cycle which is in phase with the equivalent curve of secondary production. This is caused by the biological activity of the suspension feeders responding to the increased phytoplankton productivity. Below the barrage, in the Bristol Channel, levels of dissolved organics are reduced, due to increased retention time in the modified estuary. The biomass per unit area of diatoms, which grow on the mud-flats, is unchanged by the presence of a barrage because their growth depends upon the light intensity on the exposed mud flats, which is not modified in any way by the lower turbidity of the water column. Total diatom production would be halved due to the decreased area of tidal mud flats in the region above the barrage. The levels and fluctuations of all of the state variables in GEMBASE are modified to some degree by the imposition of a tidal barrage scheme but the most important features of these changes concern those variables already discussed. Further appreciations of the changes induced by a barrage may be gained by consideration of the processes via the rate variables.

Effects of a tidal power scheme on processes

Primary production

The predicted transient (1973) and the more permanent (1974) effects of a barrage on rates of primary production by phytoplankton are given in table 4.

Table 4.—Simulated annual rates of primary production (g C m^{-2} y^{-1}) of phytoplankton for six regions of the Bristol Channel and Severn Estuary, both with and without a barrage for the years 1973 and 1974.

| | BARRAGE | | NO BARRAGE | |
	1973	1974	1973	1974
OUTER ESTUARY	81.4	43.8	0.86	0.86
INNER CHANNEL	4.3	3.6	3.5	4.1
CENTRAL NORTH	36.9	37.9	42.1	44.9
CENTRAL SOUTH	35.2	33.6	36.8	39.6
OUTER NORTH	52.3	57.9	55.0	63.0
OUTER SOUTH	43.9	48.9	47.0	52.8

In the year immediately following the introduction of a barrage (1973) the annual rate of primary production would be increased by two orders of magnitude in the Outer Estuary region (above the barrage). Subsequently (1974), this rate would moderate to a figure very similar to that which occurs in the Central Channel North in the unmodified estuary. Just below the barrage in the Inner Channel Region in 1973, there would be a slight increase in primary production which could be attributed to the growth of the extra phytoplankton, produced in the Outer Estuary and advected across the barrage. Throughout the remainder of the Bristol Channel rates of primary production would be reduced by between 5% and 15%, because of the lower levels of nutrients there, caused by the combined effects of increased primary production and increased retention time above the barrage.

Other processes

In the process flow diagrams (fig. 8 and 9) the predicted annual average biomass of the ten organic state variables (g C m^{-2}) of GEMBASE are represented as rectangles of appropriate size, linked by arrows of widths which represent the relative magnitudes of the average flows of carbon (g C m^{-2} d^{-1}) through the system. Figures 9 and 8 respectively, present these data for 1974, for the Outer Estuary Region, with and without the introduction of a barrage.

The most outstanding feature of the post-barrage, relative to the pre-barrage flow diagram (fig. 9 v. fig. 8) is the large increase in the total respiration loss of the system. This is indicative of the large general increase in biological productivity which is predicted to follow the increased primary production above a barrage. The state variables which contribute most to this effect are the primary producers (PI and AI), and the zoobenthos (BI) all of which assume a higher biomass in the presence of a barrage. The zoobenthos (BI) show the greatest increase in average biomass and a commensurate increase in their rate of intake of each component of their diet (PI, DW and HW). It is, in fact, the benthic suspension feeders (B2) that most readily exploit the changed environment brought about by the introduction of a barrage. The reduced concentration of total particulate matter (TP) in the water column enables them to filter at a greater rate for a longer period of time, and the increased primary production provides a more energetic food source.

The change in dispersion coefficients to one quarter of their natural values to represent a direct effect of a barrage has a very minor impact on the magnitude of transfers of material between the Outer Estuary and Inner Channel Regions compared with the other biological changes already discussed. It is therefore interesting to note that the net transfer of zooplankton (ZI) out of the Outer Estuary is halved, whilst its average biomass is virtually unchanged. Before the introduction of a barrage the net transfer of planktonic heterotrophs (HW) into the Outer Estuary was positive but a barrage causes the daily rate of transfer to be of the same order of magnitude but in the reverse direction. This process is the

OUTER ESTUARY FOR 1974 WITHOUT BARRAGE

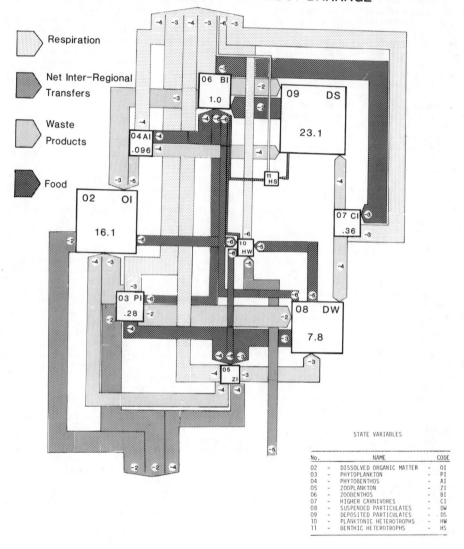

Figure 8.--Mean annual levels of state variables and processes during 1974, predicted for the unmodified Outer Estuary. The width of the squares and the width of the arrows, representing the state variables and processes respectively, have been drawn in proportion to the logarithms of their magnitude (in arbitrary units of carbon m^{-2} d^{-1}) and are labelled with the truncated exponent (e.g. 5.3 10^{-3} as -3).

OUTER ESTUARY FOR 1974 WITH BARRAGE

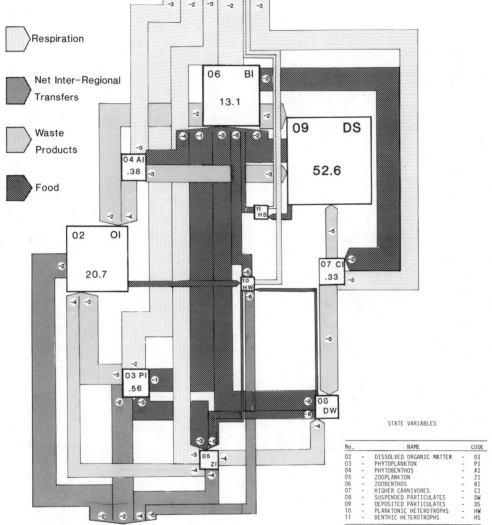

STATE VARIABLES

No.		NAME		CODE
02	–	DISSOLVED ORGANIC MATTER	–	OI
03	–	PHYTOPLANKTON	–	PI
04	–	PHYTOBENTHOS	–	AI
05	–	ZOOPLANKTON	–	ZI
06	–	ZOOBENTHOS	–	BI
07	–	HIGHER CARNIVORES	–	CI
08	–	SUSPENDED PARTICULATES	–	DW
09	–	DEPOSITED PARTICULATES	–	DS
10	–	PLANKTONIC HETEROTROPHS	–	HW
11	–	BENTHIC HETEROTROPHS	–	HS

Figure 9.--Mean annual levels of state variables and processes during 1974, predicted for the Outer Estuary assuming that a barrage had been introduced instantaneously on 1st November 1972. The width of the squares and the width of the arrows, representing the state variables and processes respectively have been drawn in proportion to the logarithm of their magnitude (in arbitrary units of carbon $m^{-2} d^{-1}$) and are labelled with the truncated exponent (e.g. 5.3 10^{-3} as -3).

most important factor in reducing the average biomass of planktonic heterotrophs above the barrage. The net transfer of all other planktonic components of the Outer Estuary is increased because their greater biomass, in each case, more than compensates for the lower dispersion caused by a barrage.

CONCLUSIONS

Although GEMBASE has not yet been validated it may be used to provide qualitative indications of the types of changes which might follow the construction of a barrage in the Severn Estuary. The effect of a barrage on the salinity of the Bristol Channel, below the structure, would be very small (< 1%) except in the region immediately adjacent to it where levels would be elevated by as much as 3%. It is not thought that these salinity changes would have significant effects on ecological processes downstream of the barrage. In the Outer Estuary, upstream of the barrage the salinity would be decreased by as much as 13% on average, which could result in a modification of the boundaries of certain Estuarine-Marine and the Euryhaline-Marine communities (Williams and Collins 1980). However, this would have a relatively slight effect on the overall distribution of these communities and no effect on the breeding populations. In general, GEMBASE predicts that a barrage would have no significant ecological effect in the Outer Bristol Channel (Regions 6 and 7 in figure 1). The higher productivity of the system above the barrage, combined with the increased retention time would lead to a decline in levels of plant nutrients in the Outer Channel but this would not result in any significant decline in primary and secondary production there The greatest impact of a barrage would be upestuary of the structure. Here primary production would stablize (fig. 3), resulting in a seasonal cycle of phytoplankton similar to that presently observable in the Central Channel North. This represents concentrations and rates of production at least an order of magnitude higher than in the unmodified Outer Estuary Region. The model predicts that the very large phytoplankton blooms which could occur in the year following barrage construction would, in subsequent years, be controlled by a combination of increased zooplankton and zoobenthos grazing. Eutrophic conditions would ,therefore, seem unlikely within the main body of the Estuary. Populations of omnivorous zooplankton (fig. 5) would respond very quickly to the increased phytoplankton but, after the first year following barrage construction, settle down to a seasonal cycle about three times higher than the present populations in the Outer Estuary Regions. Even this elevated level is much smaller than the present levels in the Central Channel North. Above the barrage, carnivorous zooplankton, in turn would respond quickly to the increase in omnivores but the system would settle down to a seasonal cycle similar to pre-barrage levels. This is rather surprising because the conditions up-estuary of a barrage are assumed to be rather similar to those in the pre-barrage Central

Channel North Region where levels of carnivorous zooplankton are several times higher than these predictions. The model is therefore suggesting that the similar phytoplankton levels are being controlled in different ways in the two different environments. The fundamental difference between the predicted ecosystem above a barrage, and the pre-barrage Central Channel North system is caused by the changes in the populations of benthic suspension feeders (figure 6). They would exploit the increased primary productivity above the barrage and build populations three or four times more dense than those now found in Swansea Bay, for example. The most consistent prediction of GEMBASE is that a barrage would encourage increased primary and secondary product on up-estuary. The model prediction that the increased secondary production would be benthic rather than planktonic, should be examined through a more detailed study including carefully planned sensitivity analysis of the factors which cause GEMBASE to make this selection. GEMBASE also predicts a gradual increase in the biomass per unit area of deposit feeders above a barrage (fig. 7). It seems unlikely that this trend would reach equilibrium within a three-year period following barrage construction. The increase is sustained by the waste products of secondary production in the water column which, together with faeces produced by the high population of suspension feeders, would enrich the carbon content of the substrate. This latter trend is confirmed in GEMBASE which predicts a small but sustained increase in the concentration of deposited particulate organic carbon above the barrage.

GEMBASE, even in its unvalidated form, has proved to be a useful tool in assessing certain aspects of the environmental impact of a tidal power generating scheme in the Severn Estuary. It can be used in conjunction with other models which provide independent predictions of modifications to dispersion coefficients and concentrations of suspended particulate matter. Interactions between biological and chemical changes should not be ignored but in the Severn Tidal Power study it has been assumed that any potential increase in levels of pollutants caused by a barrage would be counteracted by equivalent reductions in pollutant discharges to the estuary to maintain the status quo.

ACKNOWLEDGEMENTS

The General Ecosystem Model of the Bristol Channel and the Severn Estuary (GEMBASE) is a product of the Estuarine and Near Shore Systems Ecology Group of the Institute for Marine Environmental Research, a component of the UK Natural Environment Research Council. The multidisciplinary team includes Dr. I.R. Joint (microbiology and model development), Dr. P.H. Burkill (plankton), Mr. N.R. Collins (plankton), Dr. J.M. Gee (fish), Dr. A.W. Morris (inorganic chemistry), Dr. R.F.C. Mantoura (organic chemistry), Dr. R.J. Uncles (physics) and Dr. R.M. Warwick (benthos). The author of this paper who acted as systems analyst for the group would like to thank Mrs. K. M.E. Young for valuable assistance in computing and Miss C. Hawke for the graphical presentations. The work received partial financial support from the UK Department of the Environment under contract number DGR 480/48, and from the Department of Energy under contract number E/5A/CON/SB/1598/51/036.

LITERATURE CITED

Department of Energy (U.K.) 1978. The Government reply to the third and fourth reports from the Select Committee on Science and Technology. CMND 7236. H.M. Stationery Office, London.

Financial Times. 1981. European Energy Report, Financial Times Publications, London.

Hoare, A.G. and P. Haggett. 1979. Tidal power and estuary management - a geographical perspective. p. 14-25. In Tidal power and estuary management (eds., R.T. Severn, D. Dineley, and L.E. Hawker). The thirtieth sympsoium fo the Colston Research Society. Scientechnica.

Joint, I.R., and A.P. Pomroy. (in press) Primary production in a turbid estuary. Est. Coast, Shelf Sci. (in press).

Radford, P.J. 1979. Some aspects of an estuarine ecosystem model-GEMBASE. In State-of-the-art in ecological modelling (Ed. S.E. Jorgensen). Proceedings of the ISEM conference on ecological modelling, Copenhagen, 1978. International Society for Ecological Modelling, publ. p. 301-322.

Radford, P.J. 1980. SEVERN TIDAL POWER. Predicted effects of proposed tidal power schemes upon the Severn Estuary Ecosystem. Vol. 1, Water Quality. Institute for Marine Environmental Research, Plymouth. pp. 21. (Department of Energy Contract Agreement No. E/5A/CON/SB/1598/51/036).

Radford, P.J., and I.R. Joint. 1980. The application of an ecosystem model to the Bristol Channel and Severn Estuary. Institute of Water Pollution Control. Annual Conference. Water Pollution Control. Water Pollution Control, 79, 244-254.

Radford, P.J., R.J. Uncles, and A.W. Morris. 1981. Simulating the impact of technological changes on dissolved cadmium distributions in the Severn Estuary. Water Research. (in press).

Radford, P.J., and K.M.E. Young. 1980. SEVERN TIDAL POWER. Predicted effects of proposed tidal power schemes upon the Severn Estuary Ecosystem. Vol 2, Ecosystem Effects. Institute for Marine Environmental Research, Plymouth. pp. 45. (Department of Energy Contract No. E/5A/CON/SB/1598/51/036).

Uncles, R.J., and P.J. Radford. 1980. Seasonal and spring-neap tidal dependence of axial dispersion coefficients in the Severn--a wide vertically mixed estuary. J. Fluid Mechanics, 98 : 703-726.

Warwick, R.M., and R.J. Uncles. 1980. Distribution of Benthic Macrofauna Associations in the Bristol Channel in Relation to Tidal Stress. Mar. Ecol. Prog. Ser. 3: 97-103.

Williams, R., and N.R. Collins. 1980. SEVERN TIDAL POWER. The implications to benthic and planktonic faunas of presumed alterations to salinity regime in relation to the proposed Severn Barrage. Part 2. Institute for Marine Environmental REsearch, Plymouth. pp. 33. (Department of Energy Contract No. E/5A/4012/51/072).

THE IMPACT OF RESERVOIR FLOODING ON A FRESHWATER BENTHIC COMMUNITY[1]

Normand Thérien, Ken A. Morrison, and Bernard Coupal[2]

Abstract.--A dynamic model is presented of the benthos in a sub-arctic region. Two compartments are used: insects and non-insects. The negative effects of flooding are demonstrated by the model, especially the effect of reduced oxygen. Several additional tests are used to validate the model.

INTRODUCTION

Within aquatic ecosystems, the benthic organisms represent a major food source for fish. This group, composed primarily of aquatic insect larvae, mollusks and aquatic worms, inhabits the sediments and may also colonize any submerged surfaces, (e.g., submerged plants and trees). The adults of many important fish species, notably trout, feed heavily on these organisms.

Benthic organisms for the most part have limited mobility, and the major mechanism of movement in a water body is through drifting in water currents. Because of this lower mobility, changes in habitat affect the benthos directly, especially in the case of the creation of a reservoir when flows are greatly reduced as compared to those of a river. In the upper reaches of a reservoir, the habitat will be under continuously varying conditions due to changes in water levels. In the deeper parts of reservoirs, seasonal oxygen depletion may occur where there had been previously an ample supply. Sedimentation in the reservoir can also have a significant effect.

The benthos are difficult and expensive to evaluate quantitatively due to heterogeneous distributions and sampling difficulties, and the identification of the collected organisms is tedious and time-consuming. These problems are undoubtedly among the reasons that lake models reported in the literature have for the most part either avoided the benthos or have given benthic compartments less emphasis than other compartments. In their overview of northern lake models Fox et al.(1979) did not even examine whether various models included benthic compartments or not, while Chaluneau et al.(1978) found that only 5 of the 22 lake models they reviewed had benthic compartments.

The structure of the ecosystem(s) under consideration is also a factor in the desireability of examining benthic dynamics. In the region examined in this paper, the benthos play a crucial role in the aquatic food chain (Boucher 1980).

Because of the importance of the benthos to the fish, and the influences that habitat modifications have on them, an idea of the effect of reservoir creation on the benthos is an important aspect of impact assessment. This paper presents the structuring of a mathematical model of the dynamics of the benthic community residing in the sediments of a newly-created reservoir and lakes in the James Bay region of Quebec. Data are compared to simulated results, and a complete sensitivity analysis of the model is presented. Due to their dominance in the communities examined, most discussion will relate to the insects.

STUDY AREA

The data used in this research have come from the James Bay region of Quebec, Canada (approx. 53°N, 76°W) (fig.1). This area is currently being developed for hydro-electric generation by the Société d'Energie de la Baie James (SEBJ).

This region falls within what is commonly called the taiga, characterized by stunted coniferous trees, bogs, and oligotrophic lakes. It has an annual average temperature of -4°C, with a range in monthly averages of -22°C in January to $+14^\circ$C in July. The growing season is approximately 90 days, starting near the end of May (Magnin 1977).

Three sampling stations are examined in this paper. For the purposes of the data bank of SEBJ these stations have the titles G2129, G2403 and G2405. These names are kept in the text of this paper.

The stations G2403 and G2405 were located in different oligotrophic lakes in the region. These

[1] Paper presented at the international symposium Energy and Ecological Modelling, sponsored by the International Society for Ecological Modelling. (Louisville, Kentucky, April 20-23, 1981).

[2] Normand Thérien and Bernard Coupal are Professors of Chemical Engineering, and Ken Morrison is a Research Assistant in the Department of Chemical Engineering, University of Sherbrooke, Sherbrooke, Québec, Canada.

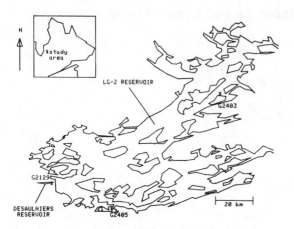

Figure 1.--Map showing location of study area and
sampling stations.

lakes were flooded over in late 1978 by the impoundment of the LG-2 hydro-electric reservoir. The data
that are discussed for these stations were collected
before impoundment of the reservoir.

The station G2129 was located in the Desaulniers River. This station was flooded in May 1977
due to the creation of the small Desaulniers Reservoir. Since this was the first reservoir
created in the region, monitoring was intensive.
The data for this station come from the year prior
to innundation and the two years after innundation.

DATA ANALYSIS

Data were collected in the reservoir from 1976
to 1978 by personnel of the James Bay Energy Corporation (SEBJ). Data from the lake stations were
collected in 1977 and 1978. Although several different instruments were used for aquatic sampling,
we have included only the data collected with
Eckman dredges. This technique retreives a known
surface area of sediment. The other techniques
provide an artificial substrate and indicate the
colonization of this substrate. Further study is
currently in progress on quantitative interpretation of the results of these other techniques.
Emergence data were collected with emergence traps.

Results retreived from the SEBJ environment
data base were reported in collected numbers of
individuals per species. In temperate waters,
numbers are not considered that reliable a measure
to use since the distribution of organisms is
extremely heterogeneous (Edmondson and Winberg
1971), while biomass is far more consistent. An

examination of the results from the reservoir
(sub-arctic) showed the distributions to the extremely heterogeneous. We therefore transformed the numbers into biomass estimates, using mean biomass values from Perron (1979 and Jorgensen et al.(1979).
Total biomass per unit area were then reported as:
B_1: insect biomass, (g m^{-2})

B_2: non-insect biomass, (g m^{-2})

Physico-chemical measures were also taken at
the same sites by the SEBJ personnel, and we obtained
the results from the data base. The data used in
this study were:
OX : dissolved oxygen, (mg ℓ^{-1})
TEMP: temperature, (^{0}C)
CARB: organic carbon (dissolved+suspended), (mg ℓ^{-1})

Analyses and discussion of the experimental
data used in this study can be found in Bachand et
al.(1979) and Cloutier and Dufort (1980).

BIOLOGY OF THE ORGANISMS

Life cycle of aquatic insects

The major insect class found at the stations
examined in this paper is the true insects, Diptera.
Of these, the two major families are Chironimidae
(midges) and Simulidae (black flies). The discussion that follows comes mostly from Oliver (1971)
and applies mainly to the former, as they are the
most numerous in the study area.

There are four distinct phases in the life
cycle, these being egg, larva, pupa, and adult.
The eggs are laid on or in the water by female
adults. The duration of the egg stage is variable,
and seems to be inversely related to temperature.
The larvae hatch from the eggs, and are planktonic
(i.e.,swim in the water) until suitable habitat is
found. They then settle onto or into the substrate,
their penetration being affected by the dissolved
oxygen in the mud. Some species regularly leave
the mud at night and swim in the water. This may
also happen when conditions become anoxic. The duration of the larval stage is usually one year in
temperate and northern latitudes although in some
conditions it may be two years and is affected by
temperature. During this phase of life, most species feed on algae and detritus obtained either by
filtration or directly from the mud, while a few
may be carnivorous.

The pupal stage is relatively short, ranging
from a few hours to a few days. The pupae float
to the surface of the water, where they break the
pupal case and emerge as adults. The adult stage
lasts at most two weeks, during which time they do
not eat, but are occupied mostly with reproduction.
Once the eggs of the female are fertilized, she
lays them on or in the water, and the cycle starts
again.

Life cycles of non-insect benthos

The non-insect benthos found at the stations examined are dominated by aquatic norms (Oligochaetes), clams (Pelecypods), snails (Gasteropods), scuds (Amphipods) and leeches (Hirudines). The distributions of these organisms tend to be extremely patchy.

In contrast to the insects, the members of this group spend their lives entirely in the aquatic habitat, and there is no emergence. Reproduction occurs in most species over the summer, and is strongly influenced by temperature. Most of the species lay eggs, while in the clams the female retains the fertilized eggs until they batch. Development and growth are affected by temperature.

For a more complete discussion of the biology of the various non-insect benthos the reader is refered to Wetzel (1976).

MODEL STRUCTURE

The benthic community is divided into 2 different compartments to recognize the fundamental differences between the insects and non-insects. The biomass of each of these compartments is affected by the biological processes of growth, recruitment, respiration, mortality, and predation. The insects are also affected by emergence (fig. 2).

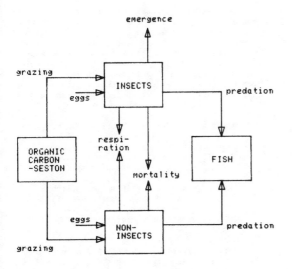

Figure 2.--Flow diagram of the model.

The state variables B_1 and B_2 are the biomass of the insects and non-insects respectively, and the two resulting state equations are:

$$\frac{dB_1}{dt} = G_1 + RE_1 - R_1 - M_1 - P_1 - E \quad (1)$$

$$\frac{dB_2}{dt} = G_2 + RE_2 - R_2 - M_2 - P_2 \quad (2)$$

where,

B_i = biomass of the compartment (i = 1 for insects, i = 2 for non-insects)

G_i = growth rate of biomass of compartment "i"

RE_i = recruitment rate of biomass into compartment "i"

R_i = respiration rate of biomass into compartment "i"

M_i = mortality rate of biomass into compartment "i"

P_i = predation rate on biomass into compartment "i"

E = emergence rate of insects

Biological processes

Growth

Biomass growth is expressed as a linear function of the biomass present. The effects of temperature dissolved oxygen, and organic carbon are considered. Also, non-insects (especially clams) improve the food supply for insects since the faeces of non-insects are more digestible than raw seston (Sephton et al. 1980). Thus the growth of the two compartments is expressed as:

$$G_i = GMAX_i * FOX1_i * FCAR1_i * FTEMP1_i * B_i \quad (3)$$

where,

G_i = growth rate of biomass of compartment "i", ($g^{-2}day^{-1}$)

$GMAX_i$ = maximum relative growth rate of compartment "i", (day^{-1})

$FOX1_i$ = relative effect of oxygen on the growth rate of compartment "i", expressed as:

$$FOX1_i = \frac{OX}{OX + KMO1_i} \quad (4)$$

where,

$KMO1_i$ = half-saturation constant for the effect of oxygen on the growth rate

$FCAR1_i$ = relative effect of organic carbon on the growth rate of compartment "i", expressed as:

$$FCAR1_i = \frac{CARB}{CARB + KMC1_i} * \frac{B_2}{B_2 + KMB1_i} \quad (5)$$

where,

$KMC1_i$ = half-saturation constant for the effect of carbon on the growth rate

$KMB1_i$ = half-saturation constant expressing the effect of the density of the non-insects on the digestibility of the organic carbon (g $^{-2}$)

$FTEMP1_i$ = relative effect of temperature on the growth rate of compartment "i", with the form :

$$FTEMP1_i = \alpha 1_i + \beta 1_i * TEMP \qquad (6)$$

Recruitment

Recruitment is considered as independent from the biomass. For the insects this is a realistic assumption since adults disperse after emergence, and thus egg-laying is fairly homogeneous. This assumption was also used for the non-insects, although for this group it may not be realistic.

Temperature and oxygen both affect development rates for eggs and larvae, and these effects are incorporated. The resulting equations are of the form :

$$RE_i = REMAX_i * FTEMP2_i * FOX2_i \qquad (7)$$

where,

RE_i = rate of recruitment into compartment "i", (g $^{-2}day^{-1}$)

$REMAX_i$ = maximum rate of recruitment into compartment "i" (g $^{-2}day^{-1}$)

$FTEMP2_i$ = relative effect of temperature on recruitment rate with the form:

$$FTEMP2_i = \alpha 2_i + \beta 2_i * TEMP \qquad (8)$$

where,

$FOX2_i$ = relative effect of oxygen on recruitment rate with the form:

$$FOX2_i = \frac{OX}{OX + KMO2_i} \qquad (9)$$

where,

$KMO2_i$ = half-saturation constant for the effect of oxygen on recruitment, (mg ℓ^{-1})

note: $FOX1_i$ and $FOX2_i$ are the same functions.

Respiration

Respiration is taken as a simple linear function of biomass and temperature. It is therefore expressed as:

$$R_i = RMAX_i * FTEMP3_i * B_i \qquad (10)$$

where,

R_i = respiration rate of compartment "i", (g $^{-2}day^{-1}$)

$RMAX_i$ = maximum relative respiration rate of compartment "i", (day^{-1})

$FTEMP3_i$ = relative effect of temperature on the respiration rate of compartment "i", with the form:

$$FTEMP3_i = \alpha 3_i + \beta 3_i * TEMP \qquad (11)$$

Mortality

Two aspects of mortality are considered. The first is natural mortality which is a function of temperature. The second aspect is mortality resulting from anoxic conditions. Thus, mortality, which is a linear function of biomass, is expressed as:

$$M_i = (MT_i + MA_i) * B_i \qquad (12)$$

where,

M_i = mortality rate of compartment "i", (g $^{-2}day^{-1}$)

MT_i = relative effect of temperature on mortality rate of compartment "i", (day^{-1})

MA_i = relative effect of anoxia on the mortality rate of compartment "i", (day^{-1})

The effect of temperature on mortality can be seen as coming from two causes. At very low temperatures, thermal stress will cause some individuals to die. At higher temperatures, bacteria and parasites will be more active and can cause increased mortality. At even higher temperatures, thermal stress could again be a problem, but such high temperature would never occur in the James Bay region. Therefore, a parabolic expression was used for the temperature related mortality as:

$$MT_i = MTMAX_i * |(TEMP - TEMPOPT_i)/(TEMPOPT_i)|^2 \quad (13)$$

where,

MT_i = relative mortality rate due to temperature of compartment "i", (day^{-1})

$MTMAX_i$ = maximum relative mortality rate due to temperature of compartment "i", (day^{-1})

$TEMPOPT_i$ = optimum temperature where $MT_i = 0$

Benthic organisms have a certain tolerance to low oxygen levels, especially chironomids and oligochaetes who have hemoglobin in their blood. However, at very low levels of oxygen (<2 mg ℓ^{-1}) some mortality will result. We therefore used OXMIN as a threshold value, and the resulting function is:

$$MA_i = MAMAX_i * (OXMIN - OX)/OXMIN \qquad (14)$$

when $0 < OX < OXMIN$

otherwise $MA_i = 0$

where,

MA_i = relative mortality rate due to anoxia (day^{-1})

$MAMAX_i$ = maximum relative mortality rate due to anoxia (day^{-1})

$OXMIN$ = threshold oxygen level $(mg\ \ell^{-1})$

Predation

The most important predators are fish, and predation is expressed as a donor-controlled process by the equation:

$$P_i = PMAX_i * FTEMP5_i * FB5_i * B_i \qquad (15)$$

where,

P_i = rate of fish predation on compartment "i"

$PMAX_i$ = maximum relative rate of fish predation on compartment "i", $(gm^{-2}day)$

$FTEMP5_i$ = relative effect of temperature on fish predation rate with the form:

$$FTEMP5_i = \alpha5_i + \beta5_i * TEMP \qquad (16)$$

$FB5_i$ = effect of biomass density of compartment "i" on fish predation with the form:

$$FB5_i = \frac{B_i}{B_i + KM5_i} \qquad (17)$$

where,

$KM5_i$ = half-saturation constant for the predation of fish on compartment "i", (gm^{-2})

Emergence

One of the major factors influencing emergence is temperature (Corbet 1964). Data on emergence were available for station G2129. Using regression and the appropriate transformation, the following equation resulted:

$$E = B_1 * EXP(-10.03+0.49*TEMP) * FTEMP6 \qquad (18)$$

where,

E = rate of emergence $(g^{-2}day^{-1})$

B_1 = biomass of insects, (g^{-2})

$FTEMP6$ = switching function that turns emergence off when the temperature falls below 4^0C

While FTEMP6 causes a discontinuity at 4^0C, the rate calculated at this temperature is quite low and the discontinuity does not constitute a significant numerical perturbation.

Model calibration

Most of the initial parameter estimates were taken predominantly from Patten et al (1975) and Jorgensen et al.(1979). The parameters for emergence were obtained from regression analysis of experimental data. Some parameters were chosen arbitrarily on the basis of qualitative discussions on benthic ecology found in the literature. Initial estimates for the parameters for the temperature functions (α_{ij}, β_{ij}) were taken from linearization of the exponential temperature functions in Patten et al.(1975). Recalibration of some of these parameters was necessary, but this was considered consistent with the adaptations necessary for the benthic community to survive in the hostile northern environments, since most of the parameter estimates came from temperate areas.

Model validation

There is no standard procedure for the validation of simulation models, although much discussion has been directed towards the problem of validation (see McCleod 1980). One approach for validation that is frequently cited is the application of the model to a data set independent of the data to which the model has been calibrated. It is in this direction that we have worked.

Two approaches were used to examine the behaviour of the model under conditions other than those to which it was calibrated. The data taken at station G2129 in the year prior to innundation were recycled for a 3-year simulation to examine what the model would yield in the absence of reservoir impoundment. The model was also applied to the physico-chemical data taken at stations G2403 and G2405 without recalibration. The results of the simulations for these two stations were compared to the limited experimental data available. Also, as a further test two researchers familiar with aquatic invertebrates in northern waters were asked to comment on the realism of the results for these stations.

Sensitivity analysis

Sensitivity analysis is a valuable aid in determining the importance of individual coefficients to the results of a model, and in examining the effects of assumptions used. We used the technique described by Thornton et al. (1979) to examine the sensitivity of the model presented here. The initial values of the compartments and each rate constant in the model were perturbed separately $\pm 10\%$, and the results after perturbation were compared to the results without perturbation over a 3-year period. For emergence, the entire rate was varied by $\pm 10\%$. This comparison was made with the sensitivity of a compartment expressed as:

$$S_{ij} = \frac{RP_{ij}}{RBASE_i} \qquad (19)$$

where,

S_{ij} = sensitivity of compartment "i" to a 10% change in parameter "j"

RP_{ij} = perturbed response of compartment "i" given a 10% change in parameter "j"

$RBASE_i$ = unperturbed response of compartment "i"

If a 10% change in a parameter did not result in at least a 10% change in one of the compartments (ie $0.9 < S_{ij} < 1.1$) then the model was not considered sensitive of this parameter. If the parameter perturbation caused more than a 10% change in one of the compartments (i.e., $S_{ij} < 0.9$ or $S_{ij} > 1.1$) then the model was considered sensitive to this parameter.

RESULTS

Baseline simulation

Station G2129

The results of the simulation are compared to the experimental results in figure 3.

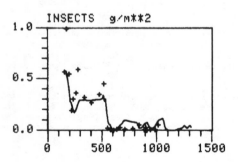

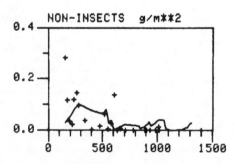

Figure 3.--Observed and simulated results for station G2129. Day 1 is 1/1/76.

The essential trends of the insect biomass compartment are aptly reproduced by the model and the impact of the reservoir impoundment can be seen. The model does not perform as well for the non-insect compartment. Since there are no apparent trends in the biomass of this compartment, variability of the data probably overshadows the dynamics of this compartment, and little more can be said. The temperature, carbon, and oxygen data used as input into the model are shown in figure 4.

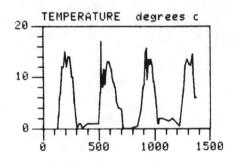

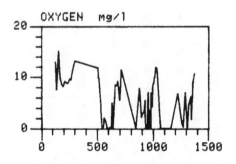

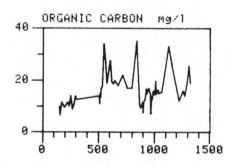

Figure 4.--Driving variables at station G2129. Day 1 is 1/1/76.

Station G2129

When the model was run cycling the physico-chemical data from the year prior to inundation the model gave cyclic response for both compartments as can be seen in figure 5. This is consistent with what should be occuring in unperturbed conditions

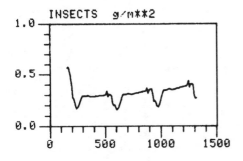

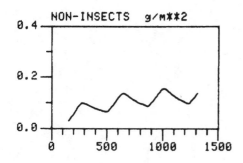

Figure 5.--Simulated results at station G2129 recycling the observed driving variables for the year prior to inundation.

Stations G2403 and G2405

The model gave cyclic responses for both of these stations when run with the two years of pre-data (fig. 6). Since the observations are only from the summer period, the major comment that can be made about the results is that they are in the same order of magnitude as the experimental results in the summer. The winter results

were considered consistent with the life cycles of the aquatic insects in nothern oligotrophic lakes [1,2].

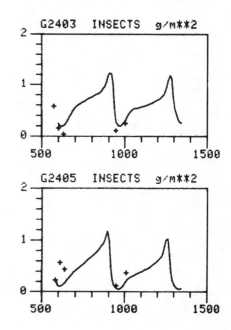

Figure 6.--Observed and simulated results for insects for the two lake stations, G2403 and G2405. Day 1 is 1/1/76.

Sensitivity analysis

Because the two compartments are connected only through detritus processing by the non-insects before ingestion by insects, neither compartment was sensitive to changes in the other compartment. Changes in the non-insect compartment did effect the insect compartment, but not to a significant extent. Therefore, the sensitivity of the model to a parameter was limited to the compartment on which that parameter had a direct effect, and the sensitivities were examined by compartment. The sensitivity trajectories over 3 years are shown in figure 7 for the parameters to which the model was sensitive. The plots commence at 0^0 (the intercept of the circle and the horizonal axis on the right). The plots proceed counter-clockwise

[1]Johnston, Murray. February 1981. Personal communication. Environment Canada, Canada Center for Inland Waters, Burlington, Ontario, Canada.

[2]Harper, Pierre-Paul. February 1981. Personal communication. University of Montreal, Dept. of Biology, Montreal, Quebec, Canada.

RESPONSE

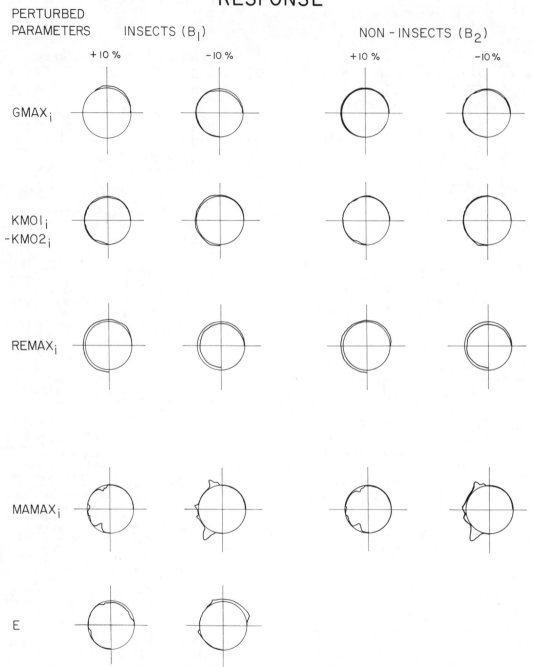

Figure 7.--Sensitivity trajectories for parameters to which the model is sensitive.
Departure from the unit circle indicates sensitivity.
Each 90° is one year, and the trajectories are plotted counterclockwise starting at 0°.

for 270^0, each 90^0 corresponding to 1 year. The circle is a unit circle, and when the line is on the circle, this indicates no difference between the perturbed and baseline values for that compartment.

The model was insensitive to the majority of the rate constants, and was also insensitive to initial values. Except for the recruitment terms, the parameters to which the model was sensitive did not cause a permanent change in the state variables, but instead caused short-term deviations from the baseline values. The sensitivity to KMO1-KMO2 was undoubtedly through recruitment. The recruitment terms caused permanent divergence of perturbed values from the baseline values.

Parameters relating to oxygen effects had consistently important effects. Of these the parameters to which the model was most sensitive were the rate constants for mortality resulting from anoxia. These caused very noticeable deviations from the baseline response, but as already stated, these deviations were only short term.

The model was sensitive to emergence mostly in the first year before inundation. This is of importance in terms of using the model in unperturbed situations.

DISCUSSION

The initial run of the model reproduces the basic trends shown by the experimental data at station G2129. The summer of the flooding of the reservoir the biomass plunges to an extremely low value, from which it never really recovers. If one examines the input variables (fig. 4) it becomes clear that oxygen is the variable really influencing the drop in the insect and benthic compartments. Temperature values are not noticeably altered by the innundation, while the amount of carbon actually increases due to the suspension of plant litter. Oxygen is however greatly decreased by the inundation, and although it does recover to higher values, it can be seen that anoxia becomes a common occurrence.

The run for this station recycling the pre-physico-chemical data shows more or less cyclic behaviour. The biomass drops from the initial value, but thereafter cycles with an increasing trend. This demonstrates that the inundation did have a negative effect on the insect population. It also demonstrates that the results obtained from the initial run of the model are not just artifacts of the parameter values or the model structure. The results for the non-insect compartment are less satisfying, but extreme heterogeneity of the population distributions make the experimental data somewhat unreliable.

In this region the major component of the insect community is chironomids. Aside from being able to withstand lower oxygen levels, many species have the ability to leave the sediments if condi-

tions become too hostile. They take up a planktonic life until the conditions again become favorable, or until they drift to a different area where conditions are favorable.

The oxygen data used for this model were taken just above the sediment surface. In the first summer after inundation, the entire water column went through anoxic periods. The qualitative date taken at this time showed little colonization of surfaces in the water column. During the second summer after inundation anoxia was limited to the lower part of the water column and some limited colonization in flooded areas was noted. These trends increased noticeably in 1979. Thus it must be noted that the model only reflects the community in the sediments. Also, part of the effect that has been taken as mortality due to anoxia is in fact just the migration of some chironomids out of the sediments and into the column.

The model results are still important in terms of food supply for some species of fish, notably trout. In the Desaulniers reservoir trout are limited to the areas above what used to be the old riverbed (Boucher 1980), and the same phenomenon has been noted for various species in Russian reservoirs (Gordeev and Permitin 1968). Thus, colonization of inundated surfaces any distance from the old riverbed will not represent food available for trout. Since the riverbed is the deepest part of the reservoir, it is there where anoxia will occur. Thus, the benthic populations will remain low in that part of the reservoir.

The results of the model from the other two stations G2403 and G2405 are cyclic much as for the second run for station G2129. There are differences, however. Growth continues at these other stations throughout winter, although the growth rate is lower than in the fall or spring. This is most probably due to temperature, since the bottom temperature in these two lakes does not go much below $4^0 C$, while in the Desaulniers reservoir the temperature can go as low as $0.5^0 C$. The results are qualitatively reasonable if no migrations are occuring over winter. Such continued growth under ice cover has been found in an arctic lake (Welch 1976).

The model results have implications for the eventual impact of the flooding of the much larger LG-2 reservoir in the same region. This reservoir was flooded over the winter, and as a result the water is extremely cold. Decomposition of inundated material is very slow, and anoxia has not been observed at all. In fact, lake trout have moved into the reservoir, pointing out that the reservoir has the major characteristics of an oligotrophic lake. Thus it is unlikely that the benthic community will be severely influenced by the inundation Qualitative data on colonization in the LG-2 reservoir has not shown the same depression as was seen in Desaulniers.

The model was in general sensitive to the effects of oxygen. This is consistent with what one would expect in northern waters. The natural unperturbed conditions are oxygen-rich running waters and oxygen-rich oligotrophic lakes, and it is to these conditions that the communities have adapted. Anoxic conditions after innundation present a situation to which the natural communities are not adapted, and strong effects can be expected. Given that all biological processes are slow in northern conditions, adaptation to occasional anoxia could take a long time.

Although the steps we have taken to validate the model may not be as vigorous as one might wish we have shown that the model is robust and gives reasonable results in several different situations. Thus we consider that the model has acceptable validity for sub-arctic fresh water bodies. Some recalibration would undoubtedly be necessary to apply the model to other regions where climatic conditions and species composition are different.

CONCLUSIONS

The model that has been constructed gives a reasonable picture of the biomass dynamics of the benthic community, more so for the insects than for the non-insects. The model indicates that in the Desaulniers reservoir a reduction in the biomass of the benthic community resulted from anoxia associated with the flooding of the reservoir. However, the scope of the model is limited to the organisms actually in the sediments. Complementary data have suggested that colonization of other flooded surfaces (e.g., inundated trees) commenced the year after flooding, although these data are only qualitative. However, this colonization does not represent a food source for trout since these fish stay in the areas of the reservoir above the old riverbed. Thus the reduction in the biomass of the benthic community means a significant decrease in the food supply for the trout.

The drop in oxygen that occured in the Desaulniers reservoir after flooding has not yet been observed in the much larger LG-2 reservoir, which was flooded over winter. Thus the immediate impact on the benthic community is probably reduced, although no data are presently available. It remains to be seen if the reservoir does develop anoxic periods after it has had several years to warm up.

ACKNOWLEDGEMENTS

Funding for this research was provided by a grant from the Direction Environment of the Société d'Energie de la Baie James (SEBJ). Helpful suggestions, fruitful discussions, and help with data analysis came from Roger Lemire, Dominique Roy, Alain Bachand, and Haider Zaidi, all of SEBJ, and we are grateful for their assistance.

LITERATURE CITED

Bachand, C.A. et al. 1979. Réseau de surveillance écologique du Complexe La Grande. Direction Environnement, Société d'Energie de la Baie James, Montreal. 250 p.

Boucher, R. 1980. Etudes écologique - Réservoir Désaulniers: Poisson. Direction Environnement Société d'Energie de la Baie James, Montreal, 135 p.

Chaluneau, F., S. des Clers, and J.A. Meyer. 1978. Les modèles de simulation en écologie lacustre - présentation des différentes approches et analyse des modèles existants, p. 355-389. In Comptes rendus, colloque sur les lacs naturels, 18-21 Sept., 1978. Assoc. Franç. de Limnologie and Assoc. Franç. pour la Protection des eaux. Paris, France. 421 p.

Cloutier, L. and F. Dufort. 1980. Etudes écologique -Réservoir Désaulniers. Communautés benthique et insectes adultes. Direction Environnement, Société d'Energie de la Baie James, Montréal, 200 p.

Corbet, P. 1964. Temporal patterns of emergence in aquatic insects. Can. Entomol. 96:264-279.

Edmondson, W. and G. Winberg. 1971. A Manual on Methods for the Assessment of Secondary Productivity in Fresh Waters. I.B.P. Handbook no. 17, Blackwell Scientific Pub., Oxford.

Fox, P., J. LaPierre, and R. Carlson, 1979. Northern lake modeling: a literature review. Water Resour. Res., 15(5):1065-1072.

Gordeev, N. and I. Permitin. 1968. On the dynamics of species composition and numerical strength of fishes in typical biotopes of Rybinsk Reservoir, p. 146-169. In Kuzin, B. (ed.), Biological and Hydrological Factors of Local Movements of Fish in Reservoirs, U.S.S.R. Academy of Sciences, Leningrad (Trans. from Russian, 1974, Amerind Publ. Co., New Delhi, 389 p.)

Jorgensen, S. et al (ed.). 1979. Handbook of Environmental Data and Ecological Parameters. I.S.E.M., Copenhagen. 1162 p.

Magnin, E. 1977. Ecologie des eaux douces du territoire de la Baie James. Service Environnement, Société d'Energie de la Baie James, Montreal. 454 p.

Mcleod, J. 1980. All about validation. Simulation 35(1):vii-x.

Oliver, D. 1971. Life history of the chironomidae. Ann. Rev. Entomol. 16:211-230.

Patten, B. et al. 1975. Total ecosystem model for a cove in Lake Texoma, p 206-421. In, Patten, B. (ed.), Systems Analysis and Simulation in Ecology -Vol III. Academic Press, New York. 601 p.

Perron, F. 1979. Détermination de poids moyens pour les organismes benthiques et les insectes adultes du territoire de la Baie James. Direction Environnement, Société d'Energie de la Baie James, Montreal. 14 p.

Sephton, T., C. Paterson and C. Fernando. 1980. Spatial interrelationships of bivalves and nonbivalve benthos in a small reservoir in New Brunswick, Canada. Can. J. Zool. 58: 852-859.

Thornton, K., A. Lessem. D. Ford, and C. Stirgus. 1979. Improving ecological simulation through sensitivity analysis. Simulation, 32(5): 155-166.

Welch, H. 1976. Ecology of Chironomidae (Diptera) in a polar lake. J. Fish. Res. Board Can. 33:227-247.

Wetzel, R. 1976. Limnology. W.B. Saunders, Toronto. 743 p.

APPENDIX

Table 1.--Coefficient values

COEFFICIENT	VALUE	UNITS
$GMAX_1$	00.065	day^{-1}
$GMAX_2$	00.025	day^{-1}
$KMO1_1$	03.50	$mg\ \ell^{-1}$
$KMO1_2$	00.500	$mg\ \ell^{-1}$
$KMC1_1$	16.0	$mg\ \ell^{-1}$
$KMC1_2$	30.0	$mg\ \ell^{-1}$
$KMB1_1$	00.150	gm^{-2}
$KMB1_2$	00.000	gm^{-2}
$REMAX_1$	00.005	day^{-1}
$REMAX_2$	00.001	day^{-1}
$KMO2_1$	03.50	$mg\ \ell^{-1}$
$KMO2_2$	00.500	$mg\ \ell^{-1}$
$RMAX_1$	00.004	day^{-1}
$RMAX_2$	00.004	day^{-1}
$MTMAX_1$	00.002	day^{-1}
$MTMAX_2$	00.002	day^{-1}
$TEMPOPT_1$	15.0	^0C
$TEMPOPT_2$	10.0	^0C
$MAMAX_1$	00.100	day^{-1}
$MAMAX_2$	00.100	day^{-1}
$OXMIN$	02.00	$mg\ \ell^{-1}$
$PMAX_1$	00.004	day^{-1}
$PMAX_2$	00.004	day^{-1}
$KM5_1$	00.05	g^{-2}
$KM5_2$	00.05	g^{-2}

Table 2.--Coefficients for dimensionless linear temperature functions of the form: $FTEMP = \alpha + \beta * TEMP$

FUNCTION	α_{ij} (dimensionless)	β_{ij} $(^0C^{-1})$
$FTEMP1_1$	0.10	0.05
$FTEMP1_2$	0.0	0.06
$FTEMP2_1$	0.0	0.05
$FTEMP2_2$	0.0	0.05
$FTEMP3_1$	0.18	0.05
$FTEMP3_2$	0.22	0.04
$FTEMP5_1$	0.04	0.06
$FTEMP5_2$	0.04	0.06

WATER QUALITY MODELING IN A SMALL STREAM: PROBLEMS, CONSIDERATIONS, AND RESULTS[1]

Lalit K. Sinha[2]

Abstract.-- QUAL II calibrated well to the measured values of dissolved oxygen, carbonaceous biochemical oxygen demand and chlorophyll a. A combination of reaeration equations and engineering judgements were necessary to achieve calibration. A way to incorporate the effects of diurnal variations on dissolved oxygen was developed. The model did not calibrate well to the measured values of nutrients such as ammonia and nitrate nitrogen and dissolved phosphorus. Twenty-one days were found inadequate to determine ultimate biochemical oxygen demand in the laboratory using unseeded stream water samples. Also, these curves did not follow the classical exponential type BOD decay equation. The calibrated QUAL II was applied to predict the dissolved oxygen profiles resulting from different effluent loadings from a sewage treatment plant in the study area.

INTRODUCTION

Quite frequently in Illinois sewage treatment plants requiring upgrading to provide advance secondary treatment levels discharge to small streams, where effluents are the primary source of stream flow under low flow conditions. Unless this level of treatment is provided these small streams may either have substandard water quality or as tributaries, they may be main cause of substandard water quality in a portion of major streams. To adequately address these problems and other water quality issues in the State emanating from complex inter-relationships among man, environment, energy, and cost of control; the Illinois Environmental Protection Agency is currently accessing through a remote terminal one water quality modeling package, QUAL II, which is available on the United States Environmental Protection Agency's (USEPA) TYMNET system in North Carolina.

The purpose of this paper is to discuss, with a case study conducted in a small stream in Central Illinois, the problems,

considerations and findings of a water quality modeling effort. Problems in instrumentations, considerations in resolving conflicting data and a sensitivity analysis of major model parameters affecting dissolved oxygen in the stream are discussed.

THE MODEL

The model refers to QUAL II/SEMCOG VERSION, developed by the Water Resources Engineers, a Camp Dresser and McKee firm, available on the USEPA TYMNET system in North Carolina. The theoretical details, computer programs, and operating instructions are documented in USEPA (1977, 1977). The complete equation for dissolved oxygen computation in this model is as follows:

$$\frac{dO}{dt} = K2\ (OSAT-O) + (L3m - L4\ p)A - K1L - K4/AX - L5B1N1 - L6B2N2 \qquad (1)$$

where,

O = the concentration of dissolved oxygen

OSAT = the saturation concentration of dissolved oxygen at the local temperature and pressure

L3 = the rate of oxygen production per unit of algae (photosynthesis)

L4 = the rate of oxygen uptake per unit of algae respired

L5 = the rate of oxygen uptake per unit of ammonia oxidation

[1]Paper presented at the International Symposium on Energy and Ecological Modelling. (Louisville, Kentucky, U.S.A. April 20-23, 1981)

[2]Lalit K. Sinha, Illinois Environmental Protection Agency, Springfield, Illinois

L6 = the rate of oxygen uptake per unit of nitrite nitrogen oxidation

K2 = the reaeration rate in accordance with Fickian diffusion analysis

K4 = constant benthic uptake rate

K1 = the rate of decay of carbonaceous biochemical oxygen demand

B1 = the rate constant for the biological oxidation of ammonia nitrogen

B2 = the rate constant for the oxidation of nitrite nitrogen

m = the local specific growth of algae

p = the local respiration rate of algae

A = algal biomas concentration

L = the concentration of carbonaceous biochemical oxygen demand

N1 = the concentration of ammonia nitrogen as nitrogen

N2 = the concentration of nitrite nitrogen as nitrogen

AX = average cross-sectional area

t = time

STUDY AREA

The study area was located in Champaign County in Central Illinois and consisted of 1.2 river miles of Copper Slough and 10.1 river miles of Kaskaskia River. These two streams, which flow through one of the most productive agricultural areas in the State, were channelized and characterized by steep grass and weed banks, sand and gravel bottoms, numerous field tile discharges and virtually no trees. Upstream from the Champaign-Urbana southwest sewage treatment plant, Copper Slough received stormwater runoff from the west side of the City of Champaign. This area was mostly medium to high income residential and included an eighteen-hole golf course. Under dry weather flow conditions almost entire flow in Copper Slough is made up of effluent discharged from the Champaign-Urbana southwest sewage treatment plant (STP). The STP was designed for a capacity of five million gallons per day (mgd). The seven-day ten-year low flow in Kaskaskia river upstream from its confluence with Copper slough was 0.05 cubic feet per second (cfs).

Figure 1 is a stick diagram, not drawn to the scale, showing the two streams, STP, 17 sampling locations and other pertinent information. Of the 17 sampling sites, eight were located in Copper Slough and nine were in Kaskaskia River. One sampling location was at a bridge site in Copper Slough while all sampling locations in Kaskaskia River were at bridge sites one to two miles apart. The

sampling site 6 was in Copper Slough just upstream from its confluence with Kaskaskia River.

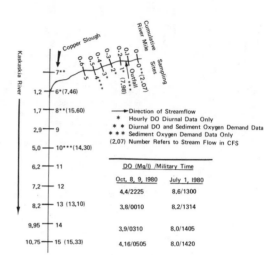

Figure 1. -- Stick diagram and pertinent information.

FIELD DATA COLLECTION

The field data collection program was conducted during the summer and early fall of 1980. Three separate sampling trips were made to the project area. However, the one conducted on October 8-9, 1980 was most comprehensive. On this day, dye (rhodamine WT) was dumped in Copper Slough immediately below the STP outfall. At every sampling location, water quality samples and in situ measurements of pertinent constituents were taken when the peak concentration of dye was reached. The underlying assumption was that the peak concentration of the dye at every site represented the average time of travel for the reach through that site and thereby an effluent plug was being followed. The dye was dumped at 11:15 a.m. on October 8, 1980 immediately below the STP outfall and sampling was finished at 5:05 a.m. on October 9, 1980 at the last sampling location in Kaskaskia River. This involved travelling 10.65 miles in 17 hours and 50 minutes or at an average velocity of 0.875 foot per second (ft/sec).

Water samples collected were analyzed for various chemical constituents including long term (21 day) biochemical oxygen demand, chlorophyll a, and phytoplankton. The 21 days biochemical oxygen demand analysis of unseeded water samples were performed with a hope to get some reliable estimates of the carbonaceous and nitrogenous decay coefficients. Two biochemical oxygen demand bottles for each sampling site were set up; one with N-Serve as an inhibitor for nitrogen demanding bacteria and another without N-Serve. The oxygen used in both bottles were read on a daily basis for 21 days. The nitrogenous curve was obtained by taking a difference between the total and carbonaceous biochemical oxygen demand readings every day.

Also, diurnal data at one-hour intervals for dissolved oxygen, temperatue, pH, conductivity and redox potential were taken on July 23-24, 1980 at six sites. Three sites were in each Copper Slough and Kaskaskia River. These sites were identified in figure 1.

The sediment oxygen demand (SOD) data were collected at one location in Copper Slough on July 1, 1980. On July 24, 1980, SOD data were collected at three locations in Copper Slough and at two locations in Kaskaskia River. SOD data collected at two sites in Copper Slough on October 8, 1980 were inconclusive. The SOD chamber at any site in the stream was kept for a period of approximately four hours. The sites for SOD measurements were shown in figure 1. 1.

PROBLEMS, CONSIDERATIONS AND THEIR RESOLUTION

The problems faced were numerous, some of which might not be uncommon when dealing with small streams. Therefore, instead of discussing each of the problems, a few major ones were highlighted under three categories; stream features and characteristics, instrumentations, and data.

Stream Features and Characteristics

Common to both the streams, Copper Slough and Kaskaskia River, was that they were losing streams (fig. 1). As measured on October 8, 1980, Copper Slough lost 0.52 cfs within a stretch of 1 mile, while Kaskaskia River lost 2.5 cfs within a stretch of 6.5 miles. However, Kaskaskia River gained 2.23 cfs immediately within next 2.55 miles. This trend was confirmed by remeasuring the stream flows on October 27, 1980. Such a trend was in agreement with findings of Singh and Stall (1973).

In Copper Slough, a considerable amount of weed outcroppings and islands were observed in the middle of the stream while walking it on foot and driving along its bank. At one location the island was of the size and nature, which created constriction to almost cause a discontinuity in flow. Also, there were a considerable number of pools and riffles.

The Kaskaskia River was neither walked on foot nor did it lend itself to windshield survey due to lack of driving path along its banks. However, a visual observation up- and down-stream from the bridge sites suggested that, though channelized, it contained some pools and riffles.

Instrumentations

Major problems arose in two areas; hydraulic characteristics such as measurement of stream flow, velocity and cross-sections, and in situ measurements of sediment oxygen demand (SOD).

Hydraulic Characteristics

The cross-sections were taken at two occasions; July 1 and October 8, 1980. At each occasion the measurements were taken by different field crews. On July 1, 1980, the cross-sections were taken by measuring the depth of water in the streams at five feet intervals. On October 8, 1980, they were taken by measuring the depth of water in the streams at one foot intervals. These two measurements resulted in substantially different cross-sections at most of the sites. As an example, the cross-sections taken at Site 13 in Kaskaskia River were illustrated in figure 2. Such a great variation in cross-section measurements greatly impact flow and velocity through the reach; especially in small streams under low flow conditions.

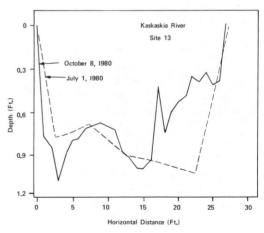

Figure 2. - - Variations in channel cross-sections.

Sediment Oxygen Demand

Measuring sediment oxygen demand in these small streams was constrained by depth of stream flow and the height of SOD chamber. The SOD chamber, which was approximately 10 inches high, must be fully submerged in water to operate properly. As a result most of the measurements had to be taken in pooled areas. The shallow depth coupled with sand and gravel in the stream bed caused substantial problems in creating a good seal between stream bed and the SOD chamber. The SOD chamber built and used was that designed by Butts and Evans (1978).

Data

Major problems arose in two areas. One was in determination of decay coefficients from long term biochemical oxygen demand data (BOD) obtained from the laboratory analysis. Another was the insufficiency of data caused by inability to pinpoint the locations for SOD and diurnal measurements.

Determination of BOD Decay Coefficients

Figure 3 is a plot of total, carbonaceous and nitrogenous BOD for 21 days using unseeded stream water samples on site 11 in Kaskaskia River. Any of these three curves in its entirety did not follow the general equation used to determine the decay coefficients. The equation is:

$$Y = L(1-e^{-kt}) \qquad (2)$$

where,

Y	= oxygen used
L	= ultimate oxygen demand
e	= exponential base, e
k	= decay coefficient
t	= time, days

However, equation 2 was fitted to the total, carbonaceous and nitrogenous BOD for the time periods starting from 5th, 11th and 4th day, respectively. The curve fitting was done using a nonlinear regression program available in the Statistical Analysis System User's Guide (1979). This nonlinear program is based on Marquardt method. The detail theory underlying Marquardt method is given by Draper and Smith (1967). The correlation coefficient, R^2, value associated with the portions fitted for total, carbonaceous, and nitrogenous BOD was 0.99 in each case. The decay coefficients thus obtained for sites, where the curves could be

fitted, were assumed to be valid for entire curve.

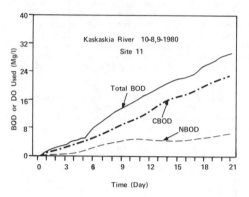

Figure 3. - - Biochemical oxygen demand versus time at 20° centigrade.

Insufficiency of Data

A few preliminary data collection efforts were made prior to the intensive water quality sampling conducted on October 8 and 9, 1980. During the preliminary efforts random measurements of dissolved oxygen at all sites were taken on July 1, 1980. Also, as mentioned in previous sections, SOD and diurnal dissolved oxygen (DO) data were collected at several locations in Copper Slough and Kaskaskia River. A major underlying objective of the preliminary efforts was that these data would help explain some of the variations that might be observed during intensive sampling program.

The intensive sampling program conducted on October 8 and 9, 1980 revealed a sudden drop in measured stream DO values (fig. 1) at sites 12 through 15. Such a phenomenon could not be explained with data collected during the preliminary and intensive sampling. However, the model, during the calibration process, suggested that this drop could be due to a heavy SOD sink in that area, sites 12 through 15. This suggestion by the model would need to be verified by making field measurements of SOD.

Also, the stream DO values taken during intensive sampling program at sites 12 through 15 were adjusted for diurnal variations from the measurements taken at site 8, which is 5.5 to 9.05 miles upstream from the sites in question. Whether adjustments were appropriate could be justified only from field measurements of diurnal DO variations at one or more of the sites 12 through 15.

RESULTS

The model was used in a steady state mode and the results were discussed under three major categories; the model calibration, the model application, and sensitivity analysis.

The Model Calibration

Calibration consisted of a process in which a very close scrutiny of the model results were performed by people involved in the field sampling programs. Also, a considerable amount of educated and engineering judgements, based on personal knowledge of the streams, were employed. The overall goals of the model calibration process were:

1. To use all measured values.

2. To place heavy reliance on the measured values.

3. To approximate as closely as possible the stream DO values measured on October 8 and 9, 1980.

4. To get a good feel of the stream reaeration.

5. To approximate as closely as possible the other measured parameters like long term carbonaceous biochemical oxygen demand, ammonia nitrogen, nitrate nitrogen, phosphorus and chlorophyll a.

6. To use the calibrated model to predict DO profiles under seven-day ten-year low flow conditions.

In a stepwise calibration process, it was felt that the measured dissolved oxygen values need to be adjusted for diurnal variations at the site upstream from the STP outfall and at all sites in Kaskaskia River. Based upon the hourly measurements of diurnal dissolved oxygen values, three adjustment curves were developed. They are presented in figure 4.

From observed hourly diurnal data, it was assumed that a constant peak or valley occurred between 11 a.m. and 5 p.m and between 10 p.m. and 6 a.m. A linear drop or rise was assumed between 5 p.m. and 10 p.m. and between 6 a.m. and 11 a.m. Also, between 8:30 a.m. and 7:30 p.m., the sunlight effect was assumed predominant. Therefore, the observed DO values during this period were adjusted downward. The observed DO values between 7:30 p.m. and 8:30 a.m. period were adjusted upward assuming predominantly darkness effects. The maximum variations in diurnal DO values were 8.8, 1.2 and 2.0 mg/l, at sites 0, 7 and average of sites 8 and 10, respectively. One-half of these values was used to adjust measured DO values during the daylight effects and another half for the darkness effects. Average of sites 8 and 10 was used to adjust all sites in Kaskaskia River downstream from its confluence with Copper Slough. No adjustments in measured DO values in Copper Slough at sites STP through 6 were made. This was because of an assumption that stream flow in this portion of Copper Slough under seven-day ten-year low flow condition would consist primarily of effluent from the STP.

O'Connor and Dobbins (1958) reaeration equation was used in Copper Slough while Tsivoglou and Wallace (1972) reaeration equation was used in Kaskaskia River with an exception in the last stream reach, where the reaeration coefficient value was set by hand at 3.2. Figure 5 contains a plot of measured, adjusted and calibrated values of dissolved oxygen. The calibrated dissolved oxygen values, with an exception at site 9 in Kaskaskia River, approximated the adjusted values quite well.

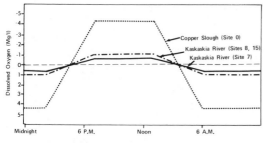

Figure 4. - - Adjustment curves for diurnal variations.

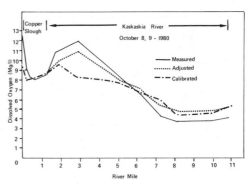

Figure 5. - - Comparison of measured, adjusted, and calibrated dissolved oxygen.

The calibration results for carbonaceous BOD, ammonia nitrogen, nitrate nitrogen, and chlorophyll a were presented in table 1. The calibration for 21 day carbonaceious biochemical oxygen demand was within 21 percent of the measured values. Approximately 60 percent of the values were within 10 percent difference and remaining 40 percent differed by more than 10 percent. In the case of ammonia and nitrate nitrogen, the calibration was good in Copper Slough. In Kaskaskia River, dramatic drops in ammonia nitrogen values could not be explained with data in hand as well as the

knowledge of existing field conditions. The calibrated and measured values of chlorphyll a were mostly within 20 percent with a few exceptions between 30 and 40 percent. At sites 1 and 3 in Copper Slough, the stream concentration of chlorphyll a was considered to be in error.

The Model Application

The calibrated model was applied to predict receiving stream DO profiles under seven-day ten-year low flow conditions within the study

Table 1.—Comparison of measured and simulated values of 21 day carbonaceous BOD, ammonia nitrogen, nitrate nitrogen and chlorophyll a

Site Number	21 day C BOD, mg/l		Ammonia nitrogen mg/l		Nitrate nitrogen mg/l		Chlorophyll a mg/l	
	Measured	Simulated	Measured	Simulated	Measured	Simulated	Measured	Simulated
COPPER SLOUGH								
0	7.62	7.62	<0.1	0.03	0.3	0.31	1.5	1.52
STP	28.9	--	3.8	--	4.8	--	2.8	--
1	28.1	23.35	2.9	2.81	4.0	3.63	1.4	2.48
2	--	23.32	--	2.81	--	3.63	--	2.50
3	26.2	23.28	2.8	2.80	4.4	3.63	1.4	2.52
4	24.4	23.25	2.1	2.80	4.3	3.63	2.8	2.57
5	24.0	23.22	2.7	2.79	4.4	3.63	2.9	2.64
6	25.2	23.08	2.3	2.77	4.6	3.63	5.6	2.95
KASKASKIA RIVER								
7	18.3	18.29	0.8	0.8	<0.1	0.10	1.4	1.41
8	21.6	20.43	1.3	1.72	2.9	1.79	4.0	2.34
9	18.7	20.24	0.9	1.69	3.1	1.80	4.2	2.84
10	16.7	19.64	0.5	1.63	3.2	1.82	5.9	3.89
11	23.1	19.48	0.4	1.60	3.3	1.83	--	4.49
12	23.5	19.33	0.3	1.59	3.2	1.84	4.4	5.10
13	21.3	19.17	0.2	1.57	3.2	1.86	6.1	5.94
14	21.8	17.24	0.1	1.41	3.1	1.72	6.1	6.79
15	16.8	16.20	0.1	1.33	3.1	1.64	7.3	7.00

area by assuming various effluent loadings from the Champaign-Urbana southwest sewage treatment plant (STP). The STP was assumed to have a design maximum flow of 8.0 cfs with the effluent dissolved oxygen concentration and temperature of 6 mg/l and 30°C, respectively. Dissolved oxygen profiles for five alternatives, as shown in figure 6, were predicted. With the exceptions of alternatives 4 and 5, there appeared to be a severe drop in the stream dissolved oxygen concentration at the lower end of the study area. This drop was closely related to a dissolved oxygen sink created in reaches 11 and 12 due to placement of heavy sediment oxygen demand. In alternative four, this sink was reduced by half and the dissolved oxygen concentration improved substantially. However, the DO concentrations at the lower end of the study area were still below Illinois general use water quality standard of 6 mg/l and at site 13 the DO value was 2.17 mg/l. When SOD sinks were completely eliminated in alternative five, the Illinois general use water quality standard for DO was marginally violated at few sites in Kaskaskia River. However, in alternative four, at site 13, the DO concentration of 2.17 mg/l created two-fold concerns. One concern was that if SOD sink really existed, can the stream DO be improved in that area? A second concern was that if the SOD sink did not exist, the current cali-bration of the model would need to be updated.

Sensitivity Analysis

A sensitivity analysis was conducted by changing the parameter values. The parameters, one at a time, were first doubled then halved. The impacts of such changes on various aspects of model outputs were studied. Sensitivity analysis for several major model parameters were performed. As an example the effects of changes in headwaters flow and sediment oxygen demand on dissolved oxygen profiles in the study area were presented in figure 7. All others are available in the Illinois Environmental Protection Agency files. It could be seen that effects of these parameters on DO profiles were more substantial in Kaskaskia River than in Copper Slough. Also, it appeared from Curves 2 and 5 in figure 7 that the effects on DO profile of doubling the headwaters flow and of reducing the sediment oxygen demand in half were almost the same in lower end of the study area. Among all parameters studied, an increase or decrease in headwaters flow appeared to be dominant in increasing or decreasing every water quality constituent evaluated under sensitivity analysis.

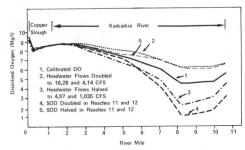

Figure 7. - - Effects of headwaters flow and sediment oxygen demand on dissolved oxygen.

1. Calibrated DO
2. Headwater Flows Doubled to 16,28 and 4,14 CFS
3. Headwater Flows Halved to 4,07 and 1,035 CFS
4. SOD Doubled in Reaches 11 and 12
5. SOD Halved in Reaches 11 and 12

CONCLUSIONS

QUAL II package, as available on the USEPA TYMNET system, is a good comprehensive steady state water quality model. It can take into account sediment oxygen demand, algae influences, nutrient cycles and other factors to produce satisfactory to excellent results under low flow conditions. The existing package needs to be modified for better adaptation to the Illinois conditions. Concerns regarding the presence of heavy sediment oxygen demand sink in the Kaskaskia River need to be verified with _in situ_ measurements. Also, the way to incorporate the impacts of diurnal variations on dissolved oxygen needs to be refined.

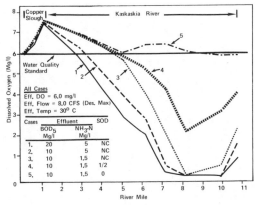

Figure 6. - - DO profiles under different effluent loadings and stream conditions.

All Cases
Eff. DO = 6.0 mg/l
Eff. Flow = 8.0 CFS (Des. Max)
Eff. Temp = 30° C

Cases	Effluent BOD$_5$ Mg/l	NH$_3$-N Mg/l	SOD
1.	20	5	NC
2.	10	5	NC
3.	10	1.5	NC
4.	10	1.5	1/2
5.	10	1.5	0

ACKNOWLEDGEMENTS

The author deeply appreciates Mr. Roger A. Kanerva, Manager, Division of Water Pollution Control, for his sincere encouragements and support to the water quality modeling efforts. Also, the author appreciates Mr. James B. Park for his critical review of this manuscript as well as for his participation in the calibration process.

LITERATURE CITED

Butts, T.A. and R.L. Evans, 1978. Sediment oxygen demand of selected northeastern Illinois streams. Circular 129, State of Illinois Department of Registration and Education, Illinois State Water Survey, Urbana, Illinois.

Draper, N.R. and H. Smith, 1967. Applied regression analysis. John Wiley and Sons, Inc., New York. Third Printing pp. 267-275.

O'Connor, D.J. and W.E. Dobbins. 1958. Mechanism of reaeration in natural streams. Transaction American Society of Civil Engineers, Volume 123.

Singh, K.P. and J.B. Stall, 1973. The 7-day 10-year low flows of Illinois streams. Bulletin 57, State of Illinois Department of Registration and Education, Illinois State Water Survey, Urbana, Illinois.

Statistical Analysis System. 1979. User's Guide. Raleigh, North Carolina, pp. 317-330.

Tsivoglou, E.C. and J.R. Wallace. 1972. Characteristics of stream reaeration capacity. Prepared for the Environmental Protection Agency, Office of Research and Monitoring, Washington, D.C.

U.S. Environmental Protection Agency, Center for Water Quality Modeling. 1977. Computer Program documentation for stream quality model (QUAL II).

U.S. Environmental Protection Agency, Center for Water Quality Modeling. 1977. User's manual for stream quality model (QUAL II).

LS:mad/sp/5557b/1-15

AN OPTIMAL PLANNING MODEL FOR HYDROPOWER DEVELOPMENT AND ANADROMOUS FISH RESTORATION[1]

Hisashi Ogawa[2] and James W. Male[3]

Abstract.--This paper presents an optimal screening model to determine the sequencing of construction activities for development of hydropower and restoration of anadromous fish. The model is applied to preliminary data for American shad in the lower Farmington River of the Connecticut River Basin.

INTRODUCTION

The energy crisis of the 1970's has generated renewed interest in using all available sources of energy. Hydropower has received increasing attention as the costs of generating energy from other conventional sources continue to rise. There has also been an initiative in major river basins in the United States to restore anadromous fish species whose populations declined as a result of human activities. The adverse impact on anadromous fish resulted primarily from poor water quality and dams barring fish from migrating to spawning grounds. During the last twenty years water quality has improved, allowing fish to survive. The problems caused by dams can also be alleviated by installing fish passageways.

In order to continue to restore anadromous fish to upstream river segments, the planning of future hydropower development must include the consideration of fish migration. This paper presents a procedure to aid in the long range planning of hydropower development and restoration of fish populations.

Hydropower

Hydroelectric power facilities provide approximately 15 percent of the energy used in the United States. These facilities are often

[1]Paper presented at the international symposium Energy and Ecological Modelling, sponsored by the International Society for Ecological Modelling (Louisville, Kentucky, April 20-23, 1981).

[2]Research Assistant, Environmental Engineering Program, Department of Civil Engineering, University of Massachusetts/Amherst.

[3]Associate Professor of Civil Engineering, University of Massachusetts/Amherst.

used to provide power at times of peak demand. In addition to conventional hydropower dams, there has been recent interest in low-head hydro and pumped storage facilities.

Low-head hydro is no different than conventional hydropower, except that the size of the dam and therefore the head is smaller. The U.S. Department of Energy (DOE) defines low-head hydro to be anything less than 20 meters of head. The term small hydro is also used and includes facilities that generate less than 15 MW (Gladwell 1980). Low-head hydro has seen a recent rejuvenation, especially in New England, because of federal and state efforts to retrofit existing dams with turbines for hydroelectric power.

Pumped storage facilities generate peaking power by pumping water to an elevated reservoir during times when the demand for electricity is low, and then releasing water through turbines for power production during high demand periods. Generally, an impoundment is required on the stream from which the water is taken.

Anadromous Fish

The presence of hydropower dams on a river severely affects the ability of anadromous fish to migrate to upstream spawning grounds. The degree to which migration is hindered depends on the size of the reservoir and height of the dam (turbine capacity). The presence of fishways helps to reduce the adverse effects considerably, however the energy expended by fish in surmounting the fishway often decreases body weight and may affect the ability to lay eggs. In some cases the fishways may also cause mortality.

Hydropower facilities affect fish populations in a number of ways (Ruggles 1980):

1. Impoundments may contribute to increased populations of predatory fish, which in turn contribute to the mortality of juvenile anadromous fish.

2. Spillways and falls contribute to mortality in a number of ways, including: rapid pressure change, rapid deceleration, shearing effects, excessive turbulence, the force of impact, and scraping and abrasion. A number of studies have been performed to determine the height from which fish can fall without excessive mortality.

3. Turbine induced mortality during downstream migration is caused by: mechanical damage due to contact with moving equipment, damage induced by low pressure areas, shearing action due to areas of extreme turbulence cavitation damage due to exposure to areas of partial vacuum, and the unknown effects of stress on surviving fish.

4. Passage through artificial impoundments may present a variety of problems during downstream migration ranging from delays to the loss of the urge to migrate.

5. Gas-bubble disease is caused where water is supersaturated with a gas due to entrapment and changes in temperature and pressure. The formation of nitrogen bubbles under the skin of affected fish is one cause of mortality.

6. Sublethal stress, although more difficult to assess, may be a significant factor.

The Conflict

The obvious conflict between hydropower development and the restoration of anadromous fish has most commonly been resolved by constructing some means for fish to pass a dam. The most common means are (1) fishways (or fishladders), consisting of a series of pools, (2) fish locks, utilizing a chamber which is raised over the dam, and (3) fish lifts (or fish elevators), which are similar to fish locks but use a water filled hopper (Hildebrand et al. 1980).

Fish passageways have been constructed on existing dams and are included in the construction of new facilities. In addition, whenever existing dams are retrofit for power generation fish passageways are generally included in the construction.

The problem that still exists, involves the planning of the construction of hydropower facilities in order to meet the future demand for hydropower and at the same time attain goals set for the restoration of anadromous fish. Obviously these goals should be achieved using the least cost alternative. In many cases this problem is approached on a site by site basis rather than looking at the river basin or region as a whole. The problems associated with hydropower generation and fish restoration are very site specific; however, they can be studied in a comprehensive manner using systems analysis techniques.

This paper develops a technique capable of addressing hydropower/fish restoration planning on a river basin. The technique considers decisions on the sequencing of (1) new dam construction, (2) hydropower retrofit of existing dams, and (3) construction of fish passageways, in light of energy supply requirements and fish population goals. The objective of the procedure is to minimize construction cost.

This procedure would aid in river basin planning with particular emphasis on hydropower. The New England River Basins Commission (NERBC) is currently involved in making recommendations for such decisions for several rivers in the New England area.

PLANNING MODEL

Two basic approaches can be taken when formulating a mathematical planning model: simulation and optimization. Simulation procedures are usually more accurate in their representation of a real world situation and they determine responses resulting from certain actions. However, simulation may not be as useful in the initial phases of planning when a large number of actions are possible. In the case at hand the actions would be decisions on when and where facilities should be constructed. The results would be the hydropower available, the number of fish restored, and the cost of facility construction.

On the other hand, optimization procedures determine what actions should be taken to achieve a desired result, while optimizing a specified objective. In this case the appropriate facility construction would be determined which minimized total construction costs yet met projected hydropower needs and desired fish restoration goals. One of the weak aspects of optimization is that simplifying assumptions often must be made in order to use appropriate solution techniques.

Since the initial stages of planning are often rough, optimization formulations are appropriate because they (1) provide insight into the modelled process, (2) are quickly solved, (3) allow the decision maker to analyze the sensitivity of the model, and (4) provide a range of good solutions when key assumptions are varied. Often the range of good solutions can be used as input to more detailed simulations which are employed in the second, more detailed, planning phase.

Model Formulation

An optimization model involves two parts: an objective function and a set of constraints. The former represents the quantity to be optimized, based on an equation incorporating different potential decisions. The constraint set is composed of relationships that must be satisfied in order for the solution of the problem to be feasible.

The problem at hand is to set a schedule for construction of dams, hydropower plants and fishways. Decisions must be made as to when and at which site(s) to construct these facilities. The planning time horizon is divided into T periods, and N sites are included. The objective of the problem is to minimize the capital construction costs of dams, hydropower plants (turbine generators) and fishways. The constraints involve the demand for hydropower and the goals for anadromous fish restoration for each period, in addition to sequencing and various conditional constraints.

The decision variables (unknowns) in the formulation represent decisions on whether to construct a facility or not. Hence they take on integer values of either zero (0) or one (1). Two subscripts are used in the model; one for site i and the other for period j in the planning time horizon. The decision variables are defined as follows:

x_{ij} = 1, if a new dam is constructed at site i during period j

= 0, otherwise

y_{ij} = 1, if a hydropower plant is constructed at site i during period j

= 0, otherwise

z_{ij} = 1, if a fishway is constructed at site i during period j

= 0, otherwise

One of the major assumptions in this and other sequencing models is that all facilities are built at the beginning of the period and operate thereafter. It also should be noted that the following formulation does not attempt to determine the sizes of the facilities. They are assumed to be pre-determined by a site specific analysis of their technical and economic feasibility. Although this assumption can be relaxed, for simplicity of illustration the problem has been formulated without continuous variables.

Demand Constraints

Demand for hydroelectric power usually increases over time. Turbine generators must be installed at dam sites to supply this demand. Let D_j denote the demand in terms of required hydropower capacity in mega-watts (MW) during period j, and let A_i be the installed hydropower capacity in MW at site i. The demand constraints are:

$$\sum_{i=1}^{N} \sum_{k=1}^{j} A_i y_{ik} \geq D_j \quad \text{for } j = 1,\ldots,T \quad (1)$$

where k is a temporary subscript used to sum over j time periods. The inner summation subscripted with k is necessary since once a hydropower plant is constructed, it is operated over the remaining periods of analysis.

Fish Population Constraints

Factors affecting changes in anadromous fish population are associated with the presence of dams, impoundments, hydropower turbines, and fishways. The effect that these facilities have on fish population can be grouped into three categories:

1. the positive effect of the availability of upstream spawning grounds.

2. the detrimental effect of (a) flooded spawning grounds due to new dam construction, (b) some mortality associated with fishway use, and (c) mortality caused by dam spillways during downstream migration, and

3. the effect of turbine operation.

The impact of turbines on fish mortality is considered separately from the other detrimental factors because the existence of a dam does not automatically imply the existence of turbine generators. Furthermore some studies have shown that the mortality of juvenile fish due to turbine operation can be a significant portion of fish mortality during downstream migration (Ruggles 1980).

In order to include the changes in fish population in the model, three additional integer variables are introduced:

u_{ij} = 1, if the spawning grounds immediately above site i are available during period j

= 0, otherwise

v_{ij} = 1, if the dam at site i reduces the potential population during period j due to flooded spawning grounds, mortality at the fishway and downstream migration losses, other than those of turbine operation.

= 0, otherwise

w_{ij} = 1, if the operation of turbines at site i reduces the potential population during period j.

= 0, otherwise.

These variables are used to replace the roles of x_{ij}, y_{ij}, and z_{ij}, by taking into account the downstream conditions preventing the upstream use of spawning grounds. For instance, the spawning grounds immediately above site i are available only if all dams downstream from site i are equipped with fishways. Similarly, the potential population losses due to turbine and fishway operations, flooded spawning grounds and downstream spillway migration depend on the existence of a dam and the availability of spawning grounds at and above site i. If no dam exists at site i or if the spawning grounds above site i are not available, these losses will never occur. Therefore, conditional relationships must be formulated between the variables, in order to avoid impossible solutions. The formulations of these and other conditional constraints are provided later in this section. In the following fish population balance constraints are formulated.

Each site must be evaluated to determine what its effect will be on the fish population. These effects are introduced as parameters:

S_i = the increase in population due to spawning ground availability immediately above site i,

M_i = the loss in population at dam site i due to flooded spawning grounds, fish mortality during upstream migration through fishways and downstream migration mortality other than that caused by turbine operation, and

Q_i = the loss in population due to turbine operation at site i.

Since the model includes existing dams as well as potential development sites, a distinction is drawn between an upstream reach with spawning grounds flooded by new dam construction and essentially no change in the presence of spawning grounds. S_i represents the latter and the effect of the former is included in the parameter M_i.

Combining all factors into one equation by summing over all sites yields:

$$\sum_{i=1}^{N}(S_i u_{ij} - M_i v_{ij} - Q_i w_{ij}) \geq P_j \qquad \text{for } j=1,\ldots,T \quad (2)$$

where P_j is the desired fish population for period j.

The activity of commercial and sport fishing may alter the fish population levels, although this last factor is not considered in this model. The analysis also excludes the consideration of benefits gained from fishing. It also should be noted that many fish population models consider the age of fish and its effects on population changes. These factors are not considered in this model.

Construction Constraints

These constraints guarantee that facilities are built no more than once for the entire planning time horizon. They are formulated as follows:

$$\sum_{j=1}^{T} x_{ij} \leq 1 \qquad \text{for } i = 1,\ldots,N \quad (3)$$

$$\sum_{j=1}^{T} y_{ij} \leq 1 \qquad \text{for } i = 1,\ldots,N \quad (4)$$

$$\sum_{j=1}^{T} z_{ij} \leq 1 \qquad \text{for } i = 1,\ldots,N \quad (5)$$

Conditional Constraints

Conditional constraints are formulated when one project depends on others. There are two types of such constraints. First, construction of a dam must be prior to or during the same period as construction of a fishway and/or a hydropower plant at any site. The second type involves conditional relationships used for the fish population constraints.

The first type is formulated as:

$$y_{ij} \leq \sum_{k=1}^{j} x_{ik} \qquad \begin{array}{l} \text{for } i = 1,\ldots,N \\ \text{and } j = 1,\ldots,T \end{array} \quad (6)$$

and

$$z_{ij} \leq \sum_{k=1}^{j} x_{ik} \qquad \begin{array}{l} \text{for } i = 1,\ldots,N \\ \text{and } j = 1,\ldots,T \end{array} \quad (7)$$

The second type requires some development. First, in order for spawning grounds above site i to be available for any period j ($u_{ij} = 1$), either no dam can exist at site i or a fishway must be present. This means that

$$\sum_{k=1}^{j} x_{ik} - \sum_{k=1}^{j} z_{ik} = 0 \qquad \begin{array}{l} \text{at site i} \\ \text{and for any} \\ \text{period j.} \end{array} \quad (8)$$

The condition $u_{ij} = 1$, also requires that all downstream sites meet the same conditions, i.e.,

$$\sum_{k=1}^{j} x_{\ell k} - \sum_{k=1}^{j} z_{\ell k} = 0 \quad (9)$$

where ℓ is any site downstream of site i. One way to formulate this is to state:

$$u_{ij} = 1, \text{ if } \sum_{\ell \in L_i} \{ \sum_{k=1}^{j} x_{\ell k} - \sum_{k=1}^{j} z_{\ell k} \} = 0 \quad (10)$$

$$= 0, \text{ otherwise}$$

where the first summation is taken over all downstream sites, i.e. L_i is the set of all sites downstream from site i.

Equation (10) involves a cumulative summation; an alternative approach is to address pairs of adjacent upstream-downstream sites in a recursive manner. In order to accomplish this and to simplify notation, an artificial parameter, p_{ij}, is introduced:

$$P_{ij} = (1-u_{mj}) + (\sum_{k=1}^{j} x_{ik} - \sum_{k=1}^{j} z_{ik}) \qquad (11)$$

where m represents the site immediately downstream from site i. Equation (11) can be expressed as: $u_{ij} = 1$ if and only if $p_{ij} = 0$, and
$u_{ij} = 0$ if and only if $1 \le p_{ij} \le 2$. However, the
recursive relationships must be formulated as
integer programming constraints. In so doing, a
theorem (provided in the Appendix) is applied to
yield linear constraints:

$$P_{ij} \le 2(1-u_{ij}) \quad \text{for i=1,...,N; j=1,...,T} \qquad (12)$$

$$P_{ij} \ge (1-u_{ij}) \quad \text{for i=1,...,N; j=1,...,T} \qquad (13)$$

Note that the upper bound of p_{ij} is 2 when

$$u_{mj} = 0, \quad \sum_{k=1}^{j} z_{ik} = 0 \quad \text{and} \quad \sum_{k=1}^{j} x_{ik} = 1.$$

Equations (12) and (13) represent constraints which define u_{ij} in terms of the
conditions at downstream sites (represented in
p_{ij}). Similar relationships can be developed
for the effects of flooded spawning grounds and
possible fishway and spillway mortality;

$$q_{ij} - 2 \le -(1-v_{ij}) \quad \text{for i=1,...,N; j=1,...,T} \qquad (14)$$

$$q_{ij} - 2 \ge -2(1-v_{ij}) \quad \text{for i=1,...,N; j=1,...,T} \qquad (15)$$

where

$$q_{ij} = u_{ij} + \sum_{k=1}^{j} x_{ik} \qquad (16)$$

hence, $\quad v_{ij} = 1, \text{ if } q_{ij} = 2$

$$= 0, \text{ if } 0 \le q_{ij} \le 1.$$

Equations (14) and (15) state that the impact of
flooded spawning grounds, and fishway and spillway mortality occur only when the spawning
grounds above site i are available ($u_{ij} = 1$)
and a dam is at the site, i.e. $\sum_{k=1}^{j} x_{ik} = 1$.

Similarly, relationships can be developed
for turbine mortality. Fish mortality occurs
only when the upstream spawning grounds are
available and turbines are operating at the site.
The integer programming constraint formulations
are:

$$r_{ij} - 2 \le -(1-w_{ij}) \quad \text{for i=1,...,N; j=1,...,T} \qquad (17)$$

$$r_{ij} - 2 \ge -2(1-w_{ij}) \quad \text{for i=1,...,N; j=1,...,T} \qquad (18)$$

where

$$r_{ij} = u_{ij} + \sum_{k=1}^{j} y_{ik} \qquad (19)$$

hence, $\quad w_{ij} = 1, \text{ if } r_{ij} = 2$

$$= 0, \text{ if } 0 \le r_{ij} \le 1.$$

Objective Function

The objective function includes the construction costs of dams, hydropower plants and
fishways. The present values of these costs are
to be minimized. The same discount rate is used
over the periods, without considering inflation.
The OM costs and other fixed annual costs are not
included. Let C_i, K_i and F_i be the construction
costs of a dam, a hydropower plant and a fishway
at site i respectively. Let h denote the discount rate for the period. The present values of
those construction costs are calculated as

$$C_{ij} = C_i (1+h)^{-j+1} \qquad (20)$$

$$K_{ij} = K_i (1+h)^{-j+1} \qquad (21)$$

$$F_{ij} = F_i (1+h)^{-j+1} \qquad (22)$$

where C_{ij}, K_{ij} and F_{ij} are the present values of
the construction costs of a dam, a hydropower
plant and a fishway at site i respectively. The
objective function is

$$\text{minimize} \quad \sum_{i=1}^{N} \sum_{j=1}^{T} C_{ij}x_{ij} + \sum_{i=1}^{N} \sum_{j=1}^{T} K_{ij}y_{ij}$$

$$+ \sum_{i=1}^{N} \sum_{j=1}^{T} F_{ij}z_{ij} \qquad (23)$$

The overall formulation of the model is given in
Table 1.

Hypothetical Example

To illustrate the use of the model in its
general form a hypothetical example is provided
which includes construction of new dams as well
as retrofitting for hydropower and installation
of fishways at existing dams. Two existing dams
and three potential dam sites are considered
(fig. 1); the upper existing dam is installed
with hydropower. Since the dam at site 1 does
not have a fishway, no spawning area is available
upstream.

The data for the hypothetical example are
given in table 2 arranged for periods and sites.
A 30 year planning horizon, with three ten-year
periods, is considered. Values of M_i and Q_i
are assumed for each site to be 5 percent and 10

Table 1.--Hydropower Development and Fish Restoration Planning Model

Objective function:

$$\text{minimize} \quad \sum_{i=1}^{N} \sum_{j=1}^{T} (C_{ij}x_{ij} + K_{ij}y_{ij} + F_{ij}z_{ij})$$

Subject to:

Demand Constraints:

$$\sum_{i=1}^{N} \sum_{k=1}^{j} A_i y_{ik} \geq D_j \quad \text{for } j=1,\ldots,T$$

Anadromous Fish Population Balance Constraints:

$$\sum_{i=1}^{N} (S_i u_{ij} - M_i v_{ij} - Q_i w_{ij}) \geq P_j \quad \text{for } j=1,\ldots,T$$

Construction Constraints:

$$\sum_{j=1}^{T} x_{ij} \leq 1 \quad \text{for } i = 1,\ldots,N$$

$$\sum_{j=1}^{T} y_{ij} \leq 1 \quad \text{for } i = 1,\ldots,N$$

$$\sum_{j=1}^{T} z_{ij} \leq 1 \quad \text{for } i = 1,\ldots,N$$

Conditional Constraints:

$$y_{ij} \leq \sum_{k=1}^{j} x_{ik} \quad \text{for } i=1,\ldots,N; \ j=1,\ldots,T$$

$$z_{ij} \leq \sum_{k=1}^{j} x_{ik} \quad \text{for } i=1,\ldots,N; \ j=1,\ldots,T$$

$$p_{ij} \leq 2(1-u_{ij}) \quad \text{for } i=1,\ldots,N; \ j=1,\ldots,T$$

$$p_{ij} \geq (1-u_{ij}) \quad \text{for } i=1,\ldots,N; \ j=1,\ldots,T$$

$$q_{ij} -2 \leq -(1-v_{ij}) \quad \text{for } i=1,\ldots,N; \ j=1,\ldots,T$$

$$q_{ij} -2 \geq -2(1-v_{ij}) \quad \text{for } i=1,\ldots,N; \ j=1,\ldots,T$$

$$r_{ij} -2 \leq -(1-w_{ij}) \quad \text{for } i=1,\ldots,N; \ j=1,\ldots,T$$

$$r_{ij} -2 \geq -2(1-w_{ij}) \quad \text{for } i=1,\ldots,N; \ j=1,\ldots,T$$

Defining Equations:

$$p_{ij} = (1-u_{mj}) + (\sum_{k=1}^{j} x_{ik} - \sum_{k=1}^{j} z_{ik}) \quad \text{for all possible neighboring pairs of upstream site (i) and downstream site (m)}$$

$$q_{ij} = u_{ij} + \sum_{k=1}^{j} x_{ik} \quad \text{for } i=1,\ldots,N; \ j=1,\ldots,T$$

$$r_{ij} = u_{ij} + \sum_{k=1}^{j} y_{ik} \quad \text{for } i=1,\ldots,N; \ j=1,\ldots,T$$

Variable	Description	Value
x_{ij}	dam construction at site i during period j	(0,1)
y_{ij}	turbine installation at site i during period j	(0,1)
z_{ij}	fishway construction at site i during period j	(0,1)
u_{ij}	spawning ground availability above site i during period j	(0,1)
v_{ij}	potential impact at site i during period j of flooded spawning grounds, mortality at fishways and downstream migration losses, other than those of turbine operations	(0,1)
w_{ij}	turbine operation at site i during period j causing reduction in fish population	(0,1)

Parameter	Description	Unit
A_i	hydropower capacity to be developed at site i	MW
S_i	population increase due to the spawning ground availability above site i excluding flooding effects	# of fish
M_i	population decrease at site i due to flooded spawning grounds, fishway mortality and downstream migration mortality other than those of turbine operations	# of fish
Q_i	population losses at site i due to turbine operations	# of fish
D_j	hydropower capacity development demand during period j	MW
P_j	desired goal for fish population restoration for period j	# of fish
C_{ij}	present value of dam construction cost at site i during period j	$\times 10^6$
K_{ij}	present value of turbine construction cost at site i during period j	$\times 10^6$
F_{ij}	present value of fishway construction cost at site i during period j	$\times 10^6$

percent of S_i respectively. A discount rate of 7 percent is used to obtain the present value of construction costs, C_{ij}, K_{ij} and F_{ij} from C_i, K_i and F_i.

The problem was solved by a direct search 0-1 integer programming algorithm. The results, showing hydropower development and fish restoration, are plotted in figure 2. Dashed lines indicate the solution with numbers showing at which site and during which period the installations of facilities take place. The solid lines indicate the hydropower capacity demands and desired population goals that must be met for each period. The difference between the solution and the desired population goals show that more than enough spawning grounds are available (above the sites indicated) to maintain the desired level of population. The hydropower development will just meet the demand after period 2.

MODEL APPLICATION AND SENSITIVITY ANALYSIS

In order to examine the applicability of the model to a real world setting, a case study was conducted with data available for the Farmington River, a tributary in the lower Connecticut River Basin (fig. 3). The mouth of the Farmington is located approximately 93 km from Long Island Sound. An anadromous fish species, American shad (Alosa sapidissima), is reported to migrate above the lowermost dam, through a fish ladder installed at the dam in 1975 (Moffitt 1979).

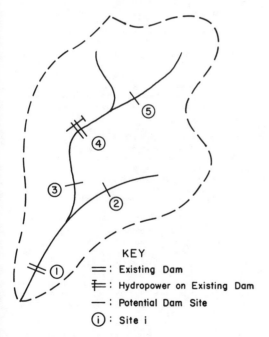

KEY

≡ : Existing Dam

⇶ : Hydropower on Existing Dam

— : Potential Dam Site

ⓘ : Site i

Figure 1.--Hypothetical Example Problem.

Table 2.--Data for the Hypothetical Example

Period, j	D_j (MW)	P_j (10^3 fish)
1	10	10
2	20	15
3	25	30

Site, i	A_i (MW)	S_i	M_i	Q_i	C_i	K_i	F_i
		(1000 fish)			(million \$)		
1	15	10	0.5	1.0	0	10	4
2	5	5	0.25	0.5	20	4	2
3	5	5	0.25	0.5	15	3	3
4	0	15	0.75	1.5	0	0	4
5	5	5	0.25	0.5	15	3	2

American shad is one of the two anadromous species considered in the Cooperative Fishery Restoration Program for the Connecticut River Basin, established in 1966 (Moffitt 1979). Atlantic salmon (Salmo salar) is the other species being considered by the program. The goal is to restore 2 million shad and 40,000 salmon at the mouth of the Connecticut River (Lanse 1977). For this to be accomplished upstream spawning areas must be available.

The importance of analyzing the entire Farmington basin or even the entire Connecticut River should not be ignored. However, this application was limited because data for the larger applications were not available, and the integer programming solution procedure is limited to moderately sized problems. In addition, for simplicity, new dam construction was not considered and fish population goals were set only for American shad.

In order to apply the procedure to very large areas, an iterative scheme can be employed. The approach would involve the use of a procedure to allocate levels of fish population and hydropower to various subbasins in the area. The optimization model could then be used with the subbasin goals. Once sequencing was determined for subbasins, the allocation scheme could be reapplied using the information derived from the solutions of the optimization scheme. Successive iterations could be performed until a desired level of accuracy is attained.

Data and Sensitivity Analysis

The data for the Farmington River case study are shown in table 3. Hydropower development data were obtained from NERBC (1980). The estimates of increases in fish population due to spawning ground availability (S_i) are calculated from a model proposed by Northrop et al. (1975). The estimates of population losses due to the existence of a dam (M_i) are assumed to be 2 percent of S_i. Mortality over spillways is reported to range from 1 to 2 percent (Fowler 1978, Ruggles 1980 and Ferguson 1981). Mortality of juvenile salmon through turbines is much higher, averaging 15 percent (Salo and Stober 1977, Fowler 1978, Ruggles 1980, Ferguson 1981). In the analyses that follow, the estimates of population losses due to turbine operation (Q_i) were varied, assuming 10, 30 and 50 percent of S_i. Estimates of fishway construction costs are calculated by using \$15,000/ft of gross head (Hildebrand et al. 1980). The present values of future construction costs are calculated using a 7 percent discount rate.

The planning horizon of the case study is divided into five periods with 10 year intervals from 1980 to 2030. Since no values are available for the desired restored population of American shad and the demand for hydropower, for the lower Farmington River as a separate entity, these values are varied in the sensitivity analysis. The goals for the restoration of American shad

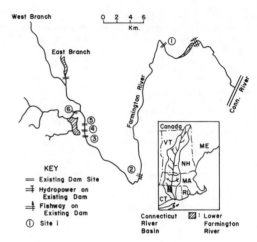

KEY
= Existing Dam Site
⫢ Hydropower on Existing Dam
⫦ Fishway on Existing Dam
① Site i

Connecticut River Basin   ▨ : Lower Farmington River

Figure 3.--Lower Farmington River.

and the demand for hydropower were assumed to be equal incremental increases for each ten year period. As a result, by specifying values at the end of the time horizon, the periodic values were simply 20, 40, 60 and 80 percent of the final value. Analyses were performed for American shad population goals of 50,000, 75,000, and 100,000 fish and hydropower requirements of 7.5, 8.0, 9.0, 10.0 and 11.0 MW. Sensitivity analyses were also performed with respect to changes in the estimates of the impact of turbine operation on fish mortality.

RESULTS AND DISCUSSION

The results of the sensitivity analyses are presented in table 4. The values listed under the hydropower capacity demand (D_j) and the American shad population goal (P_j) are those for the last period. For each of the three values of Q_i, the model was solved 15 times (5 values of D_j times 3 values of P_j). Thus table 4 presents the total present value of cost (the value of the objective function) for 45 solutions. For example, assuming that the fish population loss due to turbine operation accounts for 10 percent of the population gain due to the spawning ground availability upstream, the present value cost of 2.66 million dollars is necessary for restoration of 50,000 American shad and supply of 7.5 MW from hydropower.

Of the 45 planning schemes, 14 combinations yielded infeasible solutions as indicated by inf in table 4. As one would expect, the greater the fish restoration requirements and hydropower demands, the higher the construction costs. The results also indicate that efforts to reduce the losses of fish through turbines would be economically attractive if upstream fish goals are to be met.

Figure 4 illustrates a typical solution to the sequencing problem, showing when and at which

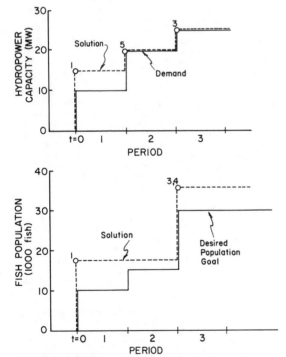

Figure 2.--Results of the hypothetical example. (Numbers on the solution line indicate sites of the solution.)

site retrofitting of hydropower and construction of fishway(s) must take place. In this solution it is not necessary to retrofit hydropower at site 1 whereas fishway construction is required at all sites. It should also be noted that fishways are constructed from downstream sites first while retrofitting of turbine generators generally takes place at upstream sites. Although the fishway construction sequence is logical, the solution for the hydropower development is more often problem specific. Inspection of other solutions indicates that the need for fishway construction is influenced significantly by changes in hydropower demands, while retrofitting for hydropower is generally insensitive to changes in fish restoration goals.

The development of multiobjective programming (e.g., Cohon 1978) allows us to accommodate several objectives in the optimization formulation. Other objectives pertinent to the problem

Table 3.--Data for the Farmington River Case Study

Site, i	A_i^1 (MW)	S_i^2	M_i^3	Q_i^4	K_i^1	F_i^5
		(1000 fish)			(million $)	
1	3.5	40.6	0.8	4.1	6.2	0.30
2	0^6	22.4	0.4	2.2	0^6	0.68
3	2.0	0.6	0	0.1	1.0	0.27
4	2.2	1.4	0	0.1	1.7	0.30
5	2.2	31.0	0.6	3.1	1.7	0.30
6	1.1	21.0	0.4	2.1	1.2	1.70

[1]The estimates A_i and K_i were obtained from NERBC (1980).

[2]The estimates of S_i are calculated from a formula proposed by Northrop et al. (1975) for the Connecticut River American shad. The formula is: (maximum sustainable spawning shad population) = 0.025 x (total bottom area in square meters).

[3]The estimates of M_i are assumed to be 2% of S_i.

[4]The estimates of Q_i are assumed to be 10% of S_i. In the sensitivity analyses, 30% and 50% of S_i are also considered for the estimates of Q_i.

[5]The values of F_i are calculated using $15,000/ft of gross head, a typical value given by Hildebrand et al. (1980). The gross head values are obtained from NERBC (1980).

[6]No new hydropower development is assumed at site 2.

Table 4.--Sensitivity analyses of the model applied to American shad in Farmington River with respect to turbine mortality rate, hydropower capacity and the fish restoration goal.

$(Q_i/S_i) \times 100^1$ (%)	D_j^2 (MW)	Present Value Cost (million $) P_j^3 (1000 fish)		
		50	75	100
10	7.5	2.66	2.82	3.18
	8.0	3.10	3.26	3.62
	9.0	3.26	3.42	3.78
	10.0	3.58	3.78	4.10
	11.0	4.33	4.53	4.85
30	7.5	2.66	2.86	inf[4]
	8.0	3.16	3.41	inf
	9.0	3.32	3.57	inf
	10.0	3.64	3.89	inf
	11.0	4.39	4.64	inf
50	7.5	2.66	2.97	inf
	8.0	3.37	inf	inf
	9.0	3.43	inf	inf
	10.0	3.81	inf	inf
	11.0	4.56	inf	inf

[1]Values of Q_i are expressed as a percentage of S_i. Constant percentages are assumed for all sites.

[2]Each value indicates the hydropower capacity demand during the 5th period. The demand is assumed to increase linearly over the planning horizon.

[3]Each value indicates the American shad restoration requirement during the last period. The requirement curve is assumed to be linear.

[4]Inf indicates an infeasible solution.

at hand are (1) to maximize available spawning grounds, and (2) to maximize hydropower generation. Obviously these objectives are usually in conflict since the quest for one is often at the expense of the other.

In figure 5 several value paths are drawn to interpret the results in a multiobjective planning scenario. The planning problem considered in this case assumed three objectives: minimizing the present value of cost, maximizing the hydropower development and maximizing the fish population. These three objectives are scaled on vertical lines as shown in figure 5. The direction of the arrows indicates that of optimizing the objectives. Figure 5 shows five value paths connecting the three objective axes. Each path represents an alternative combination of a hydropower demand, a fish population restoration

278

and the resulting present value of cost, as taken from table 4.

Path A represents a case in which high fish restoration and high hydropower development are desired. This results in the most expensive construction scheme. In path B the hydropower requirement is reduced by 3 MW relative to path A, saving more than 1 million dollars in construction cost. These indicate the trade-offs between hydropower development and least construction cost objectives. Path C indicates moderate requirements for hydropower development and fish restoration with a moderate investment. Path D shows that relative to path B, giving up 50,000 fish is equivalent to obtaining 2 MW of hydropower at the same construction cost. In path E the opposite consequence of path A is presented. Low requirements for hydropower development and fish restoration result in the least costly alternatives. Value paths like the ones shown in figure 5 may prove useful to the decision maker when deciding which alternative to pursue further.

SUMMARY AND CONCLUSIONS

This paper presents an optimization procedure capable of determining the optimal sequence of construction of hydroelectric power facilities and fish passageways. The formulation uses construction cost as the optimization criterion,

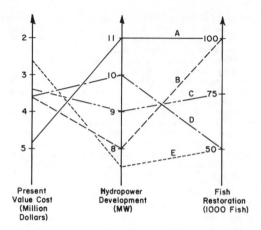

Figure 5.-- Multiobjective value paths for the Farmington River case study.

subject to meeting goals for anadromous fish restoration and satisfying future hydropower requirements. Changes in fish populations are incorporated by estimating increases due to increased availability of spawning grounds due to construction of fishways; and decreases as a result of fish mortality due to spillways and turbines, and as a result of flooded spawning grounds. The model allows for new dam construction, hydropower retrofit of existing dams and fish passageway construction. Construction costs are amortized and compared on a present value basis. The model is applied to the lower Farmington River Basin in Northern Connecticut.

The optimization procedure can be useful as an aid in river basin planning. New England River Basins Commission is currently performing similar analyses on several river basins in the region. The procedure presented in this paper is best used in the initial phase of planning as a screening model; it should be used to identify courses of action for further, more detailed study.

This study demonstrates the applicability of the planning model to two different settings: the case in which construction of new dams for hydropower development is allowed, and the case where only existing dam sites are considered for hydropower capacity expansion. The former case is applicable to unregulated streams such as in developing countries. In the latter case the model can be used as a planning tool for revitalization of existing dam sites for hydropower expansion as shown in the application example of this paper.

The approach taken in this paper allows considerable flexibility in the way the formulation is applied. Sensitivity analyses can easily be performed to study the effect on the solution of varying key assumptions and values of input

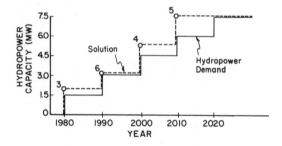

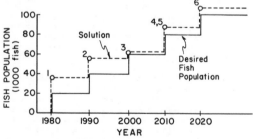

Figure 4.--Solution for the sequencing problem with ultimate hydropower capacity of 7.5 MW and American shad goal of 100,000 fish, assuming $Q_i/S_i = 0.1$. (Numbers on the solution line indicate sites in the solution.)

parameters. This has been done in the paper, in the context of different fish population goals and hydropower requirements. In addition, the results have been presented in a multiobjective context, to study the trade-offs associated with different aspects of the problem.

The major shortcoming of the formulation is its inability to address a river basin with a large number of dam sites (either potential or existing). Consequently, the model must be used in conjunction with a scheme to allocate fish population goals and hydropower demand to subbasins within the overall river basin. A number of different methods may be used to circumvent this problem. A decomposition approach (Haimes 1977) could be used to analyze such a large basin system by decomposing the basin into a number of subbasins, each of which may then be analyzed individually, and coordinating these subsolutions to form a solution in the original large-scale basin. Further research is needed on this topic.

The model can also be used as a submodel in a different context; that of planning for overall energy supply. Such an approach may include siting of fossil fuel and nuclear power plants, development of other energy alternatives, consideration of water quality and aquatic habitat affected by plant operations, and entrainment and impingement of aquatic organisms at steam generating power plants.

A number of extensions to the formulation are possible. In the application all sizing of facilities is done individually and on a preliminary basis to determine what the best hydropower capacity is. The model then determines whether the facility should be constructed at this size, or not at all. The size of the facility could be a variable in the problem, within a feasible range. This would mean that the size of hydropower facilities would be determined with other facilities in mind. Other extensions include the incorporation of annual operation and maintenance costs, benefits gained from commercial and sport fishing, considerations of changing discount rates and of cost escalation over the planning horizon, and impacts of policy changes on planning, such as energy and environmental conservations. These factors can be relatively easily incorporated into the model where data are available.

LITERATURE CITED

Cohon, J. L. 1978. Multiobjective Programming and Planning. Academic Press, New York, NY. 328 p.

Ferguson, J. 1981. Salmon: A second look. Water Spectrum, winter 1980-81. p. 39-47.

Fowler, J. A. 1978. Effects of a reservoir upon fish. p. 51-64. In Environmental Effects of Large Dams, W. V. Binger et al. ed. American Society of Civil Engineering.

Gladwell, J. S. 1980. Small Hydro - Some Practical Planning and Design Considerations. Idaho Water Resources Research Institute, University of Idaho, Moscow, Idaho. 75 p.

Haimes, Y. Y. 1977. Hierarchical Analyses of Water Resources Systems: Modeling and Optimization of Large-Scale Systems. McGraw-Hill, New York, NY. 470 p.

Hildebrand, S. G., M. C. Bell, E. D. Richey, J. J. Anderson, and Z. E. Parkhurst. 1980. Analysis of Environmental Issues Related to Small Scale Hydroelectric Development, II: Design Considerations for Passing Fish Upstream Around Dams. Environmental Sciences Division Publication No. 1567, ORNL/TM-7396. Oak Ridge National Laboratory, Oak Ridge, Tennessee. 78 p.

Lanse, R. I. 1977. Biennial Report Connecticut River Anadromous Fish Restoration Program. U.S. Fish Wildl. Serv. 38 p.

Moffitt, C. M. 1979. Recolonization of American Shad, Alosa Sapidissima (Wilson) above Rainbow Dam Fishladder, Farmington River, Connecticut. PhD dissertation submitted to the University of Massachusetts, Amherst, Massachusetts. 128 p.

New England River Basins Commission. 1980. Potential for Hydropower Development at Existing Dams in New England. Vol. 3, State of Connecticut. New England River Basin Commission, Hanover, NH. variously paged.

Northrop, G. M., W. V. McGuinness, J. C. Reidy and H. M. Katz. 1975. Connecticut River Basin: A Framework for Environmental Impact Evaluation for Electric Power Systems in a River Basin. Report prepared by the Center for the Environment and Man Inc. for the Office of Energy Systems, Federal Power Commission under Contract FP-17792. 323 p.

Ruggles, C. P. 1980. A Review of the Downstream Migration of Atlantic Salmon. Canadian Technological Report of Fisheries and Aquatic Sciences. No. 952. Freshwater and Anadromous Division Resource Branch, Department of Fisheries and Oceans, Halifax, Nova Scotia, Canada. 39 p.

Salo, E. O. and Q. J. Stober. 1977. Man's impact on the Columbia River stocks of salmon. p. 36-45. In Proceedings of the Conference on Assessing the Effects of Power-Plant-Induced Mortality on Fish Populations, W. Van Winkle, ed.

APPENDIX - THEOREM

Consider a variable defined on a set

$$Y = \{y \varepsilon R : a \leq y \leq b \text{ or } y = c\} \text{ with } c \notin [a,b]$$

and a zero-one variable $x = 0$ or 1. Then the following statements are equivalent:

(A) $x = 1$ if and only if $y = c$, and
 $x = 0$ if and only if $a \leq y \leq b$.

(B) x and y satisfy the following relations:

$$y - c \leq (b-c)(1-x) \quad (1)$$
$$y - c \geq (a-c)(1-x). \quad (2)$$

Proof. Starting with four possible assumptions, we verify the statement (A) through the relations in (B) as follows:

(i) Let $x = 1$. From (1) $y \leq c$ and from (2) $y \geq c$. This implies $y = c$ which is the condition in (A).

(ii) Let $x = 0$. From (1) $y \leq b$ and from (2) $y \geq a$. Or $a \leq y \leq b$ which is the condition in (A).

(iii) Let $y = c$. From (1) $(b-c)(1-x) \geq 0$ and from (2) $(a-c)(1-x) \leq 0$. Since $c \notin [a,b]$, either $\{(b-c) > 0$ and $(a-c) > 0\}$ or $\{(b-c) < 0$ and $(a-c) < 0\}$. Thus $x = 1$ which is the stated condition in (A).

(iv) Let $a \leq y \leq b$.

Case 1: $\{(b-c)<0$ and $(a-c)<0\}$. Let $y_1 = (y-c)/(b-c)$ and $y_2 = (y-c)/(a-c)$. Then $0<y_1<1$ and $y_2>1$. From (1) $y_1<1-x$ and from (2) $y_2>1-x$. Using the intervals of y_1 and y_2, $x<1$ and $-\infty<x<\infty$ respectively. These relations imply $x=0$ since $x=0,1$.

Case 2: $\{(b-c)<0$ and $(a-c)<0\}$. Then $y>1$ and $0<y_2<1$. From (1), $y_1>1-x$ and from (2) $y_2<1-x$. Or, $x>1-y_1$, and $x<1-y_2$. By the same argument as in case 1, $-\infty<x<\infty$ and $x<1$ respectively and thus $x=0$.

WATER QUALITY MODELING IN THE MIDDLE REACHES OF THE LOIRE RIVER[1]

Philippe Gosse[2]

The dynamic model calculates every three hours the evolution of temperature and DO, BOD, phosphorus, phytoplankton concentrations in 1977 and 1978. Measured field data at five stations are used to calibrate or verify the model. The response of the model to the implantation of power plants is studied.

INTRODUCTION

In order to assess the economic impact of future hydraulic works, nuclear power plants and waste-treatment plants upon the mid-Loire valley management, a water quality model, capable of predicting the effect of these possible installations upon the main physical and biological parameters of the river, first had to be constructed. The next objective will be to couple this model to a management model as shown in figure 1. A first stage consists in developing the water quality model in a stretch of the mid-river where measured field data allow its calibration and verification. Therefore, the studied area which is 250 kms long extends from Belleville down to Chinon (fig. 2). It will be widened later so as to simulate the evolution of the water quality from Grangent reservoir to Nantes.

The aim of the study led us to opt for a model which could be easily handled by the optimization model, but sufficiently detailed to permit a good representation of the main biochemical interactions taking place in the river. That explains the choice of six state variables for simulation and the calculation time step.

CHOICE OF THE WATER QUALITY VARIABLES

A parameter of major importance is the tem-

[1] Paper presented at the international symposium Energy and Ecological Modelling, sponsored by the International Society for Ecological Modelling (Louisville, Kentucky, April 20-23, 1981).

[2] Philippe Gosse is hydraulic engineer, Electricité de France, Direction des Etudes et Recherches, Département Environnement Aquatique et Atmosphérique 78400 - CHATOU - FRANCE.

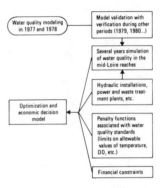

Figure 1.-- MODELS COUPLING

-perature which influences most of the biochemical and biological reactions. Another significant factor in the evaluation of water quality is the dissolved oxygen concentration. In order to represent its diurnal variation due to the temperature and photosynthetic activity, a three-hourly calculation step has been chosen. In this way, low values occuring at night, which are likely to be harmful to the fish or may lead to anaerobic processes, can be distinguished: from the management viewpoint, this short time step can be of great interest. Phytoplankton content, which is responsible for diurnal oxygen variations, is also simulated. Though we have at our disposal information (Champ 1980) on the type of phytoplankton found in the river (mainly diatoms and green algae), we do not distinguish one species from another so as to keep the model from becoming difficult to handle. But the main reason lies in the fact that most of the available field data are chlorophyll a measurements. This prevents us from being too ambitious because the phases of model calibration and verification would have been problematic. A weakness of the model consists in using chlorophyll a which is a rather poor biomass indicator because it also measures some deteriorated phytoplankton; and,furthermore, the variability of the carbon to chlorophyll a ratio is a major handicap. But in our case, measured field data were used to assess these two factors. Unlike nitrate or silicon, phosphorus can play a role in limiting algal growth in the Loire river and is modeled accordingly. Two types of phosphorus are differentiated : phosphorus (PO_4) which is available for algal growth and phosphorus which is unavailable. Lack of knowledge about phosphorus concentrations at Belleville except for PO_4 was an obstacle. However, the unavailable phosphorus concentration at the input of the model at Belleville was determined from field data at Orléans. Despite this weakness, taking into account the phosphorus cycle appeared essential in order to correctly describe the dynamics of phytoplankton population under certain management conditions. Finally, the evolution of total BOD is calculated. The only measured field data enabling its estimation were BOD_5, NH_4^+ and NO_2^- concentrations. The observed values (for instance NH_4^+ values smaller than 0.05 mg/1) proved that the nitrogenous oxidation was insignificant beside the carbonaceous oxidation in the simulation stretch. Therefore, it appeared that the simulation of both CBOD and NBOD was not necessary. After extension of the calculation area, this simplification might be reconsidered.

The model disregards the action of aquatic plants which are scarce in the Loire river. The unstable and sandy state of the river bed is probably the explanation for this phenomenon.

Erosion and sedimentation processes which can affect BOD and phosphorus contents, are not shown directly in the mass balances of these variables but are taken into account in the model when calculating the wastewater discharges mixing with the river water.

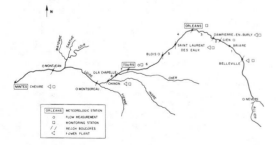

Figure 2.-- SITUATION DIAGRAM.

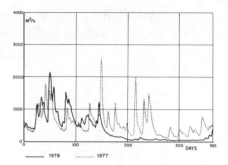

Figure 3.-- Daily averages of flow at Belleville.

INPUTS TO THE MODEL

Hydraulic data

Daily averages of flow are available on the Loire at Belleville, Gien, Orléans, Tours and La Chapelle so that the supply from the tributaries can be accurately assessed (fig. 3).

The mean velocity U (m/s) of the current and the hydraulic depth H (m) are calculated from the rate of flow Q (m^3/s) by two types of experimental laws depending on the place :

$$U = 0.165 \ Q^{0.275}$$
$$H = 0.134 \ Q^{0.4125}$$
from Belleville down to Orléans (1

$$U = 0.141 \ Q^{0.295}$$
$$H = 0.205 \ Q^{0.4425}$$
from Orléans down to Chinon (2

Tributaries

In the simulation area, the main water supply is from the Cher river whose flow sometimes amounts to 25% of the Loire flow. As no PO_4, DO, BOD and chla measurements are available on this river, we assume its only effect to be a flow increase. However, a certain error may creep in at some periods when the water quality of the Cher river is different from the Loire water quality. In particular, Khalanski and Renon (1977) has shown that this discrepancy of quality between the two rivers could influence phytoplankton content, with, at some periods, a 20% decrease in the algal density after the confluence.

Catchment area and wastewater discharges

BOD and phosphorus contents emanating from the catchment area and wastewater discharges have been calculated from data collected at the Agence Financière de Bassin Loire-Bretagne. This organization takes an inventory of the agricultural, industrial and urban discharges flowing into the Loire river. Unfortunately, there is very little information on the variability of these discharges over short periods.

Pollution injected into the stretch represents, on the average, (Q = 200 m³/s), a 500,000
population equivalent.

Heat discharges

In the mid-river, they result from two
nuclear power sites at St-Laurent-des-Eaux and
Chinon. In the stretch under consideration, the
output of the model is located just upstream
from the Chinon power plants. Consequently, only
the heat discharge from the St-Laurent once-
through-cooling power plants (2 x 400 MW) is
taken into account (fig. 4). When there are
low flows, the increase of the river temperature
can reach 6°C after mixing. Since neither phytoplankton mortality increase, or production de-

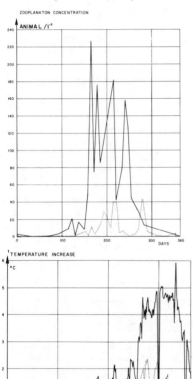

Figure 4.-- Saint-Laurent-des-Eaux :
Zooplankton concentration and river
temperature increase due to the power
plants.

crease have been shown during the phytoplankton
transit into St-Laurent power plants (Khalanski
and Renon 1977), the power plants are assumed to
act only on temperature.

Zooplankton population (mainly rotifers)

Weekly measurements at St-Laurent-des-Eaux
are used to define the grazing term in the phytoplankton equation (fig. 4). Since no other data
are available elsewhere in 1977 and 1978, the
model assumes the zooplankton population to be
stable from Belleville down to Chinon. Its concentration is assumed to be constant when proceeding along the stretch. This hypothesis is
not unrealistic ; Lair (1980) did not notice any
significant changes in the evolution of zooplankton concentration from Belleville down to Chinon
in 1978. Although Lair's study was carried out
in lentic areas, we assume this to be true in
running water as well.

Initial distribution of the 6 variables at Belleville

The initialization of the 6 variables is
defined from the following data measured at the
Belleville monitoring station :

- water temperature every three hours,

- twenty eight series of DO, BOD_5, chlorophyll a,
  and phosphorus (PO_4) measurements in a year.

MODEL FORMULATION

Main features of the model

We assume there is enough turbulence in the
river for the variables to be constant in a section. This hypothesis is realistic since the
Loire river does not usually show thermal stratification (however, a certain error is made just
downstream of the Saint-Laurent power plants
since we assume a rapid mixing). The problem
under consideration is therefore one-dimensional,
the values of the variables only depending on
the position of the section along the river. The
longitudinal dispersion is disregarded.

To make the numerical computation easier :

. the river is segmented in the direction of the
  flow into reaches in such a way that the following physical characteristics ; flow, velocity, hydraulic depth and meteorological
  conditions are assumed constant in the whole
  reach during one time step.

. External sources of pollution are included in
  the model in two ways :

  - continuous injections of phosphorus and
    BOD which are included in the mass bal-
    ance equations in the form q U/Q (mg
    l^{-1} day^{-1}) with q injection flow (g m^{-1}
    day^{-1}).

- discrete injections : they include polluting irjections from urban and industrial areas (Orléans, Tours and Blois), and temperature changes due to nuclear power plants (Saint-Laurent). These injections are taken into account as a change in the variable values in the inflowing water at the beginning of the water reach.

The simulation stretch is therefore divided into eight reaches (fig. 2). Information on the numerical method used in the model can be found in Gras (1969) and Gosse (1981).

The basic mechanisms of the model are given hereafter. An initial sensitivity study of the model led us to simplify the mass balances of certain variables in comparison with a previous version : some factors (for instance, algae settling which couldn't be correctly explained by the form σ/H with σ as calibration coefficient, or BOD decrease due to bacteria attached to the bottom) are neglected in the model and, in this way, the number of calibration parameters are limited.

Basic mechanisms

From the initial conditions defined at Belleville the mathematical scheme described hereafter is computed along the river reach-by reach. At the end of each reach, except the eighth, the calculated values of state variables become the inputs of the next reach added to the possible discrete injections.

Let us consider an elementary volume of water, located at position x_0 at time t_0 (day) with values T_0, P_0, N_0^1, N_0^2, C_0, L_0 representing the variables, temperature ($^\circ$C), phytoplankton (mg chla/m^3), available phosphorus PO$_4$ unavailable phosphorus DO (mg O$_2$/l) and BOD (mg O$_2$/l). At time $t = t_0 + \Delta t$ (with $\Delta t = 0.125$ day), this elementary volume will be at the position $x = x_0 + U\Delta t$ (the chosen time step is small enough for U to remain fairly constant during Δt). The new values of the variables are:

$$T = T_0 + \frac{\Delta t}{C_p H} \left[\text{solar radiation + atmospheric radiation − back radiation from water − latent heat flux − sensible heat flux} \right]. \quad (3)$$

$$P = P_0 + \Delta t \ (\text{GROW − RESP − MORT − GRAZ − EX}) \quad (4)$$

$$N^1 = N_0^1 + \Delta t \ (-\alpha \ \text{GROW} + K_3 \ N_0^2 + q_1 \frac{U}{Q}) \quad (5)$$

$$N^2 = N_0^2 + \Delta t \left[\alpha(\text{MORT + GRAZ + EX}) - K_3 \ N_0^2 + q_2 \frac{U}{Q} \right] \quad (6)$$

$$C = C_0 + \Delta t \left[\beta \ (\text{GROW − RESP}) + K_2(C_s - C_0) - K_1 \ L_0 - \frac{\text{BEN}}{H} \right] \quad (7)$$

$$L = L_0 + \Delta t \left[q_3 \frac{U}{Q} + \beta' \ (\text{MORT + GRAZ + EX}) - K_1 \ L_0 \right] \quad (8)$$

In each equation, the expression in brackets represents internal and external sinks or sources relating to the variable during the three hours in which we follow the elementary volume. All the parameters used in the model are listed in 1. As temperature simulation has already shown its accuracy in the Loire river (Gras and Gilbert 1978), emphasis is put on the other variables.

Temperature

Heat exchanges in brackets are expressed in $W.m^{-2}$. ρ is the water specific mass (kg.m^{-3}) and C_p is the specific heat of water at constant pressure (J.kg^{-1}.$^\circ$C^{-1}).

Table 1 - SUMMARY OF PARAMETERS DEFINITION.

a	coefficient	0.4	$^\circ C^{-1}$	determined from in situ measurements
a'	coefficient	0.35	$^\circ C^{-1}$	evaluated
T_{opt}	optimal temperature for phytoplankton growth	22	$^\circ C$	measured in situ
T'_{opt}	optimal temperature for zooplankton growth	24	$^\circ C$	measured in situ
G_{max}	maximum growth rate	2.5	day^{-1}	calibrated
I_s	light saturation intensity for phytoplankton	150	$W.m^{-2}$	measured in situ
I_0	light intensity at the water surface		$W.m^{2}$	input data
k'_e	extinction coefficient on the average	0.3	m^{-1}	measured in situ
$T_{max} = T'_{max}$	temperature above wich no growth is possible	37	$^\circ C$	literature
K_5	michaelis constant for PO$_4$	0.02	mgP/l	calibrated
K_4	respiration coefficient	0.03	day^{-1}	literature
K_7	mortality coefficient	0.0004	m3.mg chla^{-1}.day^{-1}	literature
γ	zooplankton grazing rate	0.18×10^{-2}	l.animal^{-1}.day^{-1}	calibrated
Z_0	zooplankton concentration		animal.l^{-1}	input data
K_6	excretion coefficient	0.2	dimensionless	literature
α	phosphorus-to-phytoplankton biomass ratio	0.001	$\frac{\text{mg P/l}}{\text{mg chla/m3}}$	literature
β	produced oxygen to produced carbon ratio	0.13	$\frac{\text{mg O}_2\text{/l}}{\text{mg chla/m3}}$	calibrated
β'	Edible oxygen to excreted carbon ratio	0.06	$\frac{\text{mg O}_2\text{/l}}{\text{mg chla/m3}}$	calibrated
Rate constants at 20°C				
$K_{3,20}$	phosphorus decay	0.2	day^{-1}	literature
$K_{1,20}$	BOD decay	0.22	day^{-1}	calibrated
BEN_{20}	benthal oxygen demand	0.1	g.m^{-2} day^{-1}	evaluated
Temperature constants θ in $K_T = K_{20} \theta^{(T-20)}$				
	benthal oxygen demand	1.04		literature
	phosphorus decay	1.02		literature
	BOD decay	1.047		literature
	Reaeration	1.025		literature

Phytoplankton

Growth equation

$$GROW = G_{max} \; RAD \; g \; (T_o) \; \frac{N_o^1}{N_o^1 + K_5} \; P_o \qquad (9)$$

. growth rate RAD due to insolation uses Steele's (1962) law. It is integrated and averaged along the vertical

$$RAD = \frac{e}{k_e \, H} \; (e^{-\frac{I_o}{I_s} e^{-k_e H}} - e^{-\frac{I_o}{I_s}}) \qquad (10)$$

extinction coefficient k_e is calculated by Riley's (1946) formula $k_e = k_c + 0.0088 \, P_o + \qquad (11)$
$0.054 \, P_o^{2/3}$.

. $g \; (T_o)$ represents the temperature effect on phytoplankton growth. Lassiter and Kearns (1975) formula which accurately represents algae behaviour observed in the Loire river is used:

$$g \; (T_o) = e^{a(T_o - T_{opt})} \; (\frac{T_{max} - T_o}{T_{max} - T_{opt}})^{a(T_{max} - T_{opt})} \qquad (12)$$

. $\dfrac{N_o^1}{N_o^1 + K_5}$ is the phosphorus action represented $\quad (13)$ by Monod growth kinetics.

Sink equations

. $RESP = K_4 \; e^{0.06 \, T_o} \; P_o$ represents the endogeneous $\quad (14)$ respiration term

It is now widely accepted that the endogeneous respiration variation *vs.* temperature follows an exponential law and is maximum for temperatures greater than 32°C. As the temperature does not reach this value in the model applications, a Lassiter and Kearns law is not necessary and a simple exponential law is sufficient.

. $EX = K_6 \times RESP$ is the excretion factor $\quad (15)$
(Larimore et al. approach

. $GRAZ = \gamma \; g'(T_o) \; Z_o \; P_o$ expresses the grazing $\quad (16)$ of phytoplankton by zooplankton

For the determination of GRAZ, we adopt the Di Toro et al. (1975) viewpoint and consider this death term to be proportional to the phytoplankton concentration.

$g' \; (T_o)$ represents the temperature effect on grazing (Lassiter and Kearns 1975 law). It leads to the definition of a', T'_{opt}, T'_{max}.

. MORT is the mortality term due to causes other

than grazing. Bierman's (1980) formula is used:

$$MORT = K_7 \; (1.08)^{T_o - 20} \; (P_o)^2 \qquad (17)$$

We assume that the main factors which may cause algal mortality are temperature and algal supersaturation. This mortality term which is small under normal conditions becomes significant for high temperatures and high phytoplankton concentrations.

Unavailable phosphorus

q_2 is the continuous injection flow ($g \; P \; m^{-1}$ day^{-1}) the value of which depends on the reach. Other sources are phosphorus from dead phytoplankton and excreted phosphorus. As zooplankton population is not modeled, phosphorus contained in grazed phytoplankton is assumed to compensate phosphorus excreted by zooplankton. That amounts to assuming the zooplankton population to be steady. Conversion of unavailable phosphorus into available phosphorus is supposed to be a first order reaction.

Available phosphorus

q_1 is the continuous injection flow ($g \; P \; m^{-1}$ day^{-1}) whose value depends on the reach. Another source term comes from phosphorus mineralization. Losses are due to phytoplankton assimilation.

Dissolved oxygen

Sources are photosynthetic production ; sinks are phytoplankton respiration, reaeration process, benthal demand and BOD decay.

The diffusion of oxygen into the air at the water surface is represented by the widely accepted relationship : $K_2 \; (C_s - C_o)$. The saturation value of oxygen is given by Truesdale's law : $C_s = 14.161 - 0.3943 \; T + 0.007714 \; T^2 - \qquad (18)$
$0.0000646 \; T^3$. Churchill's (1962) formula is used to calculate K_2 at 20°C :

$$K_{2,20} = 5.02 \; U^{0.969} \; H^{-1.673} \qquad (19)$$

A rough estimation of the quantity of oxygen created by photosynthesis can be derived from the basic equation expressing glucose sugar production.

$$6 \; CO_2 + 6 \; H_2O + light \longrightarrow C_6 H_{12} O_6 + 6 \; O_2 \qquad (20)$$

Therefore, the ratio of produced oxygen to produced carbon is $2.67 \; mg.O_2.l^{-1}$. If the phytoplankton carbon to total chlorophyll a ratio is chosen inside the usually accepted 30 - 50 mg C/mg chla range, the value should be between 0.08 and 0.13 Field data at Saint-Laurent-des-Eaux showed that in summer 1977 and 1978 the phytoplankton carbon to total chlorophyll a ratio had values for the most part between 30 and 50.

. The respiration process like benthal demand is assumed to have a rapid effect on DO concentration. Consequently, it is assimilated to an oxygen sink rather than a BOD sink. On the other hand, dead and excreted phytoplankton which undergo a slower decay are assimilated to a BOD sink.

B.O.D.

q_3 is the continuous injection flow (converted into a oxidizable potential mg.O_2/l) whose value depends on the reach. In order to accurately define the DO balance, the L_0 variable, appearing in the DO sink - $K_1 L_0$ due to oxidation processes, must represent the quantity of oxygen that would be necessary to oxidize all the organic matters contained in a water volume unit with the exception of live algae. So as to determine the values of this modeled BOD at the beginning of the stretch (Belleville) from the BOD_5 and chlorophyll a measurements, the following formula is used :

$$L = \frac{BOD_5}{1 - e^{-5 K_{1,20}}} - \beta ' P \qquad (21)$$

Equation (21) and the $K_1 L$ term in equations (7) and (8) assume that the oxidation processes follow a first order chemical reaction according to the widely accepted Streeter and Phelps law. In equation (21) $\beta ' P$ is a correction term taking into account the fact that phytoplankton which was alive during sample taking was dead during the BOD_5 measurement.

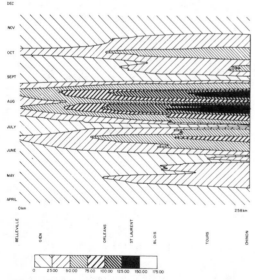

DEC
NOV
OCT
SEPT
AUG
JULY
JUNE
MAY
APRIL

0km   258km

BELLEVILLE   GIEN   ORLEANS   ST LAURENT   BLOIS   TOURS   CHINON

0   25.00   50.00   75.00   100.00   125.00   150.00   175.00

Figure 5.-- Calculated spatiotemporal distribution of the phytoplankton concentration in 1977.

Equation (21) is used in the opposite direction in order to enable, at the monitoring stations, the comparison between measured BOD_5 values and calculated values.

RESULTS OF THE MODEL

Two different types of computer output were obtained : one showing concentration vs. position (kms) and time (fig. 5) and the other showing the concentration vs. time at places where monitoring stations are situated. The s.

type of output is used to calibrate and verify the model. Apart from data at Belleville which are the inputs of the model, chlorophyll a, temperature, DO, BOD_5 and PO_4 measurements were collected at Dampierre (future power plant, position n° 1), upstream from the St-Laurent-des-Eaux power plant (position n° 3) and upstream from the Chinon power plant (output of the model at the present time, position n° 5) at the average rate of two measurements per month (measurement series n°1). Moreover, DO measurements taken once a week and daily temperature measurements were available at the water intake of the St-Laurent and Chinon power plants (positions n°3 and n°6 and measurements series n°2). Finally, PO_4, DO, BOD5 concentrations were measured once a month at Orléans (position n°2) and Tours (position n°4) by l'Agence Financière de Bassin Loire-Bretagne (series n° 3).

Model calibration

Saint-Laurent-des-Eaux (position n° 3) was chosen as the calibration station because of its measurement richness and because of its location in the middle of the stretch. Comparison between model results and field data is presented in figure 6. DO concentrations (series n° 2) collected at nine o'clock in the morning once a week are compared with values calculated by the model at nine o'clock every day. For PO_4, chlorophyll a, BOD_5 and temperature whose diurnal variations are less significant, daily average values are chosen for the model outputs. Calculated temperatures are compared with temperature measurements carried out every morning at ten o'clock (series n° 2). Calculated PO_4, chlorophyll a and BOD_5 concentrations are compared with the field data of series n° 1.

Model verification

The model is then verified at stations 1, 2, 4, 5 and 6. A good agreement is achieved between model results and available data at stations 1, 2 and 4 (fig. 7). However a significant, and at first sight inexplicable, gap was observed between the computed values and field data at the monitoring station n°5 especially in summer. Many hypotheses have been put forward to explain such a marked discrepancy for chlorophyll a ($\frac{\text{observed concentrations}}{\text{calculated concentrations}} = \frac{1}{4}$ in summer 1977) : Cher river influence, toxic pollutants, grazing increase below St-Laurent, turbidity, etc ...

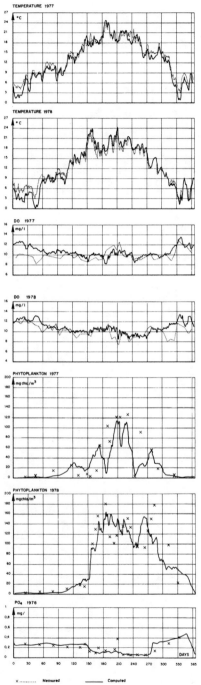

Figure 6.-- Some calibration results at
Saint-Laurent-des-Eaux.

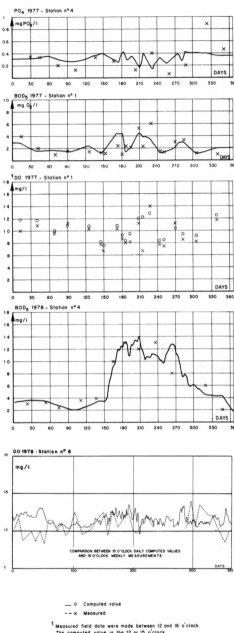

Figure 7.-- Some verifications of the model.

But from all the available physico-chemical and biological data, none of the advanced reasons could explain such a difference. Finally, a closer look at the location of the monitoring station which was said to be representative of Loire water has shown that, in fact, this station was badly situated and had measured the water quality of a Loire – Indre mixture mainly composed of Indre water. This unfortunately prevented us from verifying the model as far as Chinon for BOD_5, chla and PO_4 concentrations.

Another difficulty was to verify the diurnal outputs of oxygen concentration since all the available data had been collected during the day and no continuous measurements were taken of the levels reached at night in 1977 and 1978. However continuous measurements of oxygen concentration have been made during other periods. They show that the model results are realistic. In particular, they confirm the model is right in displaying that very often in summer DO concentration can be greater than the DO saturation concentration during the whole of the day when chlorophyll a concentration exceeds 40 mg/m3 (fig. 8).

Analysis of the model sensitivity

Analysis of sensitivity to coefficients has been carried out. That gives rise to the following salient remarks (fig. 9).

. In summer, growth dynamics of the phytoplankton and consequently DO and BOD_5 contents are more sensitive to the zooplankton grazing rate in 1978 by comparison with 1977. The reason is that zooplankton concentration in the Loire river was about four times greater in 1978 than in 1977, whereas phytoplankton concentrations were almost similar. From this, we can see how important information on zooplankton concentration may be when constructing a water quality model.

. Sensitivity of phytoplankton growth dynamics, DO, phosphorus and BOD_5 contents to the Michaelis constant for phosphorus concentration was more significant in the summer of 1978 than in the summer of 1977. This is due to the low values attained by PO_4 concentration in 1978. This phenomenon is visible in August and September 1978 where PO_4 concentrations do not exceed 0.05 mg PO_4/l and have a limiting action on algal growth.

Model application

The water quality model can be applied in numerous ways in order to study management strategies or reach some understanding of the Loire river ecosystem (effect of flow reduction or

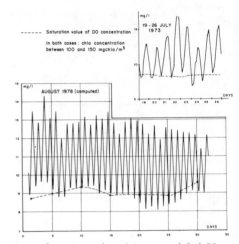

Figure 8.-- Comparison between modeled DO concentrations and continuous measurements at Saint-Laurent-des-Eaux.

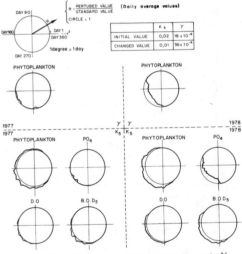

Figure 9.-- Model sensitivity to K_5 and γ. Results at Saint-Laurent-des-Eaux.

increase, impact of plants implantation, effect of any change in effluent discharges, effect of reducing concentration of phosphorus in overland flows, etc...).

Two examples of model application are discussed here :

- Impact of the 1990 nuclear programme (power plants with cooling towers) ;

- Impact of two fictive 900 MW once-through-cooling power plants (Belleville and Dampierre)

In 1990, 8 power plants will be installed
along the stretch under consideration (2 x 1300
MW at Belleville 4 x 900 MW at Dampierre and
2 x 900 MW at Saint-Laurent). They are all power
plants with cooling towers. Moreover the 2 x 400
MW Saint-Laurent power plants which are working
at the present time will be closed down before
1990. The response of the model to this 1990
programme is studied for the years 1977 and 1978
and is compared to the reference situation.
From Belleville down to Saint-Laurent, the heating of the water due to these installations, by
comparison with the original situation, is
always lower than 1°C. Owing to the low flows
at the end of 1978 and because the difference
between river temperature and the temperature
of the drainage discharge from the cooling towers
(which depends on air temperature and humidity)
is at a maximum when air and water temperatures
are low (winter period), the maximum increase
in the river temperature is reached in autumn
1978 **(fig. 10)**. This low increase of temperature has very little effect on the other variables. From Saint-Laurent down to Chinon the
replacement at Saint-Laurent of the once-through
cooling power plants by plants with cooling
towers leads to a slight cooling of the river
temperature (< 5°C) by comparison with the original situation. The model shows that the other
variables are not affected very much **(fig. 11)**.

The model is then studied under more severe
conditions (river temperature increase up to
6°C, river temperature up to 32°C). The studied
scenario is : two 900 MW once-through-cooling
power plants working continuously except when
the flow is too low (no day in 1977 and in 1978);
one at Belleville, the other at Dampierre. The
model results are compared to the reference state
at Orleans **(fig. 12)**.

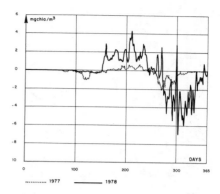

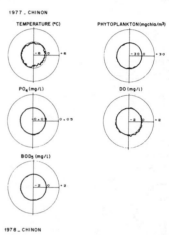

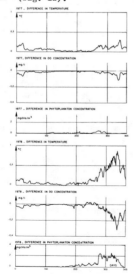

Figure 10.-- Results of the 1990 program
compared with the real figures at Orleans.

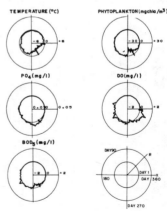

Figure 11.-- Difference in photoplankton
concentration between 1990 program and
reference state at Blois. General compar-
son at Chinon upstream from the power
plants (Daily average values).

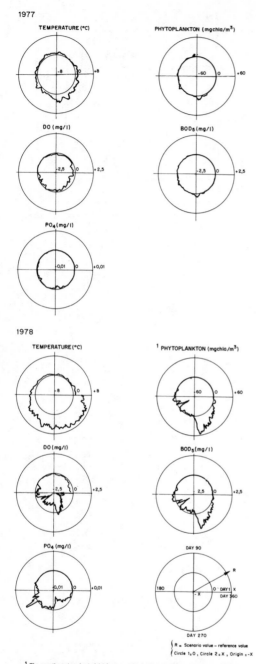

1977

TEMPERATURE (°C)   PHYTOPLANKTON (mgchla/m³)

DO (mg/l)   BOD₅(mg/l)

PO₄(mg/l)

1978

TEMPERATURE(°C)   ¹ PHYTOPLANKTON (mgchla/m³)

DO(mg/l)   BOD₅(mg/l)

PO₄ (mg/l)   DAY 90

DAY 270

R = Scenario value – reference value
Circle 1 = O , Circle 2 = X , Origin = -X

¹ The negative values (end of july) are associated with temperatures getting near 30°C

Figure 12.-- Results of the second scenario
compared with the real figures at Orleans.

Special attention is now paid to the phyto-plankton equation :
- For temperatures lower then T'_{opt}, a temperature increase due to these installations does not provoke any growth of zooplankton population since zooplankton is not modeled. Though $g'(T_0)$ is affected by the temperature increase, the grazing term GRAZ might however be slightly underestimated.

- On the other hand, for river temperatures greater than T'_{opt}, Z_0 value is not affected by this increase and GRAZ may be overestimated.

Therefore, developing a zooplankton equation seems essential to improve the modelization. Another criticism of the phytoplankton equation can come from the fact that the model does not differentiate between one species and another. This may be a weakness under certain conditions. From a biological point of view, it is probable for instance that our approach is pessimistic if high temperatures are continuously maintained because it does not take into account the fact that certain algal species could appear and bear these thermal conditions.

That is why the response of the model to this second scenario (2 x 900 MW) must be regarded as indicative of a probable trend rather than an exact solution. Lack of knowledge of phytoplankton behaviour to these high temperatures (mortality, substitution of a species for another, etc...), lack of zooplankton equation prevent us from getting very close to the numerical values in such conditions.

CONCLUSION

- The model developed is capable of representing the evolution of DO, BOD, temperature, PO₄ and phytoplankton along the river from Belleville to Tours in 1977 and 1978.

- The model shows that reaeration processes and photosynthesis activity (in summer and autumn) are sufficient to maintain a high level for the oxygen concentration because they remain more important than oxidation processes (DO concentration above 6 mg/l in 1977 and 1978).

- The model confirms that phosphorus can be limiting for algal growth in the Loire river.

- The model shows that field data measurements of zooplankton are necessary to enable a good representation of chlorophyll a content.

- The model was applied to show how the different variables would respond if the nuclear

in 1977 and 1978. The variables are hardly affected by these installations.

- The final application of the model was a simulation of a more severe condition in which two
fictive once-through-cooling power plants
were installed. In the autumn of 1978, a period
during which there were low waters, phytoplankton content would have increased by 20 % compared with the real figures.

LITERATURE CITED

Bierman, V.J., Dolan, Stoermer, Gannon and Smith.
1980 : The development and calibration of a
spatially simplified multiclass phytoplankton
model for Saginaw Bay, Lake Huron. Contribution n° 33. Environmental Protection
Agency Ecological Research Series. GLEPS,
126 p.

Champ, P. 1980. Biomasse et production primaire
du Phytoplankton d'un fleuve : La Loire au
niveau de la centrale nucleaire de Siant-
Laurent-des-Eaux. Acta Oecologica. Oecol.
Gener. Vol. 1 n° 2 p. 111-130.

Churchill, M., M. Elmore and R. Buckingham. 1962.
The prediction of stream rearation rates.
J. Sanitary Eng. Div. July 1962, n° SA4, Vol.
88.

Di Toro, D., D.J. O'Connor, R.V. Thomann, J.L.
Mancini. 1975. Phytoplankton-zooplankton-
nutrient interaction model for western lake
Erie. In B.C. Patten (Ed.) Systems analysis
and simulation in ecology, Vol. 3, Academic
Press. New York. pp. 423-474.

Gosse, P. 1981. Un modele de qualite d'eau de
riviere. Application a la Loire moyenne.
Rapport n° 1. Electricite de France. Direction des Etudes et Recherches. 78400 CHATOU
Rapport EDF/DER. E31-78 n° 7.

Gras, R. 1969. Simulation thermique d'un cours
d'eau a partir des donnees d'un reseau mete-
orologique Classique. Congres de l'AIRH,
Tokio 1969.

Gras, R. and A. Gilbert. 1978. Etude de l'impact
du site de Billeville sur le regime thermique
de la Loire. Electricite de France. Direction
des Etudes et Recherches. 78400 CHATOU.
Rapport EDF/DER-E31-78/n° 7.

Khalanski, M. and Renon. 1977. Evolution de la
teneur en pigments planctoniques dans la Loire
entre Dampierre et Montsoreau. Cahiers du
Laboratoire d'Hydrobiologie de Montereau n° 5.

Lair, N. 1980. The rotifer fauna of the Loire
river. Hydrobiologia 73, p. 153-160.

Larimore, R.W., Mac Nurney et al. 1979. Evaluation of a Cooling Lake Fishery. Vol. 2:
Lake Sangchris Ecosystem Modeling. Illinois
Natural History Survey Urbana. EPRI. EA
1148. Research Project 573.

Lassiter, R. and D. Kearns. 1975. Phytoplankton
population changes and nutrient fluctuations
in a simple aquatic ecosystem model. In
E.J Middlebrooks, D.H. Falkenborg, T.E.
Maloney (Eds.). Modeling the eutrophication
process. pp. 131-138.

Riley, G. 1946. Factors controlling phytoplankton
population on Georges Bank. J. Mar. Res.,
6 (1).

Steele, J. 1962. Environment control of photosynthesis in the sea. Limnol. Oceanogr.
n° 7.

MODELLING OF GRANGENT RESERVOIR ECOSYSTEM
Véronique Garçon[2]

Abstract.-- A numerical, one-dimensional, vertical, unsteady,
internal nutrient pool, phytoplankton simulation model is developed
and applied to field data acquired in 1978 on GRANGENT reservoir (FRANCE).
The coupled partial differential equations predict the local changes
in the variables due to vertical turbulent mixing, advection and
biological processes.

INTRODUCTION

Accelerated rates of eutrophication in reservoirs have become a major water quality issue. Excessive phytoplankton growth interferes with the uses of these water bodies as sources for drinking water and water recreation. Eutrophication is caused by excessive nutrient enrichment due to, for instance, agricultural runoff, wastewater and sewage discharges. The relationships between physical, chemical and biological processes occuring in deep stratified reservoirs are so complex that passive storage of field data in order to establish inventories is now insufficient to have a good understanding of the ecological behaviour of a reservoir ecosystem and to help formulate effective management strategies to improve water quality of the reservoir and of the outflow. To develop management strategies for dealing with the problem of eutrophication, mathematical simulation models have been developed in order to describe the cause-effect connection between external nutrient loading and phytoplankton growth in the water body (Di Toro et al. 1971, Chen and Orlob 1975, Larsen et al. 1974, Thomann et al. 1975, Canale et al. 1976, Park et al. 1974, Desormeau 1980, Bierman et al. 1980, Scavia 1980).

The objectives of the present research are twofold :
- develop a mathematical model able to describe the ecological behaviour of GRANGENT reservoir located on the upper course of the river Loire in the Eastern section of the Massif Central, in FRANCE (fig. 1);

- use the model as a tool for predicting the effect of reservoir management through a multiple outlet structure on the water quality.

Actually, only the first objective has been achieved.

It is well known that primary productivity in reservoirs is dependent upon the distribution of essential algal nutrients, the availability of light, and the water temperature. Each of these factors depends upon turbulent mixing characteristics which are determined, in turn, by local meteorological conditions and depends upon the inflow - outflow process which contributes, for a great part, to the vertical structure of the reservoir.

The work presented here can be roughly divided into two parts :
- the formulation of a model as accurate as possible representing the thermal annual cycle of a reservoir ;
- the formulation of a biologically realistic model involving the following dependent state variables : phytoplankton (diatoms and green algae), internal nutrients (nitrogen and phosphorus), external nutrients (nitrogen, phosphorus and silicon), herbivorous zooplankton and organic matter.

We set up the general equations governing the evolution of a reservoir ecosystem in the next section . Section 3 is devoted to a very brief description of the modelling of transport mechanisms and thermal structure evolution. In section 4, we give the biological model structure (choice of state variables, forcing functions, biological processes, analytical formulation of the biological interactions, numerical procedure and sensitivity analysis) and the comparison between field data and simulation results for the year 1978. By way of conclusion, we'll develop the further use of the model for reservoir water quality management optimization.

[1] Paper presented at the International Symposium Energy and Ecological Modelling, sponsored by the International Society for Ecological Modelling. (Louisville, Kentucky, April 20-23, 1981).
[2] Véronique Garçon is a Predoctoral Student in Energy and Pollution at University Paris VII, Paris, France. This study was supported by a grant from the Electricité de France, Chatou, France.

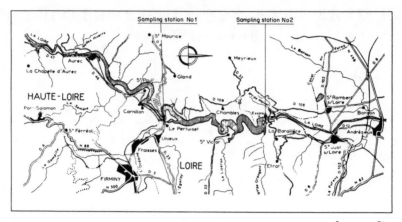

Figure 1.-- Map of GRANGENT reservoir.

0   3 km

THE MODEL GENERAL EQUATIONS

The variables temperature θ, velocity $\vec{U}$, radiative heat flux $\vec{R}$, density P are chosen to describe the dynamic and thermodynamic evolution of the waterbody. The variable X^{α} describes the concentration evolution of a biological component, with $\alpha = 1, \ldots, \mathcal{N}$, $\mathcal{N}$ being the number of biological state variables selected for an accurate representation of the ecosystem.

The hypotheses employed in this study concern the existence of a "spectral gap" in the energy spectrum of the variables, the Boussinesq approximation (Coantic 1978) and the physical existence of a multilayered structure of the waterbody ; we assume also the mean variables $\bar{\theta}$ and $\bar{X}^{\alpha}$ to be homogenous in each physical layer of elementary volume of water δV.

The local equations are the following :

$$\bar{P} = P_r \left[1 - \beta (\bar{\theta} - \theta_r) \right]$$

$$\text{div } \vec{U} = 0 \qquad (1)$$

$$\frac{\partial \bar{\theta}}{\partial t} + \text{div } (\bar{\theta}.\vec{U}) = -\text{div } (\overline{\theta'\vec{U'}}) - \frac{1}{P_r \, C_p} \text{ div } \vec{R} \quad (2)$$

$$\frac{\partial \bar{X}^{\alpha}}{\partial t} + \text{div } (\bar{X}^{\alpha}.\vec{U}) = -\text{div } (\overline{X'^{\alpha}\vec{U'}}) + \bar{F}\alpha \quad (3)$$

where the subscript r denotes constant reference values of the water density and temperature, θ' and $\vec{U'}$ denote the fluctuating parts of θ and $\vec{U}$, β is the thermal expansion coefficient, C_p is the specific heat and $F\alpha$ is the biological sources and sinks term.

As a solution technique to integrate this set of partial differential equations, we treat separately the heat transfer at the air-water interface and mixing due to wind action and convective processes, the advective processes and the biological processes, as they should occur successively within a time step, namely :

- the first part, relative to atmospheric exchanges effects:

$$\left(\frac{\partial \bar{\theta}}{\partial t}\right)_1 = -\text{ div } (\overline{\theta'\vec{U'}}) - \frac{1}{P_r \, C_p} \text{ div } \bar{\bar{R}} \quad (4)$$

$$\left(\frac{\partial \bar{X}^{\alpha}}{\partial t}\right)_1 = -\text{ div } (\overline{X'^{\alpha} \, \vec{U'}}) \quad (5)$$

- the second part, relative to inflow mixing and outflow withdrawal :

$$\left(\frac{\partial \bar{\theta}}{\partial t}\right)_2 = -\text{ div } (\bar{\theta} . \vec{U}) \quad (6)$$

$$\left(\frac{\partial \bar{X}^{\alpha}}{\partial t}\right)_2 = -\text{ div } (\bar{X}^{\alpha} . \vec{U}) \quad (7)$$

- the third part, relative to ecological interactions :

$$\left(\frac{\partial \bar{X}^{\alpha}}{\partial t}\right)_3 = \bar{F}\alpha \quad (8)$$

PHYSICAL MODEL STRUCTURE

Parameterization of the transport mechanisms induced by atmospheric exchanges

We follow here the Stefan and Ford's approach (1975): a one dimensional (vertical), unsteady, "integral" model based on energy balance scheme.

The basic assumption of an "integral" model is that temperature is uniform within the mixed layer. The two unknowns introduced are temperature θ and depth h of the mixed layer; the equation considered is the conservation equation of internal energy (4). A further equation necessary to close the system is the vertical integral of the turbulence kinetic energy equation in its balanced form (Niiler and Kraus 1977) :

$$\underbrace{\int_{-h}^{o} g\,\beta\overline{\theta'w'}dz}_{P} = \underbrace{\int_{-h}^{o} \overline{w'\vec{U'}} . \frac{\partial \vec{\bar{U}}}{\partial z} \, dz}_{M} + \underbrace{\int_{-h}^{o} \bar{\varepsilon} .dz}_{-D} \quad (9)$$

where,

P represents the rate of working of the buoyancy force,

M represents the rate of turbulent kinetic energy production due to the mean shearing flow,

D represents the rate of dissipation of turbulent kinetic energy into heat.

Surface heat transfer and convective mixing.

At each time step, the surface flux function - turbulent and radiative fluxes at the air-water interface - is evaluated. When cooling occurs at the surface creating density instability, the surface water sinks and convective mixing takes place. The depth of the mixed layer is determined from the balance between the potential (buoyant) energy deficit of the mixed layer relative to the layer immediately below it and a term proportional to the surface flux function. Knowing the new depth of the mixed layer, we get the new water temperature profile from the internal energy equation.

In case of heat input in the surface layer, the water density profile is stable ; so no entrainment or layer deepening occurs. The new water temperature profile is computed using the internal energy equation.

Mixing due to wind action.

To decide whether wind energy (work) applied at the water surface will change the accumulated potential (buoyant) energy or will be dissipated by viscosity, resulting in no additional mixing, an integral stability criterion R_i in the form of an energy ratio will be used ; this Richardson number determined from the turbulent kinetic energy equation expresses the balance between potential and kinetic energy of the mixed layer, namely :

$$R_i \equiv \frac{g \beta h \Delta \theta}{|\Delta \vec{U}|^2} \qquad (10)$$

In the case of R_i smaller than a critical value R_{ic}, entrainment occurs, resulting in an increase of the buoyant potential energy. The process is continued until R_i reaches R_{ic} (in agreement with R_i as an increasing function of h). The depth of the mixed layer is the value of h at which R_i equals R_{ic}. We may therefore calculate the temperature profile using the internal energy equation.

In the case of R_i larger than R_{ic}, no entrainment is possible, the thickness of the mixed layer is not varied.

Stefan and Ford parameterize $\frac{1}{2} h |\Delta \vec{U}|^2$ as follows :

$$\frac{1}{2} h |\Delta \vec{U}|^2 = u_*^3 \cdot \Delta t \qquad (11)$$

where,

u_* is the water shear velocity,

Δt is the accumulation time of wind energy input.

It must be kept in mind that whenever a mixed layer is induced at the surface of the reservoir not only the temperature, but every biological variable must be averaged over it (equation (5)) the mixed layer being homogeneous in every respect.

Parameterization of the transport mechanisms due to advection transport

The one dimensional transport model chosen implemented by Gaillard (1981) is very similar to the one developed by Markofsky and Harleman (1971).

The basic equations relative to the advective processes are the equations (1), (6) and (7). They are integrated over an elementary volume δV and then discretized in space and in time.

The main assumptions are the following :

the inflow enters the reservoir water column at its own density level,

the outflow is centered about the level of the outlet,

the inflow and outflow velocity fields are normally distributed (Huber and Harleman 1968, Koh 1964); the inflow and outflow standard deviations of the Gaussian curves are constant values.

We determine the vertical flows from the discretized mass conservation equation and we solve the system of N linear equations with N unknowns being the new values of temperature (or of concentrations of biological components) for the N layers of the reservoir (Gaillard 1981).

BIOLOGICAL MODEL STRUCTURE

Structural complexity

We define here the structural complexity as the choice of compartments in the model, of the processes modeled and the forcing functions for these processes.

The state variables

Phytoplankton

Phytoplankton biomass, on the basis of dry weight, is resolved into two functional groups : diatoms and green algae. Both of them require phosphorus and nitrogen as they are major components of proteins, nucleic acids and lipids ; in addition, diatoms require silicon for their test constitution. They are the main groups represented in GRANGENT reservoir.

Zooplankton

The herbivorous zooplankton is considered homogeneous. Although no data are available, we choose to simulate the zooplankton biomass dynamics in order to get a better estimation of the grazing term all over the year.

Nutrients

The nutrients included in the model are phosphorus, nitrogen and silicon because of their role as potential limiting nutrients to phytoplankton growth.

We consider algal growth to be a two-step process involving separate nutrient uptake and cell synthesis mechanisms (Rhee 1973, 1974, 1978, Droop 1973, 1974, Fuhs et al. 1972, Azad and Borchardt 1970). That permits simulation of storage of surplus nutrients (free amino-acids and eserve proteins for nitrogen and polyphosphate fractions for phosphorus). No internal nutrient pool is included actually for silicon.

We deliberately choose a variable stoichiometry, internal pool model, thus adding complexity in terms of computation and degrees of freedom though we have no data for phytoplankton internal nutrients. Indeed the experimentally observed modes of phytoplankton nutrient uptake and growth can be described over the full range of dynamic conditions

Organic matter

We mean by organic matter detritus, that is, excretion products, dead algal cells, food ingested by zooplankton but not assimilated, therefore particulate and dissolved organic matter.

Forcing functions

They include water temperature, incident solar radiation and external loadings. Water temperature is computed internally by the model (see previous section). A time-series for incident solar radiation is developed using Perrin de Brichambaut's (1975) method and is supplied as input to the model. Phosphorus, nitrogen and chlorophyll a loadings from the river Loire are determined using the data of the sampling station No 1 (see fig. 1). We estimate internal nutrient biomass loadings from algal biomass loading according to Bierman (1976), Rhee (1973, 1978) and Strickland (1965) indications.

Loadings of organic matter, silicon and zooplankton are unknown, we choose probable values constant all over the year.

Figure 2 illustrates how the state variables are linked by the biological processes considered in the model.

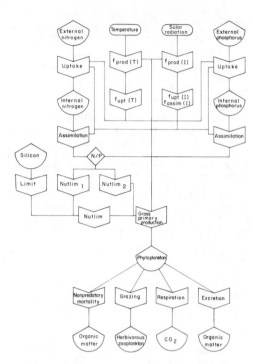

Figure 2.-- Flow diagram for the biological model showing relationships among the major processes and compartments.

Functional complexity

We define here the functional complexity as the choice of the mathematical formulations of the ecological interactions (equation (8)). The state variable equations, for each physical layer of the reservoir, are given in table 1. The state variable equations, for each physical layer of the reservoir, are given in table. 1. The definition for state variables and the definition, value, unit and reference for the coefficients included in these formulations and this paragraph are summarized in table 2.

Phytoplankton kinetics
Nutrient uptake : UPT

The nutrient uptake rate is a function of both external and internal nutrient levels, it includes appropriate feedback based on inhibitory mechanisms. As suggested by Rhee (1973, 1974, 1978), phosphorus and nitrogen uptake can be described by an equation for non-competitive enzyme inhibition.

Assimilation of internal nutrients : ASSIM

A Monod-type equation represents internal nutrient assimilation.

Table 1.-- State variable equations for each physical layer of the reservoir.

Diatoms

$$\frac{dx^1}{dt} = \left[\underbrace{K_p.1.f_{prod}(T).f_{prod}(I).NUTLIM_1}_{\text{gross primary production}} - \underbrace{\sigma\,e^{bT}}_{\text{respiration}} - \underbrace{Km_1.1.T.(x^1+x^2)}_{\text{non predatory mortality}} - \underbrace{\frac{Cm\,w_1\,x^7}{Kg+\sum\limits_{i=1}^{2}W_i\,x^i}}_{\text{grazing}} - \underbrace{K_{ex_1}.(PPROD_1-RESP_1)}_{\text{excretion}} \right].x^1$$

$$\qquad\qquad PPROD_1 \qquad\qquad RESP_1 \qquad MOR_1 \qquad BROUT_1 \qquad EXC_1$$

Green algae

$$\frac{dx^2}{dt} = \left[PPROD_2 \qquad - RESP_2 \quad - MOR_2 \quad - BROUT_2 \qquad - EXC_2 \right].x^2$$

Internal nitrogen diatoms

$$\frac{dx^3}{dt} = \left[\underbrace{-\frac{V_{max,N}}{K_{SN}+x^3}.f_{assim}(I).x^1}_{\substack{\text{assimilation of}\\\text{internal nitrogen}}} - \underbrace{MOR_1 - BROUT_1 - EXC_1}_{\substack{\text{loss of internal nitrogen}\\\text{due to non predatory mortality,}\\\text{grazing and excretion.}}} \right].x^3 + \left[\underbrace{V_{max,N}.f_{upt}(T).f_{upt}(I)\left(\frac{1}{1+x_{mN}/x^9}\right)\left(\frac{1}{1+x_{inh,N}/K_{inh,N}}\right)}_{\substack{\text{uptake of external nitrogen with appropriate feedback}\\\text{based on inhibitory mechanisms}}} \right].x^1$$

$$\qquad\qquad\qquad ASSIM_{N1} \qquad\qquad\qquad\qquad\qquad\qquad\qquad UPT_{N1}$$

Internal phosphorus diatoms

$$\frac{dx^4}{dt} = \left[- ASSIM_{P1} \quad - MOR_1 - BROUT_1 - EXC_1 \right].x^4 + UPT_{P1}.x^1$$

Internal nitrogen green algae

$$\frac{dx^5}{dt} = \left[- ASSIM_{N2} \quad - MOR_2 - BROUT_2 - EXC_2 \right].x^5 + UPT_{N2}.x^2$$

Internal phosphorus green algae

$$\frac{dx^6}{dt} = \left[- ASSIM_{P2} \quad - MOR_2 - BROUT_2 - EXC_2 \right].x^6 + UPT_{P2}.x^2$$

Herbivorous zooplankton

$$\frac{dx^7}{dt} = \left[\underbrace{\frac{a_z.Cm\sum\limits_{i=1}^{2}W_i\,x^i}{Kg+\sum\limits_{i=1}^{2}W_i\,x^i}}_{\text{zooplankton feeding}} - \underbrace{c\,e^{dT}}_{\text{respiration}} - \underbrace{Kz,m}_{\substack{\text{predation by the}\\\text{upper trophic level}}} \right].x^7$$

Organic matter

$$\frac{dx^8}{dt} = \underbrace{- D.x^8}_{\substack{\text{organic matter}\\\text{decay}}} + \underbrace{a_1.(MOR_1+EXC_1).x^1}_{\substack{\text{excretion products,}\\\text{dead algal cells.}}} + \underbrace{a_2.(MOR_2+EXC_2).x^2}_{\substack{\text{excretion products,}\\\text{dead algal cells}}} + \underbrace{(1-a_z).Cm\frac{\sum\limits_{i=1}^{2}W_i\,x^i}{Kg+\sum\limits_{i=1}^{2}W_i\,x^i}.x^7}_{\substack{\text{food ingested by zooplankton}\\\text{but not assimilated}}}$$

External nitrogen

$$\frac{dx^9}{dt} = \underbrace{- UPT_{N1}.x^1 - UPT_{N2}.x^2}_{\substack{\text{uptake of external nitrogen}\\\text{by diatoms and green algae}}} + \underbrace{\beta_N.D.x^8}_{\text{remineralization}}$$

External phosphorus

$$\frac{dx^{10}}{dt} = \underbrace{- UPT_{P1}.x^1 - UPT_{P2}.x^2}_{\substack{\text{uptake of external phosphorus}\\\text{by diatoms and green algae}}} + \underbrace{\beta_p.D.x^8}_{\text{remineralization}}$$

Silicon

$$\frac{dx^{11}}{dt} = \underbrace{K_{11}.MOR_1.x^1}_{\substack{\text{production of}\\\text{silicon by diatom}\\\text{frustules dissolution}}} - \underbrace{K_{11}.PPROD_1.x^1}_{\substack{\text{uptake for}\\\text{primary production}}}$$

Primary production : PPROD

Growth rate depends on internal nutrient levels and not on external nutrient levels (Droop 1974, 1975 , Rhee 1973, 1974, 1978). We represent the multiple-nutrient limitation by the threshold hypothesis as shown by these authors : when several nutrients are limiting, growth is limited only by the internal nutrient in shortest supply. The limiting nutrient is determined dynamically by calculating the ratio N/P_{cal} (by atoms) for each algae species. Some observed species-specific threshold ratios N/P_{obs} (by atoms) are known . If $N/P_{cal} < N/P_{obs}$, nitrogen is the limiting nutrient, otherwise phosphorus is limiting (Desormeau 1980). The nutrient limitation factor on growth is :

$$NUTLIM = \frac{SQ}{K_{SQ} + SQ} \qquad (13)$$

where,

SQ is the subsistence quota (internal nutrient limiting per unit algal biomass (Droop 1974)).

Silicon kinetics LIMIT is described using the Monod hyperbolic equation for growth rate as a function of external silicon concentration (Davis 1976) :

$$LIMIT = \frac{x^{11}}{K_{Si} + x^{11}} \qquad (14)$$

So for diatoms, we have :

$$NUTLIM = Min (NUTLIM, LIMIT)$$

The effects of light on uptake, assimilation and growth $f_{upt}(I)$, $f_{assim}(I)$ and $f_{prod}(I)$ are based on Steele's formulation (1962) :

$$f(I) = \frac{I_{ave}}{I_s} \exp\left(1 - \frac{I_{ave}}{I_s}\right) \qquad (15)$$

with :

$$I_{ave} = \frac{1}{\delta V} \int_{\delta V} I_o\,e^{-K.z}\,dV \qquad (16)$$

I_{ave} = average light intensity in a mixed layer of volume δV

K = light extinction coefficient described by a linear relationship $K = 0.5\,(1 + x^1 + x^2)$ (m^{-1}).

We choose a linear temperature dependence of nutrient uptake $f_{upt}(T)$ and an exponential one of growth $f_{prod}(T)$, namely :

$$f_{upt}(T) = 0.1\,T \qquad (17)$$

$$f_{prod}(T) = \begin{array}{ll} 1.07^{T-20} & \text{diatoms} \\ 1.08^{T-20} & \text{green algae} \end{array}$$

Respiration : RESP

We use an equation calibrated by Riley (1963), based on experimental data, to evaluate the endogeneous respiration rate as a function of temperature.

Non predatory mortality : MOR

The model includes a temperature-dependent, second order phytoplankton decay mechanism to simulate the effects of microbial decomposition (DePinto et al. 1975).

Grazing : BROUT

The selective grazing expression can be divided in two parts : a Michaelis-Menten term that relates grazing to the concentration of total food and a factor used to partition the total food eaten into each prey category. BROUT is set equal to zero if the concentration of food drops below a critical threshold value.

Table 2.-- State variables, forcing functions and biological coefficients.

State variable	Definition	Unit
x^1	diatoms biomass	g dry wt/m^3
x^2	green algae biomass	g dry wt/m3
x^3, x^5	free amino-acids and reserve proteins concentration for each algae species	g$_{surplus}$ N/m3
x^4, x^6	inorganic polyphosphates concentration for each algae species	g$_{surplus}$ P/m3
x^7	herbivorous zooplankton biomass	mg C/l
x^8	organic matter concentration	mg C/l
x^9	nitrogen concentration	mg N/l
x^{10}	phosphorus concentration	mg P/l
x^{11}	silicon concentration	mg SiO$_2$/l

Forcing function	Definition	Unit
T	Water temperature	°C
I_o	incident solar radiation at the reservoir surface	W m-2

Abbreviation	Definition	Unit
$f_{prod(T)}$	reduction factor for temperature limitation on photosynthesis	unitless
$f_{prod(I)}$	reduction factor for light limitation on photosynthesis	unitless
NUTLIM	reduction factor for internal nutrient limitation for each algae species	unitless
$f_{assim(I)}$	dependence of internal nutrient assimilation on light intensity	unitless
$f_{upt(T)}$	dependence of nutrient uptake on temperature	unitless
$f_{upt(I)}$	dependence of nutrient uptake on light intensity	unitless

Coefficient	definition	Value	Reference
K_{11}	% of silicon by dry weight	50	Bierman 1976
K_{Si}	silicon half saturation constant	0.05 mg SiO$_2$/l	Lehman et al. 1975
Cm	zooplankton maximum ingestion rate	1.8 mgC/mgC/day	Scavia 1980
K_g	half saturation concentration of total phytoplankton for grazing by zooplankton	0.2 mgC/l	Scavia 1980
a_z	zooplankton assimilation efficiency	0.8	Scavia 1980
$K_{z,m}$	predation rate of zooplankton by the upper trophic level	0.2 day^{-1}	Pichot 1980
c, d	zooplankton respiration rates	0.01 day^{-1},0.0693°C^{-1}	Kremer and Nixon 1978
D	decay rate of organic matter	0.015 day^{-1}	Scavia 1980

Coefficient	definition	diatoms	green algae	unit	Reference
K'_p	maximum growth rate	2.1	1.9	day^{-1}	Canale et al. 1976
K_m	non predatory mortality rate	0.0015	0.0015	|day°C.g dry wt/m3|$^{-1}$	Bierman 1976
K_{ex}	proportion of net primary production excreted	0.056	0.056		Desormeau 1978
I_s	optimal light intensity for photosynthesis for uptake	140 112	120 77	W.m^{-2} W.m^{-2}	Larsen et al. 1974 Canale et al. 1976
N/P$_{obs}$	observed threshold ratio (by atoms)	15	30		Desormeau 1979 Rhee 1978
W	selectivity constant for feeding on each algae species	1.0	0.7		Scavia 1980
α	% intracellular carbon by dry weight for each algae species	33	43		Strickland 1965
a, b	phytoplankton respiration rates	0.0175,0.069	0.0175,0.069	day^{-1}, °C^{-1}	Groden 1977

Coefficient	definition	nitrogen	phosphorus	Reference
U$_{max}$	maximum specific growth rate for nutrients	0.2 g$_{surplus}$N/g dry wt/day	0.1 g$_{surplus}$ P/g dry wt/day	Desormeau 1978
K_s	half saturation constant for growth by nutrients	0.005 g$_{surplus}$N/m3	0.05 g$_{surplus}$P/m3	Desormeau 1978
V$_{max}$	maximum uptake rate of nutrients	0.14 g$_N$/g dry wt/day	0.10 g$_P$/g dry wt/day	Desormeau 1978
K_m	half saturation constant for uptake of nutrients	0.05 g$_N$/m3	0.07 g$_P$/m3	Desormeau 1978
X$_{inh}$	concentration of inhibitor for nutrients uptake	g$_N$/m3	g$_P$/m3	calculated
K$_{inh}$	half saturation constant for inhibition by free amino-acids (nitrogen) and by acid soluble polyphosphate fraction (phosphorus)	0.0005 g$_N$/m3	0.0001 g$_P$/m3	Desormeau 1978
K$_{SQ}$	half saturation constant for nutrient limitation by either internal nitrogen or internal phosphorus	0.05 g$_{surplus}$N/g dry wt	0.005 g$_{surplus}$P/g dry wt	Desormeau 1978
β	variable stoichiometric ratio N : C and P : C (by weight)	mg N/mg C	mg P/mg C	calculated

Excretion : EXC

Excretion is considered as proportional to net photosynthesis (Park et al. 1974). EXC is set equal to zero if gross primary production rate is smaller than respiration rate.

When phytoplankton die, is grazed or excretes, proportional amounts of internal nutrient biomass must also be lost in order to maintain mass balance in the system.

Nutrient kinetics

Available phosphorus is considered to be phosphorus of orthophosphates PO_4-P (only PO_4^{3-} ions), available nitrogen is considered to be the sum of ammonia, nitrite and nitrate nitrogen and available silicon is dissolved reactive silicon. Unavailable forms of nutrients are not included, the detritus pool acts the role of these 3 compartments. We assume that organic matter is decomposed into inorganic nutrients through first-order decay kinetics and that decomposition of organic matter is accompanied by stoichiometric remineralization of nitrogen and phosphorus, this stoichiometry is taken equal to the variable one of phytoplankton.

Zooplankton kinetics

Zooplankton specific growth rate is a function of maximum ingestion rate, assimilation efficiency and phytoplankton concentration. Zooplankton loss rate has 2 components : the first one represents zooplankton respiration; i.e., population losses due to self-maintenance requirements and the second one a predatory death mechanism. Such a mechanism is a substitute for explicitly including an additional tropic level above herbivorous zooplankton.

Phytoplankton sinking losses are described by assigning a sinking velocity modulated as a function of water temperature (Calcagno 1979) to each phytoplankton group and therefore to each internal nutrient pool. Organic matter sedimentation description follows the same approach.

Numerical procedure

Each evolution equation (table 1) of the state variables takes the form :

$$\frac{dX^{\alpha}}{dt} = \sum_{\beta \leq \alpha} a_{\alpha\beta} X^{\beta} \qquad (18)$$

with α, β, ϵ $\left[1, \mathscr{N}\right]$, $\mathscr{N} = 11$

We use an implicit time differencing scheme. The system of 11 coupled, ordinary, non-linear differential equations written under a matrix form becomes :

$$\left[A(X)\right]\left\{X\right\} = \left\{B\right\} \qquad (19)$$

So we associate with each vector $V \epsilon \mathbb{R}^{\mathscr{N}}$ a linear mapping $A(V) : \mathbb{R}^{\mathscr{N}} \longrightarrow \mathbb{R}^{\mathscr{N}}$ and we consider this non-linear problem :

$$b \text{ given} \epsilon \mathbb{R}^{\mathscr{N}}, \text{ find } u \epsilon \mathbb{R}^{\mathscr{N}} / A(u) \cdot u = b \qquad (20)$$

We demonstrate that this problem is similar to the research of fixed points of an operator. We use a successive approximations method to find these fixed points as developed in other scientific areas (Klein 1980). The solution algorithm is iterated until convergence is achieved. The time step used is 3 hours for both the transport mechanisms and biological processes. A typical 365-day run required approximately 13 minutes of CPU time using an IBM 3033 computer with virtual storage.

Sensitivity analysis

The model has been calibrated using the data collected in 1978 at the sampling station No 2 (see fig. 1). Sensitivity analysis runs were conducted on the calibration coefficients in order to determine the values fitting best the data (within the range reported in the literature). The first step in the calibration of the model concerns the temperature calibration (Gaillard 1981). The second step concerns the biological model calibration. To allow a compact representation of the state variables sensitivity we use the particular

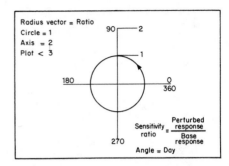

Figure 3.-- A sensitivity plot explaining the significance of the model trajectory in polar coordinates.

type of display as explained in figure 3. Figures 4, 5, 6 and 7 show the effects due to variations in certain coefficients (maximum algae growth rate, phytoplankton respiration rate) on the biological variables as seen at the surface of the reservoir and at a depth of 21 meters. We used as basic values the standard values reported in the literature, the values we finally adopted are summarized in table 2.

Results

Some results of the calibration phase are shown in figure 8. In general the thermocline depth and the shape of the temperature profile have been well predicted ; it means that the hydraulic conditions within the water body depending both on the energy transfer at the air-water interface and the advective process have been correctly represented. During the early spring, the net production remains small, mostly due to low temperature, short day length and relatively large mixing which removes the organisms from the euphotic zone for an appreciable amount of time. The nutrient concentrations are uniform over the entire waterbody. This quasi-stationary state reached by the system with chlorophyll and nutrients concentrations essentially constant in time is then perturbed when initiation of stratification together with increasing temperature in the epilimnion and available light allow a larger primary production rate and produce a moderate algal bloom. During the all summer, although nutrients levels are relatively low, the chlorophyll concentration remains nearly constant. This nutrient depletion is brought about by stratification of the water body and the fact that epilimnion is essentially uncoupled from the hypolimnion. The phytoplankton may use its internal nutrients pools to maintain the bloom at a modest level.

One feature has to be noted in the data on the vertical profile of phosphorus : the increase in phosphorus near the bottom ; it may be due to release of phosphorus at the sediment-water interface, the hypolimnion being in anaerobic conditions during the summer period. With regard to

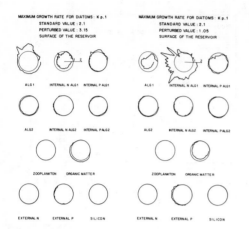

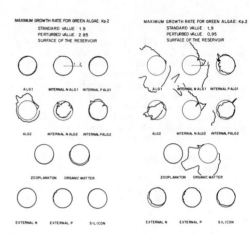

Figure 4.-- Sensitivity of model variables to a $\pm$ 50 % change in the maximum growth rate for diatoms.

Figure 5.-- Sensitivity of model variables to a $\pm$ 50 % change in the maximum growth rate for green algae.

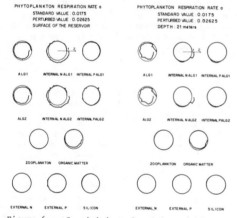

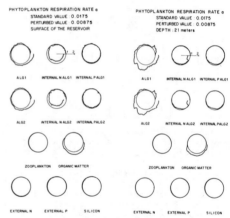

Figure 6.-- Sensitivity of model variables to a + 50 % change in the algal respiration rate.

Figure 7.-- Sensitivity of model variables to a - 50 % change in the algal respiration rate.

the vertical profiles of nitrogen, the fit is not very satisfactory. We must stress that we consider available nitrogen as the sum of NH_4^+-N, NO_2^--N and NO_3^--N and therefore don't take into account the selective uptake of nitrogen under a particular inorganic form; this could cause the relative discrepancy between results and field data.

In the fall overturn period (results not shown), conditions for growth are not favorable due to declining temperatures and decreasing cell residence time in the euphotic zone. The response of both the model and the data indicate that the dominant process is mixing. We must, however ,be very cautious in interpreting the results since very few data are available. As already mentioned, measurements of zooplankton biomass, internal nutrients, organic matter and silicon concentra-

tions are lacking and external loadings have been roughly estimated.

Once the model calibrated on the year 1978, we have then validated it, without changing parameter values, using data of the year 1979. Some of the results are presented in figure 9. Agreement between predicted and observed data is not as good as for 1978 but still remains acceptable.

SUMMARY AND CONCLUSIONS

An ecological model was developed in order to predict the spatial and temporal distribution of temperature, phytoplankton, zooplankton, nutrients and detritus in a reservoir. A wind mixing "integral" model coupled with a transport model was used for modelling of tranport mechanisms and thermal structure evolution. Several recent devel-

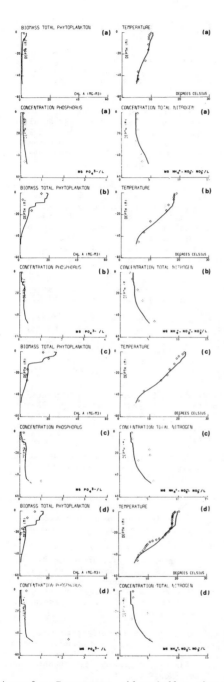

Figure 8.-- Temperature, chlorophyll a, phosphorus and total nitrogen profiles at
selected days for the year of calibration.
Simulation——, data◇.(a) 5/9/1978;
(b) 7/6/1978; (c) 8/23/1978; (d) 9/11/1978.

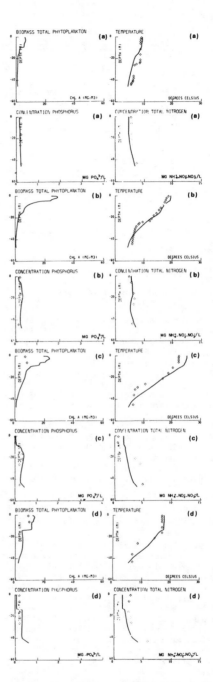

Figure 9.-- Temperature, chlorophyll a, phosphorus and total nitrogen profiles at
selected days for the year of validation.
Simulation——, data◇.(a) 4/19/1979;
(b) 6/6/1979; (c) 8/7/1979; (d) 9/25/1979.

302

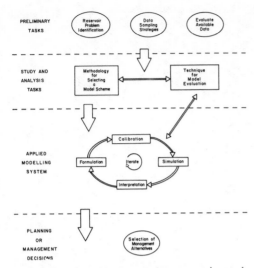

Figure 10.-- Overall context for reservoir study and applied ecosystem modelling.

opments in the modelling of aquatic ecosystems are taken into account: 1) The internal concentrations of major nutrients in algae and detritus are variable. 2) The uptake of external nutrients includes appropriate feedback based on inhibitory mechanisms. 3) The production rate of algae is controlled by the intracellular concentrations of nutrients. 4) The multiple-nutrient limitation is based upon the threshold hypothesis.

The model is both theoretically sound and accurate. Of course, further research in order to improve it would be necessary. However, despite the simplifications and weaknesses of the model, simulation results compare favorably with the few measured distributions in GRANGENT reservoir.

Figure 10 gives a representation of the definite sequence of tasks necessary in developing a modelling system that is useful for analyzing a reservoir ecosystem.

We see that the ultimate goal is now to provide a choice from among the alternative management strategies that will offer the best solution to the reservoir given problem. The choice of multiple outlet structures, and, therefore, of a selective withdrawal scheme, depends upon the type of reservoir, the hydrologic conditions, the inflow water quality and, of course, the required water quality objectives. The determination of the optimal management strategy will require the development of an optimization algorithm in order to optimize the use of a multiple outlet structure for improving the reservoir and outflow water quality and satisfying the required water quality criteria.

LITERATURE CITED

Azad, H.S. and J.A. Borchardt. 1970. Variations in phosphorus uptake by algae. Environmental Science and Technology. 4(9) : 737-743.

Bierman, V. J., Jr. 1976. Mathematical model of the selective enhancement of blue-green algae by nutrient enrichment. p. 1-31. In Modeling Biochemical Processes in Aquatic Ecosystems, (R.P. Canale, Ed.) Ann Arbor Science Publishers, Inc. Ann Arbor, Mich.

Bierman, V. J., Jr., D. M. Dolan, E. F. Stoermer, J. E. Gannon and V. E. Smith. 1980. The development and calibration of a spatially simplified multi-class phytoplankton model from Saginaw Bay, Lake Huron. Contribution No. 33. Environmental Protection Agency Ecological Research Series, Great Lakes Environmental Planning Study GLEPS. 126 p.

Calcagno, A.T.J. 1979. A wind-mixing dissolved oxygen model for reservoirs. M.I.T. Department of Civil Engineering. M.Sc.Thesis. 141 p.

Canale, R. P., L. M. DePalma and A. H. Vogel. 1976. A plankton-based food web model for Lake Michigan. p. 33-74. In Modeling Biochemical Processes in Aquatic Ecosystems, (R. P. Canale, Ed.) Ann Arbor Science Publishers, Inc. Ann Arbor, Mich.

Chen, C. W. and G. T. Orlob. 1975. Ecologic simulation for aquatic environments. p. 475-588. In Systems Analysis and Simulation in Ecology, Vol. III, (B. C. Patten, Ed.) Academic Press. New York, N. Y.

Coantic, M. 1978. An introduction to turbulence in geophysics and air-sea interactions. AGARDograph. No.232.

Davis, C.O. 1976. Continuous culture of marine diatoms under silicate limitation. II. Effect of light intensity on growth and nutrient uptake of Skeletonema costatum. Journal of Phycology. 12 : 291-300.

DePinto, J.V., Bierman, V.J., Jr. and F.H. Verhoff. 1975. Seasonal phytoplankton succession as a function of phosphorus and nitrogen levels. Proceedings of the 169th Meeting of the American Chemical Society, Philadelphia, Pennsylvania, April 6-11.

Desormeau, C.J. 1978. Mathematical modeling of phytoplankton kinetics with application to two alpine lakes. Report No.4. Center for Ecological Modeling. Rensselaer Polytechnic Institute, Troy, New York. 21 p.

_____. 1979. Personal conversation. Center for Ecological Modeling. Rensselaer Polytechnic Institute, Troy, New York.

_____. 1980. Formulation and validation of a mathematical model of phytoplankton growth. Ecology. 61(3) : 639-649.

DiToro, D. M., D. J. O'Connor and R. V. Thomann. 1971. A dynamic model of phytoplankton population in the Sacramento-San Joaquin Delta. In Advances in Chemistry Series 106 : Non Equilibrium Systems in Natural Water Chemistry, R.F. Gould, Ed. (Washington, DC : American Chemical Society) p. 131-180.

Droop, M.R. 1973. Some thoughts on nutrient limitation in algae. Journal of Phycology. 9 : 264-272.

_____. 1974. The nutrient status of algal cells in continuous culture. Journal of the Marine Biological Association of the United Kingdom. 54 : 825-855.

_____. 1975. The nutrient status of algal cells in batch culture. Journal of the Marine Biological Association of the United Kingdom. 55 : 541-555.

Fuhs, G. W., S. D. Demmerle, E. Canelli and M. Chen. M. 1972. Characterization of phosphorus-limited planton algae with reflections on limiting-nutrient concept. p. 113-133. In Nutrients and Eutrophication: The Limiting Nutrient Controversy. Proceedings of a Symposium held at the W. K. Kellogg Biological Station. Michigan State University, February 11-12.

Gaillard, J. 1981. A predictive model for water quality in reservoirs and its application to selective withdrawal. Internal Report. Electricité de France, Chatou, France.

Groden, T.W. 1977. Modeling temperature and light adaptation of phytoplankton. Report No.2. Center for Ecological Modeling. Rensselaer Polytechnic Institute, Troy, New York.

Huber, W.C. and D.R.F. Harleman. 1968. Laboratory and analytical studies of the thermal stratification of reservoirs. M.I.T. Hydrodynamics Laboratory Technical Report No.112.

Klein, J.P. 1980. Modélisation des mécanismes turbulents dans les couches marines superticielles (couche mélangée et thermocline). Thèse de doctorat d'Etat ès-Sciences Physiques. Université Aix-Marseille II, France.

Koh, R.C.Y. 1964. Viscous stratified flow towards a line sink. W.M. Keck Laboratory, Report No KH-R-6, California Institute of Technology.

Kremer, J.N. and S.W. Nixon. 1978. A coastal marine ecosystem. Springer-Verlag, Berlin Heidelberg, New York, N. Y. 217 p.

Larson, D. P., H. T. Mercier and K. W. Malueg. 1974. Modeling algal growth dynamics in Shagawa Lake, Minnesota, with comments concerning projected restoration of the lake. p. 15-31. In Modeling the Eutrophication Process, E. J. Middlebrooks, (D. H. Falkenborg and T. E. Maloney, Eds.). Ann Arbor Science Publishers, Inc. Ann Arbor, Mich.

Lehmann, J. T., D. B. Botkin, And G. E. Likens. 1975. The assumptions and rationales of a computer model of phytoplankton population dynamics. Limnology and Oceanography. 20 : 343-364.

Markofsky, M. and D.R.F. Harleman. 1971. A predictive model for thermal stratification and water quality in reservoirs. M.I.T. Hydrodynamics Laboratory Technical Report No.134.

Niiler, P. P. and E. B. Kraus. 1977. One-dimensional models of the upper ocean. p. 143-172. In Modelling and Prediction of the Upper Layers of the Ocean, (E. B. Kraus, Ed.). Pergamon Press, Inc. New York, N. Y.

Park, R.A., R.V. O'Neill, J.A. Bloomfield, H.H. Shugart, Jr., R.S. Booth, R.A. Goldstein, J.B. Mankin, J.S. Koonce, D. Scavia, M.S. Adams, L.S. Clesceri, E.M. Colon, E.H. Dettman, J.A. Hoopes, O.D. Huff, S. Katz, J.F. Kitchell, R.C. Kohberger, E.J. Larow, D.C. McNaught, J.L. Peterson, J.E. Titus, P.R. Weiler, J.W. Wilkinsin and C.S. Zahorcak. 1974. A generalized model for simulating lake ecosystems. Simulation. 23:33-50.

Perrin de Brichambaut, C. 1975. Estimation des ressources énergétiques solaires en France. Supplément aux Cahiers A.F.E.D.E.S. No.1.

Pichot, G. 1980. Simulation du cycle de l'azote à travers l'écosystème pélagique de la baie Sud de la mer du Nord. Thèse de doctorat en Océanologie. Faculté des Sciences, Université de Liège, Belgique.

Rhee, G- Yull. 1973. A continuous culture study of phosphate uptake, growth rate and polyphosphate in Scenedesmus sp. Journal of Phycology. 9 : 495-506.

_____. 1974. Phosphate uptake under nitrate limitation by Scenedesmus sp. and its ecological implications. Journal of Phycology. 10 : 470-475.

_____. 1978. Effects of N : P atomic ratios and nitrate limitation on algal growth, cell composition, and nitrate uptake. Limnology and Oceanography. 23 : 10-25.

Riley, G.A. 1963. Theory of food-chain relations in the ocean. In The Sea, Vol. II.

Scavia, D. 1980. An ecological model of Lake Ontario. Ecological Modelling. 8 : 49-78.

Steele, J.H. 1962. Environmental control of
photosynthesis in the sea. Limnology and Oceanography. 7 : 137-150.

Stefan, H. and D.C. Ford. 1975. Temperature
dynamics in dimictic lakes. Journal of the
Hydraulics Division, ASCE, Vol. 101, No.HY1,
Proc. Paper 11058. p. 97-114.

Strickland, J.D.H. 1965. Chemical Oceanography,
Vol. I. 503 p.

Thomann, R. V., D. M. DiToro, R. P. Winfield
and D.J. O'Connor. 1975. Mathematical modeling
of phytoplankton in lake Ontario. I. Model
development and verification. Environmental
Protection Agency Ecological Research Series.
EPA-660/3-75-005. 177 p.

LOTIC PERIPHYTON SIMULATION MODEL[1]

H. M. Runke[2], J. E. Shultz[3] and D. B. Porcella[4]

Abstract.--A Lotic Periphyton Simulation Model was constructed to investigate the ecological effects of various stream discharge and water temperature regimes on the algae communities of two mountain streams in Utah. Simulation of periphyton biomass under conditions where stream dewatering caused by energy development was hypothesized, showed that large increases in periphyton biomass would result from flow reductions greater than 25 percent of normal flow. Such increases were magnified by concomitant elevation of water temperatures.

INTRODUCTION

Since energy development in the Western United States will consume large amounts of water, it is of interest to learn how dewatering of streams will affect stream ecosystems. A Lotic Periphyton Simulation Model (LPSM) was constructed to investigate the ecological effects of stream discharge and water temperature on the periphyton communities of two Utah mountain streams. The model was developed to simulate conditions at Spawn Creek in Northern Utah. A tributary of Temple Fork Creek and the Logan River, Spawn Creek is located 24 km northeast of Logan, Utah in the Wasatch Mountains. Average annual discharge from the watershed is approximately 10 cubic feet per second. Maximum flow normally occurs during the snowmelt period of May and June. Upon completion of snowmelt, the stream discharge drops to a low level and remains low through the remainder of the year. The lotic periphyton were investigated as part of a larger watershed ecosystem research project. The research was aimed at the inter-related components: Meteorology, hydrology, periphyton, benthic macroinvertebrates, and fish. Continuous measurements of meteorologic and hydrologic variables were made at selected locations on the watershed. Periodic measurements of biological variables were made of the periphyton community and included standing crop and net growth rates.

The LPSM model, originally a submodel of a larger Mountain Ecosystem Simulation Package (MESP), was extracted from the overall watershed model to evaluate the effects of flow alterations on the periphyton community of a second stream. That stream, the Upper Strawberry River, is located in northeastern Utah in the Uinta Mountains. It is similar in size and character to Spawn Creek.

The objective of the original Spawn Creek modelling project was to predict the ecological impacts of weather modification on the stream. Snowmaking activities were hypothesized to have been conducted on the Spawn Creek watershed, and impacts of the resultant increased flows and lower water temperatures during periods of snowmelt were assessed. Modelling indicated that periphyton biomass in that high gradient, cold water stream would be relatively unaffected by increases in stream discharge and current velocity. Increases in watershed snowpack up to 30 percent, resulted in a ten percent (maximum) reduction in peak periphyton biomass during the following summer. The details of that study, and the formulation of the MESP model were reported by Israelsen et al. (1975). Details of supporting field and laboratory experimentation can also be found in that report. The LPSM model was applied to the Strawberry River to simulate the effects of various flow reductions in the stream on periphyton biomass. Those simulations are the principal subject of this paper.

MODEL DEVELOPMENT

Overview

A computer simulation model of the dynamics of the periphyton community of a small mountain stream

[1] Paper presented at the international symposium Energy and Ecological Modelling, sponsored by the International Society for Ecological Modelling. (Louisville, Kentuck, April 20-23, 1981)
[2] H. M. Runke is Manager of Environmental Assessment Department, Environmental Research Group, Inc., St. Paul, Minnesota U.S.A.
[3] J. E. Shultz is Reclamation Engineer, Consolidation Coal Company, Stanton, North Dakota U.S.A.
[4] D. B. Porcella is Principal Scientist, Tetra Tech, Inc., Lafayette, California U.S.A.

306

was developed from data and relationships from the literature, and from the results and impressions gathered from field experimentation at Spawn Creek. The model includes one state variable, the biomass of the periphyton community, and three rate variables: primary production, endogenous respiration and the removal processes that result in the export of biomass from the system. Realizing that the structure and function of the lotic periphyton system is not yet fully understood, this model has been constructed with what was judged to be the best information presently available. The basic modelling approach used here is similar to that of McIntire (1973) in his simple model of periphyton assemblages. Using his conceptual framework as a point of departure, expansions upon and additions to his model were made.

The periphyton assemblages of most streams are comprised of many different algal taxa. This model is based on the assumption that this diversity can be subordinated and that the periphyton community can be treated as a unit without quantitative concern for the dynamics of its many constituent populations (McIntire 1973). A similar approach to modelling phytoplankton populations has been followed (DiToro et al. 1971; Thomann et al. 1975).

Growth can be defined as the orderly increase of all chemical components in a biological system. In an adequate medium to which organisms are adapted, balanced growth occurs. During a balanced growth period, a doubling of the biomass is accompanied by a doubling of all other measurable properties of the population. This model assumes the periphyton community to be in a state of balanced growth. The model considers the environmental variables: water temperature, current velocity, light intensity and nutrient concentrations. It makes maximum (saturated) specific growth rates a function of water temperatures alone. This maximum rate is subsequently reduced according to the intensity of the environmental variables to which the periphyton is exposed. Using normalized reduction factors, including one that accounts for spatial limitation, the maximum is reduced and the resultant, adjusted rate is applied to the photosynthesizing biomass during the simulation.

In this model it is assumed that the importation of periphyton biomass from outside the system such as by drift from upstream reaches is negligible and that the rate of periphyton production controls the addition of periphyton biomass at any point within the system. The losses of periphyton biomass from the system are controlled by the rate of endogenous respiration and the rate at which biomass is removed from the community and exported from the system.

Temperature

The variation in specific growth rate with water temperature has been represented in models by exponential equations. The justification for that approach is based on the observation that enzymatic reaction rates tend to increase exponentially with temperature. A plot of saturated specific growth rate versus temperature has been constructed from data from a number of sources (DiToro, et al. 1971). These data were widely scattered, however, and only a slight functional relationship was apparent.

The variation in growth rate with temperature of marine phytoplankton was investigated by Eppley (1972). Analysis of data from laboratory cultures suggested that an equation could be written to describe the maximum expected growth rate for temperatures less than 40°C. The equation arrived at was:

$$\mu_{max} = 0.851 \times (1.066)^T \qquad (1)$$

where,

μ = the maximum expected specific growth rate (in doublings per day)

T = water temperature in degrees centigrade

Basic to this approach is the assumption that such an equation would describe the growth of algae from a wide variety of taxonomic groups and that an algal community is a diverse entity comprised of many taxa with differing temperature optima. Figure 1 shows results of equation (1) superimposed over observed specific growth rates of Spawn Creek periphyton. Those growth rates were calculated by periodic placement, removal and analysis of artificial substrates upon which periphyton colonized and subsequently grew.

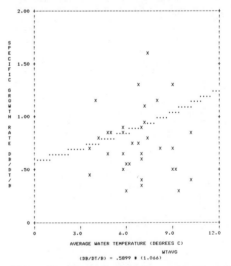

Figure 1.--Observed specific growth rates of Spawn Creek Periphyton as compared with calculated rates (Eppley 1972).

The specific growth rates calculated for Spawn Creek periphyton algae closely agree with the rates calculated from the Eppley equation within the temperature ranges encountered. Therefore, this equation was selected as the basic growth equation of the Spawn Creek periphyton model.

Growth, as described in that equation, represents organisms growing at optimal levels of light and nutrients in the laboratory. In the natural environment, however, variations in these and other environmental parameters will cause significant variations in algal growth rate.

Light

Light provides the ultimate source of energy for algae and, therefore, affects the rate of growth. Experimental evidence indicates that photosynthetic rate increases with light intensity and will reach a maximum for a particular range of intensities. At light intensities above this optimum range the photosynthetic rate is inhibited. Some mathematical models have proposed linear variation of photosynthesis with light intensity; some have proposed a hyperbolic function and others have proposed an exponential function. For the model under discussion the latter algorithm was used.

The light limitation relationship permits growth (photosynthesis) to increase exponentially to a maximum at a saturated light intensity, beyond which growth decreases with increased light intensity. The relationship is expressed:

$$\frac{I_o}{I_s} \exp\left(1 - \frac{I_o}{I_s}\right) \qquad (2)$$

where,

I_o = incident light intensity

I_s = saturated light intensity

Reported values for I_s range from 300 ly/day to 350 ly/day (DiToro et al. 1971).

The attenuation of light with depth was approximated by an exponential function:

$$I = I_o \exp(-kZ) \qquad (3)$$

where,

I = light intensity at depth Z

I_o = intensity at the surface

For present considerations the light just penetrating the water has been taken as 95 percent of that incident to the surface. Only 50 percent of the total incident solar radiation was considered photosynthetically active.

During this study light energy was measured in Langleys using an actinograph. This instrument records an accumulated daily quantity of solar energy that gives no indication of light intensity at any given sun position. During a 24 hour period the light intensities are not represented in the model. It is necessary to estimate an average value of a light-dependent growth rate reduction factor (FL) from an average daily value of light intensity. The diurnal variation in light intensity on the earth's surface can be approximated by the cosine law:

$$\ell = L \cos\left(\frac{\pi}{T_\ell} T\right), \quad -\frac{1}{2} T_\ell \le T \le \frac{1}{2} T_\ell \quad (4)$$

$$\ell = 0, \quad \text{otherwise.}$$

where,

ℓ = instantaneous light intensity

T = time measured from local noon (days)

T_ℓ = length of daylight period (days)

L = light flux at noon (langleys per day)

Photoperiod (T_ℓ) is introduced into the model as a trigonometric function of day number.

The value of L can be estimated from the following equation:

$$L = \frac{L_D \pi}{2 T_\ell} \qquad (5)$$

where,

L_D = total Langleys on the surface during the day.

Consequently, the average daily value of FL is determined numerically using a fourth order Runge - Kutta integration algorithm:

$$FL = \int_{-\pi/2}^{\pi/2} \frac{L \cos \theta}{I_s} \exp\left(1 - \frac{L \cos \theta}{I_s}\right) d\theta \quad (6)$$

where,

$$\theta = \frac{\pi}{T_\ell} T$$

The resulting values of FL were used in the model to represent the influence of light on the periphyton community.

Nutrients

Like other organisms the periphyton community must be nourished in an environment that is capable of sustaining and promoting growth. The curves relating growth rate to nutrient concentration are typically hyperbolic and fit the equation:

$$\mu = \mu_{max} \frac{S}{S + K_s} \qquad (7)$$

where,

μ = specific growth rate at limiting nutrient concentration S

μ_{max} = growth rate at saturating concentration of nutrient

K_s = a constant analogous to the Michaelis-Menten constant of enzyme kinetics, being numerically equal to the substrate concentration supporting a growth rate equal to $\frac{1}{2} \mu_{max}$.

Three nutrients were considered to be potentially limiting in the Spawn Creek periphyton simulation model: phosphorus, nitrogen and silica. The corresponding expressions (FNP, FNN and FNSI) representing the effect of suboptimal nutrient concentration on growth rate are:

$$FNP = \frac{P}{P + K_P} \qquad (8)$$

$$FNN = \frac{N}{N + K_N} \qquad (9)$$

$$FNSI = \frac{S_i}{S_i + K_{Si}} \qquad (10)$$

Nutrient availability is increased by current velocity. The effect of current velocity on growth rate is to facilitate the uptake of nutrients from water by algae. In quiet water a film deficient in nutrients forms at the surface of an algal cell, while in swift water this film is swept away, decreasing the diffusion gradient between the cell surface and the bulk liquid. Laboratory experiments have shown repeatedly that algae in currents grow better and take up nutrients in greater amounts.

The effect of suboptimal current velocity upon growth rate μ is represented by an expression FV, analogous to Michaelis-Menten enzyme kinetics. It is reasonable that this parameter should employ the same type of expression as does the nutrient limitation factor, FN, since it has been shown that the effect of current velocity on algal growth is due to increasing the exchange of materials (nutrients) between organism and stream water. The expression is:

$$FN = \frac{V}{K_V + V} \qquad (11)$$

where,

V = current velocity in feet per second

K_V = half saturation constant for algal growth with respect to current velocity in feet per second.

In the preceding discussion of environmental influences on the growth rate it was assumed that the level of only one parameter was limiting and all other environmental parameters were optimal. This may sometimes be the case in bodies of water, however, it is more probable that the level of more than one environmental parameter simultaneously limits the algal growth rate.

What has been termed Liebig's Law of the Minimum was the original means of expressing the importance of limiting factors. The element (light, energy, N, P, heat, whatever) that is in shortest supply determines the growth rate. This expression seems to be suitable for pure cultures but not appropriate for natural assemblages with varying adaptations because light, nutrients, etc. are probably never available in the optimal concentrations for all organisms present. The combined effect of simultaneous growth rate limitation by nutrients, light intensity and current velocity according to Liebig's Law may be expressed by the construct:

$$FACT = AMIN1 \ (FNP, FNN, FNSI, FL, FV) \qquad (12)$$

A second method commonly used in combining of effects of several variables is that of analyzing the effect of each factor on the growth rate and then multiplying these effects together. This expression is extreme in that even under near optimal conditions the factors have fractional values which, when multiplied, can produce a severe reduction effect in the simulation. This multiplicative combination of effects assumes that the effects are independent of each other when in reality they are interactive.

A third approach for combining effects of several variables on a growth rate treats the several effects as analogous to parallel resistance in an electrical circuit. While resistances in series are additive, resistances in parallel combine in a way indicated by the formula:

$$FACT = 5/ \ (1/FNP + 1/FNSI + 1/FL + 1/FV) \qquad (13)$$

or generally:

$$FACT = \frac{1}{N} \sum_{i=1}^{N} \frac{1}{\mu_i} \qquad (14)$$

The primary disadvantage of this formulation is that biological mechanisms seldom appear to operate as resistors in parallel and the results of such calculations are difficult to interpret in biological terms.

The LPSM was used to evaluate the three methods for combining the effects of several variables. Also, the model has the advantage that any of the three methods can be used.

Effect of Periphyton Density on Production

Since periphyton are attached algae, spatial limitation can be a factor that influences the daily production. McIntire (1973) found that the periphyton community in artifical streams reached a maximum specific rate of primary production at low levels of biomass. The same inverse relationship between periphyton density and primary production per unit biomass has been demonstrated in natural streams (Pfeifer and McDiffett 1975). As the standing crop increases, the periphyton community is composed of an outer layer of photosynthetically active cells and inner layers of scenescent and decomposing cells. Because only the outer layer is autotrophic, maximum specific productivity occurs at a relatively low biomass when the community is essentially a monolayer. The relationship between standing crop and production rate, based on the data presented by Pfeifer and McDiffett (1975), is expressed:

$$SPEC = \frac{B + 0.18B}{B + K_B} \qquad (15)$$

where,

SPEC = normalized, density dependent growth rate reduction factor

B = biomass in grams dry weight/ square meter

K_B = biomass at which specific productivity is one half its maximum

Respiration

The rate of endogenous respiration is the rate at which the algae oxidize their organic carbon to carbon dioxide. Respiration rate (RP) is calculated in this simulation model using an equation proposed by Di Toro et al. (1971):

$$RP = 0.005 * T * FV \qquad (16)$$

where,

T = water temperature (^{o}C)

FV = current velocity factor as previously defined

Current Velocity

Current velocity affects the respiration rate in a manner similar to the way it affects photosynthesis. Increased current velocity facilitates an increased endogenous respiration rate and as a first approximation of this effect, the same term (FV) as used in the growth equation was used to model this effect on the respiration rate.

Current velocity is particularly significant in the constant erosion of periphyton biomass and its export downstream. McIntire (1968) reported that export was promoted by current velocity. He chose to express that rate of export as a function of current velocity. Since herbivory is not considered as a distinct process in the LPSM model, a simplified representation of removal (slough, scour and herbivory) processes has been inserted to eliminate the requirement for modelling the herbivore community also. Removal rate (RMVL) is calculated in this simulation model using an equation statistically fit to field observations at Spawn Creek:

$$RMVL = k_o + k_1 * V * T^2 \qquad (17)$$

where,

V = current velocity

T = water temperature (^{o}C)

The functional form of the preceeding relationship is based on the recognition that current velocity (through tractive force) determines scour rate, and that herbivory is a temperature dependent process whose rate is roughly proportional to the square of water temperature. The constants k_o and k_1 are regression constants. Herbivore biomass could follow a temperature dependent pattern also.

According to the previous equations, the temperature-dependent growth rate of algae and the effects of various environmental variables on that rate have been defined. It remains now to assemble these parts into a single predictive equation. The maximum daily, specific growth rate is altered by the normalized reduction factors FNP, FNN, FNSI, FV, FL and SPEC to account for the effects of nutrient limitation, non-optimal light intensity, sub-optimal current velocity and space limitation respectively. The net daily specific growth rate is also affected by the specific rates of endogenous respiration (RP) and removal (RMVL), the rate at which biomass is removed from the substrate and exported from the system. Net, daily, specific growth rate, μ, for any day is then:

$$\mu = \mu_{max} *f \, (FNP, FNN, FNSI, FL, FV) * SPEC - RP - RMVL$$

$$= \left[\mu_{max} * FACT * SPEC \right] - RP - RMVL$$

$$= PS - RP - RMVL \qquad (18)$$

The basic growth equation of the Spawn Creek periphyton model predicts that the periphyton biomass (B) at time t_1 will be:

$$B_1 = B_0 \exp (\mu) \qquad (19)$$

when the simulation time step is one day.

Figure 2 is a flow diagram representing the instantaneous dynamics of the periphyton community in the model LPSM.

The LPSM model consists of a main program and 9 subroutines. The model employes an optimization algorithrm subroutine that performs a sequential search over a feasible region and selects model parameter values which result in a minimum value of a predefined objective function (sum of error squared in this case). The model calculations are performed on a daily time step to simulate the dynamics of the lotic periphyton.

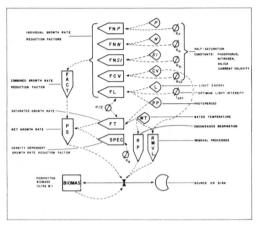

Figure 2.--A systems diagram of the LPSM model of lotic periphyton dynamics.

MODEL SIMULATIONS

The following figure 3 illustrates the prediction of the LPSM simulation model in relation to the observed field data for Spawn Creek. Relatively close agreement between observed and calculated biomass was obtained. While predicted peak biomasses agree reasonably well with corresponding observed values, the occurrence of the former seems to lag behind the latter. This seems to be related to the expression used for the removal rate constant. For the Spawn Creek simulation, the lag was approximately 60 days. A much better fit to observed data was obtained when the LPSM was applied to data from another Utah mountain stream.

The LPSM model was developed with the possibility of other applications in mind. An effort was made to construct a model general enough in nature to be applicable to other lotic systems. Only minimal changes and the appropriate data inputs would be required to use the simulation model elsewhere.

The LPSM model was applied to data from the Upper Strawberry River to assess the impacts of stream dewatering. In figure 4, a continuous curve representing daily predicted periphyton biomass is shown in comparison to measured

values. The correlation coefficient between the observed and predicted data was 0.68. The greatest deviation of predicted values from observed values appears to occur during the late summer and autumn months.

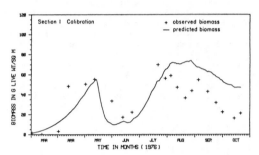

Figure 4.--Calibration run of the LPSM model for Strawberry River, Utah.

Following model calibration, a series of abritrary flow reductions such as those which may result from flow diversion of streams for energy development projects were evaluated. Input flow data to the model were reduced by a quarter, a half, and three quarters. Examples of predicted biomass resulting from such altered flows are shown in figure 5. Maximum standing crops occurred in late August or early September in all cases. The predicted maximum biomass (99.6 g/m^2) for a flow reduction of 25 percent was not much greater than the observed maximum biomass (69.8 g/m^2). However, flow reduction greater than 25 percent resulted in predicted values a great deal higher than observed biomass, i.e., 155.2 g/m^2 for 50 percent reduction and 393.8 g/m^2 for a 75 percent flow reduction.

Since reduced flows will most likely result in increased water temperatures, the model was also used to predict the effects of a combination of reduced flow and increased temperature on periphyton biomass. The magnitude of temperature increases which might occur as a result of actual diversions were not known. (However, a temperature model could be incorporated easily using one of several approaches.) Therefore, a maximum temperature increase of $7.5°C$ above observed temperatures was used in steps of $2.5C$. A series of nine additional runs of the model were made at the three levels of flow reductions and three levels of temperature increases. Resulting predicted periphyton biomass increased with increasing temperature, as well as with flow reductions.

Maximum predicted periphyton biomass for each level of flow and temperature are shown in table 1 and these should be viewed relative to the observed maximum biomass of 69.8 g/m^2 at observed discharge and no increase in temperature. Most significant changes occured with flow reductions

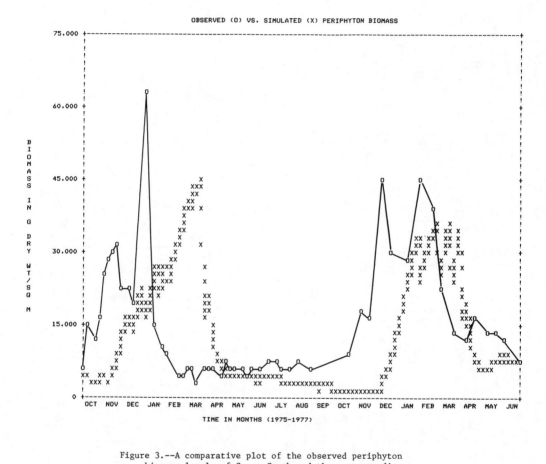

OBSERVED (O) VS. SIMULATED (X) PERIPHYTON BIOMASS

Figure 3.--A comparative plot of the observed periphyton
biomass levels of Spawn Creek and the corresponding
simulated biomass levels from LPSM.

312

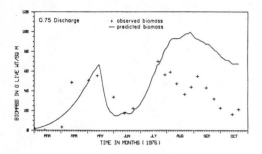

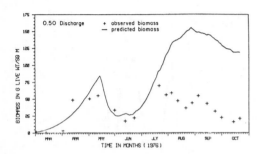

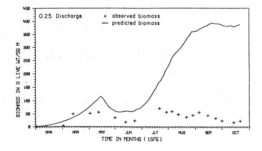

Figure 5.--Predicted periphyton biomass with
decreased flows and increased temperatures,
Strawberry River, Utah.

of 75 percent and with temperature increases
greater than 2.5C. By plotting predicted
maximum periphyton biomass against the assumed
temperature increases, a family of curves was
developed for the various levels of flow reductions
as is shown in figure 6. With a known level of
flow reduction known and a predicted temperature
increase, the resulting periphyton biomass can
be predicted.

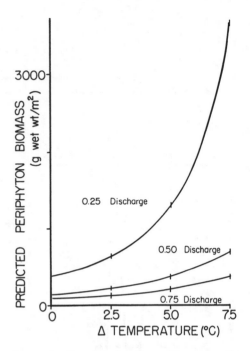

Figure 6.--Example of management runs for flow
reduction and temperature increase effects,
Strawberry River, Utah.

Table 1.--Predicted maximum periphyton biomass
with decreased flows and increased water
temperatures (grams/square meter)

+ Δ Temperature	Discharge		
	.75 Discharge	.50 Discharge	.25 Discharge
+ 0 C	99.6	155.2	393.8
+ 2.5 C	137.3	223.8	650.8
+ 5.0 C	205.5	359.9	1317.9
+ 7.5 C	349.6	681.1	3657.7

LITERATURE CITED

DiToro, D.M., D. J. O'Connor, and R. V. Thomann. 1971. A dynamic model of phytoplankton population in the Sacramento-San Joaquin delta. *In*: R. F. Gould (ed.), Advances in Chemistry. No. 106. Non equilibrium systems in natural water chemistry. American Chemical Society.

Eppley, R. W. 1972. Temperature and phytoplankton growth in the sea. Fishery Bulletin 70(4):1063-1085.

Israelsen, E.K., D. R. Bernard, T. M. Twedt, and H. M. Runke. 1975. A technique for predicting the aquatic ecosystem response to weather modification. Utah Water Research Laboratory, College of Engineering, Utah State University, Logan. PRWG 138-1.

McIntire, C. D. 1973. Periphyton dynamics in laboratory streams - simulation model and its implications. *Ecol. Monogr.* 43:399-420.

Pfeifer, R. F., and W. F. McDiffet. 1975. Some factors affecting primary productivity of stream riffle communities. *Arch. Hydrob.* 75:306-317.

Thomann, R. V., D. M. DiToro, R. P. Winfield, D. J. O'Conner. 1970. Mathematical modeling of phytoplankton in Lake Ontario. I. Model development and verification. U.S.E.P.A. 660/3-75-005 Ecological Research Series. Nat. Environ. Res. Center. Corvallis, Oregon.

PRELIMINARY ANALYSIS OF ENERGY FLOW IMPACTS
OF A RIVER REDIVERSION[1, 2]

H. McKellar, M. Homer, L. Pearlstine[3]
and
W. Kitchens[4]

Abstract.--The rediversion of a major river in South
Carolina will affect riverine, floodplain and estuarine
ecosystems in addition to power generation, harbor dredging
and industrial water supplies. The projected impacts are
discussed and compared in a preliminary energy analysis indicating the importance of ecological productivities in
region-wide water resource planning. The response of floodplain productivity to flood patterns may be very critical
in formulating water release alternatives for hydropower
generation.

INTRODUCTION

Natural water resource patterns in many regions have been radically modified by projects
which were justified on the basis of regional water
needs, flood control, navigation concerns, and/or
hydropower development. Such changes may affect
the economics of urban and industrial sectors of
these regions and also impact the ecological functioning of natural systems such as riverine habitats, flood plains and downstream coastal environments. The effective management of regional water
resources thus, requires an analytical overview of
the many complex interactions between man and his
natural surroundings.

Energy analysis has recently emerged as a valuable tool in achieving such an analytical overview
(Odum and Odum 1976; Bayley, Odum, and Kemp 1976;
Odum 1978; Gilliland 1978; Jansson and Zuchetto
1978, and others). Using this type of analysis
flow diagrams of major energy exchanges and interactions are first constructed to help identify critical relationships that may be affected by the
proposed environmental alterations. Since energy
flow is causative for all work processes in man's
fuel powered, economic system as well as in ecological systems, a broad spectrum of energy/eco-
nomic impacts of proposed water management plans
can be examined through energy analysis using
comparable units.

In addition to gaining a quantitative and functional overview, a major objective of energy analysis is to identify those alternative plans of resource management which tend to maximize the net
energy flow into productive work processes. The
basis of this objective derives from the principle
that those management plans which enhance the total,
regional energy flow will be more likely to continue
in long-term regional developments than those which
do not (Odum and Odum 1976; Odum 1978).

In this paper we use energy flow analysis in an
initial evaluation of the regional impacts of a major river rediversion. Issues concerning hydropower
cooling towers, industrial and municipal water supply, habor dredging, floodplain and estuarine productivity, and fisheries are summarized for a major
water resource modification in the coastal plain of
South Carolina.

SITE DESCRIPTION AND ISSUES OF RIVER FLOW

River Flow Modifications

With headwaters in the Carolina piedmont (fig.
1), the Santee River originally had an annual average discharge of approximately 17,400 cfs and was the
4th largest river on the east coast of the U.S. In
1941 the South Carolina Public Service Authority
(PSA) significantly altered this flow regime in order to develop hydroelectric power potential in the

[1] Paper presented at the International symposium of Energy and Ecological Modelling, sponsored by
the International Society for Ecological Modelling
(Louisville, Kentucky, April 20-23, 1981).
[2] This work was supported by the National
Coastal Ecosystems team of the U.S. Fish and Wildlife Service (Cooperative Agreement No. 14-16-0009-
80-994).
[3] Dept. Environmental Health Sciences, University of South Carolina, Columbia, S.C. 29208
[4] U.S. Fish & Wildlife Service, National Coastal
Ecosystem Team , Slidell, Louisiana 70458.

coastal plain. With the formation of 2 large impoundments, approximately 88% of the Santee discharge was diverted into the adjacent Cooper River, which was formerly a small, coastal plain stream with a mean annual flow of only 100 cfs (fig. 2). The elevation difference between the Santee and the Cooper Rivers provides a 75 ft hydraulic head behind the dam at Lake Moultrie which drives a 130 Mw hydropower generation facility. The power plant releases water directly into the upper Cooper River which now has an annual mean discharge of approximately 15,500 cfs.

Among other, less documented impacts, the Santee-Cooper project apparently led to severe shoaling problems in Charleston Harbor. The escalated shoaling rate has been attributed mainly to the heavy load of fine-grained sediments transported into the harbor by the Cooper River (US Army Corps of Engineers 1975). Maintenance of the main navigation channels in the harbor requires intensive, year-round dredging.

To alleviate this shoaling problem and associated dredging costs, the Army Corps of Engineers has been authorized to redivert most of the water flow back into the Santee River. Rediversion, which is now almost complete, involves the construction of an 18.5 km long canal from Lake Moultrie back into the Santee River (fig. 2). An additional 84 Mw hydropower plant will be constructed on the rediversion canal near St. Stephens to take advantage of some of the hydrostatic head gained through the original diversion. Thus, an annual average of 12,500 cfs will be diverted from Lake Moultrie, through the new hydroplant, and into the Santee River.

The seasonal patterns of river flow for present conditions and for post-rediversion conditions (fig. 3) include a winter-spring period of high flow and a summer-fall period of low flow. Presently large winter-spring flood releases from the Lake Marion dam into the Santee River increase the annual mean flow in the Santee up to 2000 cfs although the normal controlled flow is 500-600 cfs.

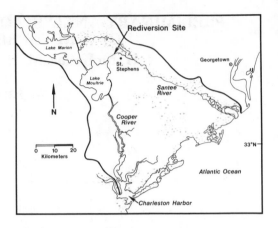

Figure 2.— Details of the Santee-Cooper River system in the coastal plain of South Carolina.

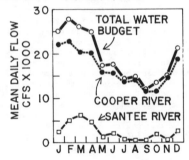

Figure 1.—The Santee-Cooper watershed. The total drainage basic covers 41,000 km^2 (16,000 mi^2).

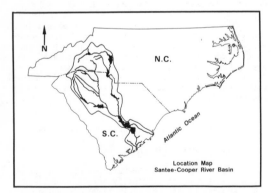

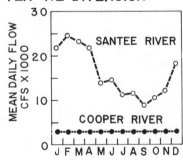

Figure 3.—Seasonal discharge patterns for the Santee-Cooper Rivers. Each points represents a monthly mean based on 9 yrs. of flow data from 1970-1978. Data Sources: U.S. Geological Survey and the S.C. Public Service Authority.

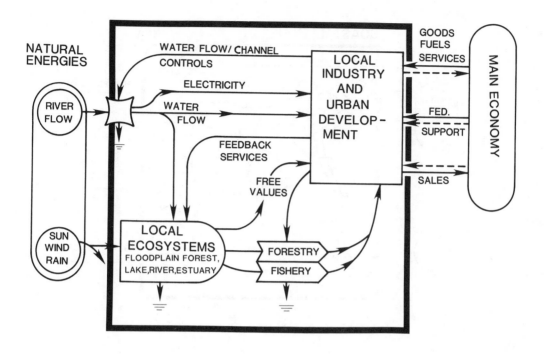

Figure 4.—Regional overview model of issues involved with river flow control. Solid lines indicate energy flows and dotted lines show economic transactions. General energy flow symbols are from Odum and Odum (1976).

After rediversion, the Santee River will be returned to within 80% of its original (pre-diversion) flow rate. Seasonal patterns will reflect both flood stages and periods of low flow (fig. 3) but, in addition, daily and weekly fluctuations will be caused by peak-demand operation patterns of the new hydroplant.

For the Cooper River, the US Army Corps of Engineers has stipulated (preliminary) that a weekly average flow of 3000 cfs must not be exceeded if shoaling in Charleston Harbor is to be effectively reduced (US Army Corps of Eng. 1975). Although figure 3 illustrates a constant monthly average flow for the Cooper River, large daily fluctuations will also probably occur due to hydropower generation patterns at the Lake Moultrie dam.

A long-range goal of our project is to identify alternative schemes of water release from the new hydroplant on the rediversion canal. By considering impacts on the ecological systems of the floodplains and estuarine marshes, fishery potential as well as hydropower generation, industrial/urban water supply, and dredging costs, small changes in water release patterns could help enhance the posi-

tive benefits of the project as well as help mitigate negative impacts downstream from the project. In this paper, we present a preliminary energy analysis in an effort to identify some of the most sensitive impacts which cou_d be critical in formulating future alternatives for water release.

Framework for Energy Analysis

An overview of energy flow relationships in which the rediversion project is embedded is given in figure 4. River flow, as a major natural driving force, is emphasized since it is the major control variable of the rediversion issue. Human controls on river flow (diversions, rediversions, channel maintenance) represent a large energy cost from society, requiring substantial purchase of goods, fuels, and services as well as federal support from the main economy. Hydroelectric generation is also indicated as a major energy flow concern of river flow controls. In addition, the resultant patterns of water flow may affect major functions of the local aquatic ecosystems as well as industrial/urban water supplies.

Ecosystem effects are reflected in their di-

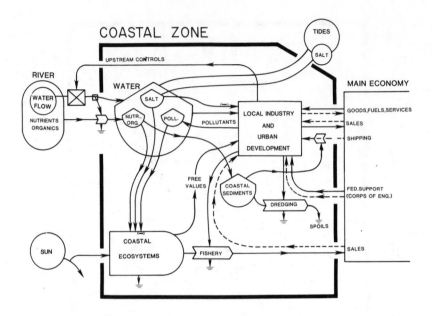

Figure 5.--Energy flow issues in the Coastal Zone.

rect support of forestry and fishery harvests which
are often enhanced by human feedback services such
as silviculture and aquaculture practices. The
harvestable products and the human feedbacks are
usually accounted for in economic transactions with
the main economy (purchase of goods, fuels, and services, and sales of forestry and fishery products).
Ecological functions also provide "free values" to
the local system which include the productive work
contributing to natural flood control, protection
from storms, water quality control, and micro-cli-
mate modifications. These values are usually not
accounted for in traditional economic evaluations,
but are major considerations of an energy analysis.
A measure of the total productive work processes of
an ecosystem is its gross community production.

To compare units of ecological energy flow with
purchased energies in the economic sector, transformation ratios are needed to convert all energies
to equivalent units. In this analysis we convert
all energies to solar equivalents using the following ratios:

25×10^7 kcal/\$ (Wang and Miller 1980), and

920 kcal/kcal gross production (Odum 1978).

In this discussion we examine some detail interactions for the coastal zone, the river-flood-
plain environments, and for the specific issues
concerning anadromous fish populations in the rivers and lakes. A tabular summary of impacts is
presented (table 3) which provides a quantitative
comparison of projected effects in equivalent units
of energy flow.

ISSUES IN THE COASTAL ZONE

Area Description

The coastal features of the Santee and Cooper
Rivers vary considerably in physical form (fig. 2)
and in degree of human development. Figure 5 summarizes some major relationships between incoming
river flow and energy exchanges in both the Cooper
and Santee coastal zones.

The Santee River mouth is a fluvial delta with
a high salinity aquatic environment. During normal
flow conditions (i.e.,when there are no flood releases from Lake Marion) the 10 ppt isohaline extends 8-10 km upstream along the North and South
Santee River branches (Kjerfve and Greer 1978).
Vegetation in the delta is dominated by salt and
brackish marshes (Spartina alterniflora, S. cyno-
suroides and Juncus roemarianus). The Santee
coastal area is relatively undeveloped although
about 40% of the marshes have been impounded and
managed for game preserves and waterfowl production (Sandifer et al 1980). In the managed brackish impoundments the dominant vegetation is usually
widgeon grass (Ruppia maritima) salt marsh bulrush
(Scirpus robustas) and dwarf spikerush (Eleocharis
parvula).

A significant economic factor in the Santee
coastal area which has apparently developed only
since the 1941 diversion is the hardclam (Mercenaria
mercenaria) fishery. There are presently almost
3 km^2 of productive clam beds within the first few
km upstream from the Santee river mouth. These

beds yield an average of about 100,000 pounds of clam meat per year with a retail value of $800,000 (Theiling 1981)[1].

The coastal zone of the Cooper River is dominated by the open water habitat of Charleston Harbor (fig. 2), although salt marshes occur along the borders of the harbor as well as the lower reaches of the river (Nelson 1974). During normal flows of the Cooper River, the 10 ppt isohaline extends 3 to 5 km upstream from its mouth at Charleston Harbor. Further upstream, salt and brackish marshes grade into typical tidal freshwater marshes with rushes (Eleocharis, Scirpus) cattails (Typha) and pickerelweed (Pontederia) along with giant reed (Phragmites communis), water millet (Zizaniopsis miliacea) and wild rice (Zizania aquatica) (Barry 1980).

The Charleston Harbor area and the lower reaches of the Cooper River support a dense industrial/ urban development. Because of the present high rates of river flow, providing both water supply and effluent dilution, much industrial development has occurred along the mid and lower portions of the Cooper River.

Charleston, with its protected harbor, is a major port city with a total population of more than 120,000. The 1941 diversion of water from the Santee drainage basin into the Cooper River has led to severe shoaling problems in the Charleston Harbor. The inflowing river water transports a load of dissolved and suspended material into the coastal water. The suspended silts and clays (and other agglomerates formed in the estuarine zone) settle out of the water column and become unconsolidated benthic sediments. These sediments are shown in figure 5 to represent a negative influence on shipping interests (specifically for Charleston Harbor) and must be removed by intensive dredging activities. To maintain critical navigational channels, the U.S. Army Corps of Engineers, the S.C. Port Authority, and the U.S. Navy dredge an average total volume of about 10.2 million cubic yards of sediments from the harbor per year (U.S. A.Corps of Eng. 1966). Dredging is a fuel intensive operation and current total costs average about $1/yd^3 (Sadler 1981)[2], or $10.2 million a year in maintaining Charleston Harbor navigation channels. This cost has significantly increased due to rising fuel costs since an earlier economic analysis (U.S. A. Corps of Eng. 1966). If present dredging trends continue, the cost could rise considerably because available spoil disposal areas are nearly full and it may be necessary to barge dredge material out to sea for disposal. Clearly, there is substantial incentive to solve the shoaling problem in Charleston Harbor.

---

[1]Theiling, D. 1981. Personal communication. S.C. Wildlife and Marine Resources. Columbia.

[2]Sadler, J. 1981. Personal communication. S.C. Ports Authority. Charleston.

Table 1.--Direct Economic Costs of the Rediversion Project (U.S. A. Corps of Eng. 1981)

ITEM	COST (10^6 dollars)
New Power Plant	85.8
Channels and Canals	35.7
Relocations	15.1
Engineering and Design	10.1
Supervision and Administration	6.3
Lands and Damages	5.0
Other*	1.1
	159.1

*roads, railroads, bridges, buildings, grounds, utilities, and permanent operating equipment

Effects of Rediversion

Direct Costs and Benefits

Although the physical rediversion will actually be outside the coastal zone, it is useful at this point to compare the direct costs of the rediversion with the major benefits due to projected decreases in harbor maintenance. The direct costs of rediversion (including the cost of the new hydropower plant) amount to $160 million (table 1). Over an estimated 50-yr life of the project, this sum yields an annual cost of $3.2 million.

The rediversion of most of the Cooper River flow back to the Santee (upstream controls, fig. 5) will reduce the sediment load to Charleston Harbor. This will, according to Army Corps engineers, reduce the present cost of dredging by 70%, yielding an annual savings of $7.1 million. Although there are other costs and benefits of the project (as discussed below) these annual savings compared to the annual direct costs provide the basic economic justification for the project.

Industrial/Urban Water Supply

Some predicted changes in salinity up the Cooper River are discussed by McKellar, Homer and Pearlstine (1981). With present flow conditions in the Cooper River, salinity drops from about 20 ppt in Charleston Harbor to less than 1 ppt by river mile 17. A verified dispersion model, developed by CH2M Hill and Betz (Environ. Eng. consultants) predicts that the 1 ppt isohaline will extend approximately 30 miles upriver with rediversion.

At river mile 33, a canal from the Cooper River now feeds the Back River Reservoir providing a water supply for several major industries and the City of Charleston. The dispersion model indicated that salinity will increase in the reservoir to levels above the maximum tolerance limits for industrial and municipal use.

Although the Corps of Engineers does not accept the validity of this dispersion model, they are making plans for a new canal leading into the reservoir from a point further upstream on the Cooper River. Independent estimates of the cost of such a canal amounted to a total of $12 million, or $240,000/yr. for a project life of 50 yr (McKellar et al. 1981). This cost is relatively small considering the estimated savings gained by the rediversion project.

Ecological Production

The vegetation changes following the 1941 diversion were poorly documented although the remnants of dead trees standing among flourishing salt marshes in the lower Santee give evidence of salinity-induced community changes (Kjerfve 1976). The largest areas involved in the coastal zone are the present expanses of salt, brackish, and freshwater marsh lands. With the drastic drop in salinity expected for the Santee coastal delta, the present salt and brackish marshes and impoundments will possibly develop into tidal freshwater marshes along the present Cooper River will probably develop more toward brackish and saltwater dominants. Differences in productivity per unit area between fresh and salt marshes may be relatively small (Whigham et al. 1978), although changes in areal distribution may be important with respect to regional productivity. Present distributional patterns are currently being mapped (Harper 1981)[1] to better evaluate such changes.

Fisheries and Wildife

Although the rediversion impacts on total ecological productivity in the coastal zones are still unclear, the physiological responses of individual organisms to salinity changes are well documented. The hard clams (M. mercenaria) which have developed in the high salinity conditions of the Santee delta may experience significant physiological stress at salinities below 25 ppt (Kinne 1971). With the 7-fold increase in discharge due to the rediversion, salinities in the Santee delta will probably plummet well below 25 ppt for most of the year, and the decline of the lucrative clam fishery is likely (Kjerfve and Greer 1978). Thus, the annual loss of the $800,000 retail income to the region for Santee clams should be considered a definite cost of the rediversion project.

An additional benefit of the project relates to the attraction of migratory waterfowl to coastal marshes. As described earlier, much of the marsh in the Santee delta is presently impounded and managed for waterfowl habitats. These management practices are more successful in fresh and brackish impoundments where the dominant vegetation is more attractive to waterfowl. The U.S. Fish and Wildlife Service originally estimated that an additional 150 km[2] could be managed for waterfowl habitat

---

[1]Harper, K. 1981. Personal communication. U.S. Fish and Wildlife Service. Charleston, S.C.

in the Santee River after rediversion. Thus the increase in hunting benefits, amounting to about $150,000/yr. would be an added benefit to the project (U.S. Fish and Wildlife Ser. 1966).

ISSUES IN THE RIVER/FLOODPLAIN HABITAT

Area Description

The Santee Floodplain

An obvious major feature of the Santee River is the extensive forested floodplain which spans 2 to 8 km wide along the 70 km distance from Lake Marion to the coastal zone (fig. 2). Although the mean flow in the Santee River is now only 10% of its original flow, there is still periodic inundation of the entire floodplain due to annual winter-spring flood releases from Lake Marion.

The vegetation of the floodplain is a mix of bottomland hardwoods and cypress-tupelo associations plus various stages of succession in logged areas. The cypress-tupelo associations typically occur in areas where the water table is fairly shallow and where extended periods of flooding are common. The bottomland hardwoods include sweetgum, green ash, American elm, water hickory and others, (Harper 1981) which require deeper water tables and shorter flood durations. Preliminary results from a detailed vegetation mapping program indicates that bottomland hardwoods may cover as much as 75% of the entire floodplain (Harper 1981).[1]

The Upper Cooper River

In contrast to the Santee, the Cooper River conspicuously lacks a forested floodplain. With high water flows from Lake Moultrie, water in the Cooper River remains in a relatively well-defined upstream channel or filters out through downstream bordering freshwater marshes.

Also in contrast to the Santee, there is much more industrial development along the banks of the Cooper. A 330 Mw coal fired steam-electric power plant is currently in full operation just downstream from Lake Moultrie. The powerhouse uses the present high discharge of the Cooper River for once-through cooling. The operation of this power plant is clearly of concern since the Cooper River flow will be drastically reduced with rediversion.

Impacts of Rediversion

Floodplain Productivity

Intuitively, the effects of the rediversion project on the Santee River floodplain could enhance ecological productivities since the total river flow will be restored to 80% of its original, pre-diversion flow (conditions under which the floodplain originally developed). However, the hydro-regimes imposed by the power plant on the re-diversion canal may be significantly different from the original, pre-diversion conditions. We suggest the possibility that the rediversion may indeed represent a negative impact on Santee floodplain productivity.

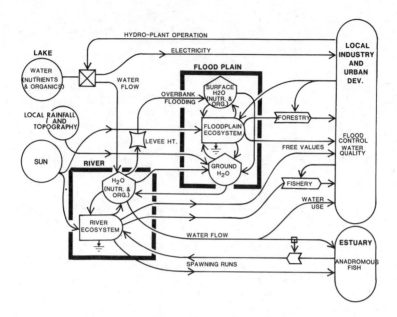

Figure 6.— Energy flow for river/floodplain interactions.

Figure 6 illustrates some critical energy flow relationships between the river/floodplain habitat and total water flow. In the Santee river, water flow is now (and will be) controlled largely by hydropowerplant operation. When the water level in the river exceeds the height of the natural levee, overbank flooding occurs and the surface of the floodplain is inundated. The corresponding nutrient and organic cycles among the surface floodwater, ground water and the floodplain vegetation constitute major functional attributes of floodplain forest ecosystems (Wharton 1970).

The frequency and duration of flooding and the depth of the water table during the growing season are critical controls on species distributions and total primary productivity. (Phipps 1979). The detailed mechanisms of these controls are related to soil redox potentials, micronutrient chemistry, root physiology and seedling survival (Gosselink and Turner 1978, Teskey and Hinkley 1977). The depth to the water table over most of the floodplain area is probably controlled more by local rainfall and topography than by river stage (Williams 1981)[1]. Therefore, a more direct effect of rediversion may be the resultant changes in the frequency and duration of flooding. Characteristic range of flood duration tolerance for bottomland hardwood species is about 5-30% of the time; for

cypress-tupelo associations the range is 40-50% (Gosselink et al. in press; Phipps 1979).

As mentioned earlier, the Santee River now floods periodically in response to winter-spring flood releases from Lake Marion. Stage-discharge relationships for the Santee River (USA Corps of Eng. 1976) and past records of Santee discharge (US Geol. Survey Annual Rep. 1973-1979) indicate that, at present conditions, the Santee floodplain is inundated for approximately 10% of the time. Even though the total annual discharge of the Santee is well below its original level, flood durations are still within the tolerance limits for the general bottomland hardwood associations which cover most of the floodplain.

After the rediversion, flow in the Santee will be controlled primarily by the daily operation of the new hydroplant which is determined by power demand. A typical week-day operation schedule for the hydroplant includes 15 hours of operation at full capacity (24,500 cfs) and 9 hr period with zero discharge (USA Corps of Eng. 1976). Simulated stage curves (USA Corps of Eng. 1976) indicate that this pulsing discharge would cause overbank flooding approximately 50% of the time in the upstream reaches of the river. At the lower reaches of the river, the same release patterns would cause overbank flooding essentially 100% of the time. Prorating these stage-discharge relationships over an entire year (and considering seasonal changes in water availability, fig. 3) indicates that after rediversion, the Santee flood-

[1]Williams, T. 1981. Personnal communication. Clemson Univ. Forest Science Institute. Georgetown, S.C.

Table 2.--Estimated Changes in Total Gross Primary
Productivity in the Santee Floodplain. I. Before
Rediversion, II. After Rediversion

	Association	% Cover*	TOTAL AREA (km^2)	GROSS PRODUCTIVITY**	
				gC/m^2.yr	gC/yr x 10^{10}
I.	Bottomland Hardwood	75	82.5	1944	19.7
	Cypress-Tupelo	25	27.5	1399	4.7
					24.4
II.	Bottomland Hardwood	37.5	41.3	1944	9.9
	Cypress-Tupelo	12.5	68.7	1399	11.8
					21.7
				change	2.7

*Preliminary estimate (Harper 1981)

**Net Production values from Day et al.(1977)
Conversion factors: 2.2g dwt/gc, and
2.7gC gross productivity/gC net prod.
(Whittaker 1975)

plain may be flooded for about 50% of the time over
a typical annual cycle (McKellar et al.1981).

These flood conditions are outside the tolerance limits of the typical bottomland hardwoods
which dominate most of the floodplain. The hydro-
regime imposed by the power plant discharges is
more typical of areas dominated by cypress-tupelo
associations. If we assume, that a certain fraction of the present bottomland hardwoods will be
taken over by cypress-tupelo dominants in long-
term succession then the change in floodplain productivity may be estimated from the literature.

Table 2 gives the corresponding calculations,
assuming that 1/2 of the bottomland hardwoods
would succeed to cypress-tupelo associations (this
partial effect was tentatively chosen to account
for upstream-downstream and cross-sectional variations in flood duration). Since typical cypress-
tupelo associations may be somewhat less productive
that bottomland hardwoods in the same area (Day et
al. 1977) there could be a net decrease in total
floodplain productivity amounting to approximately
3 x 10^{10} gC/yr (table 2). These calculations are
highly conservative since they do not include the
possible widespread loss of production during the
successional change over. Although this estimated
change represents only about a 10% decrease in total
floodplain productivity, this amount represents a
highly significant energy cost compared to other
energy costs and benefits due to rediversion (table
3). These calculations clearly illustrate the importance of floodplain production in overall regional energy flows and the possible implications of
alternative plans of water release in controlling
floodplain productivity.

[1]Byrn, B. 1981. Personnal communication. Stan-
ley Assoc. Atlanta, Georgia.

Power Plant Cooling System

When the Cooper River discharge is reduced to
a weekly mean of 3000 cfs, the 330 Mw steam-electric powerhouse downstream from Lake Moultrie will
have to resort to alternative means of waste heat
disposal (other than once-through cooling with Cooper River water). Apparently the decision has been
been made for the Corps of Engineers to construct
a cooling tower adjacent to the power plant. Although much evidence has been accumulated over the
past 10 years indicating that cooling towers are
probably not competitive alternatives for waste
heat disposal (Odum 1977, Kemp et al 1977, Reynolds
1980) the Corps of Engineers is presently proceeding in that direction at the insistence of state
environmental agencies.

While the exact design of the cooling tower
has not yet been confirmed, the total costs of
building and maintaining a concrete tower with
forced air circulation are estimated at approximately $1.4 million/yr over a 30-yr life of the
tower (McKellar et al. 1981, Byrn 1981[1]). In
comparison, the total cost of the cooling tower
represents a substantial energy involvement of the
overall rediversion project (table 3).

STRIPED BASS AND BLUEBACK HERRING

One of the better known features of the Santee
Cooper system is the landlocked striped bass (Morone
saxatilis) sport fishery. This fishery was (inci-
dentally) created during the 1940's when the Santee
River diversion caused the formation of two impoundments, Lakes Moultrie and Marion. Striped bass,
trapped within the impoundments. adapted to this
restriction on their seaward movements and established a reproducing, apparently stable, and rel-

Table 3.--Net Energy Costs and Benefits of the Santee-Cooper
Rediversion.(+ indicates net benefits or savings, - indicates costs)

ITEM	Annual Change due to Rediversion	TRANSFORMATION RATIO (kcal solar eq./unit change)	Energy Flow in Solar Equivalents (10^{14} kcal/yr)
Dredging Charleston Harbor	+\$7.1 x 10^6	2.5 x 10^7	+1.78
Direct Rediversion Costs	-\$3.2 x 10^6	"	-0.80
Operation + Maintenance of Power Plant	-\$0.2 x 10^6	"	-0.05
New Canals for Industrial/ Urban Water Supply	-\$0.2 x 10^6	"	-0.05
Cooling Tower for Steam Plant	-\$1.4 x 10^6	"	-0.35
Fish Lift and Hatchery	-\$0.07 x 10^6	"	-0.02
COASTAL PRODUCTIVITY			
Waterfowl Benefits	+\$0.15 x 10^6	"	+0.04
Clam Harvest	-\$0.80 x 10^6	"	-0.20
FLOODPLAIN PRODUCTIVITY			
Forest Metabolism (Gross Productivity)	-2.7 x 10^{11} kcal*	920	-2.48

*Assuming 10 kcal/gC (see table 3)

atively large population (Stevens 1958; 1961). The establishment of this landlocked population was heavily publicized, fish camps opened, and a substantial, 22 million dollar a year (estimate) tourist trade developed (Laurie 1977). However, during 1959-60 a sudden, severe decline in the abundance of landlocked striped bass occurred and was subsequently attributed to a significant decrease in the abundance of gizzard shad (Dorosoma cepedianum), the major prey of landlocked stripers (Stevens 1962; 1963; Laurie 1977). Coincident with the decline in gizzard shad abundance was an increase in importance of blueback herring (Alosa aestivalis) as a striped bass prey item (Stevens 1958; 1963). In the early 1960's, it was decided to schedule specific operations of the navigational lock at Pinopolis Dam to permit the passage of spawning-run herring into the impounded areas. These operations, designed to provide a food source for the landlocked striped bass, are thought to have been responsible for subsequent increases in and stabilization of the landlocked population (Laurie 1977).

[1]Curtis, T.A. 1980. Personal communication. S.C. Wildlife and Marine Resources Dept., Bonneau, S.C.

In addition to the diversion-related sport fishery, increased water flows into the Cooper River were apparently responsible for the development or enhancement of a substantial spawning run of Cooper River striped bass. This circumstance resulted in the development of a fish hatchery near Pinopolis Dam, which currently produces upwards of 50 million striped bass (Harrell 1979). These juveniles are used to stock other lakes in South Carolina and a number of other states.

With the completion of the Rediversion Project (1983), expressed concern regarding the striped bass/blueback herring sport fishery has focused on two aspects: 1) the effect on the magnitude of the herring spawning runs, and 2) the effect on riverine striped bass populations (Laurie 1977; Bulak and Curtis 1977).

Based on data collected during 1975-1979, upwards of 6 million herring are passed through the lock at Pinopolis Dam (associated cost: about \$6,000/yr) every spring into Lake Moultrie (Curtis 1978; 1979). Researchers feel that the current magnitude of herring passage is necessary to maintain present landlocked striped bass stocks (Bulak and Tuten 1979; Curtis 1980[1]). However, because

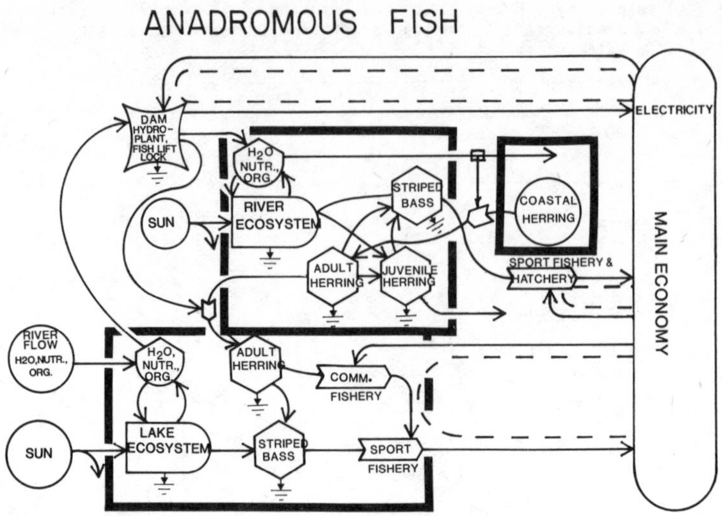

ANADROMOUS FISH

Figure 7.--Energy flows associated with striped bass and
blueback herring interactions, with emphasis on the
commerical and sport fisheries.

it appears that spawning stock size of anadromous
fish and stream discharge may be positively correlated (Setzler et al. 1980; Dwyer et al. 1980),
there does exist the possibility of a significant
decline in the magnitude of the Cooper River herring run. In anticipation of this, and the possibility of a larger run in the Santee River, a
$3,000,000 fish lift located on the rediversion
canal will be used to pass herring into Lake Moultrie.

It has been predicted that the reduction in
flow in the Cooper River will result in a significant decrease in the number of striped bass
broodfish available for hatchery purposes (Harrell
1979). As a result, the Army Corps of Engineers
is building a duplicate hatchery facility (est.
cost: about $425,000) on the rediversion canal in
anticipation of increased Santee River broodfish
availability. In addition, attempts are currently
being made to enhance Santee River striped bass
populations through the introduction of Cooper River
stock (Harrell 1978; 1979).

A flow diagram depicting the Santee Cooper system, emphasizing certain aspects of its associated
fishery is given in figure 7. Briefly, adult herring enter the river systems during the spring, stimulated by temperature, photoperiod, and water flow.
Spawning takes place in flooded, upstream areas,
after which the spent adults leave the river systems. During the spawning run, adult herring are
commercially harvested and later sold as bait in
the landlocked striped bass fishery (annual retail
value of about $70,000). In addition, a large pro-

portion of herring moving upstream in the Cooper
River, are allowed to pass through the navigational
lock into Lake Moultrie. These fish spawn either
in the lakes or in upstream rivers, and are heavily
preyed upon by striped bass and harvested by commercial fisherman (annual retail value about $130,000)
These fish are then used as bait by sport fishermen
attracted to the area by the abundance of landlocked
striped bass. Cooper River striped bass are
used as broodfish for the hatchery production of
juveniles, stimulating sport fisheries elsewhere
in South Carolina and other states. Controlling,
or significantly influencing, many of the biological processes in the Santee Cooper is stream discharge, which is, in turn, controlled by South Carolina Public Service Authority water release schedules.

For the purposes of this paper, based on preliminary findings, we are assuming that the measures taken by the Army Corps of Engineers and the
South Carolina Wildlife and Marine Resources Dept.
will both allow sufficient numbers of herring (to
maintain landlocked stripers) to be passed into the
impounded areas, and the continuation of current
hatchery production levels. Based upon this assumption, costs associated with the passage of blueback
herring and striped bass hatchery operations will
increase by about $68,500 per year over current
costs (table 3).

There does remain the possibility, however,
that the passage of minimal (about 5,000,000)
numbers of herring, as a food source for landlocked striped bass, will not occur after rediver-

sion. Some possible reasons for this would be related to: 1) homing, that is, will the expected "flip-flop" of herring populations between river systems occur; 2) fish lift operations - will herring move into the rediversion canal in sufficient numbers, how high will the lift associated mortality rate be; 3) will the increased discharge into the Santee River cause herring to spawn downstream from the rediversion canal; 4) will extended periods of low or no flow conditions in the Cooper River cause massive fish kills, as have low flow conditions in the Susquehanna River (Md.) (ERM 1979). A significant reduction in the numbers of herring currently passed into the impounded areas could result in substantial monetary losses to the local and state economies.

In addition, if the Santee River striped bass population does not respond positively to increased flows (Harrell 1979), hatchery production could decrease. Since juveniles from this hatchery are currently used to provide stock in other lakes, thus enhancing sport fisheries elsewhere, a decline in hatchery production could lead to monetary losses in areas that depend on these juveniles.

SUMMARY

In this study, we compare the magnitudes of several predicted impacts for the Santee-Cooper River rediversion. The present shoaling of Charleston Harbor and the costs of dredging apparently justify a region-wide modification in river flow. The predictable economic costs of river rediversion includes the construction and mainenance of new channels and canals, a new hydropower plant, a cooling tower in addition to the possible loss of a major clam fishery. These costs are apparently less than the accrued savings due to decreased costs of dredging in Charleston Harbor (table 3).

However, several important issues with respect to other biological/ecological impacts are not as easily predicted and quantified. If the new fish lift is successful in maintaining herring populations in the lake, this additional cost is relatviely small (table 3). However if it is unsuccessful, the possible decline in the lucrative striped bass fishery in the lakes would certainly represent a major economic loss for the state.

Whereas a small additional benefit may be realized due to enhanced fresh and brackish marsh management for waterfowl in the Santee delta, the area-wide impact on total coastal marsh productivity is still uncertain.

Region-wide impacts on the ecological productivity of the Santee floodplain may be highly significant, especially with respect to water flow and flooding patterns of the river. In this paper, we present a tentative but conservative estimate of possible energy flow impacts on the floodplain. These calculations indicate that the energy involvement of these expansive forested wetlands may be at least as significant as the energy savings due to decreased dredging (table 3). Since ecological

community composition and systems productivity of the floodplain is clearly related to water flow and flood patterns of the river, these relationships should be examined more closely in formulating alternative water release plans for the new hydropower plant.

LITERATURE CITED

Bayley, S., H.T. Odum, and W.M. Kemp. 1976. Energy evaluation and management for Florida's east coast. Trans. 41st N.A. Wildl. Mar. Res. Conf., Wildl. Manag. Inst., Wash. D.C.

Bulak, J.S. and T.A. Curtis. 1977. Santee Cooper rediversion project. Ann. Prog. Rep. SCR 1-1. S.C. Wildl. Mar. Res. Dept.

Bulak, J.S. and J.S. Tuten. 1979. Santee-Cooper blueback herring studies. Ann. Prog. Rep. SCR 1-3. S.C. Wildl. Mar. Res. Dept.

Curtis, T.A. 1978. Anadromous fish survey of the Santee and Cooper River system. Ann. Prog. Rep. AFS 3-8. S.C. Wildl. Mar. Res. Dept.

Curtis, T.A. 1979. Anadromous fish survey of the Santee and Cooper River system. Ann. Prog. Rep. AFS 3-9. S.C. Wildl. Mar. Res. Dept.

Day, J.W., Jr., T.J. Butler, and W.H. Conner. 1977. Productivity and nutrient export studies in a cypress swamp and lake system in Louisiana. Estuarine Processes, Academic Press, N.Y. Vol. 2, pages 255-269.

Dwyer, R.L., G.F. Johnson, and T.T. Polgar. 1980. Historieal relationships between anadromous fish landings in the Upper Chesapeake Bay and stock-dependent and environmental factors. Martin-Marietta/Environ. Cent. Rep.

Environmental Resources Management, Inc. 1979. Anadromous fish studies. Conowingo Dam. Draft Rep. to Maryland Power Plant Siting Program. West Chester, Pa.

Gilliland, M.W. (ed). 1978. Energy Analysis: a new public policy tool. AAAS selected symposium No.9. Westview Press, Inc.

Gosselink, J.G. and R.E. Turner. 1978. The role of hydrology in freshwater wetland ecosystems. p. 63-78. In R.E. Good, D.F. Whigham, and R.L. Simpson (eds.) Freshwater wetlands: ecological processes and management potential. Acad. Press. N.Y.

Gosselink, J.E., S.E. Bayley, W.H. Conner, R.E. Turner. in press. Ecological factors in the determination of riparian wetland boundaries. Proc. of a workshop on Riparian Wetlands Lake Lanier, Georgia. Elsevier Sci. Publ. Co., N.Y.

Harrell, R.M. 1978. Enhancement of striped bass population in Santee River. Job Prog. Rep. AFS 8-1. S.C. Wildl. Mar. Res. Dept.

Harrell, R.M. 1979. Enhancement of striped bass population in Santee River. Job Prog. Rep. AFS 8-2. S.C. Wildl. Mar. Res. Dept.

Jansson, A.M. and J. Zuchetto. 1978. Man, nature, and energy flow on the island of Gotland. Ambio 7:140-149.

Kemp, W.M., W.H.B. Smith, H.N. McKellar, M.E. Lehman, M. Homer, D.L. Young, and H.T. Odum. 1977. Energy cost-benefit analysis applied to power plants near Crystal River, Florida,

p. 507-543. In C.A.S. Hall and J.W. Day, Jr. (eds.), Ecosystem Modelling in theory and practice. John Wiley and Sons, N.Y.

Kjerfve, B. 1976. The Santee-Cooper: A study of estuarine manipulations. p. 44-56. In Estuarine processes: Vol. 1. Uses, stresses, and adaptation to the estuary. Aca. Press, N.Y.

Kjerfve, B. and J.E. Greer. 1978. Hydrography of the Santee River during moderate discharge conditions. Estuaries 1:111-119.

Kinne, O. 1971. Marine Ecology I. Environmental factors, part 2. Wiley Interscience. 1244 p.

Laurie, P. 1977. The 100 million dollar question. S.C. Wildl. Mag., Sept.-Oct., p. 9-15. S.C. Wild. Mar. Res. Dept.

McKellar, H., M. Homer, and L. Pearlstine. 1981. Conceptual models and a preliminary energy flow analysis of the Santee-Cooper rediversion project. Report to the U.S. Fish and Wildlife Ser. May 1981

Nelson, F.P. (ed.) 1974. Cooper river environmental study. Rept. No. 117. Estuarine Studies. S.C. Wat. Res. Comm. Columbia, S.C.

Odum, H.T. 1977. Energy analysis, energy quality, and environment. p. 1-39. In M.W. Gilliland (ed.), Energy analysis: a new public policy tool. Amer. Assoc. Advancement Sci., Wash., D.C.

Odum, H.T. 1978. Energy analysis, energy quality, and environment. p. 55-87. In M.W. Gilliland (ed.), Energy analysis: a new public policy tool. AAAS Selected Symposium No. 9. Westview Press.

Odum, H.T. and E.C. Odum. 1976. Energy basis for man and nature. 296 p. McGraw-Hill. N.Y.

Phipps, R.L. 1979. Simulation of wetlands forest vegetation dynamics. Ecol. Modelling 7:257-288.

Reynolds, J.Z. 1980. Power plant cooling systems: policy alternatives. Sci. 207:367-372.

Setzler, E.M. (ed.) 1980. Synopsis of biological data on striped bass, Morone saxatilis. NOAA TECHN. REP. NMFS CIRC. 433.

Stevens, R.E. 1958. The striped bass of the Santee-Cooper Reservoir. Proc. 11th Ann. Conf. S.E. Assoc. Game Fish Comm. 1957:253-264.

Stevens, R.E. 1961. Investigation of fish populations in Lakes Moultrie, Marion, Wateree, and Catawba. S.C. Wildl. Mar. Res. Dept., Proj. F-1-R-9.

Stevens, R.E. 1962. Investigation of fish populations in Lakes Moultrie, Marion, Wateree, and Catawba. S.C. Wildl. Mar. Res. Dept., Proj. F-1-R-10.

Stevens, R.E. 1963. Investigation of fish populations in Lakes Moultrie, Marion, Wateree, and Catawba. S.C. Wildl. Mar. Res. Dept., Proj. F-1-R-11.

Tesky, R.O. and T.M. Hinkley. 1977. Plant and soil responses to flooding. Vol. I. In Impact of water level changes on woody riparian and wetland communities. U.S. Fish and Wildlife Ser. Publ. Washington.

U.S. Army Corps of Engineers. 1966. Survey rept. on Cooper River, S.C. (Shoaling in Charleston Harbor) Charleston District, Charleston, S.C.

U. S. Army Corps of Engineers. 1975. The final environmental statement, Cooper River rediversion project. Charleston District, Charleston, S. C.

U.S. Army Corps of Engineers. 1976. Design memorandum No. 9. Intake and Tailrace Canals. The Cooper River Rediversion Project. Charleston District, Charleston, S. C.

U.S. Army Corps of Engineers. 1981. Summary Construction program (PB-1) Cooper River, Charleston Harbor, S. C.

U.S. Fish and Wildlife Ser. 1966. A fish and wildlife evaluation report on proposed project to reduce shoaling in Charleston Harbor, S.C. In U.S. Army Corps of Eng. Survey Rept. on Cooper Riv., S.C. Appendix G. Charleston District, Charleston. S.C.

Wang, F.C. and M.A. Miller. 1980. Comparison of energy analysis and economic cost-benefit procedure; Case Study-LaSalle nuclear power plant. Report to the Nuclear Regulatory Commission. Energy Analysis Workshop, Center for Wetlands. Univ. Florida, Gainesville.

Wharton, C.H. 1970. The southern river swamp-a multiple-use environment. Bureau of Business Research, School of Business Adm. Georgia State Univ. Statesboro.

Whigham, D.F., J.McCormick, R.H. Good, and R.L. Simpson. 1978. Biomass and primary production in freshwater tidal wetlands of the middle Atlantic coast. p. 3-20. In R.E. Good, D.F. Whigham, and R.L. Simpson (eds.). Freshwater wetlands: Ecological processes and management potential. Aca. Press. N.Y.

Whittaker, R.H. 1975. Communities and ecosystem (2nd ed.) MacMillian Publ. Co., Inc. N.Y. 387 p.

SEDIMENT TRANSPORT AND HYDROELECTRIC POWER GENERATION IN THE HIGH MOUNTAINS OF VENEZUELA[1]

Miguel F. Acevedo L.[2]

Abstract.- A mathematical model of sediment generation, transport and accumulation is developed to analyze the important interaction between environmental processes and hydro-electrical energy generation in the high mountain river basins of the Venezuelan Andes. Particular reference is made to the Santo Domingo river dam.

INTRODUCTION

In the early seventies, the hydroelectric power generation facilities on the Santo Domingo river in the high mountains of Venezuela went into operation. It was soon recognized that this engineering complex was subject to an intimate relationship with environmental processes in the watershed, and that the energy generation capacity was limited by some of these processes, especially the accumulation of sediment transported to the dam from the areas of critical erosion. Such recognition had been made by the designers, but it came as a surprise to the general public, at a time when Venezuelans felt that obtaining the adequate infrastructure for economic "development" only needed a proper budgetary output for financing engineering feats.

Now, in the early eighties, when severe rationing in electrical energy consumption is being imposed on the areas serviced by the plant, due to drought and presumably to an inefficient sediment new interest in the sediment deposition problem has arisen. Moreover, a new hydroelectric power generation complex is under construction in the neighboring Uribante-Caparo watershed, with the ambition of serving a large portion of western Venezuela. The relationships

of this new complex with the environmental processes of the area have not yet come under scrutiny, and public awareness has not yet focused attention on the possible shortcomings of the future complex. Sediment transport may again remind the development anxious public that the environment poses constraints on the energetic output of hydroelectric plants.

A complete analysis of the sediment deposition problem at the Santo Domingo river dam and its possible solutions have been reported by Yanes and Guevara (1975). Their work covers a wide range of aspects of the sediment problem, such as: technical details of erosion, agricultural and other human activities upstream from the dam, alternative solutions, and recommendations with regard to energy generation. They proposed the schematic model of sediment generation, transport and deposition shown in figure 1, where the storage compartment represents the sediment in transit, i.e. temporarily deposited on the water course upstream from the reservoir. By elaborating on this scheme, I wish to produce, in this paper, an analytical treatment of the problem using the theory of shot noise or filtered Poisson processes (Papoulis 1965; Davenport 1970).

[1] Paper presented at the international symposium on Energy and Ecological Modelling, sponsored by ISEM. Louisville, Kentucky, April 20-23, 1981.

[2] Escuela de Ingeniería de Sistemas, Universidad de Los Andes, Mérida, Venezuela

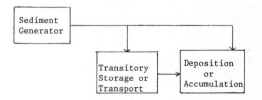

Figure 1.- Schematic model of sediment generation, transport and deposition proposed by Yanes and Guevara (1975).

THE SEDIMENT GENERATOR MODEL

Due to the characteristics of the erosion process in the area, the sediment generator output can be conceived as a sequence of sudden bursts of sediment load. Sediment generation could be modelled , then, as a train of impulses or Dirac delta functions

$$\sum_k g_i \delta(t-t_k) \qquad (1)$$

occurring at times t_k with intensity g_i which will be the total sediment volume generated at times t_k. Although some may question the realism of the following assumption, a simple way of modelling the times of occurrence of these impulses is to assume them to be random and originated by a Poisson process. In this manner, impulses of intensity g_i constitute a train of non-uniform or non-homogeneous Poisson impulses with rate parameter $\lambda_i(t)$ varying with time t. Eliminating the time dependence of this parameter reduces the model to the special case of a uniform or homogeneous Poisson impulses train. Since there will be different rates for different intensities, the sediment generator output $g(t)$ would be a sum of several trains of Poisson impulses, that is to say

$$g(t) = \sum_i g_i \sum_k \delta(t-t_k) \qquad (2)$$

A high value of g_i would be associated with a low rate $\lambda_i(t)$, whereas a low value of g_i corresponds to a high rate $\lambda_i(t)$.

It is interesting to model the rate parameters $\lambda_i(t)$ as seasonally modulated using a Fourier series expansion

$$\lambda_i(t) = \sum_n \lambda_{in} \exp(jnw_o t) \qquad (3)$$

with main frequency $w_o = 2\pi/year$. The coefficient λ_{io} will be aproximated by the inverse of the return period T_i of the flood value g_i , which are available data (Yanes and Guevara 1975).

THE TRANSIT MODEL

Now, the trains of Poisson impulses are the inputs to the storage or transit compartment in figure 1. As shown in this figure, some material can arrive at the reservoir almost instantly after the occurrence of a sediment impulse, i.e. bypassing the transit storage box which could be seen as a transport lag. The simplest assumption for the distribution of time elapsed for a generator impulse to arrive at the reservoir is then an exponential distribution

$$h(t) = ae^{-at} u(t) \qquad (4)$$

where $a < 1$ gives both the fraction of the impulse intensity arriving instantaneously at the reservoir as well as the rate of arrival of the delayed material, and where $u(t)$ is the Heaviside step

function. Note that

$$\int_0^\infty h(t)dt = 1 \qquad (5)$$

represents the concept that there is 100% efficiency in sediment transport, i.e. all sediment generated eventually reaches the reservoir. Application of the model to a different situation where the efficiency is less than 100% requires adjusting (4) so that the right side of (5) is a fraction of one. Likewise, applying the model to a situation where there is not instantaneous arrival of material requires a different distribution in (4), e.g. a gamma distribution.

Now, $h(t)$ can be conceived as the impulse response of a linear system whose dynamics represent the transport process and whose transfer function is

$$H(p) = a/(p+a) \qquad (6)$$

Of interest, then, is the volume of sediment reaching the reservoir which is given by the response of this linear system to the generator output $g(t)$. Because the system is assumed to be linear, the response is a summation of several non-uniform shot noise (Papoulis 1965) or filtered Poisson (Davenport 1970) processes.

Each one of these shot noise processes $s_i(t)$ corresponds to one of the trains of Poisson impulses, and consequently has an expected value of (Papoulis 1965)

$$E\{s_i(t)\} = g_i \int_0^\infty \lambda_i(t-\tau)h(\tau)d\tau \quad . \qquad (7)$$

This convolution integral could be evaluated via transform techniques. Denoting the magnitude and angle of $H(jw)$ by $A(w)$ and $\phi(w)$ respectively, we can write the steady-state expected value of each shot noise process as

$$E\{s_i(t)\} = g_i \sum_n \lambda_{in} A(nw_o) \exp\{jnw_o t + j\phi(nw_o)\} \qquad (8)$$

Adding over all i's, we get the total output of the transit storage compartment as

$$E\{s(t)\} = \sum_i E\{s_i(t)\} \qquad (9)$$

Of course, not only the expected value is of interest. Next, we determine the variance of $s_i(t)$ by (Papoulis 1965)

$$\sigma_{s_i}^2(t) = g_i^2 \int_0^\infty \lambda_i(\tau) \, h^2(t-\tau)d\tau \qquad (10)$$

Again we evaluate a convolution integral via transform techniques; let $\tilde{H}(jw)$ be the transform of

$$h^2(t) = a^2 e^{-2at} u(t) \qquad (11)$$

It is easy to see that

$$\hat{H}(jw) = a^2 /(jw+2a) .\qquad(12)$$

Denoting by $\hat{A}(w)$, $\hat{\phi}(w)$ its magnitude and phase respectively, we can write

$$\sigma^2_{s_i}(t)=g_i^2\sum_n \lambda_{in}\hat{A}(nw_o)exp\{jnw_ot+j\hat{\phi}(nw_o)\}\quad(13)$$

and then

$$\sigma^2_s(t) = \sum_i \sigma^2_{s_i}(t) .\qquad(14)$$

If we are willing to ignore the intra-annual variations, and use only the year to year dynamics, simpler expressions are possible:

$$E\{s(t)\}=\sum_i g_i\lambda_{io} , \quad \sigma^2_s(t) =\sum_i g_i^2 \lambda_{io}a/2\quad(15)$$

because

$$A(0)=1 \text{ and } \hat{A}(0)=a/2 .\qquad(16)$$

THE ACCUMULATION MODEL

Next, accumulation at the reservoir represented by variables $r_i(t)$ denoting the total volume of material accumulated by time t due to shot noise process $s_i(t)$. The dynamics of this last compartment are simply given by the integral

$$r_i(t) = \int_0^t s_i(\tau)d\tau\qquad(17)$$

and of course

$$r(t) = \sum_i r_i(t) .\qquad(18)$$

If we are interested in the year to year variations, t is a multiple of $2\pi/w_o= 1$ year (say m years); then simple expressions for $E\{r_i(t)\}$ can be derived because integrating the sinusoidal components of the Fourier series over an exact number of cycles cancel them out, yielding

$$E\{r_i(m)\} = \int_0^m E\{s_i(t)\}dt = g_i \int_0^m \lambda_{io}A(0)dt\quad(19)$$

and upon integration and using (16)

$$E\{r_i(m)\} =g_i\lambda_{io}m\qquad(20)$$

and therefore

$$E\{r(m)\} = m\sum_i g_i\lambda_{io} = m\sum_i g_i T_i\qquad(21)$$

which constitutes an expression useful to calculate the number of years that should elapse for the expected value of accumulated sediment to reach a critical level.

Estimating the variance of $r_i(t)$ is not so easy. We would need to calculate

$$\sigma^2_{r_i}(t) = \int_0^t\int_0^t C(t_1,t_2)dt_1dt_2\qquad(22)$$

where $C(t_1,t_2)$ is the covariance of $s_i(t)$ which in turn is given by

$$C(t_1,t_2) = g_i^2\int_0^\infty \lambda_i(\tau)h(t_1-\tau)h(t_2-\tau)d\tau\quad(23)$$

The special case of uniform shot noise allows an easier calculation because the process is stationary. Ignoring the intra-annual variations and using only a constant rate λ_{io}, we get the covariance

$$C(t_1-t_2) = \lambda_{io}ag_i^2 exp(-a|t_1-t_2|)/2\qquad(24)$$

and therefore (22) reduces to

$$\sigma^2_{r_i}(t)= 0.5\lambda_{io}ag_i^2\int_0^t\int_0^t exp(-a|t_1-t_2|)dt_1dt_2\quad(25)$$

Making $t_1-t_2=\tau$ and because $exp(-a|\tau|)$ is even in τ, (25) can be rewritten as

$$\sigma^2_{r_i}(t) =\lambda_{io}ag_i^2\int_0^t (t-\tau)e^{-a\tau} d\tau\qquad(26)$$

and therefore

$$\sigma^2_{r_i}(t) = \lambda_{io}g^2 \{t(1-2e^{-at})+(1-e^{-at})/a\}\quad(27)$$

In this special case of uniform shot noise, the expected value of $r_i(t)$ becomes $g_i\lambda_{io}t$ as it can be easily seen from (20). We conclude, then, that after several time constants a^{-1}, when the term e^{-at} goes to zero, both the mean and variance of r_i increase linearly with time.

DISCUSSION

In this paper, the details of a possible model of sediment generation, transport and deposition have been presented for the special situation of hydroelectric facilities in the Venezuelan Andes. The model is based on the strong assumption of Poissonian sediment bursts and exponential or first-order transport lag. Seasonal variations on the rate of bursts can be accounted for, but the results are harder to interpret and derive (as in the case of the variance of the volume of accumulated material at the reservoir). Simpler results can be obtained if the seasonal variation is ignored and uniform Poisson impulses are considered. In this case, the expected value of accumulated sediment load increases linearly with time, but so does the variance, suggesting a non very reliable prediction for a long time horizon. The derived results for the seasonal variation of sediment transported are steady-state results, due to the approach followed in evaluating the convolution integrals in (7) and (10).

Further work is necessary to be able to evaluate the transients , as well as to relate them and the results derived here to the policies of electrical energy generation and sediment discharge.

ACKNOWLEDGEMENTS

I wish to thank Dr. Adolfo Yanes for providing the stimulus to work on this problem.

REFERENCES

Davenport,W.B. 1970.*Probability and Random Processes.* McGraw-Hill, N.Y.

Papoulis,A. 1965, *Probability, Random Variables and Stochastic Processes.* McGraw-Hill, N.Y.

Yanes,A and Guevara J. 1975, *Sedimentos en el Embalse.* CADAFE, Caracas.

5. Modelling of Air Pollution Impact

SOME CONSEQUENCES FOR GROUND LEVEL CONCENTRATIONS OF SULPHUR DIOXIDE RESULTING FROM INCREASING POWER GENERATION CAPACITY[1]

Brian Henderson-Sellers[2]

A numerical model is described which simulates the rise and spreading of an effluent containing any percentage of a dense gas (such as SO_2). Spread rates and ground level concentrations are calculated for several different atmospheric stability classes.

INTRODUCTION

The majority of atmospheric pollution is generated by combustion. In the U.K., in 1976, power generation was responsible for sulphur dioxide emissions of 2.7 million tonnes out of a total emission of 4.1 million tonnes (Weatherley 1977). The currently increasing emissions from fossil fuelled power stations are probably a consequence of two factors: increasing demand and varying sulphur content in the fossil fuel (as prime, first grade sources become depleted). The possible reversion to a coal economy both in Europe (for instance, the decisions taken in the Venice Summit by EEC members in Summer 1980) and North America as oil reserves dwindle could be anticipated to lead to further increases in emissions. Unless source control measures are implemented this will result in a greater atmospheric loading of SO_2 and the consequent increase in acid rain, damage to human health, etc.

Many countries adopt a "tall stack" policy as well as or in place of waste gas cleaning. The dilution afforded by the atmosphere ought to be sufficient (in all except unusual meteorological conditions) to ensure that ground level concentrations are kept below an "acceptable level" (although some tall stack calculations based on flat terrain assumptions may not be accurate for a heavily industrialized, urban environment). However, if the sulphur content of the effluent were to increase, the density of the waste gas would also become greater. Alternatively, scrubbing gases, although removing sulphur dioxide, tends to cool the gas. In both these cases, the result is a less buoyant effluent whose restricted rise must

be recalculated. Many nomograms and empirical formulae are unable to take into account these and other factors, including non-planar urban topography and developing urban heat domes. The numerical model described here can readily incorporate such characteristics (of both plume and environment). Preliminary results are presented and discussed.

PLUME RISE MODEL

Numerical plume rise models may be based upon the set of conservation equations originally presented by Morton et al. (1956) and later analyzed by Morton (1971). The mixing and dilution of the plume material is characterized in terms of an entrainment velocity. The mixing rate thus determined leads both to a dilution rate and, through its effect on total plume buoyancy, possibly to a limit to the plume rise (dependent also upon the prevailing atmosphere stratification). Slawson and Csanady (1967, 1971) identified three characteristic mixing phases and this "3-phase" model was then later extended by Henderson-Sellers (1980) by relating the second and third phase entrainment velocities to readily observable parameters in the urban environment (e.g. topography and building heights, friction, velocity, etc.). In fact, the third phase is easily shown to be of identical formulation to the so-called "Gaussian plume model" (e.g. Melli and Runca 1979) for non-buoyant sources. In these models the buoyancy is expressed in terms of density differences induced by temperature differences, i.e. it is assumed that the molecular weight of the effluent is of the same order as the molecular weight of air. For plumes containing a non-negligible percentage of a dense gas such as sulphur dioxide or carbon dioxide, this assumption may no longer be strictly valid. The inclusion of such direct density effects is discussed here and the results of the 3-phase model compared with laboratory experiments (from the literature) and one previous model.

The conservation equations are given (Henderson-Sellers 1980) by

[1] Paper presented at the international symposium Energy and Ecological Modelling, sponsored by the International Society for Ecological Modelling. (Louisville, Kentucky, April 20-23, 1981).

[2] Dr. B. Henderson-Sellers is lecturer in the Department of Civil Engineering, University of Salford, Salford, U.K.

$$dV/ds = 2^{1-a}(V_e/I_1) \ (V/M)^{2a/(a+1)} \qquad (1)$$

$$dM^4/ds = 2 \ (I_3/I_2) \ B \ V \sin \phi \qquad (2)$$

$$dB/ds = - (I_1/I_4) \ N^2 \ V \sin \phi \qquad (3)$$

$$\cos \phi = \{(B_0 V_0/M_0^2) \cos \phi_0\} \ M^2/BV \qquad (4)$$

where

$$V = b^{a+1}W; \ M = b^{(a+1)/2}W; \ B = b^{a+1}W\Delta \qquad (5)$$

b is the radius length scale, Δ and W the axial values of buoyancy and velocity respectively, ϕ is the angle of the plume to the horizontal and subscript zero indicates initial values. The values of the constants I_i are given in Henderson-Sellers (1981) and a has a value of 1 for a point source and zero for a line source. N^2 is the Brunt-Väisälä frequency and V_e the entrainment velocity.

Dense Plume Model

The axial buoyancy parameter Δ is related to the density, d, at any spatial location (curvilinear coordinates q, s) by

$$d(q,s) = \Delta(s) \ G(1) \qquad (6)$$

where q is the lateral distance from the axis and s is the longitudinal distance along the trajectory. G(1) is a shape function given by

$$\text{Top hat:} \quad G(1) = 1 \ \text{for} \ |q| < \lambda \ b \qquad (7a)$$

$$\text{Gaussian:} \quad G(1) = \exp(-1^2/\lambda^2) \qquad (7b)$$

where $1 = q/b$ and λ is the turbulent Schmidt number. d is related to the density difference by

$$d = g \ (\rho_e - \rho)/\rho_1 \qquad (8)$$

where ρ_e is the atmospheric density and ρ_1 is a reference density taken as the atmospheric density at the point of emission. (Alternatively, this can be expressed in terms of the difference of potential temperature, θ).

Integrating over the cross-section gives

$$(2\pi)^a \int_{-}^{\infty} u(q,s)d(q,s)q^a dq = (2\pi)^a W(s)\Delta(s)b^{a+1}$$

$$\times \int_L^{\infty} F(1)G(1)1^a d1$$

i.e. $\int_L^{\infty} u(q,s)d(q,s)q^a dq = Wb^{a+1}\Delta(s)I_4 = BI_4 \qquad (9)$

For an emission of molecular weight not equal to that of air, the following modifications will be implemented.

Taking an effluent consisting of 100c% (by volume at atmospheric temperature and pressure) of pollutant with molecular weight μ_p and 100(1-c)% of air (molecular weight μ_a), then assuming the pollutant is well-mixed (i.e. using a "top hat" model) the plume density is given by

$$\rho(T_e) \simeq (c\mu_p + (1-c)\mu_a) \times 0.0423 \ \text{kg m}^{3} \qquad (10)$$

Using the ideal gas law

$$\rho(T_s) = T_e \ \rho(T_e)/T_s$$

$$= (c\mu_p +(1-c) \ \mu_a) \times 0.0423 \times T_e/T_s \qquad (11)$$

Hence equation 6 is written

$$d = g(\rho_e-0.043T_e(c\mu_p+(1-c)\mu_a)/T_s)/\rho_1 \qquad (12)$$

and since $\rho_e \simeq \rho_1 \simeq 0.0423 \ \mu_a$ we have

$$d = g(1-T_e/T_s) \ (1-c(1-\mu_p/\mu_a)) + gc(1-\mu_p/\mu_a) \qquad (13)$$

The buoyancy flux B in equations 1 to 4 must thus be replaced by B' given by

$$B' = B(1-c(1-\mu_p/\mu_a)) + Vgc(1-\mu_p/\mu_a)I_1/I_4 \qquad (14)$$

RESULTS

As partial verification of this model, the simulations are compared with observations of Bodurtha (1961). These are summarized by Ooms et al. (1974) who also compare their own model with these observations. These latter results were expressed in terms of specific gravity. For comparison purposes only, the special case of a stack emission at ambient temperature is considered for the calculations using the modified 3-phase model described here. The buoyancy flux B' is thus dependent only on the concentration c and the species involved. Taking μ_p=64 (SO_2) and μ_a = 29

$$B'_0 = V_0 \ gc(1-64/29) \ I_1/I_4 \qquad (15)$$

For a stack effluent of radius R_0, effluent velocity W_0 we have (assuming a "top hat" cross-sectional profile and a turbulent Schmidt number of unity)

$$B'_0 = R_0^2 \ W_0 \ g \ c \ (1-64/29) \qquad (16)$$

As a result of the use of the top hat profile, this value is in fact a "model equivalent" value (Henderson-Sellers 1981). However Ooms et al. (1974) express the buoyancy as $-g\rho_0^*/\rho_{a0}$ which is a Gaussian value (Δ_G). This is related to the top hat model equivalent value $\Delta' = B'/(b^{a+1} \ W)$ by

$$\Delta_G = 2 \ \Delta' \qquad (17)$$

Evaluation of the value of c corresponding to the value of Δ_G used by Ooms et al. (1974) is accomplished by simple substitution into this equation viz.

$$- g\, \rho_0^*/\rho_{ao} = 2R_0^2 W_0 gc(1-64/29)/R_0^2 W_0$$

Hence $\qquad c = 0.4143\, \rho_0^*/\rho_{ao}$ (18)

(Numerical values are given in table 1).

The top hat dense plume model will now satisfactorily simulate any stack emissions - the relations between the top hat and Gaussian profiles which may be assumed for the effluent at the point of emission must be taken into account (Henderson-Sellers 1981).

Only Case I and II (Ooms et al, 1974) can be simulated with the present model (using a variable percentage of sulphur dioxide). The results are displayed graphically in figure 1. Case I is a plume of zero excess buoyancy emitted at a velocity of 6.10 m s^{-1}. The rise is thus totally momentum dominated. The 3-phase simulation is in good qualitative agreement with both Ooms et al. (1974) and Bodurtha (1961). Discrepancies may be partly due to the character of the observational data; (Ooms et al comment that the flow visualization could be unrealistic for several reasons:

Table 1.--Values for ρ_0^*/ρ_{ao} and c for three dense plumes specified by Ooms et al. (1974).

Ooms et al	ρ_0^*/ρ_{ao}	c
Case I	0.0	0.0
Case II	0.52	0.2154
Case III	4.17	1.7276

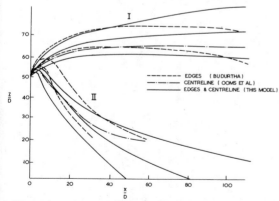

Figure 1.-- Observed (Case I and Case II) plumes (Bodurtha, 1961); calculated plumes (Ooms et al, 1974) and calculations using the modified three-phase model described here.

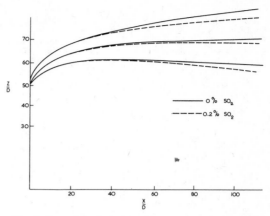

Figure 2.-- Calculated trajectory for an emission containing 0.2% SO$_2$ (cf. emission containing no dense gas).

the use of dense oil as an aerosol; presentation of results as photographs; unrealistic Reynolds numbers) as well as to dissimilarities in model construction. Better agreement exists between the two models for Case II (s.g. = 1.52) - although Ooms et al. (1974) give no indication of calculated spread rates for comparisons here. (Case III cannot be simulated using an SO$_2$ plume).

For the specific comparative study it would appear that the modified 3-phase model, developed originally for positively buoyant plumes, is equally applicable to the case of plumes denser than air.

For a more typical SO$_2$ power plant plume, however, the results are less "dramatic". Figure 2 shows a plume containing 0.2% SO$_2$ (c = 0.002). This corresponds (with the emission radius and velocity as before) to an emission of the order of 0.01 kg s^{-1}. The decrease in the calculated plume rise at a distance x/D = 100 downstream (D is the stack radius) is over 6%. The results plotted in figure 1 and 2 on non-dimensional axes indicate that there is acceptable agreement between this modified three-phase model and previous models and observations.

For concerns of health, acid rain, etc., it is of more use to characterise the plume by both trajectory and the ground level concentrations. These should be discussed at distances of up to several kilometres downwind of the source.

Ground Level Concentrations

In calculating ground level concentrations, it is usually assumed that the background pollution is negligible and thus all the pollutant is contained within the plume. If it is further assumed that the pollutant is either gaseous or an aerosol with negligible reaction rate or fall out[3] rate, then the concept of continuity can be invoked. Although a top hat profile has been used to calculate the trajectory, it is now necessary (and easily undertaken) for a Gaussian distribution of pollutant to be invoked. This can be assumed to be identical with that of density. Then conservation of the pollutant implies that

$$\int n \, \overset{*}{w} \, dA = \text{constant} = Q_S \qquad (19)$$

where Q_S is the source strength in kg s^{-1}, A is the cross-sectional area of the plume, n is the ambient concentration and $\overset{*}{w}$ the longitudinal velocity. In terms of centreline values N_G, $\overset{*}{W}$ given by

$$n(s,q) = N_G(s) \, G(l)$$
$$\overset{*}{w}(s,q) = \overset{*}{W}(s) \, F(l) \qquad (20)$$

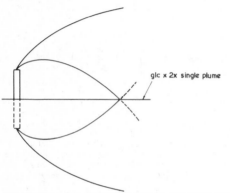

Figure 3.-- The method of images for calculating ground level concentrations.

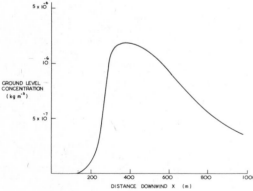

GROUND LEVEL CONCENTRATION (kg m^{-3})

DISTANCE DOWNWIND X (m)

Figure 4.-- Ground level concentrations $n(s,z_c)$ for an 0. 2% SO$_2$ plume (as shown in figure 2.

---

[3]One method of including this effect is given by Moore (1980).

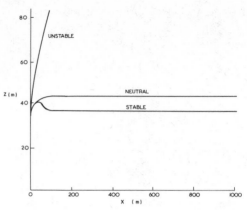

Figure 5.-- Trajectories for an 0. 2% SO$_2$ plume under neutral, stable and unstable conditions.

where F(1), G(1) are the shape functions for momentum and density respectively and $l = q/b$; equation 19 is written (for the 3-D case only) as

$$2 \pi N_G(s) \, \overset{*}{W}(s) \, b^{a+1} \, I_4 = \text{constant} = Q_S \qquad (21)$$

$$N_G(s) = Q_S/(2\pi \overset{*}{W}_G(s) \, b_G^2 \, I_{4G}) \qquad (22)$$

$$\text{or } Q_S/(\pi \overset{*}{W}_T(s) \, b_T^2 \, I_{4T})$$

$$\text{and } n(s,q) = N_G(s) \, G(l) \qquad (23)$$

where $\overset{*}{W}_T$, b_T, I_{4T} are the values that have been employed in the top hat model calculations.

Ground level concentrations (over flat terrain here) are given by setting $q = z_c$ where $z_c(x)$ is the height of the plume centreline at distance x downstream of the source. The method of images is used (fig.3) which leads to the ground level concentration being double that calculated from equations 22 and 23; (this assumes total reflection of pollutant at the solid boundary). The function $n(s,z_c)$ is plotted in figure 4 for the plume shown in figure 2 to a downwind distance of 1000 metres. The maximum occurs at a distance of 400 m - in good agreement with classical analyses of the Gaussian plume model.

The effects of the meteorological conditions can be demonstrated by altering the ambient stratification. Figure 5 shows the corresponding plume trajectories for an unstable case and for the case of an inversion at ground level (although in this case it must be stressed that the dampening effect on the cross-sectional shape has not been taken into account - this is currently being introduced into the model (paper in preparation)). In neither case does the plume touch ground within 1 km of the source. The unstable plume rises rapidly, whilst both the rise of the stable plume and its spread rate are severely damped (in comparison with figure 2). Enhanced ground level concentrations are most likely when an inversion exists just above the stack top when the fanning nature of the plume becomes important (paper in preparation).

CONCLUSIONS

The three-phase model has been successfully modified to include dense gases. Using the (approximate) top hat model, the trajectory has been calculated for various classes of atmospheric stability and, by transforming to a Gaussian cross-sectional shape, ground level concentrations have been deduced. These reflect the expected maximum at a distance of 10-15 stack heights downstream. Furthermore the model is designed to allow for non-planar natural and urban topography thus making this numerical approach useful for a wide variety of urban configurations with any prevailing meteorological conditions.

LITERATURE CITED

Bodurtha, F.T. 1961. The behaviour of dense stack gases. J. Air Pollut. Control Assoc. 11: 431-437.

Henderson-Sellers, B. 1980. The behaviour of marginally buoyant plumes in an urban environment. Ecological Modelling 9:43-56.

Henderson-Sellers, B. 1981. Shape constants for plume models. Boundary-Layer Meteorology (in press).

Melli, P. and Runca, E. 1979. Gaussian plume model parameters for ground level and elevated sources derived from the atmospheric diffusion equation in a neutral case. J. Appl. Meteor. 18:1216-1221.

Moore, D.J. 1980. Atmospheric and water pollution from power plants. p290-325. In Geophysical Aspects of the Energy Problem ed. A. Rapolla, G.V. Keller and D.J. Moore, Elsevier, Amsterdam.

Morton, B.R. 1971. The choice of conservation equations for plume models. J. Geophys. Res. 76: 7409-7416.

Morton, B.R., Taylor, G.I. and Turner, J.S. 1956. Turbulent gravitational convection from maintained and instantaneous sources. Procs. Roy. Soc. Series A 234; 1-23.

Ooms, G., Mahieu, A.P. and Zelis, F. 1974. The plume path of vent gases heavier than air, p211-219. In Proceedings 1st International Loss Prevention Symposium, the Hague/Delft, Netherlands, Elsevier, Amsterdam.

Slawson, P.R. and Csanady, G.T. 1967. On the mean path of buoyant bent-over chimney plumes. J. Fluid Mech. 28:311-327.

Slawson, P.R. and Csanady, G.T. 1971. The effect of atmospheric conditions on plume rise. J. Fluid Mech. 47:33-48.

Weatherley, M-L.P.M. 1977. Fuel consumption, and smoke and sulphur dioxide emissions, in the United Kingdom up to 1976. Warren Spring Laboratory LR258(AP).

AN APPROACH TO MODELLING VEGETATION YIELD REDUCTION FROM POINT SOURCE EMISSIONS OF SULFUR DIOXIDE[1]

T. Veselka,[2] S. W. Ballou,[3] P. A. Dauzvardis[4]

During the past decade, a great amount of research has been directed toward understanding the dose-response effects of gases emitted from industrial sources on vegetation. The results of this research, while still far from complete, can be translated into predictive models that will give estimates of damage to vegetation in terms of decreased yields and the amount of geographic areas exposed. This paper presents one such model, developed at Argonne National Laboratory, which examines the possible SO_2 - induced reduction in soybean yield located in the vicinity of a coal-fired power plant. Two hypothetical 1000 MW power plants located in Greensboro, N.C. were modeled using this technique, one with an emission rate of 1500 g/s and the other with a rate of 2250 g/s. Integrated SO_2 exposures were calculated to simulate reported dose-response data for soybeans. The results show that there was an insufficient dosage to cause detectable chronically induced yield reduction for the low emission case but a 7-8% reduction in yield was estimated for the high case. The total area which would experience a yield reduction of 5% or more was estimated to be about 1000 acres for the high emission rate case.

INTRODUCTION

Many researchers have attempted to mathematically simulate the responses of sensitive vegetation to atmospheric pollutants. Developing functional relationships, however, has proved difficult because of the complexity of the situation to be modeled. Actual plant response depends on many factors, such as climate, soil type, cultivation, sensitivity within species, agricultural practices, and temporal and spatial variations in concentrations of a mixture of air pollutants. The number of these independent variables has been reduced by studying exposures to single pollutants under controlled laboratory or field fumigation conditions. But any dose-response relationship developed for a given species under experimental conditions is, at best, an approximation. Likewise, damage functions developed from field observations also have difficulty accounting for all factors.

Most methods of assessing the potential effects of isolated point source emissions are limited to comparing a predicted air pollutant concentration to some threshold level and estimating the area exposed to concentrations above that level. Air pollutant exposure levels are generally estimated by Gaussian dispersion models or empirical techniques. Each technique has its advantages and drawbacks. Whichever technique is used, however, it must be consistent with the vegetation damage function to be utilized in the damage assessment. That is, if the technique for assessing vegetation damage requires an estimate of 3-hr average concentrations at a given location, the air quality model should provide that estimate. Techniques to determine air pollutant doses integrated over time for use in chronic vegetation damage assessments have been particularly lacking.

This report outlines some of the techniques available for vegetation damage assessments. A

[1] Paper presented at the international symposium Energy and Ecological Modelling, sponsored by the International Society for Ecological Modelling. (Louisville, Kentucky, April 20-23, 1981).

[2] Thomas Veselka, Graduate student in residence, Argonne National Laboratory, Argonne, Illinois, U.S.A.

[3] Stephen W. Ballou, Director Midwest Regional Studies, Argonne National Laboratory, Argonne, Illinois, U.S.A.

[4] Peter A. Dauzvardis, Assistant Environmental Engineeer, Argonne National Laboratory, Argonne, Illinois, U.S.A.

new method of assessing potential chronic damage has been developed and is described here. An example is presented in which this method is used to estimate the chronic damage in terms of yield reduction for a crop grown near coal-fired power plants.

AIR QUALITY MODELING METHODS

Empirical Techniques

Empirical techniques for predicting potential air pollutant concentrations around point sources are based upon observations of air quality. A relationship between emissions and ground-level concentration or impact area is established by identifying trends over time. To develop an accurate relationship, a comprehensive set of monitored air quality data is necessary. The particular source to be studied must have an established monitoring network that gives regularly reported air quality data for a large area surrounding the source. This network should be designed to determine background air quality as well as the contributions from the emissions source. Also, the data should include emissions and meteorological information.

Empirical techniques that predict air quality impacts for crop damage assessments have the distinct advantage of being based on actual occurrences. As such, they can account for climatological conditions such as fumigation and extended periods of stagnation, which are not reliably predicted by conventional dispersion models. Also, once the exposure area for a given threshold is related to the emission rate, curves can be drawn that make the use of this method quite simple.

Only a few sources have an air quality monitoring system large enough to accurately predict areas of air pollutant exposure for vegetation. Often, monitors are located close to each other to determine the peak concentrations, but do not provide enough data to determine lower exposure levels around the source. Topographic features around sources often may present anomalous results that make it difficult to empirically determine the area of impact. Also, an empirical relationship developed for emitting sources in one geographic area may not be fully applicable to other regions if the regions differ widely in climatological factors (mixing height, occurrence of stagnation, etc.).

The Institute of Ecology (TIE) of Butler University developed an empirical relationship between sulfur dioxide (SO_2) vegetation exposure areas and emissions from coal-fired power plants for its work in the Ohio River Basin Energy Study (ORBES) (Loucks et al, 1980). Data were collected from the monitoring network developed for three power plants in the American Electric Power

System. The mean was computed for the second highest 1-hr average SO_2 concentration occurring at each of 11 monitors for the months of June, July, and August 1975, 1976, and 1977. (The second highest value was used on the assumption that two episodes per month could induce damage to sensitive vegetation during the growing season.) These data were used to determine areas of SO_2 exposure for thresholds of 0.1, 0.2, and 0.05 parts per million (ppm). Results showed that the shapes of the exposure isopleths approximated ellipses. The 0.05-ppm isopleth was extrapolated by assuming a Poisson distribution of concentration in the downwind direction and a Gaussian distribution in the crosswind direction. Major and minor ellipsoid radii were plotted as a function of SO_2 emission rates (figs. 1 and 2).

Gaussian Plume Models -- Peak Level Impacts

Gaussian plume dispersion models are widely used to estimate ground-level concentrations of air pollutants emitted from point sources. The Gaussian equation takes the form (Turner 1974).

$$\chi(x,y,z,H) = \frac{Q}{2\pi u \sigma_y \sigma_z} \exp\left[-\frac{1}{2}\left(\frac{y}{\sigma_y}\right)^2\right] \quad (1)$$

$$\left\{\exp\left[-\frac{1}{2}\left(\frac{z-H}{\sigma_z}\right)^2\right] + \exp\left[-\frac{1}{2}\left(\frac{z+H}{\sigma_z}\right)^2\right]\right\}$$

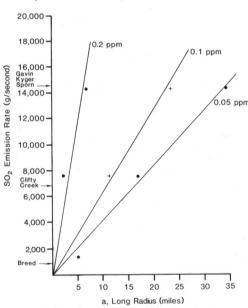

Figure 1.-- Empirical Relationship between Power Plant Emission Rate and Major Radii of Three Concentration Isopleths as Developed by Loucks et al.(1980)

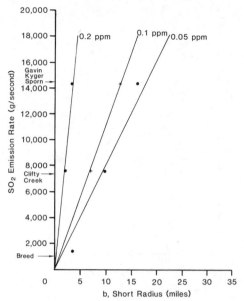

Figure 2.—Empirical Relationship between Power
Plant Emission Rate and Minor Radii of Three
Concentration Isopleths as Developed by
Loucks et al. (1980)

where:

χ = pollutant concentration, g/m^3

x = downwind distance from the
source, km

y = crosswind distance from the
plume centerline, m

z = vertical distance above the
ground, m

H = effective stack height, m

Q = source emission rate, g/s

u = mean wind speed, m/s

σ_y = standard deviation of plume
concentration distribution
in the horizontal at distance
x, m

σ_z = standard deviation of plume
concentration distribution
in the vertical at distance
x, m

The dispersion coefficients, σ_y and σ_z, have been
empirically determined from field observations.
Their magnitudes depend on the stability conditions in the lower atmosphere. Although several
sets of dispersion coefficients have been developed, those of Pasquill and Gifford provide
reasonable estimates for 10-min average concentrations over a range of several kilometers (Turner

1974). These coefficients can be represented in
the form of the power law as follows:

$$\sigma_z = ax^b \qquad (2)$$
$$\sigma_y = cx^d$$

where a, b, c, and d are empirically derived
coefficients that depend on the atmospheric
stability. Relationships for estimating the
rise of thermally buoyant plumes also have been
determined empirically.

Advantages and Disadvantages

Gaussian plume dispersion modeling is a
relatively simple technique for analyzing many
different source variables; emission rates,
stack heights, etc. are easily substituted into
the model. Isopleths of pollutant concentration
can be prepared to determine impact area. The
Gaussian model has an added benefit for crop
damage assessment because it is widely recognized by air quality analysts and thus makes
reproduction of results possible.

A major drawback of Gaussian plume models is
that the relationships for the plume rise and the
dispersion coefficients have been determined for
only a few cases. Problems can occur when
results are extrapolated to averaging times
longer than about 1 hr using the Pasquill-Gifford
dispersion coefficients, since the empirical data
from which those coefficients were derived were
based on 10-min tests (Pasquill 1974). Moreover, in most cases, the dispersion coefficients
are accurate to only about 10 km downwind of the
source (Turner 1974). Gaussian plume models
have limited value in analyzing impacts that may
occur during periods of fumigation or stagnation.

Peak Concentrations from Screening Models

Although inaccuracies exist in the Gaussian
plume model, it remains useful for assessing crop
damage. Short-term models that provide peak
ground-level concentrations for screening studies
would be especially useful for comparing the
anticipated peak concentration with a threshold
to determine if vegetation damage may occur.

Several screening models exist that can
determine peak ground-level concentrations from
point source emissions. These include the models
recommended by the U.S. Environmental Protection
Agency (EPA) such as PTMAX, PTDIS, and PTMTP
(Turner and Busse 1973) and methods developed by
Science Applications Inc. (Latimer 1979), Brookhaven National Laboratory (BNL) (Lipfert 1979),
and Los Alamos Scientific Laboratory (LASL)
(Nochumson 1980). The last two are similar
methods that can be used to estimate maximum
anticipated concentrations of relatively conservative pollutants emitted from isolated
point sources that have elevated stacks. Both
methods determine the critical meteorological
conditions under which peak concentrations may
occur and then determine the corresponding

concentration. The principal difference between the two methods is that the BNL model uses dispersion coefficients developed by Smith (1973) and the LASL model uses the Pasquill-Gifford coefficients. The latter was utilized in this study. Briefly, the following procedure is used to determine peak concentrations:

For neutral to unstable atmospheric conditions, the critical wind speed, u_c, at which maximum ground-level concentrations will occur is:

$$u_c = u\Delta h/h_s \tag{3}$$

where h_s is the stack height and Δh is the plume rise. If the critical wind speed is determined to be greater than the maximum wind speed that one could reasonably expect at plume height under a given stability class, the maximum wind speed value is assumed. The downwind distance, x_{max}, at which the maximum concentration will occur has been determined through differentiation of the Gaussian equation as follows:

$$x_{max} = \left\{ H/[a(1 + d/b)^{\frac{1}{2}}] \right\}^{1/b} \tag{4}$$

The computed value of x_{max} is substituted into the equations for σ_y and σ_z and into the dispersion equation to give the peak concentration x_{max}.

Figure 3 plots x_{max}/Q versus power plant capacity in megawatts (MW) for atmospheric stability classes A and B. These curves assumed Pasquill-Gifford dispersion coefficients and the plume rise equation developed by Briggs (1969). The stack height for each plant was determined using the following relationship (Lipfert 1979):

$$h_s = 23 \ MW^{0.326} \tag{5}$$

which was developed empirically to relate power plant building height with generating unit size. This relationship also assumes that good engineering practice for stack height is two and one-half times the building height.

To determine the peak 10-min average concentration for a coal-fired power plant under these two stability classes, one only needs to know the emission rate in grams/second (g/s). For example, given a 500-MW power plant with an SO_2 emission rate (Q) of 1000 g/s, the maximum 10-min concentration is estimated from figure 3 to be about 2350 $\mu g/m^3$ for stability class A and about 1150 $\mu g/m^3$ for stability class B. This value can then be compared with the damage threshold value for a particular plant species.

Areas of Impact

The screening model described above can also be used to determine the area exposed to ground-level concentrations higher than a given threshold, should the model results indicate that such concentrations could occur. Turner (1974) has indicated that when a Gaussian plume contacts the ground in flat terrain, the boundaries of any

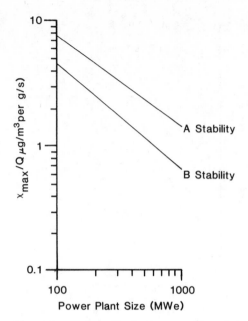

Figure 3.—Maximum 10-min Concentrations from Coal- Fired Power Plants under Unstable Conditions as a Function of Plant Capacity as Determined by LASL Screening Method

given isopleth can be approximated by an ellipse with its major (long) axis oriented in the downwind direction and minor axis in the crosswind direction.

Rather than determine exposure areas for specific concentrations, it may be more advantageous to determine a set of isopleths defined as some fraction of the maximum ground-level concentration. As an exercise, these areas were determined for the power plant sizes and peak concentrations covered in figure 3. Areas of exposure were determined for elliptical isopleths defined by 50%, 70%, 80%, 90%, and 95% of x_{max}/Q by the following technique:

The lengths of the major axes for the various isopleths were determined by plotting x_{max}/Q versus downwind distance for several power plant sizes and for the two stability classes. The following relationship can be used to determine the length of each minor axis, y, from the crosswind distance from the plume centerline at which the desired concentration will occur:

$$y = \sigma_y \left[2 \ \ln \frac{X/Q \ (plume \ centerline)}{ax_{max}/Q} \right]^{1/2} \tag{6}$$

where ax_{max}/Q is some fraction of the maximum ground-level concentration. From the set of y values for a given isopleth, the largest value was assumed to represent the minor axis. Figures

4 and 5 shows estimates of the areas exposed to some fraction of the peak concentration.

To estimate areas exposed to acute levels of air pollutants around power plants, first determine from figure 3 the maximum concentration that will occur for the specified power plant size and emission rate. If the peak concentration (from fig. 3) is greater than the damage threshold, then determine the ratio of threshold/X_{max}. The area can then be determined from the corresponding line in figures 4 or 5.

For example, assume the same conditions previously given, that is, a 500-MW power plant that emits 1000 g/s of SO_2. From figure 3, $X_{max} = 2350$ $\mu g/m^3$ for stability A. If the damage threshold level of concern is 1800 $\mu g/m^3$, this corresponds to about 77% of X_{max}. From figure 4, the area within the 1800-$\mu g/m^3$ isopleth will be about 60,000 m^2.

Only the most unstable conditions were assumed in this exercise. Other stability classes could be considered but for identifying peak, short-term concentrations, the results probably would not be meaningful because of the large distance the plume would have to travel under these conditions before it hit the ground.

Chronic-Damage Air Quality Modeling

In some situations, simple modeling of peak air pollutant concentrations is insufficient. A case in point is when functional relationships for chronic vegetation damage have been developed on the basis of long-term exposures to lower pollutant levels. To estimate the potential for vegetation damage around a point source, it may be necessary to estimate the total integrated pollu-

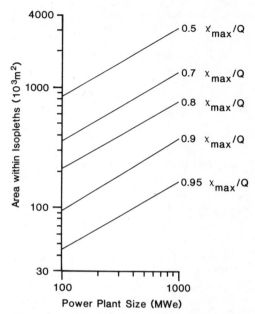

Figure 5.-- Surface Area within Isopleths Defined as a Fraction of Peak Concentration - Stability Class B

tant dose over the entire area surrounding the source. One such method, developed at Argonne, computes the dose for many smaller discrete time periods and totals the dose for some longer time period, e.g., a growing season.

Crop Damage Dose-Response Function

A study by Miller et al (1977) at Argonne examined the potential for yield reduction in soybeans with repeated exposures to moderate levels of SO_2. Test plots of soybeans were fumigated with SO_2 so that the average concentration per fumigation was more than 234 $\mu g/m^3$ over a 4-hr period. These exposures were applied 18-20 times during July and August. During this period the seed pods are developing and are most susceptible to damage. Sulfur dioxide applications were performed by using an open air fumigation technique, resulting in continual fluctuations in concentrations due to changes in wind speed and turbulence. The fluctuations in concentrations were very similar to that of actual fluctuations of SO_2 near a point source.

The results revealed that exposures to concentrations higher than 234 $\mu g/m^3$ over a 4-hr period would significantly contribute to reduction in crop yield. Exposures of 234 $\mu g/m^3$ for less than 4 hrs. were not considered damaging but each additional consecutive hour beyond a 4-hr event would contribute to yield reduction. Approximately 18-20 of these exposures were needed to induce some degree of yield reduction. An

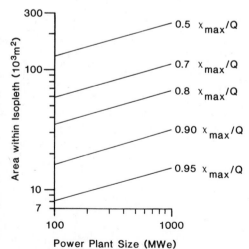

Figure 4.--Surface Area within Isopleths Defined as a Fraction of Peak Concentration - Stability Class A

equation to estimate the degree of reduction based on the integrated SO_2 exposure was developed by Sprugel et al (1978):

$$Y = 0.803D - 0.0034D^2 \qquad (7)$$

where

$$Y = \text{yield reduction in } \% \qquad (8)$$
$$D = \text{dose in ppm-hr}.$$

Method to Determine Integrated Exposure

To model Miller's work, the total air pollutant exposure must be calculated for those periods in which the 4-hr average was greater than 234 $\mu g/m^3$ over a two-month period. Also, because wind direction shifts, exposure to a pollutant will vary in all directions around the emitting source. The model must account for these variations.

ANL has investigated a method that considers variations in exposure and meteorological conditions. This method, based on the EPA Single Source (CRSTER) model (U.S. Environmental Protection Agency 1977), can also be applied to other damage thresholds or averaging times. This Gaussian plume model uses Pasquill-Gifford dispersion coefficients to compute 1-hr, 3-hr, daily, and annual average ground-level concentrations at 180 points located radially around a point source. Hourly surface meteorological observations (wind speed and direction, stability class, temperature) and twice-daily observations of upper air conditions are also incorporated. Since CRSTER is a widely-documented model, it will not be discussed further here.

Figure 6 outlines the procedure to determine integrated pollutant exposures. First, CRSTER is run for the desired emission source parameters for one set of meteorological observations. Hourly ground-level concentrations are computed for each receptor point for the daylight hours (0500 to 2200 hours) of July and August. Only daylight hours are considered because that is when plant stomata are open and most susceptible to SO_2 exposure. All of these 1-hr concentrations for each point are then stored for further analysis (about 1000 1-hr values for each receptor point).

The next step identifies periods when the average concentration is above some desired level, which, in the case of Miller's study, was 234 $\mu g/m^3$. Beginning at 0500 hours, the hourly concentrations at each receptor point for a 4-hr period are averaged to determine if the 4-hr average concentration exceeds the triggering level. If the average is not high enough, the next 4-hr period (e.g., 0600-1000 hours) is examined. If a 4-hr period exceeds the threshold, then each consecutive hour exceeding 234 $\mu g/m^3$ is counted to determine the total hours that exceed the threshold. Periods of 8-hr average

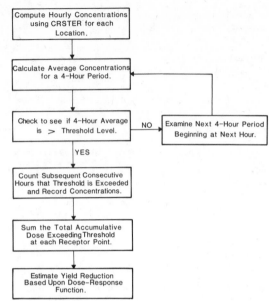

Figure 6.—Method for Determining Integrated Air Pollutant Exposures from Point Sources

concentrations greater than the threshold are considered to be two separate 4-hr periods for accounting purposes.

Exposures at each receptor may be substituted into a damaged function to determine yield losses in the vicinity of the power plant. In the first modeling exercise conducted at ANL, Sprugel's equation was used (Sprugel 1978). Sprugel's work is based on a fumigation regime that produced a series 4-hr of average SO_2 values of 234 $\mu g/m^3$. Consequently, the reduction in yield observed by Miller et al (1978) were based upon this consistent SO_2 concentration. The modelling procedure described above may, however, result in a much higher average SO_2 concentration of all tallied periods than the 234 $\mu g/m^3$ cited in Miller's work. This higher level may underestimate yield reduction. To minimize this problem, the ANL model lowers the average SO_2 value in successive steps until the total average of all tallied observations reach or approach 234 $\mu g/m^3$. This process is done by automatically adding in other observations that were not previously recorded because they did not average 234 $\mu g/m^3$.

This method is potentially useful for assessing possible yield reduction in crops near major air pollutant emitting sources. It uses observed meteorological data for specific locations and can consider spatial distributions of sensitive vegetation. It is also quite flexible in the emission source characteristics, exposure averaging times, and threshold levels that it can accommodate.

CASE STUDY: CROP DAMAGE FROM CHRONIC EXPOSURES

This report discusses methods for determining point source air quality impacts on vegetation. As such, it is not usually appropriate to present study results based on these methods, particularly ones that have been recently developed. However, a case study will be presented here to illustrate the type of problems the integrated exposure modeling technique can address.

As a test, the ANL method was used to examine the possible SO_2-induced reduction in yield that might occur in soybean crops located near a coal-fired power plant. Integrated exposures were calculated to simulate the study conducted by Miller et al (1977). For this test, meteorological observations from Greensboro, N.C., were used. Two hypothetical 1000-MW power plants were modeled, identical except for their emission rates. One plant had an assumed emission rate of 1500 g/s, corresponding to the upper limit of the New Source Performance Standard (about 1.2 lb/10⁶ Btu). The other plant had an assumed emission rate of 2250 g/s; this level represents the emission from an existing facility. Stack heights of 125 m and a stack gas exit temperature of 180°F were assumed. Modeling, using CRSTER, indicated that the peak 3-hr SO^2 concentration from the 1500-g/s plant was about 500 µg/m³, or nearly the PSD Class II increment.

Figures 7 and 8 are plots of the total integrated dose of SO_2 which accounted for each 4-hr period with an average concentration of 234 µg/m³ or more at each receptor point. The receptor grid rings are spaced at 1-km intervals, with the outermost ring located 5 km from the

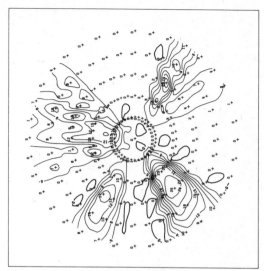

Figure 8.-- Contours of Integrated SO_2 Dosage in 10³ µg/m³-hr around 1000-MW Power Plant with Emission Rate of 2250 g/s

source. The model indicates areas of highest dosage in the northeast and southwest sectors for both cases. For damage to soybeans to occur from chronic exposure to SO_2, Sprugel's model indicates that integrated doses of 18,000-20,000 µg/m³-hr must be observed. Figure 7 shows a peak dose level of only about 10,000 µg/m³-hr for the new plant. Thus, there is an insufficient dose to cause detectable chronic yield reduction at this facility. For the existing plant case shown in figure 8, the dose reached levels as high as 25,000 µg/m³-hr, which is high enough to cause damage. Figure 9 plots the percentage reduction, based on Sprugel's equation, in soybean yield near the plant with the high-emission-rate. In some locations the model predicts reduction as high as 7%-8%.

The area affected by high SO_2 exposures was determined graphically. About 1000 acres were estimated to experience a yield reduction of 5% or more for the high-emission-rate case. Assuming a normal crop yields 25 bushels/ acre, the total loss would be about 1250 bushels. (A reduction of 5% was not considered significant, because of the accuracy limitations of the model.)

These results agree for the most part with other vegetation damage studies. The Institute of Ecology indicated that SO_2 emissions less than about 2000 g/s will have little impact on yield reduction. A study by Jones and Weatherford (1979) indicated that emission rates of 2000 g/s or more would be needed before significant damage would result. As figures 7 and 8 show, detectable levels of crop yield reduction would not occur in the case of the lower emission rate (1000 g/s).

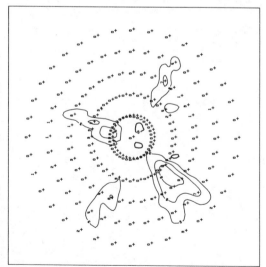

Figure 7.--Contours of Integrated SO_2 Dosage in 10³ µg/m³-hr around 1000-MW Power Plant with Emission Rate of 1500 g/s

346

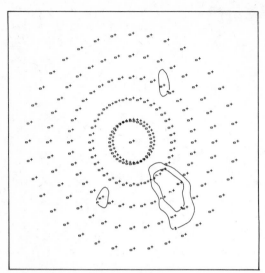

Figure 9.— Contours of Percentage Reducation of
Soybean Yield around 1000-MW Power Plant
with Emission Rate of 2250 g/s

REFERENCES

Briggs, G. A. 1969. Plume rise: a critical
survey. U.S. Atomic Energy Commission.TID
250D75 - 1969. Oak Ridge National Laboratory.
Air Resources and Atmospheric Turbulence
Laboratory. Oak Ridge, Tenn.

Jones, H. D. and F. P. Weatherford. 1979. Power
Plant Siting: Assessing Risks of Sulfur Dioxide Efforts on Agriculture, presented at
72nd Annual Meeting of the Air Pollution
Control Ass., Cincinnati, Oh.

Latimer, D. A. 1979. Power Plant Impacts on Air
Quality and Visibility: Siting and Emission
Control Implications EPA/400/02.

Lipfert, F. W. 1979. A Methodology for Analyzing
the Short-Term Air Quality Impacts of New
Power Plants, Brookhaven National Laboratory
Report BNL51292.

Loucks, O. L. et al., 1980. Crop and forest losses
due to current and projected emissions from
Coal-Fired Power Plants in the Ohio River Basin,
The Institute of Ecology, Indianapolis, final
draft report.

Miller, J. E. et al., 1977. An Open Air Fumigation
System for the Study of Sulfur Dioxide Effects
on Crop Plants, Argonne National Laboratory
Report ANL-77-65-III.

Nochumson, D. 1980. A Screening Methodology for
Calculating Worst Case Concentrations Downwind
from a Large Isolated Air Pollutant Emission
Source, Los Alamos Scientific Laboratory Draft
Report.

Pasquill, F. 1974 Atmospheric Diffusion (2nd ed.),
Halsted Press, New York City, N. Y.

Smith, M. E. 1973. Recommended Guide for the Prediction of the Dispersion of Airborne Effluents,
2nd Ed., American Society of Mechanical Engineers, New York City, N. Y.

Sprugel, D. G. et al., Effects of Chronic Sulfur
Dioxide Fumigation on Developments, Yield,
and Seed Quality of Field Grown Soybeans:
Summary of 1977 and 1978 Experiments, Argonne
National Laboratory Report ANL-78-65-III.

Turner, D. B. and A. D. Busse. 1973. Busse User's
Guides for PTMAX, PTDIS, and PTMTP Point Source
Dispersion Programs, U. S. Environmental Pro-
tection Agency.

Turner, D. B. 1974. Workbook of Atmospheric Dispersion Estimates, U. S. Environmental Protection Agency Report, AP-26.

U.S. Environmental Protection Agency User's Manual
for Single-Source (CRSTER) Model. 1977.
U. S. Environmental Protection Agency Report
EPA-450/2-77-013.

SYSTEMS ANALYSIS OF POTENTIAL AIR POLLUTION IMPACTS ON GRASSLAND ECOSYSTEMS[1]

J. E. Heasley, W. K. Lauenroth and J. L. Dodd[2]

Abstract.--A high resolution, ecosystem level, simulation model was developed to address questions about the long-term impacts of air pollution on grasslands in the Northern Great Plains of North America.

Thirty-year simulations were conducted utilizing three-hour SO_2 concentrations which resulted in annual average concentrations approximately equal to the U.S. air quality standard.

INTRODUCTION

Impacts of sulfur dioxide on individual plants has been thoroughly documented for high concentration short duration exposures as well as low concentration longer term exposures. Analogous impacts at the ecosystem level are rare. A previous analysis which relied upon application of a conceptual ecosystem model to results from a five year field experiment on a grassland ecosystem yielded the following scenario for long-term SO_2 impacts (Lauenroth and Heasley 1980).

Results of our experiments indicated that excess sulfur entering the ecosystem as SO_2 resulted in the reduction in chlorophyll in the dominant primary producer species. This decrease in chlorophyll will result in reduced energy capture by individual leaves. Exposure to SO_2 also resulted in phenological changes, producing a shift toward younger leaf-age classes. Because of the higher efficiency of energy capture of younger plant tissue, these changes will offset the effects of reduced chlorophyll on total biomass production. The energy cost of young leaf growth is large compared to the maintenance

lost of older leaf tissue. Shifts in the age structure towards younger age classes will place greater demands upon stored energy as well as reduce the amount of energy available for storage. Measurements of rhizome biomass over the five-year experiment support this hypothesis. Reductions in stored carbon may result in increased susceptibility to perturbations such as fire, drought, or grazing. Phenological changes in primary producers may also disrupt consumer life cycles, resulting in a reduction in consumer populations or changes in species composition.

The immediate impact of exposure to SO_2 will be the increase in sulfur content in all of the ecosystem components. Changes in primary-producer sulfur content may reduce the palatability of certain plants to herbivores, resulting in decreased herbivore numbers or shifts in grazing pressure to more palatable species. Such was observed for grasshoppers in this study. Changes in grazing pressure together with differential tolerance for excess sulfur could result in a change in species composition at all trophic levels of the grassland ecosystem.

Perhaps the most significant impact of SO_2 exposure in the long term will be the impact upon decomposition processes. Sulfur dioxide reduced decomposition directly through reductions in decomposer populations and decomposer activity or through shifts in decomposer species composition to species that are more tolerant of sulfur but less efficient in decomposing plant litter. Decreased nutrient cycling rates will reduce the size of the soil nutrient pools available for uptake by plants. As a result, the size of the

[1]Paper presented at the International Symposium Energy and Ecological Modelling, sponsored by the International Society for Ecological Modelling. (Louisville, Kentucky, April 20-23, 1981)

[2]J. E. Heasley, W. K. Lauenroth, and J. L. Dodd, Research Ecologist, Natural Resource Ecology Laboratory, Colorado State University, Fort Collins, Colo. U.S.A.

primary-producer and associated consumer components of the ecosystem will be reduced. Carbon and nutrients will accumulate in the soil organic matter pool. This accumulation will continue until nutrient inputs from various sources balance that released by decomposition. The sulfur content of all ecosystem components will come into equilibrium at a higher value. The soil nutrient pool will have a large sulfur component compared with non-sulfur nutrients. The turnover rate of non-sulfur nutrients will be rapid but low in volume due to lower decomposition rates.

In the long term, we expect a decrease in primary-producer carbon, largely as a result of decreased nutrient availability. Because of reduced decomposition, a larger portion of the potentially available nutrients will be tied up in organic matter. This will not apply to sulfur, which will constitute a larger portion of the nutrients in each compartment. Consumer carbon will decrease as a result of decreased primary production. Because of the very large size of the soil organic matter compartment, it will not measurably change. Decomposer carbon will decrease as a result of SO_2 input.

From the scenario we selected four hypotheses to investigate via long-term (30-year) simulation experiments.

Exposure of a grassland to SO_2 concentrations from a point source which just meets the U.S. ambient air quality standard will result in:

1. Reduced primary production.
2. Reduced cattle production.
3. Increased sensitivity to other perturbations such as heavy grazing or drought.
4. Reduced inorganic nitrogen availability.

MODEL DESCRIPTION

Our grassland model, (SAGE-Systems Analysis of Grassland Ecosystems), is a difference equation, flow oriented, ecosystem, simulation model. This model simulates the flow of carbon, nitrogen, and sulfur through the major components of grassland ecosystems (see fig. 1). These components include:

1. Abiotic subsystem
2. Primary producer subsystem
3. Soil subsystem
4. Ruminant consumer subsystem

The primary driving variables are solar radiation, air temperature, precipitation, relative humidity, and wind speed. The model simulates the response of a "typical" square meter of grassland. The difference equation representations of subsystem processes are implemented in FORTRAN utilizing MODAID, a collection of simulation programs developed at the Natural Resource Ecology Lab. The general time resolution for the model is one day with a few processes being simulated within the day. Atmospheric sulfur dioxide (SO_2) is input through the use of an SO_2 deposition submodel. The effects of SO_2 are simulated at the process level.

Abiotic Submodel

The abiotic model is structured to simulate the soil and canopy abiotic variables which influence a grassland ecosystem and is separated into two parts; (1) a water flow submodel and (2) a temperature profile submodel. The water flow submodel simulates the flow of water through the plant canopy and several soil layers. The partitioning of rainfall into evaporation and transpiration are the important processes considered. The model is generalized to handle a number of layers for soil water in which the depth and soil type are specified. The water flow model is structured to include the important feedback mechanisms between the biotic and abiotic state variables.

The temperature profile submodel simulates the daily solar radiation, maximum canopy air temperature, and soil temperature at 13 points in the soil profile. The soil temperature is determined by using a modified finite difference solution for the one-dimensional Fourier heat conduction equation in which the temperature at the upper (0 cm) and lower (180 cm) interfaces are specified. Temperature at the upper interface is calculated as a function of the air temperature at 2 m, the potential evapotranspiration rate, and the standing crop biomass while the monthly average temperature at 180 cm is used for the lower interface. Solar radiation is simulated as a function of cloud cover and the time of year, while the maximum canopy air temperature is a function of solar radiation and the maximum air temperature at 2 m.

The water flow submodel calculates the flow of water through the plant canopy into the soil water layers. Intercepted water loss by standing crop and litter interception are subtracted from daily rainfall, with remaining rainfall being infiltrated into the soil. Litter and standing crop interception is lost due to evaporation, while water loss from the soil is a function of bare soil evaporation rate and the transpiration water loss.

The temperature profile and water flow submodels use daily rainfall, cloud cover, wind speed (2 m), maximum and minimum air temperatures (2 m), and relative humidity (2 m) as the driving variables. These variables are determined either by using an observed time series of daily weather observations or a stochastic weather simulator.

Primary Producer Submodel

The primary producer submodel simulates the carbon (C), nitrogen (N), and sulfur (S) dynamics of three general categories of grassland plants.

Systems Analysis of Grassland Ecosystems

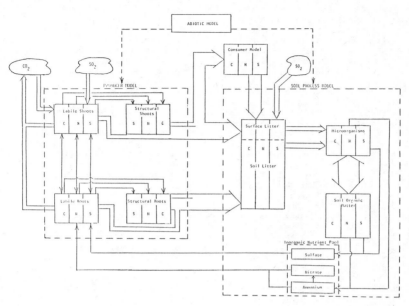

Figure 1.--System diagram of SAGE a grassland ecosystem model.

These include: cool-season grasses, warm-season grasses and cool-season forbs. Both the structural and labile forms of the C, N and S are represented. Aboveground plant parts are divided into young tissue (actively growing) and mature tissue (non-growing but photosynthetically active). Below-ground plant parts are separated into crowns or rhizomes and roots. Roots are further distin-guished according to soil depth (four soil layers: 0-1 cm, 1-5 cm, 5-20 cm, 20-60 cm).

Carbon enters the primary producer subsystem through photosynthesis. This process is repre-sented by a CO_2 model which combines a diffusion resistance network with a double Michaelis-Menton representation of carboxylation. Carbohydrate production responds to leaf temperature, atmo-spheric CO_2 concentration, leaf water potential, leaf age, leaf temperature acclimation, and light intensity. Various parameters in the model reflect the differences in the photosynthetic pathways (C_3 or C_4) of the four plant categories.

In the absence of air pollutants, nitrogen and sulfur enter the primary producer subsystem via root uptake from soil nutrient pools. Sulfur is taken up as sulfate while nitrogen is taken up as ammonium and nitrate. The dynamics of root uptake follow the Machaelis-Menton form. (In this form, the rate of uptake is determined by the soil solution concentrations of sulfate, ammonium, or nitrate.) This basic mechanism is

further modified by the effects of soil tempera-ture, soil moisture, and the relative levels of labile sulfur or nitrogen in the roots.

Growth of roots and shoots is represented as an exponential function of young tissue carbon. This function is in turn modified by temperature and moisture. Structural tissue is assumed to be synthesized with fixed C/N and C/S ratios. The labile elements associated with structural material are also assumed to occur with fixed C/N and C/S ratios. Therefore, carbon growth requirements dictate nitrogen and sulfur growth requirements. If either nitrogen or sulfur levels in the plant are less than required, carbon growth is reduced accordingly.

Nutrient allocation is a function of the demands for carbon, nitrogen, and sulfur (sink strengths), the supply of carbon, nitrogen, and sulfur (source strengths), and the age structure of the plant tissue. Young tissue is assumed to have a certain amount of excess labile material associated with it to support growth. As young tissue approaches maturity, the amount of labile support material approaches zero. Therefore the sulfur and nitrogen concentration of the plant reflects the age structure of the tissue in that plant. Nutrients are allocated in perennial plants in a hierarchical fashion with mainte-nance functions (respiration) being met first, followed by growth and reproduction. Any shoot requirements that cannot be satisfied by nutrient levels contained in the shoots may be met through translocation from belowground organs. The amount which can be translocated depends upon the relative amounts of labile nutrients contained in the crowns and roots and the maximum translocation rates for roots to shoots. Excess shoot nutrients may be translocated to belowground plant parts subject to the constraints placed upon the translocation process (relative labile levels and translocation rates).

Carbon, sulfur, and nitrogen leave the primary producer subsystem via respiration, root exudation, senescence, and litter fall. Respiration of roots and shoots is simulated as an exponential function of carbon levels, temperature, and moisture. Root exudation represents the leakage of nutrients back across the membranes into the soil pools. It is primarily a function of the concentration of labile carbon, nitrogen, and sulfur contained in the roots. Senescence is simulated as a function of tissue age, temperature, and soil moisture levels. Standing dead material enters the soil process subsystem via litter fall. Litter fall is estimated as a function of precipitation and standing dead carbon.

Soil Process Submodel

The soil submodel simulates processes restricted to the soil subsystem. These include inorganic nutrient transformations, litter decomposition, microbial processes (uptake, growth, death), fractionation of soil organic matter into humic (humads) and highly recalcitrant forms, and the movement of nutrients between soil layers (leaching). Two forms of inorganic nitrogen (ammonium and nitrate) and one form of inorganic sulfur (sulfate) are simulated. In the majority of processes, carbon, nitrogen, and sulfur flow through the soil subsystem in a parallel fashion (see fig. 1). When either plants or microbes die, they are partitioned into slowly decomposing, low N content substrate (mainly cell walls) and rapidly decomposing, high N content substrate (mainly cytoplasm and organelles). The rate of use of these substrates by bacteria and fungi depends on the physiological state of the microbes, as reflected by their C/N, C/S, or N/S ratios, and on soil temperature and moisture. Humads are fairly resistant materials formed directly from high-N substrate by adsorption onto humic material and clay minerals, or left as a residuum of resistant material, such as lignins, after the decomposition of low-N substrate. Resistant soil organic matter is the very recalcitrant material left after the decomposition of humads. It is important to recognize these various classes of substrates because their relative abundance determines the nutritive status of microbes, and whether microbes mineralize or immobilize nitrogen. Bacteria and fungi are distinguished in the model as they differ in their responses to abiotic factors, and in their abilities to break down the various classes of substrate. Usually microbes are net mineralizers

of nitrogen, but they may compete with plants for inorganic nitrogen if low-N substrate becomes abundant.

Ruminant Consumer Submodel

The ruminant consumer submodel simulates the energy, nitrogen, and sulfur requirements of ruminant grazers; their consumption and metabolism of energy, nitrogen, and sulfur, consequent periods of positive and negative energy, and nitrogen balance; and their resultant weight gains or losses.

The ruminant submodel consists of three component submodels: nitrogen and sulfur ingestion, metabolism, and loss; energy ingestion, metabolism, and loss; and a simple representation of ruminal microbial protein. The fate of sulfur in the ruminant's body is tied to the fate of nitrogen because both are involved primarily in protein metabolism.

This submodel is a straightforward representation of ruminant energy and nitrogen balance, but a few points are worthy of separate mention. Ingestion of forage by the animal is regulated by physical capacity of the rumen and the rate of passage of ingesta through the digestive tract. Passage rate is controlled by forage digestibility and rumen microbial biomass. The rate of microbial protein synthesis in the rumen is controlled in turn by the amount of energy released by digestion during the previous time step and by the amount of nitrogen in the rumen. A capability for recycling non-protein nitrogen to the rumen from the blood exists and is triggered by low dietary nitrogen concentration.

Energy and nitrogen are metabolized jointly in the model. This permits the nitrogen status of the animal to influence the manner in which energy is partitioned within the animal, and vice versa. The pool of labile energy in the body consists of amino acids as well as nonnitrogenous energy-bearing compounds contributed by rumen fermentation and by a normal mobilization of a portion of the fat stores (Koong and Lucas 1973). The amount of energy in this pool, the proportion of amino acids in it, and the energetic and growth demands placed on it control the partitioning of the energy pool and the fate of the nitrogenous compounds.

The highest priority requirement to be met from the energy pool is the maintenance requirement (basal metabolic requirement + activity + thermoregulatory requirement). This is met by the non-N component of the pool if that component is sufficiently large. If necessary, the nitrogenous components can meet a residual requirement by deamination, yielding energy, and nonprotein N. If a maintenance energy requirement persists after exhaustion of the labile pool, fat and then lean body tissue are catabolized to meet it. Catabolism of lean body tissues again results in the production of nonprotein nitrogen.

If energy remains in the labile pool after the maintenance requirement has been met, that energy can be used for some "productive" purpose. The purpose to which the energy will be put depends upon the physiological status of the animal (pregnant, lactating, underweight) and the ratio of nitrogenous to nonnitrogenous forms remaining in the energy pool.

The priorities associated with the various productive requirements are not clear. We assume that reproductive purposes are higher priorities than lean body growth which in turn is a higher priority purpose than fattening. Energy is allocated to the purposes in that order, provided the nitrogenous component of the energy pool is not limiting. Milk and lean body have characteristic ratios of nitrogenous to nonnitrogenous energy. If the ratio of N to non-N forms in the pool is higher than that in the material to be produced, some amino acids can deaminate to non-N forms, again yielding nonprotein nitrogen. If, however, the N to non-N ratio is lower in the pool than in the product to be formed, production must cease when N forms are exhausted with any remaining non-N energy converted to fat. If all N-requiring purposes are satisfied, amino acids remaining in the energy pool will deaminate yielding nonprotein nitrogen and carbon skeletons which, with any non-N components remaining in the original pool, are converted to fat.

Although sulfur flows parallel those of nitrogen, there are three departures from this stoichiometric scheme. They are 1) recyling of non-protein sulfur (NPS) to and from the rumen, which is modeled independently from NPN recycling; 2) synthesis of microbial protein in which both N and S concentrations control synthesis rates and; 3) adjustments to the amino acid pool to account for the fact that absorbed amino acids have a biological value of less than 1.

Recycling of non-protein sulfur is predicted from dietary sulfur concentration. If an addition to the rumen is predicted, it is drawing from the animal's non-protein sulfur (NPS) pool. If a net loss is indicated, a flow from ruminal sulfur to the NPS pool occurs. Nitrogen to sulfur ratio controls the efficiency of micro-protein production in terms of grams protein/K cal digested. Adjustments are made in the amino acid N pool and amino acid S pool via deamination or desulfation such that the ratio of N pool to S pool corresponds to the N/S ratio of ruminant lean body tissue.

SO$_2$ Deposition and Effects

The deposition of sulfur dioxide is simulated utilizing a diffusion resistance scheme. Wind speed is predicted in the canopy utilizing the wind speed at reference height and the equations of Ripley and Redman (1976). The canopy is divided into six layers in which the aerodynamic and boundary layer resistances are predicted for the midpoint of each layer. The resistance to diffusion of SO$_2$ is represented by a parallel resistance network from the reference height to the soil surface. Diffusion of SO$_2$ branches off at the midpoint of each canopy layer. Resistance to diffusion of SO$_2$ from each canopy midpoint to the leaf surface or the interior of the leaf consists of a parallel/series combination of stomatal resistance, leaf surface resistance and boundary layer resistance. The total flux of SO$_2$ to the canopy is determined by dividing the SO$_2$ concentration at the reference height by the total diffusion resistance from the soil surface up to the reference height. This is calculated by the parallel combination of all of the resistances. Total flux is then split among the canopy layers utilizing electrical analogue computation techniques. This method allows the prediction of SO$_2$ concentration at the midpoint of each of the canopy layers. The uptake of SO$_2$ in each of the layers is estimated as a function of the flux into that layer and the total leaf area contained in it. It is therefore necessary to provide estimates of canopy height, and leaf area distribution.

Sulfur dioxide enters the leaf sulfur pool via diffusion through the stomates. In the leaf, SO$_2$ is converted to sulfite which in turn is converted to sulfate. It is the rate of conversion from SO$_2$ to sulfate that controls the entry rate of atmospheric sulfur into the leaf sulfur pool. Any sulfur which enters the leaf sulfur pool via the stomata and is converted to sulfate is immediately available for use in building new leaf tissue. Excess sulfur in the leaf sulfur pool may be translocated to the roots depending upon the relative levels of the shoot and root sulfur pool. If SO$_2$ enters the leaf at a rate higher than the rate of conversion of sulfite to sulfate, sulfite will build up in the leaf. The destruction of leaf tissue is directly related to the level of excess sulfite in the leaf.

Sulfur interacts directly or indirectly with various plant physiological processes. These include stomatal mechanics, enzyme production, phenology, carbohydrate storage and respiration. Exposure to very low levels of SO$_2$ is assumed to stimulate stomatal opening while higher levels are assumed to stimulate stomatal closure (Majernik and Mansfield 1971; Unsworth et al. 1972; Schramel 1975; Black and Unsworth 1980). The stimulation or depression of stomatal opening is dependent upon the vapor pressure deficit.

Sulfur may influence photosynthesis through the stimulation of enzyme production or the inhibition of photosynthetic enzymes. Various researchers have shown stimulatory effects of SO$_2$ on photosynthesis beyond those effects attributed to stomatal opening (White et al. 1974; Bennett and Hill 1973; Koziol and Jordan 1978). This effect is implemented in the model through slight increases in the enzyme levels which drive the carboxylation reactions in the CO$_2$ diffusion model. At higher concentrations, SO$_2$ has been found to inhibit RuDP carboxylase, a primary photosynthetic enzyme. Inhibition is implemented in the model through reduction in the relative enzyme level utilized in the CO_2 diffusion model. Leaf respiration is increased in direct proportion to the amount of SO_2 which enters the leaf (Guderian and Van Haut 1970). The effects of sulfur on plant phenology and plant carbohydrate storage are implemented indirectly through its effect on carbohydrate production.

Decomposition rates are also influenced by sulfur dioxide. The uptake of carbon from surface litter by microbes is reduced as a function of SO_2 concentration at the soil surface. In addition to direct effects, the current model structure will allow the investigation of the indirect effects of SO_2 on phenological patterns (e.g., senescence), overall decomposition rates, the cycling rates of sulfur through the ecosystem, changes in carbohydrate storage levels, forage quality, primary and secondary productivity, and the relative abundance of species.

MODEL VALIDATION

Three years of experimental data were utilized to make model parameter adjustments and to validate or compare model output with field data. Data collected during the 1975 growing season were used to make adjustments in model parameters while 1976 and 1977 data were utilized for validation. The model was initialized for 1975 conditions and run for a three-year simulation period. Abiotic driving data were derived from data collected on the study site (for the growing season) and from data generated at the nearest U. S. Weather Station (for the remainder of the year). SO_2 concentrations measured on the study plots were input to the SO_2 deposition submodel.

The 1976 and 1977 growing seasons were quite different with respect to growing conditions. Nineteen seventy-six was a wet year while 1977 was a dry year. These years were chosen to ensure that the model responded properly to extremes of the driving variables. A three-year simulation was conducted, rather than reinitializing each year, to allow errors that accumulated during one simulation year to propagate into the following year. This approach was taken in order to provide confidence that the model could be applied to long-term scenarios.

Model output was compared to field data for aboveground biomass production, sulfur concentration in live leaves, and soil water content in the soil profile. Figures 2, 3, and 4 depict these comparisons for the 1976 and 1977 growing seasons. In most cases model output fell within the statistical variation of the field data (standard errors). The shape and timing of the responses were consistent with that observed in the field.

MODEL EXPERIMENTS

Our objective was to investigate the impact of long-term exposure to SO_2 on a grassland ecosystem. To meet this objective we conducted two 31-year simulations. One under no SO_2 and one under SO_2 concentrations at the levels of the Federal Standard (Ludwick et al. 1980). Abiotic driving data were obtained from the Billings WB AP (U. S. Climatological Survey 1958-1978). Because we considered precipitation to be the most important abiotic variable controlling the response of the system to SO_2 we subdivided our abiotic data into categories of wet, dry or normal. We partioned the year into four seasons based upon major ecological events and each were categorized with respect to the long-term mean precipitation. Representative seasons were combined to form a 20-year sequence of weather years corresponding to that experienced in Montana during 1958-1978. Weather years ran from November 1 to October 31. The last eleven weather years of the simulation were chosen to provide information on system response to perturbation.

Three-hour average SO_2 concentrations were generated with the Texas Episodic Model (Christiansen 1975-76). This model is a steadystate Gaussian plume model. The meteorological data used to drive the ecological model was utilized to generate SO_2 concentrations.

LONG TERM SYSTEM RESPONSE

The response of the primary producer component of the ecosystem was qualitatively as we had predicted. However, it was quantitatively less than expected. Over the simulation period the standing crop of primary producers was reduced by 10%. Gross primary production (fig. 5) was generally greater as a result of exposure to low level SO_2 concentrations due to increased senescence effects of SO_2. The resulting shift in age structure toward younger leaf tissue resulted in higher photosynthetic rates. During wet growing seasons, however, greater SO_2 uptake occurred resulting in lower gross primary production. Respiration (fig. 6) was generally higher in plants exposed to SO_2 due to the increased cost of dispersing the excess sulfur taken up. Net primary production was lower in most years for primary producers growing in an SO_2 environment (see fig. 7). Peak aboveground biomass was greater for plants grown in the absence of SO_2. The difference was greatest during wet years. Under drought conditions peak aboveground biomass for both treatments were similar. Sulfur uptake from the atmosphere was five to six times greater during wet years than dry years. Root uptake of sulfur was substantially reduced for those plants exposed to SO_2.

Consumption of primary producers by domestic ruminants was on the average 3% greater for the system exposed to SO_2. SO_2 increased the senescence rate of the plants resulting in a higher

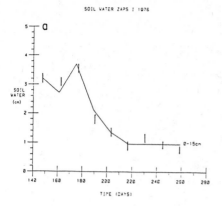

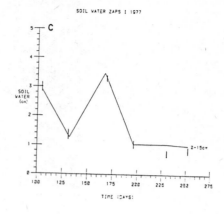

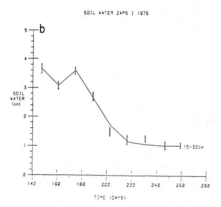

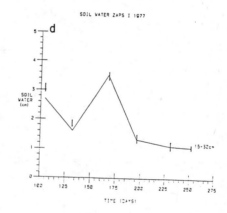

Figure 2.--Model validation predictions of soil water content compared to the standard error bars of the field
data for the 1976 and 1977 growing seasons.

proportion of dead material in the leaves. This
contributed to a slight depression in digestibility requiring an offsetting increase in intake
in order to meet energy needs. Cattle production
was higher for the system exposed to SO_2 (fig. 8).
Cattle weights at the end of the grazing season
were 1 to 14% higher than those utilizing plants
not exposed to SO_2. Much has been written about
the beneficial effects of supplementing cattle
diets with sulfur (Tisdale 1977, Garrigus 1970).
Increasing the sulfur content of the diet increases the retention rate for nitrogen. This
occurs as the N/S ratio of the forage more closely
approaches that for body proteins.

Because respiration rates were higher for
primary producers exposed to SO_2, less labile
carbon was available for storage. Plants in an
SO_2 environment stored 10 to 20 percent less
labile carbon than those in a SO_2 free environment (fig. 9). The difference in peak labile
carbon stores clearly increased with exposure
time as is illustrated in figure 10. Our original hypothesis was that lower carbohydrate
stores would increase system susceptibility to
perturbations such as drought or heavy grazing.
This, however, was not found to be the case.
Toward the end of the simulation run we imposed
two years of severe drought (dry spring and
summer). Measures of system susceptibility were

354

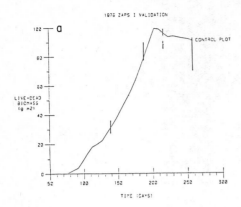

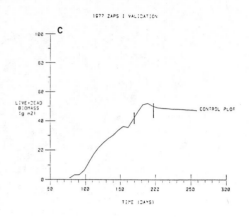

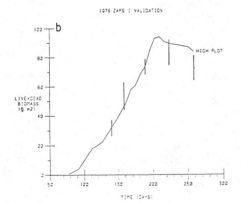

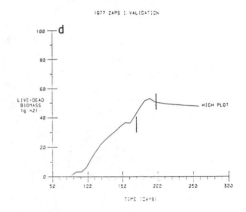

Figure 3.--Model validation predictions of live plus dead
aboveground biomass compared to the standard error bars
of the field data for the 1976 and 1977 growing season.

defined as the fraction of carbohydrate stores lost and the ratio of net primary production during the perturbation to net primary production in a normal year. In both cases there was less than 2% difference between the susceptibilities of the two systems. The apparent lack of effect on the susceptibility of the system to perturbations is partially related due to the survival mechanisms of the plants. Mechanisms which reduce respiration, shed leaves, etc. in response to stress minimize the effect of reduced carbohydrate stores during short periods of drought or heavy grazing.

Mineralization of inorganic nitrogen was 2 to 20% greater in the SO_2 free system than in the system exposed to SO_2. During the first ten years of the simulation most uptake of mineral nitrogen was slightly higher in the SO_2 exposed system. This was due to the higher nitrogen demands of the younger age structured plant populations. The inorganic nitrogen pool decreased during this period under SO_2 exposure (fig. 11). As the standing crop of primary producers decreased under SO_2 exposure the demand for nitrogen decreased with it. Root uptake of mineral nitrogen was 7 to 10 percent less than that taken up by SO_2 free plants. Therefore the inorganic nitrogen pool for the SO_2 exposed system increased to a level above that of the SO_2 free system. In addition, the increased availability of nitrogen resulted in microbial N/S ratios

1976 ZAPS I VALIDATION

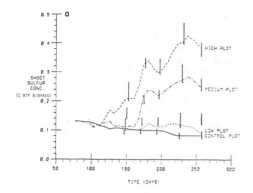

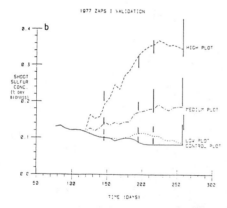

Figure 4.--Model validation prediction of shoot sulfur concentrations compared to the standard error bars of the 1976 and 1977 growing season field data.

which were more favorable for mineralization. At the end of the thirty-year simulation the inorganic nitrogen pool was greater in the system exposed to SO_2.

As predicted, sulfur concentrations increased in the system exposed to SO_2. However, the increases in sulfur were not evenly distributed among system compartments. Sulfur content of primary producers increased by 10%. The soil organic and decomposer sulfur showed little change. Over the thirty-year period 15 g m^{-2} of sulfur were added to the system from the atmosphere. Almost 90% of it ended up as inorganic sulfate. The inorganic sulfur pool increased by

65% over the thirty-year simulation. Root uptake of mineral sulfur was substantially reduced.

CONCLUSIONS

The results of the long term simulations appear to be consistent with our speculations in some areas while inconsistent in others. Simulation results indicated that primary production will be reduced and that the primary producer compartment of the ecosystem should be smaller. Secondary production will most likely be unaffected. In the case of cattle production it may even be stimulated. Carbohydrate stores will be reduced. However, because of the resilience of grasslands, this will probably have little effect on the response of the system to perturbations such as heavy grazing or drought. System response to these perturbations should be similar whether SO_2 is present or not.

Inorganic nitrogen pool levels may increase as the system reaches a new equilibrium. The greatest impact of SO_2 will be to increase the inorganic sulfur pool level. The soils in semi-arid grasslands are highly buffered. However, long term exposure to SO_2 may eventually alter the pH of the soils resulting in a reduction in decomposition rates. If this occurs, the system may more closely resemble the speculatory projections made previously.

The use of simulation models to investigate the long term effects of air pollution has proven to be very useful. It has allowed us to integrate the myriad of air pollution effects into a more comprehensible form and has pointed out weaknesses in our knowledge of grassland ecosystems and how they respond to air pollution. The model provides information relative to system behavior and identifies areas for future research.

LITERATURE CITED

Bennett, Jesse H., and A. Clyde Hill. 1973. Inhibition of apparent photosynthesis by air pollutants. J. Environ. Qual. 2:526-530.

Black, V. S., and M. H. Unsworth. 1980. Stomatal responses to sulphur dioxide and vapour pressure deficit. J. Exp. Bot. 31:667-677.

Christiansen, J. H. 1975-1976. Users guide to the Texas episodic model. Data processing section. Texas Air Control Board.

Garrigus, U. S. 1970. The need for sulfur in the diet of ruminants. Pages 126-153 in D. H. Math and J. E. Oldfield (eds.) Symp. Sulf. Nutri., 1979. Oregon State Univ., Corvallis.

Guderian, R., and H. Van Haut. 1970. Detection of SO_2 effects upon plants. Staub. 30:22-35.

Koong, L. J., and H. L. Lucas. 1973. A mathematical model for the joint metabolism of nitrogen and energy. Inst. Stat. Mimeo Ser. No. 882. North Carolina State Univ., Raleigh. 116 pp.

Koziol, M. J., and C. F. Jordan. 1978. Changes in carbohydrate levels in red kidney bean

356

(Phaseolus vulgaris L.) exposed to sulphur dioxide. J. Exp. Bot. 29:1037-1043.

Lauenroth, W. K., and J. E. Heasley. 1980. Impact of atmospheric sulfur deposition on grassland ecosystems. Pages 417-430 in D. S. Shriver, C. R. Richmond, and S. E. Lindberg (eds.) Atmospheric sulfur deposition--environmental impact and health effects. Ann Arbor Science Publishers, Inc., Ann Arbor, Michigan.

Ludwick, J. D., D. B. Weber, K. B. Olsen, and S. R. Garcia. 1980. Air quality measurements in the coal fired power plant environment of Colstrip, Montana. Atmos. Environ. 14:523-532.

Majernik, O., and T. A. Mansfield. 1971. Effects of SO₂ pollution on stomatal movements in Vicia faba. Phytopath. Z. 71:123-128.

Schramel, M. 1975. Influence of sulfur dioxide on stomatal apertures and diffusive resistance of leaves in various species of cultivated plants under optimum soil moisture and drought conditions. Bull. de l'Academie Polonaise des Sciences, Serie des science biologique 23:57-66.

Tisdale, T. S. 1977. Sulphur in forage quality and ruminant nutrition. Tech. Bull. 22. The Sulphur Institute, Washington, D.C. 13 p.

Unsworth, M. H., P. V. Biscoe, and H. R. Pinckney. 1972. Stomatal responses to sulfur dioxide. Nature 239:458-459.

White, Kenneth L., A. Clyde Hill, and Jesse H. Bennett. 1974. Synergistic inhibition of apparent photosynthesis rate of alfalfa by combinations of sulfur dioxide and nitrogen dioxide. Environ. Sci. Tech. 8(6):574-576.

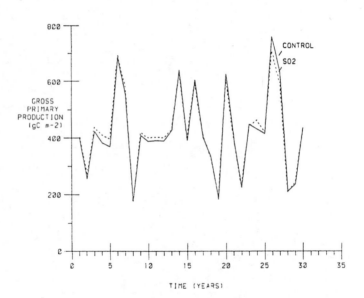

Figure 5.--Simulated gross primary production of a grassland exposed to control (no SO₂) and SO₂ (SO₂ concentrations at the Federal Standard level) conditions.

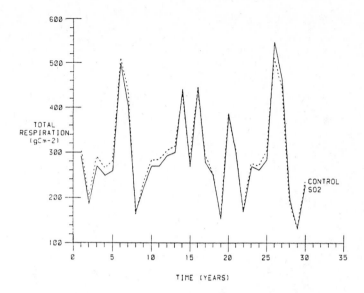

Figure 6.--Simulated total respiration for primary producers exposed to control (no SO_2) and SO_2 (SO_2 concentrations at the Federal Standard level) conditions.

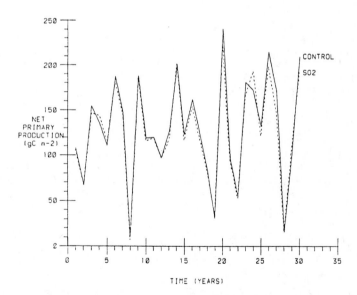

Figure 7.--Total net primary production for a grassland exposed to control (no SO_2) and SO_2 (SO_2 concentrations at the Federal Standard level) conditions.

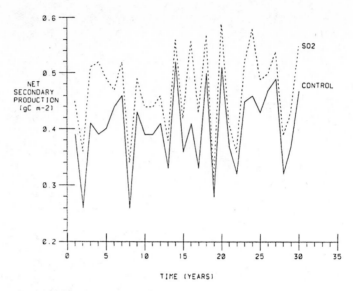

Figure 8.--Simulated net secondary production (cattle) for a grassland exposed to control (no SO_2) and SO_2 (SO_2 concentrations at the Federal Standard level) conditions.

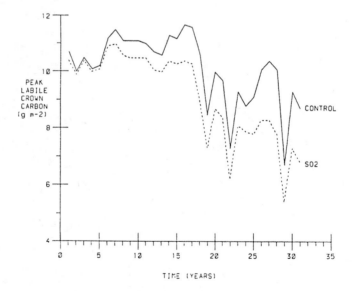

Figure 9.--Simulated peak labile crown carbon for a grassland exposed to control (no SO_2) and SO_2 (SO_2 concentrations at the Federal Standard level) conditions.

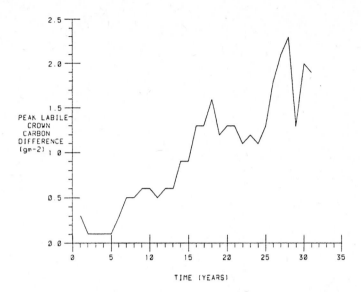

Figure 10.--The difference in peak labile crown carbon for
grassland ecosystems exposed to control (no SO$_2$) and
SO$_2$ (SO$_2$ concentrations at the Federal Standard level)
conditions.

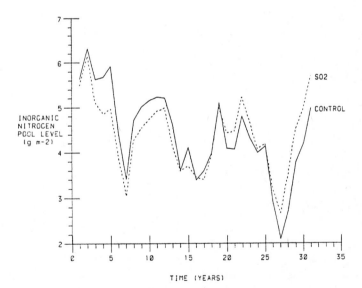

Figure 11.--Simulated inorganic nitrogen pool levels for
grasslands exposed to control (no SO$_2$) and SO$_2$ (SO$_2$
concentrations at the Federal Standard level) conditions.

A MODEL OF CARBON FLOW IN LICHENS[1]

Thomas B. Kirchner, J. E. Heasley, and W. K. Lauenroth[2]

Abstract.--We present a simulation model of lichens which may have utility in anticipating and explaining impacts of air pollution. The model represents carbon flow and water relations in lichens in a mechanistic manner. The results of simulations compare favorably with results from several different laboratory experiments.

INTRODUCTION

Lichens are widely distributed throughout a variety of environments ranging from the Arctic to the tropics. Most lichens are terrestrial and are found on soil and rocks, or as epiphytes. While the role of lichens in the ecological systems in which they occur is poorly understood they are potentially important as colonizers of extreme microsites and as nitrogen fixers (Syers and Iskander 1973; Hitch and Stewart 1973).

In its simplest form a lichen may be thought of as an ecosystem having two trophic levels, a primary producer (an alga) and a consumer (a fungi) (Farrar 1976). Most interactions between these levels can be reduced to mechanisms at a physiological level because of the closeness of the symbiotic relationship. However, the exact nature of many of these physiological processes have not yet been fully resolved using experimental approaches. Simulation modeling can provide an alternate, often productive approach to understanding the operation of a system.

Carbon flow in lichens was previously modeled by fitting polynomials to the response curves of net photosynthesis and dark respiration to light intensity and water levels (Kershaw and Harris 1971a,b). Although some erroneous conclusions were drawn from the results of simulations due to improper formulations, the model

[1]Paper presented at the international symposium Energy and Ecological Modelling, sponsored by the International Society for Ecological Modelling. (Louisville, Kentucky, April 20-23, 1981)

[2]Thomas B. Kirchner, J. E. Heasley, and W. K. Lauenroth, Research Ecologists, Natural Resource Ecology Laboratory, Colorado State University, Fort Collins, Colorado, U.S.A.

nevertheless demonstrated that the level and temporal distribution of water contents are important determinants of net assimilation in lichens (Harris 1976). Our objectives here are: (1) to describe a simulation model of carbon flow in lichens and (2) present the results of simulation experiments conducted to determine the response of the model to exposure to SO_2.

MODEL OVERVIEW

The system is represented at a level of complexity sufficient to show adaptive growth, i.e., the proportions of the biomasses of the mycobiont and phycobiont can change through time in response to environmental conditions. Thus, the model may permit more detailed analyses of the interactions which are important to carbon flow in lichens. The model also may provide the means of testing alternate hypotheses about physiological processes, indicate areas where knowledge about the physiology of lichens is lacking, and give insight into how perturbations, such as air pollutants, affect the functioning of lichens.

The physiological status of a lichen is intimately related to its water content. There may be no measurable metabolic activity in lichens when they are dry (<10% of saturation) (Harris 1976), and photosynthesis may be inhibited by high water contents. Thus, accurate representations of the processes of water loss and uptake are of primary importance in the development of a simulation model for lichens. Therefore, the model is constructed with a water relations submodel as well as a carbon flow submodel.

The model is formulated as a system of differential equations which are solved using a fourth order Runge-Kutta algorithm. The model is coded in FORTRAN.

362

WATER RELATIONS SUBMODEL

The water relations submodel consists of a single compartment, WATER, an atmospheric water vapor source/sink, and a liquid water source (fig. 1). The water contents of lichens may vary from as little as 2% of the dry weight under drought conditions (Haynes 1964) to as much as 400% when fully saturated (Blum 1973). Water uptake and loss in lichens is apparently a purely physical process. Much of the water absorbed by lichen thalli is stored in extracellular regions. This extracellular water may be in the form of capillary water within the hyphal network of the thallus, or as "water of swelling" (adsorbed water) within the gelatinous sheaths of the phycobiont and the thickened hyphal walls of the mycobiont (Ried 1960b). Most of the water in a nonsaturated lichen is contained within the algal layer (Showman and Rudolph 1972).

Differences in the water loss and absorption characteristics between species or ecotypes within species may be due primarily to morphological differences (Larson and Kershaw 1975; Larson 1979b). Morphology can directly affect the proportions of capillary and adsorbed water

in lichens (Ried 1960a,b). Capillary water evaporates at a relatively constant rate in conditions of constant temperature and humidity because there is little or no reduction in the vapor pressure of water at the evaporating surface. However, the evaporation of adsorbed water is expected to reduce the vapor pressure over the thallus in proportion to the reduction in water content of the thallus (Scofield and Yarman 1943). Many authors, e.g., Hale (1974), Blum (1973), Harris (1976), and Showman and Rudolph (1972), have noted a resemblence between water exchange in lichens and in hydrophillic gels. The resemblance is due undoubtedly to the presence of the adsorbed water in the thallus. However, Larson (1979b) points out that drying and absorption curves for gels are usually determined for solid masses, whereas lichens have a form more like a filamentous mass. Thus, such resemblance may be limited to conditions where atmospheric turbulance can be ignored.

Evaporation

Water vapor exchange in transpiring plants has often been described using a resistance analog equation of the form

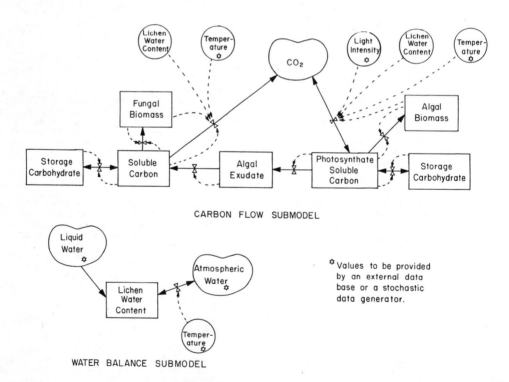

CARBON FLOW SUBMODEL

WATER BALANCE SUBMODEL

✿ Values to be provided by an external data base or a stochastic data generator.

Figure 1.--The carbon flow and water balance submodels of the lichen model.

$$evap = \frac{\rho c'}{\lambda \gamma} \frac{(e_s(T_o) - e_a)}{r_b + r_1} \quad \text{(Monteith 1965)} \quad (1)$$

where r_b = boundary layer resistance of the air; r_1 = leaf resistance to vapor exchange; ρ = density of air; c' = heat capacity of air; λ = latent heat of vaporization; γ = psychrometric constant; $e_s(T_o)$ = saturation vapor pressure at the temperature of the evaporating surface; e_a = vapor pressure of the air at the ambient temperature; and evap = flux density of water exchange (g cm^{-2} min^{-1}). The boundary layer resistance is a function of wind velocity. r_b decreases in proportion to an increase in the square root of wind velocity, $V^{\frac{1}{2}}$, until the wind flow becomes turbulent; there is little or no reduction in r_b after this velocity is reached. In practice, r_b usually cannot be distinguished from r_1. Therefore, a term representing the total evaporative resistance, r_t, is substituted for $r_b + r_1$. Thus, at low windspeeds r_t is expected to be related to windspeed by

$$r_t = aV^{\frac{1}{2}} + b \quad (2)$$

A regression using using the data for the effect of windspeed on the evaporative resistance of Clodina stellaris from Larson (1979b) shows good agreement with this model (r = 0.985, n = 5). Given the windspeed at some height, d, above the ground, the equations of Ripley and Redmann (1976) were used to estimate a wind speed profile. Wind speed above the canopy is given by:

$$u(z) = u(h) \ Ln[(z-d)/Z_o]/Ln[(h-d)/z_o] \ ; \ z{\geq}h$$

Wind speed within the canopy is given by:

$$u(z) = u(h)exp \ -[0.385(z/h) + 0.72]LAI(z) \ ; \ z{\leq}h$$

where u(i) is the windspeed at height i (m/sec), d is the zero plane displacement (cm), h is the canopy height (cm), z is the height above the ground, z_o is the aerodynamic roughness length (cm), and LAI(Z) is the total leaf area index above height z.

An implicit assumption of equation (1) is that the evaporating surface behaves as a pool of free water in terms of its vapor pressure. This assumption may be reasonable for higher plants since the range of water contents of biological interest is relatively high (>90%). However, this assumption may be invalid when considering evaporation of the adsorbed water of lichens (Scofield and Yarman 1943). Therefore, the evaporation of water from the lichen is modeled as a two step process. Equation (1) is assumed to apply only to the evaporation of capillary water. Evaporation of adsorbed water occurs only after all capillary water is gone. An equation for the vapor pressure of water over the lichen as a function of the vapor pressure of the air was developed using data for the equilibrial water contents at various humidities for several species of lichens from Ried (1960b).

The inverse of water content, WATER^{-1}, was found to be linearly related to humidity over the range of values given. Thus, the vapor pressure for the adsorbed component of water in the thallus, e_t, is related to WATER by

$$e_t = (c_1 - c_2 \cdot WATER^{-1}) \ e_s(T_o) \quad (3)$$

where c_1 and c_2 are empirically determined constants. The equilibrium water content of the thallus, W_H, for a given relative humidity, H, is given by

$$W_H = \frac{c_2}{c_1 - H/100} \quad (4)$$

Thus, the maximum amount of adsorbed water, W_{sat}, is

$$W_{sat} = \frac{c_2}{c_1 - 1} \quad$$

The difference between W_{sat} and the maximum water content of the lichen, W_{max}, is assumed to be the maximum mass of capillary water that the lichen can hold. Therefore, the water content of a "fully saturated" thallus corresponds to W_{max}.

When the lichen is fully saturated, the area of the evaporating surface is probably proportional to the dry mass of the thallus. However, the fibrous nature of lichens may tend to complicate the relationship between the area of the evaporating surface and the dry mass of the plant. The area of the evaporating surface for water adhering to cylindrical fibers is roughly proportional to WATER$^{\frac{1}{2}}$. The surface area of water adhering to spheres is proportional to WATER$^{2/3}$. In general

$$Surface \ area = c \cdot WATER^{\alpha} \quad (5)$$

where α is a parameter relating surface area to mass of water and c is a proportionality constant relating the area of the evaporating surface to the dry mass of the thallus; α may approach 0 for a fully saturated thallus.

Uptake of Water

Lichens can absorb water from either a liquid or gaseous state. The uptake of liquid water is usually rapid. The curve of absorption of water against time is similar to that for absorption by a hydrophillic gel (Smith 1962), i.e., an initially rapid rate of uptake which declines as the equilibrial level is approached. Complete saturation can be achieved in two minutes or less (Stocker 1927; Smith 1962; Blum 1973; Ried 1960b). Because saturation is achieved in a small fraction of the integration step used in the simulation, the uptake of liquid water is simulated by a simple step function; the addition of liquid water results in an immediate increase in the water content of the thallus to W_{max}.

The process of uptake of water vapor from the atmosphere is much slower than that for the uptake of liquid water. It may take several hours, or even days, for the thallus to approach its equilibrial water content (Stocker 1927; Smyth 1934; Quispel 1943). However, the general form of the absorption curve is similar to that for the uptake of liquid water. The equation developed to represent the evaporation of adsorbed water also predicts the uptake of water vapor when the water content of the lichen is below its expected equilibrial water content, i.e., when the vapor pressure of the atmosphere exceeds that of the thallus. The predicted rate of absorption is high initially and declines as the equilibrium is approached, as has been observed. Therefore, the differential equations representing uptake and loss of water vapor are

$$\frac{d\ WATER}{dt} = \begin{cases} \dfrac{WLRATE \cdot WATER^{\alpha}}{r_t}\ [e_s(T_o) - e_a];\ WATER > W_{sat} \\[2em] \dfrac{WLRATE \cdot WATER^{\alpha}}{r_t}\ [(c_1 - c_2\ WATER^{-1})e_s(T_o) - e_a];\ WATER \le W_{sat} \end{cases} \tag{6}$$

where $WLRATE = \dfrac{\rho c' c}{\lambda \gamma}$

CARBON FLOW SUBMODEL

The carbon submodel consists of seven compartments and an atmospheric carbon source/sink (CO_2) (fig. 1). The phycobiont and mycobiont are each represented by three compartments: (1) a soluble carbon compartment, (2) a storage carbohydrate compartment, and (3) a structural carbon compartment. In addition, there is a compartment which represents the pool of carbon released by the mycobiont.

The carbon that is fixed by photosynthesis enters the soluble carbon compartment of the phycobiont, GLUCOS. Carbon may leave this compartment via four pathways: (1) carbon may return to CO_2 through respiration; (2) carbon may pass into the storage compartment, ACARB; (3) carbon may pass via growth into the structural carbon compartment, ASTRCT; and (4) carbon may be released from the phycobiont to enter the extracellular carbon pool, EXCARB. Carbon may also enter GLUCOS by three other pathways: (1) a degradation of the long chain carbohydrates of ACARB into short chain, soluble carbohydrates; (2) catabolism of structural carbon; and (3) uptake of carbon from EXCARB.

Carbon enters the mycobiont through uptake from EXCARB. Carbon in the soluble carbon pool, POLYOL, is partitioned in the mycobiont in about the same manner as in the phycobiont, i.e., to storage carbohydrate (FCARB), to structural

carbon (FSTRCT), and to CO_2 through respiration. However, there is no flow of carbon to EXCARB from POLYOL. Carbon may also enter POLYOL from FSTRCT and FCARB due to catabolism.

The flow of carbon between these compartments is primarily controlled by three factors: substrate levels, water content, and temperature. Threshold response curves are used to express the effects of substrate levels and water content on many of the rates of flow used in the model. These curves can be represented by the equation

$$R/R_{max} = \frac{X^n}{K^n + X^n} \qquad \text{(Thornley 1976)} \quad (7)$$

where R is the rate observed at some level of the factor X, K is the value of X which gives half-maximal response, R_{max} is the maximum rate, and n is a constant. The equation reduces to a Michaelis-Menton response curve for n = 1. A higher value of n produces a more sigmoid response curve, with the curve approaching a step function as $n \to \infty$. Where the level of more than one factor controls the rate process, the response surface is calculated from the product of the response curves for each factor. Thus, these factors are assumed to act independently.

Photosynthesis

The flow of carbon from CO_2 to GLUCOS represents gross photosynthesis. Photosynthesis is a function of temperature and the levels of light, water, and carbon dioxide. The temperature dependence of photosynthesis may be described by two processes: (1) a rate limiting enzymatic step which is controlled by an energy of activation of an enzyme catalyzed reaction, and (2) denaturation of the rate limiting enzyme (Tenhunen et al. 1976). These two processes have been described by Johnson et al. (1942) using a single equation

$$P/P_{max} = \frac{C \cdot T_K \cdot e^{-\Delta H_2/R \cdot T_K}}{1 + e^{-\Delta H_1/R \cdot T_K} \cdot e^{\Delta S/R}} \tag{8}$$

where c = a proportionality constant; T_K = the thallus temperature in °K; ΔH_1 = the energy of activation for the denaturation of the enzyme;

ΔH_2 = the energy of activation for the enzyme catalyzed reaction; ΔS = the entropy of the denaturation equilibrium; and R = the ideal gas constant. This equation is used to simulate the response of photosynthesis to temperature.

Response curves for the effect of water content on photosynthesis in lichens typically show an optimum net assimilation rate (NAR) at water contents between 30 and 90% of the maximum thallus water content (Kershaw 1972; Ried 1960b). The NAR approaches 0 at low water content, presumably because water becomes limiting as a substrate in some enzymatic reactions affecting carbon fixation. The level of intracellular water may not be a linear function of the total water content of the lichen because of the complex nature of storage of water in lichens. Thus, physiological functions which depend upon water are likely to show threshold responses in relation to water content. Therefore the response of photosynthesis to low water content is described in the model by a generalized threshold response curve [equation (7)].

The NAR of many lichens also declines at high water levels. Two hypotheses have been suggested for this decline in NAR: (1) that the water in excess of the optimal level decreases the transmission of light to the phycobiont, and (2) that the additional water reduces the rate of diffusion of carbon dioxide into the photosynthetic sites and thus causes photosynthesis to become substrate limited.

The relationship between water content and the transmissivity of the cortex is assumed to be a function of the water content above a threshold value. Beer's law

$$I = I_o \, e^{-ks} \qquad (9)$$

is used to describe the attenuation of light as it passes through a medium, where k is a constant, s is the path length, I_o is the level of light incident on the surface, and I is the level of transmitted light. In a complex medium, the constant k can incorporate factors relating to transmission, scattering, and a non-random spatial distribution of the medium (cf. Acock et al. 1970). Although s is expected to increase with increasing water content of the lichen, the relationship between s and water content is probably nonlinear. In addition, water contents below a threshold value, W*, may have little or no effect on the transmission of light to the algal chloroplasts. Therefore, the relationship between water content and transmitted light is assumed to have the form

$$I = \begin{array}{l} I_o e^{-k_1 T} , \ \text{WATER} \leq W^* \\[2ex] I_o e^{-k_1 T} \, e^{-k_2 (\text{WATER} - W^*)^\beta} , \ W \geq W^* \end{array} \qquad (10)$$

where k_1 and k_2 are constants for the transmission of light below and above the threshold water

content, respectively; T is a constant relative to the transmissivity of the cortex at water levels below W*; and β is a constant relating water content to the thickness of the water layer through which light must pass.

A Michaelis-Menton response curve for the effect of carbon dioxide in photosynthesis is predicted by the biochemistry of the photosynthetic reaction. Response curves showing the relationship between NAR and ambient CO_2 levels for Umbilicaria lichens from Larson (1979a) and for Alectoria ochroleuca and Parmelia caperata from Larson and Kershaw (1975) appear to agree with the prediction. However, the level of CO_2 at the photosynthetic centers is assumed to be the factor affecting NAR, not the ambient levels of CO_2. Several previous models (e.g., Rabenowitch 1951; Chartier 1970; Tenhunen et al. 1976) have employed three steady state assumptions about carbon dioxide exchange in photosynthesizing tissues to simplify the estimation of CO_2 levels at the photosynthetic centers. These assumptions are (1) that the diffusion resistance between the photosynthetic centers and the atmosphere remains constant, (2) that the system is homogeneous in terms of light and CO_2 levels, and (3) respiration occurs at a constant rate and is evolved at the photosynthetic centers. Such assumptions are clearly violated in considering photosynthesis in lichens because the diffusion resistance is undoubtedly related to water content, because lichens are very non-homogeneous and because the proportion of photosynthetic tissue is small compared to the amount of respiring tissue. Thus, estimation of the levels of CO_2 in the phycobiont requires specific information about the morphology (distribution of algal cells) as well as the relationship between diffusion resistance and water content. However, the level of complexity at which the model was designed precludes consideration of specific morphological relationships, and there is currently no data on which to base assumptions about the relationship between diffusion resistance and water content (see Lange 1980). Therefore we employ the simplifying assumption that CO_2 levels at the photosynthetic centers are linearly related to water content above a threshold value. Carbon dioxide levels are expressed as a proportion of the optimal level of CO_2. Thus,

$$CO_{2p}/CO_{2opt} = \text{MIN} \, (1, \ a - b \cdot \text{WATER}) \qquad (11)$$

where CO_{2p} is the level of CO_2 at the photosynthetic centers, CO_{2opt} is the optimal level, and a and b are constants. The photosynthetic rate is related to CO_{2p} by a Michaelis-Menton function.

Respiration

Respiration in both symbionts is assumed to be a function of the level of substrate, water content of the lichen, and temperature. Two components of temperature are simulated. The first component has a relatively low maximal rate but this rate remains near maximal until extremely low levels of substrate are reached. The second component has a relatively high maximal rate which is achieved only at high substrate levels. Functionally, these two components correspond to maintenance respiration and constructive respiration, respectively. Both components are assumed to have Michaelis-Menton response curves with respect to substrate levels.

Response curves for the effect of water content on respiration in lichens have been determined for many species. The majority of these curves show a typical threshold response to water content, i.e., a sigmoid response curve which is asymptotic towards a maximum respiration rate at high water content. Measurements of the effects of water content on respiration were determined for lichens in the dark. Thus, these measurements may represent primarily the maintenance component of respiration. However, the same type of response function is also assumed to affect the constructive component of respiration. These responses are represented in the model by equation (7).

The effects of temperature on respiration in lichens are similar to those found in tissues of other plants (Smyth 1934; Stalfelt 1939; Scholander et al. 1952); respiration rates show a typical Arrhenius type response to temperature. Therefore, a Q_{10} response function is used to adjust respiration rates for the effects of temperature.

Anabolism and Catabolism of Structural Tissues

The flow from soluble carbon to structural carbon in each symbiont is proportional to the total respiration of the symbiont times a yield factor, YIELD. Thus, anabolism is a function of substrate levels, water content, and temperature. On the other hand, catabolism of structural carbon is assumed to be a function of temperature and water content, but not of substrate levels. The products of catabolism reenter the pool of soluble carbon. The response of catabolism to temperature and water parallels that of respiration. The rate of catabolism is assumed to equal YIELD times the maximum rate of maintenance respiration. Thus, anabolism due to maintenance respiration is sufficient to replace the structural carbon lost through catabolism at all but the most limited substrate levels.

In the event that substrate levels are reduced to the extent that maintenance respiration falls below the maximum rate, there will be a net loss of carbon from ASTRCT and/or FSTRCT. Extended periods of such degradation of the structural components of a cell expected to lead to the mortality of the cell. However, all of the cells of the symbiont probably would not be affected in exactly the same way; cell to cell variation in storage supplies or resistance to structural damage would produce variation in the time the cells could withstand severe substrate limitation. Therefore, a Gaussian distribution was assumed for the frequencies of these survival times. An event which may lead to the loss of lichen biomass due to mortality is initiated when there is a net loss in structural carbon. The proportion of cell mass which survives after t hours under conditions of severe substrate limitation is calculated from

$$P_t = \begin{cases} 1 - CDF & , \quad t < \bar{t} \\ CDF & , \quad t \geq \bar{t} \end{cases}$$

where P_t is the proportion surviving t hours, $\bar{t}$ is the mean time of survival, and CDF is an approximation to the cumulative distribution function of a normal distribution. CDF is equivalent to

$$\cos\left(-.4322\frac{(t - \bar{t})^2}{\sigma^2}\right)^{\frac{1}{2}} + 1.024$$

where σ^2 is the variance of the distribution. Carbon is removed from all of the compartments of the affected symbiont in proportion to P_t. Mortality due to severe substrate limitation ceases once a net gain in structural carbon is realized.

Partitioning of Carbon between Soluble and Storage Carbohydrate

The rates of flow of carbon from GLUCOS to ACARB and from POLYOL to FCARB is assumed to be Michaelis-Menton functions of GLUCOS and POLYOL, respectively, adjusted for temperature by a Q_{10} relationship. Although water does not directly affect the rate of flow into the storage compartment, there is an indirect effect of water since the levels of GLUCOS and POLYOL are expected to vary with water content through its effect on respiration and photosynthesis. This assumption agrees with the results of Harris and Kershaw (1971) which show that starch accumulated in algal cells at low water contents (3-5% of dry weight) in the dark.

Carbon may also flow from ACARB and FCARB to GLUCOS and POLYOL, respectively. Two independent mechanisms of flow are defined for each storage compartment. A flow of low rate occurs at all but the most severe levels of dryness. The maximum rate of this flow supplies sufficient carbon to support maintenance respiration. A second flow may be stimulated by higher water contents to permit constructive respiration to proceed under light limited conditions. These flows are assumed to be Michaelis-Menton functions of substrate levels adjusted for temperature effects with a Q_{10} relationship.

Carbohydrate Release by the Alga

There are three main hypotheses concerning the mechanism of carbohydrate release by the alga, (1) diffusion through the membrane, (2) hydrolysis of extracellular polysaccharide produced by the cell, and (3) active transport of carbohydrates across the membrane (Hill 1976). Diffusion is probably the most common mechanism of carbohydrate transport across membranes and may account for the release of carbohydrates by phytoplankton (Watt and Fogg 1966). Follmann (1960) showed that the permeability of Trebouxia algae to glucose was enhanced in those recently isolated from lichens. Thus the mycobiont may be able to facilitate diffusion through modification of the algal cell membrane.

Drew and Smith (1967a,b) postulated that the transport of carbohydrate from the phycobiont to the mycobiont may result from a modification of the synthesis of polysaccharides used in cell walls. Richardson et al. (1968) found that short chain glucose polymers accompanied the release of glucose in Lichina, and Hill (1972) found that ^{14}C transferred to the mycobiont in Peltigera polydactyla passed through an insoluble glucose polymer. However, Peveling and Hill (1974) found that this insoluble polymer was contained within the cell membrane of the alga. Thus, there is no definitive evidence to support the second hypothesis.

Active transport of carbohydrate across membranes occurs infrequently (Hill 1976). However, Hill and Smith (1972) found that glucose release was inhibited by darkness. This inhibition may have been due to a reduction in the level of photosynthate in the alga or to a lack of sufficient ATP to drive the active transport mechanism. Swindlhurst, Jordan, and Hill (unpublished, cited in Hill 1976) attempted to identify the mechanism of the inhibition with the use of DCMU, which stops CO_2 fixation but apparently does not interfere with ATP formation. Release of labeled carbon in the light by lichens affected by DCMU was similar to release rates of control lichens in the dark. These findings raise doubts about the possibility of active transport of carbohydrate in lichens.

Diffusion appears to be the most satisfactory mechanism postulated to explain carbohydrate release by the alga. Therefore, the rate of flow of carbon from GLUCOS to EXCARB was assumed to be proportional to the difference in the concentrations of the two pools. We assume that the mass of the intracellular water of the phycobiont is approximately three times the mass of the algal structural carbon in order to estimate the concentration of intracellular soluble carbohydrate, c_i. The concentration of extracellular carbohydrate, c_e, is assumed to equal EXCARB divided by the water content of the lichen. The rate of diffusion is then calculated by the resistance analog equation

$$R_{diff} = \frac{(c_i - c_e) \, a \cdot ASTRCT^{2/3}}{r}$$

where R_{diff} is the flux of carbon, r is the diffusion resistance, and a is a proportionality factor relating ASTRCT to the surface area of the algae. Although c_i is expected to normally be greater than c_e, the equation does permit the diffusion of carbon from EXCARB to GLUCOS in the event that the concentration gradient is reversed.

Uptake of Carbon by the Mycobiont

The carbohydrate released by the phycobiont does not accumulate in the extracellular pool and is taken up rapidly by the mycobiont. The active uptake of carbohydrate by the mycobiont from EXCARB is assumed to involve an enzymatic process. Therefore, the rate of uptake is calculated using a Michaelis-Menton function of substrate level. A Q_{10} relationship is used to calculate the response of the uptake rate to temperature. The response of uptake to water content is assumed to parallel that of respiration.

VALIDATION OF THE MODEL

Validation of the model is achieved by showing that the assumptions of the model, i.e. the mathematical descriptions of individual processes, produce results which are consistent with experimental observation and theory. Validation is particularly important where one of several alternate hypotheses is used or where mathematical representations of processes must be derived for the first time, as in the case of water exchange with the atmosphere. Model responses which are particularly important to consider in validating a model are those which result from interactions which are not explicitly specified as model assumptions. Tests of model validity are useful only when constrained by the objectives under which the model was built; in this case, the objective was to represent the major physiological processes related to carbon flow in lichens. Therefore, the model is validated by simulating several experiments involving carbon flow and comparing results obtained from the model with those obtained experimentally.

Parameterization of the Model

We could not find data representing a single species which was sufficient to completely parameterize the model. Therefore we used data from several sources and for several species. Data from Ried (1960b) for Umbilicaria polyphylla were used to determine constants used in the water exchange equation, c_1 and c_2 of equation (3) were determined using a linear regression between humidity and the inverse of the equilibrial water contents, and the rate constant

WLRATE was estimated using data for the rate of drying at a relative humidity of 60%. Parameters required to estimate the relative change in the total evaporative resistance with wind velocity were calculated using the data from Larson (1979b). However, wind velocity was assumed to be zero in the following experiments.

Data required to parameterize the majority of the carbon flow submodel were obtained from Kershaw and Harris (1971a,b) for Parmelia caperata and from Kershaw (1972) for Zanthoria fallax. These data include the response of NAR to water content at 800 foot candles light intensity and the addition of a metabolizable substrate. A Q_{10} of 2.0 was used to estimate the response of respiration and associated processes to temperature. However, parameters for the response of photosynthesis to temperature were estimated using the data for Ramalina maciformis (Del.) Bory from Lange (1969). Light intensity at the photosynthetic centers was assumed to be independent of water content in these simulations. The maximum rate of uptake of carbon by the myobiont was set sufficiently high as to assure a nearly complete and rapid uptake of the carbohydrate released by the phycobiont.

Results and Discussion of Model Validation Simulations

The water exchange equation was tested using drying curves for U. polyphylla at humidities of 80% and 95%. The model predictions for the rates of water loss agree well with the data of Ried (1960b) with no further adjustment of

parameter values (fig. 2). Although data for the absorption of water vapor by U. polyphylla are not available, absorption curves predicted by the model agree at least quantitatively with such curves reported for other species; absorption of water occurs rapidly at first but declines to a low rate as the water content approaches its equilibrial value. The water content remains below the maximum even when water is absorbed at relative humidities of 100%. Thus, the simulation of water exchange with the atmosphere appears to agree well with experimental observations.

The response of the rate of respiration to water content in P. caperata may be approximated by a Michaelis-Menton function of maintenance respiration at low water levels. The increase in respiration rate at high water contents was assumed to represent a threshold type of response to water content by constructive respiration. Thus, the curve representing total respiration as a function of water content is similar to that found by Kershaw and Harris (1971) (fig. 3).

The model also represents the response of NAR to light intensity in a realistic manner (fig. 4). Respiration rates at low light intensities are lower than at high light intensities due to substrate limitation. Kershaw and Harris (1971b) demonstrated that the mechanism involved in the suppression of respiration at low light levels was due to substrate limitation rather than a photorespiratory response by an experiment in which additional substrate (mannitol) was added to lichens at various light intensities. The net assimilation rates of the lichens

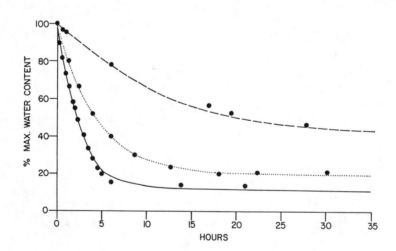

Figure 2.--Curves showing the loss of water by evaporation at relative humidities of 60% (solid line), 80% (dotted line), and 95% (broken line). Data from the drying curves of U. polyphylla (dots) were taken from Ried (1960b).

declined after the addition of the substrate. The magnitude of the change was negatively correlated with light intensity. A similar experiment was simulated using the model. The resulting changes in net assimilation rates predicted by the model are similar to those observed by Kershaw and Harris (1971b) (fig. 5). Thus, the interactions between carbon fixation, allocation of substrate, and respiration parallel those observed in P. caperata.

The theoretically derived equation relating gross photosynthesis to temperature from Tenhunen et al. (1976) was parameterized to approximate the response curve of R. maciformis (Del) Bury, a lichen from the Negev Desert. The parameters of the fitted curve indicate that the lichen has a lower energy of activation and a higher energy of denaturation than found in the red kidney bean Phaseolus vulgaris by Tenhunen et al. (1976). Lange et al. (1970) has shown that the maximum rate of net photosynthesis usually

occurs in the early morning hours while the lichen is still moist from dew. Thus, the photosynthetic enzyme systems of R. maciformis appear to be well adapted to a desert environment.

Parameters relating the effects of water content to photosynthesis were adjusted to fit the data values for the response of NAR to water content in P. caperata at a light intensity of 800 foot candles (fig. 3). Comparison of the predictions of the model for a light intensity of 1600 foot candles shows a reasonably close correlation with the observed response. However, the data show a greater reduction in NAR at high water content at 1600 foot candles than at 800 footcandles; the model does not predict such a response. Assuming that the decline in NAR at high water content is the result of an increase in the resistance to diffusion of CO_2, a proportionally greater reduction in NAR would be expected when gross photosynthetic rates are higher. However, the assumptions used in the

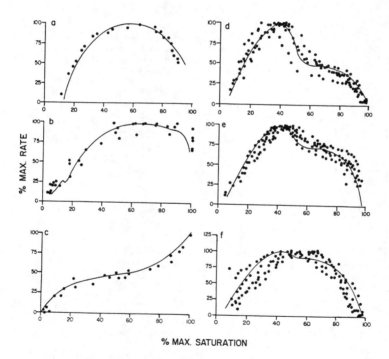

% MAX. RATE

% MAX. SATURATION

Figure 3.--Net assimilation rates as functions of water content for (a) P. caperata at 1600 footcandles, (b) P. caperata at 800 footcandles, (c) P. caperata in the dark (respiration), (d) X. fallax from very dry sties, (e) X. fallax from dry sites, and (f) X. fallax from mesic sites. Results of model simulations are shown as solid lines. Observed rates, shown as dots, were taken from Kershaw and Harris (1971a,b) (a, b, and c), and from Kershaw (1971) (d, e, and f). Both the rates (ordinates) and the water contents (abscissas) are shown as percents of maxima.

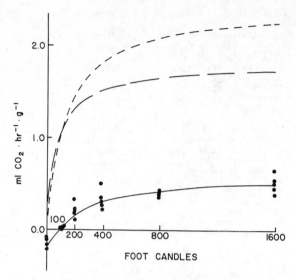

Figure 4.--Net assimilation rate of P. caperata as a function
of light intensity. Simulated values are shown by a
solid line. Measured values of N.A.R., taken from
Kershaw and Harris (1971a), are shown as dots. Also
shown are the gross photosynthetic rate (small dashes)
and the respiration rate (large dashes).

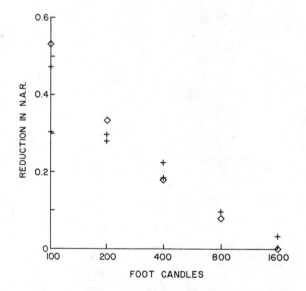

Figure 5.--Reduction in net assimilation rate as a function
of light intensity due to the addition of an external
source of carbon. Observed values (plus signs) are
taken from Kershaw and Harris (1971b). Results of
simulations of these experiments are shown as diamonds.

model to relate the reduction in CO_2 concentrations at the photosynthetic centers to water content disregard the rate of CO_2 utilization. Thus, the model does not predict this interaction.

The relationships between NAR and water content of some species have more complex response curves than those of P. caperata. For example, Kershaw (1972) found that the response curves of X. fallax from three sites of different moisture regimes had greatly different shapes (fig. 3). The model was parameterized to approximate the response curve of X. fallax from very dry sites. The curves for the dry and mesic sites were then approximated by adjusting a single parameter, the half saturation constant relating light intensity to photosynthesis. Thus, the model appears to be general enough to permit the simulation of a wide variety of lichen species.

IMPACT OF SULFUR DIOXIDE ON LICHENS

Lichens and brophytes are among the plants most sensitive to exposure to sulfur dioxide. The sensitivity of lichens to sulfur dioxide air pollution is probably the result of their inability to control the uptake and loss of materials through the thallus. Substances in solution can be absorbed over their entire surface area. Good relationships are often found between the distributions of lichens and concentrations of SO_2 in polluted areas (Johnsen and Sochting 1973). The constancy of this relationship has resulted in the widespread usage of lichens as bioindicators of air pollution (Heck et al. 1979).

Experiments have been done in the laboratory to identify some of the physiological mechanisms affected by SO_2. At atmospheric concentrations of 5 ppm over a 24 h. exposure, Rao and LeBlanc (1965) found that SO_2 caused bleaching of the algal cells and the breakdown of chlorophyll to phaeophytin. Chemical oxidation of chlorophyll may be induced at low pH by SO_2 (Puckett et al. 1973). However, ^{14}C fixation may be impaired by SO_2 even without bleaching of chlorophyll. Puckett et al. (1973) suggest that such inhibition may be due to interference by SO_3 and HSO_3 of the flow of electrons to NADP$^+$ in the electron transport chain involved with the fixation of CO_2. A slight enhancement in CO_2 fixation has been observed in some studies. Such enhancement may also represent interference by SO_3 and HSO_3 in the electron transport chain, since these species of ions can act as electron donors at near neutral pH values. Finally, bisulphite and sulfate ions may compete with phosphate for a bonding site critical to phosphorylation in chloroplasts (Ryrie and Jagendorf 1971; Hall and Telfer 1969). However, Hill (1971) found no inhibition of ^{14}C fixation in lichens supplied with excess sulfate.

Sulfur dioxide is known to affect other important processes in lichens. Baddeley,

Ferry, and Finegan (1971, 1972), Eversman (1978), and Showman (1972) have found that exposure to SO_2 can reduce the respiration rates of lichens. Permeability of cell membranes may also be increased in response to SO_2 (Puckett et al. 1974; Malholtra and Hocking 1976).

Simulation modeling of lichen response to air pollution as an indication of potential air pollution damage has not previously been attempted. Our objective was to incorporate as mechanistically as possible the effects of sulfur dioxide on the dynamics of carbon in lichens. Implementation of the effects of SO_2 on lichens is based upon results of experiments which were presented previously in the literature. However, many of these experiments were carried out at concentrations of SO_2 which are well above those that are likely to be found in field situations. Thus, much of this information is of limited usefulness in trying to assess the impact that realistic concentrations of SO_2 might have on lichen growth and metabolism.

THE SIMULATION OF EXPOSURE TO SULFUR DIOXIDE

We chose to simulate terricolous lichens exposed to concentrations of SO_2 consistent with that delivered to four treatments of an experimental area in Montana. These treatments included a control plot plus three plots to which approximately 20, 50, and 100 ppb SO_2 were delivered by means of a network of pipes. Actual measurements of air concentrations at .3 m. made at hourly intervals during the month of June, 1976 were used as data for the simulations. The concentration of SO_2 near ground level was estimated by

$$C(z) = C(h)[1- \frac{LAI(h)-LAI(z)}{LAI(h)} \cdot \frac{U(h)-U(z)}{U(h)}$$

where $C(i)$ is the concentration at height i ($\mu g/m^3$), z is the height at which the SO_2 concentration is desired, h is the canopy height (cm), LAI is leaf area index, and U is wind speed (Wong et al. 1980).

In our simulation of the impact of sulfur dioxide on the lichens, we assumed that only the gross photosynthetic rate was affected. The response was determined using the curve for ^{14}C fixation by U. muhlenbergii after one hours exposure to an aqueous solution of SO_2 (data from Richardson et al. 1979). Linear interpolation of the data presented by Hocking and Hocking (1977) was used to estimate from the concentration in air the concentration of SO_2 in solution to which the algae were exposed. The rate of photosynthesis was then reduced in proportion to the reduction in net assimilation expected at this concentration of aqueous SO_2.

Responses of the model to each of the four treatments were measured using the change in mass of structural carbon of the phycobiont and mycobiont over the 30 day period. Growth was not reduced in any of the four simulations. This result was not surprising, since the levels of SO_2 in solution predicted by the model were quite low compared to the level at which Richardson et al. found even a moderate (15%) decline in total ^{14}C fixation. These results clearly indicate that more research needs to be done concerning the effects of levels of SO_2 more likely to be found in polluted areas. Effects other than the reduction in NAR or respiration rate also should be considered. For example, although sulfur accumulation through time might be responsible for the effects observed in field populations of lichens, very little is known about the rates of such accumulation, let alone its effect on lichens. Lichens undergo major physiological changes during their frequent periods of drying and rewetting, including the leakage of important cations (Richardson and Nieboer 1980). Thus any effects of SO_2 or accumulated sulfur on the permeability of the cell membranes or on the repair of such membranes during rewetting could have significant impacts upon the functioning of the lichen. It is our belief that models, such as that presented here, can be useful tools in helping the researcher identify those processes which are most critical to the functioning of lichens, or about which little is known.

LITERATURE CITED

Acock, B., J.H.M. Thornley, and J. Warren Wilson. 1970. Spatial variation of light in the canopy. p. 91-102. In (I. Setlik, ed.) Prediction and measurement of photosynthetic productivity. Pudoc, Wageningen.

Baddeley, M.S., B.W. Ferry, and E.J. Finegan. 1971. A new method of measuring lichen respiration: response of selected species to temperature, pH and sulphur dioxide. The Lichenologist 5:18-25.

Blum, O.B. 1973. Water relations. p. 381-400. In (V. Ahmadjian and M.E. Hale, eds.) The lichens. Academic Press, New York.

Chartier, P. 1970. A model of CO_2 assimilation in the leaf. p. 307-315. In (I. Setlik, ed.) Prediction and measurement of photosynthetic productivity. Pudoc, Wageningen.

Drew, E.A., and D.C. Smith. 1967a. Studies in the physiology of lichens. VII. The physiology of the Nostoc symbiont of Peltigera polydactyla compared with cultured and free-living forms. New Phytol. 66:379-388.

Drew, E.P., and D.C. Smith. 1967b. Studies in the physiology of lichens. VIII. Movement of glucose from alga to fungus during photosynthesis in the thallus of Peltigera polydactyla. New Phytol. 66:389-400.

Eversman, S. 1978. Effects of low-level SO_2 on Usnea hirta and Parmelia chlorochroa. The Bryologist 81:368-377.

Farrar, J.F. 1976. The lichen as an ecosystem: observation and experiment. p. 385-406. In (D.H. Brown, D.L. Hawksworth, and R.H. Bailey, eds.) Lichenology: Progress and problems Systematics Association Special Volume No.8. Academic Press, New York.

Follmann, G. 1960. Die durchlassig keitseigen-schaften der photoplasten von phycobionten aus Cladonia furcata (Huds.) Schrad. Naturwissenschaften 47:405-406.

Hale, M.E. 1974. The biology of lichens. Arnold, London.

Hall, D.O., and A. Telfer. 1969. The effect of sulphate and sulphite on photophosphorylation by spinach chloroplasts. Prog. Photosyn. Res. 2:1281.

Harris, G. P. 1976. Water content and produc-tivity of lichens. p. 452-468. In (O.L. Lange, L. Kepper, and E.-D. Schulze, eds.) Water and plant life-problems and modern approaches. Springer-verlag, Berlin.

Harris, G.P., and K.A. Kershaw. 1971. Thallus growth and the distribution of stored metabolites in the phycobionts of the lichen Parmelia sulcata and P. physodes. Can. J. Bot. 49:1367-1372.

Haynes, F.M. 1964. Lichens. Viewpoints Biol. 3:64-115.

Heck, W.W., S.V. Krupa, and S.N. Linzon. 1979. Methodology for the assessment of air pollucion effects on vegetation. Air Pollution Control Association.

Hill, D.J. 1971. Experimental study of the effects of sulphite on lichens with reference to atmospheric pollution. New Phytol. 70:831.

Hill, D.J. 1972. The movement of carbohydrate from the alga to the fungus in the lichen Peltigera polyductyla. New Phytol. 71:31-39.

Hill, D.J. 1976. The physiology of lichen symbiosis. p. 457-496. In (D.H. Brown, D.L. Hawksworth, and R.H. Bailey, eds.) Lichenology: Progress and problems. Systematics Association Special Volume No.8. Academic Press, New York.

Hill, D.J., and D.C. Smith. 1972. Lichen physiology. XII. The inhibition technique. New Phytol. 71:15-30.

Hitch, C., and W.D.P. Stewart. 1973. Nitrogen fixation by lichens in Scotland. New Phytol. 72:509-524.

Hocking, D., and M.B. Hocking. 1977. Equilibrium solubility of trace atmospheric sulphur dioxide in water and its bearing on air pollution injury to plants. Environ. Pollut. 13:57-64.

Johnsen, I., and U. Sochting. 1973. Influence of air pollution on the epiphytic lichen vegetation and bark properties of deciduous trees in the Copenhagen area. Oikos 24:344-351.

Johnson, F., H. Eyring, and R. Williams. 1942. The nature of enzyme inhibitions in bacterial luminescence: Sulfanilamide, urethane, temperature, and pressure. J. Cell. Comp. Physiol. 20:247-268.

Kershaw, K.A. 1972. The relationship between moisture content and net assimilation rate of lichen thalli and its ecological significance. Can. J. Bot. 50:543-555.

Kershaw, K.A., and G.P. Harris. 1971a. Simulation studies and ecology: A simple defined system model. p. 1-21. In (G.P. Patil, E.C. Pielou, and W.E. Waters, eds.) Statistical ecology, Vol. 3. Penn State Univ. Press.

Kershaw, K.A., and G.P. Harris. 1971b. Simulation studies and ecology: Use of the model. p. 23-42. In (G.P. Patil, E.C. Pielou, and W.E. Waters, eds.) Statistical ecology, Vol. 3. Penn State Univ. Press.

Lange, O.L. 1969. Experimentell-okologische untersuchungen an flechten der Negev-Wuste. I. CO_2-Gaswechsel von Ramalina maciformis (Oel.) Bory unter Kontrollierten Bedingungen im laboratorium. Flora 158:324-359.

Lange, O.L., E.-D. Schulze, and W. Koch. 1970. Experimentell-okologische untersuchungen un flechten der Negev-Wuste. III. CO_2-Gaswechsel und wassenhaushalt von krusten-und blattflechten am natülichen standort während der sommerlichen trockenperiode. Flora 159:525-538.

Lange, O.L. 1980. Moisture content and CO_2 exchange of lichens. I. Influence of temperature on moisture-dependent net photosynthesis and dark respiration in Ramalina maciformis. Oecologia 45:82-87.

Larson, D.W., and K.A. Kershaw. 1975. Measurement of CO_2 exchange in lichens: a new method. Can. J. Bot. 53:1535-1541.

Larson, D.W. 1979a. Preliminary studies of the physiological ecology of Umbilicaria lichens. Can. J. Bot. 57:1398-1406.

Larson, D.W. 1979b. Lichen water relations under drying conditions. New Phytol. 82:713-731.

Malhotra, S.S., and D. Hocking. 1976. Biochemical and cytological effects of sulphur dioxide on plant metabolism. New Phytol. 76:227-237.

Monteith, O.L. 1965. Evaporation and environment. p. 205-234. In (O.E. Fogg, ed.) The state and movement of water in living organisms, Vol. 19. Symp. Soc. Exp. Biol., Cambridge Univ. Press, London.

Peveling, E., and D.J. Hill. 1974. The localization of an insoluable intermediate in glucose production in the lichen Peltigera polydactyla. New Phytol. 73:769-771.

Puckett, K.J., E. Nieboer, W.P. Flora, and D.H.S. Richardson. 1973. Sulphur dioxide: its effects on photosynthetic ^{14}C fixation in lichens and suggested mechanisms of phytotoxicity. New Phytol. 72:141-154.

Puckett, K.J., D.H.S. Richardson, W.P. Flora, and E. Nieboer. 1974. Photosynthetic ^{14}C fixation by the lichen Umbilicaria muhlenbergii (Ach.) Tuck. following short exposure to aqueous sulphur dioxide. New Phytol. 73:1183-1192.

Quispel, A. 1943. The mutual relations between algae and fungi in lichens. Rec. Trav. Bot. Neerl. 40:413-541.

Rabinowitch, E. 1951. Photosynthesis and related processes. Vol. II, Part 1. New York, Interscience.

Rao, D.M., and F. LeBlanc. 1965. Effects of sulfur dioxide on the lichen alga, with special reference to chlorophyll. Bryologist 69:69-75.

Richardson, D.H.S., D. Jackson Hill, and D.C. Smith. 1968. Lichen physiology. XI. The role of the alga in determining the pattern of carbohydrate movement between lichen symbionts. New Phytol. 67:469-486.

Richardson, D.H.S., E. Niebur, P. Lavoie, and D. Padouah. 1979. The role of metal ion binding in modifying the toxic effects of sulfur dioxide on the lichen Umbilicaria muhlenbergii II. [14]C fixation studies. New Phytol. 82:633-643.

Richardson, D.H.S., and E. Nieboer. 1980. Surface finding and accumulation of metals in lichens. p.75-94. In (C.B. Cook, P.W. Pappas, and E.D. Rudolph, eds.) Cellular interactions in symbiosis and parasitism. Ohio State Univ. Press, Columbus.

Ried, A. 1960a. Thallusbau und assimilationshaushalt von laub-und krusten flechten. Biologisches Zentralblatt 79:129-151.

Ried, A. 1960b. Stoffwechsel und verbreitung-sgrenzen von flechten. II. Wasser- und assimilationschaushalt, entquellungs-und submersionsresistenz von krustenflechten benachbarter standorte. Flora (Jena) 149:345-385.

Ripley, E.A., and R.E. Redmann. 1976. Grassland. p. 349-398. In (J.L. Monteith, ed.) Vegetation and the atmosphere, Vol. 2, Case studies. Academic Press, London.

Ryrie, J.J., and A.T. Jabendorf. 1971. Inhibition of photophosphorylation in spinach chloroplasts by inorganic sulphate. J. Biol. Chem. 246:582.

Scholander, P.F., W. Flagg, V. Walton, and L. Irving. 1952. Respiration in some artic and tropical lichens in relation to temperature. Amer. J. Bot. 39:701-713.

Scofield, H.T., and L.E. Yarman. 1943. Some investigations of the water relations of lichens. Ohio J. Sci. 3:139-146.

Showman, R.E. 1972. Residual effects of sulfur dioxide on the net photosynthetic and respiratory rates of lichen thalli and cultured lichen symbionts. The Bryologist 75:335-341.

Showman, R.E., and E.D. Rudolph. 1972. Water relations of living, dead, and cellulose models of the lichen Umbilicaria papulosa. The Bryologist 74:444-450.

Smith, D.C. 1962. The biology of lichen Thalli. Biol. Rev. 37:537-570.

Smyth, E.S. 1934. A contribution to the physiology and ecology of Peltigera canina and Peltigera polydactyla. Ann. Bot., Lond. O.S. 48:781-817.

Stalfelt, M.G. 1939. Der gasaustausch der flechten. Planta 29:11-31.

Stocker, O. 1927. Physiologische und ökologische unter suchungen an Laub-und strauch flechten. Flora (Jena) 121:334-415.

Syers, J., and I. K. Iskander. 1973. Pedogenic significance of lichens. p. 225-248. In (V. Ahmadjian and M.E. Hale, eds.) The lichens. Academic Press, New York.

Tenhunen, S.D., C.S. Yocum, and D.M. Gates. 1976. Development of a photosynthesis model with an emphasis on ecological applications. I. Theory. Oecologia 26:89-100.

Thornley, J.H.M. 1976. Mathematical models in plant physiology. 318 p. Academic Press, New York.

Watt, W.D., and G.E. Fogg. 1966. The kinetics of extra cellular glycollate production by Chlorella pyrenoidosa. J. Exp. Bot. 17: 117-134.

Wong, L.T.K., E.M. Preston, and T.L. Gullett. 1980. Vertical SO_2 concentration profile on ZAPS during the 1978 field season. p. 108-119. In (Eric M. Preston and David W. O'Guinn, eds.) The bioenvironmental impact of a coal fired power plant. Fifth Interim Report, Colstrip, Montana.

A MATHEMATICAL MODEL FOR CALCULATING OPTIMAL OZONE CONTROL STRATEGIES[1]

Gregory A. Holton[2]

Abstract.--This paper presents a simple deterministic mathematical model that calculates optimal ozone control strategies using representative control cost functions. Results of application indicate that, depending on the regional distribution of ozone-precursor concentrations, the most economical ozone control strategy may be to control both nonmethane hydrocarbon and oxides of nitrogen rather than nonmethane hydrocarbon only.

INTRODUCTION

The U.S. Environmental Protection Agency (EPA) is placing increased emphasis on modeling to support ozone (O_3) control strategy decisions and policies (USEPA, 1979a-c). Reliance on models is growing because they apply to a wide variety of locations through the use of region-specific data, appropriate assumptions, and analysis of model input and output uncertainties. Applications are designed, among other reasons, to identify the key elements of the complex ozone pollution problems of a region without an expensive, time-consuming field study.

There are a variety of pertinent modeling approaches and photochemical models (Whitten and Hogo 1978; Whitten et al. 1979; Jeffries and Sexton 1980). Regardless of the photochemical model employed, application usually involves the solution of a system of coupled, first-order, nonlinear differential equations which are the time-dependent representation of the photochemical mechanism.

The term "photochemical mechanism" denotes a chemical kinetics description of atmospheric photochemical processes by compiling pertinent chemical reactions and reaction rates. Computations employing such photochemical mechanisms produce relationships between ozone and its precursors -- nonmethane hydrocarbons (HC) and oxides of nitrogen (NO_x) -- that can be graphically represented by a plot called an ozone-precursor surface (fig. 1). The photochemical mechanism (Durbin et al. 1975) that was used to generate figure 1 has become outdated (Falls and Seinfeld 1978) but its use in determining chemical relationships between ozone and its

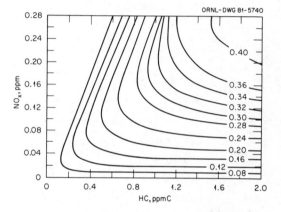

Figure 1.--Standard EKMA ozone-precursor surface with indicated ozone isopleth values for precursor concentrations of NO_x and HC.

precursors in EPA's Standard Empirical Kinetics Modelling Approach (Standard EKMA) still makes it important (USEPA 1978b).[3]

Although compilation of photochemical mechanisms is designed to provide an understandable simplification of the photochemical processes, such mechanisms are extremely detailed. Such detail is necessary to include all the important chemical reactions known to occur in mixtures of hydrocarbons and oxides of nitrogen irradiated by light. As detail increases, understanding the governing processes becomes more difficult. One technique used to achieve a better understanding of such governing processes is to develop a predictive model employing only a few parameters that adequately embodies the ozone-precursor relationships contained in the photochemical mechanism.

[1]Operated by Union Carbide Corporation under contract W-7405-eng-26 with the U.S. Department of Energy.

[2]Gregory A. Holton is a Research Associate in the Health and Safety Research Division, Oak Ridge National Laboratory, Oak Ridge, Tennessee U.S.A.

[3]Standard EKMA is being used by EPA to establish ozone strategies in certain regions of the United States (USEPA, 1979c).

A mathematical model of only a few parameters has been developed for various ozone-precursor surfaces (Holton and Jeffries 1979). This model is expressed by

$$[O_3] = (CL^N)\left[1.00 - \left(\frac{D}{L\tan\theta}\right)^G\right]^H \quad \text{for } D \geq 0 , \tag{1}$$

and

$$[O_3] = (CL^N)e^{-Q\left(\frac{|D|}{L}\right)^R} \quad \text{for } D < 0 , \tag{2}$$

where

C, N, θ, G, H, Q, and R are mechanism-dependent constants;

$|D|$ is the absolute value of the distance to the left (negative) or right (positive) of the ridge line in ppm units; and

L is the distance up to the ridge line in ppm units.

The ridge line is defined as that perpendicular line which connects all points of maximum ozone produced by parallel lines of a fixed hydrocarbon concentration ([HC]) to oxides of nitrogen concentration ([NO$_x$]) ratio. Each ozone-precursor surface has a ridge line angle θ, meaning that angle described by the intersection of the ridge line and the HC axis. Because different ozone-precursor surfaces have unique ridge line angles, each ozone-precursor surface used to develop equations (1) and (2) was rotated through the angle (90−θ) degrees for comparison. With

$$\theta = \text{Arc tan } ([NO_x]/[HC]) , \tag{3}$$

L and D can be expressed

$$L = [HC] \cos\theta + [NO_x] \sin\theta , \tag{4}$$

and

$$D = [HC] \sin\theta - [NO_x] \cos\theta . \tag{5}$$

Equations (1) and (2) have the same general form for all ozone-precursor surfaces studied, regardless of the photochemical mechanism used to develop them. Only the parameters vary with initial conditions and mechanisms. For an ozone-precursor surface similar to figure 1, the mathematical model is for D ≥ 0 ppm:

$$O_3 = 0.282 \ L^{0.683}[1.00 - (5.49 \ D/L)^{1.97}]^{0.518} , \tag{6}$$

and for D < 0 ppm:

$$O_3 = 0.282 \ L^{0.683} \ e^{-9.11(|D|/L)^{1.68}} . \tag{7}$$

The application of an ozone-precursor surface under EKMA requires two pieces of information from observed ambient air quality data. The first is the ozone design value in or near the city. The design value is defined by EPA to be the second highest hourly ozone concentration observed during the monitoring period. The other ambient data requirement is the [HC] to

[NO$_x$] ratio. According to EKMA user guidelines, the ideal ratio occurs between 6:00-9:00 A.M. local daylight time near the city center, upwind on the same day from the O$_3$ design value observation site. These monitoring values are to be obtained, if possible, by employment of standard measurement techniques (USEPA 1977).

The criteria document for ozone indicates that EKMA can be used in either a relative or absolute sense (USEPA, 1978a). For example, given an ozone-precursor surface such as figure 2 and initial concentrations of NO$_x$ and HC of about 0.10 ppm and 0.90 ppm C, respectively, one would predict an ozone concentration of about 0.25 ppm (Point A). Such a prediction represents use of EKMA in an absolute sense. If, however, one wanted to determine by what percent the maximum ozone concentration is reduced in response to a given percent reduction in HC and/or NO$_x$, then such an application would be termed relative. It should be noted that the ozone-precursor surface is generated using specific assumptions about dilution, light, reactivity, etc. Thus, an absolute prediction of O$_3$ concentration would be accurate only if: (a) all physical conditions such as atmospheric dilution, light intensity, etc., were representative of the region under study; and (b) the generating photochemical mechanisms were absolutely accurate instead of an approximation (Falls and Seinfeld 1979; Turner 1979). For these reasons, the use of EKMA in a relative sense seems more meaningful than absolute EKMA use in the development of oxidant control strategies.

RESULTS AND DISCUSSION

Use of mathematical expressions for ozone-precursor surfaces affords the incorporation of economic considerations and a simultaneous analysis of all control options through non-linear optimization techniques. These techniques are now illustrated by the formulation of an ozone control optimization problem which includes typical economic cost functions.

The National Ambient Air Quality Standard of 0.12 ppm for O$_3$ can be achieved in non-compliance regions by controlling only initial hydrocarbon, controlling only initial oxides of nitrogen, or controlling both initial HC and NO$_x$. Any ozone control strategy should be based on an analysis of these three options. Furthermore, an optimum control policy strategy must include economic information (Bilger 1978). It is realized that the following functions for control costs are only representative of typical United States control costs and are not rigorously accurate. It is also realized that because different geographical regions contain different industries, cost functions are also region-specific. For purposes of this problem formulation, however, it is assumed that the cost functions are not radically different from region to region and that the following representative functions can be universally applied without loss of meaning.

Because NO_x emissions come from a variety of both stationary and mobile sources, attainment of National Ambient Air Quality Standard (NAAQS) levels can probably be achieved without controlling all sources to the best available technology. A ranking of possible control strategies could, therefore, be determined based upon cost-effectiveness. Table 1 shows one such ranking by Tuerk. Past EPA policy concerning incremental control implementation has generally been based on such considerations (Tuerk 1976).

Establishing incremental costs for various levels of control for several source categories considered to be representative within both the mobile and stationary source framework, Tuerk combined these costs with projected emission reductions to develop cost effectiveness ratios. He then applied these ratios, coupled with projections of the air quality associated with the various control strategies for the year 2000, to provide the most cost-effective approach to each given level of control.

Tuerk considered the following categories: (1) the cost versus tons of NO_x removed for mobile sources; (2) the cost versus tons of NO_x removed for stationary sources; and (3) the cost versus tons of NO_x removed for combined mobile and stationary sources. Such a treatment by Tuerk is advantageous to this study in that by using predicted baseline emissions for the year 2000, annual cost as a function of fraction NO_x removed can be calculated. Regression of NO_x control cost in millions of dollars versus fraction NO_x removed (R_N) yields:

Table 1.--Incremental NO_x control strategies ranked by cost effectiveness

Control strategy	Incremental control	Emission Reduction (10^6 tons)	Cost Effectiveness ($/ton)
New utility boilers	25 to 50%	3.52	100
Industrial boilers	25 to 65%	2.46	150
Existing utility boilers	25 to 50%	0.62	225
Stationary internal combustion engines	25 to 75%	2.87	340
Mobile sources other than cars	1.0 to 0.8 g/mile	2.90	450
Light duty vehicles (cars)	2.0 to 1.0 g/mile	1.91	450
Utility boilers	50 to 90%	6.62	1200
Stationary internal combustion engines	75 to 90%	0.86	1700
Light duty vehicles (cars)	1.0 to 0.4 g/mile	1.15	2300

$$\text{Annual } NO_x \text{ Control Cost} = 436\ e^{5.37\ R_N}, \quad r^2 = 0.993 \qquad (8)$$

with

$$R_N = \frac{[NO_x]_0 - [NO_x]}{[NO_x]_0} \qquad (9)$$

where $[NO_x]_0$ is the initial NO_x concentration and $[NO_x]$ is the sought final NO_x value.

Tuerk performed a similar analysis with respect to the cost of removing hydrocarbon emissions based upon the most cost-effective approach to each given level of control. Table 2 shows a ranking by Tuerk of possible control strategies for hydrocarbon, based upon the most economically cost-effective control costs. Converting Tuerk's data to a form more appropriate to this study, regression of HC control costs in millions of dollars versus fraction HC removed R_H yields:

$$\text{Annual HC Control Cost} = 5.30\ e^{13.3\ R_H}, \quad r^2 = 0.981 \qquad (10)$$

with

$$R_H = \frac{[HC]_0 - [HC]}{[HC]_0} \qquad (11)$$

where $[HC]_0$ is the initial HC concentration and $[HC]$ is the sought final HC value. Equations (3) through (11) can be used to formulate the following nonlinear optimization problem (NLP).

$$\text{NLP: Minimize Cost} = 5.30\ e^{13.3\left(\frac{[HC]_0 - [HC]}{[HC]_0}\right)} +$$

$$436\ e^{5.37\left(\frac{[NO_x]_0 - [NO_x]}{[NO_x]_0}\right)} \qquad (12)$$

subject to

$$[HC] \leq [HC]_0$$

$$[NO_x] \leq [NO_x]_0$$

$$[O_3] = 0.282\ L^{0.683}[1.00 - (5.49\ D/L)^{1.97}]^{0.518} = 0.12 \text{ for } D \geq 0 \ (13)$$

$$[O_3] = 0.282\ L^{0.683}\ e^{-9.11(|D|/L)^{1.68}} = 0.12 \text{ for } D < 0 \qquad (14)$$

with

$$L = 0.984\ [HC] + 0.180\ [NO_x] ,$$

and

$$D = 0.180\ [HC] - 0.984\ [NO_x] .$$

The ozone equality constraints are based on the achievement of the ozone standard (0.12 ppm). The coefficients of $[HC]$ and $[NO_x]$ in the L and D equations are the sine and cosine of the ridge line angle $\theta = 10.3°$ from the ozone-precursor surface used to derive equations (6) and (7). Solution of this nonlinear optimization problem provides the final NO_x and HC values sought. When these values are substituted in the objective function of the NLP (minimize cost), the optimum cost of control is found.

Table 2.--Ranking of incremental hydrocarbon
control strategies

Control strategy	Incremental control	Emission reduction (10⁶ tons)	Cost effectiveness ($/ton)
Dry cleaning	0 to 80%	0.48	10
Industrial finishing	0 to 75%	1.82	10
Carbon black	0 to 95%	0.14	10
Formaldehyde	0 to 95%	0.06	10
Refining	0 to 67%	1.53	13
Architectural coatings	0 to 100%	1.00	15
Heavy-duty vehicle evaporation	5.8 to 0.5 g/mile	0.58	18
Open burning	0 to 25%	0.30	20
Ethylene oxide	0 to 95%	0.44	20
Acrylonitrile	36 to 99%	0.17	37
Ethylene dichloride	0 to 95%	0.16	40
Paint & varnish	0 to 70%	0.09	50
Degreasing	41 to 90%	0.36	85
Industrial finishing	76 to 97%	0.55	100
Gasoline handling	16 to 50%	0.57	100
Cyclohexanene	0 to 90%	0.19	120
Metal decorating	0 to 90%	0.19	120
Miscellaneous chemicals	35 to 53%	0.30	200
Gasoline distribution	67 to 99%	0.43	270
Coke ovens	0 to 80%	0.24	435
Light-duty truck exhaust	0.9 to 0.41 g/mile	0.93	470
Light-duty truck & heavy-duty vehicle exhaust	2.9 to 0.8 g/mile	0.84	470
Foundries	0 to 60%	0.15	500
Letterpress & lighography	0 to 90%	0.27	700
Gas handling	51 to 91%	0.68	700

the same, Point A´ (fig. 2), every time (although
some optimal costs vary) while the optimal
solutions for Point B (Points B_1, B_2, B_3, B_4,
and B_5) of figure 2 are different. Differences in
optimal costs are also evident for Point B.

The significance of the solutions presented
is two-fold. First, the strategy of HC control
only is not an optimal strategy throughout the
surface. Secondly, the strategy is insensitive
to control costs at [HC] to [NO_x] ratios less
than about nine to one. For atmospheric concentrations with ratios less than this value, all optimal strategies
involve HC control only. This result can be
explained from figure 1. In these regions,
control of NO_x requires an additional amount of
HC control because of ozone isopleth curvature.
NO_x control is effective at reducing ozone and
required HC control only for ozone-precursor
surface locations of high [HC] to [NO_x] values.

CONCLUSIONS

A mathematical model for ozone allows
deterministic solution of ozone control problems.
Because net ozone production dependence on
precursors can be expressed by a simple mathematical model and ozone precursor control costs
can be approximated by cost functions, nonlinear
optimization techniques allow meaningful sensitivity testing. Presented in this paper is one
sensitivity test. It indicates that while HC
control alone is cost effective for most conditions, there are potential precursor conditions
for which a combined HC-NO_x control program is
optimal. These results are found from study of
two initial concentration points that are representative of real ozone-precursor conditions.

The formulation of the NLP and computer programs which solve it to determine the minimum
cost for a control strategy decision allows different types of testing. By varying each individual cost parameter in equations (8) and (10),
the sensitivity of the optimal solution to these
parameters can be studied. By selecting different points on the ozone-precursor surface
(fig. 2), the efficacy of a given surface-wide
control strategy can be analyzed. Furthermore,
by knowing how the parameters of the ozone model
[eqs. (1) and (2)] vary with initial condition
and chemistry, the effects of chemistry changes
on a given control strategy decision can be
assessed. Solutions to the NLP thus provide
meaningful sensitivity testing data.

Summarized in table 3 are data produced
from a sensitivity test using two ozone-precursor
surface points. In this test, each parameter in
the cost functions of equations (8) and (10) was
individually reduced by 10%. Results show that
the optimal solutions for Point A (fig. 2) are

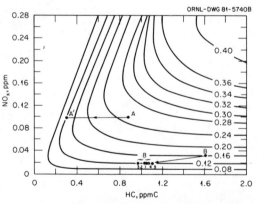

Figure 2.--Standard EKMA ozone-precursor surface
with pre-control points A and B and optimal
post-control points A´ and B, through B5.

Table 3.--Optimal ozone control strategy data for two
points on the ozone-precursor surface

Cost Function	Point A[1] Optimal [HC] (ppm C)	Point [NOx] (ppm)	Optimal R_H[3]	Control R_N[4]	Optimal Cost (10^6 \$)	Point B[2] Optimal [HC] (ppm C)	Point [NOx] (ppm)	Optimal R_H[3]	Control R_N[4]	Optimal Cost (10^6 \$)
1[5]	0.353	0.099	0.606	0.000	1.71×10^4	1.05	0.0189	0.344	0.427	4.83×10^3
2[6]	0.353	0.099	0.606	0.000	1.55×10^4	1.04	0.0190	0.350	0.424	4.77×10^3
3[7]	0.353	0.099	0.606	0.000	8.04×10^3	0.972	0.0194	0.393	0.412	4.54×10^3
4[8]	0.353	0.099	0.606	0.000	1.71×10^4	1.07	0.0188	0.331	0.430	4.39×10^3
5[9]	0.353	0.099	0.606	0.000	1.71×10^4	1.10	0.0186	0.313	0.436	3.90×10^3

[1] Initial hydrocarbon concentration $[HC]_o$ = 0.896 ppm C, initial oxides of nitrogen concentration $[NO_x]_o$ = 0.099 ppm.

[2] Initial hydrocarbon concentration $[HC]_o$ = 1.600 ppm C, initial oxides of nitrogen concentration $[NO_x]$ = 0.033 ppm.

[3] Fraction hydrocarbon removed $R_H = ([HC]_o - [HC])/[HC]_o$.

[4] Fraction oxides of nitrogen removed $R_N = ([NO_x]_o - [NO_x])/[NO_x]_o$.

[5] Objective cost function = $5.30\ e^{13.3\ R_H} + 436\ e^{5.37\ R_N}$.

[6] Objective cost function = $4.77\ e^{13.3\ R_H} + 436\ e^{5.37\ R_N}$.

[7] Objective cost function = $5.30\ e^{12.0\ R_H} + 436\ e^{5.37\ R_N}$.

[8] Objective cost function = $5.30\ e^{13.3\ R_H} + 392\ e^{4.37\ R_N}$.

[9] Objective cost function = $5.30\ e^{13.3\ R_H} + 436\ e^{4.83\ R_N}$.

What these two points do not show, however, is the frequency with which such conditions occur for a given geographic region. Before control strategy decisions can be finalized, an analysis must be made to determine such frequencies so that control strategies are governed not only by optimal solutions to individual points but by a frequency-dependent solution for a geographic region.

LITERATURE CITED

Bilger, R. W. 1978. Optimum Control Strategy for Photochemical Oxidants, Environmental Science and Technology 12(8): 937-40

Dodge, M. C. 1977. Combined Use of Modelling Techniques and Smog Chamber Data to Derive Ozone-Precursor Relationships, International Conference on Photochemical Oxidant Pollution and Its Control, Proceedings: Volume II, EPA-600/3-77-pp1b, U. S. Environmental Protection Agency, Research Triangle Park, N. C.

Durbin, P. A., T. A. Hecht, and G. Z. Whitten. 1975. Mathematical Modeling of Simulated Photochemical Smog, EPA-650/4-75-026, U. S. Environmental Protection Agency, Research Triangle Park, N. C.

Falls, A. H., and J. H. Seinfeld. 1978. Continued Development of a Kinetic Mechanism for Photochemical Smog, Environmental Science and Technology 12(13): 1398-406 (December 1978).

Holton, G. A., and H. E. Jeffries. 1979. Mathematical Analysis of Ozone Isopleth Surfaces, Proceedings Second Ozone Specialty Conference of the Air Pollution Control Association, Houston, Tex.

Jeffries, H. E., and K. G. Sexton. 1980. Comparison of Nitrogen Transformation Processes in Various Photochemical Kinetics Models. Department of Environmental Sciences and Engineering. School of Public Health, University of North Carolina at Chapel Hill.

Tuerk, E. 1976. Air Quality, Noise, and Health, Interagency Task Force on Motor Vehicle Goals Beyond 1980, TST-8, U. S. Department of Transportation, Washington, D. C.

Turner, D. B. 1979. Atmospheric Dispersion Modeling, A Critical Review, Journal of the Air Pollution Control Association 29(5):502-19.

U. S. Environmental Protection Agency. 1978a. Air Quality Criteria for Ozone and Other Photochemical Oxidants, Volume I, EPA-600/8-78-004, Washington, D. C.

U. S. Environmental Protection Agency. 1979a. Data Collection for 1982 Ozone Implementation Plan Submittals. Federal Register 44(21): 65667-70.

U. S. Environmental Protection Agency. 1979b. National Primary and Secondary Ambient Air Quality Standards. Federal Register 44(28): 8202-37

U. S. Environmental Protection Agency. 1979b. Revisions to the Ambient Air Quality Standards for Photochemical Oxidants.

U. S. Environmental Protection Agency. 1978b. User's Manual for Kinetics Model and Ozone Isopleth Plotting Package, EPA-600/8-78-014a.

U.S. Environmental Protection Agency. 1977. Uses, Limitations and Technical Basis of Procedures for Quantifying Relationships Between Photochemical Oxidants and Precursors, EPA-450/2-77-21a.

Whitten, G. Z., and Hogo, H. 1978. User's Manual for Ozone Isopleth Plotting with Optional Mechanisms, EF 78-30, Contract No. 68-02-2426, Systems Applications, Incorporated, San Rafael, Cal.

Whitten, G. Z., H. Hogo, M. J. Meldgin, J. P. Killus, and P. J. Bekowies. 1979. Modeling of Simulated Photochemical Smog with Kinetic Mechanisms, Volume 1, EPA-600-3-79-001a, Systems Applications, Incorporated, San Rafael, Cal.

MANIPULATION AND DISPLAY OF AIR QUALITY DATA USING IMAGE PROCESSING TECHNIQUES[1]

Kevin J. Hussey[2] and Richard J. Blackwell[3]

Abstract.--Output from urban airshed models can be effectively displayed and analyzed by utilizing digital image processing techniques. This was demonstrated by the production of a time lapse digital movie which shows the changes in reactive hydrocarbon emissions and ozone concentrations over the California South Coast Air Basin during a 24 hour period.

INTRODUCTION

Improved air pollution dispersion models in the air quality planning process were called for by the Clean Air Amendments of 1977 (Public Law 95-95). A research program at the Environmental Quality Laboratory of the California Institute of Technology has produced an advanced model for predicting air pollution concentrations within an urban airshed. Simu-

lated output of this model in conjunction with the VICAR/IBIS image processing system[4] have been used to show how modeled air pollution data can be numerically and graphically assembled to clearly demonstrate the spatial and temporal nature of air pollution. This was accomplished through the production of a synthetic time lapse digital movie which shows the changes in reactive hydrocarbons emissions and ozone concentrations of the California South Coast Air Basin during a 24 hour period.

The purpose of this paper is to describe the image processing techniques used to produce the time lapse sequences of reactive hydrocarbon emissions and ozone concentrations seen in the film "Pollution and Pixels"[5] and to outline further applications of these techniques. The Image Processing Laboratory (IPL) of the Jet Propulsion Laboratory (JPL) became involved with the display and manipulation of air quality data by the desire of the California Institute of Technology's (Caltech's) Environmental Quality Laboratory (EQL) to more clearly represent the dynamics of air pollution in the South Coast Air Basin (SCAB) of California.

Overview of the Caltech Urban Scale Airshed Model

The model used in data generation for the movie was developed at the California Institute of Technology under the sponsorship of the California Air Resources Board, Research Division. The principal research objective was "...to develop an improved modeling capability to assess the effectiveness of air pollution control measures in reducing photochemical smog" (McRae et al. 1979). A simplified view of the factors and processes incorporated into the Caltech model are shown in figure 1. Source emissions, meteorology and chemical conversion processes are coupled together as part of the mathematical model which, in essence, describes the formation and transport of chemically-reacting species in the turbulent planetary boundary layer" (McRae et al. 1979). A detailed discussion of this model along

---

[1]Paper presented at the international symposium Energy and Ecological Modelling, sponsored by the International Society for Ecological Modelling. (Louisville, Kentucky, April 20-23, 1981). This paper presents one phase of research conducted at the Jet Propulsion Laboratory, California Institute of Technology under NAS7-100, sponsored by the National Aeronautics and Space Adminstration.

[2]Kevin J. Hussey is an Engineer and Geographer at the Jet Propulsion Laboratory's Image Processing Laboratory, Pasadena, California, U.S.A.

[3]Richard J. Blackwell is the Group Supervisor of the Earth Resources Applications Group at the Jet Propulsion Laboratory's Image Processing Laboratory, Pasadena, California, U.S.A.

[4]The image processing techniques used throughout this study are part of the VICAR/IBIS System, developed at the Image Processing Laboratory at the Jet Propulsion Laboratory. VICAR (Video Image Communication and Retrieval) is a set of computer programs designed to facilitate the acquisition, processing, and recording of digital image data. IBIS (Image Based Information System) is a computer based approach for the analysis of geographical situations.

[5]The film "Pollution and Pixels" was produced at the Jet Propulsion Laboratory (JPL) by a joint effort between the Image Processing Laboratory at JPL and the Enviornmental Quality Laboratory (EQL) of the California Institute of Technology. It was written and directed by Kevin J. Hussey of JPL and Gregory J. McRae of EQL. The film was shown at the 1981 ISEM Symposium in Louisville, Kentucky as a portion of the presentation, "Manipulation and Display of Air Quality Data Using Image Processing Techniques" by Kevin J. Hussey and Richard J. Blackwell.

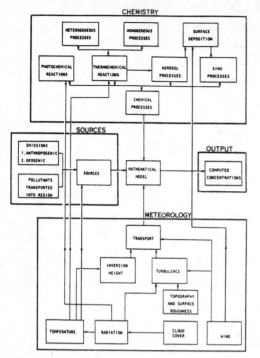

Source: Greg McRae - CALTECH

Figure 1.--Simplified view of the factors involved
in relating emissions to atmospheric air quality.

with validation, and specific applications are
presented by McRae (1979) and Goodin et al.
(1976).

The Study Area

The South Coast Air Basin is commonly the
focus of air pollution studies due to the severe
nature of the problem. It is a natural basin
bounded on three sides by elevated regions, suffers frequent temperature inversions, light winds,
and numerous sources of pollution, both mobile and
stationary. The basin covers an area of approximately 8,600 square miles (14,000 square kilometers) and had an estimated 1973 population of
9,899,000 (Horie et al. 1979).

During the summer and early fall the dynamically induced anticyclone of the North Pacific
is centered off the coast of California. The subsiding warm air associated with the eastern side
of the high "puts a lid" on the cooler marine air
found near the surface. Pollutants which are
emitted into the cooler marine air rise to the base
of the inversion and collect. The lack of vertical
mixing between the cooler and warmer air concen-
trates the pollutants between the ground and the
inversion layer, greatly increasing the occurrence
of unhealthy air. Changes in the elevation of the

inversion are in response to varying meteorological
conditions (Hagevik 1970). Meteorology and the
distribution of emissions are factors which control
areal differences in air pollution at any specific
time. The effects of meteorology on a group of
primary pollutants like reactive hydrocarbons become clearly evident when examining the resultant
distribution of ozone concentration levels.

DATA PREPARATION

This portion of the paper describes the history of the data before they were received by IPL.
Ozone was chosen as the pollutant to study because
it is the principal component of modern smog and
has a pronounced diurnal variation in both area
and intensity (U.S. EPA 1979). Reactive hydrocarbon emissions were selected because they are emitted by mobile and stationary sources and are photochemically related to ozone. The particular day
chosen for the demonstration was June 26, 1974, a
day of very high pollutant levels.

Data Acquisition and Pre-Processing

Ozone Concentration Data

The ozone concentration data used in the
movie were provided by EQL. They acquired hourly
average ozone concentrations, taken at approximately
60 monitoring stations throughout the SCAB, from
the California Air Resources Board (CARB). The
station measured data were also used by EQL to
calibrate their model and check its results.

The requirements of the EQL model necessitated
the development of continuous ozone concentration
fields from the point measured data. A number of
interpolation methods were considered by the EQL
staff which finally chose a method based on a
second-degree polynomial fit with r^{-2} weighting
(r is the distance from a grid point to a measuring
station (Goodin et al. 1979). Ozone concentration
were interpolated into the five kilometer Universal
Transverse Mercator grid indicated in figure 2.
These gridded concentration fields produced from
the interpolation process are in the exact same
format as the model's output. This provided a
standardized input matrix of 30 rows by 80 columns
and established a preliminary geographic base
which aided the subsequent image processing. The
gridded data were then made available on a magnetic tape containing 24 files, one file for each
hour of the day. These files provided ozone concentration by 25 square kilometer grid cells over
the SCAB for June 26, 1974.

Reactive Hydrocarbon Emissions Data

The reactive hydrocarbon emissions data was
also provided by EQL in model output format (30 x
80 matrix of real numbers). EQL compiled the
gridded emissions data from various sources of information. The California Transportation Department, California Air Resources Board, and the U.S.

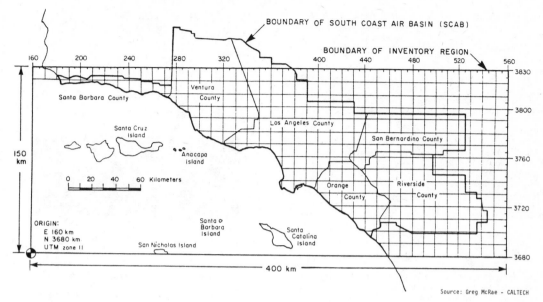

Figure 2.--Boundary of the South Coast Air Basin of Southern
California together with the U.T.M. grid which received
the interpolated data.

Environmental Protection Agency are all contributors
of the emissions data needed by EQL as input to
their model. JPL received the emissions data to
produce the time sequence on a magnetic tape containing 24 files, one for each hour of June 26,
1974. These files contained the hourly emissions
from the 25 square kilometer grid cells indicated
in figure 2. With the ozone concentration and
reactive hydrocarbon emissions data available for
use, the initial image processing could proceed.

DIGITAL IMAGE PROCESSING

An understanding of what is meant by a
"Digital Image" as well as "Image Processing" is
essential for full understanding of the techniques
used in this paper. Photographs, paintings and
maps are examples of images. A digital image is
simply a numerical representation of a scene. A
digital image could be produced from a black and
white photograph by overlaying an imaginary rectilinear grid and assigning each grid cell a number which would indicate the absolute brightness
of that area on the picture. If 256 shades of
gray were to be used, a brightness value of 0
would represent black and a value of 255 would be
white (fig. 3). The resultant matrix of brightness values (commonly refered to as picture elements
or pixels) is a digital image. A digital image can
be stored in any available data format and, with
the use of a display device, can be viewed at future times. If a cathode ray tube is chosen as the
display device, each pixel's brightness value is
used to determine the gray level of the corresponding point on the display screen

Figure 3.--Grey scale produced by assigning
visual brightness to digital brightness
values.

(Castleman 1979). The amount of detail retained
from the original photograph is mainly a function
of the rectilinear grid density used to assign the
initial brightness values. The term "image processing," as used in this paper, is the computer manipulation of digital images to achieve a desired outcome, which in the case is an improved method of
handling and displaying large volumes of air quality
data.

The image processing techniques utilized in the
production of the film were essentially the same for
the ozone and reactive hydrocarbon data. Therefore,
the procedure described below was performed on both
data types. Differences in processing the two data
will be pointed out where they occur. For demonstration purposes primarily ozone data will be used to
illustrate the various operations performed.

Image Transformation

From the definitions given above, the hourly

ozone concentration and reactive hydrocarbon data
were received in a series of digital images, 30
rows by 80 columns in size. Values in each grid
cell are numerical representations of the pollutant level at a particular place in space and time.
The raw ozone data are floating point numbers with
a range from 0.04 to approximately 0.33 parts per
million (ppm). The raw reactive hydrocarbon data
are floating point numbers ranging from zero to
approximately 1500 kilograms per hour (kg/hr).

These raw images needed to be transformed
into another format to begin further processing.
At JPL when dealing with earth resource data, it
is common to use 256 gray levels, zero being black
and 255 being white. This allows each datum to be
stored in one 8 bit byte, greatly reducing the
amount of storage space and computation time needed
to process digital images. The raw floating point
data were linearly transformed to fit within the 0
to 255 gray scale. This gave an original ozone concentration value of 0.33 ppm a gray shade integer
of 255 or white, and assigned an original reactive
hydrocarbon emissions value of 1500 kg/hr the integer 255. Figure 4 is the 1500 hours ozone concentration image before any significant image processing has been applied, in which the brightness
is a measure of the ozone concentration in the grid
cells. The transformation of the floating point
data into byte integers reduces the amount of information storage and processing required by a
factor of three.

Figure 4.--The 1500 hours ozone concentration
image before any significant image processing has been applied.

Image Expansion

The 30 by 80 image (fig. 4) was too small
for effective visualization on an electronic
screen. To correct this, each picture element
(pixel) was expanded by a factor of five to maximize hardware display capabilities used in the
final movie making process. Figure 5 demonstrates
the expansion process. Each pixel in the original
image becomes 50 pixels, with the same brightness
value, in the expanded version. Two adjacent
pixels become two adjacent groups of 50 pixels
each. This enlarges the size of the original
image to 150 lines by 500 columns. Figure 6 shows

50 PIXELS

7	7	7	7	7	2	2	2	2	2
7	7	7	7	7	2	2	2	2	2
7	7	7	7	7	2	2	2	2	2
7	7	7	7	7	2	2	2	2	2
7	7	7	7	7	2	2	2	2	2

BECOME

TWO PIXELS

7	2

Figure 5.--An example of the 5x image expansion process. Two adjacent pixels become two adjacent
groups of 25 pixels each.

BEFORE EXPANSION

AFTER EXPANSION

Figure 6.--The 1500 hours ozone concentration
image before and after 5x expansion.

the 1500 hours image before and after the expansion
process. Recall that on an ozone image black represents a concentration of less than 0.04 ppm and
white corresponds with a concentration of approximately 0.33 ppm. Intermediate shades of gray
represent intermediate concentrations of ozone.
In order to more closely represent the spatial
variation over the entire basin, an interpolation
(filtering) process was employed.

Image Filtering

Since abrupt ozone concentration changes at
grid cell boundaries (as depicted in the bottom
image of fig. 6) do not exist in reality, a
digital filter was applied to the expanded image
to smooth the transitions between cells. The two
dimensional averaging filter used is a linear image
enhancement technique which reassigns a picture
element as a function of a specified number of
surrounding pixels. The degree of filtering is
controlled by the number of surrounding pixels
that are included in the filtering process.
Equation 1 defines the specific filter used.

Let r be the radius of the filter from the
center, in pixels, not including the center pixel.
If (X_n, Y_m) is the center or filtered pixel then
its value can be determined by:

$$(X_n, Y_m) = \frac{\sum_{k=n-r}^{n+r} \sum_{i=m-r}^{m+r} (X_k, Y_i)}{(2r + 1)^2} \quad (1)$$

This operation is repeated for each pixel in the
image (Selzer 1968). Figure 7 shows the expanded

BEFORE FILTERING

AFTER FILTERING

Figure 7.--Expanded 1500 hours ozone image before and after a two dimensional averaging filter was applied.

image before and after a two dimensional averaging filters was applied. The filter radius chosen for this example was two pixels. This makes each pixel in the filtered image the average of its original value and the other 24 pixels that surround it in the unfiltered image (fig. 8). By using an expansion factor of five and a filter radius of two, the center pixel in each grid cell will retain its original concentration value. This preserves the dynamic range of the original data while still smoothing the cell to cell transition. Vertical cross-sections through the 1500 hours image before and after filtering show the linear nature of the filtering process. Figure 9 contains such cross-sections along the same line from west to east across the SCAB. In figure 9 the stair-stepped line, symbolized with the square, would be a profile along A-A', while the other line, symbolized with the octagon, is a cross-section along B-B'. Note that the profile of the filtered image (B-B') demonstrates the data retained its full range, even after interpolating from grid cell to grid cell along each row and column. The average gray level, standard deviation, beginning and ending coordinates for each profile have also been provided. The filter described in this section is just one of the many types of filters which are used in image processing (Castleman 1979). Image expansion and filtering were the techniques used to smooth cell to cell transitions, and synthetically increase the effective resolution of the ozone concentration and reactive hydrocarbon emission data from 25 km^2 to 1 km^2. The range of the data was not altered by these processes. However, before any meaningful geographic data comparisons can be made, a common geographic base must be established.

IMAGE Z BEFORE FILTERING

0	0	0	0	0	0	0	0	0	0	0
0	0	0	0	0	0	0	0	0	0	0
0	0	0	0	0	0	0	0	0	0	0
0	0	0	100	100	100	100	100	0	0	0
0	0	0	100	100	100	100	100	0	0	0
0	0	0	100	100	100	100	100	0	0	0
0	0	0	100	100	100	100	100	0	0	0
0	0	0	100	100	100	100	100	0	0	0
0	0	0	0	0	0	0	0	0	0	0
0	0	0	0	0	0	0	0	0	0	0
0	0	0	0	0	0	0	0	0	0	0

IMAGE Z AFTER FILTERING

0	0	0	0	0	0	0	0	0	0	0
0	4	8	12	16	20	16	12	8	4	0
0	8	16	24	32	40	32	24	16	8	0
0	12	24	36	48	60	48	36	24	12	0
0	16	32	48	64	80	64	48	32	16	0
0	20	40	60	80	100	80	60	40	20	0
0	16	32	48	64	80	64	48	32	16	0
0	12	24	36	48	60	48	36	24	12	0
0	8	16	24	32	40	32	24	16	8	0
0	4	8	12	16	20	16	12	8	4	0
0	0	0	0	0	0	0	0	0	0	0

Figure 8.--This figure shows a hypothetical image (Z) before and after the application of a two-dimensional averaging filter with a radius of two pixels. Note the value of the shaded pixel in the unfiltered Image Z. A box has been scribed around all of the pixels that fall within a radius of two from the shaded value. The average value of all pixels within the box is thirty-six (900 ÷ 25)= 36. This average value was then assigned to the corresponding (shaded) pixel in filtered Image Z. Each value in filtered Image Z was acquired by this averaging process.

Image to Geo-Reference Base Registration

Accurate locational reference for each pixel in all the images can be determined once they are registered to a standard map projection. Since the data were supplied within a Universal Transverse Mercator (UTM) grid, a UTM projection was used in this study for convenience. The capability exists to employ other projections.

Portions of the UTM map shown in figure 2 were digitized and registered to the model output grid with an affine transformation. This creates

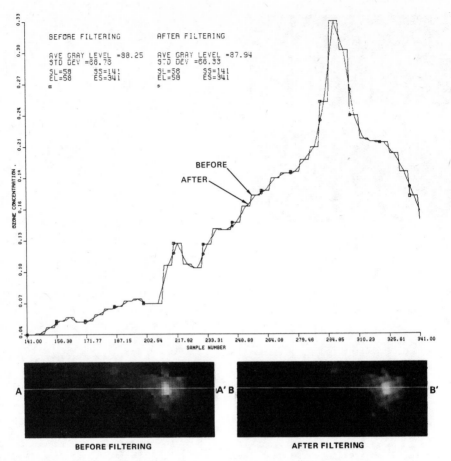

Figure 9.--Vertical cross-section through the 1500 hours
ozone concentration image. The line labeled 'Before'
and symbolized by a square in the above graph, corresponds to the line A-A' in the bottom image. The
'After' line, symbolized on the graph by an octagon,
corresponds to line B-B' in the image at the bottom.

a geo-reference image which can be combined with
the data to provide the needed geographic base.
A resulting composite image (fig. 10) depicts

Figure 10.--This image depicts the 1500 hours
reactive hydrocarbon emissions data complete
with a geo-reference base (the contour effect
was performed to visually enhance this
figure only).

the 1500 hours reactive hydrocarbon data complete
with the geo-reference base. The procedure used
to create figure 10 was then repeated 23 times
using each of the remaining data files.

Figure 11 is an example of a conventional
display technique which is used to show the spatial
variation of nitrogen oxides emissions. The rec-
tangular grid seen at the bottom of the figure delineates the 30 x 80 grid cell pattern used to
represent the SCAB. The magnitude of the nitrogen
oxides concentrations for each cell are depicted
in the perspective view as a series of lines or
vectors, the height being proportional to the
magnitude. Above the grid data is an outline map
of the geographic region included by the SCAB.

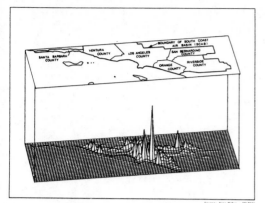

Figure 11.--Diurnal variation of nitrogen oxides emissions displayed by an X-Y plotter.

The line plot display method seen in figure 11, is suitable for cursory visual analysis of a few matrices. However, due to the nature of this type of display and the elevation angle selected for perspective, it is difficult to see relationships beyond or behind the high intensity spikes rising along the coast line. Geographic reference is also difficult to discern. Conversion of airshed model data to a digital image format eliminates these problems and has other advantages which prove useful to the scientist or interpreter. An example of increased data utilization is demonstrated by the creation of images which depict ozone concentrations at any specific time of day. This was done to portray the temporally continuous nature of pollutant emissions and concentrations. Synthetic images were created by interpolating between each hour to produce a field every 12 minutes. This greatly enhances the visual appearance of the time lapse movie without sacrificing accuracy.

Production of Synthetic Between-Hour Frames

Each of the frames that were synthesized are linear functions of two "on the hour" frames. The change in concentration between any two hours can be found by subtracting the earlier image from the later and placing the differences into another matrix. This will produce an image of the differences between the two hours. If we multiply the difference image by a constant, for example .2, we will have created a 20% of the difference image. By adding the 20% difference image to the earlier of the hourly images, a 12 minute past the hour frame has been produced. This type of synthetic frame was produced for 12, 24, 36 and 48 minutes after each hour. This process is a simple example of linear interpolation as shown by equations 2 thru 6.

Let $(I_1, I_2, \ldots I_{24})$ represent the hourly emission of concentration images with I_1 being 0100 hours, I_2 being 0200 hours, and so forth.

A general explanation of any 12 minute interval, between-hour sequence, is given. ΔI_n is the difference image of I_n and I_{n+1}.

$$\Delta I_n = I_{n+1} - I_n \qquad (2)$$

where

$$n = \text{hour of image}$$

$I_{n(12)}$ represents the 12 minute after hour n image.

$$I_{n\ (12)} = (\Delta I_n \times .2) + I_n \qquad (3)$$

Similarly, the remaining three between-hour frames would be,

$$I_{n\ (24)} = (\Delta I_n \times .4) + I_n \qquad (4)$$

$$I_{n\ (36)} = (\Delta I_n \times .6) + I_n \qquad (5)$$

$$I_{n\ (48)} = (\Delta I_n \times .8) + I_n \qquad (6)$$

The above process was repeated for all hourly frames resulting in 120 black and white images of reactive hdyrocarbon emission rates and ozone concentrations at 12 minute intervals for June 26, 1974. The next step was to visually enhance the images so that pollutant changes from frame to frame would be easily detected.

Further Image Enhancement

Image enhancement techniques are numerous and vary considerably in form and cost. In producing the time lapse movie only a few of the available enhancement techniques were utilized. For further reading on image enhancement see Gonzalez (1977), Blackwell et al. (1975). Blackwell et al. (1979) and Castleman (1979).

Due to the dynamic nature of the movie format, the need to rapidly differentiate between changing pollutant levels was the primary criteria for choosing a pseudo-color enhancement technique. The psuedo color process transforms a monochromatic image into three primary color images which when combined, in either a photographic process or on a color television, produce a color image. This was accomplished by classifying or grouping the pollution levels into 12 classes and then assigning each class three brightness values, one each in the red, green and blue portions of the spectrum. The classification used in this case was devised solely for visual interpretation enhancement. The colors were chosen with the idea that the cool hues (blues) indicate low pollutant levels and the warmer tones (red) express high levels of pollution. Figure 12 is a classified psuedo-color image of the 1500 hours ozone concentration data. The key provided at the bottom of the figure numerically relates the colors to ozone concentration levels.

The main drawback in utilizing psuedo color in digital imagery is the increased cost of display. The higher cost of color display is offset by the

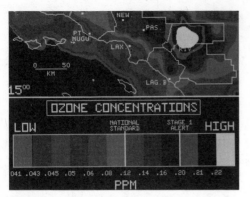

Figure 12.--This classified pseudo color image
depicts the 1500 hours ozone concentration
data with a key to the color classification.

greatly improved appearance and ease in interpretation. Once the pseudo color images were prepared the final step in making the time lapse
sequence was to transform the images from the
numerical realm into the world of visual moving
pitures. This was done by filming the images
directly off a color television monitor with a
16 mm movie camera. Twenty-four frames of the
time lapse sequence are shown in figure 13.
Each frame is a pair of images. The top image
in the pair is the reactive hydrocarbon emissions
while the bottom image depicts the measured ozone
concentrations for the indicated hour. The nu-
merals in the lower left corner of each sub-image
in figure 13, indicate the starting hour of the
average for that image.

The photochemical nature of ozone production
can be witnessed in figure 13. Note the rapid
buildup of ozone during the daylight hours and
the dissipation of the same as the sun sets. It
is also interesting to observe the eastward migration and enlargement of the highest ozone concentrations. This migration and enlargement can be
attributed to a predominant westerly wind which
carries pollutants, originating in Los Angeles,
up against the San Gabriel and San Bernardino
mountains where they photochemically react to
produce large quantities of ozone. A detailed
study of the varying spatial and temporal distributions shown in this figure yields a wealth
of cultural and environmental phenomena.

APPLICATIONS

The production of this synthetic time
lapse movie is the first step in an envisioned
process of multi-source data integration. We
believe that the unique and powerful capability
offered by digital image processing can bring new
insights into the study of photochemical air
pollution and other environmentally related problems. The ability to organize, assemble and
process massive amounts of in-situ, model,
instrument and observational data exists

(Bryant 1977). Figure 14 illustrates this
ability to integrate digital data from various
sources and provides an example of how model
data could be coupled with census tract locations and population statistics to provide
estimates of exposure levels to air pollution
for each of the approximately 1900 census
tracts in Los Angeles County.

Still another example, illustrated in figure
14, is the incorporation of automobile vehicle
registration data from the Department of Motor
Vehicles. Through a process of vehicle owner address matching to the census data, auto counts and
locations can be related to traffic zone maps and
then to the air quality model data for use in
evaluating control strategies.

The potential applications of this technology
to air quality problems appear to be very promising
as demonstrated by the examples. In addition to
the two possible examples cited above, the capability also exists to routinely acquire additional
data for use in air quality studies. These data
could be derived from processed NASA Landsat Earth
Resources Satellite data and would include the
development of land cover maps which describe land
cover types such as: forest, desert, water, and
basic land classifications of residential, commer-
cial/industrial, open space and agricultural. Additional information to improve estimates of surface
roughness and topographic expression can be taken
from digital terrain data, these estimates could
be computed for each SCAB grid cell or for each
census tract.

The integration of other available data sources
with this time lapse information is currently being
pursued within the Earth Resources Applications
Group at the Jet Propulsion Laboratory. Further
application of image processing algorithms to these
data can result in the ability to examine, historic,
current and future population exposure to various
pollutants over time.

CONCLUSION

It has been stated by the principle author of
the Caltech Urban Scale Airshed Model that "having
a better model is but one component of the task to
solve air pollution, ... we need a means of organizing the information in a region so complex as Los
Angeles..."[6]. Image processing systems, like the
one developed at JPL (VICAR/IBIS), can provide the
needed data handling capabilities required for many
types of applications. Some attributes of image
processing systems which facilitate the handling of
enviornmental data are: 1) digital format, 2) geographically referenced - map projections and coor-
dinate systems can be changed, 3) allows integration
and comparison of other data types such as census

---

[6]Gregory J. McRae, Quoted in the L.A. Times
Article by Sandra Glakeslee, "Computer Graphs 'Figure Out Smog' in Los Angeles". L.A. Times, 22 July
1979, Part 2, p. 1.

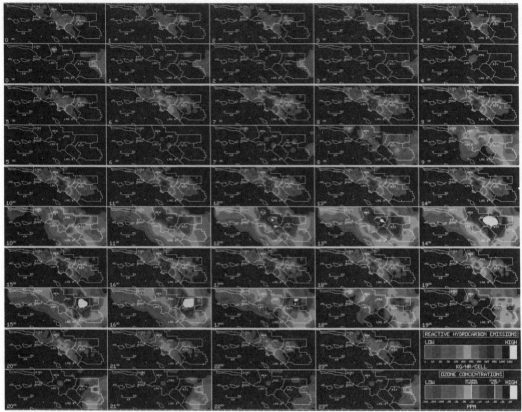

Figure 13.--The 24 "on the hours" frames from the film
"Pollution and Pixels". Note the change in pollutants
over time and space.

tracts, traffic zones, land use information and
terrain data, 4) four dimensional quality (x, y,
z and time), 5) cost and time efficient in comparison to conventional methods, 6) use of the final
digital products require no special computing resources.

While producing the movie that provided the
impetus for this paper, a majority of the applications development work was completed. This fact
greatly simplifies future system utilization from
a user's standpoint. With these features in mind,
the need for digital image processing and data base
management techniques in conjunction with air pollution models will increase greatly as modeled data
is applied to the planning process.

This paper has shown how digital image processing techniques can be used on Urban Scale Air
Pollution Model output to increase the utility of
modeled data. It is believed that the methodology
and technology described here can be used to great
advantage in gaining new insights into understanding and controlling air quality.

SELECTED BIBLIOGRAPHY

Blackwell, R.J. and W.J. Crisci. 1975. Image
Processing Technology and Its Application in
Forensic Sciences, Journal of Forensic Sciences.
20:283-304.

Blackwell, R.J. and D.H.P. Boland. 1979. Trophic
Classification of Selected Colorado Lakes,
JPL Technical Publication 78-100, EPA Publica-
tion 600/4-79-005.

Bryant, N.A. and A.L. Zobrist. 1977. IBIS: A
Geographic Information System Based on Digital
Processing and Image Raster Data Type. Geo-
science Electronics. 15:152-159.

Castleman, K.R. 1979 Digital Image Processing.
Prentice-Hall. Englewood Cliffs, N.J.

Goodin, W.R., G.J. McRae and J.H. Seinfeld. 1976.
Validity and Accuracy of Atmospheric Air
Quality Models, p. 366-373. In Proceedings
Third Symposium on Atmospheric Turbulence,
Diffusion and Air Quality. (Raleigh, N.C.,
October-19-22, 1976). American Meteorological
Society.

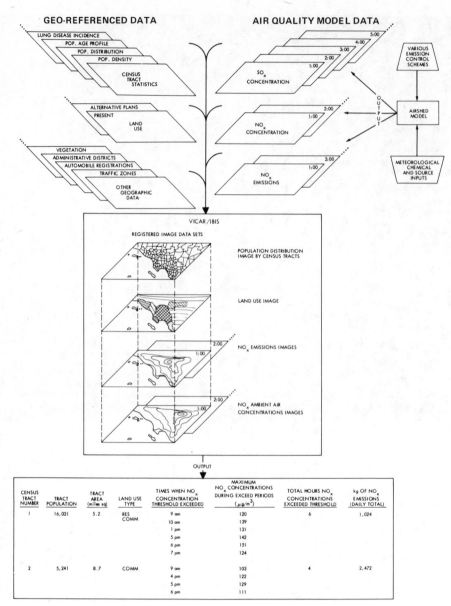

CENSUS TRACT NUMBER	TRACT POPULATION	TRACT AREA (miles sq)	LAND USE TYPE	TIMES WHEN NO$_x$ CONCENTRATION THRESHOLD EXCEEDED	MAXIMUM NO$_x$ CONCENTRATIONS DURING EXCEED PERIODS ($\mu g/m^3$)	TOTAL HOURS NO$_x$ CONCENTRATIONS EXCEEDED THRESHOLD	kg OF NO$_x$ EMISSIONS (DAILY TOTAL)
1	16,031	5.2	RES COMM	9 am	120	6	1,024
				10 am	139		
				1 pm	131		
				5 pm	142		
				6 pm	151		
				7 pm	124		
2	5,241	8.7	COMM	9 am	103	4	2,472
				4 pm	122		
				5 pm	129		
				6 pm	111		

Figure 14.--Air Quality and Geo-based Data Integration.
This figure illustrates the multisource data integration
potential of digital image processing techniques. Relationships between population density, land use, NO$_x$
emissions and NO$_x$ ambient air concentrations are analyzed
as user specified cross-tabulations and statistics.

Goodin, W.R., G.J. McRae and J.H. Seinfeld. 1979.
A Comparison of Interpolation Methods for
Sparse Data: Application to Wind and Concentration Fields. Journal of Applied Meteorology. 18:761-771.

Hagevik, G.H. Decision Making in Air Pollution
Control. 1970. Praeger Publishers. New
York, N. Y.

Horie, Y., A.S. Chaplin, N.H. Frank and W.F. Hunt, Jr. 1977. Population Exposure to Oxidants and Nitrogen Dioxide in Los Angeles. Vol. 1: Executive Summary, Publication EPA-450/3-77-004a. Environmental Protection Agency. Triangle Park, N.C.

McRae, G.J., W.R. Goodin and J.H. Seinfeld. 1979. Development of a Second-Generation Airshed Model for Photochemical Air Pollution. Fourth Symposium on Turbulence, Diffusion, and Air Pollution. Reno, Nev. (January15-18, 1979).

Selzer, R.H. The Use of Computers to Improve Biomedical Image Quality. p. 817-834. In Proceedings of the Fall Joint Computer Conference 33 1968.

U.S. Environmental Protection Agency. 1979. Air Pollution and Your Health. Publication OPS 54/8 Washington, D. C. Environmental Protection Agency, 1979.

IMPACTS OF ACID PRECIPITATION ON WATERSHED ECOSYSTEMS:
AN APPLICATION OF THE ADAPTIVE ENVIRONMENTAL ASSESSMENT PROCESS[1]

Austin K. Andrews[2], Gregor T. Auble[2], Richard A. Ellison[2], David B. Hamilton[2], James E. Roelle[2], David R. Marmorek[3], and Orie L. Loucks[4]

Abstract.--A structured workshop modelling process, Adaptive Environmental Assessment, is being used to facilitate interdisciplinary and interagency interaction in synthesizing existing knowledge and identifying research needs on the mechanisms by which acid precipitation impacts freshwater aquatic resources. The focus of this effort is on construction and iterative refinement of an ecosystem simulation model. The current model is described and evaluated in terms of performance, limitations, utility for identifying research needs, and potential for further refinement and application.

INTRODUCTION

This paper describes the ongoing application of a workshop modelling approach, Adaptive Environmental Assessment, to the analysis of mechanisms by which acid precipitation alters stream and lake chemistry resulting in impacts on fish populations and other aquatic resources. The emphasis in this application has been on the integration of existing information about mechanisms and processes and the identification of research needs and priorities rather than immediate quantitative prediction.

An initial workshop involving many scientists currently engaged in acid rain research was held in August, 1980 in Ann Arbor, Michigan. The five-day workshop could not fully address the objective of establishing research priorities for so complex a subject as acid precipitation effects on aquatic resources. However, workshop participants did develop a preliminary model and research assessment which significantly contributed toward achieving this objective (Andrews et al. 1980).

The initial workshop utilized four submodels in order to examine the relationships and research needs among the linked components of terrestrial and aquatic systems: 1) key fishery populations; 2) food-chain components, including algae, zooplankton, and benthic organisms, some of which are themselves sensitive to low pH; 3) the altered chemistry of aquatic environments, particularly relatively infrequent abrupt pH depressions; and 4) the watershed system, which receives atmospheric inputs and yields chemically altered surface and groundwaters.

A second workshop to review, evaluate, and refine the preliminary model was held in Fort Collins, Colorado, in March, 1981 (US Fish and Wildlife Service in prep.). Calibrated watershed data bases were available for partial model calibration and evaluation. Simulation runs indicated the feasibility of using the model to assist in the evaluation of resource impacts, provided that specified research needs can be met and additional regional validation can be undertaken.

[1] Paper read by Kathy E. Lowry at the international symposium Energy and Ecological Modelling, sponsored by the International Society for Ecological Modelling. (Louisville, Kentucky, April 20-23, 1981).

[2] Austin K. Andrews, Gregor T. Auble, Richard A. Ellison, David B. Hamilton, and James E. Roelle are staff members of the Adaptive Environmental Assessment Group, Western Energy and Land Use Team, Office of Biological Services, US Fish and Wildlife Service, 2625 Redwing Road, Fort Collins, Colorado 80526.
[3] David R. Marmorek is a Systems Ecologist, Environmental and Social Systems Analysts, Ltd., Vancouver, B. C. Canada.
[4] Orie L. Loucks is Science Director, The Institute of Ecology, 4600 Sunset Boulevard, Indianapolis, Indiana 46208

A section on the general philosophy and procedures of the Adaptive Environmental Assessment process is followed by a description of the specific application of the process to development of a model expressing effects of acid precipitation. The concluding section evaluates the application of this structured workshop modelling process in terms of the general performance and limitations of the current model, its utility for identification of research needs, and the potential for further development and application.

ADAPTIVE ENVIRONMENTAL ASSESSMENT PROCESS

Adaptive Environmental Assessment workshop modelling (Holling 1978) combines decision-makers, planners, scientists, and other interested parties in a highly structured environment. The workshops are organized around the construction and refinement of a simulation model, in this case representing the aquatic resource consequences of acid precipitation.

Workshop Activities

The modelling workshops are facilitated by a workshop staff trained in techniques of systems analysis, computer modelling, policy analysis, and group dynamics. The staff has four primary functions during the course of a workshop: 1) to moderate the workshop; 2) to assist participants in formulating and constructing a computerized simulation model of the resource system under consideration; 3) to assist participants in the interpretation of model output; and 4) to aid in the incorporation of workshop results into relevant research and management activities.

The initial modelling workshop starts with problem definition and assessment of the problem's space, time, and scientific bounds. Considerable effort is devoted to identifying quantities (indicators) that describe the resource system and actions or societal demands that management must consider. The resource system is usually divided into a number of logical components for detailed consideration by subgroups. Before breaking into subgroups, participants go through an exercise, termed Looking Outward, that defines interactions among the primary subsystems.

Participants divide into small working groups according to their areas of expertise and interest. Each subgroup defines and tries to quantify the functional relationships that describe the behavior of a particular subsystem.

After submodels have been programmed by the staff, workshop participants begin a process of model evaluation. This evaluation can focus on the identification of additional research needs and is accomplished by examining model response to a variety of input sequences and initial conditions (scenarios). Participants may suggest changes in data or functional relationships which the workshop staff can try to incorporate into the model, although this may lead to additional data needs. Scenarios are rerun to examine the implications and sensitivity of the model to various alterations. The model that emerges represents a group concensus concerning resource behavior and a strategy for improving the level of understanding and predictive potential of the model.

Workshop Expectations

The Adaptive Environmental Assessment process is an exercise in interdisciplinary modelling and can contribute to several aspects of environmental analysis.

Interdisciplinary Communication and Public Participation

Scientists and policymakers from Federal, State, and local agencies, as well as public interest groups, can participate in, and contribute to, the environmental assessment through an integrated systems approach. This integrated approach facilitates communication among a diverse group of participants and provides a mechanism for testing potential consequences of various actions. Each participant's perspective of the resource issue is broadened and his understanding of the resource system improved.

Problem Definition and Relevancy of Analysis

Environmental analysis is usually constrained by time and funding. The scope of impacts to be addressed must be comprehensive in scale, yet prudent in detail. The Adaptive Environmental Assessment modelling process yields a workable compromise between breadth and depth, producing crisp problem definition. The participation of scientists and policymakers from Federal, State, and local agencies, as well as public interest groups, helps ensure that data collection and analysis address relevant issues, focus on key questions and variables, and provide information responsive to policymakers' needs.

Interdisciplinary and Interagency Coordination and Research Design

Although the initial model may be incomplete, the model building and testing process helps identify gaps in available data and understanding. Research priorities can be established to fill those data gaps. A clear definition of interdisciplinary and interagency data requirements is explicit in building a workshop model. Since workshop participants represent various disciplines and affected government and private interests, the definition of data requirements can form a basis for inter-disciplinary and interagency coordination of data collection and analysis.

Impact Analysis and Evaluation of Analysis Assumptions

Environmental assessments always involve a risk that one or more of the fundamental assumptions on which predictions are based are wrong. The modelling process forces participants to state and evaluate assumptions about relationships. It also provides a mechanism for testing consequences of model assumptions that are based on little or no information. In addition, environmental studies are rarely conducted over a long enough time span to differentiate between natural environmental fluctutations and responses to cultural perturbation. Even preliminary modelling is a useful tool to explore different types of development-related changes and their potential consequences.

Integration and Synthesis

Initial models provide a framework for integrating existing information about a resource issue and for identifying what additional information is required. The Adaptive Environmental Assessment process is iterative with the periods between workshops used for research, data collection, and model refinement. Each subsequent workshop incorporates new information and produces a more credible model for use evaluating management alternatives. This iterative application provides ongoing synthesis and communication among disiciplies, agencies, and interest groups, as well as the integrated understanding on which sound resource management recommendations can be based.

MODEL STRUCTURE AND DEVELOPMENT

The acid precipitation simulation model provides a framework for expressing the dynamic interaction of a large number of variables from an ecosystem perspective. This perspective is essential in evaluating the way in which acid precipitation inputs are propagated through coupled aquatic and terrestrial systems to result in effects on aquatic organisms.

Actions and Indicators

The modelling process used in the acid precipitation workshops approached the problem of defining system components and appropriate spatial and temporal scales from a group discussion of actions (table 1) and indicators (table 2). The actions and indicators were organized into groupings associated with the major model subsystems and edited to eliminate duplication.

The focus of the acid precipitation workshops was on identification of research priorities. The list of actions is, therefore, somewhat shorter than would generally result from a workshop designed to explore potential consequences of alternative management strategies. The various hypotheses concerning system function and mechanisms by which acid precipitation impacts the components of the ecosystem are of greater interest, in this case, than system responses to management strategies.

Space and Time

The model is defined around a small watershed in a region of potential sensitivity to acidic inputs and is based on the identified actions and indicators. The watershed, by definition, contains a lake, an outflow stream, a tributary stream available for fish spawning, a network of additional tributary streamlets, and a drainage area covered by evergreen and hardwood forests, as well as bare ground. The characteristics chosen for quantification in a specific watershed can be varied to simulate conditions in other watersheds.

Finite difference equations, with a basic time step of one week, are used as a compromise between short-term events, such as plankton growth and the impact of intense summer storms, and longer term phenomena, such as lake flushing. Simulation commences on October 1 and encompasses a variable time span of 10 to 20 years. Each submodel is programmed in FORTRAN; control of the computational sequence is governed by the simulation language SIMCON (Hilborn, 1973). SIMCON controls the sequence of model computations, facilitates data input and output, and provides library functions for specialized simulation requirements.

SUBMODELS AND THEIR INTERACTIONS

The model is composed of four linked submodels: 1) fish, 2) plankton-benthos, 3) water chemistry (lake); and 4) watershed. This disaggregation into subsystems facilitates consideration of cause and effect and efficiently allocates effort and expertise to the respective submodels.

The linkages (fig. 1) were identified through construction of an interaction or Looking Outward matrix. The submodels are arrayed as both row and column headings. Where a matrix element, m_{ij}, represents an interaction between two components, the question is asked, "What information is needed from the submodel i in order to represent the dynamics of the submodel j?" The process of constructing this matrix not only identifies linkages between submodel components, but is also extremely useful in promoting interdisciplinary communication and understanding among workshop participants. Participants are

Table 1.--Management actions identified at
the initial acid precipitation
workshop, August, 1980.

Model Component	Action
Watershed	- Regulate chemical composition of incoming precipitation - Alter canopy cover and composition (evergreen, deciduous, bare ground)
Aquatic chemistry	- Alter pH of lake or stream
Fish	- Manipulate harvest - Stock species already present - Introduce acid-tolerant species - Introduce metal-tolerant species

Table 2.--Indicators (performance measures)
identified at the initial acid
precipitation workshop, August, 1980.

Model Component	Action
Watershed	- Chemical composition of water leaving watershed
Aquatic Chemistry	- Chemical composition of lake and stream
Plankton-benthos	- Productivity and standing crop - Decomposition rate - Water transparency
Fish	- Species composition - Age and size structure of populations - Harvest or numbers available for harvest - Reproductive success - Heavy metal concentrations[1] - Physical deformities[1] - Behavioral pathologies[1]
Other	- Avian populations[1] - Herp populations[1] - Ratio of bacteria to fungi in substrate[1]

[1] Indicates a performance measure identified but
not included in the model due limited time
or information.

forced to look carefully at the kinds of information they can reasonably expect to obtain from other disciplines (i.e., how their submodel dynamics are influenced by other submodels) and the kinds of information other disciplines expect from them (i.e., how their submodel influences the dynamics of other submodels).

The matrix constructed at the August workshop was edited to eliminate linkages that were originally identified, but never incorporated into the model (fig. 1). For example, the Fish Submodel originally requested data on mercury (Hg) and copper (Cu) concentrations in the lake and stream from the Aquatic Chemistry Submodel. It was subsequently discovered that insufficient information was available to relate these concentrations to specific impacts on fish growth or survival. Mercury and copper concentrations are, therefore, not shown as linkages from the Water Chemistry Submodel to the Fish Submodel, even though they may be important. Identification of items originally requested, but subsequently omitted because of lack of information, is an important step in identifying data gaps, research priorities, and needs for model refinement.

Model revisions incorporated at the March, 1981 workshop primarily dealt with internal aspects of the respective submodels. Only small changes, such as a separation of dissolved reactive phosphorus from total phosphorus, were necessary in the submodel connections expressed in fig. 1.

Following the Looking Outward exercise, participants formed subgroups to construct submodels representing the internal dynamics of each component. The basic charge of each of these subgroups was to describe the mechanisms and processes ("rules for change") that occur in that subsystem. State-dependent relationships were developed for each submodel expressing indicators and variables required by other submodels as functions of actions and inputs received from other submodels. The general logic of the "rules for change" for each submodel is described below. Technical reports summarize the state of the model after each of the two workshops in more detail (Andrews et al. 1980, US Fish and Wildlife Service in prep.).

The Fish Submodel describes fisheries dynamics in terms of the number of individuals, average weights, and lengths by life stage for brook trout and white sucker populations. The submodel also has the capability of comparing spawning activities in the lake with spawning activities in streams. The Fish Submodel requires information on a variety of parameters by life stage, including those related to growth, survival, reproduction, the effects of pH and aluminum, and relationships between growth rate and food. The state variables in the Fish Submodel define a minimum data set from which changes in fish biomass and production can be computed. Changes

Figure 1. Looking Outward Matrix constructed at the initial Acid Precipitation Workshop, August, 1980. No information transfer was identified for the single hatched off-diagonal elements. No inter-component transfer is possible for the double hatched main-diagonal elements.

	THIS SUBMODEL NEEDS INFORMATION CONCERNING:			
FROM THIS SUBMODEL:	Watershed	Aquatic Chemistry	Plankton/Benthos	Fish
Watershed	▨	Inflows to lake and stream, concentrations of alkalinity, H^+, inorganic Al, total Al, Ca, Mg, Na, K, NO_3, SO_4, PO_4, dissolved organic C, and Cl in inflows	▨	▨
Aquatic Chemistry	▨	▨	Temperature of lake, concentrations of NO_3 and PO_4 in lake and stream	pH of lake and stream, concentrations of inorganic Al and Ca in lake and stream
Plankton/Benthos	▨	Net NO_3 uptake	▨	Available standing crop of benthos (by size class) in lake and stream Available standing crop of plankton (by size class) in lake
Fish	▨	▨	Amount of plankton and benthos consumed	▨

in these variables are governed by the supply of invertebrate food items and the toxic effects of pH and aluminum concentration on reproduction and mortality.

The Plankton-benthos Submodel represents the major facets of lower trophic levels. The biological compartments are phytoplankton, zooplankton, and benthos by size class. Maximum phytoplankton growth rates are determined by light and available phosphorus.

The Aquatic Chemistry Submodel calculates the chemical composition and temperature of the stream and lake. Lake concentrations of water quality parameters (Ca^{++}, Mg^{++}, Na^+, K^+, $SO_4^=$, NO_3^-, Cl^-, DOC, NH_4^+, and forms of P) are computed from streamflow and direct precipitation loadings (from Watershed Submodel) and initial lake concentrations. An iterative ionic electro-neutrality computation is used to determine lake (stream) pH based on lake (stream) concentrations of water quality parameters, thermodynamic considerations of the carbonate buffering system and aluminum speciation, algal metabolism of nutrients, and addition of base (e.g., lime) or nutrients (e.g., phosphorous) associated with stream or lake restoration efforts.

The Watershed Submodel characteristics are instrumental in determining how susceptible a lake is to the possible adverse effects of anthropogenic or natural inputs of acidity. The Watershed Submodel utilizes precipitation patterns and associated chemical constituent loadings as its basic inputs. Water from precipitation and snowmelt is routed through a two-layer soil system to the stream. Chemical constituent concentrations in the stream are calculated from empirical relationships with stream discharge.

RESULTS AND PROSPECTUS

General Performance and Limitations

The model structure resulting from revisions incorporated at the second workshop requires a relatively comprehensive ecosystem data base for adequate calibration or validation. Initial parameter estimates were based on a composite of information from a number of geographic areas.

The current acid precipitation model has been only cursorily compared with a single data set from Harp Lake, Ontario. Much of the Harp Lake data is in unpublished form based on the work of Dillon, Harvey, Jeffries, Nichols, Scheider, and Yan. These preliminary results have been provided by P. Dillon and W. Scheider (Ontario Ministry of the Environment, Water Resources Branch-Limnology Unit, P. O. Box 213, Rexdale, Ontario, M9W 5L1), who should be contacted for further information.

The data set considered for the Harp Lake site included information on geology and general site description (Dillon et al. 1978, Scheider et al. 1979a, and Jeffries and Snyder unpubl.), hydrology (Scheider et al. 1979a, 1979b, Scheider 1981, and Jeffries and Snyder 1981), precipitation chemistry (Scheider et al. 1979a and Jeffries and Snyder 1981), stream and lake chemistry (Scheider et al. 1979a), and phytoplankton, zooplankton and fish populations. One-year model runs were compared to data for the period from October 1977 to September 1978; 10-year runs were made to explore longer term model behavior.

These model runs helped in the identification of data limitations and conceptual inadequacies. Areas where further model refinements are needed include the parameter estimates for the Watershed Submodel, primarily obtained from studies in the Adirondack Mountains which failed to yield streamflow concentrations consistent with those observed at Harp Lake; the treatment of net nitrate uptake by phytoplankton; and the representation of the relative timing of fish spawning and stream spring pH depression. These inadequacies are currently being addressed and can probably be rectified with further model refinement based on existing information from Harp Lake.

A second set of limitations identified by workshop participants involves more basic reconceptualization of certain processes and assumptions necessary before the model can be broadly applied, without serious qualification, to the variety of sites at which acid precipitation is a potentially important problem. These include the assumption of no significant net groundwater flux, the relative lack of attention to stream biological processes, an over-aggregation of the benthos components, inadequate treatment of processes occurring at the sediment-water interface, the implications of behavioral adaptations and synergistic effects in toxicity relationships, and the assumed unimportance of pH-related species replacements in species-aggregated components. The model, despite inadequacies, demonstrated utility for tracing effects of acid precipitation through a watershed ecosystem. The primary factor limiting the model's utility, after appropriate modifications are made, may be the relative lack of comprehensive data bases needed to validate the model independently for new geographical locations.

Research Needs

One of the major objectives of the modelling workshops was to provide a structured environment in which scientists could interact to consider mechanisms of acid precipitation impacts on living aquatic resources and to define priority research needs.

Some of the more important areas where the need for additional information and research was identified include the following:

1) The nature and kinetics of soil leaching and exchange reactions, including metals and metaloids, as influenced by precipitation quantity and chemistry;

2) Factors which determine susceptibility of a watershed to acidic inputs (e.g., soil depth, cation exchange capacity, porosity, and bedrock);

3) Ligands and speciation of metals at low pH;

4) Direct (pH) and indirect (e.g., cumulative metal toxicity) effects of acidification on plankton-benthos (including cation exchange at water-sediment interface);

5) Trace metal toxicity and synergistic-antagonistic effects (lethal and sublethal) of pH, Al, Cu, Hg, Zn, Pb, Fe, Cd, Mn, and Ca on different life stages of important fish species;

6) More complete baseline data for fixed study sites;

7) Differences between NO_x and SO_x driven acidification.

The participants at the two workshops recognized that conventional water quality standards, at least those relating to nonhuman health considerations, are usually set to maintain and protect "fishable" waters. The workshop results indicate that the chemical alteration of water by acid inputs is a complex process and that no simple relationship can ensure water quality protection. Data on most components of the system are available at only a few sites. The major conclusion reached at the workshops is that a holistic, interdisciplinary approach to data collection and analysis is required. The single most limiting factor in evaluating the model was that the calibrating information for several submodels had to come from different geographic regions. Carefully managed research programs covering hydrologic, biological, geochemical, and aquatic response processes are essential to advancing predictive understanding of integrated whole system responses.

Prospectus

The effect of acid deposition on aquatic resources is an ecosystem response to spatially complex discharge patterns. The problem dictates an approach capable of spanning traditional disciplinary and governmental divisions. The benefits of Adaptive Environmental Assessment modelling workshops in facilitating communication and coordination among such diverse specialties and interests with different perspectives on a complex, natural system or resource issue are much more difficult to measure, but potentially of much greater ultimate importance, than the performance of the model itself. Initial objectives and results have focused on the area of identification of information deficiencies and research priorities.

Workshop participants specifically recommended that a number of sites, for which necessary data exist, be selected for between-system comparisons to provide an interdisciplinary, interagency opportunity to explore regional differences in the reaction and susceptibility of aquatic resources to acid precipitation. The overall extent of the acidification problem and the relative jeopardy of natural systems to further acidification can be assessed only by conducting some type of regional analysis. Technical revisions and trial parameterizations of the model are currently in progress in order to evaluate the model's ability to support such regional analyses.

ACKNOWLEDGEMENTS

Application of Adaptive Environmental Assessment to acid precipitation has been supported by the National Power Plant Team of the US Fish and Wildlife Service in cooperation with the US Environmental Protection Agency. The workshops were conducted by the Adaptive Environmental Assessment Group of the US Fish and Wildlife Service's Western Energy and Land Use Team, and in the case of the second workshop, in cooperation with The Institute of Ecology.

Successful application of the Adaptive Environmental Assessment process provides a mechanism for facilitating and expressing the results of interaction among a set of specialists, managers, and interest groups with different knowledge and perspectives. Credit for the overall progress to date resides with the workshop participants, several of whom have authored sections of a report (US Fish and Wildlife Service in prep.) summarizing the second workshop. The participants in the first workshop were: L. Barnthouse, J. Bassin, J. Bennett, K. Biesinger, D. Brakke, E. Cowling, P. Daye, C. Driscoll, R. Eisler, R. Foley, R. Friedman, E. Fritz, T. Haines, D. Hoffman, A. Johannes, R. Johnson, A. Julin, W. Kovalak, D. Lenhart, R. Linthurst, R. Morison, G. Omen, C. Powers, P. Rago, T. Roush, K. Schreiber, S. Smith, M. Staley, J. Weiner, and M. Zengerle. The second workshop participants were: C. Driscoll, J. Eilers, J. Hargis, P. Rago, P. Rodgers, W. Scheider, J. Schnoor, and C. Schofield.

LITERATURE CITED

Andrews, A. K., R. A. Ellison, D. B. Hamilton, J. E. Roelle and D. Marmorek. 1980. Results of a modeling workshop concerning acid precipitation. Western Energy and Land Use Team, US Fish and Wildlife Service, Fort Collins, CO. 101pp.

Dillon, P. J., D. S. Jeffries, W. Snyder, R. Reid, N. D. Yan, D. Evans, J. Moss, and W. A. Scheider. 1978. Acidic precipitation in south-central Ontario: recent observations. Journal of the Fisheries Research Board of Canada 35: 809-815.

Hilborn, R. 1973. A control system for FORTRAN simulation programming. Simulation 20:172-175.

Holling, C.S., ed. 1978. Adaptive environmental assessment and management. John Wiley and sons, New York. 377pp.

Jeffries, D. S., and W. R. Snyder. 1981.
Atmospheric deposition of heavy metals in
central Ontario. Water, Air and Soil
Pollution 14: 133-157.

Scheider, W. A. 1981. Hydrology and water
Balance. In Studies of lakes and water-
sheds near Sudbury, Ontario. Ontario
Ministry of Environment.

Scheider, W. A., W. R. Snyder, and B. Clark.
1979a. Deposition of nutrients and major
ions by precipitation in south-central
Ontario. Water, Air, and Soil Pollution
12: 171-185.

Scheider, W. A., J. J. Moss, and P. J. Dillon.
1979b. Measurement and uses of hydraulic
and nutrient budgets. Proceedings of a
National Conference on Lake Restoration,
Minneapolis, MN. EPA 440/5-79-001.

US Fish and Wildlife Service. In prep. Results
of a modelling workshop concerning acid
precipitation. US Fish and Wildlife
Service, National Power Plant Team and
Western Energy and Land Use Team.

HIERARCHICAL SET OF MODELS FOR ESTIMATING
THE EFFECTS OF AIR POLLUTION ON VEGETATION[1]
J. R. Kercher, M. C. Axelrod, and G. E. Bingham[2]

Abstract.--Three models have been developed to estimate the effects of air pollutants on vegetation at the photosynthetic process (PHOTO), plant (GROW1), and community (SILVA) levels of resolution. PHOTO simulates the enhancement of photosynthesis at low H_2S levels, depression of photsynthesis at high H_2S levels, and the threshold effects for sulfur pollutants. GROW1 simulates the growth and development of a plant during a growing season. GROW1 has been used to assess the effects on sugar beets of geothermal energy development in the Imperial Valley, California. SILVA is a community-level model simulating the effects of SO_2 on growth, species composition, and succession, for the mixed conifer forest types of the Sierra Nevada, California.

INTRODUCTION

It is recognized (Goldstein and Mankin 1972, Goodall 1976) that ecosystems are intrinsically hierarchical in nature with levels of organization of interest occurring from relatively small objects (e.g. leaves) to relatively large objects (e.g. watersheds). Thus, a wide range of levels of resolution confront the ecological modeller and the choice of which level he chooses to model depends on the objective of his project. Typically the modeller constructs a model at one level of organization treating the phenomena at the next finer level of resolution empirically. That is, the modeller is often attempting to explain the phenomena at one level of resolution using the next finer level of resolution to construct the dynamics of the model. Goldstein and Mankin (1972) point out that the levels of

resolution are both in the spatial scales and in the temporal scales. The time scale extends from one end where one might consider fractions of an hour for photosynthesis to the other end where one might consider centuries for forest succession or even longer time units for some evolution processes. Goldstein and Mankin (1972) also suggest that the spatial and temporal scales of resolution are not independent, rather, that small things are best modelled at short time scales and large things at large time scales. Goodall (1976) points out that if different time scales are mixed in the same model, the natural consequence is a model with stiff differential equations.

Our objective is to understand and be able to predict the effects of pollutants on plant yield in the case of crops and community composition, productivity, and succession in the case of forests. These objectives have led us to three models at three different levels of resolution. In the case of crops, our belief is that a predictive capability for crop yield is best achieved with a physiologically-based model of crop growth and development. Here the plant is divided into subsystems: leaves, roots, etc. We developed a plant-level model, GROW1 (Kercher 1977), a general crop model which has been parameterized specifically for sugar beets (Beta vulgaris) for use in an assessment of the impact of geothermal energy development in the Imperial Valley, California. Because of the importance of photosynthesis to plant development and

[1]Paper presented at the international symposium Energy and Ecological Modelling, sponsored by the International Society for Ecological Modelling. (Louisville, Kentucky, April 20 - 23, 1981). Work performed under the auspices of the U.S. Department of Energy by the UCLLNL under contract number W-7405-ENG-48 and funded in part by the Environmental Protection Agency and in part by the U.S. Fish and Wildlife Service under Interagency Agreement FWS-14-16-009-78-969.

[2]J.R. Kercher, M.C. Axelrod, and G.E. Bingham are Environmental Scientist, Electrical Engineer, and Environmental Scientist, respectively, at Lawrence Livermore National Laboratory, Livermore, California, U.S.A.

because of the secondary objective of being able
to predict pollutant effects mechanistically, we
developed a photosynthesis process model, PHOTO,
at the leaf level. PHOTO (Kercher 1978) is thus
a finer resolution model than GROW1. Our objective in the case of forests led us to a stand-
level model of forest dynamics, SILVA. SILVA
does not simulate the internal dynamics of the
tree, but rather the tree is treated as a black
box with certain responses for certain environmental conditions. The dynamics between the
trees are emphasized in SILVA. SILVA (Kercher
and Axelrod 1981) is a stand-level model which
comprises many trees, and operates at annual time
steps with simulations that span decades or
centuries.

PHOTO: LEAF PHOTOSYNTHESIS MODEL

The first-generation plant growth simulator,
GROW1, incorporates PHOTO as a submodel. PHOTO
is sensitive to the pollutant level inside the
leaf. This level is calculated by a sulfur uptake submodel, SULFUR, which is also an instantaneous, steady state model formulated at the
same level as PHOTO. Hence they will both be
discussed in this section.

Figure 1 summarizes the general approach
which is considered here for modelling pollutant
efects. The plant is seen as a box with its own
internal physiology and is acted on by the
environment with such variables as CO_2 concentration, light, etc. Pollutants may act on the
plant-environment interface, e. g., the stomata,
and thereby indirectly effect the behavior of
the plant. Also, the pollutants may enter the
plant itself either by leaf or root uptake.
Once inside the plant, the pollutants may interact directly with the plant's physiology and
directly affect growth and yield.

SULFUR Submodel Description

Ziegler (1975) discusses the metabolism of
sulfur dioxide in plant cells. We model the
metabolism by assuming Michaelis-Menten reaction

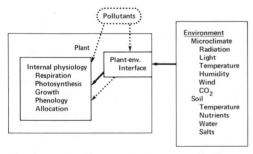

Figure 1.--Conceptualization of plant response
to environmental conditions and pollutant
level.

kinetics in the sulfur submodel. The absorption
of SO_2 into the leaf is modelled by the typical
diffusion equations dominated by the diffusion
resistances of the stomata , cell wall, etc.
Combining these equations leads to an expression
for the $SO_2^=$ concentration in the leaf as a
function of the SO_2 concentration in air.
This equation produces a threshold in the relationship of the cellular concentration of
$SO_2^=$ to the air concentration of SO_2. Below
the threshold, large changes in the air concentration produce small changes in concentration
in the mesophyll. Above the threshold, increases
in the SO_2 concentration in air produce comparable changes in the concentration in the cell.

The sulfur submodel requires a parameter
expressing the change in the internal sulfur
concentration for a change in the external concentration at low sulfur concentrations, the sum
of the leaf resistances to diffusion of SO_2,
and the threshold discussed above. The model can
be parameterized from experiments measuring
SO_2 uptake rates over a wide range of external
concentrations.

PHOTO Submodel Description

The photosynthesis submodel calculates the
rates of net and gross photosynthesis of the
leaf under the influence of elevated levels of
SO_2. This is done by equating the diffusion
rate of CO_2 into the leaf with the biochemical
assimilation rate. The diffusion rate into the
leaf is a function of the external CO_2 concentration and the diffusion resistances between
the atmosphere and the chloroplast. The biochemical assimilation rate depends on the light
reaction of photosynthesis and the dark reaction.
The light reaction is a function of the incident
photon flux and the kinetic parameters governing
the reduction of NADP to NADPH (nicotinamide
adenine dinucleotide). The dark reaction of
carboxylation is a function of the CO_2 concentration at the chloroplast and parameters of the
chemical kinetics for the conversion between
ribulose diphosphate and phosphoglyceric acid.
In the forward direction, CO_2 is taken up and
in the back conversion, the energy is supplied
by the reduced NADPH from the light reaction.
The preceeding portions of the photosynthesis
model are similar to the work of Sinclair (1972)
and Hall (1971). The temperature effects are
also modelled biochemically. This uses the temperature dependence of rate reactions of enzyme
kinetics and assumes that enzymes are inactivated
by being raised to higher internal energy states.

The photosynthesis submodel calculates the
effects of sulfur pollution on the photosynthesis
rate. First, the action of the sulfur compound
on the stomata is used in the photosynthesis
calculation. Also, the effect of sulfur is calculated on the internal biochemistry of the light
and dark reactions. The biochemical effects are
calculated in two different time frames: a

short-term model valid over fractions of an hour for reversible effects and a long-term model valid for a growing season which incorporates irreversible effects. The short-term model uses the observations of Ziegler (1975) that the Hill reaction is enhanced in the presence of sulfite and the dark reaction is inhibited by sulfite competing for the same uptake sites as the CO_2. The long-term model deals with the change over the growing season of the actual photosynthetic capacity of the plant, namely, changes in the number of reaction sites, the concentration of enzymes, or the amount of viable leaf tissue. For low levels of SO_2 or H_2S, the photosynthesis model calculates the fertilization effect of sulfur on the photosynthetic capacity. For high levels, it calculates the depressant effect. The fertilization effect is modelled by increasing the creation rate of the photosynthetic capacity at low sulfur levels and the depressant effect by increasing the destruction rate at high sulfur levels.

PHOTO requires the experimental determination of the leaf boundary-layer resistance and stomatal resistance. The biochemical parameters are found from experiments relating photosynthesis to temperature, light, CO_2, and the pollutant (SO_2) level. Photosynthesis is a multivariate function of these variables, so the best fits are obtained if these variables are varied independently over a wide range of values. In figure 2 we show an example of fitting net photosynthesis of sugar beets leaves to light level and temperature. Data are from Hall (1970). It is important to realize that all variables, e.g., stomatal resistance, SO_2, CO_2, etc., must be simultaneously measured and if possible, controlled. Because the SO_2 or H_2S response is a more complicated response than, say, the light response, correspondingly more measurements will be required to accurately parameterize this relationship. We stress that the response curve measurement should be repeated for several light, CO_2, and temperature levels.

The minimum data necessary to parameterize the long-term model is the photosynthetic capacity measured during the season for crops raised at constant pollutant levels. The data is necessary for several levels of pollutant load. These levels should bracket any complicated response such as fertilization, threshold, and depression. In addition, the CO_2 level should be held at elevated levels for some experiments to determine the long-term CO_2-sulfur interaction. See figure 3 as an example. The solid line is the short-term photosynthesis model adjusted to agree with Shinn's (unpublished) photosynthesis data for sugar beets. The triangles on the dashed line are the results of the long-term model adjusted to agree with Thompson and Kats' (1978) yield data (circles). The dark triangle and circle were model results and data for yields of plants grown at 0.3 ppm H_2S and CO_2 values of ambient plus 50 ppm.

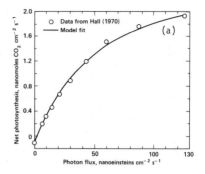

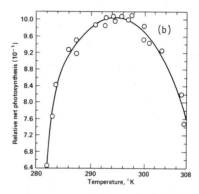

Figure 2.--Model fits of net photosynthesis of sugar beet leaves compared to data (a) light response, (b) temperature response. These fits parameterized the photosynthesis model with no pollutants present.

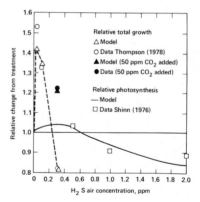

Figure 3.--Effect of H_2S treatment on photosynthesis (short-term) and growth (long-term) of sugar beets. Comparison of models and experiments. The comparisons are used to parameterize short-term and long-term photosynthesis models.

Results of PHOTO

Because PHOTO is constructed from mechanistic process considerations, it offers a high potential for producing information beyond the original data set. These results are testable experimentally. For example, in figure 4 we plot net photosynthesis against temperature for various levels of external CO_2. One sees that at low temperatures, net photosynthesis is very sensitive at CO_2 level; whereas at high temperatures, net photosynthesis is much less sensitive to CO_2 level. This is because at low temperature, there is a sufficiency of enzymes to conduct photosynthesis and CO_2 is therefore a limiting factor; at high temperatures so much enzyme inactivation has occurred that the enzyme level is limiting and CO_2 is not.

As another example, using PHOTO, Shinn and Kercher (1979) point out that estimates of effects H_2S on photosynthesis can change drastically depending on the CO_2 concentration in air. These changes are quantified by PHOTO (fig. 5) and thus the user can determine what the photosynthesis response would be after the addition of specific H_2S-CO_2 mixtures to the plant's environment. Such results can have an important application in the setting of standards.

GROW1: PLANT GROWTH MODEL

Figure 6 is a schematic showing the conceptual modules in GROW1. The core of GROW1 is a photosynthate generator which is labeled Photosynthesis Submodel and a Photosynthate Allocator which grows the plant. The Water Submodel tells the Photosynthesis Submodel the effects of soil water on stomatal resistance. The function of the Sulfur Submodel was discussed above. The Canopy Submodel calculates the light profile in canopy for use by the Photosynthesis Submodel. The Photosynthesis Submodel passes photosynthate to the Photosynthate Allocator which grows the plant.

Canopy Submodel

The canopy submodel calculates the LAI from biomass in the leaf compartment; the diurnal course of the angle of the sun above the horizon; and the light penetration in the canopy. The light penetration calculation uses a Beer's Law type formulation and is dependent on the angle of sun and the angle of the leaf.

Photosynthate Allocator

At present the allocation submodel of GROW1 does not depend on the pollutant level directly but only indirectly through the pollutant's effect on the photosynthate available for allocation. As more experimental work is done, the

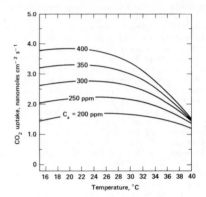

Figure 4.--Short-term prediction of net photosynthesis as a function of temperature for levels of CO_2 from 200 to 400 ppm in 25 ppm increments.

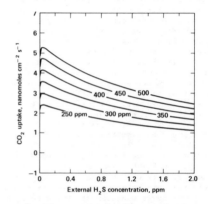

Figure 5.--Photosynthesis of sugar beets as a function of H_2S fumigation for various levels of external CO_2 concentration.

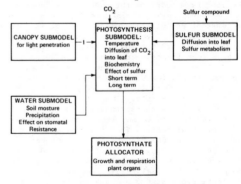

Figure 6.--Schematic structure of GROW1 showing modules.

Photosynthate Allocator may require revision to incorporate direct effects of pollutants on allocation.

The sugar beet version of GROW1 compartmentalizes the plant into leaves (tops), the structural component of the beet, the sugar compartment of the beet, fibrous roots, and a reserve storage compartment which is labile and available for use by all plant compartments. This version of GROW1 is based in part on the model of Fick (1971). Photosynthate is deposited into the reserve compartment. Growth is defined as the transfer of material from the reserve compartment to one of the structural compartments. Maintenance respiration is proportional to the weight of each compartment and is "paid" from the contents of the reserve compartment. Growth respiration is proportional to the transfer from the reserve compartment to the structural compartments and is "paid" from the reserve compartment. A transfer is expressed as a function of the size of the sink compartment. Also, it is a function of the size of reserve available for use by the plant. If reserves are low, transfers decrease linearly with decreasing reserve size. As reserves increase, transfers saturate as a function of reserve size. Priorities between plant parts are determined by relative placement of thresholds for reserve mobilization. Thus, the rate of change of the sizes of plant compartments is given by a coupled set of ordinary, nonlinear differential equations in the compartment sizes.

All rates use the same temperature dependence as respiration discussed in the photosynthesis section above.

Water Submodel

The primary importance of the water submodel is that it supplies the photosynthesis submodel with that portion of stomatal resistance due to water stress. Stomatal resistance is calculated as a function of soil water potential following Curry (1971). Soil water potential is found from soil water content which is governed by a differential equation with a precipitation input term and a transpiration output term. Transpiration is calculated using a modified Penman equation and is a function of solar radiation, relative humidity, temperature, and diffusion resistances of the canopy and stomates.

GROW1 Results

The minimum data set required by the allocation submodel is the time series of the development of the plant compartments. These measurements should be taken on the same population that the photosynthesis measurements are recorded from. The time development is necessary to fix the sink strengths, transfer thresholds, etc.,

of the allocation submodel. In figure 7, we see the leaves and beet development data of Fick (1971) compared to the adjusted model output. The model was adjusted so that the leaf results agree with experiment. The remainder of photosynthate must go to beet and fibrous root. In effect, figure 7 is a validation of the photosynthesis model. In the case of sugar beets, sugar levels must be measured to parameterize the partitioning between sugar and beet structure.

Kercher and Layton (1980) used GROW1 to forecast the effects of geothermal development of sugar beet production in the Imperial Valley, California. They were able to show that in the most highly fumigated position in the valley, the anticipated emission rate was a factor of 13 below that required to reach the threshold of production loss. In relatively remote or favorable areas the emission rate was of the order of several hundred times below the rate required to reach the damage threshold. The assessment was performed by using real time meteorological data as model input and using the time series of the simulations of the expected ground level concentrations of H_2S at twenty-two locations in the

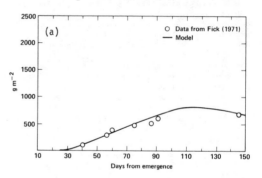

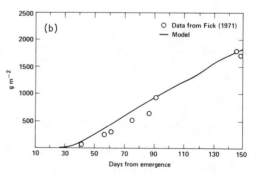

Figure 7.--Comparison of model and experiment of growth of sugar beet parts. (a) leaves (b) beet. These comparisons are used to parameterize photosynthate allocation in growth models.

Valley. In figure 8, we see the results for the five stations anticipated to see the largest concentrations. These simulations suggest that models at the plant level of resolution have the potential for being useful assessment and energy-siting tools.

SILVA: FOREST STAND MODEL

SILVA simulates the effects of SO_2 at the population and community levels given the results of the effects at the whole plant level. SILVA uses an empirical dose-response relationship for the effects of pollutants at the tree level. By virtue of the ecological interactions contained in the model, the effects at the tree level are translated into effects at the community level.

We have followed the modelling approach developed by Botkin et al. (1972) who developed JABOWA, a simulator of forests of the north-eastern USA. For a case study, SILVA has been applied to the ponderosa pine and mixed conifer forest types of the Sierra Nevada, California, USA. The associated species in these forests are ponderosa pine (Pinus ponderosa), white fir (Abies concolor), Douglas-fir (Pseudotsuga menziesii), sugar pine (Pinus lambertiana), incense-cedar (Libocedrus or Calocedrus decurrens), California black oak (Quercus kelloggii), and Jeffrey pine (Pinus jeffreyi). The model simulates recruitment, growth, and death of each tree similar to JABOWA but with extensive modifications. These modifications include the introduction of fire ecology, temporal seed crop patterns unique to the Sierra, water stress, and pollutant stress.

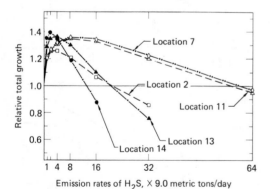

Figure 8.--Response of total growth of sugar beets compared to growth with zero emission rate (H_2S) as a function of emission rate of H_2S. Emission rates are in units of the expected emission rate (EER, 9 metric tons H_2S/day). Lines are for five most potentially fumigated sites of twenty-two studied in Imperial Valley, Calif.

Model Description

In figure 9 we show a structure diagram of the coding of SILVA. SILVA calculates environmental parameters (CALCNT, SITE) of the stand and initializes number and sizes (DIST, START) of the trees from environmental and control data respectively (TREDTA, SITDTA, CNTRL). A table of good and bad seed crop years (CYCLES) and a list of fire years (RINGS) is generated. The effect of pollution on trees is calculated (SEASO2, POLLUT, EPISOD). Then the number of new seedlings (BIRTH), the growth of each tree (GROW), and mortality (KILL) is determined for each year. Growth is modelled as a difference equation in the tree dbh and as a function of environmental variables. The killing is done stochastically depending on the probability of death as determined by ecological risk, lack of growth, and fire damage (FIRE). The dynamics of fuel accumulation (FUEL) [litter (LOAD) and brush (BRUSH)] are also modelled. Basal area is calculated (BASAL), averaged over runs (AVG), and outputted (OUTPUT).

Temporal Seed Crop Patterns

For the conifers of the Sierra Nevada, there can be significant temporal variations in the annual cone production. We modelled the phenomenon of high and low yield seed years as a Bernoulli random process with blocking.

Fire Ecology

Fire is a critical factor in the population dynamics of western forests. The most important aspect of fire is fire-induced mortality. The occurence of fire was also modelled as a Bernoulli random process with blocking parameterized by fire incidence data. The blocking in this case arises from time required for fuel to build back up to levels capable of supporting

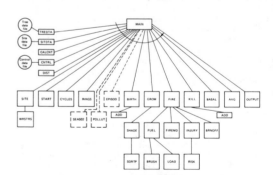

Figure 9.-- Structure of computer model of forest growth. Subroutines that communicate are connected with the calling subroutines displayed above those called.

fire propagation. Fire kills by raising the
temperature inside the tree and damaging the
crown. Fire intensity is calculated in kilo-
watts/meter of fireline length using FIREMD
(Albini 1976) and probability of death is determined as a function of dbh, bark thickness, and
scorch height (RISK). Scorch height is calculated from fire intensity, ambient temperature,
and windspeed (INJURY).

Moisture Stress

The effects of moisture stress are modelled
by multiplying the difference equation for growth
by a moisture stress factor. The moisture stress
factor is a function of the ratio of actual
evapotranspiration to potential evapotranspiration.

Modelling SO_2 - Pollutant Effects

There are two versions of the pollutant
effect section of SILVA. The seasonal average
model (SEASO2, POLLUT) calculates the pollutant
effect on the basis of the seasonal average concentration of pollution by assuming that growth
reduction is a simple function of the SO_2 concentration averaged over the growing season, or
equivalently, of the integral of SO_2 concentration over time. We use a dose-response function in which growth decreases linearly with
increasing accumulated dose based on the preliminary study of the tree-ring data of Lathe and
McCallum (1939) for ponderosa pine grown near
the smelter at Trail, B.C. The utility of the
seasonal average approach is in those cases for
which the accumulated dose or the seasonal average is the only data available or in those cases
where the pollution is of the chronic type.

Larsen and Heck (1976) have found that the
response of foliar injury to pollutant exposure
is highly non-linear in the concentration of the
episode averaged over the time-duration.
Guderian (1977) has suggested that in most cases
involving a single point source, chronic injury
results from the "short-term action of relatively
high concentration peaks". These two considerations suggest that in cases for which high concentrations occur for short time periods in
episodes, the accumulated effect should be calculated from the episodes. The successive episode model calculates a generalized dose from
each episode using the type of non-linear response found by Larsen and Heck (1976). The
cumulative generalized dose from all episodes is
used to calcuate the effect pollution on tree
productivity.

Simulation Results

Pollution Simulations

As an example of effects of pollution, we
set the SO_2-level to produce 10% growth reduction in ponderosa pine. We scaled the response

of the remaining species according to their published relative sensitivities. These calculations used the seasonal average model. The
results for ponderosa pine and white fir (fig.
10) indicates that while white fir undergoes a
nominal growth reduction of about 1 to 2% per
tree with pollution, total basal area actually
shows a dramatic increase. This is due to the
much greater growth retardation that the dominant
species experiences. We can summarize (fig. 11)

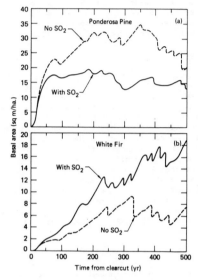

Figure 10.--Average of basal area from 25 simulations showing effects of pollution.
(a) Ponderosa pine (b) White fir.

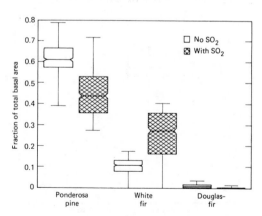

Figure 11.--Boxplots of polluted and unpolluted
cases. Median is line at notches. Top of
box is 75th percentile; bottom of box is
25th. Range is vertical line. Non-
overlapping notches indicate significance
at 95% level.

the results for ponderosa pine, white fir, and Douglas-fir by using boxplots of the distributions of the 500 annual data points of each species' fraction of the total basal area. Note the decrease in pine and the increase in fir with pollution. The basal area of Douglas-fir is extremely reduced. The environmental conditions were poor for Douglas-fir even in the absence of SO_2. The competitive disadvantage for Douglas-fir is made worse by pollution because Douglas-fir is sensitive to SO_2 and carries its needles longer than ponderosa pine. Thus, the growth reduction for an individual tree (greater than that for ponderosa pine) translates into a much larger effect on basal area.

Fire Ecology

Figure 12 summarizes the response of ponderosa pine and white fir with fire occurring at the natural frequency and with complete fire suppression. One can see that ponderosa pine improves in performace with fire and white fir is adversely affected by fire. Ponderosa pine is well adapted to fire and dominates where undergrowth is thinned by fire. The model reproduces this result and indicates white fir would eventually outcompete ponderosa pine in the absence of fire. The model suggests that a significant factor in the fire adaptation of ponderosa pine is the growth rate and growth form which allows it to evade fire by minimizing the time that the crown is exposed to fire.

CONCLUSIONS

Each of the three models discussed here has a unique capability in estimating the effects of air pollutants on vegetation. Each model operates at a spatial and temporal scale different from the others. Modified versions of PHOTO

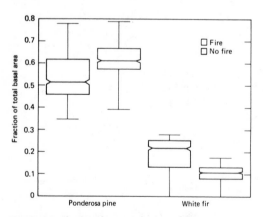

Figure 12.--Boxplots comparing results of simulation with fire suppression to those allowing fire to occur at natural frequency.

and GROW1 are being developed for use by the EPA project, National Crop Loss Assessment Network (NCLAN). The goal of NCLAN is to assess the impacts of current and projected pollutant levels on yield and economic loss. Physiological models of pollutant impacts are an integral part of future NCLAN assessments.

SILVA can be used as an aid in assessing the effects of SO_2 on western forests. For example, it could be useful in determing critical factors in site selection of fossil energy facilities, forecasting probable shifts in species composition of communities, and providing judgements of probable effects of SO_2 on communities already existing in stressed conditions.

LITERATURE CITED

Albini, F.A. 1976. Computer based models of wildland fire behavior: a user's manual. 68 p. USDA For. Serv., Intermt. For. and Range Stn., Ogden, Utah.

Botkin, D.B., J.F. Janak, and J.R. Wallis. 1972. Some ecological consequences of a computer model of forest growth. J. Ecol. 60:849-872.

Curry, R.B. 1971. Dynamic simulation of plant growth. I. Development of a model. Trans. ASAE 14:946-959.

Fick, G.W. 1971. Analysis and simulation of the growth of sugar beet (Beta vulgaris), Ph.D. dissertaion, University of California, Davis. 71-20384 University Microfilm, Ann Arbor.

Goldstein, R.A. and J.B. Mankin. 1972. Space-time considerations in modeling the development of vegetation. p. 87-97. In Modeling the growth of trees: Proceedings of the workshop. [Duke University, Oct. 11-12, 1971]. EDFB-IBP-72-11, Oak Ridge National Laboratory, Oak Ridge, Tenn.

Goodall, David W. 1976. The hierarchical approach to model building. p. 11-21. In Critical evaluation of systems analysis in ecosystems research and management. G.W. Arnold and C.T. de Wit (eds.) Center for Ag. Publishing and Documentation, Wageningen.

Guderian R. 1977. Air pollution. 127 p. Springer-Verlag, New York.

Hall, A.E. 1970. Photosynthetic capabilities of healthy and beet yellow virus infected sugar beets (Beta vulgaris L.). Ph.D. dissertation, University of California, Davis. 71-20384 University Microfilms, Ann Arbor.

Hall, A.E. 1971 A model of leaf photosynthesis and respiration. Carnegie Inst. Wash. Yearbook 70:530-540.

Kercher, J.R. 1977. GROW1: A crop growth model for assessing impacts of gaseous pollutants from geothermal technologies. UCRL-52247, 33 p. Lawrence Livermore Natl. Laboratory, Livermore, Calif.

Kercher, J.R. 1978. A model of leaf photosynthesis and the effects of simple gaseous sulfur compounds (H_2S and SO_2). UCRL-52643, 37 p. Lawrence Livermore Natl. Laboratory, Livermore, Calif.

Kercher, J.R. and M.C. Axelrod. 1981. SILVA: A model for forecasting the effects of SO_2 pollution on growth and succession in a western coniferous forest. UCRL-53109, 72 p. Lawrence Livermore Natl. Laboratory, Livermore Calif.

Kercher, J.R. and D.L. Layton. 1980. Impacts on agricultural resources. p. 9-1-9-20. In An assessment of geothermal development in the Imperial Valley of California. Vol. 1 - Environment, Health and Socioeconomics. DOE/EV-0092. Washington, D.C.

Larson, R.I. and W.W. Heck. 1976. An air quality data analysis system for interrelating effects, standards, and needed source reductions: Part 3. Vegetation injury. J. Air Poll. Control Assoc. 26:325-333.

Lathe, F.E. and A.W. McCallum. 1939. The effect of sulphur dioxide on the diameter increment of conifers. In Effect of sulphur dioxide on vegetation. National Research Council of Canada. p. 174-206. N.R.C. No. 815. Ottawa, Canada.

Shinn, J.H. and J.R. Kercher. 1979. Ecological effects of atmospheric releases from synthetic fuels processes. p. 199-203. In Proc. Symp. on Potential Health and Environmental Effects of Synthetic Fossil Fuel Technologies. DOE Conf-780903, NTIS, Springfield, Virginia.

Sinclair, T.R. 1972. A leaf photosynthesis submodel for use in general growth models. Memo Rept. 72-14. Triangle Research Site, Eastern Deciduous Forest Biome, U.S. International Biological Program, Duke University, Durham, N.C.

Thompson, C.R. and G. Kats. 1978. Effects of continuous H_2S fumigation on crop and forest plants. Environ. Sci. Technol. 12:550-553.

Ziegler, I. 1975. The effect of SO_2 pollution on plant metabolism. Residue Rev. 56:79-105.

ENERGY AND SYSTEMS

6. Regional Systems

THE ENERGY PATTERN OF DEVELOPMENT IN COASTAL LOUISIANA: PURCHASED AND NATURAL ENERGIES[1]

Charles S. Hopkinson[2] and John W. Day[3]

Abstract.--A conceptual system based on energy flow was used to analyze interrelations and trade-offs among environmental and economic processes in Barataria Basin in coastal Louisiana, U.S.A. The energy basis and ecological interactions of the region were developed for the period of 1900 through the present and projected to 1995. Barataria Basin is a distinctly unique environment of wetlands and uplands in the United States, as it is greatly influenced by the Mississippi River which defines one of the regional boundaries. Natural energy inputs to the region are more than two times greater on an areal basis than inputs to other Gulf state coastal areas. The Mississippi River supplies 47 percent of the natural energy inputs to the region. The area is also unique with respect to flows of purchased energy--witness one of the largest petrochemical complexes in the country and a per capita fuel usage that is 1.7 x that of the U.S. as a whole. During the course of development there have been substantial changes in land use and in the magnitude of both natural and purchased energy flows. We estimate that in 1900 purchased and natural energy flows were 875 and 6200 CE kcal/$m^2 \cdot$yr respectively and by 1995 the magnitudes are projected to be 48,000 and 3600 CE kcal/$m^2 \cdot$yr. The increased utilization of energy subsidies at the expense of natural energies is a pattern that has repeatedly occurred throughout history. With the probability of a declining energy budget on a worldwide basis, it would appear wise to protect and use the free work subsidy of natural energies more effectively.

INTRODUCTION

Because all life represents a continual battle against entropy, the study of material and energy flows through an organism, population or ecosystem provides useful information on ecological structure, function, and over time, patterns of succession. As dictated by the first and second laws of thermodynamics, the level of work in and by a system is totally controlled by the input of energy to that system. Some of the most complex and least understood ecosystems are those that include highly developed urban centers. The principles of energetics which have been explored in such great depth for individual organisms are also applicable for understanding the dynamics of the complex systems of man. This paper summarizes the results of a regional systems analysis of energy flows and work processes in the ecological and economic systems of a coastal region in Louisiana, U.S.A.

Human social development has been made possible by progressive acquisition of control over external sources of energy beyond the realm of wild food. Figure 1 illustrates the energy flows and interrelations that have been important in varying degrees while human social structure developed. When a hunter-gather, primary energy flows were from natural energies (sun, wind, tides, waves, rivers, rain) to the environment and thence to man.

[1]Paper presented at the International symposium Energy and Ecological Modelling, sponsored by the International Society for Ecological Modelling (Louisville, Kentucky, April 20-23, 1981).

[2]Charles S. Hopkinson, Research Associate-Marine Institute and Adjunct Assistant Professor-Zoology, University of Georgia, Sapelo Island, Georgia 31327 U.S.A.
[3]John W. Day, Professor and Director, Coastal Ecology Laboratory, Center for Wetland Resources, Louisiana State University, Baton Rouge, Louisiana 70803 U.S.A.
Contribution No. 428 from the University of Georgia Marine Institute.

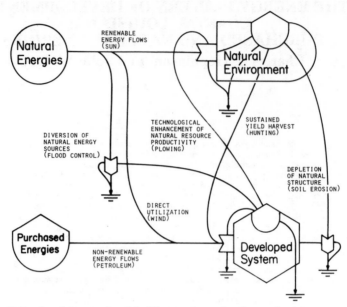

Figure 1.--Conceptual model illustrating major interrelations
and flows between man and the environment during the
social development of man.

There was no substantial buildup of population or
capital structure and little feedback between
people and the environment. With the advent of
swidden agriculture, capital structure was formed
and some high quality energy was fed back in manipulation of the natural environment. Resultant
declines in productivity led to community migration.
Throughout history, populations grew in surges
related to major acquisitions of energy. Plateaus
and declines were periods when growth was limited
by food or habitat (i.e. energy).

The rise of Greco-Roman civilization was
largely driven by the stored, fuel energy source
provided by forests (Hughes 1978). Wood was
reduced to charcoal to provide for space heating,
and heat for metallurgical and ceramic kilns. The
acquisition of this energy allowed a surplus upon
which capital structure and population increases
were built. To continually build structure, more
and more energy was expended in collecting wood
from greater distances. Massive deforestation
occurred, the result of which was catastrophic.
Unprotected by plant cover, soil was exposed to
powerful erosive forces. Within a short time
rain runoff increased, erosion increased, soil
fertility dropped, food production dropped and
rivers silted up. With a declining energy source,
capital formation slowed and the subsidy from the
natural environment lessened. Cook (1976) and
Hughes (1978) speculated that it was this pattern
of events that led to the fall of the Roman
empire.

For twelve hundred years the population of
China oscillated as the result of introduction of
new food plants, expansion of cropped land, increases in food supply, and subsequent overpopulation, soil erosion, famine and warfare (Cook 1976).
During periods of agricultural plenty, the population
rapidly increased and little energy surplus was
used for capital structure. Surplus was ultimately
transferred to deficit which encouraged overplanting,
soil erosion and reduction of the subsidy from the
natural environment.

Bowlus (1978) showed a similar pattern of
events for Medieval Europe. Through the joint
action of a series of technological innovations
(feedback from capital structure) and a rich input
of natural energies, agricultural surpluses were
created. Between 800 and 1200 there was steady
population growth (Cook 1976). By 1300, however,
the Europeans reached the limits of their available
energy resources and the natural ecosystem was
badly damaged. Deforestation was widespread and
the resultant erosion and decline in soil fertility
diminished the agricultural production base.
Bowlus (1978) hypothesized that the lack of further
technological breakthroughs and the continuing
population pressure against resources led to famine
and wars. Finally the plague struck.

From the rise and fall of these various
societies a clear pattern emerges where society
continually tries to utilize increasing amounts of
energy. The embodied energy of the natural system

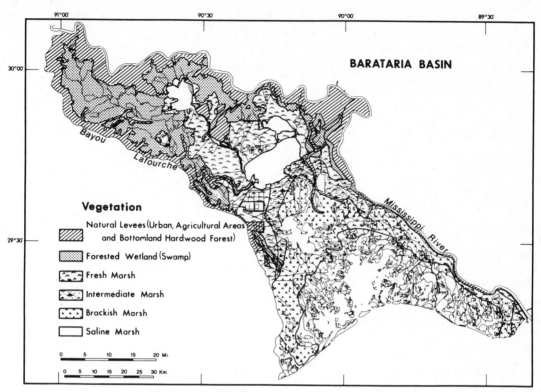

Figure 2.--Ecological zones and urban centers of Barataria
Basin.

is used to increase human productivity and to build
capital structure which is then usually used to
acquire still more energy from the environment.
In the process of energy acquisition, there is a
general lack of stewardship of the natural energies
and environment upon which society's production
is based. As environmental stresses increase, the
level and utility of the natural energies decrease.
Soils are depleted of nutrients, land is subjected
to accelerated erosion, floods increase in virulence
and agricultural productivity declines. Without
new technological advances or additional energy
inputs to counteract the decline in natural energies,
total energy capture decreases and society falters.
From the standpoint of cultural anthropology, this
is a deterministic view of society (Harriss 1977).

It appears that the world is again approaching
the declining portion of this pattern today. The
fossil fuel energies, upon which present industrial
societies are based, are finite. Environmental
degradation is widespread and becoming less of an
energy subsidy. It is in this light that the
Barataria Basin in coastal Louisiana is analyzed.
This article focuses on the energy flows in the
region and their interactions between the ecological and economic systems.

METHODS

A single conceptual system based on energy
flow was used to analyze the interrelations and
tradeoffs among environmental and economic processes. Energy is the common denominator of the
analysis because it is required by all activities
and both limits and governs the structure of human
economics (Costanza 1980). This type of energy flow
analysis is called energy analysis (Odum et al.
1979; Odum and Odum 1976; Odum et al. 1977;
Gilliland 1975). The following procedure was
carried out for the energy analysis of Barataria
Basin.

System boundaries were selected and an energy
diagram of the region was developed including all
external natural and human driving forces and the
interactions and components that these generated
within. Natural energy flows include the energies
involved in photosynthesis, the hydrologic cycle,
winds, waves and tides. Human energy flows include
agriculture, mining, manufacturing, trade, government subsidies, transportation, tourism, services
and direct fuel consumption. With the exception of
fuel production and consumption, which were quantified directly, all purchased energies were

calculated using the strong relation between
embodied energy and dollar value for the U.S.
economy (Costanza 1980). A ratio of 20,000 coal
equivalent kcal per dollar economic flow was used.
All energy flows were converted to coal equivalent
kilocalories (CE kcal) using energy quality factors.
The energy flows of the region were analyzed to
provide an understanding of the interdependency of
man's system and the natural environment. To
examine how the relationship between these two
systems has varied through time, the energy analysis
was developed for 1900 and projected to 1995.

ECOLOGICAL DESCRIPTION OF THE AREA

Bataria Basin (BBHU) is an interdistributary
basin in the Mississippi River deltaic plain
(fig. 2). It is bordered by the Gulf of Mexico,
the Mississippi River, and Bayou Lafourche, an
abandoned river distributary. Natural levees and
river crevasses are along the inland perimeter of
the basin. A chain of barrier islands separates
the basin from the Gulf of Mexico and in addition
to natural levees are the only regions not regularly
flooded. These relatively high lands slope from
up to 9 m above MSL along the river to about 0.3 m
towards the central axis of the basin. The basin
interior is wetland and waterbodies. Bordering
the Gulf of Mexico there is a band of saline marshes.
Moving inland, there are successive zones of brackish and freshwater marshes and swamp forest. The
wetland-to-water ratio is highest in the swamp
forest and lowest in the saline zone. The basin
has an area of 628,619 ha.

The integrating parameter that interconnects
and makes the entire system into a functional whole
is the hydrologic regime. The movement of water
is highly important. Before construction of flood
control levees along the Mississippi River, fresh
water was introduced into the interdistributary
basin via spring overbank flooding. This water
and rainfall would normally flow slowly through
the wetlands and be collected by bayous and waterbodies which tend to be located centrally in the
basin.

Wetlands of Bataria Basin serve as water
storage reservoirs, nursery areas, sites of chemical
transformation, and sources of organic matter and
nutrients. Bayous and waterbodies are the con-
nectors of the entire basin, providing conduits
for movements of water, materials, and migrating
organisms as well as areas of growth and feeding
of consumer organisms.

Bataria Basin serves as a major natural
resource area. It is estimated that this basin is
responsible for about 500 million pounds of commercial fishery harvest per year; about 45 percent
of the state total. This includes menhaden, trout,
croaker, crab, shrimp, oyster, catfish, and craw-
fish. There are also substantial fur and timber
harvests. There are extensive hunting and sport
fishing activities. The economic value of these
natural resources is in excess of $100 million

annually. The area is rich in non-renewable resources with numerous oil and gas fields and
several sulfur mines.

Natural geological cycles associated with the
Mississippi River and human development of the
region keep the areas of the various geomorphic
and ecological zones in a constant state of flux.
Artificial levees along the Mississippi River have
stopped all overbank flooding into the basin.
Channelization in the delta causes most riverine
sediments to be discharged over the continental
shelf break and prevents new land building. Ecological zonation is changing in response to saltwater intrusion, land loss, and urban development
(see Cleveland et al., this volume). Wetlands are
being reclaimed for urban and agricultural activities. Detailed geological and ecological discussions of Bataria Basin can be found in Gagliano
and van Beek (1970), Day et al. (1973), Hopkinson
et al. (1978), Hopkinson and Day (1977, 1979),
Craig and Day (1977), and Gael and Hopkinson (1979).

In the past couple of hundred years, considerable urban development has occurred on the natural
levees and surrounding wetlands of Bataria Basin.
Two factors were important in the urbanization:
the Mississippi River and the tremendous fossil
fuel reserves in the region. The river provided
an important navigation route between the interior
of the U.S. and foreign markets. It provided
drinking water and water for industrial cooling and
processes. With the discovery of the vast oil and
gas reserves in the area, the entire strip running
along the natural levees of the river rapidly
developed. The region is presently one of the
largest oil and natural gas producing areas in the
country, is one of the most productive petrochemical-
industrial bases in the country, and is the site of
a portion of the country's second busiest port
(New Orleans).

ENERGY FLOWS

Natural Energies

Natural energy flows of Bataria Basin were
first calculated for 1975 conditions. A small
fraction of the solar energy impinging on the basin
is captured by photosynthetic systems including
swamp, fresh marsh, brackish marsh, saline marsh,
bays and bayous, and uplands. Upland systems include forest, agricultural, industrial, commercial
and residential environments. The gross photosynthetic production of all these systems was
163.81 x 10^{12} kcal/yr or 8.3 CE kcal/yr (table 1).
To facilitate comparison, all energies were converted to coal equivalent kilocalories (CE kcal).
In this manner, diverse energies can be compared
on the basis of their ability to do work.

In addition to photosynthesis, there are
several other forms of energy inputs, the most
important of which is the Mississippi River.
Energy input in the form of hydrostatic head and
chemical free energy of fresh water relative to

Table 1.--Summary of system areas and productivities in
Barataria Basin in 1975.

Production unit	Area $(m^2 \times 10^8)$	Gross primary production $(kcal \cdot m^{-2} yr^{-1})$	Regional gross production $(kcal \cdot yr^{-1} \times 10^{12})$
Swamp	9.795	22,149	21.70
Swamp aquatic	.220	2,502	0.05
Fresh marsh	8.779	48,528	42.60
Fresh aquatic	6.724	9,252	6.22
Brackish marsh	5.504	70,308	38.7
Brackish aquatic	4.826	5,130	2.48
Saline marsh	6.336	47,947	30.38
Saline aquatic	8.214	6,673	5.48
Upland	12.235	12,998	16.20
Total	62.9		163.81

Details and a breakdown of upland systems in Hopkinson (1979).

salt water (ΔG) was calculated for the Mississippi River, Bayou Lafourche, and rainfall. This energy molds the land surface, carries sediment, builds land, distributes nutrients, and inhibits salt water intrusion. The potential energy of elevated water is 2.8 x 10^{12} CE kcal/yr, and the chemical potential energy (ΔG) is more than two times greater at 7.3 x 10^{12} CE kcal/yr.

Tidal and wave energy is greatest at the southern end of the basin near the Gulf of Mexico. Beneficial works of tides are waste removal, sediment transport, channel maintenance, organism and nutrient transport and gradient setup. Waves form the gulfward barrier islands which protect the interior basin from excessively high erosional forces and provide high land for urban activity. Tides and waves had an energy input of 0.6 x 10^{12} CE kcal.

Winds are important in evapotranspiration, removing pollutants, creating water currents and moving sand on beaches. Power associated with winds was calculated assuming a logarithmic wind profile and an average wind speed of 447 cm/sec. Winds are a substantial energy input to the region (5.6 x 10^{12} CE kcal/yr).

Annual natural energy inputs to the Barataria region totaled 24.5 CE kcal in 1975 (table 2).

Table 2.--Annual natural energy inputs to Barataria Basin
in 1975 (see Hopkinson 1979).

Energy input	Energy quality factor	Regional work CE kcal x 10^{12}/yr	Percent of total
Metabolism of natural ecosystems	.05	8.19	33.5
Tides	1.5	0.042	<1
Waves	1.2	0.590	2.4
Hydrostatic head	1.7		
Mississippi River		2.75	16.9
Bayou Lafourche		0.001	<1
Rain		0.005	<1
Chemical free energy	3.1		
Mississippi River		7.29	29.8
Bayou Lafourche		0.001	<1
Rain		0.038	<1
Wind	0.1	5.57	22.8
Total		24.48	100

The input of energy to the region was dominated by the Mississippi River. The energy potential of hydrostatic head and chemical potential energy represented 47 percent of all natural energy inputs to the basin. Almost one-third of the natural energy inputs to the region came from solar radiation used in photosynthesis. Winds provided slightly less than one-fourth of the total inputs. Due to the low wave energy along the coast and the small tidal range, these energies constituted only 2.5 percent of the total natural inputs to the region.

Purchased Energies

Major categories of purchased energies for the Barataria Basin region are fuels, electricity, goods and government tax transfers. Each of these is considered in this section.

Fuels

Production and consumption of fuel are the major purchased energy flows in the basin. Production of oil and natural gas began after 1900 in the state of Louisiana. Proved reserves of crude oil reached a maximum of 5.7×10^9 barrels in 1970 and have declined each year since at a rate of 14.8 percent per year. Maximum production was also reached in 1970. The pattern for natural gas reserves and production is the same. Maximum reserves and production were in 1970. The depletion rate was 12.1 percent per year. Since 1956, an increasingly larger percentage of crude and natural gas production has been from federally controlled offshore zones. In 1976, 52 percent of crude and 50 percent of natural gas production for the state of Louisiana was from the offshore region. Total production was 372.5×10^{12} CE kcal/yr in 1975.

Fuel consumption was calculated by summing natural gas, gasoline, LPG and electricity usage (table 3). Total natural gas consumption in 1976 was broken down as follows: major industry 57%, electricity generation 31%, gas company utility sales 12%. Electric utilities and most industry purchase their gas directly from intrastate gas pipeline companies. Fifty-five percent of gas utility company sales was to the residential and commercial sectors and 45% was to industry. There was no pattern of increased industrial usage. Consumption by residential and commercial sectors increased at rates varying from 0 to 8 percent per year.

Total electricity consumption has increased dramatically, with industry accounting for the largest increase, as well as the greatest total consumption. Industry consumed almost 69 percent of the total electricity used in Barataria Basin in 1976. Industrial consumption increased at about 15 percent per year; commercial at about 12 percent; residential at about 13 percent. This increase is not due to increasing numbers of customers alone. The total number of customers is increasing at about 7.5 percent per year, and

Table 3.--Total fuel consumption in Barataria Basin (1976).

Type of flow	Energy flow (CE kcal x 10^{12}/yr)
Electricity sold by utilities[1]	
Residential	1.07
Commercial	0.43
Industrial	3.41
Natural gas	
Electricity generation	14.6
Major industry	27.6
Utility sales	6.14
Gasoline	
Nonfarm	5.7
Farm, aircraft, fishing	0.04
LPG	0.02
Total[2]	54.0

[1]Total electricity generated is greater than that sold by utility companies because there is a significant generation by major utilities for their own consumption.

[2]Total energy sold by electric utilities is included in the total for natural gas, therefore not double counted.

total consumption is increasing at about 14 percent per year. Consequently, the consumption per customer has doubled in 16 years.

Of the remaining direct energy consumption, only nonfarm gasoline use was significant. Gasoline consumption rose from 50.7×10^6 gallon/yr to 178.3×10^6 gallon/yr between 1955 and 1975. Household use of LPG by homes scattered remotely in the basin was estimated at 7×10^5 gallon/yr.

In summary, natural gas supplied more than 89 percent of the total fossil fuel energies consumed, and gasoline supplied 10 percent. Only 15 percent of the fossil fuel energies produced in the basin are locally consumed. The remainder is exported to the rest of the state and country. In fact, natural gas is the only fuel produced in the area that is also consumed there. All crude is exported. Crude that is refined in the area (Venice is imported from Texas.

Economic Energy Flows

The remaining energy inputs that drive the basin are the economic flows of dollars or goods and materials (table 4). Industrial materials are separated from wholesale and retail goods as are industrial and agricultural products. Whereas data on the export of products was directly available, it was not for the import of goods. Therefore, these flows were calculated from gross accountings of total goods sales with assumptions about the fraction of goods originating from outside the region, and subtracting fuels and payroll. Wholesale and retail goods accounted for 59 percent of

Table 4.--Economic energy flows driving Barataria Basin in
addition to fossil fuels ('+' = in; '-' = out).[1]

Name of flow	Energy flow	
	$ x 10^6	CE kcal x 10^{12}
Goods and materials		
Industrial materials	-1040.0	+20.8
Wholesale and retail goods	-1469.0	+29.38
Product markets	+1620.9	-32.42
Exported renewable products	+0.09	-0.002
Taxes		
Federal taxes	-226.0	-4.52
State taxes	-102.0	-2.04
Federal payments	>+38.6	+0.77
State payments	+13.6	-0.27
Tourism		
Total trade	+102.	+2.04
Port services		
Cost	+125.0	+2.5
Gross inland tonnage	(44.7 x 10^6 tons)	
Gross ocean tonnage	(23.25 x 10^6 tons)	
Navigation and drainage subsidies		
State	+1.4	+0.03
Federal	+11.1	+0.22
Mining		
Total services	+63.0	+1.26
Total	4812.0	96.25
Payments out	-2837.0	+57.27
Payments in	+1975.0	-38.98
Net	-862.0	+18.29

[1]Details of accounting were presented in Hopkinson (1979).

the total imported value, which is not surprising since most of the industries are petrochemical-based and the majority of their raw material was counted earlier in the section dealing with fuel. The export of agricultural products is insignificant compared to the export of industrial products.

The balance of payments to state and federal governments (table 4) showed that the BBHU received less money from government than it paid. Tourism and outdoor recreation brought in $102.0 x 10^6 to the region annually.

An enormous tonnage of material passes through the region via the Mississippi River. The port of New Orleans, the second largest in the U.S., is partially within Barataria Basin. The value of the port to the region is in the wages paid personnel to run the port and in the goods and materials imported to build port structure. The latter were counted under goods and materials. In 1975 about $125 x 10^6 were paid to personnel to operate the port. 68 x 10^6 tons of material passed through the port; 18 x 10^6 gons were imported from overseas and 5.2 x 10^6 tons were exported. Along inland routes 44.7 x 10^6 tons were transferred.

The industrial work of producing oil and gas was quantified by determining wages paid to

personnel in the operation and dollar transfers for materials needed to produce pipelines, drilling rigs, boats, etc. Materials were included previously under goods and materials. Personnel services were $63 x 10^6 annually.

Drainage, flood control and navigation projects are the remaining major energy flows of the region. $5.75 x 10^6 were spent on drainage and flood control projects in 1975. Of this amount 51 percent came from the U.S. Army Corps of Engineers. This energy input was a direct subsidy for basin agricultural production and urban expansion (wetland reclamation, etc.). To facilitate navigation in the Mississippi River, the Army Corps spent $19.9 x 10^6 in 1975 on channel maintenance and construction, structures, and revetments and wavewash protection.

The energy flows of the urban system are overwhelmingly dominated by direct fuel use (75 percent of the total) and by the petrochemical and refinery industries in particular. These industries are both capital and energy intensive enterprises that have little labor involvement. Natural gas is their dominant raw material as well as their fuel of choice. Industry consumed 75 percent of all the fuel burned in Barataria Basin in 1975; consumption of natural gas alone represented 52 percent of the total flow of purchased energies. Despite the good growing seasons, the tremendous production of the natural system and the large fishery base of the region, product markets based on agriculture or fisheries represented less than 1 percent of the purchased energies of the region.

Total Energy Flow Summary – Uniqueness of the Area

A synoptic overview of Barataria Basin energy inputs indicates the uniqueness of this assemblage of wetlands, uplands and urban centers in the United States. The major energy flows in order of decreasing importance in 1975 were 1) fuel production, 2) fuel consumption, 3) goods and materials, 4) Mississippi River, and 5) regional gross primary production (table 5). The combination of river energies and the tremendous past, regional reserves of fossil fuel molded the development of southeast Louisiana social systems. The oil production, refinery, petrochemical, and port character of the resultant urban centers is witness to this combination. In Barataria Basin 86 percent of fossil fuel

production was exported in 1975. All crude oil was exported. Per capita total fuel consumption is 1.6 x that of the U.S. as a whole. Industrial per capita fuel consumption is 3 x that of the U.S. Natural gas makes up 80 percent of total fuel consumption while in the U.S. as a whole it contributes only 32 percent. The second busiest port in the country is in the region.

A comparison of energy signatures for a variety of regions in the U.S. and abroad reveals again the uniqueness of the energy flows that sustain Barataria Basin (fig. 3). The magnitude of natural energy inflows to the basin are second only to Gotland. Gotland, a Swedish island in the Baltic, has a tremendously large natural energy input in the form of waves and winds. Compared to the low level of primary production on the island, however, it would appear that wave and tide energies are so strong and concentrated that they are more of a stress than a subsidy. With the application of appropriate wind technology some of that energy might become available to the urban system (Jansson and Zucchetto 1978).

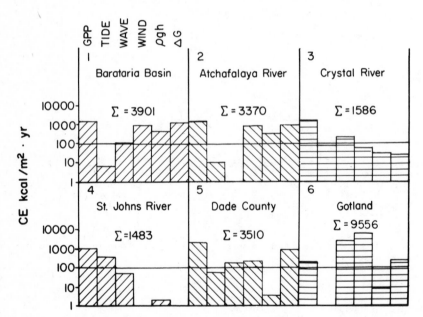

Figure 3.--Energy signatures of a variety of regions around the United States and Europe.[1]

[1]Information for Atchafalaya from Young et al. 1974; Crystal River from Kemp 1977; St. Johns from Bayley et al. 1976; Dade County from Zucchetto 1975; Gotland from Jansson and Zucchetto 1978. Method of primary production calculation was the same in all studies. Data from Young et al., Bayley et al., Zucchetto, and Jansson and Zucchetto were standardized by using equations and energy quality factors similar to those used for Barataria Basin and Crystal River. GPP = gross primary production; ρgh = hydrostatic equation; ΔG = Gibbs free energy.

Table 5.-- Temporal pattern of energy flows driving Barataria
Basin.

	Energy input by year (CE kcal x 10^{12}/yr)		
	1900	1975	1995
Fuel production[1]	0	372.5	140.4
Purchased energy			
Fuel consumption[2]	3.73	54.0	204.0
Other subsidies[3]	1.77	18.29	97.05
Natural energy			
Regional primary production[4]	17.3	8.19	6.87
Tides[5]	0.042	0.042	0.042
Waves[5]	0.590	0.590	0.590
Hydrostatic head			
Mississippi River[6]	3.93	2.75	2.75
Bayou Lafourche[5]	0.001	0.001	0.001
Rainwater[5]	0.005	0.005	0.005
Chemical free energy			
Mississippi River[6]	10.32	7.29	7.29
Bayou Lafourche[5]	0.001	0.001	0.001
Rainwater[5]	0.038	0.038	0.038
Wind[7]	6.37	5.60	5.59
Total purchased energies	5.5	72.69	301.05
Total natural energies	38.56	24.51	23.18
Total input energies	44.06	96.80	324.23
Ratio purchased: Natural	0.14:1	2.95:1	12.99:1

[1] Production 1900 - 0 -- Before petroleum discovery in Louisiana.
1995   -- From Cole (1974).

[2] Fuel consumption - 1900 -- Based on per capita energy use in 1900
(Cole 1974) and past population (USDI,
Bureau of the Census, 1899).
1995 -- Projected per capita energy use (Cole 1974)
and population (SCORP 1971).

[3] Other subsidies include Goods and Materials, Tax Payments Balance,
Tourism and Recreation, Mining Services, Port Services, Drainage and Navigation Subsidies. Extrapolations based on the past and future energy use to
present energy use ratio and 1975 values. See Table 4.

[4] Productivities calculated by multiplying present productivity per unit
area by area of each land use in the past and future. For 1900 productivities
were multipled by 1.5 to account for the suboptimum production in 1975 attributable to impoundment and oil extraction activities. Areas of various
ecological zones were calculated from measurements of land loss, salt water
intrusion, land : water changes, wetland reclamation, and farm acreage
(Gagliano and van Beek 1970; Adams et al. 1976; Craig et al. 1979; McIntire
et al. 1975; Van Sickle et al. 1976; USDI Census Reports Vol. VI-Agriculture
and Population-1899). Future urban land areas were obtained from Louisiana
State Planning office projections. (See Hopkinson 1979 for more detail.)

[5] Assume constant with time.

[6] 1900 value based on no Atchafalaya-Mississippi River connection.

[7] Based on changing area of BBHU through time (see Hopkinson 1979).

Natural energy inputs to Barataria Basin are 245 percent higher than they are to the Crystal River powershed region in Florida (Kemp 1977). Crystal River probably has an energy signature typical of most Gulf coast regions. That Barataria has a much richer energy endowment indicates the uniqueness of the region compared to adjacent Gulf coast regions. The most unique aspect of the Barataria signature is the profound influence of the energy associated with the Mississippi River. Over 47 percent of the natural energies originate with the river. Energy associated with the chemical potential of fresh water is also high in Dade County, Florida (Zucchetto 1975), but this water is mainly underground. The relative lack of importance of river energy in the St. Johns River region is due to low discharge.

The overall high level of energy inputs to the Atchafalaya River Basin is interesting because this region of Louisiana is relatively undeveloped. As with Barataria, the Atchafalaya system is largely river driven (the Atchafalaya is a distributary of the Mississippi River). The signature for the Atchafalaya, however, is young. It has only been in the past 50 years that a significant portion of the Mississippi flowed in the Atchafalaya. Until recently there was little dry land and no river flow. Presently the river is doing a considerable amount of work modifying and building land surface. Sometime in the future the Atchafalaya Basin will be a prime area for human development. The U.S. government is meeting considerable opposition by landowners in its plan to purchase the basin in order to maintain the ecological primitiveness and usefulness of the river basin as a floodway which possibly indicates the foresight of the landowners.

TEMPORAL PATTERN OF ENERGY FLOW

Analysis through time of the changing inputs of purchased and natural energies to Barataria Basin can provide insight into the interrelations and trade-offs among environmental and economic processes in the region. For this reason the energy basis and ecological interactions of the region were developed for 1900 and projected for 1995.

Clearly evident from the temporal analysis is the enormous change in the relative importance of natural to purchased energy inputs to Barataria Basin (table 5). In 1900 it is estimated that natural energies exceeded purchased energy flow by a factor greater than 6. Presently the situation is reversed and by 1995 purchased energy flows are projected to exceed natural energies by a factor of 12. Purchased energies of man's social system are expected to change from 5.5 to 301 x 10^{12} CE kcal/yr between 1900 and 1995.

As expected the increase in purchased energy inputs to the region is not without a decrease in environmental energies. Between 1900 and 1995, we predict that natural energies will have decreased 40 percent (38.6 to 23.2 x 10^{12} CE kcal/yr). The decrease is primarily due to declines in total

primary production and the energies associated with the Mississippi River. In 1900 the entire volume of the Mississippi River took the main river course past Barataria Basin. Overbank flooding still occurred during spring runoff. Presently, 30 percent of the river goes down the Atchafalaya River and, were it not for the Army Corps of Engineers, the percentage probably would be closer to 100 percent.

We predict that primary production of the Barataria region will have declined from 17.3 to 6.9 x 10^{12} CE kcal/yr from 1900 to 1995. The major causes are 1) conversion of productive wetlands to less productive waterbodies, 2) conversion from natural ecosystem to urban land use, and 3) wetland impoundment resulting largely from canaling associated with petroleum and natural gas extraction.

The decline in natural energies should be considered as conservative because little attempt was made to include an impact of urban activities on the environment beyond direct land-use changes.

CONCLUSIONS

For Barataria Basin as a whole, the ratio of natural energies to purchased energies has been declining rapidly. This ratio indicates the dependence of the region on outside sources of energy and the increasing susceptibility of the basin to fuel supply disruptions. The historical pattern of energy acquisition by society at the expense of natural energies and the environment appears to be repeating in Barataria Basin. The urban system increased its capture of purchased energies by a factor of 10 between 1900 and 1995. Projections indicate an additional quadrupling by 1995.

In the process of energy acquisition there has been a general lack of stewardship of the environmental energies that also support the region. This decline has been accompanied by an overall drop in environmental quality. Waterbodies have been becoming increasingly eutrophic as canals and levees shunted nutrient-rich, storm water runoff from uplands without permitting sheet flow across wetlands (Gael and Hopkinson 1979; Craig et al. 1979; Seaton and Day 1979). Total dissolved solids discharged by industry along the lower Mississippi increased 600 percent between 1955 and 1969. Mutagenic activity of river water samples was extremely high (Pelon and Whitman 1979) and may contribute to the high incidence of urinary and gastrointestinal cancer in the regional population. Ambient air quality was considerably below primary standards in 1975 and is probably related to the tremendous concentrations of lung cancer. Renewable products from the region have also been declining recently. Timber production in the predominantly impounded swamps is 50 percent lower than in adjacent flowing water areas (Conner and Day 1976). Potential commercial fishery harvest is expected to decline more than 17 percent by 1995. Trapping is expected to decline 23 percent.

The rapid growth of technology and population in the region has produced a strong drive to convert natural systems into developed ones. The strong positive feedback for producing more capital structure to further stimulate production is perhaps causing overdevelopment or an overreliance on outside energies. An interesting relationship to examine in the future is the proportionality between rate of purchased energy increase and natural energy decrease for a region through time. Unfortunately, we do not believe enough information currently exists to fully explore this relation.

As the remaining rich energies of fossil fuels are exhausted and their net energy declines, the importance of natural energies will rise. More profitable and efficient use of natural energies would appear wise to insure the carrying capacity of the biosphere in the future. Odum (1971) presented a lichen model that is symbolic for man. Presently man acts as a parasite on his environment with little regard to the welfare of his host. Without additional massive energy inputs to the social system, man will have to evolve to a mutualistic relation with his host since he will be an increasingly dependent heterotroph.

Several low cost opportunities now exist for man to improve his relation with nature. By allowing overbank flooding of the Mississippi into the adjacent Barataria wetlands (Gagliano 1981), by directing upland runoff into wetlands (Hopkinson and Day 1980a; Kemp 1978), by removing sheet flow barriers in wetlands (levees) (Craig et al. 1979; Cleveland et al. this volume; Hopkinson and Day 1980a, 1980b), and by halting additional canaling (Craig et al. 1979), the natural energy base of Barataria Basin could probably be restored to 1900 levels. Whether Louisiana decides to use the remaining rich oil to plan a transition to a lower energy state remains to be seen.

LITERATURE CITED

Adams, R.D., B.B. Barrett, J.H. Blackmon, B.W. Gane, and W.G. McIntire. 1976. Barataria basin: geological processes and framework. 117 p. Louisiana State University, Center for Wetland Resources, Baton Rouge, La. Sea Grant Publ. No. LSU-T-76-006.

Bayley, S.E., H.T. Odum, and W.M. Kemp. 1976. Energy evaluation and management alternatives for Florida's east coast, p. 87-104. In Transactions 41st North American Wildlife and Natural Resources Conference. Wildlife Management Institute, Washington, D.C.

Bowlus, C.R. 1978. Energy and ecology in medieval Europe (1000-1350). 8 p. Paper prepared for the Conservation Foundation, Washington, D.C.

Cleveland, C.J., J.W. Day, and C. Neill. 1981. A model of petroleum recovery operation on coastal ecosystems in Louisiana. (this volume)

Cole, N.A. 1974. Louisiana energy study. 257 p. Report to the Ozarks Regional Commission. Contract No. TA 7411.

Cook, E. 1976. Man, energy, society. 478 p. W.H. Freeman and Company, San Francisco, Calif.

Conner, W.H., and J.W. Day. 1976. Productivity and composition of a bald-cypresswater tupelo site and a bottomland hardwood site in a Louisiana swamp. Amer. J. Bot. 63(10):1354-1364.

Costanza, R. 1980. Embodies energy and economic valuation. Science 320:1219-1224.

Craig, N.J., and J.W. Day. 1977. Cumulative impact studies in the Louisiana coastal zone: eutrophication; land loss. Final report to Louisiana Planning Office, Baton Rouge, La.

Craig, N.J., R.E. Turner, and J.W. Day. 1979. Land loss in coastal Louisiana, p. 227-254. In J.W. Day, D.D. Culley, R.E. Turner, and A. Mumphrey (eds.), Proceedings of the Third Coastal Marsh and Estuary Management Symposium. 511 p. Louisiana State University, Division of Continuing Education, Baton Rouge, La.

Day, J.W., W.G. Smith, P.R. Wagner, and W.C. Stowe. 1973. Community structure and carbon budget of a salt marsh and shallow bay estuarine system in Louisiana. 80 p. Louisiana State University, Center for Wetland Resources, Sea Grant Publ. No. LSU-SG-73-04.

Gael, B.T., and C.S. Hopkinson. 1979. Drainage density, land use and eutrophication in Barataria Bay, Louisiana, p. 147-165. In J.W. Day, D.D. Culley, R.E. Turner, and A. Mumphrey (eds.), Proceedings of the Third Coastal Marsh and Estuary Management Symposium. 511 p. Louisiana State University, Division of Continuing Education, Baton Rouge, La.

Gagliano, S.M. 1981. Special report on marsh deterioration and land loss in the deltaic plain of coastal Louisiana. 12 p. Report to Louisiana Department of Natural Resources and Department of Wildlife and Fisheries. Feb. 24, 1981.

Gagliano, S.M., and J.L. van Beek. 1970. Hydrologic and geologic studies of coastal Louisiana. Part 1, vol. 1: Geologic and geomorphic aspects of deltaic processes, Mississippi Delta System. Coastal Resources Unit, Center for Wetland Resources, Louisiana State University, Baton Rouge La.

Gilliland, M.W. 1975. Energy analysis and public policy. Science 189:1051-1056.

Harriss, M. 1977. Cannibals and kings: the origins of cultures. Random House, New York, N.Y.

Hopkinson, C.S. 1979. The relation of man and nature in Barataria Basin, Louisiana. 236 p. Ph.D. dissertation. Louisiana State University, Baton Rouge, La.

Hopkinson, C.S., and J.W. Day. 1977. A model of Barataria Bay salt marsh ecosystem, p. 236-265. In C.A.S. Hall and J.W. Day (eds.), Ecosystem modeling in theory and practice: an introduction. Wiley-Interscience, New York.

Hopkinson, C.S., and J.W. Day. 1979. Aquatic productivity and water quality at the upland-estuary interface in Barataria basin, Louisiana, p. 291-314. In Livingston (ed.), Ecological processes in coastal and marine systems. Plenum Press, New York, N.Y.

Hopkinson, C.S., and J.W. Day. 1980a. Modeling the relationship between development and stormwater and nutrient runoff. Environmental Management 4(4):315-324.

Hopkinson, C.S., and J.W. Day. 1980b. Modeling hydrology and eutrophication in a Louisiana swamp forest ecosystem. Environmental Management 4(4):325-335.

Hopkinson, C.S., J.G. Gosselink, and R.T. Parrondo. 1978. Above-ground production of seven marsh plant species in coastal Louisiana. Ecology 59(4):760-769.

Hughes, J.D. 1978. Deforestation and erosion in Greece and Rome. 45 p. Paper prepared for the Conservation Foundation, Washington, D.C.

Jansson, A-M., and J. Zucchetto. 1978. Man, nature and energy flow on the island of Gotland. Ambio 7(4):140-149.

Kemp, G.P. 1978. Agricultural runoff and nutrient dynamics of a swamp forest in Louisiana. 57 p. M.S. thesis. Louisiana State University, Baton Rouge, La.

Kemp, W.M. 1977. Energy analysis and ecological evaluation of a coastal power plant. 560 p. Ph.D. dissertation. University of Florida, Gainesville, Fla.

McIntire, W.G., M.J. Hershman, R.D. Adams, K.D. Midboe, and B.B. Barrett. 1975. A rationale for determining Louisiana's coastal zone. 91 p. Sea Grant Publ. No. LSU-T-75-006. Louisiana State University, Center for Wetland Resources, Baton Rouge, La.

Odum, E.P. 1971. Fundamentals of ecology. 574 p. W.H. Saunders Company, Philadelphia, Pennsylvania.

Odum, H.T., and E.C. Odum. 1976. Energy basis for man and nature. 287 p. McGraw-Hill, New York.

Odum, H.T., W.M. Kemp, M. Sell, W. Boynton, and M. Lehman. 1977. Energy analysis and the coupling of man and estuaries. Environmental Management 1(4):297-315.

Odum, H.T. et al. 1979. Energy basis for the United States. 444 p. Report to Department of Energy, Contract EY-76-S-05-4398. University of Florida, Gainesville, Fl.

Pelon, W., and B. Whitman. 1979. Detection of mutagens and/or carcinogens in municipal water supplies and sources in southeastern Louisiana using bacterial and mammalian cell monitors, p. 89-105. In J.W. Day, D.D. Culley, R.E. Turner, and A. Mumphrey (eds.), Proceedings of the Third Coastal Marsh and Estuary Management Symposium. 511 p. Louisiana State University, Division of Continuing Education, Baton Rouge, La.

SCORP. 1971. Louisiana statewide comprehensive outdoor recreation plan for 1970-1975. 350 p. Prepared by State Parks and Recreation Commission, Baton Rouge, Louisiana.

Seaton, A., and J.W. Day. 1979. Development of a trophic state index for the quantification of eutrophication in Barataria Basin, p. 113-126. In J.W. Day, D.D. Culley, R.E. Turner, and A. Mumphrey (eds.). Proceedings of the Third Coastal Marsh and Estuary Management Symposium. 511 p. Louisiana State University, Division of Continuing Education, Baton Rouge, La.

van Sickle, V.R., B.B. Barrett, T.B. Ford, and L.J. Gulick. 1976. Barataria Basin: Salinity changes and oyster distribution. Coastal Zone Management Series, Publ. No. LSU-T-76-02. 22 p. Louisiana State University, Center for Wetland Resources, Baton Rouge, La.

Young, D., H.T. Odum, J.W. Day, and T.J. Butler. 1974. Evaluation of regional models for the alternatives in management of the Atchafalaya basin. U.S. Department of the Interior, Bureau of Sport Fisheries and Wildlife. Contract No. 14-16-0004-424.

Zucchetto, J.J. 1975. Energy basis for Miami, Florida, and other urban systems. 247 p. Ph.D. dissertation. University of Florida, Gainesville, Fla.

THE IMPACT OF ARTIFICIAL CANALS ON LAND LOSS IN THE BARATARIA BASIN, LOUISIANA[1]

Cutler J. Cleveland, Jr.,[2] Christopher Neill,[3] John W. Day, Jr.[4]

Abstract.--A simulation model was developed to address the
impacts that artificial canals have upon natural land loss rates in
the Barataria Basin, Louisiana. The effects of canals on hydrology,
sedimentation rates, and interconversions between inland and streamside marsh were examined. Results of the model indicate that canals
do alter the mechanisms underlying land loss and gain and therefore
accelerate naturally occurring land loss rates.

INTRODUCTION

This paper is about a model of land loss and
land gain in the salt marshes of the Barataria
Basin, a coastal basin in the Mississippi deltaic
plain. Over the past several thousand years
there has been a gradual growth of the delta to
about 1.4 million hectares today. For the past
several decades, however, there has been an
overall net loss of coastal wetlands. This has
also been a period of intense human activity in
the coastal zone. The loss of wetlands is
generally considered detrimental (Craig et al.
1979) and there is controversy about the effects
of different human activities upon the loss of
coastal wetlands. In this paper we investigate
the effects of a specific impact, canals, on
land changes. The process of land loss and gain
is very complex involving both natural and human
factors. In this introduction we will describe
the natural cycles and show how human impacts
affect these processes.

The deltaic plain is a large area of dynamic
geomorphic change. Since sea level stabilized
after the last glaciation, sediments from the
Mississippi River have formed a broad, coastal
plain of over 4 million hectares. About 50% is
inland water such as bays, lakes, and bayous;
about 40% is swamp and marsh; and the remainder
is terrestrial (primarily natural levee ridges
adjacent to Mississippi River distributaries).

Although there has been an overall gradual
net growth, there are large scale cycles of
growth and decay of land. Since sea level
stabilization, the river has occupied seven
different major courses and many minor ones and
built seven major deltaic lobes. The river is
now in the early stages of the eighth deltaic
lobe and a new delta is forming in Atchafalaya
Bay. During the growth phase of a lobe there is
rapid progradation and land building at the
mouth of the channel. Land is also built by
overbank flooding and infilling of older deteriorating marshes. As the channel lengthens and
the gradient becomes smaller, the river will
seek a new, shorter course to the Gulf. When a
lobe is abandoned, active land building ceases
and there is a net loss of land due to erosion
and subsidence. There has always been, therefore,
growth in part of the deltaic plain and decay in
the remainder. Historically, however, there has
been an overall net growth over the entire
coastal plain.

The process of land loss is a complex one.
There are three major natural mechanisms: Gulf
of Mexico beach retreat; lateral erosion of
streamside marsh shores; and gradual sinking of
inland marshes. Wave attack is the primary
cause of shoreline retreat and erosion. Lack of
sufficient sediments to offset apparent sea
level rise causes inland marsh loss. Apparent

[1]Paper presented at the International
Symposium for Energy and Ecological Modelling,
sponsored by the International Society for Ecological Modelling (Louisville, Kentucky, April
20-23, 1981).

[2]Cutler J. Cleveland is graduate research
assistant, Department of Marine Sciences, Louisiana
State University, Baton Rouge, Louisiana.

[3]Christopher Neill is research associate,
Coastal Ecology Laboratory, Louisiana State
University, Baton Rouge, Louisiana.

[4]John W. Day, Jr. is Professor of Marine
Sciences, Louisiana State University, Baton Rouge,
Louisiana.

sea level rise in Louisiana is due to both regional subsidence and eustatic sea level change.

Regional subsidence, a lowering of marsh elevation relative to sea level, has occurred steadily over the past 130 years over the entire western Gulf of Mexico coast, and has shown a significant rate of increase over the past 25 years. From 1848 to 1959, the measured rate of subsidence for two locations in or adjacent to Barataria Basin was 0.27 and 0.83 cm per year, respectively (Swanson and Thurlow 1973). From 1959 to 1971, these rates increased to 1.29 and 1.12 cm per year, respectively. In addition, a third location in the nearshore Gulf experienced a subsidence rate of 1.51 cm per year during the latter time period. In the most recent analysis of subsidence in the Barataria Basin, Baumann (1980) found that from 1954 to 1979, subsidence in the lower Barataria Basin was 1.30 cm per year.

The most important process affected by the deprivation of riverine sediment input are the rates of sedimentation and net marsh accretion of both streamside and inland marsh types. DeLaune and Patrick (1978) found that marsh sites nearer natural streams were accreting at 1.35 cm per year, while inland marsh sites in the same area were accreting at only 0.75 cm per year. They observed that much of the inland marsh in the study sites were accreting at rates less than the rate of apparent sea level rise. When compared to the subsidence rates mentioned above of 1.29 and 1.12 cm per year, it is evident that only the streamside marsh is accreting fast enough to offset the effects of subsidence. This is because suspended sediment must first pass through the streamside marsh before it reaches the inland marsh. Boto and Patrick (1978) also found that steamside marsh has greater stem density which facilitates the removal of sediment which in turn provides nutrients that help the marsh maintain higher levels of net primary production.

Baumann (1980) observed similar patterns of accretion in his analysis of sedimentation rates from 1975 to 1979. Streamside marsh accretion rates of 1.52 cm per year was significantly higher than inland marsh accretion rates of 0.91 cm per year. These sedimentation rates were related to the observed rates of land loss in the study area. For all inland and streamside sites that had an "aggradation deficit" (i.e., sedimentation rate less than apparent sea level rise), the weighted average sedimentation rate was 1.12 cm per year. Compared to the subsidence rate of 1.30 cm per year, this rate of sedimentation leaves a mean aggradation deficit of 0.18 cm per year for 80% of the marsh in the Barataria Basin. Baumann proposed that this deficit indicates that the lack of adequate external sediment supply may be responsible for a large part of the marsh currently being lost in the Basin.

Along with different rates of sedimentation, a shift in the source and seasonality of sedimentation in saline marshes also plays a major role influencing marsh accretion rates. Over the past 4600 years, direct introduction of sediments during the spring flood supplied marshes in the active delta lobe with enough sediment to counter the effects of subsidence, and provided a surplus rate of sediment that enabled marsh progradation to occur. Following a shift in the course of the Mississippi River, a gradual decrease in direct sedimentation takes place over the next several hundred years in the now abandoned delta lobe. The Barataria Basin is currently in such a state, with the construction of levees along the Mississippi River accelerating the loss of direct sediment supply.

Rather than the high rates of sedimentation associated with deltaic processes during flood season, the Barataria Basin now receives a relatively low rate of sediment input that arrives during high physical energy events such as storms and hurricanes. These events do not introduce any new external sediment into the basin, but instead rework and redistribute bay bottom sediments. Baumann (1980) found that 41% of all sedimentation from 1975 to 1979 in saline marsh areas occurred during the winter months of those years, 38% was due to two tropical storms that passed through the area, and 21% occurred during the remaining months of the year.

The key to the high sedimentation rates during the winter months is storm activity. Southerly winds increase as a cold front approaches from the northwest, leading to the flooding of marshes and an increase in the suspended sediment load of the water. This suspended material is deposited on the marsh surface due to a decrease in turbulence that occurs as the front passes through the area. When the front passes, winds swing to a northerly direction, leading to a rapid drop in water level (Baumann 1980).

Sedimentation, then, is the opposing force to apparent sea level rise in the overall struggle between land loss and gain in the saline marshes of Barataria Basin. On a broad, long-term scale both wetlands and uplands are built and decay. However, from a shorter range, management point of view, the "problem" of land loss almost always refers to wetland loss. "Losses occur in three basic ways: (1) Wetlands become open water because of natural or artifical processes; loss of this type may be caused by erosion, subsidence, or dredging to form canals, harbors, etc. (2) Wetlands are covered by fill material and altered to terrestial habitat. Placement of spoil from dredging is the most common example. (3) Wetlands can be wholly or partly isolated by spoil banks" (Craig et al. 1979).

Craig et al. (1979) defined land loss "as the substantial removal of land from its ecologic role under natural conditions." Therefore, impoundment or filling of wetlands does not

eliminate an area of land (i.e., by converting
it to water), but it does eliminate the ecological
function of the wetland. In this paper we
consider wetland loss to include both conversion
to water or dry land.

As indicated earlier, there has been a
dramatic shift from a net gain of land to a net
loss (Gagliano and van Beek 1970; Craig et al.
1979; Adams et al. 1976; and Gagliano 1980).
The rate of loss is accelerating and the latest
estimate is 39.4 mi^2 (102 km^2) annually for the
deltaic plain (see Figure 1). Construction of
artificial levees along the Mississippi River
and closure of several distributaries have
eliminated most new sediment input to the deltaic
plain. Large areas of older deteriorating
marshes which would have infilled to some extent
with riverine sediments are now deficient in
sediments. There are, however, other direct
impacts on wetlands which affect land loss.

In this paper we consider specifically the
impacts of canals on wetland loss. Canals are a
common feature of the coastal zone. The area of
canals (without the area of spoil deposition) is
about 1.4% of the coastal zone or about 20,200
hectares (Craig et al. 1979). Canals have been
constructed for a variety of purposes including
navigation, pipeline, oil access, drainage, and
borrow pits. Canals directly affect wetlands by
excavation and spoil deposition. They indirectly
impact wetlands by changing hydrology, sedimentation, and productivity. Canals have led to more
rapid salinity intrusion leading to death of
vegetation (Van Sickle et al. 1976). Canal
spoil banks limit water exhange over wetlands,
and thus decrease deposition of suspended sediments.
This exasterbates the existing sediment deficit
in inland marshes. Sedimentation is also the
single most important source of "new" nutrients
to the marshes (DeLaune and Patrick 1980).
A lack of sediments, therefore, leads to lower

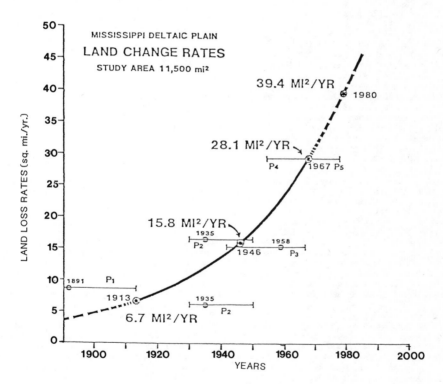

Figure 1. Land loss rates in coastal Louisiana (from Gagliano, 1980)

productivity. Craig et al. (1979) estimated that almost 40% of the total annual loss of 39.4 square miles (102 km²) is due to canals.

Once established, canals do not stabilize, but instead tend to widen over time through usage, especially as a result of wave action and altered hydrologic patterns (Craig et al. 1979). The annual increase in canal width varies from 2 to 14% for a doubling time of 5 to 50 years. As the area of canals increases, the area of natural channels decreases, indicating altered hydrologic patterns and the conversion of marsh to open water.

Spoil banks also lead to direct marsh destruction. Spoil, the material excavated by the dredging of canals, is most often deposited alongside the dredged area, a practice which destroys the adjacent marsh and leads to higher elevations which prevent water exchange with the adjacent marsh. The ratio of canal to spoil area has been estimated to be 1:2.5 (Craig et al. 1979), indicating the importance of spoil deposition to land loss. If this area is accounted for, the portion of total land loss due to canals and spoil is 69%.

Canals also have less obvious indirect effects due to such factors as salt water intrusion, altered sediment transport and inter-conversions between inland and streamside marsh. One of the most subtle yet pervasive indirect effect of canals is a decrease in sedimentation which leads to the conversion of one marsh type to another. For instance, when a canal is dredged through a marsh containing natural channels, the canal replaces channels as the main agent of transport through the marsh since it is deeper and faster flowing (Craig et al. 1979). The streamside marsh adjacent to the natural channel is steadily converted to inland marsh since the natural channel no longer supplies the sediments and nutrients necessary to maintain it. Once converted to the inland type, the marsh is then subjected to the aggredation deficit described earlier, and will eventually be converted to open water due to subsidence effects. Craig et al. (1979) estimated that the area of streamside marsh converted to inland marsh by the indirect mechanism is three to four times the area of streamside marsh converted directly to open water by canal building.

Our objective in this paper was to construct a model of the land change process and to incorporate the impacts of canals. We included the different mechanisms of streamside and inland marsh loss and the effects of canals on both of these. Overall we wanted to know the quantitative and qualitative roles of canals in land loss.

THE MODEL

The conceptual model (Figure 2) addresses the relationship between natural land change mechanisms due to subsidence and lateral erosion, riverine input of sediment, the effects of canals on sedimentation and erosion rates and how these three factors affect changes in the areas of streamside and inland marsh.

Barataria Basin was selected as the site or our study because data on the area has been collected in continuing studies at the Center for Wetland Resources at Louisiana State University. It is also one of the areas of coastal Louisiana most heavily affected by land loss. The year 1970 was used as the starting point for all simulations (see Table 1) and the model was run for 100 years.

Table 1. Initial conditions: 1970 marsh and canal area in saline zone of Barataria Basin, Louisiana

Variable	Area (ha)*	
Streamside marsh area	17,509	(30% total marsh area)
Inland marsh area	40,854	(70% total marsh area)
Total marsh area	58,363	
Canal area	2,429	

*Marsh area from Chabreck (1972). Canal area from Adams et al. (1976).

The first simulation involved determining changes in marsh area resulting from natural sedimentation and erosion rates without including riverine input or the effects of canals.

The second simulation represented the factors currently influencing land loss in the saline marshes. Riverine inputs were excluded, and direct and indirect impacts of canals were included. Direct effects of canals included marsh area loss due to conversion to open water and spoil, indirect effects are those due to interconversions between inland and steamside marsh, and altered sedimentation patterns.

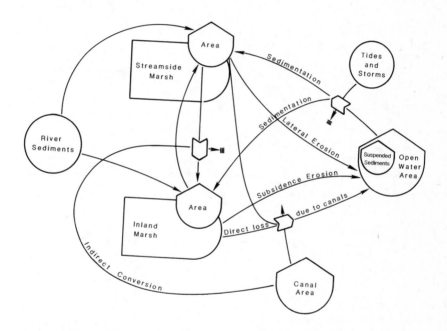

Figure 2. Conceptual model of natural and artifical land loss.

The third simulation included riverine input of new sediments without canals to determine how marsh areas would be altered if they were in fact receiving an external source of sediment.

The model does not explicitly deal with important factors such as nutrient inputs into the marshes being directly related to sedimentation rates (DeLaune and Patrick 1980). Primary productivity in the marshes is therefore influenced by any changes in sedimentation rates resulting from canals or riverine input. The problem of saltwater intrusion is also not addressed in the model. These mechanisms, however, are implicitly included since their effects are embodied in the observed changes in marsh area used in the model.

Natural decline in marsh area in a senescent delta lobe was modeled as a function of two processes: (1) erosion of streamside marsh at the edges of natural tidal channels, and (2) sinking of inland marsh as a result of lack of sediment deposition and the general subsidence and sea level rise of the Mississippi Deltaic Plain region. Equations for these natural land loss process are shown in Table 2.

$$(1) \quad 0.8 \ (58{,}363 \ \text{ha}) = C \ \text{ha}\cdot\text{cm}^{-1} \left[100 \ y \ x \right.$$

$$\left. (1.3 \ \text{cm}\cdot\text{y}^{-1} - 1.12 \ \text{cm}\cdot\text{y}^{-1}) \right]$$

The constant for the land area lost per centimeter of vertical sedimentation deficit was divided by the total area of inland marsh to obtain a second constant (K) which represents the percentage of inland marsh lost per centimeter of sediment deficit (see equation 2). It is this K that appears in the second equation in Table 2.

$$(2) \quad K = \frac{2593.9 \ \text{ha}\cdot\text{cm}^{-1}}{40{,}854 \ \text{ha}} = 0.063\%\cdot\text{cm}^{-1}$$

All losses of marsh due to subsidence and sea level rise were losses of inland marsh, since only on inland marsh does accretion fail to counter the effects of the general sinking of land with respect to water in the Mississippi Deltaic Plain.

Table 2. Equations for first simulation: Wetland loss without the influence of canals or riverine sediment input

Effect	Equation
Loss of streamside marsh by erosion along natural channels	$AREA_{ss}(y + 1) = AREA_{ss}(y) - 0.003*AREA_{ss}(y)$
Loss of inland marsh by natural subsidence due to sediment deficit	$AREA_{in}(y+ 1) = AREA_{in}(y)-k*(Subsidence - sedimentation)*AREA_{in}(y)$

Lateral erosion of streamside marsh was calculated from observations of shoreline retreat. Streamside marsh with an average width about 50 meters near Airplane Lake in the Barataria Basin showed a shoreline retreat of about 1.5 meters over a ten year period, yielding an annual erosion rate of 0.3% per year.

The relationship between the sedimentation deficit on inland marsh and areal decline was based on the estimate by Baumann (1980) that 80% of the marsh in the Barataria Basin will be innundated after 100 years. The sedimentation deficit was calculated as the difference between sea level rise (1.3 cm·yr^{-1}) and the average sedimentation on both marsh types from 1975 to 1979 (1.2 cm·yr^{-1}, Baumann 1980). A constant (C) relating sediment deficit to areal loss was calculated from equation (1) and has the value 2593.9 ha per centimeter of deficit.

Studies (Baumann 1980, DeLaune et al. 1978) have shown that, unlike on inland marsh, accretion on streamside marsh is keeping pace with subsidence and sea level rise. Currently, the elevation of streamside marsh does not appear to be increasing with respect to water level, and it was assumed that any sediment available to streamside marsh above what is necessary to counteract subsidence and maintain the present marsh area is not deposited on streamside marsh but is available to be deposited elsewhere in the marsh.

The effects of canals on wetland loss were added to the model for a second simulation, which corresponds to current conditions in the Barataria Basin. The effects of canals were judged to be either (1) direct (marsh area directly converted to open water by dredging), or (2) indirect (creation of spoil, conversion of streamside marsh to inland marsh by exclusion of overland flow of sediment carrying water). The equations for the effects of canals on wetland loss are shown in Table 3.

Craig et al. (1979) found the area of marsh affected by a canal may be five to six times the area of the canal. The indirect effects in this case would be changes in marsh hydrology that

Table 3. Equations for second simulation: wetland loss with the influence of canals but without riverine sediment input

Effect	Equation
Annual increment in canal area	$AREA_{new\ canals}(y + 1) = AREA_{canal}(y)*0.02$
Yearly increase in total canal area	$AREA_{canal}(y + 1) = AREA_{canal}(y) + AREA_{new\ canal}(y + 1)$
Direct loss of streamside marsh due to canals	$AREA\ SS_{direct}(y) = AREA_{new\ canals}(y)* \%SS(y - 1)$
Direct loss of inland marsh due to canals	$AREA\ IN_{direct}(y) = AREA_{new\ canals}(y)* \%IN(y)$
Indirect loss of streamside marsh to spoil	$AREA\ SS_{indirect}(y) = AREA\ SS_{direct}(y)*2.5$
Indirect loss of inland marsh to spoil	$AREA\ IN_{indirect}(y) = AREA\ IN_{direct}(y)*2.5$
Streamside marsh converted to inland marsh by exclusion of overland flow	$AREA\ SS_{converted}(y) = AREA\ SS_{direct}(y)*3$
Total loss of streamside marsh due to canals	$AREA_{ss}(y + 1) = AREA_{ss}(y) - AREA\ SS_{direct}(y) - AREA\ SS_{indirect}(y)$ $- AREA\ SS_{converted}(y)$
Total loss of inland marsh due to canals	$AREA_{in}(y + 1) = AREA_{in}(y) - AREA\ IN_{direct}(y) - AREA\ IN_{indirect}(y)$ $+ AREA\ SS_{converted}(y)$

The growth in canal area in the saline zone of Barataria Basin has historically been about 2% per year. A canal:spoil ratio of 1:2.5 (Craig et al. 1979) was used in the model to calculate conversion of marsh to spoil. Canals were assumed to affect streamside and inland marsh in proportion to their relative areas in the year the canal was constructed.

exclude the flow of water and sediment to inland marsh, thus resulting in loss of inland marsh. We used an estimate of three times the area of streamside marsh directly converted to canals as an estimate of the area of streamside marsh converted to inland marsh via additional indirect effects of canals.

The third simulation of the gain of marsh area in Barataria Basin with renewed sediment input to the wetland is an attempt at projecting how land loss could be slowed or reversed by the reintroduction of natural sediment flow into the region.

Much of the new land built in a prograding delta complex is formed when a break in the natural levee captures a portion of river flow and forms a new subdelta. Before the leveeing of the Mississippi River, growth of subdelta lobes was an important land building mechanism in the Barataria Basin, but currently breaks in the man-made levee confining the river are actively prevented.

To model land building in Barataria, we assumed that if the Mississippi River was not confined as it passes Barataria Bay, subdeltas would form at approximately the same rate they did in the previous century, before construction of the river levee. Information on the formation of Mississippi subdelta lobes was obtained from Morgan (1977). We estimated that an average of 2 subdeltas would form in Barataria Basin every hundred years. The areal extent and growth pattern of such subdelta lobes were derived from information on the Garden Island Bay subdelta (Morgan 1977). The increase in area of a subdelta was modeled as a logistic growth curve, with a carrying capacity (K) equal to the maximum extent of the subdelta. For the Garden Island Bay subdelta, K equals 7770 ha and the intrinsic rate of increase (r) equals 0.15. In our 100 year simulation, subdelta lobes were introduced in the first and fiftieth year. Equations for the third simulation are shown in Table 4. Marsh types were not differentiated. Each subdelta captured about 5% of the total Mississippi River flow. We assumed that in the depositional environment of a rapidly prograding subdelta complex, subdeltas, once laid down, do not rapidly decay.

RESULTS AND DISCUSSION

The results of the three simulations are compared in Figure 3. Year zero in the simulations corresponds to 1970.

Streamside and inland marsh area show natural decline, in the first simulation with sediment-deficient inland marsh disappearing faster. Ninety-two percent of the inland marsh, 26% of the streamside and 72% of the total marsh disappears after 100 years. As inland marsh sinks, new natural channels form, creating new streamside marsh. The relative area of streamside marsh increases from 30% in 1970 to 79% after 100 years, as only sedimentation on streamside marsh is able to keep pace with subsidence and sea level rise.

The results of the second simulation illustrate the accelerated land loss caused by canals. Streamside marsh disappears after 71 years; inland marsh disappears one year later. The effect is most dramatic on streamside marsh. The difference between natural loss of inland marsh and inland marsh loss due to canals is lessened by the indirect effect of canals converting streamside marsh to inland marsh.

The land loss rate for total salt marsh area in Barataria Basin for 1980 projected by the model was 4.3 square miles per year. Adams et al. (1976) estimated about 29% of the total land lost in the Barataria Basin was in the saline zone, the remainder occurring in fresh and brackish marshes. Assuming the same relationship between marsh type and land loss for 1980, our results indicate a total land loss in the Barataria Basin of about 15 square miles, more than 37% of the total land loss of 39.4 square miles in coastal Louisiana estimated by Gagliano (1981).

Table 4. Equations for the third simulation: wetland gain with the influence of riverine sediment input

Effects	Equation
Annual increment in area of subdelta lobe	$\text{NEW AREA}_{\text{lobe}_n}(y + 1) = \text{AREA}_{\text{total}}(y) * 0.15 * \dfrac{\left[7700 - \text{AREA}_{\text{total}}(y)\right]}{7700}$
Yearly increase in area of subdelta lobe	$\text{AREA}_{\text{total}}(y + 1) = \text{AREA}_{\text{total}}(y) + \text{NEW AREA}_{\text{lobe}_n}(y)$
Annual increase in marsh area	$\text{AREA}_{\text{marsh}}(y + 1) = \text{AREA}_{\text{marsh}}(y) + \text{NEW AREA}_{\text{lobe}_1}(y) + \text{NEW AREA}_{\text{lobe}_2}(y) +$ $\quad \ldots \; \text{NEW AREA}_{\text{lobe}_n}(y)$

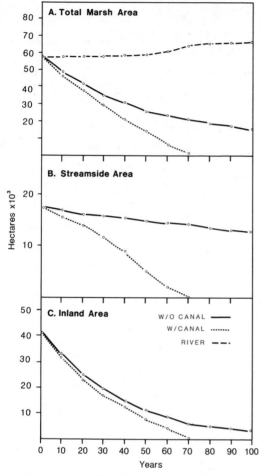

Figure 3. Results of simulation model for A) total marsh area, B) streamside marsh area, C) inland marsh area

With the influence of river sediments forming new subdelta lobes, the third simulation shows total marsh area increases from 58,363 ha to 67,077 ha after 100 years. Increase in area is most rapid in year 65, when the first subdelta is in the middle of its roughly 100 year cycle and the growth of a second lobe is beginning to contribute to areal growth.

SUMMARY AND CONCLUSIONS

Results of a simulation model of the processes leading to wetland loss in the Barataria Basin of Louisiana show the natural decline of wetland area in a senescent delta complex and the dramatic influence of canals in accelerating the functional removal of wetland habitat by conversion of marsh to open water and dredge spoil. The results of our model suggest that artificial canals, dredged primarily for the purpose of navigation and oil and natural gas extraction, significantly contribute to the decline in wetland area. We can also postulate from the results that detrimental effects of artificial canals might be lessened by the elimination of existing spoil banks or regulations limiting the height of new spoil banks. This would have the effect of allowing more sediment-carrying water to reach critical areas of inland marsh and thereby reduce the sediment deficit and slow areal decline.

The results of our model also indicate that reintroduction of riverine sediment from the Mississippi and natural progradation of subdelta lobes result in a net land gain in the Barataria Basin. A relatively small portion of the Mississippi River flow and its sediment load diverted into the Basin, might slow or even reverse the current process of wetland loss. It is unlikely that the rate of land loss in the Barataria Basin can be slowed without alteration of the current levee system, or without changes in the manner in which man-made canals are dredged and maintained. Without such changes a continued increase in the rate of wetland loss in coastal Louisiana can be expected.

ACKNOWLEDGMENTS

This paper is contribution No. LSU-CEL-81-09 of the Coastal Ecology Laboratory, LSU Center for Wetland Resources, Baton Rouge, LA. We thank R. H. Baumann for many helpful comments and suggestions.

LITERATURE CITED

Adams, R. D., B. B. Barrett, J. H. Blackmon, B. W. Gane, and W. G. McIntire. 1976. Barataria Basin: Geologic processes and framework. Louisiana State University, Center for Wetland Resources, Baton Rouge, La. Sea Grant Publ. No. LSU-T-76-006.

Baumann, R. H. 1980. Mechanisms of maintaining marsh elevation in a subsiding environment. M.S. thesis, Louisiana State University, Baton Rouge, La.

Boto, K. G., W. H. Patrick, Jr. 1978. Role of wetlands in removal of suspended sediments. In Wetlands Functions and Values: The State of Our Understanding, P. E. Greeson, J. R. Clark, and J. E. Clark, (eds.), 1978. Am. Water Resources Assoc., Minneapolis.

Chabreck, R. H. 1972. Vegetation, water, and soil chracteristics of the Louisiana coastal region. Louisiana State University Agricultural Experiment Station Bulletin No. 664.

Craig, N. J., R. E. Turner, J. W. Day, Jr. 1979. Land loss in Coastal Louisiana. Environmental Management 3:2,133-144.

DeLaune, R. D., W. H. Patrick, Jr., R. J. Buresh. 1978. Sedimentation rates determined by ^{137}Cs dating in a rapidly accreting salt marsh. Nature 275:532-533.

DeLaune, R. D., W. H. Patrick, Jr. 1980. Rate of sedimentation and its role in nutrient cycling in a Louisiana salt marsh. In Estuarine and Wetland Processes, P. Hamilton, K. B. McDonald, (eds.), Plenum Press, N. Y.

Gagliano, S. M., and J. L. van Beek. 1970. Geologic and geomorophic aspects of deltaic process, Mississippi Delta system. Hydrologic and Geologic Studies of Coastal Louisiana, Rept. No. 1. Coastal Resources Unit, Center for Wetland Resources, Louisiana State Univ., Baton Rouge, La.

Gagliano, S. M. 1981. Special report on marsh deterioration and land loss in the Deltaic Plain of Coastal Louisiana. Rept. to Louisiana Dept. of Natural Resources and Louisiana Fish and Wildlife Services, Feb. 24, 1981.

Morgan, D. J. 1977. The Mississippi River Delta: Legal-geomorphic evaluation of historic shoreline changes. Geoscience and Man, vol. XVI.

Swanson, R. L., and C. I. Thurlow. 1973. Recent rates along the Texas and Louisiana coasts as determined from tide measurements. J. Geophys. Res. 78:2665-2671.

Van Sickle, V. R., B. B. Barrett, and T. B. Ford. 1976. Barataria Basin: Salinity changes and oyster distribution. Louisiana State Univ., Center for Wetland Resources, Baton Rouge. Sea Grant Publ. No. LSU-T-76-002.

AN OVERVIEW OF THE OHIO RIVER BASIN ENERGY STUDY (ORBES) AQUATIC IMPACT METHODOLOGY[1]
Hugh T. Spencer and Clara A. Leuthart[2]

Abstract.--The ORBES project, (an in-depth study of power plant development in the Ohio River Valley), was initiated in August, 1976 at the request of Congress. Subsequently, a core team of thirteen researchers drawn from nine major universities in the midwest spent four years evaluating year - 2000 scenarios for the region. In the final analysis parameters defined through through strict versus lax environmental controls turned out to be major distinguishing factors.

INTRODUCTION

The Ohio River Basin Energy Study (ORBES) was initiated in August 1976 at the request of Congress for the purpose of providing information for the public and for government officials to assist in the decision making processes related to power plant development in the region. The final main ORBES report was issued in January 1981. Responsibility for the study rested with the U.S. EPA; however, the study was designed and effected by researchers from nine universities in the states bordering the Ohio River (U.S. EPA 1981).

A core team of researchers developed the study around 24 regional scenarios of energy development through the year 2000. The scenarios varies according to fuel emphasis, degree of environmental control, economic growth, and siting of power generating facilities. After construction of the scenarios it then became the responsibility of the core team researchers and their support researchers to evaluate the impacts of each scenario.

ORBES was conceived as an interdisciplinary study and was to address various environmental, social, economic, and engineering perspectives of energy development in the region. This paper addresses the problems encountered and the solutions attempted in such an interdisciplinary study from the perspective of the aquatic scientists.

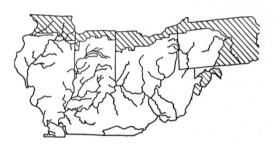

Figure 1.--Boundaries of the Ohio River Basin Energy Study Region.

AQUATIC ANALYSIS

Early in the study the researchers established boundaries for the ORBES region (fig. 1). The aquatic scientists had their problems with these boundaries, as did others of the research team, but the boundaries had to address the concerns of researchers from a variety of disciplines and thus at times were somewhat anomalous.

Aquatic systems in the region are quite varied and include mountain streams, impounded and free flowing rivers, numerous wetlands and sloughs in both the mountain valleys in the east and along major river basins of the west. Complete watershed systems within the region presented no special analysis problems. However, the impacts on systems divided by ORBES boundary lines proved difficult to evaluate. For example, scenario siting patterns extended up only the eastern half of the Mississippi, a phenomenon which came about by virtue of the Mississippi forming the state boundary of Illinois.

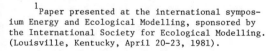

[1] Paper presented at the international symposium Energy and Ecological Modelling, sponsored by the International Society for Ecological Modelling. (Louisville, Kentucky, April 20-23, 1981).

[2] Hugh T. Spencer is Professor of Environmental Engineering in the Speed Scientific School, and Clara A. Leuthart is Assistant Professor of Liberal Studies in the University College, University of Louisville, Louisville, Kentucky, U.S.A.

The aquatic analysis centered around the fish species in the major streams and rivers of the region as well as the projected water consumption ratios for each scenario (Leuthart and Spencer 1980).

The recent occurrence of 258 species of fish in 70 of the major rivers and streams of the region was found, though only 24 of these streams were chosen for detailed analysis (fig. 2). Each stream was then assigned a protection level according to the following criteria:

Protection Level (A)--denoted a stream having a rich fauna in so far as the region was concerned; essentially a regionally unique stream.

Protection Level (B)--denoted a stream having a fauna not so rich as (A) and possibly working its way toward a more ubiquitous status. A more common place system, but one capable of moving to (A) status.

Protection Level (C)--denoted a stream with some quality but working its way toward a degraded system; a situation still found locally along some reaches of both (A) and (B) level streams.

Protection Level (D)--denoted a degraded stream having only pollutant tolerant fauna; a situation once common in some parts of the Ohio River Basin but rare at the present.

The key to the impact analysis was to develop a method which would permit a determination of the change in protection level under the conditions expressed for each scenario.

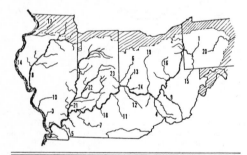

RIVER	MAP NO.	RIVER	MAP NO.
Allegheny	1	Little Miami	13
Beaver	2	Mississippi	14
Big Muddy	3	Monongahela	15
Big Sandy	4	Muskingum	16
Cumberland	5	Rock	17
Great Miami	6	Salt	18
Green	7	Scioto	19
Illinois	8	Susquehanna	20
Kanawha	9	Wabash	21
Kaskaskia	10	White	22
Kentucky	11	Whitewater	23
Licking	12	Ohio Main Stem	24

Figure 2.--Rivers selected for detailed analysis

A water and effluent component mass balance was developed (Brill et al. 1980) for the twenty-four test streams under conditions of 7 day-10 year low flow (7-Q-10). This was then applied in calculating concentrations for the twenty parameters listed in table 1, each being treated as a conservative agent and applied in tables 2 and 3.

These twenty components are all to some degree toxic to biological systems. The actual toxicity of each would depend upon both the system in question and the levels of all other agents in the immediate environment. However, the reference concentrations listed in table 1 are generally accepted statute values and do reflect current knowledge as to their relative toxicity.

Violations of reference concentrations were determined for each scenario and were tabulated according to the identifiers X_{10}, X_{25}, and X_{50} where:

X_{10} = the number of instances per scenario in which existing background concentrations plus incremental additions due to power plant loading and consumption was $\geq 10\%$ but $< 25\%$ of parameter reference concentrations.

X_{25} = the number of instances in which violations were $\geq 25\%$ but $< 50\%$ of parameter reference concentrations.

X_{50} = the number of instances in which these violations were $\geq 50\%$ of parameter reference concentrations.

These water quality events (sums for X_{10}, X_{25}, and X_{50}) were further weighted to yield a water quality impact index (WQII) where:

$$WQII = 5(X_{50}) + 2.5(X_{25}) + (X_{10})$$

The WQII is of a form easily evaluated by multiple linear regression for determination of coefficients, and was found suitable for making internal comparisons among scenarios. Aquatic impacts according to the WQII values were defined as follows.

Light--WQII < 10%. Under these conditions impacts on a system's biota would likely not be detectable except locally in the vicinity of out-falls. No change in protection level expected.

Moderate--WQII $\geq 10\% < 25\%$. Minor eutrophication with some loss of embryonic fishes existent at the time of low flow but recovery over the next several seasons would also be expected. One grade drop in protection level for the period of recovery.

Heavy--WQII $\geq 25\% < 50\%$. Eutrophication, the concentrating of heavy metal salts and possible stream dessication would be expected to combine to have a marked effect on the stream's biota. The effects would be noticeable immediately with possible local fish kills. A longer period of time would be required for recovery. Two grade drops in protection level.

Table 1.--ORBES reference concentrations and water quality criteria and standards in effect in the ORBES region (from US EPA 1981, p. 90 and 91,)

Constituent	ORBES Reference Concentrations (mg/l)	U.S. EPA Criteria (mg/l)		ORSANCO Criteria (Ohio River Main Stem) (mg/l)	State Standards (mg/l)							
		Domestic	Aquatic		IL Domestic	IL General	IN	KY	OH Domestic	OH Aquatic	PA	WVA
TDS	500	250[3]	N	500	500	1000	750	500	500 (150)	1500 (150)	500	500
TSS	50	5	N	N	N	N	N	N	N	N	N	N
Sulfate	250	[3]	N	250	250	500	250	N	250	N[7]	250	N
Ammonia	1.10	1.1[6]	N	N	N	1.5	N	N	N	1.4[7]	.5	N
Arsenic	.05	.05[8]	N	.05	.1	1.0	N	.05	.05	N	.05	.01
Barium	1.0	1.0	N[9]	1.0	1.0	5.0	N	1.0	1.0	N	N	.50
Cadmium	.01	.01		.01	.01	.05	N	.01	.01	.012	N	.01
Chloride	250	[3]	N	250[10]	250	500	250	N[10]	250	N	150	100[10]
Chromium	.01	.05	.01	.05[10]	.05	1.05	N	.05[10]	.05	.1	.05	.05[10]
Phosphorus	.05	N	N	N	N	.05	N	N	N	N	N	N
Selenium	.01	.01	.01L	.01	.01	1.0	.01L	.01	.01	.01L	N	.01
Silver	.005	.05	.01L	.05	N	.005	.01L	.05	.05	.01L[11]	N	.05
Copper	.06	1.0	.1L	.1L	N	.02	.1L	.1L	1.0	.06[11]	.1L	.1L
Iron	.30	.30	1.0	N	N	1.0	N	N	.3	1.0	1.5	N
Lead	.05	.05	.01L	.05	.05	.1	.01L	.05	.05	.03	.05	.05
Manganese	.05	.05	.1	N	.05	1.0	N	N	.05	N	1.0	N
Mercury (ug/l)[3]	.05	2.0	.05	.2	N	.5	N	N	2.0	.05	N	N
Nickel	1.00	N	.01L	N	N	1.0	.01L	.1L	N	.01L[11]	.01L	.1L
Zinc	.205	5.0[12]	.01L	.01L	N	1.0	.01L	.1L	5.0	.205[11]	.01L	.1L
Boron	1.00			N	N	1.0	N	N	N	N	N	N

N: No standard
L: 96-hour LC_{50}

[1] Toxic substance 0.1(96 LC_{50}) or 0.1 (96-hour median tolerance limit)

[2] Assuming Kanawha River criteria and all toxic substances 0.1(96-hour median tolerance limit)

[3] For chlorides and sulfates

[4] may exceed either 1500 mg/l or 150 mg/l attributable to human activities

[5] Does not reduce depth of compensation point for photosynthetic activity by more than 10% from norm.

[6] Based on 0.02 mg/l un-ionized NH_3 with pH 7.5, T = 25°C.

[7] Based on 0.05 mg/l un-ionized NH_3 with pH 7.5, T + 25°C.

[8] 0.100 for irrigation

[9] Standard (mg/l)

soft water	hard water	type of aquatic life
0.0004	0.0012	cladocerans, salmonid fish
0.004	0.012	less sensitive aquatic life

[10] As hexavalent chromium

[11] Based on total hardness = 260 - 280 mg/l as $CaCO_3$ and 0.1 (96LC_{50}) if Cu and 0.01 (96LC_{50}) if Zn

[12] 0.75 for irrigation on sensitive crops

Table 2.—Rivers studied in detail: Protection levels, pollutants violating ORBES reference concentrations at 7 day-10 year low flow, and flow per second at 7 day-10 year low flow.

River	Normal Flow Protection Level	TDS	TSS	SO$_4$	NH$_3$	As	Ba	Cd	Cl	Cr	P	Se	Ag	Cu	Fe	Pb	Mn	Hg	Ni	Zn	B	7day – 10 year Low Flow (m^3 sec^{-1})
Allegheny	A	X				X			X	X				X	X	X	X					13.3-31.3
Beaver	B		X						X	X			X	X	X	X	X			X		2.4-6.6
Big Muddy	A									X				X		X						1.05
Big Sandy	A	X							X	X			X	X	X	X	X	X			X	0.7-1.7
Cumberland	A							X	X				X			X	X					0.2-116
Great Miami	A	X						X	X	X		X	X	X	X	X	X			X		1.6-7.9
Green	A	X							X	X		X	.	X	X	X	X					8.7-14.2
Illinois	A			X						X			X	X	X	X	X					12.8-102
Kanawha	A								X	X				X	X	X	X	X				31.3-36.4
Kaskaskia	A									X				X		X						2.8
Kentucky	A																					3.2-4.6
Licking	A	X							X	X	X			X	X		X	X				0.3-2.0
Little Miami	B	X						X	X	X				X	X	X	X	X				0.8
Mississippi	A									X			X	X			X	X				446.1-1342.7
Monongahela	A								X	X				X			X	X	X			10.9-13.0
Muskingum	B	X						X	X	X				X	X	X	X					15.6-16.0
Ohio Main Stem	A	X						X	X	X		X	X	X	X	X	X	X	X			186-1286
Rock	B																					37
Salt	A																					.6
Scioto	A	X		X					X	X	X	X		X	X	X	X					1.4-9.5
Susquehanna	A																					3.9
Wabash	A	X								X			X	X			X	X				3.7-70.7
White	B	X		X						X				X	X	X	X					4.4-19.4
Whitewater	B	X								X												2.3

X = violation of ORBES reference concentration at 7 day-10 year low flow

Table 3.—Aquatic habitat impacts and protection level at low flow on rivers studied in detail for strict environmental controls and moderate growth versus lax environmental controls and high electrical energy growth.

River	Strict Impact	Strict Protection Level	Lax Impact	Lax Protection Level
Allegheny	Heavy	C	Heavy	C
Beaver	Heavy	C	Heavy	C
Big Muddy	Moderate	B	Drastic	D
Big Sandy	Heavy	C	Drastic	D
Cumberland	Heavy	C	Heavy	C
Great Miami	Heavy	C	Drastic	D
Green	Heavy	C	Heavy	C
Illinois	Moderate	B	Heavy	C
Kanawha	Heavy	C	Heavy	C
Kaskaskia	Moderate	B	Moderate	B
Kentucky	*	*	*	*
Licking	Moderate	B	Heavy	C
Little Miami	Heavy	D	Drastic	D
Mississippi	Moderate	B	Moderate	B
Monongahela	Heavy	C	Heavy	C
Muskingum	Heavy	C	Heavy	C
Ohio Main Stem	Heavy	C	Heavy	C
Rock	*	*	*	*
Salt	*	*	*	*
Scioto	Heavy	C	Heavy	C
Susquehanna	*	*	*	*
Wabash	Moderate	B	Heavy	C
White	Moderate	C	Heavy	D
Whitewater	*	*	*	*

* Background data incomplete; analysis could not be conducted

Drastic--WQII ≥ 50%. Eutrophication, heavy metal salts, dissolved oxygen depletion, siltation and stream dessication would all combine to essentially destroy the system existent at the time. Extensive fish kills would be expected all along the waterway with near complete loss of embryonic fishes. The period of recovery would be long depending on the final condition of the watershed. Three grade drops in protection level for an extensive period.

DISCUSSION OF SCENARIO IMPACTS

The 1976 pre-scenario concentrations at 7-Q-10 of the components listed in table 1 were of major concern to the aquatic scientists. These were high (see table 2) and thus tended to dominate analysis of the scenarios. Subsequently the only scenario to show an improvement in stream conditions was the one which included strict environmental controls imposed in 1975 and thence onward through the year 2000. A comparison of the scenario with strict environmental controls versus impacts anticipated as a result of high growth and lax environmental controls from 1976 through 2000 is given in table 3.

Specific and distinctive scenario effects (as differing from those due solely to pre-existing high backgrounds) were observed in all cases in which power plants were sited on streams having 7-Q-10 values less than 2.832 m^3 sec^{-1}. Though these effects could always be relieved by shifting the facilities to the main stem of the Ohio River, it was found that such a concentrating of power plants then greatly exaccerbated the regional air quality problems.

LITERATURE CITED

Brill, E.D., Jr., Shoou-Yuh Chang, Robert W. Fuessle, and Randolph M. Lyon. 1980. Potential Water Quantity and Water Quality Impacts of Power Development Scenarios on Major Rivers in the Ohio Basin. Subcontract under Prime Contract EPA R805588. University of Illinois, Urbana-Champaign.

Leuthart, Clara A., and Hugh T. Spencer. 1980. Fish Resources and Aquatic Habitat Impact Assessment Methodology for the Ohio River Basin Energy Study Region. Grant No. EPA R804816. 446 p. University of Louisville, Louisville, KY.

U.S. EPA. 1981. Ohio River Basin Energy Study (ORBES) Main Report. 311 p. U.S. EPA 600/7-81-008. 311 p. Ofice of Research and Development, U. S. EPA, Washington, D. C.

AN ANALYSIS OF ENERGETIC AND ECONOMIC VALUES ASSOCIATED WITH THE DECLINE OF SUBMERGED MACROPHYTIC COMMUNITIES IN CHESAPEAKE BAY[1]

W. R. Boynton[2], W. M. Kemp[3], A. J. Hermann[3], J. R. Kahn[4,5], T. R. Schueler[3], S. Bollinger[3], S. C. Lonergan[6,7], J. C. Stevenson[3], R. Twilley[3], K. Staver[3], and J. J. Zucchetto[6]

Abstract.--A methodology was developed for investigating the regional implications of a major perturbation to an estuarine system. This analysis coupled a suite of simulation models with the concepts of embodied energy and economic surplus to estimate cost/benefit ratios for several watershed management scenarios. Preliminary results suggest that energetic and economic evaluations tend to converge.

INTRODUCTION

In most developed regions of the world, conflicts have arisen concerning competing uses of various natural resources. For example, in many estuarine environments the aquatic resources are used both for disposal of anthropogenic wastes and for extraction of fisheries, despite the fact that the former may cause deleterious effects on the latter. The resoution of such conflicts is a matter of great concern, but it is unclear as to the appropriate methodologies for judicious allocation of competing uses. Several generic problems complicate these issues, including: 1) understanding cause/effect relationships between specific human activities and ecosystem responses; 2) evaluating quantitatively the impact of observed ecological changes on overall functioning of natural systems; and 3) estimating the loss of value resulting from any such changes in ecosystem functions. The last of these problems is, perhaps, the most vexing, and a number of approaches have been utilized for resource evaluation.

One of the oldest techniques for quantitative measurement of values associated with resource utilization is the traditional economic Benefit/Cost Analysis (BCA). Over the last several decades this method has been applied to a variety of issues, many of which involved water resource projects (e.g., Howe 1971, James and Lee 1971). A major criticism of the BCA as it has been used traditionally for environmental decision-making is its inability to quantify environmental costs which do not have a defined market-place. In an effort to deal with this problem of unequal quantification of environmental amenities, several subjectively quantitative schemes were developed, including the matrix approach of Leopold et al. (1971), the "Delphi process" of Dee et al. (1973), and the mapping procedures of McHarg (1969). With the recent emergence of "resource economics" as a reputable discipline, a new approach has been developed for estimating economic "shadow prices" to measure the value of non-market environmental goods and services (Krutilla and Fisher 1975). This method uses the concepts of marginal utility and consumer and producer surplus to evaluate environmental externalities (Kahn 1981).

More recently, an alternative approach, referred to as "energy analysis", has been developed, utilizing the basic principles of thermodynamics for evaluation of costs and benefits in energetic terms (Odum 1971). In this approach, energy flow is viewed as the ultimate limiting and driving factor in all processes, so that an activity in any system, whether or not it involves economic markets, can be measured in terms of energy. Energy analysis has been applied to a variety of issues of resource utilization (e.g., Gilliland 1975); however, it has been widely criticized by economists for violating the fundamental economic concepts of utility (Huettner 1976, Shabman and Baite 1978, Langham and McPherson 1976). Many aspects of energy analysis are discussed in this volume (e.g., Hannon, Richardson and Odum,

[1]Paper presented at the international symposium Energy and Ecological Modelling, sponsored by the International Society for Ecological Modelling. (Louisville, Kentucky, April 20-23, 1981).
[2]University of Maryland, Center for Environmental and Estuarine Studies (UMCEES), Chesapeake Biological Laboratory, Box 38, Solomons, MD. 20688
[3]UMCEES, Horn Point Environmental Laboratories, Box 775, Cambridge, MD. 21613
[4]Dept. of Economics and Bureau of Business and Economic Research, University of Maryland, College Park, MD. 20742
[5]Dept. of Economics, State University of New York, Binghampton, N.Y. 13901
[6]University of Pennsylvania, Dept. of Regional Science, Philadelphia, PA.
[7]Dept. of Geography, McMaster University, Hamilton, Ontario, Canada

Costanza and Neill, Lavine and Butler, Herendeen and others).

One environmental system whose resources are in continual conflicting demand is the Chesapeake Bay. With major urban centers on its shores, agricultural land-uses dominating its watershed, and fisheries and boating comprising important cultural and economic activities in its waters, the Chesapeake region is characterized by a myriad of legitimate but competitive uses of its estuary. This estuarine ecosystem has undergone many changes in its biological resources over the last several decades, including the drastic decline of submerged aquatic vegetation (SAV) communities which once dominated its littoral zone (Stevenson and Confer 1978, Bayley et al. 1978, Brush et al. 1980). Coincident with this loss of aquatic vegetation there have been significant changes in water quality, including increases in nutrients, phytoplankton chlorophyll, and turbidity, as well as the introduction of agricultural herbicides (Heinle et al. 1980, Mihursky and Boynton 1978, Stevenson and Confer 1978).

We have postulated (Stevenson et al. 1978) that changes in water quality are generally attributable to increased waste loading to the Bay, and that changing water quality has led to the loss of SAV. Moreover, we hypothesized that the decline in SAV has contributed to detrimental changes in fisheries production as well as a loss of waste assimilation capacity. In addition to changes in commercial fishing, other modifications more difficult to quantify may also be occurring. These include such changes as recreational fishing and hunting, aesthetic factors which attract people to the area, and the lifestyles of rural fishing communities. Hence, we suggest that this SAV issue represents a well focused case-study for examining the problem of managing and balancing the conflicting uses of a large natural resource system.

The purpose of this paper, then, is to present our preliminary analysis of the economic and ecological values associated with the loss of SAV in Chesapeake Bay, and to compare the relative magnitudes of changes in these values resulting from modifications of watershed activities. In this analysis we utilize both economic and energetic techniques which are representative of the current status of their respective methodological disciplines. We compare the outcomes of these two different approaches so as to glean insights into their relative strengths and weaknesses. The data upon which this analysis is based are very preliminary in nature, and we have chosen to use a "worst case" perspective which may tend to exaggerate certain resource values, particularly in the economic approach. For the sake of argument, we have emphasized agriculture as a major source of materials which affect water quality (nutrients, turbidity, herbicides); however, in many situations other sources will far outweigh those from farming. In any case our intent here is not

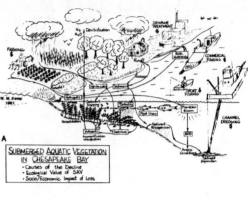

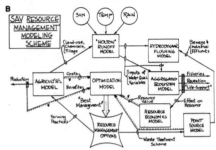

Fig. 1. Conceptual models of estuarine resource conflicts related to submersed macrophytes: a) pictorial representation; b) analytical modeling scheme.

so much to suggest what caused the SAV decline or the resulting loss of value in precise terms, as it is to present a methodological perspective on how such large-scale environmental issues can be treated with consistent and objective logic.

APPROACH TO PROBLEM

To clarify our approach to this conflict in resource use, we illustrate the web of interacting activities which relate to SAV in the pictorial fashion of Fig. 1a. Here, SAV are shown to act as natural nutrient sinks and sediment traps, both processes having economic analogs in terms of sewage treatment plants and channel dredging operations, respectively. Additionally, SAV communities are suggested to be important sources of food and habitat promoting growth of fish, shellfish and waterfowl stocks which are harvested in commercial and recreational endeavors. Various watershed activities are shown to influence estuarine water quality (nutrient, sediment and herbicide additions) via direct discharges and run-off which are in turn regulated by rainfall and other factors. Throughout this cycle some economic enterprises (e.g., farming) influence SAV while others (e.g., fishing and dredging) are affected by losses of SAV. While this presentation is useful to overview the issues, it is not sufficient to provide an operational format in which causes of SAV decline can be related quantitatively to the health and production of SAV ecosystems and finally to the relative costs and benefits associated with improving estuarine water quality.

The overall modeling framework used in this study is shown in Fig. 1b, which is a flow chart formalizing some of the concepts indicated in Fig. 1a. Briefly, at the top of the diagram sunlight, temperature, rain and agricultural chemicals are shown as inputs to a run-off model (Holton Model) which routes water, nutrients, sediments and herbicides from fields to the estuary. A simplified hydrodynamic model, based largely on steady-state flushing rates, receives agricultural runoff, sewage, and industrial inputs and transports water and materials through the estuary providing appropriate water quality concentrations to which SAV communities are exposed. These materials, along with direct agricultural runoff, provide inputs to the aggregated SAV ecosystem model (ASEM) which is a non-linear simulation model emphasizing the effects of water quality on SAV growth as well as the overall yield of fishery products. The outputs from the ASEM become input functions to a resource economic model which estimates equivalent economic value associated with changes in fish populations. Similarly, the changes in embodied energy resulting from changes in SAV also are calculated. Thus, alterations in diffuse and point-source loadings can be traced through a causal chain where they affect SAV and, in turn, fishery yields.

On the left side of Fig. 1b, an agricultural model estimates the marginal costs and benefits associated with different agricultural practices which in turn result in different loadings to the estuary. We also indicate in the diagram an optimization model which does not exist at present but could be devised to maximize any of a number of objective functions. Finally, the output of this analysis becomes a highly synthesized set of resource management options which managers, law makers and voters could consider. In the following sections the results and implications of this study are presented for the Choptank River basin, a large tributary of Chesapeake Bay.

RUNOFF OF NUTRIENTS, HERBICIDES, AND SEDIMENTS

Runoff rates for both nitrogen and sediment from various land uses in the Choptank watershed were computed using a modified version of the USDAHL runoff model developed by the Department of Agricultural Engineering at the University of Maryland (Holtan and Yaramanglou 1979). This model utilizes a revised form of the Universal Soil Loss Equation coupled with a detailed hydrologic model applicable to coastal plain areas (Fig 2a), as an esti-

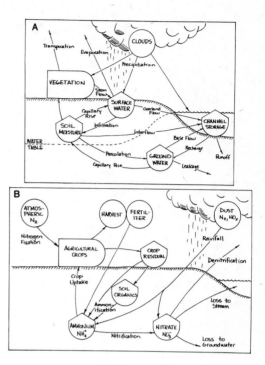

Fig. 2. Simplified representations of inputs, storages, and fluxes related to a) water and b) nitrogen export from cropland in the Choptank River watershed.

mator of sediment yield during storm events. This revised formulation considers sediment transport as proportional to peak flow of runoff waters during any runoff event.

We modified the nitrogen elements of this model to include greater biological detail. Here, nitrogen (primarily nitrate) is routed passively along most of the major pathways of the hydrologic cycle: infiltration, interflow, runoff, and evapotranspiration by plants. Biologically-mediated transformations of nitrogen, including ammonification, nitrogen-fixation, nitrification, and denitrification, were altered for the sake of both realism and precision (Fig. 2b).

Dates of tillage and mass of applied fertilizer were entered as forcing functions to the runoff model along with break-point segmented precipitation data. Soil types and land use parameters were likewise entered for each major category of land use. The model was calibrated using data collected from a flumed, predominantly agricultural sub-watershed of the Choptank River. Precipitation data from 1979 were employed for calibration of this model. This was a moderately wet year, as indicated by a thirty year record of

annual precipitation in this region. Major land
uses in the Choptank watershed considered in this
study include: cropping in corn, soybeans and
small grains using conventional and limited tillage ("no-till") practices; pasture; residen-
tial/commercial areas; forested lands; and construction sites. Simulations for each major category were performed, and the resultant loading
factors were multiplied by estimated acerage
(USDA 1978) to determine the total yearly loading
for the watershed.

The seasonal patterns of nitrate yield were
driven especially by the seasonal patterns of
rainfall and temperature, and by the timing of
fertilizer applications. Highest rates of nitrogen runoff commonly occur in the winter and
spring in the Choptank watershed; the predicted
runoff for 1979 fit this general pattern. Total
yearly runoff was calibrated using the mass estimates of Christy (1980) for the flumed subwater-
shed during that year. Changing the rate, but not
the timing, of fertilization altered the predicted nitrogen yield for the spring, without
shifting the overall seasonal runoff pattern.

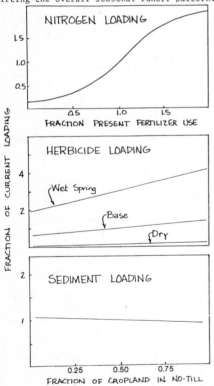

Fig. 3. Effects of tillage and fertilizer management on watershed runoff from cropland to
the Choptank River estuary. Currently, 50%
of cropland is in no-till.

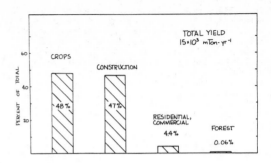

Fig. 4. Model-assisted sediment budget for the
Choptank River Basin

Conventional tillage involves yearly plowing
of cropped land; "no-till" methods do not, and
hence result in considerably lower sediment
yields from that land. Currently, about half of
the cropped land in the Choptank Basin is tilled
using conventional methods; the other half by
no-till (Ron Wade, Cooperative Extension Service,
pers. comm.). Using the runoff model output, we
related a percentage change in no-till usage for
current cropland to a percentage change in total
watershed sediment yield, and the slope of this
function appears to be slight (Fig. 3). This results because cropland is not the only major
source of sediment in this watershed. In fact, a
model assisted sediment budget for the region
(Fig. 4). suggests that construction losses are
potentially on a par with agricultural losses.

Herbicides were not explicitly a part of the
runoff model as it currently exists. Hence, we
chose a different approach for estimating their
annual delivery. Herbicides commonly used in the
Choptank Basin include atrazine, simazine, para-
quat, alachlor, 2-4 D and trifluralin. An extensive literature search yielded a range of values
for observed runoff of each chemical in coastal
plain or Piedmont soils, expressed as a percentage of applied mass (Schueler 1979). We used the
median percent yield for each herbicide as an
estimator for that compound in the Choptank Basin
under average conditions of rainfall and timing
of application. The highest percentage derived
from the literature was taken to reflect maximal
yield, presumably in a "wet" year where a major
storm event takes place just after the compound
is applied. The lowest percentage was taken as
representing a "dry" year, or best possible
timing of application between rainfall events.
For three rainfall/timing scenarios, then, we had
an estimate of yield for each chemical. Actual
use of each herbicide under recent land-uses was
derived from data presented in Stevenson and
Confer (1978). Applied mass for each was multiplied by the appropriate loading factor, and the
yields summed over all chemicals for each sce-

nario. For "dry", "average", and "wet" years, we estimated 0.44, 7.68, and 23.34 mg (m² cropland)⁻¹ yr⁻¹, respectively. The derived total yield in each case was presumed to enter the macrophyte beds and the estuarine waters at the beginning of May, dispersed evenly over one day's runoff driven by a major storm event at that time. Tillage practice affects at least the quantity, and possibly the timing, of herbicide applications to cropland. The input—output data of Stevens et al. (1981) for Maryland agriculture suggest that a typical acre of cropland in conventional tillage receives 4.5 kg herbicide compared to 9.5 kg for a typical hectare with no-till methods. These data were also employed in our estimate of herbicide runoff.

The combined runoff of nitrogen, sediment, and herbicides flows into the macrophyte beds by two routes: 1) direct impact and 2) exchange with open water. Direct impact of runoff on macrophytes occurs especially in coves, where the beds receive waters from the drainage. To estimate the loading of such shallow areas with agrochemicals, we compared the total area of watershed with the total area of shallow (0–2 m mean low water, MLW) waters in the Choptank. The ratio of these two areas is 16.8. Hence, as an estimate of direct loading, the runoff per m² of watershed was multiplied by 16.8 to yield runoff per m² of macrophyte bed.

Indirect loading of nitrogen, sediment, and herbicide to the macrophyte beds depends on the flushing rate of bed water with the river water, and the concentration gradient between those two pools. The flushing rate was calculated using a 1.0 m tidal range and a 33% flushing efficiency. For shallow (0–1 m MLW) areas, the flushing rate was 0.22 day⁻¹; for deeper (1–2 m MLW) areas, that rate was 0.33 day⁻¹. The ambient estuarine concentration of nitrogen was computed using the current seasonal pattern, the ratio of projected to current nonpoint loading, and the ratio of nonpoint to point source nitrogen loading to the Choptank. Present dissolved inorganic nitrogen (DIN) levels in the Choptank range from 2–100 µg-at; such high levels are presumed due almost entirely to the nonpoint plus point source load experienced by the river. As a first approximation we assumed that a given percentage increase in the current nitrogen load effects an equivalent percentage increase in the ambient concentrations throughout the year.

Herbicide levels vary only slightly with season in the deeper, open waters of the Choptank. A constant mean concentration was employed for modelling purposes throughout the year. As with DIN, increased watershed loading effected an equivalent increased mean concentration.

Sediment deposited in the Choptank River is subject to resuspension by wind and tide-induced turbulence. At the same time, natural processes act to consolidate loose bottom sediments. Eventually, heightened sediment loads to the river should raise the mass of the unconsolidated sediment pool. Insofar as sediment deposition, resuspension, and consolidation are linear, donor-controlled processes a doubling of sediment to the river should, in the long run, result in a doubling of ambient suspended sediment concentrations. Turbulence sets the resuspension rate of available sediment, and is itself driven by a seasonal pattern of wind and tidal range.

ESTUARINE CONCENTRATIONS AND EFFECTS ON SAV

Nitrogen affects the growth of estuarine macrophytes through two mechanisms. First, so long as DIN is the primary limiting nutrient for these plants, a rise in the nitrogen level effects a rise in macrophyte productivity. At the same time, heightened levels of DIN spur productivity of plankton and epiflora, plus the epifauna which consume these components. Together, plankton and epiphytes shade the macrophyte beds. We postulate that in the Choptank River, macrophytes are presently light limited, in part due to this mechanism.

Turbidity caused by suspended sediment likewise blocks the light reaching submersed macrophytes. In calm waters, sediment particles may settle directly on the plants; in that case shading is especially severe. Epiphytes may also trap more sediment than would be the case with an uncolonized plant; hence they exacerbate the sediment shading problem. Shading by the above mechanisms is especially problematic in deeper waters. For the purposes of this study, we assumed all present or projected macrophyte growth occurrred only in the 0–2 m mean low water range.

Herbicides stress macrophytic production by interfering with the photosynthetic pathway in those plants. For example, triazine herbicides block the Hill Reaction in photosynthesis. Toxicity varies with plant species and chemical. At least partial, and possibly complete, recovery ensues once the interfering compound has degraded or been flushed out of the surrounding waters. Experiments conducted by Jones et al. (1981) using waters and sediments from the Choptank River indicate degradation of dissolved or adsorbed atrazine (a widely used triazine compound) proceeds with a half-life of one month or less. Coupled with a considerable flushing rate for even deeper beds, and the low herbicide content of ambient waters, we expect a rapid exponential decay of this toxicant in the vicinity of the macrophytes.

We incorporated these ecological processes into an aggregated SAV ecosystem model (ASEM) to estimate the effects of projected runoff on macrophytic growth in both shallow (0–1 m) and deeper (1–2 m) habitats. An overall view of model components and design is given by Fig. 5. Flows among fifteen state variables are driven by eleven seasonally varying forcing functions in this scheme. The interactive effects of light/turbidity, dis-

446

solved inorganic nitrogen, and herbicides (atrazine) were included in the growth and mortality
functions for macrophytes, plankton, and epiphytes. Each of the dissolved or suspended components of this system are subject to exchange with
the open waters adjacent to the macrophyte bed.
The combined influence of wind, tidal exchange,
and macrophyte density yield a turbulence level
which controls diffusion, resuspension, and deposition processes within the bed. Temperature affects gross production and respiration of autotrophic components, grazing and respiration of
heterotrophs, and the seasonal dieback of the
macrophytes. By the reasoning outlined earlier,
forcing functions of direct runoff and ambient
concentrations of nitrogen, sediment, and herbicide were multiplied by a range of factors. What
emerged was a series of graphs relating potential
macrophytic productivity per unit area in each
depth range to each of the three runoff-derived
components (Fig. 6).According to the ASEM nitrogen and especially sediment exerted a strong
influence on long term macrophyte viability. The
effect of herbicides, while probably important in
some areas, was not too significant for the river
as a whole. For the present, we have ignored the
synergistic effects of two or more factors
varying simultaneously.

EFFECTS OF SAV PRODUCTION
ON FISH POPULATIONS

At the outset of our research, it was hypothesized that SAV communities possess several functions which are of considerable importance to the
overall estuarine ecosystem. Among these was the
notion that SAV communities are more productive,
in both a primary and secondary sense, than similar but non-vegetated littoral zones. Secondly,
numerous literature sources suggested that SAV

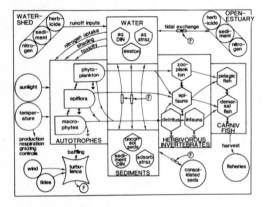

Fig. 5. Overall design of the aggregated SAV ecosystem model (ASEM). Details of component
interactions are given in Kemp et al. (1980)
and Kemp et al. (1981). See Odum and Odum
(1976) for a description of symbols.

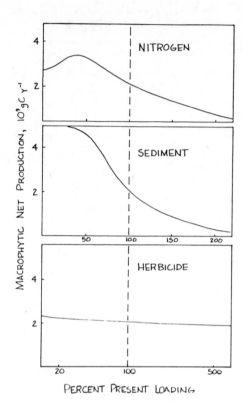

Fig. 6. Effects of watershed nitrogen, sediment,
and herbicide loading on submerged macrophyte production in the Choptank River
estuary.

provide a beneficial and in some cases necessary
habitat for some species, particularly the
younger stages. Thus, the first attribute pro-
motes the growth via food availability of heterotrophic stocks, some of which are of direct value
to man, while the second acts to reduce mortality
rates and in that way increase potential yields.

Space precludes an exhaustive review of
these features, but a few examples from our research in Chesapeake Bay will place the concepts
in perspective. In terms of organic matter production, Kaumeyer et al. (1981) reported that
rates of primary production were 2 to 5 times
higher in vegetated as opposed to non-vegetated
areas. Living plant biomass was up to 100 times
higher. Similarly, Lubbers et al. (1981) reported
that nekton stocks were from 1.5 to 10 times
higher. In terms of habitat, it appeared that at
least one SAV community served as a refuge for
juvenile fish; juveniles less than 1 g indiv.[-1]
were extremely abundant (10–20 g m[-2]) in SAV

areas. Furthermore, grass shrimp (Paleomonetes sp.), which are thought to be an important food item for various fishes, were rare in non-vegetated zones while being very abundant in SAV areas (1-3 g m^{-2}). Additional examples are available in Kemp et al. (1981).

The combined effects of habitat and high primary and secondary production were incorporated into the ASEM and used to project SAV influence on finfish production. Though the model considers both pelagic and demersal types of secondary production, we focused on demersal fishes since these are most dependent on SAV production. Growth of demersal nekton plus macroninvertebrates is proportional to the detrital plus infaunal mass in the current version of the ASEM. To determine an approximate relation between macrophyte and nektonic productivity, we varied the rate of macrophyte gross production by a series of constants. Successive model runs produced pairs of values for total annual macrophytic net production and peak demersal nekton biomass, which were used in developing the function shown in Fig. 7.

COMMERCIAL AND SPORT FISHERIES VALUES ASSOCIATED WITH SAV

We have thus far alluded to the value of macrophytes as reflected by their effect on fisheries. Here we attempt to quantify that fisheries-related value in dollar terms. We employed the Schaefer model of stock/yield relations in our analysis (Schaefer 1954); hence we assumed that fish stocks in the river determined the effort required to catch fish at any time, as well as the sustainable yield through time. The sustainable yield is roughly equal to the rate of growth of the fish population, which in turn is proportional to the stock.

The economic value of a particular yield is here defined using the traditional notion of con-

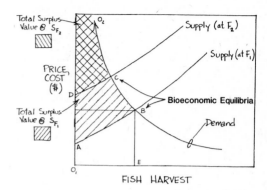

Fig. 8. Illustration of fisheries surplus value concepts with varying fish stocks. See text for explanation of figure.

sumer and producer surplus. The absence of economic rent of the fish stock is due to the open access character of the fishery, as this rent is dissipated through overfishing. Surplus value or "utility" experienced by consumers of commercially-caught fish is defined as the difference between what they are willing to pay for an additional fish and what they actually pay, integrated over the number of fish they actually purchase (Fig. 8). Presumably they find the first fish purchased more valuable than the second, and so on up to point B or C, where perceived value is just equal to the amount paid. The integrated difference is the area under the demand curve, down to the current price. In a similar fashion, producer surplus is the area between the supply curve and the current price. The supply curve is upward-sloping because, among other factors, less productive areas must generally be fished to raise the total harvest.

The current price is set by the intersection of supply and demand functions (points B and C in Fig. 8). These points are termed the "bioeconomic equilibria", since they embody not only market forces, but biological forces as well. Actually, only one point on each supply "curve" corresponds to an achievable, sustained yield of fish, given the fishing pressure set by the demand function for each level of stock. At this bioeconomic equilibrium, demand for fish per unit time is just balanced by the growth rate of those fish. Macrophytes affect the fishery by influencing the specific growth rate of certain species. A decline in macrophytic production thus effects a decline in stock density of those species. For instance, suppose stocks decline from a level F_1 to a new level F_2; this shifts the supply curve upwards from AB to DC since more effort is now required to catch fish. Therefore, the total area between

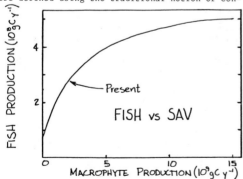

Fig. 7. Predicted effects of submerged macrophyte production on demersal fish production.

supply and demand curves is diminished; "utility"
or "welfare" equal to the area ABCD in Fig. 8 is
lost by users of the commercial fishery.

Sport fishing was treated in much the same
way, except that the welfare associated with the
sportfishing experience had to be calculated in
addition to the welfare associated with landings
of fish. The household production function technique, where the individual is viewed as both
consumer and supplier (producer) of the sportfishing experience, was used to derive demand and
supply curves from survey data on trip costs,
trip characteristics and individual socioeconomic
characteristics.

The results for total recreational plus commercial surplus benefits of major finfish species
plus blue crabs are shown in Fig. 9. In fact,
sufficient data for computing bioeconomic equilibria existed only for striped bass in the Choptank River. However, the analysis was extended to
other demersal finfish and crabs using factors of
proportionality based on relative weights of
catch of the various species. According to this
curve, losses of up to one million dollars per
year were possible, with SAV ranging from maximum
to minimum estimated abundance.

It must be noted that the above analaysis
does not include the recreational value of waterfowl as supported by macrophytes. Further, the
"existence value" of macrophyte-supported fisheries is not considered. This involves the satisfaction of people who do not utilize Choptank
River resources directly, but nonetheless benefit
from knowing that this ecosystem continues to
function properly. Both waterfowl and existence
values would raise the economic surplus loss
associated with a loss of macrophytes.

NON-MARKET VALUES ASSOCIATED WITH SAV

A well defined market currently exists for
the direct fishery-related values of SAV; how-

ever, other functional properties of SAV also
appear to influence human activities but lie outside the perviews of economic consideration. In
an effort to more fully address SAV value we have
applied two additional evaluation procedures, the
first of which uses concepts of energy analysis
as described by Odum and Odum (1976) while the
second ascribes dollar value to two SAV functions
in proportion to the cost of replacing these functions with conventional methods.

Embodied Energy Value of SAV

In this section we estimate SAV value employing energy analysis and, in particular, concepts relating to embodied energy. This approach
takes as its main thesis the notion that available energy both limits and regulates the activities of all systems. Within this framework the
term embodied energy is defined as the direct and
indirect energy needed to maintain any process.
Costanza (1980) has recently applied this concept
to the U.S. economic system. According to this
theory the energy needed to sustain a process
ultimately is equated with value. In the case of
a system of submerged macrophytes, the gross primary productivity (GPP) can be taken as a reasonably inclusive measure of the total amount of
energy used to support all ecological activities.
Thus, an estimate of embodied energy value for
SAV communities in the Choptank estuary was developed by multiplying average values of GPP per
unit area by areal coverage for the present condition and for expected coverages resulting from
four management schemes. The resultant values
were initially expressed as Kcal of GPP area^{-1}
yr^{-1}. However, these values represented work
which was of an inherently lower quality than
that associated with other forms of energy. The
concept implied here, namely energy quality, has
been discussed at length by Odum & Odum (1976)
and others. For consistency with other works we
converted GPP type work to coal equivalent Kcal
(CE) which involved dividing GPP by 10.0.

While GPP may be a good index of work in an
ecological system, it neglects some energy flow
if that system is subjected to exploitation. In
other words, we must expand our calculations to
include the energy expended in extracting the
harvest. In our study we estimated fisheries
catch under several different management scenarios and calculated CE energy expenditure using
energy intensity data developed by Rawitscher and
Mayer (1977).

Other Values Related to SAV Processes

There are, to be sure, additional values
associated with SAV which have yet to be evaluated. Two of the more obvious of these include
the capability of SAV communities to trap and
bind sediments and to assimilate nutrient uptake.
Boynton et al. (1981), using sediment traps and
by measuring tidal differentials in suspended

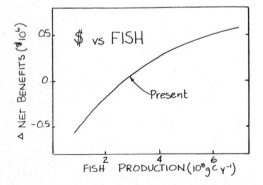

Fig. 9. Loss or gain of consumer-producer surplus
as fish production changes (1978 dollars).

sediments over SAV beds, estimated that between
0.4-0.7 cm of sediments were deposited in an SAV
bed in the Choptank River during the growing season (6 mo.). Whether the deposited material remained throughout the year and became progressively more consolidated is still open to question. We postulate that much of this sediment,
confined to the SAV littoral zone, would otherwise be transported into deep channels, thus contributing to the continuing need for maintainance
dredging of navigation channels.

SAV remove dissolved nitrogen and phosphorus
from the water column (Marbury et al. 1981)
during the growing season and thus serve as a
seasonal sink for these compounds. Additionally,
macrophyte communities may be active zones of
denitrification in which case they serve as terminal nitrogen sinks. Thus, SAV play a role
analogous to tertiary sewage treatment. Dollar
costs are associated with both dredging and wastewater treatment, and while SAV have similar capabilities which might partially mitigate the need
for such economic activities, no payment is made
for these ecological functions. To estimate the
magnitude of those potential values, the cost of
dredging a volume of sediment equivalent to that
accumulation in SAV communities was calculated
using dredging costs summarized by Kemp (1977).
Similarly, the cost of removing a mass of nitrogen via tertiary treatment procedures equivalent
to the mass removed by SAV was calculated using
data developed by Smith and McMichael (1969).
Strictly speaking, this measure of economic value
is not consistent with the estimates of consumer
and producer surplus; however, it is presently
not possible to develop supply and demand curves
for tertiary treatment or maintenance dredging.
We, therefore, provide these calculations simply
as a point of departure for present discussion
and future analysis.

AGRICULTURAL COSTS AND PRODUCTION

Clearly, the above considerations indicate
there are sound economic reasons for promoting
the net production of macrophytes in the Choptank
River. However, such gains might come only at the
expense of some land-based activity such as
farming. For direct comparison with fisheries, we
utilized the same concepts of producer and consumer surplus to quantify the welfare loss in
agriculture stemming from new practices.

Farmers' demand curves were derived using
time series of price and usage of the major variable inputs to agricultural production (land,
labor, fertilizer, herbicides, fuels) combined
with a time series of average yield per acre.
Yearly observations covered the period 1970-1977
in four separate counties which lie within the
Choptank River watershed. Log-linear regressions
related usage of an individual factor in a county
to the price of that factor, the prices of other
factors, and the yield of crops harvested in that
county. Using 1977 prices, each usage relation

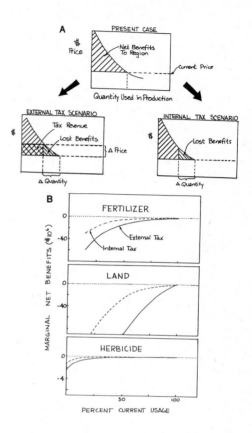

Fig. 10. Effects of management on surplus benefits
from agriculture. a) Surplus concepts for two
control strategies: "external" vs. "internal"
taxation. b) Surplus lost in the Choptank River
Basin with restricted fertilizer, land, and
herbicide usage (1978 dollars).

was then inverted to get a demand curve for that
factor. The supply curve is presumed to have zero
slope for herbicide and fertilizer, since these
are sold by the manufacturers on world markets,
which are hardly affected by the demand of one
watershed for them. As with fisheries, the utility experienced by society (in this case, by the
farmer) is the area under the demand curve and
above the supply line.

Usage of a particular item may be restricted
in several ways, broadly classed in Fig. 10 into
"external" vs. "internal" taxes. Externally, the
federal government could place a stiff tax on
herbicide or fertilizer sold, raising the supply
line and hence reducing consumption. The utility

loss is then due to both reduced yields -- say, the farmer wants to produce more because it is profitable, but cannot -- and to the tax paid out of the region. In the case of the "enlightened" farmer who voluntarily restricts his usage, no tax revenue is lost. A county tax, where revenues were returned to the farmers in some fashion, yields the same result.

We applied these techniques for land, fertilizer, and herbicides, and obtained the results shown in Fig. 10. Losses by county were area-weighted to obtain loss in the watershed. These graphs suggest that an especially high value is placed on both fertilizer and land in production by farmers. It appears that they would be more willing to part with herbicides if necessary, since the welfare loss is smaller. The external tax, as expected from Figure 10. produced a much larger welfare loss than the internal tax, especially in the vicinity of 100% of current usage. All subsequent calculations presented in this paper utilize this smaller loss of the "enlightened farmer" or farming community.

For a full comparison with the SAV values, we needed to quantify the agricultural values in embodied energy terms as well as dollars. For this purpose we considered the embodied energy throughput of current agricultural practices in the Choptank watershed. We treated embodied energy as a conservative property, presently in steady state within the agricultural sector. Here, embodied energy per unit of production was taken as the ratio of total energy input (direct and indirect) to the associated yield under the present methods of cultivation. It is assumed that the current methods are the most energy efficient ones, given the technological constraints of the times. Yields employed were total bushels for all crops; as with the fisheries analysis, energies were expressed in coal equivalent (CE) kilocalories. Energy embodied in the production of fuels, chemicals and machinery used in agriculture, as well as the embodied energy of farm

labor, were obtained from the work of Pimentel et al. (1973) and Heichel (1976). Average yields were obtained from the historical time series for corn and soybeans (MDA 1976-1980).

Yield varies with input, so that both embodied energy usage and embodied energy export changed under new scenarios of management. Projected yields were obtained using a Cobb-Douglas type production function derived by Lonergan et al. (1980) for the eastern shore of Maryland. A production surface of the type shown in Fig. 11. was used to relate the expected yield of corn (on the vertical axis) to various combinations of fertilizer and herbicides. Consistent with the surplus loss curves of Fig. 10, we see that fertilizer affects yield more strongly than herbicides. The production function was calibrated for average practice and yield for: conventionally tilled corn, no-till corn, conventionally tilled soybeans, and no-till, double-cropped soybeans with barley. Materials used in these four land uses under various management schemes were obtained from the work of Stevens et al. (1981). Acerage of each presently in the Choptank watershed was obtained from the Delmarva peninsula land use survey (USDA, 1978) and the Dorchester County agricultural extension agent (Ron Wade, pers. comm.).

ECONOMIC AND ENERGETIC BENEFITS, COSTS, AND YIELDS FOR VARIOUS MANAGEMENT STRATEGIES

The various marginal costs and benefits associated with four watershed management activities are summarized in Table 1 for both surplus economic value and embodied energy value. The primary economic analysis considers surplus losses from agriculture vs. surplus gains in fisheries, while the primary energetic analysis considers embodied energy throughput losses from cropland vs. energy throughput gain in the macrophyte fisheries system. Energy throughput changes are compared on an equivalent quality basis. Since we assumed that the present management is the most energy-efficient one for agriculture, and that embodied energy is in steady state with that sector presently, any new management option resulted in an embodied energy loss from agriculture.

Four different scenarios are considered here: two involve restricted fertilizer usage, one involves taking agricultural land out of production, and a fourth involves a conversion (actually, a reversion) back to conventional tillage techniques for all cropland. Losses or gains are shown relative to present land use, in millions of 1978 dollars for the surplus analysis, and billions of CE kcal for the embodied energy analysis. Also shown are the cost-benefit ratios, obtained by dividing agricultural loss by estuary gain. The upper ratio A:E for economic cost/benefit employs only the fisheries-related surplus changes; the lower number (A:E*) includes the equivalent cost of waste treatment and

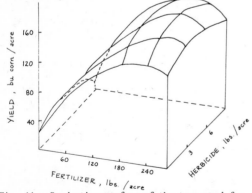

Fig. 11. Production surface of the type used for agriculture in the Choptank Basin. A Cobb-Douglas production function here relates fertilizer and herbicide usage to expected yields of corn.

Table 1. Marginal costs (−), benefits (+), and cost: benefit
ratios derived using economic and energetic analyses of
agriculture/estuary trade-offs

METHOD Sector	MANAGEMENT SCENARIOS 100% CONVENTIONAL TILLAGE	FERTILIZER USAGE 25% REDUCTION	50% REDUCTION	10% REDUCTION AGRICULTURAL LAND
CONSUMER SURPLUS ($10^6, US 1978)				
Agriculture (A)	−0.03	−0.9	−5.3	−0.2
Estuary (E)	−0.02	+0.1	+0.2	+0.05
Ratio, A:E	NA[1]	9.1	35.5	3.6
(Ratio, A:E*)[2]	(NA)	(6.5)	(23.2)	(3.1)
EMBODIED ENERGY (10^9 Kcal CE)				
Agriculture (A)	−259.0	−16.7	−152.0	NA[3]
Estuary (E)	−2.7	+4.9	+11.2	+1.3
Ratio, A:E	NA	3.4	14.0	NA
(Ratio, A:E**)[4]	(NA)	(3.0)	(12.3)	(NA)

[1]Not applicable.
[2]Includes dollar value of additional waste treatment and sediment trapping supplied by heightened macrophyte production; see text.
[3]See text.
[4]Includes coal equivalent Kcal throughput of new fisheries activity attracted by heightened macrophyte production; see text.

dredging to replace services lost by reduced macrophyte production. In a like manner, the upper ratio for energetic cost/benefit uses only the coal-equivalent loss of gross production of macrophytes, while the lower ratio (A:E**) includes the gain of embodied energy throughput of fisheries newly attracted to the region because of higher fish stocks.

For the reversion back from 50% no-till to 100% conventional tillage, the ratio is meaningless since both sectors suffer. On the other hand, a reduction by 50% in fertilizer usage incurs huge agricultural losses which are far from offset by the gained estuarine benefits and productivity. With a 10% loss of agricultural land in production, consumer surplus loss is greater than macrophyte-derived surplus gain. The low cost/benefit ratio of about 3.0 (A:E*) indicates some trade-off may be possible without losing significant total surplus. An apparent deficiency in our embodied energy method is evident in the fact that reducing cropped land by 10%, while maintaining standard cropping practice per acre, results in no net energy loss or gain. For this reason the 10% cropland reduction scenario is not included in the energy cost/benefit ratio calculation in Table 1. In the 25% fertilizer reduction scenario, the 16.7 billion Kcal loss from agriculture is slight compared to the other reduced fertilizer scenario. The 4.9 billion gain to the estuary is certainly within range of the energetic loss from agriculture, and the ratio of the two is small, especially when the attracted energy to fisheries is included

(A:E**). Given the uncertainties in the data, this cost:benefit ratio may not be significantly different from unity. Presumably then, a some degree of modification in agricultural practices may be possible without much loss of embodied energy. For example, creative parsing of farmland -- such as through planting buffer strips adjacent to drainage channels -- might boost estuarine gains and lower the cost/benefit ratio even further.

Bearing in mind the tentative nature of these calculations, we nevertheless might ask why agriculture is the dominant term for any new scenario. The Bay is currently managed to satisfy a variety of interests, both recreational and commercial. Conceivably that management focus could shift entirely to commercial concerns, e.g. towards the intensive aquaculture of the type currently practiced in the Orient. This would greatly increase the relative value of the estuarine system (in either economic or energetic terms), such that a small reduction in productivity could well swamp the computed benefits to agriculture when the two conflict. The energy throughput of the intensely managed estuarine system would be much greater; presumably so would the combined utility derived from commercial plus recreational fisheries, even though recreational use had declined. Present-day commercial use of the Bay is nowhere near as fossil-fuel intensive as agricultural use of the watershed. Well-established methods for using high-quality energy to produce high-quality carbohydrates and protein exist for land, while equivalent methods are available in only a few instances for aquatic systems. We do not necessarily advocate the development of energy-intensive aquaculture for the Choptank, but merely point out how such a technology, if widespread, would alter the current calculation.

We note that in each case the cost/benefit ratio using economic surplus is higher than the cost/benefit ratio using embodied energy. This probably reflects the more inclusive nature of embodied energy as a measurement index. Still, the results of the two analyses tend to converge for most scenarios considered. In the final analysis, considering the uncertainties in the data employed, it is surprising that the ratios for the two methods are so comparable. Perhaps, as Odum (1971) and Costanza (1980) have suggested, in the long run market processes will respond to energy constraints such that economic and energetic considerations will lead to similar conclusions.

The results of this analysis represent an objective set of quantitative measures of resource value under a variety of management scenarios. Some might argue that neither economic nor energetic techniques is inherently comprehensive, because such factors as morality, justice, aesthetics, and culture are not considered directly. Moreover, both analyses require predictions of system responses to untested management scenarios, and our ability to make these forecasts with confidence is in its infancy. Hence, such analyses should be considered as but one of several inputs to the democratic process whereby resource uses are decided. Finally, we must again emphasize the fact that the data employed in this study are very preliminary, and that no final conclusions should be drawn as to the economically or energetically "optimum" use of the Bay resources.

ACKNOWLEDGEMENTS

Valuable contributions were made by many individuals and groups in conducting this study. Many concepts used here first emerged at the University of Florida in the systems ecology group of H.T. Odum. We gratefully acknowledge the theoretical contributions of John Cumberland and Bob Costanza, as well as the conceptual counterpoints provided by Mark Sagoff. Much of the empirical underpinnings of this study are due to the efforts of L. Lubbers, K. Kaumeyer, S. Bunker, D. Marbury, J. Cunningham, T. Jones and J. Metz. This work was supported primarily by the Environmental Protection Agency's Chesapeake Bay Program and the Maryland Department of Natural Resources Tidewater Administration.

REFERENCES

Bayley, S., V.D. Stotts, P.F. Springer, and J. Steenis, 1978. Changes in submerged aquatic macrophyte populations at the head of Chesapeake Bay, 1958-1975. Estuaries 1(3):171-182.

Boynton, W.R., W.M. Kemp, L. Lubbers, III, K. Kaumeyer, S. Bunker, K. Staver, and J. Means. 1981. Influence of submerged macrophyte communities on turbidity and sedimentation in littoral zones of northern Chesapeake Bay. p. 842-875. In: Kemp et al. (eds.) Submerged aquatic vegetation in Chesapeake Bay: Its ecological role in Bay ecosystems and factors leading to its decline. Univ. of Maryland, Center for Environmental and Estuarine Studies, Cambridge, MD. Ref no. UMCEES 81-20 HPEL.

Brush, G.S., F.W. Davis, and S. Runner. 1980. Biostratigraphy of Chesapeake Bay and its tributaries. A feasibility study. Chesapeake Bay Program, U.S. Environmental Protection Agency, Annapolis, MD. EPA 600/8-80-040.

Christy, M.S. 1980. Investigations into the dynamics of water movement on a low relief landscape on the Eastern Shore of Maryland. M.S. Thesis. University of Maryland, Department of Agricultural Engineering. College Park, 156 pp.

Costanza, R. 1980. Embodied energy and economic valuation. Science 210:1219-1224.

Dee, N., J. Baker, N. Brobney, K. Duke, I. Whitman, and D. Fahringer. 1973. An environmental evaluation system for water resources planning. Water Resources Res. 9:523-535.

Gilliland, M.W. 1975. Energy analysis and public policy. Science 189:1051-1056.

Heichel, G.H. 1976. Agricultural production and energy resources. American Scientist 64:64-72.

Heinle, D.R., C.F. D'Elia, J.L. Taft, J.S. Wilson, M. Cole-Jones, A.B. Caplins, and L.E. Cronin. 1980. Historical review of water quality and climatic data from Chesapeake Bay with emphasis on effects of enrichment. Chesapeake Biol. Lab. Solomons, MD. Ref. No. UMCEES 80-15 CBL.

Howe, C.W. 1971. Benefit-cost analysis for water system planning. Water Resources Monogr. 2. Amer. Geophys. Union Publ. Press, Inc., Baltimore. 144 pp.

Holtan, H.N. and M. Yaramanglou. 1979. Procedures manual for sediment, phosphorus and nitrogen transport computations with USDAHL. Maryland Agricultural Experiment Station, University of Maryland, College Park. 32 pp.

Huettner, D.A. 1976. Net energy analysis: an economic assessment. Science 192:101-104.

James, D., and R.T. Lee. 1971. Economics of water resources planning. McGraw Hill Book Co., New York.

Jones, T.W., J.C. Means, J.C. Stevenson, W.M. Kemp, N. Kaumeyer, and J. Metz. 1981. Degradation of atrazine in estuarine water/sediment systems and selected soils. p. 231-268. In: Kemp et al. (eds.). Submerged aquatic vegetation in Chesapeake Bay: its ecological role in Bay ecosystems and factors leading to its decline. Submitted to Chesapeake Bay Program, USEPA, Annapolis, MD. Ref. No. UMCEES 81-28 HPEL.

Kahn, J.R. 1981. The economic losses associated with depletion of submerged aquatic vegetation in Chesapeake Bay. Ph.D. Thesis. Dept. of Economics, Univ. of Maryland, College Park.

Kaumeyer, K., W.R. Boynton, L. Lubbers, K. Staver, S. Bunker, W.M. Kemp and J.C. Means. 1981. Metabolism and biomass of submerged macrophyte communities in northern Chesapeake Bay. p. 353-400. In: Kemp et al. (eds.). Submerged aquatic vegetation in Chesapeake Bay: its ecological role in Bay ecosystems and factors leading to its decline. Submitted to Chesapeake Bay Program, USEPA, Annapolis, MD. Ref. No. UMCEES 81-28 HPEL.

Kemp, W.M. 1977. Energy analysis and ecological evaluation of a coastal power plant. Ph.D. Thesis. Dept. of Environmental Engineering, Univ. of Florida, Gainesville.

Kemp, W.M., A. Hermann, and W.R. Boynton. 1980. Resource dynamics and ecology of submerged aquatic vegetation in Chesapeake Bay: A model to demonstrate resource management concepts. Univ. of Maryland, Center for Environmental and Estuarirne Studies, Cambridge, MD. Ref. No. UMCEES 80-168 HPEL.

Kemp, W.M., J.C. Stevenson, W.R. Boynton, and J.C. Means (eds.). 1981. Submerged aquatic vegetation in Chesapeake Bay: Its ecological role in Bay ecosystems and factors leading to its decline. Univ. of Maryland, Center for Environmental and Estuarine Studies, Cambridge, MD. Ref. No. UMCEES 81-28 HPEL.

Krutilla, J.V., and A.C. Fisher. 1975. The economics of natural environments. The Johns Hopkins Univ. Press, Baltimore.

Langham, M.R., and W.W. McPherson. 1976. Energy analysis. Science 192:8.

Leopold, L.B., F.E. Clarke, B.B. Hanshaw, and J.W. Balsley. 1971. A procedure for evaluating environmental impact. U.S. Geol. Sur. Circ. 645. Washington, D.C. 13 pp.

Lonergan, S.C., T.R. Schueler, and W.M. Kemp. 1980. Estimating the effects of fertilizer and herbicide applications on agricultural costs and crop yields in the Choptank River Watershed: a cost and production function approach. In: Kemp, et al. 1980. Resource dynamics and ecology of submerged aquatic vegetation in Chesapeake Bay: A model to demonstrate management concepts. Univ. of Maryland, Center for Environmental and Estuarine Studies, Cambridge, MD. Ref. No. UMCEES 80-168 HPEL.

Lubbers, L., S. Bunker, K. Staver, W. Boynton, N. Burger, M. Meteyer and W. Kemp. 1981. Comparitive abundance and structure of littoral nekton communities at vegetated and non-vegetated sites in the Chesapeake Bay. p. 461-574. In: Kemp et al. (eds.). Submerged aquatic vegetation in Chesapeake Bay: its ecological role in Bay ecosystems and factors leading to its decline. Submitted to Chesapeake Bay Program, USEPA, Annapolis, MD. Ref. No. UMCEES 81-28 HPEL.

Marbury, D., J. Metz, L. Lane, J.C. Stevenson, W.M. Kemp, and R. Twilley. 1981. Nitrogen uptake kinetics for the submerged estuarine macrophyte Potamogeton perfoliatus. In: Kemp et al. (eds.). Submerged aquatic vegetation in Chesapeake Bay: its ecological role in Bay ecosystems and factors leading to its decline. Submitted to Chesapeake Bay Program, USEPA, Annapolis, MD. Ref. No. UMCEES 81-28 HPEL.

MDA, Maryland Department of Agriculture. 1976-1980. Maryland agricultural statistics: Summary for 1975-1979. Issued cooperatively by Maryland Department of Agriculture and U.S. Department of Agriculture, Publication No. 92.

McHarg, I.L. 1969. Design with nature. Natural History Press. Garden City, N.J.

Mihursky, J.A., and W.R. Boynton. 1978. Review of Patuxent Estuary data base. Chesapeake Biological Laboratory, Solomons, MD. Ref. No. UMCEES 78-157 CBL.

Odum, H.T. 1971. Environment, Power, and Society. John Wiley, N.Y. 331 pp.

Odum, H.T. and E.C. Odum. 1976. Energy Basis for Man and Nature. Wiley and Sons, New York.

Pimentel, D.P., L.E. Hurd, A.C. Bellotti, M.J. Forster, I.N. Oka, O.D. Sholes, and R.J. Whitman. 1973. Food production and the energy crisis. Science 182:443-449.

Rawitscher, M., and J. Mayer. 1977. Nutritional outputs and energy inputs in seafoods. Science 198:261-264.

Schaefer, M.B. 1954. Some aspects of the dynamics of populations important to the management of marine commercial fisheries. Inter-American Tropical Tuna Commission Bulletin Vol. 1, p. 25-56.

Schuler, T.R. 1979. Summary of runoff and loading coefficients of diffuse source pollutants: herbicides. In: Kemp et al. (eds.). Submerged aquatic vegetation in the Chesapeake Bay. Annual report to Chesapeake Bay Program, USEPA, Annapolis, MD. Vol. II. UMCEES Ref. No. 80-68 HPEL.

Shabman, L.A., and S.S. Baite. 1978. Economic value of natural coastal wetlands: a critique. Coastal Zone Management Journal 4:231-247.

Smith, R., and W.F. McMichael. 1969. Cost and perfomance estimates for tertiary wastewater treating processes. U.S. Department of the Interior, Federal Water Pollution Control Administration, Robert A. Taft Water Research Center, Cincinnati, Ohio. Report No. TWRC-9.

Stevens, G.A., Wysong, J.W., and B.V. Lessley. 1981. Farm data manual. Department of Agricultural and Resource Economics, Cooperative Extension Service, University of Maryland, College Park.

Stevenson, J.C., and N.M. Confer. 1978. Summary of available information on Chesapeake Bay submerged vegetation. U.S. Dept. of Interior, Washington, D.C. Ref. No. FWS/OBS-78/66.

Stevenson, J.C., W.M. Kemp and W.R. Boynton. 1978. Submerged aquatic vegetation in the Chesapeake Bay: Its role in the Bay ecosystem and factors leading to its decline. Univ. of Maryland Center for Environmental and Estuarine Studies, Cambridge, MD. Ref. No. UMCEES 78-32 HPEL.

USDA, United States Department of Agriculture. 1978. Delmarva river basins survey. Prepared by USDA Economics, Statistics, and Cooperative Service, Forest Service and soil Conservataion Service in cooperation with the states of Maryland, Delaware, and Virginia. Government Printing Office No. 936-349. Washington, D.C.

A THEORETICAL APPROACH TO VALUING WATER FOR ENERGY DEVELOPMENT IN THE COLORADO RIVER BASIN[1]

Martha W. Gilliland and Linda B. Fenner[2]

Abstract.--The value of water in the Colorado River Basin is derived theoretically from principles of thermodynamics; the results provide a means to quantify the water quantity and quality impacts of agricultural and energy development in the West. The calculated value is 3.9 kilocalories per liter, equivalent to $330 per million liters in 1978 dollars. The price of water rights in New Mexico now averages $25 per million liters, indicating the extent to which water is undervalued.

INTRODUCTION

Water availability and quality are among the most critical problems associated with expanded energy development in the Colorado River Basin of the U.S. The coal, oil shale, and uranium development projected for the Colorado River Basin will require large quantities of water; this water use, in turn, will cause salinity increases downstream. Consequently, the issue of how to value clean water is central to the formulation of a water policy for the West.

The water impacts of energy development in the West have been evaluated in numerous studies (See for example, White et al. 1979) where water quantity and quality impacts are treated separately and are expressed as--liters of water consumption and milligrams per liter (mg/l) increase in total dissolved solids concentrations, respectively. In fact, however, the value of a given quantity of water is a function of its quality, and quantity and quality ought to be treated simultaneously. For example, one million liters of ocean water with a total dissolved solids concentration of 35,000 mg/l is clearly less valuable then one million liters of rain water at 10 mg/l.

How does the value of ten million liters of brackish water at 3000 mg/l compare with the value of one million liters of fresh water at 300 mg/l; the former is ten times greater in quantity and ten times poorer in quality than the latter. Using a theoretical approach developed by H.T. Odum (1978), we derive a value for water that expresses quantity and quality impacts simultaneously in either energy or dollar units. That value is then utilized to express the environmental damages produced by alternative kinds of development in the West.

The theoretical approach used to derive a value for water in this paper grew out of more general research on quantifying the linkages between environmental systems and technological systems. Known, by some, simply as energy analysis and, by others, as ecoenergetics (reflecting its roots in ecology and thermodynamics), it offers a framework for examining the interactions between technological and environmental resources and measures these inputs in a common unit (unit of energy). Given the common unit of measure, energy analysis can examine the linkages among the four inputs and can identify where one input may be substituted for another. For comparison, economic analysis uses two rather than four inputs to production (capital and labor) and measures these two in dollars.

METHODS

The method for deriving a theoretical value for water is described first. Then, the water quantity and quality impacts caused by several

[1]Paper presented at the international symposium Energy and Ecological Modelling, sponsored by the International Society for Ecological Modelling. (Louisville, Kentucky, April 20-23, 1981).

[2]Martha W. Gilliland is Executive Director and Linda B. Fenner is Research Associate, Energy Policy Studies, Inc., Omaha, Nebraska, U.S.A.

kinds of development in the West are presented as the basis for evaluating the theoretical value of those impacts.

A Theoretical Value for Water

The conceptual framework for energy analysis, in general, and for deriving the value of water, more specifically, is illustrated in   1. Note that solar energy is the only energy form located outside this system boundary and that four inputs to the production sector are shown   A,B, C, and D, these are capital, labor, environmental resources, and fuels, respectively. The inclusion of environmental resources within the system boundary provides the uniqueness to energy analysis; the resources are internal to the system. Obviously, water, the concern of this paper, is an environmental resource and one of the inputs to the production sector. Based on the system boundary established by   1, the only unit of measure that is common to all four inputs is solar energy. In theory, at least, each input can be traced back to its solar origins, and the amount of solar energy required to make and maintain each input can be evaluated. Thus, using this method, all inputs to production are evaluated in terms of their "solar equivalent" value. Because people are more accustomed to thinking in terms of fuel equivalent values (e.g., Btu's of coal, barrels of oil), solar equivalent energy values are often converted to "fuel equivalents" and to dollars for ease of communication.

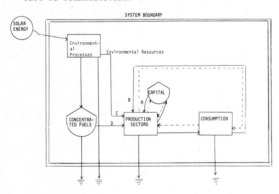

Figure 1.--System Boundary of Energy Anlysis.
Four primary factors of production are labeled A-Capital, B-Labor, C-Environmental Resources, and D-Concentrated Fuels; solid lines are energy flow and dotted lines are money flow.

In this study, the environmental resource of interest is water-fresh water which is produced in the hydrologic cycle with the use of solar energy. Table 1 shows how its energy value is .derived. The Gibbs free energy value is given in step 1, the fuel equivalent value in step 4, and the dollar value in step 6.

Fresh water, produced from ocean water by evaporation, has a free energy value relative to ocean water. More specifically, rain water has an average total dissolved solids concentration of 10 mg/1; in contrast, the ocean from which the rain was produced averages 35,000 mg/1. The free energy value of rain is a function of that quality difference. Using the standard equation for calculating the Gibbs free energy associated with a concentration gradient (step 1 in table 1) a value of -1.13 kilocalories per liter for rain water is obtained. The negative sign indicates that the change from rain water to ocean water is spontaneous or downhill thermodynamically. As the change occurs, 1.13 kcal/liter of heat are released. In reverse, +1.13 kcal/liter represents the minimum amount of energy required to desalinate sea water. It represents the amount of energy required to drive the uphill part of the cycle if the cycle operated reversibly at maximum efficiency. In fact, of course it does not operate reversibly. Consequently, the amount of energy needed to drive the uphill portion of the cycle (produce "pure" water from ocean water) must be greater than that released as pure water runs "downhill" thermodynamically to ocean water.

In nature, this larger amount is the solar energy that drives the hydrologic cycle (step 2 in table 1).[3] In energy units that are equivalent to coal or oil in quality, that amount is 3.9 kcal/ liter (step 4). In 1978, 11,891 kcal of energy, in its fuel equivalent forms, produced one dollar's worth of goods and services in the U.S. (step 5); thus, in 1978 dollars, the theoretically derived value of water is $330/million liters (step 6).

Water Use and Salinity Changes Caused by Development in the West

In order to illustrate the application of a theoretically derived value for water, the water use and salinity changes caused by several types of development in the West were evaluated for San Juan County, New Mexico. The San Juan River, flowing through the county, is the source of water for a substantial amount of existing and proposed agricultural and energy development in the county. The San Juan is part of the Colorado River System; it begins in the San Juan Mountains of Colorado and flows through the northwestern corner of New Mexico, joining the Colorado River above Lake Powell, at the Utah-Arizona border. (Fig. 2). Table 2 gives water quantity and quality information for the kinds of agricultural and energy development that either

---

[3] This method for the derivation of a solar equivalent value is thought to suffer from some double counting problems; work is underway to improve the technique (Lavine 1981).

Table 1.--Derivation of the Theoretical Value of Water in Energy and Dollar Units.

1. Calculate the Gibbs Free Energy Value of Rain

   $\Delta G = \Delta G^O + nRT\ln c_2/c_1$ where c_2 = density of rain at 10 mg/1, c_1 = density of ocean water at 35,000 mg/1, and G^O is 0 for constant temperature and pressure and no change in the ionic species.

   Therefore $\Delta G = \dfrac{1\ \text{mole}}{18\ \text{grams}}\ \dfrac{1.99 \times 10^{-3}\ \text{kcal}}{\text{mole}^O\text{K}}\ 298.15^O\text{K}\ \ln 1.00001/1.035$

   $\Delta G = -1.1336 \times 10^{-3}\ \text{kcal/g} = -1.1336\ \text{kcal/1}$

2. Calculate a Ratio for Solar Energy: Rain Energy where solar radiation = 8.57×10^{20} and continental rain quantity = $109,000 \times 10^{12}$ liters;

   $\text{Ratio} = \dfrac{8.57 \times 10^{20}\ \text{kcal}}{109,000 \times 10^{12}\ 1 \times 1.1336\ \text{kcal/1}} = 6900\!:\!1$

3. Calculate a Ratio for Fuel Equivalent Energy to Rain Energy where solar energy to fuel energy is 2000:1 (Odum 1978) and solar energy to rain energy is 6900:1 (step 2 above); therefore, fuel energy to rain energy is 3.45:1 (6900/2000).

4. Calculate the Fuel Equivalent Energy Value of Rain as 1.1336 kcal/1 X 3.45 fuel energy/rain energy equals 3.911 kcal (fuel)/1.

5. Calculate the Fuel Equivalent Energy Value of one Dollar for a Given Year.

   In 1978, the U.S. consumed 19.7×10^{15} kcal fuel + ($11.3 \times 10^{18}/2000$) kcal of solar energy measured in fuel equivalents to produce 2127.6×10^9 dollars worth of goods and services; 25.3×10^{15} kcal/ 2127.6×10^9 $ = 11,891 kcal/1978 $.

6. Calculate the Dollar Value of Rain as 3.911 kcal/1 ÷ 11,891 kcal/$ = 3.289×10^{-4} $/1 = 330 $/million 1.

exists or is expected in San Juan County; several water conservation options are included.

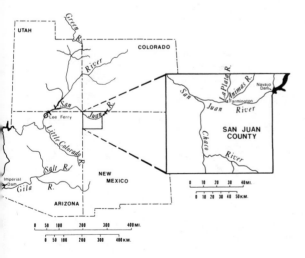

Figure 2.--Geography of the Colorado River Basin and San Juan County.

Water use was evaluated in terms of depletions or the amount of water consumed by a given kind of development. In all analyses and comparisons, depletions were distinguished from diversions and return flows. The diversion is the quantity of water withdrawn from the river for use elsewhere. For some kinds of uses, a portion of this diversion returns to the River via percolation through the soils. The return flow represents this quantity. The remainder of the diversion is consumed and represents the depletion. Return flows from agriculture are considered non-point sources of water pollution.

In these analyses, we assumed that no return flow or effluent was produced by energy development; that is, we assumed that diversions were equal to depletions and that all water diverted for an energy development either evaporated or was contained within the boundaries of the energy facility (2). This assumption is consistant with plans for energy development as well as with federal water pollution control regulations which call for zero discharge.

Salinity increases in the San Juan River and in the Colorado River Basin as a whole are the result of both salt loading and salt concentrating.

Salt loading results from a net addition of salts to the River via return flows; it occurs naturally and as a consequence of water development projects. As rainfall or irrigation water percolates through the soil, the water leaches out the natural salts; these leached salts then enter the River with return flows.

Salt concentrating results from consumptive use of water; thus all depletions produce a salt concentrating effect. For example, evaporation has a concentrating effect because the salts are left in a smaller volume of water as the water evaporates. In the case of evapotranspiration by natural vegetation or by irrigated argricultural crops, the salt content of the water transpired is left in the return flow (a smaller volume of water). In the case of other consumptive uses of water, the salt concentrating occurs with respect to conditions downstream of the diversion. The salinity of most rivers increases naturally from the headwaters to the mouth. Thus, although any upstream diversion removes both salt and water, the diversion is nearly always of higher quality water than downstream quality. Thus, less high quality water is left in the stream to dilute lower quality water downstream. Consumptive uses for energy developments are in this category.

Under traditional agricultural practices in San Juan County, water is allowed to flow to the fields via a complex and lengthy system of unlined canals and then into the fields along furrows (flood and furrow irrigation). An alternative, used by the Navajo Indian Irrigation Project in San Juan County, includes lined canals and application of the water by sprinkler systems. Because so much water percolates through the unlined canals and fields, about 2½ times as much water must be diverted from the San Juan River for a field irrigated by the flood method than for the same field irrigated by the sprinkler method and served by a lined canal (table 2). However, the amount of water actually consumed (the depletion in table 2) remains about the same regardless of the type of irrigation system. This is because the water that percolates through the unlined canals and fields eventually returns to the river. Thus, the key difference in the water impacts of the two irrigation methods is in the salinity changes they cause. In San Juan County, return flow (water returning from the canals and fields via percolation) is 3.5 times more salty than the diversion (the water removed from the River for irrigation). Water diverted for irrigation in San Juan county has a total dissolved solids concentration of about 270 mg/l; in contrast, the return flow averages 955 mg/l. For the case shown in (one 164 hectare farm) and flood irrigation, the depletion in combination with the salty return flow causes a salinity increase of 0.14 mg/l at Imperial Dam. Sprinkler irrigation of the same farm causes salinity to increase by only 0.08 mg/l at Imperial Dam. Imperial Dam was chosen as the point at which to measure and value the salinity change since the treaty between the U.S. and Mexico refers to Imperial Dam when it defines the quality of Colorado River water that is to flow from the U.S. to Mexico.

Water use and salinity changes associated with a coal-fired electric power plant (3000 Mwe capacity) and a coal gasification plant (producing 7.1 million cubic meters per day of synthetic gas) are also included in table 2. Note that the use of a combination of wet and dry cooling in a power plant, rather than all wet cooling, reduces diversions and depletions by 24,800 million liters per year (U. S. Environmental Protection Agency 1979), a 67 percent reduction for that one plant. For coal gasification plants, reductions in water use through different cooling technologies

Table 2.--Water Quantity and Quality Parameters Associated with Agricultural and Energy Development in San Juan County, New Mexico

	Diversion		Return Flow		Environmental Impact	
	Amount $(10^9 l/yr)$	Salinity (mg/l)	Amount $(10^9 l/yr)$	Salinity (mg/l)	Depletion $(10^9 l/yr)$	Salinity Increase[4] (mg/l)
Agriculture[1]						
Flood & Furrow	2.5	270	1.5	955	1.0	0.14
Sprinkler	1.0	270	0.5	955	1.0	0.08
Electric Power Plant[2]						
Wet Cooled	36.9	400	0	NA	36.9	1.97
Wet/Dry Cooled	12.1	400	0	NA	12.1	0.65
Coal Gasification Plant[3]						
Wet Cooled	9.1	400	0	NA	9.1	0.49
Wet/Dry Cooled	7.3	400	0	NA	7.3	0.39

[1] 164 hectares;
[2] 3000 megawatts of capacity.
[3] Producing 7.1 million cubic meters per day.
[4] Salinity increase measured at Imperial Dam in Arizona (see Fig. 2).

NA - Not Applicable.

are not as great as for power plants, but are
still significant. Wet/dry cooling[4] in one coal
gasification plant reduces diversions and depletions by 1840 million liters per year, a 20 percent reduction for the plant. The salinity increases, caused by the salt concentrating that is
induced by water consumption in these energy
facilities, range from 0.39 to 1.97 mg/l at
Imperial Dam (table 2). Obviously, the smaller
the water consumption is, the smaller the salinity
impact.

RESULTS

The theoretically derived values for water
with total dissolved solids concentrations ranging from 10 mg/l to 35,000 mg/l are given in
table 3. For perspective, salinity concentrations
in the headwaters of the San Juan River system in
Colorado range from 30 to 100 mg/l. By Farmington, New Mexico, salinity has increased to about
300 mg/l; when the San Juan enters the mainstem
of the Colorado River, the total dissolved solids
content is about 500 mg/l. Water leaving the
Upper Colorado River Basin at Lee Ferry (fig. 2)
has a salinity concentration of about 580 mg/l;
by Imperial Dam when the Colorado enters Mexico,
concentrations are more than 800 mg/l. Along
the course of the River, the theoretically
derived value of the water declines as its dissolved solids concentration increases (Table 3).
Its value is, of course, zero when it is finally
mixed with ocean water.

Results that utilize the values listed in
table 3 are discussed in 4 categories: (i) the
viability of desalination technology, (ii) water
rights costs in the West, (iii) the water impact
of agricultural and energy developments in San
Juan County, and (iv) choosing among water conservation options.

The Viability of Desalination Technology

In table 4, the actual cost of desalting
water in several desalination plants is compared
to the theoretically derived value for that water.
The theoretical value represents the minimum cost
of desalting regardless of technology, under the
assumption that the overall efficiency of production in nature cannot be improved upon. For
the brackish water cases in the West, desalting
costs ten times more than the theoretical minimum.
In the sea water case, desalting costs only

---

[4] Wet/dry cooling refers to the use of both
wet and dry cooling towers-dry cooling is applied
to those parts of the facility where it can most
economically be used.

Table 3.--The Theoretical Value of Water.

Dissolved Solids Concentration (mg/l)	Heat Equivalent Energy Value[1] kcal/l	Fuel Equivalent Energy Value[2] (kcal)1	Dollar Value[3] (1978$/10^6 l)
10	-1.1336	-3.9109	329
500	-1.1175	-3.8554	324
1,000	-1.1010	-3.7984	319
10,000	-0.8059	-2.7804	234
20,000	-0.4812	-1.660	140
35,000	0	0	0

[1] Calculated by substituting each dissolved
solids concentration for C_2 in the free energy
equation given in Table 1, step 1.
[2] Calculated by multiplying the heat equivalent
value by 3.45 (see Table 1, step 3).
[3] Calculated by multiplying the fuel equivalent
energy value by 1 X 10^6 liters and dividing by
11,891 kcal/1978 $ (See Table 1, step 5).

about twice as much as the theoretical minimum.
Energy costs are also shown in table 4, and, of
course, bear the same relationship. One of the
economic approaches to internalizing environmental
impacts is to calculate the cost of replacing the
environmental damage caused by some kind of development and assign that cost to the environmental
impact category. If that approach were used in
this example, water would be valued at the desalination costs (replacement costs) given in table 4.
Such a value, of course, changes with technological
improvements. In contrast, the theoretically
derived value is stable and represents "the best you
can ever do". Thus, although these results indicate that desalting technology will probably improve, at least 3.9 million kilocalories of fossil
fuel equivalent energy, costing at least 340
1978 dollars will always be required, directly and
indirectly, to produce a million liters of fresh
water from sea water. Clearly, desalting technology
will never be an economical way to obtain water
unless it is subsidized by inexpensive energy.

Water Rights Costs

Water in the West, although increasing in
cost, is still greatly undervalued. The average
cost of a water right in the San Juan River Basin
of New Mexico is about $25 per million liters.
This average cost is about ten times less than
the theoretically derived value of that water.
Knowledge of the "real" value of water does not
necessarily indicate that the user should pay
that amount. It does, however, indicate the extent

to which economic production in the West relies on free resources from nature, and it measures the minimum cost of substituting some other kind of input for water in the economic production process. For example, it indicates that the substitution of capital, labor, and fuel for water (for example, in the form of dry cooling in a power plant) would cost a minimum of $325 per million liters of water not consumed (in 1978 $).

Water Impacts of Development

The energy and dollar value of the water impacts caused by agricultural and energy development in San Juan County, New Mexico are given in table 5. In agriculture, one 164 hectare farm does about 4 billion kcal or $340,000 worth of environmental damage per year. Said differently, the farmer is receiving a 4 billion kcal or $340,000 per year subsidy from nature. Similarly, the electric power utility that uses all wet cooling is receiving a 146 billion kcal per year or $12.3 million per year subsidy. If wet/dry cooling is employed, utility costs go up and the free subsidy declines to 47.5 billion kcal ($4 million) per year. The water impact of a coal gasification plant is 28 to 35 billion kcal per year or 2.4 to 3 million 1978 dollars per year depending on whether wet or wet/dry cooling is employed.

Choosing Among Water Conservation Options

The ultimate question arising from this sort of analysis often relates to the choice of a technology. Should a technology that costs more money but consumes less water be chosen

Table 4.--Desalination: Actual vs. Theoretical Costs

	Buckeye Az[1]	Gillette Wyo[1]	Sea Water[2]
Concentration(mg/1)			
Feed Water	1500	2500	35,000
Product Water	500	500	1,000
Dollar Cost(1978$/10^6 liter)			
Actual	112	200	500-1800
Theoretical Minimum[3]	10	20	340
Energy Cost (million kcal)			
Actual	1.3	2.4	5.9-21.4
Theoretical Minimum[3]	0.1	0.2	3.9

[1]From US Dept of Interior 1974; cost data converted to 1978$ for inclusion here.

[2]Cruver and Sleigh 1976; the range in cost represents differences caused by plant size; cost data were converted to 1978$ for inclusion here.

[3] 1.136×10^{-4} kcal/mg/1 /1 X salinity difference X $10^6$1 yields the energy cost; divide by 11,891 kcal/1978 $ to obtain the dollar cost.

(or forced upon the developer) as opposed to the alternative that is cheap but consumes a lot of water? Recall from the earlier discussion on the system boundary in which these theoretical water values were derived that any one of four categories of inputs to production can substitute for another. The options listed in table 5 require all four types: water, fuel, labor, and capital. Within the systems context of fig. 1, the wet cooling option relies on more water and less fuel, labor, and capital (as embodied in the cooling towers) than the wet/dry cooling option. In the wet cooling option, water is substituted for fuel, labor, and capital. Conversely, fuel, labor, and capital substitute for water when wet/dry cooling is used. The best choice is the cheapest choice when all four inputs are counted as costs. This is in contrast to economic choices which are based on the sum of only 3 inputs.

For example, as indicated in table 6, the environmental damage that is prevented when a sprinkler irrigation system is used instead of the flood and furrow method is theoretically worth $10,000 per year; the cost to the farmer (and eventually to the consumer) of that sprinkler system is $41,000 per year. Energetically in this total systems context, flood and furrow should remain the irrigation system used. When all things are considered including the water impacts, sprinkler irrigation costs society more than flood and furrow. The same is true for the water conservation option in power and gasification plants. In a power plant, wet/dry cooling prevents 8.3 million dollars per year worth of environmental damage but costs the utility and the consumer 31.3 million dollars extra per year. In general, these results support the thesis suggested by Odum (1974) that the use of complex and expensive environmental control technology can be counterproductive.

CONCLUSION

From a global systems perspective, environmental resources are one of four kinds of inputs to economic production, and their value can be measured in energy units. Any one of the four inputs can substitute for any other, and the energy value of each is a measure of that substitutability.

In this paper, the value of water (as an environmental resource) to economic production in the West was examined in this global systems context. Its theoretical value was calculated to be 3.8 to 3.9 kilocalories per liter or $320 to $330 per million liters (1978 dollars) for a range in dissolved solids concentration of 10 to 1000 mg/1. This theoretically derived value for water was used to examine several water related issues in the West.

Table 5.-- Theoretical Value of Water Impacts Caused by Agricultural and Energy Development

| | Water Use | Salinity Increase[3] | Total | Total[4] |
	(billion kcal/yr)			(million 1978 $/yr)
Agriculture (164 hectares)				
Flood and Furrow	3.88[1]	0.15	4.03	0.34
Sprinkler	3.88[1]	0.09	3.97	0.33
Electric Power Plant (3000 Mwe)				
Wet Cooling	143.92[2]	2.16	146.08	12.28
Wet/Dry Cooling	46.83[2]	0.72	47.55	4.00
Coal Gasification Plant				
Wet Cooling	35.22[2]	0.54	35.76	3.01
Wet/Dry Cooling	28.25[2]	0.43	28.68	2.41

[1] Depletion of 1 X 10^9 liters/yr from Table 2 x 3.88 kcal/l (the replacement value of water at 270 mg/l).
[2] Depletion from Table 2 x 3.87 kcal/l (the replacement value of water at 400 mg/l).
[3] The salinity increase from Table 2 x (1.136 x 10^{-4} kcal/mg/l /l) x 9.67 x 10^{12} l/yr (flow at Imperial Dam.)
[4] Total in energy units divided by 11,891 kcal/1978 $.

Table 6.-- Value and Cost of Environmental Damage Prevented (1978 $/year)

	Theoretical Value of the Environmental Damage that is Prevented[1]	Economic Cost to the Developer of Prevention
In Agriculture (164 hectares)		
Use of Sprinkler instead of flood and furrow irrigation	10,000	41,000[2]
In Energy Development		
Use Wet/Dry Cooling instead of Wet Cooling		
in a Power Plant (3000 Mwe)	8,280,000	31,270,000[3]
in a Coal Gasification Plant (250 million cf/day)	600,000	820,000[4]

[1] Difference between the two options from Table 5.
[2] $250/hectare/year (Gilliland & Fenner, 1979, p1-27) x 164 hectares
[3] 1.7 mills/kwh (U.S. Environmental Protection Agency, 1979) x 1.8396 x 10^{10} kwh/yr (production from power plant at 70% load factor.
[4] $0.01/million Btu's (U.S. Environmental Protection Agency, 1979) x 82.125 x 10^{12} Btu's/year (production from gasification plant at 90% load factor).

First, the theoretical value indicates that water, at the current water rights cost of $25 per million liters, is greatly undervalued. Substitution of a different kind of input for water in economic production would cost at least ten times more than the water now costs.

Secondly, the theoretical derivation enables the environmental impact of agricultural development and energy development in the West to be evaluated. In San Juan County, a 164 hectare irrigated farm does about $340,000 (1978 dollars) worth of environmental damage each year in the form of water consumption and salinity increases.

The environmental damages of energy development range from $2 to 12 million per year depending on the kind of development and type of cooling system used.

Thirdly, if water inputs are counted and evaluated on an equal basis with fuel, labor, and capital inputs to energy development in the West, the use of all wet cooling systems is cheaper to society as a whole than the use of a combination of wet and dry cooling systems.

Finally, and perhaps most importantly, the theoretically derived value of water provides a sense of just how valuable this natural resource is and of just how much the U.S. economy relies on natural resources that are free. What would happen to the costs of industrial production and to household costs if this water had to be purchased from its supplier (nature)?

ACKNOWLEDGEMENTS

This work was carried out under a subcontract with the Science and Public Policy Program, the University of Oklahoma, Norman, who was contracted to the Office of Energy, Minerals, and Industry; Office of Research and Development, U.S. Environmental Protection Agency (EPA Contract No. 68-01-1916).

LITERATURE CITED

Cruver, J.E. and J.H. Sleigh. 1976. Reverse osmosis - the emerging answer to seawater desalination. Industrial Water Engineering, June/July, 1976.

Gilliland, M.W., and L.B. Fenner. 1979. Alternative water management strategies for the San Juan River Basin in New Mexico. Energy Policy Studies, Inc., Omaha, NE.

Lavine, M. J. 1981. Use of embodied energy values in the analysis of environmental/ energy/economy tradeoff decisions: an exploratory analysis: working paper. Center for Environmental Research, Cornell University, Ithaca, N.Y.

Odum, H.T. 1974. Energy cost-benefit models for evaluating thermal plumes. p 628-649. In Thermal ecology. J.W. Gibbons and R.R. Sharitz, eds. National Technical Information Service, Springfield, Va.

Odum, H.T. 1978. Energy analysis, energy quality, and environment. p. 55-88. In Energy analysis: a new public policy tool. M.W. Gilliland, ed. Westview Press, Boulder, Colo.

U.S. Department of the Interior. 1974. Saline water conversion summary report. U.S. Government Printing Office, Washington, D.C.

U.S. Environmental Protection Agency. 1979. Wet/dry cooling and cooling tower blowdown disposal in synthetic fuel and steam-electric power plants. EPA-600/7-79-085, Washington, D.C.

White, I.L., M.A. Chartock, R.L. Leonard, and others. 1979. Energy from the West: impact analysis report. U.S. Environmental Protection Agency. EPA-600/7-79-0826. Washington, D.C.

ASSESSING THE ENVIRONMENTAL IMPACTS OF ENERGY DEVELOPMENT ALTERNATIVES USING A FISH AND WILDLIFE MODEL: A SUMMARY REPORT 1977–1981[1]

C. T. Cushwa, J. M. Brown, C. W. Du Brock, D. N. Gladwin, G. R. Gravatt, W. T. Mason, Jr., and R. Nicholas Rowse[2]

Abstract.--A procedure is presented for summarizing fish and wildlife[3] species information for computer entry. Guidelines are discussed for applying this procedure in developing a statewide fish and wildlife species information system. The information system is demonstrated by developing multiple species fish and wildlife models. These models are used to assess the effects of energy development activities upon fish and wildlife populations and habitats.

INTRODUCTION

The U.S. Fish and Wildlife Service (FWS) has developed a standard methodology for summarizing fish and wildlife species information for computer entry. This methodology, "A Procedure for Describing Fish and Wildlife" (Mason et al. 1979) is being pilot tested in Pennsylvania. The Procedure provides instructions for summarizing and coding vertebrate and invertebrate species information for a specific geographic area; e.g., a State. Information on 847 species is being compiled for Pennsylvania. The standard categories of information included in the Procedure were identified by potential users of the system to aid in meeting their fish and wildlife information needs for management, planning, research, administrative, and regulatory activities (Cushwa et al. 1980). Information summarized using the Procedure is hardware and software independent. The data for Pennsylvania is stored on a Univac 1100

computer and is being manipulated by the data base management system, MANAGE (Wilcott 1978). In the paper, we will illustrate use of (1) the Procedure to develop a statewide fish and wildlife species information system, (2) the information system to develop fish and wildlife distribution, management practices, habitat, food web, niche, and other species models, and (3) these models to analyze the effects of energy development activities on fish and wildlife.

BACKGROUND ON DESIGNING FISH AND WILDLIFE SPECIES MODELS

During the past two decades in the United States, emphasis has been placed on a systems approach to resource management, planning, and assessment. Modeling has been an important part of this approach. Armentano and Loucks (1979) concluded, after a two-year study reviewing environmental data needs and availability of data to meet these needs, that existing data and models were not being used to their full potential. They cited inconsistencies in data documentation, inadequate communications between data suppliers and data users, and a lack of overall coordination of the data bases in management, research and development programs. Parzyck et al. (1980), Odum and Cooley (1980), Wagner (1980), and Hilborn et al. (1980) identified potential uses of modeling and stressed the need for further communication across disciplines to improve impact assessment capabilities at the ecosystem level. None of these authors, however, specifically addressed the complex issue of modeling the fish and wildlife resource, or using such models to analyze the effects of alternative energy development

[1]Paper presented at the international symposium Energy and Ecological Modelling, sponsored by the International Society for Ecological Modelling. (Louisville, Kentucky, April 20-23, 1981).

[2]Charles T. Cushwa, Senior Wildlife Biologist; James M. Brown, Plant Ecologist; Calvin W. Du Brock, Ecologist; Douglas N. Gladwin, Wildlife Biologist; Glenn R. Gravatt, Operations Research Analyst; William T. Mason, Jr., Aquatic Ecologist; and R. Nicholas Rowse, Wildlife Biologist; Eastern Energy and Land Use Team, U.S. Fish and Wildlife Service, Route 3, Box 44, Kearneysville, West Virginia 25430.

[3]Fish and wildlife includes vertebrate and invertebrate species.

activities on fish and wildlife populations and habitats.

Generally, detailed information on animal communities is not available in sufficient quality and quantity to address such concepts as threshold (Woodwell 1974), effects of pollutants on species and ecosystems (Levin 1974) or accurate quantitative predictions of changes in animal populations (Hoekstra et al. 1979). The state-of-the-art systems approach to animal community interrelations has, at best in the past, been limited to presence or absence of the species in a given habitat or geographic area. For example, Merriam (1892) presented the general concept of grouping animal species together on the basis of some common attributes such as distribution, and Darlington (1957) provided a detailed discussion of some basic ecological principles influencing continental patterns of animal distribution. However, both works were limited by the lack of (1) fish and wildlife species information, (2) automated data processing technology, and (3) demand for this type of information by users other than researchers and/or scholars.

During the 1970's numerous efforts were initiated to develop computerized fish and wildlife information systems. Between 1972-76, information was summarized on 721 vertebrate animals inhabiting Arizona and New Mexico (Patton 1978). These data were entered on a computerized data management system called RUN WILD. This became the first fish and wildlife species information system to combine aquatic and terrestrial vertebrate data in a common data base, which permitted analysis of relations between land/water use and the fish and wildlife resource. Also, during the mid-1970's, Thomas et al. (1976), and Thomas (1979) were developing a method of grouping terrestrial vertebrate animals by their breeding and feeding requirements. They combined information from 378 species into 16 "life-forms." The work by Patton and Thomas was useful in FWS efforts to develop statewide fish and wildlife species information systems in West Virginia, Alabama, and Pennsylvania (Cushwa et al. 1978).

Du Brock et al. (1981), Schweitzer et al. (1981), Besadny (1979), and Hirsch et al. (1979) summarize progress in the development of fish and wildlife species information systems and uses of these systems to generate species models for analyzing or assessing impacts of land use and resource development on the fish and wildlife resource. These works also provide valuable insight into the complexity of developing fish and wildlife species models and a standard methodology like the Procedure.

Some of the factors that influence the design of fish and wildlife species information systems and development of fish and wildlife models are:

1. The institutional complexity of managing the fish and wildlife resource. For example, in the United States:

   - The States own and are legally responsible for resident fish and wildlife.

   - The Federal Government has legal responsibility for the protection and management of animal species that are migratory, included in international treaties, and classified as threatened or endangered.

   - The habitats of wild animals are owned and managed by individuals, cooperatives, local, State, and Federal agencies.

2. The biological complexity of the resource.

   - The fish and wildlife resource of the United States is comprised of over 4,000 vertebrate species and tens of thousands of invertebrate species.

   - These species occupy a complex variety of aquatic and terrestrial habitats that include the surface and near-surface environments of the United States.

3. The availability, format, and completeness of information about a species or groups of species. Much of the available information is historical and scattered in a variety of files, reports, books, notes, and in the minds of experts.

4. Time and cost of coordination in developing and applying standards in collecting and compiling fish and wildlife information among private, university, State, and Federal groups.

In order to consider these factors in developing fish and wildlife species models one must first summarize data in a manner compatible with user needs. Our experience indicates that use of a standard methodology like the Procedure and an interagency team approach is an effective way in which to summarize species data for developing a statewide fish and wildlife information system. For example, it is impossible to identify an individual or agency who is an expert on all species of fish and wildlife inhabiting a state or one who knows all the institutional ramifications and information needs of managers, planners, and administrators.

Potential benefits from an interagency team approach, as opposed to a single agency or individual approach, include:

1. Reduced total cost to each organization needing basically the same fish and wildlife information.

2. A coordinated centralized information system that can be effectively managed and updated.

3. Ensured consistency and quality control of data among users within the State.

4. A more complete synthesis of fish and wildlife information to manage the resource.

5. Pooling of talent and expertise to address the biological and institutional complexity of the fish and wildlife resource.

6. Greater credibility and acceptability of fish and wildlife information included in the system.

The interagency team approach to developing and managing a statewide fish and wildlife information system, being pilot tested in Pennsylvania, Colorado, Missouri, and Minnesota, evolved through efforts to develop and manage statewide fish and wildlife species information systems in West Virginia and Alabama (Cushwa et al. 1978). From these experiences we have identified 11 major tasks which should be considered by an interagency team in implementing a statewide fish and wildlife information system. These tasks include:

1. Establish a Federal/State interagency team for guiding the development and management of the system.

2. Assess the fish and wildlife information needs of users within the State.

3. Select or design a data base format that meets user information needs identified in item 2 above.

4. Determine budget/personnel requirements, funding sources, and user fees.

5. Identify sources of existing fish and wildlife information and experts to summarize information using the design selected in item 3 above.

6. Select the hardware and software to be used and identify a data base manager.

7. Develop and implement an editing procedure for raw data.

8. Develop and implement procedures for inputting data into the data management system.

9. Develop and implement marketing and user training activities.

10. Develop and implement procedures for monitoring use and evaluating cost-effectiveness.

11. Develop and implement updating procedures.

One of the most important of these tasks is to estimate the initial cost of implementing the system. Thus far, the cost of developing a statewide fish and wildlife information system has ranged from $60-$120 per species. Cost to enter the data on a computer has ranged from $30-$50/species and cost to store data has ranged from $2-$20/species/year. Currently, the average cost to develop a 1000 species, statewide fish and wildlife information system using the Procedure is $150,000. This cost may seem high, but there are a number of benefits accrued through use of standard methods for developing and managing statewide fish and wildlife species information systems. These include: (1) eliminating redundant data collection, (2) providing a unified body of knowledge for planning, management, inventory, assessment, and other activities, (3) providing a common fish and wildlife dataset for natural resource decisions, (4) improving coordination among private organizations and Federal and State agencies, (5) making accurate, comprehensive, timely, and inexpensive fish and wildlife species information available to a wide array of users, and (6) establishing a coordinated focus for fish and wildlife data storage, exchange, and management within a State.

We believe the improved quantity and quality of fish and wildlife species data available for use in assessing land use or management practices from a statewide fish and wildlife species information system developed according to the Procedure is cost-effective. However, we will not have documentation of the use such a system receives or the effects of this information system on resource management until we complete the monitoring and evaluation of the application in Pennsylvania and Colorado.

DEVELOPING MODELS USING A STATEWIDE FISH
AND WILDLIFE SPECIES INFORMATION SYSTEM[4]

As of April 1981, information has been
summarized for 441 species according to
guidelines prepared by Mason et al. (1979).
By July 1981, there will be 847 species in the
data base. This information is being processed
and retrieved using MANAGE (Wilcott 1978) on
a Univac 1100 computer.

Odum (1971) defined a model as "a formulation
that mimics a real world phenonmenon, and by
means of which predictions can be made."
Similarly, in this report, a model is defined
as an assemblage of species information that
represents species relationships with the
environment or represents species composition
of animal communities. The information system
includes a model for each species. Each
species model contains 87 categories of
information about the following general
subjects: taxonomy, distribution, status,
food habits, niche requirements, management
practices, and references (fig. 1). We can
develop animal community models, natural and
artificial, using single or multiple criteria
from the 87 fields of information in each
species model.

These 87 fields of information can be
used to develop different models by varying
the number of fields included in a model or
by the way fields are combined. For example,
figure 2 is a model of the bald eagle (Haliaeetus
leucocephalis L.) in which all the fields of
information are listed. Field names spelled
out in figure 1 can be used to clarify
abbreviations used in this and the following
examples.

The 441 species in the information
system are distributed among the 10 major
taxonomic groups as follows:

Group	# spp.	Group	# spp.
Amphibians	33	Crustaceans	2
Birds	27	Molluscs	21
Fishes	184	Insects (aquatic)	46
Mammals	44	Insects (terrestrial)	44
Reptiles	32	Other Invertebrates	8

TOTAL: 441

[4]The system-generated information presented in
the next two sections is preliminary. It is
presented to demonstrate the "system" and the
"process" of developing models to analyze energy
development alternatives using a statewide fish
and wildlife species information system. Several
of the figures are computer printouts and are
provided to demonstrate data format and costs.

The system can generate any number of
models for each of the 87 fields of information.
Examples follow:

A. A model of species inhabiting
    Clearfield County. This model
    contains 163 species distributed
    among the following groups:

Group	# spp.	Group	# spp.
Amphibians	21	Reptiles	18
Birds	13	Insects (aquatic)	6
Fishes	37	Insects (terrestrial)	31
Mammals	33	Other Invertebrates	4

TOTAL: 163

These 163 species represent a model
of the animal community composition in
Clearfield County.

B. A model of species occurrence by water-
    shed, using Office of Water Data
    Coordination (OWDC) Hydrologic Units.
    The model for OWDC Hydrologic Unit
    #02050201 contains 191 species
    distributed among the following
    groups:

Group	# spp.	Group	# spp.
Amphibians	20	Molluscs	3
Birds	22	Insects (aquatic)	7
Fishes	51	Insects (terrestrial)	33
Mammals	54	Other Invertebrates	4
Reptiles	16		

TOTAL: 191

C. A model of species occurrence by
    Society of American Foresters (SAF)
    forest cover type. The model for SAF
    forest cover type #43 (bear oak forest
    cover type, Quercus ilicifolia Wang)
    contains 90 species distributed among
    the following groups:

Group	# spp.	Group	# spp.
Amphibians	16	Insects (terrestrial)	26
Birds	8	Other Invertebrates	2
Mammals	38		

TOTAL: 90

No.	Group	Definition
1	SCI-NAME	Species scientific name (genus and species)
2	COMMON-NAME	Species common English name
3	GROUP	Common name for species class (e.g., birds)
4	%COUNTIES	Percentage of counties in the State that the species occurs in
5	STATUS	Species legal status determined by State and Federal law
6	HABITAT	General habitat classification for species (e.g., aquatic)
7	TROPHIC	Species trophic level
8	ORIGIN	Origin of species in State (e.g., introduced)
9	PAST	Species past populaton trend for 5 years
10	FUTURE	Species expected population trend with management
11	OWNERSHIP	General ownership of land species occurs on (e.g., State land only)
12	TERRITORY	Territorial behavior of species
13	DISPERSAL	Dispersal patterns of species
14	ACRES	Home range size in acres
15	SQ.MILES	Home range size in square miles
16	PERIODICITY	Daily periodicity of species activities (e.g., nocturnal)
17	PHYLUM	
18	CLASS	
19	ORDER	
20	FAMILY	
21	GENUS	
22	SPECIES	Species taxonomic classification
23	SUBPHYLUM	
24	SUBCLASS	
25	SUBORDER	
26	SUBFAMILY	
27	SUBGENUS	
28	SUBSPECIES	
29	AUTHORITY	Taxonomic authority for species
30	ENTERED	Date species record was entered into system
31	UPDATED	Last date species record was updated
32	DISTRIB-PRES	Counties species is known to occur in
33	DISTRIB-ABS	Counties species is known not to occur in
34	DISTRIB-UNK	Counties where the species distribution is unknown
35	ABUNDANCE-HI	Counties the species occurs in high abundance
36	ABUNDANCE-M	Counties the species occurs in medium abundance
37	ABUNDANCE-LO	Counties the species occurs in low abundance
38	LATLON	Latitude and longitude of isolated population sites
39	QUAD	USGS 7½ quads that special concern species are found in
40	HYDROUNIT	Hydrologic units (river basins) species is found in
41	ECOREGION	Ecoregions (Bailey's) that species is found in
42	VEGETATION	Vegetation types (Küchler's) species occurs in
43	FTYPE	Forest cover type (SAF) species occurs in
44	FSIZE	Forest size classes species occurs in
45	RANGELAND	USGS rangeland classifications species occurs in
46	AGRICULTURE	USGS agriculture classifications species occur in
47	FOREST	USGS forest classifications species occurs in
48	URBAN	USGS urban classifications species occurs in
49	WATER	USGS water classifications species occurs in
50	BARREN	USGS barren classifications species occurs in
51	NWI	National Wetlands Inventory system classifications relative to species
52	AQUATIC	Species aquatic plant and animal associations
53	TERRESTRIAL	Species terrestrial plant and animal associations
54	FOOD-L	Species larval food habits
55	FOOD-J	Species juvenile food habits
56	FOOD-A	Species adult food habits
57	NICHE-E	Species egg niche requirements
58	NICHE-LF	Species feeding larva niche requirements
59	NICHE-LR	Species resting larva niche requirements
60	NICHE-P	Species pupa niche requirements
61	NICHE-JF	Species feeding juvenile niche requirements
62	NICHE-JR	Species resting juvenile niche requirements
63	NICHE-AB	Species breeding niche requirements

Figure 1.--The 87 fields of information included in each species model and their definitions.

No.	Group	Definition
64	NICHE-AF	Species feeding adult niche requirements
65	NICHE-AR	Species resting adult niche requirements
66	MANAGEMENT-B	Management practices beneficial to species
67	MANAGEMENT-A	Management practices adverse to species
68	References	References for species information
69	REF-CODES	Card locations various references were used
70	C-FOOD-L	Comments on larval food habits
71	C-FOOD-J	Comments on juvenile food habits
72	C-FOOD-A	Comments on adult food habits
73	C-FOOD-G	Comments on general food habits
74	C-FOOD-S	Comments on seasonal food habits
75	C-NICHE-E	Comments on egg niche requirements
76	C-NICHE-LF	Comments on feeding larva niche requirements
77	C-NICHE-LR	Comments on resting larva niche requirements
78	C-NICHE-P	Comments on pupa niche requirements
79	C-NICHE-JF	Comments on feeding juvenile niche requirements
80	C-NICHE-JR	Comments on resting juvenile niche requirements
81	C-NICHE-AB	Comments on breeding niche requirements
82	C-NICHE-AF	Comments on feeding adult niche requirements
83	C-NICHE-AR	Comments on resting adult niche requirements
84	C-NICHE-G	Comments on general niche requirements
85	C-NICHE-S	Comments on seasonal niche requirements
86	C-MANAGEMENT	Comments on beneficial and adverse management practices
87	C-OTHER	Comments on distribution and other areas

Figure 1.--The 87 fields of information included in each species model and their definitions (continued).

```
<SCI-NAME> HALIAEETUS LEUCOCEPHALIS    <COMMON-NAME> EAGLE, BALD
<GROUP> BIRDS    <%COUNTIES> <10%    <STATUS> F-ENDANGERED,S-ENDANGERED
<HABITAT> RIPARIAN    <TROPHIC> CARNIVORE    <ORIGIN> NATIVE    <PAST> UNKNOWN
<FUTURE> UNKNOWN    <OWNERSHIP> ALL    <TERRITORY> BREED/NEST
<DISPERSAL> UNKNOWN    <ACRES> UNKNOWN    <SQ.MILES> UNKNOWN
<PERIODICITY> DIURNAL,DAY ONLY    <PHYLUM> CHORDATA    <CLASS> AVES
<ORDER> FALCONIFORMES    <FAMILY> ACCIPITRIDAE    <GENUS> HALIAEETUS
<SPECIES> LEUCOCEPHALIS    <SUBPHYLUM>    <SUBCLASS>    <SUBORDER>
<SUBFAMILY>    <SUBGENUS>    <SUBSPECIES>    <AUTHORITY> LINNAEUS
<ENTERED> 80/09/06    <UPDATED>
<DISTRIB-PRES>
    39,49
<DISTRIB-ABS>
    (NO VALUE FOR THIS FIELD)
<DISTRIB-UNK>
    1,3,5,7,9,11,13,15,17,19,21,23,25,27,29,31,33,35,37,41,43,45,47,51,
    53,55,57,59,61,63,65,67,69,71,73,75,77,79,81,83,85,87,89,91,93,95,97,
    99,101,103,105,107,109,111,113,115,117,119,121,123,125,127,129,131,
    133
<ABUNDANCE-HI>
    (NO VALUE FOR THIS FIELD)
<ABUNDANCE-M>
    (NO VALUE FOR THIS FIELD)
<ABUNDANCE-LO>
    (NO VALUE FOR THIS FIELD)
<LATLON>
    412230 792230421500 792230421500 793000412230 793000
<QUAD>
    UNKNOWN
<HYDROUNIT>
    04120101,05010004,05030102,05010002,05010003,04110003,02040104,
    02040103
<ECOREGION>
    21120000,21130000
<VEGETATION>
    093,097
<FTYPE>
    055,021,094
<FSIZE>
    MATURE,OVER-MATURE
<RANGELAND>
    (NO VALUE FOR THIS FIELD)
<AGRICULTURE>
    UNKNOWN
<FOREST>
    UNKNOWN
<URBAN>
    (NO VALUE FOR THIS FIELD)
<WATER>
    (NO VALUE FOR THIS FIELD)
<BARREN>
    (NO VALUE FOR THIS FIELD)
<NWI>
    ALL
<AQUATIC>
    (NO VALUE FOR THIS FIELD)
<TERRESTRIAL>
    (NO VALUE FOR THIS FIELD)
<FOOD-L>
    (NO VALUE FOR THIS FIELD)
<FOOD-J>
    5040,5220,5310,5590
<FOOD-A>
    5050,5230,5320,5590
```

Figure 2.--A species model of the bald eagle.

```
<NICHE-E >
    05990S,05450S,06830S,05390S
<NICHE-LF>
    (NO VALUE FOR THIS FIELD)
<NICHE-LR>
    (NO VALUE FOR THIS FIELD)
<NICHE-P >
    (NO VALUE FOR THIS FIELD)
<NICHE-JF>
    06020S,05480S,06860S,05420S
<NICHE-JR>
    06020S,05480S,06860S,05420S
<NICHE-AB>
    06040S,05500S,06880S,05440S
<NICHE-AF>
    06030S,05490S,06870S,05430S
<NICHE-AR>
    06030S,05490S,06870S,05430S
<MANAGEMENT-B>
    008,013,023,027,030,043,046,052,006
<MANAGEMENT-A>
    003,029,033,036,037,015
<REFERENCES>
    01BROWN, L. AND D. AMADON.  1968.  EAGLES, HAWKS, AND FALCONS OF THE W
    01ORLD, VOL. I. MCGRAW-HILL BOOK CO., N. Y. 414PP.
    02COMMITTEE ON RARE AND ENDANGERED WILDLIFE SPECIES.  1966.  RARE AND
    02ENDANGERED WILDLIFE OF THE UNITED STATES.  BUREAU OF SPORT FISHERIES
    02 AND WILDLIFE, WASHINGTON, D. C.
    03JOHNSON.  1978.  THE STUDY OF RAPTOR POPULATIONS.  UNIVERSITY PRESS
    03OF IDAHO. 57PP.
    04LAYCOCK, G.  1973.  AUTUMN OF THE EAGLE.  CHARLES SCRIBNER'S SONS, N
    04. Y. 239PP.
    05MARSHALL, D. B. AND P. F. NICKERSON.  1976.  THE BALD EAGLE 1776-197
    056.  NAT. PARKS AND CONS. MAG. 50(7) JULY, 1976. NPCA. PP14-19.
    06TODD, W. E.  1940.  BIRDS OF WESTERN PENNSYLVANIA.  UNIV. OF PITTSBU
    06RGH PRESS. PP.150-155.
    07WARREN, B. H., M. D.  1890.  BIRDS OF PENNSYLVANIA.  COMMONWEALTH OF
    07 PENNSYLVANIA, HARRISBURG. 434PP.
    08WILDLIFE MANAGEMENT INSTITUTE TRANSACTIONS OF THE 42ND NORTHAMERICAN
    08 WILDLIFE AND NATURAL RESOURCES CONFERENCE.  MANAGEMENT STRATEGY FOR
    08 BALD EAGLES.  WASHINGTON, D. C. PP.86-92.
    09WOOD, M.  1973.  BIRDS OF PENNSYLVANIA.  PENNSYLVANIA STATE UNIV., C
    09OLLEGE OF AGR., UNIVERSITY PARK, PA. 111PP.
<REF-CODES>
    CARD# 010    REF-CODES: 09
    CARD# 050    REF-CODES: 07
    CARD# 075    REF-CODES: 091973
    CARD# 100    REF-CODES: 09
    CARD# 125    REF-CODES: 09
    CARD# 145    REF-CODES: 09
    CARD# 175    REF-CODES: 09
    CARD# 195    REF-CODES: 09
    CARD# 215    REF-CODES: 09
    CARD# 235    REF-CODES: 06
    CARD# 250    REF-CODES: 06
    CARD# 305    REF-CODES: 09
    CARD# 355    REF-CODES: 01
    CARD# 365    REF-CODES: 06
    CARD# 390    REF-CODES: 01
    CARD# 395    REF-CODES: 01
    CARD# 400    REF-CODES: 01
    CARD# 435    REF-CODES: 01
```

Figure 2.--A species model of the bald eagle (continued).

```
        CARD# 550    REF-CODES: 01,03,05
        CARD# 570    REF-CODES: 01,03,05
        CARD# 590    REF-CODES: 05
        CARD# 615    REF-CODES: 04,05,06
        CARD# 735    REF-CODES: 04,05
        CARD# 765    REF-CODES: 04,05
        CARD# 795    REF-CODES: 04,05,06
        CARD# 825    REF-CODES: 04,05
        CARD# 855    REF-CODES: 04,05
        CARD# 905    REF-CODES: 02,05,08
        CARD# 920    REF-CODES: 02,05
        CARD# 935    REF-CODES: 05
<C-FOOD-L>
    (NO VALUE FOR THIS FIELD)
<C-FOOD-J>
    (NO VALUE FOR THIS FIELD)
<C-FOOD-A>
    (NO VALUE FOR THIS FIELD)
<C-FOOD-G>
    OPPORTUNISTIC, HARASS OSPREYS TO DROP PREY*05*
<C-FOOD-S>
    (NO VALUE FOR THIS FIELD)
<C-NICHE-E >
    REQUIRES MILES OF SHORELINE WITH HIGH PERCHING AND LOOKOUT POINTS AND
    TALL, OFTEN DEAD, TREES FOR NESTS*04,05,06*
<C-NICHE-LF>
    (NO VALUE FOR THIS FIELD)
<C-NICHE-LR>
    (NO VALUE FOR THIS FIELD)
<C-NICHE-P >
    (NO VALUE FOR THIS FIELD)
<C-NICHE-JF>
    REQUIRES MILES OF SHORELINE ALONG UNPOLLUTED WATER WITH HIGH PERCHING
    AND LOOKOUT POINTS*04,05*
<C-NICHE-JR>
    REQUIRES MILES OF SHORELINE WITH HIGH PERCHING & LOOKOUT POINTS*04,05*
<C-NICHE-AB>
    REQUIRES MILES OF SHORELINE WITH HIGH PERCHING AND LOOKOUT POINTS AND
    TALL, OFTEN DEAD, TREES FOR NESTS*04,05,06*
<C-NICHE-AF>
    REQUIRES MILES OF SHORELINE ALONG UNPOLLUTED WATER WITH HIGH PERCHING
    AND LOOKOUT POINTS*04,05*
<C-NICHE-AR>
    REQUIRES MILES OF SHORELINE WITH HIGH PERCHING & LOOKOUT POINTS*04,05*
<C-NICHE-G>
    (NO VALUE FOR THIS FIELD)
<C-NICHE-S >
    (NO VALUE FOR THIS FIELD)
<C-MANAGEMENT>
    TRANSPLANTING OF EGGS AND FOSTER NESTINGS HAS BEEN SUCCESSFUL*05*
<C-OTHER>
    RARE BREEDING RESIDENT. OCCASIONAL WINTER RESIDENT*09*
    REPORTS OF NESTS IN PIKE, WAYNE COS. UNCONFIRMED*09*
    REFERS TALL, OFTEN DEAD, TREES FOR NESTS*06*
    TIMID, BUT MAY DIVE TOWARD NEST INTRUDERS*01*
    COASTS, LAKES, RIVERS; SEEN ALONG MT. RIDGES IN MIGRATION
```

Figure 2.--A species model of the bald eagle (continued).

D. A model of species adversely affected
by a management practice such as stream
channelization contains 52 species
distributed among the following
groups:

	#
Group	spp.
Amphibians	11
Fishes	13
Mammals	5
Reptiles	13
Molluscs	10

TOTAL: 52

E. A model of species whose adult life
stage requires a particular food
type. The model of adult animals
that feed on reptile eggs contains
33 species distributed among the
following groups:

	#
Group	spp.
Amphibians	2
Birds	11
Fishes	4
Mammals	9
Reptiles	7

TOTAL: 33

F. A model of species by a breeding
habitat requirement. The model of
breeding adults that require water
velocity between 1.0-1.4 FPS contains
28 species:

	#
Group	spp.
Fishes	28

Other specific fish and wildlife species
models can be developed by the user. For
example, as a specific model is developed,
the user frequently identifies other information
which is relevant and proceeds to develop a
series of submodels to meet more specific
fish and wildlife information needs. In the
next section, we demonstrate this process by
using several fish and wildlife models to
make an initial assessment of (a) impacts on
fish and wildlife of a proposed channelization
and dredge spoil deposit project, and (b) impacts
on fish and wildlife of a proposed 100-acre
surface mine.

Case Examples

I.   Permit application for a dredge disposal site and channelization project.

Location: Bucks County, Pennsylvania, SW of Croydon on Neshaminy Creek, 1/4 mile north of the junction with the Delaware River.

Proposed Action: Fill a 5-acre wetland with dredge spoil; channelize 1/2 mile of Neshaminy Creek and widen banks by 50 feet.

Questions and Responses:

1.   Are there any high-value (endangered, threatened, or sensitive) species of fish and wildlife in the general area (Bucks County)? Yes, the system identifies 12 high-value species in Bucks County.

GROUP	COMMON-NAME
AMPHIBIANS	FROG, NEW JERSEY CHORUS
AMPHIBIANS	PEEPER, NORTHERN SPRING
FISH	BASS, LARGEMOUTH
FISH	STURGEON, ATLANTIC
MOLLUSCS	(ALASMIDONTA UNDULATA)
MOLLUSCS	ELK TOE, EASTERN
MOLLUSCS	PURPLE, COMMON FLAT.
MOLLUSCS	SHELL, ALEWIFE
MOLLUSCS	SHELL, EASTERN FLOAT
REPTILES	TURTLE, BOG
REPTILES	TURTLE, EASTERN MUD
REPTILES	TURTLE, RED-BELLIED

2.   Are there any high-value species of fish and wildlife in the USGS 1:24,000 scale quadrangles covering this area? Yes, there are two species. For these two species we requested group, common name, scientific name, status, percent of counties in the State in which the species occurs, species occurrence by county, species occurrence by USGS quadrangles, general comments, comments on management practices, and information references.

```
<GROUP> REPTILES   <COMMON-NAME> TURTLE, RED-BELLIED
<SCI-NAME> CHRYSEMYS RUBRIVENTRIS   <STATUS> S-ENDANGERED   <%COUNTIES> <10%
<DISTRIB-PRES>
   17,45,55,101
<QUAD>
   407417,407427,397582,397583,397572,397573,407717
<C-OTHER>
   TWO SUBSPECIES RECOGNIZED*03*
   TINICUM MARSHES, SOUTHERN PHILADELPHIA AND ADJACENT DELAWARE COUNTIES;
   MANOR LAKE (ON SCOTT'S CREEK), BUCKS COUNTY; SILVER LAKE, BRISTOL,
   BUCKS COUNTY (LARGEST CONCENTRATION OF CHRYSEMYS R. RUBRIVENTRIS IN PA.)
   ACTIVELY THREATENED WITH EXTINCTION IN STATE. CONTINUED SUR
   VIVAL UNLIKELY WITHOUT SPECIAL PROTECTION*04*
   COULD DECREASE IF HABITATS GREW TOO SILTED*05*
   SLIGHT INCREASE MIGHT BE POSSIBLE THROUGH REINTRODUCTION TO SUI
   TABLE HABITAT. HABITAT PRESERVATION SINGLE BEST MANAGEMENT PRACTICE
   SPENDS MUCH OF THE DAY BASKING ON LOGS AND ROCKS
   APPARENTLY VENTURES ON LAND ONLY TO LAY EGGS*03*
   RELATIVELY LARGE, DEEP BODIES OF WATER PREFERRED. HAS BEEN FOUN
   D IN BRACKISH WATER*03*
<C-MANAGEMENT>
   PRESERVING AND PROTECTING THIS SPECIES' LIMITED HABITAT IN PENNSYLVANIA
   IS THE MOST IMPORTANT MANAGEMENT TOOL*05*
```

```
<GROUP> AMPHIBIANS   <COMMON-NAME> FROG, NEW JERSEY CHORUS
<SCI-NAME> PSEUDACRIS TRISERIATA   <STATUS> S-ENDANGERED   <%COUNTIES> <10%
<DISTRIB-PRES>
   17,91
<QUAD>
   407427,407534
<C-OTHER>
   INTERGRADES WITH PSEUDACRIS TRISERIATA FERIARUM
   SUBSPECIES NAMED 1955*04*LIFE HISTORY DATA PROBABLY RESEMBLE THOSE OF
   PSEUDACRIS TRISERIATA FERIARUM, WITH WHICH IT INTERGRADES
   BUCKS COUNTY-ONE MILE EAST OF TULLYTOWN. MONTGOMERY COUNTY-SUMNEYTOWN
   *04*
<C-MANAGEMENT>
   (NO VALUE FOR THIS FIELD)
```

3.   Are there any species that occur in Bucks County that are known to be adversely affected by the filling of a wetland? Yes, there are 40 species.

GROUP	COMMON-NAME
AMPHIBIANS	FROG, BULL+
AMPHIBIANS	FROG, GRAY TREE+
AMPHIBIANS	FROG, GREEN
AMPHIBIANS	FROG, NEW JERSEY CHORUS
AMPHIBIANS	FROG, UPLAND CHORUS
AMPHIBIANS	NEWT, RED-SPOTTED
AMPHIBIANS	PEEPER, NORTHERN SPRING
AMPHIBIANS	SALAMANDER, FOUR-TOED
AMPHIBIANS	SALAMANDER, MARBLED
AMPHIBIANS	SALAMANDER, NORTHERN TWO-LINED
AMPHIBIANS	TOAD, AMERICAN
AMPHIBIANS	TOAD, FOWLER'S
BIRDS	BITTERN, AMERICAN
BIRDS	BITTERN, LEAST
BIRDS	EGRET, CATTLE
BIRDS	EGRET, GREAT
BIRDS	EGRET, SNOWY
BIRDS	GOOSE, CANADA
BIRDS	GREBE, PIED-BILLED
BIRDS	HERON, BLACK-CROWNED NIGHT
BIRDS	HERON, GREAT BLUE
BIRDS	HERON, GREEN
BIRDS	HERON, YELLOW-CROWNED NIGHT
BIRDS	IBIS, GLOSSY
BIRDS	MALLARD
MAMMALS	FOX, GRAY
MAMMALS	MINK
MAMMALS	MOLE, STAR-NOSED
MAMMALS	MUSKRAT
MAMMALS	RACCOON
MAMMALS	RAT, NORWAY
MAMMALS	VOLE, MEADOW
MAMMALS	WEASEL, LONG-TAILED
REPTILES	SNAKE, NORTHERN WATER
REPTILES	SNAKE, QUEEN
REPTILES	STINKPOT
REPTILES	TURTLE, BOG
REPTILES	TURTLE, EASTERN MUD
REPTILES	TURTLE, MIDLAND PAINTED
REPTILES	TURTLE, RED-BELLIED

4. Are there any species occurring in Bucks County that are known to be adversely affected by channelization? Yes, there are 41 species.

```
        GROUP               COMMON-NAME
        AMPHIBIANS          FROG, BULL+
        AMPHIBIANS          FROG, GREEN
        AMPHIBIANS          NEWT, RED-SPOTTED
        AMPHIBIANS          SALAMANDER, FOUR-TOED
        AMPHIBIANS          SALAMANDER, LONGTAIL
        AMPHIBIANS          SALAMANDER, MARBLED
        AMPHIBIANS          SALAMANDER, NORTHERN DUSKY
        AMPHIBIANS          SALAMANDER, NORTHERN RED
        AMPHIBIANS          SALAMANDER, NORTHERN SPRING
        AMPHIBIANS          SALAMANDER, NORTHERN TWO-LINED
        BIRDS               BITTERN, AMERICAN
        BIRDS               BITTERN, LEAST
        BIRDS               EGRET, CATTLE
        BIRDS               EGRET, GREAT
        BIRDS               HERON, BLACK-CROWNED NIGHT
        BIRDS               HERON, GREAT BLUE
        BIRDS               HERON, GREEN
        BIRDS               HERON, YELLOW-CROWNED NIGHT
        BIRDS               IBIS, GLOSSY
        FISH                BASS,ROCK
        FISH                BASS,SMALLMOUTH
        FISH                EEL,AMERICAN
        FISH                PICKEREL,CHAIN
        FISH                SHAD,AMERICAN
        FISH                SUNFISH,REDBREAST
        MAMMALS             MINK
        MAMMALS             MOLE, STAR-NOSED
        MAMMALS             MUSKRAT
        MAMMALS             RAT, NORWAY
        MOLLUSCS            (ALASMIDONTA UNDULATA)
        MOLLUSCS            ELK TOE, EASTERN
        MOLLUSCS            SHELL, ALEWIFE
        MOLLUSCS            SHELL, EASTERN FLOAT
        REPTILES            SNAKE, NORTHERN WATER
        REPTILES            SNAKE, QUEEN
        REPTILES            STINKPOT
        REPTILES            TURTLE, EASTERN MUD
        REPTILES            TURTLE, MIDLAND PAINTED
        REPTILES            TURTLE, NORTHERN SNAPPING
        REPTILES            TURTLE, RED-BELLIED
        REPTILES            TURTLE, SPOTTED
```

5. Are there any species known to occur in Bucks County that are of high-value and would be adversely affected by filling a wetland and channelizing a stream? Yes, there are 9 species. For these 9 species we requested the following categories of information: group, common name, scientific name, status, species association by National Wetlands Inventory Classification, beneficial management practices, adverse management practices, and comments on management practices.

```
        <GROUP> AMPHIBIANS    <COMMON-NAME> FROG, NEW JERSEY CHORUS
        <SCI-NAME> PSEUDACRIS TRISERIATA    <STATUS> S-ENDANGERED
        <NWI>
            PALUSTRINE
        <MANAGEMENT-B>
            051
        <MANAGEMENT-A>
            001,002,003,015
        <C-MANAGEMENT>
            (NO VALUE FOR THIS FIELD)
```

```
<GROUP> AMPHIBIANS    <COMMON-NAME> PEEPER, NORTHERN SPRING
<SCI-NAME> HYLA CRUCIFER    <STATUS> SENSITIVE
<NWI>
   PALUSTRINE
<MANAGEMENT-B>
   041,051
<MANAGEMENT-A>
   001,002,003,015,032,033,034,036,037
<C-MANAGEMENT>
   DISAPPEARS RAPIDLY FROM INTENSELY CULTIVATED AREAS*03*

<GROUP> MOLLUSCS    <COMMON-NAME>
<SCI-NAME> ALASMIDONTA SAY, 1818 UNDULATA (SAY, 1817)
<STATUS> SENSITIVE,INDICATOR
<NWI>
   RIVERINE
<MANAGEMENT-B>
   004,006,027,042
<MANAGEMENT-A>
   001,002,003,004,015,020,038,040,046,047,048,049,054,057
<C-MANAGEMENT>
   THERE ARE CURRENTLY NO MANAGEMENT PRACTICES IN PENNSYLVANIA

<GROUP> MOLLUSCS    <COMMON-NAME> ELK TOE, EASTERN
<SCI-NAME> ALASMIDONTA VARICOSA    <STATUS> SENSITIVE,INDICATOR
<NWI>
   RIVERINE
<MANAGEMENT-B>
   004,006,027,042
<MANAGEMENT-A>
   001,002,003,004,015,020,038,040,046,047,048,049,054,057
<C-MANAGEMENT>
   THERE PRESENTLY ARE NO MANAGEMENT PRACTICES IN PENNSYLVANIA

<GROUP> MOLLUSCS    <COMMON-NAME> SHELL, ALEWIFE
<SCI-NAME> ANODONTA IMPLICATA    <STATUS> SENSITIVE,INDICATOR
<NWI>
   RIVERINE
<MANAGEMENT-B>
   004,006,027,042
<MANAGEMENT-A>
   001,002,003,004,015,020,038,040,046,047,048,049,054,057
<C-MANAGEMENT>
   THERE ARE CURRENTLY NO MANAGEMENT PRACTICES IN PENNSYLVANIA

<GROUP> MOLLUSCS    <COMMON-NAME> SHELL, EASTERN FLOAT
<SCI-NAME> ANODONTA CATARACTA    <STATUS> SENSITIVE,INDICATOR
<NWI>
   RIVERINE
<MANAGEMENT-B>
   004,006,027,042
<MANAGEMENT-A>
   001,002,003,004,015,020,038,040,046,047,048,049,054,057
<C-MANAGEMENT>
   THERE ARE CURRENTLY NO MANAGEMENT PRACTICES IN PENNSYLVANIA

<GROUP> REPTILES    <COMMON-NAME> TURTLE, BOG
<SCI-NAME> CLEMMYS MUHLENBERGII    <STATUS> S-ENDANGERED
<NWI>
   PALUSTRINE
<MANAGEMENT-B>
   008,013,029,042,051
<MANAGEMENT-A>
   015,030,036,037,039,040,041,049,046,048
<C-MANAGEMENT>
   MAINTAIN AND PRESERVE HABITAT. HABITAT RECLAMATION AND REINTRODUCTION
   ARE POSSIBILITIES*11*GREATEST THREAT IS HABITAT LOSS CAUSED BY MANMADE
   LAKES, DRAINING AND FILLING OF WETLAND*12*
```

```
<GROUP> REPTILES    <COMMON-NAME> TURTLE, EASTERN MUD
<SCI-NAME> KINOSTERNON SUBRUBRUM    <STATUS> S-ENDANGERED
<NWI>
    RIVERINE,LACUSTRINE,PALUSTRINE
<MANAGEMENT-B>
    042,051
<MANAGEMENT-A>
    015,036,037,040
<C-MANAGEMENT>
    (NO VALUE FOR THIS FIELD)


<GROUP> REPTILES    <COMMON-NAME> TURTLE, RED-BELLIED
<SCI-NAME> CHRYSEMYS RUBRIVENTRIS    <STATUS> S-ENDANGERED
<NWI>
    RIVERINE,LACUSTRINE,PALUSTRINE
<MANAGEMENT-B>
    008,042,048
<MANAGEMENT-A>
    002,015,037,040,054,057
<C-MANAGEMENT>
    PRESERVING AND PROTECTING THIS SPECIES' LIMITED HABITAT IN PENNSYLVANIA
    IS THE MOST IMPORTANT MANAGEMENT TOOL*05*
```

Total time and cost for building these models:

a. Computer Costs

Computer	Quantity	Cost
I/O Requests	1335	$ 8.01
I/O Words	313210	$ .32
CPU Seconds	10845	$ 1.63
Connect Minutes	86.3	$ 6.04
SUBTOTAL		$16.00

b. Operator, @$10/hr.x90 min. = $15.00

c. Telephone (long distance
    commercial telephone rate for
    90 min.)   = $25.00

        TOTAL COST   = $56.00

Interpretation:

Prior to any action on this permit request, the site should be visited to determine whether or not the 9 species identified in response to question 3, or their habitats, occur on the permit area. It appears that habitat similar to that described for the permit area is of high value to fish and wildlife.

We recommend retrieving from the system food habits and niche requirements by life stage for each of these 9 species. This information will help field personnel identify key foods or habitats during field inspection of the permit area.

II. Surface mining permit application.

Location: Clinton County, Pennsylvania, 3 miles SE of Glenn Union, on the northside of west branch of Susquehanna River.

Size: 100 acres.

Land use/land cover: deciduous forest and streams.

SAF forest cover type and size: Mature northern red oak, basswood, white ash (type #54); mature northern red oak (type #55); and mature river birch, sycamore (type #61).

Topography: less than 20% of area gently sloping with local relief (1000-3000 feet).

Reclamation planned: return to original contour and seed mined land with a mixture of 10 lb. of Festuca arundinacea (K-31 Fescue) and 2 lb. of Lespedeza sericea per acre. Convert to pasture.

Questions and Responses:

1. Are there any high-value (threatened, endangered, or sensitive) species of fish and wildlife in Clinton County? Yes, there are 7 high-value species in Clinton County.

    <GROUP> AMPHIBIANS   <COMMON-NAME> PEEPER, NORTHERN SPRING
    <STATUS> SENSITIVE
    <NWI>
        PALUSTRINE

    <GROUP> BIRDS   <COMMON-NAME> GREBE, RED-NECKED
    <STATUS> NON-CONSUMP,SENSITIVE
    <NWI>
        ESTUARINE,RIVERINE,LACUSTRINE

    <GROUP> BIRDS   <COMMON-NAME> LOON, COMMON   <STATUS> NON-CONSUMP,SENSITIVE
    <NWI>
        ESTUARINE,RIVERINE,LACUSTRINE

    <GROUP> BIRDS   <COMMON-NAME> LOON, RED-THROATED
    <STATUS> NON-CONSUMP,SENSITIVE
    <NWI>
        ESTUARINE,RIVERINE,LACUSTRINE

    <GROUP> FISH   <COMMON-NAME> BASS, LARGEMOUTH
    <STATUS> CONSUMPTIVE,SENSITIVE,INDICATOR
    <NWI>
        ESTUARINE,RIVERINE,LACUSTRINE

    <GROUP> FISH   <COMMON-NAME> TROUT, BROOK
    <STATUS> CONSUMPTIVE,SENSITIVE,INDICATOR
    <NWI>
        RIVERINE

    <GROUP> MOLLUSCS   <COMMON-NAME> SHELL, EASTERN FLOAT
    <STATUS> SENSITIVE,INDICATOR
    <NWI>
        RIVERINE

2. Are there any known threatened, endangered, sensitive, or indicator species of fish and wildlife on the permit area? No, none are referenced for that location.

3. Are there any species of fish and wildlife that will be adversely affected by the removal of mature northern red oak and white ash; mature northern red oak; and mature river birch and sycamore forest cover types? Yes, there are 28 species.

GROUP	COMMON-NAME
AMPHIBIANS	NEWT, RED-SPOTTED
BIRDS	HERON, GREAT BLUE
BIRDS	KINGFISHER, BELTED
INSECTS-TERR	CICADAS,PERIODICAL
INSECTS-TERR	CRICKET, FALL FIELD
INSECTS-TERR	GYPSY MOTH
INSECTS-TERR	SWALLOWTAIL,TIGER
INSECTS-TERR	WEBWORM,FALL
MAMMALS	BEAR, BLACK
MAMMALS	FOX, GRAY
MAMMALS	FOX, RED
MAMMALS	MOLE, HAIRY-TAILED
MAMMALS	MOUSE, WHITE-FOOTED
MAMMALS	OPOSSUM, VIRGINIA
MAMMALS	RACCOON
MAMMALS	SHREW, MASKED
MAMMALS	SHREW, SHORT-TAILED
MAMMALS	SKUNK, STRIPED

```
MAMMALS              VOLE, MEADOW
MAMMALS              VOLE, WOODLAND
MAMMALS              WEASEL, LONG-TAILED
REPTILES             COPPERHEAD, NORTHERN
REPTILES             RACER, NORTHERN BLACK
REPTILES             SKINK, FIVE-LINED
REPTILES             SNAKE, NORTHERN RIBBON
REPTILES             SNAKE, NORTHERN WATER
REPTILES             TURTLE, EASTERN BOX
REPTILES             TURTLE, WOOD
```

4. Are there any species occurring in Clinton County that may no longer be present if mature forest habitats are converted to pasture? Yes, there are 11 species.

```
GROUP                COMMON-NAME
AMPHIBIANS           NEWT, RED-SPOTTED
BIRDS                HERON, GREAT BLUE
INSECTS-TERR         CICADAS,PERIODICAL
INSECTS-TERR         GYPSY MOTH
INSECTS-TERR         SWALLOWTAIL,TIGER
INSECTS-TERR         WEBWORM,FALL
MAMMALS              BEAR, BLACK
MAMMALS              FOX, GRAY
MAMMALS              VOLE, WOODLAND
REPTILES             COPPERHEAD, NORTHERN
REPTILES             SNAKE, NORTHERN WATER
```

5. Are there any species occurring in Clinton County that are associated with the planned reclaimed habitat type (pasture)? Yes, there are 49 species.

```
GROUP                COMMON-NAME
AMPHIBIANS           FROG, GRAY TREE+
AMPHIBIANS           SALAMANDER, NORTHERN RED
AMPHIBIANS           SALAMANDER, NORTHERN TWO-LINED
AMPHIBIANS           SALAMANDER, SPOTTED
AMPHIBIANS           TOAD, AMERICAN
AMPHIBIANS           TOAD, FOWLER'S
BIRDS                EGRET, CATTLE
BIRDS                GOOSE, CANADA
BIRDS                KINGFISHER, BELTED
BIRDS                MALLARD
INSECTS-TERR         BUTTERFLY,MONARCH
INSECTS-TERR         CLOUDED SULFUR
INSECTS-TERR         CRICKET, FALL FIELD
INSECTS-TERR         CUTWORM, BLACK (OR GREASY)
INSECTS-TERR         GRASSHOPPER, REDLEGGED
INSECTS-TERR         LACEWING, COMMON GREEN
INSECTS-TERR         LACEWING, GOLDENEYE
INSECTS-TERR         LEAFHOPPER, POTATO
INSECTS-TERR         SWALLOWTAIL,BLACK
INVERT-OTHER         SOW BUG
INVERT-OTHER         WINTER TICK
INVERT-OTHER         (SPIDER) MITE, CLOVER
MAMMALS              COTTONTAIL, EASTERN
MAMMALS              COYOTE
MAMMALS              FOX, RED
MAMMALS              MINK
MAMMALS              MOLE, HAIRY-TAILED
MAMMALS              MOLE, STAR-NOSED
MAMMALS              MOUSE, DEER
MAMMALS              MOUSE, HOUSE
MAMMALS              MOUSE, MEADOW JUMPING
MAMMALS              MOUSE, WHITE-FOOTED
MAMMALS              MUSKRAT
MAMMALS              OPOSSUM, VIRGINIA
```

```
MAMMALS          RACCOON
MAMMALS          RAT, NORWAY
MAMMALS          SHREW, MASKED
MAMMALS          SHREW, SHORT-TAILED
MAMMALS          SKUNK, STRIPED
MAMMALS          VOLE, MEADOW
MAMMALS          WEASEL, LONG-TAILED
REPTILES         RACER, NORTHERN BLACK
REPTILES         SKINK, FIVE-LINED
REPTILES         SNAKE, NORTHERN BROWN
REPTILES         SNAKE, NORTHERN REDBELLY
REPTILES         SNAKE, NORTHERN RIBBON
REPTILES         TURTLE, EASTERN BOX
REPTILES         TURTLE, SPOTTED
REPTILES         TURTLE, WOOD
```

6. Are there any species occurring in Clinton County that will be adversely affected by converting mature forest to pasture? Yes, there are 11 species that will be adversely affected. For each species, the following information was requested: group, common name, scientific name, status, percent of counties in the State in which species occurs, list of counties in which species is known to occur, list of counties where species distribution is unknown, species assocation by National Wetlands Inventory Classification, adverse management practices, general comments, and comments on management practices.

```
<GROUP> AMPHIBIANS    <COMMON-NAME> NEWT, RED-SPOTTED
<SCI-NAME> NOTOPHTHALMUS VIRIDESCENS    <STATUS> NONE    <%COUNTIES> 100%
<DISTRIB-PRES>
    1,3,5,7,9,11,13,15,17,19,21,23,25,27,29,31,33,35,37,39,41,43,45,
    47,49,51,53,55,57,59,61,63,65,67,69,71,73,75,77,79,81,83,85,87,
    89,91,93,95,97,99,101,103,105,107,109,111,113,115,117,119,121,
    123,125,127,129,131,133
<DISTRIB-UNK>
    (NO VALUE FOR THIS FIELD)
<NWI>
    RIVERINE,LACUSTRINE,PALUSTRINE
<MANAGEMENT-A>
    015,033,037,040
<C-OTHER>
    FOUR SUBSPECIES ARE RECOGNIZED*05*
    LAND PHASE (EFT) MORE NUMEROUS IN DECIDUOUS THAN CONIFEROUS WOODS*02*
    LAND PHASE (EFT) ADAPTABLE TO VARYING FOREST SUCCESSIONAL
    STAGES*02*
    ADULTS AND LARVAE CLUSTER IN POOLS AND
    PONDS*06*
    AQUATIC ADULTS AND TERRESTRIAL JUVENILES
    ADULTS AND LARVAE INHABIT PONDS AND POOLS. JUVENILES
<C-MANAGEMENT>
    (NO VALUE FOR THIS FIELD)
```

```
<GROUP> BIRDS    <COMMON-NAME> HERON, GREAT BLUE    <SCI-NAME> ARDEA HERODIAS
<STATUS> NON-CONSUMP    <%COUNTIES> 100%
<DISTRIB-PRES>
    ALL
<DISTRIB-UNK>
    (NO VALUE FOR THIS FIELD)
<NWI>
    ESTUARINE,RIVERINE,LACUSTRINE,PALUSTRINE
<MANAGEMENT-A>
    002,003,015,020,033,036,044
<C-OTHER>
    REGULAR MIGRANT *10*
    REGULAR MIGRANT. BREEDS IN NORTHWESTERN COUNTIES: BUTLER, CRAWFORD,
    MCKEAN, POTTER, WARREN. *10* BREEDS MAINLY IN NORTHERN U. S., SOUTHERN
    CANADA AND ALASKA. WINTERS PRINCIPALLY ALONG COASTS AND SOUTHERN U. S.
    PLACES NEST IN TOPS OF TALLEST TREES IN AN AREA. *07*
    PROTECTED- CANNOT BE LEGALLY HUNTED OR SOLD *05*
    OUTSIDE OF BREEDING SEASON, INDIVIDUAL
    FEEDING TERRITORY IS VIGOROUSLY DEFENDED. *04*
    MORE ACTIVE AT DAWN AND DUSK.
    ONLY USE TREES AT EDGE OF FOREST
    FOUND IN SALT AND FRESH WATER ENVIRONMENTS; SHALLOW WATERS AND
    SHORES OF LAKES, PONDS, MARSHES, STREAMS, BAYS, AND ESTUARIES
<C-MANAGEMENT>
    THE CUTTING OF WOODLOTS AND DEVELOPEMENT OF REAL ESTATE CAUSES
    ABANDONMENT OF HERONRIES. *04* NEED TO CONTROL POLLUTION OF WATERS
    AND ESTABLISH SANCTUARIES. *01*

<GROUP> INSECTS-TERR    <COMMON-NAME> CICADAS,PERIODICAL
<SCI-NAME> MAGICICADA SEPTEMDECIM    <STATUS> NONE    <%COUNTIES> 50-75%
<DISTRIB-PRES>
    ALL
<DISTRIB-UNK>
    3,5,7,9,11,13,15,17,19,21,23,25,29,31,35,37,39,41,43,45,47,49,51,53,
    57,59,61,65,67,73,75,77,79,81,83,85,87,89,91,93,95,97,99,101,103,105,
    107,109,111,113,115,117,119,121,123,125,127,129,131,133
<NWI>
    (NO VALUE FOR THIS FIELD)
<MANAGEMENT-A>
    UNKNOWN
<C-OTHER>
    SYNONYMY:CICADA SEPTEMDECIM LINNAEUS;TIBICINA SEPTEMDECIM;TETTIGONIA C
    OSTALIS FABRICIUS.
    WIDELY DISTRIBUTED THROUGHOUT THE STATE DUE TO ITS DISTRIBUTION PATTER
    N.
    DISTRIBUTION DATA INCOMPLETE.
    DATA INCOMPLETE;DISTRIBUTED STATE-WIDE.
    EMERGE EVERY 17-YEAR.
<C-MANAGEMENT>
    (NO VALUE FOR THIS FIELD)

<GROUP> INSECTS-TERR    <COMMON-NAME> GYPSY MOTH    <SCI-NAME> LYMANTRIA DISPAR
<STATUS> NONE    <%COUNTIES> 100%
<DISTRIB-PRES>
    ALL
<DISTRIB-UNK>
    (NO VALUE FOR THIS FIELD)
<NWI>
    (NO VALUE FOR THIS FIELD)
<MANAGEMENT-A>
    UNKNOWN
<C-OTHER>
    THROUGHOUT THE STATES
    GENERIC SYNONYMY: LIPARIS.
    ALL COUNTIES
    INTRODUCED INTO U.S. IN 1849, BOSTON, MASS.
<C-MANAGEMENT>
    (NO VALUE FOR THIS FIELD)
```

```
<GROUP> INSECTS-TERR    <COMMON-NAME> SWALLOWTAIL,TIGER
<SCI-NAME> PAPILIO GLAUCUS    <STATUS> NONE    <%COUNTIES> 100%
<DISTRIB-PRES>
    ALL
<DISTRIB-UNK>
    1,5,9,11,13,15,19,21,23,25,31,35,37,39,41,45,47,49,51,53,55,57,59,61,
    63,65,67,69,73,75,77,79,81,83,85,87,89,91,93,95,97,99,103,105,107,111,
    113,115,117,119,121,123,127,129,131,133
<NWI>
    (NO VALUE FOR THIS FIELD)
<MANAGEMENT-A>
    UNKNOWN
<C-OTHER>
    SYNONYMY: P.TURNUS LINN.,P.FLETCHERI KEMP.
    WIDELY DISTRIBUTED; COMMON SPECIES.
    COMMON THROUGHOUT ECOREGIONS.
    OPEN SUNNY PLACES.
    BIRCH,POPLAR,ASH,CHERRY,WILLOW,AND BASSWOODS.
<C-MANAGEMENT>
    (NO VALUE FOR THIS FIELD)


<GROUP> INSECTS-TERR    <COMMON-NAME> WEBWORM,FALL
<SCI-NAME> HYPHANTRIA CUNEA    <STATUS> UNKNOWN    <%COUNTIES> 100%
<DISTRIB-PRES>
    ALL
<DISTRIB-UNK>
    (NO VALUE FOR THIS FIELD)
<NWI>
    (NO VALUE FOR THIS FIELD)
<MANAGEMENT-A>
    UNKNOWN
<C-OTHER>
    WIDELY DISTRIBUTED
    COMMONLY DISTRIBUTED.
<C-MANAGEMENT>
    (NO VALUE FOR THIS FIELD)


<GROUP> MAMMALS    <COMMON-NAME> BEAR, BLACK    <SCI-NAME> URSUS AMERICANUS
<STATUS> CONSUMPTIVE    <%COUNTIES> 50-75%
<DISTRIB-PRES>
    5,9,13,15,21,23,25,27,31,33,35,37,39,47,49,53,61,63,65,69,79,81,83,
    87,89,95,103,105,107,109,111,113,115,117,119,121,123,127,129,131
<DISTRIB-UNK>
    1,3,7,11,17,19,29,41,43,45,51,55,57,59,67,71,73,75,77,85,91,93,97,99,
    101,125,133
<NWI>
    PALUSTRINE
<MANAGEMENT-A>
    001,002,003,032,033,034,037,039
<C-OTHER>
    DAILY AND SEASON LIMIT-ONE; 1979 SEASON-ONE DAY, DEC. 17.
    *10*
    HOME RANGE VARIES BY SEASON;*09* THAT OF
    ADULT MALES IS 48432 ACRES, ADULT FEMALES WITH OFFSPRING
    LARGELY NOCTURNAL NEAR INHABITED AREAS
    RETREATS TO SWAMPS, POORLY DRAINED AREAS.*09*
<C-MANAGEMENT>
    (NO VALUE FOR THIS FIELD)
```

```
<GROUP> MAMMALS   <COMMON-NAME> FOX, GRAY
<SCI-NAME> UROCYON CINEREOARGENTEUS   <STATUS> COMMERCIAL,CONSUMPTIVE
<%COUNTIES> 100%
<DISTRIB-PRES>
    ALL
<DISTRIB-UNK>
    (NO VALUE FOR THIS FIELD)
<NWI>
    PALUSTRINE
<MANAGEMENT-A>
    001,002,003,026,032,033,034,037,039
<C-OTHER>
    HUNTING SEASON NOV. 3-JAN. 31; TRAPPED NOV. 11-JAN.31.
    NO BAG OR SEASON LIMITS.*07*
    APPEARS TO OCCUPY 0.5 TO 2 SQUARE MILES. *02*
<C-MANAGEMENT>
    (NO VALUE FOR THIS FIELD)


<GROUP> MAMMALS   <COMMON-NAME> VOLE, WOODLAND
<SCI-NAME> MICROTUS PINETORUM   <STATUS> NONE   <%COUNTIES> 100%
<DISTRIB-PRES>
    ALL
<DISTRIB-UNK>
    (NO VALUE FOR THIS FIELD)
<NWI>
    (NO VALUE FOR THIS FIELD)
<MANAGEMENT-A>
    001,003,026,032
<C-OTHER>
    FOUND MOSTLY WHERE SOILS ARE EASY FOR BURROWING: DEEP SANDY, FINE-
    GRAINED, OR HUMOUSY SOILS.*03* SIGNIFICANT POPULATIONS OCCUR ONLY IN
    SOUTH-CENTRAL AND SOUTHEASTERN COUNTIES.*04*
    FOSSORIAL HABITS PROBABLY MAKE  FOREST COVER  TYPE UNIMPORTANT.*08*
    SOIL TYPE MORE IMPORTANT THAN COVER TYPE.*02*
    WHERE ABUNDANT MAY DO CONSIDERABLE DAMAGE TO ORCHARDS BY
    GIRDLING THE TREES.*01*
    CHIEFLY FOSSORIAL.*01*
    HIGHEST POPULATIONS IN ORCHARDS.
<C-MANAGEMENT>
    IN ORCHARDS, CULTIVATION UNDER TREES, HERBICIDES, MOWING, PESTICIDES
    HAVE CONTROLLED POPULATIONS.*08,12*


<GROUP> REPTILES   <COMMON-NAME> SNAKE, NORTHERN WATER
<SCI-NAME> NERODIA SIPEDON   <STATUS> NONE   <%COUNTIES> 100%
<DISTRIB-PRES>
    1,3,5,7,9,11,13,15,17,19,21,23,25,27,29,31,33,35,37,39,41,43,45,
    47,49,51,53,55,57,59,61,63,65,67,69,71,73,75,77,79,81,83,85,87,
    89,91,93,95,97,99,101,103,105,107,109,111,113,115,117,119,121,
    123,125,127,129,131,133
<DISTRIB-UNK>
    (NO VALUE FOR THIS FIELD)
```

<GROUP> REPTILES   <COMMON-NAME> COPPERHEAD, NORTHERN
<SCI-NAME> AGKISTRODON CONTORTRIX   <STATUS> UNKNOWN   <%COUNTIES> 75-99%
<DISTRIB-PRES>
     1,3,5,7,9,11,13,17,19,21,25,27,29,31,33,35,37,41,43,45,51,55,57,59,
     61,63,65,67,69,71,73,75,77,79,81,85,87,89,91,93,95,97,99,101,103,107,
     109,111,113,115,119,121,125,127,129,131,133
<DISTRIB-UNK>
     (NO VALUE FOR THIS FIELD)
<NWI>
     (NO VALUE FOR THIS FIELD)
<MANAGEMENT-A>
     UNKNOWN
<C-OTHER>
     IN SOUTHERN TWO-THIRDS AND NORTHEAST CORNER OF STATE*05*
     GENUS INCLUDES ABOUT A DOZEN SPECIES*03*
     PENNSYLVANIA'S MOST ABUNDANT VENOMOUS SNAKE, FOUND PRIMARILY IN THE SO
     UTHERN TWO-THIRDS OF THE STATE. RARELY FOUND IN GLACIATED AREAS.
     MOST COMMON IN OR NEAR DECIDUOUS VEGETATION; COINCIDES WITH OAK
     PREFERS MOIST AREAS COVERED WITH LEAF LITTER, SHADED BY LEAF CANOPY;
     INHABITS SEVERAL TYPES OF DECIDUOUS FOREST CLIMAXES AND MANY OF
     THEIR SERAL PHASES*03*
     HOME RANGES: MALES AVERAGE 24.4 ACRES, FE
     MALES 8.5 ACRES. IN AUTUMN, AFTER BIRTH OF YOUNG
     LARGELY NOCTURNAL*05*HIBERNATES IN WINTER*03*
     MAY WANDER INTO BRUSH, GRASSLAND
     OFTEN FOUND AROUND OLD HOUSES, BARNS
     SOMETIMES FOUND IN SUBURBAN AREA
<C-MANAGEMENT>
     (NO VALUE FOR THIS FIELD)

Total time and cost for building these models:

a. Computer Costs

Computer	Quantity	Cost
I/O Requests	954	$ 5.72
I/O Words	165842	$ .17
CPU Seconds	5502	$ .82
Connect Minutes	93.4	$ 6.54
SUBTOTAL		$13.25

b. Operator, @$10/hr.x90 min. = $15.00

c. Telephone (long distance
    commercial telephone rate for
    90 min.)   = $25.00

TOTAL COST $53.25

Interpretation:

There are no known threatened or endangered species in the permit area. Converting the mature timber habitat types to pasture appears to be the major concern. Field investigation should be made by a qualified biologist to determine whether or not data presented are accurate.

SUMMARY AND CONCLUSIONS

During the late 1960's and early 1970's Congress became increasingly concerned with possible conflicts between Federal agency resource development activities and environmental quality. Accordingly, they passed legislation with an ecological perspective, and mandated assessments of the environmental consequences of major Federal land use and resource management actions. The new legislative perspectives and mandates necessitated consistent and accurate inventories and assessments of fish and wildlife populations, distributions, and habitats to meet multiple user needs. Efforts to respond to these new mandates demonstrated that required data was not available for many species, and that the existing data was widely scattered in professional journals, museum notes, research records, and other divergent sources. It became obvious that for efficient utilization in environmental analysis, and land use planning and management, that the existing data must be collected, and entered into more readily accessible data bases. Therefore, "A Procedure for Describing Fish and Wildlife" was developed to meet the information needs of multiple musers in a cost-effective manner, by providing available species information in a consistent format using standardized definitions. This Procedure is currently being pilot tested in Pennsylvania, Minnesota, Missouri, and Colorado for effectiveness in developing statewide fish and wildlife species information systems. Preliminary

results show that such systems can be used to effectively generate species models for any of the 87 fields of information or combinations of these information fields.

This report demonstrated the potential use of Procedure-based fish and wildlife information systems by developing several models to analyze some relations between the fish and wildlife resource and (a) a proposed dredged spoil disposal, channelization project, and (b) a proposed surface mining project. These examples used a preliminary 441 species dataset from the Pennsylvania statewide species information system (final system will include approximately 850 species) to identify the beneficial and adverse relations between development activities and the fish and wildlife resource. We anticipate that additional pilot testing and development will further improve our capabilities to provide fish and wildlife information in an even more cost-effective manner, and demonstrate the utility of statewide fish and wildlife information systems in supplying the needs of resource managers, planners, and other users.

ACKNOWLEDGEMENTS

Numerous agencies, State and Federal, have contributed to this effort during the past four years. Special thanks are extended to the (1) U.S. Forest Service, U.S. Bureau of Land Management, and the U.S. Fish and Wildlife Service for investing funds in the development of the project; (2) the States of Alabama, Colorado, Missouri, Minnesota, Pennsylvania, and West Virginia for being "test states"; and (3) the following individuals: Mr. Edgar A. Pash, our "Boss," for providing the opportunity to produce this effort; and Sharon Rose for the numerous hours typing proposals, study plans, and reports.

LITERATURE CITED

Armentano, T. V. and O. L. Loucks. 1979. Ecological and environmental data as under-utilized national resources: results from the TIE/Access program. 205p. The Institute of Ecology, Indianapolis, IN.

Besadny, C. D. 1979. State efforts to inventory wildlife habitat. Trans. N. Amer. Wildl. and Nat. Resour. Conf. 44:360-368.

Cushwa, C. T., D. R. Patton, W. T. Mason, Jr., and L. J. Slaski. 1978. RUN WILD EAST: a computerized data system for fish and wildlife resources. Trans. NE Fish and Wildlife Conf. 35:60-65.

_____, C. W. DuBrock, D. N. Gladwin, G. R. Gravatt, R. C. Plantico, N. R. Rowse, L. J. Slaski. 1980. A procedure for describing fish and wildlife for Pennsylvania: summary evaluation report. FWS/OBS-79/19-A U.S. Dept. of the Interior, Fish and Wildlife Service, Office of Biological Services, Eastern Energy and Land Use Team, Kearneysville, WV. 15 p.

_____ and C. W. DuBrock. 1981. Design of fish and wildlife species data bases by State and Federal agencies. In W. T. Mason, Jr., Ed. Fish and wildlife habitat protection: environmental impacts research for the decade of 1970-1980. U.S. Environmental Protection Agency, Office of Research and Development, Washington, D. C. (in preparation).

Darlington, P. J., Jr. 1957. Zoogeography: the geographic distributions of animals. 675 p. John Wiley & Sons, Inc. New York, N. Y.

DuBrock, C. W., D. N. Gladwin, W. T. Mason, Jr., and C. T. Cushwa. 1981. State-of-the-art of fish and wildlife information systems in the United States. Paper presented at the 46th N. Amer. Wildl. and Nat. Resour. Conf. 31 p.

Hilborn, R., C. S. Holling, and C. J. Walters. 1980. Managing the unknown: approaches to ecological policy design. USDI, FWS/OBS 80/26. Washington, D. C.

Hirsch, A., W. B. Krohn, D. L. Schweitzer, and C. H. Thomas. 1979. Trends and need in Federal inventories of wildlife habitat. Trans. N. Amer. Wildl. and Nat Resour. Conf. 44:340-359.

Hoekstra, T. W., D. L. Schweitzer, C. T. Cushwa, S. H. Anderson, and R. B. Barnes. 1979. Preliminary evaluation of a national wildlife and fish data base. Trans. N. Amer. Wildl. and Nat. Resour. Conf. 44:380-391.

Levin, S. A. 1974. Pollutants in ecosystems. p.4-8. In Ecosystem Analysis and Prediction. SIAM-SIMS Conf. SIAM, Philadelphia, Pa.

Mason, W. T., Jr., C. T. Cushwa, L. J. Slaski, and D. N. Gladwin. 1979. A procedure for describing fish and wildlife: coding instructions for Pennsylvania. FWS/OBS-79/19. 21 p. USDI Fish and Wildlife Service. Washington,D. C.

Merriam, C. H. 1892. The geographic distribution of life in North America. Proc. Biol. Soc. Washington. 7:1-64.

Odum, E. P. 1971. Fundamentals of ecology. 3rd edition. W. B. Saunders Co., Philadelphia, Pa.

_____ and J. J. Cooley. 1980. Ecosystem profile analysis and performance curves as tools for assessing environmental impacts. p. 94-102. In Proc. of Symp. Biol. Eval. of Environ. Impacts. USDI FWS/OBS-80/26. Washington D.C.

Parzyck, D. C., R. W. Brocksen, and W. R. Emanual. 1980. Regional analysis and environmental impact assessments. p. 114-119. In Proc. of a Symp. Biol. Eval. of Environ. Impacts. USDI FWS/OBS-80/26. Washington D. C.

Patton, D. R. 1978. RUN WILD - a storage and retrieval system for wildlife habitat information. GTR RM-51 8 p. USDA Forest Service, Rocky Mountain Forest and Range Experiment Station Fort Collins, Co.

Schweitzer, D. L., T. W. Hoekstra, and C. T. Cushwa. 1981. Lessons from the 1979 national assessment of fish and wildlife -- information and coordination needs for the future. Trans. N. Amer. Wildl. and Nat. Resour. Conf. (in press)

Thomas, J. W., J. Miller, H. Black, J. E. Rodiek, and C. Masser. 1976. Guidelines for maintaining and enhancing wildlife habitat in forest management in the Blue Mtns. of Oregon and Washington. Trans. N. Amer. Wildl. and Nat. Resour. Conf. 41:452-476.

_____. 1979. Wildlife habitats in managed forest - the Blue Mountains of Oregon and Washington. USDA Forest Service Agri. Handbook #553. 512 pp.

Wagner, F. H. 1980. Integrating and control mechanisms in arid and semiarid ecosystems -- considerations for impact assessment. 1980. p. 145-158. In Proc. of Symp. Biol. Eval. of Environ. Impacts. USDI FWS/OBS-80/26. Washington D. C.

Wilcott, J. C. 1978. MANAGE: an interactive program for data base users. U.S. Dept. of the Interior, Fish and Wildlife Service, Office of Biological Services, Western Energy and Land Use Team, Fort Collins, CO.

Woodwell, G. M. 1974. The threshold problem in ecosystems. p. 9-23. In Proc. Ecosystem Analysis and Prediction. SIAM-SIMS Conf. Siam Philadelphia, Pa.

SYSTEMS ANALYSIS OF THE PRESENT AND FUTURE ENERGY/ECONOMIC DEVELOPMENTS ON THE ISLAND OF GOTLAND, SWEDEN[1]

James Zucchetto and Ann-Mari Jansson[2]

Abstract. -- The results of a large-scale system analysis of
Gotland, Sweden are presented including quantitative measures
of economic and natural subsystems, results of modelling efforts
and potential contribution for renewable energy technologies.

Regional studies involving, as they must, a complex of interrelated activities over an extensive geographical area are complex and time consuming. They require macro-level as well as micro-level information, extensive data collection if quantitative evaluations are undertaken, and an understanding of the social and institutional arrangements which exist in order to interpret development processes which may be occurring. (Isard 1975; Odum & Brown 1975). The first step that must be undertaken in a regional study is a definition of the region: physical, political, jurisdictional, national and many other factors may lead to the definition of a "region". In our case we picked a geographical area with well-defined physical boundaries, namely, .the island of Gotland, Sweden which is located in the Baltic Sea, some 200 km south of Stockholm. Gotland offered several advantages. Import and export statistics could be obtained because of the availability of harbor statistics; the economy was not overwhelmingly large (population ca. 54,000) so that it was anticipated that a fairly detailed quantitative study could be undertaken within the constraint of our limited budget and expected length of time of the project; there was extensive dependence on photosynthetic systems such as forests, agriculture and fisheries in addition to industrial activity so that our study was anticipated to cover a variety of resource and energy analyses. Emphasis was placed on a systems framework in order to quantify the interactions between different activities, their dependence on energy and the impact of human activity on the environment and natural systems. The following is a summary of major parts of the study and ongoing

investigations. Additional information can be found in earlier publications (Jansson & Zucchetto 1978a, b; Zucchetto & Jansson 1979a, b). See appendix for units used.

OVERVIEW OF GOTLAND

Gotland has enjoyed a long history of habitation since the Late Stone Age (5000 B.C.) which has resulted in a diversity of archaeological remnants and architectural forms for the human system. The total population of the island has remained reasonably steady since 1970 at about 54,000 although it was ca. 10% higher than this in the post World War II period. Spread over an area of 3100 sq. km. the population density is 17.3 people/sq. km. in comparison to an average of 19.8/sq. km for all of Sweden. There are several small towns across the island connected by an adequate road network but the old town of Visby, on the west coast, is the dominant urban center with a population on the order of 21,000, a central, medieval walled city and the main harbor for passenger boats from the mainland.

Although recent oil drillings on Gotland have proved successful the local resources of fossil fuel are probably very limited. Consequently, the present society is dependent on the import of all fuels with oil predominating. Furthermore, there is an underwater electrical cable from the mainland, constructed in 1954, which supplies about half the required end-use electrical energy to the economy. Potential supplies of fuel are available in the form of wood, peat and other forms of biomass. Rich deposits of limestone support major exports for the extractive and cement industries. Agriculture and forestry account for major portions of the land use. Extensive drainage of marshes since 1840 has resulted in a presently stable amount of arable land while fisheries are expanding. The mix of coastal areas, natural systems, diverse architecture and maximum hours of sunshine in Sweden make Gotland an important tourist attraction.

Several developments have markedly influenced the existing system. The mechanization of agriculture in the 1950's produced a surplus of labor and a resulting migration to the towns

---

[1]Paper presented at the international symposium Energy and Ecological Modelling, sponsored by International Society for Ecological Modelling (Louisville, KY, April 20-23, 1981).

[2]James Zucchetto is at The University of Pennsylvania, Regional Science Department, Philadelphia, Pa. 19104 U.S.A. Ann-Mari Jansson is at The University of Stockholm, Asko Laboratory, Box 6801, 113 86 Stockholm, Sweden

and, in the absence of sufficient employment opportunities, to the mainland. The total population of the island dropped from 1945 to 1960 by about 10% while the population of Visby increased by 30%. This out-migration and a declining economic base were counteracted by deliberate actions of the Swedish central government such as the location of lottery headquarters in Visby, economic incentives for the location of the electrical workshop, L. M. Ericsson in 1961, and the location of military commands on Gotland. Furthermore, increasing ferry boat traffic since the mid-fifties together with extensive transportation subsidies has resulted in increased tourist activity and trade, have contributed to the economy and generated jobs. The sum total of these dynamic forces has resulted in a relatively stable population since 1970 with a somewhat positive increase since 1972 resulting from in-migration.

Energy and Economy

Oil, diesel and gasoline are the main fuel imports which in 1972 amount led to 372,400 cubic meters (13817 TJ) or ca. 34 bbl/person/yr of oil, 140 gal/person/yr of gasoline and 87 gal/person/yr of diesel fuel. About 540 TJ (150,000 MWh) of electrical energy was imported via the cable which represented about 44 % of electrical energy end-use (340,000 MWh). Although imported fossil fuels represent a much greater absolute consumption in comparison to electrical energy, the period 1972-78 has seen a steady rise in electrical energy consumption and a somewhat erratic but declining use of fossil fuels; in years when energy prices rise abruptly, consumption falls. Goods which are imported also represent embodied energy. In 1972 this was equivalent to 4,192 TJ of fossil fuel or about one third of the total energy consumption for that year on Gotland. The total gross flows of energy for Gotland are summarized in table 1.

In comparison to total Sweden, agriculture and food industries account for a much higher percentage of the labor market on Gotland and accounted for 28% of the value added. The availability of limestone and sand have contributed to a considerable impact of the energy-intensive quarry and cement industries and cement and food products are the two main exports.

There is a net subsidy to Gotland from the main economy. In 1972 this was 122 million Skr (ca. $28.7 million), compared to a gross regional product of 1158 million Skr. Expenditures by the military amounting to about 15% of the gross regional product are also a means by which the regional economy is subsidized. The per capita product for Gotland (21300 Skr) was 13% lower than for Sweden (24600 Skr) while private consumption per capita (11,700 Skr) was about 11% lower than for total Sweden (13100 kr). The ratio of direct energy

Table 1.-- Major Natural and Cultural Energy Flows for Gotland (1972).

CATEGORY	TJ/yr
Gross Photosynthesis in Natural Systems	
Terrestrial Systems	57,000
Baltic coastal systems	17,600
Managed Systems	
Agricultural	27,700
Total Gross Photosynthesis	100,000
Other Natural Energies	
Wind Energy (0-100m height)	150,000
Beach Wave Energy	56,000
Free Energy of Salt/Fresh Water Gradient	6,000
Potential Head of Water	200
Total	212,200

Cultural Energies		FFE*
Imported Fuels	13,800	(13,800)
Imported Electricity	540	(1,944)
Cost of Imported Goods	2,900	(4,192)
Total	17,240	(19,936)

*Fossil fuel equivalents.

consumption in oil equivalents to total regional product was 13.2 MJ/Skr, a much higher value than for Sweden. This reflects the impact of the very energy intensive cement industry.

SUBSYSTEMS ON GOTLAND

Natural Systems

Aside from agriculture the terrestrial system is dominated by coniferous pine forests which, in addition to Juniperus heath systems, represent two types of climax ecosystems. The coniferous forests occupy 42% of the land area and capture the most solar energy. Human activity has continuously modified the natural systems: woods have been burned and cleared for pasture, the demand for hardwood trees during several centuries diminished the oak woods, and the need for firewood for heating and lime burning as well as heavy sheep grazing impoverished the vegetation of the heath systems. Since the 19th century, 270 sq.km. of mires have been drained and converted into agricultural land. Estimates of the productivities of the various photosynthetic systems, including agriculture and the Baltic coastal waters, are presented in table 2.

The Baltic coastal waters provide a diversity of fish species which contribute to the support of the Gotland fisheries. Wave and associated turbulent energies provide work for the erosion of the sea bottom, the transport of sediments and sewage and the addition of oxygen to the surface water. In addition, wave energy

Table 2.—Annual Net and Gross Metabolic
Work for Photosynthetic Systems.

Category	Area (sq.km.)	Net (PJ)	Gross (PJ)
Coniferous Forests	1260	16.5	38.
Mixed Deciduous Forests	209	3.5	7.95
Heath with Trees	139	0.56	1.4
Grassland Heath	268	0.54	1.08
Wooded Pastures	203	2.6	5.9
Mires	31	1.4	2.1
Lakes	25	0.2	0.31
Beach Systems	62	0.62	0.6
Baltic Coastal Waters	2450		17.6
Agriculture	870		25.7
TOTAL			100.6

may be harnessed in the future with appropriate
technology. Bottom vegetation consists mainly
of Fucus on rocky bottoms and Zostera on sandy
bottoms much of which is washed up along the
coastline forming wrack strings which earlier
were used for fertilizer. Another flow from the
coastal system to the land consists of phosphorus contained in bird excretions. Impacts by
humans over the years upon the coastal system
have occurred through intense hunting of seals,
fishing activity, inputs of organic matter and
nutrient salts from sewage outlets and run-off,
oil spills, toxic chemicals, construction work
and dredging from bottom areas.

The beach systems on Gotland constitute
only 2% of the land area and have low
productivity in most areas except where high
levels of fertilizing algae accumulate. They
serve as resting and feeding grounds for a great
number of migrating birds among which can be
found some 10,000 geese. Also, these beaches
provide an extremely important attraction for
the 300,000 tourists who visit the island each
summer.

Heathlands occupy a fairly large total area
of 407 sq. km. of which one third is forested.
Their productivity is quite low as soils are
thin. The heath forests have been used as a
source of wood fuel in the 19th century as well
as for sheep grazing; ca. 1 ha. was required to
feed one ewe. Some damage of heath vegetation
is caused by military operations within the 54
sq. km. of heathland used as training fields.
Also, the exploitation of limestone is
concentrated into heathland areas and causes a
continuous decrease of the highly diverse heath
system.

Freshwater wetlands have declined from a
total area of 300 sq. km. a result of inten-
tional drainage for the purpose of argicultural
production. Vast peat deposits were associated
with the original wetlands. The energy value of
these peat layers has been estimated to be about

6900 PJ which is equivalent to several hundred
years of present fuel consumption. Drainage has
led to water quality problems with increased
nitrate values in groundwater near drained
bogs. There are also few lakes left on the
island. One large lake, Tingstade trask in the
north serves as an important freshwater
reservoir for the population in Visby.

Deciduous forests and wooded pastures
occupy a total of about 200 sq. km. A small
percentage of this area (3 sq. km.) is maintained as a parklike system of high diversity
which produced important winter fodder before
modern clover pastures and concentrated foods
were introduced.

Forests and Forestry

The economically productive forest area
amounts to a large 1260 sq.km. (40% of the land
area) dominated by pine (Pinus silvestris) and,
to a lesser degree, spruce (Picea abies). It
was estimated that tree net production is 5400
kg/ha/yr dry matter and net production of ground
vegetation 1600 kg/ha/yr dry matter which
translates into the total energy flows indicated
in table 3. The ratio of total gross production
of 38 PJ/yr to total incident sunlight of 4680
PJ/yr is ca. 0.8%. In 1972, the harvest of wood
amounted to about 108,000 cubic meters by
companies and 40,000 cubic meters by private
owners for a total energy value of 1190 TJ. The
total direct energy cost of harvesting was
approximated to be 12 TJ and the inclusion of
indirect energy costs would result in a maximum
total energy cost of around 50 TJ. Our
calculations indicate that harvest could be
sustained at higher levels than was the case in
1972.

Table 3.— Some Indicators for the
Forest System, 1972.

Economically Productive Forests	1260 sq. km.
Incident Sunlight	4800 PJ
Total Gross Production	38 PJ
Total Net Production	16 PJ
Harvested Wood 150,000 cubic meters	1190 TJ
Energy Costs of Harvest	
Direct	12 TJ
Direct and Indirect	50 TJ
Total Labor	234,000 man-hrs.
Total Sales	9.8 M Skr.

Fisheries and the Baltic

Although fisheries represented only 0.4% of total sales in economic terms, the Baltic coastal system and fisheries are interesting from the point of view of ecology and historical development. The predominating species caught include herring, cod, flounder and salmon with the protein value of 1973 fish catch sufficient to meet the protein demands of 31,000 people and with a total value of sales of 5.65 M Skr. Economic values per unit of catch were as follows: 1.53 Skr/MJ for salmon, 0.3 Skr/MJ for flounder, 0.28 Skr/MJ for cod, 0.19 Skr/MJ for assorted fish species, and 0.09 Skr/MJ for herring. Fisheries have changed substantially over time, becoming much more capital intensive but without concommitant fish yields. In 1973 a total fish catch of 18.6 TJ was supported by 3,900 PJ of sunlight at a direct and indirect cost in fuels and electricity of 67 TJ. Some measures for fisheries are summarized in Table 4.

Table 4.— Some Measures for Gotland's Fisheries, 1920 and 1973.

	1920	1973
Catch/capital (MJ/Skr)*	15.3	2.3
Catch/worker (GJ/worker)	9.6	76.2
Capital/worker (1000 Skr/ worker)	0.6	33.4
No. of people whose protein reqmt's. could be met**	20,600	31,000
Direct energy cost (TJ)	6.3	37.6
Energy Cost/Energy of Catch	0.5	2.0

*Capital expressed in 1920 prices.
**Protein required = 50 g/person/day.

Agriculture

Agricultural production is by far the photosynthetically based system which dominates the economy accounting for 15% of the total gross regional product. The amount of land under cultivation has remained reasonably constant since 1945 at 810 sq. km. while the total harvest has increased by about 60% due to mechanization and chemical additions. Primary crops consist of oil crops, potatoes, grain and sugar beets while pigs, cattle and sheep compose the livestock populations. Detailed estimates were made of direct energy costs on and off the farm, production and environmental consequences (Zucchetto and Jansson 1979b). Some of these results are presented in table 5. Energy required by farm households amounted to a significant 1380 TJ, mostly for heating.

Table 5.——Energy and Economic Measures for Gotland's Agricultural System (1972).

A.	Direct Electricity	52 TJ = 156 TJ (FFE)
B.	Direct Fuels	270 TJ
C.	Indirect Energy Costs	980 TJ (FFE)
D.	Value Added	157.5 M Skr
E.	Value of Sales	351.2 M Skr
F.	Food Production	1730 TJ
G.	Total Sunlight	3240 PJ
	(A + B)/D	2.7 MJ/Skr
	(A + B + C)/F	0.84

Industry

Detailed economic and energy data were gathered across industry sectors and through time in order to quantify the relationship between industrial activity and energy use. Increasing prices seem to have led to increases in energy efficiency in the stone and soil, chemical, quarry, graphic, wood and workshop sectors from 1973 to 1974. The industrial sectors considered in the analysis were stone and soil, food, quarries, workshops, graphics, wood, chemical and textiles. One of the main energy consumers on the island is the cement industry wich is part of stone and soil. Based on local agriculture an important food industry has developed on Gotland which gives a high contribution to the gross regional product (55 MSkr in 1972). The various industry sectors differ in their energy as well as labor requirements, both of which are important considerations in the overall economic picture. The electrical workshop in Visby provides the greatest number of jobs while the cement industry is by far the most energy consuming sector. Data has also been obtained for inter-industry flows and resource requirements. Present work is directed towards information about pollutant generation. Some measures for the industry sectors are summarized in 6.

Table 6.--Energy and Economic Measures for Industry Sectors (1973)

Sector	Direct Energy (TJ)	Energy/ Value Added (MJ/Skr)	Energy/ Wage MJ/Skr)
Stone & Soil	5088	105	329
Food	435	7.6	14
Quarries	122	7.5	16.8
Workshops	94	1.1	2.1
Graphics	28	2.8	5.5
Wood	18	2.2	2.5
Chemical	17	9.4	6.3
Textiles	1.5	1.3	1.2

Other Subsystems

Consideration of energy requirements were also made for urban-suburban, military and transportation systems. The urban-suburban system with a total population of 42,600 people had a total energy consumption of 2,500 TJ (59 GJ/person) of fuels and 163 G Wh (3.8 M Wh/person) of electricity. Total energy consumed by the military amounted to 140 MWh (0.5 TJ) of electricity and 261 TJ of fuels. Transportation energy costs were 1720 TJ (32 GJ/person) which is higher per capita than for Sweden as a whole. Some of these energy costs are summarized in table 7.

Table 7.--Some Transportation Energy Costs (1972)

	Gasoline	Diesel
Volume (1000 cu. meters)	32	20.2
Energy (PJ)	1.0	0.72
Energy/capita (GJ)	18.5	13.3

MODELING EFFORTS

Several different models have been formulated during the course of the project in order to explore some of the energy and environmental issues or questions which have been posed. For the entire economic system of the island, an input-output formulation has been chosen to link up all the major economic activities into a systematic whole. This type of model can be linked up to two primary areas of concern: environmental impact and resource availability. Knowing a given level of economic output from various sectors allows approximations of pollutant outputs to be calculated if the pollutant matrix is known which would be a matrix whose elements, P_{ij}, indicate the amount of pollutant "i" per unit of output from economic sector "j". These inputs can then serve a forcing functions to environmental models. On the resource side the economic system requires levels of inputs for a given level of activity such as energy, water, goods, labor and land. Thus, for example, the impact of a limitation of energy or water on economic activity could be analyzed. Optimization approaches have also been incorporated by stipulating some objective function, or quantity, which should be maximized subject to a given set of resource constraints. The following model has been simulated in order to assess the impact of changing energy availability on economic activity. The optimization model was as follows (Zucchetto et al. 1980):

$$\max (V_x \cdot X + V_z \cdot Z)$$

subject to

$$x \leqslant C_x$$

$$Z \leqslant C_z$$

$$B \cdot X + D \cdot Z \leqslant R$$

From the input-output model since

$$X = (I - A)^{-1} Y$$

then the above problem becomes:

$$\max V_x (I - A)^{-1} Y + V_z Z$$

subject to

$$(I - A)^{-1} Y \leqslant C_x$$

$$Z \leqslant C_z$$

$$B (I - A)^{-1} Y + DZ \leqslant R$$

where X = vector of output from economic sectors; Y = vector of final demand (including exports); Z = vector of natural and forest lands, R = vector of resource availabilities; A = technical coefficient matrix; B = resource requirements coefficient matrix for the economic sectors; D = resource requirement coefficient matrix for forest and natural land; C_x = production capacities; C_z = constraints on land use for forest and natural lands; V_x = value added coefficients for economic sectors; V_z = value added coefficients for forest and natural lands. Resource and associated constraints considered were electricity, oil, gasoline, diesel fuel, imported goods, wood, water, labor and land. The model was run using data from 1972 with some assumptions about the level of resource constraints and the flows between economic sectors having to be made.

Several scenarios were considered. For example, starting at time t = 0, the price of oil is raised 5% /yr. for the first five years and then jumps to 10% yr. When this price jump occurs oil constraints are tightened to 80% of what they were. In addition, the possibility of producing "artificial oil" from energy plantations on forest lands is considered. Initially, the price of artificial oil is 3 times imported oil. During each year of the simulation the conditions for maximizing the gross regional product with and without these energy plantations was considered. If the regional product were maximized by the inclusion of artificially produced oil then this system was retained as part of forest lands. Results show that sectors such as agriculture, food, electrical workshops, chemicals and textiles remain quite insensitive to energy restrictions and price changes. However, the energy intensive sectors such as stone and soil and quarries significantly drop in output when oil restrictions occur. An industry like wood, which generates much value added per unit of energy consumption, expands significantly. Artificial oil is selected by the model in the 8th year and this is only made possible if the labor constraint is relaxed; thus man-hrs. expand signficantly. Artificial oil reaches a level of about 10% of imported oil. This type of modeling is continuing with data for more recent years and better accuracy for intersectoral transfers and use of certain resources such as water. Additionally, more information on pollutant generation will allow us to consider pollutant constraints.

Other models which have been considered have focussed more directly on particular environmental systems. For example, high levels of nitrate in drinking water and constraints on available water supplies have fostered the development of a simple water-nitrogen simulation model (Spiller 1980) which includes the following state variables: soil water, soil nitrate, groundwater storage and groundwater nitrate. Soil nitrate is influenced by uptake by plants, decomposition, nitrification and denitrification, runoff and infiltration to groundwater. Similarly, soil water storage is influenced by precipitation, evapotranspiration and infiltration to groundwater. It is assumed that soil water runoff, evapotranspiration and infiltration to groundwater control soil nitrate runoff, nitrogen uptake by plants and soil nitrate infiltration to groundwater, respectively. Four differential equations were specified for the four state variables and simulations were conducted on a day by day basis to ascertain the influence of nitrogen inputs to soil nitrate from such anthropogenic sources as agriculture. For example, irrigation in some areas during the growing season showed significant increases in groundwater nitrate as a result. Realistically, for the complex limestone geology of the island, a spatial model should be developed to investigate this problem in more detail. Ideally, this model could then be coupled to the total economic system model; the economic model would specify a level of nitrate output which would be a forcing function to the water-nitrogen model. The environmental model would generate levels of nitrate in water supplies and, if unacceptable, this information would be used to tighten pollutant emission constraints in the economic system or induce a change in technology.

An ecological simulation model of the coastal system has also been formulated (Limburg 1980). This included primary production by phytoplankton and benthic algae as a function of sunlight, temperature and nitrogen concentration in the water; herring, cod and seals as part of the food web; recycle of nitrogen from consumers to producers; and external impacts affecting the coastal system such as nitrogen runoff from land and harvest of seals, cod and herring. This model is simplified with respect to fish populations as no age classes are considered. The existing levels of nitrogen runoff as a "pollutant" to the coastal system have little impact on the results. However, a 10-fold daily increase in allochthonous loading led to large boom-bust oscilla-

tions in biomass. Timing of a 1-month long nitrogen pulse led to interesting results: if it began in late March a second "producer bloom" was created which raised levels of consumers. Pulsing in August caused producers to oscillate while herring stocks leveled off. An October pulse creates a fall algal bloom, the system conserves more nutrients each winter and the system as a whole experiences exponential growth.

RENEWABLE ENERGY TECHNOLOGIES

Solar energy provides the fundamental energy basis in such activities as agriculture and forestry as well as supporting many of the unmanaged natural systems which perform numerous work functions and provide amenities. Tourists are attracted to Gotland because of the availability of direct sunlight and the diversity of nature which is supported by the flux of solar energy. In recent years, interest in renewable energy technologies for Sweden has arisen because of rising energy prices and the international energy situation (Lonnroth et al., 1977). For Gotland, the possibilities for renewable energy technologies may be more favorable than most places: the population density is fairly low, forest and agricultural lands are extensive, wind energy and wave energy are high and the number of days of sunlight is high for Sweden. Those facts seem to indicate that if any system might serve as a demonstration of the potential for renewable energy technologies, Gotland would be it.

At the moment there are several renewable energy technologies under consideration on Gotland and a prototype wind-electrical generator is under construction. Our assessment of the potential contribution from the renewable energy sources must depend on systems proposed in various engineering studies. Considering table 1 it is seen that Gotland has high levels of renewable energy fluxes. Next to solar energy the largest is wind which averages 6 m/sec over the year with higher values during the winter, values of velocity which are sufficient to generate economic levels of power output. The wind energy within 100 meters of the ground is 150,000 TJ/yr so that if anywhere between 1 and 10% of this wind energy can be converted into electricity, significant supplies to Gotland's economy would be realized. More specific engineering proposals call for a bank of 33 wind-electrical generators spread over a 4.5 x 16.5 sq. km. rectangle with each generator on the order of 1 or 2 MW capacity (Kjellstrom & Gustafsson 1976). Our energy calculations indicate that the output in oil equivalents would be 10 times the energy cost or conservatively as low as 3 or 4 times the energy cost. Thus, the net energy is positive, the energy yield ratio greater than one and furthermore, it is projected that the economic cost of the electrical output is comparable to existing prices. One bank of 33 generators would produce 1070 TJ/yr which is twice what is imported now and about equal to total electrical end-use consumption. There are at least 3 such areas within which these electrical generators could be constructed for a total output on the order of 3000 TJ/yr of electrical energy.

Engineering designs have also been proposed for generators which would convert wave energy in the coastal areas into electricity. (Bystrom et al. 1980.) This technology is much more tentative and uncertain but our detailed energy analysis indicates an energy yield ratio of anywhere from ca. 3 to 8 so that it is a net energy producer. Each proposed buoy would produce .45 TJ/yr. of electrical energy so that a 1000 operating buoys would generate about as much electrical energy as is presently imported; this represents a small fraction of the total wave energy in the coastal system and is perhaps a feasible technology.

Another renewable energy technology which is receiving a great deal of enthusiastic interest in Sweden is the energy plantation consisting of willow trees whose biomass would be converted into liquid fuel, burned for heat or converted into electricity. The incorporation of such a land use has already been considered in the preceeding optimization model which shows the various trade-offs involved in incorporating fast-growing trees into forest lands. These energy plantations require irrigation water as well as fertilizers, both of which may not be environmentally suitable for Gotland under constraints of water quality and supply. However, in order to generate some perspective, energy plantations are expected to generate 20 tons (dry matter)/ha/yr or about 0.38 TJ/ha/yr (Ljungbloom et al. 1978). Devoting 2 to 3% of agriculture and forest land to energy plantations would result in anywhere from 1500 to 2300 TJ/yr of energy available to be processed into fuel, heat or electricity.

Another potential energy source derived from the sun which could be tapped with emerging technologies are agricultural "wastes" such as unused parts of the crop and animal manure, both of which are returned to the soil. The energy content of these flows is on the order of 3800 TJ/yr so that even a 25% conversion rate might contribute substantial energy to the agricultural system. Various end use needs as well as the renewable energy potentials discussed above are summarized in table 8 and 9 where it can be seen that these energy sources have the potential of significant contributions. It should be emphasized that any realistic assessment of renewable energy technologies must await the operation of prototype models.

PROJECT CONTINUATION

This regional research will continue in several directions. Better data will be collected in order to increase the accuracy of

Table 8. Levels of Selected End-Use Energy Requirements on Gotland (TJ/yr) for 1972.

Activity	Fuels	Electricity
Industry		
Stone & Soil	4000	1160
Food	290	146
Quarries	12	110
Workshops	7	87
Graphics	6	22
Wood	4	14
Chemical	1	16
Textiles	1	.4
Agriculture	270	52
Agricultural Households	1300	63
Forestry	9	–
Urban-Suburban	2500	590

Table 9. Reasonable Potential Energy from Renewable Energy Technologies (TJ/yr)

Technology	Fuel	Electricity
Wind-electrical (99 generators)	––	1 - 3000
Wave-electrical (1000 buoys)	––	450
Energy Plantations	1500 to	2300
Agricultural Wastes (maximum)	3800	–––

the total system model especially with regard to resource use, pollutant emissions and intersectoral transfers in order to assess the impact of shortages on economic activity and the different levels of development on the environment. The understanding of the dynamics of subsystems such as the coastal system and the hydrological system will be pursued in order to quantify impacts more accurately. The contribution of renewable energy resources will continue to be investigated.

LITERATURE CITED

Bystrom, A., Clason, P., Linders, J., Sundvall, C-G, 1980. Vagenergi for Gotland (Wave energy for Gotland). 134 p. Chalmer's Institute of Technology, Gothenburg. (In Swedish).

Isard, W. 1975. Introduction to regional science. Prentice-Hall, Inc. Englewood Cliffs, NJ.

Jansson, A.M., and Zucchetto, J. 1978a. Energy, economic, and ecological relationships for Gotland, Sweden: a regional systems study. Ecological Bulletins 28. 155p Swedish

Natural Science Research Council. Stockholm, Sweden.

Jansson, A-M. and Zucchetto, J., 1978b. Man, Nature and energy flow on the island of Gotland. Ambio 4: 140-149.

Kjellstrom, B. & Gustafson, B. 1976. Forstudie for langsiktig energiplan for Gotlands kommun. Seminarierapporter. (Preparartory study for a long range plan for the municipality of Gotland. Seminar reports). Institute of Thermic Energy Techonology, Royal Institute of Technology, Stockholm. Mimeo. (In Swedish).

Limburg, K., 1980. Modelling the Gotland coastal ecosystem. A preliminary minimodel. Asko Laboratory. University of Stockholm. Mimeo.

Ljungblom, L., Lundberg, H., Marklund, A. & Sjoberg, S.O. 1978. Energiskog. (Energy Plantations). Royal Institute of Technology, Stockholm. (In Swedish).

Lonnroth, Steen, P. & Johansson T.B. 1977. Energy in transition. Secretariat for Future Studies. Stockholm, Sweden.

Odum, H. T. & Brown, M. T. (eds.). 1975. Carrying capacity for man and nature in S. Florida. Three Volumes. Final Report on Contract CX000130057 to National Division of State Planning. 886 p. Center for Wetlands, University of Florida, Gainesville, Florida.

Spiller, G., 1980. Modelling the nitrogen cycle on Gotland. Asko Laboratory. University of Stockholm. Mimeo.

Zucchetto, J. & Jansson, A. M. 1979a. Integrated regional energy analysis for the island of Gotland, Sweden. Environment and Planning A, 11:919-942.

Zucchetto, J. & Jansson, A-M. 1979b. Total energy analysis of Gotland's agriculture: A northern temperate zone case study. Agroecosystems. 5:329-344.

Zucchetto, J., Jansson, A-M. & Furugano, K. 1980. Optimization of economic and ecological resources for regional design. International Journal of Resource Management and Optimization. 1:111-143.

APPENDIX

Description of Units

GJ = gigajoule = 10^9 joules

TJ = terajoule = 10^{12} joules

PJ = petajoule = 10^{15} joules

MWh = megawatt-hour = 10^6 watt-hr.

Skr = Swedish Kronor

M Skr = 10^6 = Swedish Kronor

MODELLING THE EFFECTS OF REGIONAL ENERGY DEVELOPMENT ON GROUNDWATER NITRATE POLLUTION ON GOTLAND, SWEDEN[1]

G. Spiller, A. M. Jansson, and J. Zucchetto[2]

Abstract.--A mathematical model of the major nitrogen flows in the agricultural system on the island of Gotland, Sweden was developed. The model was run for a variety of crop and soil types and leaching losses of nitrate to groundwater were estimated. These agreed fairly well with measured values from lysimeter experiments on similar soil types. A sensitivity analysis showed leaching loss to be most sensitive to precipitation increasing in a nonlinear manner. The increase in leaching loss with increasing fertilizer applications was linear. The environmental effects of some energy development and land use management strategies were tested using the model.

INTRODUCTION

Regional energy development can have important environmental effects. Pollution caused by the introduction of a new technology to a region has often been accepted as a necessary side effect of development if productivity is to increase. Pollution can, however, represent a loss of energy from the system (Odum 1971). For example the loss of fertilizer nitrogen from arable land represents a loss of the energy required to produce the fertilizer.

By modelling the dynamics of the pollution process we can often evaluate the effectiveness of management strategies to reduce the detrimental impacts of regional energy development.

The present study uses a simulation of nitrogen flows to examine some environmental effects of regional energy development on the island of

[1]Paper presented at the international symposium Energy and Ecological Modelling, sponsored by the International Society for Ecological Modelling. (Louisville, Kentucky, April 20-23, 1981)
[2]Gary Spiller is a Research Associate at the Department of Chemical Engineering, University of Sherbrooke, Quebec, Canada.
Ann-Mari Jansson is Assistant Professor at the Zoological Department, University of Stockholm, Sweden.
James Zucchetto is Assistant Professor at the Regional Science Department, University of Pennsylvania, U.S.A.

Gotland, Sweden. The island has undergone considerable intensification in agricultural production, doubling from 1945 to 1974 (Jansson and Zucchetto 1978 a,b) while fertilizer use has increased by 350%.

Ninety percent of the 300 km² of peat bogs has been drained in the last 150 years, mainly for agricultural use, leaving marl soils high in nitrogen (up to 3.5%N).

There has been considerable concern expressed over the high nitrate concentrations in the groundwater on the island since most of the drinking water comes from groundwater. High nitrate concentrations can cause methaemoglobinemia in infants. If the oxygen concentration in the groundwater is low the nitrate (NO_3) is reduced to nitrite (NO_2) which can be converted to nitrosamines, known carcinogens.

Figure 1 shows the distribution of measured nitrate concentrations in the drinking water with respect to drained bogs and agricultural land. Not all the high concentrations are measured in agricultural or drained peat areas. This may de due to horizontal groundwater transport of water high in nitrogen and originating as infiltration from agricultural land.

Other sources of nitrate pollution in groundwater include precipitation, nitrogen fixation in pasture land and human sewage. However, a nitrogen budget constructed for the whole island (Spiller 1978) indicates that these flows are small relative to fertilizer input (fig. 2).

While useful in estimating the magnitudes of some inputs and outputs nutrient budgets can provide only indirect estimates of losses by leaching, denitrification and ammonia volatilization.

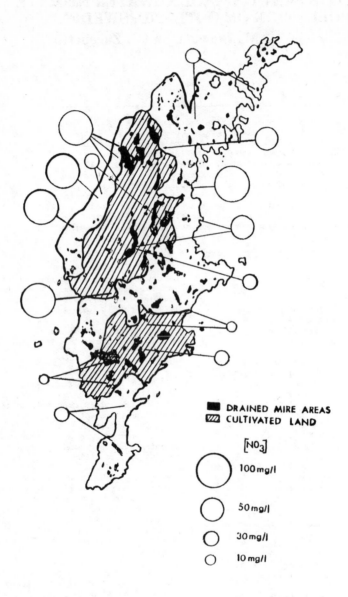

DRAINED MIRE AREAS
CULTIVATED LAND

$[NO_3]$

100 mg/l

50 mg/l

30 mg/l

10 mg/l

Figure 1.--Spatial distribution of measured groundwater nitrate
concentrations on Gotland in relation to drained peat land
and to cultivated land.

pH in the range of 6.5 to 8.5. This is the optimum range for biological processes such as decomposition and nitrification which can increase the amount of nitrate in the soil. The bedrock is limestone or marlstone so that soil water can infiltrate rapidly to the groundwater, carrying nitrate with it. These factors were integrated into a model of nitrogen flow in the agricultural ecosystem to estimate the amount of nitrate leaching to the groundwater.

DEVELOPMENT AND CALIBRATION OF THE MODEL

The model used in this study is a modified version of the one described in Spiller (1978). An attempt was made to make the model more quantitative by including more hydrological data in the calibration and by introducing equations for ammonia volatilization, denitrification and nitrogen fixation. These processes were expressed as functions of temperature and soil moisture. The assumptions made and coefficients used are described below.

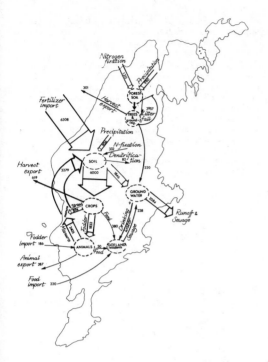

Figure 2.--Nitrogen budget for Gotland (x 10^3 kgN·yr^{-1})

There is considerable controversy over the relative importance of these processes due to the lack of experimental data in the case of denitrification and ammonia volatilization.

In this study an attempt was made to quantify these losses by means of a simulation model describing mathematically the way in which the environment affects these processes.

BACKGROUND

This project was begun as part of a study of energy flows on Gotland, a Swedish island in the Baltic sea (Jansson and Zucchetto 1978 a,b; Zucchetto and Jansson 1979). The island was modelled as a number of interacting sectors which consume resources, produce economic output and generate sets of environmental disturbances and impacts. Agriculture is a major activity on the island and the use of nitrogen fertilizers increased from an average application of 20 kgN/ha/yr in 1951 to over 70 kgN/ha/yr in 1970 (fig. 3). Nitrate concentrations in the drinking water increased from 0.2 to 10 mgNO$_3$N/l during the same period. However, the increase has not been linear and other factors may be responsible for the high nitrate concentrations in the groundwater today. Thirty percent of the agricultural land consists of drained peat soils which are high in nitrogen content (up to 3.5%). Gotland's soils are shallow and calcareous with a

Figure 3.--Average fertilizer application and average nitrate concentrations in the drinking water on Gotland 1951-1972.

DESCRIPTION OF THE MODEL

The flow chart of the model is shown in figure 4. The state variables X01 and X03 are the quantities of soil water and groundwater respectively. The state equations are:

$$\dot{X01} = \text{PRECIPITATION} - \text{EVAPOTRANSPIRATION} \quad (1)$$
$$- \text{SURFACE RUNOFF} - \text{INFILTRATION TO}$$
$$\text{GROUNDWATER}$$

$$\dot{X}03 = \text{INFILTRATION} - \text{RUNOFF} \qquad (2)$$

To this is coupled the nitrogen flow model consisting of soil nitrate (X02), soil organic nitrogen (X06), soil ammonium (X07), groundwater nitrate (X04) and plant nitrogen (X05). The equations are:

$$\dot{X}02 = \text{FERTILIZER} + \text{MANURE} + \text{NITRIFICATION} \qquad (3)$$
$$- \text{PLANT UPTAKE} - \text{INFILTRATION}$$

$$\dot{X}04 = \text{INFILTRATION} - \text{RUNOFF} \qquad (4)$$

$$\dot{X}05 = \text{UPTAKE} - \text{HARVEST} - \text{CROP RESIDUE} \qquad (5)$$

$$\dot{X}06 = \text{MANURE} + \text{CROP RESIDUE} - \text{DECOMPOSITION} \qquad (6)$$
$$+ \text{IMMOBILIZATION}$$

$$\dot{X}07 = \text{FERTILIZATION} + \text{MANURE} - \text{NITRIFICATION} \qquad (7)$$
$$- \text{IMMOBILIZATION} - \text{UPTAKE}$$

HYDROLOGICAL MODEL

This model described by Spiller (1978) and Häger & Thoms (1978) assumes that infiltration occurs only if the soil is saturated. It does not account for rapid infiltration or fissure flow through cracks in bedrock exposed to the surface. The model was calibrated to the Hemse region of Gotland, an agricultural area in the southern part of the island. About half of the 470 km² area of the region is devoted to agriculture with the remainder consisting of forests and heath. The

bedrock consists of Silurian marlstone covered by shallow soils, mainly glacial till one meter deep. The largest drained bog on the island (31 km²) is located in this region.

Groundwater recharge occurs only in the winter. This was consistent with predictions of the saturated flow model. Groundwater levels range from 1.0 to 2.7 m below the land surface. This relatively small variation in groundwater level also suggests that most of the water entering the aquifer does so as seepage from saturated soil. Assuming fissure flow is negligible and an effective porosity of 0.4% a groundwater runoff coefficient of 0.07 was calibrated from groundwater level curve and used in the equation:

$$T30 = K \cdot X03 \qquad (8)$$

where,

$T30$ = rate of groundwater runoff
$X03$ = groundwater present at time t (0-15m)
K = 0.07/day

An active mixing layer of 15 m was assumed since nitrate concentrations in wells deeper than this are very low.

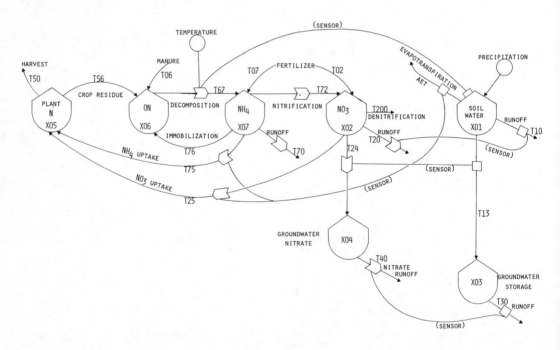

Figure 4.--Flow chart of model. Sensor denotes coupling of nitrogen flow to water flow.

Actual evapotranspiration was calculated from the formula:

$$AET = PET \cdot (S/SO)^B \qquad (9)$$

where,
PET = Potential evapotranspiration (mm)
      calculated from the Penman formula
      (Penman 1948)
   S = the amount of water in the top meter
      of the soil minus the wilting point (mm)
  SO = field capacity minus wilting point (mm)
   B = 0.5

INPUTS TO THE MODEL

Fertilization in the spring and fall and daily precipitation values are the two main inputs to the model. Initial simulations were made for 1972 since the energy study (Jansson & Zucchetto 1978a,b) was done for that year. Field capacities for the different soils range from 45 mm for sand to 210 mm for glacial till. Drained peat soil has a capacity of 100 mm (Häger & Thoms 1978).

BIOLOGICAL TRANSFORMATIONS OF NITROGEN

Decomposition

A first order decomposition model was used. The rate coefficient of ammonia production was expressed as an increasing function of temperature and soil moisture with the maximum rate occurring at $50°C$ and 60% saturation. The equation for decomposition is:

$$DECOMP = KDE \cdot ON \qquad (10)$$

where,
DECOMP = rate of NH_4–N production ($kgNH_4$–N $\cdot$ ha^{-1}
                   $\cdot$ day^{-1})
    ON = organic nitrogen in soil (kg/ha)
       (including manure and crop residue
       added)
  KDE = KDMAX $\cdot$ SMRF $\cdot$ TRF
KDMAX = decomposition rate coefficient under
       optimal temperature and soil moisture
       conditions
 SMRF = fractional reduction in decomposition
       rate at sub-optimal soil water
   TRF = fractional reduction in decomposition
       rate at sub-optimal soil temperature

If the carbon/nitrogen ratio of the added organic material exceeds 25 to 1 no net decomposition is assumed to occur. Instead all ammonia ions produced are immobilized in microbial biomass.

Nitrification

Nitrification rates were assumed to be maximum under warm temperatures and with a soil water content of 8 to 60% of the field capacity. Greater soil water content would inhibit nitrification since oxygen is necessary for the process to occur. Nitrification is probably very rapid on Gotland since the alkaline soils are within the optimum pH range for nitrification (6.5–8.5). This prediction is supported by the low concentration of ammonium found in soil drainage water (Wiklander 1977). The equation for nitrification is:

$$NITRIF = KN \cdot KNRF \cdot NH_4 \qquad (11)$$

where,
  KN = KN_{20} + C_{20}(T-20)

KNRF = soil oil reduction factor

Denitrification

Denitrification was expressed as a linear function of temperature above $12°$ and was assumed to occur only if the soil was saturated. The pH of the soils on Gotland is in the optimum range for denitrification (6.2 to 8.0) and the process can be expressed to occur wherever anaerobic conditions are present. The equation used was:

$$DENIT = KD \cdot NO_3 \qquad (12)$$
$$KD = KT_i \cdot KH \qquad \text{if XO1} \geq \text{FC (soil saturated)}$$

$$KD = 0 \text{ if XO1} < \text{FC}$$

where,
KT_1 = KT_{12} + C_{12} (T-20)
KT_2 = KT_{11} + C_{11} (T-20)
KT_1 = temperature coefficient for $T \geq 12°C$
KT_2 = temperature coefficient for $T < 12°C$
KH = pH coefficient = 1 for Gotland soils

Larsen (1977) found that both nitrification and denitrification occurred at temperatures as low as $6.5°C$.

Ammonia volatilization

This process occurs almost entirely in the surface layer (top 10 cm) of soil. It reaches its maximum in warm, moist, slightly alkaline soils. A linear temperature function was constructed for soils in the optimum pH range. Denmead et al. (1977) found a correlation between volatilization rates and evaporation rates. The equation used was:

$$NH_4VOL = KA \cdot NH_4 \qquad (13)$$
$$KA = KTV \cdot KSM$$

where,
  KTV is the temperature coefficient
  KTV = KTV_{20} + C_{20} (T-20)
  KSM is the soil moisture coefficient

Bolin and Arrhenius (1977) estimated that 60% of the nitrogen in manure would be volatilized but that 25% of this gaseous loss would be redeposited on cropland. Since volatilization losses occur on the surface layer they can be expected to be large from manure that is applied and not ploughed under or from manure piles and urine wells if the weather is warm.

Nitrogen uptake by plants

The rate of nitrogen uptake by crops was assumed to be inversely proportional to the water deficit as given by the ratio of actual to potential evapotranspiration. This model was based on one developed by Jensen (1976). The equation used was:

$$NUPTAKE = UMAX \cdot \left(\frac{AET}{PET}\right)^{D(t)} \tag{14}$$

where,
UMAX is the maximum N uptake for a given crop under optimal nutrient and soil moisture condition
$$(kg\ ha^{-1}day^{-1}) = 4\ for\ sugar\ beets,$$
1.6 for barley, 1.4 for winter wheat
$D(t)$ is a coefficient of sensitivity to the deficit
$D = 1$ for the first growth stage (8 weeks for barley)
0.5 afterwards

Crop yields were estimated from a similar equation using the same sensitivity coefficient as in nitrogen uptake.

$$YIELD = YMAX \cdot \left(\frac{AET}{PET}\right)^{D(t)} \tag{15}$$

where,
YMAX is the maximum yield for a given crop under optimal nutrient and soil moisture conditions
$$(kg\ ha^{-1}day^{-1}) = 200\ for\ sugar\ beets,$$
80 for barley, 75 for winter wheat
$D(t)$ is defined above

SENSITIVITY ANALYSIS

The sensitivity of the model to the amount of precipitation and the rate of fertilization was tested by running the model with 0.5, 0.75, 1.0, 1.25, and 1.50 times the 1972 levels of precipitation and fertilization for sugar beets growing on glacial till. The sensitivity of nitrate leaching to changes in these driving variables is shown in figure 5. Both the annual nitrogen leaching and the average annual nitrate concentration in the groundwater increased linearly with increasing fertilizer application. This relationship is similar to one found by Brink (1977) for different levels of manure application to cropland (fig. 6).

As figure 5 shows the model predicts that an increase in fertilization from 110 to 165 kgN/ha/yr would result in a leaching loss of 41 kgN/ha/yr. These losses would, however, be greater in a wet year as can be seen from figure 7.

The nonlinear relation between precipitation and nitrate leaching resembles the one presented in Kolenbrander (1969) and reproduced here (fig. 8). When the 1972 precipitation of 448 mm was reduced by 25%, the average groundwater nitrate concentra-

tion was sharply reduced from 12.5 to 1.8 mgNO$_3$-N/l due to the predicted decrease in the amount of water infiltrating to the groundwater. Increasing the level of precipitation to 672 mm resulted in an increase in the amount of nitrate leached from 31.2 to 71.6 kg/ha/yr, but a decrease in the average NO$_3$-N concentration from 12.5 to 10.0 mg/l due to a dilution effect. Soil water concentration decreased steadily as precipitation increased due to greater losses to the groundwater.

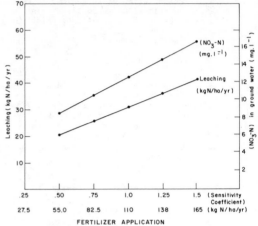

Figure 5.--Sensitivity of predicted leaching and groundwater nitrate concentrations to changes in fertilizer application.

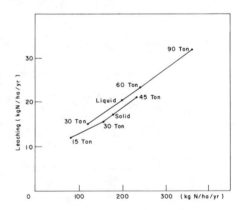

FERTILIZER APPLICATION (MANURE)

FROM BRINK (1977)

Figure 6.--Empirical relation between fertilizer application and leaching measured using lysimeters for a variety of soil types in Sweden. From Brink (1977).

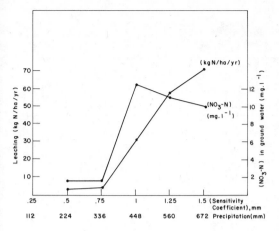

Figure 7.--Sensitivity of predicted leaching and groundwater nitrate concentrations to changes in precipitation.

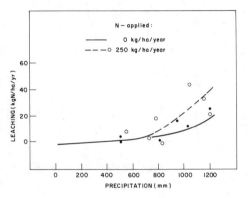

FROM KOLENBRANDER (1969)

Figure 8.--Variation of annual leaching loss with precipitation. Leaching loss was measured by lysimeter on grassland soils. From Kolenbrander (1969).

SIMULATION OF CURRENT AGRICULTURAL PRACTICES ON GOTLAND

Model simulations were made for sugar beets, barley and ley growing on different soil types (sand, glacial till and peat). The results are summarized in table 1. The maximum leaching loss predicted was 42 kgN/ha/yr from sugar beets growing on a sandy soil followed by a loss of 38 kgN/ha/yr from barley growing on peat soil. This was due primarily to high decomposition and nitrification rates in the peat (182 kgN/ha/yr decomposed in peat, 74 in till and 61 in sand).

Glacial till soils exhibited the lowest leaching of the three soils studied. This appears to be due to the larger field capacity of the soil resulting in lower nitrification rates and higher denitrification rates in early spring and fall

when the soil is saturated. A comparison of three of the runs is shown in figure 9.

Fertilizer applications were different for each soil (table 1). A low application of nitrogen fertilizer was recommended for peat soils, however, a manure application of 30 t/ha/yr (105 kgN/ha/yr) was assumed to be applied on all soils.

Denitrification rates were highest in all soils during the period of manure application. Groundwater nitrate concentrations followed the same general trend with maxima in the winter when infiltration occurs.

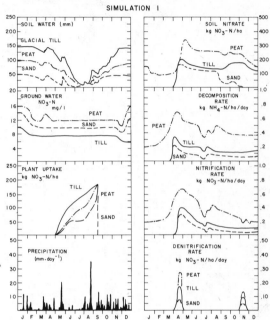

Figure 9.--Simulation of barley growing on different soil types for 1972. Fertilizer applications shown in table 1. Manure applications were 80 kgN/ha in spring and 20 kgN/ha in the fall in all runs.

VALIDATION OF THE MODEL

Predicted groundwater nitrate concentrations ranged from 2 to 16 mgNO$_3$-N/l in 1972, a year when the rainfall was below average. This is in agreement with concentrations reported by Nilsson and Rannek (1975) and by the County Health Board on Gotland (figs. 3&10). Wiklander (1977) measured very high concentrations of nitrate in drainage waters on Gotland and calculated an average of 20.1 mgNO$_3$-N/l for 53 samples.

The simulation of barley growing on peat soil in figure 5 is probably representative of a typical agricultural activity in the Hemse region of Gotland where the largest drained bog on the island

Table 1.--Simulation of leaching losses from cropland on
Gotland for 1972. (precipitation = 448 mm)

	SAND		GLACIAL TILL		PEAT	
	Fertilizer applied (kgN/ha/yr)	Leaching (kgN/ha/yr)	Fertilizer applied (kgN/ha/yr)	Leaching (kgN/ha/yr)	Fertilizer applied (kgN/ha/yr)	Leaching (kgN/ha/yr)
Sugar Beets	(110)[1]	(42)	(110)	(31)	not grown	not grown
	138	53	138	36	-	-
Barley	(75)	(34)	(75)	(26)	(30)	(38)
	62	28	62	21	0	32
Ley	(50+50)	(14)	(50+50)	(12)	(30+50)	(24)
	45	6.3	45	5.4	45	10.8

[1]Numbers in parenthesis are fertilizer applications
recommended by Supra (1976). Actual application rates are
from the Swedish agricultural year book (Anon. 1971).

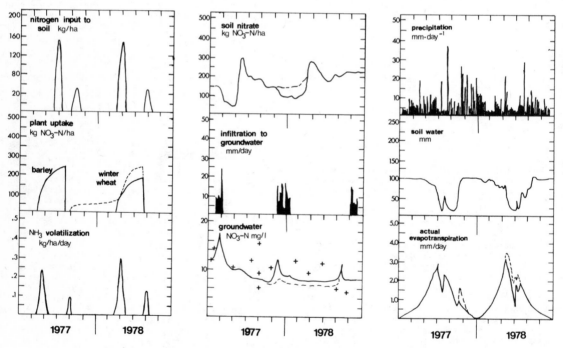

Figure 10.--A two-year simulation of barley growing on peat
soil on Martebo, the largest drained bog on Gotland.
The dashed line refers to a simulation with winter wheat
grown instead of barley in the second year. + denotes
NO_3-N concentrations in drinking water.

is under cultivation. A simulation was made for 1977 and 1978 and results were compared with measured values of NO_3-N in the drinking water (fig. 10).

The model appears to predict seasonal trends in groundwater nitrate concentrations fairly well with maximum values occurring in the winter when groundwater infiltration occurs. Both predicted and measured nitrate concentrations in the groundwater were higher for the Hemse area in 1977 than in 1978. This was probably due to the high rainfall and higher infiltration in 1977. The observed increase from 9.2 to 14.4 $mgNO_3$-N/1 in the summer of 1977 may indicate leaching from unsaturated soil which is not accounted for in the model. However, considering that the measured concentrations are a result of leaching from different crop types the model's predictions seem to be reasonably accurate. A crop of winter wheat grown in the second year decreased the average groundwater concentration from 10 to 8 $mgNO_3$-N/1.

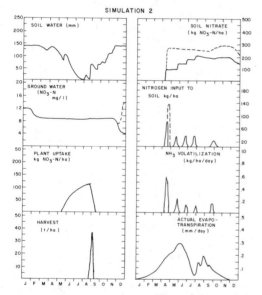

Figure 11--Simulation of the effect of increasing the frequency of fertilizer application for sugar beets on glacial till while reducing the total amount of fertilizer applied from 138 to 90 kgN/ha/yr. Applications of 30 kgN/ha were applied in May, June, and July. Manure applications were constant, 80 kgN/ha in spring and 20 kgN/ha in the fall in all runs.

MANAGEMENT SIMULATION RUNS

A simulation was made increasing the fertilizer frequency for sugar beets growing on glacial till from one large application in April of 138 kgN/ha to three small applications in May, June, and July each of 46 kgN/ha. The annual leaching loss was

reduced from 30.1 to 21.2 kgN/ha/yr. This loss was further reduced to 14 kgN/ha/yr when the three applications were reduced to 30 kgN/ha each (fig.11). The nitrogen uptake by the crop was not reduced when fertilization was decreased to 90 kgN/ha/yr suggesting that the same yields could be expected with nitrogen fertilization. Losses by ammonia volatilization were greatest from spring applications of manure when evaporation loss and ammonia applications were high.

Two irrigation runs were simulated for sugar beets growing on glacial till with current and reduced levels of fertilization (138 and 90 kgN/ha applied in the spring). Three irrigations of 35 mm were simulated on June 1, July 1, and August 1. Water was assumed to be supplied from an irrigation reservoir. When fertilizers were applied at their present levels leaching losses to the groundwater increased causing an increase in groundwater nitrate concentrations to 20 mg/l. However, nitrogen uptake from the soil also increased to 125 kgN/ha raising yields from 38 to 56 t/ha. When fertilizer applications were reduced to 90 kgN/ha spread over three applications, leaching losses decreased to 18.2 kgN/ha/yr with no decrease in yield (fig. 12).

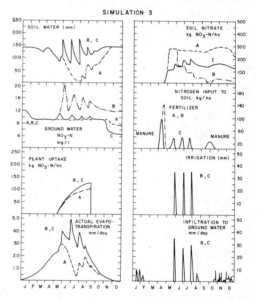

Figure 12.--Simulation results for irrigation of sugar beets on glacial till. Three applications of 35 mm were made on June 1, July 1, and August 1. Curve A is the standard run without irrigation, curve B is with irrigation and a standard fertilizer application of 138 kgN/ha. Curve C is irrigation with 90 kgN/ha fertilizer applied, distributed over three applications in May, June, and July. Manure applications were 80 kgN/ha in spring and 20 kgN/ha in the fall in all runs.

DISCUSSION

The simulation runs have predicted that yields could be increased and leaching losses reduced under a system of irrigation with lower levels of fertilization. Further runs could be made to determine an optimum fertilizer-water combination for each crop. From the simulations made to date it appears that irrigation of sugar beets with 105 mm of water and 90 kgN/ha/yr in three applications would increase yields by 47% and reduce leaching losses by 40%. The possibility of irrigating with Baltic water should be investigated since crops on Gotland are already adapted to saline soils.

Leaching losses could also be reduced by more widespread use of manure and silage collection facilities on the island.

It is significant that nitrate concentrations in lakes and streams on Gotland are considerably lower than groundwater concentrations and are even lower than concentrations in many lakes and streams on the mainland. Spiller (1978) reports values ranging from 0.06 to 1.8 mgNO$_3$-N/1(0.3 to 8 mgNO$_3$/1) for agricultural areas on Gotland. Wiklander (1977) found a range of 0.1 to 0.5 mgNO$_3$N/1 in his survey.

One explanation for the large concentration difference between surface and soil water could be high denitrification rates from the lakes and streams. Larsen (1977) reported denitrification rates up to 76 kgN/ha/yr from the sediments of some shallow Danish lakes with similar water quality characteristics to those on Gotland (shallow, high pH, high N/P ratio). Most of the denitrification occurred in the late summer when the sediments were anoxic. The importance of this loss on Gotland could be assessed by using the model to estimate the nitrate runoff from cropland to lakes and the denitrification losses from the sediments. The nitrogen lost to surface waters represents an energy loss since much of it originated from fertilizers. Some of the nitrogen and energy lost in this manner could be recovered if fast growing reeds such as Phragmites were planted in some of the surface waters on Gotland and harvested for fuel. The nitrogen removed in a good Phragmites harvest would be about 250 kgN/ha (Bjork and Graneli 1978). The feasibility of such a system for Gotland could also be tested using the model.

The maximum energy value of a good Phragmites harvest was estimated by Bjork and Graneli (1978) at 50 mWh/ha/yr or 3.45 x 10^{9}kJ/ha/yr. If applied to the lakes on Gotland (25 km^2) we get a maximum energy value of 8625 x 10^{9}kJ/yr. A more conservative estimate, assuming only half the maximum is achieved and only half the lakes are cultivated, yields an estimate of 2156 x 10^{9}kJ/yr.

CONCLUSIONS

The model may be considered to be semi-quantitative in its prediction of average nitrogen leaching rates from the soil for the major crops on Gotland. Results can be used to provide rough estimates of leaching but it must be remembered that the results are only valid if the assumptions made in formulating the model hold. More accurate predictions could probably be made by including a model for snowmelt and by dividing the soil into 10 cm layers for more precise calculation of percolation rates. These factors were neglected in this version of the model because of the mild climate, low snowfall, and shallow soils found on Gotland.

This study has shown that modelling can be used to estimate parameters such as leaching, ammonia volatilization, denitrification and decomposition which must often be guessed at in nutrient budget work. There is a need for direct measurements of these parameters where possible to test the model's predictions.

ACKNOWLEDGEMENTS

This work was funded by the Swedish Natural Science Research Council. Helpful discussions with T. Rosswall are appreciated. We would also like to thank the Swedish Coniferous Forest Project for use of their PDP-11 computer and various government agencies on Gotland for furnishing data.

LITERATURE CITED

Anon. 1971. Swedish Agricultural Yearbook. Central Bureau of Statistics, Stockholm.

Bjork, S. and W. Graneli. 1978. Energy reeds and the environment. Ambio 7 (4): 150-156

Bolin, B. and Arrhenius. 1977. Nitrogen - an essential life factor and a growing environment hazard. Report from Nobel Symposium No. 28 Ambio 6 (2-3):96-105.

Brink, N. 1977. Aspects on environmental hygiene and animal health in connection with the handling of farmyard manure. Kungl, Skogs- och Lantbruksakademiens tidskrift 116:190 (In Swedish)

Denmead, O.T., J.R. Freney and J.R. Simpson. 1977. A closed ammonia cycle within a plant canopy. Soil Biol. Biochem. 8:161-164.

Häger, S. and C. Thoms. 1978. Hydrological analysis of potential groundwater recharge. Ecological Bulletins. 28:103-106.

Jansson, A.M. and J. Zucchetto. 1978 a. Energy, economic and ecological relationships for Gotland, Sweden: a regional systems study Ecological Bulletins 28: 154.

Jansson, A.M. and J. Zucchetto. 1978 b. Man, nature and energy flow on the island of Gotland, Sweden. Ambio 7(4):140-149.

Jensen, M.E. 1976. Water consumption by agricultural plants. p. 1-22. In Kozlowski, T.T. (ed.) Water deficits and plant growth. Academic Press, New York, N. Y.

Kolenbrander, G.J. 1969. Nitrogen content and loss in drainwater. Neth. J. Agri. Aci. 17:246-255.

Larsen, V. 1977. Nitrogen transformation in lakes. Prog. Wat. Tech. 8(4/5):419-431.

Nilsson, L.Y. and J. Rannek. 1975. Nitrate in drinking water. surveying section, KTH, Stockholm.Report 3:19 (In Swedish)

Odum, H.T. 1971. Environment, Power and Society. 331 p. John Wiley, New York, N. Y.

Penman, H.L. 1948. Natural evaporation from open water, bare soil and grass. Proc. Royal Soc. London 193:120-145.

Spiller, G. 1978. Nitrogen analysis and simulation. Ecological Bulletins. 28:107-125.

Supra. 1976. Fertilization Recommendations 1975/76. Gotlands Lantman, Visby, Sweden.

Wiklander, L. 1977. Leaching of plant nutrients in soils. IV. Contents in drainage water and groundwater. Acta Agric. Scand 27:175-189.

Zucchetto, J. and A.M. Jansson. 1979. Integrated regional energy analysis for the island of Gotland, Sweden. Environment and Planning 1J: 919-942.

APPENDIX

Table Al.--Coefficient table

Decomposition
KDMAX $\quad$ 0.01 day^{-1}
SMRF $\quad$ 1 at 60% saturation,0 at 0%
TRF $\quad$ 1 at 50°C, 0 at 0°C

Nitrification
KN_{20} $\quad$ 0.03 day^{-1}
C_{20} $\quad$ 0.003 day^{-1}°C^{-1}
KNRF $\quad$ 1 at 60% saturation, 0 at 0%

Denitrification
KT_{12} $\quad$ 0.00125 day^{-1}
KT_{11} $\quad$ 0.0006 day^{-1}
C_{12} $\quad$ 0.000125 day^{-1}°C^{-1}
C_{11} $\quad$ 0.00006 day^{-1}°C^{-1}

Ammonia volatilization
KTV_{20} $\quad$ 0.003 day^{-1}
C_{20} $\quad$ 0.0003 day^{-1}°C^{-1}
KSM $\quad$ 1 at 60% saturation,0 at 0%

Table A2.--Model description

Variable

X01   Total water stored in upper meter of soil (mm)

X02 NO_3-N upper meter of soil (kg ha^{-1})

X03   Groundwater (0-15m) (mm)

X04 NO_3-N in groundwater

X05   Plant nitrogen (kg ha^{-1})

X06   Soil organic nitrogen (kg ha^{-1}) (ON)

X07   Soil ammonia (kg ha^{-1})

T01   Precipitation input (mm · day^{-1})

T10   Soil runoff SR = k · T01 if X01 > FC

$$= 0 \quad \text{otherwise}$$

where, k = .5 for soils with tile drainage

.1 for soils without drainage

FC = Field capacity of soil (maximum water storage) (mm)

T13   Infiltration to groundwater (mm day^{-1})

$$T13 = T01 - X01 \quad \text{if} \quad X01 > FC$$
$$= 0 \quad \text{otherwise}$$

T20 NO_3-N lost in soil runoff (kg ha^{-1}day^{-1})

$$T20 = (X02/X01) \cdot SR$$

T24 NO_3-N leached to groundwater (kg ha^{-1}day^{-1})

$$T24 = (X02/X01) \cdot SR$$

T40 NO_3-N lost in groundwater runoff

$$T40 = (X04/X03) \cdot T30$$

T25 NO_3-N uptake by plants

T75 NH_4-N uptake by plants

T56   Crop residue

T70 NH_4-N lost in soil runoff (kg ha^{-1}day^{-1})

$$T70 = (X07/X01) \cdot SR$$

T06   Organic N added in manure (kg ha^{-1}day^{-1})

T07 NH_4-N in fertilizer added to soil (kg ha^{-1}day^{-1})

DYNAMICS AND ENERGY FLOW IN BALTIC ECOSYSTEMS[1]

Bengt-Owe Jansson[2]

Abstract.--The Baltic ecosystem is characterized by low diversity, slow turnover and critical oxygen levels in deeper basins. Increased nutrient levels have caused increase of biomass in surface layer, increase of bluegreen algal blooms, "desertification" of 1/4 of the bottoms. Oil spills locally change the bottom systems. Toxic substances accumulate causing i.e. extinction of seal populations.

INTRODUCTION

The Baltic Sea, the largest brackish water area in the world, 365 000 km[2] and 20 000 km[3] of 6 - 10 $^O/_{oo}$S water (surface layer) is comparatively young as a system. Various organism assemblies have been tried out during different physical stages since the last glaciation ca 15 000 years ago. Switching between freshwater and fully marine conditions has resulted in a mixture of organisms which during an evolutionary short period have co-adapted and established networks processing the available potential energy.

MAIN CHARACTERISTICS OF THE BALTIC SYSTEM

Hydrodynamic patterns

The unusually stable salinity compared to other estuarine systems is maintained by efficient redistribution modes. A continuous inflow of heavy, saline North Sea water through the narrow and shallow straits is generated by the outflow of low salinity surface water originating from the big rivers in the north of Sweden and Finland. A persistent primary halocline is maintained at ca 60 m depth which effectively reduces vertical transports. The deeper basins of the shallow Baltic (mean depth 60 m) are often stagnant but are now and then flushed by meteorologically forced injections of North Sea water which temporarily oxygenates the area below the halocline. Because of the large volume and the narrow inlets the residence time of the Baltic water is long, ca 25 years (Falkenmark and Mikulski 1974). The area below the halocline acts as a huge nutrient trap fed by sedimentation of organic material and absorbing nutrients during oxygenated con-

[1]Paper presented at the international symposium Energy and Ecological Modelling, sponsored by the International Society for Ecological Modelling. (Louisville, Kentucky, April 20-23, 1981).
[2]Bengt-Owe Jansson is Professor of Marine Ecology, Askö Laboratory, University of Stockholm, Box 6801, S-113 86 Stockholm, Sweden

Natural boundaries
.... Axis for section below

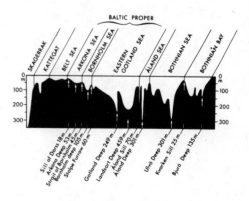

Figure 1.--The Baltic Sea with its natural divisions, the straits (Öresund and Darss) and main sills and basins (from Jansson 1978).

ditions, releasing them to the water phase during oxygen-free periods. Absence of tide in the Baltic means that one potential stablizing agent is lacking but insolation and temperature show heavy annual pulses. Light sufficient for primary production is only available down to 10 metres depth, but the highest nutrient concentrations occur in the layers below the halocline. This means that events promoting vertical upward transports are critical for the regulation of the annual primary production. In addition to the diffusive vertical exchange through the halocline (Kullenberg 1974) coastal processes such as boundary mixing and upwelling are also important. Figure 2 shows three main areas of upwelling the Baltic proper due to Ekman transport generated by south-westerly winds. The response of the plankton system to these nuclei of higher phosphate concentrations is shown in fig. 7 as areas of higher densities of bluegreen algae. North-easterly winds give downwelling water in the same areas. Both types generate strong, topographically chanelled currents, "coastal jets" (Shaffer 1975), able to transport quantities of water of 1% of the Gulf Stream. They probably act as important redistribution agents for salt and nutrients in the Baltic.

The Baltic is a cold sea with formation of ice during winter in the northern part for as much as 60% of the year. Below the temporary thermocline at 20 m depth temperatures of less than 5 – 6°C prevail during the whole year.

Biological variables

Peculiar to the Baltic system is the mixed freshwater-marine organism assembly. In the same gill-net, cod, flounder and herring can be caught together with pike, perch and bream. In the bays, brown and red algae grow together with reed and pond weed. The species diversity is low. Compared to the 1,500 macroscopic animal species off the Norwegian coast only some 90 species are found in the central Baltic. Of ca 150 macroscopic algal species in the North Sea only 24 are found off the Finnish coast. This low species diversity also means that the main functions of the system such as production decomposition and recycling are carried out by only a few "workers" and there are fewer "stand-ins" than in a fully marine system. In this respect the Baltic is more sensitive to changes than, for example, the Swedish west coast.

The smaller size of individuals compared to corresponding populations in fully marine waters is characteristic of many Baltic species such as herring and the common mussel *Mytilus edulis*. This reflects the higher costs of living in a dilute medium – higher tolerance resulting in an ability to colonize large areas has to be paid for by slower growth.

Figure 2.--Surface temperatures in September showing three main areas of upwelling, cold bottom water (from Jansson 1978). Compare this figure with fig. 7.

Subsystem properties

Pelagic system

The role of the pelagic system lies in producing organic matter through pulses during the year (fig. 3). A heavy spring bloom utilizing the rapidly increasing light and the high levels of nutrients accumulated during winter and now available due to advections from deeper systems is dominated by dinoflagellates and heavy diatoms. Of the annual primary production in the northern Baltic of ca 160 gC·m^{-2} sometimes as much as one fourth is synthesized during spring (fig. 4). In the absence of herbivores nearly half of this sinks to the bottom constituting half of the annual requirement of potential energy of the soft bottom system. The changing levels of nutrients in the free water due to biological uptake are compensated for at the systems level by switching to producers with optimal properties: small, fast, efficient green algae during summer's nutrient minima, nitrogen fixing bluegreens during early autumn's nitrogen deficit, dinoflagellates and diatoms after the decomposition of the bluegreens. In the processing of organic material, bacteria play an important role, running on excretion of dissolved carbon from the primary producers.

Typical annual estimates for the pelagic energy flow in the northern Baltic proper includes a primary production of 160 gC·m^{-2} and a bacterial production of 38 gC·m^{-2} (Larsson and Hagström in manuscript), a zooplankton production of 9 gC·m^{-2} (Larsson et al. in prep.) and a sedimentation of 30% of the primary production. Microzooplankton play an important but not yet quantified role.

Hardbottom (phytal) system

The large archipelagos of the Baltic Sea with fringes of green, brown and red algae around each rocky formation act as huge filters for land runoff. They absorb nutrients from the passing water, build organic structures for food, provide shelter for the Baltic's most diverse fauna, and recycle the respiration products. The filamentous green algae (Cladophora) of the surf zone and the brown algal belt (Fucus) with epiphytes below constitute a combination of a fast-growing, annual component and a slower, perennial, structure-building element which together effectively trap solar energy and nutrients. The annual pulse is also pronounced here with a luxurious spring bloom of sessile diatoms and filamentous brown algae. There are few herbivores present so the bulk of the produced organic matter is chan-

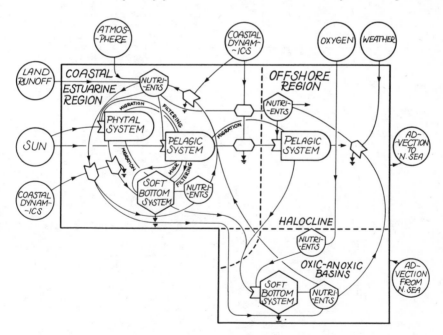

Figure 3.-- Conceptual model of the total Baltic ecosystem showing the main biological-chemical-physical couplings.
For simplicity the estuarine and coastal regions are merged and the thermocline in the surface water omitted.

neled into the detritus pool of the soft bottom system.

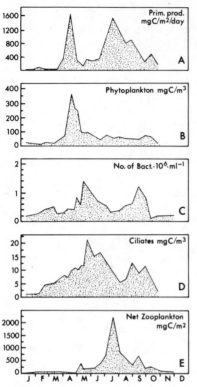

Figure 4.-- Annual dynamics of the pelagic system at the permanent monitoring station of the Askö Laboratory, northern Baltic proper.

The animal biomass although distributed over some 40 macroscopic species is totally dominated by the blue mussel *Mytilus edulis* which makes up more than 95% of the total biomass (A.M. Jansson and N. Kautsky 1977). Although small in size it effectively converts organics to inorganics through filter feeding, securing a steady nutrient supply for the primary producers even during summer thermal stratification limiting vertical advection. Production of organic material and nutrient recycling are two of the main functions of the phytal system in the total system. Another is the production of fish. Many species find food and shelter here and the commercially important herring spawns in May – June on the flourishing filamentous brown algal beds before returning to the open waters for the rest of the year.

Soft bottom system

Most of the Baltic seabed consists of sand and mud and as a whole the Baltic system can be considered as detritus-based. The imported organic material from the pelagic and the phytal systems maintain a

macrofauna dominated by two crustaceans (*Pontoporeia* spp.), a clam (*Macoma baltica*) and the blue mussel. The meiofauna is more diverse. Of the estimated annual energy requirement of the soft bottom system in the northern Baltic proper of 40 – 60 $gC \cdot m^{-2}$ (Ankar and Elmgren 1976), the macrofauna processes 10 – 50%, the bacteria about 50% and the meiofauna less than 10%. The sunlit, shallow areas are dominated by freshwater plants like reed (*Phragmites*) and pond weed (*Potamogeton*) of which the former effectively retains nutrients from land runoff, storing them in the roots during winter. The area is very productive, warmed up early in spring exposing a blooming microflora, stimulating secondary production and forming optimal conditions for spawning fish. Except for sport fishing of pike and perch, man little utilizes this productive zone the importance of which lies more in serving the total system by retaining nutrients and producing larval food for fish.

The Askö research project

Public concern for the future of the Baltic facilitated the start of the large scale project "Dynamics and energy flow of Baltic ecosystems" which was carried out by the Askö Laboratory during 1971 – 1980. Many of the already presented results and ideas were born within this project with the aim of understanding the total system dynamics. In the following some of our results of the impact of eutrophication and oil spills are merged with previous knowledge to a short summary of man-induced effects.

EFFECTS OF ENERGY USE ON THE TOTAL SYSTEM

Ability to process energy and matter

The long residence time and stable stratification of the water, the lack of flushing tidal forces, the low temperature and the low diversity of the organism assembly determine the limited capacity of the Baltic system as a processor of energy and matter. The concentration of a conservative pollutant injected in the surface water would be reduced to 50% only after 15 years (Falkenmark and Mikulski 1974). The original oligotrophic system with low concentrations and slow exchange processes has clearly reacted to man´s long use of its potential natural resources as fishing area, receiver of waste flows, transportation area and recreation center. More direct impact of man's growing use of fossil fuel may show in the effects of the hydroelectric dams in the Swedish rivers. Changes of the pulse and temperature of the river water discharging into Baltic coastal areas could affect for example migration of andromous fish. No data on this problem are available however. Another example of human impact on the Baltic may show in the future in the thermal pollution from nuclear power plants. They are few in number on the Baltic coast, however, and although long-term studies of single plants do exist (Grimås 1977) the effect on the ecosystem are local.

Changes due to the more general use of energy

in society are many. Three principle cases will be considered and discussed in the following: increased flows of nutrients from land and atmosphere, oil spill effects and accumulation of toxic substances.

Increased nutrient levels - "eutrophication"

The building of sewage treatment plants in Sweden during the last 15 years has reduced the input of phosphorus. The increased use of nitrogen fertilizers in agriculture during the same period has completely counteracted the intended effects of the treatment plants in nutrient retention. While it is easy to prove the local impact of increased nutrient concentrations the impact on a large scale system such as the Baltic Sea requires far more established data. In the surface water of the open Baltic the phosphate phosphorus concentrations have increased by a factor of 3 since the late 50´s and the nitrate level has doubled during the period 1968 – 1978 (Fonselius 1980). Present data on the total input of total phosphorus and total nitrogen from land runoff (rivers, municipal and industrial discharge) and from the atmosphere are shown in table 1. Figures for land runoff are underestimations due to incomplete data from southeastern and southern areas and the atmospheric data are rough estimates due to both few available observations and methodological difficulties. The most striking fact is the large input of nitrogen from the atmosphere.

Table 1.--Annual input in tons of total phosphorus, total nitrogen, DDT and PCB to the Baltic Sea

Source	Phosphorus	Nitrogen	DDT	PCB
[1]Land runoff	25,825	309,890	?	?
[2]Atmospheric fallout	1,400	400,000	6	8
[2]Content in the Baltic	470,000	$3,500\times10^3$		

[1]From Pawlak 1980   [2]From Rodhe *et al.* 1980

The local effects of separate discharges follow the well known patterns of marine eutrophication. More interesting are the overall effects. They can be classified as: increasing biomasses of pelagic fish and benthic macrofauna above the halocline, increased frequency of bluegreen algal blooms, decreased oxygen concentrations and benthic macrofauna below the permanent halocline.

Increase of fish and macrofauna biomasses above the halocline

Although lack of reliable and comparable primary production measurements on longterm studies prevents documentation of increased primary production in pelagic waters this has certainly occurred. This means both an increase in potential food for pelagic fish and increased sedimentation. The yield of the Baltic fisheries has increased enormously

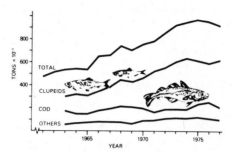

Figure 5.--Landings of fish from the Baltic Sea 1961 – 1977 (from Thurow 1980).

since the middle sixties and in 1975 reached 974 000 tons (fig. 5). This was due mainly to increased fishing by the socialist countries. Although it is not possible to differentiate between a real biomass increase of fish and the effects of increased fishing the stocks of both herring and sprat are supposed to have increased. The stocks are now overexploited. however, in spite of there being no signs of recruitment failure. A biomass increase of 60% due to successful hatching in 1971 and 1972 has been proven for cod (Thurow 1980).

By repeating Hessle´s quantitative sampling in 1920 and 1923 of 28 soft bottom stations around the islands of Öland and Gothland in the middle of the Baltic (fig. 6), Cederwall and Elmgren (1980)

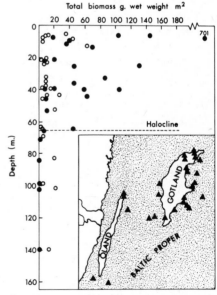

Figure 6.--Macrofauna biomasses from 1920, 1923 (unfilled circles) and from 1976, 1977 (filled circles) in the Central Baltic. Explanations in the text (from Cederwall and Elmgren 1980)

were able in 1976-1977 to prove a significant
fourfold increase in macrofauna biomass above the
halocline. As neither discrepencies in methods nor
interannual variations can explain these differences
they must have been caused by increased potential
food supply through sedimentation. This is the first
statistically confirmed biological evidence of
large-scale eutrophication of the Baltic Sea.

Increased frequency of bluegreen algal blooms

The mass occurrence of bluegreen, atmospheric
nitrogen-fixing algae, especially *Nodularia spumigena* in July-August is a well known feature of the
Baltic. The depletion of nutrients in pelagic waters
after the summer bloom makes the nitrogen-fixing
bluegreens highly competitive. Stimulated by high
temperatures and phosphorus increase (Lindahl *et
al.* 1978) they grow explosively, producing gas va-
cuoles and finally float to the surface in calm
weather forming dense patches (fig. 7). The size
and frequency of these blooms seems to have increased during the last decades (Horstmann 1975).
This ability of the system to compensate for a
lack of nitrogen by employing a medium which increases the energy trapping potential is very powerful. Over a period of only a few weeks, nitrogen in
the order of 100,000 tons can be fixed from the atmosphere in the northern and central Baltic proper
(Rinne *et al.* 1978). Calculations from remote sensing images and "groundtruth" measurements have also
shown these flows to be of the same order of magnitude as the waste flows from land (Nyqvist 1974;
Öström 1976).

"Desertification" of bottoms below the halocline

The oxygen concentrations in the water below
the primary halocline have successively decreased
since 1900 (fig. 8). Changes in the statistical characteristics of the meteorological forcings have
changed the frequency of the pulses of North Sea
water flushing the deep basins. An "oceanization"
of the Baltic is taking place with an increase of
the salinity of the deep water of roughly 1°/ooS
(Matthäus 1972) and of its temperature of 1°C.

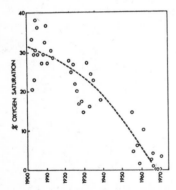

Figure 8.--The decrease in oxygen content of the
bottom water in the northern Central Baltic
(in Jansson 1978 from Fonselius 1969).

Figure 7.-- Blooms of the bluegreen, nitrogen-fixing alga *Nodularia spumigena* appear as dark
streaks off the Swedish Baltic coast from 90
km altitude (Nyqvist 1974).

Shaffer (1979) has presented hydrodynamic evidence,
however, showing that the decreasing oxygen levels
are a result of increased oxygen consumption of the
sediment rather than of changing hydrodynamic conditions. The increased oxygen consumption rate can
only partly be explained by the slight temperature
increase and must be an effect of increased sedimentation and bacterial activity.

The deeper soft bottom system is totally dominated by the processes of the sulphur cycle and
the sediment takes up or releases, in particular
phosphate, in relation to the amount of cations
(Schippel *et al.* 1973) and ammonia depending on the
quality of the organic matter (Engwall 1977). The
system switches between oxic and anoxic conditions

forced by the salt water injections and the intermittent formation of hydrogen sulphide wipes out the macrofaunal components. During the sixties and seventies these "deserts" have covered about 100,000 km[2], that is, 25% of the total Baltic Sea area (Andersin *et al.* 1978). Recolonization of these bottoms does occur but the periods of optimal oxygen conditions are too short and stochastic to allow permanent settling. The character of the community has switched from a dominance of suspension feeders to a dominance of non-selective deposit feeders (Leppäkoski 1975). The deep soft bottom system has lost its former important function as foraging area for bottom fish and spawning area for cod and offshore flounder (Lindblom 1973) and now acts more as a huge source of nutrient regeneration.

Internal nutrient cycling of the natural system

It is exceedingly interesting to compare Man´s contribution to the nutrient pool through the waste flows from land and atmosphere with the contributions within the natural system. The blue mussel is one of the most common organisms in the Baltic. Being a filter feeder it pumps the seawater through the gills, consuming the organic particles and excreting respiratory products, among them phosphorus and nitrogen. Kautsky and Wallentinus (1980) estimated the *Mytilus* biomass in the upper 25 m in the Baltic proper as being 8 million tons (dryweight including shells). Based on measurements in plastic enclosures this population can be estimated to excrete annually 250,000 tons of inorganic nitrogen, 80,000 tons of phosphorus and 100,000 tons of amino-N.

Oil spill effects

According to Coast Guard statistics, around 1,000 oil spills per year in the Baltic area inject oil in the order of 100,000 tons. Since most of this is not recovered large amounts of this carbon source must obviously be processed by the system causing more or less apparent and long-lasting changes. Oil spill effects on separate parts of the marine system such as massive bird mortality, deformation of fish larvae, decreased diversity of littoral fauna are well established whereas studies on the reactions of the total ecosystem are poorly documented. The grounding of the Soviet tanker Tsesis in October 1977 in the primary investigation area of the Askö Laboratory in the northern Baltic proper initiated Swedish-American cooperative studies from the day following the spill. Effects on both sublittoral and supralittoral systems including the chemical processes of weathering, bioaccumulation and depuration of oil (Kineman *et al.* 1980) were studied. The spill comprised ca 1,100 tons of No 5 heavy fuel oil most of which was recovered (no detergents were used in the clean-up operations) leaving ca 400 tons oiling an area of 34 km[2]. The following report generalizes the effects of only a small oil spill on the Baltic ecosystem.

The immediate effect on the plankton system was an increase in phytoplankton biomass, primary production and bacterial biomass and a decrease in zooplankton biomass (Johansson *et al.* 1980). No harmful effects on pelagic fish were noted and normal pelagic conditions were regained within one month.

The supralittoral zone showed little damage during the following summer and bird life was practically unaffected due to the low normal activity of the season. In the littoral zone however, the crustaceans were drastically reduced and *Mytilus edulis*, covering a fairly large area still contained oil a year after the spill. The algae were seemingly unaffected and the total recovery of the phytal system should be a matter of only a few years depending on the rate of immigration of species.

The soft bottom system shows the largest and most persistent effects. Sediment traps showed a transport of at least 20 tons of oil to the sediment where it could be traced in the tissues of the filter feeder *Macoma baltica*, the Baltic clam. The dominating crustaceans *Pontoporeia affinis* and P. *femorata* and the polychaete *Harmothoe sarsi* were drastically reduced whereas *Macoma* survived but contained large amounts of oil. Three years after the spill these differences are perhaps even more accentuated ([1]Elmgren *et al.* pers. comm.). Whereas *Harmothoe* has recovered, the *Pontoporeia* populations are still low and *Macoma* has greatly increased (fig. 9). This can probably be explained by decreased negative influence (including predation) of the *Pontoporeia* species on the recruitment of the clam.

A complete recovery of the soft bottom system after even a small oil spill is probably in the region of 5 - 10 years.

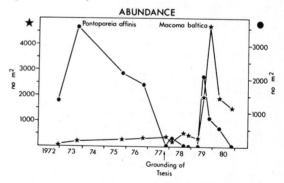

Figure 9.--Effects of the Tsesis oil spill on the soft bottom community in the northern Baltic. The previously dominating amphipods have been replaced by the suspensionfeeding clam *Macoma* ([1]Elmgren *et al*).

[1]Elmgren, R., S. Hansson and U. Larsson. 1981. Personal communication. Askö Laboratory, Institute of Marine Ecology, University of Stockholm, Stockholm, Sweden.

Accumulation of toxic substances

The presence of toxic substances in natural systems is a clear sign of man's energy processing activities leading to emissions of chemical compounds new to the system which usually lacks the mechanisms necessary to safely incorporate them. All investigated Baltic organisms contain the chlorinated hydrocarbons DDT and PCB and compared to corresponding populations in the North Sea, in concentrations ten times as high (Jensen *et al.* 1969). The White-Tailed Eagle living exclusively on fish showed concentrations of 25,000 ppm and was close to extinction. After the DDT ban in most countries, the concentrations of these compounds have decreased in, for example, eggs of guillemots while the PCBs, due to high stability, show no decline (fig. 10).

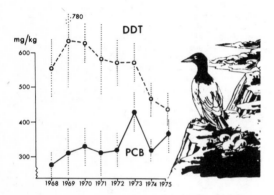

Figure 10.-- Concentrations of DDT (unfilled circles) and PCB (filled circles) in fat from guillemot eggs from Central Baltic (Olsson 1978).

Atmospheric fallout is the major source of PCB in the Baltic (table 1). Although no toxic effects of PCB have been proved in invertebrates and fish, three species of seal have greatly decreased in numbers due partly to decreasing reproductive success. At present ca 40% of the females show pathological changes in the uteri (Helle *et al.* 1976) probably caused by the high PCB concentrations. The seals in the Baltic will become extinct in the very near future. The decline in the populations of otter and porpoise may be due to the same reason.

Kihlström and Berglund (1978) have estimated the amounts of PCB in the biomass of the Baltic (table 2). Of the 2 - 3 tons incorporated in the total biomass, 1 ton is bound in fish of which 0.3 - 0.6 tons are removed yearly by fishing. The far higher input from the atmosphere indicates a continuing increase of PCBs in the Baltic Sea.

Recent determinations of heavy metals in the sediments of the Baltic have shown a threefold increase of zinc and lead, a tenfold increase over

Table 2.--Levels of PCBs in Baltic organisms. Values from various sources compiled by Kihlström and Berglund (1978).

Organisms	$mg \cdot kg^{-1}$ wet weight
Seals (fish-eating top carnivors)	34
Pelagic fish Salmon	1.3
Herring, sprat	0.6
Plankton (mixed phyto-, zooplankton)	0.1
Demersal fish (cod, flounder, plaice)	0.1 - 0.06
Phytal organisms Crustaceans (*Gammarus, Idotea* spp.)	0.1
Fucus vesiculosus	0.1
Mytilus edulis	0.03
Baltic seawater	$0.1 \cdot 10^{-6}$

previous levels of mercury and cadmium (Niemistö and Voipio 1979). Little is known of the long-term effects of many of the heavy metals. Zinc, copper and mercury seriously affect the propagation of fish (Bengtsson 1975). The increasing concentrations of mercury when moving up the food-chain are classical facts. After the mercury ban in Sweden in 1966 the levels have decreased in many areas and the mercury concentrations in the feathers of Baltic guillemots have now returned to the levels of the previous century ([2]Olsson, 1976).

A substance now much in focus is cadmium. The only present long-term experiment of its biological effects has been made by Sundelin (1981). Using the energetically important amphipod *Pontoporeia affinis* and cadmium concentrations of 5, 50 and 150 $\mu g \cdot l^{-1}$ as test system she was able to show effects of embryogenesis in all concentrations and 100% mortality of all the juveniles hatched in the three concentrations after 15 months.

THE FUTURE

Surrounded as it is by industrial countries which all use the potential natural resources with increasing intensity, the Baltic gradually changes into a huge "interface ecosystem" (sensu H.T. Odum).

Most pre-requisites for sound ecological engineering are already there. Legislative means are established through "The convention on the pollution of the marine environment of the Baltic" which is one of the most comprehensive environmental agree-

[2]Olsson, M. 1976. Personal conversation. The Swedish Museum of Natural History, Stockholm, Sweden.

ments ever ratified and was signed by all Baltic states. Understanding of the system is rapidly growing, the Baltic being one of the most investigated seas of the world. Money is potentially available thanks to the growing conviction that Nature is a sound economic investment, yielding high returns and awaits only its final allocation. Time is the one factor in short supply as man´s rapid use of auxiliary energy quickly carries the system towards the point of no return.

LITERATURE CITED

Andersin, A.-B., J. Lassig, L. Parkkonen, and H. Sandler. 1978. The decline of macrofauna in the deeper parts of the Baltic proper and the Gulf of Finland. Kieler Meeresforsch. Sonderheft 4:23-52.

Ankar, S., and R. Elmgren. 1976. The benthic macro- and meiofauna of the Askö-Landsort area (Northern Baltic Proper). A stratified random sampling survey. Contributions from the Askö Laboratory, Stockholm, Sweden, 11, 115 pp.

Bengtsson, B.-E. 1975. Some effects of zinc on different stages in the life history of the minnow, Phoxinus phoxinus L. (Pisces). Statens naturvårdsverk PM 570, 56 p.

Cederwall, H., and R. Elmgren. 1980. Biomass increase of benthic macrofauna demonstrates eutrophication of the Baltic Sea. Ophelia, Suppl. 1:287-304.

Engwall, A.-G. 1977. Nitrogen exchange at the sediment-water interface. Ambio Spec. Rep. 5:141-146.

Falkemark, M., and Z. Mikulski. 1974. Hydrology of the Baltic. Water balance of the Baltic Sea, a Regional Co-operation Project of the Baltic Countries, Project Document No. 1, 51 p.

Fonselius, S. 1969. Hydrography of the Baltic deep basins III. Fishery Board of Sweden, Series Hydrography 23, 97 p.

Fonselius, S. 1980. On long term variations of phosphorus in the Baltic surface water. ICES C.M. 1980/C:36.

Grimås, U. 1977. Some effects of heat from cooling water systems on coastal areas. Ambio Spec. Rep. 5:43-45.

Helle, E., M. Olsson, and S. Jensen. 1976. PCB levels correlated with pathological changes in seal uteri. Ambio 5:261-269.

Horstmann, U. 1975. Eutrophication and mass production of bluegreen algae in the Baltic. Merentutkimuslaitoksen Julkaisu)Havsforskningsinst. Skrift No. 239:83-90.

Jansson, A.M., and N. Kautsky. 1977. Quantitative survey of hard bottom communities in a Baltic archipelago. In: Keegan, B.S., P.O. Ceidigh and P.G.S. Boaden (eds.). Biology of benthic organisms. Plenum Press, Oxford 1977. p. 259-266.

Jansson, B.O. 1978. The Baltic - A systems analysis of a semi-enclosed sea. In: Charnock, H. and Sir G. Deacon (eds.). Advances in Oceanography. Plenum Publishing Corporation, 1978.p.131-183.

Jensen, S., and A. Johnels, M. Olsson, and G. Otterlind. 1972. DDT and PCB in herring and cod from the Baltic, the Kattegat and the Skagerrak. Ambio Spec. Rep. 1:71-85.

Johansson, S., and U. Larsson, and P.D. Boehm. 1980. The Tsesis oil spill. Impact on the pelagic ecosystem. Mar. Poll. Bull. 11:284-293.

Kautsky, N., and I. Wallentinus. 1980. Nutrient release from a Baltic Mytilus - red algal community and its role in benthic and pelagic productivity. Ophelia Suppl. 1:17-30.

Kihlström, J.E., and E. Berglund. 1978. An estimation of the amounts of polychlorinated biphenyls in the biomass of the Baltic. Ambio 7:175-178.

Kineman, J.J., and R. Elmgren, and S. Hansson (eds.). 1980. The Tsesis oil spill. - Report of the first year scientific study (October 26, 1977 to December 1978). U.S. Department of Commerce. Office of Marine Pollution Assessment. National Oceanic and Atmospheric Administration, Boulder Colorado. USA, 296 p.

Kullenberg, G.E.B. 1974. Some observations of the vertical mixing in the Baltic. In: 9th Conference of the Baltic Oceanographers, Kiel, April, 1974.

Larsson, U., and Å. Hagström. 1981. Fractionated phytoplankton primary production, exudate release, and bacterial production in a Baltic eutrophication gradient. (In communicated manuscript).

Larsson, U., and S. Johansson, and K. Skärlund. (In manuscript). Netzooplankton community structure and production in an eutrophication gradient in the northern Baltic proper.

Leppäkoski, E. 1975. Macrobenthic fauna as indicator of oceanization in the Southern Baltic. Merentutkimuslaitoksen Julkaisu/Havsforskningsinst. Skrift No. 239:280-288.

Lindahl, G., and K. Wallström, and G. Brattberg. 1978. On nitrogen fixation in a coastal area of the northern Baltic. Kieler Meeresforsch. Sonderheft 4:171-177.

Lindblom, R. 1973. Abundance and horizontal distribution of pelagic fish eggs and larvae in the Baltic Sea. Meddelande från Havsfiskelab., Lysekil, 140. 33 p.

Matthäus, W. 1972. Secular changes in oxygen conditions in the deep water of the Gotland Basin. Oikos Suppl. 15:9-13.

Niemistö, L., and A. Voipio. 1979. Notes on the sediment studies in the Finnish pollution research in the Baltic Sea. In: ICES Symp. Workshop on Sediment and Pollution Interchange in Shallow Seas and the Interpretation of Data from such Studies. Texel, the Netherlands, 1979.

Nyqvist, B. 1974. Östersjön blommar. Forskning och Framsteg. 6:1-2. (In Swedish).

Olsson, M. 1978. Bioaccumulating substances - mainly DDT and PCB - in biota in the Gulf of Bothnia. Finnish Marine Research. No. 244:227-237.

Öström, B. 1976. Fertilization of the Baltic by nitrogen fixation in the bluegreen alga Nodularia spumigena. Remote sensing of environment. 4:305-310.

Pawlak, J. 1980. Land-based inputs of some major pollutants to the Baltic Sea. Ambio 9:163-167.

Rinne, J., and T. Melvasalo, and A. Niemi, and L. Niemistö. 1978. Nitrogen fixation by bluegreen algae in the Baltic Sea. Kieler Meeres-

516

forsch. Sonderheft 4:178-187.

Rodhe, H., and R. Söderlund, and J. Ekstedt. 1980.
Deposition of airborne pollutants on the
Baltic. Ambio 9:168-173.

Schippel, F., and R.O. Hallberg, and S. Odén.
1973. Phosphate exchange at the sediment-water
interface. Oikos Suppl. 15:64-67.

Schaffer, G. 1975. Baltic coastal dynamics project
- the fall down-welling regime off Askö.
Contr. from the Askö Laboratory, Univ. of

Stockholm, Sweden. 7, 61 p.

Schaffer, G. 1979. On the phosphorus and oxygen
dynamics of the Baltic Sea. Contr. from the
Askö Laboratory, Stockholm, Sweden. 26, 90 p.

Sundelin, B. (In manuscript). Heavy metal effects
on a laboratory soft bottom ecosystem. I.
Cadmium and the life cycle of *Pontoporeia*
affinis Lindström (Crustacea, Amphipoda).

Thurow, F. 1980. The state of fish stocks in the
Baltic. Ambio 9:153-157.

ENERGY BASIS FOR HIERARCHIES IN URBAN AND REGIONAL SYSTEMS[1]

Mark T. Brown[2]

Abstract.——Understanding the hierarchical patterns of energy flow in landscapes is a major objective in the sciences of environment and human settlement. Data on regional and national patterns of landscape organization are used to test theories of energy flow control of hierarchies. Simulation models are developed to quantitatively relate ideas of mechanism and energetics to hierarchical structure, spatial pattern, and spectral distribution observed in systems of humanity and nature.

INTRODUCTION

Complex systems such as ecosystems, industrial processes, and networks of cities in the landscape appear to be organized in webs of energy flow with multiple levels of components (fig. 1a). These may be visualized in simplified form with diagrams as in figure 1b. The patterns have spatial manifestations, with many small units converging energy to a few larger ones.

Theories developed to account for these hierarchical patterns may be based, in part, on the theory that systems compete for power and survive by developing a structure of energy flows that maximizes useful power. The maximum power principle was enunciated by Lotka (1922) and additional corollaries were proposed by Odum (1967, 1971, 1975) and Odum and Odum (1976). The type and form of web adapting to different combinations of energy from the environment produces different spectral distributions and spatial patterns, which may be predictable from simple models.

Theoretical Concepts

An Energy Basis for Hierarchies

Given in figures 1a and 1b are simplified energy circuit models (Odum 1971) that depict

[1]Paper presented at the international symposium Energy and Ecological Modelling, sponsored by the International Society for Ecological Modelling. (Louisville, Kentucky, April 20—23, 1981.)
[2]Mark T. Brown is Assistant Research Scientist at the Center for Wetlands and Adjunct Assistant Professor, Department of Urban and Regional Planning, University of Florida, Gainesville, Florida U.S.A.

energy flow and storage in hierarchically organized systems. These diagrams show energy flow and control action feedbacks in five-compartment (level) hierarchies and are the basic configuration for the organization of data in this investigation of an energy theory of hierarchically organized systems.

The following concepts and theories about the relationship of energy, its spatial distribution, and resulting hierarchies are postulates and are the basis for examining data on systems of Florida and the nation.

Energy Constraints

Systems operate under the constraints of the First and Second Laws of Thermodynamics and Lotka's Maximum Power Principle (Lotka 1922) and corrollaries as proposed by Odum (1975) and Odum and Odum (1976); and are organized in a manner to remain competative and stable, increasing inflowing energy when excess energy is available.

Energy Quality and Embodied Energy

Odum (1976, 1977, 1978a, 1978b) and Odum and Odum (1976) suggest that there is a quality to energy, which is a measure of its ability to do work. Quality of energy is related to the degree to which it is concentrated; with dilute energies like sunlight, winds, waves, and other natural energies having lower quality than the more concentrated energies of fossil fuels.

Energy Quality and Frequency of Energy Sources

Recently, Odum (1981) and, in earlier studies of the cycles of order and disorder, Alexander (1978) have suggested that the quality of an energy is related to its frequency in the time domain.

Others (see Simon 1973) have suggested that frequency and place in hierarchy are related to the extent that high frequency is associated with low place in hierarchical order and low frequency with high hierarchical place.

Energy Quality and Power Density

One measure of the intensity of energy utilization in the landscape is power density (Odum, Brown, and Costanza 1976), or the rate of energy flow per unit area (Cal/acre·year). In this manner, the energy intensity of one area can be compared on a relative scale with others. In urban systems, power density is considered to be the rate of embodied energy consumption per unit area and in natural ecological systems of the landscape, power density is the rate at which energy is fixed, as measured by gross primary production.

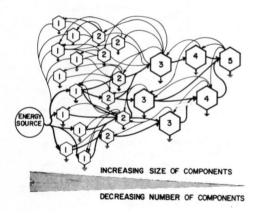

INCREASING SIZE OF COMPONENTS

DECREASING NUMBER OF COMPONENTS

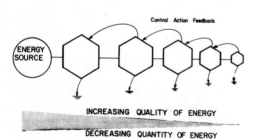

Control Action Feedback

INCREASING QUALITY OF ENERGY

DECREASING QUANTITY OF ENERGY

Figure 1. The web of energy flow and compartments of complex systems. Shown above is a hypothetical energy web, and below a simplified form organized as a hierarchy.

Previous Studies of Hierarchies

Many theories for hierarchical organization of systems and the resulting distributions of components, dating from antiquity, are prevalent in the literature. As the amount of published scientific literature has grown exponentially in the last 80 years, so has the number of scientists using the concept of hierarchy in the analysis of physical, biological, and social systems. Most notable in recent years are Woodger (1929, 1937), Whyte (1949, 1969), von Bertalanffy (1933, 1968), Simon (1962, 1973), Wilson (1969), Bunge (1969), Weiss (1971), and Laszlo (1972, 1973). Hierarchy in social systems is investigated by numerous authors (i.e., Aldrich 1979; Blau 1972, 1974; Burgess and Park 1924; Emery and Trist 1973; Glassman 1973; Landau 1969; Monane 1967; Thompson 1967; Weber 1946, 1947).

The hierarchy associated with the landscape of cities in regions was first enunciated in 1933 by Christaller (1966) and further developed by Losch (1954). Many authors have applied central place theory to regions and developed the theory further (see Dokmeci 1973; Henderson 1972; Purver 1975). Others have been critical, finding at least four specific weaknesses (see Beckman 1955; Henderson 1972; Tinbergen 1968; Van Boventer 1969). A number of authors have used gravity models and equations of diffusion for allocating regional influence of centers and calculating the spread of innovation (see Beckman 1956, 1958, 1970; Berry 1972; Hagerstrand 1966; Isard and Peck 1954; Mansfield 1963).

Zipf (1941) and later Steward (1947) have suggested there is a mathematical relationship between rank of cities and population size. MacArthur (1957) has suggested rank-abundance curves for the study of the structure of animal communities, and Odum, Cantlon, and Kornicker (1960) have postulated a hierarchical organization of ecological communities using a cumulative logarithmic species-diversity index.

Few previous studies of hierarchy have dealt with the energy control of landscape organization, but many with economic aspects and some with the physical constraints of hierarchical organization.

Plan of Study

In this study the hierarchical organization of the landscape and resulting energy spectrum of energy storage and flow were investigated at three levels: the regional level of ecosystems and urban land uses, the organization of cities in the landscape, and the organization of land uses within cities. In addition, relationship of intensity of development to the spatial area of influence was investigated at different levels of organization in the nation, the state, and within districts of the state. The specific plan for the analysis of regions, districts, and subdistricts is as follows.

First, energy spectra for many different types and sizes of systems were constructed to understand general trends of energy flow and storage in observed hierarchies.

Second, maps were made of ecosystems and land uses at two levels of study; the regions of Florida and cities within these regions.

Third, generalized models of each urban land use type were evaluated and spectra and energy storage and energy flow were calculated.

Fourth, specific analysis of the external energy requirements of areas of different sizes and an energetic evaluation of cities of different sizes were conducted.

And fifth, a series of theoretical models of hierarchical organization were simulated on analog and digital computers to explore different energy flow and storage characteristics under different organizations and pathway configurations. And then data from an aquatic ecosystem of Florida were used to test theories of hierarchical distribution and energy control.

METHODS

General Methods and Definitions

A graphic language is used throughout this paper to describe energy flow and interaction in complex systems. The language is a graphic means of depicting systems as Nth order differential equations, since each symbol represents a mathematical relationship of either energy flow, interaction, or storage relative to time. For a complete description of the language and its development see Odum (1960, 1967, 1971, 1973, 1976a).

Evaluation of Observed Hierarchies

The trends of hierarchical organization and energy spectra were graphed semilogarithmically for systems of differing scales and complexity. Data were gathered from various sources in literature and from various local, state, and federal agencies in published reports and in some cases unpublished data.

Land Use Maps of Region and Cities

Three regional areas of differing character and size were analyzed for total energy budgets, land use, and resulting hierarchic organization: the Kissimmee Everglades Basin in south Florida, a subtropical region of relatively intense urban development; the St. Johns River Basin, a region on the coast of central Florida dominated by a major river and agricultural lands with moderate urban development; and Lee County, Florida, an area in southwest Florida that is a coastal county with extensive tourism and an agriculture

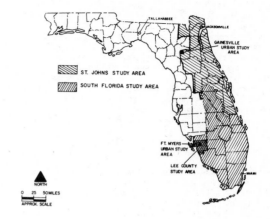

Figure 2. Map of Florida showing three regional study areas and two urban study areas.

based inland. The county has experienced recent very rapid growth (fig. 2).

Two urban areas were analyzed in detail for total energy budgets and power densities, energy budgets and power densities of land uses within the urban areas, and resulting hierarchic organization of the urban landscape: Ft. Myers, Florida, an urban area on the southwestern coast of Florida whose main economic inputs are from tourism, and is a government center; and Gainesville, Florida, an urban area in central Florida that is a governmental center and center of university education (fig. 2).

The average direct power density for each of the land use classifications was calculated by first selecting a representative sample of structures in each of the land use types (approximately 10% sample size), and then obtaining yearly energy consumption data from local utility records for each of the selected structures.

Indirect power density was calculated for the land use categories in the Ft. Myers study area only. A detailed model of energy flow between the main sectors of the local economy for 1973 was evaluated (Brown 1980) to obtain the energy embodied in goods and services that were consumed in the residential, commercial, industrial, and construction sectors of the economy. Evaluation of the flows of dollars among sectors of the economy was used and converted to embodied energies using a conversion factor of 21,000 Cal CE/$.

The structure associated with each land use was determined from property tax records of both cities where total area of structure for each of the sample structures was used. Volume of structure was calculated by multiplying area by average height of buildings.

Land Use Maps

Land use maps were constructed from false color infrared and black and white infrared photographs taken in 1973 and 1974. The land use map for the south Florida region was prepared during previous studies (Odum and Brown 1976). Areas of each land use were determined by cutting different land uses from the map and weighing them on an analytical balance. A conversion factor of grams/acre was used to convert from weight to area.

Urban Land Use Power Densities

Power density is a measure of energy flow per unit of time per unit of area. In this study, power density if expressed in units of Cal CE/acre ·year. Power density is expressed as the addition of energy consumption of fuels and electricity per unit area (referred to as direct power density), and consumption of the energy embodied in goods and services per unit area (referred to as indirect power density). Total power density results from the addition of both of these types of input energies.

Classification of Cities by Average Power Density

An average city power density was determined for all cities within each region by using averages derived in detailed studies of the two urban areas, Gainesville and Ft. Myers, Florida. The area of each city is not necessarily the actual area within legal city limits, rather it is the area that when viewed from aerial photographs is within the major concentration of urban land uses. In some cases this area may be smaller than actual city limits and in other cases, where suburban sprawl is evident, the area may be considerably larger.

Development Density and Imports/Exports

One measure of production and consumption in regional systems is gross domestic product (GDP) as determined from the total flow of dollars within in a regional economy. While domestic product is not always available for regions, it may be determined from employment data and averages for productivity per employee in each economic sector.

GDP for counties in Florida, states, and nations were determined from employment data, and "development density" was calculated by dividing GDP by land area of each county.

Export multipliers for each county and state were determined in the following manner: employment data for eight broad economic sectors (agriculture, manufacturing, wholesale and retail trade, government, services, transportation and public utilities, banking and finance, and construction) were obtained and compared to the same data from the U.S. economy. Positive departures from the U.S. percent employment were considered to indicate that portion of each economic sector that was export employment (for a detailed discussion of location quotients and methods see Hielbrun [1974]).

Exports were determined by multiplying number of export employees in each economic sector by the productivity per employee for that sector. It was assumed that local differences in employee productivity were negligible.

GDP and exports for 21 selected countries were obtained from United Nations (1978). Calculations of export multipliers and GDP were not necessary since published data are available directly.

Simulation Models of Hierarchical Organization and Energy Spectra

A series of theoretical models were simulated on both digital and analog computers to test hypothesis and evaluate structure and characteristic properties of systems organized in hierarchical fashion. As the models grew in complexity and insight gained, a final model, which was a synthesis of previous models, was simulated using data from Fontaine (1978) for an aquatic food chain.

Simulation Techniques

Models were drawn using the energy circuit language, and computer programs were written directly from the graphic model. The facilities of the Northeast Regional Data Center on the campus of the University of Florida were used for digitial computer simulation, and DYNAMO simulation language (Pugh 1970) was used. Some models were simulated on an EAI Miniac analog computer. The simulation models had one thing common to all—each is a chain of five autocatalytic components connected in series. Differences in the successive models are in the kinetics of the connections between components; with the first models having simple linear flows between components, and later models being more complex.

RESULTS

Similarities of Differing Systems and Scales

Empirical evidence of hierarchical trends in large-scale, complex systems of the landscape are presented as energy spectra in graph form, where the number of units in each level of the hierarchy is graphed on the vertical axis, and the power per unit (or power density per unit) is graphed on the horizontal axis. The energy spectra presented here are a few examples of the many systems investigated.

Figure 3 is an energy spectrum of cities in Florida. Zipf (1941), using population and rank of cities, described a frequency distribution that existed for cities in the United States and other countries. He empirically reasoned that all rank-

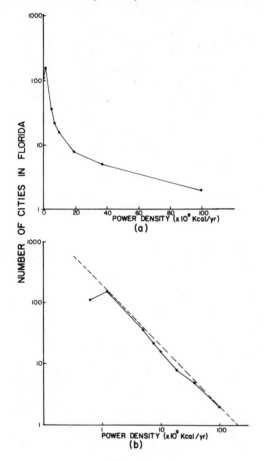

Figure 3. (a) Energy spectrum of cities in Florida graphed semilogarithmically, showing the trend of frequent occurrence of low power cities and less frequent occurrence of the very high power cities in the landscape. (b) Log-log plot of the energy spectrum of cities in Florida, after Zipf (1941), showing a negative slope of approximately 1.

Notes to Fig. 3.

Data on population of incorporated cities of the State of Florida are from the Bureau of Economic and Business Research, University of Florida (1978). The distribution of city power density was done graphically, where cities with similar power densities were grouped together and assigned a weighted average power density.

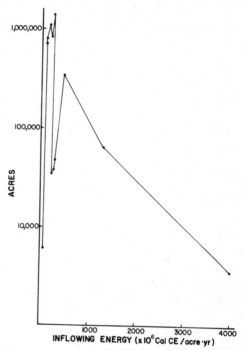

Figure 4. Energy spectrum of embodied energy in natural land production and urban land use power densities in the south Florida area, showing the spatial character of production of both natural and urban lands.

size distributions of cities should be a straight line when plotted as a log-log graph. Countries not exhibiting a straight log-log distribution were suffering from some form of disunity, and would tend toward unity (as described by his straight line log-log plot) if the forces causing disunity were removed. In a growing region like Florida, the forces of population growth may well account for the departure from a straight line distribution shown in figure 3a. The departure from the ideal distribution that occurs with the smallest cities may be a function of data, since many small cities are not incorporated and therefore not included in the statistical census of cities.

The composite energy spectra for south Florida in figure 4 is a graph of the power density of all input energies versus the number of acres of that particular power density. The graph is based on the spatial distribution of incoming energy, both natural energies and fossil fuel derived energies.

The energy spectrum in figure 5 shows the energy chain of increasing quality of energy as it flows from natural lands through the urban areas of the St. Johns River Basin and the Kissimmee-

Everglades Basin. This graph depicts the spatial character of energy chains, as there are large areas of low-energy lands to concentrate dilute low-quality energies and pass them up to the smaller yet very high-energy lands of the cities.

Regional Analysis

The Landscape of Cities within Regions

When an average power density was calculated for cities, there was a tendency for the cities to fall into the five classes of cities listed in

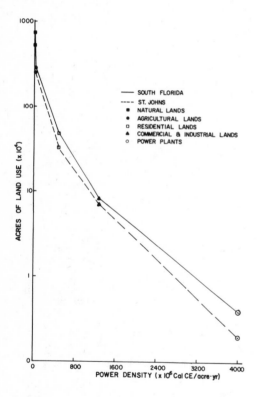

Figure 5. Energy spectrum of incoming energies to south Florida, showing the spatial distribution of both renewable and fossil fuel sources. All energies are expressed in Cal CE. Data on incoming natural energies (sunlight, rain, winds, tides, and waves) are from Costanza (1975). Data on incoming fossil fuel derived energies are from Odum and Brown (1976). The graph depicts the spatial character of inflowing energies. Because of the nature of the landscape and the inflowing energies, there are areas of concentration and areas of relatively sparse inflowing natural energy. Fossil fuel energies are somewhat point sources having very small areas of concentration. Therefore, the acreage of end use was used as the spatial distribution.

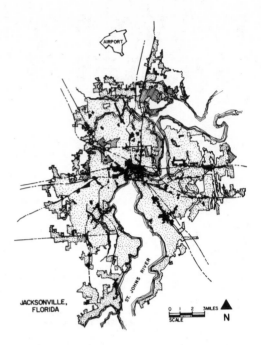

Figure 6. Generalized map of Jacksonville, Florida, a class 1 city; showing the three land use categories: commercial (black area), industrial (cross-hatched area), and residential (stippled area).

table 1. Because of the large area of wetland ecosystems in the south Florida study area, the majority of cities of all sizes were sampled from the St. Johns area. It was felt that due to the extent of this wetlands coverage, and that developable land is confined to a very narrow strip along both coasts, the south Florida region represents a unique situation from a physical standpoint, and this inhibits the development of a complete array of city sizes.

Shown in figure 6 is an example of a class 1 city (Jacksonville, Florida). Three types of urban land use are indicated and were used to calculate average embodied energy power densities.

The percentage of each land use for each of the city types were compared and are given in table 1. The extent that a city serves as a central place is indicated by the data as the percent of industrial and commercial land use increases. Thus, the class 1 cities (which includes Jacksonville and Miami) have a higher percentage of total land area in commercial and industrial uses than the other classes.

The areas of land use in each of the categories of commercial, industrial, and other uses are listed in table 2 with corresponding embodied energy power density. The percent of the total

Table 1. Urban land uses and population for 5 classes of cities in Florida.

City Class	Number of Cities	Mean Total Area (acres)	Mean Area Commercial Uses (acres)	Mean Area Industrial Uses (acres)	Mean Area All Other Uses (acres)	Mean Population
Class 1	1	988813.7	8003.9	8142.2	82667.6	504,265
Class 2	2	62877.5	4024.2	3521.1	55332.2	99,006
Class 3	5	13714.4	771.1	519.7	12423.6	37,177
Class 4	21	4584.1	244.9	151.6	4187.5	12,957
Class 5	116	692.9	29.7	9.6	653.6	1,754

Table 2. Areas of urban land use, embodied energy power density, and total embodied energy flow for five classes of cities in Florida.

City Class	Commercial Land Use		Industrial Land Use		All Other Land Uses		Total Embodied Energy Flow (x 10^{12} Cal CE/yr)
	Area (acres)	Average Embodied Energy Power Density (x 10^9 Cal CE/acre • yr)	Area (acres)	Average Embodied Energy Power Density (x 10^9 Cal CE/acre • yr)	Area (acres)	Average Embodied Energy Power Density (x 10^9 Cal CE/acre • yr)	
Class 1	8004	11.1	8142	4.4	82668	0.7	182.4
Class 2	4024	8.7	3521	4.4	55332	0.7	89.0
Class 3	771	7.5	519	4.4	12423	0.7	16.7
Class 4	245	5.4	152	4.4	4188	0.7	4.9
Class 5	30	3.9	10	4.4	654	0.7	0.6

land area of each of the uses is also listed, and when compared for each class of city, indicates the extent that each city type serves as a central place. The percent of land use in commercial and industrial uses is highest for class 1 cities and decreases with each class.

The Flows of Energy in a Regional Hierarchy: Lee County

The flows of energy through, and the storages of energy within a regional landscape, while somewhat web-like in their organization, can be grouped by quality of energy and a hierarchy emerges. Given in figure 7 is an energy model of Lee County, Florida, organized as a regional hierarchy.

Figure 7 is a "heat energy" diagram, where all flows and storages of energy are evaluated in their chemical potential energy, or heat energy equivalents.

The regional hierarchy has two energy sources inflowing from the outside. The first is

natural renewable energies that are the sum of all natural energies inflowing, including: sunlight, chemical potential energy associated with the purity of rainwater, potential energy associated with runoff of rains due to their elevation as they flow to sea level, potential energy associated with winds, potential energy in waves at the coastal margins, and the potential energy of tides over the estuarine areas. The second is the sum of fossil fuel energies inflowing and imported goods.

The renewable energies are cascaded through the regional economy and embodied in natural structure, agricultural structure, and urban structure, directly and indirectly into the higher quality components of governmental and educational structure and humans. Some of this embodied natural energy is exported in locally harvested and manufactured goods.

The inflowing fossil fuel energies and goods are the primary sources of energy for the urban structure and higher quality components. Much of this energy outflows as used heat energy, but is likewise embodied in the structure of the regional hierarchy.

524

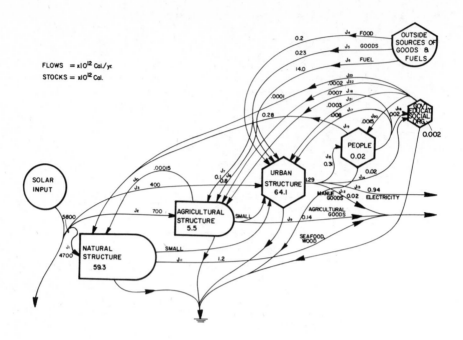

FLOWS = ×10^{12} Cal./yr.

STOCKS = ×10^{12} Cal.

Figure 7. Energy circuit model of Lee County, Florida, organized as a
regional hierarchy of components. Numbers are flows and storages of
actual energy (heat equivalents). See notes for details of calculations.

Sometimes referred to as a "first law diagram," because the inflowing energies equal the
degraded energies or heat losses from the system,
figure 7 shows the sum of inputs equal to the outflows for each component as well as for the
regional system as a whole.

Flows of energy decrease from right to left
as more and more energy is dissipated as dispersed
heat from components. In general, there are five
orders of magnitude difference in energy flows
from the inflows of natural energy to those of the
feedbacks of human work; supporting the notion
that there is a constant percent decrease from one
component to the next in hierarchically organized
systems.

When "heat" energies are converted to embodied energy Calories of coal equivalent, the values
in figure 8 result. Figure 8 is an embodied
energy diagram of the regional hierarchy, thus
there is no energy outflowing as dispersed heat,
but it is embodied in the next level components as
energy is "concentrated" through the system. Comparison of figure 7 with figure 8 shows the very
large flows of energy from natural sources, when
expressed in coal equivalent Calories of embodied
energy, as having nearly the same order of magnitude as those of fossil fuel sources.

Development Density and Imports/Exports

Development density (GDP/sq mi) was evaluated
for various counties in the State of Florida, various states in the nation, and various countries.
Then exports are evaluated using an export multiplier method for counties and states, and exports
for countries were obtained from the literature
directly. Development density was related to
exports in a series of nomographs for the three
different sized regions and are summarized in
figure 9.

The nomograph in figure 9a is a log-log plot
of development density versus export for the combined data from counties, states, and countries.
Assuming a linear relationship between development
density and exports and plotting on an arithmetic
scale gives the graphs in figure 9b. Statistical
analysis using least squares regression gives the
following equations for each set of data: (R^2 =
0.98)

Counties: Ex = 0.21 x Dev. + 5.75   (1)
States: Ex = 0.13 x Dev. - 16.04   (2)
Countries: Ex = 0.286 x Dev. - 1.86   (3)

where, Ex = Exports/sq mi, and
       Dev. = Development density (GDP/sq mi).

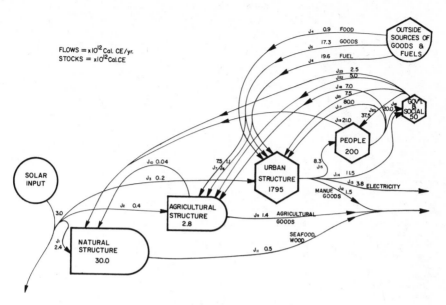

Figure 8. Energy circuit model of Lee County, Florida, evaluated in coal equivalent Calories of embodied energy.

When a regression equation is fitted to the combined set of data, the following equation is given for the line: (R^2 = 0.83)

Combined Data: Ex = 0.968 x Dev. + 23.84. (4)

Energy Flow and Structure in Urban Systems

The energy flow and structural characteristics of land uses were analyzed using 1973 data for two urban areas of Florida: Ft. Myers in southwest Florida and Gainesville in central Florida, and the data are summarized in table 3.

In table 3, the second column headed "fossil fuel power density" is defined as the power density that is from the direct use of electricity and other fossil fuels; and the column headed "power density of embodied energy in goods and services" is defined as the power density of the embodied energy that is consumed indirectly in the use of goods and services. All energy flows are expressed as a density function on a yearly basis; in this case, Cal/acre·year, rather than on a housing or commercial unit basis. The volume of enclosed space occupied by built structure.

Generally, the land uses are arranged in order of increasing power density from low density residential to the central business district (CBD). The volume of structure per acre increases with increasing power density as might

be expected, with the exception of mobile home land uses, where energy use is high as compared with the volume of structure. In this land use category, living units tend to be small (from 600 to 850 sq ft), while the energy demands of the inhabitants are approximately equivalent to those of other residential land use types.

Industrial land uses are not very energy intensive on the average in the Florida urban landscape in comparison to other industrialized areas of the nation. For example, an average value for fossil fuel power density of industrial land uses for the nation derived from the Council on Environmental Quality (1979) is equal to approximately 4,600 x 10^6 Cal/acre·yr, or about 6 times that computed for the Florida industrial land uses. This is due primarily to the "light industrial" nature of Florida industry, and also to the fact that warehouse districts were included in this classification.

Two different densities of CBD were evaluated: those areas with an average height of two stories and those with an average height of four stories. Since energy use and therefore power density is strongly related to the volume of structure associated with a land use, it seems apparent that the power densities of these land uses will differ significantly from the very urbanized areas of the nation where CBD's might have heights as much as 10 times greater than those experienced in Florida.

Table 3. Power density and total volume of structure for selected land uses in Florida.

Land Use Type	Fossil Fuel[1] Power Density (x 10^6 Cal CE/acre · yr)	Power Density[2] Embodied Energy In Goods And Services (x 10^6 Cal CE/acre · yr)	Total Power[3] Density (x 10^6 Cal CE/acre · yr)	Total Volume[4] Of Structure (x 10^3 ft^3/acre)
Single-family residential				
Low density	70	328	398	42.0
Medium density	90	411	501	70.0
High density	110	463	573	85.8
Multi-family residential				
Low rise (2 stories)	340	1557	1897	302.0
High rise (4 stories)	570	2488	3058	664.4
Mobile home				
Medium density	122	597	719	30.6
High density	230	1086	1316	54.4
Commercial strip	680	441	1121	150.3
Commercial mall	3280	2052	5332	141.6
Industrial	760	548	1308	167.2
Central business district				
Average 2 stories	2380	1525	3905	528.5
Average 4 stories	4320	2789	7109	1102.8

[1] Energy consumption data from billing records of Florida Power and Light, Ft. Myers office for 1973. In general, a 10% sample size of each land use classification was used.

[2] Goods and services consumed by each sector are from an input/output analysis of the Lee County analysis that gave total end use of goods and services by sector. Then that amount that was attributable to each separate land use within sectors was apportioned according to the same percentage of fossil fuel energies consumed by sector.

[3] Addition of column 2 and 3.

[4] Volume of structure is calculated by multiplying the square feet of structural area (obtained from property tax records) by average heights of buildings.

Simulation of Models of Hierarchical Organization and Energy Spectra

Results of the study of several hierarchically organized models are presented in this section, starting with a simple model and progressing to more complex examples. Differential equations to describe the behavior of each are presented along with time simulations of each model.

In general, the models are five compartment systems (having 5 state variables) and differ in kinetics of interaction between compartments as the models become more complex. The final model simulated is an aquatic food chain organized as a hierarchic system of energy flow, using data from Fontaine (1978) to evaluate each state variable and pathways of energy flow between variables.

Theoretical Models

Presented in figures 10, 11, and 12 are simulation results of a simple hierarchic chain of energy flow without interacting feedback pathways.

The steady state simulation results are given in figure 10 and then the results of various perturbations of the model are presented (figs. 11 and 12). In all cases, pathway coefficients are held constant in each simulation run, changing only those coefficients indicated in the models in the figures. When the initial conditions for state variables are set low and the system allowed to grow to steady state values (fig. 11), damped oscillation is exhibited by compartments Q1 and Q2, with less noticeable oscillation in "upstream" compartments.

In a final simulation of the simple chain model, pathway coefficients were adjusted so that turnover times for all compartments were equal. The simulation results are given in figure 12. Without the dampening effect of increasing turnover times for each compartment (as was the case in the simulation presented in fig. 11), increasing oscillatory behavior is exhibited by each component. The energy source has been increased twofold, acting as stimulus to the system.

A feedback pathway is added between compartments acting as a multiplicative interaction in the next simulation, which is shown in figure 13. The model has the added feature of a second energy source that is multiplicatively interacted with compartment Q3, using a switching function, so that the model first runs in a steady state and then at time = 10, the second source is turned on. The interaction of the second source changes the distribution of energy within the system, with Ql attaining a lower overall value, and Q2 and Q3 higher values. The graphs in figure 14 give the energy spectral distributions for the steady state solution, and as a result of the second source.

Aquatic Food Chain

Given in figure 15 is an evaluated aquatic food chain organized as a hierarchy of energy flow. The data used in the model given are a summary of Fontaine's (1978) data for Lake Conway

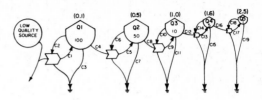

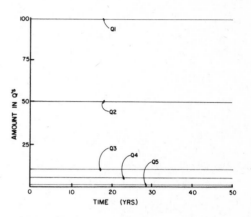

Figure 10. Model and steady state simulation results for a simple hierarchic system of compartments, with no feedback between compartments. Values in parentheses above each compartment are turnover times, values in each storage tank are steady state values, and differential equations are as follows:

$$JR = Jo/(1+KQ1)$$

$$\dot{Q}1 = C1JRQ1-C2JRQ1-C3Q1-C4Q1Q2$$

$$\dot{Q}2 = C5Q1Q2-C6Q1Q2-C7Q2-C8Q1Q2$$

$$\dot{Q}3 = C9Q2A5-C10Q2Q3-C11Q3-C12Q3Q4$$

$$\dot{Q}4 = C13Q3Q4-C14Q3Q4-C15Q4-C16Q4Q5$$

$$\dot{Q}5 = C17Q4Q5-C18Q4Q5-C19Q5$$

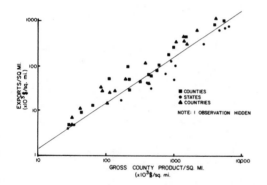

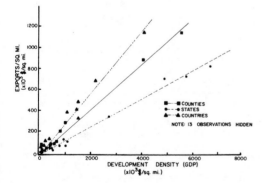

Figure 9. Graphs of the relationship of exports to gross domestic product when expressed as spatial functions for the combined data of counties in Florida, states in the United States, and countries. (a) Log-log plot as a nomograph; (b) arithmatic plot showing regression lines for each set of data.

in central Florida. Compartments were summed together into trophic levels based on primary energy source in the following manner: Ql = Phytoplankton, Macrophytes, and Epipilec Algae; Q2 = Zooplankton and Benthic Invertebrates; Q3 = Primary and Secondary Level Fish; Q4 = Tertiary Level Fish. A fifth compartment was added as a top carnivore, and values of storage and flows estimated.

The major differences between the aquatic food chain hierarchy and the previous models are the addition of a sixth compartment that represents a pool of detritus and nutrient storage that is recycled from the other five compartments, the additive pathways of energy flow up the food chain, and the additive feedback path-

528

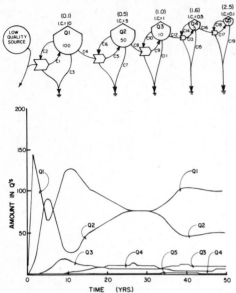

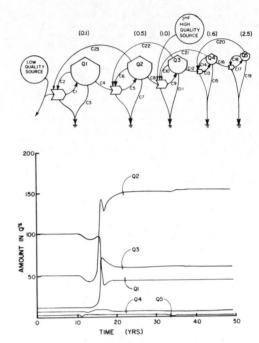

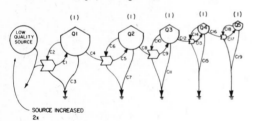

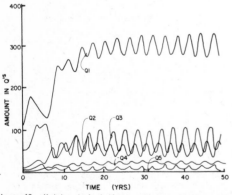

Figure 11. Simulation results of the simple chain when initial conditions are set low, as indicated above each compartment. Note that there is a difference in the vertical scale from the graph in figure 10.

Figure 12. Model and simulation results when the turnover times of each compartment are adjusted so that they are equal. Numbers in parentheses above each compartment are turnover times. Adjustment of turnover times was achieved by rescaling inflows and outflows for each compartment so that they were equal to the steady state value.

Figure 13. Model and simulation results for a hierarchical energy chain with interactive feedback and a second energy source. The simulation is first run in steady state conditions until time = 10, then the second source is switched on. Differential equations are as follows:

$$JR = Jo/[1+K(Q1+Q2)]$$

$$\dot{Q1} = C1JRQ1Q2 - C2JRQ1Q2 - C3Q1 - C4Q2Q2Q3$$

$$\dot{Q2} = C5Q1Q2Q3 - C6Q1Q2Q3 - C7Q2 - C8Q2SHQQ3Q4 - C23JRQ1Q2$$

$$\dot{Q3} = C9Q2SHQQ3Q4 - C10Q2SHQQ3Q4 - C11Q3 - C12Q3Q4Q5 - C22Q1Q2Q3$$

$$\dot{Q4} = C13Q3Q4Q5 - C14Q3Q4Q5 - C15Q4 - C16Q4Q5 - C21A2SHQQ3Q4$$

$$\dot{Q5} = C17Q4Q5 - C18Q4Q5 - C19Q5 - C20Q3Q4Q5$$

ways of control action flow from higher level compartments to lower ones.

The model given in figure 16 shows the energy source as a smooth sine wave that represents the variation in sunlight from summer to winter. The graphs in figure 16 are the steady state solution, where the effects of dampening of the fluctuations in energy source are obvious. Since Q2 draws most of its energy from the large stable pool of detritus, very little oscillation is observed. Q3, on the other hand, draws much of its energy from the first compartment (Q1) and shows oscillation, but damped from that of Q1. Yearly variation in standing crop in the higher compartments (Q4 and Q5) is relatively small due to the dampening of upstream compartments.

The model in figure 17 was simulated to test the effect of control actions by the highest level compartment. This last compartment has an additional input pathway driven by a sine wave with amplitude varying from +1 to -1, and frequency equal to the turnover time of the compartment (5 yr). The input pathway acts as both positive input and drain, causing the compartment to oscillate.

The oscillation set up in compartment Q5 is passed on to downstream compartments through feedback pathways C28 and C29, and is passed further downstream through feedback pathways C23, C24, C17, C18, and C11. Because of differences in scaling on the graph of the vertical axis for each compartmnt, the differences in the magnitude of oscillation of each compartment are not apparent. The percent change from minimum to maximum values for each compartment is given in figure 17. A general trend of decreaseing magnitude of oscillation from the highest level compartment to the lowest level compartment is observed.

In all, many simulations of the various models reported here were conducted; testing various organizations and perturbations to each model. Different kinetic organization, energy sources and magnitudes were tested for the theoretical models. The aquatic food chain was simulated testing different sources (constant and pulsing), effects of harvesting of each compartment, and the effects of a secondary energy source as a stocking function. These simulation results are reported in Brown (1980). Reported here are some of the highlights of those initial investigations.

DISCUSSION

Analysis of regions, urban systems, and smaller ecosystems showed many manifestations of hierarchy that were related to supporting energy flows. Simulated models were able to duplicate many of the observed features of hierarchical organization. This evidence supports a theory of energy control of the organization of systems of man and nature.

Pathways of energy flow were shown to be greatest, hierarchically, in the highest components of urban and regional landscapes. Here components were found to be largest in size, fewest in number, to have larger time constants, and have greatest potential control of the overall systems processes.

Hierarchical Principles of Landscape Organization

From the measurements and models, principles may be formulated for relating parts to whole landscapes and as guidelines for regional planning.

Energy Quality and Spatial Effect

While the number of components decreased in successive levels of hierarchies, the spatial area over which their effect was spread increased. The nomographs of development density versus exports suggested this relationship, for as the density of human activity increased, the total exports to other regions increased.

Energy Convergence in Landscape Hierarchies

The evaluated models of regional and ecological systems and data on regional land uses and spatial distribution of incoming energies showed the pattern of convergence of energy flows to higher and higher quality components in smaller

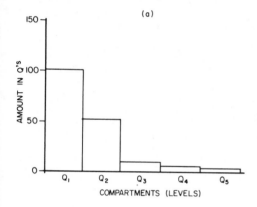

(a)

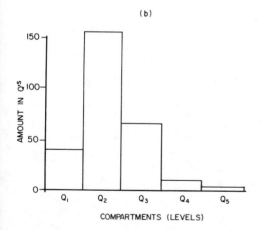

(b)

Figure 14. Graphs of the energy spectral distributions of compartments in the simulation model in figure 13 where a secondary energy source was introduced into the system. (a) The distribution achieved with single low quality source; (b) the distribution achieved with the addition of a secondary high quality energy source.

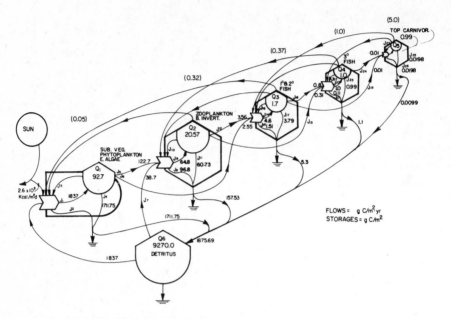

Figure 15. Model of an aquatic food web organized in a hierarchic fashion. Data are from Fontaine (1978). Numbers in parentheses are turnover times of the state variables.

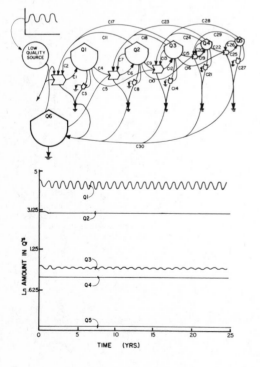

Figure 16. Model and steady state simulation results of model in figure 15, where the energy source is generated sine wave representing seasonal fluctuation of sunlight. The vertical axis on the graph is the natural log of the value in compartments. Differential equations are as follows:

JR = sine wave

$$\dot{Q}1 = C1JRQ1(Q2+Q3)Q6-C2JRQ1(Q2+Q3)Q6-C3Q1-C4(Q1+Q6)Q2(Q3+Q4)-C5(Q1+Q2)Q3(Q4+Q5)$$

$$\dot{Q}2 = C6(Q1+Q6)Q2(Q3+Q4)-C7(Q1+Q6)Q2(Q3+Q4)-C8Q2-C9(Q2+Q1)Q3(Q4+Q5)-C10(Q2+Q3)Q4Q5-C11JRQ1(Q2+Q3)Q6$$

$$\dot{Q}3 = C12(Q2+Q1)Q3(Q4+Q5)-C13(Q2+Q1)Q3(Q4+Q5)-C14Q3-C15(Q2+Q3)Q4Q5-C16(Q3+Q4)Q5-C17JRQ1(Q2+Q3)Q6-C18(Q1+Q6)Q2(Q3+Q4)$$

$$\dot{Q}4 = C19(Q2+Q3)Q4Q5-C20(Q2+Q3)Q4Q5-C21Q4-C22(Q4+Q3)Q5-C23(Q1+Q6)Q2(Q3+Q4)-C24(Q1+Q2)Q3(Q4+Q5)$$

$$\dot{Q}5 = C25(Q3+Q4)Q5-C26(Q3+Q4)Q5-C27Q5-C28(Q1+Q2)Q3(Q4+Q5)-C29(Q2+Q3)Q4Q5$$

$$\dot{Q}6 = C30(C3Q1+C8Q2+C14Q3+C21Q4+C27Q5)-C31(Q1+Q6)Q2(Q3+Q4)-C32JRQ1(Q2+Q5)Q6$$

spatial extent. Incoming energies, low in quality and spatially dilute, were transformed into higher quality energies, many as storages, as energy flows converge from many low quality components to fewer and fewer high quality ones.

Energy Divergence (Dispersion)
in Landscape Hierarchies

Energy is not only concentrated and converged in landscape processes, but much is fed back in dispersing actions of recycle and control. Spectral distributions of incoming energies

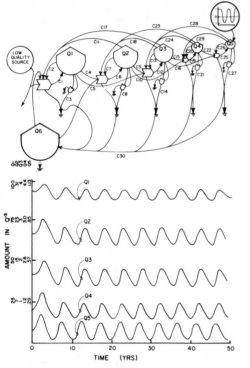

Figure 17. Model and simulation results of the aquatic food chain when compartment Q5 is caused to oscillate from an input pathway driven by a sine wave with amplitude +1 to -1. The oscillation is passed onto lower level compartments with decreasing amplitude (decreasing amplitude is not graphically apparent due to different vertical scales for each compartment on the graph). The percent change from minimum to maximum values for each compartment is as follows:

$$Q1 = 6.9\%$$
$$Q2 = 15\%$$
$$Q3 = 22\%$$
$$Q4 = 42\%$$
$$Q5 = 80\%$$

Differential equations are the same as in figure 16 except for the sine wave input to compartment Q5. The equation for Q5 is as follows:

$$\dot{Q}5 = C25(Q3+Q4)Q5+SW-C26(Q3+Q4)Q5-C27Q5-C28(Q1+Q2)Q3$$
$$(Q4+Q5)-C29(Q2+Q3)Q4Q5$$

and power density of land uses, as well as the distributions of cities within regional landscapes indicated that dispersion of high quality energy from centralized sources follows a hierarchical distribution.

Control Actions of High-Quality Components

Evaluation and simulation of hierarchically organized models suggested that the highest quality pathways were those associated with the highest level component in the hierarchy, and suggested a general principle of hierarchic organization and control action feedbacks. The greatest control effect was achieved with the highest quality pathways, since their overall cost was large and their effect must at least be equal their cost. In the simulations of the aquatic ecosystem, greatest overall effect to the system was achieved when the final compartment was perturbed.

Primary and Secondary Energy Sources
and Their Effect on Hierarchies

The landscape is a mosaic of natural lands, agricultural lands, roads, cities, and people related through pathways of energy flow and exchange. When viewed as a whole system of processes, it was seen that energy sources inflowing support the processes of the entire system. Systems that are sustained by one energy source such as the evaluated aquatic ecosystem, develop relatively smooth distributions of energy between components as energy is cascaded up the hierarchy supporting fewer and fewer components. Secondary energy sources of higher quality inflow at levels in the hierarchy where their quality nearly matches, with additional energies for support at these levels, greater structure was developed and the hierarcy was shifted somewhat in the relative distribution of energy between levels.

Stability Through the "Filtering"
Actions of Hierarchies

The term stability has been applied to a number of concepts. Orians (1975) distinguishes seven different concepts of stability. When models of hierarchies were simulated, stability was greatly enhanced if turnover times were adjusted so that turnover increased with each level in the energy chain. When the turnover times were set equal in all compartments, oscillation was exhibited, since any perturbation in one compartment was passed on to the next.

A Theory of Regional Boundaries Derived
from Place in the Landscape Hierarchy

A method that may have significance in determining the regional boundaries of systems was suggested by the results of the nomographs of exports versus development density of regions. The nomographs showed that the propensity to export (and therefore to import to maintain balance of payments) was greater as the density of development increased for regional systems of all sizes. In essence, this relationship suggests that the greater the density of development the more an area relies on external areas for sources of primary goods and energies. And, to carry it one step further, since primary goods require large areas of the landscape for their production (i.e., they are low in quality and occupy large spatial area), the greater the density of development, the greater the size of the region required for support.

ACKNOWLEDGMENTS

The author wishes to acknowledge Dr. H. T. Odum, whose reviews of the research in progress and previous drafts was invaluable. Mapping of the south Florida region and Lee County was supported under joint contract to the Division of State Planning, Department of Administration, State of Florida, and the National Park Service, U.S. Department of the Interior, H. T. Odum principal investigator. Mapping of the St. Johns region was supported under joint contract to the Jacksonville Area Planning Board and the St. Johns River Water Management District, H. T. Odum principal investigator. Initial studies of the Ft. Myers urban area were funded by Lee County Board of County Commissioners, the author and G. V. Genova principal investigators. Preliminary studies of the Gainesville urban area were funded by the Division of State Planning, Department of Administration, State of Florida, J. F. Alexander principal investigator.

Support was also received from the Department of Energy on contract EY-76-5-05-4398 entitled "Energy Models of the United States" with the Department of Environmental Engineering Sciences, H. T. Odum principal investigator.

LITERATURE CITED

Aldrich, H. E. 1979. Organization and environments. Prentice Hall, Inc., Englewood Cliffs, New Jersey.

Alexander, J. F., Jr. 1978. Energy basis of disasters and the cycles of order and disorder. Ph.D. dissertation, University of Florida, Gainesville.

Beckmann, M. 1955. Some reflections on Losch's theory of location. Papers, Regional Science Association 1:139–48.

Beckmann, M. J. 1956. Distance and the pattern of intra-European trade. Review of Economics and Statistics 38:31–40.

Beckmann, M. J. 1958. City hierarchies and the distribution of city size. Economic Development and Cultural Change 6:234–48.

Beckmann, J. 1970. The analysis of spatial diffusion processes. Papers, Regional Science Association 25:109–17.

Berry, B. J. L. 1972. Hierarchical diffusion: The basis of developmental filtering and spread in a system of growth centers. Pages 108–38 in N. M. Hanson (ed.), Growth centers in regional economic development. Free Press, New York.

Blau, P. M. 1972. Interdependence and hierarchy in organization. Social Science Research 1:1–24.

Blau, P. M. 1974. On the nature of organizations. John Wiley and Sons, New York.

Brown, M. T. 1980. Energy basis for hierarchies in urban and regional landscapes. Ph.D. dissertation, University of Florida, Gainesville.

Bunge, M. 1969. The metaphysics, epistomology, and methodology of levels in hierarchical structures. Pages 17–26 in L. L. Whyte, A. G. Wilson, and P. Wilson (eds.), Hierarchical structures. American Elsevier, New York.

Bureau of Economics and Business Research, College of Business Administration, University of Florida. 1978. Florida statistical abstract 78. University of Florida Press, Gainesville.

Burgess, E. W., and R. E. Park. 1924. Introduction to the science of sociology. University of Chicago Press, Chicago.

Christaller, W. 1966. Central places in southern Germany. (Transl. by G. C. W. Baskin.) Prentice Hall, Inc., Englewood Cliffs, New Jersey.

Costanza, R. 1975. The spatial distribution of land use subsystems, incoming energy and energy use in south Florida from 1900 to 1973. Master's terminal project. Department of Architecture, University of Florida, Gainesville.

Council on Environmental Quality. 1979. Environmental statistics, 1978. National Technical Information Service. U.S. Department of Commerce, Springfield, Virginia.

Dokmeci, V. 1973. An optimization model for a hierarchical spatial system. Journal of Regional Science 13:439–51.

Emery, F. E., and E. L. Trist. 1973. Towards a social ecology: Contextual appreciation of the future in the present. Plenum, New York.

Fontaine, T. D. 1978. Community metabolism patterns and a simulation model of a lake in central Florida. Ph.D. dissertation, University of Florida, Gainesville.

Glassman, R. 1973. Persistence and loose coupling. Behavioral Science 18(March):83–98.

Hagerstrand, T. 1966. Aspects of the spatial structure of social communication and diffusion of information. Papers, Regional Science Association 16:27–42.

Henderson, S. 1972. Hierarchy models of city size: An economic evaluation. Journal of Regional Science 12:435–41.

Hielbrun, S. 1974. Urban economics and public policy. St. Martin's Press, New York.

Isard, W., and M. Peck. 1954. Location theory and international and interregional trade. Quarterly Journal of Economics 68:97–114.

Landau, M. 1969. Redundancy, rationality, and the problem of duplication and overlap. Public Administration Review 29:346–58.

Laszlo, E. 1972. The systems view of the world. George Braziller, New York.

Laszlo, E. 1973. Introduction to systems philosophy. Harper and Row, New York.

Losch, A. 1954. The economics of location. (Transl. by U. Waglom and W. F. Stalpor.) Yale University Press, New Haven, Connecticut.

Lotka, A. J. 1922. A contribution to the energetics of evolution. Proceedings of the National Academy of Science 8:147–55.

MacArthur, R. H. 1957. On the relative abundance of bird species. Proceedings of the National Academy of Science 43:293–95.

Mansfield, E. 1963. The speed of response of firms to new techniques. Quarterly Journal of Economics 77:290–311.

Monane, J. H. 1967. A sociology of human systems. Appleton Century, Crofts, New York.

Odum, H. T. 1960. Ecological potential and analogue circuits for the ecosystem. American Scientist 48:1–8.

Odum, H. T. 1967. Biological circuits and the marine systems of Texas. Pages 99–157 in F. J. Burgess and T. A. Olson (eds.), Pollution and marine ecology. John Wiley and Sons, New York.

Odum, H. T. 1971. Environment, power, and society. John Wiley and Sons, New York.

Odum, H. T. 1973. Energy, ecology, and economics. Ambio 2(6):220–27. Swedish Royal Academy of Science, Stockholm.

Odum, H. T. 1975. Marine ecosystems with energy circuit diagrams. Pages 127–51 in J. C. J. Nihoul (ed.), Modeling of marine systems. Elsevier Oceanography Series. Elsevier Scientific Publishing Co., New York.

Odum, H. T. 1976. Energy quality and carrying capacity of the earth. Tropical Ecology 16(1):1–16.

Odum, H. T. 1976a. Macroscopic minimodels of man and nature. Pages 249–80 in B. Patten (ed.), Systems analysis and simulation in ecology, vol. 4. Academic Press, New York.

Odum, H. T. 1977. Energy, value, and money. Pages 174–96 in C. S. Hall and J. W. Day (eds.), Ecosystem modeling in theory and practice: An introduction with case histories. John Wiley and Sons, New York.

Odum, H. T. 1978a. Energy analysis, energy quality and environment. Pages 55–87 in M. W. Gilliland (ed.), Energy analysis: A new public policy tool. AAAS Selected Symposium 9.

Odum, H. T. 1978b. Net energy from the sun. Pages 196–211 in S. Lyons (ed.), Energy for a livable future comes from the sun: A handbook for the solar decade. Friends of the Earth, San Francisco.

Odum, H. T. 1981. Systems. John Wiley and Sons, New York. Forthcoming.

Odum, H. T., and M. T. Brown (eds.). 1976. Carrying capacity of man and nature in south Florida. Final contract report to the U.S. Department of Interior. National Park Service, Washington, D.C.

Odum, H. T., M. Brown, and R. Costanza. 1976. Developing a steady state for man and land: Energy procedures for regional planning. Pages 343–61 in Science for better environment. Proceedings of the International Congress on the Human Environment (HESCO), Kyoto. Asahi Evening News, Tokyo.

Odum, H. T., J. E. Cantlon, and L. S. Kornicker. 1960. An organizational hierarchy postulate for the interpretation of species-individual distribution, species entropy, ecosystem evolution, and the meaning of a species-variety index. Ecology 41(2):395–99.

Odum, H. T., and E. C. Odum. 1976. Energy basis for man and nature. McGraw-Hill, New York.

Orians, G. H. 1975. Diversity, stability, and maturity in natural ecosystems. Pages 139–50 in U. H. VanDubben and R. H. Lowe-McConnell (eds.), Unifying concepts in ecology. The University Press, Belfast, Ireland.

Pugh, A. L., III. 1970. Dynamo users manual. MIT Press, Cambridge.

Purver, D. 1975. A programming model of central place theory. Journal of Regional Science 15:307–16.

Simon, H. 1962. The architecture of complexity. Proceedings of the American Philosophical Society 106:467–82.

Simon, H. 1973. The organization of complex systems. Pages 1–28 in H. Patte (ed.), Hierarchy theory. George Braziller, New York.

Stewart, J. W. 1947. Empirical mathematical rules concerning the distribution and equilibrium of population. Geographical Review 37(3):461–85.

Thompson, S. 1967. Organizations in action. McGraw Hill, New York.

Timbergen, J. 1968. The hierarchy model of the size distribution of centers. Papers, Regional Science Association 20:65–8.

United Nations, Department of International Economic and Social Affairs. 1978. 1977 statistical yearbook (29th issue). United Nations, New York.

Van Boventer, B. 1969. Walter Christaller's central places and peripheral areas: The central place theory in retrospect. Journal of Regional Science 9:117–24.

Von Bertalanffy, L. 1933. Modern theories of development: An introduction to theoretical biology. (Transl. by J. H. Woodger.) Oxford University Press, London.

Von Bertalanffy, L. 1968. General system theory. George Braziller, New York.

Weber, M. 1946. Essays in sociology. Oxford University Press, New York.

Weber, M. 1947. The theory of social and economic organization. Free Press, Glencoe, Illinois.

Weiss, P. A. 1971. Hierarchically organized systems in theory and practice. Hafner Company, New York.

Whyte, L. L. 1949. The unitary principles in physics and biology. Henry Holt and Company, New York.

Whyte, L. L. 1969. Structural hierarchies: A challenging class of physical and biological problems. Pages 3–16 in L. L. Whyte, A. G. Wilson, and D. Wilson (eds.), Hierarchical structures. American Elsevier, New York.

Wilson, A. G. 1969. Hierarchical structures in the cosmos. In L. L. Whyte, A. G. Wilson, and D. Wilson (eds.), Hierarchical structures. American Elsevier, New York.

Woodger, J. H. 1929. Biological prinicples. Cambridge University Press, London.

Woodger, J. H. 1937. The axiomatic method in biology. Cambridge University Press, London.

Zipf, G. K. 1941. National unity and disunity. Principia Press, Bloomington, Indiana.

A REGIONAL MODELING APPROACH TO AN ENERGY-ENVIRONMENT CONFLICT[1]

J. B. Mankin, J. M. Klopatek, R. V. O'NEILL, and J. R. Krummel[2]

Abstract.--A regional analysis framework was developed to address regional environmental problems. This framework involves the use of four sets of variables: pattern, assets, and culture, as influenced by externalities. This framework was applied to the problem of resource extraction, unique ecological sites, and public opinion in the Overthrust Belt of western United States. Model results show that action groups seeking to preserve unique sites should campaign vigorously from the start. The result will be a greater amount of land preserved even though many years will pass before preservation occurs. However, if resource developers yield early to the demand for site preservation, the result will be less land preserved and a concommitant higher total extraction of resources.

INTRODUCTION

Oil and gas exploration has reached an all-time high in the United States (Tippee 1980). A key area of exploration is the Overthrust Belt which extends from the Canadian Rockies through the western United States to Mexico (fig. 1). Within a small section of this belt (6.1 x 10^6 ha in Wyoming-Utah-Idaho) reserves have been estimated at 1.2 x 10^9 m^3 of oil and 2.1 x 10^{11} m^3 of gas (Tippee 1980). These reserves represent an important national asset. In addition to oil and gas reserves, renewable timber resources are also an important asset in this region (USDA Forest Service 1979a).

The development of oil and gas resources and the harvesting of timber are complicated by the 2 x 10^6 ha that have the potential to be designated as wilderness areas (USDA Forest Service 1978). The Overthrust Belt contains important aesthetic, scenic, and recreational resources and also contains valuable wildlife habitat,

threatened and endangered animal species, unique natural areas, and other important environmental resources (Klopatek et al. 1980). Increased extraction of energy, combined with current and projected timber harvests, could threaten environmental quality in this region. In this paper we address this potential energy-environment conflict through the development of a conceptual regional analysis (ARIES) that focuses on these conflicts.

Regional Modeling

The maintenance of environmental quality should be an explicit goal in regional policy making (Nijkamp 1977). Traditionally, environmental and ecological parameters have been treated as externalities (Mäler 1974, Krutilla and Fisher 1975, Nijkamp 1977). This is because the historical basis for regional analysis lies in economics, sociology, and geography (Isard 1960, 1975; Smith 1976a,b). Only recently have ecological concerns been incorporated in regional analysis. This has largely been accomplished though the use of input-output, equilibrium materials balance, and simulation models (Nijkamp 1977).

Early attempts at including ecological variables in regional analysis used input-output models (Isard 1972) which treat ecosystems as similar to economic systems (Hannon 1973). The input-output model calculates outputs (product, pollutant, etc.) as a function of input, based on a Leontief matrix of input-output coefficients (Leontief 1966). The limitation of this type of

---

[1]Paper presented at the International Symposium Energy and Ecological Modelling. (Louisville, Kentucky, April 20-23, 1981).
[2]James B. Mankin is a Computer Scientist, Computer Sciences Division, Union Carbide Corporation Nuclear Division; and Jeffrey M. Klopatek and John R. Krummel are Research Ecologists and Robert V. O'Neill is a Senior Scientist, Environmental Sciences Division, Oak Ridge National Laboratory, Oak Ridge, Tennessee, U.S.A.

536

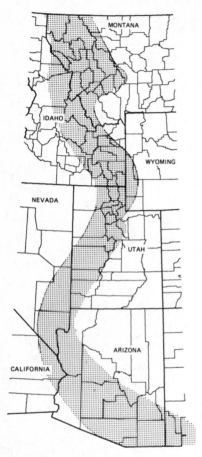

Figure 1.--Location of the Overthrust Belt in the Western United States.

Of the qualitative methods for regional analysis, simulation models place the least restrictions on problem representation (Hamilton et al. 1969). Problem definition is the most fundamental step in regional modeling, because the problem defines the boundaries of the regional system (Isard 1975). Once the system has been defined, the problem definition then determines what components or state variables should be considered in the next step of model construction. A major stumbling block in the development of any analytical framework is the definition of state variables (Mar 1976). It is often prudent to follow the advice of Watt and Wilson (1973) and Holling (1978) to build only simple models when analyzing regional systems. The validation of regional models is also a difficult if not impossible task because the model output may predict conditions never before historically experienced (Hamilton et al. 1969, Holling 1978, Mar 1974, 1976).

METHODS AND APPROACH

Conceptual Framework

In the development of a conceptual framework we chose not to utilize a disciplinary approach because of the inherent danger of viewing the problem with subjective biases. Nor did we center the framework around available data sources; only after the problem has been defined is it relevant to examine data availability.

We postulated that a regional problem requires four types of state variables, that is, the State (S) of the system is a function of the Pattern (P), the Assets (A), the Culture (C), and the Externalities (E).

Pattern refers to the spatial aspects of the environment at the regional level (e.g., hectares of forest, kilometers of undisturbed river, or distribution of air pollution). Assets can be either dynamic or static. Dynamic assets include renewable resources (e.g., timber) and nonrenewable ones (e.g., coal). Static assets (e.g., regional climate) are not, in the strict sense, a part of the system, but may impose boundary conditions. Culture includes everything related to the human population such as density, income, and value systems. Externalities are of two types: (1) demanded, i.e., those whose rate of flow is affected by something within the system, and (2) influential, i.e., those which affect the system regardless of system needs and demands. The first three categories were designated as state variables of the system, while the last, externalities, is a forcing function.

Our postulate is that this set of state variables is necessary and sufficient to address any regional environmental problem. This implies, for example, that an approach which considers economics of resource extraction and potential environmental changes, but ignores

model is that all interdependencies are presumed linear and conditions are assumed to be steady state.

A materials balance or equilibrium model provides additional insight into the relationships between economic-technological activities and ecological components. This model is based on conservation of matter/energy. All matter flowing into the system must be accounted for as storage or outflow. This model is useful because it deals with the region in a system format, and helps identify control processes and options for change. However, this type of model often requires equilibrium assumptions and is normally based on linear definitions of environmental interactions (Mäler 1974).

cultural reactions to the changes, is an incomplete approach to the problem. We also imply that these sets of variables provide all of the information needed to address regional conflicts.

Framework Application

In addition to the vast amount of oil and gas reserves in the Overthrust region, approximately 14×10^6 ha are controlled by the U.S. Forest Service with over half classified as supporting commercial forest (Green and Setzer 1974). This timber represents an important resource to local logging and wood industries (USDA Forest Service 1978). The Forest Service must, however, manage the land for a variety of purposes, and timber harvest conflicts with other uses (e.g., wildlife habitat, recreation, water quality, etc.). Management of these lands must include public discussion of timber harvest, leasing of mineral rights, and congressional mandates to preserve renewable resources (USDA Forest Service 1979a).

It is apparent that some of the potential wilderness areas contain recoverable hydrocarbon reserves as well as timber products (Voelker et al. 1979, Bell et al. 1979). Thus, increased extraction of energy resources, combined with current and projected timber harvests, could threaten the environmental integrity of the region (TIE 1977). Public response to these environmental changes, in turn, could restrict access to the needed energy and timber resources of the region (Hendee 1977).

Our approach has been to analyze the functional links between oil exploration and development, timber harvesting, environmental quality, and public reaction to environmental impacts in the form of a simple dynamic model. The model does not seek to duplicate the precise behavior of the system. Rather, the purpose is to demonstrate that dynamic linkages between state variables are necessary to address the regional conflict. The model links the renewable and nonrenewable assets with pattern changes induced by road building, well drilling, and timber harvest. As these changes affect the pattern of unique ecological sites in this region, public perception of the environmental impacts increases and general public opinion interferes with resource removal rates.

Table 1 shows the linked equations that define the model, and defines parameters and state variables. Values for state variables and parameters associated with timber harvests and oil development were based on information provided by the U.S. Forest Service, open literature, and personal communication with individuals familiar with timber harvests and mineral extraction in this region. For example, road building and timber harvests become linked in the model such that 1 km of permanent road must be built to harvest 2400 m^3 of timber (USDA Forest Service 1979b). Some parameter values were chosen simply

to represent the qualitative behavior of interactions. The constant, C_5, influences the shape of the curve for the $f(C_B)$ function. Thus, the formation of official wilderness areas decreases public opposition to resource use in the Overthrust Belt. These preserved areas are set aside when public opposition, $f(C_B)$, reaches a critical value, FCO. The designated areas protect the remaining unique ecological sites and these sites no longer enter into the dynamics of this problem.

RESULTS

Prior to the discussion of the results, let us point out again that this paper is not designed to provide a definitive solution to the problems involved with resource extraction in the Overthrust Belt. Rather, it is an initial exploration of an approach to regional environmental problems. Therefore, the reader is cautioned not to become embroiled in the details of the results.

Because many of the areas of exploration are located in commercial timber areas, oil and gas exploration in the Overthrust Belt is inextricably linked with forest harvesting. The establishment of roads for exploration will open up new areas for harvesting of timber. The harvesting rate, F (Eq. 1), is affected by three factors: (1) public opinion to associated environmental degradation; (2) the rate of oil and gas exploration; and (3) the establishment of wilderness areas that will remove timber from potential harvest. In the model, public opinion against resource extraction becomes important with increased road building and well drilling and a decrease in forested land and unique ecological sites. Public opinion becomes active interference through the $f(C_B)$ function (table 1). The model also contains the option to relieve public pressure through the establishment of reserves to protect unique sites.

The model can, perhaps, be best understood by reviewing a specific case. The case we have chosen is one in which the public is moderately concerned while the developers are very determined to resist setting aside wilderness areas. The timber reserves, oil and gas reserves, number of wells, and the timber harvest rates are shown in fig. 2. The timber harvest increases slightly due to construction of new roads for oil and gas exploration until year 5. At that point, public action brings about a decrease in both timber harvesting and the rate of oil and gas exploration. The public is mollified somewhat in year 10 when the number of wells reaches its maximum, but this is shortlived due to continued road construction for oil and gas exploration. Oil and gas exploration ceases entirely in year 33 which relieves public pressure somewhat. However, an aroused public still protests the loss of unique sites to timber harvest. This impasse continues until year 63 when wilderness reserves are set aside, thereby preserving the unique sites that

Table 1. List of linked equations and definitions for state variables and parameters used in model of the Overthrust Belt

Assets

Timber

$$A_r = -F - a_1 P_u \ (\Phi_1) \tag{1}$$

$$F = E_{D1} + \Phi_2 \ (\Phi_2)$$

$$\Phi_1 = f(C_B) - FCO$$

$$\Phi_2 = -E_{D1} + a_2 D + [(a_3 P_a + a_7 P_a)(1-e^{-a_4 A_r})] f(C_B)$$

$$(\Phi) = \begin{cases} 1, & \Phi < 0 \\ 2, & \Phi \geq 0 \end{cases}$$

Oil

$$A_N = a_5 P_w - a_6 P_u \ (\Phi_1) \ , \tag{2}$$

where:

A_r – timber (m^3)

R – regeneration (m^3/yr)

E_{D1} – timber demand (m^3)

a_1 – timber in unique ecological site (m^3/ha)

a_2 – timber per well (m^3)

a_3 – timber per road (m^3/km)

a_4 – a constant

P_u – unique ecological sites (ha)

FCO – $f(C_B)$ value to create preserves

P_a – roads (km)

D – wells drilled per year

A_N – oil (m^3)

a_5 – oil per well (m^3/yr)

a_6 – oil per unique ecological site (m^3/ha)

P_w – wells.

Pattern

Wells

$$P_w = D - p_1 P_w \tag{3}$$

$$D = [E_{D2} + \Phi_3 \ (\Phi_3)] \ (\Phi_4)$$

$$\Phi_3 = p_2 \ (1-e^{p_3(\Phi_4)}) f(C_B) - E_{D2}$$

$$\Phi_4 = A_N \ min - A_N$$

Roads

$$P_a = p_4 \ [1 - \exp \ (-p_5 \ F - p_6 \ D)] \tag{4}$$

Forests

$$P_f = p_7 \ A_R \tag{5}$$

Table 1. (continued)

Pattern (continued)

Unique Ecological Sites

$$P_u = (\Phi_5) \ [\ (\Phi_6) \ (-p_8 \ F - p_9 \ D)]f(C_B) \tag{6}$$

$$\Phi_5 = p_{10} - P_u$$

$$\Phi_6 = FCO - f(C_B) \ ,$$

where:

p_1 – well depreciation

p_2 – maximum well drilling per year

F – timber harvest per year

p_3 – a constant

A_N min – nonrecoverable oil (m^3)

p_4 – maximum road building per year (km/yr)

p_5 and p_6 – constants

P_f – forest area (ha)

p_7 – land area per amount of timber (ha/m^3)

p_8 and p_9 – constants

p_{10} – a minimum area of unique ecological sites (ha)

E_{D2} – demand for oil (m^3).

Culture

Public Opinion

$$C_B = [e^{c_1 P_w} \ e^{-c_2 P_f} + \frac{c_3}{(\Phi_5)} 2 \ (\Phi_6)] \ c_4 \ P_a \tag{7}$$

Public Action

$$f(C_B) = 1 + (\Phi_7) \ (- \frac{K \ e^{c_5(\Phi_7)}}{1 + K(e^{c_5(\Phi_7)} -1)}) \tag{8}$$

$$\Phi_7 = c_6 \ A_N + c_7 \ F + c_8 - C_B \ ,$$

where:

C_B – public opinion

$f(C_B)$ – public action

c_1 through c_7 – constants

c_8 – public inertia to action

K – a constant.

---

are still left. This results in a dramatic increase in timber harvest due to the complete relaxation of public action.

The number of wells increases dramatically in the first 10 years. This represents a maximum exploration effort consistent with the present attitude. The number of wells then decreases, but oil and gas exploration is still taking place. However, the number of new wells is less than the number of wells becoming obsolete. Exploration ceases entirely in year 33, but energy extraction continues until year 55 assuming a maintenance of present technology.

Timber reserves decrease somewhat until year 10, and they then increase until year 63 since regeneration is greater than harvest. After the establishment of wilderness reserves, the timber reserves reach a steady state.

The initial exploration rates as well as the depletion of the reserves (fig. 2) closely approximate the predictions of others for this region (Kim and Thompson 1978). This case is interesting because an examination of fig. 2 suggests that the developers would be better off by negotiating sooner. However, this model has no provisions for changing parameter values during the course of a simulation at the present time. It also has no provision for public reaction to loss of revenues due to curtailment of timber harvest and energy extraction.

The prediction of the sequence of events which might occur during an energy-environmental conflict is of little concern at the present time. The more relevant questions to which this model may be addressed concern important links between public opinion, natural resource extraction, and environmental perturbation. These

ORNL-DWG 81-10942 ESD

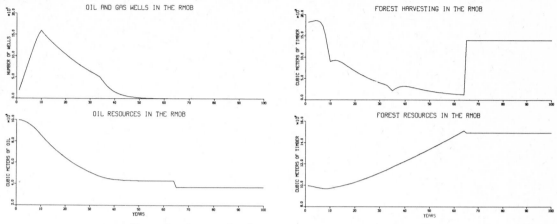

Figure 2.--An example of potential consequences of environmental conflicts in the Overthrust Belt on oil and gas
exploration and resources and timber harvesting and
resources.

become apparent in the context of this simplified
model. We can now ask relevant questions and
seek possible solutions using this simulation
model. We test our hypothesis that the framework
applied to this specific regional problem can be
a useful tool in general regional environmental
analysis.

As a test, one can analyze different policies in facilitating the extraction of resources
or the establishing of wilderness areas. This
can be analyzed in our model through the links of
public opinion to rates of resource removal (and
environmental degradation) in the Overthrust Belt
(table 1). For example, changes in the public
inertia to action represented by the public
opinion parameter (POP) (ranging from 0, no
inertia $\underline{v}$ 100, high inertia) alter the "inertia"
of public opinion or the level at which political
or legal action begins. As a strategy for
wilderness protection, an environmental group
might conduct a publicity campaign to stimulate
public opinion and thus lower the value of the
POP.

A developer would view the establishment of
wilderness areas as a potential threat to a valuable national resource. In our model, the forestry cutoff (FCO) parameter determines the value
of $f(C_B)$ at which wilderness areas become
established to protect the remaining unique ecological sites. The developer can influence the
system through lobbying. Effective lobbying
would lower the value of FCO, resulting in prevention of the formation of wilderness areas.

Figure 3 depicts the outcome of these strategies on the establishment of wilderness areas.
It is to the environmentalists' advantage to
exert immediate pressure for preservation, as

that will lead to a higher amount of land preserved. However, this figure also shows that it
is to the developers advantage to negotiate early
and agree to an early dedication of this land.
The result will be greater resource extraction.
This is exemplified in figure 4 which displays
the influence of the changes of the FCO and POP
values on the total forest harvest. Therefore,
it would seem best, according to our model, to
resolve this energy-environment conflict at the
start of resource development.

CONCLUSIONS

We have presented the results of a model
dealing with the problem of resource extraction
in the western United States. The use of the
model provided insights into the possible role of
public interaction with energy extraction, timber
harvesting, and the destruction of unique ecological sites. However, much more effort is
required before these results could be considered
as definitive. Serious applications would
require a greater level of detail and a deeper
examination of the complexities of public attitude. For instance, how would the public respond
to the loss of revenues?

The more important result in this study was
the demonstration of integrating human culture,
regional assets, and environmental pattern in
approaching an energy-environment conflict. On a
regional scale, problems are not just controlled
by economics, environmental degradation, or
energy demand. Humans pass laws, regulate utilities and interstate commerce, pass environmental
protection legislation, and countless other
activities which impact the environment and the
economy. This is the major reason we claim it is

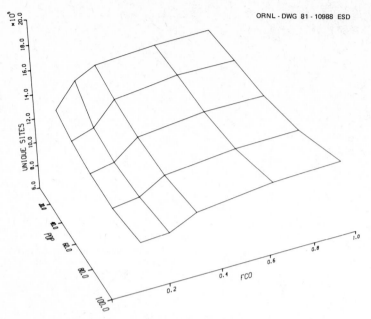

ORNL-DWG 81-10988 ESD

Figure 3.—Effect of variations in the public opinion (C_8) and the anti-wilderness pressures (FCO) on the unique sites remaining in the Overthrust Belt over a 100 year time span.

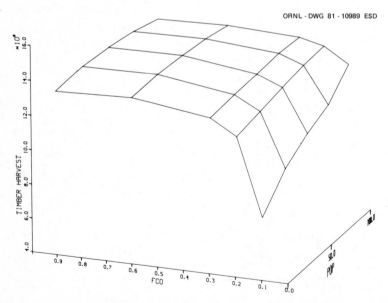

ORNL-DWG 81-10989 ESD

Figure 4.—Effect of variations in the public opinion (C_8) and the anti-wilderness pressures (FCO) on the amount of timber harvested in the Overthrust Belt over a 100 year time span.

necessary to include pattern, assets, culture, and externalities when addressing regional problems. Through application of this framework, we hope to learn how to attack regional problems.

ACKNOWLEDGEMENTS

Research sponsored by the Office of Health and Environmental Research, U.S. Department of Energy, under contract W-7405-eng-26 with Union Carbide Corporation. Publication No. 1801, Environmental Sciences Division, ORNL.

REFERENCES

Bell, E. F., K. N. Johnson, K. P. Connaughton, and R. W. Sassaman. 1979. Roadless area – intensive management trade-offs on the Bridger-Teton and Lolo National Forests. USDA Forest Service General Technical Report INT-72, 38 p. Intermountain Forest and Range Experimental Station, Ogden, Ut.

Green, Alan W., and Theodore S. Setzer. 1974. The Rocky Mountain timber situation, 1970. USDA Forest Service Resource Bulletin INT-10, 78 p. Intermountain Forest and Range Experiment Station, Ogden, Ut.

Hamilton, H. R., S. E. Goldstone, J. W. Milliman, A. L. Pugh, III, E. B. Roberts, and Z. Zellner. 1969. Systems simulation for regional analysis: an application to river-basin planning, 407 p. M.I.T. Press, Cambridge, Mass.

Hannon, B. 1973. The structure of ecosystems. J. Theor. Biol. 41:535-540.

Hendee, J. C. 1977. Public involvement in the U.S. Forest Service roadless-area review: Lessons from a case study. p. 89-103. In Sewell, D., and T. Coppock (eds.), Public Participation in Planning. John Wiley & Sons, Ltd., New York, N.Y.

Holling, C. S. (ed.). 1978. Adaptive environmental assessment and management. International Series on Applied Systems Analysis (IIASA). 377 p. John Wiley & Sons, New York, N.Y.

Isard, W. 1960. Methods of regional analysis: An introduction to regional science. 784 p. M.I.T. Press, Cambridge, Mass.

Isard, W. 1972. Ecologic-Economic Analysis for Regional Development. 270 p. The Free Press, New York, N.Y.

Isard, W. 1975. Introduction to Regional Science. 506 p. Prentice-Hall, Englewood Cliffs, N.J.

Kim, Y. V., and R. G. Thompson. 1978. An economic model-new oil and gas supplies in the lower 48 states. 110 p. Gulf Publishing Co., Houston, Tex.

Klopatek, J. M., J. T. Kitchings, R. J. Olson, D. Kumar, and L. K. Mann. 1980. An ecological analysis of the U.S. Forest Service RARE-II sites, ORNL/TM-6813, 227 p. Oak Ridge National Laboratory, Oak Ridge, Tenn.

Krutilla, J. V., and A. C. Fisher. 1975. The economics of natural environments. 292 p. The Johns Hopkins University Press, Baltimore, Ma.

Leontief, W. 1966. Input-output economics. 251 p. Oxford University Press, N.Y.

Mäler, K. 1974. Environmental Economics: a theoretical inquiry. 267 p. The Johns Hopkins University Press, Baltimore, Ma.

Mar, B. W. 1974. Problems encountered in multi-disciplinary resources and environmental simulation models development. J. Environ. Manage. 2: 83-100.

Mar, B. W. 1976. Philosophical concepts of RANN/RES projects. p. 69-75. In Regional Environmental Systems: assessments of RANN projects. National Science Foundation, NSF/ENV 76-04273, University of Washington, Seattle.

Nijkamp, P. 1977. Theory and application of environmental economics. Studies in Regional Science and Urban Economics. Volume 1. 332 p. North-Holland Publishing Co., Amsterdam.

Smith, C. A. 1976a. Regional economic systems linking geographical models and socioeconomic problems. p. 3-63. In Smith, C. A. (ed.), Regional Analysis. Economic Systems. Volume 1. Academic Press, New York, N.Y.

Smith, C. A. 1976b. Analyzing regional social systems. p. 3-20. In Smith, C. A. (ed.), Regional Analysis. Social Systems. Volume II. Academic Press, New York, N.Y.

The Institute of Ecology (TIE). 1977. Rocky Mountain environmental research: Problems and research priorities. Utah State University, Logan.

Tippee, B. 1980. Overthrust Belt action hot in and out of fairway area. p. 123-128. Oil and Gas J. July 29, 1980.

U.S. Department of Agriculture, Forest Service. 1978. RARE-II. Draft environmental statement. Roadless Area Review and Evaluation. 112 p. USDA Forest Service, Washington, D.C.

U.S. Department of Agriculture, Forest Service. 1979a. A report to Congress on the nation's renewable resources. REPA assessment and alternative program directions. 209 p. Washington, D.C. [Review Draft]

U.S. Department of Agriculture, Forest Service. 1979b. Timber management plan for final environmental statement. Bridger-Teton National Forest, Region 4, U.S. Forest Service, Jackson, Wy.

Voelker, A. H., H. Wedow, E. Oakes, and P. K. Scheffler. 1979. A systematic method for resource rating with two applications to potential wilderness areas, ORNL/TM-6739, 65 p. Oak Ridge National Laboratory, Oak Ridge, Tenn.

Watt, K. E. F., and J. L. Wilson. 1973. Regional modeling studies by the Environmental Systems Group. p. 1-36. In Land use, energy flow, and decision making in human society. Volume 3. Interdisciplinary Systems Group, University of California, Davis.

ENERGY DIVERSITY OF REGIONAL ECONOMIES[1]
by James Zucchetto[2]

Abstract. -- Several approaches are considered for incorporating the concept of energy diversity into the analysis of regional or national economies. Quantitative measures are presented for assessing a diversity vs. cost trade-off with reference to the island of Gotland, Sweden.

INTRODUCTION

Emphasis on the investigation of energy flows in a complex regional mosaic of both ecosystems and human systems leads naturally to the consideration of inputs of various kinds of energy and their diverse flows within a regional or national system (Odum, H.T. 1971, 1973; Zucchetto 1975; Browder et al. 1976; Odum and Odum 1976; Jansson and Zucchetto 1978; Zucchetto and Jansson 1979). Realistic analysis will require detailed data and complex models but there is some advantage to simplifying complex systems by defining simple indexes or measures which hopefully encapsulate important properties of the system. Both abstract and realistic approaches are considered in the present paper with regard to the energy diversity of a region. The present paper evolved out of an on-going project which has dealt with a broad energy systems investigation of the island of Gotland, Sweden.

DIVERSITY AND STABILITY

One goal in this study has been to incorporate ecological models, prinicples and concepts into the regional analysis whenever possible. The relationship between diversity and stability is addressed in much of the ecological literature (Connell and Orias 1964); Odum, E.P. 1971; May 1973; Odum, H.T. 1973; Pielou 1975; Odum and Odum 1976) and one school of thought considers high species diversity to enhance the "stability" of an ecosystem to outside perturbation: stability in this sense means the ability to recover from an external perturbation and resume its expected dynamic behavior. The more species in a system the more possibilities and strategies available to deal with changes and uncertainties in the environment, i.e., the more possibilities for establishing required feedbacks and interactions (Margalef 1968).

[1]Paper presented at the international symposium Energy and Ecological Modelling, sponsored by the International Society for Ecological Modelling. (Louisville, KY, April 20-23, 1981).

[2]University of Pennsylvania, Regional Science Department, Philadelphia, PA 19104 U.S.A.

Two particular formulations of the relationship between diversity and stability are thought to be relevant to the energy diversity of economies. Watt (1972) has discussed the diversity of food (energy) sources to an ecosystem and has attributed greater stability to the situation in which the diversity of food sources is greater. Thus, one could formulate a matrix with types of food represented by the rows and different species along the colums with any entry in the matrix, F_{ij}, representing the amount of food "i" eaten by species "j". H.T. Odum has put forth a rather different energy-diversity hypothesis (Odum, H.T. 1973; Odum and Odum 1976): the more energy made available to a system the more it will diversify and this diversification is a means for a system to capture other un- or underutilized energy resources in its tendency to "maxmimize the rate of useful energy inflows" (Lotka 1922) under conditions of competition. Diversification will cost energy and presumably the gains must outweight the costs. Several measures of diversity have been used in the ecological literature (Pielou 1975) with the following Brillouin measure being used for the diversity of "r" species among "q" habitats. In matrix notation, an element n_{ij} would represent the number of occurences of species i in habitat j. In addition,

$$\sum_{j=1}^{q} n_{ij} = \text{no. of occurences of species i}$$

in habitats of all kinds.

$$\sum_{i=1}^{r} n_{ij} = \text{no. of occurences of all species}$$

in habitat j.

$$N = \sum_{i=1}^{r} \sum_{j=1}^{q} n_{ij} = \text{total no. of occurences}$$

of all species in all kinds of habitats.

The Brillouin measure of diversity is given by:

$$H = \frac{1}{N} \log \left[\frac{N!}{\pi_i \, \pi_j \, n_{ij}!} \right] \qquad (1)$$

One could also define a food-species matrix with a similar food-species diversity index defined.

REGIONAL ENERGY DIVERSITY

In a similar fashion an energy source-activity matrix could be defined for an economic system if a matrix is constructed whose rows represent different energy types by source and the end-use activities of these energies are listed along the columns. For example,

sectors

source

		1	2	. .	m	
(1)	Oil	e_{11}	e_{12}	. .	e_{1m}	$E_{1.}$
(2)	Gasoline	e_{21}	e_{22}	. .	e_{2m}	$E_{2.}$
(3)	Coal					
(4)	Electric (from Oil)			e_{ij}		
(5)	Electric (from Wind)					
(n)	. . . .	e_{n1}	e_{n2}		e_{nm}	$E_{n.}$
		$E_{.1}$	$E_{.2}$	. .	$E_{.m}$	E

would represent a source-activity matrix. It is necessary to categorize existing energy sources as well as end-use activities. One must consider the source of the energy being used, e.g., electricity generated from nuclear, wind, coal or oil would be considered different sources. In addition, a distinction should probably be made between imported vs. internally available sources. Analogously to eq. (1) of the preceding section we define a diversity index as

$$H = \frac{1}{E} \log_e \left[\frac{E!}{\prod_i \prod_j e_{ij}!} \right] \quad (2)$$

where E = total energy used in system
e_{ij} = energy of type "i" used in activity "j".

$$\sum_{j=1}^{m} e_{ij} = E_{i.} = \text{total energy of type i.}$$

$$\sum_{i=1}^{n} e_{ij} = E_{.j} = \text{total energy use in activity j.}$$

H = diversity in nats.

This formula is not exactly analogous to eq. 1 because N is discrete and, in general, E is not. However, for all practical purposes one can consider integral units of energy use, e_{ij}, as a good approximation to the problem of distributing a given number of energy units, E, among the n x m positions of the matrix. The more evenly spread over all positions is the total energy, E, the higher will be H; conversely, if the system is dominated by a few e_{ij}'s having large values, and the rest small, the source-activity diversity will be small. The measure is usually approximated by a simpler function if E and the e_{ij}'s are large which will normally be the case for a real system. Using the Stirling approximation (log x! = x log x −

x) to evaluate the logarithms of factorials the following approximations to eq. (2) is obtained:

$$H = - \sum_{i,j} \left(x_{ij} \log x_{ij} \right) \quad (3)$$

where $x_{ij} = \frac{e_{ij}}{E}$ = fraction of total energy E

of type i used by activity j.

The diversity given by eq. 2 or 3 will depend on the size of the system, i.e., the total energy use, E. H from eq. 3 can be used to compare systems of different size by normalizing with repsect to the largest value of H, namely,

$$H_{max} = \log (n \times m).$$

This approach to energy diversity has a number of reservations. Within any given sector, j, energy types e_{1j}, e_{2j}, $\cdots$ e_{nj} are

being used and if the total energy used by this sector, $E_{.j}$ is evenly spreadout over these energy types then an interruption in any one energy source would have a small effect on sector j. But this would be true if there were substitutability between different energies; if there weren't then an interruption in an essential energy type, even if the total use of that energy were small, would be disastrous to that sector and the diversity measure loses some meaning; thus the energy quality of these different energy sources is important. Also, H as given by eqs. (2) or (3) is symmetrical in the sense that if the energy source-activity matrix is mirrored by letting $e_{ij} = e_{ji}$, H is the same but the situation is very different.

However, as with ecosystems analysis as given by equation (1), the measure may at least be used as a comparative measure between systems or as an index of change in time. Modifications to this diversity index and more realistic approaches will be considered in the following sections.

THE COSTS AND MAXIMIZATION OF DIVERSITY

So far, only brief mention has been made of the costs of diversification. In Odum's (1973) terms rich energy sources will lead to increased diversity and this diversification will, in turn, stimulate the capture of more available energy resources. However, the support of this diversity entails costs in the form of structure, organization, maintenance and operation. For example, consider the case of home heating provided by electricity supplied from a power grid as well as on-site solar energy. The electricity might be used during extremely cold and/or cloudy periods while the sun is used as a supplement whenever possible. It may turn out that the cost of supporting both systems may be greater than using the electricial system alone -- this would be an example of increased diversity facilitating the capture of an alternative energy type with consequent higher costs. One possible way to consider cost and diversity together would be to calculate the cost (energy or money) of a particular diversity pattern as

$$\text{Total Cost} = C = \sum_{i,j} P_{ij} e_{ij} \qquad (4)$$

where P_{ij} = discounted price of energy type i used in activity j (or energy cost).

If it were desired to maximize diversity subject to the cost constraint, eq. (4), then we would have

$$\max -\sum_{i,j}\left(x_{ij} \log x_{ij}\right)$$

$$\text{subject to } E \cdot \sum_{i,j} P_{ij} x_{ij} = C = \text{total cost} \qquad (5)$$

$$\sum_{i,j} x_{ij} = 1$$

If we form the Lagrangian by introducing lagrangian multipliers on the contraint equations the following problem is derived:

$$L = -\Sigma x_{ij} \log x_{ij} + \lambda_1 \cdot$$

$$(C - E \Sigma P_{ij} X_{ij}) + \lambda_2 (1 - \Sigma x_{ij}) \qquad (6)$$

The conditions for a maximum are

$$\frac{\partial L}{\partial x_{ij}} = 0; \quad \frac{\partial L}{\partial \lambda_1} = 0; \quad \frac{\partial L}{\partial \lambda_2} = 0$$

and the solution is

$$x_{ij} = e^{-(1 + \lambda_1 E P_{ij} + \lambda_2)}$$

$$C - E \Sigma P_{ij} X_{ij} = 0 \qquad (7)$$

$$1 - \Sigma x_{ij} = 0$$

and these 3 equations can be solved for the distribution of enegy, x_{ij}. It is seen that as the price, P_{ij}, goes higher x_{ij} goes down exponentially or

$$\frac{x_{ij}}{x_{\kappa\ell}} = e^{A (P_{\kappa\ell} - P_{ij})} \qquad (8)$$

$$\text{where } A = \lambda_1 E.$$

Another measure, h = H (eq. 3)/C could be defined which would be diversity per unit cost and if one tries to maximize diversity per unit cost then we have

$$\max h = \frac{H}{C} = -\frac{\Sigma x_{ij} \log x_{ij}}{E \cdot \Sigma P_{ij} X_{ij}} \qquad (9)$$

$$\text{subject to } E \cdot \Sigma P_{ij} X_{ij} = C = \text{total cost} \quad (10)$$

$$\Sigma x_{ij} = 1$$

A solution similar to the previous problem is obtained:

$$\frac{x_{ij}}{x_{\kappa\ell}} = e^{A (P_{\kappa\ell} - P_{ij})} \qquad (11)$$

$$\text{where } A = \lambda_1 C E$$

If there is no constraint on the total cost of the system or if the price of an energy use, P_{ij}, is a function of the amount of energy used, i.e., $P_{ij} = f(e_{ij})$ then the maximization of h

(eq. 9) results in a system of non-linear, simultaneous algebraic equations. Similarly, one could introduce constraints on total energy of type i or total consumption in sector j. These are probably rather realistic constraints to consider but this problem also results in a non-linear system of algebraic equations which, in principle, can be solved numerically.

A MORE REALISTIC TOTAL SYSTEM VIEW

The diversity measures presented in the preceding sections give some view of the energy structure of an economy and the way cost may enter into a consideration of energy allocation under the condition of maximum diversity. Following from the ecological approach it has been tacitly assumed that diversity is a "good" thing in that it should confer stability upon the system by spreading out the impact over sectors and that diversity indexes serve as a reasonable measure. These indexes certainly can serve as some comparative measure as to how energy is distributed within a system but they really don't address the stability of a given system: for the purposes of planning the stability of an economic system should be understood as a function of the energy source-activity matrix. Stability of an economic system could be defined as trying to maintain a given level of output or minimizing decreases in output for some anticipated interruptions in energy supplies. Thus, rather than just attempting to find a distribution of energy use which maximizes diversity or diversity per unit cost, one must also consider the likely impacts on output or production of the system.

Consider the problem in somewhat more realistic detail. There is going to be a certain level of economic output which is required or desired by an economic system. If one uses a simple input-output approach then the level of output required from economic sectors will be given by an output vector, X:

$$X = (I - A)^{-1} Y \qquad (12)$$

X = vector of output from n sectors

I = n x n identity matrix

A = n x n matrix of technical coefficients

Y = final demand (private consumption government & exports)

A planner would consider that a given level of desired consumption by consumers, government and export markets, Y, would require the output, X, given by eq. (12). It is tacitly assumed that there are not economic or environmental constraints such as energy, land, pollutant generation or levels of water quality. Constraints could be introduced if required. The activities or economic sectors which exist in a given region or economy will, for the most part, be a given: for example, on Gotland, Sweden, activities such as cement production, agriculture and forestry are there and will remain for the forseeable future. Thus, the problem becomes: given certain activities or economic sectors what is the "best" distribution of energy supply from different sources to meet the needs of these activities. Now, a criterion of "best" would be the maximization of net output for the whole system or rate of return such as

$$\max (X - C) \text{ or } \max \left(\frac{X - C}{C} \right)$$

where X = total output of the system
C = energy cost.

and if this criteria were considered over time then discounting to present value would be introduced. Now, given that a certain level of production, X_i, is required for each activity then existing technology will dictate that certain levels of energy of different kinds are required, e.g., oil of different grades, kerosene, gasoline, wood, electricity, gas, etc. Now, the question becomes what sources should supply these energies in order to maximize the criteria given or to minimize the disruption in economic output. Thus, for example, given that an activity like industry needs 10 units of electricity, what proportion of this electricity should be generated from oil, coal, wind, wood or imported. Consider that electricity will be supplied by being generated from oil and wind. We can describe the availability of each type of energy-source and its uncertainty by some form of probability distribution which would give the probabilities for various levels of energy interruption for each source so that P (a ⩽ E ⩽ b) could represent the probability that the energy supply lies between two limits a and b and each energy source probability distribution would have an expected value and variance associated with it; the more uncertain the supply the larger the variance. Now, energy is an input to each of the economic sectors and the output from the economic sectors can be found as a function of inputs, namely (Ghosh 1958):

$$x^1 = v^1 (I - \overline{A})^{-1}$$

where $x = [x_i]$ = vector of gross output from each sector; x^1 is its transpose.

$v = [v_i]$ = vector of primary inputs; v^1 is its transpose.

$\overline{A} = [\overline{a}_{ij}]$ where $X_{ij} = \overline{a}_{ij} x_i$ i.e.,

interindustry flows are proportional to the output of the producing (selling) industries.

The primary inputs, v_i, to each sector x_i can be disaggregated into component parts with different types of energy included. Thus, output x can be related to energy inputs and the probability distributions for energy can be translated into distributions for total output from the economic system. In reality, for a given region, there will probably be only a small number of energy source-activity combinations which are realistically feasible. Thus, for Gotland, Sweden, electricity may be generated from wind, oil, coal, or imported, and technological and economic considerations will eliminate all possible combinations. A finite number of different energy source combinations will result in a corresponding number of system output assessments. The net benefit or rate of return can be calculated for each combination, the one with the largest value judged to be the "best".

A SIMPLE EXAMPLE

Consider a simple economy which consists of only two sectors, industry and agriculture, and which uses electricity from two possible sources, namely, oil and wind. Approximate data for Gotland, Sweden in aggregated form is used to provide a realistic example. The input-output table 1 is as follows:

Table 1--Input-Output table for Gotland Sweden in million Swedish Kronors (1972).

From	To (1)	(2)	Final Demand	Total Output
(1) Industry	6	3	505	514
(2) Agriculture	162	16	32	210
Value Added				
I_1) elec(oil)				
	7.6	3	25	35.6
I_3) elec(wind)				
other	338.4	188	51	577.4
Total Outlay	514	210	613	1337

The value added represents inputs to the production processes in industry and agriculture, one part of which is energy.

Following the discussion in the previous section equations representing this system can be written as follows

$$x_1 = \bar{a}_{11} x_1 + \bar{a}_{21} x_2 + v_1$$

$$x_2 = \bar{a}_{12} x_1 + \bar{a}_{22} x_2 + v_2$$

where

$$\bar{a}_{ij} = \frac{x_{ij}}{x_i}$$

or

$$[x_1, x_2] = [x_1, x_2] \begin{bmatrix} a_{11} & a_{12} \\ a_{21} & a_{22} \end{bmatrix} + [v_1, v_2]$$

$$[x_1, x_2] \begin{bmatrix} 1 & 0 \\ 0 & 1 \end{bmatrix} - [x_1, x_2] \begin{bmatrix} .012 & .006 \\ .77 & .076 \end{bmatrix} = [v_1, v_2]$$

or

$$x^1 = v^1 (I-\bar{A})^{-1}$$

$$[x_1, x_2] = [v_1, v_2] \begin{bmatrix} 1.017 & .0066 \\ .848 & 1.087 \end{bmatrix}$$

Thus, knowledge of the inputs, v_1 and v_2, allows calculation of the total output $(x_1 + x_2)$. Now, suppose this system is being supplied from electricity generated from oil or wind and there is some interruption in the source of energy of a given percent. If it is assumed that all processes are controlled by energy and that a "d" percent reduction in energy results in a "d" percent reduction in all inputs then the output of the system will be given by

$$x^1 = (v - dv)^1 (I-A)^{-1}$$

In general, this reduction in energy supply will probably be described by a probability distribution with expected value and variance. The expected value of output, $E(X_1)$ and $E(X_2)$, will be given by the following equations for the problem presented above:

$$E(X_1) = 1.017 \ E(V_1) + .848 \ E(V_2)$$

$$E(X_2) = .0066 \ E(V_1) + 1.087 \ E(V_2)$$

where $E(V_1)$ and $E(V_2)$ are the expected values of

the inputs, V_1 and V_2, which is determined based

on the anticipated probabilities of interruption of the energy supply. Suppose we are given the probability distribution for the interruption of two energy sources, I_1 and I_2, as follows:

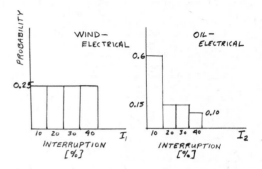

In this example, the expected interruption of electrical energy from wind is greater than from oil, and the cost of both are the same, then in order to maximize $E(X_1) + E(X_2)$ minus energy cost, the strategy will be to use all oil to minimize interruption. If there is a limitation on the amount of oil available which, along with the probability of interruption, still leaves excess capacity for production, then diversification into the wind electrical can enhance production and a more diverse system will be selected for. For the example above, all energy inputs in oil with the assumed probability distribution above leads to a total output of $E(X_1) + E(X_2) = 574$. What should also be considered is the temporal variability of the various energy sources: for example, energy for agriculture may be required for keeping livestock warm in winter, a time when wind energy is plentiful on Gotland and there is high demand for oil for the purposes of residential heating. Thus, it may make much sense to diversify because of the temporal differences, a property which is difficult to capture in the above model formulation.

DISCUSSION

An energy source-activity measure, with or without the inclusion of cost, can at least serve as some macro-measure for the comparison of economies or the evolution of a given system. Do systems diversify with time and is diversification an increasing function of total energy are questions which could be answered empirically. Other questions would include: how does energy diversity adjust to perturbations in existing energy sources. Many different types of programming or simulation approaches could be formulated in the vein of the preceding discussion: one could consider different objective functions, contraints on energy or environment, or time fluctuations of energy flows. However, in reality, a pragmatic planning approach would probably be constrained by many factors so that a choice would only need to be made between a small number of alternative source-activity matrices. In addition, external

548

economic, technical and supply considerations would probably dictate many of the entries in the matrix, namely, many of the e_{ij}'s. Thus, given a matrix, estimates would need to be made about the likelihood of interruptions in sources and consequent impacts on total system production. Thus, the choice would be one among a small set of realistically realizable energy source-activity matrices. Realistic would imply that a complex of factors have already entered the decision to generate certain levels of energy and deliver it to end-use points. In closing, then, energy diversity as defined in the above paper becomes an additional planning criteria in the evaluation of economic systems. Although much work remains to be done on the relation of this diversity to system vulnerabililty, its explicit recognition at least adds another dimension to planned organization of a total system.

LITERATURE CITED

Browder, J., Littlejohn, C. & Young, D. 1976. The South Florida Study. South Florida: Seeking a Balance of Man and Nature. Center for Wetlands, University of Florida, Gainseville, FL and Bureau of Comprehensive Planning Division of State Planning. Tallahassee, FL. 117 p.

Connell, J. H. & E. Orias. 1964. The Ecological Regulation of Species Diversity. American Natur. 98: 399-414.

Ghosh, A. 1958. Input-Output Approach in an Allocation System. Economica (February): 58-64.

Jansson, A. M. and J. Zucchetto 1978. Energy, Economic and Ecological Relationships for Gotland, Sweden: A Regional Systems Study. Ecological Bulletins (Swedish Natural Science Research Council, Stockholm), No.28.

Lotka, A. J. 1922. Contributions to the Energetics of Evolution. Proc. Nat'1. Acad. Sci. 8: 147-155.

Margalef, D. R. 1958. Information Theory in Ecology. Gen. Syst. 3:36-71.

May, R. M. 1973. Stability and Complexity in Model Ecosystems. Princeton University Press, Princeton, N.J.

Odum, E. P. 1971. Fundamentals of Ecology. W. B. Saunders Co. Philadelphia, Pa.

Odum, H. T. 1971. Environment, Power and Society. John Wiley, New York, N.Y.

Odum, H. T. 1973. Energy, Ecology and Economics. AMBIO 2: 220-227.

Odum, H. T. & E. C. Odum. 1976. Energy Basis for Man and Nature. McGraw-Hill Co. New York, N.Y.

Pielou, E. C. 1975. Ecological Diversity. Wiley-Interscience. New York, N.Y.

Watt, K.E.F. 1972. Man's Efficient Rush Toward Deadly Dullness. Natural History 81(2):74-82

Zucchetto, J. 1975. Energy-Economic Theory and Mathematical Models for Combining the Systems of Man and Nature, Case Study: The Urban Region of Miami, Florida. Ecological Modelling 1:241-268.

Zucchetto, J. and A. M. Jansson. 1979. Integrated Regional Energy Analysis of the Island of Gotland, Sweden. Environmental and Planning A, 11:919-942.

THE ECOLOGIC/SOCIOECONOMIC INTERFACE OF BIOFUEL DEVELOPMENT IN HAWAII: SILVICULTURE ON THE SLOPES OF MAUNA KEA AND MAUNA LOA[1]

Prahlad Kasturi[2]

Abstract.--This paper discusses the initiation of bio-fuel development activity in Hawaii's ecosystem; assesses its potential impact on the flow of environmental goods and amenity services and develops strategies for optimal use and control of the environments. The tradeoffs between benefits derived from preserving virgin lands and the costs associated with regulating commercial use of such lands are dealt with in an intertemporal and intergenerational context. A case study of the environmental impacts of energy tree plantations on the big island of Hawaii is presented. This is accomplished by applying a modified version of the Arizona Economic and Environmental Tradeoff Model (ATOM) to various landsites identified as being favorable for biomass production by the Biomass Energy Study team (1977). The implications for developing countries in Asia and the Pacific that are presently formulating plans for energy self sufficiency via the biomass option is noted.

INTRODUCTION

The Physiobiotic Environment

The Island of Hawaii is the biggest island in the Hawaiian Archipelago. It extends over an area of 10,480 Km^2 and is approximately eight hundred thousand years old. The island has been formed by the sequential eruptions of five volcanoes, two of which are still active. As can be expected, much of the island is mountainous in nature and the soils correspondingly have developed from volcanic ash and basaltic lava over a wide range of climatic conditions. Most of the land lies very close to the coast, less than 8 percent of the land being more than 33 km from the sea. Hence, there is a pronounced marine influence on the climate.

The remarkable altitudinal differences existing in various areas of the island show a tremendous range of environmental diversity. These extend from sub-humid, summer dry environment in the lower elevations to montane and sub-montane rain forest regions in the middle reaches. The evergreen tropical forest is replaced upward (at about 1700 m) by mountain parkland in a cool dry forest environment which is also referred to as the Heath zone. Above this zone (at about 3000 m) is the alpine tundra, a cold desert environment characterized by a sparse vegetation of moss and lichens. At the summits of these mountains (over 4000 m) there is a perennial snow. In contrast, the leeward side of the island is the rain shadow region with very little vegetation and forms the strand zone.

[1]Paper presented at the international symposium Energy and Ecological Modelling, sponsored by the International Society for Ecological Modelling. (Louisville, Kentucky, April 20-23, 1981)
[2]Prahlad Kasturi was research assistant at the Environment and Policy Institute of the East-West Center, Honolulu, Hawaii U.S.A.

The author wishes to acknowledge the helpful suggestions of Dr. Gerald Marten, EAPI; Dr. Dieter Mueller-Dombois, Dept. of Botany; and Mr. Jagdish Bahati, Ph. D. scholar, Dept. of Agr. and Res. Econ., University of Hawaii. Comments and criticism were received in the course of the study from Dr. Robert Merriam, Hawaii Division of Forestry; Craig Whitesell and Roger Skolman, U.S. Forest Service; Linda Christanty, Research intern EAPI and Susan Miyasaka, of the Bioenergy Development Corporation. The author bears responsibility for any remaining errors or deficiencies in the final product.

This paper is patterned after Norman Meyers' article on wildlife conservation in Africa appearing in the Journal of Environmental Economics and Management, Vol. 1, No. 4, December 1974.

The native terrestrial biota is estimated between 10,000 and 15,000 species. These include over 2000 species of plants, 5000 species of land arthropods, 1000 species of snails and over 30 species of birds. There is a high degree of endemism in the island's biota together with intricate coadaptation between the island's flora and fauna. The community structure is complex, though less diverse compared to tropical mainland communities (Mueller-Dombois, Bridges and Carson 1980).

Thus the island of Hawaii due to its geographic isolation, small size (relative to continental land masses), recent age, unique evolution and extra ordinary community structure forms an insulated island ecosystem (Mueller-Dombois 1975). This poses serious concerns in stability-fragility relationships stemming from pervasive human activities related to agriculture, livestock husbandry, hunting and timber growing.

Hawaii has a resident population of 80,900 people with a third of this population living in urban concentrations in and around the city of Hilo and the towns of Kailua, Papaikou and Waimea (DPED 1979a). The rest of the rural population earns and makes its living in plantation towns and villages along the Hilo-Hamakua coast and within the predominantly agricultural districts of North Kohala, Puna and Ka'u.

Employment gains in a growing tourist industry together with other perceived opportunities in diversified agriculture, aquaculture, energy development and ocean mining projects have led to a significant in-migration to the island (DPED 1979b). The annual population increase of 3.5 percent coupled with persistent efforts to stimulate growth in the island's economy has resulted in burgeoning pressure on the land resource base and led to changes in the natural environment. The pressure on environmental resources, furthermore, is likely to intensify in the future as much of the state's population growth is to be accomodated on this island and in Maui county. This has been evident from the fact that there has not only been an increase in the number of people utilizing the environment but the per capita use of resources has also grown markedly. This has been especially true of per capita water consumption and solid waste generation (DPED 1978).

Over the last hundred years, there has been a sustained use of large tracts of land for intensive sugarcane cultivation and a modest timber industry. This system of monoculture has altered considerably the character of Hawaii's shoreline, destroyed endemic flora and fauna, restricted accessibility to certain amenity services provided otherwise by a natural environment, and raised a host of environmental quality problems.

Some of the sources of environmental stress which have reduced air, water and land quality on the island are directly traceable to agricultural deforestation and livestock grazing activities (DPED 1977a). Soil, bagasse and other waste discharges which were previously not monitored, have caused siltation of coral reefs, reduction of photosynthesis and lowering of oxygen content in some in-shore areas. Intensive mechanized cultivation has caused soil erosion and sedimentation problems affecting light and water permeability. Leaching and runoff of sewage used in irrigation waters as well as agricultural chemicals has inhibited growth of certain coastal species though they could also presumably have stimulated growth in others. Agricultural burning has been another major source of air pollution but its effect remains largely unknown due to lack of knowledge regarding types of air pollutants and their impacts on total air quality. Additionally, deforestation for cultivation and construction purposes has altered wildlife habitats, fostering the spread of exotic plant species into natural areas. Water diversion for agricultural purposes and construction of irrigation channels and canals have destroyed stream fauna and affected the habitat of waterbirds some of which are endangered species. Finally, thermal effluence from sugar factories in the form of flowing water heated above ambient levels, has under some circumstances destroyed nearshore marine and estuarine ecosystems. Some of the irreversible changes occurring in the island's ecosystem, as a direct result of man's activities have significantly reduced both the quantity of amenity resources and the quality of the environment.

These changes put into perspective the current land use policies and the need for improving zoning and other land use ordinances as well as maintaining a Natural Areas Reserve System (DLNR 1975). However, an interesting issue develops in resolving conflicts between the benefits derived from conserving and preserving specific land and water areas in perpetuity for aesthetic, ecologic, geologic and educational reasons (benefits which in great measure are also captured by tourist visitors to the island) and the costs incurred (mostly by residents) through the provision for conservation districts which restrict the productive and commercial use of land-related resources in these districts for growing timber or expanding agriculture. These constraints, which result in foregone earnings and employment opportunities, bear upon the current interest in biofuels development in the context of energy self sufficiency for the island.

Economic Growth, Quality of Life
and the Energy Fix

Hawaii's population is expected to increase to 95,200 by the mid eighties and approximately 105,100 in 1990. This means that over 4700 of the 9000 new jobs needed over the next decade must be found over the next five years. Currently, agriculture on the island employs 18 percent of the total labor force of 36,500 civilian workers. Despite the low population density of 8.5 people per sq. km. as compared to 59.2 per sp. km. for the entire state, the per capita available cropland is only 1.77 acres. Tourism and corporate agriculture have borne the brunt of the economic growth process in the past. However, the annual unemployment rate averaging 9.2 percent for the last decade indicates near exhaustion of potentials for employment in these industries. It is evident that additional employment could be generated only by enhancing the economic base of the county. This implies a need to diversify and minimize dependency on a few key sectors.

Development potentials currently exist for energy production, science and research related endeavors, diversified agriculture operations, heavy industry, international commerce and commercial forestry and fishery (DPED 1977b). The constraints to rapid growth arise mainly from high costs of labor, materials, energy and transportation. Additional constraints arise from institutional factors such as zoning regulations and availability of financing. For instance, fostering industries such as energy production, diversified agriculture and commercial forestry, whilst requiring large outlays of capital, also implies far reaching changes in land use including possible additions to cropland from lands currently zoned conservation land.

Table 1 shows that nearly two thirds of all conservation and agricultural land in the state is on the island of Hawaii. A major push for zoning changes to increase crop production is expected as a result of the county's desire to attain energy self sufficiency together with a broadening of its economic base. Hawaii county now derives 40 percent of its total energy needs from electricity generated by burning bagasse, the waste from sugar processing. Geothermal energy and ocean thermal gradients are major candidates for alternate energy development in the near future, but biomass conversion to produce ethanol for gasohol, and growing silvicultural biomass farms for fuel to generate electricity, appear more imminent (DPED 1980).

In both the public and private sectors there are already several energy tree plantations

projects underway or in planning (DLNR 1979, Bioenergy Corp.1980b). They are on a relatively small scale, present plans, not amounting to more than a few thousand acres in the aggregate, and they are directed at clearly defined, small-scale needs, such as supplementation of bagasse for generating electricity and heat in sugar processing plants and supplementation of other sources of fuel for public utilities on the less populated islands. As such they are making a valuable contribution to Hawaii's energy self sufficiency efforts. Although the discussion which follows includes considerations which apply to any scale of energy farm, the energy acreages of the energy tree farms which are currently planned or underway in Hawaii are not great enough to generate significant diseconomies of the sort discussed below.

The concept of establishing large-scale silvicultural operations on the island, while attractive from energy and employment perspectives, does pose the problem of reconciling "economic growth characterized by stability and diversity fulfilling the needs and expectations of Hawaii's present and future generations" with the stated need for a "desired physical environment, characterized by beauty, cleanliness, quiet, stable natural systems, and uniqueness, that enhances the mental and physical well being of the people." These two objectives, embraced by the general framework of the Hawaii State Plan, typify the aspiration of many tropical and semi-tropical countries in Asia and the Pacific. The socioeconomic problems, population pressure, mounting unemployment and insufficient agricultural production interfaced with consequences resulting from damaged ecologies and bespoilt environments are being given a new twist by the energy dilemma. Thus the planned silvicultural operations on the lower slopes of Mauna Kea and Mauna Loa in Hawaii bring to our attention the need to deal with questions of resource use and sound environmental management that are questions of national and international concern.

Table 1.--Estimated acreage of land use districts

Land use classification	State of Hawaii	County of Hawaii	Percentage of total
Urban	151,929.6	34,457.0	23
Conservation	1,976,105.9	1,309,693.5	63
Agricultural	1,974,229.8	1,228,637.5	63
Rural	9,234.7	612.0	7
Total area	4,111,500.0	2,572,400.0	63

Source: The State of Hawaii Data Book, 1979.

Irreversibility, Externalities and
Pollution Control Strategies

As in other similar situations in Asia and
the Pacific, the possibility of growing Eucalyptus
plantations on pristine lands in Hawaii county,
raises interesting questions regarding economic
valuations of environmental resources and issues
deriving from public attitudes toward biomass
energy development. A preliminary assessment can
be made if it could be assumed that the planning
authority (or decision maker) has as its objective
to maximize the present value of net social
returns from such development. The related parameters of a biofuel development activity would
then include gains from expanded forestry,
external costs of pullution, loss of option values
and shadow price evaluations of non-market goods
such as increased economic security and reduced
dependency on outside sources for fuel. A
related approach lies in estimating the resource
savings, if any, resulting from an energy tree
farm in lieu of the lowest cost alternative
source.

Hawaii county residents have experienced a
steady improvement in living standards throughout
the past decade, resulting in a per capita income
of $6687 in 1978 (DPED 1979a). The high levels
of income have translated into tremendous
pressures for both energy and recreational
resources. It has been forecast that in the
period 1985-1990 the demand for electricity will
grow by 4.71 percent to total 5055 x 10^9 Btus and
that despite a percentage reduction in the demand
for gasoline due to adoption of conservation
measures, Hawaii county will still require 32
million gallons of that fuel. The tax receipts
on fuels alone in Hawaii county amounted to
$2,045,387 in the fiscal year 1978-79 (Dept. of
Taxation 1979). Since virtually almost all
conventional fuels on the big island are from
imported feedstock this represents a huge drain
on the county's financial and developmental
resources and reflects an adverse balance of
trade. Thus there exist cogent, convincing and
compelling reasons for the early establishment of
biofuel activities such as an expanded silviculture program on the island. On the other hand,
it has also been estimated that land-based,
resource-oriented recreation presently accounts
for 30 percent of the participation rate for
outdoor recreation activities of the residents
(Jackson 1976). These recreational activities
include hiking, back packing, wilderness
experience, environmental, historic and natural
appreciation, freshwater sports and hunting. The
increase in the per capita demand, along with
increases in resident and tourist populations, is
likely to result in growing aggregate demand for
these non-augmentable services of Hawaii's natural
environment. In the future, improving living
standards, advances in technology and associated
lifestyle changes will place increasing demands on
Hawaii's scenic and pristine environments and
create growing relative scarcity. It is interesting to note in this context, that some of the
potential sites for biomass plantations are lands

currently zoned as conservation district (Biomass
Energy Study Team 1977).

The rationale for maintaining conservation
lands appears in chapter 205-2, volume 3, Hawaii
Revised Statutes, 1968, as amended and sets its
criteria as follows:
Conservation districts shall include areas
necessary for protecting watersheds and water
resources; preserving scenic and historic sites;
providing park lands, wilderness and beach;
conserving endemic plants, fish and wildlife;
preventing floods and soil erosion; forestry;
open space areas whose existing openness,
natural condition, or present state of use, if
retained, would enhance the present or potential
value of abutting or surrounding communities, or
would maintain or enhance the conservation of
natural or scenic resources; areas of value
for recreational purposes, and other related
activities, and other permitted uses not
detrimental to a multiple use conservation
concept.

A large scale silviculture program could
conflict with many of the conservation goals
implicit in the above paragraph. Apart from
adverse modifications of the visual environment due
to construction of conversion facilities, storage
dumps and the movement of heavy trucks and other
farm vehicles, the establishment of a monocultural
biomass tree farm could have implications for
system stability (cultural or ecological homeostasis), cultural diversity and the preservation
of genetic information. Table 2 provides generic
information on the potential effects of silvicultural biomass farms on environmental quality
and the ecosystem.

There has been a pervasive tendency among
planners, administrators and legislators to regard
silviculture and agriculture as being compatible
with a multiple use conservation concept. This
essentially captures the thinking of the so called
"non purists" and has been legislated in two
separate Acts, Act 234 passed in 1957, and Act 187
passed in 1961, which have since served to change
the land use concept in Hawaii from preservation
of the forest reserve to use of land in conservation districts (DLNR 1975). The "purists" on
the other hand, recognize the implications of
technological advance associated with the above
uses and what it portends for the undisturbed
environment whose fixed supply is being compounded
by severe use intensity in recreational pursuits as
well as increase in the number of people demanding
such services (Krutilla and Fisher 1976). In the
absence of restorative technology, the establishment
of a large scale silviculture industry for harvesting biomass could involve important irreversibilities
and changes. In any case, the process of developing
virgin and pristine lands for such a scheme would
surely involve the loss of the attribute of
authenticity in the amenity services yielded by the
increasingly limited natural environment in Hawaii.
There are signs nationally and in the state of
Hawaii that this user group placing a high value on
the originality, genuineness and the primitive

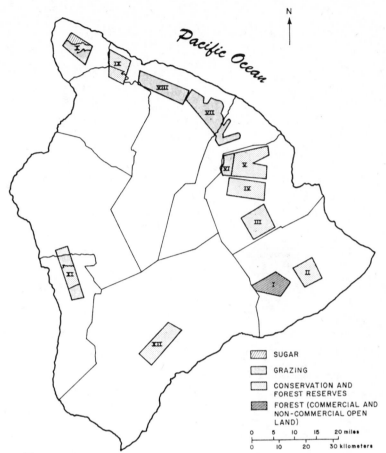

N

Pacific Ocean

SUGAR

GRAZING

CONSERVATION AND
FOREST RESERVES

FOREST (COMMERCIAL AND
NON-COMMERCIAL OPEN
LAND)

| 0 | 5 | 10 | 15 | 20 miles |

| 0 | 10 | 20 | 30 kilometers |

Figure 1.--Land Use for the Representative Site on the Island
of Hawaii

Source: Biomass Energy Study Team, Biomass Energy for Hawaii,
    Vol., IV. Institute of Energy Studies, Stanford
    University. February 1977.

nature of undisturbed and unmodifed native lands
is gaining political strength with changes in
income, education and urban composition of society.
This appears to warrant a review of existing land
use laws in light of changing public opinion in
the last two decades.

Analytical Significance

An analysis of Hawaii's biofuel development
program has implications for problems of resource
conservation and land use in virtually all
communities of the Asia and Pacific regions.
These largely rural communities depend heavily on
gathered firewood, dung and manure for their
energy requirements. The national governments of

countries within the region (both developed and
developing nations) have formulated plans in
recent years that provide for enhanced recovery
and increased utilization of fuelwood and other
crop residues (Hayes 1977, Siddiqi 1979, Asia
and Pacific Dev. Inst. 1979) (see table 3).

Biofuel development programs in these
countries stress one or more of the following
elements: 1) Increasing biomass production and
harvest from alternate agricultural and silvicultural crops; 2) Improving designs of stoves,
burners and plants that promote efficiency of
biomass utilization and 3) Developing market
structures to regulate the still largely non-
commercial uses of biomass. Although the state
of Hawaii is not economically in the same league

Table 2.--Potential effects of silvicultural biomass farms

Ecological Impact Category Potential Pollutants	Intensity of Impact		
	Site Preparation	Production, Harvesting, and Conversion	Decommissioning
Air Quality			
Diesel and Gasoline Engine Emissions	Low	Medium	Low
Particulate Production	Medium	Low	-- Low
Gaseous Emissions from Conversion Facility	-	Medium[1]	-
Surface Water			
Soil Erosion Resulting in Siltation	High	Medium	Low
Runoff Volume Increase	Medium	Low	Low
Runoff Temperature Increase	Low	-	-
Fertilizer and Soil Nutrient Enrichment of Runoff	Low	Medium	-
Herbicide Contamination of Runoff	-	-	Low
Pesticide Contamination of Runoff	-	Low	-
Effluent from Conversion and Sanitary Treatment System	-	Medium[1]	-
Ground Water			
Change in Water Table Depth	Low	-	Low
Fertilizer Accumulation	-	Low	-
Pesticide Contamination	-	Low	-
Ecological			
Soil Biota Change	Low	Low	Low
Aquatic Biota Change	Low	Medium	-
Changes in Wetland Areas	Low	Low	Low
Wildlife Disruption Due to Noise, Traffic, Gaseous Emissions	Medium	Medium	Low
Wildlife Habitat Disruption	Medium	Low	-
Vegetational Community Disruption	Medium	-	-

[1]This impact may change depending on research underway to further characterize and control conversion releases.

Source: The Mitre Corporation. Silvicultural Biomass Farms, Vol., IV. TR-7347. May 1977.

as some of the more disadvantaged nations of Asia and the Pacific, its virtual reliance on imported energy feedstock serves to highlight the vulnerability of these nations to future shortages and exorbitant pricing of conventional petroleum. Some of these countries already face severe

Table 3.--Roundwood equivalent of consumption in Escap region (million m3)

	Continuing Trend				High Assumption	
	1961	1971	1981	1991	1981	1991
ASIA FAREAST REGION						
Industrial Roundwood	148	223	310	500	320	540
Fuelwood	420	470	600	765	600	765
Total Roundwood	568	690	910	1265	920	1305

Source: FAO, Development and Forest Resources in the Asia and Far East Region, (Rome 1976), p.57.

ecological problems of unimpeded deforestation and desertification resulting from man's quest for food, fuel and fiber. These same countries have strategies for tourism development based on epitomizing the western man's dream of "getting away from it all" which fact is borne by the imagery of their travel advertisements (Dasmann et al. 1974). Thus problems of both socio-economic and politico-cultural variety could derive from externalities generated from accelerated and uncontrolled biofuel development programs (Eckholm 1976). It thus becomes important to explore the interrelationship between development and conservation based on an understand of ecology and the persistence of threatened ecosystems within the framework of the respective countries.

A decision to conserve and preserve natural ecosystems also involves a transfer of free goods to vicarious consumers, option demanders and those who may benefit from advances in the sciences through maintaining a living library of genetic information (Krutilla and Fisher 1976). The tourists who fall into the first category pay very little toward the direct and indirect costs of such decisions considering the very high social opportunity costs associated with biofuel development programs. Likewise communities throughout Asia and the Pacific may have to curtail biomass production and eventually face opportunity costs tradeoffs resulting from such decisions. (Hayes 1977; Norman 1978; Eckholm 1976). Other social costs and expenses stem from efforts designed to reduce or minimize environmental impacts such as peak flood discharges, groundwater recharge loss, groundwater consumption, soil loss, sedimentation and air pollution. A matrix method (Batelle 1973) employed to evaluate these impacts in Hawaii revealed significant costs at specific sites on the island of Hawaii (see Appendix). In the underdeveloped countries of the region conflicts may arise along the interface between ecologic and socioeconomic constraints of biofuel development since the values framework of decision makers varies widely from those of developed societies and is further compounded by the non-availability of choice.

Table 4.--Summary of plantation site characteristics

Site	Cover type	Land Use Class	Rain-fall (in/yr)	Soil Rating	Comments	Ranking
I	Commercial forest, Ohia	C	100-125	poor	Danger from volcanic activity	Preempted, native forest
II	Commercial forest, Ohia	C	125-150	poor	Danger from volcanic activity	Preempted, native forest
III	Commercial forest, Ohia-koa	C	150-225	poor	Region considered by Div.of Forestry in 1974 Planting Plan	Preempted, native forest
IV	Closed watershed, Ohia-koa forest	C	250-300	good	Unavailable, closed watershed	Preempted, native forest
V	Commercial forest, Ohia/ohia-koa	C	200-250	good	Owned by C. Brewer, Inc.	Preempted, native forest
VI	Non-forest grassland (grazing)	A	65-150	fair	Small area, owned by W.H. Shipman, Inc.	3
VII	Non-forest grassland (grazing)	A	100-150	good	Mixture of parcels & owners above sugar land--fair access	2
VIII	Non-forest grassland (grazing)	A	75-150	good	Closest to Port of Kawaihae--good access	1
IX	Mixture of non-forest grassland & poor commercial grade ohia-koa	C	100-175	very poor	Rough terrain, best suited to watershed--poor access	6
X	Mixture of intensive agriculture,	A	75-125	poor		
	non-forest grassland, &	A		good		4
	commercial forest,ohia/ohia-koa	C		poor		
XI	Non-forest grassland &	A	75-100	poor		5
	commercial forest, ohia/ohia-koa	C				
XII	Commercial forest, ohia/ohia-koa	C	100-125	good	(on hydrandept soils)	Preempted, native forest
				very poor	(on tropofolist soils)	

Source: Biomass Energy Study Team. Biomass Energy for Hawaii. Vol. IV. Stanford Institute for Energy Studies. February 1977.

Other problems relate to matters of perception and the assignment of responsibilities. Are advocates of preservation, the so called "purists" just championing meaningless esoterica or is there some inherent value to maintaining natural and unique ecosystems? Can a society faced with the threat of human survival afford preservation, the benefits of which are not quite perceived or immediately available in the form of tangibles? The press of energy related problems, which are crippling the development plans and programs of some countries, do not provide adequate time for decision makers to make informed and learned judgements regarding ecological repercussions, environmental impacts and resulting degradation of the quality of life when initiating largescale biofuel programs. In the longrun, these plans when implemented may yield suboptimal results from the viewpoint of efficiency and equity in an intertemporal and intergenerational context.

It is thus obvious that despite the similarities in Hawaii and other emerging countries of Asia and the Pacific regarding vulnerability to supplies of imported energy feedstock, reliance on tourism growth, the existence of unique ecosystems, and the ecologic/socioeconomic interface of biofuel development activity, there are also significant differences. In Hawaii, the environmental focus is on the physiobiotic environment, but it is essential to recognize that the priority in other regions of Asia and the Pacific is socioeconomic emancipation. What is needed is a framework of policy measures that transcend the divergent needs of the developed/developing countries of the region, afford protection to unique environments threatened by man's quest for newer sources of energy and provide for such needs to be reflected in the calculus of social welfare.

Asymmetry of Treatment

Bioenergy development requires substantial commitment of land and land related environmental resources for growing either agricultural or silvicultural biomass (Dowall 1980). It is thus evident that the environmental implications of biofuel development hinge significantly on issues of land preemption and tradeoffs regarding water and soil quality. Since the preservation benefits are largely regarded as intangibles or incommensurables that defy normal economic calculus, other institutional inputs are often needed to render decisions pertaining to land use. Although it may not be possible to provide an exact dollar figure for preservation benefits of keeping intact the ecological habitats of certain plant and bird species[1] that appear on the Federal List of Endangered Plants and the Federal List of Endangered Species, it is nevertheless possible for the state of Hawaii to preempt these sites since it has sufficient discretionary means to afford such luxuries. In the developing countries, it is possible, given the immense pressure on land and the very high social opportunity costs involved, that land use decisions would not take note of the intrinsic worth of such preservation benefits.

As a biofuel development program gets underway, the benefits of such a program will register
in the marketplace. The disbenefits such as loss
of genetic information on various species, noise
and visual pollution and impacts on the physiobiotic landscape on the other hand will larg ly
register as intangibles. Though an attempt ' 1s
been made in this paper (see Appendix) to measure
some of the environmental impacts in commensurable
units ($) the method is not above charges of bias
and value judgements embedded in the calculations.
It will not be reckoned as part of the solid
price-quantity data. Thus, the asymmetry of treatment in objectively estimating benefits and costs
comes heavily in favor of initiating and
developing biomass programs. The benefit-cost
ratio for growing Eucalyptus sp. trees on the big
island has been estimated at 1.53 (Khamoui 1981).
When the costs associated with the environmental
impacts are internalized, the revised benefit-cost
ratio can be revised to 1.32.

Policy Guidelines

The following broad recommendations for public
policy are made keeping in light the divergent
needs of the countries in the Asia and Pacific
region. An exercise in the social benefit-cost
estimation of biofuel development in Hawaii
revealed the need for revamping public attitudes
and requiring refinements in measuring values of
nonmarket and noncommensurable goods.

Marginal Analysis

The island of Hawaii has over 30 species of
birds and several thousand species of land
arthropods, snails and other terrestrial biota.
It is unlikely that the Hawaiian resident or the
most conscientious visitor is ever likely to
derive much additional satisfaction from knowing
that there are 15000 endemic species to choose from
instead of, say, 14000 such species. It is not known
either, that a loss of a few species would significantly affect the ecosystem stability or cause
serious loss of authenticity to an environment so
fondly cherished by the "purist." It is possible
that a percentage loss in the spectrum of species
may indeed be insignificant. And since ecologists
are at this time unprepared to speculate on the
values of various species, it suffices to save a
few rare or threatened species rather than adopt
a strategy of universal safety.

Optimal Control

A development program would entail the construction of conversion facilities, storage dumps,
movement of heavy specialized trucks and the
introduction of exotic species (e.g. Eucalyptus)
in a monocultural energy tree plantation design.
These activities potentially can modify the abiotic
base as well as change irreversibly the original
nature of the ecological habitat (Oak Ridge 1980).
Since the problem is to maximize a social welfare

function over an identifiable time horizon, a
dynamic approach can be pursued. The opportunity
costs of developing the project sites (the
irreversibly lost benefits from preservation--see
Appendix) enter directly into the expression to be
maximized. Mathematically one can find the land
use which maximizes the present value of net social
benefits over time intervals stretching from the
initial year to the terminal year of the biofuel
project. This would then involve determination of
the net present values of alternatives as well as
the threshhold value required for net preservation
benefits to equal net benefits from the biofuel
project.

Alternative Energy Technology Development

Since the preservation of unique ecosystems
in Hawaii implies the removal of competitive forms
of land use, the social opportunity costs could be
minimized if bioenergy production in the state was
shown to be noncompetitive with the preservation
benefits associated with Hawaii's natural environment. Development of solar power, geothermal power,
and ocean thermal power as alternative energy
sources might provide more compatible sources of
energy from an ecological viewpoint. The beneifts
from these alternate energy technologies could be
assessed through methodologies similar to the
alternative cost approach applied in measuring the
environmental consequences and alternatives of the
Trans Alaska Pipeline (TAP)(Cicchetti 1972).

Categories of Benefits

Benefits emerge from both developing a biomass
project and retaining a specific site in its undisturbed state for ecologic or aesthetic reasons.

The primary regional impacts from developing
large scale silvicultural programs will register
in changes in employment and the gross regional
product. It has been estimated that a Eucalyptus
farm of 1000 acres will create 38-60 new jobs in
direct employment alone (Mitre 1977). The indirect benefits stem from the fact that as payrolls
increase, the level of retail trade and personal
services will expand to readjust to consumer
demand and this in turn will generate additional
requirements of financial services, support
maintenance and transportation sectors (Stacey and
Duchi 1980).

Other benefits accrue from the maturation of
the project. Development of the industry may yield
substantial revenues to the state from taxable
income and, finally, inasmuch as it leads to fuel
substitution, the industry could help retain scarce
development funds within the state.

The benefits of preserving a specific site for
ecological, educational or aesthetic reasons may
not however seem as tangible as benefits described
above, nor be amenable to measurement. The benefits
are reflected in the enhancement of the outdoor
recreation and wilderness enjoyment experience.

This has to be viewed in the context that the demand for such experiences are growing and the opportunities for them are becoming limited (Krutilla and Fisher 1976). Retaining the sites for preservation may also ensure the protection of Hawaiian watersheds, water quality and groundwater recharge supplies. An important benefit and perhaps of paramount significance in preservation decisions is that these unique and undisturbed Hawaiian ecosystems could be of immense value to scientific research and serve as gene pools for the future.

Duration of Welfare Loss

A decision to initiate a biofuel program could lead to the escape of exotic species into the native habitat and elimination of certain endemic plants that do not compete well. This in turn may lead to elimination of habitat for certain species of fauna that depend on the original plant association. Disrupting native habitat means removing the product of ecological succession leading back to centuries of natural evolutionary processes. Technology though capable of modifying environments cannot replicate original conditions (Krutilla and Fisher 1976). Thus it is not clear that if the biofuel development program were to cease at some time in the future, the modifications in the biotic and abiotic base at the sites could be restored to their original conditions. The potential loss of authenticity is real and such irreversibilities must be recognized before any decision is made.[2] Since Hawaii is located in the tropics, the natural processes for restoration may take a shorter time, but can still be in the context of hundreds of years (Mueller-Dombois 1980).

Intergenerational Equity

Since decisions are being made which affect welfare evaluated from the point of view of present consumers, it becomes important to check for discrimination against future generations. Assumptions regarding the utility functions of unborn generations are fraught with guesswork and a great deal of uncertainty. It is perhaps possible to bind the present and the future generations in some mutually advantageous course of action under provisions of the Hawaii Revised Statutes of 1968. A legislative classification of ecologically important areas, with provisions to make reclassification more difficult might compensate for the asymmetry in decision making in an intergenerational context.

Interspecies Equity

This equity principle has to do with the rights of non-human life forms. It is an ethical question inasmuch as it is economic. Clearing and plowing a field for economic benefit is perhaps one of the most destructive operations that human beings engage in, in terms of numbers of lifeforms killed. If plants and animals have inherent value as the adage "Earth is the mother of us all" would have us believe, then human decision making has to develop the means for valuing such rights of other species (Oelhaf 1978).

The Tourist Tax

Since the tourist is a vicarious consumer and the most readily identifiable beneficiary in terms of a decision to preserve a site for either aesthetic or ecological reasons, it is conceiveable to impose a tourist tax to offset a share of the high social opportunity costs inherent in such a decision. Eventually other user groups such as academic institutions, the scientific community, wilderness seekers could also be assigned cost responsibility. However the economic aspects of ecological judgements will encounter numerous social indivisibilities and the state as a whole may end up absorbing all of the social costs.

CONCLUSIONS

This study has focused primarily on the ecologic/socioeconomic interface of biofuel development activity in Hawaii. The results of the study however have implications for a large number of countries in the Asia and Pacific region, which have plans and programs for developing biofuels in the identifiable future. The specific and general inferences drawn from the study are summarized below:

1. In Hawaii, some of the sites found suitable for silvicultural operations have competing alternative uses.
2. It is important to explore the interrelationship between development and conservation based on an understanding of ecology and danger to the stability of ecosystems.
3. Public perception of values associated with noncommensurables and intangibles is at best hazy. Better appreciation of the intrinsic worth of preservation decisions, changes in attitudes of people presently implementing land use policies and assigning responsibilities for bearing social costs and disbenefits will lead to better decisions at the margin of the ecologic/socioeconomic interface.
4. Land use decisions have to take into account value systems that differ in the developed and developing societies of Asia and the Pacific.
5. The critical constraints to development imposed by socioeconomic conditions together with the press of energy related problems in such countries have led to plans for increasing biofuel production, which when implemented might yield suboptimal results from the viewpoint of efficiency and equity in an intertemporal, intergenerational and interspecies survival context.

NOTES

[1]The following endemic plants viz.
<u>Stenogyne scrophulasioides</u> a creeping plant of the
mint family and <u>Platydesma remyii</u> a shrub and the
following birds viz. the hawaiian hawk or I'o
(<u>Buteo solitarius</u>), the hawaiian owl or Pueo
(<u>Asio flammeus sandwichensis</u>) and the marine bird
A'o (<u>Puffinus puffinus newelli</u>) appear on the
list of endangered species (Bioenergy Corporation
1980a).
[2]The effect of man's settling on the islands
is well documented in the literature. The poly-
nesians, who immigrated in waves, probably
colonized the island around 500 A.D. Since they
lived in regions close to the coast, they
exerted a particular influence on the vegetation
along the coastline. The "digstick agriculture"
they practiced in cultivating taro, bananas,
yams, breadfruit and sugarcane (all introduced
plants) resulted in the complete disappearance of
original vegetation in the areas they cultivated.

After the discovery of the Hawaiian isles by
Europeans in late eighteenth century, animals such
as cattle, horses, sheep and goats and other
exotic plant species were deliberately introduced.
Introduced plant species withstood grazing
pressure and eventually replaced indigenous
vegetation on extensive parts of the islands.
Also much of the mixed lowland Lehua forests were
cleared to plant vast acreages in sugarcane.
Coffee plantations were established on the Kona
coast by replacing Lehua and Koa forests, and
other tropical plants grown for exports such as
bananas, papayas and passion fruit resulted in the
loss of mixed mesophytic forests. As a result of
such massive disturbance of native stands,
secondary rain forests developed at lower
elevations, and in dry areas accidental or inten-
tional fires destroyed woodlands. Range pests in
the form of weeds and shrubs became common
(Fosberg 1972). Introduced mangrove trees have
converted earlier existing Hawaiian fish ponds
into swamps and the development of rural subdivisions, harbors and marinas has eliminated
many others. The marshes are now part of the
Hawaiian landscape and though currently regarded
by many as essentially wasteland, they provide
refuge, breeding and feeding grounds, for a
number of species of water and shore birds.

The loss of flora and fauna is regrettable
but more serious are the <u>irreversible</u> modifications
of aesthetically appealing environment. On the
slopes of Mauna Kea, over an area where there
once existed a scenery of "picturesque and sublime
beauty" now remains the spectre of a devastated
mamane forest destroyed by unfettered animals
that have converted the forest into a waste of
rocks, dusty subsoil and dead trees (Warner 1972).

Other <u>irrevocable</u> losses have occurred with
the disappearance of unusual faunistic assemblages
that a leisurely evolution on the islands produced.
Much of Hawaii's biota in the lower, accessible
lowlands have been destroyed due to physical changes
forced on the land, deforestation for agriculture
and competition with introduced organisms. To
remark at Hawaii's unique fauna, one must now go to
greater elevations. Many species of insects and
snails have been destroyed as native plants have
been cleared, and changing land usage and exotic
diseases have contributed heavily to the extinction
of many native Hawaiian birds which include the
Mano, O'o and Moho (Kondo 1972; Quate 1972; and
DPED 1980).

APPENDIX

Environmental Benefit-Cost
Analysis for Land Use

The principle of best use of land has often
been applied in evaluating alternative land uses.
Although this has often been interpreted to mean
the highest return on investment in simple terms of
profit, "best use" is now taking on a broader
meaning which includes the public interest. There
is an increasing public awareness of the need to
comprehend the relationship between man's economic
activities and the resulting quality of the
environment and, from an economic point of view, to
include such considerations in a broadened benefit-
cost framework.

The Matrix Method

This study has chosen a matrix model developed
by the State of Hawaii's Department of Planning and
Economic Development as a technique for demonstrating
the economic and environmental tradeoffs for
alternative land uses at three sites on the urban
fringe of Honolulu (1975). The model is a modification of the Arizona Economic and Environmental
Trade-off Model (1973), which was originated by
Batelle Laboratories to quantitatively assess the
magnitude of potential impacts of various land uses.

The Hawaiian modification overcomes certain
limitations of the original model by establishing a
common unit of cost for the impacts. It also employs
a simple three step matrix multiplication procedure
that expresses the different impacts of different
land uses at different locations as a consequence
of the different land characteristics at those
locations. The main thrust of the matrix multiplication technique has been to provide an assessment
of environmental impacts on the basis of locational
criteria and selected land uses (fig.A-1).

The matrix method yields more quantitative
information than most other methods, such as map
overlays, but it is also subject to some limitations.
One is that it can only operate in terms of linear
relationships. Another limitation is that it
requires the diverse environmental impacts with
which it is dealing to be expressed in common,
homogenous units, even though it may often be
unnatural to do so.

Environmental Impacts

The impacts of various land uses on the environment are evaluated by employing empirical data to estimate expected environmental impacts and annual costs per acre. The following impacts were considered in this study: peak discharge (flooding), ground water recharge loss, ground-water consumption, soil loss, sedimentation, and air pollution. These impacts were calculated for the following land uses:
1. conservation use (natural forest area)
2. agricultural (irrigated sugarcane)
3. silviculture (biomass tree farm)

Average impacts are associated with a specific use but do not reflect the effect of any particular location. Average impacts are then adjusted according to the land characteristics at a particular site in order to arrive at the impact on that site. The land characteristics serving to increase or decrease the impact at a specific site are slope, rainfall, permeability, erodibility, evaporation, and distance from urban centers. A scoring system of 0.5, 1 and 2 was used for each characteristic to reflect the log-linear nature of some of the land characteristic/impact relationships (TableA-1).

The average impact for each land use was determined in Figure A-2 through A-4.

Groundwater Recharge Loss

Groundwater recharge occurs primarily in forested areas in Hawaii. Assuming 55 inches of rainfall, approximately 1.4 million gallons of rainfall would be received by a forest conservation area per acre per year. With a 44 percent infiltration rate, roughly 620,000 gallons of water per acre are added annually to the groundwater supply in such areas.

In agricultural areas, this recharge would be approximately 1/2 the rate of recharge in forested areas, i.e., 310,000 gallons per acre per year. In areas devoted to energy tree plantations the infiltration rate is likely to be similar to agriculture during the year of planting and similar to conservation areas in the remaining 5 years of the production cycle. Thus the average groundwater recharge loss would be 51,600 gallons in areas devoted to energy tree plantation.

In Hawaii, the price of water is $71 per thousand gallons. This price was assigned to

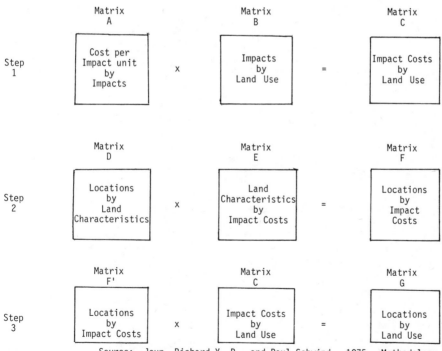

Source: Joun, Richard Y. P. and Paul Schwind. 1975. _Methodology for Evaluation of Land Use Alternatives_. Department of Planning and Economic Development. State of Hawaii.

Figure A-1.--Steps Involved in Matrix Calculation.

Table A-1.--Land Characteristics Scoring for Assessing Environmental Impact

Score	Slope	Rainfall	Permeability	Erodibility	Evaporation	Distance
.5	0-10%	0-30"	Slow	Slight	60"	0-15 mi.
1	11-30%	30-50"	Moderate	Moderate	60-80"	15-25 mi.
2	30%	50"	Rapid (0-impervious)	Severe	80"	25 mi.

measure the impact of groundwater recharge loss.

Groundwater Consumption

Groundwater consumption for forest and tree plantation areas was estimated at zero gallons per acre since no groundwater is withdrawn for consumption. The rate of groundwater consumption for agriculture in Hawaii (irrigated sugarcane) was estimated at 565 gallons per acre per year (Hawaii Water Resources Regional Study 1979). Again, a rate of $71 per thousand gallons was used as the value of groundwater consumption for agricultural use.

Soil Loss

Soil loss rates were worked out with officials of the U.S. Soil Conservation Service, using the Universal Soil Loss Equation. Average slope lengths and the soil erodibility characteristics at the sites were used to calculate the following soil loss rates: .45 tons/acre/year for conservation forest; 4 tons/acre/year for energy tree farms; and 14 tons/acre/year for agriculture.

The cost of soil loss was based on the market value of top soil ($30 per ton).

Sedimentation

Estimates of sedimentation rates were based on suspended sediment yields resulting from soil losses with the different land uses, since the sediment is transported suspended in streams to its site of deposition (Hawaii Water Resources Regional Study 1975). The sedimentation rate was determined as the percentage that suspended sediment made up of total soil loss (60 percent). The sedimentation rates are the following: .33 cubic yards for forest conservation, 2.9 cubic yards for tree plantation, 10.3 cubic yards for agriculture.

The sedimentation costs are based on current costs of dredging and ocean disposal for sediment loads of 5000 cubic yards (unit cost of $40/cubic yard) and include total cost of the operation.

Air Pollution

Air quality impacts associated with alternative land uses were estimated utilizing empirical data on SO_x emissions, particulates, CO, and NO_x, based on the Hawaii State Statistical Summary for 1979. It was assumed that forest conservation areas do not contribute to air pollution. An average of 0.16 tons/acre/year was calculated for agriculture and 0.19 tons/acre/year for tree plantations, based on reports by MITRE (1979), Oak Ridge National Laboratory (1980), and the Stanford Biomass Study (Biomass Energy Study Team 1977).

The cost per ton of emission was calculated by dividing the total damage attributable to air pollution in Hawaii (Hawaii's 0.38% share of U.S. air pollution cost $14.1 billion in 1976 and adjusted to 1980 dollars) by the total emissions from the state (urban and agricultural uses), for a rate of $90 per ton of air pollution (Council on Environmental Quality 1979).

IMPACT COSTS BY LAND USE

(DOLLARS PER ACRE)

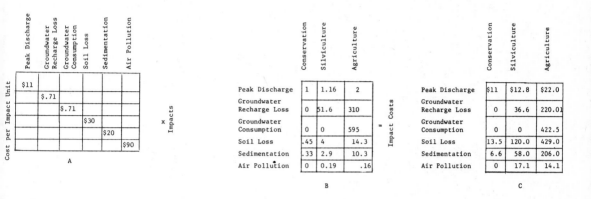

A — Cost per Impact Unit

	Peak Discharge	Groundwater Recharge Loss	Groundwater Consumption	Soil Loss	Sedimentation	Air Pollution
	$11					
		$.71				
			$.71			
				$30		
					$20	
						$90

× Impacts

B — Impacts

	Conservation	Silviculture	Agriculture
Peak Discharge	1	1.16	2
Groundwater Recharge Loss	0	51.6	310
Groundwater Consumption	0	0	595
Soil Loss	.45	4	14.3
Sedimentation	.33	2.9	10.3
Air Pollution	0	0.19	.16

= Impact Costs

C — Impact Costs

	Conservation	Silviculture	Agriculture
Peak Discharge	$11	$12.8	$22.0
Groundwater Recharge Loss	0	36.6	220.01
Groundwater Consumption	0	0	422.5
Soil Loss	13.5	120.0	429.0
Sedimentation	6.6	58.0	206.0
Air Pollution	0	17.1	14.1

LAND USES: Conservation (Natural Forest Reserves)
                 Silviculture (Eucalyptus Energy Farms)
                 Agriculture (Irrigated Sugarcane)

IMPACT UNITS: Peak Discharge (net cfs/acre) 100-year storm)
                  Groundwater Recharge Loss (1,000 gallons per acre per year)
                  Groundwater Consumption (1,000 gallons per acre per year)
                  Soil Loss (tons per acre per year)
                  Sedimentation (cubic yards per acre per year)
                  Air Pollution (tons emitted per acre per year)

Figure A-2.--Step One of Matrix Method

IMPACT MODIFICATION SCORES BY LOCATION

D — Land Characteristics (Locations)

Location	Slope	Rainfall	Permeability	Erodibility	Evaporation	Distance
1	1	2	2	.5	.5	1
2	1	2	2	.5	.5	1
3	1	2	2	.5	.5	.5
4	1	2	2	.5	.5	.5
5	1	2	2	.5	.5	.5
6	.5	2	1	.5	.5	.5
7	1	2	1	.5	.5	1
8	1	2	2	.5	.5	.5
9	1	2	2	.5	.5	.5
10 A	1	2	1	.5	.5	.5
10 B	.5	2	1	.5	.5	.5
10 C	.5	2	1	.5	.5	.5
11	1	2	2	.5	.5	.5
12	1	2	2	.5	.5	2

× Impact Costs

E — Land Characteristics × Impact Costs

	Peak Discharge	Groundwater Recharge Loss	Groundwater Consumption	Soil Loss	Sedimentation	Air Pollution
Slope	1	(0)	0	1	1	0
Rainfall	1	1	(0)	1	1	0
Permeability	(0)	1	1	0	0	0
Erodibility	0	0	0	1	1	0
Evaporation	0	(0)	1	0	0	0
Distance	0	0	0	0	0	1
	2	2	2	3	3	1

= Impact Costs (Locations)

F — Impact Costs by Location

Location	Peak Discharge	Groundwater Recharge Loss	Groundwater Consumption	Soil Loss	Sedimentation	Air Pollution
1	3	4	2.5	3.5	3.5	1
2	3	4	2.5	3.5	3.5	1
3	3	4	2.5	3.5	3.5	.5
4	3	4	2.5	3.5	3.5	.5
5	3	4	2.5	3.5	3.5	.5
6	2.5	3	1.5	3	3	.5
7	3	3	1.5	3.5	3.5	1
8	3	4	2.5	3.5	3.5	.5
9	3	4	2.5	3.5	3.5	.5
10A	3	3	1.5	3.5	3.5	.5
10B	2.5	3	1.5	3	3	.5
10C	2.5	3	1.5	3	3	.5
11	3	4	2.5	3.5	3.5	.5
12	3	4	2.5	3.5	3.5	2

Figure A-3.--Step Two of Matrix Method

COMPOSITE IMPACT COSTS BY LOCATION
(DOLLARS PER ACRE)

IMPACT COSTS

Locations	Peak Discharge	Groundwater Recharge Loss	Groundwater Consumption	Soil Loss	Sedimentation	Air Pollution
1	1.5	2.0	1.25	1.17	1.17	1.00
2	1.5	2.0	1.25	1.17	1.17	1.00
3	1.5	2.0	1.25	1.17	1.17	.50
4	1.5	2.0	1.25	1.17	1.17	.50
5	1.5	2.0	1.25	1.17	1.17	.50
6	1.25	1.5	0.75	1.00	1.00	.50
7	1.5	1.5	0.75	1.17	1.17	1.00
8	1.5	2.0	1.25	1.17	1.17	.50
9	1.5	2.0	1.25	1.17	1.17	.50
10 A	1.5	1.5	0.75	1.17	1.17	.50
10 B	1.25	1.5	0.75	1.00	1.00	.50
10 C	1.25	1.5	0.75	1.00	1.00	.50
11	1.5	2.0	1.25	1.17	1.17	.50
12	1.5	2.0	1.25	1.17	1.17	2.0

F'

Matrix F' = rescaled Matrix F (columns divided by sums of corresponding matrix E columns)

× IMPACT COSTS

LAND USE

	Conservation	Silviculture	Agriculture
Peak Discharge	$11	$12.8	$22.0
Groundwater Recharge Loss	0	36.6	220.01
Groundwater Consumption	0	0	422.5
Soil Loss	13.5	120.0	429.0
Sedimentation	6.6	58.0	206.0
Air Pollution	0	17.1	14.1

C

= LAND USE

Locations	Conservation	Silviculture	Agriculture
1	40	318	1758
2	40	318	1758
3	40	309	1751
4	40	309	1751
5	40	309	1751
6	34	258	1317
7	40	300	1437
8	40	309	1751
9	40	309	1751
10 A	40	291	1430
10 B	34	258	1317
10 C	34	258	1317
11	40	309	1751
12	40	335	1772

G

Matrix C= dollar value of environmental impacts per acre per year by land use at each location

Figure A-4.--Step Three of Matrix Method

LITERATURE CITED

Asia and Pacific Development Institute. 1979. An empirical overview of environment and development: Asia and the Pacific. United Nations Economic and Social Commission for Asia and the Pacific.

Batelle-Columbus Laboratories. 1973. Final report: Development of the Arizona Environmental and Economic Trade-off Model. Department of Economic Planning and Development, State of Arizona.

Bio Energy Development Corporation. 1980a. Bio Energy Development Corporation environmental assessment. Honolulu, Hawaii.

Bio Energy Development Corporation. 1980b. Growing energy for tomorrow. Honolulu, Hawaii.

Biomass Energy Study Team. 1977. Biomass energy for Hawaii, Vol. IV. Stanford Institute for Energy Studies, Stanford, California

Cicchetti, C. J. 1972. Alaskan oil: Alternative routes and markets. Johns Hopkins Press, Baltimore.

Council on Environmental Quality. 1979. Environmental quality. Tenth Annual Report. Washington, D. C.

Dasmann, R. F., John Milton and Peter Freeman. 1975. Ecological principles for economic development. John Wiley & Sons, New York.

Department of Land and Natural Resources. 1975. A program for the state forest lands of Hawaii. State of Hawaii.

Department of Land and Natural Resources. 1979. Energy tree farm program. Hawaii Division of Forestry (unpublished document). State of Hawaii.

Department of Planning and Economic Development. 1975. Methodology for evaluation of land use alternatives. State of Hawaii.

Department of Planning and Economic Development. 1977a. The Hawaii State plan: The environment. State of Hawaii.

Department of Planning and Economic Development. 1977b. The Hawaii State plan: The Economy. State of Hawaii.

Department of Planning and Economic Development. 1978. The Hawaii State plan. State of Hawaii.

Department of Planning and Economic Development. 1979a. The State of Hawaii data book: A statistical abstract. State of Hawaii.

Department of Planning and Economic Development. 1979b. Annual overall economic development program. State of Hawaii.

Department of Planning and Economic Development. 1980. State Energy Resources Coordinator. 1979 Annual Report.

Department of Taxation. 1979. Fiscal Year Summary. State tax collections and distribution year ending June 30, 1979. State of Hawaii.

Dowall, David F. 1980. U. S. land use and energy policy--assessing potential conflicts. Energy Policy 8(1).

Eckholm, Eric. 1976. Losing ground: Environmental stress and world food prospects. W. W. Norton & Co., New York.

Fosberg, F. R. 1972. Guide to excursion III. Tenth Pacific Science Congress. Department of Botany, University of Hawaii.

Hawaii Water Resources Regional Study. 1975. Floods study element report. Department of Land and Natural Resources. State of Hawaii.

Hawaii Water Resources Regional Study. 1979.
Hawaii water resources plan. Department of
Land and Natural Resources. State of Hawaii.

Hayes, Dennis. 1977. Energy for development:
Third world option. World Watch Institute,
Washington, D. C. Paper no. 15.

Jackson, Randal. 1976. Recreation resource and
intensive tree plantings. Industrial
Forestry for Hawaii. Proceedings of the 18th
Annual Hawaii Forestry Conference [Nov. 18-19,
Honolulu, Hawaii] DPED, State of Hawaii.

Khamoui, T. 1981. An analysis of Eucalyptus
woodchips production for export from the
island of Hawaii. Ph.D. dissertation,
Department of Agriculture and Resource
Economics, University of Hawaii.

Kondo, Y. 1972. 'Land Mullusca' Field Guide to
Excursion III (ed.) F. R. Fosberg, Department of Botany, University of Hawaii.

Krutilla, J. V. and Anthony Fisher. 1975. The
Economics of Natural Environments. Johns
Hopkins University Press, Baltimore.

MITRE Corporation. 1977. Silvicultural biomass
farms, Vol. IV. Site Specific Production
Studies and Cost Analysis. U. S. Department
of Commerce. MITRE-TR-7347. Washington, D. C.

MITRE Corporation. 1979. Environmental data for
energy technology policy analysis, Vol. 1.
Summary Washington, D. C.

Mueller-Dombois, D. 1975. Integrated island
ecosystem ecology in Hawaii. Introductory
survey. Department of Botany, University of
Hawaii.

Mueller-Dombois, D. 1980. The Ohi'a dieback phenom-
enon in the Hawaiian rain forest. In The
Recovery Process in Damaged Ecosystems. Ann
Arbor Science Publishers, Ann Arbor, Michigan.

Mueller-Dombois, D., K. W. Bridges and H. L. Carson.
(eds.) 1980. Island ecosystems: Biological
organization in selected Hawaiian communities.
Dowden, Hutchison & Ross Publishers (in press).

Norman, Colin. 1978. Soft technologies, hard
technologies. World Watch Institute,
Washington D. C., paper no. 21.

Oak Ridge National Laboratory. 1980. Environmental
Assessment Biomas Energy Systems Program.
Oak Ridge, Tennessee.

Oelhaf, Robert. 1978. Organic agriculture.
Allanheld, Osmun and Co. Publishers,
Montclair, New Jersey.

Quate, L. W. 1972. 'Hawaii's Insect Fauna' Field
Guide to Excursion III F. R. Fosberg (ed.)
Department of Botany, University of Hawaii.

Siddiqi, Toufiq A. 1979. Environmental
considerations in energy policies. East-West
Center, Honolulu, Hawaii
EAPI Reprint no. 10.

Stacy, G. S. and M. L. Duchi. 1980. Analyzing the
socioeconomic effects of large energy projects.
Environmental Impact Assessment Review, 1(3).

Warner, R. E. 1972. A Forest Dies on Mauna Kea'
Field Guide to Excursion III F. R. Fosberg
(ed.) Department of Botany, University of
Hawaii.

THE APPLICATION OF A FOREST SIMULATION MODEL TO ASSESS THE ENERGY YIELD AND ECOLOGICAL IMPACT OF FOREST UTILIZATION FOR ENERGY[1]

Thomas W. Doyle[2], Herman H. Shugart, and Darrell C. West[3]

Abstract.--This study examines the utilization and management of natural forest lands to meet growing wood-energy demands. An application of a forest simulation model is described for assessing energy returns and long-term ecological impacts of wood-energy harvesting under four general silvicultural practices. Results indicate that moderate energy yields could be expected from mild cutting operations which would neither significantly effect the commercial timber market nor the composition, structure or diversity of these forests. Forest models can provide an effective tool for determining optimal management strategies that maximize energy returns, minimize environmental detriment, and compliment existing land-use plans.

INTRODUCTION

Over the past few years the demand for fuelwood has increased substantially, along with the rising economic incentives to utilize woody biomass for energy. In many areas (e.g., east Tennessee), natural forest resources are once again being exploited to provide stovewood and, in some instances, to supplement or replace conventional fuels used in small-based industrial operations. Regional evaluations of woody biomass production confirm the potential resource from natural occuring forests as well as conversion to biomass plantations to support this growing market for wood-energy (Ranney and Cushman 1980). The development of wood-energy plantations, however, remains in its pilot stages due to many questionable factors (e.g., capital outlay, crop procurement and merchandising, species selection, site preparation and design, among others) that tend to be quite variable depending on the specific land-site and end-use.

Recent studies reveal that much of the commercial forest lands are either underutilized or that forest residues (e.g., cull trees, tops and limbs) generated from timber improvement practices are not being harvested (Curtis 1978, Howlett and Gamache 1977). In addition, there is significant acerage of marginal lands, wood lots and tree farms which at present lack any management alternatives other than a primary source of firewood. The potential role and input of these lands, particularly in the Southeast, in providing a significant wood-energy resource seemingly exists with more timely and efficient forest management.

Intensifying the use of our forests for energy purposes is of major concern, and warrants a comprehensive evaluation of the associated environmental impacts, especially with regards to indirect and long-term effects of which little is known. Equally as important to the wood-energy consumer is the reliability of the field source to sustain a relatively stable market supply over a period of years. These and related queries need to be addressed in order to fully assess the contribution and consequences of additionally managing our natural forest lands for energy production and use.

Forest simulation models can provide and have provided a necessary adjunct to field and laboratory studies, particularly where multi-use system effects and multi-year time scales are considered (Shugart and West 1980). This paper describes a preliminary application of FORET, a southern Appalachian forest simulator, to assess the energy yield and long-term ecological impact of harvesting wood-energy under four general silvicultural practices. A comparison of the resultant change in composition, structure and diversity between model

[1] Paper presented at the international symposium Energy and Ecological Modelling, sponsored by the International Society for Ecological Modelling. (Louisville, Kentucky, April 20-23, 1981).

[2] Thomas W. Doyle is a Doctoral Research Fellow in the Graduate Program in Ecology at the University of Tennessee, Knoxville, Tennessee USA
[3] Herman H. Shugart and Darrell C. West are Research Scientists in the Environmental Sciences Division at the Oak Ridge National Laboratory, Oak Ridge, Tennessee USA

simulations including wood-energy harvests and the projected pattern of natural forest succession provides the basis for this assessment.

METHODS

Model Description

FORET, forest succession model for east Tennessee, was developed by Shugart and West (1977) to simulate the forest dynamics of a southern Appalachian forest type on lower slopes in Anderson County, Tennessee. It mathematically mimics the successional pattern and competitive interrelations of individual trees for a typical 1/12 hectare plot of this forest. This stochastic model is one of a unique class of forest simulators, whose design consists of empirical formulas describing key ecological relationships of the forest community, its species and environment (Shugart and West 1980).

The major model routines, mainly tree seeding, growth and death, are computed on yearly intervals. The growth of each tree is incremented as a function of climate, light availability, total stand biomass, and the inherent growth characteristics of the individual species. The addition of new seedlings to the plot is based on the sprouting tendency and germination requirements of the species. Tree death is modeled as a stochastic process with the probability of dying inversely related to growth and longevity. For detailed documentation of the FORET model construction and validation, refer to Shugart and West (1977).

Model Application

The present version of the model considers 32 tree species. Each is assigned an average energy value of oven-dry wood expressed in Kcal/kg as shown in table 1. Heating values for some species were not found in the literature, and thus shared equal value with their congenerics. Energy harvests were then calculated by multiplying the heating value of each species times its total aboveground biomass (kg) included in the cutting.

The simulated harvests were incorporated into the model in an additional subroutine called CUT. These cutting schemes were designed from four general silvicultural practices, namely clear cut, selective cut, thinning and high-grade cut. All harvests were implemented on a discrete 60 year rotation schedule. This allowed the comparison of results among cutting selections without regards to the influence of varying harvest schedules.

The clear cut routine involved the removal of all trees regardless of species or size. The selective cut allowed the harvesting of all species, but only individuals with stem diameters exceeding 30 cm were removed. In the intense thinning simulation, all species were considered eligible for cutting, but only trees below 30 cm dbh were included in the harvest. Lastly, the high-grade cut

Table 1.--Heating values for each of the species included in the FORET model parameter list. These values represent the energy content for dry wood as compiled from many sources by Conde and Huffman (1978) and Howlett and Gamache (1977).

Species	Heat Value (Kcal/kg)
Acer rubrum	4604
Acer saccharum	4604
Aesculus octandra	4444
Carya cordiformis	4693
Carya glabra	4693
Carya ovata	4693
Carya tomentosa	4693
Cercis canadensis	4444
Cornus florida	4444
Diospyros virginiana	4444
Fagus grandifolia	4697
Fraxinus americana	4768
Juglans nigra	4444
Juniperus virginiana	5389
Liquidambar styraciflua	4563
Liriodendron tulipifera	4786
Nyssa sylvatica	4650
Oxydendron arboreum	4444
Pinus echinata	5195
Pinus strobus	5195
Pinus virginiana	5195
Prunus serotina	4790
Quercus alba	4717
Quercus coccinea	4644
Quercus falcata	4644
Quercus prinus	4644
Quercus rubra	4644
Quercus stellata	4644
Quercus velutina	4644
Robinia pseudoacacia	4444
Sassafras albidum	4444
Tilia heterophylla	4586

represented a specialized thinning where only non-commercial species (ie., non-hickory, -oak, and -poplar group) were harvested. The latter cutting schemes are considered compatible alternatives for commercial forest lands where the management plan includes timber improvement practices. Results of these simulations, however, do not include the felling of commercially harvestable trees which might otherwise be cut in actuality.

Following each cut, the model sums the total biomass and energy harvested. Model results for each simulation as a whole constitutes the average of 120 plots projected over a 500-year period. These were then compared with the output from a control simulation which represented the pattern of natural forest growth and succession.

RESULTS AND DISCUSSION

Energy returns (ie., the average expected Kcals/ha/yr) from each of the simulated harvest selections are given in table 2. As expected, considering identical harvest schedules, the clear cut reaped the highest yield, followed in order by the selective cut, intense thinning, and high-grade cut. More important than the total energy output is the required harvestable forest land necessary to maintain a wood-energy supply on a sustained-yield basis for specific conversion to electricity or space heat. Three evaluations are presented: a 10 megawatt electric plant, a 50 megawatt power facility, and a 1000 cord fuelwood supply.

It becomes evident from these findings that woodburning power facilities, even as small as 10 megawatts, would require a sizable market area in order to maintain continuous operation without directly competing for commercial timber and forest lands. It also emphasizes the importance of biomass plantations, short-rotation forestry, species selection among other alternatives, if woody biomass is to provide an energy resource for producing electricity. To read beyond the results stated here would require a specific site and source evaluation.

In contrast, relatively little land area would be necessary to generate 1000 cords of fuel-wood on a yearly basis. In this case, any of the harvesting schemes become viable alternatives regardless of the use or distribution of available lands. The primary end-use, however, would be restricted to residential space heating or indus-trial woodburning boilers. The cumulative contri-bution of these decentralized end-uses cannot be overlooked for their significance on the local energy network, particularly in regions where fragmented non-commercial forest parcels are common and commercial timber production is of prime importance.

To account for the long-term ecological con-straints of each of the simulated harvesting measures, we compared the resultant change in the composition, structure and diversity of the man-aged forest with the projected pattern of natural forest succession. Compostional differences were distinguished by evaluating species rank values

of the stand composition the year prior to each cutting, using the Spearman rank statistical test. The analysis required an ordering (ie., numerical ranking from maximum to minimum) of the relative biomass by species for both the control and harvest simulations. Table 3 lists the results of this analysis in terms of an r_s coefficient, represent-ing the degree of compostional agreement between each simulated harvest and the control. Since these values signify the sum of species composition for all plots prior to harvest, they also provide some indication of the forest's ability to recover or its resiliency to disturbance.

On the average, this test indicated no signi-ficant differences in the overall compositions resulting from any of the harvest simulations with what might otherwise be expected over 500 years of forest succession. It can be seen, however, that the more severe cutting routines impose a greater shift in the overall species array as well as a greater stress on system recovery. The model does not account for differential seeding effects that might occur with overselecting certain species or eliminating parent trees. Without this input, the true shift of forest dominance towards early suc-cessional species is probably underestimated. Although the harvest selections did not appear to significantly affect the forest composition over the long-term, this does not take into account the changes in the actual distribution of species biomass and numbers (e.g., species diversity).

Changes in forest structure due to each simu-lated harvest were analyzed by comparing the mean stem density characteristics for an average 1 ha stand. This test involved the statistical compar-ison of cumulative diameter distributions of tree size-ranges from all plots on years prior to cut-ting. In effect, only the most mature stands were included in the test sample.

Diameter distributions represent the number of trees apportioned in defined diameter size classes over the diameter size range of the stand. The size class range used for testing the model results herein was 4 cm. The smaller the size range, the more powerful this test becomes for

Table 2.--Expected energy yield and required harvestable forest land (ha) as determined for each harvest selection.

Harvest Selection	Energy Yield (Kcal/ha/yr)	Required Harvestable Forest Land (ha)		
		10-MWe	50-MWe	10^3cds
Clear cut	1.19×10^7	1414	7777	120
Selective	1.02×10^7	1649	9096	140
Thinning	2.23×10^6	7517	41343	450
High-Grade	1.46×10^6	11533	63316	700

Table 3.--Ecological index values indicating alter-ation of forest composition, structure and diversity.

Harvest Selection	Composition (r_s values)	Structure (D_k value)	Diversity (S_w value)
Clear cut	.81	.20	.66706
Selective	.88	.19	.67742
Thinning	.89	.02	.58406
High-Grade	.94	.05	.52896
Control	1.00	0.00	.59691

determining signficant differences in forest structure. From one year to the next, individual trees
may grow into larger size classes or remain in the
same one, provided tree death, disturbance or cutting does not occur. The shape and position of
these diameter distribution curves relative to
one another can give some indication of the successional maturity and dynamics of a forest stand.
Figure 1 shows the relation of each simulated
harvest with that of the control simulation.

These stem density figures were then converted into cumulative percentages of the total sample
density for successive diameter classes from
smallest to largest. This results in cumulative
frequency distributions which can be statistically
compared by using the Kolmogorov-Smirnov two-
sample test. Derived D_k values (table 3), representing the maximum differences between distributions, indicated that only the clear cut and selective cut samples were significantly different
from the control sample. The differences in D_k
values between the thinning and high-grade samples
is perhaps attributable to the unrestricted size
cut of the high-grade cutting routine.

Lastly, the simulated harvesting practices
were evaluated for their effect on forest diversity. Diversity has been defined in many terms,
but for the purposes of this text, it signifies
an evenness or equitability with which the forest
biomass is distributed among the model species.
The index values herein were derived by using the
Shannon-Weiner formula for evenness. This function
generates S_w values ranging from 0 to 1, where 0
represents complete dominance by one species as
in a monoculture plantation and 1 indicates equal
dominance by all species.

Figure 2 illustrates the contrast of diversity
through time between the simulated harvests and the
control run. The most radical silvicultural practice, clear cut, generates the highest degree of
variability in the diversity pattern, while the
remaining harvest measures appear distributed
slightly above or below the pattern for natural
succession. Table 3 lists the average diversity

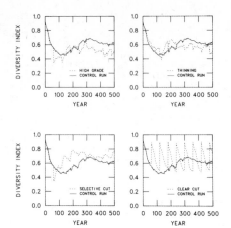

Figure 2.--Forest diversity pattern for 500 years
of model simulations, contrasting the effects
of the different harvest selections with the
expected pattern of natural forest succession.

values for all simulations and all years over a
500-year period. The repetitive clear cut and selective cut tended to increase the overall diversity
index above the expected normal. On the other
hand, the high-grade cut, selecting only cull species for harvest, had the sole distinction of decreasing the overall forest diversity. In general,
one equates increased species diversity with a
positive impact and decreased diversity with a
negative impact. However, in this case we must
consider the maximum absolute differences in diversity values as some indication of the ecological
consequence of any harvesting practice or management plan. As already pointed out, the clear cut
and selective cut seem to impose the most concern
for environmental detriment in this regard.

CONCLUSION

Results from the four simulated harvest selections indicated that the extreme silvicultural
practices, clear cut and selective cut, yielded
higher energy returns over the less severe thinning
operations. But even at the highest yield values,
considerable land area would be necessary to fuel
a 10 megawatt woodburning power facility from natural forest resources. Such natural forest expanses
are not without existing management plans or commercially valuble tree stock to allow the cutting
of timber solely for energy purposes. The alternatives then seem to narrow to less extreme practices that could be incorporated into already implemented land-use plans and/or more direct uses
of the wood-energy in small industrial boilers or
residential woodstoves. The latter proposition
appears more practical and ecologically sound when
one also considers the potential environmental
impacts not previously mentioned.

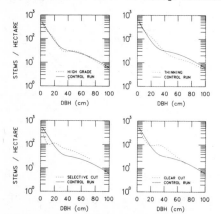

Figure 1.--Density-diameter distributions of harvest simulations compared to the control run.

Although resultant changes in the forest composition from each of the harvest selections tested with no significant differences, it was apparent that the clear cut and selective cut posed the greatest threat to changes in species composition away from the natural progression of forest succession. The structural impact of the clear cut and selective cut showed a significant shift in stem density characteristics to fewer large trees and more abundant smaller trees. The thinning and high-grade harvests demonstrated no statistical difference in forest structure with that of the control simulation. Results of the effects on forest diversity indicated similar findings to those brought out in the structure and composition tests, namely extreme harvest selections alter the species diversity much more greatly than do the milder cut operations. In addition, the reduced diversity generated in the case of the high-grade, cull removal thinning lends some consideration to the effects of overselecting preferred species.

Clearly, the ecological impacts of intensively harvesting forests for energy go beyond the implications given herein of the projected long-term effects. A more thorough treatment would also consider short-term effects, both direct and indirect, including nutrient loss, soil disturbance and compaction, stand regeneration, and even changes in wildlife habitat (van Hook et al. 1980). Though not included here, such additional considerations could be incorporated into this modeling scheme.

Forest simulation models can be effectively applied to examine potential energy yields and long-term ecological impacts of utilizing natural forests for energy needs. Where singular effects of specific forest uses (ie., timber, wildlife) are fairly well documented, at least on the short-term, the cumulative impacts of a multi-use forest plan becomes extremely difficult to determine without the aid of ecological forest models. By employing a modeling approach such as presented here, one can determine optimal or recommended harvesting strategies which maximize the energy return while minimizing any ecological detriment to acceptable standards. Also, models of this sort can be used to compliment existing management plans by simulating the expected harvest selections and schedules, and thereby compute the potential wood-energy return in forest residue.

In the years ahead, the demand for energy from woody biomass will almost certainly intensify. These needs will likely be met through more intensive harvesting of existing forests and/or the advent of biomass energy farms. New energy-use alternatives for woodlots, tree farms and marginal lands will undoubtedly develop with the rising energy demands. The increased attention drawn to woody biomass for energy will obviously lead to more careful consideration as to how we manage our forest resources. In conclusion, forest models can provide an effective tool for testing and determining forest-energy management alternatives which maximize energy return, minimize environmental detriment, and compliment existing land-use plans.

LITERATURE CITED

Conde, Louis F., and J. B. Huffman. 1978. Energy utilization from biomass - fuel plantations. p. 43-64. In Energy in forestry - production and use: Proceedings from the 10th Spring symposium for the Florida section, Society of American Foresters. (Gainesville, Fla., May 30-31, 1978) University of Florida Resources Report 5, 155 p. School of Forest Resources and Conservation, Gainesville, Fla.

Curtis, A. B. 1978. Forest residues in the South. p. 21-33. In Energy in forestry - production and use: Proceedings from the 10th Spring symposium for the Florida section, Society of American Foresters. (Gainesville, Fla., May 30-31, 1978) University of Florida Resources Report 5, 155 p. School of Forest Resources and Conservation, Gainesville, Fla.

Howlett, K., and A. Gamache. 1977. Silvicultural biomass farms: Forest and mill residues as potential sources of biomass. MITRE Tech. Rep. 7347, Vol. VI., 124 p. MITRE Corp., McLean, Va.

Ranney, J. W., and J. H. Cushman. 1980. Regional evaluation of woody biomass production for fuels in the Southeast. p. 109-120. In Biotechnology in energy production and conservation: Proceedings of the 10th symposium on biotechnology and bioengineering. (Gatlinburg, TN., October 3-5, 1979) 354 p. John Wiley and Sons, New York, N.Y.

Shugart, H. H., and D. C. West. 1977. Development of an Appalachian deciduous forest succession model and its application to assessment of the impact of the chestnut blight. J. Eviron. Mgmt. 5: 161-179.

Shugart, H. H., and D. C. West. 1980. Forest succession models. BioScience 30: 308-313.

van Hook, R. I., D. W. Johnson, D. C. West, and L. K. Mann. 1980. Environmental effects of harvesting forests for energy. p. 537-541. In Proceedings of the Bio-energy '80 world congress and exposition. (Atlanta, Ga. April 21-24, 1980) 587 p. The Bio-Energy Council, Washington, D.C.

USE OF AN ENERGY FLOW MODEL TO EVALUATE ALTERNATIVE HARVESTING STRATEGIES IN A MULTISPECIES FISHERY[1]

Joan A. Browder[2]

An energy flow model was designed and quantified to help examine the ecological implications of the large discarded fish bycatch of the Gulf of Mexico shrimp fleet and to evaluate potential impacts of reducing or utilizing the bycatch. Comparison of the relative magnitudes of relevant flows suggests the following: (1) The rate of nitrogen regeneration from discards is four orders of magnitude lower than that from other forms of organic matter in the environment and three orders of magnitude smaller than that expected from animal excretions or the nitrate-nitrite-nitrogen input of rivers; therefore, nitrogen regeneration by discards is inconsequential in stimulating primary productivity and the availability of phytoplankton-based detritus as food for shrimp. (2) The biomass of discards is several orders of magnitude lower than other potential food sources for shrimp, such as living benthos or plant-derived detritus; therefore, discards probably do not appreciably stimulate the growth of shrimp, even if heavily favored by shrimp as a food source. (3) Unless potential predators select for groundfish, the rate of kill by shrimpers may approximately equal the rate of natural predation. We cannot, therefore, discount the possibility that shrimpers affect groundfish biomass and directed groundfish harvests. (4) Unless groundfish are selecting against shrimp relative to their biomass in the environment, groundfish predation on shrimp probably is on the order of three times their harvest rate. We, therefore, cannot discount the possibility that predation by groundfish affects shrimp biomass and shrimp harvests. On the other hand, environmental effects on both shrimp and groundfish populations may be so great that the effects of predation and harvest pressures are overshadowed. Model simulations will provide additional information on the sensitivity of shrimp and groundfish biomass to harvesting and predation ratios. Field studies are needed to quantify selectivity of prey, including commercial penaeid shrimp, by groundfish species.

INTRODUCTION

Penaeid shrimp support productive and economically valuable fisheries throughout the world. In the United States, the ex-vessel value for the penaeid shrimp catch is the highest for any fishery. The commercial harvest of penaeids is by bottom trawls, which are non-selective and catch large quantities of demersal fish that share the shrimp habitat. This bycatch, which usually is discarded, is made up of many species but primarily consists of sciaenids, pomadaysids, sparids, synodotids, serranids, and bothids, which commonly are referred to collectively as groundfish or bottomfish. In the

[1]Paper presented at international symposium Energy and Ecological Modelling, sponsored by International Society for Ecological Modelling (Louisville, Kentucky, April 20-23, 1981).

[2]Joan A. Browder is Operations Research Analyst at the Southeast Fisheries Center, National Marine Fisheries Service, NOAA, 75 Virginia Beach Drive, Miami, Florida 33149. Contribution No. 81-30M.

northern Gulf of Mexico, where more than half of the U.S. shrimp catch is taken (NMFS 1979), the weight of discards averages fourteen times the weight of shrimp landings. The ratio of fish discards to shrimp landings probably is as high or higher elsewhere in the tropics and subtropics, but has not been measured in totality. These species also are harvested by directed fisheries. In the north-central Gulf of Mexico, the combined catch of these fisheries amounts to one-tenth the quantity discarded by the shrimp fleet.

Because edible species form the bulk of the discards, social pressure in a protein-starved world is mounting to either utilize the bycatch or to reduce its size by means of selective gear in order to stimulate its harvest by directed fisheries. Economic conditions have thus far prevented shrimp vessels from landing more than minor amounts of their fish bycatch. The difference between the prevailing price of shrimp and that of associated fish species is very great ($2.92 per pound for shrimp as opposed to $0.14 per pound for fish in Alabama in 1977 (Griffin and Warren 1978)), and the refrigerated space on shrimp vessels is limited; space taken up by fish cannot be utilized for shrimp.

For fishery managers interested in promoting optimum yields of both shrimp and fish, ecological as well as social and economic questions are important. The quantities involved appear at first glance to be very large. What is the effect on an ecosystem of converting such a large quantity of biomass from living to dead? Are groundfish populations diminished or are landings of the directed fleet depressed by killing so many fish? Is shrimp production stimulated by either a reduction in the number of living fish or the creation of a biomass of dead fish? If a significant portion of this bycatch were landed or eliminated how would the system change and what would be the effect on shrimp and fish populations and landings?

Answering these questions requires an understanding of the role of the living fish in the ecosystem, the role of discards in the ecosystem, and the interrelationships between shrimp and demersal fish. As part of the research to develop this understanding, energy-flow analysis of a shrimp-demersal fish system is being conducted. In the first phase of the study, a simple energy-flow model has been designed and quantified on the basis of the most applicable information in the literature, and the relative magnitudes of the stocks and flows have been compared. Presented in this report are results of this analysis.

The setting for the model is the north central Gulf of Mexico, the coastal region most directly under the influence of the Mississippi River discharge. It also receives the inflow of many coastal rivers and runoff from local rainfall. Known as "the fertile crescent" (Gunter 1963), this area, which lies along 4% of

the nation's linear shoreline (Louisiana, Mississippi, and Alabama), was responsible for 35% of the total U.S. marine landings in 1978 (NMFS 1979). A large percentage of north central Gulf landings were menhaden. Penaeid shrimp also were important. The discarded fish bycatch of the shrimp fleet was more than half the size of the menhaden landings.

The initial objectives of the modeling effort were to look at four specific questions.

1) Does predation by groundfish have an appreciable effect on commercial shrimp stocks?

2) Does the harvest of groundfish by shrimp vessels have an appreciable effect on either groundfish stocks or the groundfish landings of the directed fishery?

3) Does the regeneration of nutrients from discarded groundfish stimulate the growth rate of shrimp by appreciably increasing the food supply of the shrimp?

4) Do discards provide a direct food source that stimulates the growth of shrimp?

Considerable perspective on these questions was gained by comparison of flows. Additional perspective and insight on possible ways by which changes made in harvesting strategies could affect shrimp, groundfish, or the system will be examined through computer simulation of the model, to be conducted in the next phase of this ongoing study.

MODEL DESIGN

The first stage in model development was the design of a diagram in energy-circuit language (Odum In press) that shows the major storages of energy (biomass) of the system and the significant flows of energy and/or materials between them, as well as major sources of energy and materials to the system from outside. This design represents the structure of the system (fig. 1).

As indicated in the diagram, the high fishery productivity of the study region is fed by freshwater runoff, which provides an abundance of organic material to detrital food chains and stimulates primary productivity in the water column with land-derived nutrients such as nitrogen. Nitrogen was selected as the nutrient on which to focus attention because nitrogen is thought to be the principal limiting factor in coastal waters (Ryther and Dunstan 1972).

Photosynthesis is concentrated in the upper layers of the water column and light is prevented from reaching the bottom by the dense populations of phytoplankton and by the large quantity of dissolved and particulate carbon compounds entering from the river, so benthic vegetation contributes little to primary productivity.

The phytoplankton is an abundant source of food for zooplankton and pelagic herbivorous fishes such as menhaden. Depending on the quantity utilized by these populations, some phytoplankton sinks to the bottom, where it becomes part of the detritus pool. Zooplankton also contribute to the detrital pool -- primarily through their fecal pellets, which represent unutilized ingested phytoplankton that has been processed by gut microbes and is enriched in nitrogen, compared to the pre-ingested material (Johannes and Satomi 1966). Fish feces are a further contribution to the detrital pool.

The detritus pool is divided into two compartments in the model. Organic material from different sources varies in its physical characteristics and chemical composition and therefore in its relative digestibility and food value to feeding organisms. Dead plant material and dead animal material probably represent two extremes, with animal material more easily digestible and of higher food value than plant material because of (a) the higher concentration of nitrogen in animal material and (b) the higher proportion of labile as opposed to refractory compounds in animal material. Refractory compounds are difficult for even microorganisms to break down and are not readily available to higher organisms. For the sake of simplicity, the two detrital pools of the model are labeled "high-nitrogen" and "low-nitrogen."

Intimately associated with both organic pools are bacteria, which convert a portion of the organic mass to carbon dioxide gas through their own metabolism. In the process, they make the low-nitrogen organic material more easily utilizable by marine animals and increase its food value by increasing its protein concentration. It is

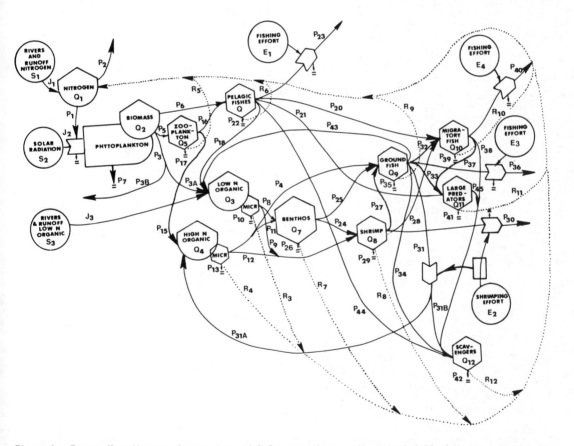

Figure 1.--Energy-flow diagram of ecosystem model for near-shore north-central Gulf of Mexico. Circles are external sources, bullet is light energy to biomass converter, tanks are storage compartments, hexagons are animal biomass compartments, double arrows are symbols indicating interactions (influence of more than one flow), solid lines with arrows are flows of energy in form of biomass, and dashed lines are nitrogen flows.

doubtful that they increase the food value of high-nitrogen organic matter; and, given similar environmental conditions, they might be expected to metabolize this material much more rapidly than the low-nitrogen material.

Benthic organisms are the primary utilizers of both high- and low-nitrogen organic material. In the model, we have included both meio- and macro-benthos in the same compartment. Meio-benthic biomass appears to be an extremely small proportion of the benthos in north central Gulf coastal habitats (Parker et al. 1980).

Low-nitrogen organic material and the benthos are the principal food sources of the commercial penaeid shrimp, which undoubtedly utilize high-nitrogen organic material to the extent it is available. Groundfish may utilize high-nitrogen organic material when it is available but probably receive little energy form low-nitrogen organic material. Their primary food source is the benthos, although the larger individuals also feed to some extent on commercial shrimp.

The model distinguishes between commercial shrimp and other crustaceans, including noncommercial penaeids, because of our specific interest in the commercial penaeids. All crustaceans except commercial shrimp are included in the benthos compartment. Polychaetes are the principal components of that compartment.

Energy flows to higher trophic levels of the system primarily through herbivorous pelagic fish and demersal fish. In addition to these food sources, migratory pelagic fish such as king and Spanish mackerel also feed on shrimp. Larger predators such as dolphins and other marine mammals feed on the migratory pelagics as well as on herbivorous pelagics and groundfish. Sharks, the principal large scavengers of the system, feed on fish at all trophic levels. They also feed on discarded groundfish. Other scavengers in the system that feed on the discards are included in the benthos.

Nutrient recycling occurs in the form of bacterial exudates and with the excrement of animals at all trophic levels in association with respiration.

Four stocks of the system -- coastal pelagics (menhaden), groundfish, migratory pelagics (mackerels), and certain penaeid shrimp -- are harvested commercially in abundance. Groundfish catches are separated into two flows, the first, the catch of the directed fleet, is removed from the system; the second, the bycatch of the shrimp fleet, is released back into the system, where what is not immediately tapped by large scavengers becomes a part of the high-nitrogen compartment, to be utilized by bacteria, the benthos, and, to some extent, by both shrimp and groundfish.

Based on the diagram, a set of simultaneous differential equations was written. This set of equations forms the mathematical structure of the system. The basic formulation of the equations is intrinsic in the symbols of the energy circuit language utilized in the diagram. The set of differential equations becomes the basis for a computer program that can be used to simulate the dynamics of the system.

This model is of necessity a gross simplification of the system. The art of model development

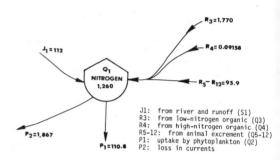

Figure 2.--Nitrogen compartment (Q1) from model, showing initial condition and estimated magnitudes of inflows and outflows for steady-state situation.
Units. Q1: mg nitrogen (N)/m^2, Others: mg N/m^2 day
Sources of values. Q1[a], J1=(2000mg NO$_3$-N/m^3)[b] (6.9 x 10^{11}m^3/yr)[c]/(3.42 x 10^{10}m^2)[d], P1=0.06[e] (P3 + P5 + P6)[f],(P2=(J1 + R3-12 - P1)[g], R3=(P10 x 0.057)[h], R4=(P13 x 0.099)[i], R5=P17[j] x 0.081[k], R6=P22[j] x 0.134[l], R7=P26[j] x 0.12[m], R8=P29[j] x 0.12[n], R9= P35[j] x 0.134[m], R10=P39[j] x 0.14[p], R11= P41[j] x 0.15[p], R12=P42[j] x 0.15[p].

[a]Sklar (1976); [b]Nitrate-nitrogen concentration in Mississippi River (from Sackett 1972); [c]Freshwater flow of Mississippi and Atchafalaya Rivers (from Sackett 1972); [d]Primary area inside 50 fathoms (from Allen et al. 1976); [e]based on C/N ratio of 6 (Strickland 1960); [f]Set equal to sum of phytoplankton predation and sinking rates in steady-state condition; [g]Set equal to sum of inflows minus sum of other outflows in steady-state condition; [h]Bacterial oxidation of low-nitrogen biomass, multiplied by N/OM ratio (from Parsons et al. 1977); [i]Bacterial oxidation of high-nitrogen biomass, multiplied by N/OM ratio (from Parsons et al. 1977); [j]Animal respiration; [k]N/OM ratio for zooplankton from Parsons et al. (1977); [l]N/OM ratio for pelagic fish based on Darnell and Wissing (1975) for pinfish; [m]N/OM ratio for benthos from Darnell and Wissing (1975) for polychaetes; [n]N/OM ratio for shrimp assumed; [o]N/OM ratio for groundfish based on Darnell and Wissing (1975) for pinfish; [p]N/OM ratio for migratory fish, large predators, and scavengers assumed.

is in including the characteristics of the system that determine its operation and its response to variations in outside forces. Considerable testing and evaluation of a model is needed to determine its reliability from this standpoint.

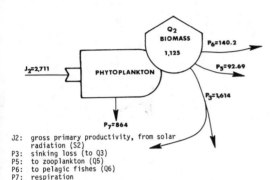

J2: gross primary productivity, from solar radiation (S2)
P3: sinking loss (to Q3)
P5: to zooplankton (Q5)
P6: to pelagic fishes (Q6)
P7: respiration

Figure 3.--Phytoplankton compartment (Q2) from model, showing initial condition and esti-mated magnitude of inflows and outflows for steady-state.
Units. Q2: mg dry organic matter (DOM)/m^2, Others: mg DOM/m^2 · day
Source of values. Q2[a], J2=(1.642[b] + 0.768[c])Q2, P3=(J2 - P5 -P6 - P7)Q2, P5[d], P6[d], P7=Q2 x RQ2[e]

---

[a]Sklar (1976), [b]Net primary productivity (from Sklar 1976), [c]Respiration coefficient (from Ryther and Guillard 1962), [d]Calculated in flow-balancing procedure, [e]Respiration coefficient for a 4µ eurythermal diatom, Cyclotella nana clone 3H (from Ryther and Guillard 1962).

MODEL QUANTIFICATION PROCEDURES

The model was quantified for steady state conditions in a series of steps. In the steady state condition, the sum of the inflows to each compartment equals the sum of the outflows, and there is no change in energy, or mass. The steps were as follows:

Step 1. Biomass, respiration rates, harvesting rates, and assimilation rates were esti-mated on the basis of the most applicable literature or data available.

Step 2. Inflows were calculated on the basis of outflows with an iterative top-down flow-balancing method, starting with the highest tropic levels.

Step 3. A few values at the bottom (front end) of the model were set in such a way that they both satisfied the steady state con-dition and seemed realistic.

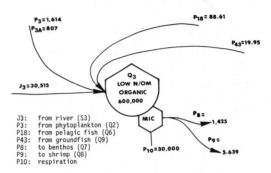

J3: from river (S3)
P3: from phytoplankton (Q2)
P18: from pelagic fish (Q6)
P43: from groundfish (Q9)
P8: to benthos (Q7)
P9: to shrimp (Q8)
P10: respiration

Figure 4.--Low-nitrogen organic compartment from model, showing initial condition and esti-mated magnitudes of inflows and outflows for steady-state situation.
Units. Q3: mg DOM/m^2, Others: mg DOM/m^2 · day
Sources of values. Q3=(15 mg DOM/g sediment)[a](1 x 10^4 cm^3/cm · m^2)(2 cm)[b](4 g sediment/cm^3)[c](0.5 labile)[d], J3[e], P3[f], P3A=P3 x0.5[g], P18=P6(1-0.42[h]) + P16(1-0.6[i]), P43=P4 (1-0.86[j]) + P25(1-0.7[k]) + P27(1-0.7[l]), P8[e], P9[e], P10=Q3 x RQ3[n]

---

[a]Hausknecht (1980), [b]Assumed depth of biolog-ically active sediment, [c]Approximated from CRC Handbook of Chemistry and Physics, [d]Assumed, [e]Calculated in flow balancing procedure, [f]See figure 3, [g]Fraction of sinking phytoplankton reaching bottom in study area, [h]Assimilation efficiency of phy-toplankton to pelagic fish (assumed), [i]Assimilation efficiency of zooplankton to pelagic fish (assumed), [j]Assimilation effi-ciency of high-nitrogen organic to groundfish (for pinfish from Darnell and Wissing 1975), [k]Assimilation efficiency of benthos to groundfish (assumed), [l]Assimilation effi-ciency of shrimp to groundfish (assumed), [m]Calculated by flow-balancing procedure, [n]Decomposition rate (for phytoplankton detritus, from Ogura 1972).

Data and literature sources for initial con-ditions and other starting values are given with excerpts from the model diagram in figures 2 through 13. The common currencies of the model are "milligrams dry weight of organic matter" (DOM) and "milligrams of inorganic nitrogen" per square meter. Conversions used to standardize the units found in the literature were from Parsons et al. (1977) and are as follows: O_2 to DOM, x 1; Carbon to DOM, x 2.5; Chlorophyll to Carbon, x 2.5; wet weight to dry weight, x 0.2. Rates are in terms of approximate days (1 yr/360) for ease of computation. I assumed an effective water depth (or photic zone) of 10 meters where necessary to convert concentration values to a unit area basis.

Source Flows

Source flows to the model were river inorganic nitrogen, river organic material, and solar radiation. Mississippi River nitrate-nitrogen multiplied by average daily Mississippi River flow serves as the river input value and was divided by the area estimated to be the primary receptor of the input (primary area). The river input of organic material was calculated for steady state conditions to balance outflows with inflows.

Stock Initial Conditions

Nitrogen (NO_3, NO_2, and NH_3-N) and phytoplankton initial stock values both were based on Sklar (1976). Both values are averages for the four seasons. Surprisingly, the two values are approximately equal, the nitrogen value being slightly greater.

Gross primary productivity for initial conditions was estimated as the sum of net primary productivity (Sklar 1976) and respiration (Ryther and Guillard 1962) for the initial phytoplankton biomass. The respiration rate that was used was for a eurythermal diatom at 25° C.

Three types of data were used to estimate initial biomass. These were fisheries landings (or bycatch estimates), research field calculations, and census data. Fisheries landings and bycatch data were divided by estimated fishing mortality rates to yield total stock biomass, using the relationship $N = C/F$ (where N is number or biomass, C is catch in terms of number or weight, and F is fishing mortality). Fishing mortality was assumed to be 1.0 for menhaden, 1.0 for shrimp, and 0.67 for groundfish. Values of 0.41 for king mackerel and 0.28 for Spanish mackerel were estimated based on the Fishery Management Plan for Coastal Migratory Pelagic Resources (GMSAFMC 1980).A multiplication factor was used to grossly estimate total stocks in the given trophic group from the biomass of commercial species. The multiplication factor to account for similar non-commercial species was 2.0 for menhaden, 2.0 for the mackerels, 1.0 for groundfish, 1.0 one for shrimp. Biomass was divided by estimated area from which the landings came, using measurements from Allen et al. (1976), to express it on a unit area basis. All areas are inside 50 fathoms (91 meters). Weights were converted to dry weight.

The biomass of large predators was estimated from census information on bottlenose dolphins (Leatherwood et al. 1978), which are considered to be the most abundant marine mammal in the study area. Summer average density figures from three

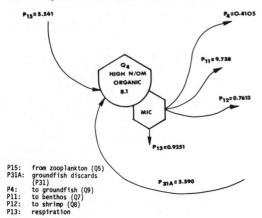

P15: from zooplankton (Q5)
P31A: groundfish discards (P31)
P4: to groundfish (Q9)
P11: to benthos (Q7)
P12: to shrimp (Q8)
P13: respiration

Figure 5.--High-nitrogen organic compartment from model, showing initial condition and estimated magnitudes of inflows and outflows for steady-state situation.
Units. Q4: mg DOM/m^2, Others: mg DOM/m^2 · day
Source of values. Q4[a], P15=P5(1 - 0.7[b])0.2[c], P31A=P31[d] - P31B[e], P4[f], P11[f], P12[f], P13=Q4 x RQ4[g]

---

[a]Assumed, [b]Assimilation efficiency of phytoplankton to zooplankton, [c]Efficiency of zooplankton gut bacteria in converting ingested phytoplankton to feces, [d]See figure 10, [e]See figure 13, [f]Calculated in flow-balancing procedure, [g]RQ4=(P15 + P31A - P4 - P11 - P12)/Q4

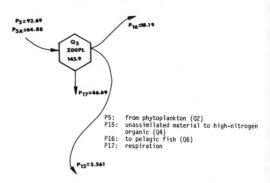

P5: from phytoplankton (Q2)
P15: unassimilated material to high-nitrogen organic (Q4)
P16: to pelagic fish (Q6)
P17: respiration

Figure 6.--Zooplankton compartment (Q5) from model, showing initial condition and estimated magnitudes of inflows and outflows for steady-state situation.
Units. Q5: mg/m^2, Others: mg/m^2 · day
Sources of values. Q5=(2360ind/m^3)[a](6 x 10^{-6}g DOM/ind)[b](10^3/m^2)[c], P5A=P5[d] x 0.7[e], P15=P5(1 - 0.7[e]), P16[d], P17=Q5 x 0.32[e]

---

[a]Density (from Reitsema 1980), [b]Individual weight (from Conover 1959), [c]Assumed depth of biologically active zone, [d]Calculated in flow-balancing procedure, [d]Assimilation efficiency of phytoplankton to copepods (from Parsons et al. 1977), [e]Respiration rate coefficient for a 6 μg D.W. copepod (from Conover 1959).

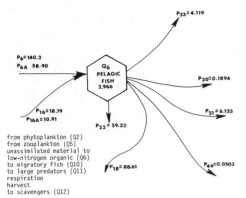

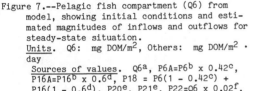

P6: from phytoplankton (Q2)
P16: from zooplankton (Q5)
P18: unassimilated material to
low-nitrogen organic (Q6)
P20: to migratory fish (Q10)
P21: to large predators (Q11)
P22: respiration
P23: harvest
P44: to scavengers (Q12)

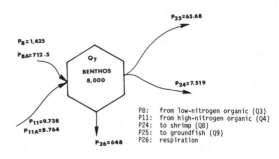

P8: from low-nitrogen organic (Q3)
P11: from high-nitrogen organic (Q4)
P24: to shrimp (Q8)
P25: to groundfish (Q9)
P26: respiration

Figure 7.--Pelagic fish compartment (Q6) from
model, showing initial conditions and esti-
mated magnitudes of inflows and outflows for
steady-state situation.
Units. Q6: mg DOM/m^2, Others: mg DOM/m^2 ·
day
Sources of values. Q6[a], P6A=P6[b] x 0.42[c],
P16A=P16[b] x 0.6[d], P18 = P6(1 - 0.42[c]) +
P16(1 - 0.6[d]), P20[e], P21[e], P22=Q6 x 0.02[f],
P23=(4.5 x 10^{14} mg)[a](0.2 D.W./W.W)[b]/(6.096 x
10m^2)[d], P44[e,j], Q6=(4.5 x 10^{14} mg)[a](0.2
D.W.W)[b](2)[c]/(6.096 x 10m^2)[d]

---

[a]Approximate Louisiana menhaden landings for
1975 (from NMFS 1978), [b]Dry weight/wet weight
conversion (from Parsons et al. 1977),
[c]Louisiana coastal area inside 50 fathoms
(from Allen et al. 1976), [d]Calculated in
flow-balancing procedure, [e]Assimilation efficiency of phytoplankton to pelagic fish
(assumed), [f]Assimilation efficiency of
zooplankton to pelagic fish (assumed),
[g]Calculated in flow-balancing procedure,
[h]Respiration rate coefficient for a 226 g
W.W. menhaden (from Hettler 1976), [i]See
figure 13 for weighting factor.

coastal Louisiana areas were divided in half to
take into consideration that either local or long-
distance migrations of dolphin into these areas
had occurred, as indicated for areas in Texas
(Shane 1980). The original value, 0.44
dolphins/km^2, seemed too large to be an average
for U.S Gulf of Mexico coastal waters as a whole.
Multiplication by the area of coastal waters north
of latitude 28° N on the east coast and 26° N on
the west coast yields a population of more than
70,000 dolphin, which probably is greater then the
actual Gulf population. To be on the conservative
side, particularly in view of quantification
results, dolphin biomass was not increased by a
factor to account for other marine mammals.

Reported field collection results from the
Strategic Petroleum Reserve Texoma/Capline studies

Figure 8.--Benthos compartment (Q7) from model,
showing initial condition and estimated
magnitudes of inflows and outflows for
steady-state situation.
Units. Q7: mg DOM/m^2, Others: mg DOM/m^2 ·
day
Sources of values: Q7[a], P8A=P8[b,c] x 0.5[d],
P11A=P11[b,e] x 0.9[f], P24[g,h], P25[g], P26=Q7 x
0.081[i]

---

[a]Parker et al. 1980, [b]Calculated in flow-
balancing procedure, [c]Weighting factor of 1
(assumed), [d]Assimilation efficiency of low-
nitrogen organic to benthos (assumed),
[e]Weighting factor of 500 (assumed),
[f]Assimilation efficiency of high-nitrogen
organic to benthos (assumed, based on Darnell
and Wissing 1975), [g]Calculated by flow-
balancing procedure, [h]See figure 9 for
weighting factor, [i]Respiration rate coefficient for 1 g W.W. benthic organism (20
times the anaerobic rate measured by Pamatmat
1980, according to the relationship suggested
by Pamatmat).

off the Louisiana coast in the Weeks Island and
West Hackberry areas were used to estimate stocks
(DOM) per unit area of zooplankton, low-nitrogen
organic material and benthos. No information was
available from which to estimate high-nitrogen
organic matter, so it was assumed to be approximately equal to one day's inflows.

Respiration Rates

Respiration rates were estimated on the basis
of weight-specific measurements in the literature
for the particular species or a similar species.
The average weight assumed for each calculation is
given in the figure legends. Benthic respiration
rates are Pamatmat's (1980) anaerobic rates for a
number of species of polychaetes and bivalves
multiplied by 20 to estimate aerobic respiration,
suggested by Pamatmat. Respiration rates are proportional to stock biomass.

Harvesting Rates

Harvesting rates are landings data and bycatch divided by the estimated area from which the landings were taken. Areas are inside 50 fathoms and come from Allen et al. (1976).

Ratio of Nitrogen in Excrement to DOM Loss in Respiration

A basic assumption in quantifying the model was that the waste nitrogen released in the excrement of animals is linearly related to the energy (expressed as DOM loss) released in the respiration of the animals and can be estimated as the N/DOM ratio of the body tissues of the animal.

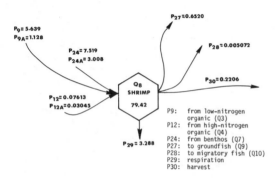

Figure 9.--Commercial shrimp compartment (Q8) from model, showing initial condition and estimated magnitudes of inflows and outflows for steady-state situation.
Units: Q8: mg DOM/m^2, Others: mg DOM/m^2 · day
Sources of values: Q8=(2.41 x 10^{13} mg)^a(0.2 D.W./W.W)^b/(6.069 x 10^{10}m^2)^c, P9A = P9^{d,e} x 0.2^f, P12A=P12^{d,g} x 0.4^h, P24A=P24^{d,i} x 0.4^j, P27^d, P28^{d,k}, P29=Q8 x 0.0414^l, P30^m

---

^{a}Approximate Louisiana shrimp landings for 1975 (from NMFS 1978), ^{b}Dry weight to wet weight conversion (from Parsons et al. 1977), ^{c}Louisiana coastal area inside 50 fathoms (from Allen et al. (1976), ^{d}Calculated in flow-balancing procedure, ^{e}Weighting factor of 1 (assumed), ^{f}Assimilation efficiency of low-nitrogen organic to shrimp (inferred from Jones 1973), ^{g}Weighting factor of 1000 (assumed), ^{h}Assimilation efficiency of high-nitrogen organic to shrimp (inferred from Jones 1973), ^{i}Weighting factor of 10 (assumed), ^{j}Assimilation efficiency of benthos to shrimp (inferred from Jones 1973),^{k}See figure 9 for weighting factor, ^{l}Respiration rate coefficient for 6.7 g brown shrimp, assuming 12 active hours (from Bishop et al. 1980), ^{m}Same as Q8.

Assimilation Efficiencies

Assimilation efficiencies were used to estimate the quantity of ingested food actually contributing to the growth or maintenance of the feeding organism. Assimilation rates for the same food source may differ for different feeding species, depending on the digestive equipment of each. A few of the assimilation efficiencies used in the model were based on published values. Most are gross estimates. Where efficiencies were thought to be 0.9 or above, they were set at 1.0 for the sake of simplicity.

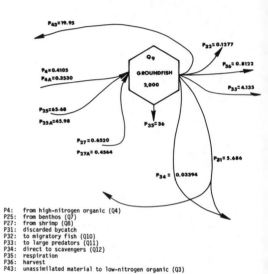

Figure 10.--Groundfish compartment (Q9) from model, showing initial condition and estimated magnitudes of inflows and outflows for steady-state situation.
Units. Q9: mg DOM/m^2, Others: mg DOM/m^2 · day
Sources of values. Q9={(4.00 x 10^{14}mg)^a(0.2 D.W./W.W.)^b/(0.67)^c}/(3.42 x 10^{10}m^2)^d, P4A=P4^e x 0.86^f, P25A=P25^e x 0.7^g, P27A=P27^e x 0.7^h, P31=(3.5 x 10^{14}mg)^i/(3.42 x 10^{10}m^2)^d, P32^e, P33^e, P34^e, P35=Q9 x 0.018^j, P36=(4.00 x 10^{14}mg)^a(0.2 D.W./W.W)^b/(3.42 x 10^{10}m^2)^d, P43=P4(1-0.86^f) + P25(1-0.7^h) + P27(1-0.7^h)

---

^{a}Groundfish catch, including discards in "Primary Area" (GMFMC 1980), ^{b}Dry weight to wet weight conversion, ^{c}Fishing mortality (assumed), ^{d}Area of "Primary Area" inside 50 fathoms (estimated from Allen et al. 1976), ^{e}Calculated by flow-balancing method, ^{f}Assimilation efficiency of high-nitrogen organic to groundfish (inferred from Darnell and Wissing 1975), ^{g}Assimilation efficiency of benthos to groundfish (assumed), ^{h}Assimilation efficiency of shrimp to groundfish (assumed), ^{i}Estimated bycatch in Primary Area, ^{j}Respiration rate coefficient for pinfish (from Hoss 1974).

Selectivity Weighting Coefficients

The selectivity weights are an important feature of the model that will be discussed in the next section.

Iterative Top-down Flow-balancing Procedure

Solving for unknown flow rate coefficients by means of simultaneous linear differential equations was not possible due to the feedback loops in the system. The iterative procedure of setting rates was a workable alternative. Starting with the highest trophic levels, inflows from all sources were calculated for each stock successively. Where more than one inflow to a

stock occurred, the relative inflows were assumed to be equal to the relative biomasses of contributing stocks. In cases where significant selectivity was thought to occur, stock weighting factors were employed.

The rationale for making relative inflows equal to relative biomasses of contributing stocks is straight-forward. Examples of the dependence of prey selection on either relative size of alternate prey or relative prey abundance are common in the literature on optimal foraging (Schoener 1971, Murdock et al. 1975). Relative biomass serves as in integration of both of these factors. Since flows are proportional to biomasses in the simple differential equations of the model, when the above method of calculating pathways is used, pathway coefficients are equal in the unweighted situation and, in the weighted situation, the same number is multiplied by the specific weighting factor to obtain the pathway coefficient. A description of the weighting factor values used in the model was postponed until

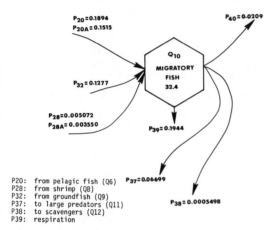

P20: from pelagic fish (Q6)
P28: from shrimp (Q8)
P32: from groundfish (Q9)
P37: to large predators (Q11)
P38: to scavengers (Q12)
P39: respiration

Figure 11.—Migratory fish compartment (Q10) from model, showing initial condition and estimated magnitudes of inflows and outflows for steady-state situation.
Units. Q10: mg DOM/m^2, Others: mg DOM/m^2 · day
Sources of values. Q10={(2.27 x 10^{12}mg)^a/0.41^b + (4.5 x 10^{12}mg)^c/0.28^d 0.67^e}/(1.81 x 10^{11}m^2)^f, P20A=P20^g x 0.8^h, P28A=P28^g x 0.7^i, P32^{g,j}, P37^g, P38^{g,k}, P39=Q10 x 0.006^l

[a]Approximate king mackerel landings of Gulf and Atlantic (NMFS 1978), [b]Estimated fishing mortality (within range considered reasonable in GMSAFMC 1980), [c]Approximate Spanish mackerel landings of Gulf and Atlantic (NMFS 1978), [d]Estimated fishing mortality (within range considered reasonable in GMSAFMC 1980), [e]Fraction of biomass in Gulf (assumed), [f]Gulf coastal area north of 28° N latitude inside 50 fathoms, [g]Calculated by flow-balancing procedure, [h]Assimilation efficiency of pelagic fish to migratory fish (assumed), [i]Assimilation efficiency of shrimp to migratory fish (assumed), [j]Assimilation efficiency of 1 (assumed), [k]See figure 13 for weighting factor, [l]Respiration rate coefficient for 1 kg skipjack tuna (from Brill 1979).

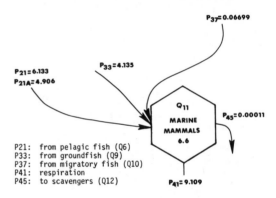

P21: from pelagic fish (Q6)
P33: from groundfish (Q9)
P37: from migratory fish (Q10)
P41: respiration
P45: to scavengers (Q12)

Figure 12.—Large predator compartment (Q11) from model showing initial condition and estimated magnitudes of inflows and outflows for steady-state situation.
Units. Q11: mg DOM/m^2, Others: mg DOM/m^2 · day
Sources of values. Q11=(0.44 x 10^{-6}/m^2)^a (0.2)^i (150 x 10^{6}mg/dolphin)^{b}0.5^c, P21A=P21^d x 0.8^e, P33^{d,f}, P37^{d,f}, P41=Q11 x 1.38^g, P45^{d,f,h}

[a]Density of bottlenose dolphins at Louisiana coastal sites (from Leatherwood et al. 1978), [b]Estimated weighted average weight of bottlenose dolphin, [c]Correction factor to account for possible higher-than-average densities in sampling area, [d]Calculated by flow-balancing procedure, [e]Assimilation efficiency of pelagic fish to large predators (assumed), [f]Assimilation efficiency of 1 (assumed), [g]Respiration rate coefficient of 150 kg bottlenose dolphin (from Irving et al. 1941), [h]See figure 13 for weighting factor, [i] dry wt./wet wt.

this portion of the discussion in order that their meaning could first be explained.

Although the weighting factors are gross assumptions and without any quantitative foundation in the literature, they were carefully chosen not only to fulfill general qualitative observations about the system but also, in some cases, to satisfy an "even-if" strategy with the objectives of the modeling effort in mind. To indicate lack of selectivity for anything other than relative biomass, weighting factors for all flows entering a compartment were set equal to 1. This was the case for all compartments, except shrimp, benthos, and scavengers.

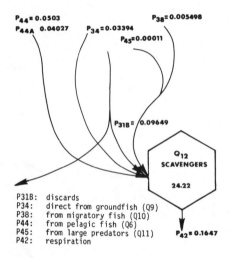

P31B: discards
P34: direct from groundfish (Q9)
P38: from migratory fish (Q10)
P44: from pelagic fish (Q6)
P45: from large predators (Q11)
P42: respiration

Figure 13.--Large scavenger compartment (Q12) from model, showing initial condition and estimated magnitudes of inflows and outflows for steady-state situation.
Units. Q12: mg DOM/m^2, Others: mg DOM/m^2 · day
Sources of Values. P31B[a], Q12=(24.29 mg/m^2)[b] (0.2 D.W./W.W.)[c] 5[d], P34[e,f], P38[e,f], P42=Q12 x 0.00684[g], P44A=P44[e,f] x 0.8[h]

[a]Calculated from flow-balancing method, using P31 rather than Q9, as biomass, with weighting factor of 1000, [b]Estimated average biomass of sharks in Gulf of Mexico (L. Rivas, unpubl. MS. National Marine Fisheries Service, Miami), [c]Dry weight to wet weight conversion, [d]Multiplication factor for near shore concentrations (assumed), [e]Calculated by flow-balancing method, [f]Weighting factor of 1 (assumed), [g]Respiration rate (routine) of 2 kg spiny dogfish (Brett and Blackburn 1978), [h]Assimilation efficiency of pelagic fish to scavengers (assumed).

Dead material is more accessible to scavengers than live material because of their feeding behavior, and the concentration of dead material in the discarding practice may increase the availability of this material to scavengers; therefore the selectivity factor for discards against other food sources of scavengers was set at 1000:1. This would tend to maximize the amount of discards immediately removed from the system by large scavengers and minimize that received by the detrital pool.

Using weighting factors, shrimp were given a selectivity of 1000 to 1 for high-nitrogen organic material (which includes the fish discards) over low-nitrogen organic material. They were given a selectivity of 10 to 1 for high-nitrogen organic material over the benthos, assuming that the concentration of discards might make it easier for shrimp to feed on them than on the benthos. A selectivity of 100 to 1 for benthos over low-nitrogen organic material follows from the above relationships and seems reasonable. The benthos was given a selectivity of 500 to 1 for high-nitrogen organic material over low-nitrogen organic material. The various foods of groundfish were not weighted for selectivity by groundfish.

In the context we have used it in this paper, the term "selectivity" refers to the occurrence in an animal's diet of one food over another in greater mass proportions than would be predicted from the relative biomasses of the alternative foods in the environment.

QUANTIFICATION RESULTS

The flow rates estimated for the model are presented in figures 2 through 13. Because of the use of assimilation efficiencies in the model, it is necessary in some cases to present a pair of numbers for the same pathway. The "A" flow is the flow of assimilated material to the recipient. At steady state, total assimilated inflows and total outflows are equal (excepting differences due to rounding error).

The objectives of this modeling effort require that we focus attention on those flows to and from nitrogen (Q1), to high-nitrogen organics (Q4), to shrimp (Q8), from shrimp, and from groundfish (Q9). Figures 2, 5, 9, and 10 facilitate this exercise.

The nitrogen compartment (Q1) received inflows from river and runoff (J1), from the breakdown (microbial respiration) of low-nitrogen organic material (R3), from the breakdown of high-nitrogen organic material (R4), and in animal excrement (R5-12). Outflows are uptake by phytoplankton associated with their growth and losses to the system in currents (figure 2).

The major contribution is from the mineralization associated with the breakdown (microbial oxidation) of low-nitrogen organic material, most of which enters the system in rivers and as coastal runoff. This contribution is more than five orders of magnitude higher than the contribution from the breakdown of high-nitrogen organic material, approximately half of which appears to be made up of discards (fig. 5). The combined contribution from animal excrement is three orders of magnitude greater than that from high-nitrogen organics.

Recycling from animals alone is almost sufficient to satisfy the daily phytoplankton uptake rate, although it might not be adequate to promote maximum photosynthesis. Nitrogen in river and runoff waters also is sufficient by itself to replenish nitrogen taken up in daily growth of phytoplankton at the calculated rate. Furthermore the quantity of nitrogen released from discards and other high-nitrogen sources is a minute fraction (0.00007) of the stock of inorganic nitrogen in the environment.

Even given the possibility of very large errors in quantifying the stocks and flows of the model, it seems safe to conclude that the discard practice does not stimulate primary production and in that way increase the quantity of food available to shrimp.

In the model shrimp receive energy from three sources: benthos (P24), low-nitrogen organic material (P9), and high-nitrogen organic material (P12) (fig. 9). Even when flows of only assimilated food are counted, the other two flows are almost two orders of magnitude higher than the flow that includes discards. This result is in spite of the food selectivity factor being weighted 1000 to 1 for high-nitrogen organics over low-nitrogen organics and 10 to 1 for high-nitrogen organics over benthos. An appreciable stimulation of the production of shrimp biomass by food in the form of discards is not likely. Improvement in the accuracy of model input numbers or results of computer simulation of the model are not likely to alter this conclusion.

Energy flows from the shrimp compartment by four routes: predation by groundfish (P27), predation by migratory pelagics (P28), harvesting by man (P30), and respiration (P29). Of these flows, respiration is by far the largest, 4% of biomass. Predation by groundfish is one fifth the size of respiration and three times the size of harvesting. Its relative magnitude suggests that variation in this flow possibly could have an appreciable effect on shrimp biomass and also on harvesting rates.

The importance of groundfish as a predatory force on shrimp is highly dependent upon the weighting factors used to estimate the flow of energy to groundfish from the several sources. In quantification of the model, equal weights were given to shrimp and other groundfish foods. This allowed the proportion from each food source to be governed entirely by the relative biomasses of the sources, as has been described. If groundfish are feeding less on shrimp than would be predicated from the biomass of shrimp relative to other

groundfish foods in the environment, then the flow of shrimp to groundfish will be lower than calculated. That the extent of the flow reduction is almost exactly inversely proportional to the weighting ratio against shrimp can be seen in table 1.

It is not unrealistic to think that selection by groundfish may be weighted against the commercial penaeids, because groundfish of the size and species that make up the bulk of the groundfish biomass may be too small to eat the commerical species of penaeids that occur with them. Our understanding of this basic aspect of shrimp-groundfish interaction might greatly benefit if future field and laboratory work, as well as analytical work, concentrated on quantifying selectivity relative to the proportions of biomasses of the various prey in the environment.

If only a small fraction of benthic biomass consisted of animals that occur in groundfish diets, then the way the model has been quantified might not be realistic; however, polychaetes, a principal food of groundfish, form a large portion of the benthic biomass in typical shrimp and groundfish habitats of the northern Gulf of Mexico (Parker et al. 1980).

Although the flow of shrimp to groundfish may be of consequence to shrimp and possibly to their harvesters, it does not appear to be of much significance to groundfish, based on its magnitude relative to the flow of energy to groundfish from the benthos (P25), which is two orders of magnitude higher (fig. 8) even with no weighting factors. The flow of energy from high-nitrogen organics (the discards) to groundfish (P4) is only slightly less than that from shrimp to groundfish (P27).

Energy is lost from the groundfish compartment through six routes (not counting unassimilated ingested material (P43)). These are predation by migratory pelagics (P32), predation

Table 1.--Effect of the weighting factors for selectivity by groundfish on the energy flow from shrimp to groundfish.

	Weighting ratios	Flow of shrimp to groundfish ($mg/m^2 \cdot day$)
Base case	1:1	0.6520
Minor selection against shrimp	10:1	0.06579
Moderate selection against shrimp	100:1	0.006586
Intense selection against shrimp	1000:1	0.0006585

by large predators (P33), predation by large scavengers (P34), respiration (P35), harvesting by directed fisheries (P36), and harvesting by the shrimp fleet (P31), which returns dead fish material to the high-nitrogen organics compartment. The largest flow, 2% of the biomass, is respiration. Two other flows may be large enough to have a significant effect on stock biomass. These are the discard catch and predation by large predators, the dolphins. That these two flows are substantial and of approximately the same size tells us two things about this system: (1) Predation by marine mammals could be a significant force in this sytem, and (2) "Predation" by man has doubled the rate of predation on the groundfish stock.

DISCUSSION

In energy-flow analysis, model quantification is merely an early stage in model development and utilization; yet, in this initial stage, we have gained considerable perspective on the questions that must be addressed. Conclusions regarding nitrogen recycling from discards and discard material as food for shrimp indicate that the return of the bycatch to the environment probably is of no consequence in providing food to promote the growth of shrimp. We can therefore conclude that the utilization of major portions of the bycatch, which would mean the removal of this material from the system, would not have a detrimental effect on shrimp biomass or harvests.

All we can conclude at this point with regard to the effect on groundfish populations of harvesting pressure by shrimpers is that this energy loss from the groundfish compartment is of sufficient magnitude to have a significant effect on groundfish biomass. It may also have an effect on predators of groundfish, particularly the marine mammals.

With regard to the importance of predation on shrimp by groundfish, the model suggests that, if groundfish prey on shrimp in proportion to their biomass relative to that of other groundfish prey (the unweighted condition), then groundfish are significant predators of shrimp, causing a rate of energy loss that may be three times greater than that resulting from shrimp harvests.

A simple test conducted by varying the weighting factors to represent selection against shrimp by groundfish indicated that the energy flow from shrimp to groundfish is very sensitive to weighting. The differences were inconsequential from the point of view of the groundfish but very important to shrimp.

These initial model results suggest that quantification of selectivity by groundfish is the key to understanding shrimp-groundfish interactions and the potential impact on shrimp of reducing the groundfish catch by shrimp vessels. The approach to quantifying selectivity that was introduced by the model may be applicable to field investigation and may be particularly useful.

This modeling work has not yet examined the role of environmental factors in influencing the biomass of shrimp and groundfish populations. Despite their magnitudes, predation and harvesting rates may be unimportant in controlling shrimp and groundfish biomass in comparison to enviromental factors.

ACKNOWLEDGMENTS

I wish to thank Mark Farber and Scott Nichols, my colleagues at the Southeast Fisheries Center in Miami, for helpful discussions. I also thank Grady Reinert for preparing the drawings.

LITERATURE CITED

Allen, D. M., J. E. Tashiro, and A. C. Jones. 1976. The present status of U.S. fisheries off Mexico (Gulf of Mexico and Caribbean Sea). A briefing paper. Southeast Fisheries Center, National Marine Fisheries Service, Miami, Florida.

Bishop J. M., J. G. Gosselink, and J. H. Stone. 1980. Oxygen cosumption and hemolymph osmolality of brown shrimp, Penaeus aztecus. Fish. Bull. 78: 741-757.

Brett, J. R. and J. M. Blackburn. 1978. Metabolic rate and energy expenditure in the spiny dogfish Squalus acanthias. J. Fish. Res. Bd. Canada 35: 816-821.

Brill, R. W. 1979. The effect of body size on the standard metabolic rate of skipjack tuna, Katsuwonus pelamis. Fish. Bull. 77: 494-498.

Conover, R. J. 1959. Regional and seasonal variation in the respiratory rate of marine copepods. Limnol. Oceanogr. 4: 259-269.

Darnell, R. M. and T. E. Wissing. 1975. Nitrogen turn-over and food relationships of the pinfish Lagodon rhomboides in a North Carolina estuary. p. 81-110 In F. J. Vernberg (ed.). Physiological Ecology of Estuarine Organisms. University of South Carolina Press, Columbia, South Carolina.

Griffin, W. L. and J. P. Warren. 1978. Costs and return data; groundfish trawlers of northern Gulf of Mexico. Unpublished report to Gulf of Mexico Fishery Managment Council, Tampa, Florida.

Gulf of Mexico Fishery Management Council (GMFMC). 1980. Draft fishery management plan for groundfish in the Gulf of Mexico. Gulf of Mexico Fishery Management Council, Lincoln Center, Suite 881, 5401 West Kennedy Blvd., Tampa, Florida. 233 p.

Gulf of Mexico and South Atlantic Fishery Management Council (GMSAFMC). 1980. Fishery Management Plan, Final Environmental Impact Statement, and Regulatory Analysis for the Coastal Migratory Pelagic Resources.

Gunter, G. 1963. The fertile fisheries crescent. J. Miss. Acad. Sci. 9:286-290.

Hausknecht, K. A. 1980. Describe surficial sediments and suspended particulate matter. Vol. V. In W. B. Jackson and G. M. Faw (eds.). Biological/chemical survey of Texoma and Capline sector salt dome brine disposal sites off Louisiana, 1978-1979. NOAA Tech. Mem. NMFS-SEFC-20, 56 p. Available from NTIS, Springfield, Virginia.

Hettler, W. F. 1976. Influence of temperature and salinity on routine metabolic rate and growth of young Altantic menhaden. J. Fish. Biol. 8: 55-65.

Hoss, D. E. 1974. Energy requirements of a population of pinfish Lagodon rhomboides (Linnaeus). Ecology 55: 848-855.

Irving, L., P. F. Scholander, and S. W. Grinnell. 1941. The respiration of the porpoise Tursiops truncatus. J. Cell. Comp. Physiol. 17: 145-168.

Johannes, R. E. and M. Satomi. 1966. Composition and nutritive value of fecal pellets of a marine crustacean. Limnol Oceangr. 11: 191-197.

Jones, R. R. 1973. Utilization of Louisiana estuarine sediments as a source of nutrition for the brown shrimp Penaeus aztecus. Ph.D. Dissertation. Louisiana State University. Baton Rouge, Louisiana. 130 p.

Leatherwood, S., J. R. Gilbert, and D. G. Chapman. 1978. An evaluation of some techniques for aerial census of Bottlenosed dolphins. J. Wildl. Manage. 42: 239-250.

Murdock, W. W., S. Avery, and M.E.B. Smyth. 1975. Switching in predatory fish. Ecology 56: 1094-1105.

NMFS (National Marine Fisheries Service). 1978. Fishery Statistics of the United States 1975. Statistical Digest No. 69. U. S. Dept. Comm. NOAA. Washington, D. C. 418 p.

NMFS (National Marine Fisheries Service). 1979. Fisheries of the United States, 1978. Current Fishery Statistics No. 7800. U. S. Dept. Comm. NOAA. Washington, D. C. 120 p.

Odum, H. T. In press. System. J. Wiley, New York.

Ogura, N. 1972. Decomposition of dissolved organic matter derived from dead phytoplankton. p. 508-515 In A. Y. Takenouti (ed.). Biological Oceanography of the Northern Pacific Ocean. Publ. Idemitsu Shoten (Tokyo).

Pamatmat, M. M. 1980. Facultative anaerobosis of benthos. p. 69-92 In K. R. Tenore and B. C. Cull (eds.). Marine Benthic Dynamics. University of South Carolian Press. Columbia, South Carolina.

Parker, R. H., A. L. Crowe, and L. S. Bohme. 1980. Describe living and dead benthic (macro- and meio-) communities. Vol. I. In W. B. Jackson and G. M. Faw (eds.). Biological/chemcial survey of Texoma and Capline sector salt dome brine disposal sites off Louisiana, 1978-1979. NOAA Tech. Mem. NMFS-SEFC-25, 103 p. Available from: NTIS, Springfield, Virginia.

Parsons, T. R., M. Takahashi, and B. Hargrave. 1977. Biological Oceanographic Processes. Pergamon Press. New York. 332 p.

Reitsema, L. A. 1980. Determine seasonal abundance, distribution, and community composition of zooplankton. Vol. II. in W. B. Jackson and G. M. Faw (eds.). Biological/chemical survey of Texoma and Capline sector salt dome brine disposal sites off Louisiana, 1978-1979. NOAA Tech. Mem. NMFS-SEFC-26, 133 p. Available from: NTIS, Springfield, Virginia.

Rivas, L. R. (Unpubl. Ms.) Estimates of biomass for pelagic and coastal sharks in the Gulf of Mexico. Southeast Fisheries Center, National Marine Fisheries Service, Miami, Florida.

Ryther, J. H. and W. M. Dunstan. 1972. Nitrogen, phosphorus and the eutrophication of the coastal marine environment. p. 375-380 In R. F. Ford and W. E. Hagen (eds.). Readings in Aquatic Ecology. W. B. Saunders Co., Philadelphia.

Ryther, J. H. and R. R. L. Guillard. 1962. Studies of marine planktonic diatoms. III. Some effects of temperature on respiration of five species. Can. J. Microbiol. 8: 447-453.

Sackett, W. M. 1972. Chemistry. In S. Z. El-Sayed et al. (eds.). Chemistry, primary productivity, and benthic algae of the Gulf of Mexico. Folio 22, Serial Atlas of the Marine Environment. American Geographical Society, New York.

Schoener, T. W. 1971. Theory of feeding strategies. Ann. Rev. Ecol. Syst. 2: 369-404.

Shane, S. H. 1980. Occurrence, movements, and distribution of bottlenose dolphin, Tursiops truncatus, in southern Texas. Fish. Bull. 78: 593-602.

Sklar, F. H. 1976. Primary productivity in the Mississippi delta bight near a shallow bay estuarine system in Louisiana. Ph.D. Dissertation. Louisiana State University. Baton Rouge, Louisiana. 96 p.

Strickland, J. D. H. 1960. Measuring the production of marine phytoplankton. Fish. Res. Bd. Canada Bull. 122: 172.

7. Energy and Ecological Theory

EXERGY AS KEY FUNCTION IN ECOLOGICAL MODELS[1]

S. E. Jørgensen[2] and H. F. Mejer[3]

Abstract. Previous investigations on the reactions of ecosystems
to perturbations have demonstrated that an ecosystem seems to develop towards increasing exergy, computed as:

$$Ex-P = RT\Sigma a_j (P_j \ln(P_j^{eq}) - (P_j - P_j^{eq})) \text{ KJm}^{-3}$$

Where Ex-P is the exergy related to P (parallel expressions for
other elements), a_j the relative size of compartment j, P_j is the
P concentration in compartment j and P_j^{eq} the corresponding thermodynamic equilibrium concentration. This hypothesis can be used to
explain changes in ecosystem structures due to higher or lower
flow of nutrients through the system or due to changes in the annual temperature pattern. Recent results have demonstrated how it
is possible to use the concept exergy to select preferable structure of ecosystems. Furthermore, it is shown that values of parameters can be determined by use of the above mentioned principle.
It is a characteristic feature of this examination that parameters
in accordance with the literature give higher exergy than unrealistic low or high values. A current adjustment of max. growth rate of
phytoplankton has been applied to account for changes caused by increased or decreased nutrient loadings.

INTRODUCTION

Many ecological models are used to make predictions
on the responses to changes in the external factors. Ecosystems are, however, soft systems, which
are able to meet changes in external factors with
only minor changes in the function of the ecosystem
due to a high flexibility in the structure.

These considerations imply that a model constructed from an observed structure valid under a
given set of external factors, might be insufficient to make predictions, if another set of external factors prevails. Impact on the system will
imply that a new structure is developed.

It has been shown that the changes in structure observed as a result of another set of external factors, is accompanied by an increase in the
thermodynamic function exergy. (Jørgensen & Mejer
1979, 1981 and Jørgensen 1981).

This principle can be used for parameter estimation to account for changes in ecological structure, as it will be discussed and demonstrated below; but before that it is necessary to show how
the exergy can be used and computed in ecological
models.

THE USE OF EXERGY IN ECOLOGICAL MODELLING

Evans et al. (1966) defined exergy as:

$$Ex = T \cdot I \qquad (1)$$

If the ecosystems were in thermodynamic equilibrium,
the entropy S^{eq} would be higher than in non-equilibrium. The excess entropy is the thermodynamic information

$$I = S^{eq} - S \qquad (2)$$

It is an old fact that I also equals the Kullbach
measure of information (Brillouin (1956):

$$I = k \sum_j p_j^* \cdot \ln\frac{p_j^*}{p_j} \qquad (3)$$

where $\{p_j^*\}$ and $\{p_j\}$ are probability distributions,
a posteriori and a priori to an observation of the
molecular details of the system, and k is Boltzmann's
constant.

It may be shown (see Evans 1966) that (2)
leads to:

[1]Paper presented at the international symposium Energy and Ecological Modelling, sponsored by
the International Society for Ecological Modelling.
(Louisville, Kentucky, April 20-23, 1981)

[2]Sven Erik Jørgensen, Professor, Dr.Ing.,
M.Sc., Royal Danish School of Pharmacy, Dept. of
Chemistry AD, 2 Universitetsparken, Copenhagen, DK.
[3]Henning F.Mejer, Processor, M.Sc., Copenhagen Engineering School, Copenhagen, Denmark.

$$I = \frac{U + PV - TS - \Sigma_j X_j n_j}{T} \qquad (4)$$

where P, T and X_j are intensive properties of <u>re-servoirs</u> that the system is assumed to interact with. n_j are mole numbers.

So from (4):

$$x = U + PV - TS - \sum_j X_j n_j \qquad (5)$$

It is easily seen that Ex degenerates to well known thermodynamic potentials at special circum-stances. If e.g. a chemical inert system interacts with a heat reservoir only, (5) becomes E=U-TS, i.e the Helmholtz free energy, since V and n_j remain constant. Similarly, E=U+PV-TS equates the Gibb's potential, if the system is coupled to heat and work reservoirs only. Ex will equate the enthalpy U+PV, if only volume can be exchanged with the surround-ings.

One of the postulates in equilibrium thermody-namics is, that entropy may be expressed by U, V and N_j in the form

$$S = \frac{U}{T} + \frac{P}{T} V - \frac{1}{T} \sum_j X_j n_j \quad \underline{\text{at equilibrium}} \qquad (6)$$

It then follows from (5) that

$$Ex = 0 \qquad \underline{\text{at equilibrium}} \qquad (7)$$

This also follows from (1) and (2) which further shows that

$$Ex \gtrless 0 \qquad \underline{\text{in general}} \qquad (8)$$

It should be stressed once more that the in-tensive properties (P,T,X_j) in (5) are <u>reservoir</u> properties. So exergy depends on the surroundings. An evacuated vessel normally has a positive exergy, but Ex = 0 if it is brought to outer space. Exergy as a free energy concept - is a measure of the maximum useful (entropy free) work that may be ex-tracted from the system on its way to that thermo-dynamic equilibrium state which is compatible with reservoir properties

The exergy balance is found by differentia-tion of (5)

$$\frac{dEx}{dt} = - \int_{\text{surface}} \bar{J}_E \cdot d\bar{a} - T \int_{\text{volume}} \sigma dV \qquad (9)$$

where $\bar{J}_E$ is the exergy outflux:

$$\bar{J}_E = (1 - \frac{T^R}{T}) \bar{J}_u + T^R \sum (\frac{X_c}{T} - \frac{X_c^R}{T^R}) \bar{J}_c \qquad (10)$$

(R denotes reservoir properties).
(Ex cannot be defined as a local property, since depends on its surroundings, i.e. sum of subsystem xergy will not equal the exergy of the combined system.)

To exemplify the use of exergy in ecomodelling, let us calculate Ex for a well mixed lake which alone communicates with heat and work reservoirs. Only soluble phosphorus (P_s) and algal bound phos-phorus (P_a) are considered. Since $P_s+P_a=P_t$ is con-stant, we have only one independent state variable. It is practical to let this be the 'reaction coor-dinate' ξ defined through

$$P_a = P_a^{eq} + \xi$$
$$P_s = P_s^{eq} - \xi \qquad (11)$$

The unsophisticated phosphorus point model is used to exemplify:

$$\frac{dP_s}{dt} = (P_{in} - P_s) \cdot \frac{Q}{V} - (G-M) \cdot P_a \qquad (12)$$

$$\frac{dP_a}{dt} = (G-M - Q/V) \cdot P_a \qquad (13)$$

where $G = G' \cdot \frac{P_s}{K + P_s} \qquad (14)$

The state variables P_s and P_a are soluble and algal bound P in concentration units. P_{in} and Q are driving variables (conc. of P and flow of water). V is system volume and G and M designate rates of algal growth (or rather: nutrient uptake) and re-mineralization.
Obvious constraints obeyed by this model are:

$$P_s > 0 \quad , \quad P_a > 0 \qquad (15)$$

$$P_s + P_a = P_{total} \qquad (16)$$

where P_{total} is a solution to

$$\frac{dP_{total}}{dt} = (P_{in} - P_{total}) \cdot \frac{Q}{V} \qquad (17)$$

(15) - (17) are fulfilled provided

$$G' > (M+Q/V) \cdot (1+K/P_{in}) > 0 \qquad (18)$$

(15) expresses feasibility of model solutions, (16) and (17) denote conservation of an element, while (12) and (13) are mass balances for components P_s and P_a.
By differentiation we find:

$$\dot{P}_a = (G-M) \cdot P_a = - \dot{P}_s \quad , \quad G=G' \cdot \frac{P_s}{K + P_s} \qquad (19)$$

Since $dP_a = -dP_s = d\xi$ the change in exergy (Gibb's potential) is

$$dEx = \mu_s dP_s + \mu_a dP_a = (\mu_a - \mu_s) d\xi \qquad (20)$$

where the chemical potentials are assumed to follow:

$$\mu_a = RT \ln \frac{P_a}{P_a^{eq}} \quad , \quad \mu_s = RT \ln \frac{P_s}{P_a^{eq}} \qquad (21)$$

From (11), (20) and (21):

$$\frac{dEx}{dt} = RT \, \ell n \, \frac{(P_a^{eq}+\xi)\,P_s^{eq}}{P_a^{eq}(P_s^{eq}-\xi)} \qquad (22)$$

By integration from $\xi = 0$ (equilibrium) to ξ we find the exergy

$$Ex = RT\left[(P_a^{eq}+\xi)\,\ell n\,\frac{P_a^{eq}+\xi}{P_a^{eq}} + (P_s^{eq}-\;)\,\ell n\,\frac{P_s^{eq}-\xi}{P_s^{eq}}\right] \quad (23)$$

Notice that $P_s^{eq} \simeq P_t$ and that P_a^{eq} is of the order of 10^{-50} gP/m^3 (based on heat combustion and specific entropy for algal biomass, Morowitz 1968). Then 23 to simplify to

$$Ex = RT\xi \, \ell n \, \frac{\xi}{P_a^{eq}} \qquad (24)$$

provided $\xi \gg P_a^{eq}$. (In steady state: $\xi = P_t - \frac{KM}{G'-M}$).

(24) tells how far the lake is from equilibrium.
 To examine whether Ex depends heavily on topology (food web intricacies, etc.) exactly the same arguments could be performed for the following expanded models:

Zooplankton-model:

$$\dot P_s = (M-G)P_a + M_z P_z$$
$$\dot P_a = (G-M)P_a - HP_z$$
$$\dot P_z = HP_z - M_z Y_z$$

Detritus model:

$$\dot P_s = K_d P_d - GP_a$$
$$\dot P_a = (G-M)P_a \qquad (25)$$
$$\dot P_d = MP_a - K_d P_d$$

New symbols:

Zooplankton-P: P_z

Mortality of zooplankton: M_z

Grazing rate: $H = \left(\frac{Max \; P_a - P_a,min}{K_z + P_a}, 0\right)$

Detritus-P: P_d

Decay rate: K_d

New variables $(\bar\xi)$ are defined by:

$$P_a = P_a^{eq} + \xi_a \qquad\qquad P_a = P_a^{eq} + \xi_a$$
$$P_z = P_z^{eq} + \xi_z \qquad\qquad P_d = P_d^{eq} + \xi_d \qquad (26)$$
$$P_s = P_t - \xi_a - \xi_z \qquad P_s = P_s^{eq} - \xi_a - \xi_d$$

The results are:

Zooplankton case:

$$Ex = RT\left[(P_a^{eq}+\xi_a)\,\ell n\,\frac{P_a^{eq}+\xi_a}{P_a^{eq}} + (P_z^{eq}+\;_z)\,\ell n\,\frac{P_z^{eq}+\xi_z}{P_z^{eq}} \right. \qquad (27)$$
$$\left. + (P_t-\xi_a-\xi_z)\,\ell n\,\frac{P_t-\xi_a-\xi_z}{P_t}\right]$$

Since $P_a^{eq} \ll P_t$, $P_z^{eq} \ll P_t$ we have far from eq.
$(\xi_a \gg P_a^{eq}, \; \xi_z \gg P_z^{eq})$:

$$Ex = RT\left[\xi_a \, \ell n\,\frac{\xi_a}{P_a^{eq}} + \xi_z \, \ell n\,\frac{\xi_z}{P_z^{eq}}\right] \qquad (28)$$

Detritus case:

$$Ex = RT\left[(P_a^{eq}+\xi_a)\,\ell n\,\frac{P_a^{eq}+\xi_a}{P_a^{eq}} + (P_d^{eq}+\xi_d)\,\ell n\,\frac{P_d^{eq}+\xi_d}{P_d^{eq}}\right. \qquad (29)$$
$$\left. + (P_t-\xi_a-\xi_d)\,\ell n\,\frac{P_t-\xi_a-\xi_d}{P_t}\right]$$

Approximation:

$$Ex = RT\left[\xi_a \, \ell n\,\frac{\xi_a}{P_a^{eq}} + \xi_z \, \ell n\,\frac{\xi_z}{P_z^{eq}}\right] \qquad (30)$$

i.e. a result identical to the zooplankton case. To summarize: Exergy expressions are independent of model details such as food web topology and rate formulas. In models with n biological state variables $(\xi_c \gg p_c^{eq}, \, c=1,\ldots,n)$ we have:

$$Ex = RT \sum_{c=1}^{n} \xi_c \, \ell n\,\frac{\xi_c}{P_c^{eq}} \qquad (31)$$

For systems with inorganic net inflow and passive organic outflow (31) becomes:

$$Ex = RT \sum_{c=o}^{n}\left[\xi_c \, \ell n\,\frac{\xi_c}{P_c^{eq}} - (\xi_c - P_c^{eq})\right] \qquad (32)$$

where ξ_o is concentration of the inorganic compound.

PARAMETER ESTIMATION BY USE OF EXERGY

 Most models have a rather simple structure compared with the real system. If, for instance, a food web or food chain is used as model structure this coarse description would be valid under all conditions, whether stress factors are relieved or increased.
 However, the change in the structure of the ecosystem, would be reflected in the parameters. New species are taken over and they might have other maximum growth rates, respiration rates etc. Parameter estimation is often the most crucial step in application of system analysis on environmental problems. Most parameters have a realistic upper and lower limit, which is often a condition for a feasible calibration. The results of many environmental models are, however, very sensitive to some parameters and quite often those available are not of sufficient quality to allow a sufficiently accurate calibration. Furthermore, most parameters will not be constant but several feed-back mechanisms are changing their values currently. Here, exergy could be used for a current modification of some crucial parameters, using the maximum exergy principle as an additional constraint on the parameter optimization.

A current modification of the maximum growth of phytoplankton in our eutrophication model (Jørgensen 1976 and Jørgensen et al. 1978) was attempted. The exergy for a wide range of values for the maximum growth rate of phytoplankton was computed, and the value which gave the highest exergy was selected. The model was applied on a hypereutrophic lake and a 99 % reduction of the phosphorus input was simulated. It was found that along with decreased phosphorus concentration and eutrophication, the selected maximum growth rate increased, see table 1.

Table 1.

Case	Max. Growth Rate at highest exergy
Oligotrophic Lake $P_{total} \leqslant 0.05$ mg ℓ^{-1}	3.3 day^{-1}
Eutrophic Lake $P_{total} \sim 0.5$ mg ℓ^{-1}	2.2 day^{-1}
Hypereutrophic Lake $P_{total} \sim 1.5$ mg ℓ^{-1}	1.6 day^{-1}

This is in accordance with the observations that phytoplankton species in oligotrophic lakes are generally smaller (the specific surface is higher giving a higher growth rate) than in eutrophic lakes.

These results have encouraged us to develop this parameter estimation method further. We use an automatic calibration method (recently developed software called PSI). The method allows introduction of any type of optimization criteria. In addition to the generally applied procedure, which is based on best possible fit to observed data by use of

$$S = \left(\frac{\sum_{i=1}^{i=n} \sum_{j=1}^{j=m} (x_{j,i} - x'_{j,i})^2}{m \cdot n - 1} \right)^{\frac{1}{2}} \quad (33)$$

as objective function ($x_{j,i}$ are model results and $x'_{j,i}$ are observed data, n data points of m state variables, we will select a certain number of crucial parameters in accordance with the maximum exergy principle. This will possibly give varying parameters, hopefully in accordance with the succession of species throughout the year. Further examination will be undertaken during the coming year.

CONCLUSIONS

It seems necessary in the coming years to introduce additional constraints on ecological models and to take the structure changes of ecosystems into consideration. As ecological models have a simple structure, these ecological changes should be reflected in the selection of parameters. Here it might be appropriate to use exergy to determine current changes in parameters, if the hypothesis that ecosystems developed towards higher exergy is valid.

The method introduced here for current change of parameters is simple. If further investigations on the method confirms its applicability, it would be rather easy to account for changes in structure of ecological models and thereby improve the predictive capability of ecological models.

LITERATURE CITED

Brillouin, L. 1956. Science and Information Theory. McGraw-Hill, New York, N.Y.

Evans, R.B. et al. 1966. Principles of Desalination. In Spiegler, K.S. (ed), New York, N.Y.

Jørgensen, S.E. 1976. A eutrophication model for a lake. Ecol. Modelling, 2: 147-165.

Jørgensen, S.E. 1981. A holistic approach to ecological modelling by application of thermodynamics. Paper presented at the conference on Energy and Ecological Modelling in invited session on Energetics, Louisville, Kentucky, April 20-23, 1981.

Jørgensen, S.E. and H. Mejer. 1979. A holistic approach to ecological modelling. Ecol. Modelling, 7: 169-189.

Jørgensen, S.E. and H. Mejer. 1981. Application of exergy in ecological models. In Progress in ecological modelling, edited by D. Dubois, Liege. (In press.)

Jørgensen, S.E., H. Mejer and M. Friis. 1978. Examination of a lake model. Ecol. Modelling, 4: 253-278.

THE RESIDENCE TIME OF ENERGY AS A MEASURE OF ECOLOGICAL ORGANIZATION[1]

Edward F. Cheslak[2] and Vincent A. Lamarra[3]

Abstract.--The hypothesis that the residence time of energy in a system is sensitive to ecological organization is tested. Succession terminates in a steady-state system of maximum organization. Using carbon as a tracer the measurement of energy residence time during a laboratory algal succession provides for a critical test of the hypothesis. The simplicity of the system enables one to determine when ecological mechanisms (e.g. competition) begin influencing community structure. Thus, the quantitative importance of molecular versus ecological contributions can be evaluated. The results indicate that ecological processes are of major importance in determining the residence time of energy. This implies that the degree of molecular organization is strongly influenced by ecological processes.

INTRODUCTION

Ecological organization can be defined as the degree of interdependence existing between a set of populations and their environment (Collier et al. 1973; Whittaker 1975). If the occurrence of each population is primarily the result of adaptations to abiotic conditions, then the resulting ecological organization is relatively low. Conversely, if the presence and importance of populations within the assemblage is determined by biotic interactions (e.g., the result of competition, predation, symbiosis, etc.), the degree of organization can be relatively high (Collier et al. 1973).

Despite our ability to define organization this concept has primarily been limited to heuristic and pedagogical applications. This is a consequence of our inability to measure this fundamental property of species associations. Many measures have been suggested; among these are the composition and abundance of species (MacArthur 1957; Hairston 1959), species diversity (MacArthur 1972; Murdoch et al. 1972; Harger and Tustin 1973a, 1973b), foliage height diversity (MacArthur and MacArthur 1961; Recher

1969), trophic structure (Heatwole and Levins 1972; Pianka 1974; Simberloff 1976), food web complexity (MacArthur 1955; Pain 1966; Watt 1968) and guild structure (Landres and MacMahon 1979). The principal problems associated with the use of these and similar indices are: 1) for large complex systems they are imprecise and unmeasurable; 2) they are usually applicable to only a small subset of the system; and 3) their relationship to organization is implied and usually untested. To advance the theory of ecosystem organization through the testing of well defined hypotheses an index of organization which overcomes these difficulties is imperative.

Morowitz (1968) addresses the question of biological organization from the perspective of thermodynamics. The thesis of his monograph is that energy flow through a system acts to organize that system. He demonstrated in a very general way that molecular order arises in systems undergoing an energy flow from a source to a sink. Since biological systems are characterized by energy flow, it follows that biological systems are organized (i.e. ordered).

Morowitz (1968) develops an order measure based on the distance a system is from thermodynamic equilibrium. He uses as his reference function the Helmholtz free energy of the system, which is given by

$$A = U - TS \tag{1}$$

where,

U = internal energy of the system
T = kinetic temperature
S = entropy of the system

This energy function is minimized when entropy is at a maximum (i.e. equilibrium). Since all

[1]Paper presented at the international symposium Energy and Ecological Modelling, sponsored by the International Society for Ecological Modelling (Louisville, Kentucky, April 20-23, 1981).

[2]Edward F. Cheslak is a doctoral candidate of the Ecology Center, Utah State University, Logan, Utah.
[3]Vincent A. Lamarra is Assistant Professor of Wildlife Science, Utah State University, Logan, Utah.

other states have higher values for the Helmholtz free energy, the difference between the non-equilibrium and equilibrium value is a measure of the distance from equilibrium. Thus,

$$\Delta A(T') = A(T')_{nonequilib.} - A(T')_{equilib.} \qquad (2)$$

where the equilibrium system used for comparison is equilibrated with a reservoir at the kinetic temperature (T') of the nonequilibrium system. Morowitz then normalizes this measure and introduces his order function,

$$L = \frac{\Delta A(T')}{kT'} \qquad (3)$$

where,
$$k = \text{Boltzman constant}$$

For large complex biological systems L is difficult, if not impossible to measure. Theoretically, however, A(T') exists and could be measured by adiabatically isolating the system and allowing it to degrade to equilibrium.

Morowitz (1968) modifies L slightly to incorporate energy storage and energy flow into the concept of order. His new function is,

$$L' = hL = \frac{\Delta A(T')}{(kT'/h)} \qquad (4)$$

where,
$$h = \text{Planck's constant}$$

Of this function Morowitz (1968, p. 143) has written: "The order L' represents a ratio of an energy to a rate; the energy stored in the system is potential and entropic modes divided by a sort of universal rate constant (Glasstone et al. 1941) depicting the decay to equilibrium. Order represents a kind of tension between storing energy and the decay of energy into the most random possible distribution." This implies that the residence time of energy in a system is a measure of its organization. The more organized the system, the longer the residence time. Normally the dynamics of energy is difficult to measure, but in biological systems the storage and flow of energy is closely linked with carbon. This important observation could make ecological organization measurable in terms of the residence time of carbon within the system.

Morowitz (1968) and Morrison (1964; quoted in Morowitz) contend that the thermodynamic properties of order are primarily the consequence of molecular level events. From this perspective ecological organization is viewed as a "small secondary property of a fundamentally [Molecular] thermodynamic system" (Morrison 1964). However, in complex communities the very existence of the diverse molecular forms (i.e. species) is the result of ecological processes. The examples of how these processes give rise to and maintain community structure (i.e. the abundance and

distribution of various molecular forms) is extensive. From a pragmatic viewpoint such examples do not allow a rejection of Morowitz's conclusion (unless each potential case were evaluated). Thus, it becomes necessary to test the relationship between ecological organization and energy residence time. To perform this test let us identify an ecological process which is known to be organizational and measure the residence time of energy as the process proceeds.

Current ecological theory views succession as a process which results in a highly organized community (i.e. high in information, content, low in entropy, with developed internal symbiosis; Cooke 1967; Margalef 1968; Odum 1969, 1971; Harger and Tustin 1973b) consider the statement by Odum (1971; quoted in Ricklefs 1973) that "the time involved in an uninterrupted succession allows for increasingly intimate associations with reciprocal adaptations between plants and animals." This description coincides with the definition of ecological organization.

Returning to our pragmatic approach, this theory would be more convincing if an independent measure of organization were available. Harger and Tustin (1973b) developed such a measure, termed the "organizational component," based on species diversity. They calculated their measure for three developmental seres; an aquatic periphyton succession, a rotting wood community succession, and a Hemlock-White Pine succession. In each case the measured component was found to increase as succession proceeded (all regressions were significant, $p < 0.001$). These results support the premise that succession is an organizing process, but do not preclude the present study because of the limitations identified for diversity type approaches.

This leads directly to the following hypothesis set:

Ia) Energy residence time is a holistic property that increases to a maximum during succession as a result of increasing ecological organization.

Ib) Energy residence time is not related to ecological organization in that it is not an increasing function throughout succession.

To test these hypotheses one need only determine the residence time of energy as measured by carbon as it changes through a successional sere.

Suppose that energy residence time is related to organization. It is still possible that ecological processes are "quantitatively unimportant." That is, most of the change in residence time could be due simply to density independent increases in the populations. We need either a) an independent measure of the degree of interdependence between the populations, or b) a way of determining when interactions start affecting population sizes.

Consider a simple algal succession in a batch culture microcosm (as in Cooke 1967). In the absence of grazers and predators, ecological organization results primarily from competitive interactions between the primary producers (ignoring for the moment the role of decomposition). If we knew when competition began we could evaluate the changes in energy residence time directly related to these interactions.

The present theory of competition requires as a premise the existence of some limiting resource. It then attempts to explain the occurrence and abundance of specific forms based on their differential ability to acquire these resources (Cody 1974; Connell 1975). The implication of this theory is that competition cannot be used to explain system organization during periods of resource abundance. If we follow the change in major nutrients and net community production in an algal succession we can determine when resource limitations begin. The net primary production of an algal community is partially determined by the availability of nutrients. When nutrients are abundant and production is increasing we can infer that nutrient limitations are not operative--all populations are in exponential growth phases (Fogg 1965). As the nutrients are depleted a time is reached where limitations are imposed. Some of the populations, due to species specific uptake kinetics, will be unable to obtain nutrients at a rate sufficient to maintain maximum production (Fogg 1965). Consequently, their production rate declines, as well as their population growth rates. This has the effect of depressing the net community production (Fogg 1965; Whittaker 1975; Wetzel 1975). Fogg (1965) notes that high production rates and exponential population growth can be extended with further additions of the limiting nutrient. Because the point of resource limitation can be identified, the associated residence time of energy becomes defined and can be compared with the maximum value. This allows us to test the following hypotheses:

IIa) Ecological organization is a quantitatively important determinant of energy residence time (accounting for 60% or more of the maximal level observed).

IIb) Energy residence time is primarily a measure of molecular, not ecological, organization (the latter accounting for 40% or less of the maximal residence time observed).

IIc) Ecological and molecular organization are of relatively equal importance in determining energy residence time.

METHODS AND MATERIALS

Central to this project is the determination of the carbon residence time of an ecosystem.

Figure 1 shows a generalized carbon cycle for such a system (Ricklefs 1973). As shown an ecosystem is composed of primary producers, consumers, and dead organic matter (C_2, C_3, and C_4 in figure 1, respectively; see Reichle et al. 1975 for similar definition and justification). We shall assume that the production of coal, oil and methane is insignificant. Under steady state conditions, the residence time of carbon is given by

$$t_c = \frac{C}{F_r} \tag{5}$$

where

$$C = C_2 + C_3 + C_4 \quad \text{(in Kg.), and}$$

$$F_r = F_2 + F_7 \quad \text{(Kg./time)}$$

The assumption of steady state conditions is violated in ecosystems undergoing succession. To remove that assumption from the determination of t_c define:

$$t'(t_i, t_{i+1}) = \frac{1}{t_{i+1} - t_i} \int_{t_i}^{t_{i+1}} \frac{C(t)}{F_r(t)} \, dt \tag{6}$$

with $t_{i+1} > t_i$. The index $t'(t_i, t_{i+1})$ is now seen to represent the average residence time of carbon for the interval $t_{i+1} - t_i$. This means that the concept of residence time when applied

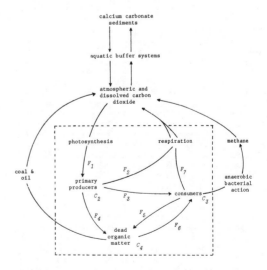

Figure 1.--The carbon cycle (based on Ricklefs 1973). The ecosystem, as defined, is contained within the dotted lines (after Reichle et al. 1975).

to nonsteady state systems is only defined with respect to intervals of time. Note that when $t_{i+1} = t_i$, t' is undefined and if $(dC/dt) = 0$ (e.g., $F_1=F_2+F_7$, fig. 1) then (6) reduces to (5) for that time interval over which steady state assumptions apply.

Moreover, the calculation of t' also assumes that the carbon is homogeneously distributed throughout the system. To eliminate this constraint define $\bar{t}$ as:

$$\bar{t}(t_i, t_{i+1}, x, y, z) = \frac{1}{t_{i+1}-t_i}$$

$$\cdot \frac{1}{x-x_0} \cdot \frac{1}{y-y_0} \cdot \frac{1}{z-z_0} \int_{t_i}^{t_{i+1}} \int_{x_0}^{x} \int_{y_0}^{y} \int_{z_0}^{z}$$

$$\frac{C(t,x,y,z)}{F_r(t,x,y,z)} \quad dz \; dy \; dx \; dt \qquad (7)$$

This states that the residence time of carbon as calculated by (7) represents the average over both time and volume of the ratio $C(t,x,y,z)/F_r(t,x,y,z)$. This equation has the advantage that averages for flow rates (F_r) and state variables (C) are experimentally measurable, whereas instantaneous values (especially for rates) are usually unmeasurable. This greatly facilitates the interpretation of experimental results.

Experimental Approach

The batch culture microcosm approach of Cooke (1967) was used as the experimental framework for testing the hypotheses. A synthetic medium (1/2 strength #36; Taub & Dollar 1964) served as the nutrient source. A "mother cosm" was obtained from a sewage lagoon and left for several months to undergo succession and reorganization. The experiment was initiated by inoculating 5 mls of the well stirred mother cosm into 295 ml of media contained in a 400 ml beaker. One hundred eight microcosms were inoculated, covered with a watch glass and placed in a 21° ± 2°C water bath. Incubation occurred under six Sylvania standard Gro-lux lamps yielding an intensity of 150-180 ft candles (simulating Cooke's methodology); a 12-hour light-dark cycle was employed. On the 20th day, the microcosms were cross inoculated; no further additions were made other than sterile distilled water to replace evaporative losses.

As the succession proceeded a destructive sampling scheme was used to determine the following properties: total organic carbon (TOC), dissolved organic carbon (DOC), PO_4-P, NH_3-N, NO_3-N, and NO_2-N concentration. Standard techniques were used in all nutrient determinations (Cheslak 1981). At each sampling date three randomly selected microcosms were chosen for measurement and sacrifice. The criterion for termination was three successive sampling dates

exhibiting decreased levels of TOC. This resulted in 187 days of observation with 34 separate sampling dates.

The Measurement of Carbon

The pH method of Beyers and Odum (1959) was used to determine the CO_2 fixed and respired by the system. pH was measured at 0900, 2100, and 0900 then converted to mg CO_2/L/time by an empirical titration curve. After the final 0900 pH measurement each system was scraped and vigorously stirred. A subsample was taken for determination of total organic carbon (TOC); an additional subsample was filtered with GFC fiber filters to obtain a dissolved organic fraction (DOC). The TOC sample was then sonified to break up the large algae mats and cells. The organic carbon in each fraction was determined using an Oceanographic International Model 05243B Total Carbon System.

This methodology of using whole water pH changes along with scraped and stirred subsamples for TOC determination experimentally reduces the complexity of the carbon residence time (t'($\Delta t, \Delta v$)) calculation. The averages for total carbon and respiration as a function of volume are handled methodologically. pH is an average property of the whole medium, maintained through chemical and physical equilibrium relationships; thus, changes in the community respiration rate as a function of volume are minimized. The destructive sampling technique described above for organic carbon also provides averages for the entire microcosm. Therefore, the integral over volume for C(t) and $F_r(t)$ in equation 7 are no longer considered.

Finally, the diffusional contribution to the observed pH change of the media was measured. Three "respirometers" were designed so that mass balances of CO_2 in the gas phase could be monitored (Cheslak 1981). Mass balances on these respirometers and the actual experimental systems allow one to determine net CO_2 flux into or out of the medium as a result of diffusional processes. This allows us to correct the production and respiration curves for errors associated with diffusion.

RESULTS

Carbon Dynamics

The change in total organic carbon (TOC) during succession followed a logistic curve (fig. 2). The a priori termination criteria of three consecutive decreasing values in TOC led to a 187 day experiment. The first of these decreasing values (day 163) was chosen as the end point for the "stationary phase" of the batch culture "limited volume" experiment. This stationary phase is assumed to represent the steady state community of a typical successional climax. This assumption appears reasonable since the stationary phase can be maintained indefinitely in semi

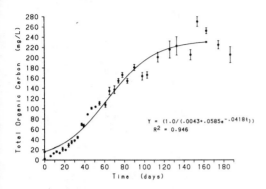

Figure 2.--Total organic carbon in mg C/L as a function of time. The ranges shown represent standard errors about the means. The equation shown is for the logistic fit drawn through the means.

or continuous cultures where nutrient additions match production requirements (Fogg 1965).

The net daytime primary production and nighttime respiration curves corrected for diffusion are shown in figures 3 and 4. The basic pattern for both processes is similar--an early bimodal peak, followed by a second peak, then a reduced level for the later part of the successional sere.

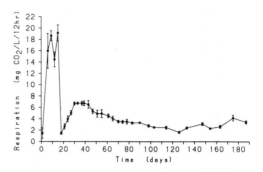

Figure 4.--Nighttime respiration in mg $CO_2/L/12$ hr. This curve has been corrected for diffusion errors. The ranges shown are standard errors about each mean.

Determination of Energy Residence Time

Given the functions for TOC and community respiration (figs. 2 and 4) it is possible to determine the residence time of energy as measured by carbon for the sampling intervals specified. Because changes in t' related to spacial distributions were handled experimentally, equation 6 gives the appropriate mathematical formulation for t'. To determine $C(t)$ and $F_r(t)$ linearity from t_i to t_{i+1} was assumed, this made the integration and evaluation of equation 6 straightforward. The resulting solution provided a direct expression for evaluating $t'(t_i, t_{i+1})$ from the available data ($F_r(t)$ = 2 x night respiration; $C(t)$ = TOC at sample time). Figure 5 shows the results of these calculations for all subintervals sampled. A logistic equation was

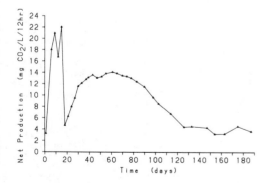

Figure 3.--Net daytime primary production in mg $CO_2/L/12$ hr as a function of time. This curve has been corrected for diffusional errors via a mass balance (Cheslak 1981).

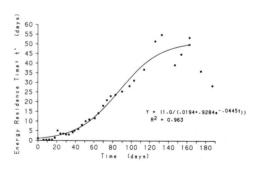

Figure 5.--The residence time of energy (t') in days as a function of time. The equation is for the logistic curve fit through the data.

fit to these data by a program which minimized the total squared error (C. Fowler, personal communication). This fit resulted in a very high r^2 value (0.96). The points at days 175 and 187 were not included in this fit because the a priori termination criteria had established day 163 as the end of the stationary phase, i.e., the attainment of a "climax" community.

Nutrient Dynamics

The concentration of NH_3-N, NO_3-N, NO_2-N, and PO_4-P (figs. 6-8) were followed during succession. The NH_3-N curve (fig. 7) has three peaks, two early in the sere and one near the end. The NO_3-N concentration (fig. 6) drops steadily, reaching below detectable levels on day 33, where it remains with the exception of two minor peaks on days 55 and 74. The nitrite curve is more varied; several peaks occur (days 6, 15, 27, 55, 83) before it falls below detectable levels. The change in ortho-phosphate concentration is shown in figure 8 (Log_{10} concentration). Its behavior is similar to the NO_3-N curve, i.e., a curve that rapidly drops to very low levels by day 39 (16 μg/l).

DISCUSSION

Ecological theory describes succession as a self-organizing process terminating in a maximally organized system (Margalef 1968; Odum 1971; Cooke 1971; Harger and Tustin 1973b; Kitchens 1978). When we examine the trajectory of energy residence time (fig. 5) we see this property also increases to a maximum. This allows a rejection of the hypothesis that the residence time of energy is unrelated to ecological organization (Ib), leaving the conclusion that as one property increases so does the other.

The fact remains that ecological organization may be a "small secondary property." We

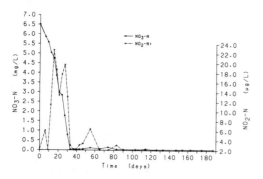

Figure 6.--Nitrate (NO_3-N) and nitrite (NO_2-N) concentrations through time. The points represent means for all microcosms sampled that day.

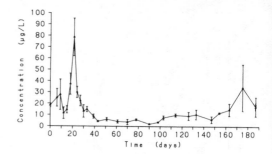

Figure 7.--Ammonia (NH_3-N) concentrations in mg/l as a function of time. Ranges shown are standard errors about the mean.

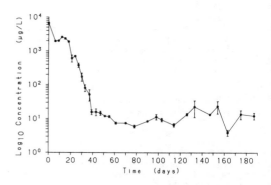

Figure 8.--Ortho-phosphate (PO_4-P) concentrations in mg/l transformed by Log_{10} as a function of time. Ranges are standard errors about the means.

need to evaluate the role of ecological processes in determining the degree of molecular order. Because this is a simple community composed only of algal species, some ciliates and bacteria (Cheslak 1981) we know that the principal processes influencing species abundance will be decomposition and competition. The inflection point in the total organic carbon curve (fig 2; day 62) indicates that some or all of the species comprising the community are experiencing a constraint on the rate at which they can grow and/or reproduce (Fogg 1965). If nutrient limitation is causing the termination of exponential growth then we have identified an important point in the successional sere; the point at which the axiom for competitive exclusion is fulfilled. To explain the dynamics in biomass and species abundance beyond this point requires the inclusion of competitive interrelationships. Those species whose uptake kinetics for the limiting nutrient enable them to efficiently extract quantities capable of maintaining growth and reproduction will increase in relative abundance (Eppley et al. 1969; MacIsaac and Dugdale 1969). Less efficient species will decrease causing a change in community structure (Fogg 1965). Furthermore, as the total biomass and abundance of algae increases these interrelations will become more, not less important, since the demand on these limited resources steadily increases. As a result, the role of decomposition in providing available nutrients also increase.

Fogg (1965) identifies several possible factors which could cause the cessation of exponential growth. These are: "1. Exhaustion of nutrients; 2. Rate of supply of carbon dioxide or oxygen; 3. Alteration of pH of the medium as a result of preferential absorption of particular constituents of the medium; 4. Reduction of the light intensity by self-shading; and 5. Autoinhibition." During autoinhibition one usually observes "an abrupt transition from the exponential to the stationary phase" (Fogg 1965); the prolonged phase of declining relative growth rates (day 62 - day 163) is inconsistent with this explanation and indicates that autoinhibition is not the factor accounting for the inflection point. The maximum surface area covered by dense algal mats in the microcosms was estimated at only 23 percent of the total area available (Cheslak 1981) and occurred on day 83. Reduction in light intensity for the entire community by self-shading also appears to be an inadequate explanation and can be rejected. The highest pH value (10.25) was observed on day 30, after which this property continuously decreased for the remainder of the experiment. Again, the cessation of exponential growth is uncorrelated with elevated pH and this alternative can be rejected as an adequate explanation of the observed dynamics. This leaves only two explanations which can be grouped into the general category of resource limitation.

The medium is a carbon and nitrogen limited mineral solution with the original molar ratios of C:N:P being 0.61:2.178:1. Comparing these initial conditions with the ratios found in an average algal cell (106:16:1; D. Porcella personal communication) we see that additional carbon and nitrogen is required from atmospheric sources to fix all the phosphorus into biomass. If all the phosphorus was fixed then the expected concentration of carbon (assuming the above ratios) would be 272.0 mg C/L. The highest concentration observed was 270.18 mg C/L (day 154; fig. 2); we can conclude that in this system phosphorus was the ultimate limiting factor. According to Fogg (1965) exhaustion of the ultimate limiting nutrient usually causes systems to fix all the phosphorus into biomass. As with autoinhibition, the observed behavior is inconsistent with our expectations if phosphorus limitation is the cause of the inflection point. It is noteworthy that the external concentration of PO_4-P hits a very low value on day 61 (7.5 µg/l). However, continued exponential growth in the presence of inadequate external

concentrations has been observed and explained frequently by the process of "luxury consumption" (Fogg 1965; Grenney et al. 1973, 1974; Lehman et al. 1975). This casts reasonable suspicion on the interpretation of low external PO_4-P concentrations as evidence for nutrient limitation. The macroscopic behavior of an attenuated approach to stationary phase suggests that some other nutrient is acting to constrain production.

Resource limitation implies that supply is insufficient to meet demand. The potential processes which supply nitrogen to this system are atmospheric diffusion of NH_3-N, nitrification, nitrogen fixation, and original additions of NO_3-N. As early as day 33 we see that the external concentration of nitrate is below detectable limits (0.04 mg/l), yet the exponential phase of growth is still occurring. Is this "luxury consumption" in its classical form? If we calculate the amount of carbon the original concentration of NO_3-N and NH_3-N would produce, 37.0 mg C/L is expected; by day 33 47.8 mg C/L was observed as particulate organics. The observed production had thus exceeded the expected levels if all the original nitrogen had been converted to algal biomass. The exponential growth from day 33 to 62 cannot be accounted for by luxury consumption and utilization of stored nitrogen. Some external source must be providing nitrogen at a rate sufficient to meet demands.

Nitrogen fixation is known to be an effective process for providing nitrogen in aquatic systems (Campbell 1977). However, the heterocystic blue-green bacteria known to fix nitrogen in oxygen plentiful environments were absent in this simple system. This leaves nitrification and ammonia diffusion and production as the only potential sources of nitrogen. The dynamics of nitrate and nitrite from day 33 to 62 are indicative of active nitrification; note the relatively high NO_2-N peak on day 55. After day 62 these oxidized forms of nitrogen drop off implying reduced nitrification rates. Similarly, ammonia decreases during the interval indicating that uptake and nitrification exceed diffusional and decompositional sources. Since this is a nitrogen limited medium, it is not unreasonable to suggest that the inflection point is a result of rate limitations imposed by nitrogen shortages. The macroscopic behavior beyond day 62 is consistent with a diffusion rate limited system; the atmosphere operating as an infinite source of free NH_3-N. This interpretation is strengthened by considering the actual data points in figure 2 rather than the statistical curve fit. Note the abrupt change in the points after day 43; it changes rapidly from an exponential to a more linear curve. This corresponds with the low point in the NH_3-N curve (fig. 6, day 43) implying that the abrupt transition is a result of nitrogen limitation; the decreased growth rates being constrained by diffusion and decomposition rates after this point in the sere.

An equally likely explanation is that CO_2 becomes rate limiting. Fraleigh and Dibert

(1980) found that CO_2 diffusion rates influenced biomass growth rates in systems similar to those reported here. Comparison of results beyond this general statement is difficult because of differences in nutrient concentrations, techniques, and objectives. The declining growth rates beyond day 62 are consistent with what Fogg (1965) and Pearsall and Bengry (1940; cited in Fogg (1965)) describes for stagnant cultures where diffusion is dependent upon surface area "presented by the culture." This leads us to conclude that the system is either CO_2 or nitrogen limited at the inflection point of the total organic carbon curve. This resource limitation is influencing community production (as seen in fig. 3) such that decreasing rates of organic carbon production is occurring. Demands are now exceeding supplies. The axiom of competitive exclusion is fulfilled (i.e. limited resources) and this process influences community organization. In addition, decomposition becomes more important because of its role in providing nitrogen to the system via recycling.

It is now possible to determine the importance of ecological interrelationships in determining the degree of molecular order. Let $t'(t_{e-1}, t_e)$ represent the residence time of energy when decomposition and competition begin to effect species abundance, and t'_s the steady state value obtained from the logistic curve fit in figure 2. The fraction of the steady state residence time attributable to the operation of ecological processes (E) becomes:

$$E = 1 - \frac{t'(t_{e-1}, t_e)}{t_s'} \qquad (8)$$

With t_e equal to 61 (the sample date closest to the inflection point in the TOC curve), E = 0.775. That is 77.5 percent of the final residence time of energy is attributable to ecological processes.

This represents a minimum estimate of E. The macroscopic behavior of the system in conjunction with the nutrient dynamics allows us to unequivocally state that competition is operating to structure the system after day 62. However, it is possible that conditions conducive to competition occur at a much earlier time. Campbell (1977) notes that "nitrification is favored at neutral or slightly alkaline pH's but above pH 8 Nitrobacter does not convert much nitrite to nitrate." The pH in these systems rose to above 8.80 on the 21st day. The peaks in nitrite after this date cannot be interpreted as indicating rapid nitrification; instead they result from the inability of this form to be converted to nitrate and made available for uptake. The disappearance of nitrate on day 33 now has a special significance; that nitrogen species becomes limited. Eppley et al. (1969) have demonstrated that the different uptake kinetics for various forms of inorganic nitrogen control the outcome of competition between

algae by affecting their relative growth rates. Microcosm species who can utilize NH_3-N directly now have a competitive advantage. Note how the concentration of ammonia in the water drops rapidly after day 33 (fig. 6). Accepting the premise that nitrate is limiting we can again invoke, if you will, the axiom for competitive exclusion. The corresponding value of E is 0.940; for t_e = 33, 94 percent of the final energy residence time results from ecological processes that influence the degree of molecular order. This value is the maximum estimate of E.

Depending upon which scenario we accept the fraction of the steady-state residence time attributable to ecological processes ranges from 0.775 to 0.940. Either of these values allows us to reject the hypotheses (IIb & e) that this macroscopic property is insensitive to ecological interrelationships. This implies that the degree of molecular order is to a large part determined by interactions among the species comprising the system. To an ecologist this is not surprising in that we view thermodynamic properties as constraints that operate on a fundamentally interactive, ecological system.

CONCLUSIONS

As ecological organization increases the residence time of energy in that system increases. In addition, the role of ecological processes in determining the steady state value of residence time is a major one; 78 to 94 percent of the final value was attributable to the influence of interactions between the species (competition and decomposition). This conclusion was based on the behavior of the total organic carbon (TOC) curve (fig. 2) and the observed nutrient dynamics (figs. 6-8). The inflection point observed in the TOC curve on day 62 was interpreted as evidence that some constraint was operating to decrease growth in the system. An analysis of the potential factors accounting for this decrease revealed that resource limitation was the most plausible alternative. This allowed us to invoke the theory of competition to explain system behavior beyond this point in the sere; giving the lower limit on E (0.775). The lack of available nitrate on day 33 was used as an alternative date in which resource limitation, and therefore competition, could be operative; this provided the upper limit of E (0.94). Both of these values supported the conclusion that ecological processes are a major influence on the expression of molecular order in these far from equilibrium, dissipative structure, thermodynamic systems.

ACKNOWLEDGMENTS

We wish to thank the Ecology Center, the Wildlife Science Dept., and the Vice-President for Research at Utah State University for funding this research. The thought, time, support, and patience of many friends have made this possible--to all of them our special thanks. The senior author wishes to thank his daughter, Shelly Anne, for her many hours of help in the lab; also he dedicates this work to Paramahansa Yogananda whose constant guidance, discipline, and understanding have brought this work to fruition.

LITERATURE CITED

Beyers, R. J., and H. T. Odum. 1959. The use of carbon dioxide to construct pH curves for the measurement of productivity. Limnol. and Oceanogr. 4:499-502.

Campbell, R. 1977. Microbial Ecology 148 p. John Wiley & Sons, New York, N. Y.

Cheslak, E. F. 1981. Energy residence time as a measure of ecological organization. PhD dissertation, Utah State University, Logan, Utah.

Cody, M. L. 1974. Competition and the Structure of Bird Communities. Monogr. in Popul. Bio. (ed. R. M. May). No. 7. 318 p. Princeton Univ. Press, Princeton, New Jersey

Collier, B. D., G. W. Cox, A. W. Johnson, and P. C. Miller. 1973. Dynamic Ecology. 563 p. Prentice Hall, Inc., Englewood Cliffs, New Jersey.

Connell, J. H. 1975. Some mechanisms producing structure in natural   A model and evidence from field experiments. Pages 460-490 In M. Cody and J. Diamond, eds. Ecology and Evolution of Communities. Harvard Univ. Press, Cambridge, Mass.

Cooke, G. D. 1967. The pattern of autotrophic succession in laboratory microcosms. Bioscience 17:717-721.

Eppley, R. W., J. N. Rogers, and J. J. McCarthy. 1969. Half-saturation constants for uptake of nitrate and ammonium by marine phytoplankton. Limnol. and Oceanogr. 14:912-920.

Fogg, G. E. 1965. Algal Cultures and Phytoplankton Ecology. 126 p. The Univ. of Wisconsin Press Madison, Wis.

Fraleigh, P. C., and P. C. Dibert. 1980. Inorganic carbon limitation during ecological succession in aquatic microcosms. Pages 369-401 In J. P. Giesy, Jr., ed. Microcosms, in Ecological Research.   DOE Symposium Series 52, CONF-781101.

Glasstone, S., K. J. Laidler, and H. Eyring. 1941. 611 p. The Theory of Rate Processes. McGraw-Hill, New York, N. Y.

Grenney, W. J., D. A. Bella, and H. C. Curl, Jr. 1973. A theoretical approach to interspecific competition in phytoplankton communities. Amer. Natur. 107:405-425.

Grenney, W. J., D. A. Bella, and H. C. Curl, Jr. 1974. Effects of intracellular nutrient pools on growth dynamics of phytoplankton. Jour. of the Water Poll. Cont. Fed. 46:1751-1760.

Hairston, N. G. 1959. Species abundance and community organization. Ecology 40:404-416.

Harger, J. R. E., and K. Tustin. 1973a. Succession and stability in biological communities. Part 1: Diversity. Intern. J. Environ. Studies 5:117-130.

Harger, J. R. E., and K. Tustin. 1973b. Succession and stability in biological communities. Part 2: Organization. Intern. J. Environ. Studies 5:183-192.

Heatwole, H., and R. Levins. 1972. Trophic structure stability and faunal change during recolonization. Ecology 53:531-534.

Kitchens, W. M., Jr. 1978. Succession in laboratory microecosystems subjected to thermal and nutrient addition stresses. PhD dissertation, North Carolina State University at Raleigh, Raleigh, N.C.

Landres, P. B., and J. A. MacMahon. 1980. Guilds and community organization: Analysis of an oak woodland auifauna in Sonora, Mexico. The Auk 97:351-365.

Lehman, J. T., D. B. Botkin, and G. E. Likens. 1975. The assumptions and rationales of a computer model of phytoplankton growth. Limnol. and Oceanogr. 20:343-364.

MacArthur, R. H. 1955. Fluctuations of animal populations, and a measure of community stability. Ecology 36:533-536.

MacArthur, R. H. 1957. On the relative abundance of bird species. Proc. Nat. Acad. Sci. 43:293-295, Washington, D.C.

MacArthur, R. H. 1972. Geographical Ecology: Patterns in the Distribution of Species. 269 p. Harper and Row, New York, N. Y.

MacArthur, R. H., and J. W. MacArthur. 1961. On bird species diversity. Ecology 42:594-598.

MacIsaac, J. J., and R. C. Dugdale. 1969. The kinetics of nitrate and ammonia uptake by natural populations of marine phytoplankton. Deep-Sea Res. 16:45-57.

Margalef, R. 1968. Perspectives in ecological theory. 111 p. Univ. of Chicago Press, Chicago, Ill.

Morrison, P. 1964. A thermodynamic characterization of self-reproduction. Rev. Mod.
Phys. 36:517-524.

Morowitz, H. J. 1968. Energy Flow in Biology:
Biological Organization as a Problem in
Thermal Physics. 179 p. Academic Press,
New York, N. Y.

Murdoch, W. W., F. C. Evans, and C. H. Peterson.
1972. Diversity and pattern in plants and
insects. Ecology 53:819-829.

Odum, E. P. 1969. The strategy of ecosystem
development. Science 164:262-270.

Odum, E. P. 1971. Fundamentals of Ecology.
3rd Ed. 574 p. Saunders Co. Philadelphia, Pa.

Paine, R. T. 1966. Food web complexity and
species diversity. Amer. Nat. 100:65-75.

Pearsall, W. H., and R. P. Bengry. 1940. The
growth of Chlorella in darkness and in
glucose solution. Ann. Bot., Lond., N.S.
4:365-377.

Pianka, E. R. 1974. Evolutionary Ecology.
356 p. Harper and Row, New York, N. Y.

Recher, H. F. 1969. Bird species diversity and
habitat diversity in Australia and North
America. Amer. Natur. 103:75-80.

Reichle, D. E., R. V. O'Neill, and W. F. Harris.
1975. Principles of energy and material
exchange in ecosystems. Pages 27-42 In
W. H. Van Dobben and R. H. Lowe-McConnell,
eds. Unifying Concepts in Ecology. Dr. W.
Junk B. V. Publishers, The Hague.

Ricklefs, R. E. 1973. Ecology. 861 p. Chiron
Press, Inc., Portland, Oregon.

Simberloff, D. 1976. structure determination and equilibrium in an arthropod
community. Ecology 57:395-398.

Taub, F. B., and A. M. Dollar. 1964. A chlorella-
daphnia food-chain study: The design of a
compatible chemically defined culture
medium. Limnol. and Oceanogr. 9:61-74.

Watt, K. E. F. 1968. Ecology and Resource
Management. 450 p. McGraw-Hill, New York, N. Y.

Wetzel, R. G. 1975. Limnology. 743 p. W. B. Saunders
Co., Philadelphia, Pa.

Whittaker, R. H. 1975. Communities and Ecosystems. 385 p. MacMillan Pub. Co., Inc.,
New York, N. Y.

A CONTROL HYPOTHESIS FOR ECOSYSTEMS –
ENERGETICS AND QUANTIFICATION WITH THE
TOXIC METAL CADMIUM[1]

Robert L. Knight[2]

Abstract.--A control hypothesis is presented that predicts correlation between control effect and energy required
(embodied energy). When these two properties of a controller
are reported in embodied energy units this correlation can
be quantified. Using an energy and matter model of cadmium
(Cd) toxicity and cycling in stream microcosms, data for
control effect of Cd were predicted. The correlation between
embodied energy of Cd and its control effect was found to be
positive at low concentrations, then negative, and approaching zero at high Cd concentrations.

INTRODUCTION

Clear evidence for the existence of regulatory processes is found in any smoothly functioning, self-maintaining system, because homeostasis
is the direct result of control. The study of the
energy basis of this control and an understanding
of the constraints under which it develops is of
great importance in theoretical ecology. In addition, a method of quantification of control with
the ability to summarize diverse information is
necessary for advancement in environmental system
management. This paper presents initial development of a theoretical and quantitative technique
to evaluate, compare, and utilize control in environmental management by illustrating the approach
with one substance, the heavy metal cadmium (Cd).

The study of control and regulation in
environmental systems is as old as the study of
ecology. Elton (1927) provided a conceptual description of population regulation early in this
century. Since that time, a parallel evolution of
understanding of control has occurred in the fields
of ecology and cybernetics. Recent interest in
combining these similar bodies of knowledge is
shown by ecological modellers such as Conrad (1976),
Mulholland and Sims (1976), Straskraba (1979), and
others. However, a useful method of quantification
of control in ecological systems is still needed.

[1]Paper presented at the international symposium of Energy and Ecological Modelling, sponsored
by the International Society for Ecological Modelling (Louisville, Kentucky, April 20-23, 1981).
[2]Robert L. Knight is a Research Associate of
Environmental Engineering, University of Florida,
Gainesville, Florida U.S.A.

Energy And Control

By "controller", I mean an abiotic substance
or biological component, that has the ability to
increase, decrease, or divert energy flows that
are greater than its own energy content. Thus
control is an amplification process in which a
small energy flow regulates a large energy flow.
An herbivore controlling plant growth may in turn
be regulated by a chemical controller such as a
toxic metal. The "energy effect" of a controller
is the net change in energy flow of the controlled
process.

An hypothesis proposed by Odum (1979) and the
author is that control may be a function of the
energy "embodied" in the controlling agent. "Embodied energy" is defined as the total energy of
a system concentrated to form a component through
convergence of energy flows. In this study, the
embodied energy of the toxic controller, Cd, is
estimated. Then, using a model of energy and
material flows in streams with and without Cd, the
embodied energy of this metal is correlated with
its energy effect on system structure and function.

Maximum Power Theory And Embodied Energy

Lotka (1922) proposed a principle of thermodynamics for open systems which states that selection in the struggle for existence is based on
maximum energy flow (power). Later, Odum and
Pinkerton (1955) and Odum (1968,1971, 1979) ex-
pounded on the maximum power theory and suggested
ways control actions generate more power and thus
tend to persist in competing systems.

Given a finite available energy source, the
complexity of a system is limited by the energy
required for each energy transformation process.
The second law of thermodynamics predicts that

some available work energy will be lost in each transformation step. The degraded energy in this process may be considered to be of "lower quality" than the upgraded or "high quality" energy resulting from the process because of less ability to do useful work. The high quality energy output of the transformation step may be considered to have embodied in it the total energy entering the transformation process. Thus, embodied energy as defined by Odum (1978), is a measure of the total energy of one type necessary to generate an energy flow or storage of another type. The ratio of embodied energy to actual energy is called "energy quality factor" or "transformation ratio" (TR).

Control Hypothesis

Biological food chains represent concentrations of energy, with each level requiring energy diverted from the structure of the primary producers. In an analogous manner, control hierarchies in human systems divert energy from the primary producers of fuel energies. Given the criterion for selection based on maximum energy flow, the diverted energy in hierarchies must be compensated for by energies fed back from high quality storages to capture greater free energy for the system. Thus, a control hypothesis follows directly from the maximum power theory. In successful systems, controllers must have a positive energy effect that is at least equal to their embodied energy.

The control hypothesis suggests that controllers will have a value as a stimulant to energy flow of a system that is proportional to their energy consumption from the system for maintenance, and that natural selective processes will eliminate items that require more energy flow than they stimulate. In an immature system the two values of a controller, i.e., the embodied energy and the energy effect, might be widely different, but in an adapted system they must balance or a more productive arrangement will be selected. Thus, an adapted system requires consumers and may be able to use toxins.

General Toxicity Effect

As a natural part of evolution, biological systems have developed toxic substances which control energy flows and perhaps maximize power. Thus plants have allelopathic chemicals and insects have venom. These substances represent a concentration of energies and have a high embodied energy.

Human systems may be similar in this respect. Toxins are collected from nature or synthesized in laboratories for the purpose of controlling environmental energy flows. In addition, toxic substances often result as by-products of industrial processes. These substances also represent large energy flows and have high embodied energy.

On the other hand, toxicity is a drag on energy flow in a system if it is not a part of material cycles and regenerative processes. If the selective criterion is maximum system energy flow, the ideal use for a controlling substance is to enhance the capture and use of energy sources. Consequently, toxins that decrease a system's power must be detoxified by surviving systems or be adapted to through species selection and evolution. If possible, powerful controllers may be incorporated into productive processes within an organism as part of enzyme systems or respiratory and photosynthetic pathways. Thus, copper and zinc (Zn) are recognized essential nutrients for many plants at low concentrations, but are toxic at higher levels. This phenonmenon has been widely recognized as the Arndt-Schultz Law in physiology (Lamanna and Mallette 1953); the subsidy-stress gradient in ecology (Odum et al. 1979); or hormesis in invertebrate toxicology (Laughlin et al. 1981). I propose that this subsidy-stress effect is a general phenomena of control by any substance that has been a part of natural systems for evolutionary time, and quantification is possible through embodied energy calculations.

Control Effect Of Cadmium

As with many other substances which are generally considered toxic, the heavy metal Cd has repeatedly shown stimulatory effects at low concentrations. Thus Doyle et al. (1975) reported Cd stimulation of net growth for three out of six microorganisms tested. Klass et al. (1974) measured a small stimulation in maximum cell numbers of Scenedesmus quaricauda at the lowest Cd concentration tested. Brkovic-Popovic and Popovic (1977) reported a slight stimulation of respiration by tubificid worms at low Cd concentration. Pickering and Gast (1972) observed greatly enhanced egg production in fat-head minnows and some increased survival of adults at low Cd levels.

On the other hand, all of these studies found Cd to be extremely toxic at higher concentrations. Thus the typical subsidy-stress curve may be a reasonable descriptor of Cd effect over a wide range of concentrations. This generalized control effect of Cd is incorporated in the stream model that follows to allow quantification of Cd effect over a wide range of embodied energy values.

METHODS

Data used to calibrate the stream model below were collected by Giesy et al. (1979) from six experimental streams located on the Savannah River Plant in Aiken County, South Carolina. Research was funded by the Environmental Protection Agency to study the fate and effects of pollutants in natural water systems.

The six channels used for the Cd study measured 92 m long, 0.61 m wide, and 0.31 m deep, were located outside, and received a continous flow of deep well water at a rate of 95 $L \cdot min^{-1}$ resulting in a mean current velocity of 1.3 $cm \cdot s^{-1}$. The well water used for this study had very low concentrations of dissolved constituents reflected in the low hardness (11.1 mg·L^{-1} as CaCO$_3$) and alkalinity (9.1 mg·L^{-1} as CaCO$_3$); as well as low concentrations of plant nutrients including nitrite-nitrate nitrogen (15.8 µg N·L^{-1}) and total phosphorus (2.9 µg P·L^{-1}). Background Cd concentration was determined as 0.023 µg Cd·L^{-1}.

Complex biological communities were rapidly established from inocula saved from previous studies. After an initial successional development of the biota for four months, Cd inputs into four of the six channels were started on March 18, 1976. Cadmium (as CdCl$_2$) was input to establish initial concentrations of 5 µg Cd·L^{-1} in two channels and 10 µg Cd·L^{-1} in another two, with the remaining two channels serving as controls. Cadmium inputs were discontinued on March 18, 1977, one full year after they were started, and data were collected for 5 months after that date. Details on methods of biological and chemical sampling and analysis may be found in Giesy et al. (1979).

Stream Model

An energy and matter flow model of the Cd streams was constructed as a diagram using the techniques and energy circuit language of Odum (1971,1972). The model is of intermediate complexity, combining storages of nutrients, algae, macrophytes, consumers, and detritus-microbes; and their uptake of and response to Cd at different concentrations.

The major biomass storages and their Cd content were monitored throughout the 2-yr study and have been used directly to calibrate the model when possible. However, few of the data were for average levels throughout the streams, but rather were for concentrations on replicable substrates. Thus, to calibrate the model to whole-system averages, extrapolations from a few measurements were made to the other data.

Also, few of the rates were actually measured during the project other than Cd uptake and release, system production and respiration, and export; and therefore, some parameterization was done by simulating the model and comparing results to the measured storages over time. Specific rates from the literature were used when available.

Computer simulations were made in BASIC computer language with a desk-top microcomputer. Integration was by means of simple difference equations with variable time steps. A copy of the program used is available upon request.

Energy Relationships

Embodied Energy

Calculations of embodied energy have been made according to the conventions set forth by Odum

(1978). Basically, embodied energy is the energy of one type necessary to produce an energy flow or storage of another type. The units of embodied energy are reported in terms of some reference energy type chosen for convenience in a given situation. For the comparisons made in this report, average insolation has been chosen as the reference energy (solar equivalent kilocalories, S.E. Cal).

In order to calculate the energy embodied in a particular energy flow or storage, a model is prepared showing all of the major energy flows responsible for generating the flow of interest. Figure 1 presents such a model for a simplified production process. Production (C) produces biomass (Q) with maintenance energy loss (E) and feedback maintenance (D). For example, this model could represent production in a pond where A is direct solar energy and B is the energy associated with all other inputs such as nutrients in rain and runoff from the land. These other energy flows must also be calculated in terms of S.E. Cal. This calculation may be made with the use of estimates of energy content (free energy of chemicals, kinetic energy of rain, etc.) and values for TR's (quality factors) published by Odum and Odum (1980), Odum et al. (1980), and Wang et al. (1980).

As defined by Odum (1978) each energy flow within the system boundary has equal embodied energy flow. Thus the embodied energy in flows C, D, and E are equal although their heat equivalent energies may be very different. The embodied energy of a storage is equal to the integrated input energy flows during one turnover time of the storage.

If the actual energies in C, D, and E (Cal) are known, then TR's may be calculated from this model by dividing the total input energy (A + B)

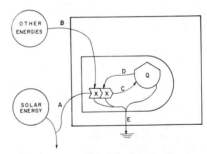

Figure 1.--Aggregated model of a production process with two energy inputs (A and B), gross production (C) of a stored product (Q), feedback maintenance (D) and maintenance energy loss from the system (E). The embodied energies of flows C, D, and E are equivalent and are equal to the sum of input embodied energies (A + B). The embodied energy in Q is equal to the total embodied energy input (A + B) multiplied by the turnover time of Q.

by the actual energy in each resulting flow. Thus
$(A + B)/C$ represents the TR for gross productivity
in figure 1, with units of S.E. $Cal \cdot Cal^{-1}$. These
TR values may then be used to evaluate some other
situation.

Values for the TR of a particular energy flow
may be calculated from several different models and
compared. There may be a theoretical minimum value
for each TR that represents the optimal efficiency
for the given energy transformation. Precise TR
values of many energy types and flows are not yet
known, so all calculations are assumed to be preliminary.

Energy Effect

The energy effect of a toxin or controller is
measured as an amplification (either positive or
negative) of an energy flow or storage expressed
in embodied energy units. This amplification or
perturbation may be measured relative to a control
with zero concentration of the toxin or other
controller.

For example, assume that a Cd input of 1 µg
$Cd \cdot m^{-2} \cdot d^{-1}$ to a steady state microcosm is found to
decrease primary productivity relative to a control
by 10 $Cal \cdot m^{-2} \cdot d^{-1}$. From other estimates we find
that the TR for primary productivity is 100 S.E.
$Cal \cdot Cal^{-1}$ and thus the energy effect of the given
Cd input is -1000 S.E. $Cal \cdot m^{-2} \cdot d^{-1}$. This toxic
effect may be divided by the Cd input to give
-1000 S.E. $Cal \cdot µg$ Cd^{-1} as the energy effect ratio.
The energy effect of a controlling substance may
be a function of its concentration and therefore
the concentration must be specified for comparisons.
As with embodied energy calculations, energy effect
calculations are in a preliminary stage and subject
to revision. The importance of the energy calculations made in this paper is as much in presenting
new, untested , but promising methods as in the
actual comparisons of embodied energy and energy
effect.

RESULTS AND DISCUSSION

Embodied Energy Of Cadmium

In order to evaluate the embodied energy of Cd
added to the artificial streams, data reported for
the industrial concentration of Cd were used.
Cadmium metal is produced commercially from by-
products of Zn production and therefore, the energy
inputs to the combined process of Zn and Cd production had to be evaluated.

Figure 2 presents a simplified model of this
process showing the evaluated actual energy and
material flows. Table 1 lists the flows and their
equivalent values in embodied energy units (S.E. Cal).
As a baseline for calculation, the embodied energy
of the unrefined Zn ore was assumed to be equal to
zero. The industrial embodied energy inputs in the
Cd purification process represent the minimum

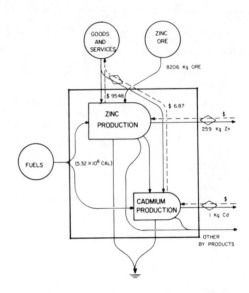

Figure 2.--Model of Zn and Cd production by the
electrolytic process with actual energy and
dollar flows evaluated. Cadmium production
is entirely a by-product recovery of Zn
purification.

amplification ability (energy effect) that the Cd
must have in human systems in order not to decrease
total power. Calculated from the data presented
in figure 2 and table 1, the embodied energy
content of pure Cd is 4.6×10^{10} S.E. $Cal \cdot kg$ Cd^{-1}.

Stream Model

The aggregated model of a stream ecosystem
used for experimentation with toxin manipulations
is presented in figure 3. This figure diagrams
the relationships between the forcing functions,
sunlight and water inflow containing nitrogen (N)
and Cd, and the complex stream community. Primary
producers include macrophytes (rooted plants, Q3)
and their associated periphytic algae (Q2). Consumers are aggregated in one unit (Q4) and all
unassimilated and dead material is cycled through
the detrital-microbial storage (Q5). The major
outputs of the stream model are export of water with
associated particulate and dissolved constituents
and degraded heat energy.

The model presented in figure 3 represents a
simplification of the actual streams and yet is a
very complex, nonlinear model. Since the purpose
of the model simulations was to provide approximate
data for the effects of Cd concentrations not
actually studied to be used in example correlations
of embodied energy and energy effect, exact fitting
of model output to actual data was not attempted.

Table 1.--Actual and embodied energy flows in the industrial purification of Zn and Cd from Zn ore resulting in 1 kg of pure Cd as illustrated in figure 2.

Type	Actual Quantity	Transformation Ratio	Embodied Energy
Zn ore	8206 kg Zn ore[1]	assumed zero	assumed zero
Fuels	5.31×10^6 Elec. Cal[2]	8000 S.E. Cal/Elec. Cal[3]	4.25×10^{10} S.E. Cal
Goods and Services	$95.48[4]	37×10^6 S.E. Cal/$[5]	3.53×10^9 S.E. Cal
Fuels	2597 Elec. Cal[6]	8000 S.E. Cal/Elec. Cal	2.08×10^7 S.E. Cal
Goods and Services	$ 6.87[7]	37×10^6 S.E. Cal/$	2.54×10^8 S.E. Cal
Purified Zn	259 kg[8]	1.79×10^8 S.E. Cal/kg	4.63×10^{10} S.E. Cal
Purified Cd	1 kg	4.63×10^{10} S.E. Cal/kg	4.63×10^{10} S.E. Cal

[1]From Petrick et al. (1979), 492 kg Zn concentrate with 60% Zn; 90% recovery from ore with 4% Zn content.

[2]Battelle Columbus Laboratories (BCL) (1975) total energy costs in Zn production converted to electrical Cal.

[3]From Odum and Odum (1980).

[4]From BCL (1975), $36.34 materials and reagents; from Cammarota (1978), $36.34 labor and $22.80 capital assuming 20-yr life for plant.

[5]Odum et al. (1980).

[6]Petrick et al. (1979).

[7]Ibid.

[8]79% efficiency of recovery from ore (Cammarota and Lucas 1977).

Instead, the model was simplified whenever possible, while retaining both fate and effects of the toxin.

Parameterization

The model illustrated in figure 3 contains 39 constants that had to be estimated from data collected during the Cd study or from published reports. Many of these constants were not known exactly and in fact many may not have been constant as species dominance changed during the 2-yr successional period. A discussion of the basis for the choice of the model parameters is presented below.

Units.--Flows in the model were in the following units: pure energy, Cal (kilogram calories); biomass, grams of dry weight (g dw); nitrogen, mg N; and Cd, μg Cd. Rates are all on a per square meter per day ($m^{-2} \cdot d^{-1}$) basis.

Solar Input.--As seen in figure 3, solar energy is one of the two primary driving forces included in the stream model. For simplicity on the microcomputer, solar input (J∅) was simulated by a sine function with maximum and minimum values of 6000 and 2720 $Cal \cdot m^{-2} \cdot d^{-1}$, respectively. Actual solar energy at the channels was taken as 70% of these maximum values from average cloud cover data published for Columbia, South Carolina (NOAA 1976, 1977).

In the channels, remaining solar input (JR) was calculated for use in productivity formulations. JR was equal to J∅ minus solar energy absorbed by algae (FA), by macrophytes (FB), and by the detritus-microbes (FC). The constant of photosynthetic efficiency used for algae and macrophytes was 2.8% measured at Silver Springs, Florida (Knight 1980). Using a conversion factor of 4 $Cal \cdot g \ dw^{-1}$ at 2.8% efficiency, 143 Cal absorbed per g dw produced was calculated. Thus the constants KA and KB were set equal to 143 $Cal \cdot g \ dw^{-1}$.

Water Inflow.--Incoming water flow (JW) was constant and equal to 136,800 $L \cdot d^{-1}$. This water contained N (N1 = 0.015 mg $N \cdot L^{-1}$) and Cd (C1 = variable with 0.023 μg $Cd \cdot L^{-1}$ as the background concentration).

Nitrogen Dynamics.--The total nitrogen concentration measured in the channel input water was 15.8 μg $N \cdot L^{-1}$ while the total phosphorus concentration was measured as 2.9 μg $P \cdot L^{-1}$ giving a ratio by atoms of about 13:1. Since the optimal ratio is generally considered to be 16:1, nitrogen may be more limiting and was chosen as the nutrient to monitor in the model. A few calculations show the importance of this nutrient in the streams and consequently in the model simulations. With a water input of 136,800 $L \cdot d^{-1}$ to each channel with 15.8 μg $N \cdot L^{-1}$ and 56 m^2 surface area, we calculate a nitrogen flow of 38.6 mg $N \cdot m^{-2} \cdot d^{-1}$. Actual net productivities measured in the channels during the second year of study were about 2.5 g $dw \cdot m^{-2} \cdot d^{-1}$.

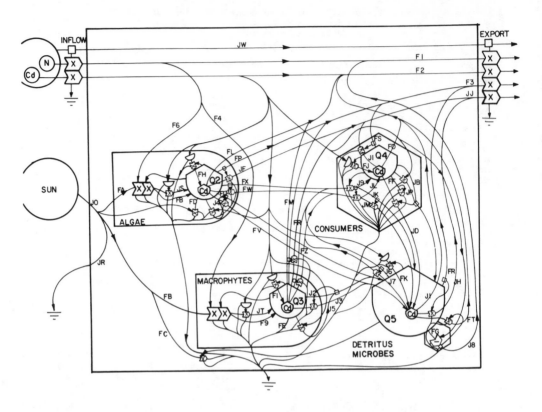

Figure 3.--Overall system model of stream microcosms receiving
Cd inputs. Sunlight interacts with dissolved chemicals
in input water to maintain complicated biological systems
and Cd cycling. Alphanumeric variable names indicate
flows and storages discussed in the text.

This means , if all of the N was taken up in biomass during each 24-hr period, the N content on a
dry weight basis could have only been about 1.5%,
which is low for algae and plants (Odum 1973).
Luxury uptake of N in the dark is known (Ketchum
1954) but algae cannot take up N from extremely
dilute water. Therefore, the constants for N uptake, K4 for algae and K6 for macrophytes, were
assumed to be 0.5% or 5 mg N·g dw^{-1}. The importance
of Cd's effect on nutrient cycling in these channels
for maximizing productivity is obvious from the
above discussion.

    Recycling of N was simplified in the model by
the assumption that all N remineralization was from
the detrital-microbial storage (Q5):

$$J8 = L8 \cdot FG \qquad (1)$$

where L8 = K4 above (0.5 mg N·g dw^{-1}) and FG is the
respiration of storage Q5, discussed below.

Overall N flow in the channel water was:

$$F1 = JN + J8 - F4 - F6 \qquad (2)$$

where JN is the N in input water:

$$JN = N1 \cdot JW \qquad (3)$$

and N1 is the N content of the inflow. Nitrogen
concentration in the stream water was calculated as:

$$N2 = F1/JW. \qquad (4)$$

    Cadmium Dynamics.--The background Cd concen-
tration (C1) was approximately 0.023 µg Cd·L^{-1}.
For the model simulations this concentation was
varied between ambient and 100 µg Cd·L^{-1}.

Cadmium inflow (JC) was calculated as:

$$JC = C1 \cdot JW. \qquad (5)$$

Cadmium uptake by the biota was modelled as a hyperbolic Michaelis-Menton function:

$$FH = \frac{KH \cdot CA \cdot Q2}{K1 + CA} \quad \text{(algae)} \qquad (6)$$

$$FI = \frac{KI \cdot CA \cdot Q3}{K1 + CA} \quad \text{(macrophytes)} \qquad (7)$$

$$FJ = \frac{KJ \cdot CA \cdot Q4}{K1 + CA} \quad \text{(consumers)} \qquad (8)$$

$$FK = \frac{KK \cdot CA \cdot Q5}{K1 + CA} \quad \text{(detritus-microbes)} \qquad (9)$$

Where the half-saturation constant (K1) was assumed to be equal to 200 µg $Cd \cdot L^{-1}$.

The Cd content (µg $Cd \cdot m^{-2}$) of each of the four biological storages was included in the model: C2 -- Cd in algae; C3 --Cd in macrophytes; C4 --Cd in consumers; and C5 --Cd in detritus-microbes.

Loss of Cd from each of these storages was in two forms: 1. dissolved Cd, modelled as a simple linear decay; and 2. particulate Cd transport to other storages. For the algae Cd decay was equal to:

$$FL = KL \cdot C2 \qquad (10)$$

where KL was measured in the streams as 0.065 d^{-1}. For the macrophytes, Cd decay equaled:

$$FM = KM \cdot C3 \qquad (11)$$

where KM was measured as 0.02 d^{-1} in the streams. For the consumers, Cd decay was modelled as:

$$FO = KO \cdot C4 \qquad (12)$$

where a value of 0.0055 d^{-1} was used for KO as measured for mosquito fish in the channels. For the detritus-microbes, Cd decay was equal to:

$$J1 = L1 \cdot C5 \qquad (13)$$

with L1 = KL = 0.065 d^{-1}. Cadmium was also remineralized in microbial respiration:

$$FT = FG \cdot CE \qquad (14)$$

where CE was the concentration of Cd in Q5 and was equal to C5/Q5.

Cadmium loss in particulate form was simply calculated as g $dw \cdot m^{-2} \cdot d^{-1}$ multiplied by the Cd concentration in the storage (µg $Cd \cdot g \ dw^{-1}$).

Overall Cd flow in the channel water was equal to:

$$\begin{aligned} F2 = JC + FL &+ FM + FO + J1 - FK - FJ \\ &- FH - FI \end{aligned} \qquad (15)$$

and Cd concentration in this water was calculated as:

$$CA = \frac{F2}{JW}. \qquad (16)$$

Algae.--The living algal component (Q2) of the periphyton was monitored separately because of its importance in the primary energy fixation in the channels. The modelled relationship for gross primary production was:

$$F8 = K8 \cdot N2 \cdot JR \cdot Q2. \qquad (17)$$

Algal respiration was equal to:

$$FD = KD \cdot Q2 \cdot JR \qquad (18)$$

where the JR term indicates photorespiration in the shallow streams. In addition, the algal storage is also drained by consumer feeding:

$$FX = KX \cdot Q2 \cdot Q4 , \qquad (19)$$

by death to detritus:

$$FU = KU \cdot Q2 , \qquad (20)$$

and by export:

$$FP = KP \cdot Q2 . \qquad (21)$$

The effect of Cd was modelled as both stimulatory at low concentrations:

$$JS = \frac{LS \cdot F8 \cdot CA}{LU + CA} \qquad (22)$$

and as a toxic drain on biomass:

$$J4 = L4 \cdot Q2 \cdot CA. \qquad (23)$$

The constants in the above relationships (K8, KD, KX, KU, KP, LS, LU, and L4) were all adjusted by an initial approximation and then simulation runs to fit model output to observed algal biomass patterns.

Macrophytes.--Gross primary production by the rooted plants (Q3) was equal to:

$$F9 = K9 \cdot N2 \cdot JR \cdot Q3 . \qquad (24)$$

Macrophyte respiration was modelled as a simple linear decay of biomass:

$$FE = KE \cdot Q3. \qquad (25)$$

Other flows were analagous to the algal flows:

$$J2 = L2 \cdot Q3 \quad \text{(death)} \qquad (26)$$

$$FY = KY \cdot Q3 \cdot Q4 \quad \text{(consumer feeding)} \qquad (27)$$

$$FQ = KQ \cdot Q3 \quad \text{(export)} \qquad (28)$$

$$JT = \frac{LT \cdot F9 \cdot CA}{LU + CA} \quad \text{(Cd stimulation)} \qquad (29)$$

$$J5 = L5 \cdot Q3 \cdot CA \quad \text{(Cd toxicity)} . \quad (30)$$

The eight constants governing macrophyte dynamics were adjusted by simulations to fit observed biomass changes in the artificial streams.

Consumers.

Consumers ($Q4$) fed on algae (eqn. 19), on macrophytes (eqn. 27), and on detritus-microbes:

$$J6 = L6 \cdot Q5 \cdot Q4 . \quad (31)$$

Assimilation efficiency was assumed to be 30%,

$$J9 = L9 \cdot (FX + FY + J6) \quad (32)$$

where $L9 = 0.3$. The remainder of the ingested food was returned to the detritus-microbes:

$$JK = FX + FY + J6 - J9. \quad (33)$$

Losses from the consumers were respiration:

$$FF = KF \cdot Q4^2 \quad (34)$$

with a square term to indicate predation or crowding effects; death to detritus:

$$JB = LB \cdot Q4 ; \quad (35)$$

and export:

$$FS = KS \cdot Q4 . \quad (36)$$

Cadmium toxicity was modelled as an interactive drain on consumer biomass:

$$JP = LP \cdot Q4 \cdot CA . \quad (37)$$

Once again the constants listed above (KX, KY, L6, KF, LB, KS, and LP) were approximated and then adjusted to give realistic output.

Detritus-Microbes.

The detritus-microbe sub-system ($Q5$) illustrated in figure 3 comprised the greatest portion of the periphyton, more than 98% by weight (with the remainder composed of living algae). Detritus and microbes were modelled as one unit because of their intimate association in the stream microcosms.

Inputs to $Q5$ were mentioned in the preceding paragraphs. The two major drains on $Q5$ were export:

$$FR = KR \cdot Q5 \quad (38)$$

and respiration:

$$FG = KG \cdot Q5 \cdot (1 - LE \cdot CA) . \quad (39)$$

As seen in equation (39), Cd toxicity to Q5 was a negative interaction with respiration and nutrient remineralization. This formulation is consistent with actual stream data for leaf litter packs (Giesy et al. 1979).

Parameter Approximation.

In order to approximate the unmeasured constants discussed above, a "steady state" period of the artificial streams succession was evaluated. Figure 4 illustrates the four biomass storages with inputs and outputs of dry weight for data from September 1976 when little net growth was observed. Biomass values, total net production and respiration, and export values were known. Gross productivity was partitioned between the algae and macrophytes as shown. The partition of respiration was also based on assumptions. Total export at this time was assumed to be twice the measured value or 3 g dw·m^{-2}·d^{-1}. This number divided by the total biomass of 210 g dw·m^{-2} gives the export constants of 0.014 d^{-1}. Consumer feeding on algae and macrophytes was assumed to be 0.2%, and 0.08% for detritus-microbes. With 30% assimilation efficiency, the return of ingested material to detritus was calculated as 0.168 g dw·m^{-2}·d^{-1}. Cadmium toxicity was taken as zero for these control stream data. By difference, death to detritus was calculated for each storage. These calculations resulted in a small net loss for Q5 of 0.011 g dw·m^{-2}·d^{-1}. The constants calculated by this procedure are also shown in figure 4. These approximate values were used in initial simulations but many were varied to give better fit to the actual time-series data from the Cd streams. The final values used in the control model simulation are listed in table 2. Table 2 also lists the differential equations for the state variables and the initial conditions used in the control simulation.

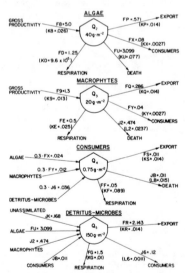

Figure 4.—Input-output diagram for four biological storages in Cd stream model of figure 3. Data from September 1976 were used to approximate steady state flows for parameter estimation. Storages are in g dw·m^{-2}; flows are in g dw·m^{-2}·d^{-1}; and constants in parentheses have variable units listed in table 2.

Control Simulation

As a baseline for calibration of the stream model, data from the control channels of the Cd streams project were used. This control model was simulated for November 1975 through August 1977. Initial conditions in the model were roughly those in the experimental streams: clean and uncolonized except for the plants and consumers that were added. Model parameters were adjusted as discussed above until simulation results were comparable to the measured data.

Table 2.--Listing of differential equations, initial conditions, and transfer coefficients used to simulate the model of stream microcosms receiving Cd inputs illustrated in figure 3.

Differential Equations

$\dot{Q}2$ = F8+JS-FD-FX-FP-FU-J4
$\dot{Q}3$ = F9+JT-FQ-FY-J2-FE-J5
$\dot{Q}4$ = J9-FS-JB-FF-JP
$\dot{Q}5$ = J2+FU+JB+JK-FR-FG+J4+JP+J5-J6
$\dot{C}2$ = FH-FL-FW-FV-JF
$\dot{C}3$ = FI-FZ-J3-FM-JG
$\dot{C}4$ = JL+FJ-FO-JD-JI
$\dot{C}5$ = J3+FV+FK-J1-JH-J7-FT+JD+JM

Initial Conditions

Q2	0.1 g dw·m^{-2}		CA	0.023 µg Cd·L^{-1}	
Q3	1.0	"	C1	0.023-100 µg Cd·L^{-1}	
Q4	0.1	"	CB	1.0 µg Cd·g dw^{-1}	
Q5	0.1	"	CC	1.0	"
C2	0.1 µg Cd·m^{-2}		CD	1.0	"
C3	1.0	"	CE	1.0	"
C4	0.1	"	JØ	3442 Cal·m^{-2}·d^{-1}	
C5	0.1	"	JR	3000	"
N1	0.015 mg N·L^{-1}		JW	2443 L·m^{-2}·d^{-1}	
N2	0.015	"			

Transfer Coefficients

K1	200 µg Cd·L^{-1}		KQ	0.008 d^{-1}
K4	5 mg N·g dw^{-1}		KR	0.008 d^{-1}
K6	5	"	KS	0.008 d^{-1}
K8	0.009 L·m^2·Cal^{-1}·mg N^{-1}		KU	0.06 d^{-1}
K9	0.0024	"	KX	0.0027 m^2·d^{-1}·g^{-1}
KA	143 Cal·g^{-1}		KY	0.0015	"
KB	143	"	L1	0.065 d^{-1}
KC	0.003 m^2·g^{-1}		L2	0.02 d^{-1}
KD	6 x 10^{-5} m^2·Cal^{-1}		L4	.003 L·d^{-1}·µg Cd^{-1}
KE	0.018 d^{-1}		L5	.0015	"
KF	0.03 d^{-1}		L6	0.0007 m^2·d^{-1}·g^{-1}
KG	0.00012 d^{-1}		L8	5 mg N·g^{-1}
KH	180 µg Cd·g^{-1}·d^{-1}		L9	0.3
KI	100	"	LP	0.008 L·d^{-1}·µg Cd^{-1}
KJ	8	"	LB	0.015 d^{-1}
KK	80	"	LE	0.01 L·µg Cd^{-1}
KL	0.045 d^{-1}		LS	0.025
KM	0.02 d^{-1}		LT	0.025
KO	0.0055 d^{-1}		LU	0.2 µg Cd·L^{-1}
KP	0.008 d^{-1}			

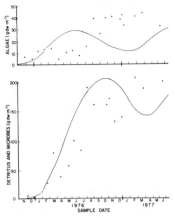

Figure 5.--Stream model simulation results for algal and detrital-microbial biomasses at control Cd concentration of 0.023 µg Cd·L^{-1}. Solid lines are simulation results and dots are measured data.

Biological Parameters.--Figure 5 illustrates a comparison between simulation results and measured algal and detrital biomass. Simulated values were in the same range as the measured ones, but the timing of peaks and troughs did not overlap. This effect is also seen in figure 6 which compares simulated output for gross production, system respiration, and export to the measured data. Figure 7

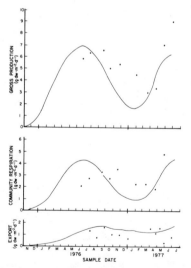

Figure 6.--Simulation results for gross production, community respiration, and export at control Cd concentration of 0.023 µg Cd·L^{-1}. Solid lines are simulation results and dots are measured data.

compares simulated and actual data for macrophyte
and macroinvertebrate biomasses. The model was not
able to predict an increase in macrophyte biomass
measured during the second summer.

For all of the above parameters the timing of
the model was approximately 1-2 months early. This
effect may have been due to the simple sinusoidal
solar input used as compared to the actual stochastic variation in solar input. A more complex
forcing function with this same model could have
perhaps provided a better fit between simulated and
actual data; however, the model behaved realistically
enough for experimentation with Cd toxicity effect.

Cadmium Dynamics.--A measured Cd input of 0.023
µg Cd·L^{-1} (control streams) resulted in concentrations in most biota of about 1 µg Cd·g dw^{-1}.
Depuration constants for Cd from the various storages
were presented earlier in this section. Uptake
rates were also estimated from stream data, and
maximum uptake rates were calculated from these
values for a half-saturation constant of K1 = 200
µg Cd·L^{-1}. These values were altered slightly to
improve the fit of model data to measured concentrations and are reported in table 2.

The control simulation run predicted all Cd
concentrations to be between 0.2 and 1.0 µg Cd·g
dw^{-1} during most of the year. These values were
slightly lower than the observed concentrations.

Cadmium Input

Cadmium input in the model was regulated by
a series of IF...THEN statements. Cadmium concentration of the input water was set as Cl.
When time reached 136 days (March 18, 1976), Cl
was set equal to ZZ, which was the elevated Cd
content of the input water, and when time reached
503 days (March 18, 1977), Cl was set back to the
control Cd concentration.

Biological Parameters.--Data from the 5 and
10 µg Cd·L^{-1} streams were used to calibrate the
toxicity constants L4, L5, LP, and LE. The
constants LS and LT allowed direct stimulatory
effect of Cd on primary production.

Simulated and measured values for algal and
detrital-microbial biomasses are presented in
figure 8. The model results show some reduction
in algal and detrital biomasses at 5 and 10 µg Cd·
L^{-1}, but the fit to the actual data was disappoint-
ing. The model also predicted a reduction in
macroinvertebrate and macrophyte biomasses with
Cd input (fig. 9).

Figure 10 shows actual and simulated values
of the system-level parameters at 5 and 10 µg Cd·
L^{-1}. The fit between actual and simulated data
for these parameters is better because they were
of primary consideration in adjustment of toxicity
constants.

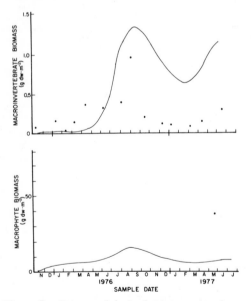

Figure 7.--Stream model simulation results for
macroinvertebrate and macrophyte biomasses at
control Cd concentration of 0.023 µg Cd·L^{-1}.
Solid lines are simulation results and dots
are measured data.

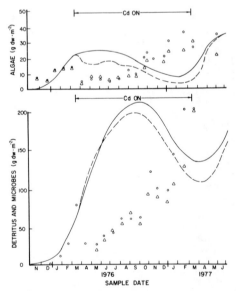

Figure 8.--Stream model simulation results for
algal and detrital-microbial biomasses at
5 and 10 µg Cd·L^{-1} input levels. Solid and
dotted lines are simulation results and
circles and triangles are measured values
for 5 and 10 µg Cd·L^{-1}.

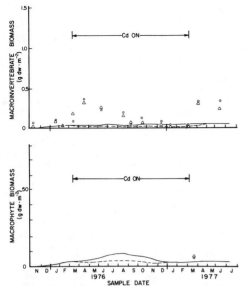

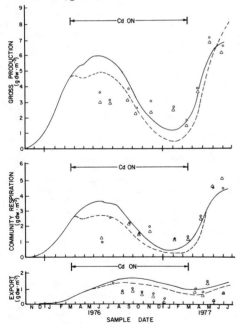

Figure 9.--Stream model simulation results
for macroinvertebrate and macrophyte biomasses
at 5 and 10 μg Cd·L^{-1} input levels. Solid
and dotted lines are simulation results and
circles and triangles are measured values for
5 and 10 μg Cd·L^{-1}.

Figure 10.--Stream model simulation results for gross
production, community respiration, and export
at 5 and 10 μg Cd·L^{-1} input levels. Solid and
dotted lines are simulation results and circles
and triangles are measured values for 5 and
10 μg Cd·L^{-1}.

Cadmium Dynamics.--The dynamics of Cd concentration in the various biological storages were
monitored in the model for comparison to the actual
measured data. Steady state concentrations of Cd
in the biota at a water concentration of 5 μg Cd·
L^{-1} were: algae, 35; macrophytes, 50; consumers,
33; and detritus-microbes, 30 μg Cd·g dw^{-1}. For
the 10 μg Cd·L^{-1} water input concentration, biotic
Cd concentrations were: algae, 60; macrophytes, 90;
consumers, 40; and detritus-microbes, 58 μg Cd·g
dw^{-1}. These compare favorably with the values
presented by Giesy et al. (1979).

The stream model provides a convenient means
of summarization of the fate of Cd in the aquatic
ecosystem because all flows of the metal are accounted for. Figure 11 illustrates the major Cd
flows and storages predicted by the model at the
three Cd concentrations actually tested (averaged
over a 1-yr period). All flows are in μg Cd·m^{-2}·
d^{-1} and storages are in μg Cd·m^{-2}. The model predicted an average lowering of water Cd concentration of 0.1 and 0.2 μg Cd·L^{-1} at 5 and 10 μg Cd·
L^{-1}, respectively. These values were slightly
lower than the measured values of 0.2 and 0.3 μg
Cd·L^{-1}. Data in figure 11 indicate that only
0.4% of the Cd output from the channels was in
particulate form. Also, the model indicated that
Cd uptake and depuration by biota was about 3% of
the dissolved Cd flow through.

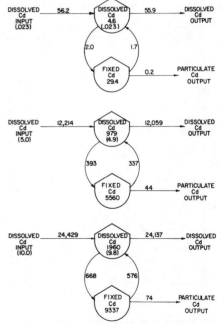

Figure 11.--Overview of Cd fate in artificial
streams from model output. Values are
averages over a 1-yr period. Storages are
in μg Cd·m^{-2}, flows are in μg Cd·m^{-2}·d^{-1}, and
concentrations (in parentheses) are in
μg Cd·L^{-1}.

Model Experimentation

The calibrated stream model presented above allows experimentation with exogenous and internal controls other than those actually tested in South Carolina. Of particular interest in this paper is the prediction of toxin effect over a range of Cd concentrations outside the ones actually studied.

The predicted response of system-level parameters to a range of Cd concentrations is presented in figure 12. The data from the three concentrations actually studied are shown for comparison. As presented in figure 3, the stream model predicts maximum enhancement of stream metabolism at a Cd concentration of 0.5 μg Cd·L^{-1}.

Embodied Energy And Control

Transformation Ratios

The calibrated stream model facilitates calculation of TR's for analysis of the energy effect of Cd on system energy flow.

The entire energy income of the Cd streams must be known in order to calculate TR values. Not only is energy received directly from the sun, but energy is also concentrated from indirect sources such as runoff from the watershed, which creates stream flow and structure. In the stream microcosms this flow and structure were added to the incoming sunlight, from human energies and fossil fuel work. The energy content of the structure

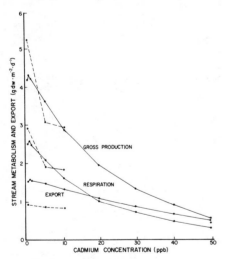

Figure 12.--Average gross productivity, respiration, and export values during 1 yr of continuous Cd input, predicted by stream model for Cd concentrations up to 50 μg Cd·L^{-1}. Measured data from Cd streams are indicated by dashed lines for comparison.

is estimated from its dollar cost. The energy content of the inflowing water is estimated by calculation of the energy content of one component, the dissolved nitrogen. Only one constituent of the water must be evaluated since it has the same embodied energy as the entire water inflow and calculation of additional components would be double counting.

The average yearly insolation used in the model was 3052 S.E. Cal·m^{-2}·d^{-1}. The total structure of the streams including plastic, sand, concrete blocks, and labor cost approximately 6.5×10^{-4} \$·m^{-2}·d^{-1}. Using the 1972 energy-to-dollar ratio of 43.2×10^6 S.E. Cal·\$$^{-1}$ from Odum et al. (1980), this expenditure is equivalent to 2.79×10^4 S.E. Cal·m^{-2}·d^{-1} for the channel area.

The other major energy input is the water flow in the streams. This flow may be evaluated by the free energy change for the essential nutrient, nitrogen. Using the concentration change measured in the control channels (10.4 to 3.6 μg N·L^{-1}) and the equation for Gibb's free energy:

$$\Delta G = nRT \ln \frac{C_2}{C_1} \qquad (40)$$

where ΔG is the change in Gibb's free energy, n is the number of moles of reactants, R is the universal gas constant = 1.99×10^{-3} Cal·$^{\circ}$K^{-1}·mol^{-1}, T is the absolute temperature in $^{\circ}$K, and C_1 and C_2 are the concentrations of nitrogen before and after the free-energy change; a decrease in free energy along the reach of the streams of 0.045 Cal·g N^{-1} was calculated, which is equivalent to 1.15×10^{-3} Cal·m^{-2}·d^{-1}. Using the TR for chemical potential of dissolved solids in stream flow of 1.17×10^6 S.E. Cal·Cal^{-1} given by Wang et al. (1980), an energy contribution from stream flow of 1346 S.E. Cal·m^{-2}·d^{-1} was calculated.

The total energy contribution to the streams is the sum of the three values listed above in equivalent embodied energy units and is equal to 32,298 S.E. Cal·m^{-2}·d^{-1}. Transformation ratios of the stream energy flows and storages have been calculated using yearly averages from the stream model and the total energy input given above. A conversion factor of 4 Cal·g dw^{-1} was used to convert the biomass units to heat energy units. These values are listed in table 3.

Embodied Energy-Energy Effect Correlation

The stream model was simulated at a series of Cd concentrations from background to 100 μg Cd·L^{-1}. Yearly averages of gross production, respiration, export, algae, macrophytes, consumers, and detritus-microbes were calculated for analysis of the effect of Cd on components of varying energy quality. The effect of Cd on a given energy flow or storage, relative to the control case, was converted from actual energy to embodied energy using the TR values in table 3. This effect could be either positive or negative.

Table 3.--Transformation ratios for major storages and flows
in stream model. Energy values are derived from dry
weight values using the factor 4 Cal = 1 g dw. Total
energy input to the streams was taken as 32,298 S.E.
$Cal \cdot m^{-2} \cdot d^{-1}$ and growth times of storages were taken
from model simulation data. All data used are from
the control simulation (0.023 µg $Cd \cdot L^{-1}$).

Flow or Storage	Average Value	Actual Energy $Cal \cdot m^{-2} \cdot d^{-1}$	TR S.E. $Cal \cdot Cal^{-1}$
Gross production	4.21 g dw·m^{-2}·d^{-1}	16.8	1923
System respiration	2.51 "	10.0	3217
Export	1.52 "	6.1	5312
Algae (Q2)	21.4 g dw·m^{-2}	2.1[1]	15,093
Macrophytes (Q3)	8.06 "	0.161[2]	2.0 x 10^5
Consumers (Q4)	0.7 "	0.014[2]	2.3 x 10^6
Detritus-Microbes (Q5)	159.0 "	3.18[2]	10,157

[1]Charge-up time to steady state biomass was estimated as 40 days.
[2]Charge-up time to steady state biomass was estimated as 200 days.

Cadmium TR was calculated for the different
concentrations simulated in the stream model. As
a baseline, the TR of pure Cd calculated for the
industrial process (4.6 x 10^7 S.E. Cal·g Cd^{-1}) was
used. Using the solubility of CdCl$_2$ in water
(1400 g CdCl$_2$·L^{-1}), a water concentration
of 8.6 x 10^8 µg Cd·L^{-1} (saturated) was
assumed to have a TR equal to pure Cd.
The lowest concentration of Cd normally found in
water is approximately 0.02 µg Cd·L^{-1}, and the free
energy change between these two Cd concentrations
was calculated using equation (40) as 0.13 Cal·g
Cd^{-1}. Dividing this value into 4.6 x 10^7 S.E.
Cal·g Cd^{-1} gives the TR of pure Cd on a Cal per
Cal basis of 3.5 x 10^8 S.E. Cal·Cal^{-1}.

In order to calculate the TR of Cd at any
other water Cd concentration, the free energy
change between the saturated solution and the given
concentration must be calculated. For example, at
10 µg Cd·L^{-1}, the free energy change is $\Delta G = -0.097$
Cal·g Cd^{-1}. This value is multiplied by 3.5 x 10^8
S.E. Cal·Cal^{-1} to give the change in TR going from
the saturated Cd solution to a less concentrated
one. For 10 µg Cd·L^{-1} this change equals -3.4 x
10^7 S.E. Cal·g Cd^{-1}. This value is subtracted
from 4.6 x 10^7 S.E. Cal·g Cd^{-1} to give the TR:

$$TR_{Cd, 10 \ \mu g \ Cd \cdot L-1} = 1.2 \times 10^7 \ S.E. \ Cal \cdot g \ Cd^{-1}.$$

These calculations of the embodied energy of
Cd were combined with the energy effect data predicted by the stream model. Figure 13 presents
the correlation diagrams for the system-level
parameters and for the biological storages. All
of the parameters except the macrophyte storage
showed first a positive, then a negative, and
finally zero correlation between embodied energy
and energy effect as Cd embodied energy increased.
Positive correlations between Cd TR and energy
effect were generally found at concentrations below

0.1 µg Cd·L^{-1}. Figure 13 also indicates that as a
biological controller, Cd has little marginal
effect above a TR of 16 x 10^6 S.E. Cal·g Cd^{-1}
(100 µg Cd·L^{-1}).

The values of the two parameters in figure 13
show an order-of-magnitude comparison of energy
effect and TR of Cd on system energy flow (gross
production and respiration). Given the assumptions
upon which these calculations were made, this

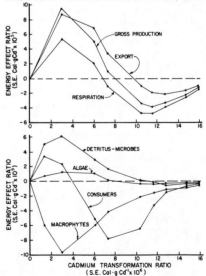

Figure 13.--Predicted correlation between Cd TR
and Cd energy effect ratio for system-level
parameters and for storages. Values are
calculated from 1-yr averages of simulation
data from the stream model.

result is encouraging. Notice that the storages have a much higher energy effect than the system parameters. The significance of this finding may be that selection is taking place on overall energy flow rather than for storage biomass as discussed earlier.

Actual stream systems that are capable of steady state growth populations may have much tighter correlations between the TR of Cd (its energy cost to the system) and the energy effect of Cd (its ability to control the system). If these correlations are found to be consistent in other systems and with other controllers, embodied energy of a controller may be calculated from its effect and vice versa. Widely different studies of energy effect could be compared using embodied energy values. Studies of several seemingly unique controllers may be comparable if their embodied energy content is known. The idea of energy effect being a direct function of energy cost may allow a needed synthesis of information in dealing with the modern world's increasing toxic wastes. The recognition of the stimulatory role of naturally-occurring chemicals in biological systems greatly broadens our theoretical understanding of processes and allows a more finely tuned control by humans of environmental systems.

The controlling influence of consumers in natural ecosystems has received attention through a series of review papers by Chew (1974), Mattson and Addy (1975), Owen and Wiegert (1976), Batzli (1978), and Kitchell et al. (1979); however, a theoretical understanding of why consumers are regulators and the energetics of their control appears to be lacking. The generality of the control effect illustrated with Cd in this paper is clearly seen in the subsidy-stress curves reported by researchers who experimentally manipulate consumer densities and measure the resultant effects on system functioning (Cooper 1973; Hargrave 1970; Flint and Goldman 1975; Knight 1980). It is hoped that the method of quantification outlined in this paper will provide a basis for comparison of control in these diverse systems.

LITERATURE CITED

Battelle Columbus Laboratories. 1975. Energy use patterns in metallurgical and nonmetallic mineral processing. (Phase 4-energy data and flow sheets, high-priority committees). U.S. Bur. Mines OFR 80-75, NTIS PB 245759/AS. 192 p.

Batzli, G.O. 1978. The role of herbivores in mineral cycling. p. 95-112. In J.P. Giesy (ed.) Environmental chemistry and cycling processes. NTIS 760429, Augusta, Ga.

Brkovic-Popovic, I. and M. Popovic. 1977. Effects of heavy metals on survival and respiration rate of tubificid worms: Pt. 2 - Effects on respiration rate. Environ. Pollut. 13:93-98.

Cammarota, V.A. 1978. Zinc. Mineral commodity profiles MCP-12. U.S. Bur. Mines. 25 p.

Cammarota, V.A. and J.M. Lucas. 1977. Zinc. U.S. Bur. Mines Minerals Yearbook,vol. 1. 34 p.

Chew, R.M. 1974. Consumers as regulators of ecosystems: an alternative to energetics. Ohio Journal of Science 74:359-370.

Conrad, M. 1976. Patterns of biological control in ecosystems. p.431-56.In B.C. Patten (ed.) Systems analysis and simulation in ecology. vol. 4, 593 p. Academic Press, New York,N.Y.

Cooper, D.C. 1973. Enhancement of net primary productivity by herbivore grazing in aquatic laboratory microcosms. Limnol. Oceanogr. 18: 31-37.

Doyle, J.J., R.T. Marshall, and P.Fander. 1975. Effects of cadmium on the growth and uptake of cadmium by microorganisms. Appl. Microbiol. 29:562-564.

Elton, C. 1927. Animal ecology. MacMillan Co., New York, N.Y.

Flint, R.W. and C.R.Goldman. 1975. The effects of a benthic grazer on the primary productivity of the littoral zone of Lake Tahoe. Limnol. Oceanogr. 20:935-944.

Giesy, J.P., H.J. Kania, J.W. Bowling, R.L. Knight, S. Mashburn, and S. Clarkin. 1979. Fate and biological effects of cadmium introduced into channel microcosms. 156 p. EPA-600/3-79-039.

Hargrave, B.T. 1970. The effect of a deposit-feeding amphipod on the metabolism of benthic microflora. Limnol Oceanogr. 15:21-30.

Ketchum, B.H. 1954. Mineral nutrition of phytoplankton. Ann. Rev. Plant Physiol. 5:55-74.

Kitchell, J.R., R.V. O'Neill, D. Webb, G.W. Gallepp, S.M. Bartell, J.F. Koonce, and B.S. Ausmus. 1979. Consumer regulation of nutrient cycling. Bioscience 29:28-34.

Klass, E., D.W. Rowe, and E.J. Massaro. 1974. The effect of cadmium on population growth of the green alga Scenedesmus quadricauda. Bull. Environ. Contam. Toxicol. 12:442-445.

Knight, R.L. 1980. Energy basis of control in aquatic ecosystems. Ph.D. dissertation, University of Florida. 198 p.

Lamanna, C. and M.F. Mallette. 1953. Basic bacteriology. Williams and Wilkins Co., Baltimore, Md.

Laughlin, R.B., J. Ng, and H.E. Guard. 1981. Hormesis: a response to low environmental concentrations of petroleum hydrocarbons. Science 211:705-707.

Lotka, A.J. 1922. Contributions to the energetics of evolution. Proc. Natl. Acad. Sci. 8:147-151.

Mattson, W.J. and N.D. Addy. 1975. Phytophagous insects as regulators of forest primary production. Science 190:515-522.

Mulholland, R.J. and C.S. Sims. 1976. Control theory and regulation of ecosystems. p. 373-388 In B.C. Patten (ed.) Systems analysis and simulation in ecology. vol. 4, 593 p. Academic Press, New York, N.Y.

National Oceanographic and Atmospheric Administration. 1976,1977. Climatological data: National summary. vols. 26 and 27. Asheville, N.C.

Odum, E.P. 1973. Fundamentals of ecology. 3rd ed. W.B.Saunders, Co. Philadelphia, Pa.

Odum, E.P., J.T. Finn, and E.H. Franz. 1979. Perturbation theory and the subsidy-stress gradient. Bioscience 29:349-352.

Odum, H.T. 1968. Work circuits and system stress. p. 81-138 In Y. Young (ed.) Primary production and mineral cycling in natural ecosystems. University of Maine Press, Orono, Me.

Odum, H.T. 1971. Environment, power, and society. Wiley-Interscience, New York, N.Y.

Odum, H.T. 1972. An energy circuit language for ecological and social systems : Its physical basis. p. 139-211 In B.C. Patten (ed.) Systems analysis and simulation in ecology. vol. 2. Academic Press, New York, N.Y.

Odum, H.T. 1978. Energy analysis, energy quality, and environment. p. 55-78 In M.W. Gilliland (ed.) Energy analysis: A new public policy tool. Westview Press, Boulder, Colo.

Odum, H.T. 1979. Energy quality control of ecosystem design. p. 221-235 In R.F. Dame (ed.) Marsh-Estuarine systems simulation. University of South Carolina Press, Columbia, S.C.

Odum, H.T., M.J. Lavine, F.C. Wang, M.A. Miller, J.F. Alexander, and T. Butler. 1980. A manual for using energy analysis for power plant siting. Final Report to the Nuclear Regulatory Commission, Center for Wetlands, University of Florida, Gainesville, Fla.

Odum, H.T. and E.C. Odum. 1980. Energy basis of New Zealand and the use of embodied energy for evaluating benefits of international trade. p. 106-167 In Proceedings of Energy Modeling Symposium, Ministry of Energy and Victoria University of Wellington, Tech. Publ. No. 7, Wellington, N.Z. November 1979.

Odum, H.T. and R.C. Pinkerton. 1955. Time's speed regulator: the optimum efficiency for maximum power output in physical and biological systems. American Scientist 43:331-343.

Owen, D.F. and R.G. Wiegert. 1976. Do consumers maximize plant fitness? Oikos 27:488-492.

Petrick, A., H.J. Bennett, K.E. Starch, and R.C. Weisner. 1979. The economics of by-product metals - Pt. 2. U.S. Bur. Mines Info. Cir. 8570, U.S. Dept. Interior.

Pickering, Q.H. and M.H. Gast. 1972. Acute and chronic toxicity of cadmium to the fathead minnow (Pimephales promelas). J. Fish. Res. Bd. Can. 29:1099-1106.

Straskraba, M. 1979. Natural control mechanisms in models of aquatic ecosystems. Ecological Modelling 6:305-321.

Wang, F.C., H.T. Odum, and P.C. Kangas. 1980. Energy analysis for environmental impact assessment. Journal Water Resources Planning and Management Division, ASCE 106:451-466.

EFFECTS OF STRUCTURAL CHANGES ON MATTER AND ENERGY MODELS OF ECOSYSTEMS[1]

Robert W. Bosserman[2]

Abstract.--Graph and matrix techniques have been used to analyze the interactive structures of ecosystems. These analyses are based on the concept that causes and effects in ecosystems are propagated through a network of energy and matter flows; therefore, both direct and indirect paths are important for assessing ecosystem structure. The connectivity structure of ecosystem models is encapsulated into a single complexity index which can be used to compare different models as well as interactions within a model. A sensitivity analysis which involves changing the relative magnitudes of flows between compartments has also been developed. The methods are applicable to Leontief-style input-output models which have been used for energy and material analyses of ecosystems.

INTRODUCTION

Like other living systems, ecosystems are self-organizing networks of matter and energy which are far from thermodynamic equilibrium and are dependent on energy and matter flows for their existence. These energy and matter flows are expressed in relatively stable configurations which generate complex patterns of behavior and organization. Ecosystems are also dynamic objects, so their behaviors and organizations change through time. However, most representations of ecosystems, as well as other living systems, are static and poorly depict their structural dynamics. Such static representations are poorly able to deal with adaptive, developmental and evolutionary processes that vary in time and space (Jantsch 1980). Indeed, much of Western science has developed around a paradigm of static structure (Prigogine 1980). Many ecosystem models, for a number of reasons, assume nearly constant structural relations between components, steady state conditions or linearity. However, ecological phenomena such as species replacement, succession, development of nutrient cycling, reaction to perturbations, and nutrient

limitations involve structural changes in the flow network, as well as simple changes in the magnitudes of fluxes.

Ecosystems change with their environments in a continual process of adaptation and evolution. Coevolutionary, competitive and cooperative relations are important in the establishment of a harmonious balance between ecosystem properties and environmental fluctuations. During the process of adaptation to the environment, the component organisms of the ecosystem orient themselves to optimize their environment (Bosserman 1979, Conrad 1976). This optimization involves the reshuffling of resources through active selection of inputs and outputs, a process which results in a stable and orderly configuration of metabolic processes (Bosserman 1979). Minor fluctuations in the environment result in small adjustments of the flow structure due to changes in inputs and changes in rate parameters; this can be accomplished by acclimatization of organisms, replacement by compensator species, or mutations which affect metabolic processes. Major fluctuations in the environment may result in more traumatic changes such as the creation or deletion of components or interactions. These traumatic changes sometimes result in the emergence of novelty in metabolic networks such as in an ecosystem (Bunge 1959, Jantsch 1980). New modes of organization and behavior develop in ecosystems on both successional and evolutionary time scales. Effects of such changes are often difficult to evaluate because they are transmitted through a network in which both direct and indirect effects are important. Therefore, effects may not be proportional

---

[1]Paper presented at the international symposium Energy and Ecological Modelling sponsored by the International Society for Ecological Modelling (Louisville, Kentucky, April 20-23, 1981).
[2]Robert W. Bosserman is Assistant Professor of Systems Science and Biology, University of Louisville, Louisville, Kentucky U.S.A.

to the initial cause but can be amplified or attenuated as they pass through the network; even small changes in the environment or in the species composition can cause large changes in the entire ecosystem. Periodic fluctuations in the environment are particularly important to the control and development of ecosystems because many natural inputs and constraints are periodic. Effects of periodic fluctuations can be complex in a network because of phase shifts, frequency modifications, and amplitude modulations. Such periodic fluctuations, which arise naturally in metabolic networks (Prigogine 1980) have been hypothesized to control many processes in living systems (Jantsch 1980, Varela 1979, Eigen and Schuster 1979).

Effects of environmental and internal changes are not expressed at only one level of organization in ecosystems. Ecosystems are hierarchical objects which simultaneously develop, adapt, and evolve at many levels of reality (Odum 1971, Patten et al. 1976, Jantsch 1980). In order to understand the structural variation of ecosystems and its relation to ecosystem behavior, several levels of an ecosystem hierarchy should be examined (Odum 1971): level below the system, level of the system and level above the system.

Graph theory and systems theory have combined in the past to produce techniques for representing and assessing the structure of system networks. However, classically defined graphs and digraphs are the most useful for representing systems with constant structure rather than systems with variable structure. Recently developed concepts in graph theory are useful for representing the variable structure of hierarchically organized systems. The way that a system is represented in a diagram is perhaps as important as the computations which can be used to analyze the behavior of the system. The diagrams used to model a system can constrain the way people conceive of the system; the edges and vertices of a graph, for example, impart the idea of a permanent structural relationship. In order to impart the concept of structural variability for systems, digraphs have been broken up into elementary components which can reform in order to reflect the active selection of inputs and outputs which occur in many living systems. The hierarchical nature of systems can also be expressed by digraphs (Bosserman and Harary 1981a,b). Matrix techniques, in conjunction with graphical methods, can be used to examine both the direct and indirect flows in a network. One technique assesses the amount of change that occurs in a system because of small changes in parameters. Another technique assesses the amount of change that occurs in a system because of the creation or deletion of components or interactions. Each technique measures different aspects of system structure.

BACKGROUND

Many different types of systems, including ecosystems, have been represented by graphs and digraphs (Roberts 1976, Lane and Levins 1977, Patten et al. 1976) so their use is not foreign to ecologists. Throughout this paper, the terminology of standard graph theories (Roberts 1976, Harary 1969, Busaker and Saaty 1965) will be used. A graph is composed of a set of points called vertices that are connected by lines called edges. In a digraph the lines are oriented by an arrowhead and are called arcs. The vertices of a digraph represent variables of interest in a system, while the arcs represent relationships between the variables. The arcs are often oriented from cause to effect or in the direction of energy and matter flow. A series of adjoining arcs is called a path (or walk). An arc which begins and ends at the same vertex is called a loop. A path which begins and ends at the same vertex is called a closed path. A closed path which includes each vertex only once is called a cycle. The length of a path refers to the number of arcs which make it up. The number of arcs making up the shortest path from vertex i to vertex j is called the distance from vertex i to vertex j.

In a system with changing structure, the relationships between vertices may change with time; i.e., the pattern of connections at one point in time may be different than at another point in time. This variability can be represented by decomposing arcs and vertices into elementary components which can dissociate and then reform into a new configuration. An arc can be decomposed into two demiarcs, a male demiarc (outarc) which leaves a vertex (figure 1a) and a female demiarc (inarc) which enters a vertex (figure 1b) (Bosserman and Harary 1981a,b). Every vertex may have a number of attached and unattached demiarcs which may be used to represent the gates through which interactions occur in living systems (Platt 1969). Compatible inarcs and outarcs mesh at a junction which is analogous to the communication channel of information theory (figure 1c) (Shannon and Weaver 1963). The vertex can also be split into elementary components, the genon and creaon (figure 1d) (Patten et al. 1976, Bosserman and Harary 1981a,b). The genon is the side of the vertex which creates and orients outputs or, in this case, outarcs. The creaon is the side of the vertex which selects and receives inputs (inarcs) from the environment. The combination of inarcs, outarcs, genons and creaons in a digraph provides a means to represent the active selection, orientation and reshuffling of inputs and outputs within a real system.

Digraphs can also be regarded as being hierarchical in that the vertices of a particular digraph can be expanded into digraphs of greater

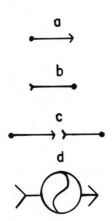

Figure 1.--Elementary components of a digraph
(a) male demiarc, (b) female demiarc,
(c) an arc formed from compatible demiarcs
(d) elementary components of a vertex.

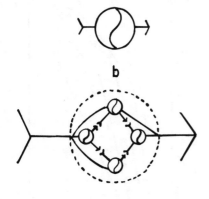

Figure 2.--The hierarchical decomposition of
(a) a level-1 digraph into (b) a level-2 digraph.

detail. Suppose that any vertex is regarded as a level-1 digraph. This vertex can be expanded into a level-2 digraph. In general, a vertex in a level-i digraph can be expanded into a level-$(i+1)$ digraph, whereas a level-i digraph can be condensed into a level-$(i-1)$ digraph. An arc can be regarded as a bundle of arcs which are split at the boundary of an expanded vertex to connect the vertices contained within. An example of an expanded vertex and demiarcs are shown in figure 2.

The pattern of connectivity in a digraph can be equivalently represented by an adjacency matrix (Harary 1969, Patten et al. 1976, Roberts

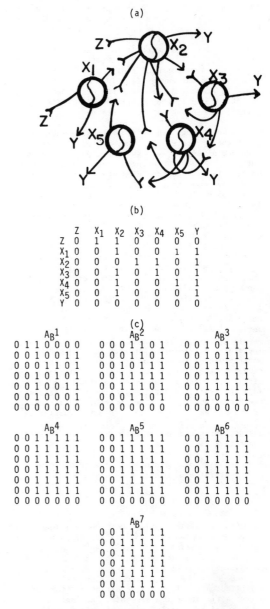

(a)

(b)

	Z	X_1	X_2	X_3	X_4	X_5	Y
Z	0	1	1	0	0	0	0
X_1	0	0	1	0	0	1	1
X_2	0	0	0	1	1	0	1
X_3	0	0	1	0	1	0	1
X_4	0	0	1	0	0	1	1
X_5	0	0	1	0	0	0	1
Y	0	0	0	0	0	0	0

(c)

A_B^1
```
0 1 1 0 0 0 0
0 0 1 0 0 1 1
0 0 0 1 1 0 1
0 0 1 0 1 0 1
0 0 1 0 0 1 1
0 0 1 0 0 0 1
0 0 0 0 0 0 0
```

A_B^2
```
0 0 0 1 1 0 1
0 0 0 1 1 0 1
0 0 1 0 1 1 1
0 0 1 1 1 1 1
0 0 1 1 1 0 1
0 0 0 1 1 0 1
0 0 0 0 0 0 0
```

A_B^3
```
0 0 1 0 1 1 1
0 0 1 0 1 1 1
0 0 1 1 1 1 1
0 0 1 1 1 1 1
0 0 1 1 1 1 1
0 0 1 0 1 1 1
0 0 0 0 0 0 0
```

A_B^4
```
0 0 1 1 1 1 1
0 0 1 1 1 1 1
0 0 1 1 1 1 1
0 0 1 1 1 1 1
0 0 1 1 1 1 1
0 0 1 1 1 1 1
0 0 0 0 0 0 0
```

A_B^5
```
0 0 1 1 1 1 1
0 0 1 1 1 1 1
0 0 1 1 1 1 1
0 0 1 1 1 1 1
0 0 1 1 1 1 1
0 0 1 1 1 1 1
0 0 0 0 0 0 0
```

A_B^6
```
0 0 1 1 1 1 1
0 0 1 1 1 1 1
0 0 1 1 1 1 1
0 0 1 1 1 1 1
0 0 1 1 1 1 1
0 0 1 1 1 1 1
0 0 0 0 0 0 0
```

A_B^7
```
0 0 1 1 1 1 1
0 0 1 1 1 1 1
0 0 1 1 1 1 1
0 0 1 1 1 1 1
0 0 1 1 1 1 1
0 0 1 1 1 1 1
0 0 0 0 0 0 0
```

Figure 3.--(a) A digraph of the Cone Springs energy model: X_1 = autotrophic plants, X_2 = detritus, X_3 = bacteria, X_4 = detritus feeders, X_5 = carnivores, Z = inputs, Y = outputs. (b) The adjacency matrix for the Cone Springs model. (c) The first n terms of the adjacency matrix series for the Cone Spring model.

1976, Gallopin 1972). Each representation provides a different perspective and answers different questions about the model. Suppose the vertices of a digraph are numbered from 1 to n. If an arc exists from vertex i to vertex j then element i,j is 1. If no arc exists then element i,j is 0. An example of a digraph and its adjacency matrix for energy flow in a cold spring is shown in figure 3. By using the adjacency matrix and its products, the position of all indirect paths in the graph can be determined.

A well known result from linear algebra and graph theory states that the adjacency matrix sequence $A^1, A^2, A^3, \ldots$, shows the position of all paths in the corresponding digraph (Harary 1969, Roberts 1976). Nonzero elements of the adjacency matrix A^j demonstrate the positions of the j-length paths in the digraph. If real arithmetic is used, and elements of A are the number of arcs in the digraph, the sequence, $A^1, A^2, A^3, \ldots$, shows the number and position of all paths in the digraph. If Boolean arithmetic is used and 1 and 0 are used to indicate the existence or nonexistence of a path, respectively, then the sequence $A_B^1, A_B^2, A_B^3, \ldots$, shows only the position of the paths. The subscript B indicates that Boolean matrices, addition, and multiplication are used ($0 \times 0 = 0$, $0 \times 1 = 0$, $1 \times 0 = 0$, $1 \times 1 = 1$; $0 + 0 = 0$, $0 + 1 = 1$, $1 + 0 = 1$, $1 + 1 = 1$). If the digraph and associated Boolean adjacency matrix are nth order, then the series $\sum_1^\infty A_B^i$ converges to the reachability matrix by the nth term (Roberts 1976, Gill 1962). Therefore, all information about the Boolean connectivity structure of the digraph is contained in the first n terms of the series. Direct connectivity can be calculated from the adjacency matrix A and the indirect connectivity can be calculated from the succeeding n-1 terms of the series $\sum_2^n A_B^i$. From this information, a structural measure of digraph complexity can be devised which depends on both direct and indirect connectivity.

Let $a^{(i)}$ be the number of 1's in the nth order partial sums matrix $A_B^{(i)}$, where

$$A_B^{(i)} = \sum_1^i A_B^j = A_B^{(i-1)} + A^i \qquad (1)$$

Elements of matrix, $A_B^{(i)}$, show the positions of all paths from length 1 to i in the digraph. The sequence of partial sums, $A_B^{(1)}, A_B^{(2)}, A_B^{(3)}, \ldots$, demonstrates the changes in connectivity as increasingly longer paths are considered. This sequence converges to a reachability matrix, which indicates all possible direct and indirect paths, after at most n elements. A complexity index c and an associated normalized index c can be defined as follows (Bosserman 1980, 1981a):

$$c = 1/n^2 \sum_1^n a(j) \qquad (2)$$

$$\bar{c} = (1/n)c \qquad (3)$$

If there are n vertices in a digraph then

the number of possible arcs in the digraph is n^2. The complexity index c, which varies from 0 to n for digraphs with n vertices, can be used to compare connectivities in digraphs of different sizes. Index $\bar{c}$ is normalized with respect to the number of vertices in the digraph and varies from 0 for a digraph with no connectivity to 1 for a digraph with complete connectivity. The complexity index is a weighted sum of the distances between vertices of the digraph. Distances of 1-length are included n times, of 2-length are included n-1 times, and, in general, distances of k paths are included n-k-1 times.

The variation of $\bar{c}$ with system size (n) and degree of direct connectivity is shown in figure 4. The degree of direct connectivity is expressed as the ratio of arc number to possible arc number (n^2). For a digraph with n vertices the value c falls with a closed curve defined by a $\bar{c}_{min}$ and $\bar{c}_{max}$ for digraphs of that size. Several of these curves are depicted in figure 4 for digraphs with 1 to 4 vertices. The number of digraphs increases greatly with the number of vertices. For digraphs without loops and with 8 indistinct vertices, there are approximately 1.79 x 10^{12} different digraphs (Harary 1969). For digraphs with loops and with 8 labelled vertices

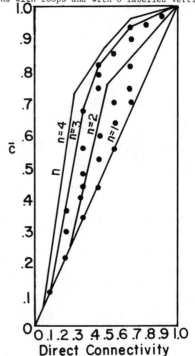

Figure 4. Variation of $\bar{c}$ with direct connectivity for digraphs of various sizes (n). Dots represent the c values and direct connectivity for 79 digraphs with 3 vertices.

there are approximately 1.84×10^{19} different digraphs ($2n^2$).

Importance values can be assigned to each arc of a digraph with respect to the complexity index. The importance values indicate how much graph connectivity changes when an arc is added or deleted. Let b_{ij} be the importance value of the arc from vertex i to vertex j, defined in the following way:

$$b_{ij} = 100 \times (1 - c'/c) \qquad (4)$$

where c is the complexity index of the original digraph and c' is the complexity index of the digraph to which the arc from vertex i to vertex j has been added or deleted. Therefore, the importance value is the percentage change in c when an arc is added to or deleted from a digraph. The importance values reflect the degree to which structural modifications affect the pattern of interactions within a graph or ecosystem network. This method addresses the problem of structural vulnerability for digraphs (Siljak 1979). The most vulnerable parts of a digraph, with respect to structural changes, can therefore be identified; removal or addition of the flows in an ecosystem may lead to the largest changes in system behavior. A similar importance value can be constructed for vertices which are removed or added.

Flow analysis

Flow analyses have been used extensively in the economic and ecological fields in order to trace patterns for material and energy flow through ecosystems. A sensitivity analysis has been devised which assesses the effects of small changes in the flow structure of an ecosystem network. Details of the analysis are presented in another paper in this volume (Bosserman 1981b).

APPLICATION OF TECHNIQUES

Structural Analysis

The structural and sensitivity analyses have been applied to an energy flow model of Cone Spring. A conceptual diagram of this model is shown in figure 5. This model has been constructed using elementary components in order to represent the structural variablity in such ecosystems. The structural and sensitivity analyses address the problems of assessing large and small changes in system structure. Cone Spring was originally studied by Tilly (1968) and was later developed as a class project at the University of Georgia (Williams and Crouthamel 1972). Compartments represent energy (kcal m^{-2}) in autotrophic plants, detritus (X_2), bacteria (X_3), detritus feeders (X_4) and carnivores (X_5). Inputs Z from the environment are from primary production of plants ($a_{z,1}$), and from allochthonous organic matter by leaffall, litterfall, and ground water

($a_{z,2}$). Outflows (kcal m^{-2} y^{-1}) occur from all compartments. Internal flows (kcal m^{-2} y^{-1}) include feeding rates ($a_{2,3}$, $a_{2,4}$, $a_{3,4}$, $a_{5,4}$) and mortality ($a_{1,2}$, $a_{3,2}$, $a_{4,2}$, $a_{5,2}$). The adjacency matrix for this model is shown in figure 5b and the first n powers of this adjacency matrix are shown in figure 5c. The direct connectivity, which is the number of direct paths/ total number of paths, of the Cone Spring model is $15/49 = 0.31$. The complexity index c = 4.04 while $\bar{c}$ =0.58, which indicates that 58% of the possible direct and indirect paths are present. Importance values for each arc in the Cone Spring model are shown in table 1. In the Cone Spring model arcs $a_{2,3}$, $a_{5,2}$, $a_{4,5}$ and $a_{1,2}$ are structurally important because if they are removed c changes between 12 and 18%. If arcs $a_{3,1}$, $a_{4,1}$ and $a_{5,1}$ were added to the Cone Spring model c would change about -14 to -16%. The addition or deletion of other arcs in the Cone Spring model would lead to smaller changes so they are regarded as structurally unimportant. Input arcs and output arcs are relatively unimportant and their removal or addition would change c between -1 and 2%. It is interesting to note that the removal of arc $a_{2,3}$ would change c 18.7% while the removal of its complement, $a_{3,2}$, would change c only 4%. Removal of arc $a_{2,3}$ isolates X_3 from the influence of other state variables, whereas the removal of arc $a_{3,2}$ would isolate no variable; therefore, it is structurally redundant in comparison to $a_{2,3}$ and its effects in the pattern of interactions can be compensated for by other arcs.

Sensitivity Analysis

A sensitivity analysis which was based on the flow structure of the Cone Spring model (Patten et al. 1976) was also done and compared with the structural analysis. The flows between compartments in the model are depicted in kcal m^{-2} yr^{-1} and are shown in table 2a. The flows within each column are normalized with respect to column sums in matrix G in table 2b. Each values g_{ij} in this matrix indicates the proportion of input into X_j coming from X_i. All columns therefore sum to 1. The structure matrix B^{-1} for the Cone Spring model is shown in table 2c, where B = $(I-G)$. Each value b^{*}_{ij} of B^{-1} indicates the amount of direct and indirect flow passing through X_j after having originated from X_i or Z. Values greater than 1.0 indicate that cycling of energy has occurred.

Table 1. Matrix showing importance values for Cone Spring model when arcs from compartments i to j were added or deleted.

	Z	I_1	I_2	I_3	I_4	I_5	Y
Z	-	3.5	2.0	-0.5	-1.0	-1.0	-
I_1	-	-0.5	12.1	-0.5	-1.0	-1.5	-0.5
I_2	-	-15.7	-0.5	18.7	4.5	-2.0	-0.5
I_3	-	-14.1	1.0	-0.5	1.0	-0.5	-0.5
I_4	-	-14.6	1.5	-0.5	-0.5	17.2	-0.5
I_5	-	-13.6	12.1	-0.5	-1.5	-0.5	-0.5
Y	-	-	-	-	-	-	-

A hierarchy of sensitivity values can be determined from this analysis. At the lowest level one can determine the sensitivity of individual elements of B^{-1} with respect to changes in b_{ij}. For each element b_{ij}, a matrix $S(i,j)$ of sensitivity values can be found (equation 10). For a network with n compartments there are n^4 sensitivity values which are in n^2 matrices. Every column k of $S(i,j)$ can be added to establish the effects of changing b_{ij} on indirect and direct inflows to X_k. Every element in $S(i,j)$ can be added together to get the total effect on the entire flow network due to changes in b_{ij}. The trace of the structure matrix, which is the sum of the main diagonal elements, indicates how sensitive the entire matrix is to changes in B. A normalized trace (Tr/n) can be used to compare matrices of different sizes (n). If there are no cycles in the network, then each element in the trace of the structure matrix is 1.0. If there are cycles, then each element is greater than or equal to 1.0. The magnitude of the elements in the trace depends on the amount of cycling in the network. Therefore, one can look at sensitivities of individual elements of the structure matrix, of inflows into compartments, and of the total network.

The sensitivity of B^{-1} with respect to changes in element $b_{2,1}$ of B is shown in table 2d. Changing $b_{2,1}$ causes the greatest changes in energy flowing from X_2 into X_1, X_2, X_3, X_4 and X_5, and causes the least changes in energy flows from X_4 and X_5. The sensitivity of B^{-1} with respect to changes in element $b_{3,z}$ of B is shown in table 2e. Changes in $b_{3,z}$ (i.e., from X_3 to the environment) cause the greatest changes in flows from X_1, X_2, and X_3.

The sensitivities of the total direct and indirect inflows to X_i in response to changes in all possible flows in the model are shown in table 3. The largest sensitivities resulted from changing $b_{5,2}$, $b_{4,2}$, $b_{4,4}$ and $b_{5,4}$. The first two flows are present in the Cone Spring model as depicted in Figure 5a; however, the latter two are not. These absent flows may represent potential points of control or management.

The sensitivity of the total structure matrix to changes in B are shown in table 4. element d_{ij} of this matrix represents the total sensitivity of the structure matrix due to changes in b_{ij}. The structure matrix is most sensitive to changes in $b_{2,1}$, $b_{2,4}$, $b_{2,5}$, $b_{3,2}$, $b_{4,2}$, $b_{5,2}$, $b_{3,2}$, $b_{4,2}$, and $b_{5,2}$.

Comparison of Structural versus Sensitivity Analyses

The results of the structural analyses can be compared to those of the sensitivity analysis. Importance values which indicate how much the complexity index c changes when arcs are added or deleted (table 1) can be compared to total sensitivity values generated by changing

elements of B (table 4). An element-wise comparison of the values generated by these two methods indicates little correspondence. In general, knowing the importance value of a flow in a network does not indicate anything about the sensitivity value. A linear regression of sensitivity values versus importance values demonstrated that no correlation existed (r = -0.20). This result indicates that the analyses are examining different aspects of network structure; they are complementary rather than redundant.

CONCLUSIONS

The representation of ecosystems as having a variable structure in which the pattern of interactions changes through time is a recent development (Prigogine 1980). Systems ecology has been constrained by its origins in other sciences which have dealt with static representations of reality. Models that have constant or static structure are easier to work with than those that do not. Many analysis techniques which developed in the engineering and mathematical sciences deal with systems which operate near a steady state or equilibrium and are thus assumed to remain nearly the same through time. Under such conditions small changes in the environment or system structure tend to give rise to small changes in effects. Such cause and effect relationships are often modelled as being linear (Bunge 1959). In many natural situations, especially in self-organizing, living systems, environmental changes generate effects which are magnified by the system network or cause structural changes in the system network. Such large changes are responsible for the emergence of novel behavior which may not be proportional to the original causes. Such novel behavior is important in the development, adaptation, and evolution of living systems in their dynamic environments.

The use of conceptual representations that have been decomposed into elementary components such as genons, creaons, and demiarcs allows one to describe the variable structure of self-organizing systems. Such conceptual techniques are useful for removing the static restrictions which are inherent in many conceptual modelling frameworks. Two analysis techniques are introduced in this paper in order to assess the effects of structural variation in ecosystem networks. The complexity analysis is based on the presence and absence of compartments and interactions. Direct and indirect paths are included in this analysis. Networks can be ranked according to their degree of complexity. Importance values can be assigned to each interaction in a network in order to evaluate how much it contributes to the complexity. The deletion or addition of interactions corresponds to changes in system structure or conceptualization. Such changes reflect drastic modification in the pattern of system interactions that occur during major disturbances, succession, or management. This analysis, although useful for assessing the

Table 2. Flow matrices for the Cone Spring Energy model. (a) Flow matrix showing direct flows (kcal/m^2/yr) to and from each compartment. (b) flow matrix normalized by column sums. (c) structure matrix $B^{-1} = (I-G)^{-1}$ (d) Matrix of sensitivity values for B^{-1} due to changes in $b_{2,1}$. (e) Matrix of sensitivity values for B^{-1} due to changes in $b_{3,6}$.

a. Matrix showing flows (kcal m^{-2} yr^{-1}) for the Cone Spring energy model.

	X_1	X_2	X_3	X_4	X_5	Z
X_1	0.0	8881.4	0.0	0.0	0.0	0.0
X_2	0.0	0.0	5204.5	2309.2	0.0	0.0
X_3	0.0	1600.2	0.0	75.1	0.0	0.0
X_4	0.0	200.0	0.0	0.0	370.1	0.0
X_5	0.0	167.0	0.0	0.0	0.0	0.0
Z	11184.0	635.0	0.0	0.0	0.0	0.0

b. Fractional inflow matrix for Cone Spring energy model.

	X_1	X_2	X_3	X_4	X_5	Z
X_1	0.0	0.773	0.0	0.0	0.0	0.0
X_2	0.0	0.0	1.0	0.969	0.0	0.0
X_3	0.0	0.139	0.0	0.031	0.0	0.0
X_4	0.0	0.017	0.0	0.0	1.0	0.0
X_5	0.0	0.015	0.0	0.0	0.0	0.0
Z	1.0	0.055	0.0	0.0	0.0	0.0

c. Structure matrix for Cone Spring Energy model.

	X_1	X_2	X_3	X_4	X_5	Z
X_1	1.0	0.933	0.933	0.933	0.933	0.0
X_2	0.0	1.207	1.207	1.207	1.207	0.0
X_3	0.0	0.169	1.169	0.200	0.200	0.0
X_4	0.0	0.039	0.039	1.039	1.039	0.0
X_5	0.0	0.018	0.018	0.018	1.018	0.0
Z	1.0	1.0	1.0	1.0	1.0	1.0

d. Matrix showing how sensitive B^{-1} is to changes in $b_{2,1}$.

	X_1	X_2	X_3	X_4	X_5	Z
X_1	-0.93	-0.07	-0.84	-0.87	-0.87	0.0
X_2	-1.21	-1.13	-1.09	-1.13	-1.13	0.0
X_3	-0.17	-0.16	-0.15	-0.16	-0.16	0.0
X_4	-0.04	-0.04	-0.04	-0.04	-0.04	0.0
X_5	-0.02	-0.02	-0.02	-0.02	-0.02	0.0
Z	-1.00	-0.93	-0.90	-0.93	-0.93	0.0

e. Matrix showing how sensitive B^{-1} is to changes in $b_{3,6}$.

	X_1	X_2	X_3	X_4	X_5	Z
X_1	-0.90	-0.90	-0.90	-0.90	-0.90	0.0
X_2	-1.21	-1.21	-1.21	-1.21	-1.21	0.0
X_3	-1.17	-1.17	-1.17	-1.17	-1.17	0.0
X_4	-0.04	-0.04	-0.04	-0.04	-0.04	0.0
X_5	-0.02	-0.02	-0.02	-0.02	-0.02	0.0
Z	-1.00	-1.00	-1.00	-1.00	-1.00	0.0

Table 3. Matrix showing sensitivity values for direct and indirect inflows into each compartment when element b_{ij} is changed.

b_{ij}	X_1	X_2	X_3	X_4	X_5	Z
1,1	-1.0	-1.0	-1.8	-1.9	-1.9	0.0
1,2	0.0	-2.4	-2.4	-2.4	-2.4	0.0
1,3	0.0	-0.3	-2.3	-0.4	-0.4	0.0
1,4	0.0	-0.1	-0.1	-2.1	-2.1	0.0
1,5	0.0	0.0	0.0	0.0	0.0	0.0
1,z	-2.0	-2.0	-2.0	-2.0	-2.0	0.0
2,1	-3.4	-3.1	-3.0	-3.1	-3.1	0.0
2,2	0.0	-4.1	-4.1	-4.1	-4.1	0.0
2,3	0.0	-0.6	-3.9	-0.7	-0.7	0.0
2,4	0.0	-0.1	-0.1	-3.5	-3.5	0.0
2,5	0.0	-0.1	-0.1	-0.1	-0.1	0.0
2,z	-3.4	-3.4	-3.4	-3.4	-3.4	0.0
3,1	-4.3	-4.0	-3.9	-4.0	-4.0	0.0
3,2	0.0	-5.2	-5.2	-5.2	-5.2	0.0
3,3	0.0	-0.7	-5.1	-0.9	-0.9	0.0
3,4	0.0	-0.2	-0.2	-4.5	-4.5	0.0
3,5	0.0	-0.1	-0.1	-0.1	-0.1	0.0
3,z	-4.3	-4.3	-4.3	-4.3	-4.3	0.0
4,1	-4.4	-4.1	-4.0	-4.1	-4.1	0.0
4,2	0.0	-5.3	-5.3	-5.3	-5.3	0.0
4,3	0.0	-0.7	-5.1	-0.9	-0.9	0.0
4,4	0.0	-0.2	-0.2	-4.6	-4.6	0.0
4,5	0.0	-0.1	-0.1	-0.1	-0.1	0.0
4,z	-4.4	-4.4	-4.4	-4.4	-4.4	0.0
5,1	-4.4	-4.1	-4.0	-4.1	-4.1	0.0
5,2	0.0	-5.3	-5.3	-5.3	-5.3	0.0
5,3	0.0	-0.7	-5.1	-0.9	-0.9	0.0
5,4	0.0	-0.2	-0.2	-4.6	-4.6	0.0
5,5	0.0	-0.1	-0.1	-0.1	-0.1	0.0
5',z	-4.4	-4.4	-4.4	-4.4	-4.4	0.0
z,1	-1.0	-0.9	-0.9	-0.9	-0.9	0.0
z,2	0.0	-1.2	-1.2	-1.2	-1.2	0.0
z,3	0.0	-0.2	-1.2	-0.2	-0.2	0.0
z,4	0.0	0.0	0.0	-1.0	-1.0	0.0
z,5	0.0	0.0	0.0	0.0	-1.0	0.0
z,z	-1.0	-1.0	-1.0	-1.0	-1.0	0.0

Table 4. Matrix showing total sensitivity values for structure matrix of Cone Spring model when flows from compartment i to compartment j were modified.

	D_1	D_2	D_3	D_4	D_5	Z
D_1	- 9.4	- 9.7	-3.5	-4.3	-0.1	-10.0
D_2	-15.8	-16.3	-5.9	-7.3	-0.2	-16.8
D_3	-20.4	-20.9	-7.5	-9.3	-0.3	-21.7
D_4	-20.7	-21.2	-2.6	-9.5	-0.3	-22.0
D_5	-22.7	-21.2	-7.6	-9.5	-0.3	-22.0
Z	0.0	0.0	0.0	0.0	0.0	0.0

effects of large changes, does not include the magnitudes of the interactions. Changing the relative magnitudes of the interactions permits infinite variety of patterns of organization.

A sensitivity analysis which examines the effect of small changes in the magnitudes of interactions in a conservative flow network has also been introduced. By examining standard input-output analyses with classic matrix techniques, one can establish how small changes in the direct flow structure can affect the indirect flows in a network. These changes can be interpreted to be the effects of error, observation, human manipulation, environmental variation, or composition alterations. Networks which have a large amount of internal recycling tend to be more sensitive to changes than those which do not. Therefore, energy models which have few cycles tend to be less sensitive than nutrient models.

These analyses have applications for ecosystem interpretation, conceptualization, and management and design. These analyses can identify the effects of interactions which were not orginally included in the model. Typically only the most important flows are included in a model while the least important flows are excluded. Any errors in conceptualization will lead to poor behavior of the model. Possible interactions which were misconceptualized but have importance because of their position in the network can be identified. Even a minor flow which was previously unmodelled can have an important impact when the entire pattern of interactions is considered.

Several aspects of stability can be examined with these analyses. The complexity analyses provide a means for examining how connectivity is related to network stability. The complexity index c is a better measure of connectivity than the direct connectivity index which has been used in many stability studies (Gardner and Ashby 1970, Bosserman 1980). It also provides better discrimination between the complexities of networks of the same size. Importance values indicate which arcs cause the greater changes in connectivity. Those arcs which close cycles are the most influential while redundant arcs are the least influential (Bosserman 1980). Structurally vulnerable systems which include arcs which high importance values will probably be the most unstable in respect to deletion or addition of arcs.

For input-output flow analyses, the magnitude of the elements of the structure matrix determine the sensitivity of system to change. Condition numbers for the structure matrix encapsulate the total sensitivity of the flow analysis. One of the most useful condition numbers is the trace of the structure matrix. The trace can be used to indicate the amount of recycling in a network (Finn 1976). It is also related nontrivially to the eigenvalues of the system, which help to determine the stability of the matrix. When the trace is large, the structure matrix is unstable with respect to changes in the direct flow structure. Networks, with a large amount of recycling are therefore most sensitive to modification; however, they also demonstrate the greatest ranges of behavior for a set of structural modifications.

Dealing with the structural variation of ecosystems is one of the most important tasks that systems ecologists will have to deal with in future modelling programs. Many phenomena, especially those in developing systems, involve changes in patterns of interactions. Techniques such as those presented here may provide some help in dealing with complex, evolving systems.

Literature Cited

Bosserman, R. W. 1979. The Hierarchical Integrity of Utricularia-periphyton Microecosystems. Ph. D. Dissertation. University of Georgia, Athens, Georgia.

Bosserman, R. W. 1980. Complexity measures for assessment of environmental impact in ecosystem networks. Pages 811-820 in Proceedings of the Pittsburg Conference on Modeling and Simulation. Pittsburg, Pennsylvania. April 20-23, 1980.

Bosserman, R.W. 1981a. Complexity measures for evaluation of ecosystem networks. ISEM Journal (in press).

Bosserman, R.W. 1981b. Sensitivity techniques for examination of input-output flow analyses. Proceedings of the International Symposium on Energy and Ecological Modelling. Elsevier, New York. (This volume).

Bosserman, R.W. and F. Harary. 1981. The hierarchical decomposition of digraphs into demiarcs, genons, and creaons. Pages 388-397 in Proceedings of the Tenth Annual Conference of the Southeastern Region for the Society for General Systems Research. Louisville, Kentucky. April 20-23, 1981.

Bosserman, R.W. and F. Harary. Demiarcs, genons, and creaons. Jour. of Theor. Bio. (in press).

Bunge, M. 1959. Causality. Harvard University Press, Cambridge, Massachusetts.

Busaker, R. G. and Saaty, T. L. 1965. Finite Graphs and Networks. McGraw-Hill, New York.

Conrad, M. 1976. Patterns of biological control in ecosystems. Pages 431-457 in B. C. Patten, eds. Systems Analysis and Simulation in Ecology, Vol. IV. Academic Press, New York.

Eigen, M. and P. Schuster. 1979. The Hypercycle: A Principle of Natural Self-Organization. Springer Verlag, Berlin.

Faddeev, D. K. and V. N. Faddeeva. 1963. Computational Methods of Linear Algebra. W. H. Freeman and Company, San Francisco.

Finn, J. T. 1976. Cycling Index: a general measure of cycling in compartmental models. Pages 138-164 in D. C. Adriana and I. L. Brisbin, eds. Environmental Chemistry and Cycling Processes Symposium, U. S. Energy Research and Development Administration, Washington, D. C.

Gallopin, G. C. 1972. Structural properties of food webs. Pages 241-283 in B. C. Patten, ed. Systems Analysis and Simulation in Ecology, Vol. II. Academic Press, New York.

Gardner, M. R., and W. R. Ashby. 1970. Connectedness of large dynamic (cybernetic) systems: critical value of stability. Nature 228:784.

Gill, A. 1962. Introduction to the Theory of Finite State Machines. McGraw Hill, New York.

Hannon, B. 1973. The structure of ecosystems. J. Theoret. Biol. 41:535-546.

Harary, F. 1969. Graph Theory. Addison-Wesley, Reading, Massachusetts.

Jantsch, E. 1980. The Self-Organizing Universe. Pergamon Press, Oxford.

Lane, P. and Levins, R. 1977. The dynamics of aquatic systems. 2. the effects of nutrient enrichment on model plankton communities. Limnol. Oceanogr. 22:454-471.

Noble, B. 1969. Applied Linear Algebra. Prentice-Hall, Inc., Englewood Cliffs, New Jersey.

Odum, E. P. 1971. Fundamentals of Ecology. W. B. Saunders Company, Philadelphia, Pennsylvania.

Patten, B. C., R. W. Bosserman, J. T. Finn, and W. G. Cale. 1976. Propagation of cause in ecosystems. Pages 457-579 in Systems Analyis and Simulation in Ecology (B. C. Patten, ed.). Academic Press, New York.

Platt, J. 1969. Theories on boundaries in hierarchical systems. Pages 201-213 in L. L. Whyte, A. G. Wilson, and D. Wilson, eds. Hierarchical Structures. Am. Elsevier, New York.

Prigogine, I. 1980. From Being to Becoming. W. H. Freeman and Company, San Francisco.

Roberts, F. S. 1976. Discrete Mathematical Models. Prentice-Hall, Inc. New York.

Shannon, C. W., and Weaver, W. 1963. The Mathematical Theory of Communication. University of Illinois Press, Urbana.

Siljak, D D. 1979. Structure and stability of model ecosystems. Pages 151-181 in E. Halfon, ed. Theoretical Systems Ecology Academic Press, New York.

Tilly, L. J. 1968. The structure and dynamics of Cone Spring. Ecol. Monogr. 38:169-197.

Varela, F. J. 1979. Principles of Biological Autonomy. North Holland, New York.

William, M. and Crouthamel, D. 1972. Systems analysis of Cone Spring. University of Georgia, Athens. Unpublished manuscript.

AN EXTENDED LOTKA-VOLTERRA MODEL
FOR POPULATION-ENERGY SYSTEMS[1]

A. A. Harms and E. M. Krenciglowa[2]

An analysis is undertaken to identify population-energy
topologies of selected national economies. Based on a Lotka-
Volterra ecological characterization of the temporal interrelationship of population and per capita energy consumption
combined with a historic record, we identify distinct national
population-energy topologies. The associated trajectories
represent characteristic "phase-portraits" of national popu-
lation-growth and energy-consumption trends and appear to
reflect intrinsic levels of economic-social development and
resource capacity.

INTRODUCTION

Energy has emerged as a profound and dominant
world-wide issue. Underlying many of the energy
related considerations is the pervasive and urgent
need to develop and implement conceptual methodology
appropriate to specific and general energy-related
problem domains. The general implication that such
approaches have an important role to play in energy
systems analysis -- recognizing that in this domain
there exists a considerable and complex interplay
between human actions and forms of inanimate matter
-- is generally recognized (Breitenecker 1980;
Grumm 1976; Haefele 1976; Roberts 1976).

For our purposes here, we address ourselves
to specific national population-energy characteristics. We consider the historic population
pattern of a nation and combine it with its record
of per capita energy consumption. Our objective
then is to formulate a rational and useful conceptual characterization of appropriate national
parameters and thereby obtain a generalized
evolutionary population-energy phase-space portrait.
The associated trajectory represents a tendency-
trend of that country's population-energy characteristics.

The conceptual model we adopt for our analysis
here belongs to the general domain of low-order
coupled rate equations which have proven to be useful and effective in the physical and biological
sciences (Haken 1977; Kemeny (1962). The describing
equations for national population-energy system are

---

[1]Paper presented at the international symposium
Energy and Ecological Modelling, sponsored by the
International Society for Ecological Modelling.
(Louisville, Kentucky, April 20-23, 1981).
[2]A.A. Harms is Professor of Engineering Physics,
McMaster University, Hamilton, Canada and E.M.
Krenciglowa is a Research Associate at the same
institution.

formulated here from a set of reasonable hypotheses
on the interplay between population and energy and
then subjected to tests involving historic records.
In adopting a highly compact, aggregate energy
systems model, the emphasis is not on the detailed
results of the model but rather on a qualitative
description of the global (geometric) structure
of the energy system which in turn can provide the
basis for establishing qualitative trends and
phase-space characterizations.

ANALYTICAL FORMULATION

We consider a society or a nation -- or other
suitable region -- characterized by a time-
dependent population $P(t)$ and a time-dependent per
capita energy consumption $q(t)$. We adopt the
premise that these two functions are coupled in a
manner which is case-specific and may vary with
time.

To establish the mathematical model of
interest here, we proceed via the following
plausibility argument applicable to a finite though
arbitrary time interval Δt. We suggest that the
fractional change in per capita energy consumption
over the time interval Δt possesses one component
which is essentially constant and another component
which is linearly dependent on the population.
That is, as a proportionality relationship using
superposition we write

$$\frac{\Delta q}{q(t)} \propto [a_1 + a_2 P(t)], \quad t \in \Delta t \qquad (1)$$

where,

a_1 and a_2 are constants.

The rationalization of Eq. (1) can be further
aided by reference to Fig. 1 where we illustrate
two distinct linear variations of $\Delta q/q$ with P over
the time interval Δt. The line with a positive
slope may, for example, be associated with the

determination and resourcefulness of a growing population to provide an increasing per capita energy accessibility; the negative slope, however, implies a diminishing per capita energy consumption with an increasing population suggesting, for example, that the total energy resource is essentially constant and must be distributed among more people. Such trends, or tendencies, can be exhibited to a varying degree by different countries, at different times in their technological development, and associated with different resource policy states.

The translation of this qualitative argument is that the proportionality relationship of Eq. (1) can be explicitly written as a finite difference equation

$$\frac{\Delta q/q(t)}{\Delta t} \sim \mu_1 + \mu_2 P(t) \qquad (2)$$

or

$$\frac{\Delta q}{\Delta t} \sim \mu_1 q(t) + \mu_2 q(t) P(t) \qquad (3)$$

where,

μ_1 and μ_2 are specific constants.

Taking the limit $\Delta t \to 0$ and allowing t to be arbitrary yields the differential equation

$$\frac{dq}{dt} = \mu_1 q(t) + \mu_2 q(t) P(t) \qquad (4)$$

While this equation is in a useful form, it represents only one differential relation for two functions q(t) and P(t). We therefore proceed to establish a comparable plausibility argument for P(t) as a function of q(t).

We consider the fractional change in the population and postulate that it is also related to the per capita energy consumption in a linear manner

$$\frac{\Delta P}{P(t)} \alpha \left[b_1 + b_2 q(t) \right], \quad t \in \Delta t \qquad (5)$$

with,

b_1 and b_2 as constants.

Referring to Fig. 2, the positive slope suggests that, for example, an increasing energy consumption leads to improved health care reducing therefore infant mortality and extending life expectancy. On the other hand, the negative slope may reflect the tendency for smaller familty units with increased standard of living equated here with increased per capita energy consumption. By the analogous arguments used to obtain Eq. (4), we obtain

$$\frac{dP}{dt} = \sigma_1 P(t) + \sigma_2 q(t) P(t) \qquad (6)$$

Eqs. (4) and (6) represent the describing equations of interest here.

Evidently, various considerations can lead to reasonable assertions concerning the signs of the coefficients μ and σ of Eqs. (4) and (6). Such sign considerations are, of course, of interest in understanding the conceptual nature of the model. However, upon recognizing the immense diversity and complexity inherent in different national economies, there is no reason a priori to expect universal restrictions on the signs of the coefficients μ and σ. Depending on the particular national or regional economy, both population and per capita energy consumption can increase or decrease with time resulting in positive and negative signs as appropriate.

Then, we note that equations (4) and (6) possess their symbolic equivalence to the so-called Lotka-Voltera formulations which have been successfully used in the analysis of ecological systems (4,5). Clearly, aside from functional similarities, there are profound interpretive differences between these "natural" processes and the "institutional-human" population-energy processes.

If Eq. (4) and Eq. (6) constitute a "reasonably" acceptable representation of "reality" then the coefficients are, in a complex way, functions of the various pertinent independent parameters; among these we list birth rates, death rates, gross national product, standard of living, cost of capital, type of economy, and many others. This would then suggest that the coefficients

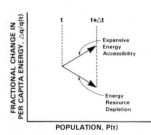

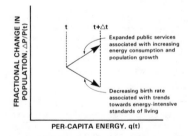

Figure 1. Schematic illustration showing hypothesized relationship of a fractional per capita energy change as a function of population during time interval Δt. Conceivable explanations for the trends are indicated.

Figure 2. Graphical depiction of hypothesized relationship between a fractional population change as a function of per capita energy consumption during time interval Δt. Plausible explanations for such linear trends are suggested.

σ_1, σ_2, μ_1 and μ_2 might best be studied by fitting Eq. (4) and Eq. (6) to specific regional data. Further, since these determining factors change with time, these coefficients could be "slow variables" of time.

The feature that the coefficients σ_1, σ_2, μ_1 and μ_2 are space and time dependent suggests that careful attention be paid to the applicability of the equations. The space dependence is easily accommodated by calculating these coefficients separately for each region of interest; the time dependence imposes the restriction that, once the coefficients are known, the dynamical equations are useful as predictive tools only if the co-efficients do not vary with time.

This latter point suggests a potential domain of application of the dynamical equations, Eq. (4) and Eq. (6), under conditions that the coefficients are calculated. It alludes to the identification of trajectory classes or phase portraits for national population-energy patterns and indicates an intrinsic geometric structure which characterizes the population-energy pattern. Our approach and interest here is directed at such global or bulk geometric features furnished by the dynamical equation rather than, and in contrast to, individual trajectories. This fundamental distinction follows the spirit of previous work (Breitenecker 1980; Haefele 1976) in which strong emphasis is placed on a qualitative characterization of the global structure and evolution of energy systems. Such an approach holds forth the prospect that the qualitative characterizations and predictions can indeed make good sense and provide useful indicators even if some of the detailed aspects of the model have shortcomings.

COMPUTATIONAL ASSESSMENT

We have undertaken to test the P-q phase-space evolutionary model represented by Eqs. (4) and (6) using historical data for the following countries thought to represent sufficiently diverse population-energy characteristics: Canada, Czechoslovakia, India, Nigeria, Sweden, U.S.A. and the U.S.S.R. The per capita energy consumption $q(t)$ and populations $P(t)$ for the chosen ten year period 1957-66 were adopted (Anon. 1975) because this period seemed to be generally free from large scale artificial perturbations of energy consumption patterns. The function $q(t)$ is in units of kilograms coal equivalent per capita and $P(t)$ is in units of capita.

To facilitate curve fittings, Eqs. (4) and (6) are re-written as

$$\frac{1}{P}\frac{dP}{dt} = \sigma_1 + \sigma_2 q \qquad (7.a)$$

$$\frac{1}{q}\frac{dq}{dt} = \mu_1 + \mu_2 P \qquad (7.b)$$

so that a standard linear form $Y = b_o + b_1 X$ for Eqs. (7.a) and (7.b) is achieved through obvious choices for the X and Y variables. Standard least

squares linear curve fitting procedures were then applied to determine the coefficients σ and μ. This procedure was carried out for the raw data and its smoothed version. For the raw data, suitable averages and differences are taken in order to treat the derivative term; linear regression is then applied. To obtain smoothed data, the raw $P(t)$ and $q(t)$ data is first fitted by a least squares quadratic polynomial in time

$$P_F(t) = a_o + a_1 t + a_2 t^2 \qquad (8)$$

and similarly for $q_F(t)$. These $P_F(t)$ and $q_F(t)$ are taken as basic input data for our analysis.

The calculated coefficients σ and μ for all countries together with related parameters are displayed in Table 1; we found that smoothing the data had a substantial effect on the correlation coefficients but little effect on the magnitude of the coefficients.

The trajectory in the (P,q)-plane for Eqs. (4) and (6) is obtained by solving

$$\frac{dq}{dP} = \frac{q(\mu_1 + \mu_2 P)}{P(\sigma_1 + \sigma_2 q)} \qquad (9)$$

which has

$$[|\frac{\sigma_2}{\sigma_1}|q]^{S(\sigma_1)} \exp[S(\sigma_2)|\frac{\sigma_2}{\sigma_1}|q]$$

$$\times \ [\{|\frac{\mu_2}{\mu_1}|P\}^{S(\mu_1)} \exp\{S(\mu_2)|\frac{\mu_2}{\mu_1}|P\}]^{-|\frac{\mu_1}{\sigma_1}|} = C \quad (10)$$

as the solution. In Eq. (10), C is the contour constant and $S(z) = \text{sign}(z) \times 1 = \pm 1$.

A computational examination of this equation together with the coordinates of the equilibrium points

$$(P_E, q_E) = (-\mu_1/\mu_2, \ -\sigma_1/\sigma_2) \qquad (11)$$

Table 1. Evaluated coefficients σ and μ based on the historical smoothed data 1957-1966.

Country	σ_1	σ_2	μ_1	μ_2	Correlation σ's	μ's
			(10^{-3})			
Canada	40.4	-0.0033	-445	25.9	-0.90	0.99
Czech.	10.6	-0.0007	1068	-74.4	-0.99	-0.99
India	20.7	0.0066	320	-0.62	0.97	-0.99
Nigeria	12.4	0.25	-929	21.9	0.82	0.95
Sweden	-3.12	0.0024	-453	67.9	0.99	0.91
U.S.A.	41.4	-0.0031	-318	1.84	-0.94	0.99
U.S.S.R.	36.0	-0.0065	-235	1.27	-0.97	0.99

provides for a categorization of the P-q surface for each country, Table 2, as well as for a graphical representation of this surface for the selected 10-year period 1957-66, Figs. 3-9.

Table 2. Geometric characterization and equilibrium point location based on the historical
smoothed data 1957–1966.

Type of Economy	Country	Type of Trajectory	Equilibrium Point $P_E(10^6)$	$q_E(10^3)$
Developed: (abundant resources)	Canada	ellipse	17.2	12.2
	U.S.A.	ellipse	173.0	13.4
	U.S.S.R.	ellipse	185.0	5.54
Developed: (limited resources)	Czech.	hyperbola	14.4	15.1
	Sweden	hyperbola	6.7	1.40
Developing:	India	hyperbola	516.0	-3.14
	Nigeria	ellipse	42.4	-0.05

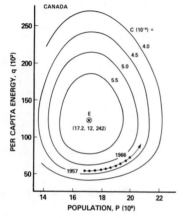

Figure 3. Population–energy surface and trajectory
for Canada, 1957–1966.

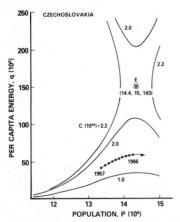

Figure 4. Population–energy surface and trajectory
for Czechoslovakia, 1957–66.

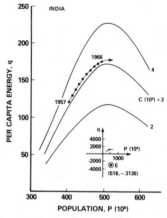

Figure 5. Population–energy surface and
trajectory for India 1957–66.

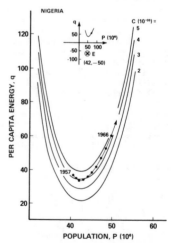

Figure 6. Population–energy surface and
trajectory for Nigeria, 1957–66.

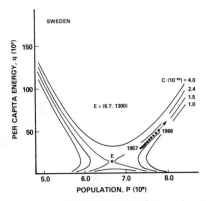

Figure 7. Population-energy surface and trajectory
for Sweden, 1957-66.

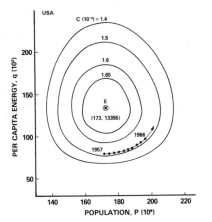

Figure 8. Population-energy surface and trajectory
for U.S.A., 1957-66.

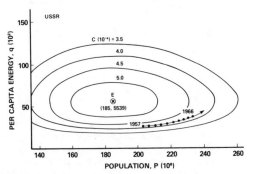

Figure 9. Population-energy surface and trajectory
for U.S.S.R., 1957-66.

The implication which follows from an examination of the P-q surfaces and the associated P-q
trajectories is that if the parameters σ_1, σ_2, μ_1
and μ_2 do not change with time then the specific
surface is stationary and the population and per
capita energy pattern is thus fully prescribed.
As depicted, these patterns are characterized by
elliptic and hyperbolic geometries about the
equilibrium point and suggest either cyclic or
asymptotic trajectory tendencies.

On examining Tables 1 and 2, several patterns
are evident. First, the σ_1 coefficient is invariably greater than zero, except in one case, so
that the expected (positive) birth-rate effect
is well reproduced by the model. Second, a
positive equilibrium population is always
obtained ($P_E = -\mu_1/\mu_2$). Third, the countries
with a highly industrialized and developed economy
(Canada, Czechoslovakia, Sweden, U.S.A., and
U.S.S.R.) have their equilibrium point in the
positive P-quadrant and positive q-quadrant
while the developing countries (India and Nigeria)
have their equilibrium point in the positive
P-quadrant but negative q-quadrant. Fourth, the
model differentiates between resource abundant
and resource limited developed economies by
assigning elliptic and hyperbolic geometric
surfaces, respectively.

TEMPORAL EVOLUTION OF THE P-q SURFACES

While the preceding provides interesting
features of the different national energy-
population patterns, the results displayed
characterize specifically the ten year period
1957-66. An assessment of the time evolution
of the P-q surfaces is clearly of interest.

We have undertaken to assess the temporal variations of the population-energy surface choosing
Canada for this purpose. The population energy data
(Anon. 1975; Anon. 1978) covered the 29 year period
1950-1978 and is displayed in Fig. 10. As an

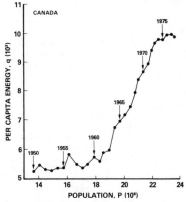

Figure 10. Historical population and per capita
energy consumption for Canada, 1950-1978.

indicator of the time variation of the P-q surface we chose to use the history of the equilibrium point (P_E, q_E) and the characteristic geometric shape -- either hyperbolic or elliptic -- for successive 10-year sliding intervals. That is, for each of the time-intervals 1950-59, 1951-60, 1952-61, 1953-62, ..., 1969-78, we calculated σ_1, σ_2, μ_1 and μ_2 from which we obtained the equilibrium point

$$E_i = (P_E, q_E)_i$$

$$= (-\frac{\mu_1}{\mu_2}, -\frac{\sigma_1}{\sigma_2})_i, \quad i = 1, 2, \ldots, 20 \quad (12)$$

and identified the geometric shape of the corresponding $(P-q)_i$ surface by

$$(\frac{\mu_1}{\sigma_1})_i \begin{cases} <0, \text{ elliptic} \\ >0, \text{ hyperbolic} \end{cases} \quad (13)$$

The polynomial data smoothing procedure used previously, Eq. (8) was also used in this part of the analysis, in addition to raw data.

The resultant trajectory of the equilibrium point $(P_E,q_E)_i$ using raw data for the 20 sliding time intervals together with the identification of whether the P-q surface is elliptic or hyperbolic was thus obtained. An examination of equilibrium point trajectory displayed some erratic features. The topology started out as a hyperbola, (1950-59) then transformed itself into an ellipse (1951-60) and returns to a hyperbola for the next two sliding time periods. From 1954-63 to 1960-69 it remains an ellipse only to transform to a hyperbola at 1961-70 and remain in that shape for the remainder of the study period. The other interesting feature was the apparent tendency to seek an extreme or resonance value of P_E or q_E for single isolated inverals.

The latter feature suggested that the use of the raw data may very well unduly weight certain data points in the determination of the functional parameters. To determine the potential effect of these data-deviants we re-examined the raw data, Fig. 10 and chose to exclude the data points for 1951, 1956 and 1975. We then found that, although not all the extremum tendencies were eliminated, the resultant equilibrium point (P_E,q_E) trajectory appeared much smoother and is shown in Fig. 11.

A noteworthy feature of Fig. 11 is the rather clear separation of the elliptic surfaces from the hyperbolic surfaces. The onset of this separation and change is the inclusion of the 1970 data point. In view of our earlier observation that developed countries with resource limitations display hyperbolic geometries, we may interpret the change in Canada's geometric surface from elliptical to hyperbolic as signalling the onset of resource restrictions. It is generally accepted that the seventies ushered in a changing economic environment which in turn reflects on energy consumption patterns in that per capita energy consumption and gross national product coupled closely. Such observations may indeed provide useful qualitative energy/population indicators.

CONCLUDING COMMENT

We have here shown that national population-energy characteristics can be effectively investigated using methodologies based on low-order coupled differential rate formulations. With such methodologies subjected to the test of historic record, our results demonstrate that useful conceptual characterizations of national population-energy trends can indeed be formulated. This is illustrated here, for example, by the models ability to differentiate between (i) developing, (ii) developed resource abundant and (iii) developed resource limited national economies. Further, for the case of Canada where an expanded historic record is utilized, such differentiation satisfactorily reflects a changing economic perspective at the beginning of the seventies. Thus, it is reasonable to conclude that the methodology of compact, global models is demonstrated here to provide meaningful tendency-indicators of characteristic population-energy patterns. The results established here suggest considerable opportunity for additional research, particularly in the areas of the application of more sophisticated statistical procedures coupled with data base extensions and the implementation of more realistic, yet highly aggregate, population/energy models.

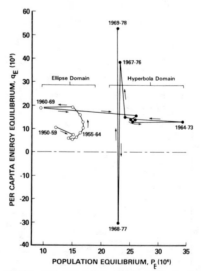

Figure 11. Illustration showing movement of equilibrium point (P_E,q_E) based on sliding 10-year periods. The years 1951, 1956 and 1975 were omitted.

ACKNOWLEDGEMENT

Financial support for the research reported here has been provided by the Natural Sciences and Engineering Research Council of Canada. We also wish to acknowledge the useful discussion with W. Haefele, J.L. Casti and M. Breitenecker, (International Institute for Applied Systems Analysis, Laxenburg, Austria), on an early draft of this research.

REFERENCES

Breitenecker, M., Grümm, H.R. 1980. Economic Evolutions and their Resilience: A Model. (Draft Paper), International Institute for Applied Systems Analysis, Laxenburg, Austria.

Grümm, H.R. and Schrattenholzer, L. 1976. Economy Phase Portraits, RM-76. International Institute for Applied Systems Analsyis, Laxenburg, Austria.

Haefele, W. et al. 1976. Second Status Reference Report of the IIASA Project on Energy Systems, RR-76-1. International Institute for Applied Systems Analysis, 2361 Laxenburg, Austria.

Haken, H. 1977. Synergetics. Springer-Verlag, Berlin.

Kemeny, J.G. and Snell, J.L. 1962. Mathematical Models in the Social Sciences. Blaisdell Publishing Co., Waltham, MA.

Roberts, F.S. ed. 1976. Energy: Mathematics and Models. Society for Industrial and Applied Mathematics, Philadelphia, Pa.

World Energy Supplies 1950-1974. Statistical Papers, Series J. No. 19, United Nations, N.Y.

World Energy Supplies 1973-1978. Statistical Papers, Series J, No. 22, United Nations, N.Y.

THERMO-CHEMICAL OPTIMIZATION OF ECOLOGICAL PROCESSES[1]

Alician V. Quinlan[2]

Abstract.-- A theory has been developed which accounts for the thermal sensitivity of several ecological processes whose optimum temperature varies linearly with the logarithm of the level of rate-limiting resource. Such processes span the first three trophic levels and include photosynthetic carbon fixation and subsequent growth by plants, ingestion of phytoplankton and subsequent growth by zooplankton, and food-intake and subsequent growth by fish. Because the levels of the rate-limiting resources for each of these processes (e.g., light intensity for plants, phytoplankton concentration for zooplankton, and food ration for fish) roughly track seasonal variations in habitat temperature, it has been conjectured that the optimum temperatures and, hence, the rate maxima of these processes may be thermo-chemically tuned to track seasonal variations in habitat temperature. Such tuning would tend to optimize process performance as habitat conditions vary throughout the year.

INTRODUCTION

Ecological processes mediated by primary producers, herbivores, and carnivores are commonly thought to be quite different; however, striking similarities exist within and among diverse sets of experimental data that show how:

(1) temperature and light intensity jointly affect the rates of photosynthetic carbon fixation and plant growth (e.g., Sorokin and Krauss 1952; Pisek and Winkler 1959; Aruga 1965; Idso 1971; Webb et al. 1974; Marks and Taylor 1978);

(2) temperature and phytoplankton concentration jointly influence the rates at which zooplankton ingest phytoplankton and grow (e.g., Vidal 1980); and

(3) temperature and food ration jointly affect the rates of food intake and growth by fish (e.g., Brett et al. 1969; Elliott 1975).

In each case, the data describing the influence of temperature and resource availability on rate are presented either as a family of rate-isotherms (for which rate is plotted versus resource level with temperature serving as a parameter) or, equivalently, as a family of

rate-isoconcentrates (for which rate is plotted versus temperature with resource level serving as a parameter). In general, as resource availability increases, each rate-isotherm at first rises almost linearly, eventually saturates, and thereafter declines when resource levels become inhibiting. On the other hand, as temperature increases, each rate-isoconcentrate non-linearly climbs toward a thermal rate maximum and thereafter falls, often sharply (Figs. 1-3 on the following page).

The temperature at which a thermal rate maximum occurs is called the optimum temperature. From Figures 1-3, it can be seen that the optimum temperature is a function of resource level. This function constitutes a thermo-chemical constraint between habitat temperature and resource availability that must be satisfied to optimize the performance of ecological processes with thermal sensitivities like those illustrated in Figures 1-3.

In this paper, it will be shown that this constraint can be derived from two sets of assumptions which are broadly applicable to primary producers and to poikilothermal herbivores and carnivores. The first assumption is simply that at any instant of time a population may be split into two cyclically connected portions, the one prepared to incorporate matter and energy, the other not. The second assumption is that all the processes involved in population growth individually and collectively obey Arrhenius temperature laws (Arrhenius 1915). The consequences of these two sets of assumptions will be pursued in light of the properties of the experimental data. As a result, it will be shown specifically that the optimum temperature of ecological processes whose rate saturates in resource level varies linearly

[1]Paper presented at the international symposium Energy and Ecological Modelling, sponsored by the International Society for Ecological Modelling. (Louisville, Kentucky, April 20-23, 1981.)

[2]Alician V. Quinlan is Assistant Professor of Mechanical Engineering, Massachusetts Institute of Technology, Cambridge, Mass., U.S.A.

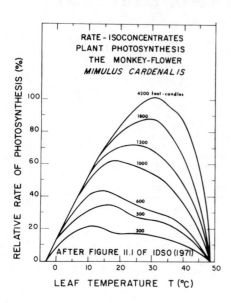

Figure 1.--An example of the thermal sensitivity of a primary producer.

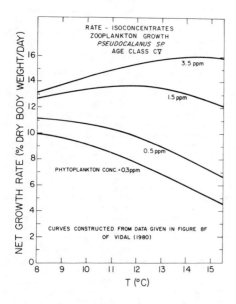

Figure 2.--An example of the thermal sensitivity of a poikilothermal herbivore.

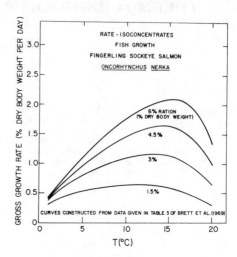

Figure 3.--An example of the thermal sensitivity of a poikilothermal carnivore (after Quinlan 1981).

with the logarithm of resource level. Then, insofar as habitat temperature tends to track resource level, it will be conjectured that the habitat temperature and the optimum temperature of many ecological processes may indeed be constrained to vary together in a way that tends to optimize process performance.

ANALYSIS

Resource Modulation of Ecological Processes

In the absence of toxic chemicals and injurious environmental conditions, producer, herbivore, and carnivore populations alike are assumed to augment their biomass according to the micromechanism shown in Figure 4. This micro-mechanism makes the following six assumptions about population growth:

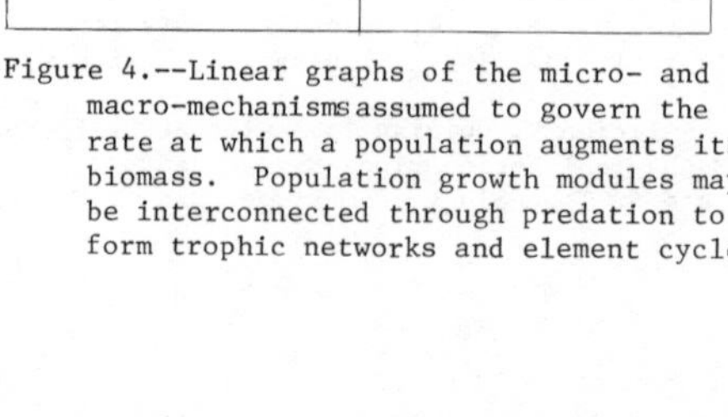

POPULATION GROWTH MODULE	
MICRO - MECHANISM	MACRO - MECHANISM

Figure 4.--Linear graphs of the micro- and macro-mechanisms assumed to govern the rate at which a population augments its biomass. Population growth modules may be interconnected through predation to form trophic networks and element cycles.

(1) the total biomass of a population (P) consists of a feeding or non-satiated portion (P_f) and a non-feeding or satiated portion (P_n);

(2) the total population gains biomass by incorporating (i) matter and energy at a rate proportional to the concentrations of the feeders (P_f) and the rate-limiting resource (R);

(3) the feeders become satiated (s) at a rate proportional to their own concentration (P_f) and that of the limiting resource (R);

(4) the non-feeders, in turn, become hungry (h) again at a rate proportional to only their own concentration (P_n);

(5) both the feeders and the non-feeders lose biomass through wastage (w_f, w_n) processes whose rate depends solely on biomass concentration (P_f, P_n, respectively), e.g., respiration, death, excretion, defecation, moulting, etc.; and

(6) the time constant for the growth of the feeding portion of the population greatly exceeds that of the non-feeding portion, or in other words, the non-feeding portion of the population achieves steady state much more rapidly than the feeding portion.

A steady-state analysis of the micro-mechanism yields the following macroscopic rate law that shows how resource level modulates the net biomass-specific rate of growth of the whole population:

$$\dot{P}/P = [VR/(K + R)] - w_f \qquad (1a)$$

where for i > s:

$$V \equiv i\,K + w_f - w_n > 0 \qquad (1b)$$

$$K \equiv (h + w_n)/s > 0 \qquad (1c)$$

The first term on the right-hand-side of equation (1a) represents a biomass-specific macroscopic incorporation rate which is equivalent to a gross biomass-specific growth rate. This rate saturates in resource level according to a generic Michaelis-Menten rate law. As a result, the macroscopic coefficients V and K may be interpreted as, respectively, the maximum-velocity and half-saturation coefficients for carbon fixation by a plant population, for phytoplankton ingestion by a zooplankton population, or for food intake by a fish population. Through their dependences on the five microscopic rate coefficients (i, h, s, w_f, w_n), the macroscopic coefficients themselves also depend on temperature and on the morphology and size of the organisms involved in the microscopic incorporation process.

The second term on the right-hand-side of equation (1a) represents the biomass-specific rate at which the population wastes away in the absence of the limiting resource. This rate identically equals the microscopic rate coefficient for wastage of the feeding or hungry portion of the population, which is reasonable because no satiated organisms should exist under starvation conditions.

The resource level (R_T) at which the macroscopic incorporation and wastage terms in equations (1) exactly balance represents a threshold concentration that must be equaled or exceeded for the survival of the population. As long as $i > [w_n/(h + w_n)]$ s, the threshold resource level is

$$R_T \equiv (w_f/i)/(1 - (s/i)\,(w_n/(h + w_n))). \qquad (2a)$$

But if i > s and h > w_n, then equation (2a) becomes approximately

$$R_T \sim w_f/i \qquad (2b)$$

In either case, the threshold resource level may be viewed as a temperature-, morphology-, and size-dependent measure of the growth efficiency of a population.

Temperature Modulation of Ecological Processes

The microscopic rate coefficients are each assumed to obey Arrhenius temperature laws (Arrhenius 1915). These laws can be approximated as follows over the temperature range supporting life processes (Quinlan 1980, 1981):

$$i \sim i \, \exp\,(T/\theta_i) \qquad (3a)$$

$$h \sim h \, \exp\,(T/\theta_h) \qquad (3b)$$

$$s \sim s \, \exp\,(T/\theta_s) \qquad (3c)$$

$$w_f \sim w_f \, \exp\,(T/\theta_{w_f}) \qquad (3d)$$

$$w_n \sim w_n \, \exp\,(T/\theta_{wn}) \qquad (3e)$$

where the coefficients i, h, s, w_f, w_n have the same units as, respectively, i, h, s, w_f, w_n; and the θ's have the same units as T. The morphology and size dependences are distributed among i, h, s, w_f, w_n and the θ's.

In turn, the temperature dependences of the microscopic rate coefficients determine those of the macroscopic coefficients V, K, R_T. Equations (1b), (1c), and (2a) suggest that these temperature dependences may be quite complicated; however, an analysis of the data used to construct Figures 1-3 showed that simple Arrhenius temperature laws also can adequately describe the temperature dependences of the macroscopic coefficients, namely,

$$V \sim V \, \exp\,(T/\theta_V) \qquad (4a)$$

$$K \sim K \, \exp\,(T/\theta_K) \qquad (4b)$$

$$R_T \sim R_T \, \exp\,(T/\theta_{RT}) \qquad (4c)$$

where V has the same units as $\dot{P}/P$, K and R_T have the same units as R, and the θ's have the same units as T. In addition, the macroscopic coefficients were all found to increase with temperature within the range of temperatures where the thermal rate maxima occur; so, the θ's must all be positive in value.

Thermo-Chemical Modulation of Ecological Processes

The properties of the families of rate-isoconcentrates shown in figures 1-3 are dominantly those of the macroscopic incorporation term in equation (1a) with equations (4a) and (4b) substituted for V and K, namely,

$$\dot{P}/P \sim [V\,R\,\exp\,(T/\theta_V)]/[K\,\exp\,(T/\theta_K) + R]. \quad (5)$$

Analysis of equation (5) shows that as long as θ_V exceeds θ_K, the rate-isoconcentrates will rise at relatively low temperatures (or high resource levels), but will fall at relatively high temperatures (or low resource levels), thereby creating a family of thermal rate maxima. Further analysis of equation (5) shows that the optimum temperatures T_0 corresponding to the rate maxima are constrained to vary as the logarithm of the ambient level of the rate-limiting resource (Quinlan 1980, 1981):

$$T_0 = \theta_K \ln R - \theta_K \ln [\theta_K/K\,(\theta_V - \theta_K)]. \quad (6)$$

Figures 5-7 depict the linear variation of optimum temperature with the logarithm of light intensity for photosynthetic carbon fixation by the monkey-flower <u>Mimulus</u> <u>cardenalis</u> (Fig. 5), with the logarithm of phytoplankton concentration for net growth by three age classes of zooplankton of the genus <u>Pseudocalanus</u> (Fig. 6), and with the logarithm of food ration for fingerling sockeye salmon <u>Oncorhynchus</u> <u>nerka</u> (Fig. 7). The

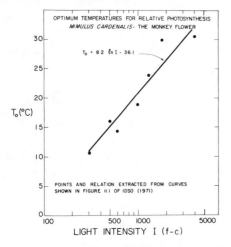

Figure 5.--An example of the variation of optimum temperature with the logarithm of light intensity for a primary producer (after Quinlan 1980).

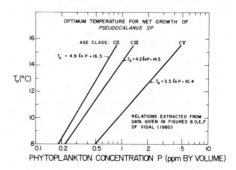

Figure 6.--An example of the variation of optimum temperature with the logarithm of prey concentration for a poikilothermal herbivore.

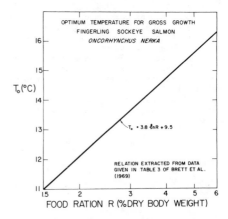

Figure 7.--An example of the variation of optimum temperature with the logarithm of prey concentration for a poikilothermal carnivore.

values of θ_K, θ_V, and K used to evaluate the slope and intercept of the lines and equations shown in these figures were estimated according to the method of Quinlan (1981) from rate-isoconcentrate data given in the cited papers. Because the optimum temperatures for the photosynthetic carbon fixation rate-isoconcentrates shown in Figure 1 could be estimated independently from this analysis, they are included in figure 5 as data points, and the correspondence between these points and the predicted linear relation appears good. Figure 6 shows how zooplankton morphology and size, as expressed through age class, can affect the linear dependence of optimum temperature on the logarithm of resource availability: both the slope and the intercept can be seen to decrease with advancing age class, with the result that for a particular prey concentration, the younger the predator, the

higher its optimum temperature, or for maintenance of a particular optimum temperature, the
older the predators, the higher its prey availability requirement.

Thermo-Chemical Optimization of Ecological
Processes

Equation (6) constitutes a thermo-chemical
constraint between habitat temperature and resource availability that must be satisfied to
optimize the performance of an ecological process whose rate saturates in resource availability.

For example, consider a situation in which
habitat temperature initially equals the optimum
temperature corresponding to the ambient level
of the resource limiting the rate of the process.
In this situation, the process should be operating at the peak of the rate-isoconcentrate
corresponding to the ambient resource level.

Now, suppose that the resource level falls
(or rises). The optimum temperature should shift
to a new lower (or higher) value determined by
equation (6). If the habitat temperature were
to remain unchanged, the process rate would be
sub-optimal. However, if habitat temperature
T_H were to vary with resource level according
to equation (6) such that $T_H \equiv T_o$, then the
process rate would migrate along the locus of
the thermal rate maxima, which is the condition
for process optimization (Quinlan 1980, 1981).

Conversely, suppose habitat temperature
falls (or rises). In order for the rate of the
process to remain optimal, the ambient level of
the rate-limiting resource would also have to
fall (or rise) according to the constraint imposed by equation (6). This constraint may be
re-written as follows to show directly how habitat
temperature determines the optimum level R_o of
the rate-limiting resource:

$$R_o = [\theta_K/K \, (\theta_V - \theta_K)] \, \exp \, (T_H/\theta_K). \qquad (7)$$

Equation (7) states that the optimum resource
level varies exponentially with habitat temperature.

CONJECTURE

Some interesting conjectures arise from
consideration of the implications of equations
(6) and (7) in relation to the variations observed in nature for habitat temperature and
for the availability of the resources limiting
the rates of key trophic and element-cycle processes.

In order for the thermo-chemical rate-op-
timization constraint stated in equations (6)
and (7) to be met in natural ecosystems, seasonal variations in rate-limiting resources would
have to track seasonal variations in habitat
temperature or vice versa. For many processes,
this seems to be the case.

For instance, light intensity and habitat
temperature show similar seasonal and spatial
variations. This similarity leads to the con-
jecture that the optimum temperature and, hence,
the thermal rate maximum of primary production
may be thermo-chemically rate-optimized on an
annual basis in natural ecosystems.

Similarly, in aquatic ecosystems phytoplankton concentrations show seasonal variations that
track both light intensity and habitat temperature; thus herbivorous predation by zooplankton
may also be thermo-chemically rate optimized on
a seasonal basis. Zooplankton concentrations
also track seasonal variations in habitat temperature; so, it is possible too that carnivorous
predation may be seasonally rate-optimized.

When interconnected through their trophic
interactions, phytoplankton, zooplankton, and
fish cascade energy and cycle elements through
trophic networks in aquatic ecosystems. Any
tendency to optimize the trophic processes mediated by these organisms would also constitute
a tendency to maximize the rates at which energy
cascades through the ecosystem and elements cycle
within the ecosystem.

ACKNOWLEDGEMENTS

The research reported in this paper was
supported by the Rockefeller Foundation through
a Fellowship Grant in Environmental Affairs
(1976-1979) and by the Henry L. and Grace Doherty
Charitable Foundation through a Henry L. Doherty
Professorship in Ocean Utilization (1978-1980).

The figures were drawn by Michele Halverson,
and the paper was formatted and typed by Anna M.
Piccolo. The author greatly appreciates the
high quality of their work.

Finally, the author thanks Professor Henry M.
Paynter for always being accessible when she
needed encouragement or a sounding board for
her ideas.

REFERENCES

Arrhenius, S. 1915. The influence of temperature
on the velocity of reactions - reactions of
cells. pp. 49-60. Quantitative Laws in
Biological Chemistry, Bell, London.

Aruga, Y. 1965. Ecological studies of photosynthesis and matter production of phytoplankton. II. Photosynthesis of algae in
relation to light intensity and temperature.
Botanical Magazine (Tokyo) 78: 360-365.

Brett, J.R., J.E. Shelbourn, and C.T. Shoop. 1969.
Growth rate and body composition of finger-
ling sockeye salmon, Oncorhynchus nerka, in
relation to temperature and ration size.
Journal of the Fisheries Research Board of
Canada 26: 2363-2394.

Elliott, J.M. 1975. The growth rate of brown
trout, Salmo trutta L., fed on reduced
rations. Journal of Animal Ecology 44:
823-842.

Idso, S.B. 1971. Potential effects of global temperature change on agriculture. pp. 184-191. In Man's Impact on Terrestrial and Oceanic Ecosystems, edited by W.H. Matthews, F.E. Smith, and E.D. Goldberg, M.I.T. Press, Cambridge, MA.

Marks, T.C. and K. Taylor. 1978. The carbon economy of Rubus chamaemorus L. I. Photosynthesis. Ann. Bot. 42: 165-179.

Pisek, A. and E. Winkler. 1959. Licht und temperaturabhängigkeit der CO_2-assimilation von fichte (Picea excelsa Link), Zirbe (Pinus cembra L.) und Sonnenblume (Helianthus annuus L.). Planta 53: 532-550.

Quinlan, Alician V. 1980. The thermal sensitivity of Michaelis-Menten kinetics as a function of substrate concentration. Journal of the Franklin Institute 310: 325-342.

Quinlan, Alician V. 1981. The thermal sensitivity of generic Michaelis-Menten processes without catalyst denaturation or inhibition. Journal of Thermal Biology 6: 103-114.

Sorokin, C. and R.W. Krauss. 1952. Effects of temperature and illuminance on Chlorella growth uncoupled from cell division. Plant Physiology 37: 37-52.

Vidal, J. 1980. Physioecology of zooplankton. I. Effects of phytoplankton concentration, temperature, and body size on the growth rate of Calanus pacificus and Pseudocalanus sp. Marine Biology 56: 111-134.

Webb, W.L., M. Newton, and D. Starr. 1974. Carbon dioxide exchange of Alnus rubra: a mathematical model. Oecologia 17: 281-291.

POWER AND A PULSING PRODUCTION MODEL[1]

John R. Richardson[2] and H. T. Odum[3]

Abstract--Simulation studies were made of the power utilization characteristics of a model with production, a self-
pulsing consumer, energy constraints, and recycle. An optimum frequency for maximum power existed over part of the
range of coefficients; nonpulsing behavior occurred at low
and very high energies. Pulses were more intense with longer
periods of production. When calibrated with biomass, soils
and woods, the interpulse period ranged from 200-300 years.
Pulses may represent surges of animal or human activity.

INTRODUCTION

In recent years, many ecosystems have been
observed to alternate between long periods of net
production followed by consumption and recycle
occurring during intensive pulses. Sprugel and
Borman (1981) reported on moving waves in balsam
fir forests. Odum (1979) suggested that pulsed
consumption may help to maximize energy use and
give a competitive advantage to systems that
pulse. While some oscillating models of
populations have been studied for a long time,
they have not been very relevant to ecosystems.
More recently, models including constraints of
energy sources, recycling of materials and
nutrients, and parallel alternate pathways by
which self organization can feed back consumer
control actions have been used to study pulsing
systems. In this paper, a pulsing model of
production and consumption with the previously
mentioned general ecosystem properties was
simulated and the relation of maximum energy
usage and pulsing behavior studied.

Verhoff and Smith (1971) simulated
producer-consumer chains without energy
constraints, but with recycling of materials and
found unstable oscillations. Quinlan and Paynter
(1976) included Michaelis-Menten modules and
found more regular oscillations. Odum (1976,
1981a) and Bayley and Odum (1976) studied
ecosystems with threshold controlled pulsing

consumer actions, in which frequencies were
internally determined by energy controlled rates
of growth rather than by frequency of external
inputs. Alexander (1978) found that consumer
unit models with a combination of linear and
non-linear input pathways had pulsing properties
that were self induced and did not require
switching actions. Ludwig, Jones, and Holling
(1978) simulated spruce-budworm ecosystems with a
three stage ecosystem model that had self pulsing
consumption. In chemical systems, Pacault
(1977), Noyes (1979), Haken (1978), and Prigogine
(1980) have found pulsing models associated with
the development of spatial structures. These
models have relationships that are expressed by
combinations of linear and non-linear pathway.
Catastrophe theory has provided an analytical
method for examining models with higher order
equations to show the sharp transitions from one
type of behavior to another. Poston and Stewart
(1978) examined the spruce-budworm behavior using
this technique.

From these studies, a hypothetical concept
emerges that natural systems may self organize
pulsing behavior to maximize energy and/or
material utilization. The natural frequencies
are self generated and may or may not be
entrained to outside controlling frequencies.
Pulsing may help maximize total resource use by
maintaining materials in readiness for production
to best capture available energy, by minimizing
the time that production is interrupted and by
using surges of high quality energies as
controlling feedback in the system. Further
review and discussion of the pulsing concept is
given in another paper at this conference (Odum
1981b).

METHODS

The model was calibrated with appropriate
values and simulated to obtain families of curves
for various parameters. The models were

[1] Paper presented at the international
symposium Energy and Ecological Modelling,
sponsored by the International Society for
Ecological Modelling, Louisville, Kentucky,
April 21-23, 1981.
[2] John R. Richardson is Graduate Research
Assistant in Environmental Engineering Sciences,
University of Florida, Gainesville, Florida.
[3] Howard T. Odum is Graduate Research
Professor in Environmental Engineering Sciences,
University of Florida, Gainesville, Florida.

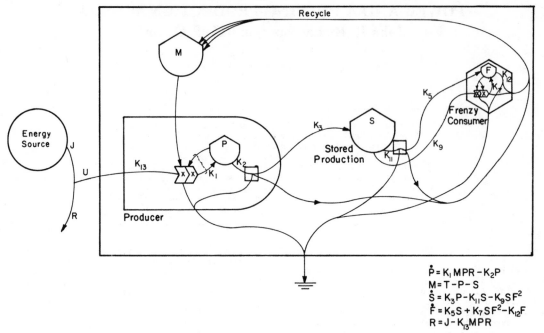

$$\dot{P} = K_1 MPR - K_2 P$$
$$M = T - P - S$$
$$\dot{S} = K_3 P - K_{11} S - K_9 SF^2$$
$$\dot{F} = K_5 S + K_7 SF^2 - K_{12} F$$
$$R = J - K_{13} MPR$$

Figure 1. Energy circuit diagram and differential equations.

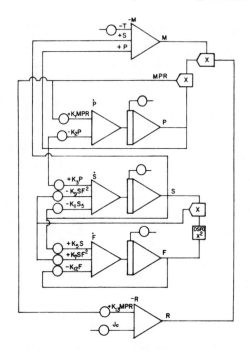

Figure 2. Analog computer diagram of model.

simulated on an EAI 2000/PDP 11 hybrid computer. Using the hybrid computer it is possible to make many runs of the model while altering various coefficients by a set amount for each run and plotting the results on an XY plotter. The rate of energy flow through the system was measured for each run and could be compared with the behavior of the model.

After the model was scaled with values appropriate to the real ecosystems, exploratory experiments were made on the following aspects:
Effect of varying the energy flow.
Effect of varying the initial conditions.
Effect of changing time constant of the consumer storages.
Effect of changing the ratio of linear to nonlinear flows.
Effect of varying the time constant of the producer storages.

The Model

The model studied is given in figure 1 as a set of differential equations and in the energy circuit language. The analog diagram is given in figure 2. It is similar to the one used by Alexander (1978) to describe catastrophes such as earthquakes and floods except that a quadratic autocatalytic feedback is part of this consumer unit. The model is similar to the Brusselator chemical reaction model of Nicolis and Prigogine (1977) except that the model in figure 1 has a

Table 1--Initial conditions and initial flow rates for world ecosystem model.

P = 1000	g Carbon/m^2
S = 4000	g Carbon/m^2
F = 50	g Carbon/m^2
M = 35600	g Carbon/m^2

R1 = 500 g C/m^2/yr	R9 = 1000 g C/m^2/yr
R2 = 500 g C/m^2/yr	R11 = 10 g C/m^2/yr
R3 = 50 g C/m^2/yr	R12 = 2.5 g C/m^2/yr
R5 = 1 g C/m^2/yr	R13 = .99 solar units
R7 = 100 g C/m^2/yr	

Source: Bolin 1970, Ryabchikov 1975, Whittaker 1975.

Table 2--Initial conditions and initial flow rates for world fossil fuel-assets.

P = 24800	g Carbon/m^2
S = 40000	g Carbon/m^2
F = $.671	/m^2
M = 40000	g Carbon/m^2

R1 = 145 g C/m^2/yr	R9 = 51 g C/m^2/yr
R2 = 145 g C/m^2/yr	R11 = .0376 g C/m^2/yr
R3 = .0376 g C/m^2/yr	R12 = $.0335 /$m^2$/yr
R5 = 3.75E-5 C/m^2/yr	R13 = .99 solar units
R7 = $.0671 /$m^2$/yr	

Source: Bolin 1970, Ryabchikov 1975, United Nations 1980, Whittaker 1975.

linear pathway from the producer to the consumer instead of an outside linear source as in the Brusselator.

The model was used to simulate two different time scales in the biosphere. First, the model was given values (table 1) of storage and flows that were appropriate for terrestrial ecosystems that develop soil and wood biomass which is then consumed in a rapid pulse by a consumer such as fire, epidemic disease or human use.

Second, the model was given values for fossil fuel reserves for the world as they are built up in the geologic process (table 2). The pulsing consumer in this time frame is our current human civilization.

RESULTS

Ecosystem Scale

Effect of Varying Energy

The result of varying the energy flow to the model with ecosystem values is shown in figure 3. The input energy was varied from 120% (bottom of graph) to 80% (top of graph) of the normal input energy. The plot is of the pulsing consumer (F), over time. The plots show a transition from a single pulse with dampened oscillation steady state at high energy values through a pulsing phase to a steady state system with no pulsing at lower energy levels. At energy levels in the range which pulsing is seen, the frequency of pulsing changed from low frequency to higher frequency as the energy level was increased until the system transitioned to the single pulse with

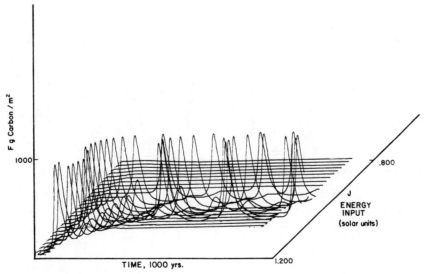

Figure 3. Results of varying energy flow. Pulsing consumer (F) graphed.

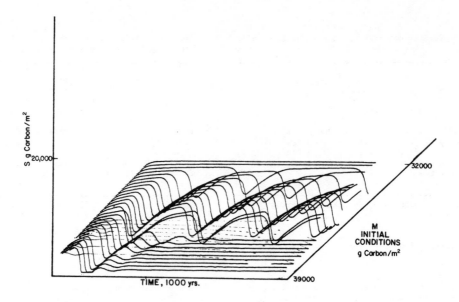

Figure 4. Results of varying initial conditions of total system carbon. Stored carbon (S) graphed.

dampened oscillation steady state. Pulsing was seen to occur at levels of energy flow that are similar to natural systems.

Effect of Varying Initial Storages

Figure 4 is a plot of the stored consumer (S) for different initial conditions on the total system carbon (M). When the initial conditions were varied on the producer (P), consumer storage (S), or the pulsing consumer (F), the only change

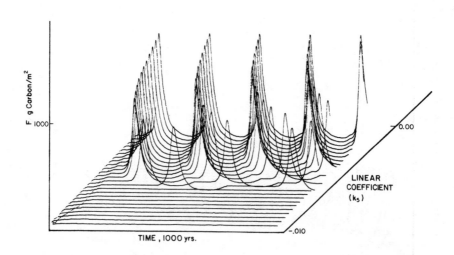

Figure 5. Results of varying linear flow K5 (K11 varied proportionately). Pulsing consumer (F) graphed.

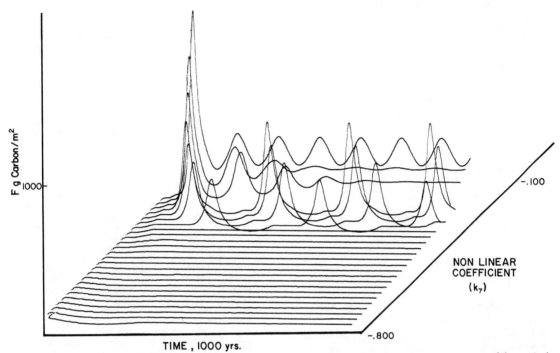

F g Carbon/m²

1000

TIME , 1000 yrs.

NON LINEAR
COEFFICIENT
(k₇)

-.100

-.800

Figure 6. Results of varying non-linear flow K7 (K9 varied proportionately). Pulsing consumer (F) graphed.

seen as a phase shift in the pulsing, with no changes in the pulsing frequency or transitions to other behavior. When the initial conditions were varied on the total carbon in the system, there was a change similar to the one seen when the energy input was varied. At 10% higher than normal levels (bottom of figure 4) the stored reserve of carbon (S) was consumed down by an initial pulse then remained at a lower steady state level. At the same time the pulsing consumer was at a high steady state level. In the middle part of the curve the frequency of pulsing decreased until it finally showed no pulsed consumption but the storage remained at a high steady state level. The pulsing consumer remained low at these values. Measurement of total power in the system showed a maximum value when the initial condition was close to the standard value.

Effect of Varying Consumer Time Constant

The effect of changing the coefficient of the linear pathway from the consumer storage to the pulsing consumer is seen in figure 5. This represents the variation of the time constant of the consumer storage. The graph depicts changing the time constant from two times the normal turnover time at the bottom of the graph to no linear flow at the top of the graph. With higher turnover time there was no pulsing behavior. The higher linear flow set up conditions where the

pulsing consumer existed at a low steady state condition but the stored biomass remained at a high steady state level. With very low or no linear flow between the two there was a pulsing behavior that was entirely dependant on the non-linear pathway. The frequency did not change after the initial transition phase.

Figure 6 shows the results of varying the nonlinear pathway coefficients. The transition here was from a low value steady state at high flow rates through a pulsing phase to a single pulse with dampened oscillation to a steady state. Finally at one half normal values a single pulse was observed with a continuous oscillation at a higher frequency than the pulsing behavior.

Effect of Varying Producer Time Constant

When the turnover time of the producer was varied, the result was similar to figure 5 with pulsing at turnover times higher than the standard run and a single pulse steady state at lower turnover times.

World Fossil Fuel Reserves-Assets

Simulation of the model with world fossil fuels-assets are given in figure 7. The producer (P) here represents the world biomass system, the

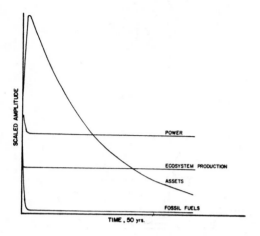

Figure 7. World fossil fuel reserves-assets. Standard model run.

stored reserves (S) represent the accumulated fossil fuels, and the pulsing consumer (F) represents the accumulated assets of the world. The standard run shown in figure 7 had a continuous increase in assets, giving a peak in 5 to 10 years followed by a decline as fossil fuels were exhausted. Various runs were made altering the coefficients and initial conditions of the model with only small changes seen in the output of the model. Because the differences in turnover times in the model are so great, realistic variations in the model show very little change in its basic behavior. If the rate of fossil fuel production were increased by a factor of 50,000 the system then goes into a pulsing behavior with peaks approximately 300 years apart.

DISCUSSION

The existence of parallel linear and nonlinear pathways in this model leads to several different types of behavior of the model. These various behaviors are determined by energy constraints, nutrient constraints and turnover times of the various compartments in the model. The rapid turnover time of the producer section of the model gives a very steady output. The slow turnover time of the stored production makes it available for a rapidly growing consumer to utilize the storage in a frenzy of consumption.

Linear Pathway

The low order, diffusion pathway, between the production storage and the consumer unit is dependent only on the production storage (S).

This flow tends to stabilize the consumer part of the system. If the linear flow is very low then there is a tendancy to generate large pulses in the system. If the flow is large enough, then the consumer will maintain itself in a steady state condition with the producer. The pulses tend to be dampened out and the system is maintained in a steady state condition.

Since the linear pathway is source dependent, the exploitation of this resource will reduce this flow. It may be best to design human systems to be receivers of this energy and to interact with it in such a way not to degrade it.

High Order Pathway

In order to grow rapidly, the consumer side of this model also has a third order autocatalytic pathway. This represents spending (or using) some of the stored structure interacting with itself to increase the flow of resources. In actuality this function may have several levels of order. Systems may use higher order feedbacks to increase energy flows and maximize their power flow. High order feedbacks may be able to process energy at higher levels while lower order pathways may be needed to process energy at lower power levels. A fundamental nature of these higher order interactions is that storages are consumed rapidly and recycled back into the resource nutrient pool thus allowing a system to diverge resources back to lower levels of the system for better use.

Maximum Power

When the energy source was held constant and the total amount of carbon was varied (figure 4), there was a change in the total power used in the system. As the carbon was increased in successive runs, the power went through a maximum in the area where pulsing was present and then power use declined as the carbon was no longer limiting. At the point where the top consumer flipped to a high level steady state, the power began to rise again. Whether or not this last transition maximizes power depends on other constraints in the model.

Perspectives

We may use the behavior of this model for insights that may apply to ecosystems and to the system of our current civilization. Perhaps pulsing may be a property of some range of energies and there may be a selection for a particular pulsing period that maximizes energy use. The question may be raised over what period is power maximized? The concept of alternation of production and pulsed consumption with a frequency that is most competitive may supply the answer.

While we deplore the excessively wasteful consumption of our natural resources by our current economic systems, it may be in the order of things that the rapid pulse of consumption returns the materials to a state of readiness for a long period of restoration and net production.

BIBLIOGRAPHY

Alexander, J. F. 1978. Energy basis of disasters and the cycles of order and disorder. Ph.D. dissertation. 232 p. Environmental Engineering Sciences. University of Florida, Gainesville.

Bayley, S. and H. T. Odum. 1976. Simulation of interrelations of everglades marsh, peat, fire, water, and phosphorus. Ecological Modelling 2:169-188.

Bolin, B. 1970. The carbon cycle. Scientific American. 223:124-132.

Haken, H. 1978. Synergetics: An introduction. 355 p. Springer-Verlag, New York, N. Y.

Ludwig, D., D. D. Jones, C. S. Holling. 1978. Qualitative analysis of insect outbreak systems: The spruce budworm and forest. Journal of Animal Ecology 47:315-322.

Nicolis, G. and I. Prigogine. 1977. Self organization in non equilibrium systems. Wiley Interscience, New York, N. Y.

Noyes, R. M. 1979. Mechanisms of checmical oscillators. p. 34-42. In Synergetics: Far from equilibrium. Ed. A. Pacault and C. Vidal. Springer Verlag, New York, N. Y.

Odum, H. T. 1979. Energy quality control of ecosystem design. p. 221-235. In Marshestuarine systems simulation. Ed. P. F. Dame, Belle W. Baruch Library in Marine Science, No. 8, University of South Carolina Press.

Odum, H. T. 1976. Macroscopic minimodels of man and nature. p. 249-288. In Systems analysis and simulation in ecology. Vol. 4. Ed. B. Patten. Academic Press, New York, N. Y.

Odum, H. T. 1981a. (in press). Systems. John Wiley, New York, N. Y.

Odum, H. T. 1981b. (in press). Pulsing, power and hierarchy. In Energetics and systems. Proceedings of the session. (Louisville, Ky., April 22, 1981) Ann Arbor Press, Mich.

Pacault, A. 1977. Chemical evolution far from equilibrium. p. 133-154. In Synergetics: A workshop. Ed. H. Haken. Springer Verlag, New York, N. Y.

Poston, T. and I. Stewart. 1978. Catastrophe theory and complexity in the physical sciences. 272 p. W. H. Freeman and Co., San Francisco, Cal.

Prigogine, I. 1980. From being to becoming: time and complexity in the physcial sciences. 272 p. W. H. Freeman and Co., San Francisco, Cal.

Quinlan, A. V. and H. M. Paynter. 1976. Some simple non-linear dynamic models of interacting element cycles in aquatic ecosystems. J. of Dynamic Systems Measurement and Control 98:1-14.

Ryabchikov, R. 1975. The changing face of the earth 205 p. Progress Publishers. Moscow, USSR.

Sprugel, D. G. and F. H. Borman. 1981. Natural disturbance and the steady state in high altitude balsam fir forests. Science 211:398-393.

United Nations Statistical Yearbook. 1980. United Nations, N. Y.

Verhoff, V. and F. J. Smith. 1971. Theoretical analysis of a conserved nutrient ecosystem. J. Theoretical Biology 33:131-147.

Whittaker, R. H. 1975. Communities and ecosystems. 358 p. Macmillan Publishing Co. Inc., New York, N. Y.

A UNIFIED THEORY OF SELF-ORGANIZATION[1]
Robert E. Ulanowicz[2]

ABSTRACT. Growth and development associated with a network of material or energy exchanges can be combined using information theory into a single measure, the ascendency. Increases in ascendency are constrained by the availability of matter or energy, dissipation, heirarchical considerations and environmental perturbations.

INTRODUCTION

Physicists are astir these days in antici-ition of a breakthrough in the search for a nified force theory. Since the turn of the entury the Holy Grail of physics has been the reconciliation of gravitational, electromag-netic, and (later) the two intranuclear forces as manifestations of a single, universal phenomenon. Events on a cosmic scale or those confined within a baryon are thought to derive from the same essence.

How different the outlook of most biolo-gists today! The cause for any event is usually assumed to lie in a mechanism evident only at a smaller scale of observation. The search for explanation cascades down a hierarchy of cate-gories until one arrives in the domain of molecular biology. It is almost as if (1814) divining angel had been rediscovered so that, knowing the sequences of nucleic acids in the gene pool of an ecosystem, the angel could tell the ecologist everything he need know about the system. Of course, in the Neo-Darwinian scenario chance can alter the course of biolog-ical events. But it appears the significant role of chance has also been relegated to the molecular level.

I submit that this wholly reductionistic attitude among many biologists is more an acci-dent of history than an accurate perception of the nature of biological phenomena. To be sure, the advances in micro and molecular biology have immensely improved scientific understanding and human well being. In contrast, to say that

[1]Paper presented at the international symposium on Energy and Ecological Modelling, sponsored by the International Society for Ecological Modelling. (Louisville, Kentucky, April 20-23, 1981)
[2]Robert E. Ulanowicz is Professor of Estuarine Science, University of Maryland, Chesapeake Biological Laboratory, Solomons, Maryland, U.S.A.

teleology, Vitalism and Lamarckism have not fared well is to speak charitably. Yet, despite this background, there appears to be no logical, apriori proscription to the existence of a principle of organization operating at, say, the level of the whole ecosystem. What does seem to be missing is a perspective of community-level phenomena which allows the articulation of development as a process which universally can be subjected to measurement.

I wish to suggest that flow analysis, the study of the network of exchanges among the components of a system, provides just such a perspective from which a useful definition of self-organization can be made. Furthermore, the key to the way in which one goes about combining flow measurements into an ensemble property reflecting the dual attributes of growth and development is to be found in the theory of quantification of knowledge, i.e., information theory.

DEVELOPMENT OF THE HYPOTHESIS

I begin the description of self-organ-zation by modifying to the efforts of MacArthur (1955) to quantify the diversity of pathways in a flow network. If T_i is the quantity of any medium (energy, material, capital, etc.) flowing through compartment i, and T is the total system throughput of the network ($T = \sum_i T_i$), then $Q_i = T_i/T$ is an estimator of the probability that at any time we shall find a given quantum of medium passing through compartment i. Using the infor-mation measure of entropy, or uncertainty, we may define the diversity of throughputs in the system as

$$C = -K \sum_i Q_i \log Q_i \qquad (1)$$

where K is an arbitrary constant.

Apter and Wolpert (1965) correctly criti-cized the use of information measures such as C to describe biological systems, saying that these indices both failed to quantify the extent, or size, of the system and did not convey any sense of how well the compartments were related one to another. To rephrase their critique, C as stated in equation (1) does not assess the attributes of network growth and development. This is a serious deficiency indeed, as we intuitively sense that, for an entity to survive vis-a-vis another system (real or putative), it requires some advantageous combination of both size and coherent structure.

Fortunately, it is possible to overcome these deficiencies in C without abandoning the use of either the flow description or information theory. For example, the lack of scale can be rectified by equating the oft-neglected constant K to the total systems throughput T. Lest this equivalence seem arbitrary or simplistic, Tribus and McIrvine (1971) note how the thermodynamic entropy function has been scaled in a similar manner since before the time of Shannon.

To assess the degree of network coherence it becomes necessary to know more than just the apportionment of throughputs. One further needs to know the probability that a quantum of throughput T_i will flow directly as input to another member j of the community. Call the estimator of this probability f_{ij}. Rutledge et al. (1976) show how these conditional probabilities reduce the uncertainty about flow partitioning by an amount equivalent to what is known in information theory as the average mutual information,

$$A = T \sum_k \sum_j f_{kj} Q_k \log [f_{kj}/(\sum_i f_{ij} Q_i)], \quad (2)$$

where I have again chosen to scale the information measure by T. McEliece (1977) shows how information measures can be superior to covariance indicies in quantifying relationships. Thus, one may look upon the average mutual information as being the best available assessment of the internal coherence of the flow network. The index A, therefore, embodies the notions of both growth and development. One might speculate that A quantifies the extent to which a given network is ascendent over similar real or putative networks. I, therefore, refer to A as the network ascendency.

The above speculation is reinforced when it is noticed that the ascendency is usually augmented by such network attributes as the tendency to internalize flows, the amount of cybernetic feedback, the degree of specialization of compartmental outputs and the number of compartments (Ulanowicz, 1980). These properties encompass all the factors which Eugene Odum (1969) cited as being typical of mature ecosystems.

If growth and development are to be identified with the optimization of network ascendency, the question immediately arises as to what, at the macroscopic level, constrains any increase in A. Here it is useful to note the inequality (McEliece, 1977),

$$C \geq A \geq 0, \quad (3)$$

that is, C serves as an upper bound on ascendency. For this reason I refer to C as the development capacity of the network.

Factors which limit the increase in capacity also place a bound on the ascendency. Two constraints on capacity seem immediately apparent. First, the total systems throughput should be constrained by the total input flows available to the system. Second, increases in the throughput diversity will be limited by how finely the total throughput can be partitioned. Extremely small compartments will be liable to chance extinctions due to arbitrary perturbations. Hence, the rigor of the environment in which the network exists also limits the degree of partitioning possible, and thereby the development capacity.

In addition to the limits on the increase in capacity one must also study the positive difference between capacity and ascendency; or, if you will, the network overhead. Overhead can be decomposed into three components. Suppose it is determined that the fraction r_i of the throughput of compartment i is dissipated, as required by the Second Law of Thermodynamics, into a form which cannot be used by any member of the community, or another similar community. Likewise, e_i is identified as the fraction of throughput exported from component i as useful medium to another community. Then the overhead can be shown algebraically to equal the sum of three non-negative quantities:

$$S = -T \sum_i r_i Q_i \log Q_i > 0, \quad (4)$$

$$E = -T \sum_i e_i Q_i \log Q_i \geq 0, \quad (5)$$

and

$$R = -T \sum_i \sum_j f_{ij} Q_i \log [f_{ij} Q_i/(\sum_k f_{kj} Q_k)] \geq 0. \quad (6)$$

The dissipation, S, is the net encumberance of the thermodynamic losses on each pass through a compartment. Odum and Pinkerton (1955) infer that r_i increases monotonically with T_i. This would make it impossible to cycle medium at an arbitrarily high rate. Thus, an upper bound would exist on how much a given amount of input flow could be translated into total systems throughput.

The export fraction, E, I have called tribute. One might expect that development would proceed in a direction so as to minimize tribute. But if exports and inputs are coupled within the context of a higher hierarchical network, it may be detrimental to the ascendency of the larger system (and hence to the original system's inputs) to reduce exports below some critical level.

The remaining fraction, R, or redundancy, was first shown by Rutledge et al. (1976) to be related to the multiplicity of pathways between two arbitrary components of the network. Again, one might expect feedback loops of greater efficiency to begin to dominate during the course of development, causing ascendency to grow at the expense of redundancy. However, chance perturbations could retard or reverse this tendency, and we thus anticipate that the redundancy would reflect the variety and magnitude of the perturbations to which a climax community is subject. Because perturbations are never absent from the real world, actual ecosystems should always possess refactory amounts of redundancy characteristic of their environments.

All of the previous speculations can be summarized by a single hypothesis: A self-organizing community flow network evolves over an adequate interval of time so as to optimize its ascendency subject to thermodynamic, hierarchical and environmental constraints. In symbols one may write

$$A = C - (S + E + R), \qquad (7)$$

noting that the development capacity of a network is characterized by the intensity and diversity of its constituent flows, and that not all of this capacity can appear as coherent structure -- a necessary amount is always encumbered for thermodynamic, hierarchical or environmental reasons. Not only does (7) bear marked similarity to the definition of Gibbs' free energy, but when combined with the inequality $S > 0$ they together subsume the two laws of thermodynamics.

IMPLICATIONS OF THE HYPOTHESIS

There are certain to be some who will look upon this hypothesis as nothing more than an exercise in epistemology (or perhaps tautology!). I would respond by noting that the theory is quantitative and can be tested. What scant data are available to test the hypothesis tend to support it (Ulanowicz 1980). Although there remain lexical problems in comparing networks of radically different ecosystems (e.g. tundra with coral reef), consistent choices for compartments should allow one to use the information variables for comparative purposes.

One exciting possibility afforded by the network variables is to give quantitative definitions to certain hitherto subjective abstractions. The common notion of a "desirable" ecosystem, for example, usually is applied to a community which is "highly-developed". Does not the network coherence (A/T) assign a number to the degree of development? "Eutrophic" systems sometimes displace more "desirable" communities. Eutrophication is seen here as an increase in ascendency corresponding to an increase in input flows. This augmented ascendency, however, is due to a disproportionate increase in total systems throughput outweighing a concomitant decrease in network

coherence. Even the "fitness" of a species vis-a-vis the community should be related to the contribution of that species to the ascendency of the ensemble... The possibilities are numerous.

Furthermore, the community variables help to resolve the conflict surrounding the earlier "diversity implies stability" hypothesis. May's (1973) remark that a benign environment allows a greater diversity to exist was echoed in my earlier statements about the limits to development capacity. At the same time a given component of this capacity, the redundancy, quantifies Odum's (1953) idea that multiple pathways buffer the system against small perturbations.

If the medium of exchange is taken to be energy, the ascendency takes on the dimensions of power. The hypothesis of maximal ascendency can then be viewed as a generalization of the Lotka maximum power principle applicable to the community as a whole (Odum and Pinkerton, 1955). If t_{kj} represents the flow of energy from k to j, the ascendency (eqn. 2) can be rewritten as

$$A = \sum_k \sum_j t_{kj} \log [f_{kj}/(\sum_i f_{ij} Q_i)]. \qquad (8)$$

That is, A can be regarded as the weighted sum of all the intramural energy exchanges. Could each weighting factor possibly be identified with the quality of its associated flow? I am not aware of any successful methodology for measuring the entropy of a living organism. Perhaps we could circumvent this difficulty by regarding bioenergetic quality as a relativistic notion. Rather than look upon quality as an intrinsic property of the medium being exchanged, might we not infer that the quality of the flowing medium is determined by the particular network of exchanges in which it is imbedded?

If future empirical evidence should lend further credence to the development hypothesis, then perhaps we need to rethink the current emphasis on reductionism in biology. Leaving aside Neo-Darwinism, we find examples of whole-ecosystem models which are implicitly reductionistic: The system is broken into compartments, and bilateral interactions between compartments are studied in isolation and given mathematical expression. These expressions are collected (usually as a set of coupled differential equations) and the behavior of the whole system is believed to correspond to the behavior of the mathematical ensemble. But, if the ecosystem is following some variational law at the community level, then who is to say that bilateral functionalities cannot change and undergo selection during community development in the same sense that genotypes are assumed to undergo selection?

Finally, and most importantly, I would hope that debate over this hypothesis will rekindle optimism concerning the search for unifying principles in biology. As this exercise demonstrates, by examining different perspectives upon the same natural phenomena, it may be possible to choose a frame of reference wherein diverse living phenomena at different temporal and spatial scales can be adequately described by a single principle. This principle might even transcend the living domain and find application wherever highly dissipative phenomena occur, e.g. economic development (E. P. Odum 1977) and meteorological systems (H. T. Odum 1971). One thing is certain, however, if we allow ourselves to be persuaded that such unity of description cannot exist, we will surely never discover it. But, if we are willing to risk the rigors and frustration of a search, there is no telling what scientific and philosophical achievements might await us!

LITERATURE CITED

Apter, M. J., and L. Wolpert. 1965. Cybernetics and development. J. Theor. Biol. 8:244-257.

LaPlace, P. S. 1814. A Philosophical Essay on Probabilities. 196p. [Translation by F. W. Truscott and F. L. Emory, copyright 1951] Dover Pubilcations, Inc., NY.

MacArthur, R. H. 1955. Fluctuations of animal populations, and a measure of community stability. Ecology 36:533-536 .

McEliece, R. J. 1977. Page 25 in The Theory of Information and Coding. 302 pp. Addison-Welsey, Reading, MA.

May, R. M. 1973. Stability and Complexity in Model Ecosystems. 235 p. Princeton University Press, Princeton, NJ.

Odum, E. P. 1953. Fundamentals of Ecology. 384 p. W. B. Saunders, Philadelphia.

Odum, E. P. 1969. The strategy of ecosystem development. Science 164:262-270.

Odum, E. P. 1977. The emergence of ecology as a new integrative discipline. Science 195:1289-1293.

Odum, H. T. 1971. Environment, Power and Society. 331 p. John Wiley & Sons, NY.

Odum, H. T. and R. C. Pinkerton. 1955. Time's speed regulator: the optimum efficiency for maximum power output in physical and biological systems. Am. Scientist 43:331-343.

Rutledge, R. W., B. L. Basore and R. J. Mullholland. 1976. Ecological stability: an information theory viewpoint. J. theor. Biol. 57:355-371.

Tribus, M. and E. C. McIrvine. 1971. Energy and Information. Sci. Am. 225(3):179-188.

Ulanowicz, R. E. 1980. An hypothesis on the development of natural communities. J. Theor. Biol. 85:223-245.

SENSITIVITY TECHNIQUES FOR EXAMINATION OF INPUT-OUTPUT FLOW ANALYSES[1]

Robert W. Bosserman[2]

Abstract.--Input-output flow analyses have been used to examine the direct and indirect pattern of conservative flows which are propagated through a causal network. In general, such studies cannot be used easily to assess the effects of structural changes in a system network. A sensitivity analysis has been developed which assesses the effects of small modifications to the flow structure of a network. This analysis is based on the 'condition' of matrices which describe sets of linear equations. Different flow structures can be compared with respect to their stabilities when changes in direct flows occur. Such techniques are also useful for examining perturbations due to measurement, observation, sampling, and manipulation.

INTRODUCTION

Flow analyses have been used extensively in the ecological and economic literature, especially in recent years with regard to energy flow (Hannon 1973, Patten et al. 1976, Finn 1976a, 1976b, Barber 1978a, 1978b, Costanza 1980). These analyses, based on the original work by Leontief (1966), represent the static structures of ecosystems and economies. However, effects of perturbations to the flow structures are rarely examined. Because of the causal propagating network which is embodied in many systems, the effects of perturbation are both direct and indirect.

In this paper, techniques are developed which allow one to examine the sensitivities of the flow structure to small changes in parameter values. Other research has examined effects of large changes in parameter values (Hannon 1973), Effects of reapportioning flows within a network or of modifying the inputs and outputs can be assessed. These results are useful for demonstrating the effects of system modification through sampling, harvesting and observation, as well as for assessing the effects of uncertainty and manipulation of inputs and outputs.

[1]Paper presented at the international symposium, Energy and Ecological Modelling, sponsored by the International Society for Ecological Modelling. (Louisville, Kentucky, April 20-23, 1981).
[2]Robert W. Bosserman is Assistant Professor of Systems Science and Biology, University of Louisville, Louisville, Kentucky 40292.

BACKGROUND

Flow Analysis

Flow analyses are initiated with the construction of a flow matrix (production matrix), A, whose elements are the amounts of matter and energy flowing from one compartment to another in an interval of time. In this paper, element a_{ij} of A will refer to the amount of flow from compartment i to compartment j. Often, the systems which are modelled are assumed to be in steady state, where the input to each compartment x_i is equal to the output from that compartment. A vector E can be formed where element e_j is the total input to compartment x_j; because the system is at steady state this value is also equal to the total output from compartment x_j. In equation form, e_j is equal to the sum of all elements in column j of matrix A:

$$e_j = \sum_{i}^{n} a_{ij} \tag{1}$$

A normalized flow matrix G is then created by dividing the elements of each column by the corresponding column sum:

$$g_{ij} = a_{ij}/e_j \tag{2}$$

In G, each input to compartment x_i is represented by a proportion of total input to x_i. Each column, therefore, sums to 1.0. One can also normalize with respect to rows, in which case each element in the normalized flow matrix G' is represented by the proportion of total output from each compartment (Patten et al. 1976).

An output vector R can be constructed

in which each element r_i is the output from compartment x_i to the environment of the system. Using the above matrices, the following relationship can be established:

$$E = GE + R = (I - G)^{-1} R$$

The matrix $(I -G)^{-1}$ has been called the structure matrix of the system (Hannon 1973) because it indicates the contributions of both direct and indirect flows. Greater insight into this matrix can be obtained by the following relationship

$$\sum_{i=0}^{\infty} G^i = (I-G)^{-1} \qquad (4)$$

Other derived concepts such as recycling indices and residence times can also be established from these analyses (Finn 1976a, Patten et al. 1976).

Sensitivity Analyses

The form of the flow analysis, as expressed in Eq. 3, makes it amenable to techniques developed in linear algebra for assessing the stability of linear equations with respect to parameter change. Eq. 3 can be rewritten in the form

$$E = B^{-1} R \qquad (5)$$

where $B = (I -G)$. Note that the elements of B are negative except for those on the main diagonal. Because of the constraints on G, B can be shown to be invertible (Hannon 1973, Patten et al. 1976). An interesting question which can be asked of B is, how do small changes in the elements of B affect B^{-1} or E? This question can be interpreted as asking how changes in the direct flows affect the indirect flows in the system. The following analysis is based on classic work by Gauss, Turing and others about the conditioning of matrices (Faddeev and Faddeeva 1963, Noble 1969).

Because B is invertible, $BB^{-1} = I$ where I is the identity matrix. When taking the partial derivative of B with respect to element b_{ij} the following equation is obtained:

$$(\delta B/\delta b_{ij})B^{-1} + B(\delta B^{-1}/\delta b_{ij}) = (\delta I/\delta b_{ij}) = 0 \qquad (6)$$

Because the derivative of a matrix is obtained by taking the derivatives of the elements of that matrix, the derivative of B with respect to b_{ij} is E_{ij}, where E_{ij} is an elementary matrix in which all elements are 0 except for the ij-th element which is 1.0. E_{ij} can be rewritten as the product of elementary vectors E_{i1} and E_{1j}. After premultiplying Eq. 6 by B^{-1} and rearranging the equation, the following equations are obtained:

$$\delta B^{-1}/\delta b_{ij} = -B^{-1}(\delta B/\delta b_{ij}) B^{-1} = \qquad (7)$$

$$-B^{-1} E_{ij} B^{-1} = - (B^{-1}E_{i1})(E_{1j} B^{-1}) = \qquad (8)$$

$$-B_{i1}^{-1} B_{1j}^{-1} = S(i,j) \qquad (9)$$

where $S(i,j)$ is a matrix of sensitivity values the elements of which describe how sensitive each element of B^{-1} is to a change in b_{ij}. Therefore, the derivative of B^{-1} with respect to b_{ij} is the product of the i-th column and the j-th row of B^{-1}. This product is a matrix $S(i,j)$ of sensitivity values, each of which indicates the relative effects of changes of b_{ij} on element b_{kl}^* of B^{-1}. One can construct one sensitivity matrix for each element of B. The effects of changing b_{ij} on any particular element of B^{-1} is the product of two elements in B^{-1}:

$$(\delta b_{kl}^*/\delta b_{ij}) = -b_{ki}^* b_{jl}^* \qquad (10)$$

These sensitivities reflect the changes due to infinitesimally small changes in b_{ij}. Absolute changes can be obtained by multiplying S by the actual change in b_{ij} (i.e., Δb_{ij})

$$(\Delta B^{-1}/\Delta b_{ij}) = \Delta b_{ij} S(i,j) = \Delta b_{ij}(\delta B^{-1}/\delta b_{ij}) \qquad (11)$$

All b_{ij}'s are restricted to the interval between 0 and 1, so the amount of possible change is limited. Because these are partial derivatives, the complete derivative is obtained by summing all the partial derivatives:

$$dB^{-1} = \sum_{ij}^{n}(\delta B^{-1}/\delta b_{ij}) \qquad (12)$$

The normalized flow matrices are constrained by the fact that the column (row) sums are 1.0. Therefore, when one element of B is changed, other elements in the same column must also be changed. These changes can counteract or amplify the changes in the element that one is interested in. There are an infinite number of ways to reapportion the flow among more than two components. The most straightforward, and most realistic, is to adjust the inputs and outputs to the system so the column (row) sums remain 1.0. These models represent open systems so they have inputs and outputs to the environment. For example, if the derivative with respect to b_{ij} is to be assessed then the input to compartment i could be readjusted to maintain the constraint $\sum_k g_{ki} = 1.0$. The derivative with respect to this change is denoted by the following equation:

$$dB^{-1} = (\delta B^{-1}/\delta b_{ij})-(\delta B^{-1}/\delta b_{zj}) \qquad (13)$$

where b_{zj} represents the normalized input from the environment. Therefore, the following relationship is obtained:

$$(\delta b_{kl}^*/\delta b_{ij})-(\delta b_{kl}^*/\delta b_{zj}) = -b_{ki}^* b_{jl}^* + b_{kz}^* b_{jl}^* \qquad (14)$$

The relevant elements of the z-th row, representing inputs to the input, are either 0 or cause the effects of modifying the inputs to cancel, so:

$$(\delta b_{kl}^*/\delta b_{ij}) - (\delta b_{kl}^*/\delta b_{zj}) = -b_{ki}^* b_{jl}^* + 0 \qquad (15)$$

The interpretation of this result is useful.

Modifying an input or output to a system is
analogous to sampling, harvesting, observing or
subsidizing a real system; such an interaction
with a system would occur through the various
inputs and outputs. With the above analysis, the
effects of small interactions with the system can
be assessed with respect to the direct and
indirect flow structure.

The sensitivity of B^{-1} to a change in b_{ij} is
an array of n^2 elements, where n is the size of
the structure matrix. It is useful to combine
these elements into several sensitivity indices
which describe the effect on indirect flows to
each state variable and on the entire system. The
sensitivity of the flows to state variable k with
respect to changes in b_{ij} is equal to $\sum_j \delta b^*_{1k}/\delta b_{ij}$,
which is the sum of all elements in column k of S.
The sensitivity of the entire network to changes
in b_{ij} is $\sum_{lk} \delta b_{1k}/\delta b_{ij}$, which is the sum of all
elements in $S(i,j)$.

The above computations also imply that one
can do the reverse analysis and assess the
changes in B (the direct flow or production
matrix subtracted from the identity matrix) by
modifying elements of B^{-1} (the structure matrix).
The following equation expresses this
relationship:

$$dB = \sum_{ij}(\delta B/\delta b^*_{ij}) \qquad (16)$$

This reverse analysis establishes how to modify
the direct flow structure in order to generate a
particular indirect system structure. Suppose,
for example, that one wished to generate a system
structure in which element b^*_{ij} has been changed
slightly (Δb^*_{ij}). The partial derivative with
respect to b^*_{ij} is multiplied by Δb^*_{ij} as shown
in the following equation:

$$\Delta b^*_{ij} (\delta B/\delta b^*_{ij}) = \Delta b^*_{ij} B_{i1}B_{1j} \qquad (17)$$

The values of the product matrix should suggest
how the direct flows should be modified to generate the indirect flow structure.

The effects of changing elements of B and
B^{-1} on the energy flow vector E and the output
vector R can be examined. The following equations
are appropriate for these analyses:

$$E = B^{-1}R \qquad (18)$$

$$B E = R \qquad (19)$$

The changes in E with respect to changes in B are
established by the following relationships:

$$(\delta E/\delta b_{ij}) = (\delta B^{-1}/\delta b_{ij}) R = -B^{-1} E_{ij}B^{-1} R = -B^{-1} E_{ij}E \qquad (20)$$

Therefore, the effect of changing b_{ij} on any
element e_k is

$$(\delta e_k/\delta b_{ij}) = -b^*_{ki} e_j \qquad (21)$$

Likewise,

$$(\delta E/\delta r_i) = B^{-1}(\delta R/\delta r_i) = B^{-1}e_{i1} \qquad (22)$$

Therefore, the effects of changing r_i on any
element e_k is

$$(\delta r_k/\delta b^*_{ij}) = b_{ki}r_j \qquad (25)$$

Essentially, this result establishes the effects
on the outflow vector due to changes in the indirect structure of the system.

APPLICATION

The preceding sensitivity analyses are applied here to two simple ecosystem models which
have been previously discussed in the literature.
Although these models are simple, the above analyses can be used easily with larger models. The
compartment model shown in figure 1a depicts the
energy flow among components in Silver Springs,
Florida. Compartments present in this model are
primary producers (X_1), herbivores (X_2), carnivores (X_3), top carnivores (X_4), and decomposers
(X_5). In table 1 are various matrices which
describe energy flow in Silver Springs. H.T.
Odum (1957) generated the original model
but flow analyses were done on the Silver Spring
model by Hannon (1973). The compartment model in
figure 1b depicts the nitrogen flow among components in a Puerto Rican tropical rainforest.
Compartments in this model are leaves (X_1), loose
litter (X_2), fibrous roots (X_3), mycorrhizae
(X_4), and wood (X_5). In table 4 are matrices
which describe nitrogen flow in the rainforest.
The original studies were done by Edmisten
(1970a, 1970b), but the model was generated by
Auble et al. (1973). Flow analyses were later
done by Patten et al. (1976).

Sensitivity of the Silver Springs Model

The flow matrix for Silver Springs is shown
in table 1a. The i,j-th element of this matrix
represents the energy flow (kcal/m^2/yr) from
compartment i to compartment j. The direct energy vector, E, which represents the sum of respiration, net production, and net export, is shown
in table 1b. Inputs are from bread thrown by
tourists and sunlight. The output vector, which
represents the output in the form of respiration
from each compartment, is shown in table 1c.
Using the values of E, the flow matrix is normalized by columns as shown in table 1d. This normalized flow matrix is designated G where B = I-G.
The structure matrix $B^{-1} = (I -G)^{-1}$ is shown in
table 1e. Using this matrix the sensitivity of
B^{-1} with respect to changes in B can be calculated. For example, suppose $b_{2,1}$ is to be modified. This coefficient describes the change in
direct flow from compartment 2 to compartment 1.
No such direct flow occurs in the original model;
although modifying this flow may be unrealistic
the analysis is still valuable because it may
represent a flow which has not been conceptual-

ized or one that could be added through management. The matrix of sensitivities calculated according to equation 10 is shown in table 1f. Sensitivity values are negative because elements of B are negative except for the diagonal elements. In contrast, elements of G are positive. A positive change in G corresponds to a negative change in B which causes a positive change in B^{-1}, the structure matrix. The most sensitive coefficients are those describing the flows from compartments X_1 and X_2. Flows from compartments X_3, X_4, and X_5 are not affected by a change in $b_{2,1}$. Effects of changing $b_{2,1}$ on elements of the direct energy vector E are shown in table 1g. Compartments X_1 and X_2 are the only ones affected; compartment X_1 is affected slightly more than X_2.

The sensitivities of flows into compartments of the Silver Springs energy model due to changes in the direct flow matrix B are shown in table 2. Total sensitivities for the entire model are also shown; total sensitivity with respect to changes in element b_{ij} is the sum of all elements in the sensitivity matrix S(i,j), whereas the sensitivities of inflows to compartment k due to changes in element b_{ij} is the sum of elements in column k of S(i,j). Modifications of $b_{4,6}$, $b_{4,1}$, $b_{3,1}$, and $b_{5,1}$ cause the largest changes in the indirect flow matrix. The largest effects tend to be for indirect flows into compartment X_1.

One also can determine which the changes which need to be made in the B matrix to generate a particular change in any element of the structure matrix. This problem is accomplished with equation 16. The amount of change required to produce a change in the indirect flow structure is a set of products of elements in the B matrix. In the Silver Springs energy model, elements $b_{5,3}$, $b_{2,3}$, $b_{5,5}$ and $b_{2,5}$ of B must be changed in order to generate changes in $b^{-1}_{1,2}$ (table 3). Element $b_{2,3}$ and $b_{7,3}$ must be changed in order to cause changes in $b^{-1}_{3,2}$. Any changes in B should be balanced by modifying inputs and outputs so the column sums of G remain 1.0.

Sensitivity of the Tropical Rainforest Nitrogen Model

The flow matrix for the tropical rainforest nitrogen model is shown in table 4a. The i,j-th element of this matrix represents the amount of nitrogen ($g/m^2/yr$) transferred from compartment i to compartment j. The flow matrix G which has been normalized with respect to columns is shown in table 4b. The structure matrix $B^{-1}=(I-G)^{-1}$ is shown in table 4e. The output vector R, shown in table 4c, reflects outputs of nitrogen to the environment. The direct flow vector E, shown in table 2b has elements e_i which are the sum of all inputs to compartment x_i. The sensitivity of B^{-1} with respect to changes in $b_{2,1}$ are demonstrated in table 2f. Flows from compartments X_1 and X_2 are over twice as sensitive to changes in this parameter as are flows from

compartments X_3, X_4, and X_5. The higher sensitivity values in the tropical rainforest nitrogen model reflect the greater amount of cycling. The cycling results in larger elements in the structure matrix. Effects of changing $b_{1,2}$ on the direct flow vector E is shown in table 2g. Flows into compartments X_1 and X_2 are most sensitive while flows into X_4 are least sensitive. All compartments are affected because they are coupled in a large cycle.

The sensitivities of flows into compartments of the Tropical Rainforest Nitrogen model due to changes in the direct flow matrix B are shown in table 5. Total sensitivities for the entire model are also shown; total sensitivity with respect to changes in element b_{ij} is the sum of all elements in the sensitivity matrix S(i,j), whereas the sensitivities of inflows to compartment i is the sum of elements in column i of S(i,j). Modifications of $b_{5,1}$, $b_{5,3}$, $b_{4,1}$, $b_{3,1}$, and $b_{2,1}$ cause the greatest changes in the indirect flow matrix. When $b_{5,1}$, $b_{3,1}$, $b_{4,1}$ and $b_{2,1}$ are changed the largest effects occur for flows into compartments X_1 and X_2. When $b_{5,3}$ is changed the largest effects occur for flows into compartments X_3, X_4, and X_5. In general, the sensitivity values of the tropical rainforest model are 4-20x those in the Silver Springs model.

As with the Silver Springs model, one also can determine which changes need to be made in the B matrix cause a particular change in any element of the structure matrix. This problem is accomplished with equation 16. In the tropical rainforest nitrogen model, element $b_{5,3}$ and $b_{2,3}$ of B must be changed in order to generate changes in $b^{-1}_{1,2}$ (table 6a). Element $b_{2,3}$ and $b_{7,3}$ must be changed in order to cause changes in $b^{-1}_{3,2}$. Any changes in B should be balanced by modifying inputs and outputs so the column sums remain 1.0.

The flow matrix for the tropical rainforest model which has been normalized by rows (G') is shown in table 7a. The structure matrix $(I-G')^{-1}=B'^{-1}$, shown in table 7b demonstrates how flows can be traced to outputs rather than to inputs. Further discussion of this matrix can be found in Patten et al. (1976). The sensitivity values for elements of B'^{-1} obtained by modifying $b'_{2,1}$ are shown in table 7c. Flows from compartments X_1, X_2, and X_5 to compartments X_1, X_2, and X_3 are most sensitive to changes in this parameter.

The tropical rain forest model is more sensitive with respect to parameter change in the flow analysis than does the Silver Springs model. This result occurs because the elements of the structure matrix of the tropical rainforest are larger than in the Silver Springs structure matrix due to the greater amount of indirect flows. In general, energy models, which tend not to have cycles, are less sensitive to changes in flow than material models, which tend to have cycles. It can be shown that the structure

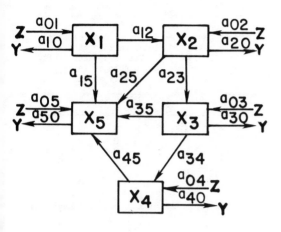

Figure 1.-- Diagram of energy storages and flows in Silver Springs energy model: X_1 = primary producers, X_2 = herbivores, X_3 = carnivores, X_4 = top carnivores, X_5 = decomposers and Z = inputs from the environment.

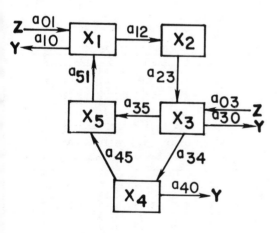

Figure 2.-- Diagram of nitrogen flows and storages in a tropical rainforest in Puerto Rico. X_1 = leaves, X_2 = loose litter, X_3 = fibrous roots, X_4 = mycorrhizae, X_5 = wood and Z = inputs from the environment.

Table 1.--Flow matrices for Silver Springs Energy Model. (a) Flow matrix showing direct flows ($kcal/m^2/yr$) to and from each compartment. (b) Vector showing direct flows ($kcal/m^2/yr$) into compartments. (c) Vector showing outputs ($kcal/m^2/yr$) from each compartment. (d) Flow matrix normalized by column. (e) Structure matrix $B^{-1}=(I-G)^{-1}$. (f) Matrix of sensitivity values for B^{-1} due to changes in $b_{1,2}$. (g) Vector for sensitivity values for E due to changes in $b_{1,2}$.

(a)
FLOW MATRIX FOR SILVER SPRINGS ENERGY MODEL

	X1	X2	X3	X4	X5	Z
X_1	0	2874	0	0	3455	0
X_2	0	0	382	0	1095	0
X_3	0	0	0	21	46	0
X_4	0	0	0	0	6	0
X_5	0	0	0	0	460	0
Z	20810	494	0	0	0	0

(b) DIRECT FLOW VECTOR, E

X1	20810
X2	3368
X3	382
X4	21
X5	5062
Z	21304

(c) OUTPUT VECTOR, R

X1	14481
X2	1891
X3	315
X4	15
X5	4602
Z	0

(d)
NORMALIZED FLOW MATRIX (BY COLUMN)

	X1	X2	X3	X4	X5	Z
X_1	0	0.853	0	0	0.683	0
X_2	0	0	1.00	0	0.216	0
X_3	0	0	0	1.00	0.009	0
X_4	0	0	0	0	0.001	0
X_5	0	0	0	0	0.090	0
Z	1.0	0.147	0	0	0	0

(e)
STRUCTURE MATRIX (B-1)

	X1	X2	X3	X4	X5	Z
X_1	1.00	0.853	0.853	0.853	0.963	0
X_2	0	1.00	1.00	1.00	0.249	0
X_3	0	0	1.00	1.00	0.011	0
X_4	0	0	0	1.00	0.001	0
X_5	0	0	0	0	1.100	0
Z	1.00	1.00	1.00	1.00	1.00	1.0

(f)
MATRIX DESCRIBING SENSITIVITY OF B^{-1} TO CHANGES IN $b_{1,2}$

	X1	X2	X3	X4	X5	Z
X_1	-0.85	-0.73	-0.73	-0.73	-0.82	0
X_2	-1.00	-0.85	-0.85	-0.85	-0.96	0
X_3	0	0	0	0	0	0
X_4	0	0	0	0	0	0
X_5	0	0	0	0	0	0
Z	0	0	0	0	0	0

(g)
VECTOR DESCRIBING SENSITIVITY OF E TO CHANGES IN $b_{1,2}$

X1	-17751
X2	-20810
X3	0
X4	0
X5	0
Z	-20810

Table 2.--Sensitivities of indirect flows into compartment X_i of the Silver Springs model due to changes in b_{ij}.

b_{ij}	X_1	X_2	X_3	X_4	X_5	Z	TOTAL
1,1	-2.0	-1.7	-1.7	-1.7	-1.9	0.0	- 9.0
1,2	0.0	-2.0	-2.0	-2.0	-0.5	0.0	- 6.5
1,3	0.0	0.0	-2.0	-2.0	0.0	0.0	- 4.0
1,4	0.0	0.0	0.0	-2.0	0.0	0.0	- 2.0
1,5	0.0	0.0	0.0	0.0	-2.2	0.0	- 2.2
1,z	-2.0	-2.0	-2.0	-2.0	-2.0	0.0	-10.0
2,1	-2.9	-2.4	-2.4	-2.4	-2.8	0.0	-12.9
2,2	0.0	-2.9	-2.9	-2.9	-0.7	0.0	- 0.0
2,3	0.0	0.0	-2.9	-2.9	0.0	0.0	- 5.7
2,4	0.0	0.0	0.0	-2.9	0.0	0.0	- 2.9
2,5	0.0	0.0	0.0	0.0	-3.1	0.0	- 3.1
2,z	-2.9	-2.9	-2.9	-2.9	-2.9	0.0	-14.3
3,1	-3.9	-3.3	-3.3	-3.3	-3.7	0.0	-17.4
3,2	0.0	-3.9	-3.9	-3.9	-1.0	0.0	-12.5
3,3	0.0	0.0	-3.9	-3.9	0.0	0.0	- 7.8
3,4	0.0	0.0	0.0	-3.9	0.0	0.0	- 3.9
3,5	0.0	0.0	0.0	0.0	-4.2	0.0	- 4.2
3,z	-3.9	-3.9	-3.9	-3.9	-3.9	0.0	- 0.0
4,1	-4.9	-4.1	-4.1	-4.1	-4.7	0.0	-22.0
4,2	0.0	-4.9	-4.9	-4.9	-1.2	0.0	-15.8
4,3	0.0	0.0	-4.9	-4.9	-0.1	0.0	- 9.8
4,4	0.0	0.0	0.0	-4.9	0.0	0.0	- 4.9
4,5	0.0	0.0	0.0	0.0	-5.3	0.0	- 5.3
4,z	-4.9	-4.9	-4.9	-4.9	-4.9	0.0	-24.3
5,1	-3.3	-2.8	-2.8	-2.8	-4.2	0.0	-15.0
5,2	0.0	-3.3	-3.3	-3.3	-0.8	0.0	-10.8
5,3	0.0	0.0	-3.3	-3.3	0.0	0.0	- 6.7
5,4	0.0	0.0	0.0	-3.3	0.0	0.0	- 3.3
5,5	0.0	0.0	0.0	0.0	-3.7	0.0	- 3.7
5,z	-3.3	-3.3	-3.3	-3.3	-3.3	0.0	-16.7
z,1	-1.0	-0.9	-0.9	-0.9	-1.0	0.0	- 4.5
z,2	0.0	-1.0	-1.0	-1.0	-0.3	0.0	- 3.3
z,3	0.0	0.0	-1.0	-1.0	-0.0	0.0	- 2.0
z,4	0.0	0.0	0.0	-1.0	-1.0	0.0	- 1.0
z,5	0.0	0.0	0.0	0.0	-1.1	0.0	- 1.1
z,z	0.0	0.0	0.0	0.0	0.0	0.0	0.0

Table 3.—(a) Elements of B matrix of the Silver Springs Energy model which must be modified to produce changes in $b^*_{1,2}$ alone (b) in $b^*_{3,2}$ alone.

(a)

	X_1	X_2	X_3	X_4	X_5	Z
X_1	0	0	0	0	0	0
X_2	0	0	0	0	0	0
X_3	0	0	0	0	0	0
X_4	0	0	0	0	0	0
X_5	0	0	-1.0	0	-0.22	0
Z	0	0	-1.0	0	-0.22	0

(b)

	X_1	X_2	X_3	X_4	X_5	Z
X1	0	0	0	0	0	0
X_2	0	0	-0.74	0	0	0
X_3	0	0	0	0	0	0
X_4	0	0	0	0	0	0
X_5	0	0	0	0	0	0
Z	0	0	-0.12	0	0	0

Table 4.--Flow matrices for Tropical Forest Nitrogen Model. (a) Flow matrix showing direct flows (g N/m^2/yr) to and from each compartment. (b) Vector showing direct flows (g N /m^2/yr) into compartments. (c) Vector showing outputs (g N /m^2/yr) from each compartment. (d) Flow matrix normalized by column. (e) Structure matrix $B^{-1}=(I-G)^{-1}$. (f). Matrix of sensitivity values for B^{-1} due to changes in $b_{1,2}$. (g) Vector for sensitivity values for E due to changes in $b_{1,2}$.

(a) FLOW MATRIX FOR TROPICAL RAIN FOREST NITROGEN MODEL

	X_1	X_2	X_3	X_4	X_5	Z
X_1	0	16	0	0	0	0
X_2	0	0	16	0	0	0
X_3	0	0	0	9.2	3.6	0
X_4	0	0	0	0	6.6	0
X_5	10.2	0	0	0	0	0
Z	7.6	0	2.7	0	0	0

(b) DIRECT FLOW VECTOR, E

X1	17.7
X2	16.0
X3	18.7
X4	9.2
X5	10.2
Z	0

c) OUTPUT VECTOR, R

X1	1.7
X2	0
X3	5.9
X4	2.6
X5	0
Z	0

(d) NORMALIZED FLOW MATRIX (BY COLUMN)

	X_1	X_2	X_3	X_4	X_5	Z
X_1	0	1.0	0	0	0	0
X_2	0	0	0.856	0	0	0
X_3	0	0	0	1.0	0.352	0
X_4	0	0	0	0	0.647	0
X_5	0.576	0	0	0	0	0
Z	0.424	0	0.144	0	0	0

(e) STRUCTURE MATRIX (B-1)

	X_1	X_2	X_3	X_4	X_5	Z
X_1	1.97	1.97	1.69	1.69	1.69	0
X_2	0.97	1.97	1.69	1.69	1.69	0
X_3	1.14	1.14	1.97	1.97	1.97	0
X_4	0.74	0.74	0.63	1.63	1.28	0
X_5	1.14	1.14	0.97	0.97	1.97	0
Z	1.00	1.00	1.00	1.00	1.00	1.00

(f) MATRIX DESCRIBING SENSITIVITY OF B-1 TO CHANGES IN $b_{1,2}$

	X_1	X_2	X_3	X_4	X_5	Z
X_1	-3.88	-3.88	-3.33	-3.33	-3.33	0
X_2	-0.94	-1.91	-1.64	-1.64	-1.64	0
X_3	-1.11	-2.25	-1.93	-1.93	-1.93	0
X_4	-0.72	-1.46	-1.25	-1.25	-1.25	0
X_5	-1.11	-2.25	-1.93	-1.93	-1.93	0
Z	-0.97	-1.97	-1.69	-1.69	-1.69	0

(g) VECTOR DESCRIBING SENSITIVITY OF E TO CHANGES IN $b_{1,2}$

X_1	-34.87
X_2	-34.87
X_3	-20.18
X_4	-13.10
X_5	-20.18
Z	-17.70

Table 5.--Sensitivities of indirect flows into compartment X_i of the Tropical Rain forest model due to changes in b_{ij}.

$b_{i,j}$	X_1	X_2	X_3	X_4	X_5	Z	TOTAL
1,1	-13.7	-13.7	-11.8	-11.8	-11.8	0.0	-62.7
1,2	- 6.8	-13.7	-11.8	-11.8	-11.8	0.0	-55.8
1,3	- 7.9	- 7.9	-13.7	-13.7	-13.7	0.0	-57.0
1,4	- 5.2	- 5.2	- 4.4	-11.4	- 9.6	0.0	-35.6
1,5	- 7.9	- 7.9	- 6.8	- 6.8	-13.7	0.0	-43.1
1,z	- 7.0	- 7.0	- 7.0	- 7.0	- 7.0	0.0	-41.8
2,1	- 15.7	-15.7	-13.5	-13.5	-13.5	0.0	-71.7
2,2	- 7.7	-15.7	-13.5	-13.5	-13.5	0.0	-63.8
2,3	- 9.1	- 9.1	-15.7	-15.7	-15.7	0.0	-65.2
2,4	- 5.9	- 5.9	- 5.0	-13.0	-11.0	0.0	-40.8
2,5	- 9.1	- 9.1	- 7.7	- 7.7	-15.7	0.0	-49.3
2,z	- 8.0	- 8.0	- 8.0	- 8.0	- 8.0	0.0	-47.8
3,1	-15.7	-15.7	-13.4	-13.4	-13.4	0.0	-71.6
3,2	- 7.7	-15.7	-13.4	-13.4	-13.4	0.0	-63.7
3,3	- 9.1	- 9.1	-15.7	-15.7	-14.7	0.0	-65.1
3,4	- 5.9	- 5.9	- 5.0	-13.0	-11.0	0.0	-40.7
3,5	- 9.1	- 9.1	- 7.7	- 7.7	-15.7	0.0	-49.2
3,z	- 8.0	- 8.0	- 8.0	- 8.0	- 8.0	0.0	-47.7
4,1	-17.5	-17.6	-15.1	-15.1	-15.1	0.0	-80.6
4,2	- 8.7	-17.6	-15.1	-15.1	-15.1	0.0	-71.7
4,3	-10.2	-10.2	-17.6	-17.6	-17.6	0.0	-73.3
4,4	- 6.6	- 6.6	- 5.6	-14.6	-12.4	0.0	-45.8
4,5	-10.2	-10.2	- 8.7	- 8.7	-17.6	0.0	-55.4
4,z	- 9.0	- 9.0	- 9.0	- 9.0	- 9.0	0.0	-53.7
5,1	-19.1	-19.1	-16.4	-16.4	-16.4	0.0	-87.4
5,2	- 9.4	-19.1	-16.4	-15.4	-16.4	0.0	-77.7
5,3	-11.1	-11.1	-19.1	-19.1	-19.1	0.0	-79.4
5,4	- 7.2	- 7.2	- 6.1	-15.8	-13.4	0.0	-49.7
5,5	-11.1	-11.1	- 9.4	- 9.4	-19.1	0.0	-60.0
5,z	- 9.7	- 9.7	- 9.7	- 9.7	- 9.7	0.0	-58.2
z,1	- 2.0	- 2.0	- 1.7	- 1.7	- 1.7	0.0	- 9.0
z,2	- 1.0	- 2.0	- 1.7	- 1.7	- 1.7	0.0	- 8.0
z,3	- 1.1	- 1.1	- 2.0	- 2.0	- 2.0	0.0	- 8.2
z,4	- 0.7	- 0.7	- 0.6	- 1.6	- 1.4	0.0	- 5.1
z,5	- 1.1	- 1.1	- 1.0	- 1.0	- 2.0	0.0	- 6.2
z,z	- 1.0	- 2.0	- 2.0	- 2.0	- 2.0	0.0	- 6.0

Table 6.--(a) Elements of B matrix of Tropical Rainforest model which must be modified to produce changes in $b_{1,2}$ alone (b) in $b_{3,2}$ alone.

(a)

	X_1	X_2	X_3	X_4	X_5	Z
X_1	0	0	0	0	0	0
X_2	0	0	0	0	0	0
X_3	0	0	0	0	0	0
X_4	0	0	0	0	0	0
X_5	0	0	-0.50	0	0	0
Z	0	0	-0.36	0	0	0

(b)

	X_1	X_2	X_3	X_4	X_5	Z
X_1	0	0	0	0	0	0
X_2	0	0	-0.74	0	0	0
X_3	0	0	0	0	0	0
X_4	0	0	0	0	0	0
X_5	0	0	0	0	0	0
Z	0	0	-0.12	0	0	0

Table 7.--Flow matrices for Tropical Rainforest Nitrogen model. (a) Flow matrix which has been normalized by rows rather than columns. (b) Structure matrix $B'^{-1} = (I-G')^{-1}$. (c) Matrix of sensitivity values for B'^{-1} due to changes in $b'_{1,2}$.

(a) NORMALIZED FLOW MATRIX FOR TROPICAL RAINFOREST (BY ROW)

	X_1	X_2	X_3	X_4	X_5	OUTPUT
X_1	0	0.904	0	0	0	0.096
X_2	0	0	1.0	0	0	0
X_3	0	0	0	0.49	0.193	0.315
X_4	0	0	0	0	0.717	0.283
X_5	1.0	0	0	0	0	0
OUTPUT	0	0	0	0	0	0

(b) STRUCTURE MATRIX (B'^{-1})

	X_1	X_2	X_3	X_4	X_5	OUTPUT
X_1	1.97	1.78	1.78	0.8	0.97	1
X_2	1.08	1.97	1.97	0.97	1.08	1
X_3	1.08	0.97	1.97	0.97	1.08	1
X_4	1.42	1.28	1.28	1.63	1.42	1
X_5	1.97	1.78	1.78	0.88	1.97	1
OUTPUT	0	0	0	0	0	1

(c) MATRIX DESCRIBING SENSITIVITY OF B'^{-1} TO CHANGES IN $b'_{1,2}$

	X_1	X_2	X_3	X_4	X_5	OUTPUT
X_1	-3.51	-3.17	-3.17	-1.42	-1.73	0
X_2	-3.88	-3.51	-3.51	-1.58	-1.91	0
X_3	-1.91	-1.73	-1.73	-0.78	-0.94	0
X_4	-2.52	-2.28	-2.28	-1.02	-1.24	0
X_5	-3.51	-3.17	-3.17	-1.42	-1.73	0
OUTPUT	0	0	0	0	0	0

matrix for a model without cycles has elements which are less than or equal to 1.0 (Patten et al. 1976). There is no such inherent limit to the structure matrix of models with cycles. As the efficiency of cycling increases, so does the sensitivity to perturbation. The amount of cycling is contained in the principal diagonal of the structure matrix (Finn 1977). Certain mathematicians have suggested that the trace (Tr A = sum of the principal diagonal elements = sum of eigenvalues) of a matrix be used as an indicator of the stability of a matrix with respect to parameter changes (Faddeev and Faddeeva 1963). Such a 'condition number' is useful for comparing the stability of matrices of the same size. A normalized trace (Tr A/n) can be used to compare matrices of different sizes. For matrices without cycles, the normalized trace is 1.0 For matrices with cycles the normalized trace would be larger than 1.0 indicating greater sensitivity to parameter change. The normalized trace of the Silver Springs structure matrix is 1.02 while the normalized trace of the Tropical rainforest structure matrix is 1.75. Other condition numbers which might be used are the Euclidean norm B^{-1} or the sum $\sum_{ij} b_{ij} b^*_{ij}$.

DISCUSSION

These techniques can be used to identify the most sensitive parts of a system with regard to the flow structure. Moreover, the techniques can provide an indication as to how stable the flow structure is when the direct flows between compartments are being changed. These results are often not intuitively obvious from examination of the flow structure, especially for large systems. Such sensitivity analyses are quicker than methods suggested by Hannon (1973) because they only involve terms of the matrices which have already been constructed for flow analysis. Results of these analyses suggest how extensively small perturbations are generated by sampling, harvesting, observation, and conscious manipulation of a system.

The stability of a structure matrix with respect to changes in the direct flow matrix is a matrix of the products of a column and a row in the structure matrix. If the elements are large, then the structure matrix is more sensitive to changes in the flow matrix. Cycles and loops in the system network tend to increase elements in the structure matrix because more of the indirect effects are propagated. In ecosystem and economic models, the presence of cycles will tend to destabilize the flow structure by increasing their sensitivity to change. This result contrasts with ecological principles which state that complexity, as evidenced by the number of components and pathways, increases stability. A system with much recycling, however, is capable of a greater range of behaviors than one without recycling, although it will be more difficult to control. Also, as long as redundant components and paths are added which do not form cycles, the flow structure remains relatively insensitive to perturbations. A sensitivity analysis, such as provided here indicates which parts of a system's structure should be modified in order to fine tune its behavior or which parts of a system need to be more heavily sampled.

CONCLUSION

Ecosystems are dynamic, self-organizing objects which are immersed in a dynamic environment. Changes in the environment or within the ecosystem cause changes in the flow structure of the ecosystem network. Such changes may be regarded as altering inputs or the parameters governing flows between compartments. Changes in inputs and outputs can be interpreted as environmental variation, manipulation and observation by an observer, internal readjustments due to species replacement or genetic change, or human impact. The effects of these changes are propagated through the network where they may be amplified or attenuated. The sensitivity technique examined here allows one the ability to assess the indirect and direct effects of small changes in the flow structure. This analysis assumes that the ecosystem has a constant structure, so the effects of traumatic modification which alters the presence and absence of flow paths cannot be examined. The most sensitive parts of a network can be identified. Such techniques relate to ecosystem stability; here, stability will be defined as how much the direct and indirect flows will change after inputs or parameters have been modified. The magnitude of the trace of the structure matrix which reflects the amount of cycling and indirect effects (Finn 1976a) provides a good estimate of the magnitude of the subsequent change in the flow network. The results of this sensitivity analysis can be applied quickly to any system in which an input-output analysis has been done. Such an analysis can provide better perspectives for how to manage a network of energy and material flows and storages.

LITERATURE CITED

Auble, G. T., J. T. Finn, D. B. Hamilton, J. Ojasti, and D. Rodman. 1973. Dynamics of nitrogen in a tropical rain forest. Unpublished manuscript. University of Georgia, Athens.

Barber, M. C. 1978a. A Markovian model of ecosystem flow analysis. Ecol. Model. 5:193-206.

Barber, M. C. 1978b. A retrospective Markovian model for ecosystem resource flow. Ecol. Model 5:125-120.

Costanza, R. 1981. Embodied energy and economic Valuation. Science 210:1219-1224.

Edmisten, J. A. 1970a. Soil studies in the El Verde rain forest. p. H79-87. In H. T. Odum, and R. F. Pigeons (Eds.). A Tropical Rainforest. U. S. Atomic Energy Commission. Oak Ridge Tennessee.

Edmisten, J. A. 1970b. Preliminary studies of the nitrogen budget of a tropical rain forest. p. H211-215. In H. T. Odum, and R. F. Pigeons (Eds.). A Tropical Rainforest. U. S. Atomic Energy Commission. Oak Ridge Tennessee.

Faddeev, D. K. and V. N. Faddeeva. 1963. Computational Methods of Linear Algebra. 621 p. W. H. Freeman and Company, San Francisco.

Finn, J. T. 1976a. Cycling Index: a general measure of cycling in compartmental models. p. 138-164. In D. C. Adriana and I. L. Brisbin (Eds.), Environmental Chemistry and Cycling Processes Symposium, U. S. Energy Research and Development Administration, Washington, D. C.

Finn, J. T. 1976b. Measure of ecosystem structure and function derived from analysis of flows. J. Theoret. Biol. 56:363-380.

Hannon, B. 1973. The structure of ecosystems. J. Theoret. Biol. 41:535-546.

Leontief, W. W. 1966. Input-Output Economics. Oxford University Press, London and New York.

Noble, B. 1969. Applied Linear Algebra. Prentice Hall, Inc., Englewood Cliffs, New Jersey.

Odum, H. T. 1957. Trophic structure and productivity of Silver Springs, Florida. Ecol. Mono., 27:55-112.

Patten, B. C., R. W. Bosserman, J. T. Finn, and W. G. Cale. 1976. Propagation of cause in ecosystems. p. 457-579. In B. C. Patten (Ed.) Systems Analysis and Simulation in Ecology, Vol. III. Academic Press, New York.

THE ENERGY EMBODIED IN THE PRODUCTS OF ECOLOGICAL SYSTEMS: A LINEAR PROGRAMMING APPROACH[1]

Robert Costanza and Christopher Neill[2]

Abstract.--Hannon (1973, 1976, 1979), Herendeen (1980), Richey et al. (1978) and others have applied input-output techniques to the analysis of energy and matter flows in ecosystems. These applications are important contributions to our understanding of interdependence in ecological systems, but they suffer from (1) reliance on "single commodity" flow matrices (i.e., enthalpy or carbon) and (2) they do not address the problem of joint products, which is particularly significant for ecological applications.

In this paper a general linear programming model is developed and used to describe and analyze interdependence in multicommodity ecosystems with joint products. The model is applied to data for Silver Springs and embodied energy intensities (that are analogous to the economist's shadow prices) for the ecosystem products are obtained. The effects of different assumptions concerning the treatment of waste heat and joint products (on the energy intensities) are also discussed, as are potential applications to resource management problems.

INTRODUCTION

The concept of value plays a major role in both ecology and economics. Both fields are involved in the study of the complex interdependence between essentially dissimilar components. The concept of value allows comparisons to be made between these dissimilar components. The "valuation problem" is the same for any system: What weighting or conversion factors can we use to compare essentially dissimilar, noncommensurable items, such as cars, swamps, and eagles? Is there a common denominator linking all systems?

The weighting factors in modern economic systems are prices. Several alternatives have been proposed as the analog of prices for ecological systems. The most common is calorimetric energy content (Odum 1971). Any proposed valuation or weighting system must satisfy the basic accounting constraint: the sum of the values (weighting factors times quantities) of inputs must equal the sum of values of outputs for all transformation processes. This value conservation constraint holds in dollars for economic entities. It also holds for calorimetric energy (by the first law of thermodynamics) and for mass (more accurately for the total mass and energy). This is one reason for the popularity of first law accounting in ecology, but it does not guarantee a "value" solution analogous with prices. The problem with first law accounting is the existence of the second law of thermodynamics, which tells us that the "usefulness" or availability of calorimetric energy decreases in any transformation. As Georgescu-Roegen (1971) has pointed out, low entropy matter and energy are necessary for economic value. Therefore, a second law accounting is more appropriate for ecological

---

[1] Paper presented at the international symposium on Energy and Ecological Modelling, sponsored by the International Society for Ecological Modelling. (Louisville, Kentucky, April 20-23, 1981).

[2] Assistant Professor and Research Associate, respectively, Coastal Ecology Laboratory, Center for Wetland Resources, Louisiana State University, Baton Rouge, Louisiana 70803.

systems, since it allows value calculations
that take into account usefulness or availability. It is the purpose of this paper to
extend the development of such an accounting.

The standard definition of entropy applies
only to reversible transformations and is usually
limited to "simple" gradients (such as temperature
and chemical concentration gradients), while the
real world is characterized by irreversible
transformations and complex gradients (such as
plants, animals, and cities). One approach to
estimating the degree of organization in real,
irreversible systems is to calculate the calori-
metric energy required <u>directly</u> <u>and</u> <u>indirectly</u>
to produce low entropy structure. This "embodied"
energy serves as an index of usefulness in the
same sense that measuring the energy required to
pump water into a reservoir is an index of the
available energy stored in the elevated water.
Embodied energy has also been shown to correlate
well with economic value given appropriate
assumptions (Costanza 1980). Using this approach
we can theoretically calculate the energy embodied
in plants, animals, soil, water, and all other
ecological goods and services once we know the
detailed web of transformations which lead to
their production.

This complex calculation can be systematized
by using matrices to represent the transformations,
and the well-known mathematical techniques of
input-output analysis to solve them. This has
been attempted by Hannon (1973, 1976, 1979) for
selected ecosystems but, as Herendeen (1980)
points out, the results were somewhat confusing.
We will attempt to show that Hannon was really
performing a first law accounting and will go on to
demonstrate how a second law accounting that
allows for irreversible processes and joint
products may be performed using the Silver
Springs ecosystem data originally published by
Odum (1957).

Joint Products

Joint products refer to the interdependent
by-products of most real transformation processes.
For the purposes of discussion we can divide all
production processes into two fundamental types:
sorting and assembly. Sorting is a fundamental
anti-entropic activity involving the splitting
of a mixed substance into its components;
assembly involves combining inputs into an
organized product with lower entropy. If all
production processes of interest were of the
pure assembly type, then joint products might not
present a problem, but the existence of sorting
implies that joint products are unavoidable.
All industries in the U.S. economy produce joint
products. The mining industry, for example,
produces refined minerals but also spent ore.
The pulp produced by the timber industry or the
sulfur removed from crude oil refining are examples
of joint products which are marketable. Other joint
products are considered waste such as air and
water pollution or spent fuel from nuclear power
plants, and have no market value.

All ecological processes also produce joint
products. Trees produce oxygen, water vapor, and
waste heat as well as tree biomass. (Waste heat
is the most ubiquitous of joint products and
requires special treatment.) Although joint
products seem to reflect the physical nature of
economic and ecological processes (it is impossible
to produce iron without also producing spent iron
ore or to produce a tree without producing
oxygen), they complicate mathematical description
and analysis and have often been ignored.

Waste heat is an unavoidable by-product of
all transformation processes. To understand its
special role we must define a third fundamental
process--dissipation of an existing gradient.
Dissipation is not a production process (since it
proceeds spontaneously in the direction increasing
entropy) but it is required to drive all production
process (which produce low entropy structures).
Waste heat is a necessary by-product of the
dissipation of a temperature gradient. The
second law of thermodynamics requires that the
amount of dissipation be greater than or equal to
the amount of sorting and assembly for irreversible
processes in an islolated system. In order to
sort or assemble at a finite rate, we must "consume"
or "dissipate" fuel, capital, labor etc. by
allowing these gradients to dissipate in a controlled way at a finite rate. (Cf., Odum and
Pinkerton 1955, Prigogine <u>et al</u>. 1977).

We can divide joint products into two
distinct groups depending on their origin as the
result of sorting or dissipation. This distinction
is critical because the products of a purely
dissipative process have no further ability to
perform work and thus no available embodied
energy and no "value," while the products of
sorting do.

For example, a temperature gradient could be
allowed to dissipate without doing any work, in
which case the only product would be uniform
heat. The available energy embodied in the
original gradient would dissapear. On the other
hand, we could insert an engine between the high
temperature source and the low temperature sink,
and the available energy embodied in the original
temperature gradient could be used to produce
another gradient (by sorting or assembly), and we
could say that the dissipation of the original
gradient was a necessary cost of production.
Thus, the original gradient could be said to be
"embodied" in the products of the engine.

In real, irreversible systems, gradients are
frequently not dissipated completely. For
example, the exhaust from an automobile engine is
above environmental temperatures and thus still
has some available energy. It is immediately
dissipated and lost without doing any work when
the exhaust is released to the environment.
Since this is a necessary cost of driving the
production process at a finite rate, we might
conclude that the energy embodied in the entire
gradient should be embodied in the outputs.
Alternatively, we could reason that only the

gradient between the engine combustion temperature and the exhaust temperature be embodied in the outputs. We will discuss the implications of these alternatives on the resulting calculations.

METHODS

The problem is to trace the flow of a primary resource (energy) through a complex web of multicommodity biological and chemical transformations to determine the amount required directly and indirectly to produce each of the components of an ecological system. This problem is analogous to the study of interdependence and the imputability of value of primary factors in economics. The economic problem was first explicitly formulated by Walras (1896) and empirical applications began with Leontief's (1941) Nobel prize-winning work. Leontief's approach has come to be known as input-output (I-O) analysis. I-O analysis has been applied to the study of energy and material flow in ecosystems (Hannon 1973, 1976, 1979, Herendeen 1980, Richey et al. 1978) and in economic systems (Bullard and Herendeen 1975, Costanza 1980) and we refer the reader to these works for detailed descriptions of the technique. I-O analysis can be shown to be a special case of the more general Linear Programming (LP) analysis (Dorfman et al. 1958). I-O analysis results when two simplifying assumptions are imposed on the LP model:

1) No joint products

2) No alternative production technologies (i.e., there is a unique combination of inputs for each output)

Since it is just these two assumptions which we wish to relax, the LP approach provides the appropriate analytical framework. We will proceed by developing the LP model, noting correspondence with the I-O model where appropriate.

A general-equilibrium model of an economy which allows joint products and alternative production technologies was first proposed by von Neumann (1945).

In the von Neumann model there are m commodities (denoted by C_1, C_2, ..., C_m) and n processes or activities for transforming commodities (denoted by Q_1, Q_2, ..., Q_n). Each transformation process can be indicated by its input commodity requirements and its jointly produced outputs, for example for process j:

$$Q_j = \begin{pmatrix} A_{1j}, A_{2j}, \ldots, A_{nj} \\ B_{1j}, B_{2j}, \ldots, B_{nj} \end{pmatrix}, \text{ or } (A_{ij}; B_{ij})$$

$$i = 1, m; \; j = 1, n$$

where A_{ij} = the input of commodity i to process j

B_{ij} = the output of commodity i from process j

and one discrete time period or accounting unit is implied. Figure 1 illustrates this general production relation.

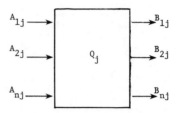

Figure 1.--General production relation.

In the Leontief system there would be only one B_{ij} for each process (j) and each commodity (i) would correspond to one process ($B_{ij} = 0$ if $i \neq j$).

We will assume a linear transformation process, or one where:

$$\begin{pmatrix} \lambda A_{1j}, \lambda A_{2j}, \ldots, \lambda A_{nj} \\ \lambda B_{1j}, \lambda B_{2j}, \ldots, \lambda B_{nj} \end{pmatrix} \text{ or } (\lambda A_{ij}; \lambda B_{ij})$$

holds for all λ.*

Next we set up a system of m linear equations (one for each commodity) which balance the total input of each commodity to all processes against the total output of each commodity from all processes. Note that m does not necessarily equal n, that is, there can be a different number of commodities and processes. Thus we have:

$$A_{11} + A_{12} + \cdots + A_{1n} + Y_1 = B_{11} + B_{12} + \cdots + B_{1n} + Z_1$$
$$A_{21} + A_{22} + \cdots + A_{2n} + Y_2 = B_{21} + B_{22} + \cdots + B_{2n} + Z_2$$
$$\vdots$$
$$A_{m1} + A_{m2} + \cdots + A_{mn} + Y_m = B_{m1} + B_{m2} + \cdots + B_{mn} + Z_m$$

(1)

---

*In practice we require that λ be small for the static equilibrium assumption. For dynamic applications we would require a more complex functional relation (probably nonlinear) relating inputs to outputs. As long as we limit applications to static accounting applications the linearity assumption is acceptable, however.

An "open" system allows for imports (commodities which are inputs from outside the system boundaries) and exports (commodities which are outputs to outside the system boundaries). We denote imports and exports of commodity i by Z_i and Y_i respectively and allow for and open system. Any change in storage of a commodity (net accumulation or depletion) is included as an "export." Also, depreciation of an already produced commodity (as distinct from losses in the process of production) is considered to be an "export" since it represents a loss of the commodity to the system. This is an important distinction, since it implies that at steady state the net input to the system is just enough to maintain all the internal stocks of commodities against depreciation.

The left side of each equation in (1) represents the demand for each commodity (as the sum of input requirements for all processes plus exports or "final demand") while the right side represents the supply of each commodity (from internal production plus imports). Note, for comparison, that in the Leontief model the right hand side of each equation is set equal to the total output of each commodity, and supply is implicitly assumed to meet or exceed demand.

Next we define two sets of technical coefficients:

$$a_{ij} = \frac{A_{ij}}{X_j} =$$ gross input coefficient, the amount of commodity i required as input for process j per unit activity of process j. These are the Leontief technical coefficients. (2)

$$b_{ij} = \frac{B_{ij}}{X_j} =$$ gross output coefficient, the amount of commodity i produced as output from process j per unit activity of process j. (3)

X_j = some measure of the activity level of process j.*

Thus we can rewrite (1) as

$$a_{11}X_1 + a_{12}X_2 + \cdots + a_{1n}X_n + Y_1 = b_{11}X_1 + b_{12}X_2 + \cdots + b_{1n}X_n + Z_1$$
$$a_{21}X_1 + a_{22}X_2 + \cdots + a_{2n}X_n + Y_2 = b_{21}X_1 + b_{22}X_2 + \cdots + b_{2n}X_n + Z_2$$
$$\vdots$$
$$a_{m1}X_1 + a_{m2}X_2 + \cdots + a_{mn}X_n + Y_m = b_{m1}X_1 + b_{m2}X_2 + \cdots + b_{mn}X_n + Z_m \quad (4)$$

in matrix notation we can indicate (4) as

$$\underline{a}X + \underline{Y} = \underline{b}X + \underline{Z} \quad (5)$$

or

$$(\underline{b} - \underline{a})X = (\underline{Y} - \underline{Z}) \quad (6)$$

where $\underline{a}$ is a n x m matrix of gross input coefficients

$\underline{b}$ is a n x m matrix of gross output coefficient

$\underline{X}$ is a m dimensional column vector of process activity levels

$\underline{Z}$ is a n dimensional column vector of commodity imports

$\underline{Y}$ is a n dimensional column vector of commodity exports (including "exports" or losses due to depreciation and any net accumulation over the time interval)

Since we do not necessarily have equal numbers of commodities and processes or a one to one correspondence between commodities and processes, the above system of equations is not necessarily solvable using standard linear algebra manipulations and is best handled as a linear programming (LP) problem. Certain assumptions can be imposed to allow input - output models to handle joint products but these either (1) can lead to negative relative weights or (2) require that we know the relative weights a priori, both of which are unacceptable for our purposes. The former (commodity-technology assumption) assumes that joint products require the same inputs when produced jointly as they do when produced as the sole output from a process. Thus, we simply transfer the inputs required for the sole output process away from the joint output process. This can lead to "negative inputs," and negative relative weights which are nonsense. The second assumption (the market shares assumption) assigns inputs to joint products based on the percentage of the total value of the output that each joint product represents (Ritz 1979). To arrive at this percentage (market share) we must already know the relative weights to apply to the joint outputs. For economic applications we can use market prices to determine market shares, but in ecological examples it is just these relative weights that we are trying to calculate and we cannot use them a priori to obtain a solution.

---

*The appropriate activity measure in I-O analysis is the total output of the one commodity produced by each sector. In the LP model the appropriate activity measure is not obvious and there are several possibilities. We could choose a "main" product output, multiply all the outputs by weighting factors and add, etc. In our examples we use the first alternative but others are certainly possible.

In the LP framework the dual of the above problem is of considerable importance since it can be used to determine the "shadow prices" or implied valuations for ecological commodities. We can formulate the generalized von Neumann model as a classic primal/dual LP pair (Dorfmann et al. 1958), with objective functions consistent with (but not identical to) Hannon's (1979) hypothesis*, and also with Oster and Wilson's (1978) hypothesis of "maximum ergonomic efficency" and Odum's (1971) "maximum power" hypothesis.

Primal

minimize $P = \Pi X$

(primary resource cost of all processes)

subject to $(\underline{b} - \underline{a})\underline{X} \geq (\underline{Y} - \underline{Z})$

$$(7)$$

(materials balance constraints one for each commodity)

and $\underline{X} \geq 0$

(non-negativity constraints)

Dual

maximize $p' = \underline{\varepsilon} (\underline{Y} - \underline{Z})$

(value of net output)

subject to $\underline{\varepsilon} (\underline{b} - \underline{a}) \leq \Pi$

$$(8)$$

(value balance constraints one for each process)

$\underline{\varepsilon} \geq 0$

(non-negativity constraints)

where $\underline{a}$ is a n x m matrix of commodity inputs per unit activity level
$\underline{b}$ is a n x m matrix of commodity outputs per unit activity level
$\underline{Y}$ is a n x 1 vector of commodity exports
$\underline{Z}$ is a n x 1 vector of commodity imports
$\underline{X}$ is a m x 1 vector of process activity levels
Π is a m x 1 vector of net primary resource inputs by process, per unit activity level (value added)
$\underline{\varepsilon}$ is a n x 1 vector of relative weights or intensities (shadow prices) per unit of commodity

*Hannon (1979) states: "In brief, the hypotheses are that while the components of an ecosystem strive to maximize their total direct and indirect energy storage within the constraints of their production characteristics, the overall system strives to minimize the metabolized energy per unit of stored biomass energy," (p. 271).

For ecological systems the primary (exogenous) resources are direct solar energy and the solar energy embodied in other primary inputs (i.e., commodities which are not also produced inside the system). The $\underline{\varepsilon}$ vector can be thus interpreted as the direct and indirect energy cost (embodied energy) per unit commodity with allowances for joint products and alternative production technologies. The system adjusts the process activity levels ($\underline{X}$) and the relative weights or embodied energy intensities ($\underline{\varepsilon}$) in order to minimize the total primary resource (energy) cost of production, which also maximizes the total value (relative weights times quantities) of the net exports from the system.

RESULTS

To demonstrate the LP model, we employ the same example system as Hannon (1973, 1976, 1979) and Herendeen (1980). Data on biomass and nitrogen flows for the Siver Springs, Florida ecosystem were obtained originally by Odum (1957).

Tables 1A and 1B show the flow data for Siver Springs. Inputs and outputs of commodities (listed along the left of the table) to and from processes (along the top of the table) are shown. Export of biomass is divided proportionally among all biomass commodities, not allocated solely to producers, as in Herendeen (1980). The flow network is shown diagramatically in Figure 2. Each commodity in the tables and figure is kept separate and we seek the conversion or weighting factors which would allow us to compare them.

The tables and figures correspond to the mathematical description outlined earlier. Table 1A lists the demand or use of each ecological commodity as the sum of inputs to internal processes and export. For example, of the total 2082 $g \cdot m^{-2} \cdot y^{-1}$ plant biomass available for use, 748.4 $g \cdot m^{-2} \cdot y^{-1}$ was used as input to herbivores, 874.2 $g \cdot m^{-2} \cdot y^{-1}$ was consumed by decomposers and 460.3 $g \cdot m^{-2} \cdot y^{-1}$ was exported. Table 1B shows the supply characteristics of each commodity in the system as the sum of internal production and imports. For example, of the total 4537.3 $gN \cdot m^{-2} \cdot y^{-1}$, 70.3 $gN \cdot m^{-2} \cdot y^{-1}$ was produced internally by decomposers and 4467.0 $gN \cdot m^{-2} \cdot y^{-1}$ was imported into the system in the inflowing water.

In Figure 2 each process is represented by a box and each commodity by a circle. Looking at the inputs and outputs to a circle is equivalent to looking at the corresponding commodity rows in Table 1A and 1B. Looking at a box in the figure is equivalent to looking at the inputs and outputs from the corresponding process column in Tables 1A and 1B. The system receives a solar energy input of 410,000 $Kcal \cdot m^{-2} \cdot yr^{-1}$. The input of 120 $g \cdot m^{-2} \cdot yr^{-1}$ of bread thrown into the system by tourists was considered a net import of plant biomass.

Table 1A. Input (use) matrix (A) for Silver Spring, Florida ecosystem ($\cdot m^{-2} \cdot yr^{-1}$)

	Primary producers	Herbivore	Carnivores	Top carnivores	Decomposers	Export	Total demand
Plant biomass (g)	0	748.4	0	0	874.2	460.3	2082.9
Herbivore biomass (g)	0	0	85.1	0	170.7	72.6	328.4
Carnivore biomass (g)	0	0	0	4.7	6.9	3.3	14.9
Top carnivore (g) −Decomposer biomass	0	0	0	0	80.6	22.9	103.5
Nitrogen (g)	121.7[1]	0	0	0	0	4415.6[2]	4537.3
Heat (kilocalories)	0	0	0	0	0	407,986	407,986
Sunlight (kilocalories)	410,000	0	0	0	0	0	0

[1] Nitrogen comprises 6.2 percent of plant biomass by weight. Net primary production in 2,082.9 $g \cdot m^{-2} \cdot yr$

[2] Net ouput of N obtained by difference (Total input − uptake).

Table 1B. Output (make) matrix (B) for Silver Springs, Florida ecosystem ($\cdot m^{-2} \cdot yr^{-1}$)

	Primary producers	Herbivore	Carnivores	Top Carnivores	Decomposers	Import	Total supply
Plant biomass (g)	1962.9	0	0	0	0	120	2,082.9
Herbivore biomass (g)	0	328.4	0	0	0	0	328.4
Carnivore biomass (g)	0	0	14.9	0	0	0	14.9
Top carnivore (g) −Decomposer biomass	0	0	0	1.3	102.2	0	103.5
Nitrogen (g)	0	0	0	0	70.3[3]	4,467.0[4]	4,537.3
Respiratory heat (kilocalories)	40,1.167	1,890	316	13	4,600	0	407,986

[3] Nitrogen produced by decomposers is 6.2 percent of biomass input to that sector. Assume all N in decomposing biomasss is regenerated.

[4] 3.395 x 10^8 $g \cdot yr^{-1}$ inflow in spring water distributed over an area of 7.6 x 10^4 m^2.

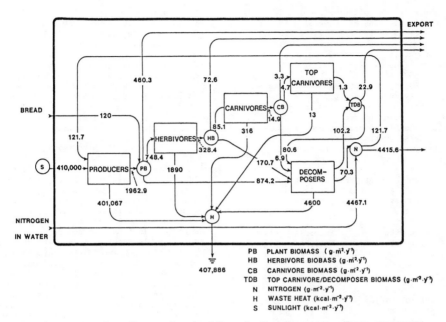

Figure 2.   Transactions diagram for the Silver Springs ecosystem.   Boxes represent
production processes or transformations.   Circles represent commodities.

Table 2.   Energy intensities (in Kcal/g) for the Silver Springs ecosystem for various values
of the quality factor for waste heat

Quality factor for waste heat*	Plant biomass	Herbivore biomass	Carnivore biomass	Top Carnivore Decomposer Biomass	Nutrients
0.0	297.05	676.95	3,866.32	13,978.24	1,422.20
0.01	294.16	670.44	3,829.35	13,844.66	1,408.60
.0428	284.68	694.02	3,707.73	13,405.31	1,363.91
0.1	268.17	611.71	3,495.85	12,639,84	1,286.07
0.5	152.67	350.80	2,014.19	7,287.07	741.78
1.0	4.50	4.50	4.50	4.53	0.0

*This is operationally equivalent to Herendeen's (1980) α.

The essence of the valuation problem is finding a set of weighting or conversion factors which would allow us to add and balance process inputs and outputs (boxes and columns) while maintaining our ability to add and balance commodities (rows and circles).

Table 1B and figure 2 show waste heat as a secondary joint product of all processes. Decomposers produce their own biomass, nutrients and waste heat as joint products. Other joint products such as other nutrients, or even non-material "services" such as feeding or nesting sites may easily be included in the model as data permit.

The treatment of waste heat is critical since it determines whether a first law, a reversible second law, or a irreversible second law analysis results.

Table 2 shows the energy intensities (ϵ's) which are analogous to shadow prices, that result from application of the LP Model.[1] The

---

[1]The ZX3LP subroutine of the IMSL Library was used to perform the analysis on the IBM 370/3033 at LSU. The subroutine uses a revised simplex technique. While the nonnegativty constraints in the LP model prevent negative energy intensities, and unequal number of commodities and process invariably leads to either zero activity levels for some processes or zero energy intensities. These results are interpretable within the theory of linear programming as indicating an uncompetive process or the existence of "slack" or unused capacity in the system (cf. Dorfman, et al 1958). In our example we avoided this result by aggregating top carnivore and decomposer biomass (since they are both at the "top" of the food chain). This allowed an equal number of commodities (4 biomass types and nitrogen) and processes (5 trophic levels) and allowed positive non-zero activity levels and energy intensities for all components. Note that an equal number of commodities and processes does not guarantee all non-zero activity levels and energy intensities. The structure of the data must be such that all processes are included in the optimal solution in order to obtain all non-zero results. This is particularly important in the analysis of joint products since unless production is distributed properly the program "decides" that all system demands for a particular commodity can be met with one process. For example, in the Silver Springs system if too much top carnivore/decomposer biomass (TDB) is produced by decomposers (relative to top carnivores) then the optimization model decides it doesn't "need" the top carnivore process and assigns it a zero activity level. See Costanza and Neill (elsewhere in this volume) for a more complete discussion of the conditions for all positive non-zero activity levels and energy intensities.

quality factor or "value" of waste heat was varied over the range from 0 to 1.[2] A quality factor of 1 for waste heat implies a first law accounting since no distinction is made between dissipation and production. A quality factor of 0 for waste heat implies a reversible second law accounting since heat outflow is assumed to be completely dissipated and thus valueless. An intermediate quality factor implies an irreversible second law accounting since waste heat is assumed to be incompletely dissipated and to still contain some available energy and some "value."

DISCUSSION

This paper demonstrates application of linear programming techniques to the valuation problem in ecological systems. Joint products, which are unavoidable in any system that contains sorting or dissipation activity, can be dealt with in the LP format, but simply casting the data in this formal structure does not guarantee intelligible results. The distinction between dissipation (change in structure in the direction of increasing entropy) and production (sorting and assembly, or change in structure in the direction of decreasing entropy) is critical. If dissipation products (such as waste heat) are not distinguished from production products then a first law accounting results. If they are distinguished, then a more useful second law accounting results. Hannon (1973, 1976, 1979), following established conventions in ecology, practiced the former and his constant energy intensities are therefore to be expected. Herendeen (1980) attempted to draw the distinction between dissipation and production by eliminating respiratory heat flow from the production matrix (see also Ulanowicz, 1972). His results show increasing energy intensities moving up the food chain as we might expect in a linear production sequence. Our results extend Herendeen's analysis by allowing for non-dissipative joint products. As a first step we include nitrogen cycling through the system. Contrary to Herendeen's expectation the inclusion of nitrogen did not tend to decrease the variance in energy intensities but rather increased it.

Herendeen's α factor is operationally equivalent to the quality factor for waste heat. A value of $\alpha = 1$ implies a first law acounting and energy intensities equal to the bomb calorimeter values for the commodities. A value of $\alpha = 0$ implies a reversible second law accounting. An intermediate value for α may be more appropriate for real irreversible systems since it implies that the dissipation products are not completely degraded or that all the dissipated potential is not embodied in the products.

When waste heat is the main dissipation product the Carnot ratio using the source-sink temperature gradient may be an appropriate independent measure of α. For example, we might define an index of α as:

---

[2]Different quality factors for heat were handled by converting it to solar equivalents (by multiplying by the chosen α) and decreasing the input, by that amount.

$$\alpha' = 1 - \left(\frac{T_2 - T_1}{T_2}\right) \qquad (9)$$

where T_2 = temperature of the source

          T_1 = temperature of the sink

The implication of this formulation is that waste heat (at sink temperature T_1) still has work potential relative to the universe which is at a still lower temperature, ($\sim 3°K$). For our Silver Springs example, we calculated a value for α' based on $T_2 = 7,000°K$ (effective temperature of sunlight) and $T_1 = 300°K$ (average temperature of the earth) as

$$\alpha' = 1 - \left(\frac{7000 - 300}{7000}\right) = .0428$$

The energy intensities based on this α value are included in Table 2. These show an average variation from the intensities calculated with $\alpha = 0$ of 0.96%. Thus, assuming that $\alpha = 0$ does not appear to lead to significant numerical error in this example since more than 95% of the available energy in sunlight is degraded by the time it emerges as 300°K waste heat.

The problem with this formulation is that Carnot ratios are based on reversible (infinitely slow) reactions. The question of what percentage of the dissipated gradient is embodied in the products is still not adequately addressed in this formulation (100% embodiment is assumed implicity). Odum and Pinkerton (1955) have suggested and defended with theoretical arguments an embodyment ratio closer to 50%. In our example, an $\alpha = .5$ would lead to energy intensities with an average variation of 48.1% compared to $\alpha = 0$.

Inspection of Table 2 shows that, except for α values close to 0 (second law case), the numerical value of α merely changes the numerical values of the energy intensities and not their distribution. Thus, questions about relative energy intensities are not as dependent on the choice of α as are questions of absolute energy intensities. This may prove useful in resource management applications, since we could "scale" the model using known resource prices without changing the relative energy intensities significantly. It also means that the choice of α does not significantly affect conclusions about an embodied energy theory of economic value (Costanza 1980) since changing α affects mainly the absolute, not the relative, energy intensities.

Resource Management Applications

Input-output type models which include ecological transformations have been suggested as potentially useful to determine "shadow prices" for non-marketed ecological commodities (Daly 1968, Isard 1972, Victor 1972, Ayres

1978, Costanza 1980). This approach to shadow pricing has several advantages over the currently more popular "willingness-to-pay" formulations (Freeman 1979). Most importantly, one does not have to interject humans into transactions in which they play no "direct" role and therefore exhibit irrelevant preferences. For example, people's willingness-to-pay for detritus is not meaningful since they do not use it directly. We must either survey detritivores about their preferences or carry the chain of production relations forward to the point where economicaly important commodities with well defined markets are produced.

Linear programming models represent a well developed, comprehensive way of detailing a complex set of production relations. As we have shown, they are applicable to ecological systems with joint products. The "shadow prices" which they produce are also interpretable as the direct and indirect primary resource cost of each product. Since sunlight is the major net input to the biosphere we may be able to use embodied solar energy as a common denominator linking ecological and economic systems (Costanza 1980, see also Costanza and Neill elsewhere in this volume). Such a common denominator would obviously be very useful for a broad range of resource management applications.

For realistic applications we would require detailed transactions tables for the ecological systems of the U.S. and the world. While these tables would always be inaccurate and incomplete (just as economic I-O tables always are) they would at least serve as a framework in which to summarize the extent of our current knowledge on ecological transactions. They would also be valuable in pinpointing areas for further research while providing "best estimates" for embodied solar energy intensities.

ACKNOWLEDGMENTS

This work was supported, in part, by NSF Project No. PRA-8003845 titled "Use of embodied energy values in the analysis of environment/ energy/economy trade-off decisions: an exploratory analysis. M. J. Lavine and R. Costanza Principal Investigators. We wish to thank B. Hannon, R. Herendeen, R. Ulanowicz and R. E. Turner for reviewing earlier drafts and providing useful comments. All remaining errors and omissions are, of course, the responsibility of the authors. This is contribution #LSU-CEL-81-06 from the Coastal Ecology Laboratory, Louisiana State University, Baton Rouge, LA 70803.

LITERATURE CITED

Ayres, R. N. 1978. Resources, environment, and economics: applications of the materials/ energy balance principle. Wiley Interscience, N. Y. 297 pp.

670

Bullard, C. W. and R. A. Herendeen. 1975. The energy costs of goods and services. Energy Policy. Dec. 1975; pp. 268–278.

Costanza, R. 1980. Embodied energy and economic valuation. Science 210:1219–1224.

Daly, H. E. 1968. On economics as a life science. J. Pol. Econ. 76:392–406.

Dorfman, R., P. A. Samuelson, and R. M. Solow. 1958. Linear programming and economic analysis. McGraw Hill, New York. 527 pp.

Freeman, A. M. 1979. The benefits of environmental improvement. Johns Hopkins. University Press, Baltimore, MD.

Georgescu-Roegen, N. 1971. The entropy law and the economic process. Harvard University Press, Cambridge, MA. 457 pp.

Hannon, B. 1973. The structure of ecosystems. J. Theor. Biol. 41:535–546.

Hannon. B. 1976. Marginal product pricing in the ecosystem. J. Theor. Biol. 56:256–267.

Hannon, B. 1979. Total energy costs in ecosystems. J. Theor. Biol. 80:271–293.

Herendeen, R. A. 1980. On the concept of energy intensity in ecological systems. ERG Document No. 271. Energy Research Group. University of Illinois, Champaign-Urbana.

Isard, W. 1972. Ecologic-economic analysis for regional development. The Free Press, N.Y. 270 pp.

Leontief, W. W. 1941. The structure of American economy, 1919, 1929: an empirical application of equillibrium analysis. Harvard University Press, Cambridge, MA.

Odum, H. T. 1957. Trophic structure and productivi of Silver Springs, Florida. Ecol. Mono. 27:55–112.

Odum, H. T. 1971. Environment, power, and society. Wiley Interscience. N. Y.

Odum, H. T. and R. C. Pinkerton. 1955. Times's speed regulator: the optimum efficency for maximum power output in physical and biological systems. Am. Sci. 43:331–343.

Oster, G. F. and E. O. Wilson. 1978. Caste and ecology in the social insects. Monographs in population Biology No. 12. Princeton University Press, Princeton, N. J. 352 pp.

Prigogine, I., P. M. Allen and R. Herman. 1977. Long term trends and the evolution of complexity. pp. 1–63 In: Laszl and J. Bierman, (eds.), Goals in a Global Community Pergamon Press, New York. 335 pp.

Richey, J. E., R. C. Wissmar, A. H. Derol, G. E. Likens, J. S. Eaton, R. G. Wetzel, W. E. Odum, H. M. Johnson, O. L. Loucks, R. T. Prentki, and P. H. Rich. 1978. Carbon flow in four lake ecosystems: a structural approach. Science 202:1183–1186.

Ritz, P. M. 1979. The input-output structure of the U. S. economy, 1972. Survey of Current Business. Feb. 1979: pp. 34–72.

Ulanowicz, R. E. 1972. Mass and energy flow in closed ecosystems. J. Theor. Biol. 34:239–253.

Victor, P. A. 1972. Pollution, economy, and environment. University of Toronto Press, Toronto, Canada. 247 pp.

von Neumann, J. 1945. A model of general economic equilibrium. Econ. Studies. 13:1–9.

Walras, L. 1896. Éléments d'économic politique Luusanne.

8. Net Energy Analysis

PRELIMINARY ENERGY ANALYSIS OF UTILIZING WOODY BIOMASS FOR FUEL[1]

Flora C. Wang, John Richardson, Katherine Carter Ewel, and Edward T. Sullivan[2]

Abstract.--Data from both published economic analyses and an on-going field study in Florida indicate that the ratio of energy output to energy input in woody biomass plantations may vary from 5 to 10, suggesting that the use of such plantations to provide fuel may be energetically feasible. On the average, fertilization and irrigation (when necessary) are the most energy-intensive processes, and weed control is the least.

INTRODUCTION

The potential for wood to contribute to the solution of the nation's energy problems is of considerable current interest, and biomass is one of the fuels being analyzed by the Department of Energy as a possible substitute for petroleum and natural gas. The objective of this paper is to evaluate the cost of production, in terms of both dollars and energy, of woody biomass growing in intensively cultivated plantations in Florida.

As an economy, we consume energy directly in the form of gasoline, electricity, natural gas, and other petroleum products. Energy is consumed indirectly when it is used to produce goods and services that are in turn purchased by consumers. The embodied energy of a product or service is the total direct and indirect energy required to produce that product or service. Herendeen and Bullard (1974) defined the energy cost as all fossil, hydroelectric, or nuclear energy consumed along the chain of extraction-refining-fabrication-sales. Therefore, the flow of goods in an economy is also a flow of energy that is embodied in the goods.

[1] Paper presented at the international symposium Energy and Ecological Modelling, sponsored by the International Society for Ecological Modelling. (Louisville, Kentucky, April 20-23, 1981).
[2] Flora C. Wang is Associate Professor, Center for Wetland Resources, Louisiana State University, Baton Rouge, La. U.S.A.; John Richardson is Graduate Assistant in Environmental Engineering Sciences, Katherine C. Ewel is Assistant Professor of Forest Resources and Conservation, and Edward T. Sullivan is Associate Professor of Forest Resources and Conservation, University of Florida, Gainesville, Fla., U.S.A.

We use energy intensity (ε) suggested by Bullard et al. (1976) in their input-output analysis of energy flow in the U.S. economy as a common accounting unit; it is expressed in terms of Btu per dollar of final product. If the physical quantities (tons, barrels, cubic meters) of different commodities are available, their total enthalpy, or heat content (Btu per physical unit), which is approximately equal to Gibb's free energy, is used as a common accounting unit.

The energy intensity coefficient was derived from data on dollar transactions in 1967. To update this value, we used energy intensity for the entire economy (energy/GNP), which includes natural energy (Wang and Miller 1980).

We are focusing in this paper on the analysis of energy yield from biomass production. For a complete economic and energy assessment, data based on at least five years of research will be required. For this preliminary study, we compare several energy budgets to reflect the range of data that may be obtained from this type of research. Since much of the information obtained from our own field experience is currently limited to relatively small experiments, we consider the analysis presented here to be subject to change as new information becomes available. We have not at this time considered the availability of land for growing woody biomass or the suitability of various soil types for the projected productivity levels.

Energy Inputs Into Woody Biomass Production

Energy inputs into biomass production systems include land clearing and site preparation, seedling production and planting, fertilization, weed control, irrigation, and harvesting. The School of Forest Resources and Conservation at the University of Florida is currently analyzing the potential for obtaining energy and chemicals from a variety of woody species in Florida. Three exotic tree species have been planted at a 4-ha test site near LaBelle, Florida: Australian pine (Casuarina spp.), Eucalyptus grandis, and Melaleuca quinquenervia. We have based our analysis on this site since energy expenditures for site preparation have already been measured. Data from other commercial forest lands being prepared, planted, and harvested using intensive methods have provided estimates for comparison with this site.

Energy Used for Land Clearing and Site Preparation

The fossil fuel consumption and labor requirements for each operation (disking, bulldozing, rotovating, and bedding) have been converted into heat equivalents in kcal/ha using the conversion ratios given in Table 1. The total energy used for site preparation and land clearing on the La-Belle site averages about 1.65×10^6 kcal/ha (Table 2). Energy costs for site preparation and land clearing from other intensive tree farming efforts vary from $22/ha (Szego and Kemp 1973) to $692/ha (Salo et al. 1979). These estimates are based on a 5-year rotation and a 6-year rotation, respectively. Using the energy/dollar ratios in Wang and Miller (1980) gives low and high energy costs of 0.32×10^6 kcal/ha and 6.44×10^6 kcal/ha, respectively. Thus, the energy inputs calculated for our LaBelle study suggest that energy costs for site preparation there are low.

Energy Cost for Seedling Production and Planting

The Florida Division of Forestry sold seedlings for the 1979-80 season at prices ranging

Table 1. Conversion ratio of energy inputs into energy equivalents.

Category	Energy Equivalent	Reference
Diesel Fuel	34,200 kcal/gal	Handbook of Chemistry and Physics, 1972
Gasoline	36,200 kcal/gal	Handbook of Chemistry and Physics, 1972
Labor	5,912 kcal/hr	Fluck and Baird 1980
Dollars	14,400-9,300 kcal/$	Wang and Miller 1980
Nitrogen	8,400 kcal/lb	Pimentel et al. 1973
Phosphorus	1,520 kcal/lb	Pimentel et al. 1973
Potassium	1,050 kcal/lb	Pimentel et al. 1973
Water	905,600 kcal/ac-ft	Pimentel et al. 1973
Other Petroleum Products	1,622 kcal/gal	ASAE, 1977
Oven Dry Wood	3,600 kcal/kg	Tillman, 1978

from $13/1000 (sand pine and slash pine) to $81/100 (Eucalyptus), including a delivery charge. Eucalyptus seedlings planted at LaBelle in summer 1980 cost $75/1000. At 1 m x 1 m spacing, 10,000 plants are required for a 1-ha plot.

Table 2. Energy used for site preparation and land clearing.

Source	Operation	Area (ha)	Labor (hr)	Diesel (gal)	Gasoline (gal)	Total Energy (10^6 kcal)	Energy Cost ($/ha)	Energy Equivalents (10^6 kcal/ha)
LaBelle Site	Disking	3.0	7.3	60	--	2.13	--	0.71
1980	Bulldozing	4.0	--	50	--	1.74	--	0.44
	Rotovating	5.1	11.0	52	--	1.88	--	0.38
	Bedding	4.4	--	--	15	0.54	--	0.12
	Subtotal							1.65
Szego and Kemp 1973[1]	Disking	--	--	--	--	--	22	0.32
Dutrow and Saucier 1976[1]	Site prep.	--	--	--	--	--	99	1.09
	Land clearing	--	--	--	--	--	86	0.95
	Subtotal						185	2.04
Inman et al. 1977[1]	Preparation and Clearing	--	--	--	--	--	457	4.66
Salo et al. 1979[2]	Site preparation	--	--	--	--	--	494	4.59
	Land clearing	--	--	--	--	--	198	1.85
	Subtotal						692	6.44

[1] Based on a 5-year rotation and a yield of 5 DTE/acre·yr (12.4 DTE/ha·yr).
[2] Based on a 6-year rotation and a projected productivity of 8 DTE/acre·yr on a 1000-acre biomass farm.

Other studies have shown that the cost of planting varies from \$18.50/ha (Dutrow and Saucier 1976) to \$135/ha (Salo et al 1979). In our analysis, a \$50/ha planting cost is assumed. The total energy costs of seedlings and planting vary from \$132/ha to \$800/ha (Table 3): 1.90×10^6 kcal/ha to 7.45 to 10^6 kcal/ha in energy equivalents. The projected energy input of 7.45 to 10^6 kcal/ha for our LaBelle site is the highest estimate because of the high cost of the exotic seedlings.

Energy Cost for Fertilization

Fertilization is necessary on short-rotation biomass farms in order to augment soil nutrients and to achieve desired productivity levels. Fertilization is not energy-efficient, as is reflected in the high cost of fertilizers.

For the LaBelle site, fertilizer application rates of 50 kg of nitrogen and 50 kg of phosphorus per hectare per year are used (Smith and

Table 3. Energy cost for seedlings and plantings.

Source	Species	Spacing	Seedling Cost (\$/1000)	Seedling Cost (\$/ha)	Planting Cost (\$/ha)	Total Cost (\$/ha)	Energy Equivalents (10^6 kcal/ha)
LaBelle Site 1980	Eucalyptus	1m x 1m	75	750	50	800	7.45
Szego and Kemp 1973[1]	---	1m x 1m	10	100	32	132	1.90
Dutrow and Saucier 1976[1]	Sycamore	4ft x 4ft	22.7	159	18.5	178	1.96
Inman et al. 1977[1]	Populus spp.	---	---	445	---	445	4.54
Salo et al. 1979[2]	Sycamore	4ft x 4ft	30	210	135	345	3.21

[1]Based on a 5-year rotation and a yield of 5 DTE/acre·year.
[2]Based on a 6-year rotation and a projected productivity of 8 DTE/acre·year on a 1000-acre biomass farm.

Table 4. Energy cost of fertilization.

Source	Operation	Application Rate (kg/ha·year) N	Application Rate (kg/ha·year) P	Energy Cost (\$/ha·yr)	Energy Equivalents (10^6 kcal/ha·rotation)
LaBelle Site 1980	First fertilization	–	–	–	0.07
	Yearly	50	50	–	5.56
	Subtotal				5.63
Szego and Kemp 1973[1]	First year	–	–	27	0.39
	2nd & 3rd year	–	–	69	1.00
	Subtotal			96	1.39
Dutrow and Saucier 1976[1]	Yearly	–	–	123.5	6.79
Inman et al. 1977[1]	Yearly	–	–	54.3	2.77
Salo et al. 1979[2]	Yearly	–	–	116	6.47

[1]Based on a 5-year rotation and a yield of 5 DTE/acre·year.
[2]Based on a 6-year rotation and a projected productivity of 8 DTE/acre·year on a 1000-acre biomass farm.

Conde 1980). Energy costs of 8,400 kcal/lb (18,800 kcal/kg) of nitrogen and 1,520 kcal/lb (3,400 kcal/kg) of phosphorus (Pimentel et al. 1973) are used in our analysis (Table 1). The total energy inputs of fertilization over one 5-year rotation are about 5.63×10^6 kcal/ha (Table 4).

Fertilizer costs for other intensive forest farms vary from \$96/ha rotation to \$123.50/ha year, or 1.39×10^6 kcal/ha·rotation to 6.79×10^6 kcal/ha·rotation. The estimated fertilizer requirement for our LaBelle site of 5.63×10^6 kcal/ha·rotation is approaching an energy-intensive level.

Energy Cost for Weed Control

Biomass farms will require weed control until trees are large enough so that competition by weeds will not inhibit their growth. Chemical weed control has shown promise, especially in combination with leguminous cover crops that not only control the weeds but also fix atmospheric nitrogen (Zavitkovski 1978).

At the LaBelle site, initial observations reveal that Melaleuca can tolerate weed competition, but that Casuarina requires weed control, either through an intensive level of site preparation or through application of herbicides (Smith and Conde 1980). The weed control requirement for Eucalyptus is not known.

Table 5. Energy cost for weed control.

Source	Frequency	Energy Cost ($/hr·yr)	Energy Equivalents (10^6 kcal/ ha·rotation)
LaBelle Site 1980	Yearly	25	1.16
Szego and Kemp 1973[1]	First Three Years	10	0.43
Dutrow and Saucier 1976[1]	Yearly	62	3.41
Inman et al. 1977[1]	Yearly	37	1.89
Salo et al. 1979[2]	First Year Second Year Subtotal	173 124.5 297.5	1.61 1.16 2.77

[1] Based on a 5-year rotation and a yield of 5 DTE/acre·year.
[2] Based on a 60-year rotation and a projected productivity of 8 DTE/acre·year on a 1000-acre biomass farm.

Other studies have shown that the cost of weed control varies from \$10/ha·year (Szego and Kemp 1973) to \$62/ha·year (Dutrow and Saucier 1976), or 0.43×10^6 kcal/ha·rotation to 3.41×10^6 kcal/ha·rotation (Table 5). In our analysis, a cost of \$25/ha·year is assumed to be spent for cultivation. This represents 1.16×10^6 kcal/ha rotation, which is a very low weed control level.

Energy Used for Irrigation System

Rapid crown closure of short-rotation fuelwood plantations will increase water use as compared to standard pulpwood silviculture (Smith and Conde 1980). Consequently, irrigation, which is costly in terms of energy demand, may be necessary. Salo et al. (1979) estimated that the average installation cost of an irrigation system, exclusive of wells and pumps, is about \$530/acre; replacement cost after every harvest averages about \$230/acre; and the average cost of the wells and pumps is about \$480/acre over a life of 30 years. This represents a very intensive energy level of 9.84×10^6 kcal/ha·rotation.

Dyson (1974) concluded that yields from Eucalyptus are variable, depending in part on the rainfall received during the growing period. At the LaBelle site, it is assumed that 25 cm of irrigation water per year in excess of rainfall is required. The energy cost of 905,600 kcal/acre·ft (736,300 kcal/hectare·meter) of irrigation water (Pimentel et al. 1973) is used in this analysis. The estimated energy used for the irrigation system at LaBelle will be about 0.92×10^6 kcal/ha·rotation, which represents a relatively low energy level (Table 6).

Energy Used for Harvesting

The harvesting operation is usually performed during the winter. Trees are felled, bunched, skidded, and chipped by different machines. The harvest sequence will depend on the characteristics of the equipment. Salo et al. (1979) suggested using a mobile feller-chipper, which will fill a trailing wagon that will then be pulled to the chip storage area by a tractor when it is full.

For the LaBelle study, the fossil fuel energy and labor input data collected from the shortwood and longwood pulpwood logging operations at Bradford Forest (Ewel and Sullivan 1979) are used (Table 7). The energy conversion of petroleum products is taken as 1,622 kcal·gal (ASAE, 1977) (Table 1). The total energy equivalents for the low-intensity and high-intensity operations do not include the manufacturing cost.

Tillman (1978) estimated the operating costs of various pieces of harvesting equipment. These varied from 33,000 Btu/ovendry ton (0.10×10^6 kcal/ha) for a chain saw to 373,000 Btu/ovendry ton (1.13×10^6 kcal/ha) for a small truck, assuming a 50-mile round trip. The energy expenditures

Table 6. Energy used for irrigation system.

Source	Irrigation System	Energy Cost ($/ha)	Energy Equivalents (10^6 kcal/ha·rotation)
LaBelle Site 1980	---	--	0.92
Szego and Kemp 1973[1]	---	--	--
Dutrow and Saucier 1976[1]	---	--	--
Inman et al. 1977[1]	Capital cost (30 years)	605	1.03
	Year cost	45	2.30
	Subtotal		3.33
Salo et al. 1979[2]	Installation cost (30 years)	1309	2.43
	Replacement cost (per rotation)	560	5.20
	Well and pump costs (30 years)	1186	2.21
	Subtotal		9.84

[1] Based on a 5-year rotation and a yield of 5 DTE/acre·year.
[2] Based on a 6-year rotation and a projected productivity of 8 DTE/acre·year on a 1000-acre biomass farm.

Table 7. Energy used for harvesting operation.

Source	Operation	Total Energy (10^6 kcal)	Energy Cost ($/ha)	Energy Equivalent (10^6 kcal/ha·rotation)
Ewel and Sullivan 1979	Shortwood Logging	96.7	--	2.54
	Longwood Logging	123.7	--	3.43
Szego and Kemp 1973[1]	Harvesting ($3.60/ green ton, 50 gt/acre)	--	445	6.40
Dutrow and Saucier 1976[1]	Harvesting/Chipping ($2.24/green ton, 50 gt/acre)	--	278	3.06
Inman et al. 1977[1]	Fuel Cost (16 gal/ hr, 1 hr/ha)	16		2.90
Salo et al. 1979[2]	Harvesting Cost	--	262	2.44
	Fuel Cost (15 gal/ hr, 1 hr/ha)	15		3.12
	Subtotal			5.56

[1] Based on a 5-year rotation and a yield of 5 DTE/acre·year.
[2] Based on a 6-year rotation and a project productivity of 8 DTE/acre·year on a 1000-acre biomass farm.

for other intensive biomass farming operations range from 2.90 x 10^6 kcal/ha (Inman et al 1977). to 6.40 x 10^6 kcal/ha (Szego and Kemp 1973). Our projected harvest ing operation costs (2.54 x 10^6 kcal/ha to 3.43 to 10^6 kcal/ha) for this analysis fall within this range of values found.

Energy Outputs from Biomass Production System

Heat value, defined as heat of combustion, is a measure of the thermochemical property of an organic substance. Chemical composition of a particular kind of wood determines both its heat value and how it produces useful energy.

Table 8. Summary of energy inputs and outputs (10^6 kcal/ha·rotation) of biomass production systems.

Source	Species	Site Prep	Seeding and Planting	Fertil-ization	Weed Clearing	Irrigation	Harvest	Total Energy Input	Energy Energy Output	Net (Available) Energy	Energy Output to Energy Input
LaBelle Site	Casuarina, Eucalyptus, Melaleuca	1.65	7.45	5.63	1.16	0.92	3.00	19.81	111	91	5.6
Szego and Kemp 1973[1]	---	0.32	1.90	1.39	0.43	--	6.40	10.44	111	101	10.6
Dutrow and Saucier 1976[1]	Sycamore	2.04	1.96	6.79	3.41	--	3.06	17.26	111	94	6.4
Inman et al. 1977[1]	Populus spp.	4.66	4.54	2.77	1.89	3.33	2.90	20.09	111	91	5.5
Salo et al. 1979[1]	Sycamore	6.44	3.21	6.47	2.77	9.84	5.56	34.29	213	179	6.2

[1]Based on a 5-year rotation and a yield of 5 DTE/acre·year.
[2]Based on a 6-year rotation and a projected productivity of 8 DTE/acre·year on a 1000-acre biomass farm.

In this analysis, the heat value of wood will be used as a primary measure of fuel value. The caloric yields for wood, bark, branch, and foliage of Melaleuca are 4.0, 6.16, 4.61, and 4.81 kcal/g, respectively (Smith and Conde 1980).

Estimates of potential yields of sand pine and slash pine are 9.0 dry tons/ha·year and 9.4 dry tons/ha·year, respectively (Smith and Conde 1980). Eucalyptus yields and heat values have not been assessed.

For our preliminary study, the average yields for a 5-year rotation of 5 DTE/acre·year and for a 6-year rotation of 8 DTE/acre·year are assumed. The heat content of air-dry wood is assumed to be 3,600 kcal/kg (Tillman 1978) in order to facilitate and standardize the calculations. Thus, the yields of 5 and 8 DTE/acre·year would give total energy yields of 222 x 10^6 kcal/ha·rotation and 427 x 10^6 kcal/ha·rotation, respectively.

Comparison of Energy Inputs and Outputs

Zavitkovski (1979) suggested evaluating the energy investment in an intensively cultivated forest plantation by calculating a ratio of energy output to energy input, and by determining the amount of energy available. The first method measures the effectiveness of the plantations and the second measures the energy yield. The available energy, the so-called net energy, or the difference between energy output and energy input, serves as a more meaningful measure of the success, and may also be compared with energy input.

For a complete net energy analysis, all of the energy inputs should be evaluated, including both the direct and indirect energy required to produce goods or services. The energy output obtainable from the forest biomass should also take into account the conversion efficiencies in combustion systems.

In our preliminary study, neither solar insolation nor any other environmental energy source such as rain, wind, and soil has been included. All values represent input of fossil fuel energies to the output of wood production in the same unit of energy equivalents.

The total energy inputs into biomass production systems include land clearing and site preparation, seedlings and plantings, fertilization, weed control, irrigation, and harvesting. Based on our preliminary analysis, these amount to 10.44 x 10^6 kcal/ha rotation for a low energy budget and 34.29 x 10^6 kcal/ha·rotation for a high energy budget (Table 8). A total of 19.81 x 10^6 kcal/ha·

rotation is consumed at the LaBelle site, reflecting a medium-intensity operation.

The net yield of energy from wood is in a different form than coal or oil, which are more concentrated energy sources. Since the inputs were measured in kilocalories of coal-equivalent energy, we multiplied the output of wood energy by 0.5. This accounts for the fact that 2 kcal of wood energy are equivalent to 1 kcal of coal energy (Odum 1978). Actual energy obtained from a biomass plantation may vary from a low of 111 x 10^6 kcal/ha·rotation to a high of 213 x 10^6 kcal/ha·rotation. Available energy, expressed as coal equivalents, ranges from 91 x 10^6 kcal/ha·rotation to 179 x 10^6 kcal/ha·rotation. The ratio of energy output to energy input ranges from 5.5 to 10.6, and averages 6.9. Combining highest energy costs for each of the inputs gives a ratio of 3.6; combining lowest energy costs gives a ratio of 14.1. Because transportation costs for any type of fuel are high, utilization should be close to the plantations where the wood is produced.

It will eventually be necessary to analyze the relationship among the chemical composition, heating value, and impurities of wood as they exist in the other combustible fuels. Experimental results have revealed that removal of silvichemicals causes a significant reduction in the heat of combustion of wood; but some of these extractives may have a marginal profit over the heat they lose (Smith and Conde 1980).

Our preliminary analysis has not included the indirect energy costs of biomass production such as the energy used to manufacture equipment. We have used the energy intensity value (Btu per dollar's worth of the final product) to circumvent this shortcoming and provide a complete (if tentative) picture of total energy cost to the biomass production system.

Other factors will also affect the pattern of wood utilization from biomass plantations. As net yields of fossil fuels continue to decrease, the need for low energy farming techniques, with lower net crop yields per hectare, will increase the demand for agricultural land. Also, the demand for wood products, which is presently close to the limit of the net annual yield from our national forest resources (Jahn and Preston 1976; Spurr and Vaux 1976), will increase because the energy cost of wood construction materials is much less than that of aluminum and steel.

We conclude from this preliminary analysis that energy invested in producing biomass energy will yield net energy, but the amounts may be small in relation to our national needs, and utilization of wood products for other uses may yield more net energy in the long run.

LITERATURE CITED

ASAE. 1977. Agricultural machinery data, pp. 326-333 in Agricultural Engineering Yearbook, American Society of Agricultural Engineers, St. Joseph. Mich.

Bullard, C.W., P.B. Penner, and D.A. Pilaiti. 1976. Energy analysis handbook. CAC Document No. 214, Center for Advanced Computation, University of Illinois, Urbana, Ill.

Dutrow, G.F., and J.R. Saucier. 1976. Economics of short rotation sycamore. USDA Forest Research Paper SO 114. 16 PP.

Dyson, W.G. 1974. Experiments on growing Eucalyptus wood fuel in the semi-deciduous forest zone in Kenya. East Africa Agricultural and Forestry Journal (April):349-355.

Ewel, K.C. and E.T. Sullivan. 1979. pp. 1-72 IN Intensive Management Practices Assessment Center, Progresss Report, School of Forest Resources and Conservation, Univ. of Fla., Gainesville, Florida.

Fluck, R.C., and C.D. Baird. 1980. Agricultural energetics. AVI Publishing Co., Westport, Connecticut.

Handbook of Chemistry & Physics. 1972. Chemical Rubber Company, Cleveland.

Herendeen, R.A., and C.W. Bullard. 1974. Energy cost of goods and services, 1963 and 1967. CAC Document No. 140, Center of Advanced Computation, University of Illinois, Urbana, Illinois.

Inman, R.E., D.J. Salo, and B.J. McGurk. 1977. Site specific production studies and cost analysis. Silvicultural Biomass Farms, Vol. IV, MITRE Technical Report 7347.

Jahn, E.C. and J.B. Preston. 1976. Timber: More effective utilization. Science 191:757-761.

Odum, H.T. 1978. Energy analysis, energy quality, and environment. p. 55-88 IN M.W. Gilliland (ed.). Energy analysis: A new public policy tool. AAAS Selected Symposium 9. Westview Press, Boulder, Colo.

Pimental, D., L.E. Hurd, A.C. Bellotti, M.J. Forster, I.H. Oka, O.D. Sholes, and R.J. Whitman. 1973. Food production and the energy crisis. Science 182:443-449.

Salo, D.J., J.F. Henry, and R.E. Inman. 1979. Design of pilot silvicultural biomass farm at the Savannah River Plant. Report to U.S. Dept. of Energy, HCP/T4101-01, The MITRE Corp., McLean, Va., March, 1979.

Smith, W.H., and L.F. Conde. 1980. Energy and
chemicals from woody species in Florida.
Annual Report to U.S. Dept. of Energy,
Grant No. ET-78-6-01-3040, School of Forest
Resources and Conservation, University of
Florida, Gainesville, Fla.

Spurr, S.T. and H.T. Vaux. 1976. Timber:
Biological and economic potential. Science
191:752-756.

Szego, G.C., and C.C. Kemp. 1973. Energy forests
and fuel plantations. Chemtech 3:275-285.

Tillman, D.A. 1978. Wood as an energy resource.
Academic Press, New York. 252 pp.

Wang, R.C. and M. Miller. 1980. Comparison of
energy analysis and economic cost-benefit
procedure: A case study - LaSalle County
Regulatory Commission, Contract #NRC-04-77-
123, Center for Wetlands, University of
Florida, Gainesville, Florida.

Zavitkovski, J. 1978. Biomass farms for energy
production-biological consideration, pp.
132-137. In North American Forests - Gate-
way to Opportunity, Society of American
Forestry, North Conference.

Zavitkovski, J. 1979. Energy production in irrigated intensively cultivated plantation
of Populus 'Tristis #1' and Jack Pine.
Forest Science 25:383-392.

NET ENERGY AND TRUE SUBSIDIES TO NEW ENERGY TECHNOLOGY[1]

Robert A. Herendeen[2]

Abstract.--For an energy technology with a poor net energy balance, the net subsidy ($/unit made available to the rest of the economy) can greatly exceed the apparent subsidy ($/unit). This is demonstrated for ethanol-from-grain for gasohol. Including various "energy opportunity costs" (equivalent to shifting the system boundary) influences results greatly.

INTRODUCTION

The production of ethanol-from-grain is a growing endeavor in the Midwest. Yet, as of February, 1981, anhydrous fuel grade ethanol is selling for $1.85/gallon; this is to be blended 10:90 with unleaded gasoline to make gasohol to compete with pure gasoline at a pump price of $1.35/gallon. Why is this a paying operation? Because in this region gasohol receives tax "breaks" of around $0.10/gallon, or, in effect, $1.00/gallon of ethanol. Subtracting this "incentive"[3] of $1.00/gallon from the $1.85/gallon price indeed makes ethanol worth producing.

Subsidies have their justification, among which are:

1. The infant industry argument: such industries will become profitable as they mature.

2. The social good argument: for example, mass transit is worth having.

---

[1]Paper presented at the international symposium Energy and Ecological Modelling, sponsored by the International Society for Ecological Modelling. (Louisville, Kentucky, April 20-23, 1981).

[2]Robert A. Herendeen is Assistant Research Professor in the Office of Vice Chancellor for Research, University of Illinois, Urbana, IL, 61801, U.S.A.

[3]Strictly speaking this is not a subsidy; it is an incentive. It has the same effect in stimulating production, I will use the two terms interchangeably. A list of state incentives is found in "State Initiatives on Alcohol Fuels," U.S. National Alcohol Fuels Commission, 412 First St., S.E., Washington, D.C. 20003.

3. The vulnerability argument: it is worth money not to depend on foreign sources of energy.

The question which I want to ask is: what is the net subsidy -- the subsidy per unit of energy made available to the economy at large? If an energy supply technology has a poor net energy balance, i.e., requires a significant amount of energy for development and operation, one suspects that the usual statement of subsidy ($1.00/gallon, etc.) is an under-estimate of the net subsidy.

This variation on the net energy theme is to some extent a rehash of the standard problems of net energy analysis, Herendeen, et al (1979) and Chambers et al (1979). In addition, I include some discussion of recent work by Baumol and Wolff (1980) which attempts to combine economic analysis and net energy analysis. To indicate one result: the effective subsidy to ethanol-from-grain ranges from +2 to -4 times the $1.00/gallon gross subsidy. The negative sign indicates an energy sink.

NET SUBSIDY vs. SUBSIDY

This is easily illustrated by reference to Fig. 1.

Two "energy ratios" can characterize the energy supply facility depicted in the figure.

1. Energy ratio without feedstock:

$$ER_{wof} = \frac{E_{out}}{E_{in}}$$

2. Energy ratio with feedstock:

$$ER_{wf} = \frac{E_{out}}{E_{in} + E_{res}}$$

Energy ratio with feedstock explicitly includes the resource commitment; therefore it is always

$\leq 1.$ [4] Energy ratio without feedstock excludes the resource commitment and reflects a traditional economic viewpoint: a resource's only cost is what the economy must use to extract it, E_{in}. For a positive net energy payoff, $ER_{wof} > 1$. For the purpose of discussing subsidies, ER_{wof} is the relevant energy ratio, and

$$S_{net} = \frac{SE_{out}}{E_{out} - E_{in}} = \frac{S}{1 - \frac{1}{ER_{wof}}} \qquad (1)$$

where,

S_{net} = net subsidy (\$/physical unit)

S = "normal" (gross) subsidy (\$/physical unit)

Equation 1 is nothing new. The question is whether subsidies are now being applied to technologies for which ER_{wof} is small enough to have an effect. Some typical figures are given in Table 1.

The wide range of ER_{wof} of ethanol-from-grain reflects first, whether the end product is ethanol or gasohol, and whether one makes energy corrections for the grain by-product and for miles-per-gallon in use for transportation fuel. This is discussed more under opportunity cost, below.

Table 1.—Energy ratios for some subsidized energy technologies. Energy ratios without feedstock generally correspond to "base case" in Tables 2 and 3, i.e., without compensation for the various opportunity costs.

ENERGY TECHNOLOGY	ER_{wof}	$1 - \frac{1}{ER_{wof}}$
Ethanol from grain (total energy), Chambers, et al, (1979) and Dovring, et al, (1980).	0.8	-4.0
Ethanol from grain (petroleum-substitutable fuels only), Chambers, et al, (1979) and Dovring, et al, (1980).	2.0	2.0
Light water reactors, Pilati (1977).	≈4	1.3
Residential energy conservation, ceiling insulation, Ford and Hannon (1980).	≈50	1.02
Residential solar, Sherwood (1978).	3.5	1.4

[4] If solar energy is assigned a resource energy cost of zero (which is usual), ER_{wf} can exceed 1.

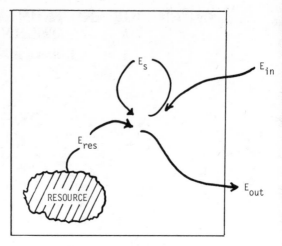

Figure 1.—Generic energy flow diagram for net energy analysis. All quantities are measured over facility's lifetime. E_s = self use, E_{res} = feedstock (resource) energy. E_s may be small, or large, relative to E_{out}. E_{in} = all energy needed to produce capital inputs and operating materials and energy.

OPPORTUNITY COST - GENERAL COMMENTS

The energy flow diagram in Fig. 1 contains a system boundary around the energy facility and its resource. It neglects other changes which may be thought of as necessary consequences of the decision to build an energy facility. For example, the capital equipment for an energy facility may be less energy intensive (BTU per dollar) than the average mix of capital equipment produced today. If (by assumption) total capital expenditures in the economy are held constant, a shift to the energy facility has a net negative impact on energy for capital equipment. Then an additional term should be added algebraically to E_{in}; the term is

$\left(\epsilon_{general\ capital} \cdot Y_{capital} \right)$, where $\epsilon_{general\ capital}$ is the energy intensity of average capital expenditures (BTU/\$) and $Y_{capital}$ is the capital expenditure (\$) for the energy facility. This approach is referred to as "respending," Hannon (1975).

A second type of opportunity cost has already been hinted at regarding the end use of the energy product produced. In the case of ethanol-from-grain, most of it is aimed at the gasohol market to be mixed 10:90 with unleaded gasoline. Since ethanol has a volumetric heat of combustion of about 2/3 of that of gasoline, one would expect the miles-per-gallon for the mixture to be about 96% that of pure gasoline -- if miles per gallon depends only on heat of combustion. The energetic consequences of other miles-per-gallon figure could be credited to the ethanol, although it is really a consequence of the fact that ethanol is used for gasoline. To illustrate, suppose gasohol gets a factor m times the miles-per-gallon of pure gasoline. Then energetically this is the same as if ethanol had a heat of combustion of (10m - 9) (heat of combustion of gasoline). If m = 1 (no change in miles-per-gallon), the ethanol has an "effective" heat of combustion equal to that of gasoline. If m = 1.07 (claimed by some researchers Scheller and Mohr (1977) but disputed), the effective heat of combustion is 1.7 times the heat of combustion of gasoline or over twice that of the ethanol. This multiplier is a consequence of the specific application in gasohol. To assume that this <u>will</u> be the application is an economic judgment; any use of opportunity cost then represents the admixing of economic criteria, Chambers <u>et al</u> (1979).

A third type of opportunity cost has to do with the disposition of process by-products. For ethanol the by-product is the grain residue after fermentation, a rather high protein material suitable for cattle feed, where it competes with soybean meal. If the number of cattle is held fixed the use of the by-product results in a savings of the energy that would have been otherwise used to produce the soybeans, Chambers <u>et al</u> (1979).

A fourth type of energy cost is a more general type discussed by Baumol and Wolff (1980). This is based upon the substitutability of labor and capital for energy in the production of all goods and services. If (say) capital is used to build an energy facility and therefore not used to produce another good, energy can be substituted into the latter process to assume unchanged production of that good. This substitution should not be confused with the respending effect, which has been discussed above. A similar comment applies to labor, and this substitution should not be confused with the energy to "maintain" the labor force, which is a controversial topic (Most standard energy analysis has ignored the latter, IFIAS (1974)). Then the Baumol and Wolff (1980) contribution (which is added to E_{in}) is expressed as

$$\left| \frac{\partial E}{\partial capital} \right|_c \cdot (\text{capital required}) +$$

$$\left| \frac{\partial E}{\partial labor} \right|_c \cdot (\text{labor required})$$

where the subscript c indicates that the derivatives are evaluated while holding the final demand for everything but energy constant.[5] It is assumed

of course, that capital and energy are actually complementary!

In any case, the Baumol and Wolff (1980) result increases E_{in} and reduces ER, thereby increasing net subsidy. I should say that the Baumol and Wolff (1980) work is interesting in a more general context because it offers hope for tying together net energy analysis and economic analysis, so that a result of "unprofitability" in energy allows strong inferences about unprofitability in dollars, or vice versa. This is a continuing discussion.

ENERGY OPPORTUNITY COSTS --
HOW LARGE ARE THEY: AN EXAMPLE FROM ETHANOL

It is of interest to determine the magnitudes of the opportunity costs mentioned so far. I illustrate for ethanol/gasohol. Recall that each of these is a correction to E_{in} (or, strictly speaking, may be so considered[6]).

a. <u>Capital respending</u>. In order to obtain this we need to know enough about the capital equipment to differentiate it from general capital equipment in an energy requiring sense. Here I only estimate the difference. All capital equipment together had an energy intensity roughly 35% higher than the average energy to GNP ratio (in 1967), Herendeen, et al., (1975) I assume that ratio holds today, as well as the 1967 variation of capital energy intensity from producing sector to sector. All this leads to a potential difference of approximately 10 kBTU per dollar (1980).

Capital requirements for an ethanol or gasohol plant are about $0.15(1980)/gal, assuming approximately a 10-year lifetime and no discounting, Dovring <u>et al</u> (1980). Capital respending then could contribute ±1.5 kBTU/gal to E_{in}. E_{in} is already around 100 kBTU/gal, Chambers <u>et al</u>, (1979). The respending effect is thus negligible.

b. <u>End use</u>. As mentioned, this is a potentially large factor because of the use of ethanol

---

[6]The use of ratios is always ambiguous, since one often has the opportunity to subtract from the denominator or add to the numerator, which can have dramatically different effects on the ratios. Since this report is concerned with subsidies in terms of the gross output of primary product, all opportunity costs are applied to the input (denominator) side.

---

[5]A subtle point arises if the amount of capital and labor is large enough so that differentials instead of derivatives are used. Then it is likely that the values of $\left| \frac{\Delta E}{\Delta capital} \right|$ differ under the two following assumptions:

a. To build the energy facility capital is shifted from other industries.

b. An excess amount of capital is available, and it is used for the energy facility instead of being used elsewhere to upgrade production, where it presumably substitutes for energy (See Baumol and Wolff (1980) pp. 8-10). A similar statement applies to labor. I do not discuss this further in this report.

in gasohol. I believe the claim of a 7% improvement in miles-per-gallon of gasohol vs. ethanol is not scientifically supported, so I will more conservatively assume a 0% change. This still results in an energy credit to the ethanol of

$(10m - 9)$ (heat of combustion of gasoline) (1.20)
  $-$ (energy to produce ethanol)

$= (10(1) - 9)$ $(125$ kBTU/gal$)(1.20)$ $-$ 100 kBTU/gal

$= 150$ kBTU/gal $- 100$ kBTU/gal $= 50$ kBTU/gal.

The factor 1.20 corrects for the energy cost of energy, occasioned by the extraction and refining necessary to deliver 1 gallon of gasoline, Bullard et al. (1978). 50 kBTU/gal is large compared with E_{in}.

c. By-product credit. Exact determination for this is very difficult, since there are questions of transportation distances, feed equivalents, and the like. An estimate is given in Chambers et al. (1979); ∿20 kBTU/gal.

d. Opportunity cost of capital and labor. This is also rather uncertain; in fact, there is some controversy about whether capital and energy are actually complements at all, Berndt and Wood (1977). There have been some estimates of elasticity of substitution of capital and energy, Berndt and Wood (1977), or of all inputs (capital plus labor) and energy, Hogan and Manne (1977). In the Appendix I use these to estimate the energy opportunity cost of labor and capital together,

$$\left.\frac{\Delta E}{\Delta(\text{labor} + \text{capital})}\right|_{\text{net output} = \text{constant}};$$

I obtain

$$\left|\frac{\Delta E}{\Delta(\text{labor} + \text{capital})}\right| \approx 10 \times 10^3 \text{ BTU/\$(1980)}.$$

This is quite low compared with the average energy to manufacture a good (the energy/GNP ratio), which is about 30×10^3 BTU/\$(1980). The reason is the dominance of the term $\frac{G}{E}$ in Eq. A4 (in the Appendix), which merely says that even today energy costs are a relatively small fraction of all input costs.

However, the amount of capital and labor "diverted" is large since both direct and indirect capital and labor are involved; i.e., even the capital needed to make the fertilizer to grow the grain must be included. Wages and capital expenditures were about 76% of the GNP in 1972, Ritz et al (1979). Detailed direct and indirect capital and employment impacts of producing ethanol can in principle be calculated, but for the present purposes I will again approximate: The total cost of a gasohol operation is about \$1.68/gal (from Dovring, et al (1980) Table 4-1, with interest on loan subtracted). Of this, approximately 3/4,

i.e., \$1.25 (1980)/gal, is for direct plus indirect capital and labor. This greatly exceeds direct capital and labor, which amount to about \$0.38 (1980)/gal. The energy opportunity cost is then

$(\$1.25/\text{gal})$ $(16 \times 10^3 \text{kBTU/\$})$ $\cong 20$ kBTU/gal

We can now summarize the effect of all these opportunity costs in ethanol from grain in Table 2. We see in the case of ethanol that some form of opportunity cost allowance is needed to give $ER_{wof} > 1$ (as long as the the other parts of E_{in}, for example, the agricultural energy inputs to produce the grain are not reduced by technological improvement). Miles-per-gallon is the single most sensitive factor. The energy opportunity cost of diverted capital and labor is relatively minor.

A fifth type of "opportunity cost" consideration involves differentiating energy types. We know that ethanol-from-grain produces a petroleum-substitutable fuel; one can ask to what extent the inputs could be provided by fuels other than petroleum. In this case, the normal value of E_{in} can be reduced to approximately 40 kBTU/gal; the most persistent petroleum use is through agricultural chemicals and field operations. A distillery can in principle run on coal. Then $ER_{wof\text{-oil only}}$ is approximately 2 before any opportunity costs are applied. In this case, the miles-per-gallon credit can reduce the effective E_{in} to

Table 2.--A summary of the magnitudes of the various opportunity costs in the net energy analysis of ethanol from grain. These represent the changes to E_{in}, which otherwise is approximately 100 kBTU/gallon of ethanol, Chambers, et al, (1979). The heat of combustion of ethanol is 80 kBTU/gallon.

EFFECT	Approximate Effect on E_{in} (kBTU/gal)	$ER_{wof} (=ER_{wf})$
Base Case	--	0.80
a. Capital respending	± 1.5	0.79 – 0.81
b. Miles-per-gallon when used in gasohol.	– 50	1.6
c. By-product credit; if substitute for soybean meal	– 20	1.0
d. Energy opportunity cost of capital and labor	20	0.67

Table 3.--Net subsidy for ethanol from grain (in gasohol) with the inclusion of various opportunity costs. The gross gasohol subsidy is assumed to be $0.10/gal, or $1.00/gal of ethanol.

OPPORTUNITY COST CONSIDERED		NET SUBSIDY	
		All Fuels ($/gal)	Petroleum Only ($/gal)
Base Case	E_{out} = 80 kBTU/gal E_{in} = 100 kBTU/gal, all fuels 40 kBTU/gal, petroleum only	- 4.0	2.0
Miles-per-gallon in gasohol	Very sensitive factor. Assume mpg for gasohol and gasoline are equal. E_{in} = - 10 kBTU/gal, petroleum only	2.7	0.9
By-product credit	E_{in} ≈ 80 kBTU/gal, all fuels E_{in} ≈ 30 kBTU/gal, petroleum only	∞	1.6
Energy Opportunity cost of capital and labor	E_{in} = 50 kBTU/gal, petroleum only	- 2.0	2.7

to zero, producing an ER of infinity. In Table 3 I combine the oil-only results with the all-types-of-energy results in calculating the net subsidy for a $0.10/gal gasohol gross subsidy.

Table 3 represents several ways of looking at the same process. We see that in terms of total energy, the net subsidy to gasohol is at best twice the gross, this for the case of correcting for miles-per-gallon of gasohol. For most of the opportunity cost "viewpoints," the subsidy is infinite or negative, restating the old claim that ethanol is not a net energy producer. For petroleum fuels only, the net subsidy is always positive and finite. However, except for the inclusion of the miles-per-gallon effect, the net subsidy still exceeds the gross.

In all cases the energy opportunity cost of capital and labor is, as calculated, relatively small. The different effects in Table 3 can, of course, be combined. Whether, and when, to do so is again the classic question of system boundary.

The results here for gasohol will probably not transfer well to so called capital intensive energy industries such as synthetic liquid fuels from coal. In addition, future conditions may complicate the analysis since 1) $\frac{E}{G}$ will not be as small as now, 2) future energy technologies probably will be capital intensive. The "capital crunch" will result in a larger effect of capital shifts.

I acknowledge Ernst Habicht's article in the Wall Street Journal, which stimulated this work. He used the same arguments, but different and less substantiated data, to obtain very large net subsidies for ethanol, Habicht (1980).

APPENDIX

SUBSTITUTABILITY OF ENERGY FOR CAPITAL AND LABOR

This is difficult to evaluate rigorously. Define

$$\sigma = \frac{\partial \ln (E/G)}{\partial \ln (P_E/P_G)} \qquad (A1)$$

where

E = energy and G = other inputs.

The derivative is evaluated with production of everything but energy held constant. σ is a long term elasticity; some estimates have been given:

SOURCE:	Hogan and Manne (1977)
DATA:	pre-1975
SECTORS COVERED:	All producing sectors except energy sectors
INPUTS:	Capital and labor
σ:	-0.5 (very approximate)

SOURCE: Berndt and Wood (1977)
DATA: 1965
SECTOR COVERED: U.S. manufacturing
INPUTS: Capital only
σ: -0.31
SECTOR COVERED: Canadian manufacturing
INPUTS: Capital only
σ: -0.46

I want to use σ to calculate a marginal substitutability of energy for other inputs, $\frac{\Delta E}{\Delta G}$ (measured in units of BTU per dollar).

Equation A1 can be integrated to yield

$$\frac{E}{G} = \frac{E_o}{G_o} \left(\frac{P_e}{P_G}\right)^{\sigma} \qquad (A2)$$

where E_o and G_o are the original values. It is permissible to express these in dollars.

Assume that G is reduced (for example, capital shunted away from its usual use and used to help build an energy facility). The price of G will rise as a consequence. Assume that the price of energy does not change; this is an approximation representative of this partial equilibrium approach. It is justified by the fact that $\frac{E_o}{G_o} \approx 0.1$ in the U.S. economy. Total expenditures for inputs is constant, so that the change in expenditures for G equals the negative of the change in expenditures for energy:

$$G \Delta p_G + p_G \Delta G \cong - p_E \Delta E \qquad (A3)$$

(the 2nd order term is neglected).

With the price change, Eq. A2 is (dropping the subscript for the values before the change):

$$\frac{E + \Delta E}{G + \Delta G} = \frac{E}{G} \left(\frac{P_E}{P_G - P_E \frac{\Delta E}{G} - P_G \frac{\Delta G}{G}} \right)^{\sigma}$$

$\frac{E}{\sigma} \ll 1$; therefore the term ΔG can be neglected from the denominator on the left hand side. Recalling that p_E and $p_G = 1$, and taking logs of both sides

$$\ln \frac{E}{G} + \ln \left(1 + \frac{\Delta E}{E}\right) = \ln \frac{E}{G} - \sigma \ln \left(1 - \frac{\Delta E}{G} - \frac{\Delta G}{G}\right)$$

Assuming $\frac{\Delta E}{E}$ and $\frac{\Delta G}{G} \ll 1$ (and hence that $\frac{\Delta E}{G} \ll 1$),

$$\frac{\Delta E}{E} = \frac{\Delta E}{G} + \frac{\sigma \Delta G}{G}$$

or

$$\frac{\Delta E}{\Delta G} = \frac{\sigma}{\frac{G}{E} - \sigma} \qquad (A4)$$

We want to express $\frac{\Delta E}{\Delta G}$ in BTU/\$, so we need an average price of energy as purchased by producing sectors of the economy. Recent Department of Energy figures indicates that the 1980 price is about \$3/10^6 BTU, Energy Information Administration (1980). The value of $\frac{E}{G}$ is around 0.1. (In the last year for which detailed input output economic data are available, 1972, $\frac{E}{G}$ was about 0.05, Scientific American (1981).) Using the estimated 1980 figure,

$$\frac{\Delta E}{\Delta G} \text{ (BTU/\$1980))} = \frac{\sigma}{\frac{3}{10^6} (10 - \sigma)}.$$

σ, from the data quoted above (based on data before the 1973 embargo) is ≈ -0.5. Substituting,

$$\frac{\Delta E}{\Delta G} = \frac{-0.5}{\frac{3}{10^6} (10 + 0.3)} \approx -16 \text{ kBTU/\$(1980)}.$$

REFERENCES

Baumol, W., and E. Wolff. 1980. Subsidies to New Energy Sources: Do They Add to Energy Stocks? Manuscript, New York University.

Berndt, E., and D. Wood. 1977. Engineering and Econometric Approaches to Industrial Energy Conservation and Capital Formation: A Reconciliation. Working Paper No. MIT-EL 77-040WP, Energy Laboratory, Massachusetts Institute of Technology, Cambridge, MA.

Bullard, C.W., P.S. Penner and David S. Pilati. 1978. Net Energy Analysis: Handbook for Combining Process and Input-Output Analysis. Resources and Energy 1: 267-313.

Chambers, R.S., R.A. Herendeen, J.J. Joyce and P.S. Penner. 1979. Gasohol: Does It or Doesn't It Produce Positive Net Energy? Science, 206: 789-795.

Dovring, Folke, Robert Herendeen, Randall Plant and Mary Ann Ross. 1980. Fuel Alcohol From Grain: Energy and dollar balances of small ethanol distilleries and their economies of size and scale. 80E-151 Agricultural Economics, ERG Doc. No. 313, Energy Research Group, University of Illinois, Urbana, IL. December.

Energy Information Administration. 1980. Annual Report to Congress, 1979. Vol. 3, DOE/EIA-0173(79)13, See Table 4.3.

Ford, C., and B. Hannon,. 1980. Labor and Net Energy Effects of Retrofitting Ceiling Insulation in Single-Family Homes. Energy Systems and Policy, 4: 217-237. Figure in Table 1 estimated from their Fig. 3, and an assumption of 20-year lifetime.

Habicht, E., Jr. 1980. Don't be Fuelish: Gasohol Production Will Be Expensive. Wall Street Journal, May 1, 1980.

Hannon, B. 1975. Energy Conservation and the Consumer. Science, 189: 95-102.

Herendeen, R., B. Segal and D. Amado. 1975.
Energy and Labor Impact of Final Demand
Expenditures, 1963 and 1967. ERG Tech. Memo
No. 62, Energy Research Group, University of
Illinois, Urbana, IL 61801, September.

Herendeen, R.A., T. Kary, and J. Rebitzer, 1979.
Energy Analysis of the Solar Power Satellite.
Science, 205: 451-454.

Hogan, W. and A. Manne. 1977. Energy-Economy
Interactions: The Fable of the Elephant and
the Rabbit? Working Paper EMF 1.3, Energy
Modeling Forum, Stanford University, Stanford,
CA 94305.

IFIAS. 1974. International Federation of Institutes for Advanced Study, Energy Analysis
(Workshop Report S-10246), Stockholm, Sweden.

Pilati, D. 1977. Energy Analysis of Electricity
Supply and Energy Conservation Options.
Energy 2: 1-7. Figure is for electricity
out divided by primary energy in over 25-year
lifetime.

Ritz, P., E. Roberts and P. Young. 1979. Dollar
Values for the 1972 Input-Output Study.
Survey of Current Business, 59(4),
February.

Scheller, W., and B. Mohr. 1977. Gasoline Does,
Too, Mix with Alcohol. Chemtech, 616-623,
October.

Scientific American. 1981. The Input-Output
Structure of the United States Economy.
Wall Chart.

Sherwood, L. 1978. Total energy Use of Home
Heating Systems. In F. Roberts, ed.,
Symposium Papers: Energy Modeling and Net
Energy Analysis, Colorado Springs, 21-25
August, 1978. The figures used represent
the heating energy saved divided by the energy
cost of the solar system. A natural gas
system is needed before installation, and
after as a backup, since solar does not
provide all energy.

A NET ENERGY ANALYSIS INCLUDING ENVIRONMENTAL COST OF OIL SHALE DEVELOPMENT IN KENTUCKY[1]

Christopher G. Lind and William J. Mitsch[2]

Abstract.--Oil shale deposits in the State of Kentucky are estimated to contain as much as 190 billion barrels of oil, but have only recently been cited as being economically feasible. The only workable method to mine oil shale in Kentucky using current technologies is strip mining, while two major processes for retrieving the shale are presently feasible. A net energy analysis was conducted to determine the energy yielded over and above the total energy amount used in mining and processing. Energy costs for environmental control were estimated separately. Yield ratios ranged from 2.8:1 to 10:1, depending on output and process choice. The assumption that oil shale will be more economically attractive as oil prices rise is disputed.

INTRODUCTION

There has been considerable interest recently in the development of oil shale in the eastern United States as an energy source. The U.S. Geological Survey has estimated the amount of known Devonian shale resources in the Eastern United States to be 400 billion barrels with probable extensions of known resources projected to 2600 billion barrels of oil (Janka and Dennison 1979).

The shales in the Western United States have higher oil yields per ton (20-30 gal/ton) than the eastern shales (10-20 gal/ton). The western shales contain more hydrogen, but both groups have similar organic carbon content (table 1). Even though the Fischer Assay oil yields are lower in the eastern shales, the overall energy content of both shales is similar. The eastern shales also have higher amounts of by-products such as coke and gas (Vyas et al. 1981).

The Institute of Gas Technology (IGT 1979) estimates that approximately 423 billion barrels of oil could be recoverable from eastern shales by their Hytort process. Forty-five percent of this amount is estimated to be recoverable from the Commonwealth of Kentucky (table 2). The energy content in the stripable eastern Devonian

[1]Paper presented at the international symposium Energy and Ecological Modelling, sponsored by the International Society for Ecological Modelling. (Louisville, Kentucky, April 20-23, 1981). Research was partially funded by the Graduate School, University of Louisville.
[2]C. G. Lind is Graduate Research Assistant in Systems Science and Environmental Engineering, University of Louisville, Louisville, Kentucky. W. J. Mitsch is Associate Professor of Systems Science and Environmental Engineering, University of Louisville, Louisville, Kentucky, U.S.A.

Table 1.--Comparison of Eastern and Western United States oil shales[1]

Parameter	Eastern	Western
Ultimate Analysis, weight %		
Organic Carbon	13.7	13.6
Hydrogen	1.6	2.1
Sulfur	4.7	0.5
Carbon Dioxide	0.5	15.9
Ash	78.3	66.8
Fischer Assay		
Oil Yield, wt %	4.6	11.4
Water Yield, wt %	2.3	1.6
Loss + Gas, wt %	2.4	2.6
Assay, gal/ton (liters/metric ton)	10.3 (43)	29.8 (124)

[1]from Feldkirchner and Janka (1979)

shales exceeds that of reported coal resources in Indiana, Kentucky, Ohio, and Tennessee (Weil et al. 1979). Estimates from a study prepared for Buffalo Trace Area Development District in north central Kentucky (Vyas et al. 1981) show that 4.4 billion barrels of oil are recoverable from shale in Lewis and Fleming Counties, Kentucky.

Eastern Shale Geology

During the middle and late Devonian and earliest Mississippian (330 to 360 million years B.P.) a majority of the Eastern United States was covered by the inland Chattanooga Sea. A few major rivers flowed into the shallow sea. The prevailing stagnant conditions and prolific algae and kelp growth were most likely what led to the

Table 2.--Estimated recoverable oil from Devonian shale deposits in Kentucky.[1]

General Area	Rock unit	Average thickness, ft (meters)	Area, sq.km	Yield, gal/ton (liter/met. ton)	Recoverable resource billion BBL
Eastern Kentucky	Sunbury	30 (9)	890	25 (104)	30
	Cleveland	30 (9)	890	30 (125)	30
	Lower Huron	30 (9)	890	20 (83)	20
Western Kentucky	New Albany	70 (21)	800	25 (104)	60
Southern Kentucky	Chattanooga	50 (15)	960	25 (104)	50
				TOTAL	190

[1]from Janka and Dennison (1979)

deposition of organic-rich (10-20 gal/ton) black shales throughout the basin. "Along the eastern edge of the sea, the organic-rich material was substantially diluted with river born sediments (clastics), leading to relatively lower concentrations of hydrogen carbon-rich material in very thick rock formations" (Janka and Dennison 1979).

The oil shales of the Green River area in Colorado are marlstones deposited from a lacustrine brackish water environment during the Eocene epoch approximately 50 million years ago, whereas the eastern Devonian black shales are true marine shales. The environment in which the shales were deposited provides valuable information on the primary factors determining their chemical properties and suitability for retorting. The sources of the organic matter in Devonian black shales have been divided into two broad groups. The humic type of organic matter was derived mainly from the woody parts of plants. This humic material contains only a few percent hydrogen by weight. Due to its low hydrogen content, the shales with a predominance of humic material test out relatively low in Fischer Assay oil yields. The sapropelic organic matter in shales is mainly from algae, spores, and pollen. This type has 10% hydrogen by weight, so it tests out higher in Fischer Assay oil yields.

Oil Shale Development in Kentucky

Figure 1 shows the extent of Devonian shale deposits at or near the surface in Kentucky. The deposits are near the surface in a semi-circular pattern of hills, known locally as the Knobs, which surround the bluegrass region of Kentucky. The New Albany shale, a major deposit in the western section of Kentucky has an average oil shale thickness of 70 feet (21 m). The thickness of the New Albany shale increases to the northwest along the outcrop from 40 to 60 feet (12-18 m) in Marion County, Kentucky. It increases to 70 to 100 feet (21-30 m) in Jefferson County, Kentucky. Kentucky has usable resources all along the outcrop areas (see fig. 1), with the exception of an area along the far eastern thrust fault. The

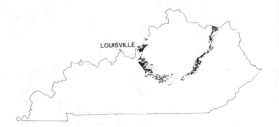

Figure 1.--Devonian shale deposits at or near the surface in Kentucky.

maximum overburden thickness was estimated at 175 feet (53 m) in the west, 200 feet (61 m) in the east, and 125 feet (33 m) in the south (Janka and Dennison 1979).

Currently accepted mining methods show that strip mining will probably be used to extract Devonian oil shale deposits in Kentucky. A recent study for the Buffalo Trace Area Development District in eastern Kentucky (Vyas et al. 1981) used a ridge-top strip mining scenario with an overall stripping ratio of 2:1. Various above ground retorting methods for recovering oil from oil shale have been analyzed with two processes found to be currently feasible for Kentucky. The Paraho method is an internal combustion retorting method, while Hytort process uses externally supplied hydrogen to heat the shale being retorted. Both processes extract oil from the black shales, leaving spent shale that has a weight ratio of 1:1 with the raw shale before processing. The spent shale will be backfilled into the strip mined area and also in the head of hollows. The spent shale will be covered with the overburden from the mining process and topped with a layer of top soil. There have not been adequate revegetation studies in the eastern United States on shale deposits to determine the feasibility of this practice in Kentucky.

The trace element concentration for the Devonian shales was predicted to be within the range of concentration found for coal preparation wastes from western Kentucky (Vyas et al. 1981), although no field studies have been done on this potential problem. The disposal regulatory guidelines used for coal mining (from the Surface Mining Control and Reclamation Act of 1977) may be sufficient for spent oil shales (Vyas et al. 1981) although additional regulations may be required. The Commonwealth of Kentucky is presently reviewing its procedures to determine if additional regulations are necessary for the protection of the environment from oil shale development.

Previous Studies On Net Energy Of Oil Shale

Net energy studies have primarily been from oil shale development in the Green River formation in Colorado. Several are summarized by Gardner (1977). Studies of oil shale development in the eastern states has only recently been cited.

Our energy yield ratios are expressed as the gross output of product divided by the external energy subsidies used in the production of the product (fig. 2). There have been numerous net energy yield ratios cited in western studies. Table 3 compares yield ratios of the western studies and one eastern study for oil shale development. The yield ratio varied from 0.68 to 15.99 using the same yield ratio formula. There were no consistent assumptions among the eleven western studies and one eastern study which makes comparison of the numbers difficult.

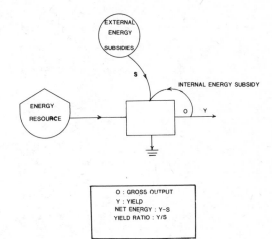

Figure 2.--General net energy model (Odum et al. 1976)

The Oregon Transition Study of Colorado oil shale was one of the first studies of net energy of oil shale (Gardner 1977). The data used in the study were for a 100,000 bbl/day (5.8×10^6 BTU/bbl) facility. Total capital costs for mining equipment, road maintenance equipment,

Table 3.--Previous net energy studies of oil shale

Name of Study Area	Final Product	Yield Ratio, Y/S*	Reference
Western Oil Shale			
Colorado(S)**	Gasoline	6.07	CERI (1976)
Colorado(U)***	Gasoline	6.32	CERI (1976)
Colorado(S)**	Electricity	3.60	CERI (1976)
Colorado(U)***	Electricity	3.89	CERI (1976)
Green River, Colorado	Syncrude	8.60	Marland et al. (1978)
Anvil Points, Colorado	Syncrude	13.30	Gardner (1977)
Colorado (Ore. Trans. Study)	Syncrude	1.42	Gardner (1977)
Colorado	Fuel Oil	15.99	Frabetti et al. (1976)
Colorado	Fuel Oil	7.62	Gardner (1977)
Rifle, Colorado	Unrefined Crude Oil	0.68	Gardner (1977)
Eastern Oil Shale			
Kentucky	Syncrude	13.70	IGT (1978)

  * Yield Ratio=Final Output/Energy Subsidy (See fig. 2)
 ** S=Surface Mining
*** U=Underground Mining

retorts, prerefining equipment, and crushing operation equipment were estimated from Bureau of Mines data. The direct operating energies and consumption of resources were estimated from either engineering projections or existing operations. The external energy subsidies were 705 BTU for 1,000 BTU of output, giving a yield ratio of 1.42 (Gardner 1977).

A net energy study in 1977 of an in situ oil shale processing operation in the Green River formation of Colorado showed an energy yield of 8.6 times the energy subsidy (Marland et al. 1978). This was for a 50,000 bbl/day operation using 20-gallon-per-ton Green River shale. The oil burned in place did not enter the energy calculations (Gardner 1977).

Frabetti et al. (1976) did a net energy analysis of a 100,000 bbl/day (30 gallon per ton) oil shale operation. The analysis used data from an economic evaluation of shale oil production by the U.S. Department of the Interior. The study used both input-output theory and process analysis. A complete energy pathway was developed using both direct and indirect energy inputs. The external energy was energy supplied outside the plants boundaries (Frabetti et al. 1976). A yield ratio of 16:1 was found with that study.

The Colorado Energy Research Institute (CERI 1976) study of oil shale development in Colorado focused on three oil shale retorting processes: the Tosco Process, the Union Oil Process, and the Bureau of Mines Combustion Process. The steps involved were mining, crushing, retorting, spent shale disposal, and refining. Energy costs for research and development were included. All the external materials, capital, and supplies were based on material costs; labor for on-site construction and labor for maintenance were not included (Gardner 1977).

A study entitled Net Energy and Oil Shale by Atlantic Richfield Co. and cited by Gardner (1977) used a 100,000 bbl/day plant in their scenario. The retorting process was the Tosco II using 35-gal-per-ton shale rock. Excluded in the analysis is the commercial plant design engineering cost of 10 million. Also excluded were natural energy flows and labor.

A net energy analysis of a Paraho Retort operation in Rifle, Colorado, claims to use data from actual operating yield and costs of operation. Fifteen thousand, eight hundred tons of shale were retorted into 10,000 barrels of unrefined crude oil during an operating period of 56 days. The reported cost was $500,000. No net energy was realized from this process with a yield ratio less than one. Capital costs of building the retort are omitted and when included should significiently reduce the low yield ratio obtained even further (Gardner 1977).

The net energy study of the Hytort process by the Institute of Gas Technology (IGT 1978)

used the net energy methodology of Melcher (1976). The yield ratio for the Hytort process was 13.7. Using the ratio of the net product to external losses plus internal energy consumption, a much lower yield ratio of 1.64 was calculated (IGT 1978).

METHODS

Our primary purpose in the analysis was to determine the net energy yield ratios for oil shale development in Kentucky using two retorting methods--the Hytort and Paraho processes. Economic data relating to both processes were obtained primarily from the Buffalo Trace Development District study of oil shale development in Kentucky (Vyas et al. 1981). The specific study sites for development in that study were in Lewis and Fleming Counties, Kentucky.

Studies of the energetics of world food production and energy sources by Odum (1967) and Odum et al. (1976) advanced the use of energetics and net energy analysis as analytical tools. In such analyses an energy diagram is drawn showing all processes, and output is compared with all internal and external feedbacks required (Odum and Odum 1976). The definition of net energy in this analysis is depicted in fig. 2.

Money and energy costs were divided into three categories; mining, processing and environmental (fig. 3). We were interested in finding out what percentage of external energy subsidies were channeled into environmental costs. Both mining and processing categories had environmental costs extracted from them. The money costs were taken from economic cost information from the Buffalo Trace Study and related estimates of reclamation costs for Kentucky coal mining. Labor and supplies used in environmental cost estimates were calculated by taking the

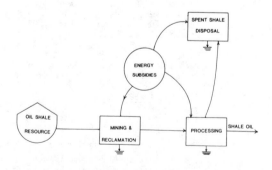

Figure 3.--Money and energy costs divided into three categories; mining, processing and environmental.

percentage of capital costs associated with environmental costs and using the same ratio (processing capital costs/environmental capital costs) to derive the operating costs associated with labor and supplies used in environmental cost estimates.

Mine reclamation costs were calculated from the estimate of $3,000 to $12,000 per acre per year for mine reclamation (W. Russo pers. comm.) by taking the average of the two figures. The average depth of seam of oil shale deposits was estimated at 30 feet (9 m).

Overall energy yield for both processes was taken from data from the Buffalo Trace Study. Most of the energy costs were calculated by obtaining estimated dollar costs and multiplying by the GNP/energy ratio for the United States (table 4). This method and the validity of using the ratio are discussed in further detail by Costanza (1980).

Table 4.--Gross national product and energy consumption for the United States (U.S. Bureau of Census 1980).

YEAR	GNP, current $ X 10^9	Energy consumption, 10^{15} BTU/year	Energy ratio, KBTU/$
1970	976	66.9	68.6
1971	1058	68.5	64.7
1972	1158	72.4	62.5
1973	1307	74.6	57.1
1974	1413	72.4	52.7
1975	1529	70.7	46.2
1976	1700	74.2	43.6
1977	1887	76.6	40.6
1978	2108	78.0	37.0
1979	2369[1]	80.0[2]	34.0
1980	2629[1]	76.3[2]	29.0

[1]Statistical Abstract of the United States (1980)
[2]From National Energy Information Center, U.S. DOE

Conceptual models of the energy flows expressed as BTU/day fossil fuel equivalents (FFE) were developed to illustrate the yield ratios for each process. Net energy yield ratios were calculated both with by-products and without by-products as gross output. The yield ratio expressed the ratio of output to external energy subsidies used to produce the output (fig. 2).

RESULTS AND DISCUSSION

Process Energy Flow

The energy balances performed by Vyas et al. (1981) for the Hytort and Paraho processes in Kentucky are summarized in figure 4 in energy diagrams. These represent only the direct energies involved in the energy analysis and are based on a 30,000 tons/day oil shale operation. The Paraho process has a greater proportion of its gross output in usable products and by-products than the Hytort process. The Paraho process results in 20 percent less shale oil but produces several other by-products including light oil, high BTU gas and electricity.

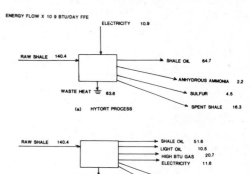

Figure 4.--Energy flow x 10^9 BTU/day FFE.

External Energy Subsidies

Money and energy costs for the mining operation for a 30,000 ton/day oil shale operation are given in table 5. The overall energy costs were 2.8 x 10^9 BTU/day. Mining costs were assumed to be independent of the retort process used.

Table 5.--Money and energy costs of oil shale mining for a 30,000 tons/day oil shale operation in Kentucky.

Account	Capital costs, $ Million[1]	Operating costs, $10^6/Year[1]	Energy costs, 10^6 BTU/day
Mining & Development	183		727
Labor & Supplies		26	2070
TOTAL ENERGY COSTS			2797

[1]Vyas et al. (1981) in 4th Quarter 1980 Dollars. These figures exclude environmental control costs.

The processing costs for both retorting methods are shown in table 6. The total energy costs for the Paraho retorting process were 7.5 x 10^9 BTU/day, and 20.2 x 10^9 BTU/day for the Hytort process. The large difference in energy costs was due to the imported electricity for the Hytort process (fig. 4). The Paraho processing method manufactured electricity over and above the electricity needed for the plant's needs.

Table 6.--Money and energy costs of processing for a 30,000 tons/day oil shale operation in Kentucky.

Account	$Million[1]		Energy Costs 10^6 BTU/day	
	Paraho	Hytort	Paraho	Hytort
Capital Costs				
Initial & Deferred	730	936	2900	3720
Operating Costs				
Annual Except Electricity	58.2	70.4	4620	5590
Electricity				
Annual	0	12.6	0	10900
TOTAL ENERGY COSTS			7520	20210

[1]Vyas et al. (1981) in 4th Quarter 1980 Dollars. These figures exclude environmental control costs.

Capital and operating costs related to environmental control are shown in table 7. Also included in the table are energy costs attributed to environmental control for both the Paraho and Hytort processes. The difference in the operating costs is attributed to the cost of limestone used in the sulfur dioxide scrubber process for the Paraho plant.

Preliminary Net Energy Ratios

Summary diagrams showing the pertinent direct and indirect energy flows for the two processes are given in figures 5 and 6. Net energy yield ratios for the two processes with and without by products are given in table 8. Yield ratios for the Paraho process were 10.0:1 with by-products and 6.0:1 without by-products.

Table 7. -- Environmental costs in money and energy for 30,000 tons/day oil shale operation in Kentucky.

Account	$Million[1]		Energy Costs 10^6 Btu/day	
	Paraho	Hytort	Paraho	Hytort
Capital Costs	43.6	51.6	173	205
Sulfur Recovery & Tail Gas Cleanup	8	15		
Waste Water Treatment	8	22		
Spent Shale Disposal	14	14		
Sulfur Dioxide Scrubber	13	--		
Sediment Ponds	0.6	0.6		
Operating Costs (Annual)	6.3	4.5	500	357
Mine Reclamation[2]	1.5	1.5		
Labor and Supplies[3]	3	3		
Limestone	1.8	--		
TOTAL ENERGY COSTS			673	562

[1]Vyas et al. (1981)
[2]based on range of $3,000-$12,000/Acre for Kentucky for coal mine reclamation (Wayne Rosso, personal communication)
[3]same percentage as capital costs

The Hytort process had a yield ratio of 3.1:1 with by-products and 2.8:1 without by-products. The results of the analysis show that both processes have a yield ratio greater than 1, and therefore both have a net yield of energy and could prove to be energetically feasible.

Although the ratio is positive for both processes, environmental costs could substantially reduce the yield ratio in actual operation. The reclamation costs for spent shale disposal have not been sufficiently researched to accurately predict the costs involved in future spent shale reclamation in Kentucky. Further research on

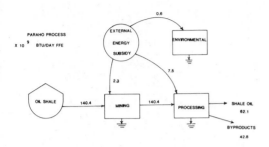

Figure 5.--Direct and indirect energy flows for Paraho process.

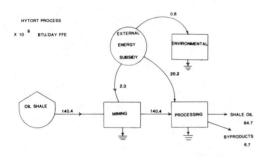

Figure 6.--Direct and indirect energy flows for Hytort process.

Table 8. --Net energy yield ratios for 30,000 tons/day oil shale development in Kentucky.

| Process | Yield ratios[1] | |
	with by-products	without by-products
Paraho	10.00:1	5.97:1
Hytort	3.09:1	2.80:1

[1]Yield Ratio=Gross Energy Yield/External Energy Subsidy

reclamation costs will be necessary to accurately predict the environmental costs. In addition, because no pilot plant has yet been built for the processes, it is difficult to determine if the capital and operating costs which were used to calculate the energy costs are accurate predictions of actual costs.

Energy Source Competition

If U.S. reserves are used now to gain independence, there would be neither independence nor means to control world prices later (Odum et al. 1976). The yield ratio of a barrel of crude oil is approximately 5:1 based on the current price of $40 per barrel on the spot market. This ratio is greater than the ratios determined for the Hytort process, and closely parallels the yield ratio of the Paraho process without considering by-products. If environmental costs were underestimated in both these processes, the yield ratio would be reduced making investment in oil shale development an energetically unwise choice for private investors and the U.S. government. If the United States decides to invest in projects with low yield ratios to gain independence, while the rest of the world has access to a higher yield ratio of imported oil, the U.S. economy would find it difficult to compete in the world market, and energy per capita and productivity in the United States might decline (Odum et al. 1976). A further look at the net energy ratio of oil shale in Kentucky in relation to other energy sources is needed. As presently analyzed, the shale represents a marginally competitive energy source.

The general assumption that energy sources become competitive when oil prices rise is disputed by energy analysis. The subsidies for oil shale development are based primarily on an oil (domestic and foreign) economy. As the cost of a barrel of oil rises, the dollar cost of the subsidies will increase proportionally and the cost of oil shale development will rise. An implicit assumption of the energy analysis method is that the energy costs shown in figures 5 and 6 would not change in such a changing economy. Energy analysis therefore has an added advantage of not changing for several years as long as the general technologies and processes remain the same. Economic estimates must now be updated almost every year because of the high rate of inflation.

CONCLUSION

There is still more research to be done in the estimation of energy costs, particularly for mining, spent shale reclamation and other environmental controls. Our analysis shows that both the Paraho and Hytort processes for oil shale recovery yield net energy above that which is required for production. However, the energy yield may not be competitive with existing energy resources. Based on the tenuous nature of the data for our calculations, oil shale development in Kentucky needs much more complete economic and energetic evaluation. The economic competitiveness of oil shale will probably not improve drastically even as oil prices rise.

LITERATURE CITED

Colorado Energy Research Institute (CERI). 1976. Net energy analysis: an energy balance study of fossil fuel resources. 217 p. Report prepared for the Department of Interior, Office of Minerals Policy and Research Analysis. National Technical Information Service, Washington, D.C.

696

Costanza, R. 1980. Embodied energy and economic valuation. Science. 210:1219-1224.

Feldkirchner, H.L., and J.C. Janka. 1979. The Hytort process. 38 p. Paper presented at synthetic fuels from oil shale symposium. [Atlanta, Georgia, December 3-6, 1979] Institute of Gas Technology, Chicago, Ill.

Frabetti, A.J., Jr., C.N. Flinkstrom, M. Gorden, R.A. Lovell, Jr., C. Sheldon, II, J.D. Westfield, and P.F. Way. 1976. A study to develop energy estimates of merit for selected fuel technologies. 376 p. Report to U.S. Department of Interior, Office of Research and Development. Development Sciences, Inc., East Sandwich, Mass.

Gardner, G.M. 1977. Comparison of energy analysis of oil shale. p. 196-232. In Energy analysis of models of the United Stated (H.T. Odum and J. Alexander, eds.). Report to U.S. Department of Energy. University of Florida, Gainesville, Fl.

Institute of Gas Technology. 1978. Net energy analysis of the Hytort process. 14 p. Institute of Gas Technology, Chicago, Ill. (Unpublished report).

Institute of Gas Technology (IGT). 1979. Eastern oil shale -a new resource for clean fuels. 22 p. Institute of Gas Technology, Chicago, Ill. (Unpublished report).

Janka, J.C., and J.M. Dennison. 1979. Devonian oil shale -a major American energy resource. 96 p. Paper presented at synthetic fuels for oil shale symposium. [Atlantic, Georgia, December 3-6, 1979] Institute of Gas Technology, Chicago, Ill.

Marland, G., A.M. Perry, and D.B. Reister. 1978. Net energy analysis of in situ oil shale processing. Energy 3:31-41.

Melcher, A.G. 1976. Net energy analysis: an energy balance study of western fossil fuels. Proceedings of the environmental oil shale symposium. [Golden, Colorado, April 29-30, 1976] Quarterly of the Colorado School of Mines 71(4):11-31.

Odum H.T. 1967. Energetics of world food production. In the world food problems, Vol. 3, report of president's science advisory committee on world supply, White House, Washington, D.C.

_____, and E.C. Odum. 1976. Energy basis for man and nature. 297p. McGraw-Hill Co., New York, N.Y.

_____, and C. Kylstra, J. Alexander, N. Sipe et al. 1976. Net energy analysis of alternatives for the United States. p. 253-302 E. In Hearings before the subcommittee on energy and power of the committee on interstate and foreign commerce, U.S. House of Representatives. [Washington, D.C., March 25-26, 1976]. Serial No. 94-63. U.S. Government Printing Office, Washington, D.C.

U.S. Bureau of the Census. 1980. Statistical abstract of the United States: 1980, 101st ed., U.S. Government Printing Office, Washington, D.C.

Vyas, K.C., G.D. Aho, and T.L. Robl. 1981. Synthetic fuels from eastern oil shale. Vol. 1. 308 p. Report prepared for Buffalo Trace Area Development District, Maysville, Kentucky. Davy McKee Corp., Cleveland, Ohio.

Weil, S.A., H.L. Feldkirchner, D.V. Punwani, and J.C. Janka. 1979. The IGT Hytort process for hydrogen retorting of Devonian oil shales. 7 p. Paper presented at 6th national conference on energy and the environment. [Pittsburgh, Pa., May 21-24, 1979] Institute of Gas Technology, Chicago, Ill.

THE ENERGY RETURN ON INVESTMENT FOR IMPORTED PETROLEUM[1]

Robert K. Kaufmann[2] and Charles A. S. Hall[3]

Abstract.--Four methods are used to calculate the energy return on investment for importing petroleum between 1963 and 1980. This ratio peaked at 22:1 in 1971 and reached its lowest value of 3.5:1 in 1980. Based upon an index that compares the economic and energetic costs of petroleum imports, and a comparison of this net energy ratio with the ratio for domestic alternatives, an explanation is offered for the apparent failure of economic analysis to predict substitution.

The United States relies on two sources of petroleum fuels; domestic resources and imports. To obtain petroleum from either source requires an investment of energy, a fact which makes energy analysis useful for calculating the energy return on investment for each. The concept of "net energy ratio" is the same for both sources in terms of the net gain of each resource to society: the quantity of energy recovered divided by the energy costs of recovery. The energy flows counted as recovery and cost of recovery differ. Fuel imports are recovered by an exchange of energy between economic systems, whereas domestic fuels are recovered by an internal expenditure of energy. An energy analysis of importing petroleum examines the energy balance of an economic exchange as opposed to the energy balance of a physical process.[4]

The energy balance of this economic exchange is important because the U.S. has become increasingly dependent upon imported fuels over the past twenty years. Annual domestic

production has, as a rule, diminished since 1970 and annual imports have increased, until the past year, in spite of considerable rhetoric and federal programs aimed at reducing them. Several additions to the domestic resource base are unlikely, and furthermore, that the net energy yield of domestic exploratory effort is likely to reach the break point within the next 20 years (Hall and Cleaveland 1981).

Politicians and economists have frequently advised the federal government to reduce American dependence on imports by intensifying efforts to find new domestic supplies and by developing domestic alternatives such as synfuels, oil shale alcohol fuels, and solar-powered satellites. Yet, despite its large and increasing economic and political costs, dependence on imports continues. This paper presents calculations of the historical and current energy return on investment for imported petroleum fuels; compares the changes in the economic and energetic costs of imports; and offers an explanation for the apparent failure of economic analysis to evaluate the productivity of different fuel sources.

METHODS

Energy return on investment is derived by dividing the quantity of energy obtained from a process or industry by the quantity of energy it uses. For imported oil and natural gas, the numerator of this ratio is the quantity of energy imported; the denominator is the quantity of energy used to produce the goods exchanged for the imported petroleum. Since petroleum fuels are not the only goods the U.S. imports in exchange for its goods, absolute quantities of imported petroleum and exported goods cannot be

---

[1]Paper presented at the international symposium Energy and Ecological Modelling, sponsored by the International Society for Ecological Modelling. (Louisville, Kentucky, April 20-23, 1981)
[2]Robert K. Kaufmann is a Research Scientist, Complex Systems Research Center, University of New Hampshire.
[3]Charles A.S. Hall is Assistant Professor, Department of Ecology and Systematics, Cornell University, Ithaca, New York.
[4]The quality of the energy flows used by the energy producing sector is at least as great as the energy flows used by the industrial sectors that manufacture the goods exported. We can, therefore, compare the net energy ratios for the different sources.

used to calculate this ratio. Instead, the energy return on investment for importing petroleum is derived by dividing the quantity of energy imported per dollar by the quantity of energy used to produce an average dollar's worth of exported good.

The levels of petroleum imports and their dollar value are available in Statistical Abstracts of the U.S. (U.S. Department of Commerce, various years-a), Highlights of U.S. Export & Import Trade, (U.S. Department of Commerce, various years-b) and Monthly Energy Review (U.S. DOE, various years) These data provide the basis for calculating the quantity of energy (in Kcals) purchased by each dollar spent on imported fuels between 1963 and 1980. This yearly figure--representing all petroleum fuels--are broken down further into liquid fuels (both crude and refined oil products) and natural gas. In order to calculate the cost of one million kcals in real dollars, the nominal cost is divided by the GNP deflator (base year 1972).

The U.S. trades a wide variety of goods for imported energy, and each dollar's worth of goods represents a different quantity of energy used in manufacturing. This amount is termed the "embodied energy" associated with that good. The energy embodied in an average dollar's worth of export is the sum of the embodied energy of all exports divided by their total dollar value. The dollar value for each type of exported good is published in Statistical Abstracts of the U.S. (U.S. Department of Commerce, various years-a).

The data we use to calculate the embodied energy in the industrial goods traded come from the Annual Survey of Manufacturers (U.S. Department of Commerce, various years-c) and Census of Manufacturers (U.S. Department of Commerce, various years-d). These sources give dollar values for the value added, value of shipments, cost of materials, and the cost and quantity of fuel used by 167 industrial sectors in the United States. We used data for two-digit industrial sectors because this is the most detailed level of aggregation available for energy use data during years prior to 1972. Similarly, our analysis begins in 1963 because of the lack of complete data for previous years. Since our data sources do not record fuel use in physical units before 1974, we used the prices for fuels paid by the industrial sector, published in Gilliker (1979), to convert data on the aggregate expenses for fuels into physical units. This conversion method produces an average absolute error of 5% in net energy ratios for years in which figures for both the cost of fuel and the physical quantity of fuel consumed are available.

We adapted two methods developed for previous energy analyses to calculate embodied energy (Herendeen and Bullard 1974; Hall et al. 1980). Both methods share one simplification. They condense industrial production into two basic steps: extraction, in which raw materials

are recovered from the environment, and benefaction, in which the recovered materials are manufactured into finished goods. Based upon this simplification, no recovered materials are used during extraction, and the energy embodied in the materials recovered is estimated as the energy used at the site of extraction divided by the value added. Without this truncation of the industrial process, the energy embodied in materials could be traced back ad infinitum. The energy embodied in finished goods is calculated for both methods by summing the energy embodied in the recovered materials bought and the energy used at the site of manufacture, then dividing this total by the value of shipments produced.

Although both methods include the energy embodied in the recovered materials as part of the energy embodied in the final good, they differ in the way they include this energy. The Annual Survey of Manufacturers and the Census of Manufacturers provide the dollar value of the materials used, but do not specify the sectors from which the materials originate. To trace the origins of these materials, we adapted a procedure developed by Herendeen and Bullard (1974) that uses input-output tables for the U.S. economy. Using these tables, we identify the sources of materials bought by each sector and the total embodied energy in this mix to calculate the energy embodied in the materials purchased. We used tables for 1963, 1967, and 1972, the last year for which they are available. We interpolated linearly between years and extropolated 1972 values to 1977. The tables are published in the Survey of Current Business (U.S. Department of Commerce 1969, 1974, 1979).

In order to buttress our analysis and to determine the importance of possible error caused by this extrapolation, we also used the procedure developed by Hall et al. (1980). This method estimates the energy embodied in materials three ways. One estimates it as the energy used per dollar value added by extractive industries, such as the timber and primary metal sectors. Since the input-output tables showed that all industrial sectors receive materials from nearly all other industrial sectors, another method sets it as the energy used per dollar value added for all industrial sectors. In order to include energy used in other parts of the economy to support the industrial sector, the final method establishes the energy embodied in materials as the energy used per dollar of GNP. The nearly standard deviation, as a percentage of the mean net energy ratio calculated by these four methods is 5.4%.

The most recent relevant industrial data published are for 1977. In order to extend the analysis to 1980, we used the change in the ratio of total U.S. energy consumption to GNP to estimate the yearly change in the energy embodied in an average dollar's worth of export. We regard this as a very conservative method, one that underestimates the energy return on investment because the ratio of energy consumption to GNP decreased at a much slower rate between 1963 and 1977 than did the energy embodied in an average dollar's worth of exported goods.

Industrial goods are not the only goods the U.S. exports. Agricultural products represent a significant portion of the foreign trade revenues, so the embodied energy in several agricultural exports, which account for approximately 65% of the total monetary value of agricultural exports, are included. The physical quantities and monetary values of these products—corn, wheat, rice, and soybeans—are published in the Statistical Abstracts of the U.S. ((U.S. Department of Commerce, various years-a). These physical quantities are converted to values for embodied energy from coefficients (Pimentel et al. 1974; Pimentel et al. 1975).

RESULTS

Figure 1 gives the average net energy ratio calculated by each of the four methods, for imported liquid fuels, natural gas, and all petroleum fuels between 1963 and 1980. The energy return on investment for imported liquid fuels peaks at approximately 22:1 in 1971. The gradual increase from 1963 to 1971 corresponds to the decreasing cost for these fuels as measured in real dollars. The rapid decline between 1973 and 1974 reflects the tripling of real price following the oil embargo. Between 1974 and 1978, the net energy ratio for imported fuels decreases gradually from 8:1 to 6:1. The cost increases that followed the 1979 Iranian cutback depressed the energy return on investment, which then levels off between 3:1 and 4:1. The historical behavior of the net energy ratio for total petroleum imports mirrors the behavior of the ratio for importing liquid fuels primarily because liquid fuels constitute the major proportion of petroleum fuel imports.

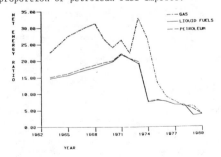

Figure 1.--Energy return on investment for imports.

Liquid fuels are such a large part of total petroleum imports that their presence masks a very different pattern displayed by the historical energy return on investment for imported natural gas. This net energy ratio shows two peaks 31:1 in 1968 and 33:1 in 1973, after which it declines rapidly. Unlike the energy return on investment for liquid fuels, which decreases gradually following the 1973-1974 embargo, the net energy ratio for imported natural gas continues to decline until it comes within range of the ratio for liquid fuels.

Using these ratios, we calculate the net quantity of liquid fuel and natural gas imported between 1963 and November 1980 (fig. 2). Until 1973, the net energy quantity of energy imported in these fuels remains very close to the gross quantity of energy imported. After 1973, when the net energy ratio for all fuels decreases, the lines for net and gross energy imports diverge. For example, although the gross quantity of liquid fuels imported between 1973 and 1975 remains fairly constant, the net quantity of energy decreases approximately 13%.

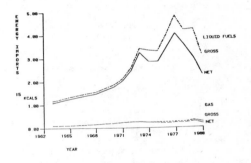

Figure 2.--Gross vs. net energy imports.

DISCUSSION

Monetary values are the most common metric for economic systems, but they may be the most useful for evaluating the productivity of various fuel sources. Because of numerous government subsidies, tax structures, and regulations, the results of monetary cost-benefit analyses do not always reflect accurately the results of energy analyses. Despite this decoupling, many economic analyses have attempted to predict the prices for imports that would trigger the substitution of domestic alternatives using only the economic cost of producing these alternatives. Based upon the economic and energetic cost of imports, and a comparison of the energy return on investment for imports and domestic alternatives, we offer an explanation for the apparent failure of economic analysis to evaluate the productivity of alternative domestic sources of energy.

Using the net energy ratio for imported liquid petroleum fuels and their cost per million kilocalories, as measured in 1972 dollars, we can compare the yearly changes in real energetic and economic costs for this source of fuel. Using 1972 as the base year (index = 1.0), real energetic cost is measured by dividing the net energy ratio in 1972, 20.9:1, by its yearly value. A similar index measures economic costs: the yearly cost of one million Kcals in 1972 dollars is divided by its cost in 1972, $1.65. The values of the comparative indices are shown in figure 3.

Figure 3.--Economic vs. energetic costs for imported petroleum.

Through 1973, the real energetic and economic costs for imported liquid fuels are fairly close. After 1973 however, they diverge sharply. The real economic cost increases by a factor of 3.6 between 1972 and 1974, and rises to 4.0 by 1978, whereas real energetic cost increases by a factor of 2.7, and rises to 3.2 by 1978. In 1979, the real economic cost rises to 7.5 then drops back to 7.0 in 1980, whereas the real energetic cost rises to 6.1 in 1979 and drops back to 5.6 in 1980. In short, the economic cost of imported liquid fuels increases more rapidly and to higher levels than does real energetic cost.

The major cause of this decoupling of economic and energetic costs is not a shift in the type of goods exchanged to less energy intensive products, but a general decrease in the energy embodied in all industrial exports. This decoupling may partly explain the apparent inability of economic analysis to predict the price levels at which domestic alternatives could compete with, and substitute for, imports.

As an example, a 1974 MIT study predicted the price of liquid fuel imports would have to reach $10.00 a barrel (1979 dollars) in order for oil shale to become an economically competitive fuel source (Adelman et al. 1974). A recent study by the Rand Corporation suggests that imports would have to reach $29.00 (1979 dollars) for the same substitution to occur (Merrow 1978). Similar predictions of substitution price levels for other fuel sources have been made (Adelman et al. 1974), yet in all cases, as prices for imported liquid fuels have exceeded these predictions, the anticipated substitutions have not taken place.

In order for domestic alternatives to substitute for imports in a competitive market, the energy return on investment for imports may have to fall to levels within range of these alternatives. The net energy ratio for importing liquid petroleum fuels, however, has not dropped to levels equal to--let alone below--the range for many domestic alternatives. Estimates of the net energy return for alcohol production range between 0.9:1 and 2.2:1 (Hopkinson and Day 1980; Jenkins et al. 1979), and the net energy ratio for the solar-powered satellite falls within a similar range (Herendeen et al. 1979). Regardless of the measurable economic costs and noncalculable political risks of dependence on imported liquid fuels, many domestic sources are unable to produce the same quantity of energy per unit of investment.

The net energy ratio for importing liquid fuels does fall within the range of one estimate of the net energy ratio for oil production from shale (Neal in prep.), but substitution is proceeding at very slow rates. The net energy ratio for importing liquid fuels may have to decline to levels well below those of domestic alternatives for any significant substitution to occur.

This possible delay in substitution may be caused by the limited availability of energy and the economic efficiency associated with producing goods to trade for imported petroleum. Producing many types of goods at levels in excess of domestic demand allows many sectors of U.S. industry to operate at increased levels of capacity and, therefore of efficiency. The world oil market has been extremely tight, and growth of excess supplies has generally prompted production cutbacks. Any significant effort by the U.S. to bring alternative domestic liquid fuel supplies on line (e.g. building and operating retorts) would require tremendous quantities of fuel. Because this energy would not be available through increased production, it would have to be diverted from other sectors of the economy. If the U.S. were to pursue a policy of reducing imports by producing domestic alternatives, one source of the energy used to produce these alternatives may be the fuel that was used previously to produce the goods traded for petroleum. This diversion of energy from many industrial sectors to a few would cause a large production cutback in the affected sectors. The resulting decreased efficiency would raise the prices of these goods in the domestic market. In addition, a sudden large increase in U.S. energy demand to support production of alternatives would increase world energy prices. Unless the U.S. is prepared to put up with increased energy costs, increased costs for industrial goods, and a decrease in the quantity of net energy available, the quantity of net energy available, imported liquid fuels will probably remain a significant and relatively productive source of U.S. petroleum supplies.

ACKNOWLEDGMENTS

We thank Linda Gauger for collecting much of the data and the librarians at the U.S. Government Repository at the University of New Hampshire for their help locating obscure sources. We also thank Steven Wolf and Berrien Moore for their critical comments.

LITERATURE CITED

Adelman, M.A., B.E. Hall, K.F. Hansen, J.H. Halloman, H.D. Jacoby, P.L. Jaskow, P.W. MacAvoy, H.P. Meissner, D.C. White, and M.B. Zimmerman. 1974. Energy self sufficiency; economic evaluation. Tech Rev., 76(4):23-58.

Gilliker, J.P. 1979. State energy prices by major economic sector from 1960 through 1977. 136 p. U.S. Department of Energy, Washington, D.C.

Hall, C.A.S., M. Lavine, and J. Sloane. 1980. Efficiency of energy delivery systems; II estimating energy costs and captial equipment. Envir. Manag., 3:505-509.

Hall, C.A.S. and C.J. Cleaveland. 1981. Petroleum drilling and production in the United States: yield per effort & net energy analysis. Science, 211:576-579.

Herendeen, R.A. and C.W. Bullard. 1974. Energy costs for goods and services 1963 and 1967. Document 140, Center for Advanced Computation, University of Illinois at Urbana, Champaign.

Herendeen, R.A., T. Korg, and J. Rebitzer. 1979. Energy analysis of the solar power satellite. Science, 205:451-454.

Hopkinson, Jr., C.S. and J.W. Day. 1980. Net energy analysis of alcohol production from sugar cane. Science, 207:302-303.

Jenkins, D.M., T.A. McClure, and T.S. Reedy. 1979. Net energy analysis of alcohol fuels. American Petroleum Institute, Publication Number 4312, Washington D.C.

Merrow, E.W. 1978. Constraints on the commercialization of oil shale. Rand Corporation, Santa Monica, California.

Neal, C. In Preparation. Manuscript In Energy, Economics, and Resource Quality, C.A.S. Hall. J. Wiley and Sons, New York, NY.

Pimentel, D., W.R. Lynn, W. Macreynolds, M.T. Hewes, and S. Rusch. 1974. Workshop on Research Methodologies for studies of Energy, Food, Man, and Environment, Phase 1. Cornell University, Ithaca, New York (June 18-20, 1974).

Pimentel, D., W. Dritschilo, J. Krammel, and J. Kutzman. 1975. Energy and land constraints in food protein production. Science, 190:754-761.

U.S. Department of Commerce. Various years-a. Statistical abstracts of the United States. Bureau of the Census, U.S.G.P.O, Washington, D.C.

U.S. Department of Commerce. Various years-b. Highlights of U.S. export and import trade. Bureau of the Census, U.S.G.P.O., Washington, D.C.

U.S. Department of Commerce. Various years-c. Annual survey of manufacturers. Bureau of the Census, U.S.G.P.O., Washington, D.C.

U.S. Department of Commerce. Various years-d. Census of manufacturers. Bureau of the Census, U.S.G.P.O., Washington, D.C.

U.S. Department of Commerce. 1969,1974, 1979. Survey of current Business. Bureau of Census, Nov. 1969, Feb. 1974, Feb. 1979, U.S.P.G.O., Washington, D.C.

U.S. Department of Energy. Various years. Monthly Energy Review. U.S.G.P.O., Washington, D.C.

A METHODOLOGY FOR ASSESSING NET ENERGY AND ABUNDANCE OF ENERGY RESOURCES[1]

Michael S. Burnett[2]

Abstract.--A methodology for assessing energy resources on the basis of the net energy yield ratio and gross resource abundance is developed. The methodology is demonstrated for a spectrum of nonrenewable and renewable resources. Tables, graphs, and a summary chart categorizing resources as to positive or negative net energy yield, relative abundance or rarity, and nonrenewable or renewable are presented.

INTRODUCTION

When attempting to quantify the true utility of an energy resource to society's end users, consideration of two aspects of that resource is necessary. First, what is the quantity of the resource? Second, what is the quality of the resource? Estimates of the gross resource abundance are most often used to measure the quantity of resources, while measurement of the quality of resources is the goal of net energy analysis.

In this paper, a methodology is presented which may be used to assess both the quantity and the quality of energy resources. Two inputs are used in the methodology. The net energy yield ratio is used as a measure of resource quality. Gross resource abundance is used as a measure of resource quantity.

To demonstrate the methodology, it has been applied to a broad spectrum of nonrenewable and renewable resources which produce electricity and/ or liquid, solid, or gaseous fuels. The results are of a preliminary nature. They are not the results of a systematic and detailed application of the methodology. The data points have been derived from various publications, or in some cases, calculated by the author. In order to draw consistent and valid conclusions for energy policy-makers, such a systematic and detailed analysis would be necessary. Thus, the purpose of this paper is not to develop conclusive recommendations on the relative merits of energy resources, but rather to demonstrate a methodology which may be useful to do so.

---

[1]Paper presented at the international symposium Energy and Ecological Modelling, sponsored by the International Society for Ecological Modelling. (Louisville, Kentucky, April 20-23, 1981)
[2]Michael S. Burnett is Energy Systems Analyst, Western Solar Utilization Network, Portland, Oregon, USA.

METHODOLOGY

As mentioned previously, the inputs to the methodology are the net energy yield ratio and the gross resource abundance. The outputs of the methodology are: a set of tables quantifying the inputs for nonrenewable and for renewable resources; a set of graphs depicting the same information; and a summary chart.

First, establishment of definitions is in order. Figure 1 shows the basic net energy model. The feedbacks F are necessary to the production of the energy resource. The gross energy output is indicated by G. The net energy N is defined as the difference between the gross output and the feedbacks: $N = G - F$. The energy yield ratio Y is defined as the ratio of the gross output to the feedbacks: $Y = G/F$. This is thus a measure of the number of units of energy yielded per unit of feedback energy. It is the yield ratio which is utilized in this methodology.

The gross resource abundance is indicated by the storage module for nonrenewables, or by a flow line for renewables. The abundances quantified here are global totals. The units are in trillions of gigajoules (GJ) for nonrenewables, or in trillions of GJ per year for renewables. Current global annual energy consumption is around 0.25 trillion GJ.

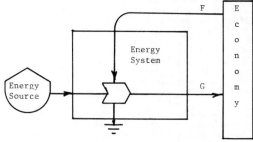

Figure 1.--Basic net energy model.

RESULTS

Table 1 shows the energy yield ratio and gross resource abundance for a spectrum of nonrenewable resources: fossil, nuclear, and geothermal.

Natural gas, oil, and coal are all rich net energy yielders. Natural gas and oil have similar abundances, and each if relied upon exclusively to supply current total global energy consumption would be exhausted in less than fifty years. The coal resource is more than twenty times as abundant. However, that which is considered economically recoverable is of the same approximate abundance as natural gas or oil resources.

Tar sands, due to its capital intensity, is apparently not a net yielder, although the resource abundance exceeds that of gas and oil combined. Oil shale is in a similar situation to tar sands, although the total resource is far greater.

The conventional fission reactor is a slight net yielder, with an abundance on the order of natural gas or oil reserves. The breeder reactor, although greatly magnifying resource abundance, may not yield net energy due to its greater capital intensity per unit of output in comparison with a conventional fission reactor. The situation for fusion is similar: highly abundant, but even more capital intensive per unit of output.

Geothermal resources of sufficient quality to produce electricity are relativiely rare, although rich net yielders.

Figure 2 is a graph portraying the data points just discussed. The yield ratio appears on the horizontal axis, and the gross resource abundance appears on the vertical axis. Both axes are log scales.

The vertical dashed line indicates a yield ratio of one. It divides the negative net yielders on the left from the positive net yielders on the right. The horizontal dashed line indicates the current global annual energy consumption of 0,25 trillion GJ per year. It divides the relatively rare resources below from the relatively abundant resources above.

Table 2 shows the energy yield ratio and gross resource abundance for a spectrum of renewable resources: solar, biomass, hydroelectric, wind, and tidal.

Although the total solar resource is very abundant, the amount which might be converted to electricity is far smaller. Both photovoltaics and power towers are not net energy yielders due to high capital and energy intensities.

The world terrestrial standing crop of biomass has an energy content exceeding combined natural gas and oil resources. The annual net productivity is seven times current global energy consumption.

Of this, approximately 20% may be considered as potentially utilizable. Present biomass consumption for all uses is approaching the theoretical potential, while the biomass actually used for productive purposes is much less. Technologies which convert biomass to liquid or gaseous fuels do not yield net energy, however. Wood as a solid fuel is a rich yielder with a potential abundance of 40% of current consumption levels.

Hydroelectric power is very rich net yielder with a potential abundance of about one fifth of current total consumption levels.

The situation for wind energy is similar to that of solar energy. The total resource is very abundant. The usable resource is much smaller, and low speed wind is not a net yielder. High speed wind does yield net energy, but such sites are relatively rare.

Tidal energy is a rich yielder in the special instances where it is viable.

Figure 3 is a graph portraying the data points just discussed. This is similar in layout to the graph for nonrenewables shown in figure 2. Remember that for renewables, abundance is in terms of annual flow, while for nonrenewables it was in terms of total stock.

CONCLUSIONS

Figure 4 is a summary chart classifying resources as to nonrenewable or renewable, positive or negative net energy, and relative resource abundance or rarity. This chart points out two important conclusions, although they are, as previously stated, of a preliminary nature.

o   First, there is no abundant positive net yielding renewable resource which could take up the slack from the abundant nonrenewables as they reach depletion.

o   Second, many of the proposed technological solutions to the energy crisis may not yield net energy, and thus may not be wise investments for society.

To conclude, I would like to suggest some possible directions for net energy analysis in the eighties.

First, we need a systematic and rigorous quantification of all potential primary resources in terms of energy yield ratio and gross resource abundance.

Second, we need to apply net energy analysis to energy end uses in a systematic fashion, including analysis of conservation and end use renewables.

Third, we need to assess the economic costs of energy resources in terms of net, rather than gross, output. Thus, quantification in terms of net BTU per dollar would flag expensive, negative net yielders as economice drains.

Table 1.--Energy yield ratio and gross resource abundance for
various nonrenewable energy resources.

Energy Resource	Energy Yield Ratio	Gross Resource Abundance $(10^{12}$ GJ)	Footnote
Fossil fuels			
Natural gas	6		a
Reserves		2.35	b
Resources		12.2	c
Oil	6		d
Reserves		4.72	e
Resources		10.3	f
Coal (including lignite)	6		g
Economically recoverable		13.8	h
Resources		246.	i
Tar sands	0.4	32.2	j
Oil shale	0.33		k
Presently recoverable		1.23	l
Reserves		19.4	m
Resources		2,070.	n
Nuclear			
Conventional	1.7	3.52	o
Breeder	0.9	211.	p
Fusion	0.4		q
Li 6, land		500.	r
Li 6, oceans		22.5×10^6	s
Deuterium, oceans		17.3×10^9	t
Geothermal	11.8	0.16	u

a Odum and Odum (1976).
b Grenon (1977) cites world natural gas reserves of 58,796 x 10^9 m³.
c Brown (1976) cites world natural gas liquids resource as 320 x 10^9 bbl oil equivalent and natural gas resource as 1680 x 10^9 bbl oil equivalent, for a total natural gas resource of 2000 x 10^9 bbl oil equivalent.
d Odum and Odum (1976)
e Grenon (1977) cites prospective and proven reserves of 107.2 x 10^9 metric tons.
f Grenon (1977) cites estimated remaining ultimate crude oil recovery of 233.4 x 10^9 metric tons.
g Odum and Odum (1976)
h Grenon (1977) cites economically recoverable coal reserves of 4.70 x 10 metric tons of coal equivalent at 7000 kcal/kg.
i Grenon (1977) cites total coal resources of 8.41 x 10^{12} metric tons of coal equivalent at 7000 kcal/kg.
j Maugh (1978) cites a capital investment of $20,000 to $24,000 per daily barrel of oil produced for two Canadian tar sands projects, with a comparable figure of $325 for Saudi Arabian oil. Tar sands processing is thus approximately seventy times as energy intensive as Saudi Arabian oil production. Gilliland (1975) cites thirty as the yield ratio

of imported crude. The yield ratio of tar sands is 30/70 = 0.4. Grenon (1977) cites tar sands and heavy crude resources of 732 x 10^9 MT of oil equivalent.
k Lovins (1977) cites oil shale as costing ninety times as much as crude oil imports in terms of capital investment per daily barrel of capacity. Gilliland cites thirty as the yield ratio of imported crude. The yield ratio of oil shale is 30/90 = 0.33.
l Grenon (1977): 28 x 10^9 MT oil equivalent.
m Grenon (1977): 442 x 10^9 MT oil equivalent.
n Brown (1976): 340,000 x 10^9 bbl oil equivalent.
o Kylstra and Han (1975) calculated a yield ratio for a conventional nuclear reactor of 2.7. Reactor cost has risen by about one third since 1972 when measured in real dollars (Zebroski and Levenson, 1976). Reactors are one third more energy intensive per unit of output, and the yield ratio has dropped to 2. Kylstra and Han's study did not include the energy costs of plant decommissioning and waste reprocessing. Rossin and Rieck (1978) cite $20 - $40 million as the cost of decommissioning. Resnikoff (1975) cites a cost of $145 per kilogram of spent fuel for reprocessing. There are 30 tons per reactor per year of waste (Anon, 1977). For a forty year

reactor lifetime, the cost of waste reprocessing is $174 million. Decommissioning and waste reprocessing total cost is $214 million. Parkins (1978) cites $975 million as the cost of a 1000 Mwe conventional reactor. Commoner (1978) reports the distribution of the cost of nuclear electricity as 73.8% capital, 6.3% operation and maintenance, and 19.8% fuel. Relating these percentages back to Parkins' figure gives $83 million for operation and maintenance and $262 million for fuel over the lifetime of the plant. The total of capital, operation and maintenance, and fuel is $1320 million. Back end costs of $214 million constitute an additional 16% in money cost over the plant lifetime. Assuming that there is a close correlation between money cost and energy cost, this represents an additional 16% in energy cost per unit of output from the reactor. The yield ratio drops to 2/1.16, or 1.7. The Organization for Economic Co-operation and Development (OECD) Nuclear Energy Agency (1977) cites reasonably assured and estimated additional uranium resources available at $50 or less per pound U_3O_8 of $4,290 \times 10^3$ metric tons. Assuming that a conventional reactor converts 1% of the theoretically available energy to electricity, the gross resource abundance is 3.52×10^{12} GJ.

p Parkins (1978) cites as construction costs $1820/kwe for breeder reactors and $975/kwe for conventional reactors. Breeder reactors are 1.9 times as energy intensive as conventional reactors. The yield ratio of a conventional reactor is 1.7 (Footnote o), so the yield ratio for a breeder reactor is 1.7/1.9 = 0.9. Assuming that a breeder reactor converts 60% of the theoretically available energy to electricity, the gross resource abundance is sixty times that of conventional reactors.

q Parkins (1978) cites as construction costs $4450/kwe for fusion reactors and $975/kwe for conventional reactors. Fusion reactors are 4.6 times as energy intensive as conventional reactors. The yield ratio of a conventional reactor is 1.7 (Footnote o), so the yield ratio for a fusion reactor is 1.7/4.6 = 0.4.

r Ehrlich, Ehrlich, and Holdren (1977) cite 500×10^{12} GJ as the energy potentially produceabl by fusion when land based Li 6 reserves are considered as the limiting resource.

s Ehrlich, Ehrlich, and Holdren (1977) cite 22.5×10^{18} GJ as the energy potentially produceable by fusion when oceanic Li 6 reserves are consi ered as the limiting resource.

t Ehrlich, Ehrlich, and Holdren (1977) cite 17.3×10^{21} GH as the energy potentially produceable by fusion when oceanic deuterium reserves are considered as the limiting resource.

u Gilliland (1975) cites yield ratios of 12.6 for a dry steam process and 10.7 for a wet steam process. Averaging these gives 11.8 as the yield ratio for geothermal electricity. Mining Congress Journal (1976) cites geothermal resources of 0.16×10^{12} GJ electricity.

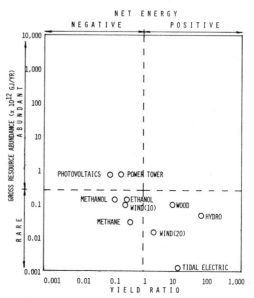

Figure 3--Net energy and abundance of renewable resources.

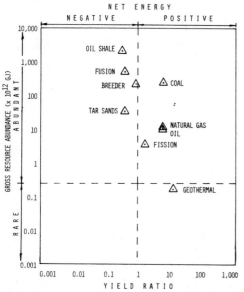

Figure 2--Net energy and abundance of nonrenewable resources.

Table 2. Energy ratios and gross resource abundances for various
renewable energy resources.

Energy Resource	Energy Yield Ratio	Gross Resource Abundance	Footnote
Solar			
Reaching out atmosphere		5,400.	a
Reaching earth's surface		3,800.	b
Reaching land surface		731.	c
Photocells	0.1	0.731	d
Power tower	0.2	0.731	e
Biomass			
World terrestrial standing crop		32.4	f
World terrestrial net productivity		1.76	g
Potential annual biomass consumption		0.352	h
Present biomass consumption		0.308	i
Present biomass utilization		0.0573	j
Methane	0.40	0.028	k
Methanol	0.14	0.126	l
Ethanol	0.31	0.126	m
Wood, transported and air dried	8	0.4	n
Hydroelectric	60	0.0450	o
Wind			
Energy available over globe		57.0	p
Energy available over land		6.01	q
Electric, 10 mph average	0.28	0.947	r
Electric, 20 mph average	2.2	0.015	s
Tidal	12	0.00480	u
Maximum electric production		0.00120	v

a IEEE Transactions on Power Apparatus and Systems, November – December 1975 (p. 1927) cites solar radiation reaching the outer atmosphere as $5,400 \times 10^{12}$ GJ.

b IEEE Transactions on Power Apparatus and Systems, November – December 1975 (p. 1927) cites solar radiation reaching the earth's surface as $38,000 \times 10^{12}$ GJ.

c IEEE Transactions on Power Apparatus and Systems, November – December 1975 (p. 1927) cites solar radiation reaching land surface as 731×10^{12} GJ.

d Kelly (1978) cites a cost of $11 per peak watt of output for a photovoltaic system. The common relationship for these systems is that each dollar per peak watt results in electricity costing $0.20/kwe. Thus, this electricity costs $2.20/kwe. The current ratio of primary energy flow to dollar flow in the U.S. economy is 13500 kcal/$. For a high quality manufactured item the price may be multiplied by this ratio to calculate an approximate embodied value. The yield ratio of a photovoltaic system exclusive of storage is 0.1. Ehrlich, Ehrlich, and Holdren (1977) cites a figure of 3.397×10^{4} GJ for solar operating at 10% efficiency over 1% of the earth's surface.

e Yield ratio from Odum (1978). Gross resource abundance is as for Footnote d.

f Whittaker and Likens (1975) cite 32.4×10^{12} GJ as the energy of biomass of terrestrial ecosystems.

g Leith (1975) cites 1.76 and 10^{12} GJ as the net primary production of terrestrial excluding lakes and streams.

h Author estimate. Assumes potential annual biomass resource is 20% of net primary production.

i Woodwell, Whittaker, Reiners, Likens, Delwiche, and Botkin (1978) cite a total release of carbon to the atmosphere from forest clearing and harvest and agriculture of 7.4×10^{15} g. An additional 5% is assumed to enter permanent storage. This gives a total of 7.8×10^{15} g C. Assuming 4.25 kcal/gm organic matter and 0.45 gm Carbon per gm organic matter gives 0.308×10^{12} GJ.

j Whittaker and Likens (1975) cite 1.22×10^{9} MT as annual agricultural harvest and 2×10^{9} MT as annual forestry harvest, for a total of 3.22×10^{9} MT biomass harvest. Assuming 4.25×10^{6} kcal per MT this equals 0.0573×10^{12} GJ.

k McCann and Sadler (1976) cite 125.9 MJ/kg as the energy cost of methane. Assuming 239 kcal/MJ and 12,000 kcal/kg gives a yield ratio of 0.40. Author estimates potential farm waste biomass as 2×10^{12} watt. yr. Biomass to methane conversion cited by McCann and Sadler was 44% efficient, or

0.028×10^{12} GJ.

l McCann and Sadler (1976) cite 147 MJ/kg as the energy cost of methanol. Assuming 239 kcal/MJ and 4,800 kcal/kg gives a yield radio of 0.14. Author estimates as potential photosynthesis fuel 8×10^{9} kw-yr. Assuming that one-half of this is methanol and one-half is ethanol, this is 0.126×10^{12} GJ.

m McCann and Sadler (1976) cite 92.4 MJ/kg as the energy cost of ethanol. Assuming 239 kcal/MJ and 6800 kcal/kg gives a yield ratio of 0.31. Author estimates as potential photosynthesis fuel 8×10^{9} kw-yr. Assuming that one-half of this is methanol and one-half is ethanol, the gross resource abundance is as calculated in Footnote 1.

n Blankenhorn, Bowersox, Hillebrand, and Murphey (1977) give date on an energy plantation which allow the calculation of a yield ratio of 8.0, if the energy cost of chipping and drying is excluded, and the muscular energy for labor instead of its embodied energy is utilized. If chipping and drying are included, the yield ratio is 3.2, excluding the energy cost of the chipping facility. Author estimates fuel wood energy 0.085×10^{12} GJ.

o Yield ratio from Odum and Odum (1976). Grenon (1977) cites hydroelectric potential of 0.0450×10^{12} GJ.

p IEEE Transactions on Power Apparatus and Systems, November – December 1975 (p. 1927) cites energy available in winds over the globe as 57.0×10^{12} GJ.

q IEEE Transactions on Power Apparatus and Systems, November – December 1975 (p. 1927) cites energy available in winds over the globe as 57.0×10^{12} GJ.

r Yield ratio from Odum and Odum (1976), for a wind generator operating under 10 mph average wind-speed. Ehrlich, Ehrlich, and Holdren (1977) cite an estimated exploitable electric production from the wind of 0.0947×10^{12} GJ.

s Since power output from the wind varies as the cube of windspeed, the same wind generator as Footnote r will yield eight times as much energy operating at 20 mph average. The yield ratio is (8) (0.28) = 2.2. Merriam (1978) states that about 2% of wind recording stations in the western United States have averages exceeding 15.7 mph. If high speed areas are 2% of the estimated exploitable electric production cited in Footnote 4, and this is at eight times the power, the gross abundance is 0.0152×10^{12} GJ.

t Odum and Odum (1976).

u Merriam (1978) cites as hydraulic energy at potential tidal power cites 0.00480×10^{12} GJ.

v Merriam (1978) states that a maximum of 25% of the hydraulic energy available at tidal power sites is convertable to electricity. Gross resource abundance is 0.00120×10^{12} GJ.

	POSITIVE NET ENERGY		NEGATIVE NET ENERGY	
ABUNDANCE:	ABUNDANT	RARE	ABUNDANT	RARE
CLASS:	I	II	III	IV
NONRENEWABLE:	COAL FISSION NATURAL GAS OIL	GEOTHERMAL	BREEDER FUSION OIL SHALE TAR SANDS	
RENEWABLE:		HYDROELECTRIC TIDAL ELECTRIC WIND ELECTRIC (20 MPH) WOOD	PHOTOVOLTAICS POWER TOWER	ETHANOL METHANE METHANOL WIND ELECTRIC (10 MPH)

Figure 4--Classification of energy resources as to nonrenewable or renewable, positive or negative net energy, and relative resource abundance or rarity.

Fourth, we need to assess the multiplicative effects of declining quality and quantity of energy materials, and water.

Fifth, we need to investigate the opposing effects of declining embodied energy in materials due to conservation, and the increase in embodied energy caused by declining net energy of primary resources.

Sixth, we need to analyze the effects of declining embodied energy in labor due to conservation and lifestyle changes resulting from high prices.

Finally, we need to evaluate net energy through a time sequence, looking at time lags and the energy costs of reindustrialization during an era of declining net energy.

LITERATURE CITED

Anonymous. 1977. New spent fuel policy unveiled. Science 199:591.

Blankenhorn, P. R., T. W. Bowersox, J. Hildebrand, and W. K. Murphey. 1977. Forest biomass evaluation procedure for consideration as a fuel source for a 100 Megawatt Electric Generating facility. School of Forest Resources, Pennsylvania State University, University Park.

Brown, H. 1976. Energy in our Future. Annual Review of Energy. 1:1-36.

Commoner, B. 1978. Freedom and the ecological imperative: beyond the poverty of power. p. 11-49. In appropriate visions, R. C. Dorf and Y. L. Hunter (eds.) Boyd and Fraser, San Francisco, Cal.

Ehrlich, P. R., A. H. Ehrlich, and J. P. Holdren. 1977. Ecoscience: population, resources, environment. 1051 p. W.H. Freeman, San Francisco, Cal.

Gilliland, M. W. 1975. Energy analysis and public policy. Science 189:1050-1056.

Grenon, M. 1977. Global energy resources. Annual Review of Energy 2:67-94.

IEEE Transactions on Power Apparatus and Systems. November -December, 1975. P. 1927.

Kelly, H. 1978. Photovoltaic power systems: a tour through the alternatives. Science 199: 634-643.

Kylstra, C. and K. Han. 1975. Energy analysis of the U. S. nuclear power system. Section IV-F, p. 138-205. In Energy models for environment, power and society. Report to U. S. ERDA, Department of Environmental Engineering Sciences, University of Florida, Gainesville.

Leith, H. 1975. Primary production of the major vegetation units of the world. p. 203-215 In Primary productivity of the biosphere, H. Leith and R. H. Whittaker, Eds. Springer-Verlag, New York. N. Y.

Lovins, A. B. 1977. Soft energy paths. Harper & Row, New York. 231 pp.

Maugh, T. H., II. 1978. Tar sands: a new fuels industry takes shape. Science 199:756-760.

McCann, D. and H. Sadler. 1976. Photobiological energy conversion in Australia. Search 7: 12:17-23.

Merriam, M. F. 1978. Wind, waves, and tides. Annual Review of Energy. Volume 3

Mining Congress Journal. March 1976. P. 20.

Odum, H. T. 1978. Net energy from the sun. p.126-211 In Sun! : A Handbook for the Solar Decade. Friends of the Earth. San Francisco, Cal.

Odum, H. T. and E. C. Odum. 1976. Energy Basis for Man and Nature. McGraw-Hill, New York.

OECD Nuclear Energy Agency. 1977. Uranium: resources, production, and demand. Organization for Economic Co-operation and Development, Paris. Pp. 20-21.

Parkins, W. E. 1978. Engineering limitations of fusion power plants. Science 199:1403-1408.

Resnikoff, M. 1975. Expensive enrichment. Environment 17(5):28-35.

Rossin, A. D. and T. A. Rieck. 1978. Economics of nuclear power. Science 201:582-589.

Whittaker, R. H. and G. E. Likens. 1975. The biospher and man. Pp. 305-328 In Primary productivity of the biosphere, H. Leith and R. H. Whittaker, Eds. Springer-Verlag, New York, N. Y.

Woodwell, G. M., R. H. Whittaker, W. A. Reiners, G. E. Likens, and D. Botkin. 1978. The biota and the world carbon budget. Science 199:141-146.

Zebroski, E. and M. Levenson. 1976. The nuclear fuel cycle. Annual Review of Energy. 1:101-130.

A METHODOLOGY FOR ASSESSING NET ENERGY YIELD AND CARBON DIOXIDE PRODUCTION OF FOSSIL FUEL RESOURCES[1]

Michael S. Burnett[2]

Abstract.--A methodology for evaluating fossil fuel resources on the basis of net energy yield per unit of carbon dioxide produced is developed. The net energy yield ranges from 24.3 x 10^3 BTU to 66.6 x 10^3 BTU per kilogram of atmospheric carbon produced.

INTRODUCTION

Two inputs critical to the development of sustainable energy policies at the national and global levels are the net energy yield and the carbon dioxide production for various energy options. The former forms the basis for the maintenance and growth of technological societies, while the latter through its climatic implications might limit the use of fossil fuels. Comparative analyses of fossil energy system have been performed independently for each of these measures of resource viability. The purpose of this paper is to synthesize the results of the comparisons into one measure of energy benefit to society: net energy yielded per unit of atmospheric carbon produced.

Net energy is defined as the energy yielded after the energy utilized to produce the energy is subtracted. It is the net energy yield of the fully developed resource, not the gross energy values of the geological reserve, which measure the true value of an energy resource to society (Odum 1973). The net yields of our primary fossil fuels have been dropping as reserves have become deeper, more remote, and more dilute (Hall and Cleveland 1981). This has resulted in increasing importance of net energy analysis to national energy policymakers.

Concern about atmospheric carbon dioxide levels derives from its ability to stimulate a greenhouse effect. The global average temp-erature is very sensitive to the abundance of carbon dioxide in the atmosphere, and the atmospheric pool of carbon is much smaller than the pool of recoverable carbon contained in fossil fuel resources (Baes, et al. 1977). Atmospheric carbon dioxide levels have been rising at a rate proportional to the exponential growth in fossil fuel consumption fostered by the industrial revolution (Rotty 1978).

Rotty (1978) has shown that the production of CO_2 per unit of energy output ranges from 14.4 kg C/million BTU for natural gas to 33.8 kg C/million BTU for synthetic fuel. These figures are based upon the calculated amounts of CO_2 produced by energy conversion of a unit weight of fossil fuels (Keeling 1973). It is apparent that if a policy decision is made to limit the annual CO_2 production to a given amount, then the total energy available to society will depend upon the mix of fossil resources utilized.

The range in the above figures was calculated in terms of gross energy output, not net energy output. Since it is net energy which is the true social benefit of a resource, the actual energy available to society's end users in the carbon dioxide limiting situation must be measured in terms of net energy. The purpose of this paper is to develop such measure: net energy yielded per unit of atmospheric carbon produced.

METHODOLOGY

For an energy production process, the energy yield ratio divides the gross energy output by the total amount of energy consumed in the production process. These latter energy feedbacks include both energy directly consumed in the production process, and that embodied in materials and equipment used in the production process.

[1] Paper presented at the international symposium Energy and Ecological Modelling, sponsored by the International Society for Ecological Modelling. (Louisville, Kentucky, April 20-23, 1981)

[2] Michael S. Burnett is Energy Systems Analyst, Western Solar Utilization Network, Portland, Oregon, USA.

712

(a) Energy flows

(b) Feedback as internal loss

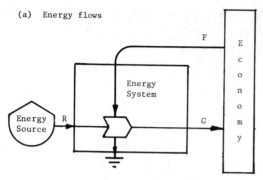

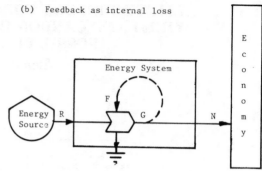

Figure 1.-- Basic net energy model

In figure 1.a., the feedbacks F are necessary to develop the resource R and produce a gross yield of G. The net yield is the difference between the two: N = G - F. The energy yield ratio equals G/F: this quantifies the number of units of gross energy produced per unit of energy feedback to the process.

Figure 1 b. shows an equivalent model where the feedback energy F is quantified as a loss internal to the production process, leaving net energy as the only output from the process. This is obviously unrealistic, and is used as an approximation for analytical purposes.

The energy yield ratios for four representative fossil energy production systems--natural gas, petroleum, coal/electric, and synfuel from coal--were derived from Melcher et al. (1976).

Since Rotty's work was quantified in terms of gross energy, it was necessary to calculate net/gross energy ratios in order to convert his results into net energy terms. These ratios are related to the energy yield ratio.

Rotty calculated carbon output per unit of energy produced. In this paper, energy output per unit of carbon produced is examined. This necessitates conversion of Rotty's figures of kg $C/10^6$ BTU into the units of 10^3 BTU/kg C.

These figures were then multiplied by the net/gross ratios in order to obtain the units desired for use in comparison of fossil systems: Net energy output per unit of atmospheric carbon produced.

Figure 2 shows a conceptual model of the processes under study. The economy is driven by two inputs: the net energy output of fossil fuel production systems and the outputs of the biosphere and climatic system. Fossil energy production processes have two outputs: net energy and carbon dioxide. The net energy has beneficial effect on the economy, while the carbon dioxide, through its adverse effects upon climate and the biosphere, has detrimental effect.

$$\frac{N}{G} = \frac{G-F}{G} = 1 - \frac{F}{G} = 1 - \frac{1}{X} \qquad (1)$$

where X is the yield ratio.

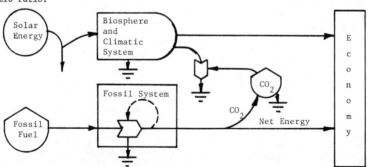

Figure 2.--Energy circuit diagram of net energy yield and
carbon dioxide production of fossil energy
systems, and their interactions with the economy.

RESULTS

Tables 1 and 2 summarize the results of the analysis. In table 1, the energy yield ratios range from 23.2 for natural gas to 5.7 for synfuel from coal. The corresponding range in net/gross ratios is from 0.96 to 0.82.

Table 2 shows both the gross and the net yield per unit of carbon produced. The latter is calculated by multiplying the former by the net/gross ratio. These results vary by a factor of nearly three: from 66.6×10^3 BTU/kg C for natural gas to 24.3×10^3 BTU/kg C for synfuel from coal. As compared with natural gas, the synthetic fuel yields only 36% as much net energy per unit of atmospheric carbon produced.

A graph summarizing these results appears in figure 3. There is apparently a correlation between the gross energy yield unit of atmospheric carbon produced and the energy yield ratio. The lease squares linear regression fit is

$$y = 2.1x + 21.8, \quad r^2 = 0.93. \qquad (2)$$

Table 1. Energy yield ratios and net/gross ratios for four fossil energy systems.

Energy System	Energy Yield Ratio	Net/Gross Ratio
Natural Gas	23.2	0.96
Petroleum	10.0	0.90
Coal-Electric	8.6	0.88
Synfuel (from coal)	5.7	0.82

Source: Melcher et al. (1976).

Table 2. -- Net energy yield per unit of atmospheric carbon dioxide produced for four fossil energy systems.

Energy System	10^3 Gross BTU / Kg C [1]	10^3 Net BTU / Kg C	Ratio To Natural Gas
Natural Gas	69.4	66.6	1.00
Petroleum	49.1	44.2	0.66
Coal	39.4	34.7	0.52
Synfuel (from coal)	29.6	24.3	0.36

1. Source: Adapted from Rotty (1978).

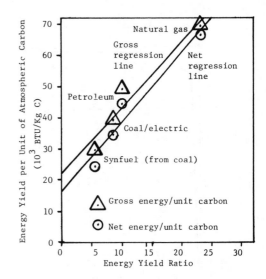

Figure 3.--Atmospheric carbon production and energy yield ratio for four fossil energy systems.

DISCUSSION

Two general conclusions may be drawn from these results:

o The gross energy yield per unit of atmospheric carbon produced is directly related to the energy yield ratio for the fossil energy systems investigated.

o As the energy yield ratio declines, there are increases in both the absolute and the proportional differences between the gross energy produced per unit of atmospheric carbon produced and the net energy produced per unit of atmospheric carbon produced.

If it becomes necessary to establish a policy limiting the total carbon delivered to the atmosphere by fossil fuel consumption, the first conclusion has dramatic ramifications. Suppose that the current average energy yield ratio for all fossil fuels is 20, and that it corresponds to a figure of 60×10^3 BTU/kg C. Suppose also that shortly after the turn of the century, natural gas and oil resources are exhausted (Hubbert 1973) forcing a dependency on sources with an energy yield ratio of 10. This would correspond to a 40×10^3 BTU/kg C figure. Thus, in the event that a carbon-limiting policy becomes necessary, it is conceivable that the total gross energy production may be cut by one third early in the next century.

The effect of the second conclusion is graphically illustrated in figure 4. For energy

714

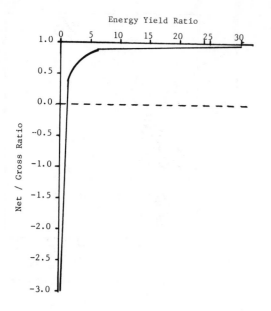

Figure 4.--Net/gross ratio vs. energy yield ratio.

yield ratios above 10, the net/gross ratios are greater than 0.9, and the difference between net and gross energy may be considered insignificant. Below an energy yield ratio of 10, the net/gross ratio drops precipitously and becomes a significant factor. While an energy yield ratio of 10 corresponds to a net/gross ratio of 0.90, 5 corresponds to 0.80, 2 corresponds to 0.50, 1 corresponds to zero (no net energy), and 0.1 corresponds to -9.0 (negative net energy).

Most of our remaining energy options lie in this very sensitive range of energy yield ratios of 10 or below. If we develop a major synfuels capacity with an energy yield ratio of 5, this would correspond to an approximate figure of 30 x 10^3 gross BTU/kg C. However, since the net/gross ratio is 0.80, this actually represents 24 x 10^3 net BTU/kg C. Compare this with the example baseline of 60 x 10^3 gross BTU/kg C at a 0.95 net/gross ratio, or 57 x 10^3 net BTU/kg C. While the gross yield has been cut in half, the net yield has decreased by nearly three fifths (58%).

In this paper a methodology synthesizing net energy analysis with an environmental parameter is developed in order to evaluate the relative performance of energy policy options. While atmospheric carbon was the pollutant studied here, this methodology may be generalized to other environmental problems, such as acid rain, radiation, and carcinogens.

This paper has two significant results for policy-makers should it become necessary to limit carbon dioxide production from fossil fuel conversion. First, the gross energy available to society may decline dramatically. Second, the net energy which sustains the economy may decline more dramatically. Since most of our future policy options lie in the range which is very sensitive to net energy considerations, net energy analysis should provide a necessary complement to economic analysis when selecting among policy options.

LITERATURE CITED

Baes, C.F., Jr., H.E. Goeller, J.S. Olson, and R.M. Rotty. 1977. Carbon dioxide and the climate: The uncontrolled experiment. American Scientist 65: 310-320.

Hall, C.A.S. and C.J. Cleveland. 1981. Petroleum drilling and production in the United States: Yield per effort and net energy analysis. Science 211: 576-579.

Hubbert, M. K. 1973. Energy resources. Pp. 157-242 in Resources and man, National Academy of Sciences-National Research Council. W. H. Freeman, San Francisco.

Keeling, C.D. 1973. Industrial production of carbon dioxide from fossil fuels and limestone. Tellus 25: 174-198.

Melcher, A.G., K. Maddox, C. Prien, T. Nevens, V. Yesavage, P. Dickson, J. Fuller, W. Loehr, R. Baldwin, and R. Bain. 1976. Net Energy Analysis: An Energy Balance Study of Fossil Resources. Colorado Energy Research Institute, Golden, CO.

Odum, H.T. 1973. Energy, ecology, and economics. Ambio 2:220.

Rotty, R.M. 1978. Atmospheric carbon dioxide: Possible consequences of future fossil fuel use. Resources and Energy 1:231-249.

YIELD PER EFFORT AS A FUNCTION OF TIME AND EFFORT FOR UNITED STATES PETROLEUM, URANIUM AND COAL[1]

Charles A. S. Hall[2], Cutler J. Cleveland[3], and Mithell Berger[4]

Abstract.--We have examined the behavior of several energy industries over time using concepts borrowed from fisheries analysis. Petroleum, coal and uranium show similar trends in that there is a secular trend of decrease in energy obtained per effort as well as an inverse relation to the instantaneous rate of exploitation at any point in time.

The phrase "net energy analysis" has been used in enough different ways to lose its utility for research or policy. It seems important to us to define net energy analysis so that unambiguous policy criteria can be met. For example, it would be useful to establish a set of conditions under which a net energy assessment provides information that would allow us to reject a technology because it does not have a positive contribution to our economic system, or provide a means by which different technologies could be ranked based upon their net energy contribution to the system of interest. Later refinements can help assess more ambiguous situations and we may choose, of course, to select one energy technology over another for reasons other than relative rankings of net energy assessments. This paper presents one final (petroleum) and two preliminary (nuclear and coal) analyses of major energy systems for the United States. We also include two interrelated refinements of the net energy return approach: an examination of trends observed for specific industries over time and an analysis of the effect, both cumulatively over time as a

resource is depleted and at any one point in time--showing how increased effort decreases yield per effort. Since the data bases available for these analyses are varied it is not always possible to make precise intercomparisons or to provide the data in the same format, but we believe that the results indicate several clear and disturbing trends.

We prefer the term "energy return on investment" rather than the more commonly used "net energy analysis" because it is analogous to the economic term "return on investment" both conceptually and with respect to the goals of economic investment. For our purposes here:

$$\text{EROI} = E_o * Q/(E_c + E_f + E_r + E_d + E_m) \quad (1)$$

where,

EROI = energy return on investment
E_o = kilocalories output of the project that is delivered to final demand
Q = quality factor (1 for fossil fuels, or 3 for electricity)
E_c = energy cost of capital equipment including replacement
E_f = energy cost of fuel necessary to run the capital equipment
E_r = energy required for repairs
E_d = energy for transporting the end product
E_m = energy cost of maintenance of all required equipment.

Further corrections for labor etc. are given in Hall et al. (1979a; 1979b). All terms must be defined for the same time period, usually the nominal life span of the project.

We define "energy return on investment 1" (EROI1) as the assessment of energy return on energy invested where numerator and denominator

[1]Paper presented at the international symposium Energy and Ecological Modelling, sponsored by the International Society for Ecological Modelling. (Louisville, Kentucky, April 20-23, 1981).

[2]Charles A. S. Hall is Assistant Professor of Ecology, Cornell University, Ithaca, New York U.S.A.

[3]Cutler J. Cleveland is a graduate student, Department of Marine Sciences, Louisiana State University, Baton Rouge, Louisiana U.S.A.

[4]Mitchell Berger is a graduate (B.S. with Honors) of Cornell University, Ithaca, New York U.S.A.

Financed by funds from the New York College of Agriculture and Life Science.

716

both are measured in similar units (e.g. fluid hydrocarbons or kcal) and where both on-site and off-site industrial energy is used, i.e. E_c, E_f, E_r and E_m. We do not include in this stage E_d or the extremely important but more difficult to quantify labor and environmental energies. The only "quality" corrections used in this analysis, in both numerator and denominator, are those for converting fluid hydrocarbons or coal into electricity, since a unit of electricity is "worth" (i.e. requires) the consumption of about 3 times its thermal equivalent of primary fuel. EROI1 gives an unambiguous means of rejecting an energy technology unless less desirable fuels are used to produce more desirable fuels, i.e. coal used to run oil-field equipment.

Energy return on investment 2 (EROI2) includes the energy cost of delivery of the energy to the consumer (E_d); energy return on investment 3 (EROI3) also includes an assessment of the fossil energy embodied in labor (E_e); and energy return on investment 4 (EROI4) includes an energy assessment of environmental losses and gains (E_e). While we think it important to include all appropriate energy costs, we choose to exclude environmental costs in this paper since the methodologies for assessing them are not yet well-developed. The analyses presented in this paper are determined in large part by what data were available and are in general not sufficiently comprehensive to do all of the above analyses. Where possible we include sensitivity analyses, but all of our estimates of total energy costs are conservative, making a given energy technology look more favorable than if we included other components.

U.S. PETROLEUM

This analysis and its limitations are detailed in Hall and Cleveland (1981), although we include here a brief summary and some additional commentary and graphs.

The analysis is essentially that of Schaefer (1957), who examined the effect of fishing effort on catch per effort for tuna fish. Our earlier attempts to analyze raw data on the barrels found per foot of drilling in the oil industry appeared not to make sense--until it became obvious that the catch per effort for oil, like that for fish, was very responsive to effort. We include figures 1 and 2 from our original paper as a summary and include figures 3, 4, 5 and 6 as a more comprehensive demonstration of the very strong correlation between "catch per effort" and effort for oil during this 3 decade period.

As indicated in the earlier paper, the data for 1979 indicated a higher yield per effort than our statistical model predicted based on the previous three decades. The yield per effort for 1980 also was higher than predicted but lower than for 1979 (fig. 4). Possible reasons for this include: 1) a statistical abnormality, 2) that

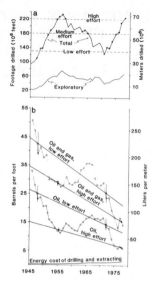

Figure 1.--(a) Rates of drilling for the U.S. domestic petroleum industry. Open circles represent years of low effort, half-filled circles years of medium effort, and filled circles years of high effort. (b) Gains and costs of petroleum exploration and development for the United States, 1946 through 1978. The energy content of gas used and found was converted to barrels of oil equivalents (5680 cubic feet of gas = 1 barrel of petroleum). The topmost irregular line is annual yield per effort for oil and gas, the middle irregular line is annual yield per effort data for oil alone, and the bottom line is the mean energy cost of exploration and development per foot drilled that year. Vertical bars and associated points are the variance (where possible) and mean of years of low and high effort as per (a). The four straight sloping lines are the least-squares fits of all low-effort and all high-effort data to a linear regression (from Hall and Cleveland 1981).

our rate of becoming more clever at finding petroleum has increased relative to the rate of increase from 1946 to 1978, 3) that we may be drilling a larger proportion of our wells in genuine "new-province" regions that presumably would have, on average, a higher return per foot drilled, although not necessarily a higher return per dollar or calorie invested, 4) that we may be reclassifying a large number of known but previously uneconomic fields (such as happened with the large but relatively low-grade Kern River field in 1979) due to the economic incentives resulting from the increase in the economic value of oil relative to the economic cost of obtaining it, and 5) the same incentives may be causing petroleum companies to exploit previously known but previously ignored small fields that are "sure bet" strikes—but of small size. It is important to realize that the modest gains in yield made during 1979 and 1980 compared to 1976-1978 were accomplished only by very large increases in drilling effort. The annual percentage increases in drilling effort in 1979 and 1980 were consecutively by far the largest in the history of the petroleum industry.

Thus it is possible that the increase in the inflation-corrected price of petroleum and the response by the industry to these increased incentives may be adding to our drilling success without substantively changing our "known" oil reserves—we may be simply classifying as reserves small and low-grade reservoirs that have actually been known for many years, a practice that artificially inflates our "finding" rates for at least a few years. Alternately, the intense exploration of new regions such as the overthrust belt of the Rockies, the Tuscaloosa trend and various deep sediments may be paying off—it is too early to tell. Our feeling since writing Hall and Cleveland (1981) is that the future finding rates per foot might be a little higher than predicted in figure 2—but that so will the costs per foot—both in response to the increased real price of petroleum. Until the increased energy costs of drilling and extraction can be assessed empirically, it is impossible to determine whether or not the small increases in yield per effort in 1979 and 1980 are in fact adding a positive net amount of oil to the nation's proved reserves.

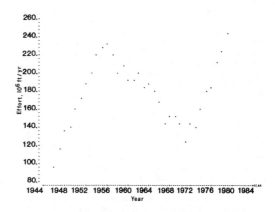

Figure 3.—Effort in United States petroleum industry as per figure 1a (computer-plotter output).

NET ENERGY FROM PRESENT-DAY U.S. NUCLEAR POWER

The objective of the next study was to access the feasibility of nuclear power from the perspective of energy return to society compared to the energy invested in certain parts of the nuclear fuel cycle (e.g. EROI1). This paper does not address certain potentially important questions such as environmental impacts or the energy costs of a large scale accident. Thus our analysis is only the beginning of a total assessment of nuclear power but we believe that some of the more important aspects are covered well.

Seven earlier studies have compared the electric output generated by a nuclear power plant with the energy required as inputs in an attempt to determine the energy return on investment for nuclear power. Such studies were suggested by the fact that the nuclear power research and fabrication industries are both very dollar- and energy-intensive, and apocryphal reports in the late 1960's that while nuclear power generated 4% of the nation's energy it required some 4% of our electrical output just to run building K-9, the gaseous diffusion plant (and the nation's largest building) at Oak Ridge Tennessee. Clearly, if nuclear power required as much energy as it produced there would be no particular reason to trade coal-generated electricity for an equal quantity of nuclear-generated electricity. On the other hand, uranium may be an effective "coal extender," increasing the quantity of electricity that can be obtained from a ton of coal. It should be noted that even energy technologies that have a positive EROI may be poor choices when compared with alternative systems that have higher ratios.

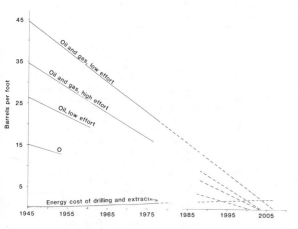

Figure 2.—Linear extrapolations (dashed lines) of energy gains and energy costs of figure 1 for high and low drilling intensity. Solid lines are from figure 1b. The inclusion of Prudhoe Bay finds into this extrapolation would extend the time of intersection by about 6 years.

The results of these earlier studies normally were expressed as the ratio of energy produced to energy invested (E_p/E_i) for the lifetime of the plant. But these studies differed greatly in their conclusions and produced confusion in what should be a more or less unambiguous analysis. Some studies stated that nuclear power will yield a large return on investment (e.g. Chapman and Mortimer 1974; E_p/E_i=15.6), others found that nuclear power yielded a much smaller return on investment (e.g. Lem, Odum and Bolch 1974: E_p/E_i=3.23), and the rest ranged all over the interval (table 1).

We converted all the energy requirement values given in these studies for each process used to produce nuclear power (i.e. mining, milling, conversion to gaseous uranium, enrichment, fuel rod fabrication, reprocessing, waste storage, plant construction, etc.) from their various original units into units of kcal of heat. Ideally this included the energy content of fuels used or purchased for carrying out a process and the energy required to produce the materials and capital equipment required for the process, including the energy requirements for production of raw materials (EROI1). In practice some studies did and some did not include energy estimates of capital equipment and this accounts for a small part of the variation among the results of the different studies. We do not include the energy input as uranium ore but assume our investment energy of interest is fossil fuel. If the total energy content of the uranium fuel were counted the energy return on energy invested would be much lower.

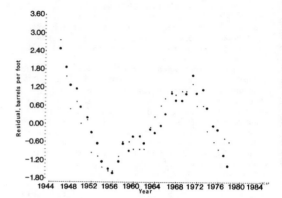

Figure 5.--Residual from regression of figure 4, i.e. difference between actual data and linear model. Solid dots are inverse of effort (i.e. figure 4, flipped over).

No estimates for costs of transporting electricity to consumers were included in the original paper but this presumably is about 3% transmission losses (Hall et al. 1979a). An inclusion of this factor would produce an EROI2 estimate 3% less than EROI1.

The energy used by labor was not included explicitly in any analysis studied.[†] There is disagreement among energy analysts as to whether to include, for example, the energy it takes to get a worker to the factory, the energy required to manufacture his or her automobile, to educate the worker, to feed his or her family, etc. One could argue that if the energy requiring recompense was not included, labor would not be available. There is some risk of double counting, however, for some small part of a worker's salary will be purchasing energy or energy-derived products from the operating power plant.

The most important results of our comparison of these seven studies was that nearly all gave much more similar results (5 ± 1.5 kcal of electricity returned per kcal invested) when standardized to similar assumptions, operating parameters, fuel qualities, and boundary assumptions (table 1). In other words, the results of these energy return-on-investment analyses is much more a function of the assumptions used by the different investigators than properties inherent in different power systems or the data base. In particular, the most important assumption was whether or not a quality factor was given to the electrical output when compared to fossil energy inputs. Weighting the output by a factor of three for the quality of electricity gives an energy return on investment of

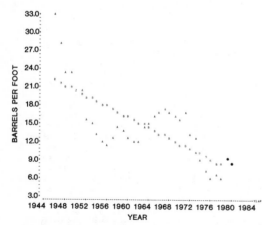

Figure 4.--Yield per effort over time (as per figure 1b) including a best least-squares linear fit (a) to the data (b) from 1946 to 1978 ($r^2 = 0.52$). Solid circles are for 1979 and 1980 which are used for a separate analysis (see text).

[†]But see Kylstra and Han (1975).

approximately 15 to 1. Other important considerations were the load factors assumed, boundaries
selected for analysis (e.g. whether an energy cost
was given for waste storage or potential accidents)
and variations in the estimate of energy cost for
some critical components such as the fabrication
of fuel rods. Details are given in Berger (1981).

Projections of Future EROI for Fission Energy

A preliminary assessment follows of fission
EROI scenarios for the future as a function of
the continuing depletion of higher grades of
uranium ore. Generally, when mining any reasonably
well-explored resource, the best grades (i.e.
highest percent mineral per ton of ore) are
exploited first. As these higher grades are
depleted, lower and lower grades of ore are
utilized. Such has been the case for the copper
industry (Lovering 1969) and also holds for the
uranium industry (fig. 7).

We developed a simple computer model to
simulate future energy return on investment for
nuclear power plants. This model was made with
the assumption that present technology would be
used for future fuel processing. It also assumed
that mean EROI values for presently operating
plants and processes would be characteristic of
future plants and processes except that the energy
cost of mining and milling was linearly proportional
to present day costs and directly, but inversely,
proportional to ore grade. A different analysis
would have to be made if reprocessing or fuel
breeding were used, but neither is being used
presently.

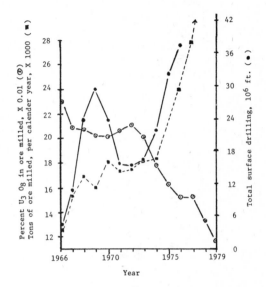

Figure 7.--Drilling rates (combined exploratory
and development, solid circles), production
(of U_3O_8 concentrate, squares) and grade of
ore milled for the domestic uranium industry
(sources: Zimmerman (1978) (drilling rates);
and USDOE GJO-100 (1980)) respectively. The
general decline in ore grade is also inversely
proportional to drilling rates and perhaps
to instantaneous production rates as well.
See also Facer (1977).

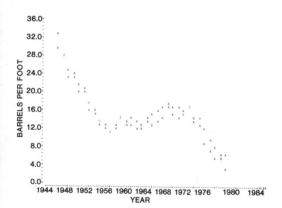

Figure 6.--Multiple regression model similar to
figure 4 but including an inverse linear
function for effort ($r^2 = 0.92$).

The model results indicated that in 1961,
mining and milling represented approximately 2.5%
of the total energy investment for a nuclear
power system. But as lower quality ores were
used, simulated costs for mining and milling
increased by as much as 20 fold (i.e. to 45%)
when low grade Chattanooga Shales were utilized.
Consequently, as time progresses, the energy cost
of mining and milling will become more significant
in determining net energy yield. The energy
return on investment (EROI1 with a quality factor
of 3) decreased from about 15 to 1 to about 3 to 1
when all now known high and moderate grades (i.e.
deposits greater than 0.01% U_3O_8, and available
at less than $100 (1980) dollars per pound U_3O_8)
of ore were depleted. Future research will refine
these values and assess the energy cost and gains
of new technologies such as centrifugal enrichment
and the impact of recycling fissionable materials
in waste.

Therefore, even with the use of low grades
of ore, nuclear power cannot be eliminated as
having a <1.0 EROI within the limited criteria
used in this analysis if we give the electrical
output a weighting of three. On the other hand,
factors not included in this analysis, such as the

increasing safety-related capital costs of plants, public fear of real or potential accidents, and/or the difficulty in producing an effective waste storage system may be more important in restricting the growth of nuclear power. Alternately, reprocessing and breeder reactors might make EROI1 more positive.

YIELD PER EFFORT AND NET ENERGY ANALYSIS OF THE U.S. BITUMINOUS COAL MINING INDUSTRY

The relationship between effort and energy yield-per-effort EROI for the domestic coal industry is not a consistent one as has been the case for the oil and gas industry, where the two are consistently and inversely related to each other. Technological advances in coal extraction techniques may explain some of the observed fluctuations in the relationship between effort and EROI. In general, the EROI for coal at the mine mouth is higher than it is for oil and gas at the well head. For example, in 1977, the EROI for coal, taking into account only those direct and indirect energy costs necessary to

bring the coal to the surface (i.e. not including labor and transportation) was 42.4 kcals returned per kcal invested. When transportation and an energy assessment for labor was included (EROI3) the ratio was 20 to 1. The same figure for oil and gas in 1977 was about 10 to 1 (fig. 1).

Based upon the trends of effort and EROI over time, it is convenient to divide the history of the coal industry into three phases: 1929 to 1954, 1954 to 1969 and 1969 to 1977. The numbers given below for EROI include not only direct and indirect energy costs (EROI1), but labor and transportation costs as well (EROI3). This analysis was an attempt, therefore, to include the majority of energy costs of producing and delivering an average ton of bituminous or lignite coal and to explicitly compare EROI1 with more comprehensive analyses. This more comprehensive analysis was not possible with the data we had available for the petroleum or nuclear industry.

From 1929 to 1954, EROI remained relatively constant at about 30 kcals returned per kcal

Table 1.--Energy return on investment for eleven studies investigating energy return on investment for nuclear power plants.

Source	energy return on investment: original numbers	energy return on investment: (standardized)[a]
Rotty et al.[b]	9.63	4.98
Rotty et al.[b]	8.93	4.49
Rotty et al.[b]	12.11	6.52
Rotty et al.[c]	3.40	2.31
Rotty et al.[c]	4.89	3.31
Rombough & Koen	6.31	5.39
Lem, Odum & Bolch	3.23	2.74
Chapman & Mortimer	10.9±2	3.54
Chapman & Mortimer	15.6±3	6.08
Chapman & Mortimer	12.9±2	5.13
Chapman & Mortimer	16.5±3	6.75

[a]Standardized by assuming the same quality factors (only for electrical inputs), fuel types and load factors (.75).

[b]using 0.176% uranium ore

[c]using 0.006% uranium ore (i.e. Chattanooga Shale)

invested (see fig. 8b) while total energy costs declined slightly (see fig. 8c). During the same time period effort, as measured by production, fluctuated greatly, and peaked at over 600 million tons in the mid 1940's, due mainly to increased war-time energy needs (see fig. 8a). Periods of low effort occurred during the depression and also after World War II due to the expansion of oil and gas use. Effort in the coal industry also is influenced greatly by factors such as labor-management relations, which historically have had great impacts on the behavior of the industry as a whole. During this time period, for example, the number of man days lost due to strikes averaged over 5 million per year.

The consistency of the EROI from 1929 to 1954 may be due to the static nature of coal production technology during this period relative to later years. Conventional underground mining methods produced the vast majority of coal from the early 1900's until well into the 1950's. Unchanging extraction techniques could lead to the observed constant EROI for two reasons. First, in the early stages of the life cycle of a large nonrenewable resource, the energy-intensifying effects of declining resource quality and availability do not yet have a significant impact. Second, the coal industry is unlike the oil industry in that only a small percentage of

its total effort is expended in exploration activities. Coal deposits of roughly equal quality and availability, whose location is already known and which are extracted with similar technologies, could produce a relatively constant EROI.

During the second time period of interest, 1954 to 1969, EROI steadily increased to 36 kcal returned per kcal invested, while production increased from 390 million tons in 1954 to 560 million tons in 1969. The reason for the simultaneous increase of both EROI and production probably is due to technological advances in coal extraction techniques that greatly increased the efficiency of coal production. Continuous mining machines, which incorporate several of the individual stages of conventional mining into one process done by one machine, were not introduced into U.S. coal mines until the late 1940's, and by 1954 accounted for only 6% of total production. By 1969, however, continuous mining machines were producing 50% of total output (National Coal Association 1979). Employment in the coal industry decreased 55% during this period, indicating that increased mechanization was able to replace human labor while also increasing production levels by over 40%. This displacement is also reflected in the changes in the energy costs of production (fig. 8c). From 1954 to 1969 indirect energy costs increased slightly, reflecting the increase in mechanization, while labor costs declined due to a sharp drop in employment. This data indicates that the increased total energy costs incurred due to increased mechanization were more than offset by increases in extraction efficiency, resulting in an increase in the EROI.

Another reason for increased production during this period was the relative stability of labor-management relations. Man days lost due to strikes averaged about 220 thousand from 1954 to 1969, a sharp decrease from the 5 million man days per year from 1929 to 1954.

During the final period of interest, 1969 to 1977, EROI3 and effort exhibit the same type of inverse relationship previously shown for oil and gas and also for uranium. EROI3 decreased from about 36 kcal per kcal in 1969 to about 20 in 1977, while at the same time production increased steadily to about 690 million tons in 1977, the highest output up to that time in the history of the coal industry. It is interesting to note that production levels in the 1940's were about equal to those in the mid 1970's (about 600 million tons), but that the EROI3 for the latter time period was only two-thirds the level it was during the mid 1940's (see fig. 8b).

There are several reasons behind the rapid increase in energy costs during this time. Continued increases in mechanization of both surface and underground mining methods resulted in a sharp increase in direct energy costs. Surface mining in particular, which increased from 38% of total production in 1969 to over 60% in 1977,

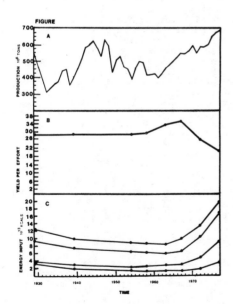

FIGURE

Figure 8.--a) Effort, as measured by production, for U.S. coal industry, b) yield per effort for EROI4, c) total energy used in coal mining.

Table 2.--Energy return on investment for domestic coal.

YEAR	Energy return on investment, kcals out/kcals in			
	Direct Energy Costs Only	Direct & Indirect (EROI1)	Direct & Indirect & Labor	Direct & Indirect & Labor & Transportation (EROI3)
1929	108	94	37	28
1939	133	88	34	29
1954	196	105	39	29
1958	212	103	42	30
1963	217	102	48	34
1967	270	175	62	36
1972	202	76	34	26
1977	112	42	23	20

experienced rapid increases in energy costs due to the development of sophisticated mechanical equipment such as immense draglines and other earth and coal-moving machines. Labor costs also increased due to a growth in employment and real wages. But the important reason for the rapid drop in EROI is probably due to changes in the quality of the resource itself. Especially in the underground mines of Appalachia, lower quality coal seams were being exploited. Decreasing seam thickness, deeper mines and exploitation of seams at more severe angles all led to increased energy costs per ton. At the same time there was a steady decrease in the heat content of the average ton of bituminous coal produced in the U.S. From 1969 to 1977 the heating value of a ton of bituminoius coal decreased about 10% from about 25 million BTU/ton to 22.7 million BTU/ton (U.S. Department of Energy 1979). Unlike the period from 1954 to 1969, increases in direct and indirect energy costs in the form of increased mechanization resulted in decreases in the efficiency of energy delivery by the industry, since these advances did not result in large enough increases in output to counter their increased energy costs. The overall decline in EROI can probably be attributed to declines in the quality and availability of the resource itself.

CONCLUSION

Two energy industries examined here (petroleum and coal), as well as imported oil, examined in Kaufmann and Hall (1981) all show decreasing energy return on investment in recent decades. The fourth (nuclear) shows a decline in grade of fuel that is probably associated with a declining return on investment although that cannot be proven by our present analysis. Additionally, all four industries show a decline in yield per effort (or in resource grade in the case of uranium) as a function of increasing effort at any point in time, although in the case of coal the relationship is true only for the past decade. Such declines in energy returns on investment probably are important contributors to national inflation and if they continue will substantially diminish our future energy options. Only coal appears to have an EROI that is competitive with a large scale program of installing regional insulation (EROI3=18 to 23.3; EROI4=12.9: Sloane et al. 1979) which may continue to represent one of our best national strategies for adjusting to depleting quality and quantity of fuel resources (Stobaugh and Yergin 1979).

LITERATURE CITED

Berger, M. 1981. Energy return on energy invested for presently operating and future nuclear power systems. Honors Thesis, Division of Biological Sciences, Cornell University, Ithaca, N.Y. 23 pp + appendices.

Chapman, P. F., and N. Mortimer. Energy inputs and outputs for nuclear power stations. ERG 005, Energy Research Group, Open University Group, Milton Keynes, Bucks., September 1974, revised December 1974.

Facer, J. F. 1977. Uranium production trends. Supply Branch, Supply Analysis Division. U.S. Department of Energy.

Hall, C. A. S., and C. Cleveland. 1981. Petroleum drilling and production in the United States: Yield per effort and net energy analysis. Science 211:576–579.

Hall, C. A. S., M. Lavine, and J. Sloane. 1979a. Efficiency of energy delivery systems: Part I. An economic and energy analysis. Environmental Management 3(6):493–504.

Hall, C. A. S., E. Kaufman, S. Walker, and D. Yen. 1979b. Efficiency of energy delivery systems: Part II. Estimating energy costs of capital equipment. Environmental Management 3(6): 505–510.

Kaufmann, R. and C. A. S. Hall. 1981 (This Volume).

Kylstra, C., and K. Han. 1975. Energy analysis of the U.S. nuclear power system. Report to Energy Research and Development Administration. ERDA Grant E(40-1)-4398. University of Florida, Gainesville.

Lem, P. N., H. T. Odum, and W. E. Bolch.1974. Some considerations that affect the net yield from nuclear power. Health Physics Society, 19th Annual Meeting, Houston, Texas. July 1974.

Lovering, T. 1969. Mineral resources from the land. p. 109–134. In Resources and Management. National Academy of Sciences. W. H. Freeman and Co., San Francisco, Ca.

National Coal Association. 1979. Coal facts N.C.A. Washington, D. C.

Rombough, C. T., and B. V. Koen. 1975. Total energy investment in nuclear power plants. Nuclear Technology 26:5–11.

Rotty, R. M., A. M. Perry, and D. B. Reister. 1975. Net energy from nuclear power. Institute for Energy Analysis (IEA) Report, Oak Ridge Associated Universities, Oak Ridge, Tenn.

Schaefer, M. B. 1957. A study of the dynamics of the fishery for yellowfin tuna in the eastern tropical Pacific Ocean. Bulletin International American Tropical Tuna Commission 2:245–285.

Sloane, J., C. A. S. Hall, and L. Fisher. 1979. Efficiency of energy delivery systems: Part III. Assessing potential energy savings through a comprehensive regional insulation program. Environmental Management 3(6): 511–515.

Stobaugh, R., and D. Yergin. 1979. Energy future. 353 p. Random House, New York, N. Y.

U. S. Department of Energy. 1979. Monthly Energy Review. Washington, D. C.

U. S. Department of Energy. 1980. Statistical data on the uranium industry. GJO-100. Grand Junction Office, Colo.

Zimmerman, C. F. 1978. Uranium resources on federal lands: an evaluation of policy alternatives. Doctoral Dissertation, Cornell University, Ithaca, N.Y.

THE SOLAR ENERGY MACHINE IS GREEN[1]
Ane D. Merriam[2] and Martha Gilliland[3]

Abstract.--The green plant is the oldest solar collector known. It is a finely tuned solar machine not requiring R&D expenditures to develop. This fact, in conjunction with results of an analysis of energy flow in Capitol Park, Sacramento, California, leads us to conclude that, in cities, a partial transition to the solar age can be accomplished through an aggressive urban forestry program. The challenge is in designing interface systems which directly and consciously couple native vegetation within native ecosystems with the by-products of technology and urban activity.

The Capitol Park complex is comprised of approximately 40 acres, five of which are buildings housing the Legislature and Governor's Office. The remaining 35 acres constitute an open botanical garden of 350 different species of mature trees. The annual value of services provided by the Capitol Park natural ecosystem was evaluated in energy and dollar units using two different techniques: energy analysis was used to measure the energy flow within the system, and traditional economic methods were used to account the cost of providing the services by technological means.

Results of the energy analysis indicate that the Capitol Park ecosystem produces between 0.9 and 4.7 billion BTU's of energy per year (equivalent to fossil fuel in quality), and is worth between $14,000 and $70,000 per year (1975 dollars). Results of the economic analysis indicate that the Capitol Park ecosystem saves between 2.7 and 19.3 billion BTU's of fossil fuel valued between $10,000 and $137,000 per year in services that would otherwise have to be provided if the ecosystem were not present.

The energy analysis indicates a value for the services provided of between $40 and $200 per tree per year; the economic analysis indicates between $30 and $400 per tree per year. The results of the two types of analyses are remarkably similar: energy analysis in fact yields the more conservative values. A comparsion technique is presented to better quantify the societal value of natural systems technologies.

INTRODUCTION

May 3, 1978 was proclaimed nationally as SUN DAY--a day set aside to tout the benefits, wonders and hopes of solar energy for the U.S. The quest for alternative kinds of energy since the 1974 oil embargo has emerged amidst the spirit of a scientific Don Quixote pursuit. Solar energy use on SUN DAY and since has been praised by some (Commoner 1978; Metz 1978; CEQ 1978) as a panacea, while others have continuously questioned the feasibility of using a dispersed, diluted energy source for a concentrated industrial society (Odum 1973; Odum and Odum 1976).

Working in sunny California, there was substantial professional pressure to come up with a striking SUN DAY display for the Governor and his Cabinet to use, with the expressed interest in maximizing the publicity, which is a subset of power in political systems. Yet the competition for press coverage was intense featuring every possible solar gimmick imaginable, ranging from sophisticated photoelectric cells to bath tubs painted black for solar hot tubs.

Our surface challenge was to design a solar display from which the Brown administration could preach an alternative energy message. Yet being somewhat more motivated than the situation called for, we decided to couple this project with some education in the hope that

[1] Paper presented at the International Symposium Energy and Ecological Modelling, sponsored by the International Society for Ecological Modelling (Louisville, Kentucky, April 20-23, 1981).
[2] Water Resources Director, Institute of Science and Public Affairs, Florida State University, Tallahassee, Florida.
[3] Executive Director, Energy Policy Studies, Inc., Omaha, Nebraska.

when SUN DAY was over there would be something to carry over in the aftermath.

Therefore we added some further goals to our assignment including:

- not wanting to preach solar energy use blindly;
- wanting others to understand the energetic implications of solar energy use;
- wanting to tell the story in such a way that key decision makers would use the knowledge in future considerations;
- wanting to develop a new evaluation technique upon which to base some comparisons of alternative technologies, namely solar applications.

With these additional goals in mind, coupled with our assignment, and without much hesitation we decided to turn to and learn from the oldest solar collector known: green plants.

Concepts of urban forestry as it is developing in conjunction with urban renewal programs across the country have lifted the perception of trees from one of aesthetic components in our cities to that of working components as well (Lamphear 1971; SMUD 1979; Bernatzsky 1966). Trees collect, upgrade, and store solar energy. The trees and the forests of which they are a part can and do use that solar energy in ways that directly benefit cities. However, those benefits and associated costs are rarely measured primarily because of the inadequacy of measurement methodologies. This paper describes some of those benefits and presents and uses a theory and methodology for quantifying them. The theory was developed by Professor H.T. Odum (1971, 1978) and the method is being applied and tested in research supported by the National Research Council (Lavine and Meyburg 1976), the Nuclear Regulatory Commission (Center for Wetlands, ongoing), the Environmental Protection Agency (Bayley and Odum 1973) and the Department of Energy (Odum 1976).

The California Resources Agency, under the leadership of Secretary Huey D. Johnson and Governor Jerry Brown, sought to apply these techniques in a situation familiar to most people in California. We set out to investigate and quantify the work (in the form of essential services) being performed by the 35 acres of natural landscape surrounding the State of California Capitol Buildings in Sacramento. This paper includes the results of that investigation. The first part summarizes descriptively and qualitatively some of the known functions of trees and urban forests in cities. The second part describes the energetic theory and methods which allow those functions to be quantified. Our analytical results for Capitol Park, Sacramento, California are then presented, along with some implications of the results for city planning. A final section discusses the energy quality and end use considerations and the need for additional theoretical and applied work in this area.

NATURAL COMPONENTS IN URBAN ECOSYSTEMS: THEIR FUNCTION

Natural components within or around urban sys-

tems provide a host of services of value to the city. Since the energy source for these natural work processes is the sun, no economic cost is associated with them. The exception is, of course, any fertilizing, watering or maintenance costs incurred to sustain nonnative species. Pollution abatement, hydrological control, and climate control have been recognized widely as functions which can be carried out, in part, by trees.

Trees or greenbelts in general absorb noise, air pollutants, and water pollutants. Trees are widely used along highways and surrounding industrial parks as barriers to noise pollution. Along highways they also absorb pollutants from automobile exhaust systems. One study found that a 500-meter wide greenbelt surrounding a factory reduced nitric oxide concentrations in the air by 27%, sulfur dioxide by 26%, and hydrogen sulfide by 1% (Bach 1971). Another compared particulate concentrations in the same part of town on tree-lined and non tree-lined streets (de Albornoz 1973). Along streets without trees, the air contained 10,000 to 12,000 particulates per cc.

In addition, water pollutants in urban runoff waters are taken up by plant communities. In fact, little urban runoff from greenbelts occurs since the plants make use of both the water and pollutants. Municipal sewage effluent can be directed onto greenbelts (either natural or altered for recreational purposes such as a golf course), mitigating the need for advanced sewage treatment as well as for irrigation water normally drawn from the city's water supply system. In some cases, this water may also recharge an aquifer. Plant communities deter soil erosion which further reduces water pollution. As barriers along rivers and estuaries, they act as a buffer zone which protect the city from flood waters.

Greenbelts also exert a major influence on microclimate within cities. In all, they mitigate extremes in temperature, in wind, and in precipitation. It is widely recognized that trees and shrubs around buildings act as insulation. By absorbing wind energy, they reduce the air exchange between the inside and outside of a house. The American Physical Society (1975) found that wind dominates the dynamics of that air exchange; thus wind buffers can be an important energy conservation strategy. Farmhouses on the prairie grasslands are nearly always surrounded by cottonwoods and junipers. These farmers have known for decades that trees mitigate the wind and temperature extremes in the farmhouse. Temperature measurements in forests substantiate this affect, indicating temperature differences of 10 to 20 degrees (F) between the forest floor and the top of the canopy.

These functions are becoming widely recognized and widely discussed. There are many others that are more subtle and more difficult to define, but probably no less important. One of these more subtle functions is performed by the dimension of diversity that ecosystems add to the urban system as a whole. Social science studies as well as evidence from the biological sciences are beginning to show that a diversity of natural and technological systems in cities has value; for example, diversity may represent an element of crime control.

Technology can also perform these functions, and the trend in our cities over the past 100 years has been to engineer some kind of a device to perform them. A wall constructed of special materials will insulate people from highway noise or industrial park noise; automobile emission control systems will remove air pollutants from car exhaust; stack gas control systems will remove pollutants from industrial stacks; larger sewage treatment plants and separate storm water sewerage systems will control water pollutants; recharge wells will recharge aquifers; specially engineered barriers will stop soil erosion and protect cities from floods; consumption of large amounts of fuel energy for heating and cooling will control temperatures within buildings; and even larger police forces will be invoked to regulate dense populations. All of these technological devices replace functions that can be performed entirely or in part by plant communities and all require enormous fuel investments.

To the extent that "urban forests" can replace technology in cities providing equivalent functions, scarce fuels will be conserved and city budgets may be reduced. We only have to examine the budgets of our big cities to recognize that just the public service component is enormous and growing exponentially. Perhaps some of those services could better be accomplished by urban forests. Solar energy (as it is manifested in trees) can substitute for fuels (as they are manifested in technology) and consequently, can conserve fuels and save money.

The extent to which urban forests can substitute for technology and perform the services described depends largely on the relative density of human activity (and associated industry), forest activity and on the manner in which the by-products of technology are coupled to or interfaced with the forests. A great deal of research remains to be done on interface designs and on the choice of appropriate species of trees. Trees which are more or less isolated from each other by concrete and buildings do not represent much of an interface; much of the energy flow through the trees is uncoupled. For example, trees in isolation and surrounded by tall buildings have a limited exposure to the sun. The small volume of soil around the tree lacks humus; the salt load in urban runoff becomes highly concentrated and may reach levels which are toxic to the tree. The roots of the tree may cause the concrete to crack and falling limbs become a hazard. Similarly, some ecological communities and species types are more adapted than others to pollutant uptake and recycling. Sugar maples (<u>Acer saccharum</u>) seem to do quite well. A study of sugar maples in three sections of Montreal where air pollution levels ranged from very high to very low indicated that leaf hairs on the leaves may develop in response to the high pollution levels (Elias and Irwin 1976). It was hypothesized in the study that the leaf hairs collect the particulate matter in the air, keeping it from entering the pores.

The point is that greenbelts probably need to be complete ecological communities of a size, spacing and composition is yet largely unknown. But, more serious, is the adequacy of methods for measuring the contribution of urban forests (i.e., of solar energy) to the city's economy. The purpose of the Capitol Park study was to make such measurements, evaluating the economic and energetic benefits of the park ecosystem surrounding the Capitol complex. Both traditional economic methods and a relatively new energetic approach were used and the results of the two compared.

THEORY AND METHODS: ENERGY FLOW AS A MEASURE OF ECOSYSTEM SERVICE

The most common approach to the problem of valuing environmental services is one of evaluating the dollar cost of all the technologies and labor which, in combination, would have to be invoked to replace the services provided by the environment. This method is indirect at best. Because no money is exchanged when services are provided by trees and greenbelts, economics must consider those services as external to urban function and urban productivity. In contrast, the theory and methods for evaluating ecosystem functions which are summarized here may allow those services to be internalized and evaluated directly.

Theory

Recognizing that we do not pay a tree, greenbelts, or the sun for the services each provides, Odum (Odum 1971; Odum and Odum 1976) sought a different metric than dollars for their value. The one commodity which is stored and flows through any kind of system is energy. Energy flow occurs in both technological and natural systems and its role in both cases is similar. More specifically, a machine and a living organism are the same in that they both depend on a flux of energy in order to function (Odum and Odum 1972). The machine depends on concentrated fuels; energy is routed through the machine and is eventually dissipated as heat. In the interim, work is done by the machine; some commodity is manufactured or upgraded. Ecologists are quite familiar with the role of sun energy in ecosystems and energy flow among trophic levels. Less familiar is the concept that energy flow through an ecosystem measures the work done by nature, just as energy flow through a machine measures work done. For example, the gross primary productivity of an ecosystem (expressed as energy flow per unit of time) is a measure of the amount of sunlight captured and concentrated by the ecosystem. Portions of this energy are then used by the plants and animals to maintain metabolic processes, to reproduce and to grow. At the ecosystem level, these processes in turn enable "pollutants" to be absorbed, degraded and recycled. In the physics context, energy flow per unit of time is a measure of work performed.

In an urban system, some of the work in support of the city is carried out by technologies using fuels and some of the work is carried out by nature using the sun. In Capitol Park, the five acres of buildings, air conditions, elevators, and lights are a marked contrast to the redwoods, pines and groundcover surrounding them. Only those activities carried out by people through technology are paid for with money; but energy flow drives both the human-technological system and the park system. Energy is consumed by both and are necessary components to the whole system.

This holistic perception is shown diagrammatically in Figure 1 for the Capitol complex. As Figure 1 indicates, the sun provides the energy drive to Capitol Park and fossil fuels (as embodied in goods and services)

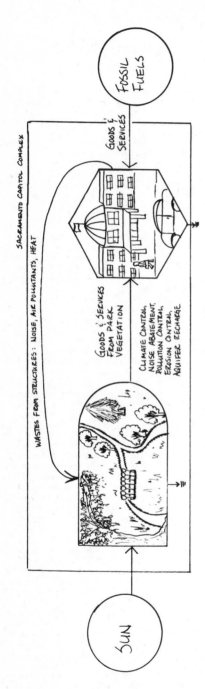

Figure 1.—Interaction of Capitol Park with Capitol Structures in Sacramento's Capitol Complex.

provide the energy drive to the Capitol structures. Certain functions from the park (e.g., climate buffering) are shown as additional inputs to the Capitol structures and certain wastes from the structures (e.g., heat) are shown being recycled to the park for processing. The conceptual framework is one of an interaction between the structures and the park. This theoretical framework suggests that a city's energy budget is the sum of the fuels used by its technologies plus the sun's energy used by its ecosystems. It suggests that both the technologies and the ecosystems contribute to the productivity of the city (Odum et al. 1977).

A city's economic productivity is usually taken to be the dollar value of the goods and services produced there for final demand, its contribution to the nation's Gross National Product. This energy based theory suggests, however, that urban economic productivity is supported by and dependent on natural environmental work as well as technological work. That is, it is dependent on solar energy as well as on fuels. Such a dependency on two primary energy sources suggests opportunities to substitute solar energy for fuels (trees for technology) and consequent opportunities to reduce costs. But the history of city development is one of reverse substitutions (technology for trees) with exponentially rising costs. For example, trees are removed to provide space for subdivisions where air conditioners are invoked to control climate. The air conditioners reject waste heat to the environment which in turn causes everyone's air conditions to run longer and produce more waste heat and so on in an exponential fashion. Retaining the trees in the subdivision, while not eliminating the need for air conditioners, greatly mitigates the exponential increase in their use. The price of electricity at the level of entire cities provides another example. The average cost of electric power to a consumer in the U.S. was $.024 per kw/hr. in 1975. In the same year in New York City, electricity was $.068 per kw/hr. The cost of power in New York City may be high because no ecosystems are available to provide any services which might otherwise subsidize the utility. All environmental controls are carried out with technology at an added economic cost. The effect becomes exponential again as, for example, taxes increase to pay for city services and the cost of heating and cooling buildings increases. At the extreme, products manufactured in cities where few services are provided by ecosystems can no longer compete with the same products manufacturing in other cities which use the environment at no cost.

In summary, this energy based theory suggests that technology and humans in cities depend on natural ecosystems. The natural ecosystems subsidize the technologies and may even be required for competitive economic production (Odum 1975). Capitol Park is an example of such a relationship where the landscape features offer energy and economic subsidies to the operation of the state government. The Capitol buildings have the advantage of the noise and climate control provided by the 35 acres of greenery and of their pollution absorption and recycling capacity.

Method

Because energy flow occurs in both the technological and natural components of the city and because energy flow per unit of time is a measure of work provided, methods for evaluating ecosystem contributions to urban productivity center in energy unit measures. But energy occurs in many forms, and it is intuitively understood that a Btu of electricity is somehow different from a Btu of coal. Similarly, a Btu of tree biomass is different from a Btu of sunlight.

Thus, methodologically, some kind of equivalency must be established among different kinds of energy. Establishing this equivalency is fundamental to measuring nature's services in energy units. In addition, communication of results requires conversion of energy units to dollar units. City planners and other interest groups are simply more comfortable planning in dollar units. The method of establishing equivalency among energy forms is discussed first, followed by the method for conversion to dollar units.

Equivalency among energy forms is established by evaluating the quantity of one kind of energy consumed to "make" another kind. The most straightforward method to perform that evaluation is to examine the natural or technological system that is designed to convert one kind of energy to another and measure the energy consumed for that system. For example, electric power plants make electricity from coal and we know that 3.6 units of coal are consumed to make one unit of electricity (Calculation 1, appendix). Solar energy is consumed to make organic biomass and we know that, on the average, 100 units of solar energy are consumed for each one unit of gross primary production and 200 units are consumed for each one unit of net primary production (Calculation 2, appendix). Organic biomass is converted or "consumed" to make coal, oil, and gas via biological and geological processes. But the conversion efficiency in nature has not been measured; thus, the straightforward method cannot be used. We do know that the heat content of organic matter is nearly equivalent to that of low-grade coal (lignite at 800 Btu's per pound), and that hardwood trees have a heat content nearly equivalent to bituminous coal. With that knowledge and concepts from theoretical thermodynamics, Odum (1973) estimates that a maximum of 5 units of organic matter are consumed to make one unit of fossil fuel on the average. We use that maximum in the calculations for Capitol Park to obtain a range of values, making our estimates of nature's services conservative.

Next, consider the relationship between energy flow and money. As has been stressed throughout this paper and shown in Figure 1, the economy and associated dollar flows depend on two fundamental types of energy: the sun and fossil fuel. Total annual economic production is measured in dollars as Gross National Product which in 1975 was $1,437 billion. The GNP measures the value of all goods and services produced. The energy consumed to produce those goods and services is the sum of fuel consumption and solar energy consumption. We have emphasized in this paper that the production of goods and services depends, not only on fuel consumption, but also on manifestations of solar energy (e.g., hydrologic, pollution and climate control). Thus, total energy consumption in 1975 was 99.6×10^{15} (quadrillion) Btu's. This is the sum of 73 quadrillion Btu's of fuel (Calculation 3, appendix) and 26.6 quadrillion Btu's of solar energy (Calculation 4, appendix). Thus, in 1975, technology using fuels contributed about 2/3 of the

power behind the GNP and environmental resources using sunlight contributed the other 1/3. Finally, conversion of energy units to dollar units is made possible using the ratio of energy consumption to dollar value of commodities produced (in 1975) or about 70,000 Btu's per dollar (99.6×10^{15} Btu's/1437×10^9).

In summary, evaluating nature's services in energy units requires establishing an equivalency among energy forms. Any energy form may be chosen as the standard to which other forms are related. For convenience, fossil fuel is chosen and used in this paper as the standard. The other energy forms considered in the Capitol Park analysis are equated to one unit of fossil fuel using the following coefficients: 2000 units of solar energy, 2 to 10 units of gross primary production, 1 to 5 units of net primary production, 1 unit of natural gas, and 0.28 units of electricity. Conversion of energy units to dollar units uses the ratio of 70,000 Btu's per dollar where the energy forms must be in fossil fuel equivalent units prior to applying the ratio.

RESULTS: ENERGY ANALYSIS
OF CAPITOL PARK

The Capitol Park complex is comprised of approximately 40 acres located in downtown Sacramento, California. Five acres are buildings which house the Governor's Office and staff and the Legislative offices and chambers. The remaining 35 acres constitute an open botanical garden of 350 different species of mature trees. The park is visited yearly by about 264,000 persons. In addition, the park has become a haven for city workers in Sacramento, a spot frequented by many at noontime.

The analysis focused on two kinds of assessments:

1. The capital asset value of the structure in the vegetation and the annual value of the services provided by the park were evaluated in energy and dollar units using the methods described above.

2. Using traditional economic methods, the cost of providing some of those services by technological means was evaluated. Both the dollar cost of and the fuel consumed by those replacement technologies were analyzed.

The first assessment computes energy values directly and money values indirectly using the energy to dollar ratio (Table 1). The second assessment computes money-values directly and fossil fuel values indirectly using traditional economic methods (Table 2).

Table 1 presents the results of the energy analysis. The capital asset value of the vegetation represents the stock of standing biomass accrued as the trees reach maturity. It is analogous to the fuel consumed to construct a facility or manufacture equipment. In either case, the capital structure must be in place for the system (the trees or the facility) to provide services or produce a commodity. For example, if the park space is allocated to air pollution control, then either trees or a control facility must be in place. The energy and dollar amounts are hypothesized to represent the capital asset value of the park to the city. The energy value (which is in fossil fuel equivalent units) is also thought to represent the quantity of fossil fuel that would have been consumed to "manufacture" Capitol Park. Similarly, the dollar amount is thought to represent the cost of "manufacturing" Capitol Park. As calculated, those values and costs are 30 to 140 billion Btu's and $0.4 billion to $2 billion. Similarly, the value of the services provided annually by the park (Table 1) are thought to represent the fossil fuel consumed (900 to 4,700 million Btu's per year) and dollar costs ($14,000 to $70,000 per year) of providing the functions with technology which Capitol Park currently provides with vegetation. The range of organic matter to its equivalent in fossil fuel units.

Table 1.--Energy and Dollar Values for Capitol Park Predicted Using Energy Analysis Methods

	Energy (Btu's)	Dollars (1975)
Capital Asset Value Total Park (5)	30-140 billion	0.4-2.0 million
Each tree in park (6)	80-400 million per tree	1140-5700 per tree
Annual Value of Services Total Park (7)	0.9-4.7 billion per year	14,000-70,000 per year
Each Tree in Park (8)	2.7-13.4 million per year	40-200 per tree per year

Table 2 presents the results of the second kind of assessment using economic methods. Four kinds of services are "broken out" and the dollar cost of and fossil fuel consumed in providing those services with a replacement technology is evaluated. For each service, a range of values is given. The actual value of the work provided depends on how well interfaced the park system is with the buildings, e.g., positioning of trees for climate and pollution controls. The low values represent a minimum and the high values represent a maximum where trees and shrubbery are planted with climate and pollution control functions in mind. It was not possible to determine how well interfaced the park vegetation and buildings are. The park is providing the four kinds of services listed in Table 2 as well as others such as mentioned before which were not evaluated.

Total fossil fuel saved for these four services alone, as calculated, ranges from 2.6 to 19.3 billion Btu's per year (Table 2). This is the equivalent of 464 to 3,450 barrels of oil per year. The savings in economic costs attributable to the work of Capitol Park vegetation range from $10,100 to $137,300 per year. In comparison, the energy analysis assessment (Table 1) suggests the savings predicted by the energy analysis of between 1 and 5 billion Btu's and $14,000 to $70,000. This dollar amount falls entirely within the low to high range calculated using economic methods. The quantity of energy given in Table 1 (1 to 5 billion Btu's per year)

overlaps, but is not entirely within the range calculated using economic methods (2.6 to 19.3 billion Btu's from Table 2). While there may be other explanations, two reasons for that difference are offered here. First, nature's designs may be, on the whole, more efficient than people's designs of systems which provide the services analyzed. Secondly, the minimum predicted value uses the conservative quality conversion factor of 5 for net primary production to fossil fuel. If that quality conversion factor is 3, then the fuel savings predicted are in the range of that calculated.

Table 2.--Energy and Dollar Values for Capitol Park Calculated Using Economic Methods as Fossil Fuel and Dollars Saved*

	Fossil fuel saved in million Btu's/year	Dollars saved in $/year
Heating/cooling (9)	2000-6000/park	4100-12,300/park
	6-19/tree	12-35/tree
Water treatment (10)	160-270/park	3000-5000/park
	0.5-0.8/tree	9-14/tree
Air purification (11)	170-5900/park	2000-7000/park
	0.5-17/tree	6-20/tree
Soil conservation (12)	130-6500/park	1000-50,000/park
	0.4-18/tree	3-140/tree
Total range of annual measurable services	2,660-19,270/park 7.4-54.8/tree	10,100-137,300/park 30-389/tree

*Calculations in Appendix, number of corresponding calculations referenced.

DISCUSSION AND CONCLUSIONS

Using a theory proposed and developed by H.T. Odum, we evaluated the energy value of the services that Capitol Park in Sacramento, California provides to the Capitol Park complex and city. Capitol Park uses solar energy to provide these services. This solar energy as embodied in biomass was expressed in fossil fuel equivalent units and in dollars. The fossil fuel equivalent energy value is believed to represent the amount of fossil fuel that would be consumed if technology were invoked to provide the same services now provided by Capitol Park. Similarly, the dollar amount is believed to represent the cost of providing those services with technology.

In order to examine the accuracy of the energy analysis method, economic techniques were also used to compute the cost of providing several kinds of services with technology. The results, in close agreement, indicate that Capitol Park has a capital asset value between 400,000 and 2 million dollars and provides annually services worth $14,000 to $70,000. While the evaluation methodology may be improved and expanded, the method is relatively easy to apply and the results provide some indication of the value of our trees and vegetation in urban settings.

The energy calculations are based on the energy requirements of producing or sustaining the trees or ecosystem. Yet this leads to another measurement problem. The value of trees is, then a function not only of the embodied energy or money in the substituted technology, but it is also a function of the value society places on the end use services. That is we need to factor in the value society has for a particular end use and adjust the thermodynamic and/or economic estimates accordingly. Of particular concern would be society's demand (or need) for the services, whereas the energy calculation focuses more on the supply aspects of the services. Economists might argue that the price of providing the alternate technology is a reflection of the value society places on the services, yet the area of environmental quality, quality of life, and economics is not that precise as yet. Many pollution control functions are subsidized through government loans, regulatory programs and the like, thereby warping the economic reflections of the value of pollution control. Additionally, the societal value may change from environment to environment, season to season, with changing levels of education in the populus, and in differing economic climates.

We would propose a general evaluation model for societal end use value of natural systems functions which would include the following components.

1. The energetics of the system (since we do not pay nature for her services), where:

   EQI   = Energy Quality Units of Input (to run the natural system)

2. The end use economic and energetic value of the services provided and the societal value of those end uses, where:

   EU   = End Use (Energy cost of replacement technology by human application)

Then for each technology where there is a substitute function by natural systems or renewable, natural based energies, the value of the natural system functions (NSV) would be calculated by:

$$\frac{EQI}{EU} = \text{Natural Systems Value} = \text{NSV}$$

Once established a sliding scale of NSV's could be used to compare the various "appropriate technologies" with each other and with the high technology options they could replace. Yet for now, an NSV of less than 1 would signify that the natural system is performing the service functions cheaper and in a less energy intensive manner than would the replacement technology.

What is needed now is a refined methodology

coupling energy, sociology and economics to quantify the components of the NSV equations, to better test the hypothesis. One such beginning is implicit in this study as the societal value of trees would include the following calculations:

1. EQI   = 2.8 billion Btu's/yr

2. EU   = mid range of energy saved by park functions

        = 10.9 billion Btu's/yr

3. Therefore, the value of the natural system (in this case trees in Capitol Park) is:

$$\frac{2.8}{10.9} = 0.26 \text{ NSV}$$

Even though the NSV represents less than unity, reflecting an efficient pollution and climate control system by Capitol Park, this value is virtually meaningless without other technologies and other values for comparison. Yet it perhaps begins to quantify some of the difficult aspects energy modellers and economists continue to grapple with in their separate and combined analyses.

Energy analysis measures physical properties and many economists do not view that measurement as indicative of the "real world" situation. Economists add some social perspectives on the energy values, yet may represent some lags and "slack" in the system. The overall system is a combination of the two, with two masters being served simultaneously. True, the laws of thermodynamics provide the baseline realms of possibilities, but the economics fine tune the perspective, and the economics change more often than do the energetics of a case study.

Despite this cautious beginning of this combined approach to couple energetics and economics to form the societal value of natural systems, much more work needs to be done to perfect the theoretical basis of the evaluation in addition to more calculations upon which to base some comparisons, and energy planning decisions. Yet one point is very clear: the value of trees in cities and urban centers is not trivial. In fact, cities are and have been running on both fossil and solar fuels through trees and greenbelts, and the expansion of this knowledge does not require a lot more calculation or debate. The additional challenges lie in other natural system uses such as wetlands for water purification, and nursery grounds for fishing economies, watersheds for water management functions, and the like. The societal value of trees is just a stepping stone from which to venture into future aggressive couplings of urban and natural systems.

LITERATURE CITED

American Physical Society. 1975. Efficient use of energy: a physics perspective. American Physical Society, New York, N.Y. 237 p.

Bach, Wilfred. 1971. Steps to better living on the urban heat island. Landscape Architecture, Vol. 61. 139 p.

Bayley, S. and Odum, H.T. 1973. Energy evaluation of water management alternatives in the upper St. Johns River Basin of Florida. Final Report to the Environmental Protection Agency, Region IV, Atlanta, Georgia.

Bernatsky, Aloys. 1966. Climatic influences of the greens and city planning. Anthos, No. 1.

Carrillo de Albornoz, J. 1973. Green space and atmospheric pollution. American Hort., 52: (3) 15.

Commoner, Barry. 1978. The solar transition. Environment, 20: (3) April.

Council on Environmental Quality (CEQ). 1978. Solar energy progress and promise.

Elias, T.S. and Irwin, H.S. 1976. Urban trees. Scientific American, 235: (5) 111-118.

Lanphear, F.O. 1971. Urban vegetation: values and stresses. Hort. Science, 6: (4).

Lavine, J.J. and Meyburg, A.H. 1976. Toward environmental benefit/cost analysis: Measurement Methodology, Final Report to the National Cooperative Highway Research Program and the National Research Council.

Metz, William D. 1978. Energy storage and solar power: an exaggerated problem," Science, Vol. 200.

Odum, H.T. 1971. Environment, power and society. 317 p. Wiley-Interscience, New York, N.Y.

_____. 1973. Energy, ecology and economics. AMBIO, No. 6.

_____. 1975. Energy quality principles for estimating the environmental carrying capacity for man. Given at the Environmental Health Conference at Pahlayi University, Tehran, Iran.

_____. 1978. Energy analysis, energy quality and environment. In Energy analysis: a new public policy tool. M.W. Gilliland, ed. Westview Press, Boulder, Colorado.

Odum, E.P. and Odum, H.T. 1972. Natural areas as necessary components of man's total environment. Transactions of the 37th North American Wildlife and Natural Resources Conference, Washington, D.C.

Odum, H.T. and Odum, E.C. 1976. Energy basis for man and nature. McGraw-Hill, New York, N.Y. 288 p.

Odum, et al., 1977. A manual of energy analysis methodology for power plant siting, the energy value of land. Research in Progress at the Center for Wetlands, University of Florida, for the Nuclear Regulatory Commission.

_____. 1976. Net energy analysis of alternatives for the United States. Prepared for the Committee on Energy and Power of the United States Congress, House of Representatives, Hearings, Serial No. 94093.

Sacramento Municipal Utility District (SMUD). 1979. How to save energy naturally. Sacramento, Ca.

APPENDIX

Calculations:

1. This 3.6 units includes 3 units of coal consumed in the power plant boiler and 0.6 units consumed indirectly in constructing and operating the power plant. (Odum et al. 1976)

2. This represents a 1% solar input to gross production efficiency and a 50% gross to net production efficiency.

3. This was 94% fossil fuel (coal, oil and gas).

4. The solar energy component is calculated as solar insulation on the U.S. divided by 2000 to convert sunlight to its equivalent fossil fuel energy form. It assumes that all solar energy is coupled to the U.S. system creating clean air, water, soils, biomass, land forms and so on.

5. Standing stock at 55,000 metric tons per km^2 or 7.8×10^9 grams for the 35 acres at 4.5 kcal/gram and 3.968 Btu/kcal is 140×10^9 Btu's. Dollar conversion at 70,000 Btu/$ (17,640 kcal/$) yield 2×10^6. All values divided by 5 using the conservative conversion to fossil fuel units, and the low end of the range.

6. Values given in note 5 divided by 350 trees.

7. Gross primary production at 16,800 $kcal/m^2$/year or 2.38×10^7 kcal/year for the 35-acre park; 3.968 Btu's/kcal yields 9.44 billion Btu's; divided by 2 to convert to its equivalent in organic matter, yields 4.7 billion Btu's. Dollar conversion at 70,000 Btu/$ yields $67,143. All values divided by 5 using the conservative conversion to fossil fuel units and the low end of the range.

8. Values in note 7 divided by 350 trees.

9. Capitol Park buildings consume 22.5×10^9 Btu's of natural gas and 47,259 kwh (5.8×10^8 Btu/th) for heating or a total of 28×10^9 Btu's. Other studies indicate greenbelts save a minimum of 7% of heating demand without greenbelts and, if well interfaced, save 25% for large commercial type buildings. Heating needs assumed to be 25% higher without park, or 30 to 35 $\times 10^9$ Btu's for a savings of 2 to 7 billion Btu's. Dollar conversion at $1.80 per cubic foot for gas saved and 3 cents per kwh for electricity saved.

10. Cost of sewage treatment capacity ranges from $64 to $106 per year for 1000 gallons of capacity. Urban runoff requiring treatment in the absence of the park is 17.1×10^6 gallons/year (18 inches over 35 acres) for 47,000 gallons/day capacity; 64 to 106 $\times$ 47 = 3000 to 5000 per year. Fuel consumed in materials for construction and manufacturing at

54,000 Btu's for each dollar of materials purchased, or 162 to 270 million Btu's.

11. Cost of sulfur removal systems for industry, while highly dependent on the degree of sulfur removal attempted, averages about $550 per year per ton of SO_2 removed. Based on a literature survey, we estimate that Capitol Park trees "take up" a minimum of 3.7 tons of SO_2 per year. If technology were to remove the equivalent, cost would be ($550 x 3.7) = $2,000 per year. If technology was also invoked to remove the particulates, CO, NO_x, and HC from automobiles, then the cost would be about $70,000 per year. The manufacture of the material and equipment needed to construct and operate air pollution control systems consumes 85,000 Btu's of fuel per dollar of cost. Thus, $2,000 to $70,000 per year corresponds to the consumption of 170 to 5900 million Btu's of fuel.

12. Low dollar value represents self fertilization with organic matter at a minimum of 11 tons per acre, replacing the need for synthetic fertilizers at $90 per ton; total dollar savings is $990. High dollar corresponds to construction of a soil catchment basin required to "save soil" if the vegetation were not present. It is based on an actual catchment system constructed in a disturbed area to collect the soil eroded until trees could be reestablished. The fossil fuel consumed by the fertilizer industries was 130,000 Btu's for each dollar of fertilizer produced in 1975. Conversion to Btu's uses 130,000 Btu's/$.

9. Energy Economics

ENERGY FLOW IN PEASANT AGRICULTURE[1]

Bruce Hannon[2]

Abstract.--It is hypothesized that the Japanese population was near the steady state population during the Tokugawa
period (1600 to about 1860) because of an energy resource
(land) shortage. This shortage gave rise first to chaos and
then peaceful existence under a strict hierarchy which, among
other regulations, taxed away more than one-third of the
peasant production. The energy flow in Japanese peasant
agriculture is modeled and criteria for optimum levels of the
population are found under conditions of food taxation. The
equations provide a means to verify the hypothesis.

Japanese history, from 1400 to about 1860 provides us with a rather well recorded example of a
human population adjusting to and attaining a steady
state level, (Reischauer 1980; Smith 1977). During
the first two to three centuries of this period, the
population growth rate slowed and was nearly zero by
the 1700's, (Reischauer 1980; Smith 1977; Nakamura
1972; Droppers 1894; Ishii 1937; Taeuber 1958. The
population was racked by almost continual warfare
before 1600. The only viable, narly self-sufficient
social unit was the extended family whose population
was about 20 to 30 persons made up of 3 generations
of blood related, adopted and indentured individuals,
(Nakamura 1972; Smith 1959). Internal trade before
1600 on any significant scale was impossible due to
the continual struggle over feudal boundaries.

From the late 1500's to about 1625, a succession of three feudal lords achieved control over
nearly one-seventh (not contiguous) of the area of
Japan, (Reischauer 1980; Ishii 1937). The
third of these became known as the first of the
Tokugawa Shoguns. It was at their insistence that
the remainder of the almost 270 lords fixed their
then existing boundaries, virtually stopped trade
with the outside world, essentially stopped internal
warfare and instituted a nearly standard arrangement of control within their domain. Over the
next century the population grew slowly to about
thirty million, possibly due to improved internal
trade which made better use of disparate land

quality. The population remained nearly steady
for the next 170 years. Nearly 85 percent were
peasants who farmed the land and lived in small
villages consisting of several families. About
6 percent of the population were Samurai warriors,
at first the military force, and later, the bureau-
cracy of each lord. Another 5 to 6 percent were
the craftsmen and merchant classes. The remainder
of the population were the priests, small business-
men and the poor who lived in the medium to large
castletowns of the lords, (Reischauer 1980; Smith
1977; Nakamura 1972; Droppers 1894; Ishii 1937;
Taeuber 1958).

The peasants provided nearly all of the food
and fiber for the Japanese population, although
some sugar and silk were imported in the early
1600's. About 30 to 35 percent of the peasant
produce was removed by the feudal lords as a tax.[3]
This tax was consumed by the lords, Samurai and the
remainder of the administrative class. Sometimes
a portion of the tax was stored against periods of
low production as a hedge against flood and drought
(Smith 1959). It is believed that this large tax
was indirectly the control mechanism responsible
for the steady peasant population. (Droppers 1894;
Benedict 1967). In the 15th-17th centuries, it
was probably to the strategic military advantage
of every feudal lord to try to maximize the number
of his Samurai warriors. Therefore, as much food
as could be practically diverted from the peasants
was distributed over the largest possible number of
warriors. Due to such practices, the actual food

---

[1] Paper presented at the international symposium Energy and Ecological Modelling, sponsored by
the International Society for Ecological Modelling.
(Louisville, Kentucky, April 20-23, 1981).

[2] Bruce Hannon is Associate Professor in the
Office of Vice Chancellor for Research, and Direc-
tor of the Energy Research Group, University of
Illinois, Urbana, IL, 61801, U.S.A.

[3] In a study of the Silver Springs ecosystem
(Hannon 1973), I find an analogy between the Tokugawa peasants and the producers in that steady state
ecosystem. The producers absorb sunlight, respire
and yield their newly created biomass to the herbivores. This new biomass is 30 percent of the captured sunlight, the rest being needed for producer
maintenance.

available to all classes was limited, and consequently, so was their population. Abortion and infanticide seemed to have been practiced by the entire population, (Smith 1977; Nakamura 1972; Droppers 1894; Ishii 1937), either to improve economic conditions in the family or to keep the population below that which followed flood and drought.

New military technologies were discouraged, peasant geographic mobility was constrained and peasant ownership of weapons was banned. The development of new agricultural technologies was carefully controlled so as to supress possible sudden population variations between neighboring domains. Inter-domain espionage was widespread and with regular census-taking, the Shogun could control changes which he felt might lead to conflict, (Smith 1959).

From the beginning of the Tokugawa period trade flourished, labor became more diverse and job specialties arose. The cities grew while the extended family dissolved into the more modern (stem) form. Trade developed rapidly to take advantage of the variation in the natural productivity of the land to produce rice, cotton or silk for example.

Some authors believe that the extremely hierarchial nature of Tokugawa Japan so denied the peasants the right to keep their produce that a steady population resulted. (Hanley and Yamamura 1977)* Smith (1959).[4] My hypothesis is the opposite. I favor the Malthusian hypothesis that the growing population pressed against the natural resource limits of an isolated island economy and produced significant strife and disorder as a result. The highly controlled structure in the country was an alternative more acceptable to the peasants than the chaos of war. The tax on the peasants very likely did constrain their population[5] as did the food distribution policies for the other classes. But in my view, the tax and distribution schemes maintained the steady state condition. They did not cause it. The

---

[4] These authors show that the tax declined slightly as a fraction of production over the Tokugawa periods. This was probably due to a cessation of fertile land surveys by the feudal lords after 1721.

[5] Because the tax was collected at the village level by the headman as an annual quota (rather than a percentage), each family would have been fully aware of the capability and actual contribution of all other famililies. Those who would not be able to make their historic contribution because of an enlarging family were ostracized by the community. Infanticide usually followed, Smith (1977). with the long-term result of a markedly steady population. Thus the Tokugawa Japanese avoided a kind of "Tragedy of the Commons" wherein each family games against society by producing many offspring to improve the probability of achieving wealth, prominence and old age security.
*(Hanley and Yamamura 1977).

hierarchy therefore, in my view, provided a workable scheme for allocating scarce resources, a no doubt unpopular solution to a nearly intractable problem. The hierarchial solution is seemingly a natural one, due in part to the nearly universal desire of the population to seek order as their highest priority in times of chaos, and the increasing plenty of those wishing to lead or control.

The tax also probably reduced the starvation during floods and famines by providing food storage, by restraining the population density (reduced contagion) and by allowing housing location to occur away from flood plains, (Droppers 1894).[6]

If this view of resource constraint on the population is correct, then i) what are the technological and physiological functions which relate the peasants to their finite resource base? and ii) what is the best measure of this resource? and finally iii) how can we test the hypothesis that the population is limited by resources rather than political constraints? These questions can be addressed by a mathematical model which is the main subject of this paper. For the Japanese, the resource measure seems to be land, or more appropriately, the ability of the land to produce food and fibre - or still more precisely - the ability of the land (and the fisheries) to absorb solar energy. Energy is a convenient measure for all the stocks and flows in the Japanese peasant system - the flows of food and fiber to both the population, the work animals and the capital equipment - the flows of waste from humans and animals and the depreciation flow of capital, and finally, the measure of the stock of humans, animals and capital equipment. That energy is such an overall appropriate measure of the stocks and flows in a peasant society is certainly an assumption. At least two authors have used it as a measure, (Rappaport 1967; Johnson, et al. 1977) My view is that physical measures of living system activity are at least appropriate when that system exists at the subsistence condition, that is when each element of the system is unable to generate a surplus of its own type over any sustained time, without reducing the amount of activity of any other element. To the extent that aggregate demand has a purely discretionary portion, that is a demand for goods which does not add to the quantity of life of a particular consumer and does not reduce the quantity of life anywhere else in the system, then that system is above the subsistence condition. Physical measures of activity may be accordingly inappropriate as the system departs from the subsistence condition. Note that certain forms of system growth may be at the subsistence condition. If each increment of surplus production is used to produce greater quantities of life and not aimed at improving the quality of life, then the growth path is at the subsistence condition. Ecosystems and perhaps agrarian human systems such as the Tokugawa Japanese may be examples which meet this definition of the subsistence condition. A simple representation of a peasant society is shown in Fig. 1.

---

[6] For a discussion and abstract of the major works on the Tokugawa period see(Sumiya and Taira 1979).

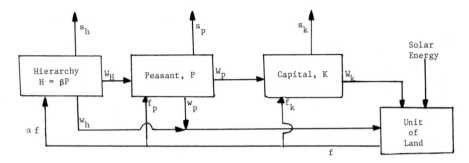

Fig. 1. -- The relationship of the administrative class,
the peasants, their capital and land.

The symbols have the following meanings:

H   is the stock of the hierarchy per unit
of land (e.g., calories per hectare).

P   is the stock or biomass of peasants
per unit of land (e.g., calories per
hectare).

K   is the stock of biomass of capital per
unit of land (e.g., calories/hectare).

β is the fraction of the peasant stock
which is Heirarchy, $0<\beta<1$.

α is the fraction of the total biomass
production which is devoted to the
heirarchy, $0<\alpha<1$.

f_p is the food consumed by peasants
(e.g., calories/hectare-year).

f_k is the biomass for capital replacement (e.g., calories/hectare-year).

f is the total biomass produced
(e.g., calories/hectare-year).

w_p is the waste (fertilizer) flow from
peasants (e.g., calories/hectare-
year).

s_p is the non-work peasant metabolism
(e.g., calories/hectare-year).

s_k is the capital depreciation rate
(e.g., calories/hectare-year).

W_k is the power exerted by the capital
on the land and is equal (100 percent
mechanical efficiency to W_p, the
power output of peasants on the
capital (e.g., calories per hectare-
year). Power is the rate at which
work is done.

$s_k = a_k K + b_k W_k$
where

a_k = constant (e.g., $\frac{1}{year}$).

b_k = dimensionless constant, $0<b_k<1$.

This last equation is an assumption which means
that the depreciation rate of the capital is a
function of the size of the capital stock and the
power passed through it.

The power input and the rate of biomass output
are assumed to be related through the production
function:

$$f = A \left(\frac{W_p}{W_o} \right)^{\delta}$$

where A and δ are constants which measure technological efficiency. The exponent is constrained:
$0<\delta \leq 1$, which means that the output rate diminishes
with increases in W_p. The power W_p must be indexed
by an arbitrary value W_o in order to make
the production function statistically tractable.
The constant   is defined in economics as the
"elasticity" of output with respect to the input.
From our definition this would be

$$\delta = \frac{W_p}{f} \frac{df}{dW_p} .$$ Thus a one percent change
in W_p will produce a
percent change in f.

W_H is the power used by the hierarchy in controlling the system. It includes, for example, the
energy used in census taking, in travel to feudal
kingdoms, and by the weavers and armorers who make
materials for the administrative class.

$W_H = \beta QP.$ The units of Q are for example,

$\frac{1}{years}$. Q is the rate that energy
is expended by all the
members of the hierarchy
in excess of their resting metabolism, plus the
rate that they consume all the energy contained in
their materials, normalized by the caloric biomass
of the hierarchy.

The following assumptions are also adopted:

α is the "tax" rate on the peasant's food production used to support the hierarchy, $(0 < \alpha < 1)$.

$f_k = Cf$ where C is a dimensionless constant, $0 < C < 1$.

$(1-\alpha) f = f_p + f_k$, $\quad w_p = b_p f_p$; where

b_p is a dimensionless constant; $0 < b_p < 1$.

$s_h = a_p \beta P$; $\quad s_p = a_p P$; $\quad a_p$ is a constant $(e.g., \frac{1}{year})$.

A final assumption is that if there exists an energetic optimal stock for the peasant and for the capital, then the peasant will strive to reach these stock levels simultaneously. This assumption is a representation of the Malthusian tendency of the population to press against the limits of its resources.

It is now possible to establish a meaningful energy balance across the peasant:

$$\dot{P} = f_p - s_p - W_p - w_p \quad . \tag{1}$$

Solving this differential equation for P gives

$$P = \frac{1}{a_p} \left[(1 - b_p)(1-\alpha-C)f - \frac{f}{A}^{1/\delta} W_o \right] \tag{2}$$

as the long term solution.

Plotting equation (2) shows that P has a maximum (see Fig. 2).

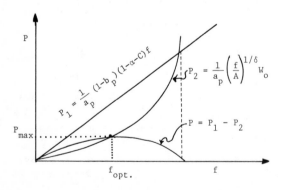

Fig. 2.--The variation of peasant biomass with land output.

Setting $\frac{dP}{df} = 0$ gives, for the peasant:

$$f_{optimum} = \left[\delta \frac{A^{1/\delta}}{W_o} (1-\alpha-C)(1-b_p) \right]^{\frac{\delta}{1-\delta}} \quad . \tag{3}$$

Therefore, the maximum possible long term peasant biomass is

$$P_{max} = \frac{1}{a_p} (1-\delta) \left[\frac{A\delta^{\delta}}{W_o^{\delta}} (1-b_p)(1-\alpha-C) \right]^{\frac{1}{1-\delta}} \tag{4}$$

By forming for the capital, $K = f_k - s_k$, solving $\frac{dK}{df} = 0$, I find that

$$f_{optimum} = \left[\frac{\delta CA^{1/\delta}}{W_o b_k} \right]^{\frac{\delta}{1-\delta}} \quad . \tag{5}$$

This result is based on the concept that if the peasants find capital at all useful, they will maximize the stock of capital subject to the technical constraints of a production function and a depreciation rate.

The food output of equations (3) and (5) would be equal in the case where peasant and capital biomass are simultaneously optimum. Doing so produces a solution for C, the portion of the energy output which is chosen by the peasants to make up for capital depreciation.

$$C = \frac{1-\alpha}{\frac{1}{b_k (1-b_p)} + 1} \quad . \tag{6}$$

Equations (4) and (6) now yield the optimum or maximum peasant population level in terms of exogenously determined constants:

$$P_{max} = \frac{1-\delta}{a_p} \left[\frac{A W_o^{-\delta} \delta^{\delta} (1-\alpha)}{\frac{1}{1-b_p} + b_k} \right]^{\frac{1}{1-\delta}} \tag{7}$$

Thus the maximum peasant population is a function of the properties of a food production function, peasant respiration and waste generation coefficients and a capital depreciation rate.

I can derive another expression for the optimum peasant population by setting up an equilibrium differential equation for the hierarchy similar to equation (1). Solving that equation for the steady state condition and substituting equations (5) and (6) yields the desired expression,

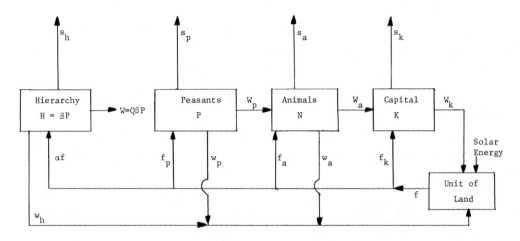

Fig. 3.--The relationship of the administrative class, the
peasants, their work animals, capital and land.

$$P_{max} = \frac{\alpha}{(a_p + Q)\,\beta} \; \left[\frac{A\,W_o^{-\delta}\,\delta^\delta(1-\ )}{\frac{1}{1-b_p} + b_k}\right]^{\frac{1}{1-\delta}} \tag{8}$$

Equating expressions (7) and (8) gives a linear
relation between α, the tax rate and β, the fraction
of the human biomass which belongs to the hierarchy,

$$\alpha = (1-\delta)(1 + \frac{Q}{a_p})\beta \tag{9}$$

The tax rate is a function of the output elasticity,
the unit human metabolism rate, the unit control
work and the hierarchial fraction. With equation
(9), I should be able to determine the value of α
given a certain value of β and experimentally derived the values for δ, Q and a_p. Assuming that
the assumptions hold and the peasants are energy
optimizers, the value of α for Tokugawa Japan should
be about 0.30 to 0.35 when the value of β is about
0.13 to 0.15. If equation (9) would produce a
significantly lower value for α then the restrictions placed on the Tokugawa Japanese peasants would
probably have been unjustified from a physical point
of view.

The foregoing discussion may not accurately
represent the more advanced peasant agriculture.
It describes a very primitive agricultural system.
While the Japanese peasants did not have work
animals, it remains to be shown why these animals
were not used. Figure 3 is a representation of a
more complex system which contains animals and capital stock particularly sutied for work animal use.
The terms are the same as in the earlier example
with the following additions:

N is the animal biomass per unit of land
(e.g., calorie/hectare)

$s_a = a_a N$ is the non work metabolism of
the animals and
a_a is a constant (e.g. $\frac{1}{year}$).

W_p, the work output of the peasants is
assumed to be directly proportional
to the work output of the animals:

$W_p = BW_a$ where B is a dimensionless
positive constant, $0<B<1$.

W_k, the work output of the capital = W_a
(100 percent mechanical efficiency)

The food output is related to the work animal
input given by:

$$f = D\left(\frac{W_a}{W_o}\right)^\eta$$ which is the production function
for the animals; and $0<\eta<1$, where
η is the elasticity of output with
respect to W_a, the workoutput of
the animals.

W_o is the indexing value for the animal work.

$$f = f_p + f_a + f_k + \alpha f; \quad f_p = c_p f, \quad f_a = c_a f$$

$$0 < c_a, \; c_p < 1; \quad w_a = b_a f_a; \quad 0 < b_a < 1 \text{ where } b_a \text{ is}$$

a dimensionless constant.

An energy balance differential equation can be
written for each component of the system and solved
for the steady state condition:

$$P = \frac{1}{a_p}\left[(1-b_p)\ c_p f - B\left(\frac{f}{D}\right)^{1/\eta} W_o\right], \qquad (10)$$

$$N = \frac{1}{a_a}\left[(1-b_a)\ c_a f - \left(\frac{f}{D}\right)^{1/\eta} W_o\right], \qquad (11)$$

$$K = \frac{1}{a_k}\left[(1-c_a-c_p-\alpha)\ f - b_k\left(\frac{f}{D}\right)^{1/\eta} W_o\right], \qquad (12)$$

By setting $\frac{dP}{dt} = \frac{dN}{df} = \frac{dK}{df} = 0$, three independent expressions for the optimal energy production level can be obtained. Under the assumption that the peasants do adjust the size and activity levels of the animals, capital and themselves to reach a simultaneous optimum, these three production level equations are equated. By solving for and equating the optimizing energy output levels for P, N and K, the value of c_p can be obtained and when this value is substituted into equation (1), the optimal value of the peasant population can be determined. These values are:

$$P_{max} = \frac{1}{a_p}\left[(1-b_p)\ c_p\ f_{opt.} - BW_o\left(\frac{f_{opt.}}{D}\right)^{1/\eta}\right]; \quad (13)$$

where:
$$f_{opt.} = \left[\frac{(1-b_p)\ c_p\ D^{1/\eta}\ \eta}{BW_o}\right]^{\eta/1-\eta}$$

and:
$$c_p = \cfrac{1}{\cfrac{(1-b_p)}{B}\ b_k + \cfrac{1}{1-b_a} + 1}$$

The result is an expression for c_p, the fraction of the total energy which is diverted into peasant food, expressed in terms of two physiological and two technological constants. With the production function and c_p, the optimal energy production rate can be determined. With this rate and equations (10), (11) and (12), the simultaneous maximum stocks of peasant, animals and capital can be determined as shown by equation (13).

If, as before, the equilibrium differential equation for the hierarchy is solved for the steady state condition and then substituted into equation (13), a result is obtained which is remarkably similar to that of equation (9),

$$\alpha = (1 - \eta)(1 + Q/a_p)\beta \qquad (14)$$

except that instead of δ we have η, the land output elasticity with respect to the animal work input. The result is independent of any capital information because of the perfect pass-through efficiency assumed for the work on the capital devices. If the resulting estimate of the maximum population value was lower than that from equation (8) for the conditions in Tokugawa Japan, the assumptions used in this paper tend to be verified. That is, the peasants strive to maximize their own population and animals are unable to add a greater food surplus than can be had without them.

A future task is to identify actual values for the coefficients in equation (4). During the Tokugawa period the urban areas grew in population and the farming population declined, (Ishii 1937). The equations in this paper are not applicable in such cases as the coefficients A, D, δ and η might change with time. The equations may apply in exceptionally steady state provinces, however, and if they are proven even reasonably accurate they may then be used to detect the effect of different tax levels on the size of population, on the frequency and severity of revolt and on the severity of death due to famine, flood and other natural disasters. At the least, the concepts presented here may suggest a methodology which could supply a better understanding of the structure and operation of resource-short economies.

Future work would include extension of the model to cover communities which trade in order to achieve optimal use of the productivity of their land. In a more fundamental sense, however, the argument of cause and effect must be resolved; did the hierarchy cause or result from the steady population in the Japanese Tokugawa period? A study of the land use during the end of the period and after, could also help reveal whether or not the ruling classes so suppressed the population growth as to leave land idle. If, however, the land used remains unchanged and yet the agricultural yield increases (presumably due to imported technology, fuel and fertilizers) and if equation (14) is valid, then the ruling classes can be viewed as having been necessary to provide order in resource scarce society, at least during the early part of the Tokugawa period. In this latter case, the ruling classes could be justifiably viewed as represeive only when such technologies and goods were known to be available and yet denied to the Japanese peasant.

REFERENCES

Benedict, Ruth. 1967. The Chrysanthemum and the Sword. New American Library, New York. This classic book, by an anthropologist, was intended to help prepare the American military for war with the Japanese.

Droppers, G. 1894. The Population of Japan in the Tokugawa Period. Transactions of the Asiatic Society of Japan, Yokahama, Vol. XII. An early review of the population statistics and the effectiveness of the Malthusian positive and Preventive checks on population growth.

Hanley, S. and K. Yamamura. 1977. Economic and Demographic Change in Preindustrial Japan, 1600–1868. Princeton University Press, Princeton, N.J. See especially the first four chapters which deal with the country as a whole. These authors argue that desire for economic improvement led the average Tokugawan family to limit its size. It is presented as an alternative to a Marxian argument of peasant enslavement and poverty as a result of extreme hierarchial control.

Hannon, B. 1973. The Structure of Ecosystems. Journal of Theoretical Biology, 41:535–456.

Ishii, R. 1937. Population Pressure and Economic Life in Japan. P.S. King, London, A good summary of the historic population levels.

Johnson, W., V. Stoltzfus, and P. Craumer 1977. Energy Conservation in Amish Agriculture. Science, 28 October 1977, pp. 373–78. A comparison of the energy use in modern and horse-agricultural systems.

Nakamura, J. 1972. Social Structure and Population Change: A comparative study of Tokugawa Japan and Ch'ing China. Japanese Quantitative Economic History Association, Osaka. A clear exposition of the effects of land scarcity on political, social and economic structure. See also: Shigeki, T. Historical Preconditions for the Meiju Restoration, in Readings on Meiji Restoration, R. Kenkyukai, editor, Society for the Study of History, Tokyo, 1958.

Rappaport, R. 1967. Pigs for Ancestors, Ritual in the Ecology of a New Guinea People. Yale University Press. A study of energy flow in a slash-and-burn agricultural system.

Reischauer, E.O. 1980. The Japanese. Harvard University Press, Cambridge, Mass. Chap. 6. An excellent and sweeping view of Japanese development.

Smith, T.C. 1959. The Agrarian Origins of Modern Japan. Princeton University Press. Princeton, N. J.

Smith, T.C. 1977. Nakahara. Stanford University Press, Stanford, California. Chapter 1 is a review of the general population trends for Japan. The remainder is a description of "Family Farming and Population in a Japanese village, 1717–1830.

Sumiya, M. and K. Taira. 1979. An Outline of Japanese History: 1603–1940. University of Tokyo Press.

Taeuber, I. 1958. The Population of Japan. Princeton University Press. The population data from primitive times to 1950.

THE ENERGY EMBODIED IN THE PRODUCTS OF THE BIOSPHERE[1]

Robert Costanza and Christopher Neill[2]

Abstract.--Solar energy is the basis for the operation
of the biosphere. It is used directly by plants to power
the photosynthetic reaction. It is also used indirectly,
since the other required inputs to plant production (nutrients, water, etc.) are the products of solar-driven biohy-
drogeologic processes. Given this we may ask: how much
solar energy is required directly and indirectly, to produce
or concentrate the "commodities" of the biosphere?

This paper develops an input-output (I-O) model of the
biosphere using recently published data on global water,
carbon, nitrogen, phosphorous, biomass, and other material
and energy flows. The model is used to estimate the direct
and indirect solar energy cost, or embodied solar energy, of
these commodities. The embodied solar energy intensities
thus derived are analogous to fossil fuel energy intensities
which have been calculated for economic systems. They may
be useful for comparing ecological systems. Recent studies
have indicated that embodied energy and economic value are
correlated when the system boundaries are drawn in an all-
inclusive way. Thus, embodied solar energy intensities may
be useful for valuing (shadow pricing) environmental goods
and services and quantifying environmental impacts in units
comparable with economic impacts.

INTRODUCTION

Modern economics has for the most part limited its field of view to exchanges which occur in
well defined markets. The "economic problem" of
allocating scarce resources extends beyond these
somewhat artifical boundaries, however. As Daly
(1968, p. 400) has noted:

Such phenomena have long been recognized
(grudgingly) in economic theory under the
heading of externalities - that is interrelations whose connecting links are
external to the economists' abstract world

of commodities but very much internal to
the world in which we live, move, and have
our being. Perhaps "non-market interdependence" is a more descriptive term.

This paper attempts to place non-market
interdependence in a global perspective by developing an input-output (I-O) model of the biosphere. The earth's hydrologic, geologic, and
nutrient cycles, and net primary productivity all
depend on the planet's only significant net input
- solar energy - to drive them. The embodied
solar energy, or the energy required directly and
indirectly to drive the cycles of the biosphere,
is a measure of their interdependence and thus
their relative "value". This measure of value is
analogous to the economists' concept of "shadow
prices" ,which are implied prices for transactions
that occur outside the market. The I-O model is
capable of handling the web of extra- and intra-
market exchanges with equal facility. Thus, the
solar energy equivalents calculated using this
comprehensive model may be useful as a common denominator for the valuation of all products
exchanged both within and without the human market
system. While the actual numerical results of
this first attempt are aggregated and crude, they

[1] Paper presented at the international symposium on Energy and Ecological Modelling, sponsored by the International Society for Ecological
Modelling. (Louisville, Kentucky, April 20-23,
1981).

[2] Assistant Professor and Research
Associate, respectively, Coastal Ecology
Laboratory, Center for Wetland Resources,
Louisiana State University, Baton Rouge, LA 70803.

746

do indicate that the approach is operational.

Joint Products

The biosphere as a whole can be considered to be a composite process producing multiple products. The only significant net input is solar energy, while the net output is the maintenance of a broad range of structures far from thermodynamic equilibrium. The problem of allocating the solar energy cost to the interdependent joint products of the biosphere is the subject of this paper.

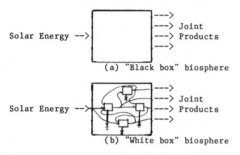

Solar Energy --> ---> ---> Joint ---> Products --->

(a) "Black box" biosphere

Solar Energy --> ---> ---> Joint ---> Products --->

(b) "White box" biosphere

Figure 1.--Alternative views of the biosphere.

Figure 1 illustrates the problem. In fig. 1a the biosphere is considered to be a "black box". This means that we know only the inputs and outputs but nothing of the internal connections. Some might argue that this is the extent of our real knowledge of the structure of the biosphere. Given this view, how could one allocate the solar energy cost to the multiple products? Odum (1978, p.61) has suggested that "Since there is only one source for all flow pathways...all pathways have the same numerical cost value." This implies that each of the joint products of the biosphere "costs" the total solar energy input. While this is one of the few possible ways of allocating costs in a "black box" system,[1] it does not possess the additivity property which is the basis for the usefulness of any accounting system. If the value accounting system is not conservative, then simple comparisons or rankings are not possible. This makes interpretion of the results of recent studies based on this methodology (and comparison with other studies) extremely difficult (cf. Odum et al. 1980; Wang and Miller this volume).

If, instead, we had perfect information on the internal connections, as illustrated by the "white box" in fig. 1b, it might be possible to trace the flow of solar energy through the system and allocate the solar energy cost to the interdependent products while maintaining the additivity property. The problem is analogous to the study of interdependence in economic systems as originally

developed by Leontief (1941), that has come to be known as input-output (I-O) analysis. I-O analysis has been applied to the study of energy flow through economic and ecological systems by Herendeen and Bullard (1974), Hannon (1973; 1976; 1979), Costanza (1980), and others. I-O analysis usually assumes that the system can be divided into internal subprocesses each of which produce only one product.

In reality, the biosphere (or any system) will always be something of a "gray box" to us since we can never hope to have perfect information. Real systems also cannot be adequately described as collections of single product subprocesses. Joint products remain even at the most detailed level of disaggregation since all processes produce "waste products" and some processes involve the sorting or separation of raw material input. Sorting is inherently a joint product activity since it always results in at least two "piles" of sorted material.

In economic applications, joint products are frequently ignored (the usual case with waste products) or dealt with using the relative market prices of the joint products to allocate costs. Neither of these approaches is acceptable for the present purpose. Solutions to the joint product problem are possible, however, with slightly more general versions of the standard I-O model. In the next section we lay out the more general model and discuss the conditions for solution.

METHODS

A general equilibrium model of an economy which allows joint products was first proposed by von Neumann (1945)[2]. To produce this model, von Neumann simply kept track of commodities and processes separately and allowed for production of several commodities by each process (as opposed to the standard I-O model which assumes each commodity is produced by only one process). This approach has now been adopted by the Bureau of Economic Analysis (BEA) in presenting and formulating the national I-O tables (Ritz 1979) and is called the "commodity-by-industry" format. Our purpose here is different from most economic I-O modeling, however, since we seek to calculate the weighting factors which are the analog of prices in economic systems. Most economic I-O modeling is not aimed at calculating prices since prices for intra-market exchanges are known[3]. In a general equilibrium framework, the weighting factors can be stated in terms of the "primary resources", that is, those resources which are necessary for production but which are not

---

[1] The only other would be to simply divide the input equally among the outputs.

---

[2] See Costanza and Neill (elsewhere this volume) for a more complete description of the von Neumann model.

[3] The notable exceptions are theoretical works (cf. Sraffa 1960, Victor 1972) and planned economies which employ some form of general equilibrium modeling to set prices.

themselves produced in the system. Our interest in modeling the biosphere partly stems from the fact that at this level the system has essentially one primary resource – sunlight. Thus, the weighting factors can all be stated as "embodied solar energy" intensities.

Figure 2 illustrates the energy balance for a

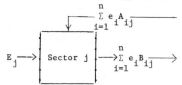

Figure 2. Single sector energy balance with joint products.

multiproduct process. This is a generalization of the single product energy balance development by Herendeen and Bullard (1974) and Hannon (1973). There are n commodities in the system and m processes. The energy balance for process j for one discrete time interval is given by

$$E_j + \sum_{i=1}^{n} e_i A_{ij} = \sum_{i=1}^{n} e_i B_{ij} \qquad (1)$$

where

E_j is the direct energy input to process j (direct solar energy in the global model)
A_{ij} is the use of commodity i by process j
B_{ij} is the output of commodity i by process j
e_i is the embodied energy intensity of commodity i

In the standard I-O model $B_{ij} = 0$ for all $i \neq j$.

In matrix notation for all m processes we have

$$E + eA = eB \qquad (2)$$

where

E is a m dimensional column vector of direct energy inputs
e is an dimensional column vector of embodied energy intensities
A is a m x n matrix of commodity inputs by process
B is a m x n matrix of commodity outputs by process

If there are an equal number of processes and commodities (m = n) we can solve this system of equations for the embodied energy intensity vector, e.

$$E = e(B-A) \qquad (3)$$

$$e = E(B-A)^{-1} \qquad (4)$$

While this system of equations is solvable, negative energy intensities can result[1]. These are difficult to interpret physically. Below we try to define the conditions under which all positive energy intensities prevail, using this model.

1) A sufficient condition for all positive energy intensities is the absence of joint products ($B_{ij} = 0$ for all $i \neq j$). In this case (B-A) will have all positive diagonal terms and all negative off-diagonal terms, leading to all positive terms in $(B-A)^{-1}$ and e. This condition is not necessary, however, for all positive energy intensities.

2) If there are joint products, then one of two general conditions can arise
   a) If the joint products are in all cases smaller than the corresponding inputs of the same commodities ($B_{ij} < A_{ij}$ for all i $\neq$ j) then the off-diagonal terms in (B-A) will still be all negative, $(B-A)^{-1}$ will be all positive and e will be all positive. For example, if a given amount of a commodity is both an input and an output of a process, it can be considered to be simply flowing through the process and is not a real joint product, even though it may have been required for some real transformation within the process.

   b) If, on the other hand, some of the joint products are larger than the corresponding inputs of the same commodities ($B_{ij} > A_{ij}$ for some i and j, i $\neq$ j) we call these "real" joint products. Negative energy intensities can arise only in the presence of real joint products. This is because (B-A) will have positive off-diagonal terms and $(B-A)^{-1}$ will have negative terms. Whether or not negative energy intensities do arise depends on the relative strengths of the positive and negative terms in $(B-A)^{-1}$ and the size and distribution of the direct input vector (E). We have not been able to determine any explicit conditions for all positive energy intensities with negative elements in $(B-A)^{-1}$. The best we can do at this point is to note that, from our experience with running models with these conditions, if the real joint products are "small" relative to production by the main production process, then all positive energy intensities will generally result. The meaning of this statement will become clearer with reference to the results of the next section.

RESULTS

The different processes of the biosphere require characteristic inputs and produce arrays

---

[1] The general conditions for solution of this system of equations are the same as those for the standard I-O equations (ie. linear independence, no zero rows or columns). Negative intensities can also be avoided by casting the problem in a linear programming (LP) format, but this introduces other problems (see Costanza and Neill, elsewhere this volume).

of different outputs which are in turn used as inputs by other processes. This interdependent nature of the biosphere can be expressed in the form of an input-output table. The magnitudes of the flows of carbon, nitrogen, phosphorus and water at the global level have been separately estimated by many researchers. We summarized published information on these flows in the input-output table shown in table 1. The data we used as input to the model is our estimate of the average values of the flows between sectors based on the literature we surveyed. Each value in table 1 has a note at the end of the paper explaining its sources and assumptions.

We divided the biosphere into an equal number of processes and commodities. Definitions of the sectoring scheme and descriptions of the commodities are shown in tables 2a and 2b. The number of processes we chose for the model was dependent on the degree of aggregation of the data and the amount of detail desired in the model output. For our first attempt at a global input-output model we selected aggregated sector definitions, but the number of sectors could later be increased to accomodate more detailed data. An equal number of processes and commodities is required for a strict input-output formulation of the model (see Costanza and Neill elsewhere in this volume). The sector boundaries differed slightly from what we would consider the ideal case, as some changes in sector definitions were required as a prerequisite for obtaining positive solar energy intensities (see Discussion).

The solar energy intensities obtained for the products of the biosphere are shown in table 3.

DISCUSSION

In order to interpret the global model's results, we first convert them into more conventional units. Table 3 lists the embodied energy intensities calculated using the global I-O model (in kcal solar/unit) and also the same intensities converted into various other units, including a "shadow price" (in $/unit). While we cannot place much credence in these numbers at present (because of uncertainties in the data and the model) we can at least interpret them for "reasonableness". Investigation of table 3 indicates that the values are indeed reasonable. For example, the intensities calculated for manufactured goods and services and fossil fuel, when combined, yield a value of about 18,000 kcal FF/$, which compares favorably with published values for the U. S. and other countries (cf. Kemp et al. this volume). The value of 10,711 kcal solar/kcal FF compares well with the value of 2000 kcal solar/ kcal FF estimated by Odum (1978), when one considers that the latter value was based on a linear sequence from sunlight to wood to electricity without adequate treatment of the embodied energy contributions of water and nutrients, and the former is based on very rough estimates of the current rate of fossil fuel formation. This rate is certainly lower than the formation rate during the main fossil fuel forming eras, and may lead to a high value for the stored fuel. The "shadow

prices" for agricultural and natural products, water, and nutrients also seem to be reasonable. For example, the calculated shadow price for fresh water of $.01/gal is reasonably higher than the market price of domestic water (which includes only the cost of collecting and treating the water) of about $.0005/gal to $.001/gal.

Negative Energy Intensities

As noted earlier, the general equilibrium system of equations for calculating energy intensities in the pressence of joint products can lead to negative values. We considered negative energy intensities to be difficult to justify and sought the conditions that would allow all positive energy intensities. While our results for the global model do show all positive intensities, we noticed that relatively minor changes in the data would produce some negatives.

We found that we could obtain positive energy intensities with the greatest regularity if we sectored the model so each process produced one "major" output, i.e., a real output greater than all the other real outputs of that commodity from all other processes. The major outputs, which correspond to sole outputs from processes in input-output formulations of systems without joint products, are lined up up along the diagonal of the output table.

To achieve this configuration with our aggregated sectoring scheme, two slight modifications were made in the sector boundaries: (1) Domestic animals were moved from the agricultural sector to the natural sector because the output of nitrogen from domestic animals would have given the agricultural sector two major outputs, nitrogen and agricultural crop biomass, and (2) The ocean was divided into two sectors to allow one to produce phosphorus as its major output.

We found that no solution to the model could be obtained if the sum of liquid water and water vapor entering each process exactly balanced the sum of liquid water and water vapor leaving (since this implies linear dependence between these rows in the matrix).

This problem is avoided by assuming that a percentage of the water taken up by plants is incorporated into biomass and leaves the sector as a major output of biomasss from the plant process. We also included a net output of liquid water, which is interpreted as a recharge of world groundwater storages.

Even with the data in this "canonical" form, we found that negative energy intensities were still possible with small changes in the data. We began to wonder if negative energy intensities might not have some reasonable interpretation. If the energy intensities represent the amount of additional sunlight required, directly and indirectly, to produce an additional unit of commodity, then a negative intensity might be interpreted as the reduction in the sunlight required (directly and indirectly) when an additional unit of commodity is produced. The next question is: does it make sense to have "negative" commodities that reduce the total energy input the more of

Table 1.--Process-by-commodity input-output table for the biosphere (all values are annual flows for ∿ 1970)*

Commodity Inputs	Urban Economy (1)	Agri-culture (2)	Natural Plants (3)	Animals (4)	Soil (5)	Deep Ocean (6)	Surface Ocean (7)	Atmos-phere (8)	Deep Geology (9)	Net Output	Totals
1. Manufactured goods and services (10^{12} $)	2.71	.08								1.19	3.98
2. Agricultural products (10^{15} g.d.w.)	1.28	4.55		3.27							9.1
3. Natural products (10^{15} g.d.w.)	1.18			27.9	103.4	34.6			0.16		167.3
4. Nitrogen (10^{12} g.)	55.0	62.4	208.0		493.6		168.0	389.5			1,376.5
5. Carbon dioxide (10^{15} g. C)		8.2	147.0			15.6	37.2	110.3			318.3
6. Phosphorous (10^{12} g.)	12.6	28.5	1,345.7		8.4		21.0	9.5	13.0		1,438.7
7. Water vapor (Km^3)								496,100			496,100
8. Fresh water (Km^3)	1,008	15,490	51,226		111,419		424,700			2,000	605,843
9. Fossil fuel (10^{15} g. C)	5										5
Sunlight (10^{18} Kcal)		23	227				606				856

Outputs	(1)	(2)	(3)	(4)	(5)	(6)	(7)	(8)	(9)	Net Input
1. Manufactured goods and services (10^{12} $)	3.98									3.98
2. Agricultural products (10^{15} g.d.w.)		9.1								9.1
3. Natural products (10^{15} g.d.w.)			163.4	3.9						167.3
4. Nitrogen (10^{12} g.)	80.0	31.0		295.0	340.5		182.0	448.0		1,376.5
5. Carbon dioxide (10^{15} g. C)	5.0	6.1	73.6	14.0	46.5	15.6	49.5	108.0		318.3
6. Phosphorus (10^{12} g.)	14.2				241.3	1,161.1		9.5	12.6	1,438.7
7. Water vapor (Km^3)	79	5,931	50,740		14,650		424,700			496,100
8. Fresh water (Km^3)	929	9,829			98,985			496,100		605,843
9. Fossil fuel (10^{15} g. C)									0.07	4.93	5.00

* See NOTES TO TABLE 1 at end of paper.

Table 2a.--Definition of processes used in
biosphere model

Process	Definition
1) Economic process	People, industries, households, and other man-made structures and institutions
2) Agriculture	Soil on cultivated land, crops, and agricultural capital and farm households
3) Natural plants	Natural primary producers, including marine producers and managed forestry crops
4) Animals	Wild and domestic animals
5) Soil and surface waters	Soil on uncultivated land, surface water and accessible groundwater
6) Deep ocean	Ocean water below the euphotic zone
7) Surface ocean	Ocean water in the euphotic zone
8) Atmosphere	Atmosphere
9) Deep geologic process	Deep geologic formations and storages of mineral resources

Table 2b.--Definitions of commodities used in
biosphere model

Commodity	Definition
1) Economic goods and services	Goods and services, measured 1972 U.S. dollars
2) Agricultural products	Agricultural crop biomass
3) Natural products	Uncultivated crop biomass (including forestry products), and animal biomass
4) Nitrogen	Nitrogen in simple inorganic nitrogenous compounds
5) Carbon dioxide	Carbon in carbon dioxide
6) Phosphorus	Phosphorus in simple inorganic phosphorus-containing compounds
7) Water vapor	Water vapor
8) Fresh water	Fresh liquid water
9) Fossil fuel	Carbon in fossil fuel

Table 3.--Embodied energy intensities from the
global model in various units.

Commodity	Embodied energy intensity*
1. Manufactured goods and services	191.2 E6 kcal solar/$ = 17,850 kcal FF/$
2. Agricultural products	13.9 E3 kcal solar/g = 6.2 E6 kcal solar/lb = $.03/lb
3. Natural products	39.2 E3 kcal solar/g = 17.7 E6 kcal solar/lb = $.09/lb
4. Nitrogen	0.63 E6 kcal solar/gN = $1.49/lbN
5. Carbon dioxide	57.1 E3 kcal solar/gC = $.13/lbC
6. Phosphorous	1.17 E6 kcal solar/gP = $2.75/lbP
7. Water vapor	.55 E18 kcal solar/cu. km = $2.87/cu. m = $.01/gal
8. Liquid water	.55 E18 kcal solar/cu. km = $2.87/cu. m = $.01/gal
9. Fossil fuel	96.4 E3 kcal solar/g FF = 10,711 kcal solar/kcal FF

*The values above were calculated using standard conversion factors (451 g/lb, 264.2 gal/cu. m, etc) and the intensities calculated by the model for the other commodities (ie. 191.2 E6 kcal solar/$). Fossil fuel was converted to kcal using 9 kcal/g FF.

them that are produced? While this notion is counter to our intuition, we cannot deny the possibility in a complex, interdependent system with joint products (like the biosphere). One example might be a product that directly blocks sunlight (like particulates), or one that reduces photosynthesis (like toxins). The problem is that none of the commodities included in the global model seem to fit in this category of "bads", if not in reality then in the way we have defined their production relations. In the model, increased inputs always lead to increased physical outputs. The "negativity" must have something to do with the way joint products behave in the system and the valuation aspects of the energy intensities.

We can gain some insight by considering the inverse matrix, $(B-A)^{-1}$. The elements of this matrix can be interpreted as the direct and indirect amount of commodity i required (or produced if the element is negative) per unit output of process j. Here, the influence of joint products can be seen directly, since real joint products lead to negative terms in the inverse matrix. This is because they are "produced" as by-products and not necessarily "required" as input to other processes. If these negative elements are large and placed properly in the matrix, they can outweigh some of the positive elements in the calculation of energy intensities. The problem is that while we can easily accept this interpretation for individual elements in the matrix, we still expect that, overall, the energy intensities should be all positive.

If negative energy intensities do not make sense physically then their presence is an indication of problems in the model or the data. While we have operated under this latter interpretation, we cannot rule out the possible validity of the former, and must leave this as an open question.

Summary and Conclusions

This paper is a first attempt at the application of input-output analysis techniques to the study of global material and energy flows. As such, it is more interesting for the methodological ground it breaks and the problems it unearths than for the numerical values it generates. The fact that it is possible to construct such a model and produce values that are at least "in the ballpark" relative to our expectations is encouraging at this point.

The methodological problems that still loom before us can be divided into two general categories: 1) data problems, and 2) the possibility of negative energy intensities. Data problems will, of course, always be with us since we can never have perfect information. The problems can be quite severe at the global level as evidenced by the contortions we sometimes went through to sector and estimate flows for the model (see NOTES TO TABLE 1). The problem of negative energy intensities is more subtle, however. On the one hand, this problem may be nothing more than a nasty manifestation of the data problem, i.e. if we really had accurate data we would find that the real system operates to avoid negative energy intensities. On the other hand, negative intensities may be perfectly valid and interpretable as the normal implications of joint production. One piece of information arguing for the latter interpretation is the existence of what amounts to "negative prices" for the (joint) waste products of our economic systems. One might argue that the more pollution produced, the less energy consumed, since energy-expensive anti-pollution devices are eliminated. On the other hand, these "negative prices" may be nothing more than a reflection of the narrow, imperfect vision of the market. If a comprehensive accounting with perfect information were done one might find that increased pollution actually increases overall energy consumption (by

increasing health costs, etc.) and should have a positive intensity.

Results like those reported in table 3 represent an attempt at "shadow pricing" in the most comprehensive sense. The potential utility to environmental managers and decision makers of defendable answers to questions like: what is the true extra-market value of rainfall? or ecosystem services? or nonrenewable resources? is enormous. The vision of this potential utility has guided us down a rocky analytical road and we can only hope that the problems we encountered are resolveable. If not, we will at least have tested the path.

ACKNOWLEDGEMENTS

This work was supported, in part, by NSF Project No. PRA-8003845 titled "Use of embodied energy values in the analysis of environment/energy/economy trade-off decisions: an exploratory analysis" M. J. Lavine and R. Costanza Principle Investigators. We thank C. Cleveland for reviewing an earlier draft and providing useful comments. This is contribution #LSU-CEL-81-07 from the Coastal Ecology Laboratory, Louisiana State University, Baton Rouge, LA 70803.

NOTES TO TABLE 1

The numbers preceeding the notes are keyed to the commodities (first number) and processes (second number) listed in Table 1. Inputs to all processes are given first, under the heading "inputs", followed by a listing of all outputs. For example 1,2 in the input section refers to the input of commodity 1 (manufactured goods and services) to process 2 (agriculture). Scientific notation is expressed using "E format" (5.00 E9 is read 5.00 times 10 to the ninth)

Inputs

1,1 World GNP for 1970 was estimated at 3.98 E 12 U.S. 1972 dollars by the U.S. State Dept. (1972). Of this total, we roughly estimated the percentage consumed (as inputs to the urban economy and agriculture) v. that invested (net output). The ratio of gross business capital formation plus net exports plus net inventory change to GNP for the U.S. in 1967 was 0.15 (U.S. Department of Commerce 1976). However, this excludes human capital formation. When human capital formation is included, the value for the gross capital formation:GNP ratio for the U.S. for 1967 is 0.64 (Kendrick 1976) World values for this ratio would be significantly lower due to lower investment in education, health, etc. We estimated a ratio of about 0.3 for the world economy. This leads to an estimate of 2.79 E 12 $U.S. as the amount consumed by urban and agriculture and 1.19 E 12 $U.S. as net output. The total consumption was distributed to urban and agriculture on the basis of these sectors' relative proportion of GNP in the U.S. economy in 1967 (0.03 agriculture, 0.97 urban).

1,2 and 1,10 (see above)

2,1 Human consumption of agricultural crops (Whittaker and Likens 1975)

2,2 Litter on agricultural land. 50% of net production (Reiners 1973)

2,4 Animal consumption on agricultural land. Production on agricultural land (see output 2,2) minus human consumption (see input 2,1) minus litter (see input 2,2)

3,1 Human consumption or use of biomass other than agricultural crops. Sum of consumption of animals (0.08, Whittaker and Likens 1975) and wood (1.1, Bolin 1977).

3,4 World animal consumption, excluding consumption on agricultural land (Ajtay et al. 1979)

3,5 Litter production on natural land. Total nonagricultural net production of plants (163.4, Whittaker and Likens 1975) and animals (3.9, Ajtay et al. 1975) minus human consumption (see input 3,1), minus animal consumption (see input 3,4), minus marine detritus (see input 3,6), minus deep sedimentation (see input 3,9)

3,6 Marine detritus production. Marine net production (55, Whittaker and Likens 1975) minus marine animal consumption (20.23, Ajtay et al. 1979) and deep sedimentation (see input 3,9)

3,9 Sedimented biomass converted to fossil fuel. (Garrels et al. 1973) Converted from grams C to grams biomass using 0.45 gC/g dry weight biomass.

4,1 Input of nitrogen to industry and households. Sum of: industrial fixation, 55; nitrogen fixed in the burning of fossil fuels, 19; (Soderlund and Svensson 1976)

4.2 Input of nitrogen to agricultural crops. Sum of: world fertilizer consumption, 33; ammonium nitrogen in precipitation and dry fallout, 13*; nitrate-nitrite nitrogen in precipitation and dry fallout, 5.4*; fixation by crops, 11**; (Soderlund and Svensson 1976)

4,3 Input of nitrogen to natural plants. Sum of: fixation by terrestrial plants (excluding agricultural crops) 128**; fixation by marine plants, 80; (Soderlund and Svensson 1976)

4,5 Input of nitrogen to soil. Sum of: ammonium nitrogen in precipitation and dry fallout, 126*; nitrate-nitrite nitrogen in precipitation and dry fallout, 52.6*; wild animal excrement (calculated as ten times the amount of nitrogen released into the atmosphere from excrement), 40; human excrement, 20; domestic animal excrement (calculated as ten times the amount of nitrogen released into the atmosphere from excrement), 255, (Soderlund and Svensson 1976)

4,7 Input of nitrogen to oceans. Sum of: nitrate-nitrite nitrogen in runoff, 8; ammonium nitrogen in runoff, 1; nitrate-nitrite in precipitation and dry fallout, 22; ammonium nitrogen in precipitation and dry fallout, 35; marine denitrification, 102; (Soderlund and Svensson 1976)

4,8 Input of nitrogen to the atmosphere. Sum of: nitrogen fixed and released in burning of fossil fuels, 19; terrestrial denitrification, 134; ammonium nitrogen released in coal burning, 8; wild animal excrement volitilized, 4; human excrement volitilized, 5; domestic animal excrement volitilized, 25.5; ammonium nitrogen from soil to balance fallout, 139; nitrate-nitrite nitrogen input from soil to balance fallout, 55; (Soderlund and Svensson 1976)

5,2 Carbon dioxide uptake by agricultural crops***. GPP asumed to be twice NPP (NPP=9.1 g. d.w., Whittaker and Likens 1975)

5,3 Carbon dioxide uptake by marine and terrestrial plants, excluding agricultural crops***. GPP assumed to be twice NPP (NPP=163.4 g. d.w., Whittaker and Lkens 1975)

5,6 Carbon dioxide input to deep ocean from biomass decomposition. All marine detritus production (equal to 34.6 g. d.w., see input 3,6) assumed to decompose.

5,7 Carbon dioxide input to ocean. Sum of: respiration of marine producers, 24.8*** (Respiration assumed to be equal to NPP, NPP=55 g. d.w., Whittker and Likens 1975); animal respiration, 10.1***. (Animal respiration assumed to be 99% ingestion. Marine animal ingestion equals 72% total animal ingestion, Ajtay et al. 1979. Total animal ingestion equals 31.2, see inputs 2,4 and 3,4); input from atmosphere, 2.3 (see output 5,8)

5,8 Carbon dioxide input to atmosphere. Sum of: natural plant respiration, 48.8*** (equal to NPP, 108.4 g. d.w., Whittaker and Likens 1975); agricultural crop respiration, 4.1*** (equal to NPP, 9.1 g. d.w., Whittaker and Likens 1975); decomposition on natural land, 46.5*** (assumed all litter decomposed, see input 3,5); decomposition of litter on agricultural land, 2.0*** (assumed all litter decomposed, see input 2,2); animal respiration, 3.9*** (Animal respiration assumed to be 99% ingestion. Terrestrial animal ingestion equals 28% total animal ingestion, Ajtay et al. 1979. Total ingestion equals 31.2, see inputs 2,4 and 3,4); fossil fuel combustion, 5 (Baes et al. 1976)

6,1 Phosphorus mined (Pierrou 1976)

6,2 Phosphorus input to agriculture. Sum of: crop uptake** (16.8, Pierrou 1976); fertilizer, 10.0; precipitation and dry fallout, 0.6*; human excrement (50% of world total is applied to agricultural land), 1.1; (Pierrou 1976)

6,5 Phosphorus input to soil. Sum of: output from households and industry, 2; human excrement (25% discharged into freshwaters), 0.55; precipitation and dry fallout, 5.8*; (Pierrou 1976)

6,7 Phosphorus input to ocean. Sum of: precipitation and dry fallout, 3.1; Runoff, 17.4; human excrement (25% discharged into coastal waters), 0.55 (Pierrou 1976)

6,8 Phosphorus input to atmosphere to balance fallout (Pierrou 1976)

6,9 Phosphorus sedimented (Pierrou 1976)

7,8 Global evapotranspiration (Baumgartner and Reichel 1975)

8,1 Liquid water input to urban economy. Sum of: water used for energy production, 502; industrial water use, 305; and domestic water use, 201; (U.S. C.E.Q. and Dept. of State 1980)

8,2 Liquid water input to agriculture. Sum of:

water taken up and transpired by crops, 4260**
(total transpiration, 55,000, Ricklefs 1979);
water taken up and incorporated into biomass,
40.8** (calculated as 1% of water transpired,
Penman 1970); water lost in transit to agriculture, 311 (17% of water withdrawn for irrigation, U.S. C.E.Q. and Dept. of State 1980);
water for irrigation that eventually runs off,
439 (24% water withdrawn for irrigation, U.S.
C.E.Q. and Dept. of State 1980) precipitation
on agricultural land 10,439* (total terrestrial
precipitation, 111,100, Baumgartner and
Reichel 1975)

8,3 Liquid water input to natural plants. Water
taken up and transpired, 50,740** (total transpiration, 55,000, Ricklefs 1979); water taken
up and incorporated into biomass, 485.8**
(Penman 1970, see input 8,2)

8,5 Liquid water input to soil. Sum of: precipitation on natural land, 102,340* (total terrestrial precipitation, 111,100, Baumgartner
and Reichel 1975); urban runoff, 929, (see
output 8,1); runoff from agriculture, 9829,
(see output 8,2)

8,7 Liquid water input to oceans. Sum of: precipitation over oceans, 385,000; runoff, 39,700;
(Baumgartner and Reichel 1975)

8,10 Net output of water. Corresponds to groundwater recharge (see text)

9,1 Human consumption of fossil fuels. Assumed
to be equal to the carbon released to the
atmosphere through the burning of fossil fuel
(5.0, Baes et al. 1976)

10,2 Sunlight falling on agricultural land. (Total
sunlight reaching the earth over land and
ocean equals 856 E 18. Agricultural area
accouts for 3% of the earth's total area,
Whittaker and Likens 1975)

10,3 Sunlight falling on nonagricultural land.
(Total sunlight reaching the earth equals 856
E 18. Nonagricultural area accounts for 26% of
the earth's total area, Whitaker and Likens
1975)

10,7 Sunlight falling on oceans. (Total sunlight
reaching the earth equals 865 E 18. Oceans
account for 71% of the earth's total area,
Whittaker and Likens 1975)

Outputs

1,1 1970 World GNP, 1972 U.S. dollars (U.S. Dept.
of State 1972)

2,2 Net primary production on agricultural land
(Whittaker and Likens 1975)

3,3 World net primary production, excluding production on agricultural land (Whittaker and
Likens 1975)

3,4 World animal production (Ajtay et al. 1979)

4,1 Nitrogen outputs from human economy. Sum of:
nitrogen fixed in fossil fuel burning, 19;
human excrement, 20; ammonium-nitrogen
released in coal burning, 8; fertilizer application, 33; (Soderlund and Svensson 1976)

4,2 Nitrogen output from agriculture. Sum of:
soil denitrification, 13*; ammonium nitrogen
output to atmosphere to balance fallout, 13*;
nitrate-nitrite nitrogen output to atmosphere
to balance fallout, 5 (Soderlund and Svensson
1976)

4,4 Nitrogen output from animals. Sum of: wild
animal excrement, 40; domestic animal excrement, 255; Excrement input to soil assumed to
be ten times the nitrogen in excrement eventually volitilized. (Soderlund and Svensson
1976)

4,5 Nitrogen output from soil. Sum of: soil denitrification, 121*; ammonium nitrogen in
runoff, 1; nitrate-nitrite nitrogen in runoff,
8; wild animal excrement volitilized, 4;
domestic animal excrement volitilized, 25.5;
human excrement volitilized, 5; ammonium
nitrogen output to atmosphere to balance fallout, 126*; nitrate-nitrite nitrogen output to
atmosphere to balance fallout, 50*; (Soderlund
and Svensson 1976)

4,7 Nitrogen output from oceans Sum of: marine
denitrification, 102; nitrogen fixed by marine
producers, 80; (Soderlund and Svensson 1976)

4,8 Nitrogen output from atmosphere. Sum of:
terrestrial fixation, 139; industrial fixation, 36; fixation in fossil fuel burning, 19;
ammonium nitrogen in precipitation and dry
fallout over oceans, 35; nitrate-nitrite precipitation and dry fallout over oceans, 22;
ammonium nitrogen in precipitation and dry
fallout over land, 139; Nitrate-nitrite in
precipitation and dry fallout over land, 58;
(Soderlund and Svensson 1976)

5,1 Carbon dioxide output from urban economy
(Baes et al. 1976)

5,2 Carbon dioxide output from agriculture. Sum
of: agricultural crop respiration, 4.1***
(equal to NPP, 9.1 g. d.w., Whittaker and
Likens 1975); decomposition of litter on agricultural land, 2.0*** (assumed all litter
decomosed, see input 2,2)

5,3 Carbon dioxide output from natural plants.
Sum of: terrestrial plant (excluding crop)
respiration, 48.8*** (assumed to be equal to
NPP, 108.4 g. d.w., Whittaker and Likens,
1975); marine plant respiration, 24.8***
(equal to NPP, 55 g. d.w, Whittaker and Likens
1975)

5,4 Carbon dioxide output from animals. Sum of:
marine animal output (see input 5,7); terrestrial animal output (see input 5,8)

5,5 Carbon dioxide output from soil (see input
5,8)

5,6 Carbon dioxide output from deep ocean
(assumed to be equal to input to deep ocean
from decomposition, see input 5,6)

5,7 Carbon dioxide output from surface ocean.
Equal to uptake by marine producers, 49.5***
(GPP assumed to be twice NPP, NPP=55 g. d.w.,
Whittaker and Likens, 1975)

5,8 Carbon dioxide output from atmosphere. Sum
of: terrestrial plant uptake, 97.5*** (GNP
assumed to be twice NPP, NPP=108.4 g. d.w.,
Whittaker and Likens 1975); agricultural plant
uptake, (see input 5,3); output to oceans, 2.3
(inputs to atmosphere assumed to exceed oututs
by 2.3. About half of the urban input of carbon dioxide to the atmosphere is an input to
the ocean)

6,1 Phosphorus output from households and industry. Sum of: human excrement, 2.2; fertilizer, 10.0; industrial and household waste,
2.0; (Pierrou 1976)

6,2 Phosphorus output from soil. Sum of: runoff, 17.4; input to atmosphere as dust, 6.4*; uptake by terrestrial and freshwater plants, 217.5; (Pierrou 1976)

6,6 Phosphorus output from the deep ocean. Sum of: phosphorus taken up by marine producers, 1145; sedimentation, 13; input to atmosphere in sea spray, 3.1; (Pierrou 1976)

6,8 Phosphorus output from atmosphere. Sum of: precipitation and dry fallout over land, 6.4; precipitation and dry fallout over oceans, 3.1; (Pierrou 1976)

7,1 Evaporation from urban economy. It was assumed that all water classified as "consumed" was evaporated. Sum of: energy production, 8 (1.5% of 502 water withdrawn consumed, U.S. C.E.Q. and Dept. of State 1980); water for industy, 34 (11% of 305 water withdrawn consumed, U.S. C.E.Q. and Dept. of State 1980); water for domestic use, 38 (18.7% of 201 water withdrawn consumed (Harte and El-Gassir 1978); All estimates of water consumed are for the United States.

7,2 Evaporation from agriculture. Sum of: transpiration by crops (4260**, total transpiration, 55,000, Ricklefs 1979); loss to evaporation in transit to agriculture, 311 (17% of water withdrqwn for irrigation, U.S. C.E.Q. and Dept. of State 1980); evaporation on agricultural land, 1360 (transpiration:evaporation ratio assumed to be the same as for total land area, 55,000:71,400 (Ricklefs 1979, Baumgartner and Reichel 1975)

7,3 Transpiration by plants, excluding agricultural crops, 50,740** (total transpiration, 55,000, Ricklefs 1979)

7,5 evaporation from soil. Global terrestrial evaporation, 71,400 (Baumgartner and Reichel 1975) minus total terrestrial transpiration (55,000) minus evaporation from urban (79) minus evaporation of water in transit to agriculture (311) minus evaporation from agricultural soil (1360, see output 7,2)

7,6 Evaporation from oceans (Baumgartner and Reichel 1975).

8,1 Liquid water output from urban economy. Water not consummed assumed to be discharged into freshwaters. Sum of water for energy production, 494.5 (998.5% of 502 water withdrawn not consumed, U.S. C.E.Q. and Dept. of State 1980); water for industry, 271.5 (89% of 305 water withdrawn not consumed, U.S. C.E.Q. and Dept. of State 1980); water for domestic use, 163.4 (81.3% of 201 water withdrawn not consumed (Harte and El-Gasseir 1978)

8,2 Agricultural runoff. Equal to the sum of liquid water inputs to agriculture (15,490) minus transpiration (4260, see output 7,2), minus evaporation 1360, see output 7,2), minus water incorporated into biomass (40.8, see input 8,2)

8,5 Liquid water outputs from soil. Sum of runoff to oceans and inputs to plants, urban, and agriculture for irrigation. Calculated by difference to balance liquid water inputs and outputs.

8,8 Global precipitation (Baumgartner and Reichel 1975)

9,9 Conversion of biomass into fossil fuel by deep geologic processes (Garrels et al. 1973).

9,10 Net depletion of fossil fuel storages. Difference between fossil fuel consumption (5.0, see input 9,1) and formation (0.07, see output 9,9)

* Distributed between natural land and agricultural land on the basis of relative terrestrial area (90.6% natural, 9.4% agricultural, Whittaker and Likens 1975)

** Distributed between natural plants and agricultural plants on the basis of relative productivity (92.3% natural, 7.7% agricultural, Whittaker and Likens 1975)

*** All biomass to carbon conversions are based on 0.45 g.C/g. biomass d.w. (Whittaker and Likens 1973)

LITERATURE CITED

Ajtay, G. L., P. Ketner, and P. Duvigneaud. 1979. Terrestrial primary production and phytomass. pp. 129-181 in: Bolin, B., E. T. Degens, S. Kempe, and P. Ketner (eds.), The global carbon cycle. SCOPE 13, John Wiley and Sons, New York.

Baes, C. F., H. E. Goeller, J. S. Olsen, and R. M. Rotty. 1976. The global carbon dioxide problem. ORNL-5194, 1-72. Oak Ridge National Laboratory, Oak Ridge, Tenn.

Baumgartner, A., and E. Reichel. 1975. The world water balance. Elsevier, New York.

Bolin, B. 1977. Changes of land biota and their importance for the carbon cycle. Science, 196:613-615.

Costanza, R. 1980. Embodied energy and economic valuation. Science 210:1219-1224.

Costanza, R. and C. Neill. 1981. The energy embodied in the products of ecological systems: a linear programming approach. Paper presented at the International Conference on Energy and Ecological Modelling, Louisville, KY, April 20-23.

Council on Environmental Quality and Department of State. 1980. The global 2000 report to the president: volume II, the technical report. U. S. Government Printing Office, Washington, D. C.

Daly, H. E. 1968. On economics as a life science. J. Pol. Econ. 76:392-406.

Freyer, H. D. 1979. Variations in the atmospheric CO 2 content. pp. 79-99 in: B. Bolin, E. T. Degens, S. Kempe, and P. Ketner (eds.), The global carbon cycle. SCOPE 13, John Wiley and Sons, New York.

Garrels, R. M., F. T. MacKenzie, and C. Hunt. 1973. Chemical cycles and the global environment. W. Kaufmann, Los Altos, CA.

Hannon, B. 1973. The structure of ecosystems. J. Theor. Biol. 41:535-546.

Hannon, B. 1976. Marginal product pricing in the ecosystem. J. Theor. Biol. 56:256-267

Hannon, B. 1979. Total energy costs in ecosystems. J. Theor. Biol. 80:271-293.

Harte, J. and M. El-Gasseir. 1978. Energy and water. Science 199:623-634.

Herendeen, R. A., and C. W. Bullard. 1974 Energy costs of goods and services, 1963 and 1967. CAC Document No. 140. Energy Research Group. University of Illinois, Champaign, IL.

Kemp, W. M., W. R. Boynton, and K. Limberg. 1981. The influence of stored and renewable energy resources on the economic production in 63 nations. Paper presented at the International Conference on Energy and Ecological Modelling, Louisville, KY, April 20-23.

Kendrick, J. W. 1976. The formation and stocks of total capital. 256 p. National Bureau of Economic Research, New York, N. Y.

Leontief, W. W. 1941. The structure of American economy, 1919, 1929: an empirical application of equilibrium analysis. Harvard University Press, Cambridge, MA.

Odum, H. T. 1978. Energy analysis, energy quality, and environment. pp. 55-88 in: M. W. Gilliland (ed.), Energy analysis: a new public policy tool. AAAS Selected Symposia 9. 110 p. Westview Press, New York, N. Y.

Odum, H. T., F. C. Wang, J. Alexander, and M. Gilliland. 1980. Energy analysis of environmental values: a manual for estimating environmental and societal values according to embodied energies. Annual Report to the Nuclear Regulatory Commission. Contract # NRC-04-77-123. Center for Wetlands, University of Florida, Gainesville. 32611.

Penman, H. L. 1970. The water cycle. Sci. Amer. 223:99-108.

Pierrou, U. 1976. The global phosphorus cycle. in: B. H. Svensson and R. Soderlund (eds.), Nitrogen, phosphorus, and sulfur--global cycles. SCOPE Report 7. Ecol. Bull. (Stockholm) 22:75-88

Reiners, W. A. 1973. Terrestrial detritus and the carbon cycle. pp. 303-327 in: G. M. Woodwell and E. V. Pecan (eds.), Carbon and the biosphere. United States Atomic Energy Commission, CONF-720510.

Ricklefs, R. E. 1979. Ecology. Chiron Press, New York.

Ritz, P. M. 1979. The input-output structure of the U. S. economy, 1972. Survey of Current Business, Feb. 1979: p. 34-72.

Sellars, W. D. 1965. Physical climatology. University of Chicago Press, Chicago, Illinois.

Soderlund, R. and B. H. Svensson. 1976. The global nitrogen cycle. in: B. H. Svensson and R. Soderlund (eds.), Nitrogen, phosphorus, and sulfur--global cycles. SCOPE Report 7. Ecol. Bull. (Stockholm) 22:23-73.

Sraffa, P. 1960. Production of commodities by means of commodities: prelude to a critique of economic theory. Cambridge University Press, Cambridge, England. 99 pp.

United States Department of Commerce. 1976. The national income and product accounts of the United States, 1929, 1974. U. S. Bureau of Economic Analysis. U. S. Government Printing Office. Washington, D. C.

United States Department of State, Bureau of Intellegence and Research. 1972. Research studies: the planetary production in 1972, systems in dissarray. U. S. Government Printing Office, Washington, D. C.

Victor, P. A. 1972. Pollution, economy, and environment. 247 p. University of Toronto Press, Toronto, Canada.

von Neumann J. 1945. A model of general economic equilibrium. Rev. Econ. Studies. 13:1-9.

Wang, F. C., and M. A. Miller. 1981. Energy analysis applied to environmental issues. Paper presented at the International Symposium on Energy and Ecological Modelling, Louisville, KY, April 20-23.

Whittaker, R. H. and G. E. Likens. 1973. Carbon in the biota. pp. 281-300 in: G. M. Woodwell and E. V. Pecan (eds.), Carbon and the biosphere. United States Atomic Energy Commission, CONF-720510.

Whittaker, R. H. and G. E. Likens. 1975. The biosphere and man. pp. 305-328 in: H. Leith and R. H. Whittaker (eds.), Primary productivity of the biosphere. Springer-Verlag, New York.

ENERGY ANALYSIS AND ECONOMIC ANALYSIS: A COMPARISON OF CONCEPTS[1]

M. J. Lavine and T. J. Butler[2]

Abstract.--Comparisons of underlying concepts reveal both important similarities and differences between embodied energy analysis and economic benefit-cost analysis. They have common goals and recognize many of the same parameters of value. However, significant differences, particularly concerning boundaries of analysis, also exist and suggest appropriate limits for comparing or combining embodied energy and economic values.

INTRODUCTION

This paper reports on a comparison of the basic concepts underlying certain embodied energy analysis and economic analysis tools. The interpretations and comparisons are based on a proposed use of embodied energy analysis (sometimes referred to as the Odum school of energy analysis or ecoenergetics) to supplement the traditional economic benefit/cost analyses practiced by a Federal agency (Odum, Lavine, et al. 1981). This comparison of energy and economic analysis concepts differs from similar previously-published comparisons (for example, see International Federation of Institutes for Advanced Study 1975) in that the form of energy analysis being considered bases valuations on the amount of energy in all factors of production, including both capital and labor, not just in fuels and certain other scarce factors. The embodied energy analysis tools have been proposed to supplement traditional economic analyses such that natural environmental factors are given consideration comparable to traditionally-evaluated economic factors.

[1]Paper presented at the international symposium Energy and Ecological Modelling, sponsored by the International Society for Ecological Modelling. (Louisville, Kentucky, April 20-23, 1981) This paper is based on research supported by the U.S. Nuclear Regulatory Commission (NRC-04-77-123) and the National Science Foundation (PRA-8003845).

[2]M.J. Lavine is Sr. Research Associate at the Center for Environmental Research and the Department of Environmental Engineering, and T.J. Butler is Research Support Specialist at the Center for Environmental Research and the Section of Ecology and Systematics, Cornell University, Ithaca, New York, U.S.A.

In embodied energy analysis each component of a system, including natural environmental resources and flows, is valued according to its embodied energy content. A system component's embodied energy content is defined as the aggregate of all direct and even remotely indirect uses of available work involved in producing and making the component available to the system when that system is operating near an optimum balance of speed and efficiency to maximize work output. Money values associated with certain components are hypothesized to be consistently related to the embodied energy values associated with those components. The comparison of concepts presented in this paper is intended to promote a clear understanding of (a) any differences in the meanings of the energy and economic analysis values and (b) the areas in which the two forms of analysis may supplement each other.

COMPARISONS

Table 1 presents some of the basic theoretical assumptions of both energy and economic analyses. Related assumptions are shown adjacent to each other so that similarities and differences between the two forms of analysis may be easily identified. The discussions below elaborate on each of the comparisons that are outlined in the table.

Goal of Analysis Procedure

When comparing alternatives, the goals of economic benefit/cost analysis and embodied energy analysis are generally very similar but have one important procedural difference. The overall end of both forms of analysis is to identify the alternative which provides the system of concern with the most physical production. However, different procedural goals are pursued as indicators of that end.

Table 1.--Comparison of theoretical assumptions.

Parameter To Be Compared	Economic Analysis	Energy Analysis
Goal of analysis procedure	Identify the alternative that has the minimum total cost (including both internal and external costs, i.e. all costs to the total system of concern) or that causes the least use of scarce resources within the system of concern.	Identify the alternative with which the system of concern accomplishes the most work (or power).
Resources of value	Scarce resources (summarized as land, labor, and capital available to the system of concern).	Embodied energy (summarized as the minimum total amount of available work required, directly and indirectly, to provide a given product when the system of concern is operating at optimum efficiency for maximum work output).
Substitutability of valuable resources in production processes	Scarce resources (land, labor, and capital) may substitute for each other.	There is no substitute for embodied energy.
Effect of time on value	There is a time component of value operating such that a resource controlled now is usually more valuable for satisfying wants than the same resource controlled later.	There is a time component of value operating such that a unit amount of embodied energy contributed to a process generates less value when used rapidly than when used slowly.
Model for valuing a resource	A system of economic markets with commodity and monetary transactions, operating with competition for satisfying wants with the least use of scarce resources.	A unified global system of man and nature with energy transformations operating under competitive conditions in which there is the best possible efficiency consistent with system-wide maximum work output.
Boundaries of the model's production system	The boundaries encompass business capital (equipment and plant) and intermediate products, all driven by net inputs of natural resources, government services, and labor services (i.e. primary factors).	The boundaries encompass a more-broadly defined set of capital (including both man and nature-- i.e. humans, human institutions, business capital, and natural structures) as well as intermediate products, all driven by net inputs of current and stored sunlight (and other presumably minor sources of available work).
Mechanism for valuation	The market-pricing mechanism, which identifies, for each transaction in the economy, the balance of a resource's marginal cost and its marginal utility.	The maximum power mechanism, which identifies, for each type of energy transformation, the systemwide direct and indirect energy requirement when the system is operating under competitive conditions in which there is the best possible efficiency consistent with maximum work output. Those conditions represent an optimum balance of efficiency and speed.

Table 1 (continued)

Parameter To Be Compared	Economic Analysis	Energy Analysis
Practical boundaries of concern for valuing a resource	A local, regional, national, international, or other economy of man and scarce resources. In practice, evaluations are often limited to the system of markets within that economy and consider effects for an un-specified but usually limited time into the future.	A global system of man and nature (i.e. large enough to account for all feedback effects), considering effects as far into the future as they may occur (i.e. allowing for full life-cycle effects).
Value relationship between the cost of a process's inputs and the effects of its outputs	In a free-market economy, the amount of money paid for the scarce-resource inputs to a process (i.e. land, labor, and/or capital used in that process) is considered a measure of the value of the cost of the process. The amount of money paid for the outputs of the process may be different and is considered a measure reflecting the value of both the cost of the inputs and the utility of the outputs.	In surviving competitive systems of energy transformations, the embodied energy in the inputs to a process is considered a measure of the value of the cost of the process. That embodied energy is, by definition, also embodied in the outputs, which may ultimately achieve a different value only if energy transformation efficiencies change in future use of that embodied energy.
Interdependence of sub-systems within the boundaries of concern	All prices and markets are interdependent.	All embodied energy values and energy transformation processes are interdependent.
Potential for double counting	The monetary value of the product of one process (e.g. sheet steel) may account for at least part of the same value that is accounted for in the monetary value of the product or another process (e.g. automobile) because both cost and utility of the two products overlap.	The embodied energy value of the product of one process (e.g. biotic production) may account for at least part of the same value that is accounted for in the embodied energy value of the product of another process (e.g. soil production) because the energy embodied in the two products may result from different transformations of energy flows from a common source.

Economic analysis seeks to identify the alternative which has the minimum total cost or the least use of scarce resources. Thus, in economics the assumption is made that physical production is maximized for the system by maximizing the efficiency of use of the system's net input resources. As discussed below, economic analysis assumes that a system's previously-achieved level of efficiency may be maintained or improved even when a resource input becomes more scarce, by changing production technology and/or substituting less-scarce resources. Thus even in times of declining resource availability, it is assumed that the system's physical production may be maintained or increased.

Energy analysis takes a different tack: it equates the most productive system with the alternative which accomplishes the most work for the system. Thus, although energy analysis also seeks maximum production for the system, it does not assume that that is necessarily promoted by maximizing efficiency. Instead, as discussed below (effect of time on value), energy analysis recognizes that efficiency must vary with speed and assumes that an optimum balance of efficiency and speed of resource use will maximize work and thus production.

The difference between these two algorithms for maximizing production can lead to considerably differing projections of future production for systems with a declining resource base. The economic analysis algorithm assumes that increasing system-wide efficiency will lead to a proportional increase of physical production, whereas the energy analysis algorithm assumes that the system's speed of production may change along with the change of efficiency, thereby allowing for a different rate of production per unit of embodied energy input. (As discussed below in the mechanism for valuation section, energy analysis assumes that production in the present macrosystem has evolved to approach the point of maximum efficiency for a given level of input.)

Resources of Value

Conceptually, both economic and energy analyses have universally-applicable classifications of resources which are assumed to carry value. In economic analysis, resources of value are called scarce resources, which may be land, labor, and/or capital. (All resources may be classified as land, labor, or capital.) Any resource may be defined as scarce depending on whether the perceived demand for the resource is limited by its availability. In economics it is conceptually possible for any resource not to be scarce and thus carry no economic value. In energy analysis, the basic resource of value is embodied energy--which, as described above, is a very comprehensive aggregate of available work. In all real systems only heat at ambient temperature has no available work and thus is assumed to carry no value.

Comparing the use of these economic and energy analysis classifications of valuable resources, it is evident that although both are universally applicable, many resources which are not economically scarce and thus carry no economic value, do have available work and thus do carry an energy value. For example, sunlight is rarely considered an economically scarce resource because economic analysts traditionally perceive it as being, for the most part, unused by the current system and even where it is used, its future availability is perceived as undiminished; however, it is assigned an energy value according to the available work associated with its input to the system. Hence, even in agricultural processes which require sunlight directly for photosynthesis, sunlight is not an economically valued input, but it is an energy-valued input.

Substitutability of Valuable Resources

There is a potentially important difference between economic and energy analyses regarding the substitutability of valuable resources. In economic analyses, land, labor, and capital may be expected to substitute for each other in efforts to minimize a given product's cost. Therefore, there may be considerable uncertainty when using economic analysis to assess future values if relative scarcities, and therefore relative values, of land, labor, and capital change in the future. In energy analysis, there is no lower-cost substitute for embodied energy because it is defined as a limiting requirement, i.e. as the amount of available work required by a system which is already operating at (or near) optimum work efficiency. Consequently, energy analysis specifies a resource with no lower-cost substitute as an indicator of value. Therefore, when assessing future values, energy analysis may

provide an extra measure of confidence. (It should be noted that in both economic and energy analyses, time has an effect on value and might, in some sense, be considered a substitute for scarce resources or embodied energy. A comparison of the treatment of that parameter in both forms of analysis is given below.)

Effect of Time on Value

Although time is considered to have an effect on value in both economic and energy analyses, calculations of that effect are based on different algorithms. In economic analysis, a resource controlled now usually is considered more valuable than the same resource controlled later. This change of resource value with respect to time is reflected in a (positive) discount rate. That rate is often quantified explicitly in economic valuations by projecting into the future, the trends of recent changes of value with time. There are both time-preference and productivity rationales for economic discounting, both apparently based on projection of the economic system's past trends.

In energy analysis, the time component of value operates such that a unit amount of embodied energy contributing to a process generates less value when used rapidly than when used slowly. In some sense this may be considered a substitution of time for energy because speeding up a production process saves time but costs energy. That is due to the phenomenon of the efficiency of energy transformation varying with speed. This algorithm explains a physical reason for time's affecting value. Future research is needed to quantify this relationship in varous systems. With such quantification, the time component may be made explicit in energy analysis valuations by identifying rates of resource use and associated efficiencies of energy transformation at maximum power loading. That is, energy transformation ratios may vary with time according to the speed with which the system draws on and uses its resource supplies. Thus, in times of increasing rates of growth, embodied energy values would be affected in a similar fashion to economic discounting, i.e. energy use would become less and less efficient resulting in the same amount of energy being capable of generating less and less value as time progresses. However, should growth rates slow toward steady state or even decline, then the traditionally positive economic discount rates determined by projection of past trends would suggest a considerably different effect on value than would be indicated by the energy analysis algorithm relating efficiency and speed. That difference may be extremely important when considering values of long-term effects, such as many environmental impacts.

Model for Valuing a Resource

There is a particularly important difference between the models used for valuing a resource in economic and energy analyses. The energy analysis model is a unified global system of man and nature with embodied energy being the common unit of all transactions. In contrast, the economic analysis model is a limited system of economic markets with money being the common unit of only scarce-resource transactions. Because there are no economic markets in the natural environment, and consequently no money transactions, natural-environmental contributions to value are often external to the model. The value of a natural-environmental resource is recognized in the model only when that resource is perceived as scarce, thus implying some relationship between the resource's availability and the market's use of the resource. Although some value of other natural resources is often recognized in theory, assessment of that value can not be governed by the market model. (Various procedures for "shadow pricing" have been proposed to supplement the traditional market model, although none are widely-accepted and used.) Consequently, the economic analysis model applies only to market resources whereas the energy analysis model may apply to any resource (except heat at ambient temperature).

This difference between economic and energy analysis models may help to better define particular areas in which the two forms of analysis might appropriately supplement each other. The obvious possibility is that of using energy analysis to evaluate resources which are outside the traditional economic model, i.e. natural resources which all markets do not perceive as being scarce. However, it must be remembered that energy analysis evaluations require new research calculations to be made, whereas traditional economic evaluation, where applicable, is accomplished with the traditionally-kept accounts derived from the market's transactions. Thus, where the market model does the job, there would appear to be no advantage in undertaking energy analysis evaluations. However, where the market model does not apply, such as for valuing many environmental flows, or where the market model is incomplete or flawed, including cases of imperfect markets and rapidly-changing resource bases, then energy analysis may provide a reasonable supplementary model, providing that the information is available with which to carry out the required embodied energy calculations.

Boundaries of the Model's Production System

The resource valuation models discussed above each include a production system which transforms net input resources into useful output. The boundaries of the system separate the internal components of the production system, which comprise the machinery for producing output, from the system's net inputs, which supply that machinery and were themselves generated without any dependence on the system's own outputs. When comparing the economic analysis and energy analysis production models, the production system boundaries are found to be in different locations. To a great extent the difference reflects differing conceptions of the roles of humans and nature in the production process. This difference has considerable bearing on the meanings of values determined by the two forms of analysis.

As conceived in economic analysis, the boundaries of the production system encompass business capital (equipment and plant) and intermediate products produced by that capital (e.g. business inventory). These internal components of the production system are assumed to be driven by the system's net inputs, which are assumed to have been generated outside the system boundaries. Those net inputs, what economists call primary factors, include natural resources, government services, and labor services. Hence, in economic analysis, humans and their institutions (e.g. government) as well as non-scarce natural resources are conceived as net inputs to the production process, meaning that their availability to the process is assumed to be completely independent of the process's output.

In contrast, the energy analysis conception places humans, their institutions, and all natural resources within the boundaries of a larger unified production system, all driven by net inputs of current and stored sunlight (and other presumably minor sources of available work). With this concept, the availability of labor, government, and natural resources is assumed to be at least partially-dependent upon the production system's own output. That is, humans, their institutions, and natural environmental components are treated as parts of the production system's stock of capital in the sense that they, like business capital, comprise part of the system's endogenous capacity for using exogenous factors (i.e. the net inputs) in the production of output.

The difference in production-system boundaries between the energy and economic models has important ramifications for the potential compatibility of the two forms of analysis. Because in economic analysis, humans, human institutions, and many natural environmental structures are not treated as endogenous components of the production system, any change of the productive capacity of those components is not necessarily attributed as a change of value of the system's current production of net output. For example, changes in levels of human eduction, training, and health care that would change labor's capacity to contribute to future production traditionally are not counted as changes in the economic value of the system's current net output. Similarly, changes in the maintenance of many "non-scarce" natural environmental structures that may change the value or availability of natural resource inputs to future production traditionally are also not counted as changes in the economic value of the system's current net output. Hence, the traditional economic concept of a nation's net output (i.e. gross national product) does not reflect any such changes of human or natural-environmental capital. In contrast, because energy analysis does treat humans, human institutions, and all of the earth's natural-environmental structure as endogenous components of the production system, (i.e. they are treated as capital), the energy concept of a nation's net output would be different from the traditionally-evaluated GNP. Therefore, a change in the value of a system's net output according to an energy analysis evaluation may be valuing a change that traditional economic accounting would not recognize until some future accounting period. (It should be noted that some economic analysts have proposed changes in traditional economic analysis conventions which would account for changes in human and government capital (for example, see Kendrick 1976 and Eisner et al. 1980). The implementation of such proposals would narrow a large part of the difference between the energy and economic accounts of net output.)

Mechanism for Valuation

The mechanisms by which energy and economic analyses assess value are different but may, in certain situations, achieve the same end. In economic analysis, the market-pricing mechanism is used to assess value of any flow by assuming that cost balances utility. In energy analysis, the maximum power mechanism is used (in energy transformation ratio calculations) to assess the value of the system's net input resources by assuming that efficiency and speed of resource use approach an optimum balance to maximize the system's production of net output. The balancing of cost and utility of resource use may well be homologous to the balancing of efficiency and speed of resource use. Therefore, the two analysis mechanisms merely may be using different means to assess the same end. However, the energy analysis (maximum power) mechanism is governed by the previously established physical laws of nature whereas the economic analysis (market-pricing) mechanism is governed by current human perceptions of resource scarcity and utility, which define economic markets. Because we are not always fast to perceive the full utility or some-times even the true global scarcity of certain resources, the energy analysis mechanism may enable us to be more rapidly responsive to certain resource changes. That is one reason why energy analysis is offered as a tool for valuing certain externalities even if, in the long run, both forms of analysis might achieve essentially the same finding.

Practical Boundaries of Concern for Valuing a Resource

Similar boundaries of concern for valuing a resource are often defined in theory by both energy and economic analysts. However, in practice, important differences are often evident regarding both spatial and temporal boundaries. The spatial boundaries of economic analyses include man, scarce resources, and the products developed from interactions of man and scarce resources and may be defined as a local, regional, national, international, or other boundary. Economic benefit-cost analyses often assume a national or global boundary; but usually only market transactions within that boundary are evaluated. Thus, often there are non-market (e.g. environmental) value changes to be considered which technically may be within the spatial boundaries of concern but nonetheless are external to the economic analysis. (See the above dis-cussions on the models and mechanisms for valuing resources and the discussion on the boundaries of the model's production system.) When considering the value of natural environmental resources, energy analysis has no such discrepancy between the conceptual and the practical spatial boundaries used. The values of both market and non-market natural environmental resources are evaluated within the bounds of a unified global system of man and nature.

When considering temporal boundaries of concern, economic analysis may again be seen to have a discrepancy between concept and practice, a discrepancy which is not evident in energy analysis. The economic market-pricing mechanism, due to its dependence on current human perceptions of future utility, often gives only minimal consideration to long-term value changes even though conceptually economic analyses, including benefit-cost analyses, have open-ended temporal boundaries. In contrast, energy analysis, with evaluations based on the limiting physical requirements of full life-cycle flows in the global system, considers and evaluates value changes regardless of how far into the future they may occur.

These practical differences in the space and time boundaries of concern have ramifications for the meanings and compatibility of energy and economic analysis findings. The calculation of economic values may be made explicit to a variety of spatial boundaries. In practice, Federal benefit/cost analyses are usually concerned with national boundaries. Energy analysis values are inherently derived from a global perspective. Hence, energy analysis findings may account for value changes, outside the national boundaries, that economic benefit/cost analyses explicitly exclude from evaluation. For example, any value changes associated with air pollution caused by activities in one country but realized as a change of some scarce resource (such as lake water quality changed by acid rain) only in an adjacent country are reflected in energy analysis findings but not in economic analysis findings. To the extent that the analysis concern is intended to be limited to the national boundaries, the two forms of analysis are incompatible.

Of course a long time boundary of concern may change one's perspective on the above statement. It is quite possible that, given a long enough time boundary, international changes of value eventually will be passed back to their country of origin. Hence, even a nationally-limited analysis with a long time boundary might appropriately account for international value changes. In that case, energy analysis findings, based on long-term global boundaries of concern, may provide a valuable supplement to traditional economic benefit/cost analyses.

Value Relationship Between Input and Output

The economic relationship between the values of a process's input costs and its output effects has both similarities to and differences from the energetic relationship of those values. In economic analysis, the value of input costs may differ from the value of output effects depending on the utility of the outputs. In energy analysis, the two values may differ only when the system-wide efficiency of energy transformations changes from what was assumed in the embodied energy evaluation of the system's net inputs. That value difference is defined as zero when the system is operating at steady state. That is, embodied energy cost equals embodied energy effect under such conditions because system-wide energy transformation efficiencies remain constant as long as the system's speed of operation, it's rate of use of net inputs, remains constant. It should be noted that embodied energy analysis as practiced often assumes steady state conditions for embodied energy evaluations of the system's net input. When the production system being analyzed, say the U.S. economy, is not operating at steady state, then such calculated embodied energy values can only be interpreted as limiting values of the system's net output.

Economic and energy analyses differ on this parameter due to another factor as well. Because the energy analysis production model uses broader boundaries, considering some non-market (e.g. environmental) processes, that model may account for the value of a process's outputs due to their use in those non-market processes. The economic analysis model does not account for such value. For example, when an industrial process's residuals released in the atmosphere cause changes in non-market environmental processes such that scarce natural resource values are changed, only the energy analysis production model accounts for such change as part of the value of the industrial process's output.

Interdependence of Subsystems

In valuing resources and products, both energy and economic analyses recognize interdependencies of processes within the system of concern. In other words, the system's complex web of interconnected processes causes at least part of the value developed in the production of any good or service to be at least indirectly dependent on the production of any other good or service. Thus, in economic analysis, all prices and markets are interdependent; and in energy analysis, all embodied energy values and energy transformation processes are interdependent. It should be noted, however, that energy and economic

analyses do not necessarily assume the same degree of such interdependence. Such interdependencies within a value system create a potential for double counting values. That potential is discussed below.

Potential for Double Counting

Both economic and energy analyses use measures of value that have considerable potential for double counting. That is, the quantification of any given monetary transaction according to price, or energy flow according to embodied energy value, often accounts for the same value that is accounted for in other monetary transactions or energy flows. Economic analysis accounts for such overlapping of value by separating out the value added portion of each transaction and by calculating system-wide value only at final demand, which is considered an aggregate of all of the system's value added components. Energy analysis accounts for such overlapping of value by calculating embodied energy value of the system's net output only for some single integrative energy flow, which embodies most or all of the energies from the other flows. For example, the embodied energy value calculated for the energy flow of a system's biotic production is often an integrative value in environmental analyses based on the assumption that most of the energy embodied in most other natural environmental processes is used either directly or indirectly in biotic production. Thus, although energy and economic analyses each use a measure of value that has considerable potential for double counting, they employ somewhat different strategies to avoid such double counting when attempting to calculate system-wide, or aggregate, net output values.

A different embodied energy accounting scheme, using input-output matrix calculations, has also been proposed (Costanza 1980). That scheme is in a sense, a hybrid of the two above forms, employing a partitioning scheme similar to that used in economic analysis to avoid double counting but otherwise being consistent with the energy analysis assumptions described above.

Both forms of embodied energy accounting and the traditional form of economic accounting avoid double counting among different parts of the system's net output. That is, in each of the three forms of accounting, the system's aggregate net output value is set exactly equal to the system's aggregate net input value. However, as noted above (in the section on boundaries of the model's production system), net input (or net output) is not defined with the same system boundaries in energy and economic analyses. Hence, although each form of analysis is internally consistent in avoiding double counting among components of net output, that in itself does not assure that net output values are comparable between energy and economic analyses. For that assurance, the boundaries that are used to define the system's net input and net output must also be comparable.

SUMMARY AND CONCLUSIONS

There are many methodological differences concerning both accounting conventions and valuation mechanisms, between energy and economic analyses as practiced. However, the comparisons presented in this paper also suggest opportunities for strengths of each form of analysis to supplement weaknesses of the other. The comparisons also make it evident that any compatibility of the two forms of analysis depends on acceptance of a common convention concerning spatial and temporal boundary conditions of the production system. The likely convention would define a globally-bounded production system in which all valuations reflect full life-cycle considerations for both man and nature.

Given common acceptance of such boundary conditions, energy and economic analyses may supplement each other in two different ways. The first way would keep all energy and economic analyses separate, the two different analytical perspectives being used as checks on each other or to provide base information for each other. The second way would combine the two forms of analysis, creating a hybrid form of analysis which might be more useful than either one alone. The hybrid form would take advantage of both energy analysis's inherent ability to evaluate natural environmental factors of production and economic analysis's long-established tools and widely-available data base for human-system accounting. Of course, the validity of any such hybrid analysis would also depend on existence of a constant proportional value relationship between embodied energy and economic values. Preliminary evidence suggests such a relationship (for example, see Costanza 1980, and Lugo and Brown 1979); further testing is underway (Lavine 1980).

Both the separate and hybrid ways of using energy and economic analyses may be useful for a variety of applications. For example, environmental impact assessment may be expanded with energy analysis, enabling a variety of impact types to be valued according to a single (embodied energy) metric. Resulting values could supplement economic considerations; or, with the hybrid approach, such impact values could be incorporated in economic benefit-cost (or other) analyses. Assessment of environmental management alternatives may also be similarly expanded with energy analysis, providing a new perspective for checking economic analyses of the alternatives; or, with the hybrid approach, the energy-derived values could be used to expand such economic analyses. Technology assessments, particularly net energy analyses of energy production or conservation proposals, may also be improved by including consideration of environmental changes in embodied energy terms and comparing with economic findings or, with the hybrid approach, using economic values as base information to be incorporated in the net energy analyses. Regional and natural resources planning efforts may also benefit from the ability to combine energy and economic analyses. Evaluation of appropriate economic discount rates, particularly when long time periods are involved, may also benefit from a compatible energy analysis physical explanation for changes of value with time (indicating that resource values change according to the speed with which they are used and that systems will change speed of resource use as the size of the available resource base changes).

Regardless of the particular approach (comparative or hybrid), appropriate use of combined energy and economic analyses requires an understanding of the differences, as well as the similarities, outlined in this paper. Energy and economic values should not be indiscriminately combined. Even given a common convention for boundary conditions of the production system and a constant proportional value relationship, embodied energy and economic values are assumed to be comparable for only net input and net output flows, not intermediate flows. That is because the energy analysis mechanism for valuation, the maximum power mechanism, operates to use the system's net input to maximize productivity at net output, not necessarily at intermediate steps. In contrast, the market-pricing mechanism (balancing cost and utility) applies to all transactions. Thus, economic prices of intermediate goods and services (i.e. not yet at final demand) may not be comparable with embodied energy values.

Other differences must also be understood so that energy and economic values are appropriately compared. For example, discounted economic values often reflect an assumption that trends of the recent past (concerning change of value with time) may be accurately projected long into the future, whereas embodied energy values reflect an assumption that future changes of value must be limited by limits to increasing rates of resource availability. For another example, economic analysis values often exclude an accounting of environmental externalities but reflect a much more detailed consideration of market flows. In contrast, energy analysis values do account for values of non-market flows but often reflect a less-detailed accounting of market flows. It is only when these and the other differences outlined in this paper are accounted for, that energy and economic analyses may appropriately be combined. Given recognition and understanding of these differences, there appear to be many applications which would benefit from a combination of embodied energy and economic analyses.

LITERATURE CITED

Costanza, Robert. 1980. Embodied energy and economic valuation. Science 210:1219-1224.

Eisner, Robert, E.R. Simons, P.J. Pieper, and S. Bender. 1981. Total incomes in the United States, 1946 to 1976. Department of Economics, Northwestern University. (Preliminary)

International Federation of Institutes for Advanced Study. 1975. Workshop report, on workshop on energy analysis and economics, Lidingo, Sweden, June 22-27, 1975. IFIAS Report No. 9, 103 p. Stockholm, Sweden.

Kendrick, John W. 1976. The formation and stocks of total capital. National Bureau of Economic Research, Number 100, General Series, 256 p. Columbia University Press, New York, N.Y.

Lavine, Mitchell J. (Principal Investigator). 1980. Use of embodied energy values in the analysis of environment/energy/economy tradeoff decisions: an exploratory analysis. National Science Foundation Project PRA 8003845. (In progress at Center for Environmental Research, Cornell University, Ithaca, N.Y.)

Lugo, Ariel E. and S. Brown. 1979. Economics, energy, ecology. Memorandum to Chairman, Council on Environmental Quality, March 1979. Washington, D. C. (Also forthcoming In S. Brown and A.E. Lugo. Management and Status of Commercial U. S. Marine Fisheries. Council on Environmental Quality. Washington, D.C.)

Odum, Howard T., M.J. Lavine, F.C. Wang, M.A. Miller, J.F. Alexander, Jr., and T. Butler. 1981. A manual for using energy analysis for plant siting. Final Draft, Report to the United States Nuclear Regulatory Commission, Contract NRC-04-77-123, Mod. 3. Energy Analysis Workshop, Center for Wetlands, University of Florida, Gainesville, Florida.

TIME-SERIES ANALYSIS OF INTERNATIONAL ENERGY-ECONOMIC RELATIONSHIPS[1]

James Zucchetto and Robert Walker[2]

Abstract. -- Several statistical models for testing energy-economic relationships among, and within countries are discussed as well as an ongoing project of international energy analysis. Cross-sectional regressions of gross domestic product (GDP) vs. energy from 1950-1974, and within country changes of GDP to energy ratios from 1950-1974 for OECD countries have remained remarkably stable over the time period. Energy/economic ratios have varied significantly within countries, some increasing and others decreasing. Future analysis of extended data sets is addressed.

INTRODUCTION

The relationship between the availability of energy for the support of economic activity is well recognized by now (Barnett 1950; Cottrell 1955; Gambel 1964; Cook 1971; Odum, 1973). Non-industrialized countries depend to a great extent on solar energy by harvesting production from forests, fields, lakes and seas and utilizing the kinetic energy of the winds. The inherent production limits of these systems results in income levels which are relatively low compared to economies with other sources of energy in the form of fuels and electricity. These additional energies can support intense industrial activity, increase the rate of production of goods as well as the material standard of living. The quantitative relationship between economic activity and energy consumption is usually expressed as total gross domestic product (GDP) equal to some function of total energy consumption (E) or GDP/capita = f (E/capita). The basic notion is that if one views a national economic systems as an "engine", whose total output is measured by the sum of goods and services produced within its boundaries, then this output is generated by the input of energy which drives the economic engine. Quantitative analysis of the relationship between economic output and energy consumption have been undertaken by various inves-

tigators for the United States economy (Barnett 1950; Cook 1971; Kylstra 1974; Darmstadter et al. 1977). Upon extending this argument, other investigators have derived regression equations for GDP = f (E) by analyzing energy and GDP data for many different countries in a given year; the basic conclusion is that higher GDP's/capita require higher energy consumption per capita, although countries vary in the efficiency with which energy is used to generate economic output. Cross-sectional regressions in the literature across a number of countries, including developed and undeveloped countries, show this relationship between GDP/capita and E/capita for 1963, 1965 and 1972 (Gambel 1964; Darmstadter et al. 1977; Perry and Landsberg 1977).

In recent years rather detailed energy analyses have been conducted for a selected number of OECD countries (U.S., Canada, France, Italy, Japan, Netherlands, Sweden, United Kingdom, W. Germany) for the purposes of comparison in order to determine why differences in energy efficiencies exist (Darmstadter et al. 1977; Schipper and Lichtenberg 1976). Details of household, commercial, transportation and industrial energy consumption relative to economic output have highlighted possibilities for energy conservation in such countries as the United States. Dunkerley et al. (1980) have concentrated on a time-series analysis of the nine OECD countries mentioned above in an effort to understand increases or decreases in the energy/GDP ratios for these industrialized countries.

The detailed study of energy/economic measures is important for several reasons. Although there are difficulties in making comparisons (Starr and Field 1978; 1979) such studies are of interest in and of themselves for the purposes of understanding how different

[1]Paper presented at the international symposium Energy and Ecological Modelling, sponsored by the International Society for Ecological Modelling. (Louisville, KY, April 20-23, 1981).

[2]University of Pennsylvania, Regional Science Department, Philadelphia, PA 19104.

kinds of economies use energy and the efficiency with which economic output is generated. Cross-sectional studies between economies generates questions as to why differences exist and time-series analysis indicates what kinds of development trends are occuring. From a more pragmatic viewpoint detailed quantitative knowledge of energy/economic measures for economies leads to estimates of energy requirements for planned levels of economic development. Furthermore, during times of energy scarcity, future economic activity may be better predicted by accurate information about energy availability and energy requirements for given levels of economic activity.

The present paper describes a study of international energy/economic relationships including a discussion of data to be analyzed, models to be tested and some very preliminary results arrived at for 24 OECD countries.

DATA BASE AND COLLECTION

A study of international energy/economic relationships requires the collection of economic, energy and geographical data. Our effort has been aided by Yeh (1981) who, in constructing a three-region world model, has collected extensive economic time-series data for many countries including population, economic consumption, gross capital formation, gross domestic product and savings for the Developed countries, Latin America, Africa, Asia and the Planned economies for varying periods between 1950 and 1974: data for the developed countries for the developed countries are available for early periods and, in general, the undeveloped or developing countries have data available back to 1960. The process of compiling these data onto computer tape is underway and so far the data for the OECD countries have been compiled with the African and South American data sets soon to follow. In addition, we have compiled energy data for many of the countries and, for now, are concentrating on the OECD countries for the period 1950-1974. There are several reasons for beginning with the OECD countries. First, data collection in these countries is more extensive and more accurate which leads to greater confidence in any results. Second, when making international comparisons of GDP, corrections must be made for purchasing power parity (PPP) (Kravis et al. 1978). These corrections are relatively small for the industrialized countries so that, for the moment, the data have not been converted to equivalent purchasing power. All our data are expressed in dollars at constant 1970 prices.

Other data which are required to accomplish a thorough analysis of energy/economic relations include the following: prices of energy, environmental variables such as temperature or degree-days of heating or cooling, measures of population concentration such as population density, level of development and structure of economy. These data are anticipated to be collected and computerized overf the next six months. The analysis presented in this paper are for the following OECD countries for the period 1950-1974: Austria, Australia, Belgium, Canada, Denmark, Finland, France, Greece, Iceland, Ireland, Israel, Italy, Japan, Netherlands, Norway, New Zealand, Poland, S. Africa, Spain, Sweden, Switzerland, Turkey, United Kingdom, United States and W. Germany.

DATA ANALYSIS

There are several regression models that have, and will be, tested throughout the conduct of this research project. For example, one model of interest is to test whether the cross-sectional relationship across countries for different years remains stable through time. Thus, models of the form

$$y = ax + b \qquad (1a)$$

$$\text{and } y = ax^b \qquad (1b)$$

where x = total energy or energy/capita/yr.
y = total GDP or GDP/capita/yr.
a,b = constants

can be tested for a given year where data given for different countries. The form of equation (1) can also be tested for data for a given country to see if the internal relationship between energy and GDP is highly correlated. Furthermore, in terms of the cross-correlation analysis among countries, it can be determined whether the slope of the equation, a, is remaining constant through time. All data for all years and countries can be pooled to estimate equation (1) also. For a given country, the energy/GDP ratio can be calculated over time to ascertain trends, anomalies and change as a function of stage of development. These energy to GDP ratios can then be used as a basis of comparison with other countries. Furthermore, for a given country, the change in GDP with respect to energy consumption can be monitored, i.e.,

$$e = \frac{y_i - y_{(i-1)}}{x_i - x_{(i-1)}} \qquad (2)$$

e = change in GDP per unit of change in energy consumption

$y_i - y_{(i-1)}$ = change in GDP from yr. (i - 1) to yr. i.

$x_i - x_{(i-1)}$ = change in energy consumption from yr. (i - 1) to yr. i.

in order to assess whether the ability of an economy to generate economic output with respect to energy fluctuates or remains relatively constant.

If one is comparing two or more different countries and energy/GDP ratios are being used as some measure of "efficiency" then corrections must be made for climate, geography or industrial mix. Thus, for example, if d_1 = degree-days in country 1 and d_2 = degree-days in country 2, then if $d_1 > d_2$ an energy consumption of e_1 in country 1 would correspond to an energy consumption equal to $e_1 (d_2/d_1)$ in country 2. A new variable modifying energy consumption by a degree-day ratio could be introduced into equation (1) for x and regressed against y. A similar approach could be taken towards different energy prices in different countries. Or one could take a multiple regression approach including those independent variables assumed to affect energy consumption. Factors which differ between countries, such as climate, must be considered in cross-sectional studies but are not of importance for time-series within country analysis because these factors would presumably remain constant. However, certain actions within countries may have impacts upon energy consumption at a later time. Thus, regressions of the form for each country are anticipated to be tested after the required data is processed:

$$E_t = a + b \ GDP_t + cP_{t-k} + dCF_{t-k}$$

where E_t = energy consumption at time t.
GDP = gross domestic product at time t.
P = energy price at time (t-k).
CF = capital formation at time (t-k).

Thus, price measures or capital investment, perhaps in new technology, may possibly lead to decreased energy consumption at some future time after a period of adjustment.

RESULTS

Cross-sectional Regressions

For each of the years 1951 to 1973 a regression of the following was performed:

$$y = ax + b \qquad (4)$$

where x = per capita energy consumption in tons coal equivalent

y = gross domestic product in $1000/capita at 1970 prices

The results for this regression are contained in table 1.

Regression of total gross domestic product as a function of total energy consumption or

$$y = ax + b \qquad (5)$$

where x = total energy consumption in million tons coal equivalents

Table 1.— Cross-correlation analysis for GDP/capita vs. Energy/capita for 1951-1973. (eq. 4).

Year	b	a	t-statistic	r^2
1951	316	372	7.4	.81
1952	290	377	7.07	.78
1953	316	365	6.61	.76
1954	439	381	6.57	.68
1955	443	381	6.85	.70
1956	467	386	7.18	.70
1957	487	390	7.38	.71
1958	513	390	7.03	.69
1959	519	395	7.5	.72
1960	558	395	7.0	.69
1961	515	406	7.57	.72
1963	529	405	8.03	.74
1964	529	426	8.03	.74
1965	514	432	8.83	.77
1966	494	441	9.13	.78
1967	489	437	9.68	.80
1968	562	425	9.36	.79
1969	596	406	9.55	.79
1970	526	434	9.72	.80
1971	657	419	8.86	.78
1972	729	420	9.59	.81
1973	744	434	8.74	.78

where, y = total GDP in millions of U.S. dollars at 1970 prices.

led to the results in table 2.

These regression results are remarkable in the sense that the slope (a) of the equations, especially for equation (5), has remained so stable over time. For the per capita plots (eq. 4), the slope of the line seemed somewhat to change from the high 300's to the low 400's from the 1950's to the 1960's. More recent data will enable an extension of these results to the late 1970's, a time of energy disruptions and price increases which should probably affect the stability of these results.

Within Country Regressions

The evolution through time of energy to economic relations for a given country is also of interest. Assuming an equation for each country of the form:

$$y = ax^b \qquad (6)$$

where x = per capita energy consumption

y = per capita GDP

and taking logarithms we get

$$\log y = \log a + b \log x$$

Table 2.--Cross-correlation analysis
results for Equation 5. All regressions*
had r² = .99.

Year	b	a
1950	-491	421
1951	-808	437
1952	-1154	445
1953	-1213	442
1954	1142	463
1955	1507	454
1956	2221	461
1957	3273	458
1958	3729	468
1959	4292	459
1960	5005	460
1961	4215	470
1962	3663	468
1963	4768	462
1964	5670	475
1965	6034	476
1966	6776	473
1967	8603	465
1968	8542	455
1970	10013	443
1971	9853	446
1972	9827	462
1973	9612	467

*All t-statistics significant.

Table 3.--Elasticities of GDP per capita
with respect to energy consumption
per capita for different countries*
(equation 7).

Country	b	r²
Australia	.89	.98
Austria	1.3	.92
Belgium	1.44	.93
Canada	0.86	.96
Denmark	0.69	.96
Finland	0.53	.96
France	1.00	.95
Iceland	1.21	.69
Ireland	0.57	.91
Israel	0.95	.96
Italy	0.63	.99
Japan	1.03	.99
Netherlands	0.68	.97
Norway	0.75	.98
New Zealand	0.65	.97
Portugal	0.92	.95
South Africa	1.22	.88
Spain	1.38	.94
Sweden	0.86	.96
Switzerland	0.48	.97
Turkey	0.57	.85
U.S.A.	1.00	.95
W. Germany	1.37	.90

*All t- statistics are significant.

$$\text{or} \quad \frac{dy}{y} = b \frac{dx}{x} \quad \text{or} \quad b = \frac{dy/y}{dx/x} \qquad (7)$$

so that a regression analysis of equation (7)
gives the percent change in GDP per percent
change in energy consumption or elasticity of
GDP with respect to energy. If b is small then
percent changes in GDP occur for larger percent
changes in energy and vice versa. Thus, countries with relatively larger elasticities will
respond more to a given percent change in energy
consumption than countries with smaller b's.
The results of this analysis are presented in
table 3.

Another parameter of interest is a measure
of the energy-economic efficiency or how much
GDP is produced per unit of energy consumption. For within country comparisons through
time one can neglect such things as climate and
culture and look for discernible trends. As of
yet a statistical analysis has not been done but
visual perusal of graphs of GDP/ton coal equivalent for different countries gives the estimates
arrived at in table 4.

Table 4.--GDP/ton coal equivalent
over time. Numbers are approximate
Units are $1970/ton coal equivalent.

Country	1954-58	1970-73	Trend
Australia	565	545	↓
Austria	550	580	↑
Belgium	380	490	↑
Canada	480	450	↓
Denmark	750	620	↓
Finland	800	525	↓
France	670	720	↑
Greece	700	460	↓
Iceland	480	600	↑
Israel	600	440	↓
Italy	960	620	↓
Japan	560	590	↑
Netherlands	590	465	↓
Sweden	780	730	↓
United Kingdom	320	430	↑
U.S.A.	440	450	-
W. Germany	450	590	↑

It can be seen from these cursory results that
some countries have increased, while others have
decreased, in this measure of efficiency. Note
that an increase in the ratio in table 4
connoted an increase in efficiency.

Considering the measure defined by equation (2), which represents the change in GDP with respect to changes in energy from year to year, our analysis has indicated great variations from year to year; possibly the economic change over such a short period of time is controlled by external events.

DISCUSSION

These very preliminary results presented above give some glimmer of the types of analyses to be undertaken for a more extensive data base. More recent data after 1973 should perhaps indicate sharp departures from historical trends such as the stability in the cross-sectional regressions, changes in within country elasticities as well as energy-economic efficiencies. Also of interest to be included will be the natural energy flows associated with these economies in order to assess their contribution: these data, however, are difficult to accurately evaluate over time. Post-1973 price changes would also be expected to impact results. Difficulties of obtaining accurate energy consumption in much of the undeveloped countries will hopefully be aided by a current Peace Corps project (Man 1981).

LITERATURE CITED

Barnett, Harold J. 1950. Energy uses and supplies U.S. Bureau of Mines Information Circular 7852. (Oct.). Washington, D. C.

Cook, Earl. 1971. The flow of energy in an industrial society. Sci. Amer. 224 (3): 135-144.

Cottrell, Fred. 1955. Energy and society McGraw-Hill Book Co., Inc. New York, N.Y.

Darmstadter, J., J. Dunkerley and J. Alterman. 1977. How industrial societies use energy (a comparative analysis). John Hopkins University Press. Baltimore, Md.

Dunkerley, J., J. Alterman and J. J. Schmag, Jr. 1980. Trends in energy use in industrial societies. Resources for the Future, Washington, DC.

Gambel, A. B. 1964. Energy R and D and national progress. Office of Science and Technology Office of the President. Washington, D.C.

Kravis, J., A. Heston, and R. Summers. 1978. Real GDP per capita for more than one hundred countries. The Economic Journal, 88 (June): 215-242.

Kylstra, C. 1974. Energy analysis as a common basis for optimally combining man's activities and nature. The national symposium on corporate social policy. Chicago, IL.

Man, Ada Joe. 1981. Peace corps energy project. Peace Corps, Washington, D.C.

Odum, H. T. 1973. Energy, ecology and economics. AMBIO. 2:220-227.

Perry, H. & H. Landsberg. 1977 Projected world energy consumption. p. 35-50. In Energy and Climate. National Academy of Sciences. Washington, D. C.

Schipper, L. and Allan Lichtenberg. 1976. Science, efficient energy use and well-being: The Swedish example. Vol. 194. No. 4269. (December 3).

Starr, Chauncey and Stanford Field. 1978. Energy use proficiency: the validity of interregional comparisons. Energy Systems & Policy, 2:211-232.

Starr, Chauncey & Stanford Field. 1979. Economic growth, employment and energy. Energy Policy. 7:2-22.

Yeh, Bijou Y. 1981. A global economic model of a three-region world. Ph.D. Dissertation, Department of Economics, University of Pennsylvania.

ECONOMIC ANALYSES OF ENERGY OPTIONS:
A CRITICAL ASSESSMENT OF SOME RECENT STUDIES[1]

K. S. Shrader-Frechette[2]

Abstract.--Focusing on five analyses of the relative cost-effectiveness of nuclear- versus coal-generated electricity, this essay discusses several methodological assumptions inherent in (1) the Inhaber Report, (2) the Schmidt-Bodansky analysis, (3) WASH-1224, (4) the NSF report, and (5) the Miller study. The main thesis is that correction of the simplifying assumptions of exclusion, range, and aggregation probably reverses the conclusion of analyses (1)-(3), that nuclear is more cost-effective, but reinforces the conclusion of studies (4)-(5), that coal power is more economical.

INTRODUCTION

It is a truism that the validity of any prediction generated by a given model is a function of the various assumptions built into the model. It is also true, according to the authors of the 1979 Harvard Business School Study, that all the major recent models (since 1973) of the US energy situation have given us predictions that have consistently been more optimistic than the facts later proved (Koreisha and Stobaugh 1979, p. 235). Some of these optimistic forecasts, issuing out of erroneous or misguided modeling assumptions, have misled both policymakers and the public. As a consequence, we have been led into numerous questionable situations, such as importing 50% of US petroleum from foreign suppliers, and we have often not become clear about the causes of, and solutions to, our energy problems.

Since "energy policy indeed has been affected to an important extent by formal models" (Koreisha and Stobaugh 1979, pp. 235-236), one important way to help clarify and improve current energy policy is to discover and evaluate the assumptions employed in influential and widely cited energy models. Scientists, policymakers, and the public, however, often have a tendency to neglect methodological evaluations, probably in part because the assumptions underlying given models are frequently neither recognized nor understood. In this essay, I will point out several highly significant methodological presuppositions in some recent energy models. My purpose is to show that, if these questionable assumptions are rejected, then the alleged conclusions they support are often no longer plausible.

Since most energy models are based primarily on cost-benefit estimates, and since economics is routinely made the final test of public policy (Koreisha and Stobaugh 1979, p. 234; Galbraith 1967, p. 408; Ravetz 1971, p. 396; and Schumacher 1973, p. 38), my focus will be on econometric assumptions. From a contemporary perspective, some of the most interesting econometric models are those which make predictions on the basis of calculations of the relative costs of coal- and nuclear-generated electricity. Both because these are the two major US short-term energy options, and because there is widespread disagreement about their relative costs, I will discuss some comparative models of the costs of these two fuel cycles.

ENERGY OPTIONS AND COST-BENEFIT ANALYSIS

Economic methodology is never value-free (Boulding 1970, p. 122), and various evaluative presuppositions enter energy models in two main ways, through the general method of cost-benefit analysis and through particular econometric premises specific to the energy case being studied. Let us look first at some of the

[1]Paper presented at the international symposium, Energy and Ecological Modelling. (Louisville, Kentucky, April 20-23, 1981)

[2]K. S. Shrader-Frechette is Professor, Philosophy of Science, University of Louisville, Louisville, Kentucky U.S.A.

general methodological assumptions characterizing cost-benefit analysis.

Although the goal of cost-benefit analysis is to select "those opportunities which yield the maximum social benefit per dollar transferred" (Mishan 1976, p. xiii), in reality this goal is almost never achieved. This is because a working assumption in most economic analyses is that one either ought not or cannot integrate social scientific and environmental data into the approaches; as a consequence, cost-benefit calculations almost always exclude the social and environmental costs of certain actions (Davoll 1978, p. 7; Congress 1976, p. 89). Some of the costs typically omitted include those associated with:

(1) long-term effects of chronic, low-level exposures to pollutants (Congress 1976, pp. 3, 77);

(2) long-term health effects of acute exposures to pollutants (Congress 1976, pp. 73-75); and

(3) long-term health effects based on cumulative, synergistic, or food-chain exposure to pollutants (Congress 1976, pp. 79-81; Congress 1979, pp. 260-261).

The effects of ignoring such long-term costs are obvious. For one thing, the analysis makes it difficult, if not impossible, to predict correctly prospective economic and environmental problems (Congress 1976, p. 40). More importantly, cost-benefit conclusions regarding the same policy may be fundamentally different, depending on whether or not long-term costs and benefits are taken into account; for this reason, ignoring long-term costs makes it likely that forthcoming generations will not have been left free to arrange their own lives (Burkhardt and Ittelson 1978, p. xiii; Davoll 1978, p. 13; Gofman and Tamplin 1971, pp. 77-78). Short-term analyses tend to militate against energy and materials conservation and the substitution of low-impact materials and technologies. In other words, "traditional economic approaches discount the future more heavily than makes sense from the standpoint of the welfare of future generations" (Congress 1976, p. 95; Shrader-Frechette 1981, pp. 59-81).

These problems with long-term costs are particularly instructive in the case of deciding whether nuclear fission- or coal-generated electricity is more economical. For many years the Atomic Energy Commission, and then the Nuclear Regulatory Commission, argued that nuclear-generated power was cheaper than that produced from coal. This conclusion, however, completely omitted consideration of the long-term cost of surveillance, storage, and control of commercial radioactive wastes, currently about $775 million per year. It also ignored the long-term health costs of normal radiation releases from storage

sites throughout the country. Once both these long-term costs are calculated and included in the comparative data, nuclear power can be shown to be clearly more expensive than that produced from coal. As one NRC engineer put it, "if the licensing board should have to factor into the cost-benefit balance...waste disposal...then...I couldn't license any reactor" (Quoted in Shrader-Frechette 1980, p. 58).

Besides long-term costs, another parameter typically excluded from cost-benefit analyses is the degree to which particular energy technologies help or hurt the employment situation. Technology, in general, causes unemployment. This is both because innovation means old skills are no longer valuable and because automation replaces people (Bowen 1966, pp. 2-7; Bauer 1969, pp. 196-201; Silk 1969, p. 52; Hoffer 1972; p. 65). In the case of energy options, solar and coal-generated electricity, for example, are much more labor intensive than that produced by nuclear fission (Horwitch 1979, p. 94; Mishan 1967, pp. 177-179; Hoffer 1972, p. 65; Novick 1976, pp. 248-249). Hence if one places a social cost on factors such as unemployment, then it is clear that the cost-benefit analyses could be said to support quite different energy policies, depending on whether unemployment was taken into account.

Perhaps the two most basic problems with cost-benefit analysis are that it takes no account of non-monetary costs and benefits, such as depletion of nonrenewable resources, and no account of distributional costs and benefits, e.g., those accruing to given individuals as a result of a given energy policy (Mishan 1976, pp. 412-415; Schumacher 1973, pp. 41-47; Emmett 1978, p. 363). Both types of parameters might be called "externalities," social costs and benefits external to the standard means of economic accounting. Because they do not enter the cost-benefit calculation, it cannot be taken for granted that assessments of energy technology necessarily reveal the actual costs and benefits of that technology to society (Mishan 1967, pp. 53-55; Mishan 1969, pp. 29-56; Mishan 1976, pp. 109-164; Lutz and Lux 1979, pp. 149-150; Stobaugh and Yergin 1979, p. 229; North 1977, p. 49ff.). Externalities are the main reason, say economists, why national resources are often misallocated (Mishan 1967, p. 55; Mayo 1972, p. 78; Davoll 1978, p. 12; Bauer 1966, p. 35; Gross 1966, pp. 222-223). In the cases of nuclear fission and coal-generated electricity, for example, some relevant externalities are the risk of a catastrophic nuclear accident and the cost of premature death and respiratory disorders because of sulfur dioxide emissions. To ignore such parameters, as indeed most cost analyses of energy policy do, is inconsistent with all allocative judgments in economics (Mishan 1976, p. 126). This means that most data on energy costs are both inaccurate and seriously deficient in methodology.

SPECIFIC ENERGY ASSESSMENTS AND THEIR SHORTCOMINGS

Apart from the general reasons why cost-benefit analysis often does not provide an adequate vehicle for determining energy policy, there are some instructive points to be gleaned from recognizing the methodological assumptions specific to several recent comparative analyses of the nuclear and coal options. Five recent studies are particularly interesting, because not all their conclusions and predictions appear to be mutually consistent.

The 1978 Inhaber report, the 1976 Schmidt and Bodansky analysis, and the 1974 US government document, WASH-1224, all tend to support the conclusion that nuclear power is cheaper per kilowatt hour than that produced by coal (Inhaber 1978; 1979, pp. 718-723; Schmidt and Bodansky 1976; US AEC 1974). The 1976 National Science Foundation study and the 1976 Miller analysis, however, conclude that coal-generated electricity is cheaper per kilowatt hour than that produced by nuclear fission (Barrager, Judd, and North 1976; Miller 1976). Which studies are more likely to be correct? Let's take a look at some of the key methodological assumptions undergirding the respective conclusions, and a clearer perspective of nuclear and coal economics will emerge.

The Inhaber Report

In his now famous report for the Canadian Atomic Energy Control Board, Herbert Inhaber concluded that total risk to humans was greater per megawatt-year net output (over lifetime of system), for coal than for nuclear-generated electricity (Inhaber 1979, p. 722). Once dollar figures are placed on these relative hazards, then Inhaber's data allow one to to conclude that the costs of nuclear power may be said to exceed those of coal. This conclusion, however, is only as valid as the methodological presuppositions upon which it is based, and there are a number of questionable assumptions in the Inhaber analysis.

For one thing, Inhaber repeatedly uses the principle of exclusion, the assumption that the influence of any factor not included in his model is unimportant in affecting his conclusions (Koreisha and Stobaugh 1979, pp. 237-238). For example, he defines "risks" as solely medical, even though medical risks are only a subset of the total class of risks borne by the public as a result of electricity generation. This omission is central to his conclusions regarding the cost-effectiveness of nuclear-generated electricity, because atomic power causes great financial risks to the public. Owing to their magnitude, these risks are also likely to affect the social and psychological well-being of numerous persons. Take, for example, a catastrophic nuclear accident which, according to US government estimates, could result in $18 billion in property damages alone, apart from other billions in medical expenses (US AEC 1957; Mulvihill et al. 1965; Berger 1977, p. 45). The US has by law (the Price-Anderson Act) limited liability in such a catastrophe to $560 million (This represents 3% of total possible property damages, apart from other losses. Although this law has been upheld by the Supreme Court, other federal courts have ruled that it violates due process and equal protection guarantees of the Fifth and Fourteenth Amendments.). This means that with nuclear technology, a majority of the medical and financial consequences of a catastrophic accident would be borne by its victims, the public, rather than by those who caused it. Hence, from the consumers' point of view, the financial risks of nuclear fission are monumental, even though they are not considered by Inhaber.

Another instance of Inhaber's use of exclusion is his omission of the social costs associated with inequitable distributions of various risks, e.g., waste management. Clearly one technology could have a smaller magnitude of risk than another, but if that risk were grossly inequitably distributed, then the overall social cost of the technology might be greater, particularly if those who bore that cost did not receive the benefit of the energy produced. In the coal versus nuclear case, the issue of the distribution of costs is important since radioactive wastes will be lethal for hundreds of thousands of years and many of its costs will be borne by future generations.

Other methodological assumptions, having to do with his data range (Koreisha and Stobaugh 1979, p. 238) also plague Inhaber's report. He claims to use nuclear risk data taken from the Rasmussen Report, but his upper-limit figures are three times smaller than those actually given in the study; once corrected for this error alone, the risk figures for nuclear energy multiply by a factor of 40 (Holdren, Smith, and Morris 1979, p. 564; Inhaber 1979, p. 722). Apart from this problem, however, the Rasmussen Report has been criticized both by the US government and by numerous independent groups of scientists for its mathematical and methodological errors (US NRC 1975; Doctor et al. 1976; Lewis et al. 1975; Nader 1976, p. 33; Shrader-Frechette 1980). If all these critics are correct, then nuclear risk data are much worse than as calculated in the Rasmussen Report; as a consequence, Inhaber may well have grossly underestimated nuclear costs relative to those of coal. Compounding the problem, he has also overestimated the costs of coal generation of electricity. His figures for occupational and public risks from the coal fuel cycle are clearly based on mining practices and powerplant SO_2 emissions which have been illegal for some time. As one scientist points out, "present dust standards imply occupational deaths from black lung disease as much as 60 times lower than the figure used by Inhaber;" his calculation of the public deaths from sulfur dioxide disease correspond not to emissions in force but to "emissions five times higher" (Holdren, Smith, and Morris 1979, p. 564; Inhaber 1979, p. 722). What all this suggests is that Inhaber has erroneously magnified coal costs and reduced nuclear costs, such that his final conclusion is highly doubtful.

The Schmidt-Bodansky Analysis

Much the same problem occurs in the Schmidt-Bodansky analysis, but for different reasons. Like Inhaber, they conclude that nuclear power is more economical than that produced by coal (Schmidt and Bodansky 1976, p. 46). Their conclusion is problematic, however, because of their simplifying assumption about the aggregation of actual and projected costs (Koreisha and Stobaugh 1979, p. 238). To obtain the cost per kilowatt of both nuclear- and coal-generated electricity, Schmidt and Bodansky appear to have used actual capital costs for the coal plants, but projected costs for the nuclear plants. Since capital outlays represent roughly 75% of the total nuclear costs and 66% of the total coal costs, substantial errors in the capital cost figures are likely to affect the validity of the Schmidt-Bodansky conclusion, that nuclear power is cheaper than coal. This is precisely what happened.

Because of massive cost overruns of 300% and 400% on nuclear plant construction, the cost of fission power increased 700% between the years of 1963 and 1973. By 1975 actual capital costs for a fission reactor were $800 per kilowatt, as compared to projected capital costs of $344 per kilowatt at the same time period. Hence, if one rejects the methodological assumption, that actual and projected costs may be aggregated, then nuclear power can easily be shown to be more expensive than that from coal (Novick 1976, p. 300; Willrich 1971, pp. 35-36; Bupp 1979, pp. 119ff; Barrager et al. 1976, p. 66).

WASH-1224

The Atomic Energy Commission Report, WASH-1224, likewise concludes that nuclear fission is cheaper than coal-generated electricity. This conclusion, however, rests on doubtful premises, as do the Inhaber, Schmidt, and Bodansky judgments. The most obvious flaw in WASH-1224 is a consequence of the assumption of exclusion (Stobaugh and Koreisha 1979, pp. 237-238). Numerous nuclear cost and risk parameters were simply omitted from the analysis, thus allowing one to conclude that nuclear power was more cost effective than coal. For example, occupational injuries and fatalities related to the effects of storage of radioactive wastes were ignored, as were injuries, fatalities, and costs associated

with nuclear transportation and contamination accidents (US AEC 1974, pp. 5-29, 5-38, 5-39). This latter omission is particularly significant, economically, because most nuclear materials are shipped by rail, and OTA authorities claim that 65 percent of all US railway cars loaded with hazardous materials are involved annually in accidental releases of these substances (Congress 1978, pp. 141-161).

As was already mentioned, WASH-1224 is also seriously deficient in failing to include the costs of waste storage in its calculations of comparative nuclear fission and coal economics. The report alleges that 1000 MWe nuclear and coal plants cost almost the same to maintain and operate per year, with coal-fired installations annually costing approximately $200,000 more. Once the data for current expenditures for commercial waste storage are added to the government tables, however, then coal power can be shown to be more than 1 million dollars cheaper, per 1000 MWe plant, per year, than that generated by nuclear energy (Shrader-Frechette 1980, pp. 54-59). Hence, because of their methodological assumptions regarding exclusion, range, and aggregation, there are reasons for believing that Inhaber, Schmidt, Bodansky, and the authors of WASH-1224 might be wrong in alleging that nuclear-generated electricity is more cost-effective than that produced by coal.

National Science Foundation Report

Let's take a look at the economic methodology of some studies whose conclusions contradict those of Inhaber, Schmidt-Bodansky, and WASH-1224. The authors of the 1976 NSF report on the costs of nuclear- versus coal-generated electricity concluded that coal power was cheaper. They calculated that, using the 1980 cost in constant 1975 dollars, coal-fired plants burning high-sulfur coal had a delivered electricity price of about $0.035 per KWh to New York area customers. Even in areas of the country where utility costs are highest, or where coal power is made more expensive by high transmission costs or by burning low-sulfur Eastern coal, the highest total 1980 cost (in 1975 dollars) for burning coal (with scrubbers) was $0.049 per KWh as compared to the average fission plant cost of $0.092 per KWh (Barrager, Judd, and North 1976, pp. 11-13; see also Congress 1979, pp. viii, 11; Horwitch 1979, p. 87).

Like the Inhaber model, the NSF report suffers from errors related to exclusion; it omits costs of injuries and deaths related to radioactive waste disposal, although it does include figures for all other types of radiation- and nonradiation-related deaths and injuries, including those caused by transport accident (Barrager, Judd, and North 1976, p. 49). Even though some nuclear costs are excluded, the NSF report easily shows coal generation to be more cost-effective than nuclear. The authors do conclude, however, that although total short-term nuclear costs double those of coal (per KWh), the short-term environmental costs associated with coal generation are greater. Note, however, that environmental costs associated with long-term radioactive waste disposal were not calculated. This suggests that, if all such long-term costs are incorporated into the data, then coal power could be shown to be more economical by an even greater margin than that originally alleged in the NSF study. If so, then even though the model's assumptions about range may be changed, it is not clear that this change reverses the final conclusion of the study.

The Miller Analysis

A similar result follows when one employs the 1976 study by the conservative economist and investment banker, Saunders Miller. Like the authors of the previous studies, Miller also falls victim to questionable assumptions regarding range and exclusion. He neither prices the long-term economic costs of waste storage, nor any social or environmental costs in his analysis. Instead he relies solely on current actual monetary outlay in order to calculate relative costs and benefits for nuclear fission and coal generation of electricity. Miller concludes that (in most areas of the country) nuclear power is not only more expensive than coal, on "a true economic basis," but also is more expensive as regards economic risk. Because an individual utility cannot control all fuel cycle risks and economic constraints, e.g., cartels, availability of yellowcake, Miller argues that atomic energy is more risky, economically speaking, and more costly, in actual dollars, than coal generation of electricity (Miller 1976, pp. 87-88).

Because of his assumptions regarding range and exclusion, Miller has clearly underestimated both coal and nuclear costs. If the data used in the NSF model is correct, however, then including these omitted parameters and widening the range (to include long-term costs) are likely to increase the actual magnitude of both coal and nuclear costs without changing their relative cost-effectiveness as predicted by Miller.

CONCLUSION

Although there has not been time for an in-depth analysis of energy models, this brief discussion of cost-benefit analysis and comparative coal and nuclear studies suggests that the validity of current predictions, especially about the relative cost-effectiveness of coal and nuclear power, is undercut by serious methodological flaws. Once the methodological assumptions based on exclusion, range, and aggregation are modified so as to include all relevant parameters, then actual numerical cost predictions made by the authors of all five studies can be shown to be in error. In the case of the Inhaber Report, the Schmidt-Bodansky analysis, and WASH-1224, correction of these errors is likely to reverse their final conclusion that nuclear power is more cost-effective than coal-generated electricity. In the case of the NSF and Miller studies, correction of these errors of exclusion and range is likely to reinforce more dramatically their final conclusion that coal is more cost-effective than is nuclear power. In any case, consideration of these econometric assumptions suggests that, once comparable methodology is employed, there may well be no inconsistency among the predictions generated by the five models. This, in turn, suggests that methodological analyses might help in directing energy policy. Too often in the past, long-term energy decisionmaking was held hostage by the claim that no consistent econometric predictions were available.

LITERATURE CITED

Barrager, S. M., B. R. Judd, and D. W. North. 1976. The economic and social costs of coal and nuclear electric generation. xi, 127 p. National Science Foundation, US Government Printing Office, Washington, D.C.

Bauer, R. A. 1966. Detection and anticipation of impact: the nature of the task. p. 1-67. In R. A. Bauer (ed) Social indicators. xxi, 357 p. MIT Press, Cambridge, Mass.

Bauer, R. A. 1969. Second-order consequences: a methodological essay on the impact of technology. xii, 240 p. MIT Press, Cambridge, Mass.

Berger, J. J. 1977. Nuclear power. Dell, New York, N.Y.

Boulding, Kenneth. 1970. Economics as a science. vi, 157 p. McGraw-Hill, New York, N.Y.

Bowen, H. R., chairperson. February 1966. Technology and the American economy. Vol. 1, xiv, 115p. National commission on technology, automation, and economic progress, US Government Printing Office, Washington, D.C.

Bupp, I. C. 1979. The nuclear stalemate. p. 108-135. In R. Stobaugh and D. Yergin (eds) Energy future: Report of the energy project at the Harvard Business School. x, 353 p. Random House, New York, N.Y.

Burkhardt, D. F., and W. H. Ittelson, editors. 1978. Environmental assessment of socio-economic systems. xv, 597 p. Plenum, New York, N.Y.

Congress of the US, Office of Technology Assessment. 1976. A review of the US Environmental Protection Agency environmental research outlook: FY 1976 through 1980. xi, 122 p. US Government Printing Office, Washington, D.C.

Congress of the US, Office of Technology Assessment. 1978. An evaluation of railroad safety. xvi, 224 p. US Government Printing Office, Washington, D.C.

Congress of the US, Office of Technology Assessment. 1979. The direct use of coal. xv, 411 p. US Government Printing Office, Washington, D.C.

Davoll, John. 1978. Systematic distortion in planning and assessment. p. 3-15. In D. F. Burkhardt and W. H. Ittelson (eds) Environmental assessment of socioeconomic systems. xv, 597 p. Plenum, New York, N.Y.

Doctor, R., et. al. 1976. The California nuclear safeguards initiative. Sierra Club Bulletin. LXI, No. 5: 4-6, 44-46.

Emmett, B. A., et al. 1978. The distribution of environmental quality: some Canadian evidence. p. 361-376. In D. F. Burkhardt and W. H. Ittelson (eds) Environmental assessmentment of socioeconomic systems. xv, 597 p. Plenum, New York, N.Y.

Galbraith, J. K. 1967. The new industrial state. xv, 427 p. Houghton Mifflin, Boston, Mass.

Gofman, J. W., and A. R. Tamplin. 1971. Poisoned power. 368 p. Rodale, Emmaus, Pa.

Gross, B. M. 1966. The state of the nation: social systems accounting. p. 154-271. In R. A. Bauer (ed) Social indicators. xxi, 357 p. MIT Press, Cambridge, Mass.

Holdren, J. P., K. R. Smith, and Gregory Morris. 1979. Energy: calculating the risks (II). Science 204, No. 4393: 564-568.

Hoffer, Eric. 1972. Automation is here to liberate us. p. 64-74. In W. E. Moore (ed) Technology and social change. vii, 236 p. Quadrangle Books, Chicago, Ill.

Horwitch, Mel. 1979. Coal: constrained abundance. p. 79-107. In R. Stobaugh and D. Yergin (eds) Energy future: report of the energy project at the Harvard Business School. x, 353 p. Random House, New York, N.Y.

Inhaber, Herbert. 1978. Risk of energy production. Report AECB 1119, Third Edition. Atomic Energy Control Board, Ottawa, Ontario.

Inhaber, Herbert. 1979. Risk with energy from conventional and nonconventional sources. Science 203, No. 4382: 718-723.

Koreisha, S., and R. Stobaugh. 1979. Appendix: limits to models. p. 234-266. In R. Stobaugh and D. Yergin (eds) Energy future: Report of the energy project at the Harvard Business School. x, 353 p. Random House, New York, N.Y.

Lewis, H. W., et al. 1975. Report to the American Physical Society by the study group on light-water reactor safety. Reviews of Modern Physics. XLVII, Supplement 1: S1-S7.

Lutz, M. A., and K. Lux. 1979. The challenge of humanistic economics. Benjamin Cummings, London, England.

Mayo, L. H. 1972. The management of technology assessment. p. 71-122. In R. G. Kasper (ed) Technology assessment: Understanding the social consequences of technological applications. x, 192 p. Praeger, New York, N.Y.

Miller, Saunders. 1976. The economics of nuclear and coal power. xv, 151 p. Praeger, New York, N.Y.

Mishan, E. J. 1976. Cost-benefit analysis. xx, 454 p. Praeger, New York, N.Y.

Mishan, E. J. 1967. The costs of economic growth. xxi, 190 p. Praeger, New York, N.Y.

Mishan, E. J. 1969. Technology and growth. xix, 193, p. Praeger, New York, N.Y.

Mulvihill, R. J., D. R. Arnold, C. E. Bloomquist, and B. Epstein. 1965. Analysis of United States power reactor accident probability. Planning Research Corporation. Los Angeles, California.

Nader, Ralph. 1976. Nuclear power: more than a technological issue. Mechanical Engineering 98: 32-37.

North, D. C. 1977. Political economy and environmental policies. Environmental Law 7, No. 3: 449-462.

Novick, S. 1976. The electric war: the fight over nuclear power. 376 p. Sierra Club Books, San Francisco, California.

Ravetz, J. R. 1971. Scientific knowledge and its social problems. xi, 449 p. Clarendon Press, Oxford, England.

Schmidt, F., and D. Bodansky. 1976. The energy controversy: the fight over nuclear power.

xiii, 154 p. Albion, San Francisco, California.

Schumacher, E. F. 1973. Small is beautiful: economics as if people mattered. 305 p. Harper, New York, N.Y.

Shrader-Frechette, K. S. 1980. Nuclear power power and public policy. xvi, 176 p. D. Reidel, Boston, Massachusetts.

Shrader-Frechette, K. S. 1981. Environmental ethics. xiv, 358 p. Boxwood, Pacific Grove, California.

Silk, L. S. 1969. Direct and indirect effects of large technology programs. p. 51-60. In P. K. Eckman (ed) Technology and social progress--synergism or conflict? xi, 158 p. American Astronautical Society, Washington, D.C.

Stobaugh, Robert, and David Yergin. 1979. Conclusion: toward a balanced energy program. p. 216-322. In R. Stobaugh and D. Yergin (eds) Energy future: report of the energy project at the Harvard Business School. x, 353 p. Random House, New York, N.Y.

US Atomic Energy Commission. 1957. Report WASH-740: Theoretical possibilities and consequences of major accidents in large nuclear power plants. ix, 105 p. US Government Printing Office, Washington, D.C.

US Atomic Energy Commission. 1974. Report WASH-1224: Comparative risk-cost-benefit study of alternative sources of electrical energy. 243 p. US Government Printing Office, Washington, D.C.

US Nuclear Regulatory Commission. 1975. Report WASH-1400: Reactor safety study--an assessment of accident risks in US commercial nuclear power plants. 198 p. US Government Printing Office, Washington, D.C.

Willrich, M. 1971. Global politics of nuclear energy. xii, 204 p. Praeger, New York, N.Y.

ENERGY ANALYSIS APPLIED TO ENVIRONMENTAL ISSUES[1]

Flora C. Wang[2] and Michael A. Miller[3]

Abstract.--A computational procedure for evaluation of environmental issues is developed by energy analysis technique. A nuclear power plant siting problem is chosen to evaluate the changes in energy flows and storages of nature and human activities. Three objectives and criteria are suggested for comparing alternatives. Results have shown that each alternative could be a preferable candidate depending upon the objective or criterion chosen.

INTRODUCTION

Various resource economists (James and Lee 1971, AEC 1972) have widely employed a dollar cost-benefit procedure whereby the positive and negative effects of a given project within the human sector are quantified with economic tools. Their methods have not measured the total work contribution of nature to man's economy. It is our intention to show that the unquantified environmental values may be given appropriate considerations in decision-making, along with economic and technical considerations.

Since flows of energy from the environment are essential to develop processes and storages, they may be considered as a basis of economic value in human system. An evaluation technique that employs an energy measurement unit that is common to all system functions and external influence has been applied by Odum (1972), Lavine et al. (1978), and Wang et al. (1980).

If a decision has already been made to build a power plant--for example, the LaSalle County Nuclear Power Plant (LSCS) near Seneca, Illinois (NRC 1978)--then the procedure for calculating changes in energy flow and storage between preplant condition and the alternatives needs to be employed. This paper describes the computational aspect of the procedure for evaluation of environmental issues using LSCS as a case study.

PROCEDURE

The procedure for evaluating environmental services in energy terms is presented in the following section. A more complete analysis is given by Odum, Wang, Alexander, and Gilliland (1980).

Energy Transformation Ratio

Researchers have recognized that different types of energy do not possess equal ability to perform work (Bullard 1975, Huettner 1976, Hannon 1978). This fact implies that energy flows have a quality factor associated with them. Odum (1978) provides a mechanism to achieve comparability of different flows by relating heat calories to embodied energies.

Embodied energy is defined as the energy of one type sequestered in a process Herendeen and Bullard 1974). Examining the embodied energy in a process necessary to convert one form of energy into another allows us to estimate the ratio of the lower quality energy to the upgraded energy. This ratio is termed energy transformation ratio.

In this study, the amount of solar energy required directly and indirectly to produce other types of energy in the biosphere is used as a scale of transformation of energy. Figure 1 is an aggregated overview model of energy flows and storages for an agricultural production process. The farm product depends on the web of natural energies, such as sunlight, wind, rain, river, and geologic energy. These natural energies provide environmental services but do not cost money, since they are supplied through solar energy, atmospheric climate, and geologic systems. The storages within the system boundary are those that take long periods to accumulate, such as wood biomass, soil structure, landscape geomorphology, and the farmland itself.

[1]Paper presented at the International Symposium Energy and Ecological Modelling, sponsored by the International Society for Ecological Modelling. (Louisville, Kentucky, April 20-23, 1981).

[2]Flora C. Wang, Associate Professor of Department of Marine Sciences, Center for Wetland Resources, Louisiana State University, Baton Rouge, Louisiana 70803 U.S.A.
[3]Michael A. Miller, Research Assistant of Center for Wetlands, University of Florida, Gainesville, Florida 32611 U.S.A.

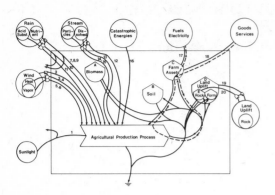

Figure 1.—Aggregated overview model of energy
flows and storages for an agricultural
production process.

The embodied energy in each of the environmental flows or storages in figure 1 is calculated
as the solar energy required to develop that
flow. That is, the energy transformation ratio,
Q, is computed as:

$$Q = \frac{\text{Calories flux of sunlight}}{\text{Calories flux of environmental flow}} \qquad (1)$$

For example, the solar energy to the biosphere not
including albedo (Sellars 1965) is:

Solar energy = (4600 Cal/m^2 day) x

(5.1 x 10^{14} m^2 area of earth) x

(365 day/year)

= 8.56 x 10^{20} Cal/yr (2)

The rate of production of atmospheric kinetic
energy (Monin 1972) is:

Wind energy = 2 x 10^{12} KW

= (2 x 10^{12} KW) x

(860 Cal/KW hr) x

(8,760 hr/yr)

= 1.51 x 10^{19} Cal/yr (3)

Thus the energy transformation ratio for wind is:

$$Q_{wind} = \frac{\text{Solar insolation}}{\text{Wind kinetic energy}}$$

$$= \frac{8.56 \times 10^{20} \text{ Cal/yr}}{1.51 \times 10^{19} \text{ Cal/yr}}$$

= 56.7 Solar Cal/Wind Cal (4)

Derivations and calculations of energy transformation ratios for other environmental flows and
storages are available elsewhere (Odum et al. 1980).
Table 1 summarizes the energy transformation ratio
in solar calories per calorie of environmental
flow or storage. The number or letter in the
table corresponds to the flow or storage displayed
in fig. 1.

Embodied Energy and Dollar Ratio

As fig. 1 shows, in a human sector (the
agricultural production process), the money and
energy (fuels, electricity, goods, and services)
flow countercurrent to one another. While one
may use an energy-to-dollar ratio, where the
energy counted is only that of concentrated
fuels, a more reasonable ratio includes all
sources, solar energy as well as fossil fuels,
since it is the use of all the energies that
drives money in circulation.

Table 2 lists the ratio of total U.S.
energy flows (fossil fuels and solar energy) to
its GNP from 1956 to 1978 (Kylstra 1974; USDA
1979). In the table, the last column represents
the ratio of embodied solar energy to dollars,
using an energy transformation ratio of 2,000
solar calories per fossil fuel calorie (Odum
1978). The various sectors of our economy are
so interwined that this ratio can be used in
giving perspective to the evaluations done in
units of embodied solar energy.

In our analysis it is assumed that there is
a proportional relationship between economic
production and the embodied energy flow of the
systems of man and nature contributing to that
production. Therefore, for given data on dollar
flow in human sectors of any year, an estimate
of the embodied energy is obtained by multiplying
by the energy/dollar ratio for that particular
year.

APPLICATION

Energy analysis, a decision-making technique
using energy flow as a common unit for both man
and nature, is used to assess the environmental
impacts. A nuclear power plant siting problem
is chosen to illustrate the evaluation of the
changes in energy flows and storages associated
with the plant siting and its cooling alternatives.

Table 1.--Energy transformation ratios in solar calories per calorie of
environmental flow and storage (Odum et al. 1980)

	Flow or storage	Transformation ratios
	(Number or letter in Fig. 1)	(Solar Cal/Cal)
1	Sunlight	1.0
2	Wind	56.7
	Heat	
3	Vertical exchange	12.9
4	Horizontal advection	5.3
	Vapor	
5	Vertical exchange	14.1
6	Horizontal advection	14.1
	Rain	
7	Kinetic energy	2.38×10^5
8	Gravitational energy	4.00×10^3
9	Chemical potential of total dissolved solids	6.90×10^3
10a	Chemical potential of nitrogen	5.88×10^8
10b	Chemical potential of phosphorus	5.28×10^9
11	Chemical potential of acid substances	1.32×10^9
	Stream	
12	Physical energy in stream flow	1.06×10^4
13	Chemical potential of total dissolved solids	1.55×10^7
14	Chemical potential energy in sediments in streams	0.88×10^6
15	Physical potential energy in materials in streams	2.33×10^7
	Catastrophic	
16a	Catastrophic energy in earthquake	3.98×10^6
16b	Catastrophic energy in tornado	2.61×10^{10}
	Fuels	
17a	Gasoline	2.00×10^3
17b	Electricity	7.20×10^3
	Goods and services	
18a	Fertilizer	0.69×10^3
18b	Machinery	2.00×10^3
18c	Labor	4.18×10^5
18d	Commodity	1.98×10^4
	Land uplift	
19	Potential energy	3.48×10^{10}
20	Chemical potential energy	2.83×10^7
A	Energy stored in biomass	0.70×10^3
B	Energy stored in soil	2.04×10^4
C	Farm assets	2.00×10^3
D	Energy stored in uplifted land	1.50×10^5
E	Energy stored in base rock	2.83×10^7

Power Plant Siting Problem

Public concerns with nuclear power plants
have concentrated on the impact of thermal
effects of nearby water bodies (Ramsey and Reed
1974), since the heat produced by a power plant
must be released following the generation of
electricity, and large quantities of water are
needed for cooling purposes.

Of the various cooling system alternatives
(cooling reservoir, once-through cooling, and
cooling tower), a cooling reservoir requires the
largest area of land. Reynolds (1980) estimated
that at least one acre of land must be inundated
for each megawatt of electric capacity. The
effect of the cooling system on land resources is
a function of individual site characteristics.
Thus the issues have evolved into the consideration
of impacts of power plant siting on surrounding
land areas.

In our analysis, it is assumed that there is
a need for the power plant and a good utilization

Table 2.--Ratio of energy flows in U.S. society to the gross national
product[1]

Year	Fossil fuels	Fossil fuels plus solar energy[2]	GNP	Solar Calories per GNP[3]
	$(10^{15}$ Cal/yr) - - - -		$(10^9$ \$)	$(10^6$ Solar Cal/\$)
1956	10.58	17.32	419.2	82.6
1957	10.56	17.30	441.1	78.4
1958	10.46	17.20	447.3	76.8
1959	10.94	17.68	483.7	73.2
1960	11.33	18.07	503.7	71.8
1961	11.52	18.26	520.1	70.2
1962	12.06	18.80	560.3	67.2
1963	12.51	19.25	590.5	65.2
1964	12.98	19.72	632.4	62.4
1965	13.60	20.34	684.9	59.4
1966	14.40	21.14	749.9	56.4
1967	14.68	21.42	793.9	54.0
1968	15.56	22.30	864.2	51.6
1969	16.37	23.11	930.3	49.6
1970	16.94	23.68	976.4	48.6
1971	17.33	24.07	1050.4	45.8
1972	18.17	24.91	1151.8	43.2
1973	18.80	25.54	1306.6	39.2
1974	18.24	24.98	1412.9	35.4
1975	17.82	24.56	1528.8	32.2
1976	18.70	25.44	1700.1	30.0
1977	19.30	26.04	1887.2	27.6
1978	19.66	26.48	2107.6	25.0

[1] Data are obtained from Kylstra (1974) and USDC (1979).

[2] Solar energy contribution to the U.S. is estimated at 6.74×10^{15} Cal fossil fuel/year.

[3] An energy transformation ratio of 2,000 solar calories per fossil fuel calorie is used (Odum 1978).

of its electrical power output. In doing so, the power plant siting problem can thus be formulated into a resource management framework, namely, the resource utilization and associated environmental impacts. The relative effects of cooling alternatives can then be quantified based on changes in embodied solar energy.

Categories for Energy Flow Evaluation

The step for evaluating the energy flows in a power plant siting problem is structured in figure 2, which shows the local system of environmental resources and the existing main economy, the new plant and cooling system to be sited, and their newly added connections to the environmental system and surrounding human economy.

The categories for calculating the changes in energy flows before and after the construction of a power plant are set forth in table 3 with reference to fig. 2. Six major categories are defined. First is the power plant productive contribution (ΔA), which is the difference between electricity generated (ΔA_1) and its feedback (ΔA_2). Second are the changes in environmental production (ΔB) because of changes in land uses, such as forest land (ΔB_1), agricultural land (ΔB_2), and other productive land (ΔB_3). Third

Table 3.—Categories for calculating the changes in energy flows due to the construction of a power plant

Flow or storage	Item of changes
(Letter in Fig. 2)	– – – – Changes in actual energy Cal/yr – – – – – – – –
ΔA	Power plant productive contribution
ΔA_1	Electric power generated
ΔA_2	Electricity fed back into use
ΔB	Changes in environmental production
ΔB_1	Gross production of forest land
ΔB_2	Agricultural productive land
ΔB_3	Other productive land
ΔS	Changes in environmental resources
ΔS_1	Wood biomass
ΔS_2	Soil
ΔS_3	Local assets
ΔC	Changes in input from main economy
ΔC_1	Goods and services to local system
ΔC_2	Goods and services to plant and cooling system
ΔF	Changes in fuel and raw materials from main economy
ΔF_1	Fuels and raw materials to local system
ΔF_2	Fuels and raw materials to plant and cooling system
ΔE	Changes related to environmental technology and waste
ΔE_1	Entrainment mortality for zooplankton
ΔE_2	Entrapment of fish population
ΔE_3	Heat impact on phytoplankton
ΔE_4	Blowdown water quality
ΔE_5	Other waste release

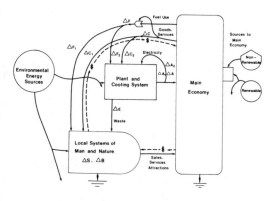

Figure 2.—Categories of changes in energy flow before and after the construction of a power plant (Odum et al. 1981).

are the changes in environmental resources (ΔS) through losses of wood biomass (ΔS_1), soil (ΔS_2), and local assets (ΔS_3).

The fourth one is the energy that is brought in by human development (ΔC) as goods and services to the local system (ΔC_1) and to the plant and cooling system (ΔC_2). Five is the embodied energy of the raw materials and fuels (ΔF) that are used by the local system (ΔF_1) and the plant (ΔF_2).

The last category concerns environmental technology and waste (ΔE)—for example, aquatic organisms entrained in the makeup water system because of thermal and mechanical shock (ΔE_1), entrapment of fish populations (ΔE_2), heat impact on phytoplankton (ΔE_3), deterioration of blowdown water quality (ΔE_4), and other waste release (ΔE_5).

Table 4.--Vegetation and land use subsystems before and after the construction of the power plant[1]

Preplant	Acreage	With Plant	Acreage
Plant site (3061)			
Soybeans	1450	Lake	2058
Corn	971	Vegetation	517
Hay	161	Old field	466
Oats	146	Structures	20
Pasture	13		
Fallow	267		
Marsh	13		
Homesites	40		
Pipeline corridor (697)			
Soybeans	75	Old field	458
Corn	111	Forest	215
Hay	16	Marsh	24
Fallow	197		
Forest	265		
Marsh	24		
Homesites	9		
Railroad spur and tower bases (127)			
Soybeans	65	Old field	98
Corn	39	Right-of-way	29
Hay	6		
Oats	6		
Fallow	11		
Total (3885)			
Agriculture	3059	Lake	2058
Fallow	475	Old field	1022
Forest	265	Vegetation	517
Homesites	49	Forest	217
Marsh	37	Structures	49
		Marsh	24

[1]Data are derived from Commonwealth Edison Company (1977).

Brief Description of Case Study

The LaSalle County nuclear power station is located four miles from the Illinois River, and about 70 miles southwest of Chicago, as shown in fig. 3 (AEC 1973). The station consists of two boiling water reactors generating a capacity of 2156 MW (NRC 1978). The exhaust steam is condensed by water circulated through an artificial cooling reservoir. The lake's intake and discharge waters are taken from and returned to the Illinois River.

The power plant was built on fertile soil. A total of 3885 acres of land were used for the LaSalle County Station. Of this, 3059 acres were agricultural land. Principal crops in the area were corn and soybeans. Other land uses

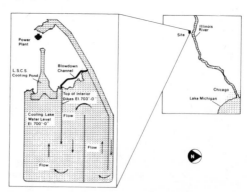

Figure 3.--Location of LaSalle County nuclear power station (AEC 1973).

were for fallow land, forest, marsh, and homesites. The vegetation and land use subsystems before and after the construction of the power plant are summarized in table 4.

An important issue at the LSCS site has been the land conversion problem, that is, whether fertile farmland should be diverted from good production to use as a site for building a large cooling lake that displaces farming activity. In an earlier paper (Wang et al. 1980), we addressed this issue in the format of land use problems, using a hypothetical example. In this paper, we further extend the land use issue into the question of whether alternative cooling systems should be considered for the LSCS case study that might be more beneficial energetically to the region.

EVALUATION

Energy evaluations involve calculating the potential energy in each category of both natural and human activities. The sequence of computation is described in the following.

Data Base for Energy Analysis

Data for energy and dollar flows into and out of the LSCS station and alternative cooling systems are primarily based on the reports from NRC (1978), CEC (1977), AEC (1973), and BCL (1972). Their data are arranged by classifying benefits and costs into four categories: economic benefits and costs and environmental benefits and costs.

Economic benefits are mainly derived from the generation and use of electricity and from the number of jobs provided. Economic costs associated with LSCS are divided into the annual fuel costs, operation and maintenance costs, and decommissioning costs. Environmental benefits are projected from the use of the cooling lakes as a recreation facility. Environmental costs are assessed by impacts on land, water, and air.

Impacts on land include a diversion of 3885 acres of productive land to industrial use at the site and a relocation of about 130 residents (AEC 1973). Impacts on water include an average water consumption of 44.2×10^6 m^3/yr (NRC 1978); a thermal plume of 4.1 acres and a total heat discharge of 745×10^6 Btu/hr to the Illinois River (NRC 1978); a blowdown water of 30,000 gpm (AEC 1973) containing dissolved and suspended solids from the cooling lake to the river; a release of radionuclides to the river; and an entrainment of aquatic organisms and fish. Impacts on air are mainly because of a formation of ice and fog, resulting in hazardous road conditions and inconvenience to ground transportation, and gaseous radwaste effluents in the atmosphere. Table 5 summarizes the essential data used in our analysis.

Energy Evaluation of All Flows and Storages

In table 5, it is noted that the data from different sources are not measured in the same units, so they are not directly additive, and no single ratio depicting the benefits and costs in dollars can be developed. It is our purpose to convert them into an equivalent amount of energy of a common type, so that they can be directly comparable among alternatives.

Energy flows and storages are calculated for the LaSalle plant site before construction with the preexisting pattern of agriculture, human settlements, and economic activity. The actual energy in heat equivalents for each of flows and storages as identified in fig. 1 is calculated first and then multiplied by energy transformation ratios from table 1, or energy/$ ratios of a particular year from table 2, to express the flow and storage in embodied solar energy. Table 6 summarizes the values of energy flow and storage before the construction of the LSCS power plant.

The evaluation of all energy flows and storages gives perspective to the basis of value. On a per-unit-area basis, the total flows of man and nature on the agricultural land at the LaSalle plant site as expressed in embodied solar energy amount to approximately 10^9 solar Cal/m^2 yr, and the embodied solar energy storage in the combined systems of man and nature is equal to 144×10^{12} solar Cal/m^2. This table is then used as a base for comparing alternatives.

Evaluation of Changes in Energy Flows and Storages

Proceeding with the assumption that building a power plant will maximize the total work accomplished in the combined system of man and nature, it is simpler to establish the preplant conditions and then to identify and calculate the changes effected by the cooling alternatives at each plant site. In estimating the changes in embodied energy, both the production and interruption of environmental and economic processes using fuels, goods and services are included.

For our LSCS case study, three cooling alternatives are considered. First, changes in energy flow and storage are estimated from the plant as built using the cooling reservoir compared with the preplant pattern of landscape and economic use; second, the differences in energy flow and storage are calculated for the plant as if it has used a cooling tower in place of the cooling reservoir; and third, the changes in energy flow and storage are computed as if the plant has been built on the shore of a large lake and uses the lake for once-through cooling.

Table 5.--Data of economic and environmental benefits and costs associated with LSCS station[1]

Classification	Magnitude and units of measurement	
Economic benefits		
Power generated	$11,300 \times 10^6$	Kwh/yr
Power capacity	$2,156 \times 10^3$	Kw
Employee payroll	2.5×10^6	$/yr
Economic costs (1980 $)		
Nuclear fuel	61×10^6	$/yr
Operation and maintenance	34×10^6	$/yr
Decommissioning	2.18×10^6	$/yr
Environmental benefits		
Recreational use of cooling lake	0.2×10^6	$/yr
Environmental costs		
Impact on land		
Land use for site	3,885	acres
Loss of 130 residents	0.72×10^6	$/yr
Impact on water		
Water consumption	44.2×10^6	m^3/yr
Area of thermal plume	4.1	acres
Heat discharge to river	745×10^6	Btu/hr
Chemical discharge to river		
Blowdown water	30,000	gpm
Total dissolved solids	1,000	mg/ℓ
Total suspended solids	220	mg/ℓ
Radionuclide contamination		
Tritium	15×10^6	μCi/yr
Others	1.9×10^5	μCi/yr
Entrainment and impingment		
Plankton organisms	_2	
Fish population	_	
Impact on air		
Effect of fogging and icing	10	day/yr
Noble gas effluent	0.06	mrem/yr
Radioiodine and particulates	0.70	mrem/yr

[1]Data are adapted from NRC (1978), CEC (1977), AEC (1973), and BCL (1972).

[2]Qualitative information on diversity only.

Referring to fig. 2, categories of changes in energy flows and storages are calculated as set forth in table 3. Table 7 shows the result of energy changes associated with the power plant as constructed using cooling reservoirs; table 8 displays the energy changes associated with the alternative using a cooling tower; and table 9 lists the changes related to the alternative of placing the plant on a large lake and using the lake for cooling. The details of each calculation are given elsewhere (Odum et al. 1980).

RECOMMENDATIONS

The following section provides the basic concepts used in forming the objectives for ranking alternatives, and explains some underlying principles in generating the criteria for comparison

Table 6.—Energy evaluation of environmental flows and storages for the LaSalle Plant site before the construction of power plant

Flow or storage	Heat equivalents	Solar equivalent
(Number or letter in Fig. 1)	$------ (Cal/m^2\ yr)\ ------$	
1 Sunlight	1.01×10^6	1.01×10^6
2 Wind	6.83×10^3	3.87×10^5
Heat		
3 Vertical exchange	1.86×10^6	2.40×10^7
4 Horizontal advection	3.03×10^6	1.61×10^7
Vapor		
5 Vertical exchange	2.34×10^5	3.30×10^6
6 Horizontal advection	2.19×10^5	3.09×10^6
Rain		
7 Kinetic energy	6.43	1.53×10^6
8 Gravitational energy	1.49×10^2	5.96×10^5
9 Chemical potential of TDS	1.05×10^3	7.25×10^6
10a Chemical potential of nitrogen	0.85×10^{-3}	0.49×10^6
10b Chemical potential of phosphorus	0.16×10^{-3}	0.85×10^6
11 Chemical potential of acid	2.13×10^{-2}	2.81×10^7
Stream		
12 Physical energy in stream flow	3.43	3.64×10^4
13 Chemical potential of TDS	0.52	0.81×10^7
14 Chemical potential in sediments	2.60	2.29×10^6
15 Physical potential in materials	1.69×10^{-3}	0.39×10^5
Catastrophic		
16a Catastrophic energy in earthquake	4.00×10^{-2}	1.59×10^5
16b Catastrophic energy in tornado	1.07×10^{-2}	2.79×10^8
Fuels		
17a Gasoline	1.97×10^2	3.94×10^5
17b Electricity	7.65×10^1	5.51×10^5
Goods and services		
18a Fertilizer	4.15×10^2	2.86×10^5
18b Machinery	1.04×10^2	2.08×10^5
18c Labor	1.21	5.06×10^5
18d Commodity	0.60×10^2	1.19×10^6
Land uplift		
19 Potential energy	2.57×10^{-6}	0.89×10^5
20 Chemical potential energy	0.23×10^2	6.51×10^8
	$----- (Cal/m^2)\ -------$	
A Energy stored in biomass	2.65×10^3	1.86×10^6
B Energy stored in soil	1.81×10^5	3.69×10^9
C Farm assets	4.14×10^3	0.83×10^7
D Energy stored in uplifted land	2.40×10^5	3.60×10^{10}
E Energy stored in base rock	5.06×10^6	1.43×10^{14}

Objectives Uses for Ranking Alternatives

One of the values of energy analysis is the recognition of many external environmental inflows that are an important part of the basis for man's economy. We suggest that the energy embodied in environmental processes be used as a measure of potential contribution to the economy. Thus the decision of power plant siting should minimize the disturbance of existing environmental production, maximize the contribution of new environmental production, and minimize the displacement of environmental resources. Such objectives are displayed for three cooling alternatives in Table 10, in which the entries are derived directly from items of ΔB and ΔS in tables 7, 8, and 9. According to this objective, alternative 3, building the plant on a large lake and using that lake for once-through cooling, is the most favorable.

Table 7.--Embodied energy evaluation of changes from the construction and operation of LaSalle County station with cooling reservoir

Flow or storage	Items of change	Change in actual energy	Change in embodied energy
(Letter in Fig. 2)		(10^9 Cal/yr)	(10^{13} solar Cal/yr)
ΔA	Power plant productive contribution		
ΔA_1	Electric power generated	+9720	+7000
ΔA_2	Electricity fed back into use	0	0
ΔB	Environmental production changes		
ΔB_1	Forest, marsh, and lake	+ 63.2	+ 5.81
ΔB_2	Agricultural production	- 311	- 36.10
ΔB_3	Other vegetation	+ 23.1	+ 2.13
ΔS	Changes in stored resources		
ΔS_1	Wood biomass	- 0.233	- 0.067
ΔS_2	Soil	- 37.7	- 132.327
ΔS_3	Farm assets	- 1.28	- 0.256
ΔC	Changes in inputs from main economy		
ΔC_1	Fertilizer	+ 5.05	+ 0.348
	Machinery	+ 1.29	+ 0.258
	Labor	+ 0.015	+ 0.627
	Commodities	+ 0.743	+ 1.47
ΔC_2	Operation and maintenance	- 436	- 87.2
	Capital investment	- 256	- 51.2
ΔF	Changes in fuels and other energy		
ΔF_1	Liquid fuel	+ 2.44	+ 0.488
	Electricity	+ 0.947	+ 0.682
ΔF_2	Nuclear fuel	- 769	- 154
ΔE	Changes related to environmental waste		
ΔE_1	Entrainment	- 0.0098	- 0.0218
ΔE_2	Entrapment	- 0.0011	- 0.0028
ΔE_3	Heat impact	- 0.0502	- 0.0046
ΔE_4	Chemical impact	- 0.4406	- 0.3040

The second goal of energy analysis is the use of energy flow in evaluating the work contribution from the main economy. This provides a linking mechanism between the environmental system and the human system. As shown in fig. 2, money and energy flow countercurrent to one another, and it is assumed that there is a consistent proportional relationship between economic production and embodied energy flows. In the LSCS example, a given power output is assumed. Therefore, feedback from the main economy, in the forms of goods and services (ΔC) and fuels and raw materials (ΔF), is made up of energy resources that are no longer available to do work in that main economy. Thus, the objective in power plant siting can be to minimize the decrease in existing economic production and minimize the diversion of fuels, goods, and services from the main economy. Table 11 summarizes the comparison of three cooling alternatives based on the data (ΔC and ΔF) obtained from tables 7, 8, and 9. With this objective in mind, alternative 1, the plant with cooling reservoir is preferable.

A third utility in energy analysis is the ability to assess environmental impacts on the combined economy of man and nature associated with each alternative (Kemp et al. 1973; Wang et al. 1980). We propose to use the Gibb's free energy as a measure of importance of each impact on the system, thus providing an indication as to the value of natural environmental flows. Therefore, a power plant should be placed where it would cause the least deterioration in environmental quality. Table 12 provides the assessment of environmental effects on the system for various alternatives. The entries in table 12 are taken from the one (ΔE) corresponding to the entries of tables 7, 8, and 9. It is interesting to note that alternative 1, building the plant using a cooling reservoir, is the best alternative here.

Table 8.— Embodied energy evaluation of changes from the construction and operation of LaSalle County station with cooling tower

Flow or storage (Letter in Fig. 2)	Items of change	Change in actual energy (10^9 Cal/yr)	Change in embodied energy (10^{13} solar Cal/yr)
ΔA	Power plant productive contribution		
ΔA_1	Electric power generated	+9720	+7000
ΔA_2	Electricity fed back into use	− 144	− 104
ΔB	Environmental production changes		
ΔB_1	Forest, marsh, and lake	− 47.4	− 4.36
ΔB_2	Agricultural production	− 63.2	− 7.33
ΔB_3	Other vegetation	+ 15.8	+ 1.45
ΔS	Changes in stored resources		
ΔS_1	Wood biomass	− 2.98	− 0.861
ΔS_2	Soil	− 1.83	− 6.423
ΔS_3	Farm assets	− 0.268	− 0.0536
ΔC	Changes in inputs from main economy		
ΔC_1	Fertilizer	+ 1.02	+ 0.0704
	Machinery	+ 0.269	+ 0.0538
	Labor	+ 0.0031	+ 0.131
	Commodities	+ 0.155	+ 0.307
ΔC_2	Operation and maintenance	− 422	− 88.4
	Capital investment	− 259	− 51.8
ΔF	Changes in fuels and other energy		
ΔF_1	Liquid fuel	+ 0.51	+ 0.102
	Electricity	+ 0.198	+ 0.143
ΔF_2	Nuclear fuel	− 769	− 154
ΔE	Changes related to environmental waste		
ΔE_1	Entrainment	− 0.0128	− 0.0284
ΔE_2	Entrapment	− 0.0021	− 0.0053
ΔE_3	Heat impact	− 0.105	− 0.0097
ΔE_4	Chemical impact	− 1.084	− 0.748

Criteria Suggested for Comparisons

Policies and pressures emanating from the 1972 Amendments to the Federal Water Pollution Control Act (FWPCA) tend to favor the installation of cooling towers in power plants. Reynolds (1980) indicates that adverse environmental effects associated with cooling towers are more likely to be irreversible than impacts associated with cooling reservoirs and cooling lakes. We suggest three criteria for comparing cooling systems.

The system from which we derive our energy includes sources from both man and nature. According to the maximum power principle (Lotka 1922), the system that will survive in the long run is the one that maximizes the use of available energies in useful work processes. The useful work process in this case is the production of electrical energy. Energy analysis seeks to identify the system that accomplishes the most work with the least diversion of energy resources. In the LSCS case study, the total diversion of energy resources, on a relative basis, can be derived from the algebraic sum of the decrease in environmental process (table 10) and the increase in diversion of economic production (table 11). Such magnitudes are displayed in column 3 of table 13. Alternative 3, a plant built on a large lake, is preferable according to this criterion of minimizing the diversion of energy resources of man and nature.

In once-through cooling systems (reservoir or lake), the pumping of large volumes of water may create problems of plankton entrainment and fish entrapment at the plant intake, and the thermal effluent may reduce metabolic activities in the discharge area. Also, cooling reservoirs

Table 9.--Embodied energy evaluation of changes from the construction and operation of LaSalle County station with once-through cooling

Flow or storage	Items of change	Change in actual energy	Change in embodied energy
(Letter in Fig. 2)		$(10^9$ Cal/yr)	$(10^{13}$ solar Cal/yr)
ΔA	Power plant productive contribution		
ΔA_1	Electric power generated	+9720	+7000
ΔA_2	Electricity fed back into use	0	0
ΔB	Environmental production changes		
ΔB_1	Forest, marsh, and lake	− 23.7	− 2.18
ΔB_2	Agricultural production	− 31.6	− 3.67
ΔB_3	Other vegetation	+ 7.9	+ 0.73
ΔS	Changes in stored resources		
ΔS_1	Wood biomass	− 1.49	− 0.431
ΔS_2	Soil	− 0.897	− 3.148
ΔS_3	Farm assets	− 0.134	− 0.0268
ΔC	Changes in inputs from main economy		
ΔC_1	Fertilizer	+ 0.510	+ 0.0352
	Machinery	+ 0.135	+ 0.0270
	Labor	+ 0.0016	+ 0.0656
	Commodities	+ 0.0777	+ 0.154
ΔC_2	Operation and maintenance	− 436	− 87.2
	Capital investment	− 259	− 50.2
ΔF	Changes in fuels and other energy		
ΔF_1	Liquid fuel	+ 0.255	+ 0.0510
	Electricity	+ 0.0991	+ 0.0714
ΔF_2	Nuclear fuel	− 769	− 154
ΔE	Changes related to environmental waste		
ΔE_1	Entrainment	− 0.148	− 0.329
ΔE_2	Entrapment	− 0.0412	− 0.103
ΔE_3	Heat impact	− 3.01	− 0.277
ΔE_4	Chemical impact	0	0

and lakes displace large amounts of valuable and productive land. Energy analysis can address the issue of whether the environmental effects resulting from the use of once-through cooling systems would justify the investment of capital energy in building a closed-cycle cooling tower to alleviate the losses. A comparison can be made by deriving the ratio of economic investment in fuels, goods, and services (table 11) to the environmental impacts on the system (table 12). The result is shown in column 5 of table 13. The plant with the cooling tower, alternative 2, is justified as a means for mitigating losses to natural systems.

According to a general principle of system evolution, among competing systems of man and nature with approximately equivalent amounts of capital energy to invest, the most economically competitive system would be the one that invests the capital energy in such a way as to produce the most productive contribution from the system. Such a criterion can be expressed in energy analysis as the ratio of power plant productive contribution (ΔA in tables 7, 8, and 9) to the diverted economic production (11). This ratio is listed in column 7 of 13. Alternative 1, the plant with a cooling reservoir, is the best candidate selected from this criterion.

SUMMARY

Emanating from the 1972 Amendments to the FWPCA, public concern about nuclear power plants have stressed the impacts of siting on surrounding land, the effects of thermal effluents on nearby water bodies, and the possibility of using the cooling tower as a means of mitigating

Table 10.-- Ranking cooling alternatives by objective A: environmental process

Objective A: Minimize the disturbance of existing environmental production
Maximize the contribution of new environmental production
Minimize the displacement of environmental resources

	Changes of environmental production			Change of environmental resources	Decrease in environmental process
Alternative	Disturbance	Contribution	Net disturbance	Displacement	Process
	— — — — — — — — — — — — (10^{13} solar Cal/yr) — — — — — — — — — — —				
1 Plant with cooling reservoir	36.10	7.94	28.16	132.65	160.81
2 Plant with cooling tower	11.69	1.45	10.24	7.34	17.58
3 Plant built on large lake	5.85	0.73	5.12*	3.61**	8.73***

*The alternative with the least disturbance of environmental production.

**The alternative with the smallest displacement of environmental resources.

***The alternative with the least diversion of environmental process.

losses to natural systems. In response to these concerns, we have proposed several objectives used in energy analysis for ranking alternatives, and then compared these alternatives by applying a variety of criteria to the consequences.

Energy analysis is an extended approach of system analysis. It is a group of decision-making techniques using energy flow as a common denominator, as opposed to the dollar unit used in economic analysis. The LaSalle County nuclear power station is chosen as a case study for applying energy analysis to environmental issues. As listed in table 5, research data for energy and dollar flows into and out of the LCSC station are gathered from different sources. No single relation depicting the dollar cost-benefit ratio can be derived. We have converted these measurements into the equivalent amount of energy, that is, the embodied solar energy via energy transformation ratios, so that different kinds of effects are directly comparable.

In our approach, we have evaluated the changes in energy flow and storage for three alternatives: the plant as built using a cooling reservoir; the plant using a cooling tower in place of a cooling reservoir; and the plant to be built on the shore of a large lake, using the lake for cooling. All are compared with the preplant pattern of landscape and economic use. Therefore, all results are in terms of the relative basis with respect to the preexisting pattern of agricultural and farming activities.

Three objectives are suggested in this study, namely, to minimize the diversion of environmental process, to minimize the diversion of economic production, and to minimize the environmental

impact on the system. Each objective is used separately as a goal in ranking alternatives. The results are summarized in tables 10, 11, and 12 and have shown that the preferable candidate is identified as alternative 3, 1, and 1, respectively according to the three objectives as proposed.

Next, we have recommended three criteria for comparing alternatives by grouping jointly the results of tables 10, 11, and 12 into an overview table. The criteria are: minimizing the diversion of energy resources of the combined system of man and nature, minimizing the capital energy investment in alleviating environmental losses, and maximizing the productive contribution in relation to the diverted production. Their consequences are displayed in table 13. Based on these criteria, alternative 3, 2, and 1 are chosen, respectively.

DISCUSSION

The foregoing results and conclusions are based on the assumed accuracy of calculations, assumptions in deriving the energy transformation ratios, and concepts in forming the objectives and criteria for comparing alternatives. As with all forms of system analysis, our energy analysis is not complete. We have not taken into consideration possible errors of double counting in embodied energy evaluation, nor have we discounted energy value with time. Additional research needs to be done in estimating the embodied energy in nuclear wastes, thereby enabling us to quantify the effects of gaseous radwaste effluents to the atmosphere. Due to

Table 11.--Ranking cooling alternatives by objective B: economic production

Objective B: Minimize the decrease in existing economy production
Minimize the diversion of economic production for the main economy

| | Changes in diversion of economic production | | | | | | Diversion of economic production |
| | Goods and services | | | Fuels and raw materials | | | |
Alternative	Decrease	Increase	Net increase	Decrease	Increase	Net increase	Diversion
	- - - - - - - - - - - - $(10^{13}$ solar Cal/yr) - - - - - - - - - - - - - -						
1 Plant with cooling reservoir	2.70	138.4	135.70	1.17	154	152.83	288.53*
2 Plant with cooling tower	0.56	140.2	139.64	0.25	154	153.75	293.39
3 Plant built on large lake	0.28	137.4	137.12	0.12	154	153.88	291.00

*The alternative with the least diversion of economic production.

these problems, our results are not exact. Nevertheless, we have described the computational procedure using LSCS as a case study in evaluating embodied energy, have quantified the environmental impacts, and have ranked the cooling alternatives by various objectives and criteria. The conclusion is that each of the alternatives can be a preferable candidate depending upon the objective or criterion chosen.

ACKNOWLEDGMENT

The work reported in this paper was supported by the United States Nuclear Regulatory Commission under the Contract No. NRC-04-77-123 Mod. 3. This is contribution #LSU-CEL-81-08 from the Coastal Ecology Laboratory, Louisiana State University, Baton Rouge, LA 70803.

LITERATURE CITED

Atomic Energy Commission. 1972. Proposed AEC guide to the preparation of benefit-cost analysis to be included in applicant's environmental reports. U.S. Government Printing Office, 483-0261159, Washington, D. C.

Atomic Energy Commission. 1973. Environmental statement: Related to the LaSalle County nuclear station. Docket Nos. 50-373, 50-374, Commonwealth Edison Company, Chicago, Illinois.

Battelle Columbus Laboratories. 1972. Environmental impact report: Supplemental information to the LaSalle County environmental report, Supplement II, Columbus, Ohio.

Bullard, C. W. III. 1975. Energy costs, benefits, and net energy. CAC Document No. 174, Center for Advanced Computation, University of Illinois at Urbana-Champaign, Urbana, Ill.

Commonwealth Edison Company. 1977. LaSalle County station: Environmental Report-Operating License Stage, Vol. 1 Chicago, Ill.

Hannon, Bruce. 1978. An energy theory of value for ecosystems. Energy Research Group, Center for Advanced Computation, University of Illinois at Urbana-Champaign, Urbana, Ill.

Herendeen, R. A., and C. W. Bullard, III. 1974. Energy costs of goods and services, 1963 and 1967. CAC Document No. 140, Center for Advanced Computation, University of Illinois at Urbana-Champaign, Urbana, Ill.

Huettner, D. A. 1976. Net energy analysis: An economic assessment. Science 192:101-104.

James, L. D., and R. R. Lee. 1971. Economics of water resources planning. 613 p. McGraw Hill, New York, N.Y.

Kemp, W. M., W. H. B. Smith, H. N. McKeller, M. E. Lehman, M. Homer, D. L. Young, and H. T. Odum. 1977. Energy cost-benefit analysis applied to power plants near Crystal River, Florida. p. 507-543. In Ecosystem modeling in theory and practice: An introduction with case studies. Hall, C. A. S., and J. W. Day (eds.), John Wiley and Sons, New York.

Table 12.--Ranking cooling alternatives by objective C: environmental impact

Objective C: Minimize the loss of natural ecosystems
Minimize the deterioration of environmental quality

Alternative	Changes related to environmental technology		Environmental impact
	Loss of natural ecosystem	Deterioration of environmental quality	Impact
	$- - - - - - (10^{13}$ solar Cal/yr) $- - - - - -$		
1 Plant with cooling reservoir	0.0292	0.304	0.333*
2 Plant with cooling tower	0.0434	0.748	0.791
3 Plant built on large lake	0.709	0	0.709

*The alternative with the least environmental impact on the system.

Table 13.-- Comparing cooling alternative by applying a variety of criteria to the consequences

Criterion A: Minimize the diversion of energy resources of combined system of man and nature
Criterion B: Maximize the capital energy investment to alleviate environmental losses
Criterion C: Maximize the productive contribution related to the diverted production

Alternative	Decrease in environmental process (Table 10) (1)	Diversion of economic production (Table 11) (2)	Diversion of energy resources (3)=(2)+(1)	Environmental impact (Table 12) (4)	Ratio of economic investment to environmental impact (5)=(2)/(4)	Power plant productive contribution (Tables 7, 8, 9) (6)	Ratio of productive contribution to diverted production (7)=(6)/(2)
	$- - - - - - - (10^{13}$ solar Cal/yr) $- - - - - - -$					$(10^{13}$ Cal/yr)	
1 Plant with cooling reservoir	160.81	288.53	449.34	0.333	866.46	7000	24.26***
2 Plant with cooling tower	17.58	293.39	310.97	0.792	370.44**	6896	23.50
3 Plant built on large lake	8.73	291.00	299.78*	0.709	410.44	7000	24.05

*The alternative with the least diversion of resources of combined system of man and nature.

**The alternative with the least capital energy in alleviating environmental losses.

***The alternative with the most productive contribution in relation to the diverted production.

Kylstra, C. D. 1974. Energy analysis as a common basis for optimally combining man's activities and nature. Energy Center, University of Florida, Gainesville, Florida.

Lavine, M. J., T. J. Butler, and A. H. Meyburg. 1978. Toward environmental benefit/cost analysis: Energy analysis. Center for Environmental Research, Cornell University, Ithaca, New York.

Lotka, A. J. 1922. Contribution to the energetics of evolution. Proceedings, National Academy of Science 8:147-155.

Monin, A. S. 1972. Weather forecasting as a problem in physics. MIT Press, Cambridge, Massachusetts.

Nuclear Regulatory Commission. 1978. Environmental statement-related to the LaSalle County nuclear power station unit Nos. 1 and 2. NUREG-0486, Commonwealth Edison Company, Chicago, Ill.

Odum, H. T. 1972. Use of energy diagrams for environmental impact statement. p. 197-213. In Tools for coastal zone management. Proceedings of the Conference [Washington, D. C., February 14-15, 1972], Marine Technology Society, Washington, D. C.

Odum, H. T. 1978. Energy analysis, energy quality, and environment. p. 55-87. In Energy analysis: A new public policy tool. Gilliland, M. W. (ed.) American Association for the Advancement of Science, Washington, D. C.

Odum, H. T., M. J. Lavine, F. C. Wang, M. A. Miller, J. F. Alexander, Jr., and T. Butler. 1981. A manual for using energy analysis for plant siting. Report to the Nuclear Regulatory Commission, Contract NRC-04-77-123 Mod. 3., Energy Analyses Workshop, Center for Wetlands, University of Florida, Gainesville, Florida.

Odum, H. T., F. C. Wang, J. Alexander, and M. Gilliland. 1980. Energy analysis of environmental values-Volume II: A manual for estimating environmental and societal values according to embodied energies. Report to Nuclear Regulatory Commission, Contract NRC-04-77-123, Center for Wetlands, University of Florida, Gainesville, Florida.

Ramsey, W., and P. R. Reed. 1974. Land use and nuclear power plants: Case studies of siting problems. Directorate of regulatory standards, U. S. Atomic Energy Commission, Washington, D. C.

Reynolds, J. Z. 1980. Power plant cooling systems: Policy alternatives. Science 207:367-372.

Sellars, W. D. 1965. Physical climatology. University of Chicago Press, Chicago, Ill.

United States Department of Commerce. 1979. United States statistical abstract. U. S. Government Printing Office, Washington, D. C.

Wang, F. C., H. T. Odum, and P. C. Kangas. 1980. Energy analysis for environmental impact assessment. Journal of the Water Resources Planning and Management Division, Proceedings of the American Society of Civil Engineers, No. WR2 106:451-466.

ENERGY AND WAR - A SURVIVAL STRATEGY FOR BOTH[1]

Ane D. Merriam[2] and Wilson Clark[3]

Abstract.--The subject of energy supply and availability is often addressed in periods of war. In fact, the flow of energy and materials at just the right time may have an overriding influence in the outcome of the war activity. Conversely, the subject of military defense and vulnerability is mentioned with respect to major, centralized energy facilities. This paper addresses these two issues simultaneously, seeking both to maximize the energy supply and availability, and to decrease military vulnerability due to energy centralization.

U.S. reliance on imported fuel and centralized systems for energy production presents problems for national security and emergency preparedness in the event of a major nuclear crisis or war. Through extensive investigations and summary energy flow models, this paper focuses on the need to link energy supply and demand planning with civil defense planning not only to insure minimal vulnerability prior to an attack but also to maximize survival and recovery capabilities. The development of decentralized and renewable energy sources such as cogeneration, wind, biomass, solar and small hydro not only provides a workable solution to the defense strategy of protecting energy supply views throughout the country, but also provides an exciting energy program complete with new jobs and economic activity for the country during peace times.

INTRODUCTION

The debate between energy experts and researchers on centralized versus decentralized systems has been elevated from the hallowed halls of academic institutions to the larger audiences of local governments, utility managers, state energy offices, League of Women Voters conferences, and in general public conversations. What's more, the once erudite relationship between energy and economics is fast becoming a regular dinner table subject. The subject of energy supply and demand has become so popularized that it has become a regular issue with the news media and even worse, a major political issue for existing and aspiring office holders.

The 1974 oil embargo sent out ripples of panic amongst modern, energy-intensive America. The notion that the well might become dry some day was not easily embraced by the American public. For some time after the embargo and to some extent even today, the public has viewed the "energy crisis" as a contrived management scheme by the oil industry to squeeze out more profits from an already over inflated citizenry. Yet amidst the heated debates about oil company profits reemerging with the innauguration of Ronald Reagan, there is an increasing percentage of the populus that is taking the subject seriously, and seeking solutions in their own backyards and attics.

With just a few months in office, however, we are faced with the emergence of the Reagan administration's top priorities: reduction of taxes; inducements to business; and defense and military capability enhancement.

The subject of energy planning and management is not expected to play the turn key role in this administration as it did in the Carter four years. Yet most political experts would agree that energy planning is just as important an issue now as it was in 1974, 1976, or 1980. Additionally energy experts and modellers continue to view flows of energy as predominant economic, social, political and cultural regulators (Hannon 1973). Yet, instead of front end policy making on energy, this administration may deal with energy in some indirect ways, at best. Therefore, a closer look into

[1] Paper presented at the International Symposium Energy and Ecological Modelling, sponsored by the International Society for Ecological Modelling (Louisville, Kentucky, April 20-23, 1981).

[2] Water Resources Director, Institute of Science and Public Affairs, Florida State University, Tallahassee, Florida.

[3] Energy Advisor, California Governor's Office, Washington, D.C.

the Reagan administration's focal points with obvious potential for coupling with energy issues is warranted.

The first topic--taxes--would be an interesting area to couple with energy matters, as the economic and equity issues of energy supply, demand and public consumption would sum up many loose ends and could provide a blueprint for a more stable economy (Richards 1911; Odum and Bayley 1974). However, there are but a few professional economists and energy planners able to agree and communicate on this subject, thereby not offering an optomistic pathway for quick assimilation and policy making at this time.

The next topic is business inducements and it is becoming obvious to many energy planners that trying to market a comprehensive energy planning message from the outside into the inside worlds of Exxon, Chrysler, and The World Bank is much like having a salmon during spawning season swim upstream with 50 pound weights tied to its back.

That leaves the area of defense. On close inspection there appears to be much opportunity to couple energy knowledge with military planning, as it has already started (Nye 1980).

Couching an energy message in defense terminology might appear inconsequential to astute students of history, but the larger message of pointing out the pervasiveness of energy planning is eased one step closer to reality if successful. Yet, even in the energy business world, the idea that the U.S. is in a transition energy phase between centralized and decentralized energy systems is still not uniformly accepted. The initial debates between centralized and decentralized approaches were reminiscent of the culturally charged 1960's Viet Nam discussions. Counter-culture "energy freaks" showed up in forums, energy fairs, energy bazaars, and Sun Day celebrations in moderately sized, but enthusiastic showings across the country. Erstwhile, the pin striped three piece suits of Westinghouse, General Electric, and others continued their showings in Legislative and Congressional Committee Hearings across the nation. It appeared that the two levels of consciousness were too far from each other to ever merge much less overlap. While the decentralized proponents espoused the natural process of transferring to a decentralized energy approach, the centralists were drawing monetary bottom lines, and kept clinging to their consulting contract powered checkbooks.

Yet here it is 1981, not even 10 years since the oil embargo, and the numbers of decentralized proponents increases daily, bringing along with it many of the bastions of conservatism. In 1979 Stobaugh and Yergin popularly legitimized the options of alternative energy systems and conservation, in their book Energy Future (Stobaugh and Yergin 1980). By couching the discussion in Harvard Business School-ese, they introduced the decentralized notions to a whole new audience. Stobaugh and Yergin's accomplishments in the enhancement of alternative energy communications were equated with George Lucas' technical and cinematographic success reflected in his movie "Star Wars". Bridging unknowns, bringing together different universes of thought were revealing themes in both ventures. And it reflected the innermost thoughts of quiet energy planners across the world. Alternative energy and conservation discussions began appearing in some of the most unusual places.

Not to lose the momentum of this attitude swell, we offer yet another focus on the discussion of the value of decentralized energy systems for the U.S. This paper features the vulnerabilities inherent in following a predominantly centralized energy supply approach. Much has been written regarding the political vulnerabilities with which we've become so familiar recently (Schurr 1979; Fesharaki 1980; Energy and Defense Project 1980). Yet, not to take away from any of these previous discussions, we opt to add another dimension to the vulnerability issue, namely the military and war implications associated with centralized energy systems.

The U.S. reliance on imported fuel and centralized systems for energy production presents problems for national security and emergency preparedness in the event of a major war or nuclear crisis. We add to the body of knowledge supporting decentralized energy systems by pointing out how energy supply and demand planning should be linked to civil defense planning in order to insure minimal vulnerability prior to an attack as well as to maximize recovery capabilities.

It is not unlikely to contemplate terrorist attacks or well sponsored enemy attacks on the U.S. energy systems. Who would have predicted the Iranian hostage crisis, the Algerian, Polish and El Salvador problems just a few years ago? Looking closer into these incidents we find deep rooted energy causes and effects, which up until very recently the U.S. has had the upper hand in controlling. Therefore, we shed another light onto the discussion of alternative energies and the value of decentralized energy systems as a civil defense and military strategy, adding to the previous discussions which have focused primarily on the supply and demand dilemma.

It is a widely recognized fact that the U.S. depends on imported pertroleum to meet almost half of its demand. Yet the numerous economical, political, social, and environmental repercussions that could result from this heavy foreign dependence are not fully realized.

Oil dominates the primary fuel scene and we have come to learn that the energy sector is vital to the industrial, agricultural, communications and other sectors of society. A failure in the system's ability to produce and distribute energy throughout the U.S. would leave the country unable to support or defend itself. Therefore, the way in which primary fuels are produced and fed into the system's activity veins not only dictates economic well-being, but also constitutes the levels of vulnerability to outside forces.

Assessing the current energy system in the U.S. we find there are two major forms of vulnerability against which the U.S. must protect itself. The first is the insecure availability of imported energy supplies and strategic materials for adequate levels of defense, economic activity and stability.

The second form of vulnerability is the centralized nature of the American energy system. Because energy is vital for maintaining the U.S. economy, an adversary would enjoy a strong strategic advantage by crippling that energy system. Centralized energy facilities add to the degree of vulnerability of the U.S. energy systems because, as energy targets, they are larger and there are fewer of them.

In our investigations we site historical accounts of wartime stategies involving energy systems, using Germany and Japan as case studies in World War II, followed by the Korean, and Middle East examples more recently. The issue of an all-out nuclear war is not the point of discussion in this investigation. That case is so unique due to the health aftereffects that we prefer to focus on the more realistic notions of targeted war attacks, sabotage and terrorism on the U.S. energy systems. We are confident in our contention that the issue of centralization versus decentralization of energy systems belongs as properly in military attaches, as it does in an Oil and Gas Department.

Furthermore, we provide a theoretical model to display a broader perspective in support of the decentralized approach for the current energy phase in which we find ourselves emersed.

METHODOLOGY

Trying to relate energy supply to civil defense can be a mindful problem, not because the relationships are obscured, but contrarily because they are so ubiquitous. It is not unlike the dilemma of linking energy supply to inflation. Those people who understand the energy-money relationship almost assuredly were not convinced of it through a single event, study, or presentation. Instead, it is more a case of adding up the parts of a picture until there are enough pieces fitting together so that the whole picture falls into place, with or without the remaining individual pieces.

Since 1973 energy planners have been trying to package the components of the whole energy picture in various ways hoping to find some universal appeal and form of communication. Besides the usual academic jealousy problems and "not invented here" syndromes, there have been two major barriers preventing universal expression.

First, is the issue of emphasis and familiarity. An energy planner or a physicist views energy as the primary force affecting the world and all its activity, through the Laws of Thermodynamics.

Yet a water engineer sees water as the major economic pump primer, causing those who have water to prosper and those who don't to dwindle and fail.

Economists see money as the baseline of social and cultural activity thus providing the frame of reference from which all activity ensues.

And a general or military strategist sees ammunition, timing, trained personnel, information and communication as the compelling forces in both preventive and actual war fare.

For some time several energy researchers have been trying to convey the energy implications and relationships of not only the water, economic and military examples above, but also the fact that all processes have an energy component to them. Not all processes have a water factor, or involve the exchange of money (for example the environmental services provided by natural systems), or requires a military option. Energy shares that limelight with no other constituent, making it difficult for nonenergy researchers to swallow the overriding significance of monitoring energy flows to explain what is happening in the natural, physical, social, economic and political world around us.

We all have a built in emphasis bias determined by what we know best and relate to most. It is much like defending the male-generated "double standard" in trying to explain objectively to nonenergy professionals how energy rises above all the prejudice and familiarity and just happens to be in the driving seat of the universe.

We do not pretend to solve this perceptional problem here, except to mention it in support of the approach used in this paper. We seek to tell the story a little differently first, by the nature of the study, relating energy to war; and secondly by employing a simplified form of energy diagramming.

The second barrier to universal information flow has been due to the dichotomous nature of decision makers and researchers. The level of information and form of disposition required to meet the rigorous standards researchers employ in judging themselves may be vastly different from the information necessary for a top decision maker to set a major policy or take substantial actions on an issue.

Beyond the obvious differences of familiarity with a subject, the dichotomous nature may be characterized with researchers most often filling the specialists' shoes, and policy makers as generalists. The fine tuned difference is one of timing. Whereas the researcher sets a time table for study and report, choosing a proper forum for presentation, the policy maker often times needs to go from a ground zero information level to complete understanding in a timeframe proportional to the speed of light by most researcher's scales.

Since we feel the subject of energy systems design, civil defense and attack vulnerabilities is sufficiently complex, we propose to maximize the measured research approach with the overview, general policy implications for decision makers. We seek to bridge the valuable detailed information level of a specialist's study and the broad brushed needs of a top decision maker.

The notion that energy is embodied in all of life as we know it is well supported by H.T. Odum and others (Odum 1971). Whether we focus on the energy implications of inflation, or the net energy considerations of mining oil and gas, the contention is clear: energy is a part of every thing we do, and is the basis of work in the universe. Sometimes we don't see the energy implications of issues we are studying, but that is not because energy is not relevant, it is often because of how we perceive things. Breaking through with a universally accepted perception tool is not easy, and perhaps no one has done it yet, but we are using Odum's energy systems diagramming techniques, modified slightly to better convey our study (Odum and Odum 1976). Odum's energese language helps relate the major variables of energy production, supply and consumption to civil defense and military planning, in a concise one page format.

We employ the Energese language in conceptual models, yet highlight some of the quantifications that would be necessary for computer simulation models. The language helps to bridge the gap between conceptual and quantitative approaches, especially in this case, where a computer model was not employed, but some pertinent values were gathered and are available for discussion purposes. Sometimes an issue merely needs to be seen clearly and holistically, for a decision maker to grasp the significance of a few single events. Once an understanding of the system's relationships is gained, then the next phase, scientific assimilation, is possible even for individuals not considered experts in the field. This study provides the system overview of the vulnerability question, and begins the quantification steps leading to the second phase.

The initial investigations of this issue providing much of the data for this report was conducted by an interdisciplinary study team funded by the Federal Emergency Management Agency (FEMA). The project entitled "Dispersed, Decentralized and Renewable Energy Sources: Alternatives to National Vulnerability and War", was performed under the direction of Wilson Clark (Energy and Defense Project 1980).

RESULTS

The results of the Energy and Defense study are reported in the 326 page project report released by FEMA in December 1980. The detailed account of this issue provides much of the basic data for this report. In this paper, we summarized the main features of the report, and have taken it one step further by relating the information in simplified conceptual models that may be used in broader or more generalized discussions.

Several conceptualized energy flow diagrams are presented in this section to:

- Provide an historical account of centralized and decentralized energy systems in times of war;
- Highlight the areas of vulnerability for the U.S. system;
- Offer alternative solutions to the vulnerabilities.

The conceptual models are displayed and discussed in the following order:

The German Case Study (fig. 1)
The Japanese Case Study (fig. 2)
The Korean Case Study (fig. 3)
The Middle East Case Study (fig. 4)
An Overview of the current U.S. Energy System (fig. 5)
The U.S. Oil System (fig. 6)
The U.S. Natural Gas System (fig. 7)
The U.S. Coal System (fig. 8)
The U.S. Nuclear Power System (fig. 9)
The U.S. Hydropower System (fig. 10)
Options available for decreasing vulnerability (fig. 11-24)

Generalized Centralized-Decentralized Theoretical Models (fig. 25, 26)

German Example

Figure 1 is an energy flow diagram of the Germany Case Study during World War II. As can be readily seen in the diagram, coal was the primary fuel for both electric power generation and synfuels used primarily for transportation. In 1941, 80.2% of the electricity produced in Germany was obtained from coal, and 20% from hydropower. Germany's heavy dependence on coal and electricity in their war effort made them extremely vulnerable to attack. Even though there were over 8,000 power plants in the country, 113 of them constituted 56% of their total power. Another 313 plants produced 26% of the power, and while there were plants throughout the country, geographical concentration also existed. Therefore bombing of 416 plants in 5 main areas could devastate the country's electrical capabilities.

The existence of the national grid system originally caused the Allies to question the vulnerability of the German power system (U.S.S.B.S. 1947). But in contrast, the German officials were concerned that the Allies would recognize the strategic vulnerability of Germany's centralized power system (U.S.S.B.S. 1947). The German consternation over this issue was expressed well by Reichminister Albert Speer (Minister for Armament and War Production):

I think the attacks on power stations, if concentrated, will undoubtedly have the swiftest effect; certainly more quickly than attacks against steel works, for the high quality steel industry, especially electro-steel, as well as the whole production of finished goods and public life, are dependent upon the supply of power . . . . The destruction of all industry can be achieved with less effort via power plants (U.S.S.B.S. 1947).

As if this level of vulnerability was not enough, German officials chose to work towards a massive output of 11 million tons annually by 1944 of synthetic fuels in large centralized production plants. These synfuel plants were also heavily dependent on electrical power and coal. The three major oil products were aviation gasoline, motor gasoline and diesel oil, all critical to the German war effort. The synfuels industry was concentrated near the major coal mines in the Ruhr Valley, and thus were susceptible to energy attack.

Because the Allies' prime targets were initially strategic military facilities, they failed to take advantage of Germany's energy vulnerability until very late in the war. When the Allies finally did destroy Germany's main synthetic fuel and electricity producing plants, the German war economy was essentially incapacitated.

The history of the Allied bombing attacks upon Germany in World War II demonstrates that with the growing interdependence of energy intensive economies, the more concentrated and centralized the energy sources, the more vulnerable the economy to an attack. The instigation of attacks upon the energy production and transportation systems brought rapid and excessively damaging results, particularly with the attacks upon the

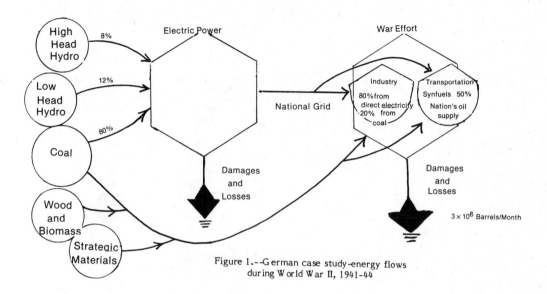

Figure 1.--German case study-energy flows
during World War II, 1941-44

means to transport coal and produce synthetic fuels.

Japanese Example

Figure 2 points out that Japan, on the other hand, had a very decentralized energy network during World War II. According to the U.S. Strategic Bombing Survey (Pacific), "the electric power system of Japan was never a primary strategic target" because most of the power requirements of Japan were "so numerous, small and inaccessible that their destruction would have been impractical, if not impossible" (U.S. Strategic Bombing Survey, 1945).

In 1944, the total generating capacity on the home islands was 10,120,000 kilowatts (10,120 MW). Generation in the peak war year, 1943, was 38.4 billion kilowatt-hours from all sources including utility, railway and industrial facilities. Water power from small hydro plants provided 78% of the total electricity in the system, with the remainder of use supplied by steam plants (mostly antiquated coal plants). During the war the largest hydroelectric plant in Japan was a 165 MW plant on the Shinanogawa River. This plant supplied only 2.7% of the annual electrical consumption (U.S.SBSP 1945).

Yet Japan had another problem. While their power system was dispersed, their total power supply was not substantial and they were unable to increase their over-all electrical sytem when it became apparent that they should do so. However, the U.S. electrical war economy grew at an annual rate of 33 percent (compared to the Japanese electrical system growth of only three percent per year). The Strategic Bombing Survey points out that "Japan could, with relative ease, have increased her production of kilowatt hours over the 1943 level --so far

as the capability of her predominantly water-driven generation system was concerned". However, supplies of necessary materials were diverted to the war effort, rather than increasing the size of the electrical system.

We are seeing renewed interests in Strategic Materials, with the USSR mounting a major effort to divert the flow of strategic materials into the U.S. It has added a new level of sophistication to political warfare.

From a military perspective however, Japan's decentralized energy system did not offer the Allies the kind of prime target that Germany did, as reflected in the formal conclusions of the Allied Bombing Survey:

> Most of the power requirements of Japan, however, come from hydro-generating plants, which are so numerous, small and inaccessible that their destruction would be impractical, if not impossible. If their supply could be eliminated or drastically curtailed by some other means, electric power supply could be reduced to a point where the shortage would assume economic importance. It has been shown that neither the transmission nor the distribution system is, of itself, vulnerable.

Korean Example

Korea's energy system during the Korean conflict is depicted in Figure 3. Much like the Japan case study before (fig. 2), Korea relied heavily on hydropower. Yet unlike Japan, Korea did have a few large plants geographically clumped for easy targeting. Since World War II, power plants and electrical facilities have be-

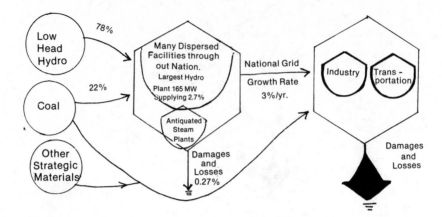

Figure 2.--Japanese case study-energy flows
during World War II, 1941-44

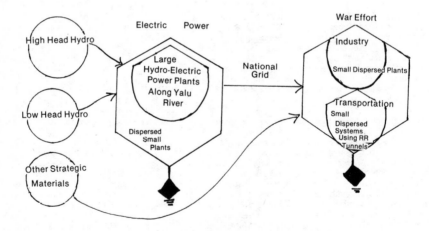

Figure 3.--Korean case study-energy flows
during Koren Conflict, 1950-52

come prime targets. During the Korean war, the U.S. made any early decision not to bomb large hydroelectric dams along the Yalu River, but reversed the decision two years later in 1952. As Bennet Ramberg points out, "the decision was reversed . . . when negotiations deadlocked and destruction of the plants seemed necessary to hasten the war's conclusion and to make more difficult the repair work the Communists were doing in small industrial establishments and railway tunnels" (Ramberg 1980).

The energy diagram (fig. 3) graphically shows how the energy flows were sufficiently strong during the first few years of conflict. Yet as soon as the U.S. decided to target the crucial facilities, centralized in nature and location, the conflict ended with the loss of power.

Middle East Example

War in the Middle East causes problems for two reasons. First is the unrest and destruction to the warring nations. Second, when the principle oil supply nations of the world are engaged in war, the importions in the rest of the world may feel the war ripples through oil supply shortfalls. Figure 4 summarizes these two points of vulnerability.

In the Middle East, during the 1973 war, Israeli warplanes bombed power stations at Damascus and Homs, Syria, "to subdue Syrian military activity and to deter other countries from entering the conflict" (Calder 1978).

Power plants and oil refineries have been targeted, most recently during the 1980 war between Iran and Iraq. The Abadan oil refinery complex at Kharg Island was bombed. The lesson in vulnerability in the Middle East affects the entire industrial world, as critical oil supplies must pass through the narrow Straits of Hormuz which is currently threatened by military actions.

In fact, the Persian Gulf war may prove to be a threatening indicator to the future, as most primary energy targets, ranging from refineries to key oil fields to the Iraqi nuclear research center, Tuwaitha, were selected for bombing forays. The September 30, 1980 attack on the French-built Osirak and Isis research reactors of the Tuwaitha facility raised the spectre of radioactive fallout from conventional bombing. Although officials of the Nuclear Regulatory Commission contend "that there (is) very little risk that bombing a research reactor would ever cause a significant fallout problem" a worst-case scenario allows for radioactive pollution to spread at least a mile or two from the reactor (Marshall 1980).

"Bombs, presumably delivered by Iranian pilots, hit the research site about ten miles from the center of Baghdad. They damaged an auxiliary building and forced the French technicians working on the project to leave. The attack did not damage the reactors, but it did shut the program down indefinitely" (Marshall 1980).

The only missing element in this Middle East duel was the presence of nuclear tipped warheads.

Current U.S. Energy System: Overview

The U.S. energy system has been the subject of discussion in research, political and economic sectors since 1973. Despite the various complexities of the system, there are a few generalities which adequately convey the essence of the system for purposes of this paper.

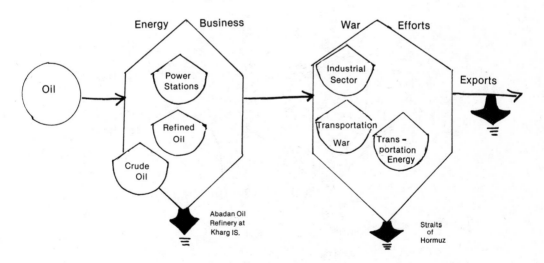

Figure 4.--Iran case study-energy flows of
Middle East War, 1979-80

First, the U.S. energy system is highly centralized, relying primarily on oil and gas for most of its energy needs, with approximately 50% supplied from foreign imports. Next, the back-up supply options, namely coal, hydropower and nuclear, are also highly centralized. The problems associated with oil supply vulnerabilities are not eliminated through the use of the other fuel supply options. The coal option requires as much as 30% of the mining and delivery energies to be supplied by diesel fuel, and the nuclear option is equally consumptive of oil and gas in the enrichment and numerous transportation phases of the uranium fuel cycle (fig. 5).

In general, the production of electricity for consumer use depends on an increasingly centralized system of large generating plants, which in turn depend on other centralized systems of fuel production, transportation, refining and storage systems.

Second, a physical aspect of centralization is that it is accompanied by energy intensity, which is further accompanied by thermodynamic efficiency losses. It takes a lot of energy just to maintain the systems in place, much less adding maintenance pressures from any energy growth. The equivalent of about 1.6 million barrels of oil per day are used for electricity generation. Of that, 1.1 million are irretrievably lost due to thermodynamic losses. This loss is an amount equal to about 17% of all oil imports to this country. With the current political situation in the Middle East countries, and the U.S. historical record of energy consumption, losses of any kind become the subject of discussion and debate in conference rooms across the country.

In this paper, we focus on the major energy contributors to the current U.S. energy scene, namely oil, gas, coal, nuclear and hydro. A summary energy diagram of the roles these fuels play in the general energy picture is depicted in figure 5. Next, each major fuel is discussed separately with accompanying energy flow diagrams to point out the major positions of vulnerability for each.

Once the baseline for vulnerability is established, we will offer several options for consideration in lessening the vulnerabilities of the U.S. energy system. We are not offering these options because we believe the U.S. took an inappropriate path in its energy system development. To the contrary, we recognize the logic in the choice of constantly replacing diluted fuels with more concentrated ones. The discussion is offered in recognition that using oil and gas was a wise path, which now needs to be followed by some other wise paths.

Oil

Oil is the predominant fuel in the U.S. and together with natural gas, supplies almost 74% of the total U.S. energy needs. It fuels transportation, is converted into electricity, heats homes, powers industry, and is also an important raw material in petrochemicals which are ubiquitously found from medicine to agriculture, space ships to baby bottles.

What the Arab oil embargo provided for the U.S. was a glimpse of the dependency we have developed on oil. The sites for energy consumption tend to remain predictable due to population, weather, industrial activity, etc., yet the sites for energy supply are not as predictable, and we have found may change rapidly over a short period of time.

The U.S. oil production system is predominantly concentrated in Texas and Louisiana, Alaska and California, and Wyoming and New Mexico. The movement of oil in the country is primarily from the East to the West.

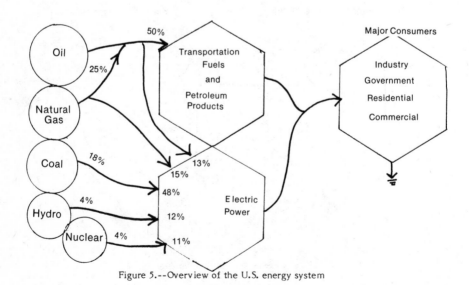

Figure 5.--Overview of the U.S. energy system

Major new sources of domestic crude oil appear to be in Alaska and on the Atlantic and Pacific Outer Continental Shelves. Despite the debate of 'how much', 'where', and 'when', the experts agree that the supply of U.S. oil is not without limits, and that we reach our limit anywhere between one and seven decades from now.

Figure 6 points out that Texas is the largest U.S. oil producer, totalling more than a billion barrels of oil in 1979. Alaska produced about half that much that year, with Louisiana running a close third, and California resting in fourth place for production. Of the top producers the greatest hope is in Alaska, and the Alaska reserves are estimated by the U.S. Geological Survey to be eleven billion barrels (recoverable). If Alaska were our sole source of oil, and we used it at current rates, it would last not quite two years. Furthermore, the offshore resource estimates are placed at about six billion barrels, about one year's supply at the current consumption rate.

It appears that the international sources of oil will assume even greater importance as domestic sources are depleted in the coming years. That alone signals vulnerabilities as we have seen in the Middle East war example. Current major contributors to the U.S. oil supply include the Middle East as the largest contributor, followed by South America, Africa and the Caribbean, Europe, Asia, Canada, Central America and Australia for crude oil, and the Caribbean (mostly Venezuela) and Europe for refined oil.

Figure 6 points out the delicate balance between domestic and foreign sources, and further points to the major levels of consumption:

1. Direct consumption of 4% minimally in the refineries;
2. Consumption due to transportation of oil by pipeline to the Southeast, North Central and California, and by water to the Northeast and to Florida. Additionally, transportation within states is accomplished by trucking, also using oil.
3. The major consumers of all the oil products in the U.S. economy are:

   --California and New York in the top consumer category
   --Texas
   --Pennsylvania
   --Illinois
4. The top consumers of gasoline include:
   --The same top 5 in the overall category with the addition of Ohio to the list.
5. The top consumers of distilled fuel for home heating include:
   --New York
   --Pennsylvania
   --New Jersey
   --Massachusetts
   --Texas
6. The top consumers of residual oil used in power plants for producing electricity include:
   --New York
   --California
   --Florida

If the recent and reoccuring economic woes of New York could be tied to any one thing, it might be the

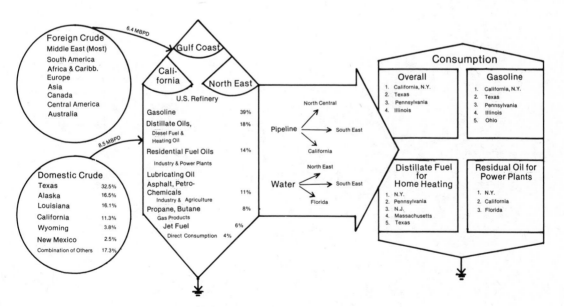

Figure 6.--Overview of the United States Oil System.

strong dependence on oil. After all New York might just be an exaggeration of the centralized and energy-intensive statement of the whole energy system.

Natural Gas

Natural gas would be the star of the energy system if it was as abundant as oil has been. Gas offers some positive things that oil does not, namely it is clean burning, and is even more efficient than oil. The main difference leading to second billing for natural gas is that it does not provide such a versatile transportation fuel.

However, natural gas provides almost 25% of the nation's energy supplies, and the U.S. production is even greater than domestic production of oil. Domestic production of both peaked in the early 1970's, and the ensuing decline has resulted in increasing curtailments of interstate natural gas commerce, even though there is still substantial interstate commerce in producing states (CRS 1977).

The major natural gas reserves in the world are found in the Soviet Union (31.8%) and the Persian Gulf (26.2%). The U.S. reserves are estimated at just under 10% of the world's total.

Figure 7 points out that as in the oil example, Texas and Louisiana provide a high percentage of the domestic supply. In fact 70% of the marketed domestic gas was supplied from Texas and Louisiana. Oklahoma, Kansas and New Mexico provide 19% of the domestic supply, and a fraction is provided through coal gasification.

To supplement the domestic production, we are importing Liquified Natural Gas (LNG) from Algeria, Lybia and Indonesia. Again we find ourselves in a similar situation as the oil dilemma, in that we are a major natural gas consumer, but we do not control a major portion of the reserves, and must seek foreign sources just to break even.

A major step in the natural gas centralized system is the elaborate storing system, primarily for economic reasons. Domestic gas is transported primarily by pipelines to underground storage facilities and underground aquifers. Then it is transported by trucks and pipelines to "peak shaving plants" or satellite facilities. Peak shaving plants store LNG that gas companies buy at low summer rates and then resell at times of peak winter demand. Storage facilities that store LNG, employ double-walled insulated tanks to keep the LNG at its required low temperature.

Figure 7 further points out that once distributed, the major consumers of natural gas include:

1. The natural gas producing states themselves, with Texas, Louisiana, Oklahoma, Kansas and New Mexico consuming 40% of the domestic supplies.
2. The major sectors of consumption ranked in descending order include:
   --Industry, with almost half of its supply needs filled by gas;
   --Residential and Commercial with states varying in their uses;
   --Utilities powering electrical production.

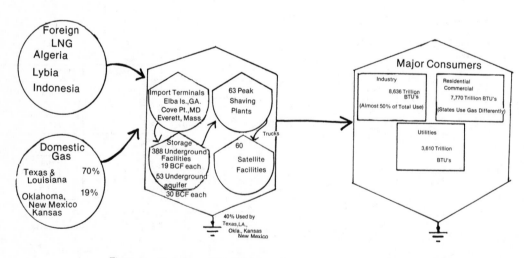

Figure 7.--Overview of the U.S. natural gas system

Coal

Coal, once the premiere fuel for U.S. consumption, has been replaced by oil and natural gas, but is touted to be the most available and abundant domestic resource we have to bail us out of the oil dilemma in which we have recently found ourselves. Energy planners across the nation are looking to the nation's tremendous domestic reserves of coal to be the primary source of energy for the nation's electric power plants--for new plants, and to replace oil in some plants. Yet it has become apparent that the increased use of coal will not be without a number of attendant environmental and health problems.

Nevertheless, the U.S. government, through several important pieces of legislation is encouraging utilities and private industry to convert from the use of oil and gas to the use of coal. The Power Plant and Industrial Fuel Use Act requires utilities to use fuels other than oil or gas in new utility boilers after 1990 (FEA 1976). This trend is not expected to change with the Reagan administration, but is expected to increase.

The notion of using coal as a way of getting away from the oil dependence may rest in faulty logic, however. Figure 8 depicts the national coal system. Although coal may be abundant, the use of coal is heavily dependent on both oil and water. Oil in the mining and transport of coal, and water in the mining primarily (Ballantine 1976). The reliance on oil and water is not a casual one, in fact it is as much a part of the coal flow system as is the coal in the ground. Because coal is concentrated geographically in the Appalachians, Pennsylvania, Wyoming and Montana, distribution of the resource is an important and elaborate step in the overall coal system. In fact, about 70% of the coal in the U.S. is located west of the Mississippi, but much of the consumption found east.

Besides the obvious vulnerabilities of oil and water dependence, the concentrated nature of coal mining due to the location of the resource, we are just beginning to grasp the environmental and health vulnerabilities of coal mining and consumption. And certainly the problems the railroad has encountered in recent years contributes to the vulnerability picture, as much of the broader distribution to states is by rail, and a sizeable portion of the coal lands are controlled by the railroads.

As would be expected, the major U.S. consumers of coal already include:

1. the electric industry
2. the steel industry
3. other metal industries
4. chemical industries.

Yet despite its problems, coal remains an important part of the future as both a domestic source, and for export potentials.

Nuclear Power

The Three Mile Island incident is still fairly fresh in the public memory, and certainly continues to be a contributor to the nightly news shows. The potential of tremendous power contributions from nuclear reactors espoused some 30 years ago has not been realized. Now with recent public sentiment somewhat skeptical about the safety and health aspects of nuclear power, the debate continues to rage between strong advocates and opponents, with little change in the percent contribution that nuclear plays in the overall U.S. energy picture.

Nevertheless nuclear power is still considered an option for the U.S., and some states rely heavily on nuclear power, such that energy expansion would most assuredly mean more nuclear. The question still remains

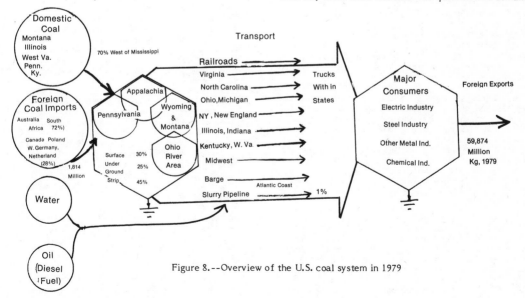

Figure 8.--Overview of the U.S. coal system in 1979

as to whether a stronger move to nuclear would decrease the U.S. dependence on oil and natural gas for electric power generation.

Figure 9 points out that there are several aspects of the nuclear power system that currently rely heavily on oil, and if the motivation toward more nuclear is to offset oil dependency, then our nuclear planners need to take a closer look into the whole system.

proponents continue to profess the efficiencies of nuclear power. Figure 9 displays some of the key energy intensive process, not all of which are included in the normal energy analyses.

The basic processes inherent in the nuclear fuel system include mining, milling, conversion, enrichment, fabrication, consumption in power plants, and waste disposal. These processes are summarized in figure 9.

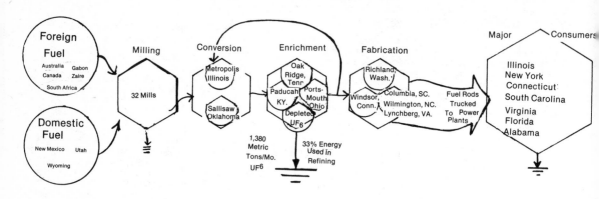

Figure 9.--Overview of the U.S
nuclear power system

First is the issue of nuclear fuel source availability. The advantage nuclear fuel provides is that it is so concentrated that one train unit of nuclear fuel would equal the fuel content of 3,000 train units of an equivalent amount of coal. That alone is an attractive feature of nuclear power. Yet as will be discussed later, that extreme concentration factor has some costs associated with it.

Despite the attractive concentrated features of the fuel, the quantity of domestic uranium is not sufficient to supply the existing power plants. The U.S. Geological Survey estimates that the domestic reserves of uranium that the U.S. now has could supply only 15% of the uranium that the U.S. plans to use between now and 2000. According to Frank C. Armstrong of the U.S.G.S., the U.S. would need between 1.6 and 2.0 million tons of uranium over the next 25 years, yet U.S. domestic production could supply less than 1% of that need (USGS 1975).

Second is the energy intensity of the nuclear fuel cycle. Various studies have indicated that it would take between 13 and 25 years to achieve a positive energy flow from a nuclear power plant, without considering the energy costs associated with decommissioning or long term waste disposal (Anon. 1975; Kylstra et al. 1975). There continues to be much debate on this issue as many

The major domestic sources of fuel are New Mexico, Wyoming, and Utah, supplemented by foreign sources from Canada, Australia, South Africa, Zaire, and Gabon. Once mined, the uranium ore is shipped to one of thirty-two milling sites in the U.S. Each mill can process 800,000 tons per month yielding 1,320 tons of yellow cake per month. Combined, the mills transport 42,240 tons per month to one of two conversion which transform the milled uranium into Uranium hexaflouride (UF6). Each of these processes require electrical power and oil products in order to properly begin the preparation of uraniun for final consumption.

The two conversion plants (Metropolis, Illinois and Sallisaw, Oklahoma) combined can produce about 1,380 metric tons of UF6 per month. The UF6 is shipped to Paduca, Kentucky for initial enrichment and then on to Oak Ridge, Tennessee (mostly) or Portsmouth, Ohio for full enrichment. These three enrichment facilities are all governmentally owned and operated gaseous diffusion plants which consume one-third of the energy that goes into producing refined uranium fuel. This highly energy intensive process is subsidized by the federal government and points out the vulnerability of the Oak Ridge facility.

Enriched UF6 is shipped to a full assembly fabricator, where it is made into pellets that are inserted into fuel rods. The first step in this process is the conversion of UF6 to uranium oxide (UO2), again an energy consuming step. There are five fabrication plants, one each located in Richland, Washington, Windsor, Connecticut, Columbia, South Carolina, Wilmington, North Carolina, and Lynchberg, Virginia, with the majority of fabrication occurring in the latter three sties.

Once fabricated, uranium fuel fully concentrated by the energy intensive preparatory steps, is shipped out to the nation's power plants for production of electricity. The states consuming the majority of the nuclear power includes Illinois, New York, Connecticut, South Carolina, Virginia, Florida and Alabama.

Besides the known energy intensities and vulnerabilities of the nuclear power system, two large unknowns still exist. The first is the monetary and energy costs associated with decommissioning nuclear power plants, and the second is the costs associated with nuclear waste disposal. Despite the lack of experience on these two additional processes, it is clear they will also be significant energy consumers.

Hydropower

Hydropower is one of the most efficient power sources and is a completely domestic resource. The combination of low head (less than 16 feet) and high head (greater than 16 feet potential) contributes 4% of the nation's energy needs, or 12% of the nation's electrical needs. This makes it equivalent with nuclear power in its contribution.

Figure 10 summarizes the features of the hydropower system, citing its localized and limited supply, primarily found in Washington, California, Oregon and New York. There have been some recent investigations into the contributions of nationally dispersed low head hydro systems, but the overall scope of net energy, environmental impacts, etc. is still not completely settled.

Alternatives

It is widely recognized that concentrated, clean energy sources are becoming harder and harder to find. The old sources are reaping lesser and lesser net yields resulting from more production energy going into digging deeper and deeper and hauling it over greater distances. Thus the quest for additional fuel sources remains a constant and pressing one.

Primarily there are two general options available to the U.S.:

- developing more centralized systems, capitalizing on the energy structure already in place, or
- development of decentralized energy systems also capitalizing on the centralized ones in place, but not in the same sense as in the expansion of those or similar energy systems.

The terms "centralized" and "decentralized" are not distinct terms, and overlap with the terms "large scale" and "small scale". For purposes of this discussion we will define centralized systems as those that depend on a relatively small number of larger components. Conversely, dispersed systems would imply a number of small components contributing to the entirety of the system.

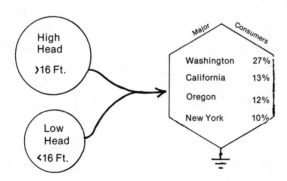

Figure 10.--Overview of the U.S.
hydropower system

Since we are investigating issues of vulnerability, and the issue enjoying almost national agreement is the supply shortfall possibilities due to either political or war events in foreign countries, or due to production problems domestically or abroad, the first options we will present are the centralized ones, most directly focusing on the systems in place.

Centralized Options

Seven centralized options are briefly presented including four using coal, and three using oil shale:

Coal Synthetic Fuels
Gasification
Natural Gas
Methanol
Oil

Oil Shale
True In-Situ
Modified In-Situ
Above Ground Retorting

Figure 11 summarizes the centralized options employing coal. It must be remembered that for a full energy picture, the processes discussed in figure 8 must be added on at the front end of Figure 11. The already energy intensive aspects of coal use are increased by each of the four options in figure 11, but coal is considered the most abundant domestic energy resource, and there is a super structure in place in the front end, upon which to capitalize.

The types of synthetic fuels being considered in the U.S. energy system, include the following types, anticipated costs and time frames.

Table 1.-- Costs of Synthetic Fuels from Coal

Technology	Commercially Available	Estimated Cost $/million BTU*
Low-Medium BTU gas	1985	4.40 - 6.60
Natural Gas	1985	5.50 - 8.80
Methanol	1986	6.92 - 7.90
Oil	1990	5.50 - 7.70

*1980 dollars; does not include transportation costs of fuels (Energy and Defense Project 1979).

Despite the substantial capital investments, the energy investments need to be considered, and judging from our initial analyses of the current energy systems, the net yield potentials of these options might be in jeopardy. We will continue our discussion of these options, keeping in mind that the net energy questions remain unanswered due to the lack of precise information about system dynamics of these new ventures.

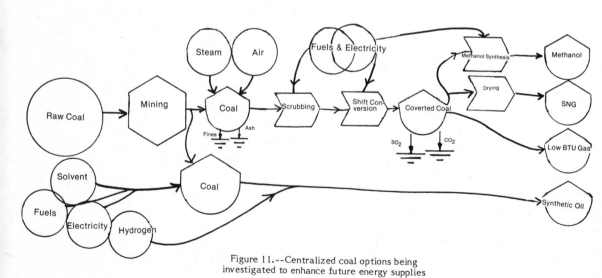

Figure 11.--Centralized coal options being investigated to enhance future energy supplies

Low and Medium BTU Gas derived from coal may take one of several technological paths: the Lurgi, Kopper-Totzek, and Winkler first generation systems; and the Stagging-Lurgi, Texaco, and U-Gas systems. In general they all start with the partial oxidation of coal in the presence of steam and oxygen. The gas produced contains combustible components including carbon monoxide, hydrogen, and methane, as well as noncombustible gases such as carbon dioxide and sulfur compounds (CEC 1978, 1980; U.S. DOE 1978a).

Before it can be used the gas must be cleaned to remove impurities such as hydrogen sulfide and carbon disulfide. Depending upon the particular processes used, carbon dioxide may be reduced, yet gasification produces solid and liquid wastes which require disposal.

There are several first generation processes in commercial operation in different parts of the world, but not in the U.S. Second generation processes are being developed to improve efficiency and operation flexibility. Commercial suppliers currently exist for the first generation gasifiers such as the Lurgi and Koppers-Totzek, but not as yet for the Texaco process.

The cost estimates shown in table 1 are conservative, in that they do not include transportation costs. As we have seen in figure 8, the transportation subsystem of the coal system is rather elaborate, and energy intensive.

Synthetic natural gas (SNG) is a potential substitute for natural gas and may be used in conventional systems. SNG is a methane-rich, high BTU gas which has had many of the impurities removed. Its heating value is approximately 1000 BTU per standard cubic foot (3,200 BTU per standard cubic meter). It can be produced from coal, agricultural and lumber residues, municipal solid waste and many other organic materials. We will focus on coal in this discussion since it is the most concentrated and available.

The major steps in processing SNG from coal are illustrated in figure 11. First the coal is gasified to produce medium BTU gas. Then the medium BTU gas must undergo a shift-conversion process to increase its hydrogen concentration in order to achieve the appropriate hydrogen-carbon ratio for producing methane (CH_4), which is the primary constituent of SNG. In the final methanation step, the gas is reacted catalytically to form methane from carbon monoxide and hydrogen. After drying, the resulting high BTU SNG is ready for on-site use or for transmission by pipeline. Pilot demonstrations of catalytic methanation to obtain SNG from medium BTU gas from coal have been performed in Scotland, South Africa and Austria.

Despite the commercial viability of plants located elsewhere in the U.S., there remains additional problems in integrating coal gasification and SNG production components into a working commercial scale system.

Methanol, a liquid fuel derived from coal and other organic materials such as wood and petroleum, could be used as a fuel in conventional utility and industrial systems as well as in the transportation sector. Methanol (CH_3OH), like SNG, can be synthesized by catalytically reacting medium BTU gas produced by any coal

gasification process which produces $CO/H2$ mixtures.

The synthesizing of methanol from medium BTU gas is a well-proven technology. At least two companies (Imperial Chemical Industries and Lurgi) offer proprietary processes with guarantees backed by multiple commercial-scale plant operating experience. A small sub-commercial-scale plant using the Koppers-Totzek gasification process and the Imperial Chemical Industries methanol process has been operating at the Modderfontein, South Africa plant for over two years (CEC 1980). Yet a commercial sized methanol from coal plant does not currently exist.

Based on conservative estimates, it would appear that a commercial sized plant would probably use at least 5,000 tons of coal per day. With that amount of coal in operation it would require additional clean ups to avoid poisoning the methanol synthesis catalyst.

America's thirst for oil has led to the investigation of producing synthetic oil from coal. The transportation fuels are particularly dependent upon a constant supply of oil to the refineries. There are several potential coal hydroliquifaction processes that could be used to produce synthetic oil for conventional applications. In the hydroliquifaction process, coal is dissolved in an appropriate solvent, then reacted with hydrogen to produce liquid fuel oils. The resulting synthetic fuels could substitute for residual fuel oil or distillate fuel in conventional systems as described in figure 6.

Although there are no plants in operation now, Germany used the Bergius process for catalytic hydrogeneration of coal to make approximately 90% of its aviation fuel in World War II. All modern hydroliquefaction process are descendants of this process.

There are significant technical differences between synthetic oil production and the gasification process which lead to questions regarding the long-term desirability of developing the liquefaction technologies. Since dried coal is required for the hydroliquefaction processes, use of western coals with higher moisture content may severely reduce the thermal efficiency. Additionally, the coal liquefaction process does not function well with western coals due to their high oxygen content, high alkalinity and low sulfur levels, and because the high oxygen content results in massive consumption of process gas. The high alkalinity interferes with catalytic reactions, and the low sulfur levels inhibit the dissolution of the coal in the solvent. These suggest that the cost of liquefying many western coals would likely be higher than for eastern coals (CEC 1980).

Synthetic fuels from oil shale may be grouped into one of three general type processes:

1. True in-situ (TIS) processes in which the oil shale is left underground and is heated by injecting hot fluids.
2. Modified in-situ (MIS) processes in which a portion of the shale deposit is mined and the rest is fractured with explosives to create a highly permeable zone through which hot fluids can be circulated.
3. Above ground retorting (AGR) processes in

which the shale is mined, crushed, and heated in vessels near the mine site (CEC 1980).

The dawning of the TIS process included Armand Hammer discussing its effectiveness on national television on the "Good Morning America Show" about a year ago. The issue in question revolved around the use of water, with particular concern for the limited water supplies in the west. The TIS process can be generally described with the following procedural steps:

1. Dewatering, if the deposit occurs in a ground water area;
2. Fracturing or rubbling, if the deposit is not already permeable to fluid flow;
3. Injection of a hot fluid or ignition of a portion of the bed to provide heat for pyrolysis; and
4. Recovery of the oil and gases through production wells.

Figure 12 summarizes these steps, and while they continue to be costly processes, the net energy implications are not known for sure. This technique is substantially less energy consuming than the original AGR system, but a closer look should include a complete net energy computation, as the AGR is known to be a significant net loser (Gardner 1978).

In the MIS process, some shale is mined from the deposit, then explosives are detonated in the remaining deposit to increase the permeability of the oil shale. This procedure creates a chimney-shaped underground retort filled with broken shale. Access tunnels are sealed and an injection/hole is drilled from the surface to the top of the fractured shale. The shale is ignited at the top by injecting air and burning fuel gas, and heat from the combustion of the top layers is carried downward in the gas stream. The bottom of the oil shale is pyrolyzed and oil vapors are swept down the retort to a sump from which they are pumped to the surface. The burning zone moves down the retort fueled by residual carbon in the retorted layers. When the zone reaches the bottom of the retort, the flow of air is stopped and combustion stops.

The MIS processes are more advanced than TIS methods. The principal advantages of MIS are that large deposits can be retorted, oil recovery ratios are high, and relatively few surface facilities are required. These would all add to the net yielding capabilities of such an operation. However, some mining and disposal of solid wastes on the surface are required and oil recovery per unit of ore processed is low relative to above ground retorting methods. In addition, the burned-out MIS retorts have the potential for ground water pollution (U.S. DOE 1978b; OTA 1978). Again the net energy implications of the total process need to be investigated with the new techniques currently being tested by the Occidental company.

The AGR process differs from the in-situ processes in that all the shale feedstock is mined. The principal advantage of AGR is the oil recovery efficiency is increased. There are four classes of operation, each varying with heating transfer technologies and efficiencies. This process has undergone substantial study both in and out of industry, and the net energy aspects continue to indicate large net losses due to the energy intensity of the mining operation.

Dispersed, Decentralized Energy Option

In addition to the centralized options reported, there are numerous dispersed and decentralized energy options available for the U.S. In this paper we will present only a few characteristic examples of decentralized options, focusing primarily on renewable resource

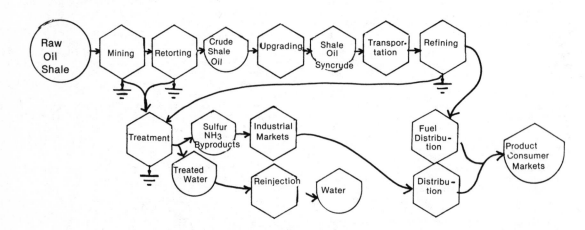

Figure 12.--Oil shale options currently being tested for future energy supplies

options. The next several energy diagrams present in some cases net energy calculations of the options, and are presented in the same general format as the other cases, with the addition of some calculations. The cases to be presented include:

conservation
cogeneration
solar water heating
solar electric
biomass
wind

Conservation was a main feature of the Carter administration's energy program, and the data is just beginning to come in on its effectiveness. Again, Stobaugh and Yergin broke the code regarding energy conservation as a new "source" of energy. Their recent summary of the significance of conservation highlights the significance of moving rapidly to implement an aggressive energy conservation program:

The telescoping of the energy emergency in 1971 has greatly increased the urgency of early action. As things stood in 1978, and given the decision now made to decontrol oil prices, we might have hoped to continue with 'business as usual' on energy conservation, allowing higher prices to work through the economy and gradually cause us to increase energy efficiency . . . .

In current circumstance, however, such a course will not be adequate. The gap between energy resources and energy demand would be closed by 'unproductive conservation'-the shutting down of factories, higher unemployment, higher inflation, offices too warm in the summer for efficient work, colder houses, a choice for some between food and fuel . . .

Far more desirable is the alternative of accelerated energy efficiency. Our whole industrial system is like a vehicle built to operate on $3 oil, puffing along with an inefficient engine and with a body leaking vast amounts of energy. Each drop wasted drives higher the price of future oil purchases . . . (Stobaugh and Yergin 1981).

This investigation of conservation will include three broad areas: residential use, industrial use, and transportation energies.

Residential energy use accounts for about 20% of the energy consumed in the U.S. (Yergin 1979). Figure 13 (a,b,c) presents three net energy diagrams which ultimately shows that the energy consumed in the production and installation of typical house insulation yields a positive net ratio of 60 units of energy saved for each one unit consumed for an average single family home for a family of four. That is the highest net yield recorded, with the high quality geothermal energy close behind it, yet conservation is already in place, and occurring more and more every day.

Table 2 below shows the percentage breakdown of residential energy use, pointing out the high proportion used for space heating (or cooling in the hotter, more tropical climates). Thus the insulation energy calculations are even more impressive considering this large space conditioning use.

Besides the more easily quantified insulation measures, there are numerous retrofit options which would stimulate the building economy, and would vary with different types of housing structures and their deficiencies.

Table 2.--Residential Energy Consumption, By Use (Yergin 1979).

Use	Percentage
Space Heating	53
Hot Water	14
Cooling	5
Air Conditioning	7
Other	21

Industry accounts for 39.5% of total U.S. energy consumption (Yergin 1979). The industrial sector has made the greatest progress in energy conservation, due to their direct profit losses from energy losses. Some examples include Lockheed with a reduction of 55% of its energy demand between 1927 - 1977 in its Los Angeles factory complex at little or no capital expense. In its U.S. refineries, Exxon reduced energy use by 21 % during the same period -- with 80% of those savings developed with little or no capital invested. Just for Exxon alone, the savings are equivalent to 11.3 million barrels of oil per year (Yergin 1979).

It is hoped that the deregulation of oil will not turn industry's efforts away from its significant conservation efforts. For it is in a highly centralized and energy intensive system such as many industries represent, that the easiest conservation efforts may occur.

The transportation sector accounts for 26% of the total U.S. energy consumption, with automobiles accounting for over half that amount. The dispersed settlement patterns characteristic of the U.S. indicate that the auto will remain a focal point for conservation efforts for some time. The most viable conservation targets can be met with reduced driving speeds and increased automobile efficiency. Some conservation might be attained through the development of efficient, small mass transit systems such as buses and van pools.

The major gains in auto efficiency have been the results of weight reduction and the importation of foreign technology. Yergin points out that "substantial technological innovation is needed in materials, engine and design; and this kind of innovation, as opposed to styling, has not been a major priority for the industry or its suppliers. Massive capital investment is needed over a decade for the four U.S. automobile companies, which will increase vehicle costs" (Yergin 1979).

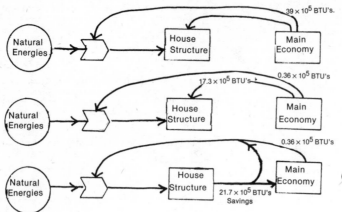

(a) Uninsulated house consumes 39×10^5 BTU's

(b) Insulated house: by adding 0.36×10^5 BTU's worth of insulation materials into a house, the fuel consumption drops down to 17.3×10^5 BTU's

(c) So for an investment of, 0.36×10^5 BTU's in insulation, society reaps a savings of 21.7×10^5 BTU's.

Figure 13.--Net energy analysis of housing
insulation, compared with fuel saved
(Brown and Arrington, 1976)

Cogeneration is a term which has two meanings: one is specific, the other general. The general definition is the use of "wasted" energy from a power producing system to be used in another process. The specific definition is:

- An efficient method of producing electricity in conjunction with process steam or heat and can be fueled by renewable resources.

A flow diagram is presented in figure 14 showing the key features of cogeneration in conjunction with electrical production. There are several applications of cogeneration in the U.S. including the following:

Industrial
Chemical plants
Oil refinery
Oil fields
Food processing plants
Wood products
Steel plants
Cement plants

Municipal
Solid waste
Sewage treatment
Hospitals
Universities and schools

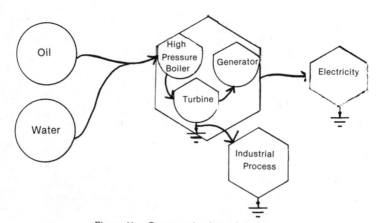

Figure 14.--Cogeneration in conjunction
with electrical production

Residential
Large apartment complexes
Modular integrated utility systems

Cogeneration is accomplished most often by using conventional steam turbines, combustion turbines, diesel engines, or other generation systems in what is known as "topping cycle," or as in the case of industrial waste heat, in a "bottoming cycle" (CEC 1978). The "topping cycle" uses various boiler-turbine configurations to generate electricity and then makes use of the valuable waste heat from steam for other processes. Another basic cogeneration system uses the back-pressure steam turbine. In a conventional steam turbine generator, steam is exhausted from the turbines into a condenser at a very low temperature and pressure.

The fuel savings derived from the cogeneration facilities are significant, and they are cropping up in many industries across the country. Additionally, there are new processes being tested using fluidized bed, combined-cycle technologies, and metal turbines.

Current savings potentials are estimated by the Department of Energy to be 6.15 quadrillion BTU per year of energy by the year 2000. Dow Chemical suggests that reduction or relaxation of governmental and institutional restraints to cogeneration would result in industry's generation of 71,105 megawatts of power by 1985. This amount is equal to 1.45 quadrillion BTU annually, or roughly the equivalent of 680,000 barrels of oil per day.

Solar energy has been the focus of much intensive debate surrounding the whole centralized versus decentralized energy systems discussion. The main features of the debate have been economics, yet there have been a few discussions regarding the potential net energy implications of the various solar options. The field continues to change rapidly with more sophisticated solar applications, but there are a few applications that have not changed significantly in over forty years of use. Solar hot water heating is one such use.

For this discussion, we focus on a simple active solar heating system. The system is summarized in Figure 15. It uses a flat plate collector to heat air or liquids which are circulated to a heat exchanger directly to the point of use or indirectly via rockbed, water or eutectic salt storage (Howell and Bereny 1979; Antolini 1978; Barnaby et al. 1978).

The maximum orientation for solar gain is due south, and so the collectors are placed either on structures or nearby on the ground with a due south face. In most systems a back-up is usually included for when storage temperature drops below room temperature.

Despite the somewhat questionable net energy aspects of solar hot water heating, the substantial displacement of oil is reason enough to consider its application in residences across the nation, not to mention the dispersed but significant economic stimulation to the building, plumbing and carpentry industries that would result.

Solar electric potentials currently include:
- Solar Thermal Power Systems
- Salt Ponds
- Photovoltaics

Of the three, the Salt Pond method is the most promising at the present time. Salt gradient ponds are natural phenomena that can be artifically created and used as sources of heat, electricity or both. A rule of thumb is that they can produce about one megawatt of electricity per 50 acres of pond area (Holland 1979). The optimal economics for electricity-producing ponds is probably between 12 and 60 megawatts, although small ponds at about 5 megawatts can be constructed without a large cost penalty. The ponds are inexpensive to construct compared to other energy sources. They are very stable under conditions of environmental stress, require relatively short lead times for constructing, can start up power generation on only a few minutes notice, are modular by nature, use a low-maintenance, proven technology to generate electricity, and present very limited environmental problems. Figure 16 shows the energy flow aspects of the solar pond.

The Israelis were the first to experiment with solar ponds and began operating a 150-kw pilot solar electrical power station in 1979 at Ein Bokek on the Dead Sea. The plant collects heat in a rubber-lined, 70,000 square-foot pond.

There are experimental saline ponds producing electricity in New Mexico, Nevada and Virginia (Anon. 1980a), but only one is being considered as precursor to commercial-scale production in the U.S.

The negative aspects of solar ponds include:
1. The energy production fluctuates seasonally, and is subject to wind influence. The wind's impact may be minimized by windbreaks, screens, and other devices. Yet, the seasonal aspects are a problem, as quite often the greatest output is at times of least consumption.

2. For solar electrical ponds, there is a difference between peaking capacity and continuous operation capacity. The pond at Ein Bokek, for example, is rated at 150 kw, but only at peak production. The pond could sustain about 35 kw in summer and fifteen in winter at continuous operation (Anon. 1980a).

Solar photovoltaics technology is characterized by the production of electricity when light strikes certain materials, and was first reported by the French physicist E. Becquerel in 1839 (Monegon 1980). Later in 1905, it was explained in a classic paper by Albert Einstein. The major features are summarized in Figure 17.

Proponents of photovoltaics maintain that the industry will ultimately reach a breakthrough similar to the transistor advancement of calculators and other devices. The current cost of photovoltaics is prohibitive, and reflects probable energy intensities in the production aspects. As long as photovoltaic development is tied to oil products and oil produced electricity, the breakthrough will have to be significant for general use.

Solar power tower technology is just being tested. There are three approaches being investigated:

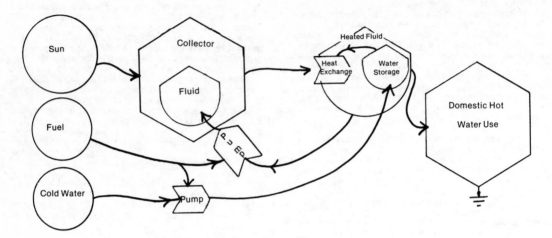

Figure 15.--Solar hot water
for domestic consumption

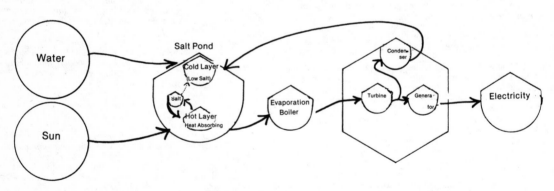

Figure 16.--Solar salt pond
generation concept

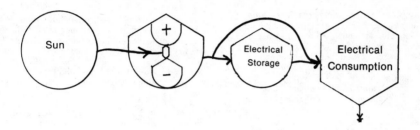

Figure 17.--Solar photovoltaic system

1. The central receiver concept which separates the collectors from the receiver. Solar energy that is collected by double-axis track-ing heliostats (flat mirrored tracking sur-faces) is directed to a focal point on a central receiving tower. Figure 18 sum-marizes the main aspects. Current systems include a 100-kwt unit at the University of Genoa in Italy, a 400 kwt system at Georgia Tech, a 5-kwt system at Sandia Labs in Alburquerque, New Mexico, and a 10 MWe system under construction near Barstow, California.

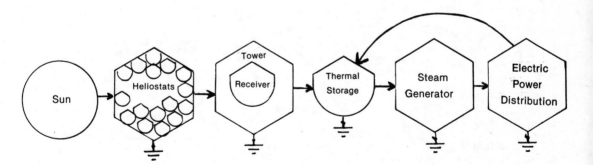

Figure 18.--Power tower concept

2. The solar "farm" concept which uses many parabolic point focus collectors or parabolic trough linear collectors to create steam in a receiver located in each collector. The steam is then piped to a central generator.

3. The distributed receiver concept integrates the collector, receiver and generator into one unit. The collector configuration is either a single-axis or double-axis tracking parabolic dish. Sunlight is focused on a receiver at the focal point of the trough or dish. The generator is located in the unit creating a thoroughly decentralized, autonomous energy source. One prototype reportedly costs about $3,000/kWe.

The use of biomass as an energy source is not new, and we have already seen that in times of war nations will burn anything with an organic content. There are however, five basic technologies which describe the uses of biomass. These five technologies fall into one of two basic conversion methods, thermochemical or biochemical.

Thermochemical includes:
*   Direct Combustion (burning in the presence of enough air to convert all the chemical energy into heat);
*   Pyrolysis (biomass is partially oxidized to give a gaseous form of energy, pyrolytic oils and char);

Biochemical includes:
*   Methane Fermentation (anaerobic digestion)
*   Alcohol Fermentation (ethanol production)
*   Biophotolysis

Figure 19 summarizes these processes.

A full statement about biomass would include the system from which the biomass is derived. In most cases this inclusion would enhance the biomass picture, as many sources are considered "wastes" to be disposed of, consuming energy in the process.

Recent work in California has shown that wood and biomass power plants are cheaper to operate and the fuel is easier to manage than conventional fuels. In fact, the forestry and agricultural industries are slowly moving into the power business. The notion of generating electricity closer to the fuel and moving the electricity to the consumer is prevalent with wood and biomass operations.

Wind is another ubiquitous source of energy, yet experience has shown that there is a threshold of wind speed that is necessary for a positive energy flow in the generation of electricity.

Windmills were used thousands of years ago in Persia to grind grain, hundreds of years ago and even today to pump water. Yet new designs are appearing on the market today involving the latest techniques in aerodynamics, electronics, structural engineering, control theory and new materials. All of these advances reduce the energy intensity of the wind systems to enhance the net yield.

The most thoroughly developed electrical generating wind turbine is the horizontal axis system. Vertical axis systems are also being tested, with the most successful being the Darrieus, where two or three slender blades resembling air foils are attached to the top and bottom of a vertical shaft or torque tube. Currently the Department of Energy is testing several wind energy systems.

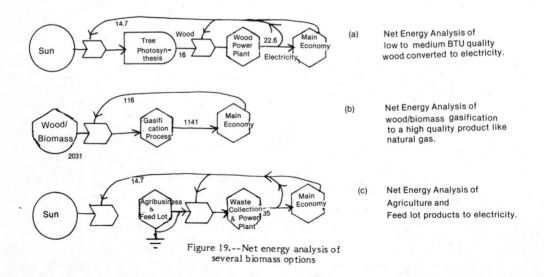

(a) Net Energy Analysis of low to medium BTU quality wood converted to electricity.

(b) Net Energy Analysis of wood/biomass gasification to a high quality product like natural gas.

(c) Net Energy Analysis of Agriculture and Feed lot products to electricity.

Figure 19.--Net energy analysis of several biomass options

DISCUSSION AND CONCLUSIONS

Vulnerability is apparent in the evolution of the U.S. energy network. With the rapid industrialization of the U.S. during the late 1800's, it became evident that the introduction of larger facilities led to profitable economies of scales. Marginal costs decreased as greater numbers of goods were produced. Centralization likewise applied to the American energy production and distribution system, and today the concentration of facilities has become an integral characteristic of the economy's energy sector. In fact, it is the energy sector that drives the U.S. economic system.

Thus the U.S. energy system is like a two sided sword. The concentration of energy sources as depicted in figure 5 allows exceptional work to take place in a short period of time. Growth became possible due to the storage and resultant concentration factors of centralization. Economic growth became one with energy centralization.

The other side of the sword also shown in figure 5 is that when energy is concentrated for exceptional work and uses, then the using systems become dependent on the concentrated quickened flows of energy, and the systems become an easier target for destruction, since it is all gathered up in very few, but large places.

The petroleum industry as shown in figure 6 is extremely centralized, energy intensive and vulnerable. From the beginning of the system pumping oil out of the ground until the distribution of oil products, the oil system follows an increasingly centralized production chain. The centralization of oil operations and continued use of energy intensive equipment and operations make it highly vulnerable to an attacker's disruptive intentions.

Domestic crude oil production is the least vulnerable step in the oil chain. Despite the concentration of oil production in only a few states, the individual wells are still too dispersed for easy attack.

Transportation of crude oil is primarily by pipelines, which are mostly found underground. However, pumping stations needed to move oil through the pipelines are located above ground 50-100 miles apart for more than 66,000 miles of crude oil pipelines. Only about one fourth of the domestic crude is not transported by pipeline, but the ships, trucks, and railroads used to transport this 25% are also vulnerable to attack due to the centralized nature of highways, port terminals and railroad stations.

Perhaps the most vulnerable points with respect to attacks in the U.S. oil system, however, are the very few, large refineries clumped along the Gulf Coast, the North East and California (fig. 6). Much like the German electric power and synfuel production the refineries of the U.S. offer a classic opportunity for energy debility. Nearly all crude oil is converted to gasoline and other products before use, and loss of the refineries' capabilities would not only devastate the U.S. economy, it would render it defenseless. Thus large refineries are considered to be a prime target in a nuclear attack because of their crucial role in the economy and because

they are the most concentrated segment of the petroleum chain.

In addition the oil industry's reliance on electric power is all too reminiscent of Germany's World War II vulnerabilities. In that regard, the federal government has suggested changes that would reduce damage done in an attack on the oil industry. These solutions include building petroleum refineries underground, maintaining separate electric power sources for each refinery, and building refineries with fallout protections. However, they are generally considered too expensive in an industry where market considerations, transportation and crude oil supply are major factors in determining site location for plants (Ford Foundation 1979).

What may not be apparent to the oil industry's managers who are focused on the money aspects of their operation, is that there may be a substantial and legitimate energetic cost to their centralized oil system. That cost would be the embodiment of necessary measures to lessen attack vulnerabilities. Figure 20 summarizes this notion highlighting the costs associated with attack vulnerabilities.

The natural gas industry exhibits vulnerabilities that are somewhat different from those already seen in the oil industry.

The production phase is concentrated in just a few states, but as in oil production the fields are fairly well dispersed. Even though the gas processing step is similar to refining in the oil system, it is not nearly as centralized, with over 750 plants in the country in 1974 (CONAES 1980).

The main area of vulnerability for natural gas is in the transmission lines carrying the gas from the processing plant to storage tanks. The greatest vulnerability rests in the fact that the pipelines and compressor stations are run by automated systems. Complex communications equipment is vital to this operation, and few people are skilled in repairing it. Thus even lightly damaged equipment could be rendered unusable if no one with the expertise survived or was able to travel to make repairs. Figure 21 highlights this vulnerability to attack with an obvious flow line activated by either an attack or non attack breakdown, indicating the subtle energetic costs associated with this centralized system.

With the 1974 oil embargo and continued wars in the Middle East, most energy vulnerability analyses have concentrated on oil, natural gas, and electric power as the prime consideration has been toward supply shortfalls. Yet this study looks beyond the supply issues and includes the attack vulnerabilities, and the lessons learned by the German World War II example are entirely applicable here.

It is certainly true that the coal industry is less complicated by comparison with oil and gas, but it nevertheless remains vulnerable because of its great dependence on electric power and transportation, with strong oil consumption overtones.

In 1979 the national averages of electric power were generated from the following sources:

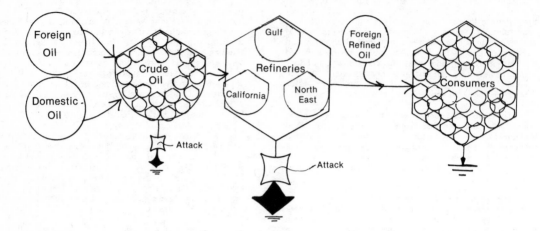

Figure 20.--Vulnerabilities of the U.S. oil system

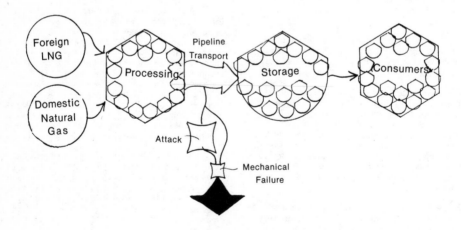

Figure 21.--Vulnerabilities of the
U.S. natural gas system

Coal	48%
Natural Gas	15%
Oil	13%
Hydro	12%
Nuclear	11%

However, these statistics vary considerably across the nation. For example, New England relied on oil for 60% of its power and coal for 7%.

Yet, the transportation sector is fueled by oil such that without movement of coal, economic and energetic productivity would only be a fraction of what it is now. Since the electric power industry is so heavily dependent on coal, the loss of transportation fuels (oil) would stifle electric power production nationwide. Figure 22 features the oil dependent vulnerabilities of the coal system.

Increasing attention is being paid to the possibility of an enemy attack upon nuclear power plants and its subsequent effects. The 11% electrical potential generated by nuclear power is contained in less than 200 potential targets. Yet they pose an even greater risk in that some are located very close to large population centers, thereby contributing to their selection as a strategic nuclear targeting, conventional bombing site or terrorist site.

A recent Princeton University study indicates several reasons nuclear plants may be chosen (Sant 1979). Foremost in the study is the fact that nuclear plants are enormous capital investment structures, and would cause quite a drain on the economic system. Additionally the emotional aspects of radioactive leaks would be prevalent, whether significant or not. The psychological effects of destroying a nuclear power plant combined with the economic sump aspects signifies additional vulnerabilities due to nuclear power centralization. A further subtlety of vulnerability is characterized by the electromagnetic pulse (EMP) which results from nuclear explosions. EMP creates a substantially higher electric field strength than an ordinary radio wave and disappears in a fraction of a second. Its main adverse affect is that it renders inoperative sensitive electronic equipment and necessary components of nuclear power systems. Figure 9 points out the oil dependency of nuclear power, such that destruction of the oil flow would bring nuclear generation to a halt. Figure 23 shows the vulnerability aspects of the U.S. nuclear system.

The notion of destroying large dams as an attack strategy is not a new idea, as it accomplishes three acts of destruction: halts electrical production, causes flooding of adjacent areas, and preempts navigation.

The obvious benefits of the very efficient hydro-generated electricity have far outweighed concerns toward attack vulnerabilites for such a small total energy contributor. Figure 24 outlines the major vulnerability of high head hydropower due to its naturally occurring concentrated nature.

The Options for the Future

With the two pronged vulnerability posture established for the existing U.S. energy system (supply shortfalls and attacks), the question then becomes which pathway of options should be pursued in our next phase of energy planning: more centralized or the dispersed, renewable, decentralized options.

The answer may seem obvious, but on closer inspection it appears to be not such an easy decision.

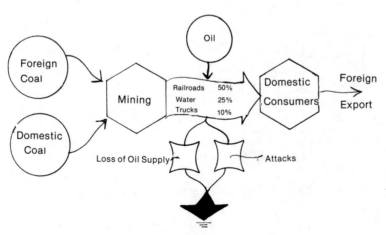

Figure 22.--Vulnerabilities of the
U.S. coal system

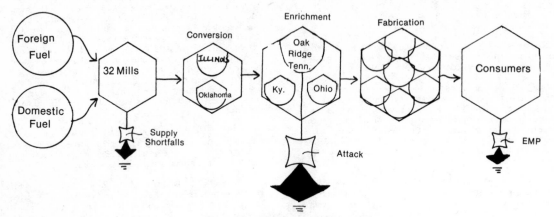

Figure 23.--Major vulnerabilities of the
U.S. nuclear power system

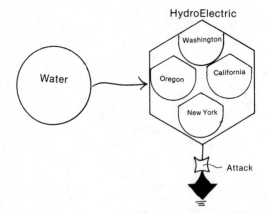

Figure 24.--Vulnerability of the
U.S. hydropower system

The first consideration is the technological feasibility of the various options. The nature of new ventures however, is that they can change in a very short period of time depending on cash flows and 'breakthroughs'. However, with all else being equal, it appears that the centralized options of coal synfuels, and oil shale are at best in a dead heat with the decentralized options of conservation, and the multi solar uses.

What's more, if a dollar to technological information curve was established, the decentralized systems would show a greater output to input ratio in almost every case (Sant 1979; CONAES 1980).

The second concern is for the long-term renewability of the options. Would choosing a particular technology reduce our short-term supply shortfall vulnerability and also contribute to the overall system vulnerabilities of supply interruptions and attack possibilities?

The German case study is a good example of over extending a good idea. The centralized efforts that led to the German strength in World War II also ultimately gave the Allies the opportunity for significant destruction and eventual ending of the war. The simple proverb of putting your eggs in one basket becomes apparent here.

The third consideration is made by way of the Japan case study (fig. 2). Japan was virtually safe from targeted attacks to large centralized energy facilities with renewability also on their side. Their highly decentralized energy system offered no such single attack options.

However, Japan had another problem. That problem was that in times of war the need for energy is quick and great. The war power as we have demonstrated is directly tied to energy availability. Japan was unable to increase its electrical power necessary for the war industry since the same strategic materials needed to increase power facilities were being consumed in the war products.

So the lesson in the Japan example is that by not having centralized systems decreases targeted power attacks while possibly opening up a new vulnerability of not being able to harness sufficient energy quickly enough to counteract the massive slugs of energy driven war attacks being scattered across the nation.

It appears we have reached a standoff. Too much centralization increases vulnerability, yet not enough lessens the defense posture. But an even closer look into this dilemma shows there is a resolution for this apparent conflict.

Perhaps the answer lies in the end use system. Figure 25 shows a two pronged approach wherein high energy needs requiring quick response times and concentrated energies including military defense, medicine, industrial processes, necessary transportation, etc. are powered by the existing centralized systems. These systems represent enormous energy concentrations by society and should continue to be used for high energy quality needs. Additionally, high quality energy "protec-

tive systems" are also funded through this high quality end of the nation's energy budget.

The second prong of the approach is where the centralized systems are also used initially to develop dispersed, decentralized systems to be used for low energy quality purposes, such as space conditioning and water heating, cooking, community electrical needs and small mass transit.

The argument for this approach is fourfold. First, the matching of energy quality (Odum 1975) with end use is a logical process supported by the Laws of Thermodynamics (APS 1975). Second, the use of alternative energy systems on a small community scale further offsets the use of oil, gas, coal and nuclear. Some of the alternatives may not be immediate net energy producers, but once established in a community could break even or become net yielding while relieving the oil and gas drain for higher quality uses, while adding to the recuperative ability of the communities. This approach provides the business and economic link necessary for a smooth national transition into an energy period of both centralized and decentralized systems, with accompanying business and economic spinoffs. And finally, from a politician's perspective, it's perfect because it supports both camps.

The chief advantage of the approach is that it offers the flexibility for the U.S. to be a dynamic economic producer, have top military response capabilities, whereas at the same time enable the dispersed U.S. communities to carry on in the event of an attack or recoup from an attack due to the decentralized energy systems working across the county. This approach would surely reduce the possibility of a targeted attack, since not all of our eggs would be in the centralized basket, and would probably reduce the likelihood of an all-out nuclear war since the U.S. would be managing its own disordering. This would disallow such a tremendous centralized energy build-up to continue to occur, which could only be dispersed by some profound holocaust.

824

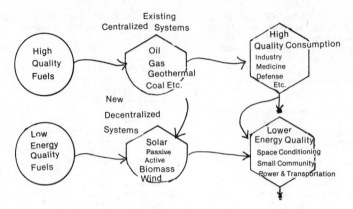

Figure 25.--Two-pronged energy consumption
maximizing centralized and decentralized options

LITERATURE CITED

American Physical Society. 1975. Efficient use of energy: a physics perspective. 237 p. American Physical Society, New York, N.Y.

Anon. 1980a. Salton sea study to determine electrical generation potential. Solar Engineering. p. 21-24.

Anon. 1980. Salton sea solar pond concept. Solar Engineering. p. 18-20.

Anon. 1975. Why atomic power dims today. Business Week (November 17, 1975) 98:106.

Antolini, H.L., (ed.). 1978. Sunset homeowner's guide to solar heating. Lane Publishing Co., Menlo Park, Ca. 63 p.

Ballantine, T. 1976. A net energy analysis of surface mined coal from the Northern Great Plains. Masters Thesis of Engineering, Department of Environmental Engineering Sciences, University of Florida, Gainesville, Florida. 78 p.

Barnaby, C.S., P. Caeser, and B. Wilcox. 1977. Solar for your present home. 57 p. California Energy Commission, Sacramento, Ca.

Brown M. and Arrington, L. 1976. Net energy analysis of housing insulation compared with fuel saved. p. 423-438. In Energy analysis of models in the United States. Report to Department of Energy Contract #EY 76-S-05-4398. Washington, D.C.

Calder, N. 1978. Nuclear nightmares, an investigation into possible wars, Viking Press, New York, N.Y.

California Energy Commission (CEC) and Arthur D. Little, Inc. 1978. California clean fuels study. 131 p. Sacramento, Ca.

CEC. 1980. Commercial status: electric generation and nongeneration technologies, State of California. 83 p. Sacramento, Ca.

Committee on Nuclear and Alternative Energy Systems (CONAES). 1980. Opportunities for technological research and development in support of energy conservation. Energy in transition: 1985-2010, W.H. Freeman and Co., 210 p., San Francisco, Ca.

Congressional Research Service (CRS). 1977. National energy transportation, Volume 1, Washington, D.C.

Energy and Defense Project. 1979. Dispersed, decentralized and renewable energy sources: alternatives to national vulnerability and war. Report to F.E.M.A. Contract No. DCPA 01-79-C-0320, Program Unit 2314-F. 326 p. Washington, D.C.

Federal Energy Administration (FEA). 1976. National energy administration. 210 p. U.S. GPO, Washington, D.C.

Fesharaki, F. 1980. Global petroleum supplies and the 1980's: prospects and problems. Wm. Freeman and Co. 310 p. San Francisco, Ca.

The Ford Foundation. 1979. Energy conservation: opportunities and obstacles. In Energy: the next twenty years, Hans H. Landsberg, Chairman, Cambridge, Ma.

Gardner, G. 1976. Net energy analysis of oil shale. 160-172. In Energy Analysis of Models of the U.S. Report to Department of Energy, Contract No. EY-76-S-05-4398. 232 p. Washington, D.C.

Hannon, B. 1973. An energy standard of value. Annals of Amer. Acad. of Polit. and Soc. Sci. 410: 139.

Holland, E. 1979. About mirrors, troughs and dishes. Solar Age, 9 p.

Howell, Y. and J.A. Bereny. 1979. Engineer's guide to solar energy. Barnes and Nobel Books, New York, N.Y.

Kylstra, C. and Ki Han. 1975. Energy analysis of the U.S. nuclear power system. In Energy models for environment, power and society. Report to E.R.D.A. Contract E-(40-1) 4398.

Marshall, E. 1980. Iraqi nuclear program halted by bombing. Science. 101:508-509.

Monegon, Ltd. 1980. The future of solar electricity, 1980-2000: developments in photovoltaics. Gaithersburg, Md. 42 p.

Nye, J. 1980. Energy and security: report of a Harvard workshop, Harvard University, Cambridge, Mass.

Odum, H.T. 1971. Environment, power and society. 319 p. Wiley-Interscience, New York, N.Y.

Odum, H.T. 1975. Energy quality principles for estimating the environmental carrying capacity for man. Given at the Environmental Health Conference at Pahlayi University, Tehran, Iran.

Odum, H.T. and S. Bayley. 1974. A model for understanding the relationships of money, energy and survival value. 358 p. In R. O'Connor and E. Lochman, eds., Economics and decision making for environmental quality. Univ. of Florida Press.

Odum, H.T. and E.C. Odum. 1976. Energy basis for man and nature. McGraw-Hill, New York, N.Y. 288 p.

Office of Technology Assessment (OTA). 1978. An assessment of oil shale technologies. 137 p. U.S. GPO, Washington, D.C.

Ramberg, B. 1980. Destruction of nuclear energy facilities in war: the problem and implications. Heath and Company, Lexington, Ma.

Richards, D. 1911. Principles of political economy and taxation. 219 p. E.P. Dutton, New York, N.Y.

Sant, R.W. 1979. The least-cost energy strategy: minimizing consumer costs through competition, Energy Productivity Center, Mellon Institute, Arlington, Va.

Schurr, S.M., ed. 1979. Energy in America's future: the choices before us. 233 p. Johns Hopkins University Press. Baltimore, Maryland.

Stobaugh, R. and D. Yergin. 1981. Energy and emergency telescoped. Foreign Affairs, 58: (3): 593-594.

Stobaugh, R. and D. Yergin (eds.) 1980. Energy futures: report of the energy project at the Harvard Business School. 353 p. New York, N.Y. Ballantine Books. 353 p.

U.S. Geological Survey. 1975. Known U.S. uranium reserves won't meet demand. USGS Press Release. Washington, D.C.

U.S. Department of Energy. 1978a. Overview of technology commercialization assessment, 57 p. Washington, D.C.

U.S. Department of Energy. 1978b. Draft commercialization strategy report for oil shale. 49 p. Washington, D.C.

U.S. Strategic Bombing Survey. 1947. The effects of strategic bombing on the German War economy. 115 p. U.S. GPO, Washington, D.C. 115 p.

U.S. Strategic Bombing Survey (Pacific). 1945. The electric power industry of Japan. Electric Power Division. Washington, D.C.

Yergin, D. 1979. Conservation: the key energy source. 136-183. In Energy future: report of the energy project at the Harvard Business Service. Robert Stobaugh and Daniel Yergin, eds., Random House, N.Y.

THE INFLUENCE OF NATURAL RESOURCES AND DEMOGRAPHIC FACTORS ON THE ECONOMIC PRODUCTION OF NATIONS[1]

W. M. Kemp[2], W. R. Boynton[3], K. Limburg[4]

Abstract.--A compilation of data describing the socio/economic, demographic, physiographic and natural resource characteristics of 63 nations is presented. Statistical correlations suggested that economic output and efficiency at the national level are influenced by population, area, fuel consumption and natural resource base.

INTRODUCTION

The relative importance of various factors influencing the structure and production of national economies is a central matter for the establishment of economic policies. Traditional views have held that cultural, social and political institutions are key elements in successful economic production. While a neo-Malthusian perspective has re-emphasized the importance of resource contraints as limits to economic growth (Hubbert 1969, Odum 1971, Forrester 1971, Meadows et al. 1972), a sense that human "resourcefulness" (i.e., technological development) and efficient institutional structures can overcome any shortages of materials or energy seems to prevail (e.g., Nordhaus 1974, Goeller and Weinberg 1976). Of the natural resources which potentially limit a nation's economic output, some have argued that materials shortages are most crucial (Cloud 1973, Cook 1976, Georgescu-Roegen 1977), while others maintain that energy availability is fundamental to all economic activities including mineral extraction and processing (Odum 1971, Cook 1971, Hayes 1976).

Perhaps, the first step toward unraveling this perplexing issue is the development of a conceptual framework which contains all generic elements in the structure of a national or world economy (Leontief 1974). Clearly, such a model must include in its internal structure: human factors of population (labor, consumers); institutional factors (industry, government); and physiographic factors (area, geography). Connections to the external world include, fuels, minerals, goods and services which are imported, exported, and/or extracted from nature within the nation's borders, as well as a broad category of renewable solar-based resources such as insolation, wind, waves, tides, and rain. In addition, internal structures such as agricultural, energy, and mining industries must be considered as the nodes where such inputs are coupled into the economic process. Some economists have argued that nature, per se, should be considered as separate from the economic sphere (this view is summarized by Simon 1980). However, a more recent view among economists has recognized a linkage between economic and natural systems via the discharge of pollutants, and has espoused the moral obligation of humans to minimize degradation of such natural systems (e.g., Ayers and Kneese 1969, 1971, Leontief 1974). Others (primarily from the natural sciences) have maintained that nature provides numerous "life-support" services for human society which are, in effect, surrogates for economic activity (Odum 1971, Gosselink et al. 1974, Biswas and Biswas 1976, Bormann 1976, Gilliland 1975, Westman 1977, Costanza 1980). Since these services of nature are outside the market system, they generally are perceived (rightly so) as being free, but they are also sometimes incorrectly considered as being valueless.

What is needed is an integrated view of the economic system which includes all factors contributing to production. We have attempted to present such a perspective in the highly-aggregated form of figure 1. Here we have defined government, commerce and industry, as well as demographic variables (such as population and area) as part of the "Main Economy", with satellite sectors (such as agriculture and mining) involved in the extraction, collection and processing of natural resources defined as direct inputs to the production process (i.e., gross domestic product). Natural ecosystems are included, not only as waste receptacles, but also

[1]Paper presented at the international symposium Energy and Ecological Modelling, sponsored by the International Society for Ecological Modelling. (Louisville, Kentucky, April 20-23, 1981)

[2]University of Maryland, Center for Environmental and Estuarine Studies (UMCEES), Horn Point Environmental Laboratories, Cambridge, MD., USA
[3]UMCEES, Chesapeake Biological Laboratory, Solomons, MD. USA
[4]University of Florida, Dept. Environmental Science and Engineering, Gainesville, FL. USA

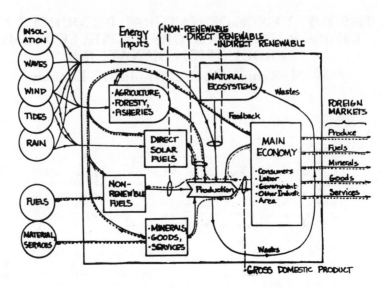

Figure 1.--Aggregated conceptualization of a national system
emphasizing the major input factors to economic
production.

as amplifiers (which have no associate monetary
flow) to economic production. Energy inputs to
the economy are grouped into three categories
(stored, direct renewable, indirect renewable)
based on their relative rate of renewal and the
directness of their entry into the economy.

A crucial issue for developing nations is
how the interrelations among various factors in
such a structure might change during the course
of economic development. Moreover, such countries
need to know how their existing resources and
social/political institutions might best be
brought to bear on economic growth. The direct
use of renewable energy resources in agriculture,
biomass or hydro-power might be one such example
(Strout 1977, Ramachandran and Gururaja 1977,
Hammond 1978, Revell 1980). However, such strategies are not without problems and, in fact, may
disrupt cultural and ecological patterns (Smil
1979, MacKillop 1980, Gorney 1980, Revell 1980).
In any case, careful selection of appropriate
technologies seems to be fundamental in the development process (Weiss 1979, Ward et al. 1980).
The concept of "filling the GNP gap" between rich
and poor nations (i.e., gross national product
per capita) has been often championed as a mis-
sion of major importance, and energy analysts
have sought, recently, to effect such change with
maximum efficiency of energy use (Bullard and
Foster 1976, Craig et al. 1976). However, others
question the appropriatness of such objectives in
that GNP per capita may not adequately measure
human welfare (Daly 1974, Morawetz 1974).

The purpose of this paper then is to present
partial results of a large compilation of data on

economic, resource, physiographic, social and
demographic variables for 63 nations which range
across a broad spectrum of economic development
levels. We present the preliminary results of
some simple statistical analyses of these data to
test the relative importance of aggregated components of a national economy (as indicated in fig.
1) on economic output and efficiency. We group
these countries into three categories of economic
development and resource potential to examine how
the interrelationships among various factors
might change during the course of development,
and to consider potential strategies for "filling
the GNP gap" between lesser and more developed
nations.

APPROACH TO THIS ANALYSIS

Sample of Countries

The statistical sample used in this analysis
includes 63 nations which represent a wide range
of factors. We selected nations randomly from
groups which were stratified according to quantitative criteria of size, latitude, continent,
rainfall, proximity to the sea, and level of development (as indicated by GNP/capita), as well
as qualitative matters of history and government.
We attempted to obtain as evenly distributed a
sample as possible, both in terms of percentage
of our sample and percentage of the world. The
actual distributions are provided in figure 2
for: a) within the sample; and b) within the
world. It can be seen that our sample represents
a reasonable balance among these criteria, although in some cases it was impossible to obtain

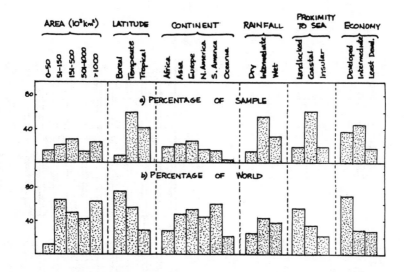

Figure 2.--Distribution of 63-nation sample among six criteria of physiography and economy, where each category
is considered as a) percentage of the total sample, and
b) percentage of total world.

even distributions both within/sample and
within/world, as indicated by the percentage of
"boreal" nations.

Definition of Energy Resources

A primary objective of this analysis was to
examine the extent to which economic output and
efficiency are influenced by fuel consumption,
population and areal scale, as well as the density of renewable resources. The wide diversity of
solar-based resources (including insolation,
wind, waves, tides and hydrostatic potential)
motivated us to utilize a scheme which considered
all such resources as a function of their thermodynamic potential. There are numerous ways in
which the potential energy of these resources
could be manifested; however, as a matter of convenience we compared them in terms of their ability to generate electricity. Of course, to the
extent that electric generating technologies have
been continually upgraded over the last century,
the energy potential calculated in this manner
will be a minimal estimate. Using this criterion
takes into account the differing thermodynamic
qualities of these various resources (Tribus and
McIrvine 1971, Odum 1971, Costanza 1980). This
approach, which has traditionally been used in
energy accounting schemes to compare, for
instance hydro-power with petroleum (Darmstadter
et al. 1971) or biomass with coal (Revelle 1980),
enabled us to maintain the aggregated view of a
national economic system (fig. 1).

While we estimated these renewable energy
resources as that available in an average year
from 1965-1975, we considered energy fuels in
terms of actual consumption during 1970 regardless of whether such fuel was obtained internal
or external to the national boundaries. Moreover,
we did not directly consider non-energy mineral
resources. In a sense, fuels consumed represents
some integration of fuels and minerals extracted
from within the national boundaries, since both
generate capital which will be used to purchase
any energy and/or materials necessary to maintain
production. This approach is consistent with the
energy analysts' notion of "embodied energy"
(e.g., Bullard and Herendeen 1975, Costanza
1980). We assumed that, on the whole, energy is
the primary limiting factor for economic production (others, e.g. Brown (1976) have made the
same assertion). The validity of this assumption
is borne-out in the remarkably significant regression between GNP and fuel consumption, which will
be discussed subsequently.

Only a small fraction of the available
solar-based energy resources is utilized directly
in a national economy. These resources are manifested in the market as agricultural, fisheries
and forestry products, and as hydroelectric
power, as well as firewood, charcoal and other
biomass fuels. Much of the potential energy from
solar-based sources is not directly utilized by
the economic system, either because it is too
dilute or too inaccessible to be of economic
value or because there is no local demand. Nevertheless, these energies continue to function in
the earth's large-scale hydrological, geological
and biogeochemical cycles, and we postulate that
the intensity of these energies will influence
indirectly the relative success of human enter-
prise.

830

Therefore, we have considered three separate categories of energy input to each national system: 1) Stored Energies (S) -- fossil fuels and nuclear; 2) Direct Renewable Energies (R_D) -- agriculture, fisheries, forestry, hydroelectric and biomass fuels (as well as wind and tidal electric power); 3) Indirect Renewable Energies (R_I) -- including all other solar. We have grouped geothermal with stored energies eventhough its time-constant for depletion may be much longer than that of fossil fuels and uranium. In calculating R_D we included only the solar inputs, and omitted contributions of S, which in some cases may be very large (Pimentel et al. 1973, Steinhart and Steinhart 1974). This procedure avoided double-counting.

Definition of Economic Output

The economic output of each nation was taken as its gross domestic product (GDP) with correction for "purchasing power parity" (GDP_P). Whereas statistics are readily available for gross national product (GNP), data for GDP and GDP_P are less accessible. Essentially, GDP differs from GNP in that it does not include economic production from activities outside the particular nation's borders, while GDP_P considers the fact that the actual purchasing power of a nation's currency internally may be different from that in the international market. Countries which are more isolated from the world economy (i.e., relatively less international trade) are likely to have GDP_P much greater than GDP. In fact, GDP_P is the most realistic measure of the output of a nation's economy, and we, therefore, utilized the detailed analysis of Kravis et al. (1978) to obtain estimates of GDP_P for 51 of our 63-country sample. We developed a multiple regression of GDP_P versus GNP and R/S which had a coefficient of determination, $r^2 = 0.978$. This compares favorably with the regressions of Kravis et al. (1978) which considered the import/export ratio as an independent variable and obtained $r^2 = 0.989$. We assumed, following the logic of Clague and Tanzi (1972), that relative dependence on renewable versus stored energies would provide an index of extent of isolation from international trade.

Data Sources

Data for this study was obtained from myriad sources which were intercalibrated among each other. To the extent possible we compiled 1970 data for each of the 63 countries for: physiographic factors (area, latitude, coastline length, number of harbors, river flow, continental shelf area, major soil types, mean, range and distribution of elevations, tidal range and frequency, oceanic wave heights, oceanic currents); climatological factors (precipitation, temperature, wind speed and frequency, humidity and vapor pressure, evapotranspiration); economic factors (imports and exports of goods and services, foreign aid given and received, inter-

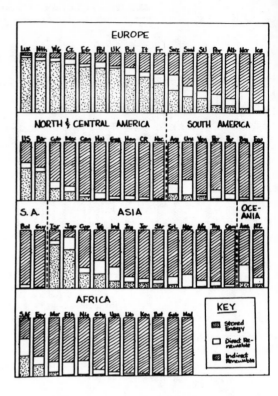

Figure 3.--Relative balance among three categories of energy resources available to each of 63 nations.

-national reserves, tourism, GNP, GDP, GDP_P); demographic factors (population size, density, and distribution, migration, birth and death rates, life-expectancy); and social factors (political system, religions, number of universities, literacy rate, food consumption rate). Much of the economic and demographic data were obtained from World Bank publications and from statistical yearbooks of either the United Nations, FAO, or a particular nation. Oceanographic and climatological data were obtained largely from US Naval Atlases and NOAA publications. Land-use and agricultural information came from USDA, UNESCO and FAO publications. Various other compendia served to provide additional sources, for example Landsberg (1969-76), Lieth and Whitaker (1975), Van Hylckama (1956), Walter (1973) and Darmstater et al. (1971). A more detailed presentation of these data is available in Limberg et al. (1977). While we attempted to cross-check all data between two or more sources, it should be noted that these kind of data must be considered first-order approximations (Darmstadter et al. 1971, Schipper and Lichtenberg 1976). This is particularly true of the information for communist-block and Third World nations.

RESULTS AND DISCUSSION

We present here some preliminary results of our data analysis and provide a few brief comments.

Social Factors and Economic Output

Various indices of social structure and standard-of-living were selected from this data set to examine possible relationships with national economic output (GDP_p and GDP_p/capita). Eight such social indices were used, including: population density and growth rate, percentage of population in urban and agricultural habitats, average rate of per capita protein consumption, birth rate, life expectancy, and literacy rate. We tested all possible combinations of arithmetic and logarithmic functions, and, remarkably, obtained only two variables which explained more than 50% of the variation in GDP_p or GDP_p/capita. These were: percentage of population in agriculture (negative slope, $r^2 = 0.53$); and literacy rate (positive slope, $r^2 = 0.56$). It seems that both of these correlations are more likely to represent effects rather than causes of economic production. Our hypothesis at the outset was that such measures of societal internal structure would be less important than external resource-related factors.

Stored Energy and Economic Output

Numerous investigators have observed strong relationships between consumption of commercial energy and gross national product for various nations (Darmstadter et al. 1971, Cook 1971, Häfele 1974a, Krenz 1977). However, most have used per-capita expressions for both x and y variables, and few have obtained r^2 values in excess of 0.90 (e.g., Darmstadter et al. reported $r^2 = 0.79$ for a sample of 49 countries). The vast majority of these published correlations used GNP/capita as the independent variable and per capita energy consumption as the dependent variable. We examined the relationship between GDP_p and S for 63 countries in log-log regressions for which the best r^2 value was 0.91 without the per-capita transformation. Throughout our analysis we have chosen to use GDP_p rather than GDP_p/capita as our dependent variable because we were interested in understanding factors affecting the output of national economic processes rather than the average per capita productivity. We, thus, viewed each national economy as an integrated system of which population is but one characteristic property.

The strength of this relationship is impressive, indeed, but our hypothesis was that other aggregated input factors must also influence economic output at the national scale. Despite the relatively small unexplained variation in this regression, several observations encouraged us to forge ahead. Darmstadter et al. (1971) noticed

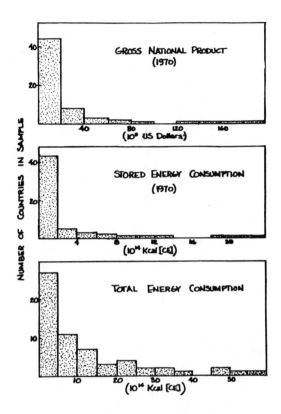

Figure 4.--Spectral distributions of GNP, fuel consumption and total energy availability for 63-nation sample.

that the regression was improved somewhat by including hydroelectric power generation, which struck us as a distinctly different kind of input, being essentially renewable through the hydrologic cycle. Moreover, in each of the previous versions of this relationship which included Third World nations, those less developed countries exhibited a higher economic output per unit energy input than did the industrialized countries. Other reports have shown that this ratio of GDP/S appears to change during the process of economic development (e.g., Häfele 1980). Before reporting the results of our search for factors affecting GDP_p, we will first present general patterns for the distribution of energy resources among our sample countries.

Distribution of Energy Resources

The relative balance among three categories of energy resources (1. stored fuels, S; 2. renewable solar-based energies directly used in economic activities, R_D; 3. other renewable energy resources not directly utilized) available to each nation is portrayed in figure 3. It is

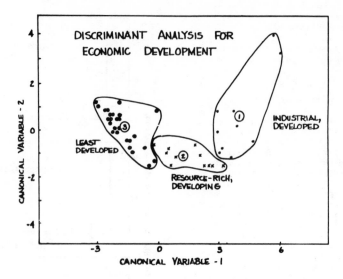

Figure 5.--Discriminant analysis for three classes of economic development, considering GDPp/capita, S/R and
literacy rate.

apparent that the European continent is generally
most dependent on S, and even those countries
such as Norway and Portugal which consume less
fuels have relatively large proportions of R_D.
Here, at the extremity is Luxembourg which is
virtually without a renewable resource base. Nations in the other continents have generally
higher fractions of R_I. Sweden and South Africa
come the closest to an even balance among the
three energy inputs. While this aggregated
presentation of data serves to provide a salient
view of the energy basis for respective economies, it may be misleading in some cases. For
example, Canada and Saudi Arabia have very similar "energy signatures" in this figure; however,
the actual nature of their resources and the per
capita values for total energy budget are dramatically different for the two countries.

Spectral distributions of world fuel consumption (S) (fig. 4, middle panel) follow the negative exponential form which typifies many power
spectra (Odum 1971). The great majority of nations is characterized by low fuel budgets. This
energy spectrum follows closely the patterns reported by Häfele and Sassin (1977) and Sassin
(1980), the latter of which used population
rather than number of nations as an abcissa. We
see, however, in the bottom panel of figure 4
that the distribution is somewhat modulated by
including renewable energies along with S. This
suggests that there may be more parity among nations than predicted by the S spectrum. The distribution of GNP (top panel) lies somewhere between the S and (S+R) distributions, perhaps
indicating that GNP may be related to both S and
R (in fact, the GDPp spectrum looks even more
like that for S+R, but this figure is not pre-

sented here for lack of space).

Grouping of Nations by Development Stage

We have postulated that the kinds of causal
relationships which would emerge from these data
might change during the process of economic development. Therefore, before proceeding into a sequence of regression analyses, we attempted to
divide the sample into three groups of roughly
defined stages of development. A discriminant
analysis was performed using criteria of:
GDPp/capita as a measure of average individual
wealth, literacy rate as an index of information
structure, and S/R as an index of renewable resource base. This scheme defined groups which
were reasonably well separated (fig. 5), and
which we have termed: 1) Industrial, Developed;
2) Resource-Rich, Developing; and 3) Least Developed. These groups roughly correspond to First,
Second and Third World nations which are traditionally defined solely on the basis of
GNP/capita. Häfele and Sassin (1977) have used
criteria of percent of GNP involved in agriculture, industry and services to separate the
world's nations into four categories of development. Obviously, numerous such classification
schemes are potentially available; however, we
selected one which was consistent with our perception of a nation as a system.

Demographic Factors and Economic Production

While the various social indices described
earlier had generally little predictive value for
economic production, we performed regression

analyses for two aggregate demo/physiographic factors (population, and area) for each development group and obtained some interesting patterns. We tested the statistical effects of population (P) and area (A) on GDP_r and on (GDP_p/S). The latter of these two dependent variables was used as an index of economic fuel-efficiency at the national scale, where higher values of (GDP_p/S) indicate greater economic output per unit of fuel consumption. Numerous investigators have discussed the possibility that this fuel efficiency (or the inverse which is termed "energy intensiveness") changes during economic growth (e.g., Summers 1971, Krenz 1977, Häfele and Sassin 1977, Goldemberg 1978). We present, here, the results of regression analysis for each of the three development groups so as to examine the relationship between level of development and fuel efficiency. We have omitted Luxembourg and Israel from the Group I regression because they both act as stastical outliers. This is to be expected in view of the fact that Luxembourg is essentially a city-state which is capable of supporting a high population in a small area with few natural resources, because it draws heavily on external resources (Day and Day 1973). Israel, while not a city-state, has been able to experience rapid development on few resources through external subsidies from the USA and other sources of private capital.

Population appears to exert a strong positive influence on economic output of a national system. This is illustrated in figure 6, where 84-95% of the variability in GDP_p is explained by P. The slope of this relationship is somewhat greater for the more developed (Group I and II) countries, suggesting that the effect of population on GDP_p increases with economic development. The relation between population and economic fuel efficiency, however, was not so simple (fig. 7). People are obviously a resource in themselves, capable of promoting economic output both through physical labor and as information storages and amplifiers (Odum 1971). A negative effect on (GDP_p/S) was observed for the entire sample of Group I countries, and for Group II and III nations beyond about 2 and 20 million persons, respectively. For the Group II and III countries (most of which are continuing to experience relatively rapid economic growth (Morawetz 1977)), lower population levels actually appear to enhance fuel efficiency. This may indicate that very large populations act as a drain on fuel resources, requiring a comparatively large proportion of the budget to be channelled into service activities which have less multiplier effect on economic production. Brown (1976) has reported a similar parabolic shaped function describing population effects on GNP/capita, to which he ascribes an analoguous interpretation. The two-fold nature of this relationship between population and economic output may account for the disparity between those who hold that population growth does not affect economic growth (e.g., Simon 1980) and those who argue that in extreme population can drain resources which would otherwise promote economic development (e.g., Erlich

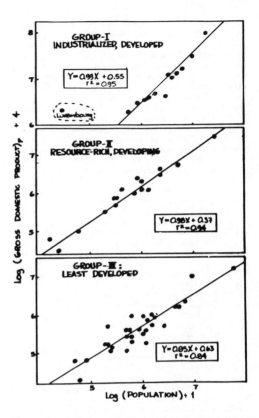

Figure 6.--Relationship between gross domestic product and population for nations in each of three groups of economic development. Note that x and y axes are scaled by factors of 10 and 10^4, respectively.

et al. 1977). These results corroborate the proposition of Sanderson and Johnston (1980) who state that, "in some circumstances rapid population growth can be quite detrimental to a country's development prospects, while in other circumstances it may even be helpful."

The areal dimensions of a nation appear to increase GDP_p for all three economic groups (fig. 8). These regressions are all significant, and the slopes tend to decrease with further economic development, indicating greater importance of area in the lesser developed nations. This is just the opposite of the relation for GDP_p vs population (fig. 6), which had slopes directly proportional to development. Apparently, less developed nations with larger area can generate more economic output per unit area, whereas more developed nations obtain more marginal effect from people. This might be explained by the fact that greater area generally captures greater solar energy (which may be more important for agrarian-based economies), while people are most important (in a production sense) as reservoirs and manipulators of information which becomes increasingly significant with the intricacies of industralized systems. The effect of A on fuel efficiency is, however, quite the opposite (fig. 9). Here, an inverse influence is indicated, the significance of which is greatest for the developed nations. Darmstadter et al. (1977) have observed that countries with larger areas tend to have lower GDP_p/S, (higher S/GDP_p) in a sample of nine highly developed industrial nations. They interpret this as a repercussion of increased transportation costs, since the more developed nations tend to devote a greater percent of their GNP toward transportation (Cook 1971).

Renewable Energy Resources and Economic Production

Simple regressions of GDP_p vs R_D and vs R_I exhibited strong positive relations for each of the three development groups. All relations were significant but highest r^2 values were obtained for Group II nations (0.76 for R_I and 0.79 for R_D). Unfortunately, we have not yet developed a figure to display these data, but in general the relationships were of similar statistical significance as was observed for GDP_p vs P and vs A (figs. 6 and 8). The slopes for these expressions were significantly greater for R_D than R_I as would be expected from the direct nature of R_D. In fact, the relation between energy consumed (S) and economic output was markedly improved by adding R_D to S. However, since S already explained most of the variation in GDP_p, there was little room for improvement (where, for example, r^2 values for Group I nations improved from 0.951 to 0.962). The implication here is that renewable solar energies contribute significantly to the economic production of nations, despite the fact that these "externalities" are not explicitly valued by economic markets. The renewable energies (R_D) which enter the economy through agriculture, fisheries, forestry and hydroelectric power are slightly better predictors of GDP_p than are the so-called indirect renewables, R_I. This is not surprising since the economic linkage is more direct (Costanza 1980), and considering the fact that, historically, they have been the basis for all agrarian societies (Rappaport 1971). However, what is most interesting is that R_I inputs to the national systems also exhibit strong, positive correlation (particularly for the more developed countries).

It is easy to speculate but difficult to pinpoint the mechanism(s) whereby R_I would act to enhance a nation's economic production. Häfele (1974a) and Darmstadter et al. (1977) have observed that among the industrialized countries of Europe and North America those with more favorable climates tend to have better economic fuel efficiencies, and their explanation is that these nations require less energy for heating. Thus, the influence of solar energies on climate appears to reduce the need for fuels. However, it can be seen in our figures 7 and 9 that the highest ratios (GDP_p/S) correspond to the least devel-

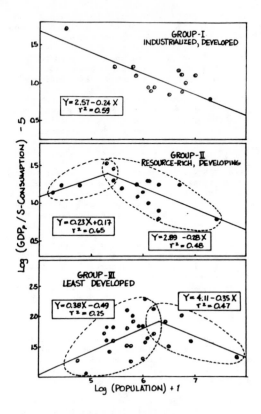

Figure 7.--Relationship between gross domestic product per unit of commercial fuel consumed and population for nations in each of three groups of economic development. See note in figure 6 caption.

oped nations, many of which are in tropical jungles, and some might argue that these climates are not conducive to economic productivity without expensive air conditioning. In fact, in many such nations, the dense and productive tropical ecosystems (which are the benefactors of rich sources of renewable energies) may even act as impediments to economic development. We know that agricultural production is often greatest in regions with bountiful rainfall and sunlight (e.g., Gross et al. 1979), and that R_D seems to stimulate GDP_p, so that the effects of R_I may act through R_D activities. Yet, we suspect that while these climatic and agricultural effects of R_I may be important, the notion of general "life-support" functions must be somehow involved here.

This issue was further investigated by examining the effect of R_I and R_D on the economic fuel efficiency of nations (GDP_p/S). Broadly speaking, we found that strong positive relationships were obtained for the most developed group of nations, with significant but weaker regres-

sion for Group II nations and little or no relationships for Group III. Moreover, the statistical fit for R_I was better for the more developed
nations while R_D showed a tighter fit with the
less developed countries. The results of regressions in which R_I and R_D were added to obtain R
(note in fig. 3 that R is dominated by R_I for
most of the nations) are presented in figure 10,
which shows a secular increase in both slope and
fit moving from least to most developed. We
interpret this relation to imply that the economic benefits of nature's "life-support" functions are most pronounced where nature's processes are most reduced (e.g., through clearing
of forests, filling of wetlands and discharge of
wastes to rivers, lakes and estuaries). Perhaps,
these services of nature are more than just a
subject of the poet's pen and the naturalist's
musing. Perhaps, they are also a source of economic advantage in the competition to generate
the most economic production out of the least
consumption of the world's dwindling fuel
resources.

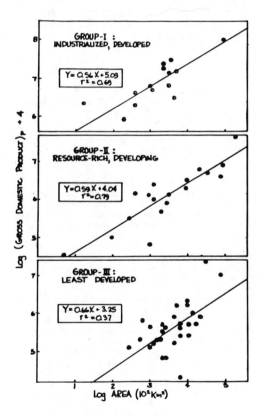

Figure 8.--Relationship between gross domestic
product and area for nations in each of three
groups of economic development. See note in
figure 6 caption.

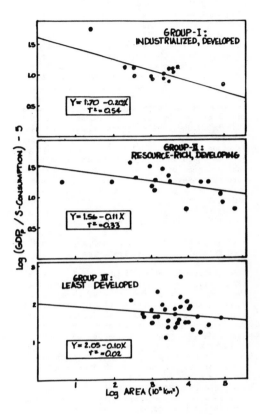

Figure 9.--Relationship between gross domestic
product per unit of commercial fuel consumed
and area for nations in each of three groups
of economic development. See note in figure
6 caption.

"Filling the GNP Gap"

If we assume that these correlations between
GDP_p and solar-based energy resources are indicative of causal relationships, we might ask how
can less developed nations exploit available renewable resources to raise their GNP/capita. This
quotient, GNP/capita, is used often as an index of
standard-of-living, and while many have argued
that it is a hollow concept (e.g., Daly 1974,
Morawetz 1977), it nonetheless provides a workable metric for planners. Since the less developed nations tend to have the best GDP_p/S ratios
(Ridker and Crosson 1975, Goldemberg 1978), it is
unlikely that they can improve economic production by achieving even greater fuel efficiency.
In fact, several investigators have suggested
that a transitional period of particularly low
economic fuel efficiency can be anticipated before full economic development is achieved (Summers 1971, Krenz 1977, Häfele and Sassin 1977,

MacKillop 1980), and our own data have corrobor-
ated this notion (see figs. 7 and 9).

Many of the Third World nations are currently utilizing renewable energy resources
directly through agriculture and fisheries (e.g.,
Scrimshaw and Taylor 1980), draft animals (Ward
et al. 1980), hydroelectric power (Strout 1977,
Hammond 1978), and biomass fuels (Ramachandran
and Gururaja 1977). Some optimists propose that
the economic gap between rich and poor nations
can be bridged by judicious use of stored fuels
(Häfele 1974b), while others suggest that such
options may not be available to many of the less
developed nations (Smil 1979), and that purchase
of such fuels would have detrimental impact on
their balance-of-payments which are already
stretched (Revelle 1980). Smil (1979) has articu-
lated an appropriate caveat for this issue, "At a
time when the most advanced industrial powers are
searching for ways to reduce their dependence on
fossil fuels by turning to renewable energy resources, it would be ironic if most of the poor
nations pursued development strategies that tied
them to non replaceable resources".

The results of our statistical analysis indicate that, indeed, GDP_p can be expanded through
the direct exploitation of solar-based resources.
Whether emphasis should be placed on agriculture,
hydroelectric or biomass fuels depends on the
particular signature of renewable energies available to a given country. One option may very well
conflict with another, for example, the conversion of truck-crop fields into sugar cane plantations for production of alcohol seems foolish in
the face of an undernourished population (Revelle
1980, Gorney 1980). The macroscopic perspective
that we have used in this analysis, while not
appropriate for addressing detailed socio/eco-
nomic isues, can provide a salient view of the
broad horizons for economic development.

We examined the percentage of R which is
coupled directly into R_D for each nation and each
type of renewable energy (insolation, hydrostatic
potential, wind and tides). For the sake of argument we assumed that any nation could, at best,
directly exploit R at rate equivalent to the highest ratio of R_D/R found in our 63-country sample
for a given type of energy. Thus, we estimate
that about 60-80% of insolation can be channelled
into agriculture, fisheries and forestry, while
only 45% of hydrostatic potential can be tapped
into electric power generation under ideal conditions of topography and precipitation. The question is, how much increase in GDP_p can be obtained by exploiting these available renewable
resources?

To illustrate the approach we use two
examples: Afghanistan, a poor nation whose renewable energies are dominated by hydrostatic potential; and Australia, a resource-rich developing
nation which possesses a fairly balanced suite of
solar-based energies. Both nations are utilizing
their insolation/photosynthetic resources nearly

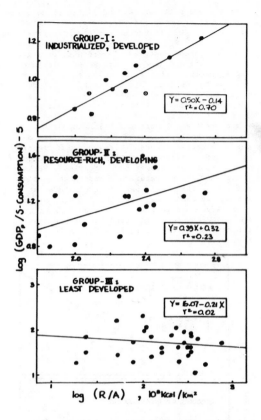

Figure 10.--Relationship between gross domestic
product per unit of commercial fuel consumed
and areal density of renewable energy resources for nations in each of three groups
of economic development. See note in figure
6 caption.

-to their potential. In figure 11 we have plotted
the log-log regression line for GDP_p versus the
sum of $S+R_D$. Two oval shapes outline the
boundaries containing "less developed" (Group
III) and "more developed" (Groups I and II) countries. The respective locations of Afghanistan
and Australia in this state-space are presented
as are the extent to which each could advance
economically by utilizing the available R as R_D.
We assume that they must move along the slope of
the line (i.e., they cannot generate more than
the average GDP_p per unit of $(S+R_D)$). This analysis suggests that Afghanistan could increase its
GNP/capita from $80 to $800 by exploiting these
resources (assuming there was no increase in population), while Australia could treble its
GNP/capaita from $2820 to $8765. Obviously, these
examples represent scenarios for maximum economic
growth through use of renewable resources, and
more than likely such growth would be accompanied
by attendant increases in population which would

reduce the per capita wealth.

SUMMARY AND CONCLUSIONS

We have presented the preliminary results of analyses examining relationships between various input factors (including population, area, and availability and utilization of energy resources) and economic output (GDP_p, corrected for purchasing power parity) as well as economic fuel efficiency (GDP_p/S) for 63 nations. We have taken a macroscopic perspective which considers the structure of a national system in a highly aggregated form. While we recognize the potential weakness in a data set compiled from so many different sources, we have made every attempt to check and correct inconsistencies. In any case, we maintain that this scale of analysis facilitates one's ability to see gross interrelationships. Since it is virtually impossible to conduct experiments at the level of a national economy, analyses of this kind of data are necessarily inductive in approach. We are limited to the use of statistical correlations, which per se are inappropriate for addressing issues of causality. We have nonetheless used these statistics to draw inferences on cause–effect couplets in reference to the hypothesized structure of national systems which we conceptualized at the outset. We cannot, however, rule out the possibility that such correlations are simply manifestations of covariances with a third factor which we did not consider.

Various indices of social welfare and standard-of-living exhibited little ability to predict GDP_p, indicating that social structure is probably not the force that drives economic production. Inversely, this absence of correlation may represent a corroboration of the fact that economic production is a poor measure of social wellbeing. Gross demographic and physiographic factors of population and area showed some interesting relationships with economic output and fuel efficiency. In general, countries (within the same economic development group) with larger populations, had larger GDP_p, suggesting the importance of human labor and information in economic production. Similarly, countries with larger areas also had larger GDP_p, perhaps implying the importance of space for expanded development, as well as area to capture solar energies. However, area had a negative effect on GDP_p/S, probably due to the increased use of fuels for transportation in larger countries. This economic fuel efficiency appears to be negatively affected by population for the more advanced nations, while poorer countries exhibited a "diminishing-returns" relation, where smaller populations enhanced efficiency while larger populations reduced it.

We observed, as have previous investigators, that consumption of stored fuels (S) provided an excellent predictor of GDP_p, and we interpreted this to represent the fact that fuels are often the limiting factor of economic production. The

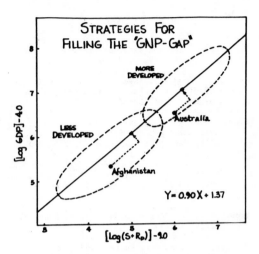

Figure 11.--Regression of gross domestic product versus the sum of commercial fuels and renewable energies utilized in the economics of 63 nations. Two ovals outlined with dashed lines delimit the regions containing less and more developed countries. See text note in figure 6 caption for further explanation.

addition of hydroelectric power, agricultural production and other renewable energies directly linked to the economy (R_D) to S improved the fit of this relation. Moreover, R_D in itself correlated well with GDP_p, all of which underscores the direct significance of solar inputs to economic processes. The often cited inverse relation between level of development and economic-output-per-energy-input ratio virtually disappears with the inclusion of R_D.

What is perhaps the most astounding outcome of our analysis is the observation that those renewable energy resources not directly involved in economic processes (R_I), nevertheless, explained up to 76% of its variability. In fact, for the most developed countries the areal density of these indirect renewable resources positively correlated with economic fuel efficiency. We infer from this relation a demonstration of the ultimate economic value of nature's "life-support" functions. Thus, the winds that blow continually over Chicago maintain the urban air relatively clear of automotive and industrial pollutants, thus reducing the need for expensive air pollution controls. Similarly, the tides which flush Boston Harbor twice per day reduce that city's need for advanced sewage treatment. So we see that, while these services of nature go unaccounted by the market system, they improve society's economic efficiency and possess indirect (but very real) economic value.

838

LITERATURE CITED

Ayers, R.U. and A.V. Kneese. 1969. Production, consumption and externalities. Amer. Econ. Rev. 59:282–297.

Ayers, R.U. and A.V. Kneese. 1971. Economic and ecological effects of a stationary economy. Ann. Rev. Ecol. Syst. 2:1–22.

Biswas, A.K. and M.R. Biswas. 1976. State of the environment and its implications to resource policy development. Bioscience 26:19–25.

Bormann, F.H. 1976. An inseparable linkage: Conservation of natural ecosystems and the conservation of fossil energy. Bioscience 26:754–760.

Brown, H. 1976. Energy in our future. Ann. Rev. Energy. 1:1–36.

Bullard, C.W. and C.Z. Foster. 1976. On decoupling energy and GNP growth. Energy 1:291–300.

Bullard, C.W. and R.A. Herendeen. 1975. The energy cost of goods and services. Energy Policy 3:268–278.

Clague, C. and V. Tanzi. 1972. Human capital, natural resources and the purchasing power parity doctrine: Some empirical results. Economia Internazionale 25:3–18.

Cloud, P. 1973. Is there intellegent life on Earth?, p. 264–280. In: G.M. Woodwell and E.V. Pecana (ed.) Carbon in the biosphere. 392 pp. ERDA Conf. Series CONF-720510, Nat. Techn. Inf. Serv., Springfield, Va.

Cook, E. 1971. The flow of energy in our industrial society, pp. 83–91. In: Scientific Amer. (ed.) Energy and power. 144 pp. Freeman, San Francisco.

Cook, E. 1976. Limits to exploitation of nonrenewable resources. Science 191:677.

Costanza, R. 1980. Embodied energy and economic valuation. Science 210:1219–1224.

Craig, P.P., J. Darmstadter, and S. Rattien. 1976. Social and institutional factors in energy conservation. Ann. Rev. Energy 1:535–551.

Daly, H.E. 1974. The world dynamics of economic growth. The economics of the steady state. J. Amer. Econ. Assoc. 64:15–21.

Darmstadter, T., P.D. Teitelbaum and T.G. Polach. 1971. Energy in the world economy: A statistical survey of trends in output, trade, and consumption since 1925. Johns Hopkins Univ. Press, Baltimore.

Darmstadter, J., J. Dunkerley and J. Atterman. 1977. How industrial societies use energy. A comparative analysis. 282 pp. Johns Hopkins Univ. Press, Baltimore.

Day, A.T. and L.H. Day. 1973. Cross-national comparison of population density. Science 181:1016.

Erlich, P., A. Erlich and J. Holdren. 1977. Ecoscience. Freeman, San Francisco.

Forrester, J. 1971. World dynamics. 142 pp. MIT Press, Cambridge, MA.

Georgescu-Roegen, N. 1977. The steady-state and ecological salvation: A thermodynamic analysis. BioScience 27:266–270.

Gilliland, M.W. 1975. Energy analysis and public policy. Science 189:1051–1056.

Goeller, H.E. and A.M. Weinberg. 1976. The age of substitutability. Science 191:683–689.

Goldemberg, J. 1978. Brazil: Energy options and current outlook. Science 200:158–164.

Gorney, C. 1980. Brazil steps up crusade for alcohol fuel. (Nov. 3, 1980). p. A23. The Washington Post, Washington, D.C.

Gosselink, J.G., E.P. Odum and R.M. Pope. 1974. The value of the tidal marsh. 30 pp. Center Wetland Resources, Louisiana State Univ., Baton Rouge, LA. Publ.No. LSU-SG-74-03.

Gross, D.R., G. Eiten, N.M. Flowers, F.M. Leoi, M.L. Ritter, D.W. Werner. 1979. Ecology and acculturation among native peoples of central Brazil. Science 206:1043–1050.

Häfele, W. 1974a. Energy choices that Europe faces: A European view of energy. Science 184:360.

Häfele, W. 1974b. A systems approach to energy. Amer. Scientist 62:438–447.

Häfele, W. and W. Sassin. 1977. The global energy system. Ann. Rev. Energy 2:1–30.

Hammond, A.L. 1978. Energy: Elements of a Latin American strategy. Science 200:753–754.

Hayes, E.T. 1976. Energy implications of materials processing. Science 191:661.

Hubbert, M. King. 1969. Energy resources, p.157–242. In: National Academy of Sci/National Res. Council. Resources and man. 240 pp. W.H. Freeman Publ., San Francisco.

Kravis, I.B., A. Heston and R. Summers. 1978. Real GDP per capita for more than one hundred countries. The Econ. Journal. 88:215–242.

Krenz, J.H. 1977. Energy and the economy: An interrelated perspective. Energy 2:115–130.

Landsberg, H.E. (ed.) 1969–1976. World Survey of climatology. Elsevier Publ., Amsterdam.

Leontief, W. 1974. Structure of the world economy: Outline of a simple input-output formulation. Amer. Econ. Rev. 64:823–834.

Lieth, H. and R.H. Whitaker (eds.). 1975. Primary productivity of the biosphere. 339 pp. Springer-Verlag, New York.

Limberg, K., W.M. Kemp, W.R. Boynton. 1977. The influence of renewable energies on national economics of the world. Ref. No. UMCEES–77–104 CBL. 62 pp. Chesapeake Biological Laboratory, Solomons, MD., USA.

MacKillip, A. 1980. Economic considerations for solar and renewable energies in developing countries. Natural Resources forum. 4:169–179.

adows, D.H. et al. 1972. Limits to growth. 185 pp. Universe Books, New York.

Morawetz, David. 1977. Twenty-five years of economic development: 1950–1975. 126 pp. The World Bank (International Bank for Reconstruction and Development), Washington, D.C.

Nordhaus, W.D. 1974. Resources as a constraint on growth. J. Amer. Econ. Assoc. 64:22–26.

Odum, H.T. 1971. Environment power and society. 331 pp. John Wiley, New York.

Pimentel, D., L.E. Hurd, A.C. Bellotti, M.J. Forster, I.N. Oka, O.D. Sholes, R.J. Whitman. 1973. Food production and the energy crisis. Science 182:443–450.

Ramachandran, A. and J. Gururaja. 1977. Perspectives on energy in India. Ann. Rev. Energy 2:365–386.

Rappaport, R.A. 1971. The flow of energy in an agricultural society, p. 69–80. In: Scientific Amer. (ed.) Energy and power. 144 pp. Freeman, San Francisco.

Revelle, R. 1980. Energy dilemmas in Asia: The needs for research and development. Science 209:164–174.

Ridker, R.G. and P.R. Crosson. 1975. Resources, environment and population, p. 202–222. In: W.C. Robinson (ed.) Population and development planning. The Population Council, New York.

Sanderson, W. and B.F. Johnston. 1980. Bad news: Is it true? Science 210:1302–1303.

Sassin, W. 1980. Energy. Scientific Amer. 243:119–132.

Schipper, L. 1976. Raising the productivity of energy utilization. Ann. Rev. Energy 1:455–517.

Schipper, L. and A.J. Lichtenberg. 1976. Efficient energy use and well-being: The Swedish example. Science 194:1001–1013.

Scrimshaw, N.S. and L. Taylor. 1980. Food. Scientific Amer. 243:78–88.

Simon, J.L. 1980. Resources, population, environment: An oversupply of false bad news. Science 208:1431–1437.

Smil, V. 1979. Energy flows in the developing world. Amer. Scientist 67:522–531.

Steinhart, J.S. and C.E. Steinhart. 1974. Energy use in the U.S. food system. Science 185:482–494.

Strout, A.M. 1977. Energy and economic growth in Central America, Ann. Rev. Energy 2:291–305.

Summers, C.M. 1971. The conversion of energy, pp. 95–106. In: Scientific Amer. (ed.) Energy and power. 144 pp. W.H. Freeman, San Francisco.

Tribus, M. and E.C. McIrvine. 1971. Energy and information, p. 121–128. In: Scientific American (ed.) Energy and power. 144 pp. W.H. Freeman, San Francisco.

Van Hylckama, T.E.A. 1956. The water balance of the earth. Publications in climatology, Vol. IX. Centerton, N.J.

Walter, H. 1973. Vegetation of the earth: In relation to climate and the eco-physical conditions. 237 pp. Springer-Verlag, New York.

Ward, G.M., T.M. Southerland, J.M. Southerland. 1980. Animals as an energy source in Thrid World agriculture. Science 208:570–574.

Weiss, C. 1979. Mobilizing technology for developing countries. Science 203:1083–1089.

Westman, W.E. 1977. How much are nature's services worth? Science 197:960–964.